TECHNICAL AMERICANA

A Checklist of Technical Publications Printed Before 1831

TECHNICAL AMERICANA

A Checklist of Technical Publications Printed Before 1831

BY

EVALD RINK

Foreword by
Eugene S. Ferguson

Sponsored by
the Eleutherian Mills Historical Library

KRAUS INTERNATIONAL PUBLICATIONS
MILLWOOD, NEW YORK

A Division of Kraus-Thomson Organization Limited

The publication of this work was supported by the National Science Foundation under Grant No. SES 79-19675. As the manuscript was completed prior to the awarding of the grant, any opinions, findings, and conclusions or recommendations are those of the author and do not necessarily reflect the views of the National Science Foundation.

First Printing

Printed in the United States of America

Library of Congress Cataloging in Publication Data

Rink, Evald, 1916-
Technical Americana.

Includes index.
1. Technology—Early works to 1850—Bibliography.
2. United States—Imprints. I. Title.
Z7912.R56 [T45] 016.6 81-4036
ISBN 0-527-75447-1 AACR2

Contents

Foreword

I RECALL my first response to this *Checklist* as I scanned the catalogue cards that Mr. Rink had so carefully prepared and assembled. It was a sobering experience to see what he had found in a period and a subject that I thought I knew quite a lot about. Every third card, it seemed, suggested fresh and unfamiliar topics and new points of view. My second response was one of delight as I realized that here was another huge segment of source materials that was being made readily available to students of American technology and, hopefully, to a much wider audience of historians who have begun to seek in technology some of the causes of social, political, and intellectual change.

The earliest works listed here were published in the seventeenth century, and many if not most of the titles before the Revolutionary War are concerned with agricultural pursuits. The bulk of the *Checklist*, however, deals with what might be called the "critical period" of American technology, when a radical qualitative change occurred not only in the way things were made but also in the outlook of technologists. An age of steam and iron replaced one of water and wood, and the promoters of change saw unlimited mechanization as an unquestionably desirable goal.

The beginning of the critical period coincided with the grand Federal Processions of 1788, which were held in seaboard cities to celebrate the adoption of the new Constitution of the United States. Militia companies and bands of music participated, of course, but the unique flavor of the parades was supplied by the dozens of floats imaginatively designed by companies of craftsmen. The shipwrights' float was a miniature ship of state; the blacksmiths manned an operating forge wherein swords were beaten into plowshares; on the printers' float was a fully equipped print shop, turning out a shower of newly printed handbills proclaiming the fruits of liberty. Hundreds of skilled craftsmen, proudly plying their ancient crafts with traditional tools, emphasized the manual, muscular aspects of American technology.

Even as the handcraftsmen marched in Philadelphia and New York and Boston, a new era was already struggling to be born. A young man in Delaware was hard at work, developing a totally new idea in manufacturing. Oliver Evans's automatic flour mill, which received wheat at one end of the building and discharged flour at the other end, was designed to operate without human intervention except to adjust the machinery. The ultimate significance of Evans's continuous-process factory, in which materials were moved from one station to the next by mechanical conveyors, was not recognized in 1788, but the American system of mass-production was implicit in that Delaware flour mill.

The critical period continued well into the nineteenth century, as one after another of the traditional crafts was mechanized. New machines called for new skills and more rigid materials, and the workbench and vise gave way to an array of machine tools, capable of working iron and other metals into intricate, precise shapes. It was nearly midcentury before waterwheels were replaced by hydraulic turbines, and it was even later that steam engines, which appeared first in river boats and railroad locomotives, took over the bulk of industrial power requirements. Yet by 1830 the enthusiasms of the progress-minded critical period were in full cry. The outlines of the mechanized future had been sketched in the books and pamphlets and broadsides that were published by that date.

The earliest publication to be listed is a Massachusetts broadside of 1683, making public a law to encourage the building of brick and stone houses, instead of wood, in the city of Boston. This, it may be noted, was less than twenty years after the great London fire of 1666. Many early titles dealt with agriculture. In 1710 was published a husbandman's guide, which had one section on planting, one on the art of the farrier, and one on distempers of men, women, and children. A number of promotional pieces before 1750 encouraged the production of hemp, flax, and "tar from trees." Manuals and encyclopaedic works on farming and rural economy were numerous after 1775; one, published in German in Lebanon, Pennsylvania, was entitled (in translation) *The American Gold Mine; or The Art of Farming.*

Lotteries were held, between 1770 and 1800, to encourage a china factory in Delaware and a glass factory in Pennsylvania, to provide navigational aids in the Connecticut River, to improve public landings in Philadelphia, to build a bridge at New Haven, and to finance the building of other factories and several canals.

Price-fixing was customary before 1830. Carpenters' charges were specified in Boston, Philadelphia, and Troy; other standard prices were published for boots and shoes, watches and clocks, cut stone and marble, and, in Boston, for printing.

Handbooks and manuals were published for the information of a wide variety of readers. Many but not all of the books were copied from English works, often without extensive editorial changes. Nevertheless, to reprint an English book required nearly the same effort in the printing shop as to produce an original American book. The type had to be set, the paper made, the sheets printed, and the whole work bound between covers.

One of the most durable books of general instruction, including lessons on spelling and reading, measuring carpenters' and masons' work, gauging, and dyeing, among other topics, was a book of 378 pages, *The American Instructor: or, Young Man's Best Companion*, first printed in Philadelphia in 1748 by Benjamin Franklin and David Hall. Reprinted in half a dozen other towns over the next sixty years, an edition was issued as late as 1810. American editions of British works were also available for the instruction of merchants, shipmasters, carpenters, weavers, and other craftsmen.

Before 1830, three great encyclopaedias had been reprinted in the United States. Beginning in 1790 and finishing in 1797, Thomas Dobson in Philadelphia reprinted the eighteen-volume *Encyclopaedia Britannica*. Abraham Rees's *The Cyclopaedia*, comprising 41 volumes of text and 6 volumes of plates (all re-engraved), was published in Philadelphia and New York between 1805 and 1824; the London edition had been issued in parts from 1802 through 1818. Finally, the American edition of the *Edinburgh Encyclopaedia* was republished in 18 volumes, 1813–1831. Although significant corrections and additions were made to these encyclopaedias by American authors, nobody, to my knowledge, has made any systematic comparisons of the British and American works.

While British technical works dominated the American scene throughout the period of this *Checklist*, a fresh stream of American handbooks was inaugurated by Oliver Evans's *Young Mill-Wright and Miller's Guide*, first published in 1795 and eventually running through fifteen editions. Evans used British books as a source of his theoretical information, but the *Guide* was an original work; in his explanation of the Evans's automatic flour mill, he was, as suggested above, laying the philosophical and technical foundation for the era of mass production. In 1805 Evans published his second original volume, *The Abortion of the Young Steam Engineer's Guide*, in which he explained the philosophy and operation of his high pressure steam engine. The title of the book reflected his frustration at being unable to find adequate time or money to complete his studies of the steam engine in the way he thought they should be done.

Science in the service of the new republic is one theme that runs through many works in American technology. Starting near the end of the eighteenth century, the works of the popular English scientific lecturers George Adams and James Ferguson were reprinted in Philadelphia. American lecturers, of whom the most famous was Jacob Bigelow, continued the tradition of adult education in explaining how science might be applied to practical pursuits. For children, an adaptation of a French juvenile work, presenting a *General View or Abstract of the Arts and Sciences; Adapted to the Capacity of Youth,* was published in Philadelphia, first in French, then within a year in English.

Books on science and practice were made easily available to willing readers after 1820 as dozens of apprentice and mercantile libraries were opened. Societies for the promotion of agriculture and the useful arts appeared by the score in the generation following the Revolutionary War. In the 1820s, a number of mechanics' institutes were organized; the Franklin Institute in Philadelphia, one of the earliest, became the most prominent. In the 1820s also, a surprising number of polytechnic and scientific colleges were proposed. By 1830 it was clear that science (then as now confused with technology) was to play an increasing rôle in American industry as it became a force to be reckoned with in world markets and world diplomacy.

The preoccupation of many nineteenth century Americans, after the War of 1812, with internal improvements—roads, bridges, canals, railroads—is reflected in the very numerous entries under those subjects: nearly one-third of this *Checklist* is devoted to internal improvements. Indeed, the task of filling the continent and subduing it was assumed to be the primary task in realizing the manifest destiny of the new republic. In one way or another, nearly every title in the pages that follow supports or promotes progress in some facet of that exciting task.

The number of periodical publications before 1830 concerned in some way with technology is quite astonishing. Over a hundred are listed, starting with *The American Magazine* of Boston, in 1743, and ending with one of the many publications entitled *Mechanics' Magazine*, again in Boston. No matter that many of the journals existed for only a year or two, the body of material is nevertheless enormous, and the finding aids are few.

A start has been made, however, by two scholars on indexing articles in periodicals. J Meredith Neil, "Architectural Comment in American Magazines 1783-1815," American Association of Architectural Bibliographers, *Papers* V (1968), 15-45, is concerned primarily with aesthetics and style, but other information is to be found in the many descriptions of towns and individual structures. Some articles describe water works and factory buildings, and Neil's section on "bridges and other pieces of engineering" lists a dozen articles on American structures.

Penrose R. Hoopes, a major collector of early technical books, serials, and pamphlets, compiled a particularly valuable index of articles pertaining to technology in American magazines through 1799. The index, entitled *Technical Americana*, has not been published, but a copy is on deposit in the Eleutherian Mills Historical Library.

In the works of George F. Dow, Rita S. Gottesman, and Alfred C. Prime* are to be found advertisements and news articles concerning the mechanical trades in long runs of early newspapers between 1704 and 1800.

Ralph R. Shaw's "Engineering Books Available in America prior to 1830," New York Public Library *Bulletin* XXXVII (1933), 38-61, 157-60, 209-22, 301-10, 539-55, is an ad hoc selection of books on "engineering," a difficult term to delimit. American imprints total 223, all of which will be found in this *Checklist*: about 252 titles were printed in England, and another 188 in France. The list is still useful for the foreign titles that it contains.

While many scholars are more interested in the social effects of the American Industrial Revolution than in the details of technical change, an acceptable explanation of social change must inevitably rest upon a solid understanding of what actually happened

*George F. Dow, *The Arts & Crafts in New England 1704-1775* (Topsfield, Mass., 1927); Rita S. Gottesman, *The Arts & Crafts in New York 1724-1776* (New York Historical Society, *Collections* LXIX 1936); Alfred C. Prime, *The Arts & Crafts in Philadelphia, Maryland, and South Carolina 1721-1800* (2 vols., Topsfield, 1929-32).

within the technical world itself. Thus far, very little has been done in any concerted way to make connections between technology and the society it served. It is to such a task that the present work can contribute so greatly.

There is still a great deal of detailed, page-by-page search to be done by individual scholars in journals, newspapers, and in manuscript collections, but the outlines of the territory are made much clearer by Mr. Rink's painstaking work. Given the impetus of such definitive segments of the total picture, scholars of the next generation should come to see the history of technology as an integral, inseparable part of the remarkable American experience.

Eugene S. Ferguson

15 May 1980

Introduction

THE PURPOSE of this checklist is to assist all who are interested in the history of American technology during its formative period. Most of the publications included here have been previously recorded in various bibliographies and checklists. Now brought together they reflect the evolution of technological knowledge and its applications. The improvements in crafts and manufacturing processes, the invention of machines and other labor saving devices to ease the burden of physical work, and the advancements in means of communication present a fascinating story which can be followed through contemporary publications.

The year 1830 has been adopted as the cut-off date for practical purposes. The number of technological publications printed prior to 1831 is still within a manageable range. Thereafter it grows very rapidly, and each year of coverage added to this checklist would have substantially increased its size and made a comprehensive listing virtually impossible.

There are general bibliographic tools, such as Evans's *American Bibliography, 1638–1800*, Shaw and Shoemaker's *American Bibliography, 1801–1819*, Shoemaker's *A checklist of American Imprints, 1820–1829*, and Cooper's *A Checklist of American Imprints for 1830* available for American publications issued before 1831, which have made the selection of relevant material easier. In addition to these the *National Union Catalog: pre 1956 Imprints* (in progress), a number of specialized bibliographies as well as the catalogs of several libraries, particularly the American Antiquarian Society, the American Philosophical Society, the Library Company of Philadelphia, the Pennsylvania Historical Society, the New York Historical Society, the Franklin Institute in Philadelphia, and the Eleutherian Mills Historical Library have been searched. In some cases, e.g. the American Antiquarian Society, access has been granted to collections of uncataloged ephemeral materials. The staff of the American Antiquarian Society added the symbol of their library to publications they hold, and suggested a number of additions not recorded in the sources to which the compiler had access, increasing the usefulness of this checklist.

Only separately published works have been included. The term "technology" has been interpreted rather broadly, and the checklist contains not only the known works on various technologies but also those which indicate the advocacy and the extent of the application of technological improvements, as well as the availability of products created by such applications.

Certain categories of publications which were not primarily meant for the dissemination of technological information contain contributions to these topics. Such publications have been included selectively on the basis of the amount and significance of the technological information they contain. Practically all journals and newspapers issued prior to 1831 carried articles on technological topics. Only the journals with an appreciable number of technological contributions have been included. All newspapers have been omitted from this checklist. These are quite adequately covered by Clarence S. Brigham's *History and Bibliography of American Newspapers, 1690–1820*, Worcester: American Antiquarian Society, 1947, and Winifred Gregory's *American Newspapers, 1821–1936, a Union List*, New York: H. W. Wilson Company, 1937. General reference works, such as encyclopedias and manuals often contain useful information on technological topics and have been included. The same is true for many travel books, gazetteers, and guides of which a considerable number have been selected.

The following types of publications, frequently containing technological information, have been omitted:

- All articles in periodicals, unless they have also been issued separately;
- Almanacs. A comprehensive guide to these publications is Milton Drake's *Almanacs of the United States,* New York: Scarecrow Press, 1962;
- City and county directories. These are adequately covered by Dorothea N. Spear's *Bibliography of American Directories through 1860,* Worcester: American Antiquarian Society, 1961. The directories themselves are available in a microfiche edition;
- Cookbooks, well covered by Eleanor Lowenstein's *Bibliography of American Cookery Books, 1742–1860,* 3rd ed., Worcester: American Antiquarian Society, 1972;
- Literature promoting industrial development and manufactures, unless such publications contain specific information on the state of an industry or manufacture;
- Literature on tariffs, trade and commerce;
- Maps, except those specifically referring to technological improvements, such as roads, canals, railroads, and industrial sites;
- Medicine and veterinary medicine, except publications concerned with techniques employing medical or pharmaceutical instruments;
- Works on military discipline and drill books for movement in formations;
- Most Federal publications on postal routes because of their repetitive nature;
- Sciences are represented selectively by samples to indicate what was available. No attempt was made to be comprehensive or to list all editions of a particular work. Its widespread availability in libraries frequently guided the selection of a particular edition. Works which contain more than the usual amount of information on scientific instruments and apparatus, and those on applied sciences are included.

The checklist is topically arranged. In order better to serve the historian's interests, the material within the several topics is chronological by year, thus presenting through contemporary publications a continuous record of technological developments. Within any one year the publications are listed alphabetically according to their authors or, in case of anonymous works, titles. In addition, an index of names, subjects and anonymous titles is provided for all entries.

The primary concern of this work is to make the early technological publications available to the users. It does not attempt to be a bibliographical study of individual items. Therefore the bibliographical description of these works, while trying to be accurate, is limited to essential features, sufficiently complete for their identification. The entries agree generally with those used in the National Union Catalog, with some exceptions due to the recent changes in cataloging rules. The names of some government agencies have been shortened. Lengthy titles have been abbreviated, if this did not interfere with their intended message, and the omissions are indicated. For certain frequently occurring words, such as "edition," "revised," and "corrected," standard abbreviations are used. If the place of publication is not known, the abbreviation "S.l." (sine loco) is used. Capitalization and punctuation follow modern library practices, but the spelling conforms to the text of the title page or to that of the bibliographical source where the publication has been previously recorded. The edition and the imprint year are always given in Arabic numerals. The size is provided only for broadsides and maps, if available. The presence of illustrations and maps in a publication is indicated.

Whenever possible, references have been

made to general bibliographical sources, such as Evans's *American Bibliography,* Shaw and Shoemaker's *American Bibliography,* Shoemaker's *A checklist of American Imprints,* Cooper's *A Checklist of American Imprints for 1830,* and frequently to *The National Union Catalog: pre 1956 Imprints.* The Evans, Shaw, and Shoemaker references are particularly important because these numbers guide the user to the microprint editions of the respective publications. References to specialized bibliographies have been made only when the publications are not listed in the general ones, or when the additional citations are useful for their identification.

The Readex Microcard Corporation's *Early American Imprints, 1638–1800*, edited by Clifford K. Shipton, has produced in microprint the full text of all works listed by Evans, and in the *Supplement* compiled by Roger P. Bristol. The symbol "Mp" is used for such works, the numeral following it being identical to the Evans number, except for publications listed in Bristol's *Supplement.* A partial list of locations for the microprint edition is available in the *National Index of American Imprints through 1800,* by Clifford K. Shipton and James E. Mooney, published by the American Antiquarian Society in 1969. The symbol "S&S" is used for publications issued from 1801-1819, "S" for 1820-1829, and "C" for 1830. The reproduction in microprint of publications listed by Shaw and Shoemaker is in progress, with copies held probably by the same institutions as listed in the *National Index* referred to above. The microprint editions make many of the publications in this checklist easily available in all sections of the country.

The assignment of locations is generally based on the data given by Evans, Bristol, Shipton, Shaw, Shoemaker, and Cooper, as well as by the *National Union Catalog: pre 1956 Imprints.* In the interest of economy the number of locations has been limited, rarely exceeding a dozen. Numerous locations have been spread to cover the whole country. If no more than five locations are known, all have been given. The locations of a number of periodicals are recorded as "Major libraries." This means that most or all large or specialized research libraries hold at least a partial run of the publication. For further information the *Union List of Serials* should be consulted. The completeness of a run held by libraries is not indicated.

The majority of entries have been personally verified; in many other cases librarians in distant libraries have graciously agreed to verify a number of items. Particular effort was made to check these items of which the number of known extant copies is very small or just one. Even so, some publications listed in various sources could not be physically located and verified.

The entries are frequently supplemented by notes. These clarify the subject matter of the work if this is not evident from its title, or supply information about the authorship or publication history of the item.

While attempting to be comprehensive, this checklist does not include every publication on technical topics issued prior to 1831. Some works have been intentionally omitted because their contributions are negligible. This is particularly true for many Federal and state bills, laws, proclamations and resolutions. Some others could not be included because the technical information they may contain is not reflected in their titles. Finally, in spite of conscientious effort, a number of relevant publications may have escaped the compiler's attention. It is hoped that this checklist will help to identify such works and will also stimulate the search for all listed but unlocated technological publications.

The help and cooperation received from many individuals and institutions have been most generous and are here thankfully acknowledged. A full listing would be far too long, but the contributions by the following institutions and their staffs deserve special gratitude: the American Antiquarian Society, the American Philosophical Society, the Baker Library of Harvard University, the Boston Athenaeum, the Boston Public Library, the Brown University Library, the Buffalo and Erie County Public Library, the Columbia University Library, the Eleutherian Mills Historical Library, the Essex Institute, the Henry E. Huntington Library, the Library Company of Philadelphia, the Library of Congress, the Louisiana State University Library, the Massachusetts Historical Society, the New York Historical Society, the Ohio Historical Society, the Pennsylvania Historical Society, the Pequot Library (Southport, Connecticut), the Rutgers University Library, the State Historical Society of Wisconsin, the University of Delaware Library, the University of North Carolina Library, the Virginia State Library, and the Western Reserve Historical Society. Their assistance has greatly increased the completeness and usefulness of the checklist; any omissions and inaccuracies that still remain are solely the responsibility of the compiler.

The compiler is particularly indebted to Professor Eugene S. Ferguson for his many useful suggestions while the material was being gathered and organized. The editorial advice of David T. Gilchrist, Director of Publications, Eleutherian Mills-Hagley Foundation, and the technical help rendered by Mrs. Nannie Brightwell, Mrs. Elaine Laudenslager and Mrs. Marilyn Siebold of the Eleutherian Mills Historical Library are here thankfully acknowledged. My wife Hildegard has spent countless evenings and weekends to ferret out inconsistencies and misprints in the manuscript, and the index owes much of its usefulness to Miss Lorraine Cheban.

Finally, without the generous support of Dr. Walter J. Heacock, General Director of the Eleutherian Mills-Hagley Foundation and Dr. Richmond D. Williams, Director of the Eleutherian Mills Historical Library, this checklist could not have been completed.

Bibliographical References

Annenberg Annenberg, Maurice. *Type Foundries of America and Their Catalogs*. Baltimore and Washington: Maran Printing Services, 1975.

B Bristol, Roger P. *Supplement to Charles Evans' American Bibliography*. Charlottesville: University Press of Virginia, 1970.

BRE Bureau of Railway Economics, Washington, D.C. *Railway Economics; a Collective Catalogue of Books in Fourteen American Libraries*. Chicago: University of Chicago Press, [1912].

Bowker Bowker, Richard R. *State Publications; a Provisional List of the Official Publications of the Several States of the United States from Their Organization*. New York: Publishers' Weekly, 1908.

C Cooper, Gayle. *A Checklist of American Imprints for 1830*. Metuchen, N.J.: Scarecrow Press, 1972.

E Evans, Charles. *American Bibliography*. New York: P. Smith, 1941-1959.

Goldsmiths London. University. Goldsmiths' Company's Library of Economic Literature. *Catalogue of the Goldsmiths' Library of Economic Literature*, compiled by Margaret Canney and David Knott. Cambridge: University Press, 1970-1975.

Graff Newberry Library, Chicago. *A Catalogue of the Everett D. Graff Collection of Western Americana*, compiled by Colton Storm. Chicago: University of Chicago Press, 1968.

Hasse (Mass.) Hasse, Adelaide R. *Index of Economic Material in Documents of the States of the United States. Massachusetts*. Washington: Carnegie Institution of Washington, 1908.

Hasse (New Jersey) __________ *New Jersey*. Washington: Carnegie Institution of Washington, 1914.

Hasse (New York) __________ *New York*. Washington: Carnegie Institution of Washington, 1907.

Hasse (Ohio) __________ *Ohio*. Washington: Carnegie Institution of Washington, 1912.

Heyl Heyl, Edgar. "Unrecorded Pre-1832 Maryland Publications." *Maryland Historical Magazine* 71 (1976): 527-547.

Hitchcock Hitchcock, Henry R. *American Architectural Books: a List of Books, Portfolios and Pamphlets on Architecture and Related Subjects Published in America Before 1895*. Minneapolis: University of Minnesota Press, 1962.

Hummel Hummel, Ray Orvin. *Southeastern Broadsides Before 1877*, edited by Ray O. Hummel, Jr. Richmond: Virginia State Library, 1971.

Kress Harvard University. Graduate School of Business Administration. Baker Library. Kress Library of Business and Economics. *Catalogue; with Data upon Cognate Items in Other Harvard Libraries.* Boston: Baker Library, 1940- .

Meyer Meyer, Balthasar H. *History of Transportation in the United States Before 1860.* Prepared under the Direction of Balthasar Henry Meyer by Caroline E. MacGill and a Staff of Collaborators. [New York]: P. Smith, 1948.

Mp Shipton, Clifford K., ed. *Early American Imprints, 1639–1800.* Worcester: American Antiquarian Society, [1956-]. (Reproduction on Readex microprint of the items listed in Evans' *American Bibliography,* and the *Supplement*, compiled by Roger P. Bristol).

N *The National Union Catalog: Pre-1956 Imprints; a Cumulative Author List Representing Library of Congress Printed Cards and Titles Reported by Other American Libraries.* London: Mansell, 1968- .

Parsons New York. Public Library. "Catalogue of the William Barclay Parsons Collection." *Bulletin of the New York Public Library* (Jan. and July 1941).

Poore Poore, Benjamin P. *A Descriptive Catalogue of the Government Publications of the United States.* Washington: 1885.

Romaine Romaine, Lawrence B. *A Guide to American Trade Catalogs*, 1744-1900. New York: R. R. Bowker Company, 1960.

S Shoemaker, Richard H. *A Checklist of American Imprints, 1820–1829.* New York: Scarecrow Press, 1964-1972.

S&S Shaw, Ralph R. *American Bibliography; a Preliminary Checklist, 1801–1819*, compiled by Ralph R. Shaw and Richard H. Shoemaker. New York: Scarecrow Press, 1958-1966.

Sabin Sabin, Joseph. *Bibliotheca Americana. A Dictionary of Books Relating to America.* New York: J. Sabin [etc.], 1868-1936.

Shaw Shaw, Ralph R. "Engineering Books Available in America Prior to 1830." *Bulletin of the New York Public Library* (Jan., Apr., June 1933).

Tapley Tapley, Harriet S. *Salem Imprints, 1768–1825; a History of the First Fifty Years of Printing in Salem, Massachusetts.* Salem: The Essex Institute, 1927.

Thomson Thomson, Thomas R. *Check List of Publications on American Railroads Before 1841.* New York: New York Public Library, 1942.

Thornton (Bibl.) Thornton, Mary L. *A Bibliography of North Carolina, 1589–1956.* Chapel Hill: University of North Carolina Press, 1958.

Thornton (Off.) ________ *Official Publications of the Colony and State of North Carolina, 1749–1939. A Bibliography.* Chapel Hill: University of North Carolina Press, 1954.

ULS *Union List of Serials in Libraries of the United States and Canada.* 3rd ed. Edited by Edna Brown Titus. New York: H. W. Wilson Company, 1965.

Wheat Wheat, James C. *Maps and Charts Published in America Before 1800; a Bibliography,* by James Clements Wheat and Christian F. Brun. New Haven: Yale University Press, 1969.

Whitford Whitford, Noble E. *History of the Canal System of the State of New York.* [Albany: Brandow Printing Co., 1906.]

Woodbury Woodbury, Charles J. H. *Bibliography of the Cotton Manufacture.* Waltham, Mass.: 1909-1910.

Key to Locations

A-Ar Alabama Department of Archives and History, Montgomery.
AAP Auburn University, Auburn.
ABS Birmingham-Southern College, Birmingham.
AFlT Florence State College, Florence.
AMaJ Judson College, Marion.
AU University of Alabama, University.
AkU University of Alaska, College.
Ar-Hi Arkansas History Commission, Department of Archives and History, Little Rock.
ArBaA Arkansas College, Batesville.
ArCH Hendrix College, Conway.
ArL Little Rock Public Library.
ArU University of Arkansas, Fayetteville.
AzU University of Arizona, Tucson.
BM British Library, London.
C California State Library, Sacramento.
C-S __________ Sutro Branch, San Francisco.
CBPac Pacific School of Religion, Berkeley.
CHi California Historical Society, San Francisco.
CL Los Angeles Public Library.
CLCM Los Angeles County Museum, Los Angeles.
CLM Los Angeles County Medical Association, Los Angeles.
CLSM Southwest Museum, Los Angeles.
CLSU University of Southern California, Los Angeles.
CLU University of California, Los Angeles.
CMont Monterey Public Library.
CSf San Francisco Public Library.
CSfA California Academy of Sciences, San Francisco.
CSfCMS San Francisco County Medical Society Library, San Francisco.
CSfCW San Francisco College for Women, San Francisco.
CSmH Henry E. Huntington Library, San Marino.
CSt Stanford University Libraries, Palo Alto.
CSt-L __________ Lane Medical Library.
CU University of California, Berkeley.
CU-A University of California, Davis.
CU-B University of California, Berkeley. Bancroft Library.
CU-L __________ Law Library, Berkeley.
CU-S University of California, San Diego.
CV Vallejo Public Library.
CoCC Colorado College, Colorado Springs.
CoD Denver Public Library.
CoFS Colorado State University, Fort Collins.
CoHi Colorado State Historical Society, Denver.
CoU University of Colorado, Boulder.
Ct Connecticut State Library, Hartford.
CtB Bridgeport Public Library.

CtHC Hartford Seminary Foundation, Hartford.
CtHT Trinity College, Hartford.
CtHT-W __________ Watkinson Library.
CtHi Connecticut Historical Society, Hartford.
CtNl New London Public Library.
CtNlCG U.S. Coast Guard Academy, New London.
CtSoP Pequot Library Association, Southport.
CtU University of Connecticut, Storrs.
CtW Wesleyan University, Middletown.
CtY Yale University, New Haven.
CtY-M __________ Medical School
DAIA American Institute of Architects.
DAU American University.
DBRE Bureau of Railway Economics Library of the Association of American Railroads.
DFo Folger Shakespeare Library.
DGU Georgetown University.
DHEW U.S. Department of Health, Education and Welfare Library.
DI U.S. Department of the Interior Library.
DI-GS __________ Geological Survey Library.
DIC U.S. Interstate Commerce Commission.
DLC U.S. Library of Congress.
DN U.S. Department of the Navy.
DNA U.S. National Archives.
DNAL U.S. National Agricultural Library.
DNLM U.S. National Library of Medicine.
DNW U.S. National War College Library, Fort McNair.
DP U.S. Patent Office.
DS U.S. Department of State Library.
DSI Smithsonian Institution.
DWP Public Library of the District of Columbia.
De-Ar Delaware Public Archives Commission.
DeGE Eleutherian Mills Historical Library, Greenville.
DeHi Historical Society of Delaware, Wilmington.
DeU University of Delaware, Newark.
DeWI Wilmington Institute and the New Castle County Free Library.
DeWint Henry Francis DuPont Winterthur Museum, Winterthur.
F Florida State Library, Tallahassee.
FMU University of Miami, Coral Gables.
FOA Albertson Public Library, Orlando.
FSaHi St. Augustine Historical Society.
FTaSU Florida State University, Tallahassee.
FU University of Florida, Gainesville.
FWpR Rollins College, Winter Park.
G Georgia State Library, Atlanta
GA Atlanta Public Library.
GAOC Oglethorpe University, Atlanta.
GAU Atlanta University, Atlanta.
GDC Columbia Theological Seminary, Decatur.
GEU Emory University, Atlanta.
GEU-M __________ A. W. Calhoun Medical Library.
GHi Georgia Historical Society, Savannah.
GMM Mercer University, Macon.
GMW Wesleyan College, Macon.
GMiW Georgia College at Milledgeville.
GU University of Georgia, Athens.
GU-De __________ DeRenne Georgia Library.
GU-M __________ University of Georgia Medical School Library, Augusta.
GWay Public Library, Waycross.
I Illinois State Library, Springfield.
I-Ar __________ Archives Division.
IC Chicago Public Library.

ICA Art Institute of Chicago.
ICHi Chicago Historical Society.
ICJ John Crerar Library, Chicago.
ICLaw Chicago Law Institute.
ICMcC McCormick Theological Seminary, Chicago.
ICN Newberry Library, Chicago.
ICRL Center for Research Libraries, Chicago. (Formerly Midwest Inter-Library Center.)
ICU University of Chicago.
ICU-M __________ Bio-Medical Libraries.
IEG Garrett Theological Seminary, Evanston.
IEN Northwestern University, Evanston.
IEN-M __________ Medical School, Chicago.
IEdS Southern Illinois University, Edwardsville Campus.
IG Galesburg Public Library.
IGK Knox College, Galesburg.
IHi Illinois State Historical Library, Springfield.
INS Illinois State University, Normal.
IObB Bethany Theological Seminary, Oak Brook.
IP Peoria Public Library.
IU University of Illinois, Urbana.
IU-M __________ Medical Sciences Library, Chicago.
Ia-HA Iowa State Department of History and Archives, Des Moines.
IaAS Iowa State University of Science and Technology, Ames.
IaB Burlington Free Public Library.
IaCrM Iowa Masonic Library, Cedar Rapids.
IaDaM Davenport Public Museum Library.
IaDuU University of Dubuque.
IaGG Grinnell College.
IaHi State Historical Society of Iowa, Iowa City.
IaU University of Iowa, Iowa City.
IaU-L __________ College of Law Library.
IdU University of Idaho, Moscow.
In Indiana State Library, Indianapolis.
InCW Wabash College, Crawfordsville.
InGrD De Pauw University, Greencastle.
InHan Hanover College.
InHi Indiana Historical Society, Indianapolis.
InI Indianapolis. Marion County Public Library, Indianapolis.
InLP Purdue University, Lafayette.
InNd University of Notre Dame.
InSNHi Northern Indiana State Historical Society, South Bend.
InTR Rose Polytechnic Institute, Terre Haute.
InU Indiana University, Bloomington.
InU-M __________ School of Medicine Library, Indianapolis.
KBB Baker University, Baldwin City.
KEmT Kansas State Teachers College, Emporia.
KHi Kansas State Historical Society, Topeka.
KMK Kansas State University, Manhattan.
KPT Kansas State College of Pittsburg.
KSalW Kansas Wesleyan University, Salina.
KU University of Kansas, Lawrence.
KWiF Friends University, Wichita.
KyBB Berea College.
KyBgW Western Kentucky State University, Bowling Green.
KyCov Kenton County Public Library, Covington.
KyDC Centre College of Kentucky, Danville.
KyHi Kentucky Historical Society, Frankfurt.
KyLo Louisville Free Public Library.

KyLoF Filson Club, Louisville.
KyLoS Southern Baptist Theological Seminary, Louisville.
KyLx Lexington Public Library.
KyLxT Transylvania College, Lexington.
KyU University of Kentucky, Lexington.
L-M Louisiana State Museum, New Orleans.
LN New Orleans Public Library.
LNA New Orleans City Archives.
LNB New Orleans Baptist Theological Seminary.
LND Dillard University Library, New Orleans.
LNFM Louisiana Masonic Grand Lodge, New Orleans.
LNHT Tulane University, New Orleans.
LNL Loyola University, New Orleans.
LNOP Orleans Parish Medical Society, New Orleans.
LU Louisiana State University, Baton Rouge.
LU-E Louisiana State University, Eunice.
M Massachusetts State Library, Boston.
MA Amherst College.
MAJ Jones Library, Amherst.
MAm Public Library, Amesbury.
MAr Robbins Public Library, Arlington.
MAtt Joseph L. Sweet Memorial Public Library, Attleborough.
MB Boston Public Library.
MBAt Boston Athenaeum.
MBBS Bostonian Society, Boston.
MBC American Congregational Association, Boston.
MBCo Countway Library of Medicine (Harvard-Boston Medical Libraries) Boston.
MBD Episcopal Diocese of Massachusetts, Boston.
MBH Massachusetts Horticultural Society, Boston.
MBL Boston Library Society.
MBMu Museum of Fine Arts, Boston.
MBNEH New England Historic Genealogical Society, Boston.
MBP Massachusetts College of Pharmacy, Boston.
MBS Social Law Library, Boston.
MBSi Simmons College, Boston.
MBU Boston University.
MBdAF U.S. Air Force Cambridge Research Center, Bedford.
MBilHi Billerica Historical Society.
MBr Public Library of Brookline.
MC Cambridge Public Library.
MCM Massachusetts Institute of Technology, Cambridge.
MChB Boston College, Chestnut Hill.
MDedHi Dedham Historical Society.
MDeeP Pocumtuck Valley Memorial Association, Deerfield.
MF Fall River Public Library.
MFiHi Fitchburg Historical Society Library.
MH Harvard University, Cambridge.
MH-A ________ Arnold Arboretum.
MH-AH ________ Andover-Harvard Theological Library.
MH-BA ________ Graduate School of Business Administration.
MH-L ________ Law School.
MH-Z ________ Museum of Comparative Zoology.
MHa Haverhill Public Library.
MHi Massachusetts Historical Society, Boston.
MLow Lowell City Library.
MLy Lynn Public Library.
MMal Malden Public Library.
MMeT Tufts University, Medford.
MMhHi Marblehead Historical Society.
MNBedf New Bedford Free Public Library.
MNF Forbes Library, Northampton.
MNS Smith College, Northampton.
MNe Newburyport Public Library.
MNoan-MV Merrimac Valley Textile Museum, North Andover.

MNt Newton Free Library.
MNtcA Andover Newton Theological School, Newton Center.
MPB Berkshire Athenaeum, Pittsfield.
MPeaHi Peabody Historical Society.
MPeaI Peabody Institute.
MPlA Antiquarian House, Plymouth.
MPlPS Pilgrim Society Library, Plymouth.
MSCV Connecticut Valley Historical Museum, Springfield.
MSa Salem Public Library.
MSaE Essex Institute, Salem.
MSaP Peabody Museum of Salem.
MSbra Thayer Public Library, South Braintree.
MSo Public Library of the City of Somerville.
MStoc Stockbridge Library.
MTaHi Old Colony Historical Society, Taunton.
MTem-NHi Narragansett Historical Society, Templeton.
MU University of Massachusetts, Amherst.
MW Worcester Public Library and Central Massachusetts Regional Library.
MWA American Antiquarian Society, Worcester.
MWCL Worcester County Law Library Association.
MWH College of the Holy Cross, Worcester.
MWHi Worcester Historical Society.
MWM Worcester Art Museum.
MWal Waltham Public Library.
MWeA Westfield Athenaeum.
MWelC Wellesley College.
MWes-tonR Regis College, Weston.
MWey Tufts Library, Weymouth.
MWeyAA Abigail Adams Historical Society, Weymouth.
MWfo J. V. Fletcher Library, Westford.
MWiW Williams College, Williamstown.
MWiW-C Williams College, Chapin Library, Williamstown.
MWo Woburn Public Library.
Md Maryland State Library, Annapolis.
Md-LR Maryland Department of Legislative Reference, Baltimore.
MdAN U.S. Naval Academy, Annapolis.
MdAS Saint John's College, Annapolis.
MdBE Enoch Pratt Free Library, Baltimore.
MdBJ Johns Hopkins University, Baltimore.
MdBJ-G ———— John Work Garrett Library.
MdBJ-W ———— William H. Welch Medical Library.
MdBL Loyola College, Baltimore.
MdBM Medical and Chirurgical Faculty of the State of Maryland, Baltimore.
MdBMA Baltimore Museum of Art.
MdBP Peabody Institute, Baltimore.
MdBS St. Mary's Seminary and University, Roland Park, Baltimore.
MdBT Towson State College, Baltimore.
MdHi Maryland Historical Society, Baltimore.
MdToH Sheppard and Enoch Pratt Hospital Library, Towson.
MdU University of Maryland, College Park.
MdU-H ———— Health Sciences Library, Baltimore.
MdW Woodstock College.
Me Maine State Library, Augusta.
Me-LR ———— Legislative Reference Bureau, Augusta.
MeB Bowdoin College, Brunswick.
MeBa Bangor Public Library.
MeBaT Bangor Theological Seminary.
MeHi Maine Historical Society, Portland.

MeL Lewiston Public Library.
MeLB Bates College, Lewiston.
MeP Portland Public Library.
MeU University of Maine, Orono.
MeWC Colby College, Waterville.
Mi Michigan State Library, Lansing.
MiD Detroit Public Library.
MiD-B __________ Burton Historical Collection.
MiDSH Sacred Heart Seminary, Detroit.
MiDU University of Detroit.
MiDW-M Wayne State University Medical Library, Detroit.
MiEM Michigan State University, East Lansing.
MiGr Grand Rapids Public Library.
MiHM Michigan College of Mining and Technology, Houghton.
MiMu Hackley Public Library, Muskegon.
MiPh Port Huron Public Library.
MiU University of Michigan, Ann Arbor.
MiU-C __________ William L. Clements Library.
MiU-L __________ Law Library.
MiU-T __________ Transportation Library.
MnHi Minnesota Historical Society, St. Paul.
MnM Minneapolis Public Library.
MnS St. Paul Public Library.
MnSH Hamline University, St. Paul.
MnSJ James Jerome Hill Reference Library, St. Paul.
MnSM Macalester College, St. Paul.
MnSS St. Paul Seminary, St. Paul.
MnU University of Minnesota, Minneapolis.
MnU-A __________ Saint Paul.
MnU-B __________ Bio-Medical Library.
MoHi Missouri State Historical Society, Columbia.
MoK Kansas City Public Library.
MoKU University of Missouri at Kansas City.
MoMM Missouri Valley College, Marshall.
MoRM University of Missouri at Rolla.
MoS St. Louis Public Library.
MoSB Missouri Botanical Garden, St. Louis.
MoSHi Missouri Historical Society, St. Louis.
MoSM Mercantile Library Association, St. Louis.
MoSMed St. Louis Medical Society.
MoSU St. Louis University.
MoSU-M __________ School of Medicine.
MoSW Washington University, St. Louis.
MoSW-M __________ Medical School Library.
MoSpD Drury College, Springfield.
MoU University of Missouri, Columbia.
Ms-Ar State Department of Archives and History, Jackson.
MsU University of Mississippi, University.
MtMis Missoula Public & Missoula County Free Library.
N New York State Library, Albany.
N-L __________ Law Library.
NAl Albany Public Library.
NAlI Albany Institute of History and Art.
NB Brooklyn Public Library.
NBC Brooklyn College.
NBLiHi Long Island Historical Society, Brooklyn.
NBM Medical Research Library of Brooklyn.
NBu Buffalo and Erie County Public Library.
NBuG __________ Grosvenor Reference Division.

NBuHi Buffalo Historical Society.

NBuU-H State University of New York at Buffalo Medical School Library.

NCH Hamilton College, Clinton.

NCaS St. Lawrence University, Canton.

NCanHi Ontario County Historical Society, Canandaigua.

NCooHi New York State Historical Association, Cooperstown.

NCorniC Corning Glass Works Library.

NGH Hobart & William Smith Colleges, Geneva.

NGeno Wadsworth Library, Geneseo.

NGlf Crandall Library, Glens Falls.

NGos Goshen Library and Historical Society.

NHerk-CHi Herkimer County Historical Society.

NHi New York Historical Society, New York.

NHu Huntington Public Library.

NIC Cornell University, Ithaca.

NICLA Cornwell Library Association, Ithaca.

NJQ Queens Borough Public Library, Jamaica.

NKi City Library, Kingston.

NN New York Public Library.

NNA American Geographical Society, New York.

NNB Association of the Bar of the City of New York.

NNBG New York Botanical Garden, Bronx Park.

NNC Columbia University, New York.

NNC-A ——— Avery Library of Architecture.

NNC-L ——— Law Library.

NNC-M ——— Medical Library.

NNC-P ——— College of Pharmacy Library.

NNC-T ——— Teachers College Library.

NNC-Typ ——— American Typefounders' Library.

NNCC Chemists' Club, New York.

NNCoCi City College of the City University, New York.

NNE Engineering Societies Library, New York.

NNF Fordham University, New York.

NNFM Grand Lodge of New York, F. & A. M. Library & Museum, New York.

NNG General Theological Seminary of the Protestant Episcopal Church, New York.

NNGr Grolier Club, New York.

NNLI New York Law Institute Library, New York.

NNM American Museum of Natural History, New York.

NNMM Metropolitan Museum of Art, New York.

NNMer Mercantile Library Association, New York.

NNMus Museum of the City of New York.

NNNAM New York Academy of Medicine.

NNPM Pierpont Morgan Library, New York.

NNR Russell Sage Foundation Library, City College of New York.

NNRT Racquet and Tennis Club Library, New York.

NNS New York Society Library, New York.

NNUT Union Theological Seminary, New York.

NOg Ogdensburg Public Library.

NP Adriance Memorial Library, Poughkeepsie.

NPSA Novitiate of Saint Andrew-on-Hudson, Poughkeepsie.

NPV Vassar College, Poughkeepsie.

NPtw Port Washington Public Library.

NR Rochester Public Library.
NRAB Samuel Colgate Baptist Historical Library of the American Baptist Historical Society, Rochester.
NRHi Rochester Historical Society.
NRSB Saint Bernard's Seminary and College, Rochester.
NRU University of Rochester.
NRU-M __________ School of Medicine and Dentistry.
NRom Jervis Library Association, Rome.
NRvS Suffolk County Historical Society, Riverhead.
NSchU Union College, Schenectady.
NSm Smithtown Public Library.
NSy Syracuse Public Library.
NSyOHi Onondaga Historical Association, Syracuse.
NSyU Syracuse University.
NT Troy Public Library.
NTR Rensselaer Polytechnic Institute, Troy.
NUt Utica Public Library.
NUtHi Oneida Historical Society, Utica.
NWM U.S. Military Academy, West Point.
NYBT Boyce Thompson Institute for Plant Research, Yonkers.
NbCrD Doane College, Crete.
NbHi Nebraska State Historical Society, Lincoln.
NbO Omaha Public Library.
NbOU University of Nebraska at Omaha.
NbU University of Nebraska, Lincoln.
NbU-M __________ College of Medicine, Omaha.
Nc North Carolina State Library, Raleigh.
NcA Pack Memorial Public Library, Asheville.
NcA-S __________ Sondley Reference Library.
NcC Public Library of Charlotte & Mecklenburg County.
NcD Duke University, Durham.
NcD-L __________ School of Law.
NcD-MC __________ Medical Center.
NcGU University of North Carolina at Greensboro.
NcGuG Guilford College.
NcHil Confederate Memorial Library, Hillsboro.
NcRS North Carolina State University, Raleigh.
NcU University of North Carolina, Chapel Hill.
NcW Wilmington Public Library.
NcWsS Salem College, Winston-Salem.
NcWsW Wake Forest College, Winston-Salem.
NdHi State Historical Society of North Dakota, Bismarck.
NdU University of North Dakota Library, Grand Forks.
Nh New Hampshire State Library, Concord.
NhD Dartmouth College, Hanover.
NhHi New Hampshire Historical Society, Concord.
NhM Manchester City Library.
NhPet Town Library, Peterborough.
NhPoA Portsmouth Atheneum.
Nj New Jersey State Library, Archives and History.
NjHi New Jersey Historical Society, Newark.
NjHoS Stevens Institute of Technology, Hoboken.
NjMo Morris County Free Library, Morristown.
NjN Newark Public Library.
NjP Princeton University.
NjPT Princeton Theological Seminary.
NjR Rutgers–The State University, New Brunswick.
O Ohio State Library, Columbus.
O-Ar Ohio State Archives, Columbus.

OAkU University of Akron.
OBg Bowling Green Public Library.
OC Public Library of Cincinnati and Hamilton County.
OCHP Cincinnati Historical Society.
OCLaw Cincinnati Law Library.
OCLloyd Lloyd Library and Museum, Cincinnati.
OCU University of Cincinnati.
OCU-M __________ School of Medicine.
OCX Xavier University, Cincinnati.
OCY Young Men's Mercantile Library Association, Cincinnati.
OCl Cleveland Public Library.
OClW Case Western Reserve University, Cleveland.
OClW-H Cleveland Health Sciences Library.
OClWHi Western Reserve Historical Society, Cleveland.
ODW Ohio Wesleyan University, Delaware.
ODa Dayton and Montgomery County Public Library.
OEac East Cleveland Public Library.
OFH Rutherford B. Hayes Library, Fremont.
OGK Kenyon College, Gambier.
OHi Ohio State Historical Society, Columbus.
OMC Marietta College.
OO Oberlin College.
OOxM Miami University, Oxford.
OSW Wittenberg University, Springfield.
OU Ohio State University, Columbus.
OUrC Urbana College.
OkS Oklahoma State University, Stillwater.
OkU University of Oklahoma, Norman.
OrCS Oregon State College, Corvallis.
OrHi Oregon Historical Society, Portland.
Or-A State Archives Library, Portland.
OrP Library Association of Portland.
OrU University of Oregon, Eugene.
OrU-M __________ Medical School, Portland.
P Pennsylvania State Library, Harrisburg.
PAtM Muhlenberg College, Allentown.
PBL Lehigh University, Bethlehem.
PCC Crozer Theological Seminary, Chester.
PDoBHi Bucks County Historical Society, Doylestown.
PHC Haverford College.
PHi Historical Society of Pennsylvania, Philadelphia.
PJA Abington Library Society, Jenkintown.
PLF Franklin and Marshall College, Lancaster.
PMA Allegheny College, Meadville.
PP Free Library of Philadelphia.
PPA Athenaeum of Philadelphia.
PPAN Academy of Natural Sciences, Philadelphia.
PPAmP American Philosophical Society, Philadelphia.
PPAmS American Sunday School Union, Philadelphia. (Inactive Library. Books not available for consultation.)
PPB Philadelphia Bar Association.
PPC College of Physicians of Philadelphia.
PPCC Carpenters' Company, Philadelphia. (Inactive Library. Books not available for consultation.)
PPF Franklin Institute, Philadelphia.
PPFHi Historical Society of Frankford, Philadelphia. (Inactive Library. Books available for consultation.)

PPFM Free and Accepted Masons of Pennsylvania, Grand Lodge Library, Philadelphia.
PPG German Society of Pennsylvania, Philadelphia.
PPHa Hahnemann Medical College and Hospital, Philadelphia.
PPHor Pennsylvania Horticultural Society, Philadelphia.
PPL Library Company of Philadelphia.
PPLT Lutheran Theological Seminary, Krauth Memorial Library, Philadelphia.
PPLas La Salle College, Philadelphia.
PPPH Pennsylvania Hospital Medical Library, Philadelphia.
PPPM Philadelphia Museum of Art.
PPPrHi Presbyterian Historical Society, Philadelphia.
PPRF Rosenbach Foundation, Philadelphia.
PPULC Union Library Catalogue of Pennsylvania, Philadelphia.
PPWI Wistar Institute of Anatomy and Biology, Philadelphia.
PPWa Wagner Free Institute of Science, Philadelphia. (Inactive Library. Books available for consultation.)
PPeSchw Schwenckfelder Historical Library, Pennsburg. (Inactive Library. Books available for consultation.)
PPi Carnegie Library of Pittsburgh.
PPiAM Pittsburgh Academy of Medicine.
PPiHi Historical Society of Western Pennsylvania, Pittsburgh.
PPiPT Pittsburgh Theological Seminary.
PPiU University of Pittsburgh.
PR Reading Public Library.
PSC Swarthmore College.
PSC-Hi ———— Friends Historical Library.
PSt Pennsylvania State University, University Park.
PU University of Pennsylvania, Philadelphia.
PU-D ———— Evans Dental Library.
PU-FA ———— School of Fine Arts.
PU-L ———— Biddle Law Library.
PU-Math ———— Mathematics-Physics Library.
PU-S ———— Edgar Fahs Smith Memorial Library.
PU-V ———— School of Veterinary Medicine Library.
PU-W ———— Wharton School of Finance and Commerce Library.
PWW Washington & Jefferson College, Washington.
PWb Osterhout Free Library, Wilkes-Barre.
PWcHi Chester County Historical Society. (Inactive Library. Books available for consultation.)
PWcS West Chester State College.
PY Martin Memorial Library, York City and County Library, York.
R Rhode Island State Library, Providence.
RHi Rhode Island Historical Society, Providence.
RJa Philomenian Library, Jamestown.
RNHi Newport Historical Society.
RNR Redwood Library and Athenaeum, Newport.
RP Providence Public Library.
RPA Providence Athenaeum.
RPB Brown University, Providence.
RPE Elmwood Public Library, Providence.
RPJCB John Carter Brown Library, Providence.
RPM Rhode Island Medical Society, Providence.
RPaw Deborah Cook Sayles Public Library, Pawtucket.

ScAr South Carolina State Archives, Columbia.

ScC Charleston Library Society.

ScCC College of Charleston.

ScCM Medical College of the State of South Carolina, Charleston.

ScClP Presbyterian College, Clinton.

ScCleU Clemson University.

ScHi South Carolina Historical Society, Charleston.

ScU University of South Carolina, Columbia.

T Tennessee State Library, Nashville.

TCU University of Tennessee, Chattanooga.

THi Tennessee Historical Society, Nashville.

TKL Public Library of Knoxville and Knox County.

TMC Cossitt Reference Library, Memphis.

TN Public Library of Nashville & Davidson County.

TNJ Joint University Libraries, Nashville.

TSewU University of the South, Sewanee.

TU University of Tennessee, Knoxville.

Tx Texas State Library and Historical Commission, Austin.

TxAuC Church Historical Society, Austin.

TxCM Texas A & M University, College Station.

TxDW Texas Woman's University, Denton.

TxDaM Southern Methodist University, Dallas.

TxH Houston Public Library.

TxHuT Sam Houston State College, Huntsville.

TxLT Texas Technological College, Lubbock.

TxSaSM Saint Mary's University, San Antonio.

TxU University of Texas, Austin.

TxU-M ________ Medical School, Galveston.

UPB Brigham Young University, Provo.

UU University of Utah, Salt Lake City.

Vi Virginia State Library, Richmond.

ViAl Alexandria Public Library.

ViFbE U.S. Army Corps of Engineers, Fort Belvoir.

ViHi Virginia Historical Society, Richmond.

ViL Jones Memorial Library, Lynchburg.

ViLxW Washington & Lee University, Lexington.

ViPet Petersburg Public Library.

ViRA Richmond Academy of Medicine.

ViRM Medical College of Virginia, Richmond.

ViRUT Union Theological Seminary Library, Richmond.

ViRVal Valentine Museum, Richmond.

ViSwC Sweet Briar College.

ViU University of Virginia, Charlottesville.

ViW College of William and Mary, Williamsburg.

ViWC Colonial Williamsburg.

ViWn Handley Library, Winchester.

Vt Vermont State Library, Montpelier.

VtBrt Brattleboro Free Library.

VtHi Vermont Historical Society, Montpelier.

VtMiM Middlebury College.

VtMiS Sheldon Art Museum, Middlebury.

VtU University of Vermont and State Agricultural College, Burlington.

VtWood Norma Williams Public Library, Woodstock.

WBB Beloit College.

WBara-Hi Sauk County Historical Society, Baraboo.

WGr Brown County Library, Green Bay.

WHi State Historical Society of Wisconsin, Madison.

WKenHi Kenosha County Historical Society.

WM Milwaukee Public Library.

WNaE Episcopal Theological Seminary, Nashotah.

WU University of Wisconsin, Madison.

WU-A ________ Agricultural Library.

WWauHi Waukesha County Historical Society, Waukesha.

WaPS Washington State University, Pullman.

WaU University of Washington, Seattle.

WvU West Virginia University, Morgantown.

WyU University of Wyoming, Laramie.

CHECKLIST

I. GENERAL WORKS

1. Periodicals

1. The American magazine and historical chronicle. v. 1–3; Sept. 1743–Dec. 1746. Boston: 1743–1746.

Contains articles on fireplaces, the Yale orrery, and new inventions made in Europe.

ULS-p. 273.

Copies: CSmH; CtY; DLC; ICN; MBAt; MH; MWA; MdBP; MiD-B; N; NN; PPL; RPB.

2. The American magazine and monthly chronicle for the British colonies. v. 1; Oct. 1757–Oct. 1758. Philadelphia: Printed and sold by William Bradford, 1757–1758. [2], 654, 6 p.

Contains articles on agriculture and new inventions.

ULS-p. 273.

Copies: CSmH; CtY; DLC; ICN; MWA; MiU-C; NN; PPL; RPJCB; TxLT.

3. The New American magazine. no. 1–23; Jan. 1758–March 1760. Woodbridge, N.J.: Printed and sold by James Parker, [etc.], 1758–1760.

Edited by Sylvanus, i.e. Samuel Nevill.

Superseded the *American magazine and monthly chronicle*, Philadelphia, 1757–1758.

Includes articles on farming, mines, and construction of roads.

ULS-p. 2981.

Copies: CtHi; DLC; MB; MHi; MWA; MiU-C; NN; NjP; PPL; WHi.

4. American magazine; or, General repository. v. 1, no. 1–9; Jan.–Sept. 1769. Philadelphia: William and Thomas Bradford, 1769. 328 p., plates.

Includes articles on technological topics: culture and dressing of flax, silk culture, and measurement of heat.

ULS-p. 273.

Copies: MB; MWA; MiU-C; MnHi; NN; PPL.

5. American Philosophical Society, Philadelphia.

Transactions of the American Philosophical Society, held at Philadelphia, for promoting useful knowledge. v. 1– ; 1769– . Philadelphia: 1771– . illus.

ULS-p. 301.

Copies: Major libraries.

6. The Royal American magazine; or, Universal repository of instruction and amusement. v. 1–2, no. 3; Jan. 1774–March 1775. Boston: 1774–1775. illus.

Contains articles on the manufacture of indigo, on the properties and use of saltpeter, and on distilling salt water.

ULS-p. 3715.

Copies: CSmH; CtY; DLC; MB; MBAt; MWA; MiD-B; MiU-C; N; NN; PPL; WHi.

7. The Pennsylvania magazine; or, American monthly museum. v. 1–2; Jan. 1775–July, 1776. Philadelphia: Printed and sold by R. Aitken, 1775–1776. illus., maps, plates.

ULS-p. 3300.

Copies: CSmH; CtY; DLC; ICN; MH; MWA; MiU-C; MnU; NN; NjP; PPAmP; PPL; PPiU; PU; RPB; WHi.

8. American Academy of Arts and Sciences, Boston.

Memoirs of the American Academy of Arts and Sciences. v. 1–4; 1780–1821; new

8 (*continued*)
ser., v. 1– ; 1826– . Boston, [etc.]: 1785– . illus.

Volume 1 contains memoirs dating from the foundation of the academy in 1780 to the end of the year 1783.

ULS-p. 161.

Copies: Major libraries.

9. The Boston magazine, containing a collection of instructive and entertaining essays, in the various branches of useful, and polite literature. v. 1–3 (no. 1–35); Nov. 1783–Nov./Dec. 1786. Boston: 1783–1786. illus.

ULS-p. 751.

Copies: CSmH; CtY; DLC; MB; MBAt; MWA; MiU-C; MnU; NN; NNMM; NhD; PHi; WHi.

10. The Columbian magazine; or, Monthly miscellany. v. 1–9; Sept. 1786–Dec. 1792. Philadelphia: 1786–1792. illus., maps, plates.

Title changed to: *The Universal asylum and Columbian magazine,* March 1790.

Superseded by *Columbian museum; or, Universal* asylum.

ULS-p. 4327.

Copies: Major libraries.

11. New Haven gazette and the Connecticut magazine. v. 1–4; Feb. 16, 1786–June 18, 1789. New Haven: 1786–1789.

ULS-p. 2997.

Copies: CtY; DLC; ICN; MB; MWA; MiU-C; NN; NhD; WHi.

12. The Worcester magazine.... Containing politicks, miscellanies, poetry and news. v. 1–4, no. 26; Apr. 1786–March 1788. Worcester, Mass.: 1786–1788.

Contains articles on bridges, mills, ironmaking, and cheesemaking.

ULS-p. 4539.

Copies: CtY; DLC; IC; MB; MBAt; MWA; MnU; NHi; NNS; RPB; WHi.

13. The American magazine, containing a miscellaneous collection of original and other valuable essays.... no. [1]–12; Dec. 1787–Nov. 1788. New York: Printed by S. Loudon, 1787–1788. 882 p., plates, plan.

Editor: Noah Webster.

Contains articles on agriculture, machines, and inventions.

ULS-p. 273.

Copies: CSmH; CtY; DLC; MB; MH; MWA; MiU-C; NN; NcU; OCl; PPL; WHi.

14. The American museum, or, Repository of ancient and modern fugitive pieces. v. 1–12; Jan. 1787–Dec. 1792. Philadelphia: Printed by Mathew Carey, [etc.], 1787–1792. illus., maps.

Title changed in 1790 to *The American museum; or, Universal magazine; containing essays on agriculture—commerce—manufactures—politics.*

Editor: Mathew Carey.

ULS-p. 288.

Copies: Major libraries.

15. The Christian's, scholar's and farmer's magazine; calculated in an eminent degree to promote religion; to disseminate useful knowledge ... and to advance the interests of agriculture. v. 1–2; April/May 1789–Feb./March 1791. Elizabethtown, N.J.: Printed by Shepard Kollock, 1789–1791.

ULS-p. 1041.

Copies: CSmH; CtY; Ia; MBAt; MH; MWA; MdBE; NN; NNS; NhD; NjP; NjR; PPAP; PPPrHi WHi.

16. Gentlemen and ladies' town and country magazine: consisting of literature, history, politics, arts, manners, and amusements, with various other matter. v. 1–2, no. 7; Feb. 1789–Aug. 1790. Boston: Nathaniel Coverly, 1789–1790. illus., plates.

ULS-p. 1691.

Copies: CSmH; DLC; MB; MWA; MiD-B; N; NN; NNC; NcD; RPJCB; TxLT; WHi.

17. The Massachusetts magazine; or, Monthly museum of knowledge and rational entertainment. v. 1–8; Jan. 1789–Dec. 1796.

Boston: Isaiah Thomas and Co., [etc.], 1789–1796. illus.

Contains articles on agriculture, sciences, arts, manufactures, and commerce.

ULS-p. 2555.

Copies: CtHW; DLC; MBAt; MH; MHi; MWA; MiU-C; NN; NjR; RPJCB; WHi.

18. The New-York magazine; or, Literary repository. v. 1–6; 1790–1795; new ser., v. 1–2; 1797. New York: 1790–1797. illus.

Contains articles on manufacturing processes, new inventions, machines, and canals.

ULS-p. 3036.

Copies: CSmH; CtY; DLC; ICN; MBAt; MH; MWA; NHi; NN; PHi; PPL; WHi.

19. The American Apollo. v. 1–3; Jan. 6, 1792–Dec. 25, 1794. Boston: 1792–1794.

Contains articles on bridges, glass works, and printing presses.

ULS-p. 169.

Copies: CSmH; CtY; DLC; MB; MWA; MiU-C; N; NN; OClWHi.

20. Society for the Promotion of Useful Arts, Albany.

Transactions of the Society for the Promotion of Useful Arts, in the state of New York. v. 1–4; 1792–1819. New York; Albany: 1792–1819. plates.

Vol. 1, entitled: *Transactions of the Society, instituted in the state of New York, for the promotion of agriculture, arts, and manufactures,* issued in 4 parts; parts 1–2 published in New York, 1792–94; parts 3–4 in Albany, 1798–99. It was reissued as 2d edition, revised, Albany, 1801. Other volumes were published in 1807, 1814, and 1816–19.

ULS-p. 3980.

Copies: Major libraries.

21. The Columbian museum; or, Universal asylum: containing essays on agriculture—commerce—manufactures—politics—morals—and manners . . . For the year 1793. Part I. From January to June. Philadelphia: From the press of John Parker, 1793. 60, 8, [4] p.

Continuation of the *Universal asylum and Columbian magazine.*

E-25324; ULS-p. 1124.

Copies: CSmH; CtHW; MWA; NN; PP; PPL; PU; TxLT.

22. The New-Hampshire magazine; or, The monthly repository of useful information. v. 1, no. 1–6; June–Nov. 1793. Concord: Printed by Elijah Russell, 1793.

ULS-p. 2996.

Copies: DLC; MWA; MnU; NHi; NhD; WHi.

23. American monthly review; or, Literary journal. v. 1–3; Jan.–Dec. 1795. Philadelphia: printed by S. H. Smith, 1795–1796.

Contains articles on dyeing and colors, on flour milling, and rigging and seamanship.

ULS-p. 287.

Copies: CtY; DLC; MH; MHi; MWA; MiU; NN; PHi; PPL; RPB.

24. The Monthly military repository. . . . v. 1–2. New York: printed by William A. Davis, 1796–1797. plates, maps.

Editor: Charles Smith.

ULS-p. 2748.

Copies: DLC; MWA; NHi; NN; RPJCB.

25. The American universal magazine. v. 1–4; Jan. 2, 1797–March 7, 1798. Philadelphia: 1797–1798. illus.

ULS-p. 340.

Copies: CoU; CtY; DLC; DeGE; ICN; MB; MWA; MiU-C; MoU; NN; NcU; NjP; OO; OrU; PPAmP; PPL; RHi; TxLT; WyU.

26. The Medical repository. v. 1–23; July 26, 1797–1824. New York: 1797–1824. illus.

Contains articles on dental and medical instruments, on the production of potash, paper, and cement, on the warming of rooms, and on the Philadelphia water supply.

ULS-p. 2583.

Copies: Major libraries.

27. Philadelphia monthly magazine; or, Universal repository of knowledge and entertainment. v. 1–2, no. 3 (no. 1–9); Jan.–Sept. 1798. Philadelphia: 1798. illus.

Contains articles on tanning, hat making, nautical inventions, salt mines, and the dressing of flax.

ULS-p. 3327.

Copies: ICN; ICU; MH; MWA; MiU; MiU-C; N; NNC; NjR; PHi; PPL; WHi.

28. The Rural magazine. v. 1; Feb. 17, 1798–Feb. 9, 1799. Newark, N.J.: 1798–1799.

Contains articles on bridges, mines, the making of bee-hives, soap, and bread.

ULS-p. 3737.

Copies: DLC; MWA; N.

29. The Weekly magazine of original essays, fugitive pieces, and interesting intelligence. v. 1–4; Feb. 3, 1798–June 1, 1799. Philadelphia: J. Watters, 1798–1799.

Publication suspended from Sept. 1798 to Jan. 1799 inclusive.

ULS-p. 4463.

Copies: CtY; DLC; DeGE; Ia; MH; MWA; NHi; NN; OClWHi; PPL.

30. The American museum; or, Annual register of fugitive pieces, ancient and modern. For the year 1798.... Philadelphia: printed for Mathew Carey by W. & R. Dickson, 1799. 416, [5] p.

Editor: Mathew Carey.

Continuation of *The American museum; or, Universal magazine.*

ULS-p. 289.

Copies: CtY; DLC; GEU; ICN; Ia; MB; MH; MWA; MdBP; MeB; MiU; MnU; MoU; NN; NNC; NcD; NhD; NjP; OCl; PPL; PPiU; PU; RPB; ViU.

31. The Monthly magazine, and American review. v. 1–3; April 1799–Dec. 1800. New York: Printed by T. & J. Swords, 1799–1800.

Contains articles on agriculture, patents and inventions.

ULS-p. 2747.

Copies: CSt; CU; CtY; DLC; GU; ICN; MB; MBAt; MH; MWA; MdBE; MnU; NN; NjP; OMC; PPL; PU; TxU; WHi.

32. American review, and literary journal. v. 1–2; Jan./March 1801–Oct./Dec. 1802. New York: Printed by T. & J. Swords, 1801–1802.

Includes descriptions of machines and inventions.

ULS-p. 313.

Copies: Major libraries.

33. National magazine; or, Cabinet of the United States. no. 1–8; Oct. 1801–Jan. 11, 1802. Washington, D.C.: 1801–1802.

Superseded the *National magazine,* 1799–1800 (Richmond).

ULS-p. 2892.

Copies: DLC; MH; MWA; N; ViW; WHi.

34. The Port folio. By Oliver Oldschool, Esq. [i.e. Joseph Dennie]. v. 1–5; 1801–1805; new ser., v. 1–6; 1806–1808; new [3d] ser., v. 1–8; 1809–1812; 3d [i.e. 4th] ser., v. 1–6; 1813–1815; 4th [i.e. 5th] ser., v. 1–20; 1816–1825; Hall's 2d ser., v. 1–2; July 1826–Dec. 1827. Philadelphia: printed by H. Maxwell, etc., 1801–1827. 47 v., maps, plans, plates.

Founded by Joseph Dennie; beginning with 1816 published by John Elihu Hall.

ULS-p. 3404.

Copies: Major libraries.

35. Society for the Promotion of Useful Arts, Albany.

Transactions of the Society, for the Promotion of Agriculture, Arts, and Manufactures, instituted in the state of New-York. Volume I. The second ed., rev. Albany: Printed by Charles R. and George Webster, 1801. viii, xiv, [15]–418 p.

S&S-1338.

Copies: DeGE; MH; MWA; NN.

36. The Balance, and Columbian reporter. v. 1–7; Jan. 1802–Dec. 1808; [new ser.], v. 1–2; Jan. 1809–Dec. 1810; [3d ser.], v. 1; Jan.–Dec. 1811. Albany: Croswell & Frary, [etc.], 1802–1811.

Title varies: 1808, *The Balance;* 1809–1810, *The Balance, and New-York state journal;* 1811, *The Balance, and state journal.*

ULS-p. 587.

Copies: CSt; CtY; DLC; ICN; MBAt; MH; MWA; MiU-C; N; NIC; NjR; OClWHi; OrU; PPL; RPB; TxU.

37. The Boston weekly magazine.... v. 1–3; Oct. 30, 1802–Oct. 19, 1805; [new ser.], v. 1 (no. 1–27); Oct. 26, 1805–Apr. 26, 1806. Boston: 1802–1806.

Issued under the title *The Boston magazine* from Oct. 26, 1805 to Apr. 26, 1806.

Contains articles on agriculture, useful arts, and inventions.

ULS-p. 751.

Copies: CSt; DLC; FU; ICN; IaU; MB; MWA; MiU-C; NN; NjP; PHi; PP.

38. The Lancaster hive; devoted to morality, literature, biography, history, poetry, agriculture, &c.&c. v. 1–2; June 22, 1803–June 12, 1805. Lancaster, [Pa.]: Printed by McDowell & Greear, [etc.], 1803–1805.

Caption title and running title: *The Hive.*

ULS-p. 2346.

Copies: DLC; MWA; PHi; WHi.

39. The Literary magazine and American register. v. 1–8; Oct. 1803–Dec. 1807. Philadelphia: John Conrad & Co., 1803–1807.

Includes articles on technological topics.

ULS-p. 2438.

Copies: C; CtY; DLC; ICU; MB; MH; MWA; MdBE; NN; NjR; OC; PPL; PPiU; PU; RPB; TxU; Vi; WHi.

40. The Monthly anthology, and Boston review. Containing sketches and reports of philosophy, religion, history, arts and manners. v. 1–10; Nov. 1803–June 1811. Boston: Munroe & Francis, [etc.], 1803–1811.

ULS-p. 2744.

Copies: Major libraries.

41. Philadelphia medical and physical journal.... v. 1–3; Nov. 1804–1808. Philadelphia: J. Conrad & Co., [etc.], 1804–1808. plates.

Also: 1st–3d supplement, March–May, 1809.

ULS-p. 3327.

Copies: Major libraries.

42. Philadelphia medical museum. Conducted by John Redman Coxe. v. 1–5; Sept. 17, 1804–1808; new ser., v. 1; 1810–1811. Philadelphia: T. Dobson, [etc.], 1804–1811.

ULS-p. 3327.

Copies: Major libraries.

43. Polyanthos. v. 1–5; Dec. 1805–July 1807. New ser., v. 1–2; Feb.–Sept. 1812. 3d ser., v. 1–4; Oct. 1812–Sept. 1814. Boston: J. T. Buckingham, 1805–1814. 11 v., illus.

ULS-p. 3392.

Copies: CL; CSmH; CtY; DLC; ICN; Ia; MB; MBAt; MWA; MiU-C; N; NhD; NjR; PPL; RPB; WHi.

44. The American register; or, General repository of history, politics and science. v. 1–7; 1806–1810. Philadelphia: C. & A. Conrad, 1806–1811.

Contains notices on roads, canals, bridges, and steam navigation.

ULS-p. 312.

Copies: Major libraries.

45. The Medical and agricultural register. v. 1, no. 1–24; Jan. 1806–Dec. 1807. Boston: Manning & Loring, [1806–1807]. 378 p.

ULS-p. 2574.

Copies: CSt; CU; CtY-M; DLC; DNLM; ICJ; InU-M; MB; MBCo; MWA; MiU; NIC; NN; NSU; NhD; NjR; OC; PPL; RPB; VtU; WU.

46. Society of Dutchess County for the Promotion of Agriculture.

Transactions of the Society of Dutchess County for the Promotion of Agriculture, with select essays on rural economy.... v. 1–2. Poughkeepsie, [N.Y.]: Printed by Bowman, Parsons & Potter, [etc.], 1807–1809.

Vol. 2 printed by Paraclete Potter.

ULS-p. 3985.

Copies: DLC; DNAL; MWA; N; NN; NP.

47. The Gleaner; or, Monthly magazine. v. 1–2, no. 3; Sept. 1808–Nov. 1809. Lancaster, Pa.: Printed by W. Greer, 1808–1809.

Sept.–Nov. 1809 issued under the title: *Monthly magazine.*

Contains articles on roads and other improvements.

ULS-p. 2747.

Copies: CSmH; DLC; DeGE (microfilm); ICU; MWA; MnU; NjR; RPB; WHi.

48. Philadelphia Society for Promoting Agriculture.

Memoirs of the Philadelphia Society for Promoting Agriculture. Containing communications on various subjects in husbandry & rural affairs. v. 1–5; 1808–1826. Philadelphia: 1808–1826. illus., plates.

ULS-p. 3328.

Copies: CSmH; DLC; DNAL; DeGE; IU; IaAS; MB; MH; N; NNC; NjR; PHC; PLF; PPF; PPL; PPiU; WU.

49. The Useful cabinet, published in monthly numbers, for the Newengland Association of Inventors and Patrons of Useful Arts. no. 1–6; Jan.–June 1808. Boston: Printed for the Association by E. C. Beals, 1808. 144p., 6 plates.

ULS-p. 4344; Kress-B5409.

Copies: DLC; MH; MHi; MWA.

50. The American magazine of wonders, and marvellous chronicle, intended as a record of accounts of the most extraordinary productions, events, and occurrences, in providence, nature, and art, that have been witnessed at any time in Europe and America.... New York: Printed by Southwick and Pelsue, 1809. 2 v., plates.

Edited by Donald Fraser.

ULS-p. 273.

Copies: CtHC; DLC; DeGE; MWA; NcU; PPL; ViU.

51. Omnium gatherum; a monthly magazine recording authentick accounts of the most remarkable productions, events, and occurrences, in providence, nature, and art. v. 1; Nov. 1809–Oct. 1810. Boston: Printed by T. Kennard, 1809–1810. 560 p., plates.

ULS-p. 3177.

Copies: DLC; InU; MB; MBAt; MH; MHi; MWA; NN; WHi.

52. The Agricultural museum: designed to be a repository of valuable information to the farmer and manufacturer.... v. 1–2, no. 11; July 4, 1810–May 1812. Georgetown, Ca.: Printed by W. A. Rind, 1810–1812.

ULS-p. 90.

Copies: DLC; DNAL; DSI; MH; MU; MWA; NcU; PU.

53. The American medical and philosophical register; or, Annals of medicine, natural history, agriculture and the arts. Conducted by David Hosack ... and John W. Francis.... v. 1–4; July 1810–April 1814. New York: C. S. Van Winkle, 1810–1814. illus., plates.

ULS-p. 278.

Copies: CLM; CtY; DLC; DNLM; ICJ; KyU; MH; MWA; MiU; NN; Nh; NjP; OO; PPAN; PPL; PU; TU; TxU-M.

54. American mineralogical journal: being a collection of facts and observations tending to elucidate the mineralogy and geology of the United States.... Conducted by Archibald Bruce.... v. 1, no. 1–4; Jan. 1810–1814. New York: Printed by Collins & Co., 1810–1814. vi, 270, [2] p., plates.

ULS-p. 285.

Copies: DI-GS; DLC; ICJ; MMeT; MWA; MWiW; MiU; NjR; PPAN; PPAmP; PPL.

55. Archives of useful knowledge, a work devoted to commerce, manufactures, rural and domestic economy, agriculture and the useful arts. v. 1–3; July 1810–Apr. 1813. Philadelphia: D. Hogan, 1810–1813. illus.

Editor: J. Mease.

ULS-p. 463.

Copies: CtY; DLC; DeGE; In; KMK; MH; MWA; NN; NcD; PP; PPAmP; PPC; PU; ViU.

56. Connecticut Academy of Arts and Sciences, New Haven.

Memoirs. v. 1, pt. 1–4; 1810–1816. New Haven: 1810–1816. 412 p., illus.

ULS-p. 1173.

Copies: Major libraries.

57. Independent mechanic. April 6, 1811–March 28, 1812. New York: 1811–1812.

Includes articles on improvements in mechanical equipment, and descriptions of new patents.

ULS-p. 1949.

Copies: N.

58. Niles' Weekly register. v. 1–76; Sept. 7, 1811–Sept. 28, 1849. Baltimore, [etc.]: Hezekiah Niles, [etc.], 1811–1849.

Title varies: *The Weekly register,* 1811–Aug., 1814; *Niles' Weekly register,* Sept. 1814–Aug. 1837; *Niles' National register,* Sept. 1837–1849.

ULS-p. 3065–66.

Copies: Major libraries.

59. The Literary and philosophical repertory: embracing discoveries and improvements in the physical sciences, the liberal and fine arts.... v. 1–2; April, 1812–May, 1817. Middlebury, Vt.: Printed for S. Swift by T. C. Strong, [etc.], 1812–1817.

ULS-p. 2436.

Copies: CtHT; DLC; ICN; MB; MWA; Mi; MnU; NN; NhD; OMC; Vt; VtU; WHi.

60. The Analectic magazine. v. 1–14; 1813–1819; new ser., v. 1–2, 1820; third ser., v. 1, 1821. Philadelphia: M. Thomas, [etc.], 1813–1821. 17 v., illus.

Third series has the title: *The Literary gazette; or, Journal of criticism, science, and the arts.*

Editors: Washington Irving and Thomas I. Wharton.

ULS-p. 354.

Copies: Major libraries.

61. Columbian Chemical Society of Philadelphia.

Memoirs of the Columbian Chemical Society of Philadelphia. Volume 1. [Philadelphia]: Isaac Peirce, 1813. xv, 221 p., illus.

ULS-p. 1124; S&S-28190.

Copies: CtW; DLC; DeGE; ICJ; MoSU; PMA; PPC; PPL; PU.

62. The Emporium of arts & sciences. v. 1–2; May 1812–Apr. 1813; new ser., v. 1–3, June 1813–Oct. 1814. Philadelphia: J. Delaplaine, 1812; Kimber & Richardson, 1813–1814. 5 v., plates, ports.

Editors: v. 1–2, J. R. Coxe; new ser., T. Cooper.

ULS-p. 1442.

Copies: Major libraries.

63. The Massachusetts agricultural repository and journal. v. [3]–10; Nov. 1813–April 1832. Boston: Wells & Lilly, [etc.], 1813–1832. illus.

Issues for 1798–1810, published in 1799–1811, consist of unnumbered papers with title: *Papers on agriculture* (1809, *Georgick papers*).

Issued by the Massachusetts Society for Promoting Agriculture; strong on agricultural inventions and improvements.

ULS-p. 2550.

Copies: CU; CtY; DNAL; DeGE; ICU; MBH; MH; MU; MWA; MeB; N; NIC-A; NjR; OU; RPB; TxU.

64. The Western gleaner; or, Repository for arts, sciences, and literature. v. 1–2, no. 4; Dec. 1813–Sept. 1814. Pittsburgh: Cramer, Spear and Eichbaum, 1813–1814.

ULS-p. 4487.

Copies: CtY; MWA; OClWHi; PMA; PPL.

65. The New-York magazine and general repository of useful knowledge. v. 1, no. 1–3; May to July, 1814. New York: 1814.
ULS-p. 3036.
Copies: DLC; MH; MWA; N; Nhi, PHi; RPB.

66. The American magazine; a monthly miscellany devoted to literature, science, history, biography, and the arts. v. 1, no. 1–12; June 1815–May 1816. Albany: Printed by E. & E. Hosford, 1815–1816. 443 p., illus.
ULS-p. 273.
Copies: Ia; MB; MH; MWA; MnU; N; NN; PMA.

67. Literary and Philosophical Society of New York.
Transactions of the Literary and Philosophical Society of New-York. Instituted in the year 1814. v. 1–2, pt. 1. New-York: Published for the Society by Van Winkle and Wiley, 1815–1825. 2 v., plates.
Includes articles on agriculture, manufactures and useful arts.
ULS-p. 2436.
Copies: CtY; ICJ; MH; MWA; NIC; NN; NNM; NjP; OC; PBL; PPAP; PPL; TNV; VtU; WHi.

68. North American review. v. 1–248; May 1815–Winter 1939/40. Boston, [etc.]: 185–1940. illus.
ULS-p. 3081.
Copies: Major libraries.

69. The National register, a weekly paper, containing a series of the important public documents, and the proceedings of Congress; statistical tables, reports and essays, original and selected, upon agriculture, manufactures, commerce and finance.... v. 1–9, no. 6; March 2, 1816–Feb. 5, 1820. Washington City: [J. K. Mead, etc.], 1816–1820.
ULS-p. 2901.
Copies: CSmH; CtY; DLC; MH; MWA; NjR; TxU.

70. Academy of Natural Sciences of Philadelphia.
Journal [1st ser.] v. 1–8; 1817–1842. Philadelphia: 1817–1842. illus.
Includes articles on mineral resources and geology.
ULS-p. 36.
Copies: Major libraries.

71. The Journal of science and the arts. Edited at the Royal Institution of Great Britain. v. 1–5; 1817–1818. New York: James Eastburn, 1817–1818. illus.
American reprint of the *Quarterly journal of science,* London.
ULS-p. 2239.
Copies: CSmH; CtHT; DLC; IU; MB; MH; MWA; N; NNN; NWM; NjP; NjR; PSt; ViU.

72. New England galaxy. v. 1–21, no. 44, Oct. 10, 1817–Nov. 1838; new ser., v. 1, no. 1–14; Dec. 28, 1838–March 29, 1839. Boston: 1817–1839.
Includes descriptions of new inventions and news of progress in the sciences and mechanic arts.
ULS-p. 2987.
Copies: CSmH; CtY; DLC; MB; MH; MWA; MiU-C; N; NIC; OCIWHi; WHi.

73. American journal of science and arts. v. 1 [July 1818]– . New Haven, [etc.]: 1818– . illus., plates, maps.
Founded by Benjamin Silliman.
Since 1880 published under the title: *American journal of science.*
ULS-p. 262.
Copies: Major libraries.

74. Essex Agricultural Society.
Transactions. [v. 1]– . Salem, Mass: 1818– . illus., plates.
The volumes contain the annual addresses delivered before the Society, reports, etc. The first volume, comprising 10 numbers, has no collective title.
ULS-p. 1479-80.
Copies: DLC; DNAL; ICJ; MBH; MHi; MWA; MiU; Nh.

75. Monthly scientific journal, containing disquisitions in natural philosophy, chemistry, and the arts. v. 1, no. 1–6; Feb.–July, 1818. New York: W. Marrat, 1818. [152] p.
ULS–P. 2750.
Copies: NHi.

76. Scientific journal. v. 1, no. 1–10; Feb. 1818–Jan. 1820. Perth Amboy, N.J.; New York: 1818–1820.
Includes articles on practical topics, such as "damp" in houses and clockmaking.
ULS-p. 3826.
Copies: CtY; MB; MdBJ; NjR.

77. The American farmer, containing original essays and selections on rural economy and internal improvements, with illustrative engravings. v. 1–15; April 2, 1819–1834. Baltimore; Washington: 1819–1834. illus.
Continued as 2d series in May 1834. Ceased publication in 1897.
ULS-p. 219.
Copies: CSmH; CtY; DLC; DNAL; ICJ; In; LU; MBH; MWA; MdBP; MoU; NIC; NN; NcD; NjR; PPL; PU; TxU.

78. The Philadelphia register, and national recorder. v. 1; Jan. 2–June 26, 1819. Philadelphia: Littell & Henry, 1819. 432 p. illus.
Continued as the *National recorder* (1819-1821) and the *Saturday magazine* (1821–1822).
ULS-p. 3785.
Copies: CU; CtW; DLC; DeGE; MB; MWA; MiU-C; PHC; PPL; PU.

79. Plough boy. v. 1–4, no. 51; June 5, 1819–July 15, 1823. Albany: 1819–1823. illus.
Title changed to *Plough boy, and journal of the Board of Agriculture,* Jan, 1820.
Includes many articles on internal improvements, roads and canals.
ULS-p. 3374.
Copies: CtY; DLC; DNAL; ICJ; IU; In; MB; MBH; MWA; MiU; MnHi; N; NN; NSU; NjR; PPL; PU; TU; VtU.

80. The Rural magazine and farmer's monthly museum, devoted to history, biography, agriculture, manufacture, miscellany.... v. 1, no. 1–6; Feb.–July 1819. Hartford: J. & W. Russell, 1819. 208 p.
ULS-p. 3738.
Copies: CtHi; CtY; DLC; Ia; MH; MWA; MiD-B; N; OClWHi; PPL; RPB.

81. Saturday magazine: being in great part a compilation from the British reviews, magazines, and scientific journals. v. 1–5; Jan. 2, 1819–June 30, 1821; [new ser.], v. 1–2, July 7, 1821–July 29, 1822. Philadelphia: 1819–1822. illus.
Title varies: Jan.–June 1819, *Philadelphia register and national recorder;* July 1819-June 1821, *National recorder.*
Emphasis on internal improvements, particularly the Pennsylvania canals, and descriptions of inventions.
ULS-p. 3785.
Copies: DLC; MWA; MiU-C; NjR; OC; PHC; PPL; PU.

82. Agricultural intelligencer, and mechanic register. v. 1, no. 1–26; Jan. 10–July 7, 1820. Boston: 1820. 208 p., illus.
ULS-p. 89.
Copies: DNAL; MU; MWA.

83. The Patron of industry. Agricultural, manufacturing, industrial. v. 1, no. 1–104, June 28, 1820–June 27, 1821. New York: 1820–21.
ULS-p. 3279.
Copies: CSmH; CtY; DLC; PPL.

84. The Farmer's and planter's friend. no. 1–7; March 26–April 24, 1821. Philadelphia: 1821.
Signed: Guatimozin [pseud.]
ULS-p. 1534.
Copies: DeGE; MBAt; MHi; N; PPAmP.

85. American museum, and repository of arts and sciences, as connected with domestic manufactures and national industry. Also, a

85 (*continued*)
list of all the patents granted by the United States up to the end of the year 1821.... Washington: Printed by S. Elliot, [etc.]; published by J. Milligan, Georgetown, D.C., 1822–1823. 60, 72, xxiv, 13, 16 p., 4 plates.

Patents continued by supplements to April 1823.

ULS-p. 289; S-7801.

Copies: DLC; MB; MiU; OU; PU.

86. The New England farmer, and horticultural register. v. 1–24; Aug. 3, 1822–June 24, 1846. Boston: J. Breck, [etc.], 1822–1846. illus.

Founded and for some years edited by T. G. Fessenden.

Superseded by the *New England farmer, devoted to agriculture, horticulture, and their kindred arts and sciences.*

ULS-p. 2987.

Copies: Major libraries.

87. Pennsylvania Agricultural Society.

Memoirs of the Pennsylvania Agricultural Society, with selections from the most approved authors.... Philadelphia: J. S. Skinner, 1823–1824. 2 v., illus.

ULS-p. 3297.

Copies: CtY; DLC; DNAL; DeGE; IU; KMK; LU; MH; MWA; MdBP; NIC; NN; Nh; NjR; PPAN; PPAmP; PPL; PU; ScU; ViU.

88. Pennsylvania Agricultural Society.

Proceedings of the Pennsylvania Agricultural Society. v. 1–3; 1823–1825. Philadelphia: 1823–1825.

ULS–p. 3297.

Copies: CtY; DLC; MHi; N; NjR; PP; PPL; RPB; WHi.

89. Washington quarterly magazine of arts, science, and literature. With illustrative engravings. Robert Little, editor. v. 1, no. 1–2; July 1823–April 1824. Washington City: P. Thompson, [etc.], 1823–1824. 156 p., 6 plates.

ULS-p. 4453.

Copies: CSmH; DLC; DP; MH; MWA; OCIWHi.

90. The Boston journal of philosophy and the arts, intended to exhibit a view of the progress of discovery in natural philosophy. mechanics, chemistry, geology and mineralogy ... and the fine and useful arts. v. 1–3; May 1823–Dec. 1826. Boston: Cummings, Hilliard & Co., 1824–1826. plates.

ULS-p. 751.

Copies: CtY; DI-GS; DLC; DNLM; DeGE; ICJ; ICN; InU; MB; MH; MWA; MdBP; NBu; NhD; NjP; PPAmP; PU; RPB; TxU; VtU.

91. Geographical, historical and statistical repository. v. 1, no. 1–2; Sept.–Oct. 1824. Philadelphia: 1824. 136 p., illus., maps.

Caption title: *Darby's monthly.*

Includes articles on canals and other internal improvements.

ULS-p. 1694.

Copies: CtY; DLC; DeGE; MB; MWA; MiU; OCIWHi; PPL. PPWa; ViU; WHi.

92. American mechanics' magazine: containing useful original matter, on subjects connected with manufactures, the arts and sciences: as well as selections from the most approved domestic and foreign journals. Conducted by associated mechanics. v. 1–2; Feb. 5, 1825–Feb. 11, 1826. New York; J. V. Seaman, [etc.], 1825–1826. illus.

Issued under title: *Mechanics magazine,* Feb. 5–19, 1825.

ULS-p. 278.

Copies: CtY; DLC; DeGE; ICN; KyU; MBdAF; MH; MWA; MiU; NN; NjR; PPAmP; RP; WHi.

93. The American traveller. v. 1–26; July 5, 1825–Dec. 10, 1864. Boston: 1825–1864.

ULS-p. 339.

Copies: DLC; M; MB; MBAt; MWA; MiU–C; MnU; N; OCIWHi; RPB; WHi.

94. Badger and Porter's stage register; containing a full account of the principal lines of stages, steamboats and canal packets in the

New England states and the state of New York ... no. 1–71; July 5, 1825–May/June 1838. Boston: Badger & Porter, 1825–1838.
Supplement to *The American traveller.*
ULS-p. 583.
Copies: DLC; ICN; MB; MH; MH-BA; MWA; NRU; ViU.

95. The Carolina journal of medicine, science and agriculture. v. 1, no. 1–4; new ser., no. 1–2; Jan. 1825–May 1826. Charleston: Gray & Ellis, 1825–1826.
ULS-p. 937.
Copies: DNLM; MBCo; NNN; NcD; OC; PPC; ScCM.

96. The Merrimack magazine, and monthly register of politics, agriculture, literature.... v. 1, no. 1–7; Jan.–July 1825. Haverill, Mass.: E. W. Reinhart, printer, 1825. 112 p.
Editor: Jeremiah Spofford.
ULS-p. 2614.
Copies: CtY; DLC; DNAL; MWA.

97. The New Jersey and Pennsylvania agricultural monthly intelligencer and farmer's magazine. v. 1, no. 1–10; May 2, 1825–Oct. 26., 1826. Camden, N.J.: S. Ellis, 1825–1826.
ULS-p. 2999.
Copies: DeGE; N; NjR.

98. The Casket, and Philadelphia monthly magazine. Embracing every department of literature.... v. 1–17; Jan. 1826–Dec. 1840. Philadelphia: Atkinson & Alexander, [etc.], 1826–1840. plates.
Consolidated with *Burton's Gentleman's magazine* in 1841 to form *Graham's Lady's and gentleman's magazine.*
ULS-p. 1753.
Copies: Major libraries.

99. Franklin Institute, Philadelphia.
Journal of the Franklin Institute. v. 1– ; 1826– . Philadelphia: 1826– . illus., plates, maps.
1826–1827 issued under title: *The Franklin journal, and American mechanics' magazine.*
ULS-p. 1629.
Copies: Major libraries.

100. The American quarterly review. v. 1–22. Mar. 1827–Dec. 1837. Philadelphia: Carey, Lea & Carey, [etc., 1827]–1837.
ULS-p. 310.
Copies: Major libraries.

101. Farmers' and mechanics' journal. v. 1, no. 1–18; Aug. 12–Dec. 15, 1827. Harrisburg, Pa.: 1827.
ULS-p. 1534.
Copies: NjR.

102. The Farmer's, mechanic's, manufacturer's, and sportsman's magazine. By George Houston.... v. 1; March 1826–February 1827. New York: Printed by Vanderpool & Cole, 1827. vii, 472 p., illus., plates.
Nos. 1–2 issued under the title: *Farmers, mechanics, and manufacturers magazine.*
ULS-p. 1536.
Copies: CtY; DLC; DNAL; DeGE; MWA; MiD; LNHT; N; NN; NjR; PWcS.

103. Maclurian Lyceum, Philadelphia.
Contributions of the Maclurian Lyceum to the arts and sciences. v. 1, no. 1–3; Jan. 1827–Jan. 1829. Philadelphia: J. Dobson, 1827–1829. plates.
ULS-p. 2503.
Copies: CU; CtY; DLC; DSI; ICJ; MWA; NN; NNM; PPAN; PPC.

104. The American journal of improvements in the useful arts, and mirror of the patent office in the United States. v. 1, no. 1–4; Jan.–Dec. 1828. Washington: Printed by W. Greer, 1828. 560 p., plates (part fold.).
Editor: I. L. Skinner.
ULS-p. 255.
Copies: AU; CtY; DGU; DLC; DeGE; MB; MH; MWA; NN; NcU; PSC; ViU.

105. Civil engineer, and herald of internal improvement. v. 1, no. 1–20; June 28–Nov. 1,

105 *(continued)*
1828. Columbus, Ohio: 1828.
ULS-p. 1064.
Copies: DLC; MHi; MWA; N; NIC; NNC; NNE; OO; PP; WHi.

106. Hazard's register of Pennsylvania, devoted to the preservation of facts and documents, and every kind of useful information respecting the state of Pennsylvania.... v. 1–16; Jan. 1828–Dec. 1835. Philadelphia: Printed by W. F. Geddes, 1828–1835. illus.
Editor: Samuel Hazard.
Vols. I–VII (1828–1831) have the title: *The Register of Pennsylvania.*
ULS-p. 1825.
Copies: Major libraries.

107. New England farmers' and mechanics' journal. v. 1, no. 1–12; Jan.–Dec. 1828. Gardiner, Me.: P. Sheldon, 1828. illus.
ULS-p.2987.
Copies: MU; MWA; Me; MeB; MeU; Nh.

108. The New-York farmer and American horticultural repository. v. 1–10; Jan. 1828–Dec. 1837. New York: 1828–1837.
Subtitle varies: *American gardeners' magazine.*
ULS-p. 3031.
Copies: CtY; DLC; DNAL; Ia: MBH; MWA; N; NIC; NN; NjP; NjR.

109. Southern agriculturist and register of rural affairs. v. 1–12, 1828–1839, 1840; new ser. v. 1–6, 1841–1846. Charleston, [S.C.]: 1828–1846.
Title varies: 1840, *Southern cabinet of agriculture, horticulture and domestic economy;* 1841–1846, *Southern agriculturist, horticulturist, and register of rural affairs.*
ULS-p. 4011.
Copies: CSmH; CtY; DLC; DNAL; GU; ICU; KyU; LU; MU; MWA; NIC; NN; NcD; NjP; ScU; TxU.

110. Western tiller; a weekly journal, devoted chiefly to agriculture and rural affairs. v. 1–4, no. 14; Aug. 23, 1828–Nov. 26, 1831. Cincinnati: 1828–1831. illus.
ULS-p. 4496.
Copies: DLC; DNAL; MWA; NN; OCHP; OCIWHi.

111. Mechanics' magazine, museum, register, journal and gazette. North American series. v. 1–6; Aug. 29, 1829–Sept. 29, 1832. Boston: 1829–1832. illus.
North American edition of the London journal with the same name, later *Iron,* established in 1823.
ULS-p. 2570.
Copies: MCM; N; NcU.

112. Western agriculturist and general intelligencer. v. 1, no. 1–52; 1829–July 3, 1830. Brookville, Ind.: 1829–1830.
ULS-p. 4481.
Copies: ICHi; In.

113. Albany Institute, Albany.
Transactions of the Albany Institute. v. 1–12; 1830–1892. Albany: 1830–1892.
Vol. 1 contains the transactions of 1828–30.
Supersedes the Society for the Promotion of Useful Arts, Albany.
ULS-p. 131.
Copies: Major libraries.

114. The Magazine of useful and entertaining knowledge. v. 1–2, no. 5; June 15, 1830–May, 1831. New York: 1830–1831. illus.
Vol. 1, no. 1–2 have the title: *Mechanics' and farmers' magazine of useful knowledge.*
ULS-p. 2514.
Copies: CtY; DLC; Ia; NN; NNC; NNS; NjR.

115. The Mechanics' magazine, and journal of public internal improvement; devoted to the useful arts, and the recording of projects, inventions, and discoveries of the

age. v. 1–7, no. 4; Feb. 1830–1836. Boston: S. N. Dickinson, 1830–1836. illus., plates.
ULS-p. 2570.
Copies: CtY; DLC; DeGE; Ia; MB; MBAt; MH; MWA; NN; NNE; NjR; PBL; PPi; WHi.

2. Encyclopedias and dictionaries

116. Encyclopaedia; or, A dictionary of arts, sciences, and miscellaneous literature.... The lst American ed., in eighteen volumes, greatly improved.... Philadelphia: Printed by Thomas Dobson, 1798 [1790–1797]. 18 v., plates, maps.

Issued in parts, 1790–1797. Title-pages, printed after completion of the work, are all dated MDCCXCVIII.

Based on the 3d ed. of the *Encyclopaedia Britannica,* considerably revised and expanded with new material.

Mp-22486, [etc.]; N-E0119237.
Copies: CtY; DLC; DeGE; ICU; KyLx; MH; MWA; MeB; MiU; NN; Nh; PHC; PPAmP; PPL; PU; RPJCB.

117. ———— Supplement to the Encyclopaedia.... Philadelphia: Printed by Budd and Bartram, for Thomas Dobson, 1803. 3 v., plates.

Issued in parts, 1800–1803.
Mp-38593; N-E0119237; S&S-4162.
Copies: CtY; DLC; DeGE; ICU; KyLx, MH; MWA; MeB; MiU; NN; Nh; PHC; PPAmP; PPL; PU; RPJCB.

118. Harris, Thaddeus Mason, 1768–1842.
The minor encyclopedia; or, Cabinet of general knowledge: being a dictionary of arts, sciences, and polite literature.... Compiled ... by Thaddeus M. Harris.... Boston: Printed for West & Greenleaf, by Manning & Loring, 1803. 4 v.

S&S-4340.
Copies: DLC; IObB; MB; MH; MWA; MeB; MiU-C; NHi; NcD; OClWHi.

119. Willich, Anthony Florian Madinger.
The domestic encyclopaedia; or, A dictionary of facts, and useful knowledge. Comprehending a concise view of the latest discoveries, inventions, and improvements.... By A. F. M. Willich ... 1st American ed., with additions ... by James Mease.... Philadelphia: William Young Birch and Abraham Small and Mathew Carey, 1803–1804. 5 v., illus.

First published in London, 1802.
S&S-5592.
Copies: CSmH; DeGE; MH; MWA; NN.

120. The new and complete American encyclopaedia: or, Universal dictionary of arts and sciences; on an improved plan: in which the respective sciences are arranged into complete systems and the arts digested into distinct treatises.... Illustrated with one hundred and seventy copper plates.... New York: John Low, 1805–1811. 7 v., illus., plates (part fold.), maps (part fold.).

From the *Encyclopaedia Perthenis,* with improvements.
S&S-7863, 8971.
Copies: DLC; DeGE; MB; MHi; NN.

121. Rees, Abraham, 1743–1825.
The cyclopaedia; or, Universal dictionary of arts, sciences and literature. By Abraham Rees ... with the assistance of eminent professional gentlemen.... 1st American ed., rev., corr., enl., and adapted to this country, by several literary and scientific characters.... Philadelphia: S. F. Bradford; New York: P. A. Mesier, [etc., 1805–24]. 41 v., illus., 6 v. of plates.

121 (*continued*)

Publication reported first in *Port folio,* August, 1805.

Held also by several libraries as originally issued (in 83 parts and plates) with cover title: *Dr. Rees's New cyclopaedia.... The new cyclopaedia, or Universal dictionary of arts and sciences.*

N-R0115316, R0115336; S&S-9234, 19907.

Copies: CtY; DLC; DeGE; ICU; MH; MWA; NcD; OHi; PPAN; PPL; ViU; VtU.

122. Smith, Thomas, 1775 or1776–1830.

The wonders of nature and art; or, A concise account of whatever is most curious and remarkable in the world; compiled from historical and geographical works of established celebrity, and illustrated with discoveries of modern travellers. By the Rev. Thomas Smith.... Rev., corr., and improved by James Mease.... Philadelphia: Printed by Robert Carr, for Birch & Small; sold also by M. Carey, Kimber, Conrad & Co. and Jacob Johnson, 1806–1807. 14 v., illus., plates.

First published in London, 1803–1804.

S&S-11386.

Copies: DeGE; MWA; MdBS; MoSW; NcU; NcWsW; PPL; RP; ViR-Val; ViW.

123. Edinburgh encyclopaedia.

The American edition of the new Edinburgh encyclopaedia. Conducted by David Brewster.... And now improved, for the greater satisfaction and better information of the people of the United States, in the ... history of their country, in American biography, and in the great discoveries in mechanics and the arts. Philadelphia: Edward Parker, [etc.], 1812–1831. 18 v., plates.

Also the "First American edition," Philadelphia: Joseph and Edward Parker, 1832, issued in 36 parts plus plates, 1813–1831.

N-E0038133; S-24929.

Copies: CSt; CtY; DLC; DeGE; ICN; IU; MWA; MWiW; MeB; MiDU; NN; OU; PPL; PU; ViU.

124. Edinburgh encyclopaedia.

Second American edition of the new Edinburgh encyclopaedia, conducted by David Brewster ... assisted by upwards of one hundred gentlemen in Europe.... And now improved for the greater satisfaction and better information of the people of the United States.... New York: Samuel Whiting and John L. Tiffany,[etc.], 1813 – 1831. 18 v., plates, maps.

Vols. 17–18 have title: *American edition of the new Edinburgh encyclopaedia.*

S&S-29277.

Copies: CtY; IEN; InLP; NN; OClWHi; PPL; ScClP; ViU.

125. Gregory, George, 1754–1808.

A dictionary of arts and sciences. By G. Gregory.... 1st American, from the 2d London ed.; considerably improved and augmented. Charlestown: S. Etheridge; Philadelphia: I. Peirce; S. Etheridge, Jun., printer, 1815–1816. 3 v., plates.

Vols. 2–3 have imprint: Philadelphia: I. Peirce; Dennis Heartt, printer, 1816.

Added engraved title pages: *A new and complete dictionary of arts and sciences, including the latest improvement and discovery and the present state of every branch of human knowledge....* 1st American ed., considerably improv'd & augmented. Philadelphia: I. Peirce, 1816.

First published in London, 1806.

S&S-34812, 34813.

Copies: CU; CtHT-W; DeGE; IEN; InU; MH; MWA; MdBL; NIC; NN; NcD; PPL; PPi; PSC; PU; ScCC; TNJ; ViU.

126. Nicholson, William, 1753–1815.

American edition of the British encyclopedia; or, Dictionary of arts and sciences, comprising an accurate and popular view of the present improved state of human knowledge.... Illustrated by upwards of 160 elegant engravings.... Philadelphia: S. A. Mitchell & H. Ames, 1816–1817. 7 v., plates.

S&S-38471.

Copies: Ar-Hi; DLC; GDC; MH; MWA; NN; NcHil; NcU; OClWHi; ScCM; ViU.

127. ———— Illustrated with upwards of 180 elegant engravings.... Philadelphia: Ames and White, 1818. 12 v., plates.

Added engraved title page: *Second American edition of Nicholson's British encyclopedia.*

S&S-45107.

Copies: DLC; ICMcC; InNea; KMK; MDeep; MoS; NIC; NRom; NcC; OUrC; RP; TCU; VtU.

128. Gregory, George, 1754–1808.

A new and complete dictionary of arts and sciences, including the latest improvement and discovery and the present state of every branch of human knowledge, by G. Gregory.... New York: Collins and Co., 1819. 3 v., plates.

S&S-48115.

Copies: CtHT; CtW; MWA; NjP.

129. Nicholson, William, 1753–1815.

American edition of the British encyclopedia; or, Dictionary of arts and sciences.... [3d American ed.]. Philadelphia: Mitchell, Ames and White, 1819–1821. 12 v., plates.

Added engraved title page: *Third American edition of Nicholson's British encyclopedia.*

S&S-48936–8.

Copies: DSI; DeGE; ICJ; MH; MWA; NN; Nj: PU.

130. Gregory, George, 1754–1808.

A dictionary of arts and sciences. By G. Gregory.... 2d American, from the last London ed., considerably improved and augmented. New-York: Published for W. T. Robinson, by Collins & Co., 1821. 3 v., plates.

Added engraved title pages: *A new and complete dictionary of arts and sciences....* New York: 1821–1822.

S-5499.

Copies: DLC; KyBB; MH; MWA; MoK; NN; NNC; Nj; OU; PP; PU; RPB; WU.

131. Willich, Anthony Florian Madinger.

The domestic encyclopedia; or, A dictionary of facts and useful knowledge ... with an appendix containing additions in domestic medicine and the veterinary and culinary arts. 2d American ed. with additions by Thomas Cooper. Philadelphia: Abraham Small, 1821. 3 v., illus. First published in London, 1802.

S-7676.

Copies: CtW; DeGE; ICJ; KyDC; MB; MdBM; NWM; NcD; OHi; PPAmP; PPL; RP; TNJ; ViL.

132. Gregory, George, 1754–1808.

A new and complete dictionary of arts and sciences.... New York: Collins and Co., 1822. 3 v., plates.

Same as no. 130.

S-8890.

Copies: MnSH; NR; NSyU.

133. Encyclopaedia Americana. A popular dictionary of arts, sciences, literature, history, politics and biography ... on the basis of the seventh edition of the German Conversations-Lexicon. Edited by Francis Lieber, assisted by E. Wigglesworth.... Philadelphia: Carey, Lea & Carey, 1829–1833. 13 v.

S-38263, 38479.

Copies: C; CoD; CtW; DLC; IU; MB; MWA; MdBL; MeB; MiDU; MnHi; MoS; Nh; Nj; OU; PPAmP; PPL; PU; RPB; Vi; WvU.

134. Crabb, George, 1773–1851.

A dictionary of general knowledge; or, An explanation of words and things connected with all the arts and sciences.... American ed., with many improvements. Boston: Gray and Bowen, 1830. iv, 368 p., illus.

C–1032.

Copies: DN; MH; MWA; MoSU; PPA; ViU.

135. ———— New York: White, Gallagher and White, 1830. iv, 368 p., illus.

C-1033.

Copies: CLCM; IC; MBAt; MH; PPL.

136. Encyclopaedia Americana. A popular dictionary of arts, sciences, literature, history,

136 (*continued*)
politics and biography ... on the basis of the seventh edition of the German Conversations-Lexicon. Edited by Francis Lieber, assisted by E. Wigglesworth.... Philadelphia: Carey & Lea, 1830–1833. 13 v.

Second issue of the first edition.

N-E0118632.

Copies: DNLM; FU; MBAt; MH; MWA; MNoanMV; NN; NcU; ViU.

3. Handbooks and manuals

137. [Prince, Thomas] 1687–1758.

The vade mecum for America; or, A companion for traders and travellers: containing I. An exact and useful table, shewing the value of any quantity of any commodity.... II. A table of simple and compound interest. III. The names of the towns and counties in the several provinces.... IV. The time of the general meetings of the Baptists and Quakers. V. A description of the principal roads from the mouth of Kennebeck-River in the north-east of New-England, to James-River in Virginia.... Boston, N.E.: Printed by S. Kneeland and T. Green, for D. Henchman, 1731. [2], iv, 220 p.

Mp-3470.

Copies: DLC; MB; MBAt; MWA; RPJCB.

138. ———— Boston, N.E.: Printed by S. Kneeland and T. Green, for D. Henchman & T. Hancock, 1732. [2], iv, [2], 220 p.

Mp-3598.

Copies: DLC; MWA.

139. Smith, Eliza.

The compleat housewife; or, Accomplished gentlewoman's companion: being a collection of upwoards of five hundred of the most approved receipts in cookery, pastry ... preserving ... made wines, cordials.... To which is added, a collection of near two hundred family receipts of medicines ... Williamsburg, [Va.]: Printed by William Parks, 1742. [12], 228, [12] p.

Mp-5061.

Copies: CSmH; MWA.

140. Fisher, George.

The American instructor; or, Young man's best companion. Containing spelling, reading.... Together with the carpenter's plain and exact rule: shewing how to measure carpenters, joyners, sawyers, bricklayers, plaisterers, plumbers, masons, glaziers and painters work. How to undertake each work, and at what price ... with Gunter's line; and Coggeshal's description of the sliding rule. Likewise the practical gauger made easy; the art of dialling ... instructions for dying, colouring, and making colours.... The whole better adapted to these American colonies.... The 9th ed., rev. and corr. Philadelphia: Printed by B. Franklin and D. Hall, 1748. v, 378 p., illus.

First published in London, 1727, under title: *The Instructor.*

Mp-6238.

Copies: CSmH; CtY; DLC; DeWI; ICN; MB; MWA; MiU-C; NN; NNC; PPAmP; PPL; PU; RNR.

141. Smith, Eliza.

The compleat housewife.... Williamsburg, [Va.]: 1752.

Also New York: Gaines, 1761, and New York: Gaines, 1764, listed as E–9010 and 9840 respectively, with no copies located.

E-6934.

No copy known.

142. Fisher, George.
The American instructor; or, Young man's best companion.... Together with the carpenter's plain and exact rule: shewing how to measure carpenters, joyners, sawyers, bricklayers, plaisterers, plumbers, masons, glaziers and painters work.... Likewise the practical gauger made easy; the art of dialling ... instructions for dying, colouring, and making colours.... The 10th ed., rev. and corr. Philadelphia: Printed by B. Franklin and D. Hall, 1753. v, 384, [2] p., illus.
Mp-7120.
Copies: CtY; DLC; ICHi; MB; MH-BA; MnU; NN; PHi; PPF.

143. ———— 12th ed., rev. and corr. New York: Printed and sold by H. Gaine, 1760. v, 378 p., illus.
Mp-8736.
Copies: DLC; MWA; NHi; NNC; NcWsW.

144. ———— 14th ed., rev. and corr. New York: Printed and sold by H. Gaine, 1770. v, 390 p., illus.
B-3181; Mp-42090.
Copies: CtY; DLC; ICU; MH; MWA; MiU-C; NN; NhD; NjP; PPL; RPJCB.

145. ———— 15th ed., rev. and corr. Philadelphia: Printed and sold by John Dunlap, 1770. v, 390 p., illus.
Mp-11859.
Copies: MWA; PPL.

146. Fisher, George.
The instructor; or, Young man's best companion.... Together with the method of measuring carpenters, joiners, sawyers, bricklayers, plasterers, plumbers, masons, glaziers and painters work.... Likewise, the practical gauger made easy.... 21st ed. Burlington, [N.J.]: Printed and sold by Isaac Collins, 1775. xii, 372 p., front.
Mp-14458.
Copies: MWA.

147. Fisher, George.
The American instructor; or, Young man's best companion.... Together with the carpenter's plain and exact rule; shewing how to measure carpenters, joiners, sawyers, brick-layers, plaisterers, plumbers, masons, glasiers and painters work.... Likewise, the practical gauger made easy; the art of dialing, and how to erect and fix any dial; with instructions for dying and colouring, and making colours.... Boston: Printed for John Boyle and J. D. M'Dougall, 1779. vi, 378 p., illus.
Mp-16524.
Copies: CSmH; CtY; IEN; MH; MWA; MiU-C; NNC; NNNAM.

148. Fisher, George.
The instructor; or, American young man's best companion.... Together with the method of measuring carpenters, joiners, sawyers ... work.... To which is added, The family's best companion: with instructions for marking on linen, how to pickle and preserve; to make excellent wine.... The 30th ed., rev., corr., enlarged and improved. Worcester, Mass.: Printed by Isaiah Thomas for I. Thomas in Worcester; J. Boyle, E. Batelle, W. Green, B. Larkin, and J. Condy, in Boston, 1785. xii, [13]–384 p., illus., plates.
Mp–19244.
Copies: CSmH; CtY; DLC; InU; MB; MH; MWA; MnU; NN; NhM; RPJCB; WU.

149. ———— 2d Worcester ed., enlarged and improved. Worcester, Mass.: Printed by I. Thomas and sold at his bookstore, 1786. xii, [13]–384 p., illus., plates.
Mp-19990.
Copies: DeGE; MWA.

150. Fisher, George.
The American instructor; or, Young man's best companion.... Together with the method of measuring carpenters, joiners ... and painters work.... Likewise, the practical gauger made easy.... Philadelphia: Printed

150 (*continued*)
and sold by Joseph Crukshank, 1787. xii, 372 p., 6 plates.

Mp-20709.

Copies: MWA; MiU-C; NN; NNC.

151. The golden cabinet: being the laboratory, or handmaid to the arts, containing such branches of useful knowledge, as nearly concerns all kinds of people.... Part the first [to third]. Philadelphia: Printed and sold by William Spotswood, and H. and P. Rice, 1793. [1], iv, 84, iv, [1], [87]–219, [1], vi p.

All three parts have separate title pages.

Mp-25551.

Copies: C-S; CtY; DLC; MBAt; MH; MWA; MiU-C; PBL; PPL; PSt.

152. Fisher, George.

The instructor; or, American young man's best companion.... Together with the method of measuring carpenters, joiners, sawyers ... work.... Likewise, the practical gauger made easy; the art of dialling, and how to erect and fix dials; with instructions for dying, colouring and making colours.... Walpole, N.H.: Printed by Isaiah Thomas and David Carlisle, for said Thomas in Worcester, Massachusetts, [etc.], 1794. xii, 384 p. plates.

Mp-27706.

Copies: DAU; MWA; MiU; NN.

153. Morse, William.

Mechanical arts, in thirty-two receipts, viz. For jacking and varnishing leather. To make oil cloth. To colour linen and cotton yarn blue. To lacker brass and tin ware. To colour feathers, fur and hair. To prepare paints, varnishes &c. To stain wood, of different colours. To colour hats green on one side &c. &c. ... Hartford: Printed by Elisha Babcock, 1795. 14 p.

B-9183; Mp-47494.

Copies: CtHi; CtY; MH.

154. One thousand valuable secrets, in the elegant and useful arts, collected from the practice of the best artists, and containing an account of the various methods of engraving on brass, copper, and steel. Of the composition of metals.... And a variety of other curious, entertaining and useful articles. lst American ed. Philadelphia: Printed for B. Davies and T. Stephens, 1795. xxxvi, 377 p.

Other editions published under the title: *Valuable secrets in arts and trades.*

Mp-29242.

Copies: DLC; DeGE; MWA; PPL; RPJCB.

155. Valuable secrets concerning arts and trades; or, Approved directions, from the best artists, for the various methods of engraving on brass, copper, or steel. Of the composition of metals and varnishes. Of mastichs and cements, sealing-wax, &c. ... Norwich, [Conn.]: Printed by Thomas Hubbard, 1795. 1 p.l., xxii, 240 p.

First published in London, 1775.

Mp-29243.

Copies: DLC; DeGE; MH; MWA; NN; PPF; PPL; PU.

156. Fisher, George.

The instructor; or, Young man's best companion.... Together with the method of measuring carpenters', joiners', sawyers', bricklayers' ... and painters' work.... Likewise, the practical gauger made easy.... Wilmington, [Del.]: Printed and sold by Peter Brynberg, 1797. xi, 300 p., plates.

Mp-32840.

Copies: CSt; CtY; DLC; DeHi; InU; MH; MWA; MiU; NNC; Nh; PPL; PU; RPJCB.

157. Valuable secrets, concerning arts and trades; or, Approved directions from the best artists, for the various methods of engraving on brass, copper or steel. Of the composition of metals and varnishes. Of mastiches and cements, sealing wax, &c. Of colours and painting, for carriage-painters.... Of colours of all sorts for oil, water, and crayons. Of the art of gilding. Of the art of dying.... Of the art of making wines.... With an appendix, containing valuable selections, in addition to, and never before published in this work. 3d American ed. Boston: From J. Bumstead's

Printing-Office, 1798. xxiv, [25]–264 p.
Mp-34913.
Copies: DLC; MWA; RPJCB.

158. Fisher, George.
The instructor; or, American young man's best companion: containing instructions for ... mensuration, gauging, the art of dialling, dying and making colours, pickling, preserving, and the art of making several sorts of wines.... Philadelphia: Printed by John Bioren, for John Conrad & Co., [etc., etc.], 1801. xii, [13]–346 [348] p., illus., plates.
S&S-498, 499.
Copies: CSmH; DGU; DLC; ICRL; MH; MWA; MiU-C; NN; PHi; PPLT; ViU.

159. Aikin, John, 1747–1822.
The arts of life: 1. Of providing food, 2. Of providing cloathing, 3. Of providing shelter; described in a series of letters, for the instruction of young persons, by Dr. Aikin. Boston: Printed by Hosea Sprague for Samuel H. Parker, 1803. 144 p.
First published in London, 1802.
S&S-3642.
Copies: CtY; DLC; DNLM; KyLx; MB; MBCo; MHi; MWA; NNNAM; ODW; PMA.

160. Ein vortrefliches Kräuter-Buch für Haus-Väter- und Mütter, nebst etlichen auserlesenen Recepten. Wie auch eine Anweisung zur Färbe-Kunst, blau, roth, gelb ... zu färben. [Ephrata, Pa.: J. Baumann?], 1803. 32 p.
Copies: PPL.

161. The book of trades; or, Library of the useful arts.... lst American ed. White Hall, [Pa.]: Jacob Johnson; Dickinson, printer, 1807. 3 v., 67 plates.
First published in London, 1804.
S&S-12173.
Copies: DLC; DeGE; MH; MPB; MWA; MiDSH; NN; PP; PPL; Vi; ViU.

162. Parker, Solomon.
Parker's American citizen's sure guide; or, Ready reckoner, measurer, and ... family companion. Sag-Harbor: Printed by Alden Spooner, for the author, 1808. iv, 287, [1] p.
S&S-15849.
Copies: CSmH; CtY; DLC; MWA; MiU; NN;

163. Valuable secrets in arts, trades, &c. Selected from the best authors, and adapted to the situation of the United States.... New York: E. Duyckinck, 1809. iv, [5]–380, [20] p.
S&S-19085.
Copies: DLC; DeGE; MNoanMV; MWA; NNS; NjP; PPL.

164. Ein vortrefliches Kräuter-Buch für Haus-Väter- und Mütter, nebst etlichen auserlesenen Recepten. Wie auch eine Anweisung zur Färbe-Kunst, blau, roth, gelb ... zu färben. Hannover, Pa.: Gedruckt bey Starck und Lange, 1809. 32 p.
N-V0239108.
Copies: DSI; MNoanMV; MWA; MiU-C.

165. Fisher, George.
The instructor; or, American young man's best companion improved ... containing instructions for ... mensuration, gauging, the art of dialling, dying and making colours, pickling, preserving, and the art of making several sorts of wines.... Philadelphia: Printed and sold by John Bioren, 1810. xi, [13]–300 p., illus., plates.
S&S-20118.
Copies: MH; MWA; NjR; OAkU; PSt.

166. Nash, Norman.
The artist's assistant, containing a choice selection of receipts from various authors, principally from the valuable Secrets of arts and trades; to which is prefixed A chymical treatise on the metals. Springfield, Mass.: 1810. 72 p.
Extracted from: *Valuable secrets in arts and trades*.
S&S-20829.
Copies: CtY; DeGE; MH; MH-BA; MSCV; MWA.

167. Fisher, George.
The instructor; or, American young man's best companion; containing instructions for ... mensuration, gauging, the art of dialing, dying and making colours, pickling, preserving, and the art of making several sorts of wines.... Philadelphia: John Bioren, 1812. 300 p., illus., plates.
S&S-25429.
Copies: CtY; FTaSU; InU; MB; MWA; MiDSH; McD; OClWHi; PHi; T.

168. Denson, Jesse.
A compendium of useful information.... Nashville, T[enn.]: Printed for the author, 1813. iv, 234 p.
S&S-28312.
Copies: CSmH; DLC; T; TMC.

169. The artist's companion, and manufacturer's guide, consisting of the most valuable secrets in arts and trades.... By a friend to American manufactures. Boston: J. Norman; E. G. House, printer, 1814. xiv, 179 p.
Also published under the titles: *Five hundred receipts in the arts and trades; Valuable secrets in arts and trades; One thousand valuable secrets.*
Sometimes attributed to John Norman.
S&S-32363.
Copies: CU; CtY; DLC; DeGE; ICJ; MB; MWA.

170. Cutbush, James, 1788–1823.
The American artist's manual; or, Dictionary of practical knowledge in the application of philosophy to the arts and manufactures. Selected from the most complete European systems with original improvements and appropriate engravings. Adapted to the use of the manufacturers of the United States.... Philadelphia: Johnson & Warner, and R. Fisher, 1814. 2 v., illus., plates (1 fold.).
S&S-31276.
Copies: DLC; DeGE; ICU; LND; MWA; NIC; NWM; NcA-S; PPL; PU; ViU.

171. The universal receipt book; or, Complete family directory; being a repository of useful knowledge in the several branches of domestic economy ... by a Society of gentlemen in New-York. New York: Published by I. Riley; Van Winkle & Wiley, printers, 1814. 284 p.
Edited by Richard Alsop.
S&S-33515.
Copies: MH; MSaE; MWA; NN; NNNAM; NjR.

172. Valuable secrets in arts and trades. Containing directions from the best artists, for calico printing.... With above five hundred valuable modern receipts ... collected from the latest European publications. By a friend to American manufactures. Boston: Published by J. Norman; E. G. House, printer, 1814. [2], xiv, 179 p.
S&S-33523.
Copies: MDedHi; MWA.

173. Blake, John.
The private insturctor, and young gentleman's pocket companion, comprising every thing necessary in arithmetic, mensuration, guaging, bookkeeping, and conveyancing.... Trenton: D. Fenton, 1815. vii, [1], 14–244 p.
S&S-34144.
Copies: CSmH; DAU; MWA; MiU.

174. Five hundred receipts in the arts and trades, containing directions from the best artists.... By a friend to American manufactures. Boston: J. Norman, [ca. 1815]. xiv, 179 p.
Running title: *Secrets in arts and trades.*
Issued between 1814 and 1817.
Copies: MWA.

175. Eine kurtzgefasste neue Sammlung, in sich haltend mehrentheils wunderseltsamen, auserlesenen, nützlichen und bewährten Recepten und Kunst-Stücke.... [S. l.]: Auf Kosten des Liebhabers herausgegeben, 1815. [5], 30 p.
S&S-35067.
Copies: MWA; NNRT; PPL.

176. Harwell, H. I.
The domestic manual; or, Family directory, containing receipts in arts, trades and domestic economy; selected from the best authors, and practical artists; and containing many processes never before published. By H. I. Harwell. New London, [Conn.]: Printed by Samuel Green, 1816. 84 p.
S&S-37812.
Copies: CtHi; CtNl; DLC; DNLM; MWA.

177. Valuable secrets in arts, trades, &c., selected from the best authors, and adapted to the situation of the United States.... New York: E. Duyckinck, 1816. iv, [5]-[317], [18] p.
Last page of text is erroneously numbered: 139.
S&S-39655.
Copies: DeGE; MH-BA; MWA; NN.

178. [Phillips, Richard, Sir] 1767–1840.
The universal preceptor; a general grammar of arts, sciences, and useful knowledge. By the Rev. David Blair [pseud.].... 1st American, from 8th London ed., with additions and improvements. Philadelphia: E. & R. Parker, 1817. vi, 316 p., illus., fold. map.
S&S-41805.
Copies: DLC; ICU; MH; MWA; MWHi; MeBa; P; PPL.

179. ———— 2d American ed., with additions and improvements. Philadelphia: E. &. R. Parker, 1817. vi, [2], 316 p., illus., fold. map.
S&S-41806.
Copies: DeGE; MWA.

180. Bingley, William, 1774–1823.
Useful knowledge; or, A familiar and explanatory account of the various productions of nature, mineral, vegetable, and animal, which are chiefly employed for the use of man. Illustrated with numerous figures, and intended as a work both of instruction and reference.... Philadelphia: A. Small, 1818. 3 v., plates.
First published in London, 1816.
S&S-43363.
Copies: CLSU; DLC; KyU; MWA; MiD; NN; NWM; OU; PPAN; RP; ScU.

181. Homespun, Priscilla, pseud.
The universal receipt book; being a compendious repository of practical information in cookery, preserving, pickling, distilling, and all the other branches of domestic economy. To which is added, some advice to farmers. By Priscilla Homespun [pseud.]. Philadelphia: Printed by J. Maxwell, for Isaac Riley, 1818. 378 p.
S&S-46618.
Copies: Nh.

182. ———— 2d ed., with great additions. Philadelphia: Isaac Riley; J. Maxwell, printer, 1818. 378 p.
S&S-44368, 46619.
Copies: DLC; ICJ; MSCV; MWA; NCH; Nh; PPL; PPi; ScU.

183. The new family receipt book, containing eight hundred truly valuable receipts in various branches of domestic economy.... A new ed., corr. Philadelphia: Collins & Croft; William Brown, printer, 1818. xxvi, 402 p.
Attributed to Maria Eliza Ketelby Rundell.
First published in London, 1810.
S&S-44990, 45596.
Copies: CSmH; MWA; NN; NNNAM; NTRPI; PP; ViRUT.

184. Curtis, Samuel, 1747–1822.
A valuable collection of recipes, medical and miscellaneous. Useful in families, and valuable to every description of persons. Amherst, N.H.: Printed by Elijah Mansur, 1819. 70, [2] p.
S&S-47762.
Copies: CtY-M; DLC; DeGE; MBCo; MWA; NhD; OC.

185. The family receipt book, containing eight hundred valuable receipts in various

185 (*continued*)
branches of domestic economy; selected from the works of the most approved writers.... 2d American ed. Pittsburgh: Randolph Barnes, 1819. xxxii, [25]–408 p.

Appeared same year at New Haven under title: *The new family receipt book....* A new edition.

S&S–47937.

Copies: ArU; CSmH; CtY; DLC; MH; MWA; NUtHi; OO; P; PPL; PPi; Vi.

186. Krauss, Johann.

Oeconomisches Haus- und Kunst-Buch, oder Sammlung ausgesuchter Vorschriften, zum Nutzen und Gebrauch für Land- und Hauswirthe, Handwerker, Künstler und Kunst-Liebhaber. Zusammengetragen aus den besten englischen und deutschen Schriften.... Allentown, [Pa.]: Gedruckt bey Henrich Ebner, 1819. xvi, 452 p.

S&S-48443.

Copies: CtY; DNLM; DeGE; MH; MWA; NN; NjR; P; PDoBHi; PHi; PPeSchw.

187. The new family receipt-book, containing eight hundred truly valuable receipts in various branches of domestic economy.... A new ed., corr. New Haven: Printed by Howe & Spalding and Samuel Wadsworth, 1819. xxxi, [37]–429 p.

Attributed to Maria Eliza Ketelby Rundell.

S&S-48845.

Copies: Ct; CtHi; DLC; MWA; MdBP; OO; ViU.

188. ———— New ed. New Haven: A. H. Maltby & Co., 1819. 429 p.

S&S-48846.

Copies: MBedf.

189. [Phillips, Richard, Sir] 1767–1840.

The universal preceptor; being a general grammar of arts, sciences and useful knowledge. By the Rev. David Blair [pseud.].... 3d American ed., with additions and improvements. Philadelphia: E. & R. Parker, 1819. iv [vi], [2], 316 p. illus., plate, map.

S&S-49102.

Copies: MB; MWA; MoKU.

190. [Preston, John] b. 1755.

Domestic economy revived; in a series of recipes.... Also a few sketches, worthy the attention of the farmer, the mechanic, and the housewife. Compiled by a philanthropist. Albany: Printed by E. and E. Hosford, 1820. 72 p.

S-2862.

Copies: MWA; N; NN.

191. A select collection of approved, genuine, secret, and modern receipts, for the preparation and execution of various valuable and curious arts, as practised by the best artists of the present age. Concord, Mass.: R. Porter, printed by J. T. Peters, [1820?]. 48 p.

Attributed to Rufus Porter.

S-3163.

Copies: DLC; MHi; MWA; NN.

192. Endless amusement; a collection of nearly 400 entertaining experiments in various branches of science, viz: acoustics, arithmetic, chemistry, electricity, hydraulics, hydrostatics, magnetism, mechanics, optics, wonders of the air pump.... To which is added, a complete system of pyrotechny.... 2d ed., with considerable alterations and improvements.... Philadelphia: M. Carey & Sons, 1821. xii, [13]–216 p., illus., 2 fold. plates.

S-5254.

Copies: MB; MH; NGos; NN.

193. ———— 3d ed., with considerable alterations and improvements.... Philadelphia: M. Carey & Sons, 1822. xii, [13]–216 p., illus., 2 fold. plates.

S-8625.

Copies: CU; DLC; DeGE; N; NcWsS; NjP.

194. Curiosities for the ingenious: selected from the most authentic treasures of nature, science and art, biography, history, and general literature. Philadelphia: H. C. Carey and I. Lea, 1822. xi, 14–192 p., plates.

First published in London, 1821.

Includes descriptions of inventions, machinery and technical equipment.

S-8488.

Copies: CSmH; MBAt; MH; MdBP; NN; NcU; OCl.

195. Hemenway, Moses.

The manufacturer's, farmer's, & mechanic's guide; being a compilation of the most valuable receipts, for manufacturers, agriculturists, and mechanics. Pittsburgh: Printed for the compiler, 1822. 175, [4] p.

S-8992.

Copies: ICN; OClWHi.

196. [Phillips, Richard, Sir] 1767–1840.

The universal preceptor; being a general grammar of arts, sciences, and useful knowledge. By the Rev. David Blair [pseud.].... 4th American ed., with additions and improvements. Philadelphia: E. Parker, 1822. vi, [2], 316 p., illus., fold. map.

S-9920.

Copies: CtHT-W; DLC; MB; MH; MiDSH; NUtHi; P; RPB.

197. Parker, M.

The arcana of arts and sciences; or, Farmers' & mechanics' manual; containing a great variety of valuable receipts and useful discoveries, in the various departments of human knowledge, many of which were never before published. By Dr. M. Parker.... Washington, Pa.: Printed by J. Grayson, 1824. 2 p.l., [7]–348 p.

S-17520.

Copies: CSmH; DNAL; DeGE; InSNHi; MWA; OClWHi; OkU.

198. Beman, David.

The mysteries of trade; or, The great source of wealth; containing receipts and patents in chemistry and manufacturing, with practical observations on the useful arts.... Boston: Printed for the author by W. Bellamy, 1825. 152 p.

S-19656.

Copies: CtHT-W; CtY; DeGE; ICJ; MB; MBP; MBSi; MH; MH-BA; MW; MWA; MiU-C; NjR.

199. Curiosities for the ingenious; selected from the most authentic treasures of nature, science and art, biography, history, and general literature. Philadelphia: Ash and Mason; R. Wright, printer, 1825. xii, [13]–192 p., plates.

First published in London, 1821.

Includes descriptions of inventions, machinery and technical equipment.

S-20237.

Copies: CtY; DLC; DeGE; Ia-HA; MB; MWA; MdBL; MoS; NCH; PPL; PU.

200. Family receipts: being a compilation from several publications, for cooking, dyeing, varnishing, painting, glueing, cementing, etc. ... Syracuse: L. H. Redfield's bookstore; T. A. Smith & Co., printers, [1825?]. 60 p.

S-20466.

Copies: MWA (59, [1] p., dated [1840?]); PPL.

201. Mackenzie, Colin.

Five thousand receipts in all the useful and domestic arts, constituting a complete and universal practical library, and operative cyclopaedia.... Philadelphia: A. Small, 1825. 670 p.

First published in London, 1823.

S-21281.

Copies: CtHT-W; DLC; GEU-M; KMK; MB; MWA; NNS; NcD; NjP; PPAmP; PPL; ViU.

202. A select collection of valuable and curious arts, and interesting experiments.... Concord, [N.H.]: Rufus Porter; J. B. Moore, printer, 1825. viii, [9]–102 p., 1 l.

Attributed to Rufus Porter.

N-P0497419, S0398282.

Copies: DeGE; MBC; MSo; MWA; NCorniC.

203. The cabinet of useful arts and manufactures, designed for the perusal of young persons. New York: Printed by James Bloomfield, 1826. 96 p., front.
S-23985a.
Copies: DeGE; MAtt; MWA; RNHi.

204. [Phillips, Richard, Sir] 1767–1840.
The universal preceptor; being a general grammar of arts, sciences, and useful knowledge. By the Rev. David Blair [pseud.]. From the 13th English ed., rev. and improved. Greenfield, Mass.: A. Phelps, 1826. 312 p., illus.
S-25764.
Copies: CtHi; DLC; MB; MDeeP; MH; NjP; OO; PU; PWCHi.

205. A select collection of valuable and curious arts, and interesting experiments.... Concord, [N.H.]: R. Porter, 1826. 102 p., 1 l.
Attributed to Rufus Porter.
S-26043.
Copies: DLC; MB; MH; MiD; Nh; NhHi; NjR.

206. ———— 2d ed. Concord, [N.H.]: R. Porter; J. B. Moore, printer, 1826. viii, [9]–132 p.
S-26044.
Copies: CtHi; CtW; MNF; NN; PPL.

207. ———— 3d ed. Concord, [N.H.]: R. Porter, 1826. 132 p.
S-26045.
Copies: CU; MB; Nh; PU-S.

208. ———— 4th ed. Concord, [N.H.]: R. Porter, 1826. viii, 172 p.
S-26046.
Copies: CSmH; CtHi; DeGE.

209. ———— 5th ed. Concord, [N.H.]: R. Porter, 1826. 172 p.
S-26047.
Copies: MWA; MiD; NhHi.

210. The cabinet of useful arts and manufactures: designed for the perusal of young persons. New York: Caleb Bartlett, 1827. 143 p., illus.
S-28355.
Copies: DLC; KU; MH; MWA; NcD; PPL.

211. Mackenzie, Colin.
Five thousand receipts in all the useful and domestic arts, constituting a complete and universal practical library, and operative cyclopaedia.... Philadelphia: Printed by J. Kay, Jun., for J. Grigg, U. Hunt, [etc.]; New York: O. A. Roorbach, 1827. 661 p., illus.
Running title: *Universal receipt book.*
S-29577.
Copies: MWA; MdBP; NN; NcRS; OrCS; PJA; PPL; PU-D; TN; ViU.

212. Roberts, Robert.
The house servant's directory; or, A monitor to private families: comprising hints on the arrangement and performance of servants' work ... and upwards of 100 useful receipts.... [1st ed.]. Boston: Munroe and Francis, [etc.], 1827. xiv, [15]–180 p.
S-30468.
Copies: ICJ; MB; MH-BA; MHi; MWA; NPV; NhHi; PPL.

213. [Shepard, John.]
The artist & tradesman's guide. Embracing some leading facts & principles of science, and a variety of matter adapted to the wants of the artist, mechanic, manufacturer, and mercantile community. To which is annexed an abstract of tonnage, duties, custom-house tares, and allowances.... Utica, [N.Y.]: Printed by W. Williams, 1827. 216 p.
Copyrighted by John Shepard.
S-30585.
Copies: CSmH; DeGE; IU; LU; MH; MH-BA; MWA; NUt; PHi.

214. Holland, Mary.
The modern family receipt book.... Philadelphia: R. Desilver, 1828. xiii, 230 p.
S-33580.
Copies: DLC; MWA; NNC; PP.

215. A new family receipt book containing all the truly valuable receipts for the various branches of cookery.... To which is added a number of valuable receipts for brewing, and several for removing spots and stains from silk, linen & woollen cloths.... Utica: Printed for the publisher, 1828. 20 p.

N-N0147766.

Copies: MWA; NBuG.

216. Roberts, Robert.

The house servant's directory; or, A monitor for private families: comprising hints on the arrangement and performance of servants' work ... and upwards of 100 useful receipts.... [2d ed.]. Boston: Munroe and Francis, [etc.], 1828. xiv, [15]–180 p.

S-35055.

Copies: CtHT-W; DLC; MBAt; MH; OCX; PPL.

217. Mackenzie, Colin.

Five thousand receipts in all the useful and domestic arts, constituting a complete and universal practical library, and operative cyclopaedia.... Philadelphia: J. J. Woodward, 1829. 546 p.

S-39367.

Copies: DeGE.

218. Mackenzie, Colin.

Mackenzie's five thousand receipts in all the useful and domestic arts: constituting a complete practical library relative to agriculture, bees, bleaching.... New American, from the latest London ed. With numerous and important additions.... By an American physician. Philadelphia: J. Kay, jun., and brother; Pittsburgh: H. C. Kay, [1829]. 456 p., illus.

S-39366.

Copies: CLCM; CSt-L; CtHT-W; DLC; IU; MH; MWA (3 variants); MdBL; MiD; MoSU; NCH; OCl; PPF; PPL; PU; TNJ; TxH; ViU; WHi.

219. [Shepard, John.]

The artist & tradesman's guide; embracing some leading facts & principles of science, and a variety of matter adapted to the wants of the artist, mechanic, manufacturer, and mercantile community. To which is annexed an abstract of tonnage, duties, custom-house tares, allowances, and useful mensuration tables.... 2d ed., with additions and improvements. New York: Printed by J. C. Johnson, 1829. 321, [3] p.

S-37508.

Copies: CtHT-W; DeGE; MWA; NNE; NhD; WWauHi.

220. ———— 3d ed., with additions and improvements. New York: Printed by J. C. Johnson, 1829. 321, [3] p.

N-S0493458.

Copies: CtY; CU; MSo.

221. [Aikin, John] 1747-1822.

The arts of life, described in a series of letters: 1. Providing food.—2. Providing clothing.—3. Providing shelter. For the instruction of young persons. By the author of Evenings at home. First[*sic*] American edition. Boston: Carter & Hendee, and Waitt & Dow, 1830. 1 p.l., [7]–158 p.

First published in London, 1802; 1st American ed., Boston, 1803.

C-41.

Copies: DLC; DeGE; MB; MH; MnU; MWA; RPB.

222. Mackenzie, Colin.

Mackenzie's five thousand receipts in all the useful and domestic arts, abridged; constituting a complete practical library, relative to agriculture, bees, bleaching.... A new American from the latest London ed.... Adapted to the convenience, necessities and means of the western states. Hamilton, Ohio: Taylor Webster, at the Telegraph office, 1830. 312 p., 2 l.

C-2332.

Copies: ICHi; InRM; MWA.

223. Mackenzie, Colin.

Mackenzie's five thousand receipts in all the useful and domestic arts; constituting a complete practical library relative to agriculture, bees, bleaching.... A new American,

223 (*continued*)
from the latest London ed. By an American physician. Philadelphia: J. Kay, Jun.; Pittsburgh: J. I. Kay & Co., 1830. 456 p., illus.
C-2331.
Copies: ICJ; MB; MWA; NjP; NjR; OC; OHi.

224. A new collection of genuine receipts, for the preparation and execution of curious arts, and interesting experiments, medical and miscellaneous, domestic and agricultural.... To which is added, a complete and much approved system of dyeing, in all its varieties. Stereotype ed. Concord, N.H.: Printed and published by Fisk & Chase, 1830. 102, 6 p.
Attributed to Rufus Porter.
N-N0144640.
Copies: MBCo; MWA.

4. Gazetteers

225. Scott, Joseph.
The United States gazetteer: containing an authentic description of the several states. Their situation, extent, boundaries, soil, produce, climate, population, trade and manufactures. Together with the extent, boundaries and population of their respective counties.... Illustrated with nineteen maps. Philadelphia: Printed by F. and R. Bailey, 1795. vi p., 1 l., [292] p., fold. maps.
The first gazetteer of the United States. cf. Pref.
Mp-29476.
Copies: DLC; DeGE; MH; MWA; NN; RPJCB.

226. Morse, Jedidiah, 1761–1826.
The American gazetteer, exhibiting, in alphabetical order, a much more full and accurate account ... of the states, provinces, counties, cities, towns, villages.... Describing the ... productions, commerce, manufactures.... Boston: Printed at the presses of S. Hall, and Thomas & Andrews, and sold by E. Larkin, [etc.], 1797. viii, [619] p., 7 maps.
Mp-32509.
Copies: CtY; DLC; ICJ; InU; MBAt; MH; MWA; NN; NjP; PPL; RPJCB; TxU; ViU.

227. Morse, Jedidiah, 1761–1826.
An abridgment of the American gazetteer .... Abridged by the author. Boston: Printed by Thomas and Andrews, 1798. 388 p., map.
Mp-34143.
Copies: CSt; CtY; DLC; InU; MBAt; MH; MHi; MWA; NN; PU; RPJCB.

228. Morse, Jedidiah, 1761–1826, comp.
The American gazetteer, exhibiting a full account of the civil divisions, rivers, harbours, Indian tribes, &c., of the American continent, also of the West India and other appendant islands; with a particular description of Louisiana. Compiled from the best authorities, by Jedidiah Morse ... Illustrated with maps. 2d ed., rev., corr. and enl. ... Charlestown, [Mass.]: Printed by and for Samuel Etheridge, and for Thomas and Andrews, Boston, 1804. vi, [628] p., 6 fold. maps (incl. front).
S&S-6825.
Copies: CLU; DLC; DeGE; MH; MWA; MWiW; NN; NjP; ViU.

229. Dean, James, 1777–1849.
An alphabetical atlas; or, Gazetteer of Vermont.... Montpelier, [Vt.]: Printed by

Samuel Goss, 1808. 43, [1] p.
S&S-14840.
Copies: CSmH; CtY; DLC; ICN; MB; MH; MWA; MiD-B; NN; OClWHi; VtHi; VtU.

230. Morse, Jedidiah, 1761–1826.
The American gazetteer, exhibiting a full account of the civil divisions, rivers, harbors.... And accompanied by a new and elegant general atlas of the world.... 3d ed., rev. and corr. Boston: Thomas & Andrews, 1810. 4 p.l., [590], 39 p., maps.
Atlas issued separately.
S&S-20788.
Copies: CSmH; DLC; DeU; MB; MH; MWA; NN; NcD; NjR; OCl; TU; ViU.

231. Spafford, Horatio Gates, 1778–1832.
A gazetteer of the state of New-York; carefully written from original and authentic materials.... Albany: H. C. Southwick, 1813. 334, ii p., fold. map., plates.
S&S-29836.
Copies: DLC; DeGE; IaDuU; KHi; MH; MWA; MWiW; MoSW; NIC; NjR.

232. Kilbourn, John, 1787–1833.
The Ohio gazetteer; or, Topographical dictionary, containing a description of the several counties, towns ... &c., in the state of Ohio.... Columbus: P. H. Olmstead & Co., July, 1816. 4 p.l., [13]–166 p.
S&S-38003.
Copies: DLC; I; MH; MiU-C; OClWHi; PPiHi; PU.

233. ———— 2d ed., improved. Columbus: Pub. by and for J. Kilbourn; Smith & Griswold, printers, Nov. 1816. 114 p.
S&S-38004.
Copies: CtY; ICHi; MWA; NHi; OC; OCHP; OClWHi; PPiU; WHi.

234. Brown, Samuel R., 1775–1817.
The western gazetteer; or, Emigrant's directory, containing a geographical description of the western states and territories, viz. the states of Kentucky, Indiana, Louisiana, Ohio, Tenessee [*sic*] and Mississippi: and the territories of Illinois, Missouri, Alabama, Michigan, and North-Western. With an appendix, containing sketches of some of the western counties of New-York, Pennsylvania and Virginia.... By Samuel R. Brown. Auburn, N. Y.: Printed by H. C. Southwick, 1817. vi, [7]–360 p.
S&S-40334.
Copies: CSmH; CtY; DLC; DeGE; ICN; InU; MB; MWA; MoS; NIC; NN; NjR; OClWHi; PPiU; TKL; WHi.

235. Kilbourn, John, 1787–1833.
The Ohio gazetteer; or, Topographical dictionary, containing a description of the several counties, towns ... &c., in the state of Ohio.... 3d ed., improved. Albany: Joshua Fish; G. J. Loomis & Co., printers, 1817. vi, [7]–180 p.
S&S-41200.
Copies: CtY; DLC; MWA; NUtHi; OClWHi; OHi; WHi.

236. ———— Columbus, [Ohio]: J. Kilbourn; printed at the office of the Intelligencer and gazette by P. H. Olmsted, 1817 v, [9]–74 p., map, plan.
S&S-41201.
Copies: CtY; DLC; ICN; MWA; N; NN; NjPT; OClWHi; OHi; PHi; PPiU; WHi.

237. Merrill, Eliphalet, 1765–1853.
Gazetteer of the state of New-Hampshire.... Compiled from the best authorities by Eliphalet Merrill and the late Phineas Merrill, Esq. Exeter, [N.H.]: Printed by C. Norris & Co., for the authors, 1817. 218, [13] p.
S&S-41414.

237 (*continued*)
Copies: CU; DLC; ICJ; InU; MHi; MWA; MWey; NRU; NjP; PC; PPL; ViW.

238. Worcester, Joseph Emerson, 1784–1865.
A geographical dictionary; or, Universal gazetteer, ancient and modern.... Andover, [Mass.]: Printed by Flagg and Gould, for the author; Salem: Henry Whipple, 1817. 2 v.
S&S-42957.
Copies: CSt; Ct; DLC; KyU; MH; MWA; MiU; NNC; NhHi; NjP; OUrC; THi; ViL.

239. Kilbourn, John, 1787–1833.
The Ohio gazetteer; or, Topographical dictionary, containing a description of the several counties, towns ... in the state of Ohio.... 5th ed., improved. Columbus: J. Kilbourn, [etc.], 1818. 144 p., illus., map.
S&S-44515.
Copies: DLC; ICN; MWA; NHi; NN; OClWHi; OO.

240. Worcester, Joseph Emerson, 1784–1865.
A gazetteer of the United States, abstracted from the Universal gazetteer of the author; with enlargement of the principal articles.... Andover, [Mass.]: Printed for the author by Flagg and Gould, 1818. viii, [360] p.
S&S-46867.
Copies: CtHi; DeGE; MWA; NICLA; NRU; NcA-S; NjP; OClWHi; RPB; ScC.

241. Kilbourn, John, 1787–1833.
The Ohio gazetteer; or, Topographical dictionary; containing a description of the several counties, towns ... in the state of Ohio.... 6th ed., improved. Columbus: Published and sold by Kilbourn. Sold also by Bailhache and Scott, Chillicothe, [etc.]; Bailhache & and Scott, printers—Chillicothe, 1819. 176 p., illus., 2 fold. maps, fold. plan.
S&S-48423.
Copies: CSmH; CtY; DLC; ICN; MWA; NHi; NN; OClWHi; PPL; PU; WHi.

242. Pease, John Chauncey, 1782–1859.
A gazetteer of the states of Connecticut and Rhode-Island.... By John C. Pease and John M. Niles. Hartford: William S. Marsh, 1819. vii, [1], 389, [1] p., fold. map.
S&S-49034.
Copies: CSt; Ct; DLC; ICN; ICU; Ia-HA; KHi; M; MB; MH; MWA; MdHi; MeHi; MiU; NN; NNC; OClWHi; RPA; WHi.

243. Kilbourn, John, 1787–1833.
The Ohio gazetteer; or, Topographical dictionary, containing a description of the several counties, towns ... in the state of Ohio.... 7th ed., carefully rev. and corr. Columbus: J. Kilbourn, 1821. 204 (i.e. 203)p., fold. maps, fold. plans.
N-K0133852; S-5766.
Copies: CtY; DLC; MWA; MiU; NIC; NN; OClWHi; OO; OOxM; PPL; PPiU; WHi.

244. Beck, Lewis Caleb, 1798–1853.
Gazetteer of the states of Illinois and Missouri, containing a general view of each state, a general view of their counties, and a particular description of their towns, villages, rivers.... Albany: C. R. & G. Webster, 1823. viii, [9]–352 p., illus., 5 maps and plans.
S-11769.
Copies: CSmH; CtY; DLC; ICJ; ICN; KU; MWA; Md; MiU-C; MnSJ; MoHi; MoKU; MH; N; NN; NNC; NWM; OCU; OClWHi; PHi; PPiU; PU; WHi.

245. Commercial directory; containing, a topographical description, extent and productions of different sections of the Union, statistical information relative to manufactures, commercial and port regulations, a list of the principal commercial houses, tables of imports and exports, foreign and domestic; tables of foreign coins, weights and measures, tariff of duties. Philadelphia: J. C. Kayser & Co., 1823. [4], [iii]–viii, 242, [8], 41 p., front. (fold. map), 6 plates.

S-12218.

Copies: CtY; DLC; DeGE; GU-De; ICU; IaHi; MH; MWA; MdBP; NNC; NcU; OCHP; PPL; PU; RPB; Vi.

246. Farmer, John, 1789–1838.

A gazetteer of the state of New-Hampshire. By John Farmer and Jacob B. Moore.... Concord: Jacob B. Moore, 1823. 276 p., illus., plates, map.

S-12494.

Copies: CtY; ICU; Ia-HA; InCW; MB; MH-BA; MWA; MdHi; MeBa; MiD-B; MnHi; NNS; Nh; NhD; NjR; OClWHi; PHi; RPB; VtU; WHi.

247. George, Noah J. T., d. 1849.

A pocket geographical and statistical gazetteer, of the state of Vermont. Embellished with diagrams.... To which is prefixed a particular description of the city of Washington, and a large number of statistical tables of the United States.... Haverhill, N.H.: S. T. Goss, 1823. vii, [9]–264 p.

S-12662.

Copies: CSmH; DLC; MB; MH; MWA; NN; NNC; Nh; Vt; VtU.

248. Morse, Jedidiah, 1761–1826.

The traveller's guide; or, Pocket gazetteer of the United States; extracted from the latest edition of Morse's Universal gazetteer. With an appendix.... New Haven: N. Whiting, 1823. 323, [1] p., map.

S-13404.

Copies: CtHi; CtY; DLC; MWA; NN; OU; PP; ViU.

249. Worcester, Joseph Emerson, 1784–1865.

Geographical dictionary; or, Universal gazetteer ... By J. E. Worcester. 2d ed. Boston: Cummings & Hilliard, University Press, Hilliard & Metcalf, 1823. 2 v.

S-14977.

Copies: CBPac; CtW; DLC; IEG; Ia; KyLx; MH; MWA; MdBL; MoS; NNC; Nh; NjP; P; PHi; RPB.

250. Spafford, Horatio Gates, 1778–1832.

A gazetteer of the state of New-York.... Albany: B.D. Packard; Troy, [N.Y.]: The author, 1824. 620 p., fold. map.

S-18063.

Copies: CL; CSmH; CtHT-W; DLC; IC; KyBB; MH; MWA; MdBP; MiD-B; MnU; MoS; NN; NhD; NjR; OClWHi; OO; PPL; PU; ViU; WHi; WaU.

251. Thompson, Zadock, 1796–1856.

A gazetteer of the state of Vermont... Montpelier: E. P. Walton, 1824. 310, [2] p., map, plates.

S-18190.

Copies: CL; CSmH; CtSoP; DLC; ICU; MB; MH; MWA; MnU; MoRM; NHi; NIC; NdU; NhD; O; OCHP; VtU; WHi.

252. Kilbourn, John, 1787–1833.

The Ohio gazetteer; or, Topographical dictionary, containing a description of the several counties, towns ... in the state of Ohio.... 8th ed., carefully rev. and corr. Columbus: J. Kilbourn, 1826. 231 p., fold. map.

S-25032.

Copies: DLC; MWA.

253. Morse, Jedidiah, 1761–1826.

Traveller's guide; or, Pocket gazetteer of the United States; extracted from the latest edition of Morse's Universal gazetteer. With an appendix, containing a description of

253 (*continued*)
some of the principal places in Canada.... 2d ed., enl., rev. and corr. New Haven: S. Wadsworth, 1826. 326 p., map.
S-25426.
Copies: CtHT-W; CtHi; CtY; DLC; ICHi; ICN; ICU; MB; MH-BA; MWA; MiD-B; NN; NWM; ViL; VtMiS.

254. Scott, John, 1793–1838.
The Indiana gazetteer; or, Topographical dictionary.... Centreville: John Scott & Wm. M. Doughty, 1826. 143 p.
S-26026.
Copies: DLC; ICN; InU; MiU-C; OHi.

255. Sherwood, Adiel, 1791–1879.
A gazetteer of the state of Georgia.... Charleston: Printed by W. Riley, 1827. iv p., 1 l., [7]–143 p.
S-30590.
Copies: DLC; G; GA; GAU; GU; NNMer; NcRS; OO.

256. Spofford, Jeremiah, 1787–1880.
A gazetteer of Massachusetts: containing a general view of the state.... With a map of the state. Newburyport: Charles Whipple, 1828. 348 p., map.
S-35374.
Copies: DLC; ICJ; MB; MBAt; MH; MH-BA; MWA; N; NN.

257. Hewett, Daniel, d. 1846.
A gazetteer of the New England. New York: Charles S. Francis, 1829. 84 p.
S-38966.
Copies: CtHT-W; MWA.

258. Kilbourn, John, 1787–1833.
The Ohio gazetteer; or, Topographical dictionary, describing the several counties, towns, villages, canals, roads, rivers, lakes, springs, mines, &c., in the state of Ohio. 9th ed., carefully rev. and corr. Columbus: J. Kilbourn, 1829. 335 p.
S-39197.
Copies: CtY; DLC; InHi; MB; MH; MWA; NBuG; NN; NjR; O; OCHP; OClWHi.

259. Sherwood, Adiel, 1791–1879.
A gazetteer of the state of Georgia.... 2d ed. Philadelphia: Printed by J. W. Martin and W. K. Boden, 1829. 1 p.l., [v]–vi, [2], [9]–300 p., illus., map.
S-40405.
Copies: Ct; DLC; DeU; GU; ICN; MB; MH; MiD; MnHi; NcD; OClWHi; P; PHi; WHi.

5. Description and travel

260. [Rawle, Francis] 1662–1727.
Ways and means for the inhabitants of Delaware to become rich; wherein the several growths and products of these countries are demonstrated to be a sufficient fund for a flourishing trade. Humbly submitted to the legislative authority of these colonies.... Philadelphia: Printed and sold by S. Keimer, 1725. 65, [1], [6] p.
The pamphlet was privately reprinted in Philadelphia, 1878.
Mp-2697.
Copies: CtY; DeWI; IC; MB; MH; MWA; MdBJ; MiU-C; PPAmP; PPL; PU; RPJCB.

261. [Cluny, Alexander.]
The American traveller: containing observations on the present state, culture and commerce of the British colonies in America, and the further improvements of which they are capable; with an account of the exports, imports and returns of each colony respectively, –and of the numbers of British ships and seamen, merchants, traders and manufacturers employed by all collectively.... In a series of letters, written originally to the Right Honourable the Earl of******** By an old experienced trader. [Philadelphia: Printed by Crukshank and Collins], 1770. 89, [1] p.
First published in London, 1769.
Mp-11603.
Copies: MWA; PPL.

262. Hutchins, Thomas, 1730–1789.
An historical narrative and topographical description of Louisiana, and West-Florida.... The climate, soil, and produce, whether animal, vegetable, or mineral; with directions for sailing into all the bays, lakes, harbours and rivers on the north side of the Gulf of Mexico, and for navigating between the islands situated along that coast.... Philadelphia: Printed for the author, and sold by Robert Aitken, 1784. iv, [5]–94, [1] p.
Mp-18532.
Copies: CU-B; CtY; DLC; FTaSU; GU; Ky; MiU-C; MWA; NcD; OCl; PPAmP; PPL; RPJCB; ViU.

263. Hutchins, Thomas, 1730–1789.
A topographical description of Virginia, Pennsylvania, Maryland, and North Carolina, comprehending the rivers Ohio, Kenhawa, Sioto, Cherokee, Wabash, Illinois, Mississippi, &c. The climate, soil, and produce ... roads.... Boston: Printed and sold by John Norman, 1787. [2], ii, 30, 2 p., 2 fold. tables.
Mp-20424.
Copies: MB; MBAt; MWA; MiU-C; NN.

264. Morse, Jedidiah, 1761–1826.
The American geography; or, A view of the present situation of the United States of America. Containing astronomical geography, geographical definitions. Discovery, and general description of America.... Summary account of ... productions; population; government; agriculture; commerce; manufactures.... Elizabeth-Town, [N.J.]: Printed by Shepard Kollock, 1789. xii, 534, [2] p., 2 maps.
Later editions issued under the title: *The American universal geography.*
Mp-21978.
Copies: DeGE; MWA; PPL.

265. An account of the soil, growing timber, and other productions of the lands in the countries situate in the back parts of the states of New-York and Pensylvania, in North America, and particularly the lands in the county of Ontario, known by the name of the

265 (*continued*)
Genesee tract.... [S.l.:] Printed in the year 1791. [2], 37 p., 2 maps.

Contains advice in farming and instructions to make maple syrup.

Now considered a London imprint.

Mp-23101.

Copies: DLC; MBAt; MWA; NHi.

266. Brissot de Warville, Jacques Pierre, 1754–1793.

New travels in the United States of America. Performed in 1788. By J. P. Brissot de Warville. Translated from the French.... New York: Printed by T. & S. Swords, for Berry & Rogers, 1792. xxvi, [27]–264, [3] p., fold. table.

First published in Paris, 1791. The present work contains the translation of the first two volumes only; it includes an account on the testing of John Fitch's steamboat.

Mp-24146.

Copies: CtY; DLC; KyU; MWA; MiU-C; NHi; NN; NcU; OCl; PPL; RPJCB.

267. [Lincoln, Benjamin] 1733–1810.

A description of the situation, climate, soil and productions of certain tracts of land in the district of Maine, and Commonwealth of Massachusetts. [Philadelphia?: 1793.] 44 p.

Attributed also to William Bingham.

Mp-25720.

Copies: MBAt; MWA; NHi; NN; RPJCB.

268. Morse, Jedidiah, 1761–1826.

The American universal geography; or, A view of the present state of all the empires, kingdoms, states, and republics in the known world, and of the United States of America in particular. In two parts.... Part I being a new edition of the American geography, corrected and greatly enlarged. Boston: Printed by Isaiah Thomas and Ebenezer T. Andrews, 1793. 2 v., maps.

Mp-25847.

Copies: CtY; DLC; MB; MBAt; MH; MWA.

269. Colles, Christopher, 1738–1816.

The geographical ledger and systemized atlas, being an united collection of topographical maps projected by one universal principle.... Containing I. Alphabetical references for pointing out the situation of lakes, islands, shoals, mills, mines, churches, iron works, forts, bridges, fords, ferries.... II. An alphabetical index referring to different parts of the map, whereby any city, town, river, creek, island, lake, &c. can be speedily found.... III. An actual survey of a number of roads.... New York: 1794. viii, [8], viii, [44] p., diagr., 8 maps.

Mp-26781.

Copies: DLC; MiU-C; NN; PPAmP.

270. Coxe, Tench, 1755–1824.

A view of the United States of America, in a series of papers, written at various times, between the years 1787 and 1794 ... interspersed with authentic documents: the whole tending to exhibit the progress and present state of civil and religious liberty, population, agriculture, exports, imports, fisheries, navigation, ship-building, manufactures, and general improvement. Philadelphia: W. Hall, and Wrigley & Berriman, 1794. v p., 1 l., [7]–13, 513 p., incl. fold. tables.

Mp-26829.

Copies: AU; CSmH; CtY; DLC; DeGE; IaU; InU; KyU; MBAt; MH; MHi; MWA; MeB; MiU-C; NN; NcU; NjP; PPL; RPJCB; ScU.

271. Griffith, Dennis.

A map of the state of Maryland, laid down from an actual survey of all the principal waters, public roads, and divisions of the counties therein; describing the situation of the cities, towns, villages ... furnaces, forges, mills, and other remarkable places ... as also a sketch of the state of Delaware, shewing the probable connexion of the Chesapeake and Delaware bays. By Dennis Griffith, May 19, 1794. Philadelphia: Thackara & Vallance, engravers, 1794. Map 82 x 140 cm.

Probably same as no. 275, cf. Wheat-511.

E-27070.

Copies: Md (not located).

272. Georgia Mississippi Company.

Grant to the Georgia Mississippi Company, the constitution thereof, and extracts relative to the situation, soil, climate, and navigation of the western territory of the state of Georgia.... Augusta: Printed by John Erdman Smith, 1795. 28 p.

Mp-28742.

Copies: DLC.

273. ——— Augusta: Printed by John Erdman Smith, and reprinted with appendix, [1795]. 39 p.

Mp-28743.

Copies: MBAt; MH; MWA; RPJCB.

274. ——— Augusta: Printed by John Erdman Smith, 1795. 40 p.

E-28744.

No copy known.

275. Griffith, Dennis.

Map of the state of Maryland, laid down from an actual survey of all the principal waters, public roads, and divisions of the counties therein; describing the situation of the cities, towns, villages, houses of worship and other public buildings. Furnaces, forges, mills, and other remarkable places ... and also a sketch of the state of Delaware shewing the probable connexion of the Chesapeake and Delaware bays; by Dennis Griffith, June 20th 1794.... Philadelphia: J. Vallance, 1795. Map 85 x 169 cm.

Mp-28772.

Copies: DLC; ICU; M; MH; MdBP; MdHi; MiU-C; NN.

276. Morse, Jedidiah, 1761–1826.

American universal geography; or, A view of the present state of all the empires, kingdoms, states, and republics ... and of the United States in particular.... 2d ed. Boston: Printed by Isaiah Thomas and Ebenezer T. Andrews, 1795–1796. 2 v., maps.

No copy of vol. 1 has been located.

Mp-29110; 30823.

Copies: MBAt; MH; MHi; MWA; RPJCB.

277. ——— 3d ed., corr. and considerably enlarged.... Boston: Printed by Isaiah Thomas and Ebenezer T. Andrews, 1796–1801. 2 v., plates, maps.

Mp-30824; S&S-957.

Copies: CtHT; DLC; MWA; MWiW; NN; NjR.

278. Whitelaw, James, 1748–1829.

A correct map of the state of Vermont from actual survey: exhibiting the county and town lines, rivers, lakes, ponds, mountains, meeting houses, mills, public roads, etc. Anno Domini 1795 ... [Rutland: S. Williams & Co., 1796.] Map.

Mp-31626.

Copies: CtY; DLC; MH; MHi; MWA; NhD; VtHi; VtU.

279. Brissot de Warville, Jacques Pierre, 1754–1793.

New travels in the United States of America. Performed in 1788, by J. P. Brissot de Warville. Translated from the French.... Boston: From the press of J. Bumstead, 1797. xxvi, [27]–276, [3] p.

Includes an account of the testing of John Fitch's steamboat.

Mp-31871.

Copies: DLC; DeGE; MB; MH; MWA; NN; PPL; RPJCB.

280. Morse, Jedidiah, 1761–1826.

A description of the soil, productions, commercial, agricultural and local advantages of the Georgia western territory.... Extracted and published in this form ... from Rev. Dr. Morse's American gazetteer.... Boston: Printed by Thomas & Andrews, 1797. 24 p., map.

Mp-32510.

Copies: DLC; MBAt; MHi; MWA; NN; RPJCB.

281. [Williamson, Charles] 1757–1808.

Description of the Genesee country, its rapidly progressive population and improvements: in a series of letters from a gentleman to his friend. Albany: Printed by Loring Andrews & Co., 1798. 37 p., 2 maps, plate.

281 (*continued*)
Mp-35033.
Copies: CSmH; DLC; MWA; MiU-C; NHi; NN; PPL; RPJCB.

282. [Williamson, Charles] 1757–1808.
Description of the settlement of the Genesee country, in the state of New-York. In a series of letters from a gentleman to his friend. New York: Printed by T. & J. Swords, 1799. 63 p., fold. map.
Includes hints on agriculture and references to mill sites and mineral resources.
Mp-36727.
Copies: CSmH; DLC; ICN; MBAt; MH; MHi; MWA; MiU-C; NN; PPL; WHi.

283. Ogden, John Cosens, 1751–1800.
An excursion into Bethlehem & Nazareth, in Pennsylvania, in the year 1799; with a succinct history of the Society of United Brethren, commonly called Moravians.... Philadelphia: Printed by Charles Cist, 1800. 1 p.l., 167, [1] p.
Mp-38149.
Copies: CSmH; CtY; DLC; DeGE; MBAt; MH; MWA; NN; NNC; NjP; P; RPJCB.

284. Drayton, John, 1766–1822.
A view of South Carolina as respects her natural and civil concerns.... Charleston: Printed by W. P. Young, 1802. iv, 252 p., maps, 6 plates.
Describes the natural resources, agriculture, particularly cotton production, manufactures, and inland navigation.
S&S-2159.
Copies: CSmH; CtY; DLC; ICU; IaU; MH; MWA; MiU-C; NN; NcD; NjP; PPL; ScU; ViU; WHi.

285. Moore, S.S.
The traveller's directory; or, A pocket companion: shewing the course of the main road from Philadelphia to New York, and from Philadelphia to Washington.... By S. S. Moore and T. W. Jones. Philadelphia: Mathew Carey, 1802. 3 p.l., 52 p., 38 maps.
S&S-2686.
Copies: CSmH; DLC; MH; MWA; MiU-C; NN; NjP; PHi; PPAmP.

286. Morse, Jedidiah, 1761–1826.
The American universal geography.... 4th ed., corr. and considerably enl. Boston: Printed by Isaiah Thomas and Ebenezer T. Andrews, 1802. 2 v., maps.
S&S-2698.
Copies: CSt; MNe; MWA; NN.

287. Ellicott, Andrew, 1754–1820.
The journal of Andrew Ellicott, late commissioner on behalf of the United States ... for determining the boundary between the United States, and the possessions of His Catholic Majesty in America, containing occasional remarks on the situation, soil, rivers, natural productions ... of the different countries on the Ohio, Mississippi, and Gulf of Mexico ... the whole of West Florida, and part of East Florida. To which is added an appendix, containing all the astronomical observations made use of for determining the boundary ... likewise a great number of thermometrical observations.... Philadelphia: Printed by Budd & Bartram for Thomas Dobson, 1803. vii, 299, 151 p., illus., 8 fold. maps, 6 fold. plates.
S&S-4147.
Copies: CU; CtY; DLC; FMU; ICJ; KyLx; MBAt; MH; MWA; MiU-C; NN; NcD; OCl; PHi; PU; TU; ViU.

288. Moore, S.S.
The traveller's directory; or, A pocket companion, shewing the course of the main road from Philadelphia to New York, and from Philadelphia to Washington.... By S.S. Moore and T. W. Jones. 2d ed. Philadelphia: Mathew Carey, 1804. 2 p.l., 37, 19 p., 38 maps.
S&S-6815.
Copies: CSmH; DLC; DeU; MH-BA; MWA; NN; NcD; NjP; OCl; ViU.

289. [Williamson, Charles] 1757–1808.
A description of the Genesee country, in the state of New-York: in which the situation, dimensions, civil divisions, soil, minerals, produce, lakes and rivers ... navigation, trade and manufactures ... are impartially described.... By Robert Munro [pseud.]. New York: Printed for the author, 1804. 16 p., map.
S&S-6836, 7770.
Copies: CSmH; CtY; DLC; MB; MWA; MiU-C; NN.

290. ——— [Canandaigua?, N.Y.]: Printed for the author, 1804. 24 p.
S&S-6835, 7769.
Copies: MWA (variant issue, incomplete); NN.

291. [Williamson, Charles] 1757–1808.
A view of the present situation of the western parts of the state of New York, called the Genesee country.... Frederick-Town, [Md.]: Printed at the Herald press for the author, 1804. 23 p.
S&S-7771.
Copies: CSmH; DLC; DeU; NNPM; NjN.

292. Harris, Thaddeus Mason, 1768–1842.
The journal of a tour into the territory northwest of the Alleghany Mountains; made in the spring of the year 1803. With a geographical and historical account of the state of Ohio.... Boston: Printed by Manning & Loring, 1805. viii, [9]–271 p., fold. map, fold. plates.
Includes observations on mineral resources and industrial enterprises.
S&S-8587.
Copies: CSmH; DLC; DeGE; ICJ; InU; KU; KyU; MB; MWA; MdBP; MiU-C; MnU; MoU; NN; NcD; OClWHi; OU; PPL; RPJCB; ViU.

293. Morse, Jedidiah, 1761–1826.
The American universal geography.... 5th ed., corr. and improved. Boston: Printed by J. T. Buckingham for Thomas & Andrews, 1805. 2 v., maps.
Part 1: 5th ed.; part 2: 4th ed.
S&S-8933.
Copies: DLC; MWA; NN; NcD.

294. Ogden, John Cosens, 1751–1800.
An excursion into Bethlehem & Nazareth in Pennsylvania, in the year 1799; with a succinct history of the Society of United Brethren, commonly called Moravians.... Philadelphia: Printed by Charles Cist, 1805. 1 p.l., 167 p.
S&S-9050.
Copies: CSmH; CtY; MB; MWA; NN; OU; PBL; PSC; ViU.

295. Scott, Joseph.
A geographical dictionary of the United States of North America. Containing a general description of each state.... With a succinct account of Indiana, Michigan, and upper and lower Louisiana territories.... Philadelphia: Printed by A. Bartram, for J. Johnson, 1805. iv, [584] p., fold. map.
S&S-9325.
Copies: CSmH; DLC; NN.

296. ——— Philadelphia: Printed by A. Bartram, for T. Armstrong, 1805. iv, [584] p., map.
S&S-9326.
Copies: DLC; MWA; NN.

297. [Blodget, Samuel] 1757–1814.
Economica: a statistical manual for the United States of America.... Washington: Printed for the author, 1806. viii, 202, xiv p.
S&S-10004.
Copies: CtY; DLC; DeGE; IaU; KU; MH; MWA; MiU; NIC; NN; NcD; Nh; PP; PPAmP; PPL; ScU; ViW; WaU.

298. Scott, Joseph.
A geographical description of Pennsylvania: also of the counties respectively.... Philadelphia: Printed by Robert Cochran, 1806. 147, [1] p., fold. map.
Includes descriptions of manufacturing

298 *(continued)*
enterprises, roads, canals, etc.
S&S-11331.
Copies: DLC; MBAt; MWA; PPL; PPi.

299. [Mitchill, Samuel Latham] 1764–1831.
The picture of New-York; or, The traveller's guide through the commercial metropolis of the United States. By a gentleman residing in this city. New York: I. Riley, 1807. viii, 223 p.
S&S-13104.
Copies: CSt; CtY; DLC; DeGE; MH-BA; MWA; NN; NjR; OMC.

300. ________ New York: 1807. [xv], 216 p.
Kress-B5230.
Copies: MH.

301. Scott, Joseph.
A geographical description of the states of Maryland and Delaware; also of the counties, towns, rivers, bays and islands.... Philadelphia: Printed by Kimber, Conrad, 1807. [2], 191 p., 2 fold. maps.
Includes observations on internal improvements and manufactures.
S&S-13555.
Copies: CSmH; DLC; DeGE; DeWi; MB; MWA; MdBE; NN; PPL.

302. Tongue, James.
A letter addressed to the people of Maryland, giving a short account of the country on the south shore of Lake Erie; including a brief description of its situation, climate, soil, production, trade, commerce, manufactures and exports.... Washington City: Printed by J. D. Westcott and R. Dinsmore, 1807. 32 p.
S&S-13718.
Copies: MBC; MWA; NHi; NN; PPL.

303. Ashe, Thomas, 1770–1835.
Travels in America, performed in 1806, for the purpose of exploring the rivers Alleghany, Monongahela, Ohio, and Mississippi, and ascertaining the produce and condition of their banks and vicinity.... London, printed. Newburyport, [Mass.]: Re-printed for W. Sawyer & Co., by E. M. Blunt, 1808. ix, [11]–366 p.
Offers data on the navigability of the rivers, also on salt springs, lead deposits, and manufacturing enterprises.
S&S-14380.
Copies: CLU; CtY; DLC; DeGE; MWA; NN; NjR; WHi.

304. ________ Pittsburgh: Reprinted by Cramer and Spear, 1808. 366 p.
S&S-14381.
Copies: MWA.

305. [Blodget, Samuel] 1757–1814.
Economica; a statistical manual for the United States of America, with additions to the beginning of the year 1810. Washington: 1810. vii, 202, xiv, 8 p.
S&S-19579.
Copies: ICU; MH; MWA; NN; NNC; NjR; PHC; PPL.

306. Cuming, Fortescue, 1762–1828.
Sketches of a tour to the western country, through the states of Ohio and Kentucky; a voyage down the Ohio and Mississippi rivers, and a trip through the Mississippi territory, and part of West Florida.... Pittsburgh: Cramer, Spear and Eichbaum, 1810. 504 p.
Includes descriptions of roads, bridges, ferries, and manufacturing enterprises.
S&S-19902.
Copies: CSmH; CtY; DLC; DeWI; GMW; IC; ICN; IaGG; Ia-HA; KyU; MH; MWA; MdHi; MiD-B; MoKU; MoS; MoSHi; MoSM: Ms-Ar; NN; OC; OClWHi; OrHi; P; PPAmP; PPL; PPiU; ScU; T; ViU.

307. Schultz, Christian.
Travels on an inland voyage through the states of New-York, Pennsylvania, Virginia, Ohio, Kentucky and Tennessee, and through the territories of Indiana, Louisiana,

Mississippi, and New-Orleans; performed in the years 1807 and 1808.... New York: Printed by Isaac Riley, 1810. 2 v., 5 maps, 2 plates.

Gives data on roads, mines, and mineral resources.

S&S-21289.

Copies: DLC; DNAL; DeU; ICN; MH; MWA; MWiW; MnHi; Ms-Ar, NN; NWM; NcU; PPL.

308. Varlé, Charles.

Topographical description of the counties of Frederick, Berkeley & Jefferson, situated in the state of Virginia.... Also, the quality of the soil, the manufactories, mills, &c.... Winchester, [Va.]: Printed by W. Heiskell, 1810. 34 p.

S&S-21888; Sabin-98630.

Copies: MWA; NHi; NcD; WHi.

309. Whitelaw, James, 1748–1829.

A correct map of the state of Vermont from actual survey; exhibiting the county and town lines, rivers, lakes, ponds, mountains, meetinghouses, mills, public roads, &c.... The 2d ed., with alterations and additions, published Septr. 1810. Engraved by Amos Doolittle, New Haven, 1796, and by James Wilson, Vermont, 1810. [Ryegate, Vt.]: 1810. Map 103 x 70 cm.

S&S-22047.

Copies: MWA; NhD.

310. Ashe, Thomas, 1770–1835.

Travels in America, performed in 1806, for the purpose of exploring the rivers Alleghany, Monongahela, Ohio, and Mississippi, and ascertaining the produce and condition of their banks and vicinity.... New York: 1811. 366 p.

S&S-51052.

Copies: CtY; MWiW; NN; PPL; PU.

311. Connecticut Academy of Arts and Sciences, New Haven.

A statistical account of the towns and parishes in the state of Connecticut. v. 1, no. 1–2. New Haven: Printed and sold by Walter and Steele, 1811–1815. xi, 83, [1], [39] p.

No more published.

No. 1 includes *A statistical account of the city of New-Haven,* by Timothy Dwight.

N-C0639247; S&S-22603, 34427.

Copies: Ct; CtHi; MH; MWA (variant issue: xi, 124 p.); NN; NhD; OO; PU; RHi.

312. Dwight, Timothy, 1752–1817.

A statistical account of the city of New-Haven. [New Haven: Printed and sold by Walter and Steele, 1811.] xi, 83, [1] p. (Connecticut Academy of Arts and Sciences. A statistical account of the towns and parishes in the state of Connecticut, v. 1, no. 1.)

Includes observations on agricultural methods and products, and on manufacturing enterprises and their production.

N-D0470641.

Copies: CtY; DLC; DeGE; ICN; MB; MH-A; NcGuG; OClWHi; PPAN; PU; ViU.

313. [Goodenow, Sterling.]

A brief topographical & statistical manual of the state of New-York: exhibiting the situation and boundaries of the several counties ... and designating the principal places.... Albany: J. Frary, 1811. 36 p.

Contains information on canals.

S&S-22928.

Copies: CSmH; CtY; DLC; MB; MH; NBu; NHi; NN; NjR; PHi; WHi.

314. Mease, James, 1771–1846.

The picture of Philadelphia, giving an account of its origin, increase and improvements in arts, sciences, manufactures, commerce and revenue. With a compendious view of its societies.... Its police—the public buildings.... Philadelphia: B. & T. Kite, 1811. xii, 376 p., fold. map.

S&S-23363.

Copies: ArBaA; CSmH; DeGE; MBCo; MH; MWA; MoSM; NIC; NNMer; NjR; PP; PPL; PU; RHi; RP.

315. ——— Philadelphia: B. & T. Kite, 1811. xii, 358 p., fold. map.

315 (*continued*)
N-M0388837.
Copies: DLC; NjP; NBu; RPJCB.

316. Melish, John, 1771–1822.
Travels in the United States of America in the years 1806 & 1807, and 1809, 1810, & 1811.... Illustrated by eight maps.... Philadelphia: The author, 1812. 2 v., maps (4 fold.), 2 fold. tables.
S&S-26062.
Copies: DeGE; Ia; MCM; MH; MWA; MWiW; MdHi; NNC; NNS; NcA-S; PR; ViAl.

317. Morse, Jedidiah, 1761–1826.
The American universal geography.... 6th ed. Boston: Thomas & Andrews, [etc.], 1812. 2 v., maps.
S&S-26142.
Copies: CSmH; CSt; DLC; DeWI; GEU; ICU; LNHT; MH; MWA; MdBP; MiD-B; MnU; MoHi; NNC; NcW; OO; RPB.

318. Dickinson, Rodolphus, 1787–1863.
A geographical and statistical view of Massachusetts proper.... Greenfield, [Mass.]: Printed by Denio and Phelps, 1813. 80 p.
Includes an account on manufactures.
S&S-28328.
Copies: CSmH; CtY; DLC; DeGE; MH-BA; MWA; MeBa; MiD-B; MnHi; NIC; NN; NcD; NjPT; NjR; O; WHi.

319. Ellicott, Andrew, 1754–1820.
The journal of Andrew Ellicott, late commissioner on behalf of the United States ... for determining the boundary between the United States, and the possessions of His Catholic Majesty in America, containing occasional remarks on the situation, soil, rivers, natural productions ... of the different countries on the Ohio, Mississippi, and Gulf of Mexico ... the whole of West Florida and part of East Florida. To which is added an appendix, containing all the astronomical observations made use of for determining the boundary ... likewise a great number of thermometrical observations.... Philadelphia: Printed by William Fry, 1814. vii, 299, 151 p., illus., 8 fold. plates, 6 fold. maps.
S&S-31400.
Copies: CU; DLC; KyU; MWA; MdBP; NBuG; NjP; OU; PHC; PPL; RPJCB.

320. Melish, John, 1771–1822.
The traveller's directory through the United States: consisting of a geographical description of the United States, with topographical tables of the counties, towns, population, &c., and a description of the roads.... Philadelphia: Printed for the author, by G. Palmer, 1815. [4], 32, 11, 82 [41], [2] p., maps.
S&S-35248.
Copies: DLC; ICU; MH; MWA (also variant copy: [4], 32, iv, 100 [44] p.); NNC; NjP; OCl; PPL; PU.

321. Melish, John, 1771–1822.
Travels through the United States of America, in the years 1806 & 1807, and 1809, 1810 & 1811.... With corrections and improvements to 1815 and a new set of coloured maps.... Philadelphia: J. Melish, 1815. 2 v., 8 maps.
S&S-35249.
Copies: DLC; GU-De; MWA; MiU; NcU; PU; RPB; ViU.

322. Greenleaf, Moses, 1777–1834.
A statistical view of the district of Maine; more especially with reference to the value and importance of its interior.... Boston: Cummings and Hilliard, 1816. viii, [9]–154 p., map.
S&S-37745.
Copies: DLC; ICJ; IaU; MB; MH; MWA; MiD; MoSU; NN; NhD; OC; PPL; PU.

323. Melish, John, 1771–1822.
A geographical description of the United States of America, with the contiguous British and Spanish possessions.... Philadel-

phia: The author, 1816. 182 p., 5 maps.
S&S-38219.
Copies: DLC; MWA; NcWsW; OkU.

324. ———— 2d ed. Philadelphia: The author; T. H. Palmer, printer, 1816. 182 p., 5 maps.
S&S-38220.
Copies: DLC; InHi; MWA; MiU-C; NjR; P; PPL; ViU.

325. Melish, John, 1771–1822.
The traveller's directory through the United States. Consisting of a geographical description of the United States, with topographical tables of the counties, towns, population &c. and a description of the roads.... Philadelphia: T. & G. Palmer, printers, 1816. 40, 37–134, [2] p., maps.
S&S-38222.
Copies: DLC; MWA (also variant copy: 40, vi, 100 p., maps); MiU-C; NNA.

326. Pitkin, Timothy, 1766–1847.
A statistical view of the commerce of the United States of America: its connection with agriculture and manufactures: and an account of the public debt, revenues, and expenditures of the United States. With a brief review of the trade, agriculture, and manufactures of the colonies, previous to their independence. Accompanied with tables, illustrative of the principles and objects of the work. Hartford: Printed by C. Hosmer, 1816. xii, 407, xix, [1] p., incl. tables.
S&S-38648.
Copies: CtHT; DLC; DeGE; GHi; ICU; IaU; InHi; KHi; KyLo; MH; MWA; MdHi; MiU; MnU; MoK; NN; NcD; NjP; OClWHi; PPL; RPB.

327. Blunt's stranger's guide to the city of New-York.... To which is prefixed, an historical sketch, general description, plan and extent of the city.... New York: Edmund M. Blunt, 1817. xv, 306 p., plan, 3 plates.
S&S-40278.
Copies: CtY; DLC; ICN; MHi; MW; MWA; MiU-C; NNA; NNMer; NNS; PPL; RNR.

328. Pitkin, Timothy, 1766–1847.
A statistical view of the commerce of the United States of America: its connection with agriculture and manufactures: and an account of the public debt; revenues, and expenditures of the United States. With a brief review of the trade, agriculture, and manufactures of the colonies, previous to their independence. Accompanied with tables, illustrative of the principles and objects of the work.... 2d ed., with additions and corrections. New York: James Eastburn & Co.; printed by Hamlen & Newton, Hartford, 1817. xii, 445, vii, [1] p.
S&S-41832.
Copies: CSt; Ct; DLC; DeGE; GU; MH; MWA; MnM; MoSW; NIC; NMW; NcD; NjR.

329. ———— New Haven: Durrie and Peck, 1817. 445 p.
S&S-41831.
Copies: MnM.

330. Bristed, John, 1778–1855.
The resources of the United States of America; or, A view of the agricultural, commercial, manufacturing, financial, political, literary, moral and religious capacity and character of the American people.... New York: J. Eastburn & Co.; Abraham Paul, printer, 1818. xvi, 505, [1] p.
S&S-43445.
Copies: AU; CLU; DLC; DeGE; ICJ; InU; MAJ; MB; MWA; MnSM; NICLA; NcU; NjP; OU; PPL; TxU; ViU; WHi.

331. Cobbett, William, 1763–1835.
A year's residence in the United States of America; treating the face of the country, the climate, the soil, the products, the mode of cultivating the land, the prices of land, of labour, of food, of raiment.... In three parts.... New York: Printed for the author by Clayton and Kingsland, 1818–1819. 3 pts. in 1 v., illus.
Part I (also issued separately in 1818) contains *Ruta Baga. Culture, mode of preserving and uses of the Ruta Baga, sometimes called the*

331 (*continued*)
Russia, and sometimes the Swedish turnip: p. [70]–134.
S&S-43654. 47652.
Copies: DLC; DeGE; MWA; NNS; NjR.

332. Darby, William, 1775–1854.
The emigrant's guide to the western and southwestern states and territories comprising a geographical and statistical description of the states of Louisiana, Mississippi, Tennessee, Kentucky and Ohio.... New York: Kirk & Mercein, 1818. [6], 311, xiii p., maps.
Includes observations on silk culture, roads and other improvements, and inland navigation, with "a complete list of the road and river routes, west of the Allegheny Mountains."
S&S-43809.
Copies: AU; CU; CtHi; DLC; DeGE; IC; KyLx; MWA; MiU-C; NN; NcWsW; NjP; OFH; PPL; RPJCB; ViU.

333. The emigrant's guide; or, Pocket geography of the western states and territories, containing a description of the several cities, towns, rivers, antiquities, population, manufactories, prices of land, soil, productions, and exports.... Cincinnati: Phillips & Speer, 1818. iv, [5]–266 p.
S&S-43943.
Copies: CoD; DLC; ICN; MWA; NHi; OClWHi; PPi;WHi.

334. Melish, John, 1771–1822.
A geographical description of the United States, with the contiguous British and Spanish possessions.... 3d ed. Philadelphia: The author, 1818. 186 p., maps.
S&S-44791.
Copies: DLC; GMiW; IHi; InHan; KyCov; MWA; MiU; MoSM; NBu; NN; OClWHi; PPAmP; PPL; Tx; ViU.

335. Melish, John, 1771–1822.
The traveller's directory through the United States, consisting of a geographical description of the United States, with topographical tables of the countries, towns, populations, and a description of the roads. Philadelphia: J. Melish & S. Harrison, 1818. 40, 35–134, [27] p., maps.
S&S-44796.
Copies: MHi; MWA; MiPh; NN.

336. [Mitchill, Samuel Latham] 1764–1831.
The picture of New-York, and stranger's guide through the commercial emporium of the United States.... New York: A. T. Goodrich & Co., 1818. 357 p.
S&S-45335.
Copies: CSmH; NbOU.

337. Seybert, Adam, 1773–1825.
Statistical annals: embracing views of the population, commerce, navigation, fisheries, public lands, post-office establishment, revenues, mint, military and naval establishments, expenditures, public debt and sinking fund, of the United States of America: founded on official documents: commencing on the fourth of March seventeen hundred and eighty-nine and ending on the twentieth of April eighteen hundred and eighteen.... Philadelphia: Thomas Dobson & Son; William Fry, printer, 1818. xxvii, [1], 803 p.
S&S-45692.
Copies: DLC; DeGE; IaU; MBAt; MH-BA; MWA; MiU; MnM: MnSM; NIC; NcWsW.

338. Waldo, Samuel Putnam, 1780–1826.
The tour of James Monroe, president of the United States, in the year 1817; through the states of Maryland, Pennsylvania, New-Jersey, New-York, Connecticut, Rhode-Island, Massachusetts, New-Hampshire, Vermont, and Ohio together with a sketch of his life, his inaugural speech and first message; and historical and geographical notices of the principal places through which he passed.... Hartford: Printed by F. D. Bolles & Co., 1818. xii, 300 [288] p., front. (port.).
S&S-46675.
Copies: A-Ar; CSmH; CSt; CtHT; CtHi; DLC; GU; ICN; IaU; KHi; LNHT; M; MH; MSaE; MWA; MdBJ; MeBa; NCaS; NjR; RPB.

339. Dana, Edmund.
Geographical sketches of the western country.... Also, a list of the principal roads. By E. Dana. Cincinnati: Looker, Reynolds & Co., 1819. iv. [5]–312 p.
S&S-47781.
Copies: CSmH; CtY; DLC; ICN; InU; KyU; MBAt; MWA; MiU-C; NIC; NcD; NjP; OC; OrU; PPL; TxU; ViU.

340. Darby, William, 1775–1854.
A tour from the city of New-York, to Detroit in the Michigan Territory, made between the 2d of May and the 22d of September, 1818.... New York: Published for the author by Kirk & Mercein; printed by E. Worthington, Brooklyn, 1819. viii, [9]–228, lix, [7], lxi–lxiii p., maps.
Includes references to steamboats, bridges, roads and canals.
S&S-47786.
Copies: CSmH; DLC; ICU; MWA; MiU-C; MnU; NNC; NcD; PPL; PU; ScU.

341. McMurtrie, Henry, 1793–1865.
Sketches of Louisville and its environs.... 1st ed. Louisville, [Ky.]: Printed by S. Penn, jun., 1819. viii, 255 p., front. (fold. plan).
"Estimate for the Ohio Canal": fold. sheet inserted at end.
S&S-48551.
Copies: C; CtY; DLC; ICJ; ICU; KyU; LU; MH; MWA, MiU; NIC; NN; NcD; OCU; OClWHi; PPiU; RPJCB; ViU; WHi.

342. Melish, John, 1771–1822.
The traveller's directory through the United States. Consisting of a geographical description of the United States, with topographical tables of the counties, towns, population &c. and a description of the roads.... 5th ed. Philadelphia: Printed for the author. T. H. Palmer, printer, 1819. 134, [14] p., maps.
S&S-48670.
Copies: CU-S; KyU; MHi; MWA; NUt; OClWHi; PHi.

343. Melish, John, 1771–1822.
The traveller's directory through the United States of America; being a complete list of the direct and cross roads, together with the conveyance by water throughout the different states and territories: including the connecting roads.... Philadelphia: Printed for J. Melish and S. Harrison, 1819. 102 p., maps.
S&S-48669.
Copies: ArU; MH; MHi; NN; OFH.

344. Morse, Jedidiah, 1761–1826.
The American universal geography.... 7th ed. Charlestown, [Mass.]: Lincoln & Edmands, [etc.], 1819. 2 v., maps.
S&S-48764.
Copies: CSt; DLC; IP; In; KyLo; LNHT; MH; MWA; MiU-C; NVP; NbHi; NhD; NjP; OClWHi; PAtM; ScC; TNJ; WHi.

345. Thomas, David, 1776–1859.
Travels through the western country in the summer of 1816. Including notices of the natural history, antiquities, topography, agriculture, commerce and manufactures: with a map of the Wabash country, now settling. Auburn, N.Y.: Printed by David Rumsey, 1819. 2 p.l., 320 p., front. (fold. map).
S&S-49585.
Copies: CSmH; CtY; DLC; DeGE; ICU; InU; MB; MWA; MiD-B; NIC; NN; OCU; OClWHi; PPiU; PU; WHi.

346. Waldo, Samuel Putnam, 1780–1826.
The tour of James Monroe, President of the United States, through the northern and eastern states, in 1817; his tour in the year 1818; together with a sketch of his life.... Hartford: S. Andrus, 1819. xii, [13]–348 p., port.
S&S-50014.
Copies: CSf; Ct; CtSoP; IP; MC; MWA; MoSW; NBLiHi; NR; NhD; OCl; OO; RPB; TSewU; ViU; VtWood.

347. Melish, John 1771–1822.
The traveller's directory through the United States. Consisting of a geographical description of the United States, with topographical tables of the counties, towns, population, &c. and a description of the roads.... Philadelphia: Printed for the author, 1820. 134, [14] p., maps.
N-M0421906; S-2225.
Copies: CtY.

348. [Tudor, William] 1779–1830.
Letters on the eastern states. New York: Kirk and Mercein, 1820. vi, [2], [5]–356 p.
Includes letters on agriculture and manufactures.
S-3497.
Copies: CtW; DLC; DeGE; IaU; InU; KyLx; LU; MH; MWA; MeHi; MiU; MnHi; NIC; NN; OCl; PPL; RPB.

349. Waldo, Samuel Putnam, 1780–1826.
The tour of James Monroe, President of the United States, through the northern and eastern states, in 1817; his tour in the year 1818; together with a sketch of his life; with descriptive and historical notices of the principal places through which he passed.... 2d ed. By S. Putnam Waldo.... Hartford: Silas Andrus, 1820. xii, [13]–348 p., front. (port.).
S-4143.
Copies: CoD; CtHi; DLC; DeGE; IHi; IaU; KU; MWA; MdBP; MiD-B; MiU; MoS; MtMiS; NN; NcD; OC; PPL; TNJ; ViU; Vt.

350. Dwight, Timothy, 1752–1817.
Travels in New England and New York. New Haven: T. Dwight; S. Converse, printer, 1821–1822. 4 v., maps.
Includes references to agricultural production, industrial enterprises, roads, bridges and public buildings.
S-5221.
Copies: CSmH; CtY; DLC; DeWI; FOA; GDC; ICU; IaU; InCW; LN; MH; MWA; MdBP; MeU; MiU; MnU; NIC; NjP; OU; PPL; RP; ScU; TxU; WHi; WaU.

351. Tudor, William, 1779–1830.
Letters on the eastern states. 2d ed. Boston: Wells and Lilly, 1821. viii, [9]–423 p.
Includes letters on agriculture and manufactures.
S-7016.
Copies: CU; Ct; DLC; GDC; IC; IaAS; InHi; KyLo; LNHT; MH; MWA; MdBJ; MeU; MiU; MnHi; NN; NhD; OC; PPL; RPB; ScU; TxU; WHi; WaU.

352. [Clinton, De Witt] 1769–1828.
Letters on the natural history and internal resources of the state of New-York. By Hibernicus [pseud.]. New York: E. Bliss & E. White, 1822. 224 p.
S-8370.
Copies: C; CSmH; CtY; DLC; IC; ICJ; MB; MH; MWA; MdBE; MiU-C; MnU; NIC; NN; NNE; NjN; OCHP; PHC; PPA; PPL; RNR; RP.

353. [Goodenow, Sterling.]
A brief topographical and statistical manual of the state of New-York.... 2d ed., enl. and improved. Containing also, an account of the grand canals.... New York: E. Bliss and E. White, 1822. 72 p.
Also issued the same year with "Second supplement, containing the alterations and additions made since 1821," 88 p.
S-8855.
Copies: CtY; DLC; DeGE; MB; MH; MWA; MdBP; MiD-B; NIC; NN; NNC; NbU; OClWHi; PPAmP; PPL; RPB; ViL; WHi.

354. Hawley, Zerah, 1781–1865.
A journal of a tour through Connecticut, Massachusetts, New-York, the north part of Pennsylvania and Ohio, including a year's residence in that part of the state of Ohio, styled New Connecticut, or Western Reserve.... New Haven: Printed by S. Converse, 1822. 158 p.
S-8974.
Copies: CtY; DLC; ICN; MB; MWA; MiU-C; N; NN; OClWHi; PPi; WHi.

355. Melish, John, 1771–1822.
A geographical description of the United States, with the contiguous countries, including Mexico and the West Indies; intended as an accompaniment to Melish's map of these countries.... A new ed., greatly improved. Philadelphia: The author, 1822. 2 p.l., viii, [9]–491, [15] p., maps.
S-9456.
Copies: CMont; DLC; DeGE; GEU; ICU; IaU; In; KyBgW; MH; MWA; Me; MiU; MoSM; NNC; NbU; NcU; NjR; OClWHi; OrP; PHi; TxU; WHi.

356. Melish, John, 1771–1822.
The traveller's directory through the United States: containing a description of all the principal roads ... with copious remarks on the rivers and other objects.... Philadelphia: The author, 1822. xix, 183 p., 2 fold. maps.
S-9460.
Copies: DLC; Ia-HA; InHi; MB; MH; MHi; MWA; MeHi; NN; NUtHi; OClWHi; PHi; PPL.

357. Shaw, Joshua.
United States directory for the use of travellers and merchants, giving an account of the principal establishments, of business and pleasure, throughout the Union.... Philadelphia: Printed by J. Maxwell, [1822]. 2 p.l., ix, 156 p., fold. map, fold. table.
S-10249.
Copies: DLC; DeGE; GU-De; MBNEH; MH-BA; P; PHi; PP; PPL.

358. Duncan, John Morison, 1759?–1825.
Travels through part of the United States and Canada in 1818 and 1819. By John M. Duncan.... New York: W. B. Gilley; New Haven: Howe & Spalding, 1823. 2 v.
Includes observations on roads and canals, bridges, public buildings, Philadelphia water-works, flour mills, and steamboats.
S-12404.
Copies: CSf; CtY; DeGE; GAU; IC; IaU; KyLxT; LNHT; MBAt; MWA; MdBE; MnSM; NNUT; NhHi; OC; PPL; RPJCB; ScU; TxU; Vi.

359. Hodgson, Adam.
Remarks during a journey through North America in the years 1819, 1820 and 1821, in a series of letters.... Collected, arranged, and published by Samuel Whiting. New York: [J. Seymour, printer], 1823. 335 p.
S-12851.
Copies: AU; CtY; DLC; GU; IU; IaU; InHi; KHi; KyU; L-M; MB; MWA; MdHi; MiU; NN; NbHi; NcD; NhHi; NjP; OU; PPL; RPB; ScU; TKL; TxU; ViU; WBB.

360. [Preston, John] b. 1755.
A statistical report of the county of Albany, for the year 1820, written by a friend of agriculture.... Albany: Packard & Van Benthuysen, 1823. 30 p.
A report on the agriculture of the county.
S-13843.
Copies: MB; MWA; N; NAl; NCH; NHi; NN; WHi.

361. Wilson, Thomas, 1768–1828?.
Picture of Philadelphia for 1824, containing the "Picture of Philadelphia, for 1811, by James Mease, M.D." With all its improvements since that period. Philadelphia: Printed by T. Town, 1823. 96, [iii]–xii, 358 p., fold. front., plates.
Includes observations on manufactures, bridges, and water-works.
S-14950.
Copies: DLC; DeGE; IC; IaCrM; MH; MWA; MdBJ; NNA; NhD; NjP; OMC; P; PPL; PU.

362. A list of streets, roads, lanes, alleys, avenues, courts, wharves, ship-yards, public buildings, manufactories, places of amusement, &c., &c.... Philadelphia: H. C. Carey & I. Lea, 1824. 33 p., fold. plan.
S-16913.
Copies: MHi.

363. Philadelphia in 1824; or, A brief account of the various institutions and public objects in this metropolis.... To which is

363 (*continued*)
prefixed, an historical and statistical account of the city. With a plan of the city, view of the water-works, and other engravings. Philadelphia: H. C. Carey & I. Lea, 1824. xi, [1], 238, 16 p., front., fold. map, plan.

Includes observations on manufactures, canals, navigation companies, steamboats, bridges, and turnpike roads.

S-17612.

Copies: DLC; DeGE; ICN; KHi; LNHT; MH; MWA; MiD-B; NIC; Nh; OCHP; PP; PPAmP; PPL; PU; WHi.

364. [Preston, John] b. 1755.

A statistical report of the county of Albany, for the year 1820. 2d ed. Albany: Reprinted by E. & E. Hosford, 1824. 32 p.

A report on the agriculture of the county.

S-17694.

Copies: CSmH; DLC; MHi; N; NN; NNS.

365. Spafford, Horatio Gates, 1778–1832.

A pocket guide for the tourist and traveller, along the line of the canals, and the interior commerce of the state of New-York.... New York: Printed by T. and J. Swords, 1824. iv, [5]–72 p.

S-18064.

Copies: DLC; DeGE; ICJ; MH; MWA; MiD; NHi; NN; NhHi; P; PHi; RHi; ViU; WHi.

366. [Dwight, Theodore] 1796–1866.

The northern traveller; containing the routes to Niagara, Quebec, and the Springs; with descriptions of the principal scenes, and useful hints to strangers.... New York: Wilder & Campbell, 1825. 213, [1] p., maps, plates.

Includes observations on internal improvements.

S-20355.

Copies: CtY; DLC; MB; MH-BA; MWA; MnHi; NIC; NN; NjP; PPL; RPB.

367. [Gilpin, Henry Dilworth] 1801–1860.

A northern tour: being a guide to Saratoga, Lake George, Niagara, Canada, Boston, &c., &c., through the states of Pennsylvania, New-Jersey, New-York, Vermont, New-Hampshire, Massachusetts, Rhode-Island, and Connecticut, embracing an account of the canals.... Philadelphia: H. C. Carey & I. Lea, 1825. [4], iv, [2], 279, [6], [4] p., fold, map.

S-20678.

Copies: CSmH; CoU; CtY; DLC; GHi; MB; MWA; MdBE; MoSM; NNC; NjP; OHi; P; PU; WHi.

368. Hewett, Daniel, d. 1846.

The American traveller; or, National directory, containing an account of all the great post roads, and most important cross roads, in the United States.... Washington: Printed by Davis & Force, 1825. 440 [464] p.

S-20872.

Copies: CtY; DLC; ICU; InU; LNHT; MB; MsU; N; NN; WHi.

369. Melish, John, 1771–1822.

The traveller's directory through the United States; containing a description of all the principal roads through the United States, with copious remarks on the rivers and other objects.... New ed., rev. and enlarged. New York: A. T. Goodrich, 1825. xix, 194 p., fold. maps.

S-21404.

Copies: DLC; InU; KyLo; MB; MWA; MiD-B; MiU; OOxM.

370. [Mitchill, Samuel Latham] 1764–1831.

The picture of New-York; or, the stranger's guide to the commercial metropolis of the United States. New York: A. T. Goodrich, 1825. xv, [2], 14–357 p.

S-21899.

Copies: MHi; MWA; PP.

371. Schoolcraft, Henry Rowe, 1793–1864.

Travels in the central portions of the Mississippi Valley, comprising observations on its mineral geography, internal resources, and aboriginal population.... New York: Collins & Hannay, 1825. iv, 459 p., maps, plates.

Schoolcraft's "Answer to a letter from Charles G. Haines, concerning lead and other mines and mineral resources," p. 446–454.

S-22204.

Copies: ArL; CSmH; DLC; ICN; IaB; MH; MWA; MdBJ; MiU; MnHi; NNS; NhHi; NjP; OC; OkU; PPL; RNR; ScU; TxU; ViU; WHi.

372. Spafford, Horatio Gates, 1778–1832.

A pocket guide for the tourist and traveler along the line of the canals, and the interior commerce of the state of New York.... 2d ed., enl. Troy, [N.Y.]: William S. Parker, 1825. 88 p.

Sabin-88849.

Copies: CtY; N.

373. Tanner, Henry Schenck, 1786–1858.

The traveller's guide map of the roads, canals and steam boat routes of the United States.... Philadelphia: H. S. Tanner, 1825. Map 43 x 51 cm.

S-22422.

Copies: DLC; ICU; IU; MH; MH-BA; MWA; MnSJ; NIC; PHi.

374. [Dwight, Theodore] 1796–1866.

The northern traveller; containing the routes to Niagara, Quebec, and the Springs, with the tour of New-England, and the route to the coal mines of Pennsylvania.... 2d ed., improved & extended. New York: A. T. Goodrich, 1826. iv, [2], 382, [4] p., maps, plates.

Includes observations on internal improvements, mines, and manufacturing enterprises.

Also a copy, with inserted pages "Additions made to The northern traveller in May, 1827," held by MWA.

S-24388.

Copies: CLSU; CSmH; CtY; DBRE; DLC; DeGE; MB; MH; MWA; MiDSH; MoSHi; NN; NhD; OClWHi; ViAl; VtU.

375. Jones, Samuel.

Pittsburgh in the year eighteen hundred and twenty-six, containing sketches topographical, historical and statistical; together with a directory of the city.... Pittsburgh: Printed by Johnston & Stockton, 1826. 152, [2] p., front.

S-25002.

Copies: DLC; ICJ; MWA; NIC; NN; NjP; OC; PHi; PPPrHi; PPi.

376. Kilbourn, John, 1787–1833.

The Ohio manual; or, Compendium of statistical information concerning the state of Ohio.... Columbus: J. Kilbourn, 1826. 16 p.

S-25033.

Copies: DLC.

377. Melish, John, 1771–1822.

A geographical description of the United States, with the contiguous countries, including Mexico and the West Indies.... A new ed., greatly improved. New York: A. T. Goodrich, 1826. viii, [9]–497, [15] p., 12 maps.

S-25301.

Copies: DLC; ICN; IaK; LNHT; M; MH; MiU; NCg; OCl.

378. ———— Philadelphia: The author, 1826. 497, [15] p., maps.

S-25302.

Copies: CtW.

379. Mills, Robert, 1781–1855.

Statistics of South Carolina, including a view of its natural, civil, and military history.... Charleston: Hurlbut and Lloyd, 1826. vii, 782 p., 48 maps.

S-25379.

Copies: A-Ar; CtSoP; DI-GS; GA; ICN; KyLoS; MH; MWA; MdBP; MiD; MoSM; NIC; NNC; NcA-S; NjP; PPL; ScC.

380. Taylor, Isaac, 1759–1829.

Scenes of wealth; or, Views and illustrations of trades—manufactures—produce and commerce, for the amusement and in-

380 *(continued)*
struction of tarry-at-home travellers.... London: Harris & Son; Hartford: Republished by Oliver D. Cooke, 1826. vi, 168 p., map, plates.
S-26177.
Copies: CSmH; CtHi; InGrD; MB; MH; MWA; MdBJ; MdBS; NN; NNC; OClW; P; PPL.

381. Chastellux, François Jean, marquis de, 1734–1788.
Travels in North America in the years 1780–81–82. By the Marquis de Chastellux.... Translated from the French, by an English gentleman who resided in America at that period. With notes by the translator.... New York: White, Gallaher & White, 1827. 416 p.
S-28446.
Copies: CSmH; CtY; DLC; GEU; ICHi; InHi; MH; MdHi; MiD-B; NN; NcD; NjP; OU; PU; TKL.

382. Drake, Benjamin, 1794–1841.
Cincinnati in 1826. By B. Drake and E. D. Mansfield. Cincinnati: Printed by Morgan, Lodge and Fisher, 1827. 100 p., 2 plates.
Two issues: the February 1827 issue lacks the "Preface" and "Contents," present in the October 1827 issue.
S-28729.
Copies: CSmH; CtHT; DLC; ICN; InHi; MH; MiD-B; MiU-C; NN; NcD; OC; PPL; RHi; WHi.

383. A sketch of the present state of commerce, agriculture and manufactures, in the United States, with some observations on the impolicy of high tariff laws. Philadelphia: E. Littell, 1827. 1 p.l., 102 p.
Signed on p. 18: Agricola.
N-S0594914.
Copies: ICU; MiU-C; PPAmP.

384. Bernhard Karl, Duke of Saxe-Weimar-Eisenach, 1792–1862.
Travels through North America, during the years 1825 and 1826. Philadelphia: Carey, Lea & Carey, sold in New York by G. & C. Carvill, 1828. 2 v. in 1.
S-32277.
Copies: CSt; CtY; DLC; GU-De; ICHi; IaU; InHi; KU; LNHT; MH; MWA; MdBE; MeB; MiU-C; MnHi; MoSHi; NNS; NbU; NcU; NhHi; NjP; OClWHi; PPL; RNR; ScU; TNJ; Vi; VtU; WHi.

385. Chastellux, François Jean, marquis de, 1734–1788.
Travels in North America, in the years 1780–81–82 by the Marquis de Chastellux.... Translated from the French by an English gentleman, who resided in America at that period. With notes by the translator. Also a biographical sketch of the author.... New York: White, Gallaher & White, 1828. 416 p.
S-32654.
Copies: CSmH; DLC; ICU; KyBgW; LNL; MHi; MWA; MdU; MiD; NRSB; OMC; RJa; Vi.

386. [Cooper, James Fenimore] 1789–1851.
Notions of the Americans: picked up by a travelling bachelor.... Philadelphia: Carey, Lea & Carey, 1828. 2 v.
S-32827.
Copies: CL; CSmH; CoU; CtHT-W; DLC; DeGE; GHi; IC; KU; KyLx; LNHT; MH; MWA; MdHi; MeB; MiU-C; MnHi; MoS; NNPM; NcU; NjP; OHi; PPL; PU; RNR; ScCC; TNJ; TxDaM.

387. Darby, William, 1775–1854.
View of the United States, historical, geographical, and statistical; exhibiting, in a convenient form, the natural and artificial features of the several states, and embracing those leading branches of history and statistics best adapted to develop the present condition of the North American union.... Philadelphia: H. S. Tanner, 1828. 2 p.l., iv, [5]–654, 11 p., fold. maps.
C-32890.

Copies: A-Ar; CtY; DBRE; DLC; DeGE; ICN; IU; InU; KHi; LNHT; MH; MWA; MdBP; MeB; MiD-B; NNC; OClWHi; PPL; RP; TNJ; ViU; WHi.

388. [Dwight, Theodore] 1796–1866.

The northern traveller (combined with The northern tour.) Containing the routes to Niagara, Quebec, and the Springs. With the tour of New England, and the route to the coal mines of Pennsylvania. Embellished with 19 maps and 11 landscapes. 3d ed., rev. and extended. New York: G. & C. Carvill; [Danforth & Penfold, printers], 1828. 403 p., illus., maps.

Includes observations on internal improvements, mines, and manufacturing enterprises.

S-33023.

Copies: CtHT-W; CtY; DLC; GMM; ICU; IU; MH; MWA; MiU-C; MoSM; NCH; NN; PPL; RPB; TxU; ViU.

389. Royall, Anne (Newport) 1769–1854.

The black book; or, A continuation of travels in the United States.... Washington City: Printed for the author, 1828–1829. 3 v.

Includes descriptions of manufacturing enterprises and machinery used in them, also references to canals, river and road transportation.

S-35075, 35076, 40322.

Copies: CSmH; CtY; DLC; ICN; MWA; MdBP; NNC; NjR; OMC; PPL; RPB; ViU; WHi.

390. [Tanner, Henry Schenck] 1786–1858.

The stranger's guide to the public buildings, places of amusement, streets ... steam-boat landings, stage offices, etc., etc., of the city of Philadelphia and adjoining districts.... Philadelphia: H. S. Tanner, 1828. 38 p., front., fold. maps.

S-35461.

Copies: DLC; MB; PHi; PPL; PU.

391. Bowen, Abel, 1790–1850.

Bowen's picture of Boston; or, The citizen's and stranger's guide to the metropolis of Massachusetts, and its environs. Boston: Abel Bowen, [etc., etc.], 1829. 252 p., fold. maps, plates.

S-37919.

Copies: CSmH; CtHT-W; DLC; M; MH; MHi; MWA; MdBP; MnHi; MoSM; NBLiHi; NNPM; PHi.

392. Greenleaf, Moses, 1777–1834.

Atlas accompanying Greenleaf's map and statistical survey of the state of Maine. Portland, [Me.]: Shirley and Hyde, [1829]. 7 maps.

S-38831.

Copies: DLC; MeB; Nj.

393. Greenleaf, Moses, 1777–1834.

A survey of the state of Maine, in reference to its geographical features, statistics and political economy; illustrated by maps.... Portland: Shirley and Hyde, 1829. viii, [9]–468 p., and atlas of 5 fold. maps.

Substantially a new edition of his *A Statistical view of the district of Maine,* Boston, 1816.

S-38832.

Copies: CU; CtY; DLC; ICJ; MH; MWA; Md; MeU; MnHi; MoSM; NIC; NcU; Nh; Nj; OC; PHi; RP; WHi.

394. Hall, Basil, 1788–1844.

Travels in North America, in the years 1827 and 1828.... Philadelphia: Carey, Lea & Carey, 1829. 2 v., map.

S-38887.

Copies: CU; CtY; DeGE; GU; ICU; InU; KyLx; LNHT; MH; MWA; MdHi; MiD; MnU; MoSM; NNS; NcU; NhD; NjP; OCIWHi; P; PPL; RPB; ScU; TxDW; VtMiM; WGr.

395. Royall, Anne (Newport) 1769–1854.

Mrs. Royall's Pennsylvania; or, Travels continued in the United States.... Washington: The author, 1829. 2 v.

The author's third series of American travel, preceded by *Sketches of history, life and*

395 (*continued*)
manners in the United States, 1826, and *The black book; or, A continuation of travels in the United States,* 1828–29.

Includes observations on mining, iron foundries and manufacturing enterprises; also, a description of a ride on the railroad at Mauch Chunk.

S-40323.

Copies: DGU; DLC; DeGE; GHi; ICN; MH; MWA; MdBP; MnU; NIC; NN; PPL; PPi; Vi.

396. [Dwight, Theodore] 1796–1866.

The northern traveller, and northern tour; with the routes to the Springs, Niagara, and Quebec, and the coal mines of Pennsylvania; also a tour of New-England. Embellished with thirty-two copperplate engravings. 4th ed., rev. and extended. New York: Printed by J. & J. Harper, sold by Collins & Hannay, [etc.], 1830. 444 p., map, plates.

Includes observations on internal improvements, mines, and manufacturing enterprises.

C-1218.

Copies: CSmH; CtY; DLC; IC; MH; MWA; MdBP; MiD-B; NIC; NN; NSyOHi; OClWHi; PPL; PSC.

397. Morrison, Thomas.

The traveller's companion. Morrison's North River traveller's companion, containing a map of the Hudson or North River, with a description of the adjoining country.... Also a table showing the distance of the principal towns in the United States from each other; the length of the principal railroads and canals, finished or in progress, in the United States. Philadelphia: [Thomas Morrison, 1830?] Fold. broadside 20 x 56 cm.

C-2638.

Copies: DBRE; MB; MWA (2 variants).

398. Tanner, Henry Schenck, 1786–1858.

The strangers' guide to the public buildings, places of amusement, street lanes ... steam boat landings, stage offices, etc., etc. of the city of Philadelphia and adjoining districts.... 2d ed. Philadelphia: H. S. Tanner, 1830. 38, 16 p.

C-3666.

Copies: DeGE; PHi; PPL.

399. The traveller's directory: exhibiting the distances on the principal land and water thoroughfares of the state of New-York: and showing the reciprocal and relative distances of each place from the chief towns on three routes between Albany and Buffalo. Rochester: Marshall, Dean & Co., 1830. 12 p.

C-3750.

Copies: DLC.

400. [Vandewater, Robert J.]

The tourist; or, Pocket manual for travellers on the Hudson River, the western canal, and stage road; comprising also the routes to Lebanon, Ballston, and Saratoga Springs. New York: Printed by J. & J. Harper, 1830. 59 p., fold. map.

C-5259.

Copies: CSt; CoD; CtSoP; DLC; DeGE; IU; MH; MWA; NIC; NN; VtU; WHi.

II. TECHNOLOGY

1. Societies

401. American Academy of Arts and Sciences, Boston.

An act to incorporate and establish a society for the cultivation and promotion of arts and sciences. Boston: Printed by Benjamin Edes and Sons, [1780]. [2] p.

Mp-16841.

Copies: MHi; MWA.

402. Alexandria Society for the Promotion of Useful Knowledge.

Constitution of the Alexandria Society for the Promotion of Useful Knowledge.... [Alexandria: Hanson and Rond, 1787.] Broadside 35 x 28.5 cm.

B-6436.

Copies: PPL.

403. Providence Association of Mechanics and Manufacturers.

The Charter, constitution and bye-laws, of the Providence Association of Mechanics and Manufacturers.... Providence: Printed by Bennet Wheeler, for the Association, [1789]. 28 p.

Mp-22085.

Copies: MH; MWA.

404. Pennsylvania Society for Mechanical Improvements and Philosophical Inquiries.

The constitution and fundamental rules of the Pennsylvania Society for Mechanical Improvements and Philosophical Inquiries. [Philadelphia: 179-?] Broadside 42 x 26 cm.

B-7270; Mp-45788.

Copies: NN (photocopy).

405. General Society of Mechanics and Tradesmen of the City of New York.

The charter and bye-laws of the General Society of Mechanics and Tradesmen of the City of New York. Also, the rules and orders with a catalogue of names of the members. Published by order of the Society. New-York: Printed by John Harrisson, 1792. 30 p.

Mp-24609.

Copies: MWA; NN.

406. New York (State). Laws.

An act to incorporate the Society of Mechanics and Tradesmen of the city of New-York, for charitable purposes. Passed the 14th of March, 1792. New York: Printed by Samuel Loudon, Jun., [1792]. 11 p.

E-24608.

No copy of a separate printing known.

407. Newport Association of Mechanics and Manufactures, Newport, R.I.

The charter, constitution and bye-laws of the Newport Association of Mechanics and Manufacturers, with a list of officers and committee of correspondence. Newport: Printed by Henry Barber, for the Association, [1792]. vi, 24 p.

Mp-24618.

Copies: MWA; RNHi; RPB.

408. Society for the Attainment of Useful Knowledge, Philadelphia.

Constitution of the Society for the Attainment of Useful Knowledge.... Philadelphia: Printed by Zachariah Poulson, junr., 1794. [12] p.

E-27513; N-S0692592.

Copies: PHi.

409. Associated Mechanics of the Town of Boston.

409 (*continued*)

Constitution of the Associated Mechanics of the Town of Boston.... [Boston: Printed and sold by John W. Folsom, 1795.] 23, [1] p.

Includes a list of members and trades to April 16th, 1795.

Mp-28315.

Copies: MB; MWA.

410. General Society of Mechanics and Tradesmen of the City of New York.

The charter and bye-laws of the General Society of Mechanics and Tradesmen of the City of New-York. Also–the rules of orders with a catalogue of members names. Published by order of the Society. New York: Printed by G. Forman, 1798. 24 p.

Mp-34224.

Copies: DeGE; MH-BA; MiU-C; NN; NcD; PBL; RPJCB.

411. Providence Association of Mechanics and Manufacturers.

The charter, article of agreement, bye-laws, rules and regulations of the Providence Association of Mechanics and Manufacturers. Also, a catalogue of names of the members.... Providence: Printed by Bennett Wheeler, 1798. 40 p.

Mp-34423.

Copies: MWA; NHi; NN; RPJCB.

412. Associated Mechanics and Manufacturers of the Commonwealth of Massachusetts.

Constitution of the Associated Mechanics and Manufacturers of the Commonwealth of Massachusetts. Boston: Printed by Russell and Cutler, 1800. 24 p.

Mp-37928.

Copies: MWA.

413. Associated Mechanics and Manufacturers of the State of New-Hampshire.

Constitution.... Portsmouth, [N.H.]: Printed by N. S. & W. Peirce, 1802. 14 p.

S&S-1774.

Copies: MWA.

414. Albany Mechanic's Society.

Charter of incorporation of the Albany Mechanic's Society, granted by the Legislature of the state of New York, 6th March, 1801. Albany: Printed by C. R. & G. Webster, 1803. 10 p.

S&S-3645.

Copies: DLC; N; NjR.

415. Associated Mechanics and Manufacturers of the State of New-Hampshire.

Bye-laws of the association. [Portsmouth?, N.H.: 1803?] [7] p.

Includes membership list, with occupations.

S&S-3696.

Copies: NN.

416. Albany Mechanic's Society.

Albany Mechanic's Society. Organized in the year 1793. Incorporated by act of the Legislature, February, 1802. Alphabetical list of members. [Albany: 1804?] Broadside.

S&S-5674.

Copies: N.

417. Massachusetts Charitable Mechanic Association.

Constitution of the Massachusetts Mechanic Association. Instituted March 15, 1795. Boston: Printed by Gilbert & Dean, 1804. 24 p.

S&S-6747.

Copies: MBAt; MBBS; MWA.

418. Albany Mechanic's Society.

Alphabetical list of members.... [Albany: 1806.] Broadside.

Dated: Tuesday, February 4, 1806.

S&S-9818.

Copies: NHi.

419. Newburyport Mechanic Association.

Constitution.... October 22, 1806. Newburyport, Mass.: 1807. 12 p.

S&S-13248.

Copies: PHi.

420. Newengland Association of Inventors and Patrons of Useful Arts.

Constitution of the New England Association of Inventors and Patrons of Useful Arts. Boston: 1807. 16 p.

S&S-13187.

Copies: DLC; InU; MWA.

421. Newengland Association of Inventors and Patrons of Useful Arts.

Rules and regulations, adopted October 22d, 1807, by the Newengland Association of Inventors and Patrons of Useful Arts. Boston: Snelling and Simons, printers, 1807. 23, [1] p.

S&S-13189.

Copies: DLC; DeGE (xerox copy); MWA; NN; PPAmP; PPF.

422. Providence Association of Mechanics and Manufacturers.

The charter, article of agreement, by-laws, rules and regulations of the Providence Association of Mechanics and Manufacturers. Providence: Printed by J. Carter, 1808. 35 p.

S&S-16022.

Copies: MH-BA; MWA.

423. Associated Mechanics and Manufacturers of the State of New-Hampshire.

Act of incorporation, constitution, by-laws &c. of the Associated Mechanics and Manufacturers of the State of New-Hampshire. Portsmouth, [N.H.]: Printed by W. Treadwell, 1810. 24 p.

S&S-19389.

Copies: MWA.

424. Master Mechanics' Benevolent Society of the City and County of Philadelphia.

The constitution and rules of the Master Mechanics' Benevolent Society of Philadelphia. Instituted March 2d, 1810. Philadelphia: Printed by John W. Scott, 1810. 24 p.

Copies: DeGE.

425. Massachusetts Charitable Mechanic Association.

Order of performance for the second triennial celebration of the Massachusetts Charitable Mechanic Association, December 17, 1812. [Boston: 1812.] Broadside 21 x 17.5 cm.

N-M0314311.

Copies: MB.

426. American Philosophical Society, Philadelphia.

Laws and regulations of the American Philosophical Society, held at Philadelphia, for promoting useful knowledge.... Philadelphia: Printed at the office of the United States gazette, 1814. 36 p.

S&S-30679.

Copies: DLC; MWA; PHi; PPAmP; PPF; PPL.

427. General Society of Mechanics and Tradesmen of the City of New York.

The charter and by-laws of the General Society of Mechanics and Tradesmen of the City of New York. New York: Printed by J. Seymour, 1814. 40 p.

S&S-31567.

Copies: MWA; MiD-B; NHi; NjR.

428. Massachusetts Charitable Mechanic Association.

Constitution of the Massachusetts Charitable Mechanic Association, instituted March 15, 1795 and incorporated March 8, 1806. Boston: Printed by J. T. Buckingham, 1814. 26 p.

S&S-32074.

Copies: CtY; MWA; MnU.

429. Society for the Promotion of Useful Arts, Albany.

Charter and bye-laws of the Society for the Promotion of Useful Arts; together with a list of the members and the standing committees. Albany: Printed by Websters and Skinners, 1815. 15 p.

S&S-35974.

Copies: MB; MBAt; MWA; N; NHi.

430. ——— Albany: Printed by Websters and Skinners, 1815. 15 p.

430 (*continued*)

"Additional by-laws, passed 7th April, 1818," 1 leaf, inserted after the title page.

Copies: MWA.

431. Mechanic Society of Staunton, Va.

Constitution & by-laws. [Staunton? Va.:] 1817?] 5 p.

N-M0390024.

Copies: Vi.

432. Salem Charitable Mechanic Association, Salem, Mass.

Constitution of the Salem Charitable Mechanic Association, instituted Oct. 1, 1817. Salem, [Mass.]: Printed by W. Palfray, Jr., 1817. 23 p.

S&S-42033.

Copies: MH; MSaE; MWA.

432a. Massachusetts Charitable Mechanic Association.

Order of performances at the fourth triennial celebration of the Massachusetts Charitable Mechanic Association.... December 18, 1818. [Boston: 1818.] Broadside 31 x 25 cm.

Copies: MWA.

433. Providence Accociation of Mechanics and Manufacturers.

Catalogue of the members of the Providence Association of Mechanicks and Manufacturers. July 13, 1818. Providence: Printed by Miller & Hutchens, 1818. 13 p.

S&S-45448.

Copies: MWA.

434. Northampton Mechanic Association.

The constitution.... January, 1819. Northampton, Mass.: Printed by T. W. Shepard, 1819. 16 p.

S&S-48950.

Copies: MNF.

435. Associated Mechanics and Manufacturers of the State of New-Hampshire.

Constitution of the Associated Mechanics and Manufacturers of the State of New-Hampshire. Portsmouth, N.H.: Beck and Foster, printers, 1820. 24 p.

S-155.

Copies: MWA; NN; NNC.

436. Providence Association of Mechanics and Manufacturers.

The charter, article of agreement, bylaws, rules and regulations of the Providence Association of Mechanics and Manufacturers. Providence: Printed by Brown & Danforth, 1820. 24 p.

S-2907.

Copies: MH-BA; MWA; RHi.

437. General Society of Mechanics and Tradesmen of the City of New York.

The charter and by-laws of the General Society of Mechanics and Tradesmen of the City of New York.... New York: William A. Mercein, printer, 1823. 48 p.

S-12653.

Copies: NHi; NjR.

438. Maryland. Laws.

An act to incorporate a Society for the Promotion of Useful Arts. [Annapolis: 1823.] 24 p. (House. Bill No. 39)

Heyl-224.

Copies: MdHi.

439. Salem Charitable Mechanic Association.

Constitution of the Salem Charitable Mechanic Association, instituted October 1, 1817. Incorporated June 14, 1822. Revised Feb. 5, 1823. Salem, [Mass.]: Printed by Warwick Palfray, Jun. 1823. 24 p.

Includes by-laws and lists of officers and members, with their trades.

S-14035.

Copies: DeGE; MH; MSaE.

440. Mechanics' Society of Detroit.

Constitution of the Mechanics' Society of Detroit.... Detroit: Printed by Sheldon & Wells, November 1825. 17 p.

Its program included the establishment of a library.

S-20301.
Copies: MiD-B.

441. Petersburg Benevolent Mechanic Accociation, Petersburg, Va.
Constitution and bye-laws of the Petersburg Benevolent Mechanic Association. Petersburg, [Va.]: 1825. 14 p.
S-21866.
Copies: Vi.

442. Providence Association of Mechanics and Manufacturers.
Charter and by-laws, rules and regulations of the Providence Association of Mechanics and Manufacturers. With list of members. Providence: Brown & Danforth, 1825. 40 p.
Sabin-66305.
No copy known.

443. General Society of Mechanics and Tradesmen of the City of New York.
Charter. An act relative to the General Society of Mechanics and Tradesmen, of the City of New-York, passed April 3, 1811. [New York: 1827.] 66 p., illus.
"By-laws," p. [17]–48; "Appendix, containing the names of the members.... On July 4th, 1827," p. [49]–64.
Copies: DeGE; NHi.

444. Massachusetts Charitable Mechanic Association.
Constitution of the Massachusetts Charitable Mechanic Association.... Boston: Printed by Monroe & Francis, 1827. 36 p.
S-29689.
Copies: MB.

445. Providence Association of Mechanics and Manufacturers.
The charter, article of agreement, by-laws, rules and regulations, of the Providence Association of Mechanics and Manufacturers. Providence: Printed by Walter R. Danforth, 1827. 24 p.
S-30380.
Copies: MWA; RPB.

446. Newark Mechanics' Association.
Constitution and by-laws of the Newark Mechanics' Association for mutual improvement in the arts and sciences. Adopted January 8, 1828. Newark, N.J.: Printed by W. Tuttle & Co., 1828. 12 p.
N-N0223570.
Copies: MH.

447. Boston Society for the Diffusion of Useful Knowledge.
[Constitution, list of members, etc. Boston: 1829.] 8 p.
N-B0677041.
Copies: DLC.

448. General Society of Mechanics and Tradesmen of the City of New York.
Reports of the library and school committees.... New York: W. A. Mercein, 1829. 16 p.
S-38736.
Copies: MH; MWA; NBu.

449. General Society of Mechanics of New Haven.
Constitution of the General Society of Mechanics of New-Haven: adopted March 23, 1807—revised, Oct. 6, 1819. Together with the Act of incorporation. New Haven: Printed for the Society, 1829. 11 p.
S-38737.
Copies: CtY.

450. Massachusetts Charitable Mechanic Association.
Constitution of the Massachusetts Charitable Mechanic Association. Instituted March 15, 1795, and incorporated March 8, 1806. Boston: Printed by John Cotton, 1829. 36 p.
List of officers and members: p. 16–36.
S-39517.
Copies: MB.

451. Boston Society for the Diffusion of Useful Knowledge.
The first annual report of the board of managers of the Boston Society for the Dif-

451 (*continued*)
fusion of Useful Knowledge. Presented at the annual meeting April 2, 1830. Boston: Office of the Daily advertiser; W. L. Lewis, Printer, 1830. 16 p.
C-614.
Copies: DLC; MBAt; MH; MHi; MWA; MoS; NIC; PPL.

452. Concord Mechanic's Association, Concord, N.H.
Report submitted to the Concord Mechanic's Association, at a quarterly meeting, April 6, 1830. [Concord, N.H.: 1830.] 8 p.
C-975.
Copies: DLC.

453. Massachusetts Charitable Mechanic Association.
Order of services at the eighth triennial festival of the Massachusetts Charitable Mechanic Association, on Thursday, October 7, 1830.... [Boston: 1830.] Broadside 30.5 x 19.5 cm.
C-2487.
Copies: MB.

2. General works

454. [Franklin, Benjamin] 1706–1790.
A proposal for promoting useful knowledge among the British plantations in America. [Philadelphia: Printed by B. Franklin, 1743.] [2] p.
Dated: Philadelphia, May 14, 1743.
Mp-5190.
Copies: CtY; DLC; MHi; PPAmP.

455. ———— [Philadelphia: 1768.] 2 p.
N-F0340717.
Copies: PPAmP.

456. Matlack, Timothy, 1736–1829.
An oration, delivered March 16, 1780, before the patron, vice-presidents and members of the American Philosophical Society, held at Philadelphia, for promoting useful knowledge.... Philadelphia: Printed by Styner and Cist, 1780. 27 p.
Advocates the application of science to agriculture and the use of new implements.
Mp-16867.
Copies: DLC; DeGE; MWA; NHi; PPAmP; PPL.

457. Vancouver, Charles.
A general compendium; or, Abstract of chemical, experimental, and natural philosophy. To which is added a complete system of commerce.... In four volumes. Vol. I. Philadelphia: Printed in the year 1785. viii, 48 p., plate.
Ends with an unfinished sentence on p. 48; no more published. The plate illustrates a machine for rescuing persons trapped in the upper floors of burning buildings.
Mp-19337.
Copies: DLC; DeGE; MBAt; MH; NN; PPL.

458. Wilson, James, 1760–1839.
An oration, delivered before the Providence Association of Mechanics and Manufacturers, at their annual election, April 14, 1794.... Providence: Printed and sold by Bennett Wheeler, 1794. 27, [4] p.
Hymns, performed at the anniversary election of the officers of the Providence Association of Mechanics and Manufacturers, on Monday, April 14, 1794, [4] p. at end.
Mp-28108.
Copies: CtY; DLC; DeGE; MH; MHi; MWA; NHi; NN; RPJCB.

459. ________ 2d ed. Providence: 1794.
E-28109.
No copy known.

460. Maxcy, Jonathan, 1768–1820.
An oration, delivered before the Providence Association of Mechanics and Manufacturers, at their annual election, April 13, 1795.... Providence: Printed and sold by Bennett Wheeler, 1795. 2 p.l., [3]–17, [4] p.
Mp-29053.
Copies: CSmH; CtY; ICN; MH; MHi; MWA; MiU-C; NcD; RHi; RPJCB.

461. Howland, John, 1757–1854.
A lecture, read before the Providence Association of Mechanics and Manufacturers, at their quarterly meeting, January 14, A.D. 1799. ... Providence: Printed by Bennett Wheeler, 1799. 14 p.
Mp-35644.
Copies: CSt; CtY; DLC; MHi; MWA; RHi; RPJCB.

462. Connecticut Academy of Arts and Sciences.
New Haven, January 1, 1800. To ________ Sir, the Connecticut Academy of Arts and Sciences, desirous of contributing to the collection and propagation of useful knowledge, and of procuring the material for a statistical history of Connecticut, request you to furnish them with ... information ... [New Haven: 1800.] 12 p.
Topics for which information was desired, included furnaces, forges and mills, descriptions of any curious machinery and improvements in arts and sciences.
Mp-37233.
Copies: CtHi; CtY; MBAt; MWA.

463. Messer, Asa, 1769–1836.
An oration, delivered before the Providence Association of Mechanics and Manufacturers, at their annual election, April 11, 1803.... Providence: Printed by John Carter, [1803.] 15 p.
S&S-4647.
Copies: DeGE; MH-BA; MWA; MiD-B; NN; RPB.

464. United States. Congress. House. Committee on the Letter of William Tatham.
Report from the Committee appointed on the tenth of February last, on the letter of William Tatham to the Speaker. April 1st, 1806.... City of Washington: A. & G. Way, printers, 1806. 23 p.
Includes William Tatham's letter on the means to improve transportation, technology and commerce, and a survey of his manuscripts, books, surveys, instruments, models, and tools, offered for sale to the government.
S&S-11682.
Copies: DLC; DeGE; MWA; N; NN.

465. Howland, John, 1757–1854.
An address delivered before the Providence Association of Mechanics and Manufacturers, April 9, 1810: being the anniversary of the choice of officers in the Association. By John Howland, Esq., Secretary of the Association. Providence: From the press of Jones & Wheeler, 1810. 25 p.
S&S-20396.
Copies: CSmH; CSt; CtY; DeGE; MH; MHi; MWA; MiU-C; NN; RHi; RNHi; RPB.

466. Russell, Benjamin, 1761–1845.
An address delivered before the Massachusetts Charitable Mechanick Association, December 21, 1809; being the anniversary of the choice of officers, and the first triennial celebration of their publick festival.... Boston: From the press of John Eliot, Jun., 1810. 45 p.
Appendix: *Past-officers, and officers of the Massachusetts Charitable Mechanick Association,* p. [35]–45.
S&S-21249.
Copies: CSt; DLC; DeGE; MH; MWA.

467. United States. Congress. House. Select Committee on the Memorial of Benjamin Tyler, and John Tyler.
Report of the Select Committee to whom was referred on the seventeenth instant, the memorial of Benjamin Tyler, Jun. and John Tyler. January 28, 1811. Read, and

467 (*continued*)
referred to the Committee of the Whole House on the bill for the encouragement of learning and the promotion of the useful arts. Washington: Printed by A. & G. Way, 1811. 4 p.

S&S-24298; 24312.

Copies: DLC; MWA.

468. United States. Laws.

An act making further provision for the Corps of Engineers. In the House of Representatives of the United States. December 5th, 1811. [Washington City: 1811.] 5 p.

S&S-24117.

Copies: DLC.

469. Fairbanks, John, 1777–1814.

An address delivered before the Massachusetts Charitable Mechanic Association, December 17, 1812.... Boston: Printed by Joseph T. Buckingham, 1812. 31 p.

S&S-25391.

Copies: DLC; InU; MB; MBr; MH-BA; MHi; MWA; Nh; PHi.

470. Fitz, Henry.

An address, delivered before the Newburyport Mechanic Association.... Oct. 28, 1812.... Newburyport, Mass.: Printed by E. H. Allen, 1812. 18 p.

S&S-25434.

Copies: DLC; MNe; MWA; MeHi.

471. Steele, Oliver.

An address, delivered before the General Society of Mechanics of New-Haven.... July 7th, 1813.... New Haven: Printed by Oliver Steele, 1813. 16 p.

Sabin-91133.

Copies: CtY; MWA.

472. Elliott, Stephen, 1771–1830.

An address to the Literary and Philosophical Society of South-Carolina; delivered in Charleston, on Wednesday, the 10th August, 1814.... Charleston: Printed by W. P. Young, 1814. 20 p.

A review of subjects to which the Society had devoted its attention, including mechanics, chemistry, mineralogy, agriculture, manufactures and internal improvements.

S&S-31405.

Copies: CtY; DLC; MHi; MWA; NN; PPAmP; PPL.

473. Clinton, De Witt, 1769–1828.

An introductory discourse, delivered before the Literary and Philosophical Society of New-York, on the fourth of May, 1814.... New York: Printed by N. Van Riper, for David Longworth, 1815. 148 p.

Includes references to developments in various sciences and techniques, particularly in "Notes and illustrations."

S&S-34371.

Copies: CU; CtY; DLC; FOA; MH; MWA; MdHi; MiU-C; MnU; NN; NcU; NjP; PPL; ScC.

474. ———— New York: Printed by Van Winkle and Wiley, 1815. 160, [1] p.

S&S-34372.

Copies: DLC; DeGE; MWA; MiD; NN; PPAmP; PPiU; RNR.

475. Train, Charles, 1783–1849.

A discourse, delivered at Hopkinton, Massachusetts, January 2, 1815, before the Worcester and Middlesex Mechanical Society.... Worcester, [Mass.]: Printed by Henry Rogers, 1815. 16 p.

Copies: MWA.

476. Buckingham, Joseph Tinker, 1779–1861.

An address before the Massachusetts Charitable Mechanic Association, at their public celebration of their triennial festival, December 21, 1815. Boston: 1816. 23 p.

S&S-37116.

Copies: DLC; InU; MH; MWA; NNUT; PHi; RPB; WHi.

477. Howland, John, 1757–1854.

An oration delivered before the Providence Association of Mechanicks and Manufacturers, April 13, 1818, being the anni-

versary of the choice of officers.... Providence: Printed by Miller and Hutchens, 1818. 24 p.
S&S-44387.
Copies: CtY; DeGE; InU; MWA; MiD-B; MiU-C; RPB.

478. Jenkins, Joseph, 1781–1851.
An address delivered before the Massachusetts Charitable Mechanick Association, December 17, 1818, being the anniversary of the choice of officers.... Boston: Printed by Munroe & Francis, 1819. 24 p.
S&S-48371.
Copies: CtHT-W; DLC; MBAt; MBr; MH; MHi; MMeT; MWA; MnHi; PHi.

479. Mercein, Thomas R., 1782–1843.
An address, delivered at the opening of the Apprentices' Library of the city of New-York, on the twenty-fifth of November, 1820.... To which are added, a reply on behalf of the apprentices, by an apprentice. An address on behalf of the scholars belonging to the Mechanic Society school, by a scholar. New York: Printed for the Society by W. A. Mercein, 1820. 22 p.
S-2240.
Copies: MWA; NN.

480. Wells, Thomas.
An address delivered before the Massachusetts Charitable Mechanic Association, October 4, 1821. Being the anniversary for the choice of officers and the fifth triennial celebration of their public festival.... Boston: J. T. Buckingham, printer, 1821. 40 p.
S-7625.
Copies: DLC; MBAt; MHi; MWA; WHi.

481. Danforth, Walter Raleigh.
An oration delivered before the Providence Association of Mechanicks and Manufacturers, April 8, 1822.... [Providence]: Miller and Hutchens, 1822. 24 p.
S-8499.
Copies: CtY; MH-BA; MWA; RPB.

482. Noah, Mordecai Manuel, 1785–1851.
An address delivered before the General Society of Mechanics and Tradesmen of the City of New-York, on the opening of the Mechanic Institution ... to which is added, the remarks made at the request of the Mechanics' Society, on laying the corner stone of that edifice, by Thomas R. Mercein.... New York: W. A. Mercein, printer, 1822. 28 p.
S-9720.
Copies: DLC; MB; NHi; NIC; NjR.

483. Jewett, Dyar Throop.
A guide to national improvement; or, Dissertation on our public regulations.... With a philosophical examination of arts & sciences, etc. New London: Printed by S. Green, for the author, 1823. 201, [1] p.
S-12955.
Copies: CtHi; ICU; MH; MWA.

484. Kean, Peter, 1788?–1828.
An address, delivered at the first anniversary meeting of the Elizabeth-Town Apprentices' Library Association, January 7, 1823.... Elizabeth-Town, [N.J.]: Printed by J. & E. Sanderson, 1823. 8 p.
Reviews the achievements of the Library in improving the competence of tradesmen.
S-12986.
Copies: NjR.

485. Cary, Alpheus.
An address, delivered before the Massachusetts Charitable Mechanic Association, October 7th, 1824, being the anniversary for the choice of officers, and the sixth triennial celebration of their public festival.... Boston: Munroe and Francis, 1824. 44 p.
Includes references to mechanical inventions and a list of members of the Massachusetts Charitable Mechanic Association.
S-15693.
Copies: CtHT-W; DLC; DeGE; MB; MH; MWA; MiD-B; RPB.

486. Gould, James.
An oration, pronounced at New-Haven,

486 (*continued*)
before the Connecticut Alpha of the Phi Beta Kappa Society, September 13, 1825. New Haven: Printed by T. G. Woodward, 1825. 31 p.

Concerns the promotion of technological education and technology.

S-20727.

Copies: CSmH; CtY; DLC; MB; MH; MWA; MiD-B; NNC; NjPT; PPL; RPB.

487. Holley, Orville Luther, 1791–1861.

The connexion between the mechanic arts and the welfare of states. Troy, [N.Y.]: 1825. 14 p.

N-H0465210.

Copies: MH; NN.

488. Appeal to the mechanic interest of the city of New York in relation to apprentices, together with an account of the Mutual Instruction Society. New York: Printed by S. Wood & Sons, 1826. 27 p.

"Originally published as numbered articles in *New York Observer and Chronicle,* signed: 'An old apprentice.' Now published together and submitted by a mechanic to the editor of the *Long Island Star.*"

N-A0357734.

Copies: CtY.

489. Greenleaf, Abner.

An address delivered before the society of Associated Mechanics & Manufacturers of the State of New-Hampshire, at the celebration of their anniversary in Portsmouth, Oct. 5, 1826.... Portsmouth, [N.H.]: T. H. Miller and C. W. Brewster, printers, 1826. 32 p.

S-24740.

Copies: DeGE; ICN; KU; MBAt; MH; MH-BA; MWA; MdBJ; MiD-B; NN; Nh; NjR; PHi.

490. Lawton, Josiah.

Oration on the importance of scientific knowledge, delivered before the Providence Association of Mechanics and Manufacturers. At their thirty seventh anniversary, April 10, 1826.... Providence: Miller & Grattan, printers, 1826. 18 p.

S-25071.

Copies: CtY; DLC; DeGE; MB; MH-BA; MWA; MiD-B; PHi; RP; RPB.

491. United States. Laws.

A bill providing for the gradual increase of the Corps of Engineers, for the organization of the Corps of Topographical Engineers, and for other purposes. January 16, 1826.... [Washington: 1826.] 3 p. (H.R. 51)

S-26690.

Copies: DNA.

492. United States. War Dept.

Augmentation of Engineer Corps. Letter from the Secretary of War, transmitting the information required by a resolution of the House of Representatives, of the 25th ult., in relation to an augmentation of the Corps of Engineers. April 12, 1826.... Washington: Printed by Gales & Seaton, 1826. 51 p., fold. tables. (19th Cong., 1st sess. House. Doc. 153)

Includes a statement from the Secretary of War, James Barbour, on the projected fortifications.

N-U0051319.

Copies: DeU; MWA; NIC.

493. Beard, Ithamar A.

An address delivered before the Middlesex Mechanic Association, at their anniversary, October 4, 1827.... Lowell: Printed at the Journal office, 1827. 10 p.

S-28078.

Copies: MB.

494. Emerson, George Barrell, 1797–1881.

An address, delivered at the opening of the Boston Mechanics' Institution, February 7, 1827. Boston: Hilliard, Gray, Little, and Wilkins, 1827. 24 p.

S-28784.

Copies: CLU; CSmH; CtHT-W; DeGE; MBAt; MH; MH-BA; MHi; MWA; MnHi; NhHi; OO; PPL; RPB.

495. Hilliard, William, 1778–1836.
An address delivered before the Massachusetts Charitable Mechanic Association, October 4, 1824, being the anniversary for the choice of officers, and the seventh triennial celebration of their festival.... Cambridge, [Mass.]: Printed by Hilliard, Metcalf, and Co., 1827. 48 p.
Includes observations on the progress of mechanic arts.
S-29217.
Copies: CSmH; DLC; DeGE; MB; MBAt; MH; MHi; MWA; MiD-B; NNC; Nh; PPL; WHi.

496. Buckingham, Joseph Tinker, 1779–1861.
An address delivered before the Massachusetts Charitable Mechanic Association, at the celebration of their eighth triennial festival, Oct. 7, 1830. Boston: The Association, 1830. 30 p.
C-688; N-B0917689.
Copies: CSmH; DLC; MB; MBAt; MH; MWA; NBu; NcD; Nh; OClWHi.

497. Burges, Tristam, 1770–1853.
Address, delivered before the American Institute of the City of New York, at their third annual fair, held at Masonic Hall, October 1830.... New York: John M. Danforth, 1830. 34 p.
C-707.
Copies: CtHT-W; CtY; MH; MWA; PHi; PPAmP; PPL.

498. Butler, Benjamin Franklin, 1795–1858.
Anniversary discourse, delivered before the Albany Institute, April 23, 1830.... Albany: Printed by Webster and Skinners, 1830. viii, [9]–88 p.
The Albany Institute was formed in 1824 by the union of the Society for the Promotion of Useful Arts with the Albany Lyceum of Natural History; one of its aims was to promote "useful arts," agriculture and physical sciences.
C-723.
Copies: CSmH; DLC; MB; MBAt; MH; MWA; MdHi; MiD-B; N; NCH; NbU; NjP; PPL; ScU.

499. Howland, John, 1757–1854.
Address delivered before the Providence Association of Mechanicks and Manufacturers, on the occasion of opening Mechanicks' Hall, January 10, A.D. 1825. By John Howland, Esq., President of the Association.... Providence: H. H. Brown, printer, 1830. 12 p.
C-1937.
Copies: DLC; DeGE; MWA; RHi; RNHi; RPB.

500. Southard, Samuel Lewis, 1787–1842.
Address delivered before the Newark Mechanics Association, July 5, 1830. Newark: Printed by W. Tuttle & Co., 1830. 27 p.
C-3554.
Copies: CSmH; DLC; IU; MB; MWA; MdHi; NHi; NjR; P; PHi; PPAmP; PPL; WHi.

3. Technical education and libraries

501. Clark, Stephen.
A catalogue of the Annapolis Circulating Library, consisting of near fifteen hundred volumes, in agriculture, arithmetic, astronomy, biography, chymistry, commerce, gardening ... husbandry ... navigation ...

501 (*continued*)
surveying.... [Annapolis: Printed by F. and S. Green, 1783.] [49] p.
Mp-17872.
Copies: MdBP; Md-LR.

502. Mechanic Library Society, New Haven.
The constitution and bye-laws of the Mechanic Library Society of New Haven, with a catalogue of books and list of the proprietors. New Haven: Printed by Abel Morse, 1793. 1 p. l., 7, [2], 11–17 p.
B-8410; Mp-46831.
Copies: CtY.

503. ———— New Haven: T. Green and Son, 1801. 27 p.
N-N0151106; S&S-1008.
Copies: CtY.

504. Library of Arts and Sciences, Salem, Mass.
Rules and regulations of the Library of Arts and Sciences. Salem: Printed by Thomas C. Cushing, 1802. 8, [8] p.
Catalog of the library: [8] p. at end.
S&S-2532.
Copies: MSaE.

505. [Barlow, Joel] 1754–1812.
Prospectus of a national institution, to be established in the United States. Washington City: Printed by Samuel H. Smith, 1806. 44 p.
Proposes a national university, which would include technical education and a museum for scientific instruments, patent models and collections of minerals.
S&S-9922.
Copies: CU; CtY; DLC; DeGE; ICJ; MH; MWA; MiU; NN; NNC; NcU; PPAmP.

506. Mechanic Library Society, New Haven.
The constitution and bye-laws of the Mechanic Library Society in New Haven; with a catalogue of books. [New Haven: 1812.] 22 p.
N-N0151107.
Copies: CtY.

507. ———— [New Haven: O. Steele & Co., 1815.] 24 p.
N-N0151108.
Copies: CtY; MWA.

508. Hanna, James.
The American instructor; or, Every one his own teacher; comprising ... arithmetic, surveying, carpenters' measurements, bookkeeping.... Trenton: The Author, 1818. x, [13]–304 p.
S&S-44253.
Copies: CSmH; DAU; DLC; MWA; NjP; OClWHi; PU.

509. Society for the Promotion of Useful Arts, Albany.
A catalogue of books in the library of the Society for the Promotion of Useful Arts. [Albany: 1818.] 4 p.
S&S-45752; Sabin-85997.
Copies: MB.

510. Colburn, Warren, 1793–1833.
Encouraged by several gentlemen, the subscriber has opened in Boston a school on the following plan. Instruction will be given in all branches of pure mathematics, and in their application to navigation, surveying, gauging, &c ... [Boston: 1820?] Broadside 24 x 18.5 cm.
S-810.
Copies: MHi.

511. General Society of Mechanics and Tradesmen of the City of New York. Library.
Catalogue of the Apprentices' Library ... with the names of donors. To which is added an address ... by Thomas R. Mercein.... New York: Printed by William A. Mercein, 1820. 48, 22 p.
S-2240, 2520.
Copies: MWA; NN.

512. Massachusetts Charitable Mechanic Association Library.

Catalogue of the Apprentices' Library, in Boston, to be loaned gratis, under superintendence of the Massachusetts Mechanic Association, to the mechanic, and all needy apprentices, with the names of donors. Boston: Printed by Munroe and Francis, 1820. 24 p.

Dated: February 22, 1820.

S-2197.

Copies: MBAt; MBBS.

513. Salem Mechanic Library.

Rules and by-laws of the Salem Mechanic Library. Instituted August, 1820. Salem, [Mass.]: Printed by Warwick Palfray, Jun. 1821. 16 p.

S-6704.

Copies: MSaE.

514. Apprentices' Library Company, Baltimore.

Catalogue of the Apprentices' Library in Baltimore, to be loaned gratis to apprentices and other young persons.... Baltimore: Printed by Thomas Murphy, 1822. 12 p.

S-7885.

Copies: MdHi.

515. Apprentices' Library Company, Baltimore.

Circular of the Apprentices' Library Company of Baltimore. Baltimore: T. Murphy, [1822]. 8 p.

A statement of purpose for the library, aiming, among other things, to better prepare the workers for particular trades, arts or professions; lists categories of books to be acquired.

S-7886.

Copies: MdBE.

516. Manufacturers' Library, Waltham, Mass.

Catalogue of books, in the Manufacturers' Library, Waltham. Boston: Printed by True & Green, 1822. [20?] p.

Copies: DeGE (imperfect: all after p. 18 wanting).

517. New York Mechanic and Scientific Institution.

Charter, constitution, and by-laws of the New-York Mechanic and Scientific Institution.... New York: W. A. Mercein, printer, 1822. 20 p.

S-9700.

Copies: DLC; NNC; NNMus; PPF.

518. Apprentices' Library Company, Philadelphia.

Catalogue of books belonging to the Apprentices' Library Company of Philadelphia. [Philadelphia]: Jesper Harding, printer, 1823. 84 p.

S-11630.

Copies: PHi.

519. Apprentices' Library Company, Philadelphia.

Report of the managers of the Apprentices' Library Company.... March 12, 1823. Philadelphia: Jesper Harding, printer, 1823. 12 p.

S-11631.

Copies: PPAmP; PPL.

520. Architectural Institute of Philadelphia.

Constitution and by-laws of the Architectural Institute of Philadelphia. Instituted 2d month 9th, 1822. Philadelphia: Printed by Solomon W. Conrad, 1823. 16 p.

Copies: DeGE.

521. Mercantile Library Company of Philadelphia.

Catalogue of the Mercantile Library of Philadelphia, April, 1823. [Philadelphia: Printed by I. Ashmead & Co., 1823.] 28 p.

"Constitution of the Mercantile Library Company," p. [3]–6.

S-13301.

Copies: PPL.

522. New York Mechanic and Scientific Institution.

Charter, constitution and by-laws of the New York Mechanic and Scientific Institu-

522 (*continued*)
tion. To which is added: The names of the officers, board of managers, and standing committees. Published by order of the committee. New York: W. A. Mercein, 1823. 20 p.

S-13567.

Copies: NN; NjR.

523. Salem Mechanic Library.

Rules & by-laws of the Salem Mechanic Library, instituted August 1820. Salem, [Mass.]: Printed by W. Palfray, Jun. 1823. 24 p.

S-14036.

Copies: MSaE; MWA.

524. Apprentices' Library Company, Burlington, N.J.

A catalogue of the books belonging to the Burlington Apprentice's Library Company. Instituted 1822. Burlington, N.J.: Printed by D. Allinson, 1824. [26] p.

S-15601.

Copies: DLC.

525. Apprentices' Library Company, Philadelphia.

Apprentices' Library, January 28, 1824. [Philadelphia: 1824.] 4 p.

S-17605.

Copies: PHi.

526. Franklin Institute, Philadelphia.

Charter of incorporation, constitution and by-laws of the Franklin Institute of the State of Pennsylvania for the Promotion of the Mechanic Arts. To which is annexed, a list of members, and of the officers for 1824. Together with the first quarterly report of the Board of Managers, made to the Institute at a meeting held April 15th, 1824. Published by order of the Institute. Philadelphia: Printed by W. Fry, 1824. 34 p., 1 l.

S-16182.

Copies: CtY; DLC; DeGE; MBMu; MWA; PPF; PPL.

527. Franklin Institute, Philadelphia.

Constitution of the Franklin Institute of the State of Pennsylvania, for the Promotion of the Mechanic Arts. [Philadelphia: 1824.] 4 p.

Copies: MWA.

528. New York Mechanic and Scientific Institution.

Report on a plan for extending & more perfectly establishing the Mechanic and Scientific Institution of New-York. [New York]: Printed by Daniel Fanshaw, 1824. 28 p.

Prepared by A committee consisting of John Griscom, Thomas Richards, and Richard E. Mount.

S-17415.

Copies: DLC; DeGE; MB; MH.

529. Franklin Institute, Philadelphia.

First annual report of the proceedings of the Franklin Institute of the State of Pennsylvania, for the Promotion of the Mechanic Arts. To which are prefixed the charter, constitution, and by-laws of the Institute, with a list of the members, and officers for 1824 and 1825, and of the standing committees for 1825. Philadelphia: J. Harding, printer, 1825. 107 p.

"Report of the exhibition of 1824," p. [64]–107.

S-20555.

Copies: CSmH; CtY; DLC; DeGE; MBAt; MH-BA; MWA; MdBJ; NCH; NbU; PP; PPAmP; PPF; PPL; PPi; PU.

530. Franklin Institute, Philadelphia.

The Franklin Institute of the State of Pennsylvania for the Promotion of Mechanic Arts.... [Philadelphia: 1825.] Broadside.

Dated at head of title: September 6, 1825.

S-20558.

Copies: DNA.

531. Mercantile Library Association, New York.

Catalogue of the books belonging to the Mercantile Library Association of the city of

New York; to which are prefixed, the constitution and the rules and regulations of the same. New York: Hopkins & Morris, 1825. 106 p.

S-21688.

Copies: ICJ; ICN; MH-BA; MWA; PPL.

532. Rensselaer Polytechnic Institute, Troy, N.Y.

The constitution and laws of Rensselaer School, in Troy, New-York; adopted by the Board of Trustees, March 11, 1825. Troy, [N.Y.]: Printed by Tuttle and Richards, 1825. 23 p.

N-R0177314.

Copies: DLC; MWA; PHi; PPAmP.

533. American Lyceum of Science and the Arts, composed of associations for mutual instruction, and designed for the general diffusion of useful and practical knowledge. Worcester, Mass.: Spy office, S. B. Manning, printer, 1826. 7 p.

Attributed to Josiah Holbrook.

S-23487.

Copies: MWA; OClW.

534. Apprentices' Library Company, Philadelphia.

Annual report of the Apprentices' Library Company.... Philadelphia: W. Fry, printer, 1826. 12 p.

S-25736.

Copies: PHi.

535. Brooklyn Apprentices' Library Association.

Catalogue of books, belonging to the Brooklyn Apprentices' Library Association. Instituted in November, 1823. Brooklyn: A. Spooner, 1826. 35, 4 p.

S-23931.

Copies: NBLiHi (not located).

536. Massachusetts. Commissioners on Establishment of Institution for Instruction in Practical Arts and Sciences.

Report of the Commissioners, appointed by a resolve of the 22 February, 1825. [Boston: 1826.] 55 p.

Report on a system for the establishment of one or more institutions for instruction in the practical arts and sciences.

Signed: Theodore Sedgwick, L. M. Parker, James Savage.

An act to establish the Massachusetts Seminary of Practical Arts and Sciences, p. 49–53.

S-25267.

Copies: DLC; MH; MWA; NIC; PPL.

537. Polytechnic and Scientific College of Philadelphia.

Address of the trustees of the Polytechnic and Scientific College to be established in Philadelphia.... [Philadelphia: 1826.] 4 p.

Dated: May 11, 1826.

S-25796.

Copies: PU.

538. Polytechnic and Scientific College of Philadelphia.

The following address, written by one of the trustees of the proposed new college ... respectfully submits it to his fellow citizens. To the citizens of the city and county of Philadelphia. [Philadelphia: 1826.] 7 p.

Dated: April 17, 1826.

Concerns technical education.

S-25797, 25800.

Copies: MWA; PPAmP; PU.

539. Polytechnic and Scientific College of Philadelphia.

The following report, which is to be submitted to a town meeting ... is now printed in order that the citizens may have fair opportunity of judging of the proposed plan. [Philadelphia: 1826.] 4 p.

Dated: March 20, 1826.

S-25798.

Copies: PU.

540. Polytechnic and Scientific College of Philadelphia.

The Polytechnic and Scientific College of Philadelphia. [Philadelphia: 1826.] 3 p.

Announcement of courses for the next term.

S-25799.

Copies: PPAmP.

541. Rensselaer Polytechnic Institute, Troy, N.Y.

Constitution and laws of Rensselaer School, adopted by the Board of Trustees, April 3, 1826. Together with a catalogue of officers and students. Albany: Printed by Packard & Van Benthuysen, 1826. 24 p.

MWA copy has 46, [2] p., including *Supplement* (p. [25]–28), and *Rensselaer School exercises, in the fall, winter and spring terms . . . to be added to the pamphlet containing the statute of incorporation,* dated on the last page: Troy, June 27, 1827 (p. [29]–[48]).

S-25922.

Copies: MH; MWA; N; PPL.

542. American Lyceum of Science and the Arts, composed of associations for mutual instruction and designed for the general diffusion of useful and practical knowledge. Worcester, [Mass.]: Printed by Samuel H. Colton and Co., 1827. 7 p.

Attributed to Josiah Holbrook.

S-27836.

Copies: DLC; MWA.

543. Apprentices' Library Company, Philadelphia.

Report of the managers of the Apprentices' Library. Philadelphia: I. Ashmead, printer, 1827. 12 p.

S-30221.

Copies: PHi.

544. Boston Mechanics Institution.

The constitution of the Boston Mechanics' Institution. Boston: Printed by Moore and Sevey, 1827. 7 p.

S-28245.

Copies: DLC; MBAt; MBBS; MH-BA; MHi.

545. Maryland Institute for the Promotion of the Mechanic Arts, Baltimore.

Proposals of the Maryland Institute for the Promotion of Mechanic Arts, for the exhibition of November, 1827, addressed to the mechanics and manufacturers of the United States. Baltimore: Printed by John D. Toy, 1827. 11 p.

S-29654.

Copies: MdHi.

546. Massachusetts. Commissioners on Establishment of Institution for Instruction in Practical Arts and Sciences.

Report of the Commissioners, by virtue of a resolution of the House of Representatives of the 22d February, 1826. The subject of the establishment of a seminary of practical arts and sciences.... [Boston: 1827.] 23 p. (H.R. 5)

S-29673.

Copies: MB; MWA.

547. Boston Mechanics Institution.

The first annual report of the board of managers of the Boston Mechanics' Institution . . . with the constitution and a list of the members. Boston: Printed by John Cotton, 1828. 24 p.

S-32439.

Copies: DLC; M; MBAt; MH; MHi; PP; WHi.

548. Brooklyn Apprentices' Library Association.

Catalogue of books, belonging to the Brooklyn Apprentices' Library Association. Brooklyn: A. Spooner, 1828. 35 p.

S-32484.
Copies: NBLiHi.

549. Mercantile Library Association, New York.

Catalogue of the books belonging to the Mercantile Library Association of the city of New York; to which are prefixed, the constitution, and the rules and regulations of the same. New York: Printed by J. & J. Harper, 1828. 188 p.
S-34545.
Copies: NNS; RPA.

550. Mercantile Library Company, Philadelphia.

Catalogue. [Philadelphia]: Printed by I. Ashmead & Co., 1828. 76 p.
S-34765.
Copies: MH; PHi; PPi; PU.

551. Oneida Institute, Whitesboro, N.Y.

First report of the trustees of Oneida Academy, March, 1828. Utica, [N.Y.]: Printed by Hastings & Tracy, 1828. 11 p.
N-00093520.
Copies: MiU-C; PPPrHi.

551a. Rensselaer Polytechnic Institute, Troy, N.Y.

Triennial catalogue of the officers and members of Rensselaer School.... Troy, N.Y.: Printed by Francis Adancourt, 1828. 15 p.
Copies: MWA.

552. Woodbine Institute.

A prospectus of Woodbine Institute, for practical education and agriculture; including some of the prominent points of Fellemberg's institutions in Switzerland. Under the superintendence of John M. Keagy, M.D. Harrisburg: Printed by John S. Wiestling, 1828. 8 p., fold. plate.
S-37225.
Copies: PPL.

553. Chittenango Polytechny.

Charter of the Polytechny, adopted October 17, 1829. Chittenango, N.Y.: Printed by Isaac Lyon, 1829. 15 p.
S-38139.
Copies: DLC;

554. Chittenango Polytechny.

A view of the course of studies and system of education pursued in the Polytechny; to which is added a catalogue of the officers & students since its commencement. Chittenango, N.Y.: From the Polytechny Press, 1829. 14 p.
S-38140.
Copies: DLC; PPL.

555. Craig, John D.

An address, delivered at a meeting of the citizens of Cincinnati, convened for the purpose of forming a mechanics' institute; and published at request of said meeting. Cincinnati: Ohio Mechanics' Institute; Wm. J. Ferris, printer, 1829. 17 p.
S-38296.
Copies: OCHP; OClWHi.

556. [Holbrook, Josiah] 1788–1854.

American Lyceum; or, Society for the Improvement of Schools, and Diffusion of Useful and Practical Knowledge. Boston: Perkins & Marvin, 1829. 24 p.
S-39008.
Copies: CtY; DLC; MWA.

557. ——— Boston: T. R. Marvin, 1829. 15 p.
S-39009.

557 (*continued*)
Copies: CtY; DLC; ICU; MBAt; MH; MWA; MeB; MiD-B; NCH; OO; PPAmP; ScC.

558. Manual Labour Academy of Pennsylvania, Philadelphia.
First annual report of the board of trustees of the Manual Labour Academy of Pennsylvania. November 10, 1829. Philadelphia: Printed by W. F. Geddes, 1829. 16 p.
S-40052.
Copies: PHi; PPAmP; PPL.

559. ________ Philadelphia: Printed by W. F. Geddes, 1829. 15 p.
S-40053.
Copies: MWA; PPL.

560. Ohio. Laws.
An act to incorporate "The Ohio Mechanics' Institute." Feb. 9, 1829. [Cincinnati?: 1829.] 3 p.
N-00053437.
Copies: OCHP.

561. Ohio Mechanics' Institute, Cincinnati.
Annual report of the board of directors of the Ohio Mechanics' Institute. Cincinnati: 1829.
Founded 1828.
S-39914.
No copy known.

562. Shrewsbury, N.J. Polytechnic Institute.
Constitution and by-laws of the Polytechnic Institute of Shrewsbury, Monmouth County, state of New-Jersey. New York: Printed by Elliott & Palmer, 1829. 12 p.
S-40423.
Copies: MB; MH; NN; NNM.

563. Apprentices' Library Company, Philadelphia.
Catalogue of books, belonging to the Apprentices' Library Company of Philadelphia, instituted in 1820. Philadelphia: Printed for the Apprentices' Library, 1830. 88 p.
C-3031.
Copies: PHi; PPL.

564. Baldwin, Henry, 1780–1844.
An address delivered before the subscribers to the Mechanic's Institute, on Thursday evening, July 22, 1830.... Pittsburgh: Printed by Joseph Snowden, 1830. 8 p.
Concerns technical education.
C-230.
Copies: NjR.

565. Boston Mechanics Institution.
The third annual report of the board of managers of the Boston Mechanics Institution; with the constitution, by-laws, and a list of members. Boston: Printed by Samuel N. Dickinson, 1830. 23 p.
C-607.
Copies: DLC; MB; MBAt; MWA.

566. Franklin Institution, New Haven.
Outline of the Franklin Institution of New Haven. [New Haven: Baldwin and Treadway, printers, 1830?] 10 p.
C-1444.
Copies: CtY; DLC; MWA.

567. Ismar, F. A.
The school of industry, at New-Harmony, state of Indiana, and Madame Maria Duclos Fretageot; a letter to Mr. William Maclure.... [New Harmony, Ind.]: Printed for the Author, 1830. 18 p.
C-2014.
Copies: NNC.

568. Maclurian Lyceum, Philadelphia.
Report of the transactions of the Maclurian Lyceum of the arts and sciences of Philadelphia, from its commencement in 1826 to January 1830. Submitted by John T. Sharpless.... Philadelphia: J. Richards, printer, 1830. 11 p.
N-M0079307.
Copies: DNLM; MWA; MiU; PHi.

569. Manual Labour Academy of Pennsylvania, Philadelphia.
Second annual report of the board of trustees of the Manual Labour Academy of Pennsylvania. November 9, 1830. Philadelphia: Printed by W. F. Geddes, 1830. 15 p.
C-3005.
Copies: PPL; PPPrHi.

570. Mercantile Library Association, New York.
Catalogue of the books belonging to the Mercantile Library Association of the city of New York; to which are prefixed the constitution and the rules and regulations of the same. New York: Printed by J. & J. Harper, 1830. 160 p.
C-2819.
Copies: MWA; NBu; NN.

571. Oneida Institute, Whitesboro, N.Y.
Second report of the trustees of the Oneida Institute of Science and Industry. Whitestown, March 20, 1830. Utica, [N.Y.]: Press of William Williams, 1830. 28 p.
C-2912.
Copies: MB; PPPrHi.

572. Providence Association of Mechanics and Manufacturers.
Catalogue of the Mechanics' and Apprentices' Library, established by the Providence Association of Mechanics and Manufacturers, in the year 1821. Providence: Cranston & Knowles, 1830. 22 p.
Sabin-66281 (note).
Copies: MWA; RPB.

573. Salem Mechanic Library.
Catalogue of the Salem Mechanic Library, together with the rules and by-laws: instituted August, 1820. Salem, [Mass.]: W. Palfray, Jr., Register office, 1830. 31 p.
C-3390.
Copies: MHi.

574. United States. Military Academy, West Point.
Catalogue of the library of the U.S. Military Academy at West-Point. May 1830. New York: Printed by J. Desnoues, 1830. 132 p.
C-5253.
Copies: DLC; MHi; NWM; PPAmP.

4. Sciences applied to technology

a. General works

575. Abel, Thomas.

Subtensial plain trigonometry, wrought with a sliding-rule, with Gunter's lines: and also arithmetically, in a very concise manner. And this method apply'd to navigation, and surveying. To which is added, I. Mensuration of masons work. II. A solution of rota, or Aristotle's wheel. III. A brief discourse upon gravity. By Thomas Abel, of Bourn in Lincolnshire, Old England. Philadelphia: Printed and sold for the author, by Andrew Steuart, 1761. [4], 86 p., 7 fold. plates.

Mp-8777.

Copies: CtHT-W; DAU; DLC; DeGE; MBAt; MH; MWA; MiU-C; NN; OHi; PHC; PPL; RPJCB.

576. Nicholson, William, 1753–1815.

An introduction to natural philosophy. Illustrated with copperplates.... The 3d ed., with improvements. Philadelphia: Printed for Thomas Dobson, 1788. xvi, [17]–570, [9] p., 25 fold. plates.

First published in London, 1782.

Many of the plates describe scientific instruments and experiments.

Mp-21333.

Copies: DLC; DeGE; KyU; MWA; MeB; NN; NjP; PPAmP; PPL; ViU; WaU.

577. Stearns, Samuel, 1741–1809.

The American oracle, comprehending an account of recent discoveries in the arts and sciences, with a variety of religious, political, physical, and philosophical subjects, necessary to be known in all families, for the promotion of their present felicity and future happiness. By the Honourable Samuel Stearns.... New York: Printed for and sold by Hodge and Campbell, Berry and Rogers, and T. Allen, 1791. 4 p.l., 627, xviii p., plate.

Mp-23795.

Copies: DLC; DeGE; MHi; MWA; NHi; NN; PPL; RPJCB.

578. Nicholson, William, 1753–1815.

An introduction to natural philosophy.... A new ed. with improvements. Philadelphia: Printed by T. Dobson, 1795. 2 v., 25 fold. plates.

First published in London, 1782.

Many of the plates describe scientific instruments and experiments.

Mp-29218.

Copies: DLC; KyU; LNHT; MWA; MiU-C; NIC; NcD; OOxM; P; PPC; PU.

579. Moreau de Saint Méry, Méderic Louis Élie, 1750–1819.

Idée générale ou abrégé des sciences et des arts à l'usage de la jeunesse.... Philadelphie: Imprimé par l'éditeur, 1796. xi, [2], 408 p.

Based on J. H. S. Formey's *Abrégé de toutes les sciences,* Paris, 1784.

Mp-30819.

Copies: DLC; DeGE; MWA; NN; PPAmP; PPL; RPJCB.

580. Moreau de Saint Méry, Méderic Louis Élie, 1750–1819.

General view or abstract of the arts and

sciences; adapted to the capacity of youth.... Translated from the French by Michael Fortune. Philadelphia: 1797. xii, 363, [7] p.

Mp-32505.

Copies: DeGE; MWA; PPAmP.

581. [Mansfield, Jared] 1759–1830.

Essays, mathematical and physical, containing new theories and illustrations of some very important and difficult subjects of the sciences.... New Haven: Printed by William W. Morse, [1801]. x, [2], 274, [46] p., 13 fold. plates.

Dated variously [1795], [1800], [1801], [1802], [1819].

Includes observations on nautical astronomy and the theory of gunnery.

N-M0185290–1; S&S-866, 2586, 48576.

Copies: CtHi; CtY; DLC; DeGE; ICJ; LU; MB; MH; MWA; MiU; MnU; NCH; NN; NjP; OCU; OkU; PPAmP; RPJCB; ViU.

582. Outlines of a course of lectures upon the application of the principles of natural philosophy and chemistry, to arts, manufactures, and the common purposes of life. Philadelphia: Printed by John Thompson, 1801. 8 p.

Copies: DeGE.

583. Cowen, Esek.

An oration, on the progress of arts and sciences; composed and delivered at the close of an exhibition, on the 19th of October 1803, in the town of Hartford, county of Washington, and state of New-York. Salem, [N.Y.]: Printed by Henry Dodd, [1803?]. 30 p.

S&S-4023.

Copies: MWA.

584. Miller, Samuel, 1769–1850.

A brief retrospect of the eighteenth century. Part first; in two volumes: containing a sketch of the revolutions and improvements in science, art, and literature, during that period.... New York: Printed by T. and J. Swords, 1803. 2 v.

No more published.

S&S-4654.

Copies: DLC; DeGE; MWA; MWiW; NN; NNC; PU.

585. Adams, George, 1750–1795.

Lectures on natural and experimental philosophy, considered in its present state of improvement.... In four volumes.... This American ed., printed from the last London ed., edited by William Jones ... is carefully rev. and corr. by Robert Patterson.... Whitehall: Printed for W. W. Woodward, Philadelphia, 1806–1807. 4 v., fold. plates.

S&S-9799, 11930.

Copies: CSt; CtW; DeGE; KyU; MH; MWA; N; NjR; PU; PWW; ViU; VtU.

586. Ferguson, James, 1710–1776.

Ferguson's Lectures on select subjects, in mechanics, hydrostatics, hydraulics, pneumatics, optics.... A new ed., corr. and enlarged. With notes and an appendix, adapted to the present state of arts and sciences. By David Brewster. This American ed., carefully rev. and corr. by Robert Patterson.... Philadelphia: M. Carey, 1806. 2 v.

First published in London, 1760.

S&S-10406.

Copies: CtY; MH; MWA; MiU; NIC; NN; NjP; OCIW; PPL; PU; RP; TU.

587. Ferguson, James, 1710–1776.

Plates illustrating a new edition of Ferguson's Lectures on mechanics, hydrostatics, hydraulics, pneumatics, optics.... Edited and enlarged by David Brewster; this American edition carefully rev. and corr. by Robert Patterson.... Philadelphia: M. Carey, 1806. [1] p., 48 plates.

S&S-10033.

Copies: GHi; IEG; MWA; MWiW; MeU; NcU; NhPoA; TU.

588. Webster, John.

Elements of natural philosophy; explaining the laws and principles of attraction, gravitation, mechanics, pneumatics, hydrostatics, hydraulics, electricity, and optics:

588 *(continued)*
with a general view of the solar system.... By John Webster. With notes and corrections by Robert Patterson.... Philadelphia: B. and T. Kite; Fry and Kammerer, printers, 1808. x [xi], [13]-231 p., illus.

S&S-16689.

Copies: DLC; DeGE; LU-E; MB; MH; MMeT; MWA; NjR; PPL.

589. Ewing, John, 1732–1802.

A plain elementary and practical system of natural experimental philosophy, including astronomy and chronology, by the late Rev. John Ewing.... The whole carefully rev. and prepared for the press with sundry explanatory and supplementary notes, by Robert Patterson.... With a biographical sketch of the author. Philadelphia: Hopkins and Earle, 1809. xxv, 538 p., plates (part fold.).

S&S-17470.

Copies: AU; CtY; DeGE; ICU; MB; MH; MWA; MWiW; MiU; NN; NjP; OSW; PPAmP; PPL; PPPrHi; PU; ViU.

590. Walker, Adam, 1731–1821.

Analysis of a course of lectures on natural and experimental philosophy. viz. 1. Properties of matter. 2. Mechanics. 3. Chymistry. 4 & 5. Pneumatics. 6. Hydrostatics. 7. Electricity.... 9. Optics.... Lancaster, [Pa.]: Printed by William Greer, 1809. 59 p.

S&S-19130.

Copies: MWA; OClW; PHi.

591. Barton, William, 1754–1817.

Memoirs of the life of David Rittenhouse, LLD., F. R. S., late president of the American Philosophical Society, &c. interspersed with various notices of many distinguished men: with an appendix, containing sundry philosophical and other papers, most of which have not hitherto been published. By William Barton.... Philadelphia: E. Parker, 1813. lxxviii, [79]–614 p., front. (port.), fold. facsim.

Includes description of Rittenhouse's orrery.

S&S-27835.

Copies: CSmH; DLC; DeGE; MBNEH; MH; MHi; MWA; MdBP; MeB; MnU; MoS; NBu; NIC; NjP; NjR; OC; PP; PU; ViU.

592. Cavallo, Tiberius, 1749–1809.

The elements of natural or experimental philosophy.... 1st American ed., with additional notes selected from various authors, by F. X. Brosius.... Philadelphia: Thomas Dobson; William Fry, printer, 1813. 2 v., 21 fold. plates.

First published in London, 1803.

S&S-28093.

Copies: CSt; DAU; DLC; KyU; MCM; MH; MWA; MWiW; MdBS; NBu; NNC; NcU; NjP; PPF; Vi; ViU.

593. De Witt, Simeon, 1756–1834.

The elements of perspective. Albany: Printed by H. C. Southwick, 1813. xxxi, [33]–69, [1] p., 12 fold. plates.

S&S-28320.

Copies: DLC; DeGE; GHi; MWA; MWiW; NSchU; NWM; NjP.

594. Farish, William, 1759–1837.

A plan of a course of lectures on arts and manufactures, more particularly such as relate to chemistry. Cambridge, Mass.: Smith, 1813. 37 p.

Probably a Cambridge, England, imprint, cf. N–F0033150.

S&S-28480.

Copies: NNC; PPF.

595. Ferguson, James, 1710–1776.

Lectures on select subjects in mechanics, hydrostatics, hydraulics, pneumatics, optics.... New ed., corr. and enlarged with notes and an appendix adapted to the present state of the arts and sciences by David Brewster. 2d American ed., carefully rev. and corr. by Robert Patterson. Philadelphia: M. Carey; A. Small, printer, 1814. 2 v.

Accompanied by a separate volume of plates.

S&S-31470.
Copies: DLC; KyU; LU; MB; MWA; MiU; NNC; NjP; PPL; PU; TU; ViU.

596. Ferguson, James, 1710–1776.
Plates illustrating a new edition of Ferguson's Lectures on mechanics, hydrostatics, hydraulics, pneumatics, optics, geography, astronomy, and dialling. Edited and enlarged by David Brewster, A.M. This American ed., carefully rev. and corr. by Robert Patterson.... Philadelphia: M. Carey, 1814. 2 p.l., 48 fold. plates.
Vol. 3 of Ferguson's *Lectures on select subjects.* 2d American ed.
S&S-31470.
Copies: DeGE; LNB; LU-E; MPIA; MWA; NNNAM; TU.

597. Griscom, John, 1774–1852.
Hints relative to the most eligible method of conducting meteorological observations.... New York: 1815. 16 p.
S&S-34824.
Copies: CtY.

598. Bigelow, Jacob, 1787–1879.
Inaugural address, delivered in the chapel of the University at Cambridge, December 11, 1816. By Jacob Bigelow ... Rumford Professor in Harvard University. Boston: Printed by Wells and Lilly, 1817. vi, [7]–24, [2] p.
Mainly on Sir Benjamin Thompson, Count Rumford, and his discoveries.
S&S-40248.
Copies: DLC; DeGE; MBAt; MBCo; MWA; MdBJ-W; MeHi; NN; NjPT; PPAmP; PPL; RPB.

599. Cutbush, Edward, 1772–1843.
An address delivered before the Columbian Institute, for the Promotion of Arts and Sciences, at the city of Washington, on the 11th January, 1817.... Washington: Printed by Gales and Seaton, 1817. 29 p.
" ... an outline of the objects and importance of the Columbian Institute, for the promotion of Arts and Sciences," cf. Prefatory remarks.
S&S-40604.
Copies: CtY; DLC; DNA; DeGE; InU; KyLxT; MBAt; MWA; MnHi; NNNAM; NcD; PPL; ViU.

600. Cavallo, Tiberius, 1749–1809.
The elements of natural or experimental philosophy.... 2d American ed., with additional notes, selected from various authors, by F. X. Brosius.... Philadelphia: Thomas Dobson & Son; T. H. Palmer, printer, 1819. 2 v., 21 fold. plates.
S&S-47529.
Copies: AU; DGU; MH; MWA; MdW; NcU; NjR; PP; PU; ViU.

601. [Farrar, John] 1779–1853.
An elementary treatise on the application of trigonometry to orthographic and stereographic projection, dialling, mensuration of heights and distances, navigation, nautical astronomy, surveying and levelling.... Cambridge, [Mass.]: Printed at the University Press, by Hilliard & Metcalf, 1822. vi, 153, [73] p., 9 fold. plates.
S-8684.
Copies: CSt; CtY; DeGE; IHi; InU; MB; MH; MdBP; MeB; MiU; NIC; NNC; NjP; RPB; TxU; ViU.

602. Keating, William Hypolitus, 1799–1840.
University of Pennsylvania. Syllabus of a course of mineralogy and chemistry, as applied to agriculture and the arts. Philadelphia: 1822. 8 p.
Copies: PPAmP.

603. [Phillips, Richard, Sir] 1767–1840.
A grammar of natural and experimental philosophy.... With one hundred engravings on wood.... By the Rev. David Blair [pseud.].... From the 12th London ed., improved and enl. Hartford: Huntington and Hopkins; Sidney's Press; New Haven, 1822. vi, [7]–216 p., illus.
S-9917.
Copies: CtHY-W; CtHi; NNE.

604. ——— Hartford: S. G. Goodrich, 1822. vi, [7]–216 p. illus.

S-9918.

Copies: DLC; MA; MDeeP; NjR; PPL; PU; VtMiM.

605. ——— New Haven: John Babcock & Son, [etc.], 1822. vi, [7]–216 p., illus.

S-9919.

Copies: CtHi; CtW; DeGE; MB; MH; NNC; OClWHi; PPL; RPB.

606. Webster, John.

Philosophical instructor; or, Webster's Elements of natural philosophy, subdivided into principles and illustrations.... By A. Eaton. Albany: Printed by Websters and Skinners, 1824. 216 p., illus.

S-16035.

Copies: GU; IaMp; KU; KyLxT; MB; MH; MWA; MiU-C; NCH; NRU; NSyU; OO; TNJ.

607. Cavallo, Tiberius, 1749–1809.

The elements of natural or experimental philosophy.... 3d American ed., with additional notes, selected from various authors, by F. X. Brosius.... Philadelphia: Towar & Hogan, 1825. 2 v., 21 fold. plates.

S-20011.

Copies: DLC; KyU; LNHT; MB; MeB; NN; NjP; TNJ; ViU.

608. Farrar, John, 1779–1853.

An elementary treatise on the application of trigonometry to orthographic and stereographic projection, dialling, mensuration of heights and distances, navigation, nautical astronomy, surveying and levelling.... 2d ed. Boston: Hilliard, Gray, Little & Wilkins, 1825. viii, 155, [73] p., 9 plates.

S-20490.

Copies: DAU; MB.

609. Genet, Edmond Charles, 1763–1834.

Address on the several subjects of science, useful knowledge, public improvements, finances, and public law, in reference to rivers, canals, navigation, and commerce; delivered on the fifth of February, 1824, in the assembly room, at the Capitol of Albany, before the members of the Legislature, and the Society for the Promotion of Useful Arts.... Albany: Printed by Packard & Van Benthuysen, 1825. 43 p.

S-20644.

Copies: CtY; DeGE; MB; MiU-C; N; NHi; NNC.

610. Genet, Edmond Charles, 1763–1834.

Memorial on the upward forces of fluids, and their applicability to several arts, sciences, and public improvements; for which a patent has been granted by the government of the United States to the author. Albany: Printed by Packard & Van Benthuysen, 1825. 112 p., diagrs., plates.

S-20645.

Copies: CtY; DLC; DeGE; ICJ; LNHT; MB; MH-BA; MWA; MiD; MiU-T; NN; NNC; NNE; NjR; OC; PPL; TxU; ViU.

611. Fenning, Daniel.

A new and easy guide to the use of the globes; to which is added, a plain method of constructing maps. Illustrated with plates. 1st American ed., rev. and corr. by Jos. Walker, Sr. and Jos. Walker, Jr. Baltimore: Fielding Lucas, Jr. [1827]. x, 244 p., plates.

First published in London, 1794.

S-28868.

Copies: CtHT-W; CtW; DGU; MWA; MWH; MdBE; MdBL; OSW.

612. [Genet, Edmond Charles] 1763–1834.

Vindication of Mr. E. C. Genet's Memorial on the upward forces of fluids: in two letters to Professor Silliman. To which are added remarks on aerostation. New Haven: Printed by T. G. Woodward, 1827. 34 p.

Letters signed: E. C. Genet.

S-29005.

Copies: DeGE; MB; N; NN; NNC; WU.

613. A treatise on the nature and effects of heat, light, electricity, and magnetism, as being only different developments of one element. Cambridge, [Mass.]: Hilliard and

Brown, 1827. [4], 91 p.
Attributed to Marshall Tufts.
S-30828.
Copies: CtY; MH; MWA; NNNAM; NjR; PPAmP.

614. Farrar, John, 1779–1853.
An elementary treatise on the application of trigonometry to orthographic and stereographic projection, dialling, mensuration of heights and distances, navigation, nautical astronomy, surveying and levelling.... 3d ed. Boston: Hilliard, Gray, and Company, 1828. viii, 155, [73] p., 9 plates.
N-F0040385; S-33150.
Copies: CtY; DI-GS; ICU; NWM; NjP; PPL; PU; RPB.

615. Arnott, Neil, 1788–1874.
Elements of physics; or, Natural philosophy, general and medical, explained independently of technical mathematics, and containing new disquisitions and practical suggestions.... 1st American from the 3d London ed., with additions by Isaac Hays.... Philadelphia: Carey, Lea & Carey, 1829. 552 p., illus.
Includes chapters on statics, dynamics, mechanics, pneumatics, hydraulics, heat and light.
S-37506.
Copies: CSt; CU; GU; ICJ; IEG; MB; MH; MWA; MiU; NNC-M; OCl; PPC; RPB.

616. Bigelow, Jacob, 1787–1879.
Elements of technology, taken chiefly from a course of lectures delivered at Cambridge, on the application of the sciences to the useful arts. Now published for the use of seminaries and students.... Boston: Hilliard, Gray, Little and Wilkins, 1829. xii, 507 p., 22 plates (part fold.).
S-37847.
Copies: CoCC; DLC; DeGE; ICJ; IU; KU; LNHT; MB; MWA; MiD; NIC; NhD; NjP; NjR; OCl; PPF; PPL; RPB.

617. Cavallo, Tiberius, 1749–1809.
The elements of natural or experimental philosophy.... 4th American ed., with additional notes, selected from various authors, by F. X. Brosius.... Philadelphia: Towar & Hogan, 1829. 2 v., 21 fold. plates.
S-38095.
Copies: CtY; DGU; ICJ; MdBE; MiD; NNC; NjP; OCU; OO; PP; PPF; ViU.

618. Comstock, John Lee, 1789–1858.
A system of natural philosophy, in which the principles of mechanics, hydrostatics, hydraulics, pneumatics, acoustics, optics, astronomy, electricity, and magnetism are familiarly explained.... Hartford: D. E. Robinson, 1830. 295 p., illus.
C–969.
Copies: CtHT-W; CtY; GU; ICU; KyDC; MBC; MH; MdU; MiU; NjR; TNJ.

619. Wilbur, Hervey, 1787–1852.
A lexicon of useful knowledge, for the use of schools and libraries, with several hundred engravings.... New York: White, Gallaher & White, 1830. 295 p., illus.
C-5465.
Copies: CSmH; DLC; MBAt; NBuG; OClW; PMA; TNJ; WHi.

b. Astronomy

620. Churchman, John, 1753–1805.

J. Churchman's address to the members of the different learned societies in Europe and America, in support of the principles of the magnetic variation, their application to determining the longitude at sea; shewing wherein what the learned Doctor Euler published on this subject at Berlin, is deficient.... [S.l.: 1789.] 4 p.

Same text was also printed in the *American Museum,* May 1789.

Copies: DLC; MBAt.

621. Churchman, John, 1753–1805.

An explanation of the magnetic atlas, or variation chart, hereunto annexed; projected on a plan entirely new, by which the magnetic variation in any part of the globe may be precisely determined ... and the variation and latitude being accurately known, the longitude is of consequence truly determined.... Philadelphia: Printed by James & Johnson, 1790. 46, 5, [1] p., fold. map, 2 fold. tables.

Mp-22406.

Copies: CtY; DLC; MBAt; MHi; MWA; MiU-C; NN; NjP; PPAmP; PPL; RPJCB; ViU.

622. Croswell, William, 1760–1834.

Tables for readily computing the longitude, by the lunar observations. Partly new, and partly taken from the requisite tables of Dr. Maskelyne. With their application, in a variety of rules and examples. By William Croswell.... Boston: Printed by I. Thomas and E. T. Andrews, 1791. v, [2], 128 p.

Mp-23300.

Copies: DLC; DeGE; MHi; MWA; NcD; PMA; RPJCB.

623. Waring, William, d. 1793.

A journal for lunar observations, by which the calculation of longitude is much expedited, the mariner being led through the operation by a regular printed form in each page.... With directions exemplified, &c. Philadelphia: 1791.

E-23954.

No copy known.

624. A geographical clock: calculated for the meridian of this city; exhibiting at one view, the time of day or night in every part of the globe; with a full description, in letter press, of the clock and their planisphere of the earth.... Philadelphia: Printed and sold by Francis Bailey, 1792.

E-24347.

No copy known.

625. Adams, George, 1750–1795.

Astronomical & geographical essays: containing, I. A ... view, on a new plan, of the general principles of astronomy. II. The use of the celestial and terrestrial globes.... III. The description and use of the most improved planetarium, tellurian, and lunarium. IV. An introduction to practical astronomy.... 4th ed. With the author's last improvements.... Whitehall: Printed for William Young, Philadelphia, 1800. xvi, 194 p., 2 l., viii, [9]–238, [9]–148 p., 16 fold. plates.

First published in London, 1789.

Mp-36756.

Copies: CtY; DeGE; MH; MWA; NN; PPL; RPJCB.

626. ——— Whitehall, [Pa.]: Printed for William young, Philadelphia, 1800 [1808]. xvi, 194 [186], viii, ii, [9]–238, [9]–142 p., plates.

The essay, on the use of the celestial and terrestrial globes.... (viii, ii, [9]–238 p.), by his

father, George Adams, has a special title page, with edition statement and imprint: 5th ed. Philadelphia: W. W. Woodward, 1808.
N-A0060390.
Copies: MWA; PPULC; PU.

627. Adams, George, 1720–1773.
An essay, on the use of the celestial and terrestrial globes ... exhibiting the general principles of dialing & navigation. By the late George Adams.... 4th ed. With the author's last improvements, illustrated with copper-plates. Whitehall: Printed for William Young, Philadelphia, 1800. viii, [9]–238 p., plates.
Mp-36757.
Copies: CtY; MH; MWA; MdBP; NN; PHi; RPJCB; ViU.

628. ———— 5th ed. ... Philadelphia: W. W. Woodward, 1808. viii, [9]–238 p., plates.
S&S-14264.
Copies: DSI; GU; MWA; NjR; OClW; PU.

629. United States. Congress. House. Committee to Whom was Referred the Memorial of William Lambert.
Report of the Committee to whom was referred, on the twenty-fifth of January, 1810, the memorial of William Lambert, accompanied with sundry papers relating to the establishment of a first meridian for the United States, at the permanent seat of their government. March 28, 1809 [1810].... Washington City: A. & G. Way, printers, 1810. 31 p., 2 fold. tables.
S&S-21840.
Copies: DLC; DNA; DeGE; MWiW.

630. Wallace, James, d. 1851.
A new treatise on the use of the globes, and practical astronomy ... In four parts. I. An extensive collection of astronomical and other definitions. II. Problems performed by the terrestrial globe, including those relative to geography, navigation, dialling.... III. Problems performed by the celestial globe, including those of finding the longitude at sea, new methods of finding the latitude.... The necessary astronomical instruments are pointed out, and the most useful tables are inserted in the work.... New York: Smith & Forman, 1812. viii, 512 p.
S&S-27419.
Copies: DLC; MB; MH; MWA; NN; NjP; PPL.

631. Bennett, Samuel.
A new explanation of the ebbing and flowing of the sea, upon the principles of gravitation. By S. Bennett.... New York: Published for the author; Gould and Van Pelt, printers, 1816. viii, [9]–79 p., fold. plate.
S&S-36930.
Copies: DLC; DeGE; MWA; NN.

632. Egypt. A brief account and description of the great circular zodiac, or planisphere of the Egyptians. Now exhibiting, for a short time, at Scudder's New-York Spectaculum, Chatham-Street. New York: Printed by D. Fanshaw, 1825. 12 p.
S-20386.
Copies: MWA.

633. United States. Congress. House. Select Committee on National Observatory.
National observatory. March 18, 1826. Mr. Mercer, from the committee to which was referred so much of the President's message as relates to the erection of an observatory, made the following report. [Washington]: 1826. 23 p. (19th Cong., 1st sess. House. Rept. 124)
Includes a report from the Engineer Department, containing a list of instruments.
S-27266.
Copies: DLC; DeU; MWA; NjP.

634. Courtenay, James Carroll, 1803–1835.
An inquiry into the propriety of establishing a national observatory. By James

634 (*continued*)
Courtenay, of Charleston.... Charleston: Printed by W. Riley, 1827. 27 p.

Includes a description of the facilities of the Greenwich Observatory.

S-28611.

Copies: CSfCW; CtY; DLC; DeGE; MB; MBAt; MH; MWA; NN; NcD; PPL; ScC.

c. Chemistry and pharmaceutical chemistry

635. A short treatise of the virtues of Dr. Bateman's pectoral drops.... To be sold only by James Wallace, in New York. [New York]: Reprinted by J. Peter Zenger, [1731]. 4, 36 p.

An abstract of the patent granted ... the inventor of Dr. Bateman's pectoral drops ... reprinted by John Peter Zenger in New-York, 1731, 4 p. preceding the text.

B-856; Mp-39981.

Copies: NN (photocopy); NNNAM.

636. Prior, Thomas, 1682?–1751.

The authentic narrative of the success of tar water ... with remarks by Thomas Prior.... To which are subjoined, two letters from the author of Siris, showing the medical properties of tar water, and the best manner of making it.... London, printed 1746. Boston, N.E.: Reprinted and sold by Rogers and Fowle, 1749. iv, [5]–80 p.

Includes instructions to make tar and tar water.

Mp-6409.

Copies: DNLM; MBAt; MH; MWA; NN; NcD.

637. Penington, John, 1768–1793.

Chemical and economical essays, designed to illustrate the connection between the theory and practice of chemistry, and the application of that science to some of the arts and manufactures of the United States of America.... Philadelphia: Printed by Joseph James, 1790. [2], iv, [2], 200 p., 2 plates.

The first three chapters appeared in the *Columbian magazine,* Aug.—Dec. 1789, illustrated with a plate.

Mp-22757.

Copies: CtY; DLC; DeU; KyU; MH; MWA; MiU-C; NN; PPAmP; PPL; PU.

638. Penington, John, 1768–1793.

An inaugural dissertation, on the phoenomena, causes and effects of fermentation: Submitted to the examination of the Revd. William Smith, S.T.P. Provost; the trustees and medical professors, of the College of Philadelphia, for the degree of Doctor of Medicine, on the second day of June A.D. 1790. Philadelphia: Printed by Joseph James, 1790. 30 p.

Mp-22758.

Copies: DLC; DeGE; MWA; PPAmP; PPF; PPL; PU.

639. Lewis, William, 1714–1781.

The Edinburgh new dispensatory: containing I. The elements of pharmaceutical chemistry. II. The materia medica.... III. The pharmaceutical preparations and medicinal compositions.... And copperplates of the most convenient furnaces, and principal pharmaceutical instruments.... A new ed., with many alterations, corrections, and additions [by Andrew Duncan]. Philadelphia: Printed by T. Dobson, 1791. xxxii, [33]–656 p., plates.

First published at Edinburgh, 1786.

Mp-23503.

Copies: CSt; DLC; IU; KyLxT; MWA; MdBJ; MeB; NIC; NN; NcD; NhD; PPC; RPJCB; ViU; WU.

640. Woodhouse, James, 1770–1809.
An inaugural dissertation on the chemical and medical properties of the persimmon tree, and the analysis of astringent vegetables; submitted to the examination of the Revd. John Ewing, S.T.P. Provost, the trustees and medical professors of the University of Pennsylvania, for the degree of Doctor of Medicine.... Philadelphia: Printed by William Woodhouse, [1792]. 34 p.
Discusses the application of the juice of unripe persimmons to tanning, the making of black ink and black dye.
Mp-25055.
Copies: DNLM; MB; MWA; PPAmP; PPL; PU.

641. Prior, Thomas, 1682?–1751.
An authentic narrative of the success of tarwater in curing a great number and variety of distempers.... To which are subjoined, two letters from the author of Siris, shewing the medicinal properties of tar-water, and the best method of preparing it. Providence: Reprinted by J. Carter, 1793. 75, [5] p.
Includes instruction to make tar and tar water. First published in London, 1746.
Mp-26038.
Copies: DLC; DNLM; MWA; RHi.

642. Woodhouse, James, 1770–1809.
Observations on the combination of acids, bitters, and astringents: being a refutation of some of the principles, contained in Dr. Percival's essay, on bitters and astringents.... Philadelphia: Printed by Jones, Hoff & Derrick, 1793. vii, [9]–20 p.
Includes experiments on making ink.
Mp-26496.
Copies: DLC; DNLM; DeGE; MWA; PPAmP; PPL; PU.

643. Prentiss, John.
The compleat toilet; or, Ladies companion. A collection of the most simple and approved methods of preparing baths, essences, pomatums, powders, perfumes and sweet scented waters; with receipts for cosmetics of every kind that can smooth and brighten the skin, give force to beauty and take off the appearance of old age.... A new ed. New York: 1794.
E-27544.
No copy known.

644. Ross, William Morrey, d. 1818.
A chemico-physiological inaugural dissertation on carbone, or charcoal. Submitted to the public examination of the faculty of physic, under the authority of the trustees of Columbia College, in the state of New-York ... for the degree of doctor of physic; on the fifth day of May, 1795.... New York: Printed by T. and J. Swords, printers to the faculty of physic of Columbia College, 1795. [2], 58 p.
Includes observations on the production and use of charcoal.
Mp-29431.
Copies: DLC; DNLM; DeGE; MBAt; MWA; NN; PPL; PU.

645. Lewis, William, 1714–1781.
The Edinburgh new dispensatory: containing I. The elements of pharmaceutical chemistry. II. The Materia medica... III. The pharmaceutical preparations.... With the additions of the most approved formulae ... and several copperplates of the most convenient furnaces, and principal pharmaceutical instruments. Being an improvement of the New dispensatory by Dr. Lewis [by John Rotheram]. The 3d American from the 4th Edinburgh ed., with many alterations, corrections and additions.... Walpole, Newhampshire: Printed by D. Carlisle, jun. for Thomas & Andrews in Boston, [etc.], 1796. xxix, 622 p., 3 plates.
Mp-30692.
Copies: CtY; DNLM; MB; MWA; MiU; NNNAM; NjR; OCU; PPC; TxU.

646. ———— The 4th ed., with many alterations, corrections and additions.... Philadelphia: Printed by Thomas Dobson, 1796. xxvii, 622 p., 6 plates.

646 (*continued*)
Mp-30693.
Copies: CoU; DNLM; IU-M; IaU; MB; MWA; MiU; NNC-M; NcU; ScCM; VtU.

647. Woodhouse, James, 1770–1809.
The young chemist's pocket companion; connected with a portable laboratory. Containing a philosophical apparatus, and a great number of chemical agents.... Philadelphia: Printed by J. H. Oswald, 1797. 56, [2] p., plate.
Mp-33245.
Copies: CSmH; CtY; DLC; MWA; NcD-MC; PPL; PU.

648. Brailsford, Edward, d. 1856.
An experimental dissertation on the chemical and medical properties of the Nicotina tabacum of Linnaeus, commonly known by the name of tobacco.... Philadelphia: Printed by John Ormrod, 1799. 77 p.
Mp-35230.
Copies: CSmH; DLC; DNLM; MH; NN; NNNAM; NcD; PPL.

649. ———— Philadelphia: Printed by John Ormrod, 1799. 78 p., fold. col. plate.
Mp-35231.
Copies: DLC; DNLM; NN; PP.

650. Lavoisier, Antoine Laurent, 1743–1794.
Elements of chemistry, in a new systematic order, containing all the modern discoveries.... Translated from the French, by Robert Kerr.... 4th ed., with notes, tables and considerable additions. Philadelphia: Mathew Carey, 1799. 1 p.l., [v]–xlii, [49]–592 p., 13 fold. plates.
Translation of his *Traité élémentaire de chemie.*
The plates depict chemical apparatus and laboratory equipment.
Mp-35718.
Copies: CtY; DLC; DeGE; ICU; MH; MWA; NN; OGK; OkU; OrU; PPAmP; PU; RPB.

651. ———— 4th ed.... New York: Swords, 1801. 592 p., fold. plates.
S&S-799.
Copies: CtHT-W.

652. Pascalis-Ouvière, Félix, 1750-1833.
Annual oration, delivered before the Chemical Society of Philadelphia, January 31st, 1801. By Felix Pascalis.... Published by order of the Society.... Philadelphia: Printed by John Bioren, 1802. 48 p.
Observations on the application of chemistry to useful arts and manufactures.
S&S-2850.
Copies: DLC; DeGE; MWA; NNNAM; PPAmP; PU-S; WU.

653. Ives, Eli, 1779–1861.
An oration, on chemistry and botany; delivered before the Phi Beta Kappa Society, at New Haven, Dec. 9, 1802.... New Haven: Printed by Joel Walter, 1803. 16 p.
Includes observations on the practical application of new discoveries in chemistry.
S&S-4439.
Copies: CtY; DeGE; MWA; NNUT; NcD; PPM.

654. Thomson, Thomas, 1773–1852.
A new system of chemistry, including mineralogy, and vegetable, animal, and dyeing substances, comprehending the lastest discoveries and improvements of the science. Philadelphia: T. Dobson, 1803. 364 p., plates.
S&S-5164.
Copies: CSmH; DLC; PU.

655. [Lewis, William] 1714–1781.
The Edinburgh new dispensatory: containing I. The elements of pharmaceutical chemistry. II. The materia medica.... III. The pharmaceutical preparations and compositions.... And several copperplates, explaining the new system of chemical characters, and representing the most useful pharmaceutical apparatus. By Andrew Duncan, jun.... 1st Worcester ed. Worcester: From the press of Isaiah Thomas, [etc.], 1805. xii, 694 p., 6 plates.
S&S-8374.

Copies: CSt-L; CtY; DLC; DNLM; ICJ; MWA; MWiW; MdBJ; NNC-M; NcA-S; OCU; OrU; PPC; PU; ViU; WU.

656. Black, Joseph, 1728–1799.

Lectures on the elements of chemistry, delivered at the University of Edinburgh; by the late Joseph Black.... Published from his manuscripts by John Robison.... 1st American ed. from the last London ed. Philadelphia: M. Carey, 1806–1807. 3 v., plates.

Includes suggestions for practical application of chemistry.

S&S-10002, 12157.

Copies: DSI; DeGE; KyU; MBCo; MH; MWA; MWiW; NN; NNC; NcD; NjR; PPL; VtU.

657. Coxe, John Redman, 1773–1864.

The American dispensatory, containing the operations of pharmacy; together with the natural, chemical, pharmaceutical and medical history of the different substances employed in medicine.... With several copperplates, exhibiting the new system of chemical characters and representing the most useful apparatus.... Philadelphia: Printed by A. Bartram for T. Dobson, 1806. xvi, 787 p., 6 plates.

S&S-10222.

Copies: CtY; DNLM; ICJ; MH-BA; MWA; MiU; NNNAM; OClWHi; PPAmP; PPL; PU; RPM.

658. Ewell, Thomas, 1785–1826.

Plain discourses on the laws or properties of matter; containing the elements or principles of modern chemistry; with more particular details of those practical parts of the science most interesting to mankind, and connected with domestic affairs.... New York: Printed for Brisban & Brannan, 1806. 469 p., plates.

S&S-10377.

Copies: CSt; CtY; DLC; DeGE; ICJ; KyLx; MB; MWA; MiU-C; NN; NcD; PPF; PU; ViU.

659. Lavoisier, Antoine Laurent, 1743–1794.

Elements of chemistry, in a new systematic order, containing all the modern discoveries.... Translated from the French by Robert Kerr.... From the 5th Edinburgh ed. With notes, tables and considerable additions. New York: E. Duyckinck, 1806. 2 v. in 1., 14 plates.

Plates describe chemical apparatus and laboratory equipment.

S&S-10706.

Copies: CU; CtY; DLC; DeGE; ICJ KyU; LU; MWA; NN; NcU; OU; OkU; PU; ViU.

660. [Marcet, Jane Haldimand] 1769–1858.

Conversations on chymistry in which the elements of that science are familiarly explained and illustrated by experiments and plates.... Philadelphia: James Humphreys, 1806. 2 vols. in 1, 11 plates.

First published in London, 1806.

S&S-10786.

Copies: CSmH; CtW; DeGE; MWA; NNN; PHi; PPAmP; PPC; PU; RPB; ViAl.

661. Parkes, Samuel, 1759–1825.

A chymical catechism; or, The application of chymistry to the arts, for the use of young people, artists, tradesmen, and the amusement of leisure hours.... Philadelphia: James Humphreys, 1807. xxviii, 29–334, [2] p., plate.

First published in London, 1806.

S&S-13317.

Copies: DNLM; MMeT; MWA; MiU; NN; NNNAM; PPL; PU-S.

662. Accum, Friedrich Christian, 1769–1838.

System of theoretical and practical chemistry ... from the improved London ed.... Philadelphia: Kimber & Conrad; Brown and Merritt, printers, 1808. 2 v., illus., 7 plates.

Plates illustrate laboratory equipment and instruments.

662 (*continued*)
First published in London, 1803.
S&S-14259.
Copies: CSt-L; CtY; DLC; DNLM; ICJ; MWA; MdBJ; PPC; RPA.

663. Henry, William, 1774–1836.
Epitome of chemistry in three parts. Part I. Intended to facilitate the acquisition of chemical knowledge.... Part II. Directions for the analysis of mineral waters; of earths and stones; of ores and metals.... Part III. Instructions for applying chemical tests and reagents to various useful purposes.... From the 4th English ed., much enlarged and illustrated with plates. To which are added notes by a professor of chemistry in this country. New York: Collins & Parkins, 1808. xl, 416, lxxxiv p., plates.
S&S-15213.
Copies: CtY; MH; MWA; MdBS; NN; Nh; PU; RNR.

664. [Marcet, Jane Haldimand] 1769–1858.
Conversations on chemistry, in which the elements of that science are familiarly explained and illustrated by experiments and plates. To which are added ... a description and plate of the pneumatic cistern of Yale College and a short account of artificial mineral waters in the United States. With an appendix, consisting of treatises on dyeing, tanning and currying. New Haven: From Sidney's Press for I. Cooke & Co., 1809. xi, 358, 17, [6] p., 12 plates.
S&S-17979.
Copies: CU; CtY; DNLM; DeGE; MWA; MWiW; NN; ViU.

665. [Marcet, Jane Haldimand] 1769–1858.
Conversations on chymistry.... 2d American ed.; enlarged by an appendix consisting of a description, with plate ... of a new hydro-pneumatic blow-pipe, invented by Mr. Joseph Cloud ... also of three disquisitions, one on dyeing, one on tanning, and one on currying. Philadelphia: J. Humphreys, 1809. 427 p., 12 plates.
S&S-17980.
Copies: AkU; DNLM; IU; MB; MH; MWA; MiD-B; NNN; OkU; PPL; ViU.

666. Coxe, John Redman, 1773–1864.
The American dispensatory, containing the operations of pharmacy; together with the natural, chemical, pharmaceutical and medical history of different substances employed in medicine.... With several copperplates, exhibiting the new system of chemical characters and representing the most useful apparatus. 2d ed.... Philadelphia: Thomas Dobson; Fry and Kammerer, printers, 1810. viii, [9]–839 p., 6 plates.
S&S-19883.
Copies: CtY; DNLM; DeGE; GWay; ICJ; MWA; MdU-H; NRU; NcU; OU; PPL; PPiAM; ViU.

667. Henry, William, 1744–1836.
An Epitome of experimental chemistry, in three parts.... The 2d American, from the 5th English ed., illustrated with plates. To which are added notes on various subjects ... by B. Silliman.... Boston: William Andrews, 1810. xii, [ii]–xviii, 432, lxxvii, [8] p., 8 plates.
S&S-20324.
Copies: CSt; GU-M; MB; MWA; MWiW; PU.

668. Parkes, Samuel, 1759–1825.
The rudiments of chemistry; illustrated by experiments, and eight copperplate engravings of chemical apparatus.... Philadelphia: Bradford & Inskeep, [etc.]; J. Maxwell, printer, 1810. xii, 291 p., plates.
S&S-20994.
Copies: MH; MNBedf; MWA; NcD.

669. Thacher, James, 1754–1844.
The American new dispensatory.... With an appendix, containing.... Medical electricity. Galvanism. An abridgment of Dr. Currie's Reports on the use of water. The cultivation of the poppy plant, and the method of preparing opium.... Boston: T. B. Wait and Co., 1810. 529 p.
S&S-21476.

Copies: DNLM; IU-P; LNFM; MBP; MWA; MeB; MiDW-M; NN.

670. Brinley, George.

A catalogue of drugs, medicines, dye stuffs, printers' colours, surgeons' instruments, &c. &c. To be sold by George Brinley, at his store, No. 3, south side of the old market house, Boston.... [Boston: 1812.] 16 p.

S&S-24935.

Copies: MWA.

671. Cooper, Thomas, 1759–1839.

The introductory lecture of Thomas Cooper, Esq., professor of chemistry at Carlisle College, Pennsylvania.... With notes and references. Carlisle, [Pa.]: Printed by Archibald Loudon, 1812. viii, 236 p.

Includes observations on various discoveries and processes. Recommends the construction of turnpikes, railroads and canals.

S&S-25164.

Copies: CtY; DLC; DeGE; ICU; KU; MB; MWA; NN; NNNAM; NjR.

672. Cutbush, James, 1788–1823.

The philosophy of experimental chemistry.... Philadelphia: Isaac Peirce; Merritt, printer, 1813. 2 v., fronts.

Offers information on the production and refinement of chemicals, particularly metals.

S&S-28261.

Copies: CtHT-W; CtY; DLC; DeGE; IU; MWA; MiD-B; PMA; PPL; ViRUT.

673. [Marcet, Jane Haldimand] 1769–1858.

Conversations on chemistry ... to which are added ... a description and plate of the pneumatic cistern of Yale College. And a short account of artifical mineral waters in the United States. With an appendix, consisting of treatises on dyeing, tanning and currying. New Haven: From Sidney's Press, for I. Cooke & Co., 1813. x, 358, 17, [6] p., 12 plates.

S&S-29047.

Copies: CLU; CtHi; DLC; MWA; MiU; NcD; NjP; OU; ViU.

674. Thacher, James, 1754–1844.

The American new dispensatory.... With an appendix, containing an account of mineral waters.... Medical electricity. Galvanism.... An abridgment of Dr. Currie's Reports on the use of water. The cultivation of the poppy plant, and the method of preparing opium.... 2d ed. Boston: Thomas B. Wait and Co., and C. Williams, [1813]. xxvii, [25]–732 p.

S&S-29937.

Copies: DLC; DNLM; IU-M; MB; MWA; NNNAM; NbU; OC; PU; RPB; WU.

675. Accum, Friedrich Christian, 1769–1838.

System of theoretical and practical chemistry.... With an appendix ... by Thomas Cooper.... [2d American ed.] Philadelphia: Kimber and Conrad, 1814. 2 v., illus., plates.

S&S-30616.

Copies: AzU; DLC; ICJ; KyU; MWA; MiU; NWM; NcU; NjR; OU PPL; PU; ViU.

676. Coxe, John Redman, 1773–1864.

The American dispensatory, containing the operations of pharmacy.... With several copperplates, exhibiting the new system of chemical characters, and representing the most useful apparatus. 3d ed., much improved.... Philadelphia: Thomas Dobson, 1814. iv, 771 p., 6 plates.

S&S-31255.

Copies: DLC; DNAL; DNLM; MWA; NcU; PPC; PU.

677. Henry, William, 1774–1836.

The elements of experimental chemistry.... The 3d American, from the 6th English ed., greatly enlarged by the author, and illustrated with nine plates.... With an additional plate, illustrating the decomposition of the fixed alkalis, by heat—and a frontispiece, exhibiting the pneumatic cistern at Yale College, by B. Silliman. Boston: Thomas & Andrews, [etc.], 1814. 2 v., 10 plates.

N-H0287877; S&S-31693.

677 (*continued*)
Copies: CtY; DNLM; IU; InU; KyU; MH; MWA; NcD; OO; PPC; RPB; ViU.

678. ———— New Haven: Printed by Joseph T. Buckingham, for Howe and Deforest, 1814. 2 v., 10 plates.
S&S-31694.
Copies: CSt; MnSM.

679. [Marcet, Jane Haldimand] 1769–1858.
Conversations on chemistry . . . to which are added . . . a description and plate of the pneumatic cistern of Yale College and a short account of artificial mineral waters in the United States. With an appendix consisting of treatises on dyeing, tanning and currying. New Haven: Sidney's Press, for Howe & DeForest, I. Cooke & Co., 1814. xi, 358, 17, [7] p., 12 plates.
S&S-32008.
Copies: MWA.

680. ———— New Haven: Printed by Sidney's Press, for Increase Cooke & Co., 1814. xi, 358, 17, [7] p., 12 plates.
S&S-32009.
Copies: CtHi; CtY; GAOC; ICU; MB; MH; MWA; MdBM; NN; NNC; ViU.

681. Davy, Humphry, Sir, bart., 1778–1829.
Elements of agricultural chemistry, in a course of lectures for the Board of Agriculture. . . . Philadelphia: John Conrad & Co; Baltimore: F. Lucas, Jr., [etc.], 1815. viii, 332, lxxiv, [6] p., plates.
First published in London, 1813.
S&S-34538.
Copies: DLC; ICJ; MWA; MiU-C; Nh; PU.

682. ———— New York: Eastburn, Kirk & Co., 1815. viii, 332, lxxiv, [6] p., plates.
S&S-34539.
Copies: MWA; PU.

683. ———— Boston: Ward, 1815. 332 p., plates.
S&S-34540.
Copies: MU; MBL; MH; MU; MWelC.

684. Parkes, Samuel, 1759–1825.
The chemical catechism, with notes, illustrations, and experiments. . . . From the 6th London ed. with emendations and considerable additions. New York: Collins & Co., 1816. xiv, 7–559, 1 p., illus., fold. plate.
Contains references to industrial application of chemistry.
S&S-38557.
Copies: CtHT; MBD; MWA; MeB; NjR; OCLloyd.

685. Accum, Friedrich Christian, 1769–1838.
A practical essay of chemical re-agents, or tests. Illustrated by a series of experiments. . . . Philadelphia: M. Carey & Son, 1817. xxiv, [1], 26–204 p., illus.
S&S-39915.
Copies: DLC; MWA; NcD; PU.

686. Henry, William, 1774–1836.
The elements of experimental chemistry, by William Henry. . . . The 4th American, from the 7th London ed., greatly enlarged; and illustrated with nine plates, engraved by Lowry. To which are added notes, and two additional plates, by John Redman Coxe. . . . Philadelphia: James Webster, 1817. xxx, 656, xcv p., 10 fold. plates.
N-H0287879; S&S-41034.
Copies: CoU; DLC; DNLM; KyU; MWA; MiU; OU; PU; ViU.

687. Thacher, James, 1754–1844.
The American new dispensatory, containing the general principles of pharmaceutic chemistry. . . . With an appendix, containing an account of mineral waters. . . . Medical electricity. Galvanism. . . . The cultivation of the poppy plant, and the method of preparing opium. . . . 3d ed. Boston: Thomas B. Wait & Sons, 1817. xxiv, [25]-724 p.
S&S-42261.

Copies: DNLM; InU-M; MB; MBCo; MWA; MnU; NBuU-M; NNNAM; PPC; RPM; ScCM; TxU.

688. Accum, Friedrich Christian, 1769–1838.

Chemical amusement, comprising a series of curious and instructive experiments in chemistry.... Philadelphia: M. Carey & Son, 1818. 217 p., illus.

Includes practical hints: how to make mill stones and invisible ink, and how to engrave on glass.

S&S-43004.

Copies: DLC; IEN-M; MBP; MWA; MiD-B; MnSS; NNS; NjR; PHi (in PPL); PU; TN.

689. ———— 2d American ed., from the 3d improved London ed. With additions by Thomas Cooper.... Philadelphia: M. Carey & Son, 1818. 3 p.l., [iii]–xxxii, [27]–406 p., illus., 2 plates.

S&S-43005.

Copies: DLC; MH; MNBedf; MWA; MiDSH; MiU; NGlf; NKi; NNNAM; PHi; PU; RPB.

690. Coxe, John Redman, 1773–1864.

The American dispensatory, containing the operations of pharmacy.... With several copperplates, exhibiting the new system of chemical characters, and representing the most useful apparatus. 4th ed., much improved.... Philadelphia: Thomas Dobson, 1818. clv, 735 p., plates.

S&S-43760.

Copies: CSmH; DNLM; ICJ; KyU; MH; MWA; NNNAM; NcD; OU; PPC; ViU.

691. Formulae selectae; or, A collection of prescriptions of eminent physicians, and the most celebrated patent medicines; together with a syllabic view of mineral waters and poisons.... New York: R. Donaldson, 1818. 202 p.

S&S-44060.

Copies: CtW; DLC; IaU; MBCo; MWA; NIC; NNNAM; NjR; WaPS.

692. [Lewis, William] 1714–1781.

The Edinburgh new dispensatory ... with numerous tables and plates ... by Andrew Duncan, Jun., from the 8th and last Edinburgh edition. With large additions, relating, principally to the vegetable productions of the United States, by Jacob Dyckman.... New York: James Eastburn & Co., 1818. 798 p., plates.

S&S-44579.

Copies: CtY; DNLM; IU; KyU; MMeT; MWA; NNC; NcU; NhD; NjP; OClW; PU; TxU; ViU; WU.

693. [Marcet, Jane Haldimand] 1769–1858.

Conversations on chemistry ... illustrated by experiments. From the 4th and latest English ed., rev., corr., and considerably enl. To which are added, notes and observations; by an American gentleman. Greenfield, Mass.: Denio and Phelps, 1818. 2 p.l., [ix]–xxii, [25]-410 p., 19 plates.

Additions by J. L. Comstock.

S&S-44687; 44688.

Copies: CtY; DLC; MB; MH; MWA; MiD-N; NNC; TU.

694. ———— Rev. and corr. by Thomas Cooper, M.D., from the 5th London ed., considerably enl.... Philadelphia: M. Carey & Son, 1818. 2 v., 15 plates.

S&S-44689.

Copies: DLC; GU-M; KyU; MNBedf; MWA; NNNAM; PPL; ViU.

695. Orfila, Mathieu Joseph Bonaventure, 1787–1853.

Practical chemistry; or, A description of the processes by which various articles of chemical research ... are procured.... Translated from the French of P. M. Orfila.... By John Redman Coxe.... To which are added, a variety of subjects of practical utility.... Philadelphia: Thomas Dobson and Son; William Fry, printer, 1818. xv, 355, lxxxvi p., 8 plates.

S&S-45184.

Copies: AzU; CSfCMS; DLC; IaU; KyLxT; MB; MWA; MoSW-M; NNNAM; NjR; OClW; PU; ViL.

696. Parkes, Samuel, 1759–1825.

The chemical catechism, with notes, illustrations, and experiments.... From the 8th London ed., greatly enlarged, and carefully adapted to the present state of chemical science. New York: Collins, 1818. xvii, [1], 7, 17–597 p., illus., fold. plate.

Additional notes at the end contain references to industrial applications such as iron manufacture, coke-ovens, apparatus to collect gas from coal, and sugar manufacture.

S&S-45205.

Copies: DLC; DeGE; MWA; NcWsW; PEaL.

697. Smith, James, 1771–1841.

Prospectus of a permanent national vaccine institution, to be established in the city of Washington, District of Columbia. Baltimore: 1818. 32 p.

S&S-45728.

Copies: DLC; DNLM; NNNAM.

698. Smith, James, 1771–1841.

Two letters relative to the vaccine institution, addressed to the members of the Hon. the General Assembly of Maryland.... Baltimore: 1818. 35 p.

S&S-45729.

Copies: DLC; DNLM; MdBM; MdBP; OClWHi.

699. Davy, Humphry, Sir, bart., 1778–1829.

Elements of agricultural chemistry, in a course of lectures for the Board of Agriculture.... 2d American ed. Hartford: Printed by Hudson & Co., 1819. 304 p.

S&S-47795.

Copies: Ct; CtY; ICU; IU; MBC; MBH; MH; MWA; MeU; Mi; NT; NjP; OCU-M; PHC; PPL; RNR; UPB; ViU; WU.

700. Gorham, John, 1783–1829.

The elements of chemical science.... Boston: Cummings and Hilliard; University Press, Hilliard & Metcalf, 1819–1820. 2 v., plates.

Observations on the practical application and illustrations of chemical apparatus.

S&S-48100.

Copies: CSfCMS; CtY; DeGE; ICJ; IGK; InU; MH; MWA; MeL; NNNAM; NhD; OO; PPAmP; PPF; PPL; PU; RPM.

701. Henry, William, 1774–1836.

The elements of experimental chemistry. 1st American from the 8th London ed.... Together with an account of Dr. Wollaston's scale of chemical equivalents. Also, A substitute for Woulfe's or Nooth's apparatus, and A new theory of galvanism, by Robert Hare. Philadelphia: R. Desilver, 1819. 2 v., 13 plates.

N-H0287881; S&S-48213.

Copies: DLC; DNLM; KyU; IU; MdU; MiU; NN; NNC; NWN; PP; PPC; PPL.

702. [Marcet, Jane Haldimand] 1769–1858.

Conversations on chemistry.... From the 5th and latest English edition, rev., corr. and considerably enlarged.... 2d ed. Greenfield, Mass.: Denio & Phelps, 1820. 420 p., 15 plates.

S-2098.

Copies: ICJ; IaHi; InU; MB; MH; MWA; MiD; N; NNC; NcD; RPB; VtU.

703. Meigs, Josiah, 1757–1822.

Memorial of Josiah Meigs, and others, for an act of incorporation of a national vaccine institution for the United States of America. January 5, 1820.... Washington: Gales & Seaton, 1820. 8 p. (16th Cong., 1st sess. House. Doc. 29)

S-3868.

Copies: CtY-M; DLC; DNLM; DeU; MBCo; MWA; MDBJ; NNNAM; NjP; O; PPL.

704. Davy, Humphry, Sir, bart., 1778–1829.

Elements of agricultural chemistry in a course of lectures for the Board of Agriculture.... To which is added A treatise on soils and manures . . . by a practical agriculturist. Philadelphia: B. Warner, [etc.]; Baltimore: F. Lucas, jr., [etc.], 1821. viii, [9]–304, viii, [9]–92 p., illus.

"Appendix: An account of the results of experiments on the produce and nutritive qualities of different grasses, and other plants, used as food for animals. Instituted by John, duke of Bedford," p. [252]–297.

A treatise on soils and manures, viii, [9]–92 p. at the end of the volume, has a separate title page.

S-5158.

Copies: CSfCMS; CtY; DLC; DeGE; KyLx; LNOP; MBP; MWA; MdBM; MeU; MoSW-M; NIC; NOg; NcD; NjP; OCLloyd; PPF; PPi; PU; ScU.

705. Parkes, Samuel, 1759–1825.

The chemical catechism, with notes, illustrations, and experiments. . . . From the 9th London ed., carefully corrected and adapted to the present state of chemical science. New York: Collins & Co., 1821. 576 p., fold. plate.

N-P0104572; S-6375.

Copies: CtY; ICJ; KU-M; MBCo; MH; MWA; NBuG; NNC: OClW; PPF.

706. Salem Laboratory Company.

Act of incorporation and bye-laws of the Salem Laboratory Company. Salem, [Mass.]: Printed by W. Palfray, Jr., 1821. 16 p.

The laboratory was going to manufacture refined saltpeter, aquafortis, muriatic acid, oil of vitriol, and alum.

S-6703.

Copies: MSaE.

707. Thacher, James, 1754–1844.

The American new dispensatory. Containing general principles of pharmaceutick chemistry. . . . With an appendix, containing an account of mineral waters. . . . Medical electricity. . . . The cultivation of the poppy plant, and the method of preparing opium. . . . 4th ed. Boston: Thomas B. Wait, 1821. 736 p.

S-6952.

Copies: DeGE; GEU; IEN-M; IU; IaU; LU; MB; MBP; MdBM; MeB; MiU; NNNAM; NRU-M; NbU; OO; PPC; WU.

708. Ure, Andrew, 1778–1857.

A dictionary of chemistry, on the basis of Mr. Nicholson's, in which the principles of the science are investigated anew and its applications to the phenomena of nature, medicine, mineralogy, agriculture, and manufactures detailed. . . . 1st American ed., with some additions, notes, and corrections by Robert Hare, assisted by Franklin Bache. Philadelphia: R. Desilver, 1821. 2 v., illus.

S-7496.

Copies: CtHT; DLC; IU; InLP; KyLx; MH; MWA; MdBL; MiU; MoS-Med; NBuU-M; NcD; NjHoS; OClM; PPL; PU; RPB; VtU.

709. Cotting, John Ruggles, 1783–1867.

An introduction to chemistry. With practical questions. . . . Boston: C. Ewer, 1822. vii, [4], 420 p., plates.

Contains references to practical aspects of chemistry and illustrations of chemical apparatus.

S-8454.

Copies: ArCH; CtY; DLC; DeGE; ICU; IaHi; KU; MB; MH; MiU-C; NNC; NNNAM; NhD; NjHoS; PU; WKenHi.

710. Coxe, John Redman, 1773–1864.

The American dispensatory, containing the natural, chemical, pharmaceutical and medical history of the different substances employed in medicine; together with the operation of pharmacy. . . . 5th ed. . . . Philadelphia: H. C. Carey & Lea, 1822. iv, 815 p., plates.

S-8458.

Copies: DLC; DNLM; NNC-M; NcD; OClW; PPC; PU; ViU.

711. Eaton, Amos, 1776–1842.

Chemical instructor: presenting a familiar method of teaching the chemical prin-

711 *(continued)*
ciples and operations of the most practical utility to farmers, mechanics, housekeepers.... Albany: Websters and Skinners, 1822. 231 p.

S-8586.

Copies: CtHT-W; CtY; DeGE; ICU; MH; MdBJ-W; N; NT; NTR.

712. Hare, Robert, 1781–1858.

Minutes of the course of chemical instruction in the Medical Department of the University of Pennsylvania.... Philadelphia: Brown, 1822–1823. 3 v. in 1.

Includes observations on the production and refining of chemicals.

N-H0116979; S-8943.

Copies: DNLM; NcU; ODW; PPAmP; PPF; PPL; PU-S; RPB.

713. Henry, William, 1774–1836.

The elements of experimental chemistry, by William Henry.... 2d American from the 8th London ed., with a supplement, containing the new matter of the 9th.... Together with an account of Dr. Wollaston's scale of chemical equivalents. Also, a substitute for Woulfe's or Nooth's apparatus, and A theory of galvanism; by Robert Hare.... Philadelphia: R. Desilver, 1822. 3 v., illus., 14 plates.

S-8999.

Copies: CtY; DLC; DeGE; GEU; ICJ; IU; KWiF; KyLxT; MH; MdBM; MiU; NNCoCi; NcU; OClW; PPC; ScCC; ViU; WaPS.

714. [Marcet, Jane Haldimand] 1769–1858.

Conversations on chemistry; in which the elements of that science are familiarly explained and illustrated by experiments ... The 8th American, from the 6th London ed., rev. and enl. ... Hartford: Oliver D. Cooke, 1822. xii, 383 p., 16 plates.

S-9359.

Copies: CSmH; CtY; DLC; GDC; MB; MH; MnU; NNC; NcD; NjP; PP; PU; RPB; TNJ; TxU; ViL; VtBrt.

715. Henry, William, 1774–1836.

The elements of experimental chemistry.... 2d American from the 8th London ed., with a supplement, containing the new matter of the 9th.... Together with an account of Dr. Wollaston's scale of chemical equivalents. Also, a substitute for Woulfe's or Nooth's apparatus, and A theory of galvanism; by Robert Hare.... Philadelphia: R. Desilver, 1823. 3 v., illus., 14 plates.

S-12814.

Copies: CU; CtY; DLC; DNLM; FU; KyU; LNHT; MB; MdBJ; NNC; OCU; OkU; PPF; ViU.

716. Parkes, Samuel, 1759–1825.

The rudiments of chemistry; illustrated by experiments and copper-plate engravings of chemical apparatus.... A new ed., carefully corr., and adapted to the present state of chemical science. Philadelphia: A. Small, 1823. xx, [13]–312 p., 8 plates.

S-13664.

Copies: CtW; DeGE; MB; MH; MdU-H; NSyU; NhPet; OSW; PPA; PU; ViU.

717. Hare, Robert, 1781–1858.

Minutes of the course of chemical instruction in the Medical Department of the University of Pennsylvania.... Philadelphia: 1824. 107, [1], 62, 69, 56 p., 5 plates.

Includes: *Supplement to Minutes* ... , 62 p.; *Lectures on electricity and galvanism,* 69 p.; and *Appendix,* 56 p.

Offers observations on the production and refining of chemical compounds, and on the use of chemical apparatus.

N-H0116980; S-16410.

Copies: CtW; CtY; DLC; NcU; PU; RPB; ViU.

718. Magendie, François, 1783–1855.

Formulary for the preparation and mode of employing several new remedies; namely, morphine, iodine, quinine, cinchonine, the hydrocyanic acid, narcotine, strychnine, nux vomica, emetine ... with an introduction, and copious notes. By the late Charles Thomas Haden, Esq. Translated

from the French of the 3d ed. of Magendie's "Formulaire." By Robley Dunglison.... Rev. and corr. by a physician of Philadelphia. With an appendix. Philadelphia: J. Webster, 1824. xxiv, [25]–268 p.
S-16983.
Copies: CSmH; DLC; ICJ; MB; MBCo; MH; MWA; MdBJ; NIC; NjP; OClW; OrU-M; PPC.

719. [Marcet, Jane Haldimand] 1769–1858.
Conversations on chemistry; in which the elements of that science are familiarly explained and illustrated by experiments, and sixteen copper-plate engravings. The 9th American from the 8th London ed., rev., corr. and enl. To which are now added, explanations of the text—questions for exercise—directions for simplifying the apparatus, and a vocabulary of terms—together with a list of interesting experiments. By J. L. Comstock, M.D. Hartford, [Conn.]: O. D. Cooke and Sons, 1824. xii, 386 p., 16 plates.
N-M0203615.
Copies: CtY; DLC; DeGE; MB; MH; NN; NjR; ViU.

720. ——— The 10th American, rev. and corr. from the 8th London ed., by William H. Keating.... Philadelphia: Thomas de Silver, [etc.]; Baltimore: J. E. Coale, 1824. xviii p., 1 l., 335, [1] p., 15 plates.
N-M0203617; S-16762.
Copies: DLC; MB; PPF; PU.

721. Parkes, Samuel, 1759–1825.
The rudiments of chemistry; illustrated by experiments, and copper-plate engravings of chemical apparatus.... From the 3d London ed., carefully corrected and adapted to the present state of chemical science. With notes by James Renwick.... New York: S. B. Collins, T. & J. Swords, Collins & Hannay, etc.; Clayton & Van Norden, printers, 1824. xi, 278 p., plates.
S-17521.
Copies: CtHT-W; LNB; LNHT; MBAt; MH; MNS; NhD; OUrC.

722. ——— New York: E. B. Clayton, 1824. xi, 278 p., plates.
N-P0104596.
Copies: DLC; NNC; OClW.

723. Hare, Robert, 1781–1858.
Essays chemical, electrical & galvanic ... [Philadelphia: R. Wright, printer, 1825.] 48 p., 5 plates.
S-20812.
Copies: CtY; DNLM; DeGE; MB; MdBP; PPL; PU.

724. Hare, Robert, 1781–1858.
Minutes of the course of chemical instruction in the Medical Department of the University of Pennsylvania.... Reprinted, with corrections and additions. Philadelphia: Clark & Raser, 1825. 96, 62, 56 p., illus.
"Supplement," 62 p. has the imprint: Philadelphia: Robert Wright, printer, 1824.
"Appendix," 56 p. at end.
Includes observations on the production and refining of chemical compounds, and on the use of chemical apparatus.
S-20813.
Copies: DLC; DeGE; GEU-M; ICU; MdBM; PP; PU; TxU-M; ViU.

725. MacNeven, William James, 1763–1841.
Introductory discourse to a few lectures on the application of chemistry to agriculture, delivered before the New-York Athenaeum in the winter of 1825.... New York: G. & C. Carvill, 1825. 40 p.
S-21289.
Copies: CtY; DGU; DLC; DeGE; IObB; IU; MB; MBAt; MBH; MH; MdHi; NN; PMA; PPAmP; WU.

726. Eaton, Amos, 1776–1842.
Chemical instructor: presenting a familiar method of teaching the chemical principles and operations of the most practical utility to farmers, mechanics, housekeepers and physicians.... 2d ed., considerably altered. Albany: Websters and Skinners, 1826. 263 p.
S-24394.

726 (*continued*) Copies: CtY; DLC; IEN-M; KU; MH; MiU; MnSM; N; NSyU; Nh; ScCC.

727. [Marcet, Jane Haldimand] 1769–1858.

Conversations on chemistry.... The 10th American, from the 8th London ed., rev., corr. and enl. To which are added, explanations of the text—directions for simplifying the apparatus, and a vocabulary of terms.... By J. L. Comstock.... Hartford: O. D. Cooke and Co., 1826. xii, [13]–348 p., illus.

S-25223.

Copies: CtY; DLC; IaHi; KU; MB; MH; MeLB; MiU; MoU; NWM; NjP; OO; PU; RPB; ScCleU; TxU.

728. Philadelphia College of Pharmacy.

The druggist's manual. Being a price current of drugs, medicines, paints, dyestuffs, glass, patent medicines, etc. ... Compiled by direction of the Philadelphia College of Pharmacy. Philadelphia: Printed for the College by S. W. Conrad, 1826. xvii, 119 p.

S-25739.

Copies: DLC; DNLM; ICJ; ICU-M; MBCo; PPL; PPiU; ViRM; WU.

729. Magendie, François, 1783–1855.

Formulary for the preparation and employment of many new medicines, such as nux vomica, salts of morphine, prussic acid, strychnine, veratrine, alkalis of the cinchona, emetine, iodine, iodine of mercury, cyanide of potassium, croton oil, salts of gold, salts of platina, &c. &c. Translated from the 5th ed., rev. and augm. by John Baxter, with notes and additions. New York: G. H. Evans, 1827. xii, 138 p.

Translation of his *Formulaire pour la préparation et l'emploi de plusieurs nouveaux médicamens,* originally published in Paris, 1821.

S-29590.

Copies: CSt; DNLM; MB; NNNAM; NRU-M; WU.

730. Eaton, Amos, 1776–1842.

Chemical instructor: presenting a familiar method of teaching the chemical principles and operations of the most practical utility to farmers, mechanics, housekeepers and physicians.... 3d ed. ... Albany: Websters and Skinners, 1828. 275 p.

S-33038.

Copies: CU; CtY; IU; MH; MiU; NSyHi; OO; PU-S; TxU.

731. Magendie, François, 1783–1855.

Formulary for the preparation and employment of many new medicines, such as nux vomica, salts of morphine, prussic acid, strychnine, veratrine, alkalis of the cinchona, emetine, iodine, iodine of mercury, cyanide of potassium, croton oil, salts of gold, salts of platina, &c. &c. Translated from the 5th ed., rev. and augm. by John Baxter, with notes and additions. 2d ed. New York: G. H. Evans, 1828. x, 138 p.

S-33972.

Copies: CSt-L; DNLM; MBCo; MeB; MnU-B; NNC; PPC; TxU; ViU; WaU.

732. [Marcet, Jane Haldimand] 1769–1858.

Conversations on chemistry.... 10th american from the 8th London ed., rev., corr. and enl. To which are now added explanations of the text; directions for simplifying the apparatus ... by J. L. Comstock.... Hartford: O. D. Cooke, 1828. 348 p., illus.

S-34010.

Copies: CSt; CoU; CtHT; DLC; MH; MWA; MeB; NNE; OClWHi; PPiU; PU-S.

733. Slack, Elijah.

A key to the technical language, and a few other difficulties of chemistry, or chemical nomenclature. Together with a concise view of the doctrine of definite proportions and chemical elements. For the use of students in the Medical College of Ohio....

Cincinnati: Printed by G. T. Williamson, 1828. 25 p.
S-35217.
Copies: NjP; OCHP.

734. Turner, Edward, 1798–1837.
Elements of chemistry, including the recent discoveries and doctrines of the science. 1st American, from the 1st London ed. Philadelphia: J. Grigg, 1828. 499, [1] p., illus.
S-35570.
Copies: CtW; MdBM; MiU; NNC-P; NcU; OClW; PPC; PU; ViU; VtMiM.

735. Bullock, William, fl. 1808–1828.
A concise and easy method of preserving objects of natural history.... From the 3d London ed., with additions by a naturalist in America. New York: Printed for the publisher, 1829. 36 p.
S-37984.
Copies: CtY; MWA; NNM; NNNAM.

736. Magendie, François, 1783–1855.
Formulary for the preparation and employment of several new remedies: namely, resin of nux vomica, strychnine, morphine, hydrocyanic acid, preparations of cinchona, emetine, iodine, piperine, chlorurets of lime and soda, salts of gold and platina, phosphorus, digitaline, &c. Translated from the 6th ed. of the Formulaire of M. Magendie published in Paris, October 1827. With an appendix ... by Joseph Houlton. ... London: T. and G. Underwood; New York: Reprinted by J. & J. Harper, for F. and N. G. Carnes, 1829. xi, [13]–145, 18 p.
S-39385.
Copies: CtY-M; DLC; ICJ; MB; MBCo; MWA; MiU; NcD-M; PPL; TxU.

737. [Marcet, Jane Haldimand] 1769–1858.
Conversations on chemistry.... The 11th American, from the 8th London ed., rev., corr. and enl. To which are added, explanations of the text—directions for simplifying the apparatus, and a list of interesting experiments. By J. L. Comstock.... Hartford: Cooke and Co., 1829. xii, [13]–348 p., illus.
S-39427.
Copies: CSmH; CtHT-W; GU-M; ICU; MB; MH; MeHi; MoSB; NRHi; NjR; OAkU; PU.

738. Turner, Edward, 1798–1837.
Elements of chemistry, including the recent discoveries and doctrines of the science. 2d American ed. with important corrections. Philadelphia: John Grigg, 1829. 496p., illus.
S-40695.
Copies: DNLM; IP; IaDaM; KyLxT; MB; NN; NNC; NjP; PPAmP; PPC; PU.

739. Gray, Samuel Frederick, fl. 1780–1836.
The chemistry of the arts; being a practical display of the arts and manufactures which depend on chemical principles. With numerous engravings. On the basis of Gray's Operative chemist, adapted to the United States; with treatises on calico printing, bleaching, and other large additions. By Arthur L. Porter.... Philadelphia: Carey & Lea, 1830. 2 v. (803, [1] p.), 70 plates.
Gray's *The operative chemist* was first published in London, 1828.
C-3137.
Copies: CtY; DLC; DSI; DeGE; ICRL; IJC; KU; MH; MiU; MnU; NIC; NN; OU; PP; RPB.

740. [Marcet, Jane Haldimand] 1769–1858.
Conversations on chemistry; in which the elements of that science are familiarly explained.... 12th American, from the last London ed., rev. corr. and enl. ... Hartford: Cooke and Packard, and Butler, 1830. 348 p., illus.
C-2388.
Copies: KU.

741. Turner, Edward, 1798–1837.

Elements of chemistry, including the recent discoveries and doctrines of the science. By Edward Turner.... 3d American from the 2d London ed., with notes and emendations by Franklin Bache.... Philadelphia: John Grigg, 1830. xix, [13]–580 p., illus.

J. Grigg's advertisements, p. 579–580.
C-3781.

Copies: DeGE; IU; MH; NN; NcD; NjP; PMA; PU.

742. Vauquelin, Louis Nicolas, 1763–1829.

Dictionary of chemistry, containing the principles and modern theories of the science, with its application to the arts, manufactures, and medicine.... Translated from Le dictionnaire de chimie, approuvé par Vauquelin.... With additions and notes by Mrs. Almira H. Lincoln.... New York: G. & C. & H. Carvill, 1830. xxviii, [29]–531, [1] p., front.

C-5261.

Copies: DLC; DeGE; IGK; MBC; MWA; NCH; NNF; PU.

d. Electricity

743. Bates, Benjamin.

Notice is hereby given to the curious, that at the house of Mrs. Abegal Smith in Providence will be exhibited and continued from day to day (the weather being fair) for a few days, a course of experiments on the newly discovered electrical fire. [Newport?: Printed by James Franklin?, 1752.] Broadside 31.5 x 19 cm.

Signed: Benjamin Bates, April 6, 1752.
E-6816.

Copies: MH.

744. Kinnersley, Ebenezer, 1711–1778.

Newport, March 16, 1752. Notice is hereby given to the curious, that at the Court-House, in the Council-Chamber, is now to be exhibited and continued from day to day, for a week or two; a course of experiments, on the newly-discovered electrical fire; containing, not only the most curious of those that have been made and published in Europe, but a considerable number of new ones lately made in Philadelphia.... [Newport: Printed by J. Franklin, 1752.] Broadside 32 x 22.5 cm.

B-1572; Mp-40620.

Copies: CtY; MBU; MWA (photocopy); PPRF; RPB.

745. Kinnersley, Ebenezer, 1711–1778.

A course of experiments, in that curious and entertaining branch of natural philosophy, called electricity; accmpanied [*sic*] with explanatory lectures: in which electricity and lightning will be proved the same thing.... [Philadelphia]: Printed by A. Armbruster, 1764. 8 p.

Mp-9708.

Copies: NN; PPL; PPRF.

746. ———— accompanied with lectures on the nature and properties of the electric fire. By William Johnson. New York: Printed by H. Gaine, 1764. 8 p.

B-2481; Mp-41454.

Copies: RPJCB.

747. ________ accompanied with lectures on the nature and properties of the electric fire. By William Johnson. New-York: H. Gaine, 1765. 8 p.
Mp-10027; N-J0131636.
Copies: DLC.

748. Mason, David, 1726–1795.
A course of experiments in that instructive and entertaining branch of natural philosophy, called elictricity[*sic*].... [Boston: 1765.] Broadside 23.5 x 18 cm.
Program of two lectures, indicating use of apparatus.
B-2594; Mp-41559.
Copies: MB; MWA (xerox copy).

749. Greenwood, Isaac, 1730–1803.
Sublime entertainment. This evening will commence at the court-house, a brilliant electrical exhibition; by I. Greenwood.... [Providence: 1793.] Broadside 33.5 x 20 cm.
Dated: Providence, February 1793.
Refers to a "machine with all its apparatus, worth upwards of 150 dollars."
B-8353; Mp-46767.
Copies: RPJCB.

750. King, William.
Rational amusements.... William King, philosophical and medical electrician ... proposes to entertain with ... a lecture upon electricity, which will be illustrated by ... philosophical experiments.... [S.l.: ca. 1800.]
N-K0145847.
Copies: CtY-M.

751. Dearborn, Benjamin, 1754–1838.
To the President and Members of the American Academy of Arts and Sciences. Gentlemen: As that season of the year is approaching in which a general apprehension of danger from lightning is most excited, a few remarks, on the various means of seeking safety.... Boston: 1807. Broadside.
S&S-12408.
Copies: MSaE; MWA (xerox copy).

752. [Green, Jacob] 1790–1841.
An epitome of electricity & galvanism. By two gentlemen of Philadelphia.... Philadelphia: Printed by Jane Aitken, 1809. [226] p. (various pagings), front.
By Jacob Green and Erskine Hazard.
S&S-17672–3.
Copies: CtY; DLC; DeGE; GU-M; IU; KU; MB; MWA MiU-C; N; NNE; Nh; NjP; OCHP; PPAmP; PPL; PU.

753. King, William.
Electrical exhibition, at the house of Joseph Wheeler, in Kingston, William King, electrician ... with lectures upon electricity, illustrated with ... philosophical experiments.... Kingston, [N.Y.]: 1810. Broadside 35 x 25 cm.
Concerns experiments with an electric battery and the display of an electrical orrery.
N-K0145846.
Copies: CtY-M; MWA.

754. Abbot, Joel, 1766–1826.
An essay on the central influence of magnetism.... Philadelphia: Printed for the author, 1814. 24 p., 1 l.
S&S-30609.
Copies: DLC; DNLM; DeGE; MB; MWA; MdBP; NNNAM; NcD; NjPT; OkU; PPAN; PPAmP; PPL.

755. Hare, Robert, 1781–1858.
A new theory of galvanism, supported by some experiments and observations made by means of the calorimotor, a new galvanic instrument. Also, a new mode of decomposing potash extemporaneously. Read before the Academy of Natural Sciences, Philadel-

755 (*continued*)
phia.... Philadelphia: M. Carey, 1819. 17, [2] p., fold. front.
S&S-40998 (mistakenly dated 1817), 48167.
Copies: DLC; DNLM; DeGE; MBCo; MHi; MWA; P; PHi; PPAmP; RPB.

756. Hare, Robert, 1781–1858.
Animadversions by Dr. Hare, on the review of his theory of galvanism, by Dr. Patterson.... [Philadelphia: 1820.] 18 p.
Reprinted from the *Philadelphia journal of the medical and physical sciences,* vol. 1, no. 1, 1820.
S-1514.
Copies: MB; MH; MWA; NNNAM; PPA; PPL; PPPH; PU.

757. Hare, Robert, 1781–1858.
A memoir of some new modifications of galvanic apparatus.... [Philadelphia?: 1820?] 17 p., plate.
S-1515, 5541.
Copies: IU; MB; MH; MWA; NN; NNNAM; PPAN; PU.

758. Hare, Robert, 1781–1858.
An essay on the question whether there be two electrical fluids, according to Du Faye, or one, according to Franklin. [Philadelphia: 1823.] 10 p., plate.
Contains also: Description of an electrical plate machine and of an improved blowpipe.
Reprinted from *Philadelphia journal of the medical and physical sciences*, v. 6, 1823.
N-H0116953.
Copies: MB; NN; PHi; PPAmP; PU.

759. Hare, Robert, 1781–1858.
An essay on the question whether there be two electric fluids, or one; also a description of an electrical plate machine, the plate mounted horizontally ... and of a blowpipe by alcohol.... [New Haven?: 1824?] 10 p., plate.
Reprint of three successive articles in the *American journal of science and arts,* v. 7, p. 103–111, 1824.
S-16408.
Copies: DLC.

760. Hare, Robert, 1781–1858.
Letter from Robert Hare ... to B. Silliman ... on some improved forms of the galvanic deflagator.... Also—An account of an improved single leaf electrometer.... Reprinted, with corrections & additions, from Silliman's Journal, no. 1, vol. VII. Philadelphia: Printed by W. Fry, 1824. 7 p., plate.
S-16409.
Copies: DLC; IEN-M; PHi; PPPH; PU-S.

761. Hare, Robert, 1781–1858.
Lectures on electricity & galvanism.... Printed for the use of his pupils. [Philadelphia: 1825.] 69, 56 p., 5 plates.
Appendix, 56 p., at end. Usually bound with his: *Minutes of the course of chemical instruction, 1825.*
S-20813.
Copies: AMaJ; DeGE; GEU-M; ICU; MdBM; OCLloyd; PP; PU; RPB; TxU-M.

762. King, William.
A manual of electricity: containing observations on the electrical phenomena, and directions for the construction of metallic conductors. Also for the making of electrical machines & Galvanic troughs, with instructions for applying their influence in aid of medicine, and in restoring suspended animation. Newbern, N.C.: 1825. iv, [5]–83 p., illus.
S-21118.
Copies: CtY; DeGE; MB; MH; MnU; NcU; RPB.

763. Cunningham, Obadiah.
New York, April 1826. Sir, I trust your

honorable station, and the magnitude of my subject, will excuse the liberty I take in making the following communication. I am confident ... that I have discovered that law in nature, or cause of the variation of the magnetic needle, and can elucidate, or demonstrate the cause, why the needle will not, at all times, point the same in the same place.... [New York: 1826.] 1 p.

S-24264.

Copies: DLC.

764. Farrar, John, 1779–1853.

Elements of electricity, magnetism, and electro-magnetism, embracing the late discoveries and improvements.... Cambridge, N.E.: Printed by Hilliard and Metcalf, 1826. vii, 395, [1] p., 6 fold. plates.

Selected from J. B. Biot's *Précis élémentaire de physique,* 3. éd., Paris, 1824.

S-24507.

Copies: CSt; CtHT; DLC; DeGE; FU; IU; KMK; MH; MdBS; MeB; MiD; MnU; MoSM; NNC; NjP; OC; PU; RP; ViU; VtU.

765. [Hare, Robert] 1781–1858.

On electricity. [Philadelphia: 1826?] 39 p.

Includes descriptions of apparatus. Contains references to his *Essays chemical, electrical & galvanic,* published in 1825.

N-H0116998.

Copies: CtY; DeGE.

766. Green, Jacob, 1790–1841.

Electro-magnetism; being an arrangement of the principal facts hitherto discovered in that science. With plates of the apparatus.... Philadelphia: Judah Dobson; Jesper Harding, printer, 1827. 216 p., 4 plates.

S-29064.

Copies: DLC; GU; KyU; MdAN; MdBS; MiD; MoU; NNE; NSchU; NjP; P; PPAmP; PPL; PU; WNaE.

767. Fansher, Sylvanus.

Concise treatise on electricity; with directions for constructing and applying lightning rods, used for the purpose of shielding houses, ships, etc., against the dangerous effects of electrical fluid from the thundercloud. New Haven: Printed for the author, 1830. 36, [2] p., illus., plates.

C-1320.

Copies: CtHi; CtY; DLC; MH; MHi; MWA; PHi.

768. Johnson, Walter Rogers, 1794–1852.

Observations on the electrical characters of caoutchouc, or gum elastic; with some applications of which they are susceptible.... (Read April 20, 1830). [Philadelphia: 1830.] 8 p.

C-2067.

Copies: DLC; MH; MWA; PP.

e. Geology and mineralogy

769. Barba, Alvaro Alonso, b. 1569.

Gruendlicher Unterricht von den Metallen, darinnen beschrieben wird, wie sie werden in der Erden generirt; und was man insgemein dabey findet. In zwei Büchern. Vormals im Spanischen beschrieben durch Albaro Alonso Barba.... Hernach in das Engländische übersetzt durch Edward, Graff von Sandwich, Anno 1669. Und nun um seiner Vortrefflichkeit willen zum erstenmal ins Hoch-Teutsche übersetzt, und zum Druck befördert durch G.R. ... Nebst einem neuen Anhang betreffend obige Materie. Ephrata, [Pa.]: Gedruckt durch J. Georg Zeisiger, 1763. 2 pts. in 1 v. ([4], 5–198, [4], 14 p.), illus.

Translation of his *Arte de los metales,* first published in Madrid, 1640.

Mp-9333.

Copies: CtY; DLC; MWA; NNE; PPAmP; PPL; RPJCB.

770. A compendious system of mineralogy & metallurgy; extracted from the American edition of the Encyclopaedia, now publishing by Thomas Dobson ... [Philadelphia]: T. Dobson, 1794. iv, 168, 171–525 p.

E-29601 (dated 1795); Mp-26801.

Copies: MWA; PPF; PPL; RPJCB.

771. A compendious system of mineralogy and metallurgy; extracted from the Encyclopaedia. Philadelphia: A. Bartram, 1798. iv, 505 p.

B-10290; Mp-48398.

Copies: NBuG; PPL.

772. Mease, James, 1771–1846.

A geological account of the United States; comprehending a short description of their animal, vegetable, and mineral productions, antiquities and curiosities.... Philadelphia: Birch & Small; R. Carr, printer, 1807. 3 p.l., 496, xiv p., plates.

S&S-13046.

Copies: C; DLC; DeGE; IC; MB; MWA; NN; NNA; PPL.

773. Accum, Friedrich Christian, 1769–1838.

A practical essay on the analysis of minerals, exemplifying the best methods of analysing ores, earths, stones, inflammable fossils, and mineral substances in general.... 1st American, from the London ed. Philadelphia: Kimber & Conrad, and B. & T. Kite, 1809. xxiv, [25]–236 p. plates.

S&S-16788.

Copies: DI-GS; DLC; DeGE; MB; MH; MWA; NN; PPAmP; PU; RNR.

774. Aikin, Arthur, 1774–1854.

A manual of mineralogy.... 1st American ed. Philadelphia: S. W. Conrad, 1815. viii, [9]–275 p.

First published in London, 1814.

S&S-33808.

Copies: DLC; MSaP; MWA; NWM; NjR; OO; OrHi; PPAmP; PPL; PU.

775. Cleaveland, Parker, 1780–1858.

Elementary treatise on mineralogy and geology, being an introduction to the study of these sciences, and designed for the use of pupils,—for persons attending lectures on these subjects,—and as a companion for travellers in the United States of America. Illustrated by six plates.... Boston: Cummings and Hilliard, 1816. xii, 668 p., fold. diagrs., fold. map.

Includes notes on the use of minerals.

S&S-37266.

Copies: CtW; DI-GS; DeGE; IEdS; InCW; KyU; MH; MWA; MdBJ; MeB; NB; NN; NcD; NjR; PPL; PU.

776. Phillips, William, 1775–1828.

Outline of mineralogy and geology.... New York: Collins, 1816. xii, 192 p., 4 plates.

S&S-38624.

Copies: CSt; DI-GS; ICU; MB; MWA; NN; NhD; NjR; OO; PPi; PU; VtMiM.

777. Maclure, William, 1763–1840.

Observations on the geology of the United States of America; with some remarks on the effect produced on the nature and fertility of soils, by the decomposition of the different classes of rocks; and an application to the fertility of every state in the Union, in reference to the accompanying geological map.... Philadelphia: Printed for the author, by Abraham Small, 1817. ix, [10]–127, [2] p., fold. map, fold. plate.

Read as a memoir before the American Philosophical Society, and included in the 1st vol., new series, of their *Transactions.*

S&S-41325.

Copies: CU; CtY; DI-GS; DLC; DeGE; LNHT; MWA; MiU-C; NIC; NN; NjP; NjR; PPL; RPJCB; ViU.

778. Dana, James Freeman, 1793–1827.

Outlines of the mineralogy and geology of Boston and its vicinity, with a geological map. By J. Freeman Dana ... and Samuel L. Dana ... Boston: Cummings and Hilliard; University Press, Hilliard and Metcalf, 1818. 108 p., front (fold. map).

"A list of the authors and editions of the works consulted ... ," p. [7].

S&S-43805.

Copies: CtW; DLC; DeGE; ICJ; MB; MWA; MiU-C; MsU; NN; NjR; OC; PPL; PU; ViU.

779. Eaton, Amos, 1776–1842.

An index to the geology of the northern states, with a transverse section from Catskill Mountain to the Atlantic. Prepared for the geological classes at Williams College.... Leicester [Mass.]: Printed by Hori Brown, sold by Websters and Skinners, Albany, [etc.], 1818. 52 p., folded map.

S&S-43906.

Copies: CSmH; CtY; DLC; LU; MWA; MWiW; NIC; Nh; PPAN; PPL; RPB; TNJ.

780. Maclure, William, 1763–1840.

Observations on the geology of the United States of North America; with remarks on the probable effects that may be produced by the different classes of rocks on the nature and fertility of soils.... [Philadelphia: A. Small, 1818.] 91 p., map.

S&S-44663.

Copies: MdBJ; NIC; NNC.

781. Phillips, William, 1775–1828.

An elementary introduction to the knowledge of mineralogy: including some account of mineral elements and constituents; explanations of terms in common use; brief accounts of minerals, and of the places and circumstances in which they are found ... by William Phillips.... With notes and additions on American articles, by Samuel L Mitchill.... New York: Collins, 1818. x, xxxiv, [9], 246, [9] p.

First published in London, 1816.

S&S-45320.

Copies: DI-GS; DeGE; MH; MWA; NUtHi; PU; ViU.

782. Akerly, Samuel, 1785–1845.

An essay on the geology of the Hudson

782 *(continued)*
River, and the adjacent regions; illustrated by a geological section of the country, from the neighborhood of Sandy-Hook, in New-Jersey, northward, through the highlands in New-York: towards the Catskill Mountains.... New York: A. T. Goodrich & Co., 1820. 69 p., fold. map.

S-50.

Copies: CSmH; CSt; CtY; DI-GS; ICJ; MH-Z; MWA; NIC; NN; NNC; NNE; NjR; PPL.

783. Albany County Agricultural Society.

A geological survey of the county of Albany, taken under the direction of the Agricultural Society of the county. [Albany: Printed by S. Southwick, 1820?] 55 p., diagrs.

Preliminary observations signed: Amos Eaton, T. Romeyn Beck.

S-60.

Copies: CtY; DI-GS; DLC; MB; MBAt; MiD; NN.

784. Eaton, Amos, 1776–1842.

An index to the geology of the northern states, with transverse sections, extending from Susquehanna to the Atlantic, crossing Catskill Mountains.... 2d ed., wholly written over and anew, and published under the direction of the Troy Lyceum.... Troy, N.Y.: W. S. Parker; Albany: Websters and Skinners, [etc., etc.],1820. 286 p., 2 plates.

S-1074.

Copies: DLC; ICJ; KU; KyU; MWA; NIC; NN; NWM; NcD; Nh; NjP; NjR; OO; PPAN; PPL; PU.

785. Albany County Agricultural Society.

A geological survey of the county of Albany, taken under the direction of the Agricultural Society of the county.... [Albany: 1822.] 55 p., illus.

Probably same as no. 783.

S-7768.

Copies: PPL.

786. Cleaveland, Parker, 1780–1858.

An elementary treatise on mineralogy and geology.... 2d ed.... Boston: Cummings and Hilliard, 1822. 2 v., col. map, plates.

Includes notes on the practical use of minerals.

S-8364.

Copies: CtY; DLC; DeGE; MH; MWA; MdBJ; MdU; MeBaT; MnHi; NNE; NcD; NjP; O; OrP; PPL; PPi; RPB; TxU; ViU; WU; WyU.

787. [Eaton, Amos] 1776–1842.

A geological and agricultural survey of Rensselaer County, in the state of New-York. To which is annexed, a geological profile, extending from Onondaga Salt Springs, across said county, to Williams College in Massachusetts. Taken under the direction of the Honourable Stephen Van Rensselaer. Albany: Printed by E. and E. Hosford, 1822. vii, [9]–70 p., plate, fold. table.

Dedication signed: Amos Eaton.

S-8587.

Copies: CSmH; CtY; DLC; DeGE; ICJ; InU; MB; MH; MWA; MdBJ; MoSHi; NhHi; NjR; OkU; PPL; ScC.

788. [Eaton, Amos] 1776–1842.

To gentlemen residing in the vicinity of the Erie Canal. [Albany?:1822.] 10 p.

Signed: Amos Eaton. Troy (N.Y.) Dec. 13, 1822.

Explains a geological and agricultural survey along the Erie Canal, planned by Stephen Van Rensselaer.

S-8589.

Copies: DeGE; MB.

789. Steel, John Honeywood, 1780–1838.

Report on the geological structure of the county of Saratoga, in the state of New-York; together with remarks on the nature and properties of the various soils and modes of culture; intended to illustrate the natural history and domestic resources of the coun-

ty.... Published by order of the Executive Committee of the Society for the Promotion of Agriculture and Domestick Manufactures. Saratoga Springs, [N.Y.]: G. M. Davison, printer, 1822. 56 p.
S-10338.
Copies: CtY; DeGE; IU; MB; MWA; NN; PPAN.

790. Varley, Delvalle Lowry.
Conversations on mineralogy.... 1st American from the last London ed. Philadelphia: U. Hunt, 1822. 332 p., plates.
S-11289.
Copies: ICU; KU; MBAt; MH; MWA; MiU; NCH; NjR; OClWHi; OO; P; PPi; RPB; WHi.

791. Hitchcock, Edward, 1793–1864.
Sketch of the geology, mineralogy and topography of Connecticut.... New Haven: S. Converse, 1823. iv, 154, iv p., illus., map.
S-12837.
Copies: MA; MH; MWA; NIC.

792. [Eaton, Amos] 1776–1842.
A geological and agricultural survey of the district adjoining the Erie Canal in the state of New York. Taken under the direction of the Hon. Stephen Van Rensselaer.... Albany: Printed by Packard & Van Benthuysen, 1824. 163 p., 2 fold. profiles.
S-16033.
Copies: CSmH; CSt; CtY; DLC; DeGE; ICJ; MH; MWA; MdBJ; MdHi; MeBa; MiHM; N; NIC; NN; NNE; OC; OClWHi; P; PPi; PPL; RPB; VtU; WU.

793. Hall, Frederick, 1780–1843.
Catalogue of minerals, found in the state of Vermont, and in the adjacent states; together with their localities.... Hartford, [Conn.]: P. B. Goodsell, printer, 1824. 44 p.
S-16381.
Copies: CtHi; DLC; MB; MWA; PPL; VtU.

794. North Carolina. Board of Agriculture.
Report on the geology of North-Carolina, conducted under the direction of the Board of Agriculture. Parts I and II, by Denison Olmsted. [Raleigh: 1824-1825.] 84, 87-152 p.
N-00080483; S-17443.
Copies: CtY; DI-GS; DLC; NcU; PPWa.

795. Webster, Matthew Henry.
A catalogue of the minerals, which have been discovered in the state of New York: arranged under the heads of the respective counties and towns in which they are found. Albany: Printed by Websters and Skinners, 1824. 32 p.
S-19232.
Copies: MB; MWA; N; NN; NNM; PPAN.

796. Robinson, Samuel.
A catalogue of American minerals, with their localities; including all which are known to exist in the United States and British provinces, and having the towns, counties, and districts in each state and province arranged alphabetically. With an appendix, containing additional localities and a tabular view.... Boston: Cummings, Hilliard, 1825. vi, [2], 316 p.
S-22115.
Copies: CSfA; DLC; DeGE; IG; KyLxT; MA; MMeT; MWA; Me; MiU-C; MsU; NNS; OkU; PPL; PU; WHi.

797. Browne, Peter Arrell, 1782–1860.
An address, intended to promote a geological and mineralogical survey of Pennsylvania, the publication of a series of geological maps and the formation of state and county geological and mineralogical collections.... Philadelphia: Printed by P. M. Lafourcade, 1826. 8 p.
S-23951.
Copies: DLC; MH-BA; MWA; PPAmP; PPF; PPL.

798. Troost, Gerard, 1776–1850.

Geological survey of the environs of Philadelphia: performed by order of the Philadelphia Society for Promoting Agriculture, by G. Troost.... Philadelphia: H. S. Tanner; Clark & Raser, printers, 1826. 39, [1] p., map.

S-26246.

Copies: DLC; DeGE; MA; MBAt; NN; NNM; P; PHC; PPAN; PPAmP; PPF; PPL; PU.

799. North Carolina. Board of Agriculture.

Report on the geology of North-Carolina, conducted under the direction of the Board of Agriculture. Part III. By Elisha Mitchell.... November, 1827. [Raleigh: 1827.] 60 p.

"Gold mines" by Charles E. Rothe, p. [29]–45; "Report on the establishment of manufactures and growing of wool ... by Mr. Fisher, from Rowan," p. [45]–60.

N-00080483.

Copies: CSmH; DLC; NcU.

800. Proposals for making a geological and mineralogical survey of Pennsylvania, for publishing a series of geological maps, and forming state and county geological and mineralogical collections. [Philadelphia: 1827.] 20 p.

S-30356.

Copies: DLC; MH; P.

801. Carpenter, George Washington, 1802–1860.

On the mineralogy of Chester County; with an account of some of the minerals of Delaware, Maryland and other localities. [Philadelphia]: 1828. 16 p.

S-32604.

Copies: DLC; DeWI; PHi; PPAN; PPAmP; PPi.

802. Eaton, Amos, 1776–1842.

A geological nomenclature for North America; founded upon geological surveys, taken under the direction of the Hon. Stephen Van Rensselaer. Albany: Printed by Packard and Van Benthuysen, 1828. 31 p., 3 col. plates.

S-33039.

Copies: CtY; DLC; KyU; MH; N; NCH; NjP; OCHP; PPF; PPL.

803. Mitchell, Elisha, 1793–1857.

Professor Mitchell's report on the geology of North-Carolina: with C. E. Rothe's remarks on the gold country: and a report on the subject of manufacturing establishments and the raising of sheep. Published by order of the Board of Agriculture. Raleigh: Printed by J. Gales & Son, 1828. 60 p. [No. 3].

S-34165.

Copies: NcU.

804. Mitchill, Samuel Latham, 1764–1831.

A lecture on some parts of the natural history of New-Jersey, delivered before the Newark Mechanic Association for Mutual Improvement in the Arts and Sciences, on Tuesday, June 3, 1828. By Samuel L. Mitchill. New York: Printed by Elliott and Palmer, 1828. 34 p.

Includes references to mineral resources, such as iron ore and copper, and the availability of water power.

S-34172.

Copies: CSmH; NBLiHi; NHi; NNNAM; NjR; PPAmP; PPL; PPPrHi.

805. Mitchell, Elisha, 1793–1857.

On the geology of the gold region of North Carolina, in a letter to the editor, dated Aug. 25, 1828. [S.l.: 1829.] 19 p.

Reprinted from the *American journal of science and arts,* vol. 16, no. 1, 1829.

N-M0650648.

Copies: NcU.

f. Mechanics

806. Greenwood, Isaac, 1702–1745.
A course of philosophical lectures, with a great variety of curious experiments, illustrating and confirming Sir Isaac Newton's laws of matter and motion. [Boston?: 1726.] 4 p.
Lecture topics include the principles of mechanics, and the laws of motion and of projectile motion.
B-715; Mp-39848.
Copies: MH; MHi.

807. Greenwood, Isaac, 1702–1745.
An experimental course of mechanical philosophy. Whereby such a competent skill in natural knowledge may be attained to (by means of various instruments and machines, with which there are above three hundred curious, and useful experiments performed) ... make themselves better acquainted with the principles of nature.... Boston: Printed [by B. Green], 1726. [2], 9 p.
Mp-2746.
Copies: MHi; MWA (photocopy).

808. [Colden, Cadwallader] 1688–1776.
An explication of the first causes of action in matter, and of the cause of gravitation. New York: Printed by James Parker, 1745. vi, 43, [1] p.
Dedication signed: Cadwallader Colden.
Mp-5564.
Copies: CSmH; CtHT-W; DeGE; MWiW; NN; NNC; RPJCB.

809. Adgate, Andrew, d. 1793.
The mechanics' lecture: showing the usefulness of mechanic arts, and who was the first mechanic; and giving a short history of taylors, masons, carpenters, ship carpenters, joiners, cabinet-makers, blacksmiths, white smiths, bakers, barbers, weavers, fullers, hatters, tanners, shoemakers, coblers, dyers, carvers, coopers, musical-instrument-makers, clock-makers, coach-makers, saddlers, sail-makers, and printers.... Philadelphia: Printed by John M'Culloch, 1789.
E-21627.
No copy known.

810. [Adgate, Andrew] d. 1793.
A lecture; containing a short history of mechanics, and of useful arts and manufactures. Reverently dedicated to the respectable supporters of liberty and property, the mechanics of the United States, but particularly addressed and delivered to the mechanics of Philadelphia, by their faithful servant and fellow labourer, Absalom Aimwell, Esquire [pseud.].... Catskill, [N.Y.]: Reprinted [by Mackay, Crosswell & Co.], 1795. 29 p.
Mp-28154.
Copies: NN.

811. Marrat, William, 1772–1852.
An introduction to the theory and practice of mechanics.... Boston: Printed and sold by J. Hellaby, 1810. xii, 455 p., 13 plates.
Probably a Boston, Eng., imprint.
S&S-20646.
Copies: CtY; DAU; MiU; NNC; NjR; PPL.

812. Cutbush, James, 1788–1823.
Hydrostatics, or a treatise on specific gravity; with a description of the different hydrometers, and their application to the purposes of philosophy and the arts, with sundry tables. Illustrated with problems.... Philadelphia: Printed for the author, 1812. 3 p.l., 49 p.

812 (*continued*)

"Originally commenced in a series of numbers, which, in part, appeared in the *Freemason's magazine.* "—Pref.

S&S-25201.

Copies: DeGE; MWA; PPAN; PPL.

813. Windship, Charles William.

An introductory discourse of the phenomena of vitality; or, Laws of mobility and motion in animal bodies.... By Charles W. Windship, M.D. Boston: Charles Callender, 1818. 16 p.

S&S-46833.

Copies: CtHT-W; DeGE; MB; MWA; NNNAM.

814. Craig, John D.

An introductory lecture, to a course on experimental philosophy, delivered Nov. 5, 1819 ... by John D. Craig.... Baltimore: Fielding Lucas, Jun., 1819. 20 p.

Observations on the practical application of mechanics and chemistry.

S&S-47745.

Copies: DeGE; MdBE.

815. Farrar, John, 1779–1853, comp.

An elementary treatise on mechanics, comprehending the doctrine of equilibrium and motion, as applied to solids and fluids, chiefly compiled, and designed for the use of the students of the university at Cambridge, New England.... Cambridge, N.E.: Printed by Hilliard and Metcalf, at the University Press. Sold by W. Hilliard, Cambridge, and by Cummings, Hilliard & Co., Boston, 1825. vii, 438 p., 10 fold. plates.

First ed., first state.

Copies: DeGE.

816. ________ Cambridge, N.E.: Printed by Hilliard and Metcalf, at the University Press. Sold by W. Hilliard, Cambridge, and by Cummings, Hilliard & Co., Boston, 1825. vii, 440 p., 10 fold. plates.

"Errata," p. [439]–440.

S-20489.

Copies: CSt; DLC; DeGE; GDE; ICJ; IaAS; InCW; MB; MH; MWA; MdBS; MiD; MoSW; NN; NNG; PPL; PU; RPB; TNJ; Vi.

817. Renwick, James, 1790–1863.

Lecture, introductory to a course of lectures on applied mechanics, delivered at the New-York Atheneum in the winter of ... 1824–25. [S.l.: 1825?] 16 p.

Sabin-69654 (note).

No copy known.

818. Richard, Gaspard.

A new system of power, as substitute of steam, by weight and lever. Invented by Gaspard Richard. New York: Printed for the author, 1825. 23 p.

N-R0242320.

Copies: InU.

819. Young, David, 1781–1852.

A lecture on the laws of motion, containing many interesting investigations, recently delivered at many places in New-Jersey. Caldwell, N.J.: Printed by Stephen Gould, 1825. 36 p.

N-Y0025013.

Copies: DeGE; NjR.

820. Franklin Institute, Philadelphia.

Report of the Committee of the Franklin Institute, on the inclined plane of Professor James Renwick. [Philadelphia: 1826.] 18 p.

From the *Franklin journal.*

S-24576.

Copies: DeGE; MWA; NN; PPAmP; PPF.

821. Richard, Gaspard.

Prospectus of a new plan to make a

large fortune by a new system of power as substitute of steam by weight and lever.... New York: The author, 1827. 50 p.
S-30448.
Copies: DLC; NN.

822. Allen, Zachariah, 1795–1882.
The science of mechanics, as applied to the present improvements in the useful arts in Europe, and in the United States of America: adapted as a manual for mechanics and manufacturers, and containing tables and calculations of general practical utility. Providence: Hutchens & Cory, 1829. v, 364, 1 p., fold. front., diagr., plates (part fold.).
S-37424.
Copies: CSmH; CtY; DLC; DeGE; GU; ICJ; MB; MH-BA; MWA; MiU; MnU; NIC; NNE; NjR; OCl; PPF; PPL; RPB; TN; WM.

823. Johnson, Walter Rogers, 1794–1852.
Address introductory to a course of lectures on mechanics and natural philosophy, delivered before the Franklin Institute, Philadelphia ... 1828. Boston: Hiram Tupper, printer, 1829. 15 p.
S-39150.
Copies: CtHT-W; MiD-B; PHi; PPAN; PPAmP.

824. Johnson, Walter Rogers, 1794–1852.
Observations on the relation between rolling and dragging friction ... October 22, 1829. [Philadelphia: 1829.] 12 p.
S-39152.
Copies: DP; MBAt; MH; MWA; NN; PPF; PPL.

825. Brunton, Robert, 1796–1852.
A compendium of mechanics; or, Text book for engineers, mill-wrights, machine-makers, founders, smiths, &c. Containing practical rules and tables connected with the steam engine, water wheel, pump, and mechanics in general.... 1st American from 4th London ed., with plates. Edited by James Renwick.... New York: G. & C. & H. Carvill, 1830. 228 p., front., 4 plates.
First published at Glasgow, 1824.
C-672.
Copies: CU; CtY; DLC; DeGE; GU; MB; MH; MWA; MnHi; NIC; NN; NNE; PPL; PU-P; VtU.

826. ——— 1st American, from the 2d English ed. Boston: Hilliard, Gray, Little, and Wilkins, 1830. 11, [3], [13]–148 p., plates.
C-673.
Copies: DLC; MNBedf; MWA (Hilliard, Gray & Co. catalog of scientific works (6 p.) at end); MnU; N.

827. Hanna, John Smith.
A lecture on mechanics: in which a principle of self-motion is advocated.... Washington: Printed by Gales & Seaton, 1830. 24 p., illus., fold. front.
C-1754.
Copies: DeGE; MdBL.

5. Scientific instruments and apparatus

828. Greenwood, Isaac, 1702–1745.

Explanatory lectures on the orrery, armillary sphere, globes and other machines, instruments, and schemes made use of by astronomers. Accompanied with a great variety of physical experiments and curious remarks.... Boston: Printed in the year 1734. [4] p.

A prospectus of lectures on scientific instruments.

Mp-3776.

Copies: CSmH; DLC; ICN; MB; MBAt; MH; MHi; MWA (xerox copy); MiU-C; RPJCB.

829. Just arrived from London, for the entertainment of the curious and others, and is now to be seen, by six or more, in a large commodious room, at the house of Mr. Vidal, in Second-Street; the solar or camera obscura microscope, invented by the ingenious Dr. Liberkhun. [Philadelphia: Printed by B. Franklin, 1744.] Broadside 30.5 x 19 cm.

The exhibition also included David Lockwood's musical clock.

Mp-5419.

Copies: PPAmP; PPL.

830. Lamb, Anthony.

Anthony Lamb, mathematical instrument maker at Sr. Isaac Newton's Head, New York.... [New York: 1755?] Broadside 31 x 21 cm.

B-1725; Mp-40759.

Copies: NHi.

831. An account of the fire at Harvard College, in Cambridge; with the loss sustained thereby.... Boston: Printed by R. and S. Draper, 1764. Broadside.

Contains descriptions of the scientific instruments lost in the fire on January 14, 1764.

Also recorded under the title: *From the Massachusetts-gazette, Thursday, February 2, 1764. An account* ... (E-9557).

Mp-9556.

Copies: MH; MWA.

832. Juliana Library Company, Lancaster, Pa.

The charter, laws, catalogue of books, list of philosophical instruments, &c. of the Juliana Library-Company, in Lancaster.... Philadelphia: Printed by D. Hall and W. Sellers, 1766. 2 p.l., 56 p.

The list of philosophical instruments includes 19 items.

Mp-10350.

Copies: PPAmP; PPL.

833. West, Benjamin, 1730–1813.

An account of the observation of Venus upon the sun, the third day of June, 1769, at Providence, in New-England. With some account of the use of those observations.... Providence: Printed by John Carter, 1769. [6], 22 p., illus.

Includes descriptions of the instruments used in the observation, and a note on Joseph Brown's electrical apparatus.

Mp-11525.

Copies: CSmH; CtY; DLC; DeGE; MB; MHi; MWA; MiU-C; NHi; NN; PPL; RPJCB.

834. Description of the scale accompanying Dycas's hydrometer. [Philadelphia?: 1792?] Broadside 41.5 x 34 cm.

With two diagrams in pen and ink.

N-D0196004.

Copies: PPRF.

835. [Scott, Joseph].

A description of the geographical clock: which contains the names and situations of the most remarkable places in the world; and exhibits at one view the time of day or night at all these places round the globe.... Philadelphia: Printed for Joseph Scott, and sold by Francis Bailey and Peter Stewart, 1792. [24] p.

A description of a cardboard diagram with a movable disc.

B-7978; Mp-46429.

Copies: PP; PPF.

836. An easy method of working by the plain and sliding rules. Collected from various authors, and designed for the use of architectors, &c. Lansingburgh: Printed by Silvester Tiffany, and sold at his office and at Spencer's bookstore, Albany, 1793. 24 p., fold. table.

B-8329; Mp-46740.

Copies: DeGE; MWA; NHi; RPJCB.

837. Moreau de Saint Méry, Médéric Louis Élie, 1750–1819.

Catalogue of books, stationary [*sic*], engravings, mathematical instruments, maps, charts, and other goods of Moreau de St. Mery & Co's store.... Philadelphia: [Printed by Moreau de Saint Méry], 1795. 76 p.

Mp-29107.

Copies: MB; MWA; PPAmP.

838. Dearborn, Benjamin, 1754–1838.

A description of the vibrating steelyard, or just balance. Invented by the subscriber, and secured to him by patent. This convenient and accurate steelyard unites more advantages for weighing with ease, dispatch and certainty, than any instrument heretofore offered to the public.... Taunton (Commonwealth of Massachusetts), Sept. 1799. [New-Bedford: Printed by John Spooner, 1799.] Broadside, illus. 44 x 25 cm.

Mp-35388.

Copies: MHi.

839. Hare, Robert, 1781–1858.

Memoir on the supply and application of the blow-pipe. Containing an account of a new method of supplying the blow-pipe either with common air, or oxygen gas: and also of the effects of the intense heat produced by the combustion of the hydrogen and oxygen gases. Illustrated by engravings. Published by order of the Chemical Society of Philadelphia; to whom it was presented by Robert Hare, Jun. ... Philadelphia: Printed for the Chemical Society, by H. Maxwell, 1802. 34 p., fold. front.

S&S-2371.

Copies: CtY; DNLM; DeGE; MWA; NNNAM; NjR; OClW; PBL; PPAmP; PPL.

840. Templeton, James J.

An explanation of the trigonometrical quadrant, a newly invented mathematical instrument, in which is included a brief system of trigonometry. Georgetown, Potomack: Printed by W. A. Rind & Co., 1802. 40, [3] p., illus., fold. plate.

Karpinski-p. 141; S&S-3148.

Copies: MWA; NjP.

841. [Dearborn, Benjamin] 1754–1838.

The patent balance, compared with other instruments for weighing. Philadelphia: Printed by R. Cochran, 1803. 20 p., illus.

S&S-4051.

Copies: DeGE (xerox copy); MBAt; P; PPAmP.

842. Dearborn, Benjamin, 1754–1838.

Description of the manner of using Dearborn's facility for casting interest.... Boston: Gilbert & Dean, printers, 1805. Broadside 46.5 x 30 cm.

Description and directions for operating a calculating instrument constructed in the form of a circular slide rule.

N-D0098160.

Copies: PPAmP.

843. Pinchbeck, William Frederick.
The expositor; or, Many mysteries unravelled.... Comprising ... acoustic temple ... penetrating spy-glasses, optical and magnetic, and various other curiosities.... Boston: Printed for the author, 1805. 100 p., illus.
Describes equipment used in magic tricks.
S&S-9149.
Copies: CtY; DLC; MB; MH; MWA; MiU-C; NjP; PPL; RPB.

844. Kneass, W.
Patent revolving interest table. Philadelphia: [181-?]. 2 circular engraved cards, with a metal attachment mounted in center.
S&S-20506.
Copies: NN.

845. Colles, Christopher, 1738–1816.
An account of the astonishing beauties and operations of nature, in the minute creation, displayed by the solar microscope. New York: Samuel Wood, 1813. 44 p., illus.
Includes instructions on the use of the microscope.
S&S-51283.
Copies: DeGE; MSaP; MWA; NN; PPL.

846. Chapman, John.
Chapman's patent sliding interest table, calculated at 7 per cent. [New York?]: 1815. Broadside.
S&S-51475a.
Copies: DLC.

847. Colles, Christopher, 1738–1816.
An account of the astonishing beauties and operations of nature, in the minute creation, displayed by the solar microscope. New York: Samuel Wood, 1815. 44 p., illus.
S&S-34392.
Copies: DLC; MWA; ViW.

848. Jelleff, Joseph.
Jelleff's patent revolving interest table.... Middlebury, Vt.: Printed by T. C. Strong for L. Walker & Co., [1815].
A revolving disc, about 20 cm. in diameter.
S&S-35013.
Copies: MH; MWA; Vt; VtHi.

849. McClure, David.
The description and application of the pedalion, in trigonometry, navigation, astronomy, &c. Philadelphia: Printed for the author, 1815. 44 p.
N-M0023120.
Copies: NjR.

850. McClure, David.
The pedalion, applicable to trigonometry, navigation, astronomy, &c. Philadelphia: Invented and published by David McClure, 1815. Broadside 34 x 34.5 cm.
S&S-51509.
Copies: MB; PPL.

851. Colles, Christopher, 1738–1816.
An account of the astonishing beauties and operations of nature, in the minute creation, displayed by the solar microscope. New York: Samuel Wood & Sons, 1816. 44 p., illus.
S&S-37296.
Copies: DeGE; MSaP; MWA; MeHi; NN.

852. Directions for using the gold standard balances, invented by Benjamin Dearborn, of Boston, Massachusetts.... [Boston?: 1817.] Broadside.
S&S-40669.
Copies: MHi.

853. Bussanich, James L.
Terrestrial pendulum.... [Wilmington, Del.: Star Printing Co., 1818.] [6] p., illus.
Erroneously dated; should be [1918], cf. N-B1003933.
S&S-43501.
Copies: DLC; DeWI.

854. Hare, Robert, 1781–1858.
Strictures by Robert Hare, M.D., professor of chemistry in the University of Pennsylvania, in reply to remarks on the

calorimotor, published at different times, in the American Medical Recorder. Philadelphia: W. Brown, printer, 1820. 28 p., plate.
S-1516.
Copies: CtY; DNLM; MB; MWA; PPAN; PPAmP; PPPH; PU; RPB.

855. Hare, Robert, 1781–1858.
Strictures on a publication entitled, Clark's gas blowpipe.... Philadelphia: Robert Desilver; T. Town, printer, 1820. 24 p., 2 plates.
Concerns the blowpipe of Edward Daniel Clarke.
S-1517.
Copies: CtY; MWA; MoSU-M; NN; OC; PPAmP; PPL; PPPH; PU.

856. Thompson, Gabriel H.
Plane sliding sector; or, Trigonometrical scale, designed for practical geometry, trigonometry, navigation, and surveying; by Gabriel H. Thompson. Boston: Printed by Henry Bowen, 1822. 10 p., diagrs.
S-10437.
Copies: DeGE; ICJ.

857. Bolles, William.
A description and practical application of Bolles's trigonometer. [Griswold, Conn.?]: 1824. [2], 32, [1] p., front. (diagr.).
S-15490.
Copies: DLC; DeGE; PPL.

858. Curtis, George.
A treatise on Gunter's scale, and the sliding rule: together with a description and use of the sector, protractor, plain scale, and line of chords; or, An easy method of finding the area of superfices; and of measuring boards; and of finding the solid contents of bodies.... To these are added several useful logarithmick tables, tables of latitude and departure.... Whitehall, N.Y.: Printed and published by E. Adams, 1824. 119 p., illus.
S-15910.
Copies: CSmH; CtY; DAU; DeGE; IU; InLP; MWA; MWHi; MiU; MoS.

859. United States. Laws.
An act authorizing the Secretary of the Treasury to adopt a new hydrometer for ascertaining the proof of liquors. December 21, 1824.... [Washington: 1824.] 1 p. (H.R. 175)
S-18276.
Copies: DNA.

860. United states. Laws.
A bill authorizing the Secretary of the Treasury to adopt a new hydrometer for ascertaining the proof of liquors. Mar. 24, 1824.... [Washington: 1824.] 1 p. (H.R. 175)
S-18397.
Copies: DNA.

861. Hare, Robert, 1781–1858.
An account of some audiometers of an improved construction.... [Philadelphia?: 1825?] 12 p., illus.
S-20810.
Copies: DLC; ICJ; MB; NN; PPL.

862. Farrar, John, 1779–1853.
An experimental treatise on optics, comprehending the leading principles of the science, and an explanation of the more important and curious optical instruments.... Cambridge, N.E.: Printed by Hilliard and Metcalf, 1826. vii, 350 p., 6 fold. plates.
Selections from J. B. Biot's *Précis élémentaire de physique experimental,* 3. éd., Paris, 1824.
S-24508.
Copies: CSt; DLC; DeGE; ICJ; MBAt; MH; MWA; MdBJ; MeB; Mi; MoS; NNC; NcD; Nh; NjR; OO; PPL; RPB; TxU; ViU; VtU.

863. Hare, Robert, 1781–1858.
Engravings and descriptions of a great part of the apparatus used in the chemical course of the University of Pennsylvania. With appropriate theoretical explanations.... Printed for the use of his pupils. In two parts.... Philadelphia: Printed by Clark & Raser, 1826. viii, 52, iv, 52 p., illus.
S-24791.

863 (*continued*)
Copies: DLC; DeGE; MA; MB; NNC; NbU; P; PBL; PP; PPAmP; PPL; RPB.

864. United States. Congress.
Joint resolution (submitted by Mr. Bradley) directing experiments to be made to ascertain the length of the pendulum, vibrating sixty times in a minute. May 16, 1826. [Washington: 1826.] 1 p.
S-27130.
Copies: DNA.

865. Catalogue of prints, oil paintings, miniatures, electrical machines, philosophical apparatus, &c.&c., to be sold at Columbian Hall (formerly Museum), Common Street, on Friday, June 13, 1828. By order of the administrator on the estate of the late William M. S. Doyle. J. L. Cunningham, auctioneer. [Boston: 1828.] 7 p.
S-32619.
Copies: DLC.

866. Holbrook, Josiah, 1788–1854.
Apparatus for schools, academies and lyceums.... Boston: Carter, Hendee & Babcock, [1830?]. Broadside.
C-1882.
Copies: MB (discarded 1944).

867. [Oliver, Henry Kemble] 1800–1885.
An elementary treatise on the construction and use of the mathematical instruments usually put into portable cases. Boston: Perkins & Marvin, 1830. 68 p., 23 diagrs. on 6 fold. plates.
Preface signed: H.K.O.
C-2898.
Copies: CU; CtY; DAU; DLC; DeGE; KyDC; MH; MWA; NNG; NcU.

868. Stansbury, Daniel.
Directions for regulating, using and taking care of chronometers, on shore and at sea. Together with formulas for working the time and applying the error and rate.... New York: E. & G. W. Blunt, 1830. 19, [49] p.
N-S0863672.
Copies: DLC.

6. Mensuration and weights and measures

869. New Hampshire (Colony). Laws.
An act for regulating the admeasurement of lumber, the gaging of oak, the making and size of bricks. [Portsmouth: 1760.] 4 p.
B-2160; Mp-41151.
Copies: MWA.

870. Workman, Benjamin.
Gauging epitomized; or, A short treatise of gauging, in which that branch is rendered familiar to the meanest capacity. To which are added, accurate tables for finding the mean-diameters of casks by inspection. Also, a comprehensive ullage table, and an accurate method of ullaging casks.... Philadelphia: Printed and sold by W. Young, 1788. [4], 120 p.
Mp-21618.
Copies: DeGE; MB; MH; MWA; NHi; NN; P; PHi; PPL; RPJCB.

871. [Bordley, John Beale] 1727–1804.
On monies, coins, weights, and measures, proposed for the United States of America. Philadelphia: Printed by Daniel Humphreys, 1789. [2], 25, [1] p.
Signed: B.
Mp-21698.
Copies: DeGE; MBAt; MH-BA; MWA; PPL; PPPH; RPJCB.

872. [Bordley, John Beale] 1727–1804.
A supplement to the essay On monies, coins, &c. proposed for the United States of America. Philadelphia: Printed by Daniel Humphreys, 1790. [2], 7 p.
Signed: B.
Mp-22368.
Copies: DeGE; MBAt; MH-BA; NN; NNC; PPL; PPPH; RPJCB.

873. Bumstead, Joseph.
Boston, January, 1790. First American treatise on mensuration. Proposals for printing by subscription ... an original work, entitled A system of mensuration. Upon a plan entirely new, by William Croswell.... Subscriptions taken by Joseph Bumstead, the publisher ... [Boston: 1790.] Broadside 39 x 31 cm.
The advertised work was not published.
B-7334; Mp-45841.
Copies: MH; MHi.

874. United States. Dept. of State.
Report of the Secretary of State, on the subject of establishing a uniformity in the weights, measures and coins of the United States. Published by order of the House of Representatives. New York: Printed by Francis Childs and John Swaine, 1790. 21 p.
Thomas Jefferson, Secretary of State.
First impression.
Mp-22994.
Copies: DLC; DeGE.

875. ———— New York: Printed by Francis Childs and John Swaine, 1790. 22 p.
Second impression.
Mp-22995.
Copies: DLC.

876. ———— New York: Printed by Francis Childs and John Swaine, 1790. 22 p.
Third impression.
Mp-22996.
Copies: DLC.

877. ———— New York: Printed by Francis Childs and John Swaine, 1790. v, [3], [9]–49 p.
Fourth impression.
Mp-22997.
Copies: DLC; DeGE; MWA; PPAmP; PPL; RPJCB.

878. ———— New York: Printed by F. Childs and J. Swaine, 1790 [1791]. 52, [1] p.
Revised edition, published in 1791 but retaining the 1790 title page. "Postscript" signed on p. 52: Thomas Jefferson, Secretary of State, January 10, 1791.
Mp-23910.
Copies: MBAt; PPAmP; RHi.

879. United States. Congress. Senate. Committee on Weights and Measures.
United States, in Senate, April 4, 1792. The committee to whom the subject of weights and measures was referred, report that it is their opinion.... [Philadelphia]: Printed by John Fenno, [1792]. 2 p.
Recommends the adoption of a decimal system of weights and measures.
Mp-24943.
Copies: DLC; NN.

880. United States. Congress. Senate. Committee on Weights and Measures.
United States. In Senate, January 29, 1793. The committee to whom was referred the motions ... relative to the measures and weights ... report.... [Philadelphia]: Printed by John Fenno, [1793]. [1], 3 p.
Mp-26348.
Copies: MWA; NN.

881. France. Convention nationale, 1792–1795. Comité de salut public.
In Senate of the United States, January 9th, 1795. Ordered, that three hundred copies of the communication from the Minister of the French Republic, of a decree of the Committee of Public Safety of the National Convention, referred to in the message of the President of the United States of the 8th instant, on the subject of weights and measures, be printed for the use of the Senate....

881 (*continued*)
[Philadelphia]: Printed by John Fenno, [1795]. [8] p.

Joseph Fauchet, Minister of France.
Mp-29722.
Copies: DLC; MH; MWA; MiU-C; PPL; RPJCB; ViU.

882. A dimension proper for an unit of measures pointed out, and an improvement in the second pendulum described; which will fit it for making the experiment recommended by the Committee of Congress on Weights and Measures in their report to that House in last session with a plan for a reform of our measures, which shews that they may be extended by a complete decimal scale, formed on the said unit of measures.... By a citizen of America. Philadelphia: Printed by Budd and Bartram, 1796. 24 p.

E-30353.
Copies: DLC(not located).

883. United States. Congress. House. Committee on Coinage, Weights and Measures.

Report of the Committee to whom were referred so much of the report of the Secretary of State, made the 13th of July, 1790; and the message from the President of the United States, of the 8th of January, 1795, as relate to weights and measures. 12th April, 1796.... [Philadelphia: 1796.] 7 p.

Mp-31390.
Copies: DeGE; MBAt; MWA.

884. United States. Laws.

An act directing certain experiments to be made to ascertain uniform standards of weights and measures for the United States.... Passed in the House of Representatives, May 19, 1796. [Philadelphia]: Printed by John Fenno, [1796.] Broadside.

Mp-31333.
Copies: DLC.

885. ________ In Senate, May the 31st, 1796. [Philadelphia]: Printed by John Fenno, [1796]. 8 p.

Contains the act passed by the House of Representatives on May 19, 1796, an amendment proposed by a committee of the Senate on May 31, 1796, and an extract from a report of a Senate committee, 18 December, 1792.

Mp-31334.
Copies: MWA.

886. Hawney, William.

Hawney's complete measurer; or, The whole art of measuring. Being a plain and comprehensive treatise on practical geometry and mensuration. Preceded by decimal and duodecimal arithmetic, and the extraction of the square and cube root.... Corr. and improved by T. Keith. Philadelphia: Printed by Cochran & M'Laughlin, for Mathew Carey and James Thackara, 1801. [10], 19–341, [1] p., diagrs.

First published in London, 1717.
S&S-636.
Copies: DLC; DeGE; MWA; NjP; PHC; PU.

887. Dilworth, Thomas, d. 1780.

The schoolmaster's assistant; being a complete system of practical arithmetic. Carefully revised and amended. To which is added, A compendious system of practical gauging, containing all the rules, fully exemplified, necessary to a perfect practical knowledge of this useful art, by Robert Patterson. Philadelphia: Printed by J. Bioren, 1802. 219 p., port., tables.

First published in London, 1744.
S&S-2147.
Copies: KyU; MH; MWA; NNC; PU.

888. Patterson, Robert, 1743–1824.

A compendious system of practical gauging.... Philadelphia: John Bioren, 1802. 23 p., illus.

S&S-2852.
Copies: DGU.

889. Dearborn, Benjamin, 1754–1838.

Dearborn's system for weighing with ease, dispatch, and precision; purchased for half the cost, and used with half the labor, of scales and weights. Boston: 1804. Broadside, illus. 51 x 31 cm.

Supplement to the *Boston Gazette,* Monday, July 2, 1804.
Copies: DeGE.

890. Goodfellow, J.
The merchant's and ship-master's ready calculator, and the complete pocket assistant, for all persons concerned in the freight of goods.... By J. Goodfellow. Carefully rev. and corr. London printed; Trenton: Reprinted by James Oram, for T. & J. Swords, [etc.], New York, 1804. 2 p.l., 193, [3] p.
S&S-6404.
Copies: CtY; MWA; NN.

891. ———— Trenton: Reprinted by James Oram for J. Conrad & Co., [etc.], Philadelphia, 1804. 2 p.l., 193, [3] p.
N-G0320385.
Copies: DLC.

892. Dilworth, Thomas, d. 1780.
The schoolmaster's assistant; being a complete system of practical arithmetic. Carefully revised and amended. To which is added, A compendious system of practical gauging; containing all the rules, fully exemplified, necessary to a perfect practical knowledge of this useful art, 2d ed., by Robert Patterson. Philadelphia: Printed and sold by John Bioren, 1805. 227 p., front.
A Compendious system of practical gauging; p. [207]–227.
S&S-8334.
Copies: DAU; DeGE; MWA; NNC.

893. Thompson, James.
A complete treatise on the mensuration of timber: containing, besides all the rules usually given on the subject, some new & interesting improvements.... With the description of the sliding rule and Gunter's scale.... Troy, N.Y.: Printed by Wright, Wilbur & Stockwell, 1805. v, [1], [7]–87 p.
S&S-9474.
Copies: DNAL; DeGE; MH; MWA; NBuG; NN; NNC; NcD.

894. Hawney, William.
Hawney's complete measurer; or, The whole art of measuring; being a plain and comprehensive treatise on practical geometry and mensuration. Corr. and improved by T. Keith. 2d American ed.... Philadelphia: Mathew Carey, 1807. iv, [6], 19–341 p., illus.
First published in London, 1717.
S&S-12732.
Copies: ICU; MH; MWA; MdW; ViU.

895. Hutton, Charles, 1737–1823.
The compendious measurer; being a brief, yet comprehensive treatise on mensuration and practical geometry.... Philadelphia: H. Maxwell, 1807. 190 p., illus.
First published in London, 1787.
S&S-12800.
Copies: CoU; DAU; DGU; MWA; MdBJ; MiU; NjP; PPL; PSC.

896. Pennsylvania. General Assembly. Senate. Committee on Weights and Measures.
Report of a committee of the Senate, on the subject of weights and measures. Mr Dorsey, from the committee ... reports. [Lancaster: W. C. Smyth, printer, 1808.] 35 p.
S&S-15880 (erroneously: Harrisburg), 18340 (erroneously dated: Philadelphia, 1809).
Copies: ICJ; PHi (at PPL); PPAmP.

897. Day, Jeremiah, 1773–1867.
A practical application of the principles of geometry to the mensuration of superficies and solids.... New Haven: O. Steele, printer, 1811. 2 p.l., 96 p., 2 fold. plates.
S&S-22667.
Copies: CSt; CtY; InU; MH; MWA; MeB; MiU; NcD; ViU.

898. Bonnycastle, John, 1750?–1821.
An introduction to mensuration and practical geometry.... To which is added, an appendix, containing a concise system of gauging. 1st American, from the 10th London ed., rev. and corr. Philadelphia: Kimber and Conrad, 1812. 262 p.
First published in London, 1782.

898 (*continued*)
S&S-24896.
Copies: MH; NcU; OC; PPL; RPB.

899. Dilworth, Thomas, d. 1780.
The schoolmaster's assistant; being a complete system of practical arithmetic, by Thomas Dilworth. Carefully revised and amended. To which is added, A compendious system of practical gauging, containing all the rules, fully exemplified, necessary to a perfect practical knowledge of this useful art. By Robert Patterson.... 3d ed. Philadelphia: Printed and sold by John Bioren, 1812. 10, [9]–211, [1] p., illus.
S&S-25270.
Copies: MWA; MiU; NjP; ViU.

900. Hawney, William.
Hawney's complete measurer; or, The whole art of measuring; being a plain and comprehensive treatise on practical geometry and mensuration. Corrected and improved by T. Keith. 2d ed. ... rev. and corr. by John D. Craig.... Baltimore: F. Lucas, Jr., [1812]. 341, [1] p., illus.
First published in London, 1717.
S&S-51206.
Copies: DLC; MiU.

901. ———— 3d ed. With an appendix containing rules and examples for finding the weight and dimensions of balls and shells, with their appropriate quantities of powder. Rev. and corr. by John D. Craig.... Baltimore: F. Lucas, Jr., and Neal, Wills & Cole, 1813. xi, 307, [7] p.
S&S-28720.
Copies: CU; CtY; DAU; GAOC; IObB; MB; MH; MWA; MdBE; MiU; NN; OC; TxH.

902. Crosbie, Thomas.
A key to Bonnycastle's Mensuration; containing as plain a solution to every problem and question left unsolved in that treatise, as the nature of each case can admit. Arranged for the American edition of 1812.... Philadelphia: M. Carey; Lydia R. Bailey, printer, 1814. 83 p., diagrs.
S&S-31269.
Copies: DAU; DeGE; MWA; MiU; PPL.

903. Goodfellow, J.
The merchant's and shipmaster's ready calculator, and the complete pocket assistant for all persons concerned in the freight of goods.... By J. Goodfellow.... Carefully rev. and corr. New York: Printed by G. Long, for Evert Duyckinck, 1816. 2 p.l., 193, [3] p.
Also a variant imprint: New York: Printed and published by George Long, 1816. Held by MWA.
S&S-37734.
Copies: CtY; MH; MWA; MiU-C; NcD; RPB.

904. Torres, Manuel.
An exposition of the commerce of Spanish America; with some observations upon its importance to the United States. To which are added, a correct analysis of the monies, weights, and measures of Spain, France, and the United States; and of the new weights and measures of England: with tables of their reciprocal reductions.... Philadelphia: G. Palmer, 1816. viii, 119, [1] p.
S&S-39104.
Copies: DLC; DeGE; MBAt; MWA; PPL.

905. Anthony, Daniel.
The gauging inspector, and measurer's assistant; shewing the most common errors in the practice of gauging, and how to correct them.... Also, tables for ascertaining the measure of grindstones, wood, bark, packages, etc.... Providence: Printed by Miller & Hutchens, 1817. 76 p., diagrs.
S&S-40009.
Copies: DAU; DeGE; MBAt; MH-BA; MWA.

906. United States. Congress. Senate. Committee on Weights and Measures.
In Senate of the United States. March 1, 1817. Mr. Dana, from the committee to whom was referred so much of the message of the President of the United States as relates to weights and measures, reported the following

resolution. [Washington: 1817.] 2 p.
S&S-42561.
Copies: DLC; MWA.

907. Bonnycastle, John, 1750?–1821.
An introduction to mensuration and practical geometry.... To which is added, an appendix, containing a concise system of gauging. The 2d American, from the 10th London ed., rev. and corr. Philadelphia: Kimber and Sharpless; T. H. Palmer, printer, 1818. 262 p.
S&S-43409.
Copies: CSmH; DLC; ICU; MWA; NjP; PPL; PU.

908. The ready reckoner, for timber and board measure. New York: Printed [by Joseph Desnoues] for Christian Brown, 1818. 20 p.
S&S-45477.
Copies: MWA; NN.

909. Stewart, Mordecai.
The youth's guide, dedicated to merchants, mechanics, students, &c. &c. containing instructions in mensuration, book-keeping, geography ... also many useful tables.... Baltimore: Published by the author; Benjamin Edes, printer, 1818. viii, [13]–288 p., plate.
S&S-45801.
Copies: MWA; MdBP.

910. White, John, 1786–1857.
The art of measuring: with a full description of the carpenter's rule and plain directions for the use thereof.... The 1st American, from the last London ed. By the Rev. J. White.... Philadelphia: Printed by Anderson and Meehan, 1818. viii, [9]–67 p.
S&S-46789.
Copies: DAU; DLC; MB; MWA.

911. Young, David, 1781–1852.
Something new and useful. The correct and easy calculator. Containing a series of logarithmic numbers, so constructed as to facilitate the computation of duodecimals, in the measurement of fire-wood, bark, timber, boards, &c.... Morris Town: Printed by Jacob Mann, 1818. 16 p., 2 fold. tables.
S&S-46898.
Copies: DeGE; MWA; NjHi; NjMo.

912. United States. Congress. House. Select Committee on Weights and Measures.
Report of a Select Committee, upon the subject of fixing a standard of weights and measures, January 25, 1819.... [Washington: 1819.] 12 p. (15th Cong., 2d sess. House. Doc. 109)
S&S-49899h.
Copies: DLC; DeGE; DeU; MWA; NN.

913. Bazeley, Charles W.
Arithmetical rules and tables of money, weights, measures and time; together with abbreviations used in writing and printing; and an explanation of French and Latin phrases.... 4th ed. Philadelphia: R. Desilver, 1820. 48 p.
S-351.
Copies: DLC; MH; MWA.

914. Fernald, Benjamin C., 1801–1868.
Tables by which the price of any quantity of staves, wood, or bark may be ascertained; with the rules for ascertaining the contents of any load of wood or bark. To which is added a table shewing the number of feet of boards any log will make before it is sawed. By B. C. Fernald & A. H. Seavy. Portland, [Me.]: Printed by Thomas Todd & Co., 1820. 24 p.
S-1223.
Copies: MWA.

915. Firor, Jacob.
The young man's monitory; containing a compendium of arithmetic.... Also, mensuration of the works of carpenters, joiners, bricklayers, masons, plasterers, painters, glaziers, &c.... Frederick-Town, [Md.]: Published by the author; George Kolb, printer, 1820. xx, [21]–245 p.
S-1231.
Copies: MWA; MdHi; ODa.

916. Hawney, William.

Hawney's complete measurer; or, The whole art of measuring. Being a plain and comprehensive treatise on practical geometry and mensuration. Corr. and improved by T. Keith. 4th ed. With an appendix containing rules and examples for finding the weight and dimensions of balls and shells, with their appropriate quantities of powder. Rev. and corr. by John D. Craig. Baltimore: F. Lucas, Jr., 1820. 312 p., illus.

First published in London, 1717.

S-1546.

Copies: CtHT-W; DAU; DLC; ICU; IEdS; MWA; MdHi; MiU; NNC-T.

917. Stewart, Mordecai.

The youth's guide, dedicated to merchants, mechanics, students, &c. containing instructions in mensuration, bookkeeping, geography ... also many useful tables. Baltimore: James Lovegrove, and the author, 1820. viii, [13]–288 p., plate.

S-3335.

Copies: DAU; MiU.

918. United States. Dept. of State.

Report of the Secretary of State, upon weights and measures, prepared in obedience to a resolution of the House of Representatives of the fourteenth of December, 1819. February 22, 1821.... Washington: Printed by Gales & Seaton, 1821. 245 p., 3 fold. tables. (16th Cong., 2d sess. House. Doc. 109)

Printed also as Senate document 119. Issued also, without document number, under the title: *Report upon weights and measures,* by John Quincy Adams, Secretary of State of the United States.... Washington: Printed by Gales & Seaton, 1821.

Papers in the *Appendix* (p. 137–241) include tables of weights and measures as used in the United States, and laws of the several states, 1734–1819.

S-7457.

Copies: DLC; DeGE; DeU; MWA.

919. United States. Dept. of State.

Report upon weights and measures, by John Quincy Adams, Secretary of State of the United States. Prepared in obedience to a resolution of the Senate of the third March, 1817. Washington: Printed by Gales & Seaton, 1821. 245 p., 3 fold. tables.

Issued also as House document no. 109, and Senate document no. 119, 16th Congress, 2d session.

Papers in the *Appendix* (p. 137–241) include tables of weights and measures as used in the United States, and laws of the several states. 1734–1819.

S-7467.

Copies: CtHT-W; DLC; DeGE; ICJ; KyDC; LNHT; MH; MWA; MdBJ; MiD; MoSW; NNCC; O; PPL.

920. Brown, Thomas.

The timber measurer's, merchant's, and shipmaster's assistant; consisting of tables to expedite the measurement of all kinds of sawed timber.... New-York: J. Seymour, 1822. 139 p.

S-8194.

Copies: DLC; MB.

921. Hawney, William.

Hawney's complete measurer; or, The whole art of measuring. Being a plain and comprehensive treatise on practical geometry and mensuration. Corrected and improved by T. Keith. 5th ed. With an appendix containing rules and examples for finding the weight and dimensions of balls and shells, with their appropriate quantities of powder. Rev. and corr. by John D. Craig.... Baltimore: F. Lucas, Jr., [1822.] xii, 312 p., illus.

First published in London, 1717.

S-8975.

Copies: DAU; DLC; MdHi; MiU; PSt.

922. Parke, Uriah.

The farmers' and mechanics' practical arithmetic; being a complete collection of the practical rules of arithmetic and mensuration.... Winchester, [Va.]: Printed by Samuel H. Davis, 1822. 6 p.l., [9]–207, [1] p.

S-9811.

Copies: DAU; DeGE; NcD; ViHi; ViWn.

923. Pennsylvania. General Assembly. Senate. Committee on Weights and Measures.

Report on the subject of weights and measures, read in the Senate, March 1, 1822. [Harrisburg: Printed by Charles Mowry, 1822.] 3 p.

S-9855.

Copies: PPAmP.

924. Stewart, Mordecai.

The youth's guide, originally dedicated to merchants, mechanics, students, &c. Now designed for the use of schools & families, being greatly improved and enl. Containing mensuration, bookkeeping, geography.... 2d ed. Baltimore: The author, 1822. 288 p.

N-S0937230; S-10354.

Copies: CtY; InU.

925. United States. Congress. House. Select Committee on Weights and Measures.

Report of the Select Committee to which was referred, on the 26th of December last, the report of the Secretary of State upon the subject of weights and measures. March 11, 1822.... [Washington: 1822.] 4 p. (17th Cong., 1st sess. House. Rept. 65)

S-11251.

Copies: DLC; DeGE; PU.

926. ———— [Washington]: Reprinted December 12, 1822. 3 p. (17th Cong., 2d sess. House. Rept. 44)

S-11252.

Copies: DeU; MWA.

927. Bonnycastle, John, 1750?–1821.

An introduction to mensuration and practical geometry.... To which is added, an appendix, containing a concise system of gauging. The 3d American, from the 10th London ed., rev. and corr. Philadelphia: Kimber & Sharpless, 1823. 256 p., illus.

S-11918.

Copies: PPF.

928. Cooper, Thomas, 1759–1839.

Two tracts: on the proposed alteration of the tariff, and on weights and measures. Submitted to the consideration of the members from South-Carolina, in the ensuing Congress of 1823–24. Charleston, [S.C.]: A. E. Miller, printer, 1823. 44 p.

S-12260.

Copies: CtY; DLC; DeGE; MB; MH; MWA; NN; NNC; NbU; PPAmP; PU; ScC; TxU.

929. Bonnycastle, John, 1750?–1821.

A key to the third American edition of Bonnycastle's Mensuration. Adapted also to the former editions.... By Benjamin Hallowell.... Philadelphia: Kimber and Sharpless, 1824. 90 p.

S-15496.

Copies: CtHT; DLC; MiD-B; PPeSchw; PSC-Hi.

930. Cooper, Thomas, 1759–1839.

Two tracts: on the proposed alteration of the tariff, and on weights & measures.... [Washington: 1824.] 32 p.

S-15868.

Copies: MH.

931. Blunt, Edmund March, 1770–1862.

The merchant and seaman's expeditious measurer; containing a set of tables which show ... the solid contents of all kinds of packages and casks according to their several lengths, breadths and depths.... Stereotype ed. [New York]: E. M. Blunt, 1825. 196 p.

Based on the *Merchant's and ship-master's ready calculator,* by J. Goodfellow, first published in London.

S-19773.

Copies: CtB; DeGE; MH; MH-BA; MWA; NjR.

932. Day, Jeremiah, 1773–1867.

A practical application of the principles of geometry to the mensuration of superficies and solids.... 2d ed. New Haven: S. Converse, 1825. 2 p.l., 96 p., plates.

S-20273.

Copies: CLU; CtHT-W; CtY; MH; MoSU; NCH; NNC; OClW.

933. Exercises in mensuration, taken chiefly from Gregory's Mechanics for practical men, Hutton's Mensuration, and Legendre's Geometry. Gardiner, Me.: Printed and published by P. Sheldon, 1826. 15 p.
S-24467.
Copies: DeGE.

934. Walsh, Michael, 1763–1840.
The mercantile arithmetic, adapted to the commerce of the United States, in its domestic and foreign relations, with an Appendix, containing practical systems of mensuration, gauging, and book-keeping. A new ed., rev. and impr. Boston: Richardson & Lord, 1826. 248, 59 p.
The previous editions lack the Appendix.
S-27542.
Copies: DAU; DeGE; MB; MH; MeHi; OMC; PPL.

935. Bonnycastle, John, 1750?–1821.
An introduction to mensuration and practical geometry.... To which is added an appendix, containing a concise system of gauging. The 4th American from the 10th London ed., enl. and corr. Philadelphia: Kimber and Sharpless, 1827. 252 p.
S-28221.
Copies: PU; UU.

936. The mechanics' assistant; or, Universal measurer, containing a collection of tables of measures, weights and powers, adapted to the use of engineers, millwrights, iron founders, smiths.... With an appendix, exhibiting the strength of beams, &c., and a correct method of calculating the horse powers of a steam engine.... Providence: John Hutchens, 1827. 58, [1] p.
S-29711.
Copies: DeGE; InLP; MWA; MiU; NN; NjR; PPF; RPB.

937. King, Andrew M.
A complete system of mensuration, in all its branches: containing many new and necessary improvements.... By Andrew M. King & J. Tinkler. Auburn, [N.Y.]: Printed by Richard Oliphant, 1828. 120 p., illus., 2 fold. plates.
N-K0140777.
Copies: CtY; N; NNNC; NRU; NjP; ViU.

938. Roberts, of Leeds, Eng.
Roberts's Mechanics' assistant; or, Universal measurer; containing a collection of tables of measures, weights & powers, adapted to the use of engineers, millwrights, iron founders, smiths.... With an appendix, exhibiting the strength of materials, &c., and a correct method of calculating the horse powers of a steam engine. 1st American, from the 4th English ed. Philadelphia: Printed for John Walshaw, engineer, by John Young, 1828. 58, [1] p.
Copies: MWA.

939. Walsh, Michael, 1763–1840.
The mercantile arithmetic, adapted to the commerce of the United States, in its domestic and foreign relations, with an Appendix, containing practical systems of mensuration, gauging, and bookkeeping. A new ed., rev. and impr. Boston: Richardson & Lord, 1828. 239, 76 p.
S-37073.
Copies: CtW; DAU; MH; NcGU; TxU.

940. Barney, William Bedford, 1781?–1838.
Par of weights; or, Table of the relative proportions actually existing between the weights of foreign commercial countries, and those of the United States: embracing other weighty information. Baltimore: Printed by Benjamin Edes, 1829. 63, [1] p.
S-37707.
Copies: DLC; MH-BA; MWA; MdHi; MnU; NN.

941. Bazeley, Charles W.
Arithmetical rules and tables of money, weights, measures and time.... 5th ed. Philadelphia: Robert Desilver; Jesper Harding, printer, 1829. 40 p.
S-37730.
Copies: DAU; NNMus; Vi.

942. Bonnycastle, John, 1750?–1821.
An introduction to mensuration and practical geometry.... To which is added, an appendix, containing a concise system of gauging. 5th American, from the 10th London ed., with corrections and additions. Philadelphia: Kimber and Sharpless, 1829. 252 p.
S-37883.
Copies: DAU; MiU; OO; PPAN; PPL; PSC; RPB; ViU.

943. Connolly, James L.
Connolly's arithmetic; or, The Ohio accomptant. Being a plain practical treatise, with a complete system of mensuration. Pittsburgh: Cramer & Spear, 1829. 250 p.
S-38259.
Copies: CSmH; DAU; MiU; OClWHi; PPi; TxU.

943a. Rules for measuring hammered granite stone, adopted April, 1829. Boston: J. Howe, printer, 1829. 8 p.
Manuscript notes, "Rules for measuring slating," tipped in at end.
N-R0509462.
Copies: NcD.

7. Inventions and patents

944. United Sates. Laws.
A bill to promote the progress of science and useful arts by securing to authors and inventors the exclusive right to their respective writings and inventions. [New York: Printed by Francis Childs and John Swaine, 1789.] 11 p.
E-22192.
No copy known.

945. United States. Laws.
An act to promote the progress of useful arts.... [Philadelphia]: Printed by John Fenno, [1790]. 3 p.
Concerns the issuing of patents to protect the interests of inventors.
B-7542; Mp-46067.
Copies: DLC; NhD.

946. United States. Laws.
A bill to promote the progress of useful arts.... [New York: Printed by Francis Childs and John Swaine, 1790.] 4 p.
Concerns the issuing of patents to protect the interests of inventors.
B-7554; Mp-46023.
Copies: DLC.

947. United States. Laws.
A bill to amend an act, intituled, "An act to promote the progress of useful arts." Philadelphia: Childs & Swaine, [1791]. [3] p.
Concerns the issuing of patents to protect the interests of inventors.
Presented in Congress on Feb. 7, 1791.
Mp-23848.
Copies: DLC; NN (photocopy).

948. Barnes, Joseph.
Treatise on the justice, policy, and utility of establishing an effectual system for promoting the progress of useful arts, by assuring property in the products of genius. To which are added, observations, on the deficiency of, and exceptions to the bill reported in March 1792. With notes, tending to demonstrate, that no property is secured in the products of genius, under the existing patent system. Also, the principles upon which a bill ought to be formed, to be effectual and equitable.... Philadelphia: Printed by Francis Bailey, 1792. 34 p.
Mp-24076.

948 (*continued*)
Copies: DLC; DeGE; MBAt; MWA; MiU; PPAmP; PPL; RPJCB; ViW.

949. United States. Laws.
An act to promote the progress of useful arts, and to repeal the act heretofore made for that purpose. [Philadelphia: Printed by Childs & Swaine, 1793.] [4] p.
Approved February 21, 1793.
Mp-26309.
Copies: DLC; MWA; NN; RPJCB.

950. United States. Laws.
An act, supplementary to the act intituled "Act to promote the progress of useful arts...." [Philadelphia: Printed by F. Childs and J. Swaine, 1794.] [2] p.
Approved June 9th, 1794.
Mp-27875.
Copies: DLC; MWA; NN; RPJCB.

951. American Philosophical Society, Philadelphia.
Premiums. The American Philosophical Society, held at Philadelphia, for promoting useful knowledge, in order the more effectually to answer the ends of their institution, have agreed to appropriate, annually, a part of their funds to be disposed of in premiums to the authors of best performances, inventions, or improvements, relative to certain specific subjects of useful knowledge.... [Philadelphia: 1796.] Broadside 41.5 x 33.5 cm.
Dated: Philadelphia, May 1796.
Mp-29977.
Copies: DLC; PPAmP.

952. United States. Congress. Senate. Committee on Patents.
Report of the committee, to whom was referred the bill, entitled "An act to extend the privilege of obtaining patents for useful discoveries and inventions, to certain persons therein mentioned, and to enlarge and define the penalties for violating the rights of patentees." [Philadelphia]: Printed by order of the Senate of the United States, 1800. 4 p.
Dated: March 28, 1800.
Mp-38883.
Copies: MWA; NN.

953. United States. Laws.
A bill for the encouragement of learning and for the promotion of useful arts. [Washington]: Duane & Son, printers, [1803]. 15 p.
Concerns the protection of the interests of inventors.
S&S-5222.
Copies: DLC.

954. United States. Dept. of State.
Letter from the Secretary of State, accompanied with a list of the names of persons who have invented any new and useful art, machine, manufacture or composition of matter, or any improvement thereon, and to whom patents have issued for the same, from the office of the Department of State, with the dates and general objects of such patents ... 22d February.... [Washington]: 1805. 48 p.
S&S-9542, 9544.
Copies: DLC; DP; DeGE; MWA; NN.

955. [Evans, Oliver] 1755–1819.
Address of the advocate of the patentees, inventors of useful improvements in the arts and sciences; petitioners to Congress, for redress of grievances. Delivered December 19th, 1806, before the Honorable Committee to whom their petition was referred, in defence of mental property.... Washington City: Printed by Duane & Son, 1806. 16 p.
S&S-9805.
Copies: CSt; DLC; DeGE; MH; MWA.

956. Newengland Association of Inventors and Patrons of Useful Arts.
Remarks on the rights of inventors, and the influence on their studies in promoting the enjoyments of life and public prosperity. Boston: Printed by E. Lincoln, 1807. 16 p.
Prepared by a committee appointed at a meeting of patentees and proprietors of patents, Boston, December 1st, 1806.
Also attributed to Paul Pilsbury.
N-P0368111; S&S-13188, 13480.
Copies: CSt; DLC; DeGE; DeU; MH-BA; MWA; NN; PPF.

957. United States. Laws.
A bill for the encouragement of learning, and for the promotion of the useful arts. January 2, 1807. [Washington: 1807.] 16 p.
Concerns the issuing of patents for the protection of inventors.
S&S-13815.
Copies: DLC.

958. Fessenden, Thomas Green, 1771–1837.
The register of arts; or, A compendious view of some of the most useful modern discoveries and inventions.... Philadelphia: C. and A. Conrad, and Co., and Byrch & Small, 1808. xi, 404 p., 2 plates.
S&S-15012.
Copies: CSmH; CtY; DLC; DeGE; ICJ; MH-BA; MWA; MiD-B; NHi; NN; NcD; NjR; PPL; PU-S; ScC; ViAl.

959. Evans, Oliver, 1755–1819.
To the honorable the senators and representatives in Congress individually. [Washington: 1809.] 2 p.
A petition concerning the rights of patentees.
Copies: DLC.

960. Fessenden, Thomas Green, 1771–1837.
An essay on the law of patents for new inventions. With an appendix, containing the French patent law, forms, &c. By Thomas G. Fessenden.... Boston: D. Mallory, [etc.]; S. T. Armstrong, printer, 1810. [viii]–xxxix, 41–229, [1] p.
S&S-20110.
Copies: DLC; DeGE; IU; KU; KyU; MBS; MH; MWA; MoU; NN; NjR; PPL; RPB.

961. United States. Laws.
A bill for the encouragement of learning, and for the promotion of the useful arts. January 18th, 1810. [Washington: 1810.] 17 p. (H.R. 35)
Concerns patents for the protection of the inventors.
S&S-21585.
Copies: DLC.

962. United States. Dept. of State.
Letter from the Secretary of State, accompanied with "A list of the names of persons who have invented any new and useful art, machine, manufacture or composition of matter, or any improvement thereon, and to whom patents have been issued for the same from the Office of the Department of State with the dates and general objects of such patents...." February 22d, 1805 ... January 28th, 1811, ordered to be reprinted. Washington: Printed by R. C. Weightman, 1811. 31 p.
S&S-24157.
Copies: DLC; DNA; DeGE; MWA; RPB.

963. United States. Dept. of State.
Letter from the Secretary of State, transmitting a list of the names of patentees, their places of residence, and the nature of their inventions or improvements.... January 14th, 1811. Ordered to lie on the table. Printed by order of the House of Representatives of the U. States. Washington: Printed by R. C. Weightman, 1811. 40 p.
S&S-24159.
Copies: DLC; DeGE; MWA; N; PMA; RPB.

964. United States. Patent Office.
Patents. Patent Office, March 5, 1811. [Washington: Jonathan Elliot, printer, 1811.] 8 p.
Signed: Wm. Thornton.
S&S-24254.
Copies: MB; MH-BA; MWA; PPAmP.

965. United States. Congress. House. Committee to Inquire into the State and Business of the Patent Office.
Report, in part, of the Committee appointed to ascertain the number of persons employed in and to inquire into the state and condition of the Patent-Office, June 12, 1812.... Washington: Printed by A. & G. Way, 1812. 32 p.

965 *(continued)*
S&S-27292.
Copies: DLC; MBAt; MWA; NN.

966. United States. Congress. House. Select Committee on Patents.
Amendments reported by the Select Committee to whom was referred the bill, from the Senate "To promote the progress of science and useful arts, by securing, for a limited time, to the inventors, the exclusive right to their dicoveries." March 9th, 1812. [Washington: 1812.] 3 p.
S&S-26949.
Copies: DLC.

967. United States. Dept. of State.
Letter from the Secretary of State, transmitting, in obedience to a resolution of the House of Representatives of the thirteenth instant, a list of the names of persons to whom patents have been issued for any new or useful invention, from the twenty-eighth December, 1810, to the first January, 1812. January 21st, 1812. Washington City: Printed by R. C. Weightman, 1812. 15 p.
S&S-27159.
Copies: DLC; MWA.

968. United States. Laws.
A bill to promote the progress of science and useful arts by securing for a limited time, to inventors, the exclusive right to their respective discoveries. (S. 19). January 22d, 1812. [Washington: 1812.] 14 p.
S&S-27111.
Copies: DLC.

969. [Evans, Oliver] 1755–1819.
Patent right oppression exposed; or, Knavery detected. In an address, to unite all good people to obtain a repeal of the patent laws. By Patrick N. I. Elisha, esq., poet laureate [pseud.], To which is added, an alarming law case; also, reflections on the patent laws. Illustrated with notes and anecdotes, by the editor. Philadelphia: R. Folwell, and G. Allchin, 1813. xi, [1], 189, [1] p.
S&S-28459.
Copies: CtHT-W; DLC; DeGE; ICJ; MWA; MdToH; MiU-T; NIC; NN; PPL; PSC; PU; ViU.

970. United States. Dept. of State.
Letter from the Secretary of State, transmitting a list of the names of persons to whom patents have been issued for any new or useful art, machine, manufacture or composition of matter, or any improvement thereon, from January 1, 1812, to January 1, 1813. January 22d, 1813. Washington: A. & G. Way, printers, 1813. 14 p.
S&S-30204.
Copies: DLC; MPeal; MWA.

971. [Evans, Oliver] 1755–1819.
Patent right oppression exposed; or, Knavery detected. In an address, to unite all good people to obtain a repeal of the patent laws. By Patrick N. I. Elisha, Esq., poet laureate [pseud.]. To which is added, an alarming law case; also, reflections on the patent laws. Illustrated with notes and anecdotes, by the editor. 2d ed. Philadelphia: Published by the booksellers, 1814. xi, [1], 189, [1] p.
S&S-31432.
Copies: CSmH; DLC; DeGE; MB; MH; MWA; NN; PPL; RPB; ViW.

971a. Thomas, John.
John Thomas's inventions. Washington: 1824. 16 p.
Dated: March 1, 1824.
N-T0167554.
Copies: DLC.

972. United States. Dept. of State.
Letter from the Secretary of State, transmitting a list of the names of persons to whom patents have been issued for any new or useful art, machine, manufacture, or composition of matter, or any improvement thereon from January 1st, 1813 to January 1st, 1814. January 31, 1814. Washington: A. & G. Way, printers, 1814. 16 p.
S&S-33257.
Copies: DLC; MWA; MoSpD; O.

973. United States. Laws.
A bill in addition to the acts concerning patents for useful discoveries and inventions. January 10th, 1814. . . . [Washington: 1814.] 2 p. (S. 7)

S&S-33121.
Copies: DLC.

974. United States. Dept. of State.
Letter from the Secretary of State, transmitting a list of persons who have made any new and useful invention, and for which patents have been obtained, from thirty-first December, 1813, to the first January, 1815. January 30, 1815. Washington City: Printed by R. C. Weightman, 1815. 13 p.
S&S-36285.
Copies: CSmH; DLC; MWA; MiD-B; MoSpD; O.

975. [Evans, Oliver] 1755–1819.
Exposition of part of the patent law. By a native born citizen of the United States. To which is added, Reflections on the patent law. [S.l.]: 1816. 78 p.
S&S-37540, 37541.
Copies: CSt; DeGE; MH-L; MHi; MWA; MiU; NN; PPAmP; PPL.

976. United States. Dept. of State.
Letter from the Secretary of State, transmitting a list of the names of persons to whom patents have been granted, for the invention of any new or useful art, or machine, manufacture, or composition of matter, or improvement thereon, from the 1st of January, 1815, to the 1st of January, 1816. February 3d, 1816. . . . Washington: Printed by William A. Davis, 1816. 13 p.
S&S-39504.
Copies: DLC; DNA; DeGE; MWA.

977. United States. President (Madison).
Message from the President of the United States, recommending alterations in the manner of conducting the Patent Office and further restrictions on the issue of patents. April 11th, 1816. Washington City: Printed by William A. Davis, 1816. 4 p.
S&S-39548.
Copies: DLC; MWA.

978. Evans, Oliver, 1755–1819.
Oliver Evans to his counsel, who are engaged in defence of his patent rights, for the improvements he has invented: containing a short account of two out of eighty of his inventions, their rise and progress in despite of all opposition and difficulties; and two of his patents, with explanations. [Philadelphia: 1817.] 55, [1] p.
S&S-40778.
Copies: CSt; DLC; DeGE; MH-BA; MWA; PPAmP; PPL.

979. United States. Dept. of State.
Letter from the Secretary of State transmitting a list of the names of persons to whom patents have been issued, for the invention of any new or useful art, or machine, manufacture, or composition of matter, or any improvement thereon, from January 1st, 1816, to January 1st, 1817. . . . Washington: Printed by William A. Davis, 1817. 12 p.
S&S-42622.
Copies: DLC; DeGE; MWA; MWiW; O.

980. Millions of dollars' worth tendered gratis, and not accepted on those terms. Submitted to the consideration of Congress, by a native citizen of the United States. Washington: Feb. 10, 1818. 19 p.
Concerns the protection of inventions; sometimes attributed to Oliver Evans.
S&S-44848.
Copies: DeGE; MH-BA; MWA; MdHi; MiU-C.

981. United States. Dept. of State.
Letter from the Secretary of State, transmitting a list of the names of persons to whom patents have been issued, for the invention of any new or useful art, or machine, manufacture, or composition of matter, or any improvement thereon, from ths[sic] 1st of January, 1817, to the 1st of January, 1818. January 13, 1818. . . . Washington: Printed by E. De Krafft, 1818. 14 p. (15th Cong., 1st sess. House. Doc. 48)
S&S-46321.
Copies: CSmH; DLC; DeGE; NjP; R.

982. United States. Dept. of State.

Letter from the Secretary of State, transmitting a list of the names of persons to whom patents have been granted, for any useful invention, from first of January, 1818, to first of January, 1819. January 7, 1819.... Washington: Printed by E. De Krafft, 1819. 13 p. (15th Cong., 2d sess. House. Doc. 78)

S&S-49812.

Copies: DLC; DeGE; DeU; G; MWA; NN; PPAmP; R.

983. A list of patents granted by the United States, for the encouragement of arts and sciences, alphabetically arranged, from 1790 to 1820: containing the names of patentees, their places of residence, and the dates of the patents; with an alphabetical list of the names of the patentees: also, all the acts passed by Congress on the subject of patents. Washington: Printed by Alfred Elliot, 1820. 72, xxiv p.

Second title page: *An Alphabetical list of all the patents granted by the United States for inventions....*

S-1972.

Copies: DLC; MWA; NIC; PPAmP; RPJCB.

984. United States. Dept. of State.

Letter from the Secretary of State, transmitting a list of the names of persons who have invented any new or useful art or machine, manufacture, or any improvement thereon; and to whom patents have been issued for the same, from the thirty-first day of December, 1818, to the first day of January, 1820 ... January 7, 1820.... Washington: Printed by Gales & Seaton, 1820. 9 p. (16th Cong., 1st sess. House. Doc. 35)

S-3776.

Copies: DLC; DeGE; DeU; MWA; O; PPL; R.

985. United States. Dept. of State.

Letter from the Secretary of State, transmitting a list of the names of persons to whom patents have been issued, for any useful invention, during the year 1820. January 5, 1821.... Washington: Printed by Gales & Seaton, 1821. 10 p. (16th Cong., 2d sess. House. Doc. 46)

S-7312.

Copies: DLC; DeGE; DeU; MWA; NjP; PPL.

986. Fessenden, Thomas Green, 1771–1837.

An essay on the law of patents for new inventions. By Thomas Green Fessenden ... 2d ed. Boston: Charles Ewer, 1822. xxxii p., 2 l., [35]–427, [1] p.

S-8698.

Copies: CU-L; Ct; DeGE; ICLaw; IaU-L; LNHT; MB; MH; MWA; Md; MeU; MiU-C; NIC; NNLI; NjR; OCl; PPAmP; PU-L; RPB; ViU.

987. United States. Dept. of State.

Letter from the Secretary of State, transmitting a list of the names of persons to whom patents have been issued for any new or useful art, or machine, manufacture, or composition of matter, or improvement thereon, for one year, prior to the 1st January, 1822. January 8, 1822.... Washington: Printed by Gales & Seaton, 1822. 11 p. (17th Cong., 1st sess. House. Doc. 18)

S-10925.

Copies: DLC; DeGE; DeU; MWA; NjP; PPL.

988. [Fessenden, Thomas Green] 1771–1837.

Reply to strictures on Thomas G. Fessenden's Essay on the law of patents for new inventions. [Boston: Phelps & Farnham, printers, 1823.] 7 p.

S-12538.

Copies: DLC; MBAt; MWA.

989. A list of patents granted by the United States, for the encouragement of arts and sciences, alphabetically arranged, from 1790 to 1820; continued by supplements to April, 1823.... Also, a list of books &c. deposited in the Department of State, for securing copyrights; from January 1796, to January 1822; continued by a supplement, to April 1823.... Washington, D.C.: Printed by S. A. Elliot,

1823. [2], 72, xxiv, 13, 6, [9]–16 p.
S-13115.
Copies: DLC; ICRL.

990. A list of patents granted by the United States, for the encouragement of arts and sciences. Supplement . . . from January 22, To April 1823 [continued by supplements to May 1826. Also Supplement to the list of books, maps, charts, &c. deposited in the Department of State, for securing copy-rights. From January 1822, to May 1825.] [Washington: 1823?–1826?.] [9]–39 p.
N-L0403173.
Copies: DLC.

991. United States. Congress. House. Committee on the Expenditures of the Department of State.
Report of the Committee on the expenditures of the Department of State on the Patent Office. February 7, 1823. . . . [Washington: 1823.] 5 p. (17th Cong., 2d sess. House. Rept. 86)
Includes a list of models in the Patent Office.
S-14713.
Copies: DLC; DeU; MWA; NjP.

992. United States. Dept. of State.
Letter from the Secretary of State, transmitting a list of the names of persons to whom patents have been granted for any new and useful invention, during the year 1822. January 22, 1823. . . . Washington: Printed by Gales & Seaton, 1823. 15 p. (17th Cong., 2d sess. House. Doc. 36)
S-14561.
Copies: DLC; DeGE; DeU; MWA; NjP.

993. United States. Dept. of State.
Letter from the Secretary of State, accompanied with a report of the names of persons to whom patents have been granted, for new and useful inventions during the year 1823. January 7, 1824. . . . Washington: Printed by Gales & Seaton, 1824. 13 p. (18th Cong., 1st sess. House. Doc. 25)
S-18783.
Copies: DLC; DeGE; DeU; MWA; NjP.

994. Franklin Institute, Philadelphia.
September 6, 1825. "The Franklin Institute of the state of Pennsylvania for the promotion of mechanic arts. . ." [Philadelphia: 1825.] Broadside.
Circular letter issued by Peter A. Browne, corresponding secretary, announcing the formation of a committee to investigate new inventions.
S-20558.
Copies: DNA; PPF (archives).

995. United States. Dept. of State.
Letter from the Secretary of State, transmitting a list of names of persons to whom patents have been issued, for the invention of any new or useful art, or machine, manufacture, or composition of matter, or any improvement thereon, from January lst, 1824, to January lst, 1825. January 3, 1825. . . . Washington: Printed by Gales & Seaton, 1825. 17 p. (18th Cong., 2d sess. House. Doc. 28)
S-22921.
Copies: DLC; DeU; MWA; NjP; PPL.

996. ——— from January 1st, 1825, to January 1st, 1826. January 3, 1826. . . . Washington: Gales & Seaton, 1826. 23 p. (19th Cong., 1st sess. House. Doc. 22)
S-27165.
Copies: DLC; DeU; MWA; NjP; PPL.

997. United States. Dept. of State.
Patent Office. Letter from the Secretary of State, transmitting the information . . . upon the subject of the organization of the Patent Office. January 13, 1827. . . . Washington: Printed by Gales & Seaton, 1827. 8 p. (19th Cong., 2d sess. House. Doc. 47)
S-31448.
Copies: DLC; DeU; MWA; NjP.

998. United States. Dept. of State.
Patents for useful inventions in 1826. Letter from the Secretary of State, transmitting a list of the names of persons to whom patents have been issued, for the invention of any new or useful art, or machine, manufacture or composition of matter, or any

998 (*continued*)
improvement thereon, from Jan. 1st, 1826, to Jan. 1st, 1827. January 8, 1827.... Washington: Printed by Gales & Seaton, 1827. 21 p. (19th Cong., 2d sess. House. Doc. 27)
S-31449.
Copies: DLC; DeGE; DeU; MWA; NjP.

999. A list of patents granted by the United States, for the encouragement of arts and sciences, alphabetically arranged, from 1790 to 1828 ... containing the names of the patentees, their places of residence, and the dates of their patents. With an alphabetical list of the names of patentees, also, the mode of applying for a patent, and all the acts passed by the Congress on the subject of patents. City of Washington, D.C.: Printed and sold by S. A. Elliot, 1828. 118, xxiv, 65–72 p.
S-36514.
Copies: DLC; DeGE; IObB.

1000. United States. Dept. of State.
Patents—1827. Letter from the Secretary of State, transmitting a list of the names of the persons to whom patents have been granted for the invention of any new or useful art, or machine ... from January 1st, 1827, to December 31, 1827. January 2, 1828.... Washington: Printed by Gales & Seaton, 1828. 23 p. (20th Cong., 1st sess. House. Doc. 34)
S-36696.
Copies: DLC; DeGE; DeU; MWA; NjR; PPL.

1001. United States. Dept. of State.
Patents—1828. Letter from the Secretary of State, transmitting a list of patents, with the names and residence of the patentees, granted for the invention of any new or useful art or machine ... from the first of January to the thirty-first of December, 1828, inclusive. January 8, 1829.... [Washington: 1829.] 20 p. (20th Cong., 2d sess. House. Doc. 59)
S-41275.
Copies: DLC; DeGE; DeU; MWA; NjP; PPL.

1002. Elliot, William.
The patentee's manual; containing a list of patents granted by the United States for the encouragement of arts & sciences, alphabetically arranged, from 1790 to 1830. (To be continued by supplements.) Also, the laws of Congress for granting patents; with a digest for all the decisions which have taken place in the courts of the United States respecting patents. Washington: Printed by S. A. Elliot, 1830. xviii, 3–118, [137]–153 p.
C-1247.
Copies: CSmH; DLC; MnU; PPL.

1003. A list of patents granted by the United States, alphabetically arranged from 1790 to 1828.... [Washington: 1830?] 110 p.
Caption title: *An Alphabetical list of all patents and patentees on record in the Patent Office.*
Supplements [1–2] to the list continued from June 1827 to January 1830. [Washington: 1830?] 2 parts.
Attributed to William Elliot.
N-E0091714.
Copies: DP (imperfect copy without title page).

1004. United States. Dept. of State.
Letter from the Secretary of State, transmitting a list of the names of persons to whom patents have been granted for the invention of any new and useful art of machine, manufacture, or composition of matter, or improvement thereon, from the 1st day of Jan. 1829, to the 1st day of Jan. 1830. January 3, 1830.... [Washington: 1830.] 33 p. (21st Cong., 1st sess. House. Doc. 16)
C-4901.
Copies: DLC; DeU; MWA; NjR; PPL.

1005. United States. Dept. of State.
Patent Office. Message from the Pres-

ident of the United States, transmitting a communication from the Secretary of State in relation to the Patent Office. January 27, 1830.... [Washington: 1830.] 14 p. (21st Cong., 1st sess. House. Doc. 38)

Includes a list of patents issued from 1790–1828.

C-5038.

Copies: DLC; DeGE; DeU; MWA; NjR.

III. AGRICULTURE

1. Societies

1006. Philadelphia Society for Promoting Agriculture.

Laws of the Philadelphia Society for Promoting Agriculture.... Philadelphia: Hall and Sellers, 1785. 8 p.

B-6148; Mp-44773.

Copies: MWA.

1007. South Carolina Society for Promoting and Improving Agriculture.

Address, and rules of the South Carolina Society for Promoting and Improving Agriculture and other rural concerns. Charleston: Timothy, 1785. 12 p.

Mp-19254.

Copies: MBAt; NcWsW.

1008. Philadelphia County Society for the Promotion of Agriculture and Domestic Manufactures.

Constitution of the Philadelphia County Society for the Promotion of Agriculture and Domestic Manufactures. [Philadelphia: 1788.] 4 p.

Copies: DeGE.

1009. Philadelphia Society for Promoting Agriculture.

Laws of the Philadelphia Society for Promoting Agriculture; as revised and enacted by the said society, January 10, 1786, with the premiums proposed, February 5, 1788. To which is prefixed, a list of members of the society.... [Philadelphia]: 1788. 24 p.

Mp-21389.

Copies: PPL.

1010. ———— February 16th, 1789, with the premiums proposed, February 3d, 1789. To which is prefixed a list of the members of the society. [Philadelphia]: 1789. 23 p.

Mp-22067.

Copies: DLC; DNAL; DeGE; PPL.

1011. Massachusetts Society for Promoting Agriculture.

Laws and regulations of the Massachusetts Society for Promoting Agriculture, containing the names of its present officers and members, list of premiums, &c. With some interesting extracts from foreign and domestick publications. By the trustees. Boston: Printed by Isaiah Thomas and Ebenezer T. Andrews, 1793. v, [7]–56 p.

Mp-25794.

Copies: DLC; DNAL; DeGE; MBAt; MH; MHi; MWA; NHi; RPJCB.

1012. [Philadelphia Society for Promoting Agriculture.]

Outlines of a plan for establishing a state society of agriculture in Pennsylvania. Philadelphia: Printed by Charles Cist, 1794. 1 p.l., 11 p.

"At a special meeting of the Philadelphia Society for Promoting Agriculture, on the 21 of January, 1794, agreed that Mr. Bordley, Mr. G. Clymer, Mr. Peters and Mr. Pickering, be a committee to prepare Outlines of a plan.... "

Attributed to John Beale Bordley.

Mp-26682, 27512.

Copies: DLC; DeGE; MH; MWA; PPL; RPJCB.

1013. Van der Kemp, Francis Adrian, 1752–1829.

Speech of Fr. Adr. Van der Kemp, at a meeting, the first of June, one thousand seven hundred and ninety-five, at Whitestown, for

1013 (*continued*)
the institution of a society of agriculture. Whitestown, [N.Y.]: Printed by Oliver P. Easton, 1795. 19 p.

Mp-29785.

Copies: CSmH; MBAt; MH; NHi.

1014. Massachusetts Society for Promoting Agriculture.

Rules and regulations of the Massachusetts Society for Promoting Agriculture: containing, the names of its present members and officers, several important communications, and interesting extracts, from foreign and domestic publications, with a list of premiums, donations, &c. by the trustees. Boston: printed by Thomas Fleet, jun., 1796. 78 p., plate.

Mp-30773.

Copies: DLC; DeGE; ICU; InU; MH; MWA; MiU-C; NHi; NN; PPF; PU; RPJCB.

1015. South Carolina Society for Promoting and Improving Agriculture.

Address and rules of the South-Carolina Society for Promoting Agriculture, and other rural concerns. Charleston: Printed by Freneau & Paine, 1798. 8 p.

Mp-34577.

Copies: DLC; DNAL; MWA.

1016. Sandy Spring Farmers' Society.

The constitution of the Farmers Society, established at Sandy Spring in Montgomery County, Maryland, the 11th of the 5th month, 1799, revised and amended the 31st of the 1st month, 1801. [Sandy Spring, Md.: 1801.] Broadside 46 x 26.5 cm.

Copies: DNAL (photocopy).

1017. Massachusetts Society for Promoting Agriculture.

Sir, a number of questions, stated by the trustees of the Agricultural Society, are inclosed.... Boston: 1803. Broadside 30 x 20 cm.

Dated: January 1803.

Concerns the publication by the Society of information on the improvement of agriculture.

Copies: MWA.

1018. Attleborough Society for the Encouragement of Agriculture, Arts, and Social Intercourse.

Constitution of the Attleborough Society for the Encouragement of Agriculture, Arts, and Social Intercourse. Providence: Printed by Heaton and Williams, [1804]. 15 p.

S&S-6069.

Copies: MAtt.

1019. Society for the Encouragement of Agriculture, Arts, and Social Intercourse, Providence.

Constitution of the Society for the Encouragement of Agriculture, Arts, and Social Intercourse. Providence: [1805]. 15 p.

S&S-9390.

No copy known.

1020. Philadelphia Society for Promoting Agriculture.

Act of incorporation ... by-laws, &c. [Philadelphia: 1807.] 16 p.

S&S-13379.

Copies: PHi; PPL.

1021. Philadelphia Society for Promoting Agriculture.

Laws of the Philadelphia Society for Promoting Agriculture, as revised and enacted at the annual meeting, held on the 14th January, 1812. To which is prefixed a list of the members of the Society. Incorporated February 14, 1809. Philadelphia: Printed by L. R. Bailey, 1812. 15 p.

S&S-26453.

Copies: DLC; DeGE; NjR; P; PPL; PPPrHi; PU-V; ScCleU.

1022. Suffolk County (N.Y.) Agricultural Society.

Constitution of the Suffolk County Agricultural Society, adopted October 6, 1813. New York: Printed by J. Gray & Co., 1813. 4 p.

S&S-29897.
Copies: NRvS.

1023. Agricultural Associates in Shrewsbury and its Vicinity.
Constitution and by laws of the Agricultural Associates in Shrewsbury and its vicinity as by them adopted Jan. 2, 1816. Worcester, [Mass.]: Printed by W. Manning, 1816. 9, [2] p.
S&S-36693.
Copies: MH; MWA.

1024. Berkshire Agricultural Society.
Bye-laws, and names of officers and committees of the Berkshire Society for the Promotion of Agriculture and the Useful Arts. Also a list of honorary and ordinary members. Pittsfield, [Mass.]: Printed by P. Allen, 1816. 16 l.
N-B0349000.
Copies: DNAL.

1025. Agriculture and Manufacturing Society of Jefferson County, Ohio.
[Constitution. Steubenville, Ohio: 1817.]
S&S-39954.
No copy known.

1026. Cheshire Agricultural Society, Cheshire, N.H.
An act establishing the Cheshire Agricultural Society. Together with the regulations and by-laws of the Society, and a list of the officers for the present year.... Windsor, [Vt.]: Printed by Jesse Cochran, by order of the Society, 1817. 8 p.
S&S-40459.
Copies: DLC; Nh.

1027. Hartford County Agricultural Society.
Articles of association and by laws of the Hartford County Agricultural Society.... Hartford: Printed by George Goodwin & Sons, 1817. 8 p.
S&S-41011.
Copies: CtHi; CtY; MWA.

1028. [Massachusetts Society for Promoting Agriculture.]
Objects of inquiry and remark. [Boston: 1817.] 4 p., illus.
N-M0316647.
Copies: MWA; NN.

1029. Agricultural Association of the Counties of Dutchess and Columbia.
Agricultural Association of Dutchess & Columbia. The Farmers' Club of the counties of Dutchess and Columbia wishing anxiously to promote a spirit of industry and improvement in agriculture, offers the following premiums for the year 1818 ... premiums for 1819.... [Hudson, N.Y.: 1818.] Broadside 32 x 19 cm.
Signed: William Wilson.
N-A0098078.
Copies: MiU-C.

1030. Agricultural Society of Albemarle, Charlottesville, Va.
Rules and regulations of the Agricultural Society of Albemarle. Richmond: Printed by Shepherd & Pollard, 1818. 8 p.
S&S-43028.
Copies: DLC; DNAL; ViU.

1031. The Agricultural Society of New-Brunswick, N.J.
Constitution.... New Brunswick, N.J.: Printed by D. Fitz Randolph, 1818. 16 p.
S&S-43029.
Copies: DNAL; MWA.

1032. Genesee Agricultural Society.
Constitution, &c. of the Genesee Agricultural Society. [Batavia?, N.Y.]: 1818. 7 p.
S&S-44140.
Copies: N.

1033. Hillsborough Society for the Promotion of Agriculture and Domestic Manufactures.
Constitution and bye-laws of the Hillsborough Society for the Promotion of Agriculture and Domestic Manufactures. Am-

1033 (*continued*)
herst, N.H.: Printed by R. Boylston, 1818. 12 p.
S&S-44318.
Copies: DNAL.

1034. Litchfield Agricultural Society.
Articles of association, by-laws, etc., of the agricultural society, established at Litchfield, February 11, 1818. Litchfield, Conn.: 1818.
S&S-44589.
Copies: NHi (not located).

1035. Society of Virginia for Promoting Agriculture.
Laws of the Virginia Society for Promoting Agriculture. [Richmond: Printed by John Warrock, 1818.] 8 p.
Sabin-100566.
Copies: DLC.

1036. Westchester Agricultural Society.
Constitution of the Westchester Agricultural Society. Mount Pleasant, [N.Y.]: Printed by Stephen Addington, for the Society, 1818. 7 p.
S&S-46766.
Copies: NHi.

1037. Agricultural Association of the Counties of Dutchess and Columbia.
Encouragement to farmers, their wives, daughters and others. Agricultural Association of the Counties of Dutchess and Columbia. The Farmers' Club of Dutchess and Columbia offer the following premiums for the year 1819.... Premiums for 1820.... [Hudson, N.Y.: 1819.] Broadside 41 x 32.5 cm.
Signed: W. Wilson, Secretary.
N-A0098079.
Copies: MiU-C.

1038. Agricultural Society of North Carolina.
Constitution of the Agricultural Society of North-Carolina, established at Raleigh, in December, 1818.... Raleigh, N.C.: Printed by J. Gales, 1819. 13 p.
S&S-46944.
Copies: NN; NcU.

1039. Albany County Agricultural Society.
Constitution and by-laws; with the address of the president and the reports of the committees, Oct. 13, 1819; also, the names of officers. Albany: Buel, 1819. 24 p.
S&S-46958.
Copies: MB.

1040. Kings County Society for Promoting Agriculture and Domestic Manufactures.
Constitution, considered and adopted at a numerous meeting of the inhabitants of the county of Kings, in the state of New-York, held at the court-house, Aug. 17, 1819. Brooklyn: Worthington, [1819.] 8 p.
S&S-48432.
Copies: NBLiHi.

1041. Maryland Agricultural Society.
Articles of association of the Maryland Agricultural Society. Baltimore: Printed by J. Robinson, 1819. [3], 4–12 p.
Heyl-58.
Copies: MdHi.

1042. Philadelphia Society for Promoting Agriculture.
Philadelphia Society for Promoting Agriculture.... [Philadelphia: 1819.] 16 p.
Contains: An act to incorporate the Philadelphia Society for Promoting Agriculture (approved 27th March, 1819); Laws of the Philadelphia Society for Promoting Agriculture; List of members.
S&S-49083.
Copies: CSmH; DeGE.

1043. Queens County Society for the Promotion of Agriculture and Domestic Manufactures.
Constitution. New York: 1819. 8 p.
S&S-49224.
Copies: NBLiHi.

1044. Rutland County Agricultural and Manufacturing Society.

Laws and regulations of the Rutland County agricultural and Manufacturing Society.... Constitution.... [Rutland?, Vt.: Fay & Burt?, 1819.] 2 l.

McCorison-2146.

Copies: VtHi.

1045. Washington Agricultural Society.

The constitution and bye-laws of the Washington Agricultural Society: together with and address, delivered before the Society, at Argyle, February 11, 1819, by Zebulon R. Shipherd, Esq. Sandy-Hill, [N.Y.]: Printed for the Society, by E. G. Storer, 1819. 16 p.

S&S-50045; Sabin-101974.

Copies: N.

1046. Watson, Elkanah, 1758–1842.

History of the rise, progress, and existing state of the Berkshire Agricultural Society, in Massachusetts, with practical directions for societies forming in North-Carolina, on the Berkshire model.... Albany: E. &. E. Hosford, 1819. 80 p., illus.

S&S-50054.

Copies: DeGE; MWA; NN; NNBG; NcAS; Nj; NjR; WHi.

1047. Windham County Agricultural Society.

Constitution and by-laws of the Windham County Agricultural Society, as amended January 18, 1819.... Worcester, Mass.: Printed by William Manning, 1819. 8 p.

S&S-50141.

Copies: MWA.

1048. Agricultural Society of New Castle County, Del.

Memoirs of the Agricultural Society of the County of New-Castle. Vol. 1. To which is prefixed the Act of incorporation, etc., etc. Wilmington, [Del.]: Printed by R. Porter, 1820. 84 p.

S-46.

Copies: DeHi.

1049. Agricultural Society of South Carolina.

Report of the curators of the Agricultural Society of South-Carolina, at their annual meeting, held in Columbia, in December, 1819; together with a list of the officers for the ensuing year. Columbia, S.C.: Printed at the Telescope Press, 1820. 48 p.

S-3281.

Copies: CtY; DLC; ScU.

1050. Albany County Agricultural Society.

Second anniversary of the Albany County Agricultural Society. Tuesday, October 10, 1820. [Albany: 1820.] 24 p.

S-61.

Copies: DLC; MB; N; NN.

1051. Pendleton Farmers' Society.

The constitution and bye-laws of the Pendleton Farmers' Society, together with the letters and papers which have been read before the society, at its various meetings. Columbia, S.C.: Printed at the Telescope Press, 1820. 112 p.

S-2676.

Copies: NcU; ScCleU.

1052. Rockingham Agricultural Society.

An act to incorporate and establish a society by the name of the Rockingham Agricultural Society, passed June 14, 1814. Also the constitution of said society, adopted October 21, 1819. Concord, N.H.: Printed by Hill and Moore, 1820. 9 p.

S-2430.

Copies: DLC; NhD.

1053. Saratoga County Society for Promoting Agriculture and Domestic Manufactures.

The constitution and by-laws of the Saratoga County Society for Promoting Agriculture and Domestic Manufactures. Ballston Spa, [N.Y.]: Printed at the Saratoga farmer office, 1820. 6 p.

S-3108.

Copies: DNAL; MWA; N; NN.

1054. [Watson, Elkanah] 1758–1842.

History of agricultural societies, on the modern Berkshire system, from the year 1807, to the establishment of the State Board of Agriculture in Albany, January 10, 1820. Albany: D. Steele, 1820. 2 p.l., [1], [111]–210, [2]p., plates.

Usually bound with his *History of the rise, progress, and existing condition of the western canals in the state of New York.*

S-1599a.

Copies: CSmH; CtW; DLC; DeGE; ICJ; MWA; MiD; NN; NjP; PPL; RPB; WHi.

1055. Agricultural Society of the County of Columbia.

The Agricultural Society of Columbia County offer the following premiums for the year 1821.... By order of the Society, W. Wilson, Secretary, January 18, 1821. [Hudson?, N.Y.: 1821.] Broadside 46 x 27 cm.

S-4423.

Copies: MiU-C.

1056. Albany County Agricultural Society.

Third anniversary of the Albany County Agricultural Society. October, 1821. [Albany: 1821.] 23 p.

Reports of committees, p. 17–23.

S-4441.

Copies: DLC; NN.

1057. Hillsborough Society for the Promotion of Agriculture and Domestic Manufactures.

Constitution and bye-laws and act of incorporation of the Hillsborough Society for the Promotion of Agriculture and Domestic Manufactures. Amherst, N.H.: Printed by Elijah Mansur, 1821. 8 p.

S-5587.

Copies: CSmH; NjR.

1058. Society for the Promotion of Agriculture and Domestic Manufactures in and for Cumberland County.

The act of incorporation and bye-laws.... Carlisle, Pa.: 1821. 8 p.

S-5116; Sabin-85910.

Copies: DLC (not located).

1059. South Carolina Society for Promoting and Improving Agriculture.

Address and rules of the South-Carolina Society, for Promoting and Improving Agriculture and other rural concerns. Charleston: Printed for the Society by A. E. Miller, 1821. 13 p.

S-6831.

Copies: ScC; ScHi.

1060. Windham County Agricultural Society.

Charter and constitution ... as adopted at their annual meeting in January, 1821. Brooklyn, Conn.: Printed by Henry Webb, 2d, [1821?] 12 p.

S-7683.

Copies: CtHi

1061. Maryland. Laws.

An act to incorporate the Maryland Agricultural Society. [Annapolis: 1822?] 18 p. (House. Bill No. 43)

Heyl-71.

Copies: MdHi.

1062. Washington Agricultural Society.

The constitution and bye-laws of the Washington Agricultural Society; together with an address, delivered before the Society, at Argyle, February 11, 1819, by Zebulon R. Shipherd, Esq. [Salem, Mass.: 1822.] 8 p.

Sabin-101974.

Copies: N.

1063. Agricultural Society of South Carolina.

The circular address of the Importing Committee of the Agricultural Society of South-Carolina. With directions for putting up seeds, to be sent from foreign countries. [Charleston: 1823.] 14 p.

Dated: Charleston, South-Carolina, September, 1823.

S-14169.

Copies: DLC.

1064. Agricultural Society of the County of Essex, N.J.

Constitution and by-laws of the Agricultural Society of the County of Essex, State of New Jersey. Newark: Printed by Kinney and Cameron, 1823. 10 p.

S-11550.

Copies: PHi.

1065. Pennsylvania Agricultural Society.

Proceedings of the Pennsylvania Agricultural Society: with the address of their President, Jonathan Roberts, Esq., delivered at their first annual meeting. Philadelphia: Printed by Clark & Raser, 1823. 39 p.

S-13718.

Copies: DeGE; MBAt; NjR; PHi; PP; PPAmP; PPL.

1066. Agricultural Society of the Valley.

Rules and regulations adopted at the several meetings of the Agricultural Society of the Valley. Winchester, Va.: Printed by S. H. Davis, 1825. 16 p.

N-A0098558.

Copies: DNAL.

1067. Merrimack Agricultural Society.

Rules and regulations for the government of the Merrimack Agricultural Society, organized February 24, A.D. 1824. Together with the act of incorporation and list of the members. Concord, [N.H.]: Printed by I. Hill, 1825. 12 p.

S-21426.

Copies: CSmH; DNAL; MWA; NhHi; NjR.

1068. Pennsylvania Agricultural Society.

Proceedings of the Pennsylvania Agricultural Society, at their annual meeting, held at Prospect Hill, in Philadelphia County, on the 19th, 20th, and 21st October, 1825: with the address, delivered by William Darlington, M.D. Philadelphia: Clark & Raser, 1825. 24 p.

S-21837.

Copies: P; PHi; PMA; RPB; WHi.

1069. Maryland Agricultural Society.

The memorial of the Maryland Agricultural Society to the General Assembly of Maryland, January, 1826.... Annapolis: Printed by Jeremiah Hughes, 1826. 5 p.

N-M0291129; S-25240.

Copies: MWA; NcD.

1070. Agricultural Society of Hamilton County, Ohio.

Constitution of the Agricultural Society of Hamilton County, Ohio: adopted at Cincinnati, June 9th, 1827. Cincinnati: Printed at the office of the Western tiller, 1827. 8 p.

S-27791.

Copies: OCHP (not located).

1071. Consolidated Association of the Planters of Louisiana.

An act to incorporate the subscribers to the Consolidated Association of the Planters of Louisiana. New Orleans: Printed by Buisson & Boimare, 1827. 15 p.

S-28583.

Copies: LNHT.

1072. Consolidated Association of the Planters of Louisiana.

Ordonnances établissant des réglemens pour l'administration de l'Association consolidée des cultivateurs de la Louisiane. Nouvelle-Orleans: 1827. 16 p.

S-28585.

Copies: LNHT (not located).

1073. Consolidated Association of the Planters of Louisiana.

Acte pour amender l'acte intitulé. "Acte pour incorporer l'Association consolidée des cultivateurs de la Louisiane." Fev. 19, 1828. [Nouvelle-Orleans: 1828.] 4 p.

S-32817.

Copies: LNHT (not located).

1074. Cumberland County (N.J.) Agricultural Society.

Constitution of the Cumberland County Agricultural Society, to which is added selections on agricultural subjects from prac-

1074 (*continued*)
tical writers ... by a member. Philadelphia: T. S. Manning, 1828. 40 p.
N-C0830726.
Copies: DNAL.

1075. Consolidated Association of the Planters of Louisiana.
An act to incorporate the subscribers to the Consolidated Association of the Planters of Louisiana.... New Orleans: Printed by Benjamin Levy, 1829. 48 p.
S-38261.
Copies: LNHT; NN.

1076. Merrimack County Agricultural Society.
A catalogue of the members of Merrimack County Agricultural Society. Concord, N.H.: Printed by L. Roby, 1829. 16 p.
S-39564.
Copies: ICU; MH; MWA; MiD-B; Nh; NhHi.

1077. Woburn Agricultural and Mechanic Association.
Charter and by-laws of the Woburn Agricultural and Mechanic Association. Boston: J. Howe, 1830. 14 p.
C-5526.
Copies: MH.

2. General works

1078. The husband-man's guide, in four parts. Part first. Containing many excellent rules for setting and planting.... Part second. Choice physical receipts for dangerous distempers in men, women and children. Part third. The experienc'd farrier.... Part fourth. Containing rare receipts to make cordial waters.... Boston: Printed by J. Allen for E. Phillips, 1710. 107 p.
Karpinski-p. 37–38; Mp-1500.
Copies: RPJCB.

1079. ———— The 2d ed., enlarged. New York: Printed by Will & Andrew Bradford, 1712. 96 p.
B-399; Mp-39544.
Copies: RPJCB.

1080. [Smith, John] ca. 1607–1673.
The husbandman's magazine. Being a treatise of horses.... Together with plain rules for improving arable & pasture lands; improving most sort of grain.... The manner and ordering of flax and hemp.... By J. S. Boston in New-England: Reprinted by John Allen for Nicholas Boone, 1718. [4], 145, [1] p., illus.
B-551; Mp-39692.
Copies: CSmH; MWA; NN.

1081. The husband-man's guide, in four parts.... New York: Printed by William Bradford, 1727.
E-2884.
No copy known.

1082. Eliot, Jared, 1685–1763.
An essay upon field-husbandry in New-England as it is, or may be ordered.... New London: Printed and sold by T. Green, 1748. [4], iv, 25, [3] p.
Mp-6132.
Copies: MHi; MWA.

1083. Eliot, Jared, 1685–1763.
A continuation of the Essay upon field-husbandry, as it is or may be ordered in New-England.... The 2d part.... New London: Printed and sold by T. Green, 1749. [4], 34, 2 p.
Mp-6313.
Copies: MHi; MWA; NHi; PPL; RPJCB.

1084. ———— The 3d part. With an appendix by Ebenezer Silliman. New London: Printed by T. Green, 1751. [2], 32 p.
Mp-6666.
Copies: CtY; MH; MWA; NHi; PPL.

1085. ———— The 4th part.... New York: Printed and sold by J. Parker and W. Weyman, 1753. [2], 33 p.
Mp-6999.
Copies: NN; PPAmP; PPL.

1086. ———— The 5th part.... New York: Printed and sold by J. Parker and W. Weyman, 1754. 1 p.l., [3]–44 p.
E-7190; N-E0085667.
Copies: DLC; NHi; PPL.

1087. Belgrove, William.
A treatise upon husbandry, or planting. By William Belgrove, a regular bred and long experienc'd planter, of the island of Barbadoes. And may be of great use to the planters of all the West India islands. Boston, N.E.: Printed by D. Fowle, 1755. [2], 86 p.
Instructions for the management of Drax-Hall and the Irish-Hope plantations, by Henry Drax, p. [51]–86.
Mp-7356.
Copies: DLC; MHi; MWA; NN; OHi; RPJCB; ViU.

1088. Eliot, Jared, 1685–1763.
The sixth essay on field-husbandry, as it is, or may be ordered in New-England.... New Haven: Printed by J. Parker, and Company, 1759. 1 p.l., [3]–34 p.
Mp-8344.
Copies: DLC; NHi; RPJCB.

1089. Eliot, Jared, 1685–1763.
Essays upon field-husbandry in New-England, as it is or may be ordered.... Boston: Printed and sold by Edes and Gill, 1760. 3 p.l., 166 p.
"The foregoing essays were first printed in New-London and in New-York; the 1st in 1748, 2d in 1749, 3d in 1751, 4th in 1753, 5th in 1754, 6th in 1759," cf. p. [158].
Appendix dated: June 1761.
E-8847; Mp-8590.
Copies: CtHT-W; CtY; DLC; DNAL; ICJ; MBAt; MH; MHi; MWA; MeB; MiU-C; NN; NcD.

1090. ———— Six essays upon field-husbandry in New-England, in the year 1747. New ed., with an appendix by another hand. New York: Printed by J. Parker & Co., 1761.
E-8848.
No copy known.

1091. Randall, John.
The semi-Virgilian husbandry, deduced from various experiments; or, An essay towards a new course of national farming.... By Mr. Randall. Philadelphia: Printed by William Bradford, 1764.
Reprint from the London ed. of the same year.
E-9808.
No copy known.

1092. Cooke, George.
The complete English farmer; or, Husbandry made perfectly easy in all its useful branches.... London printed. Boston: Reprinted by Daniel Kneeland for Nicholas

1092 *(continued)*
Bowes, [1770?]. 142 p.
Mp-11612.
Copies: MBAt; MWA.

1093. ———— Boston: Reprinted by Daniel Kneeland for Henry Knox, [1772?]. 142 p.
B-3444.
Copies: MWA.

1094. ———— Boston: Reprinted by Daniel Kneeland for Thomas Leverett, [1772?]. 142 p.
B-3445; Mp-42334.
Copies: DLC; MnU-A.

1095. Logan, Martha, 1702–1779.
A treatise on gardening. Charleston: 1772.
Not printed at this time, according to Turnbull, *Bibliography of South Carolina,* v. I, p. 184.
E-12437.

1096. [Young, Arthur] 1741–1820.
Rural economy; or, Essays on the practical parts of husbandry.... To which is added, The rural Socrates.... The 2d ed. London: Printed; Philadelphia: Reprinted and sold by James Humphreys, junr., 1776. 245 p.
"The rural Socrates.... Written in German by Mr. Hirzel," p. [155]–245.
Mp-15226.
Copies: DLC; MWA; NN; PPL.

1097. Select essays, containing: The manner of raising and dressing flax, and hemp. Also, the whole method of bleaching or whitening linen-cloth. Likewise, observations on the management of cows and sheep. The manner of raising radishes, turnips, cabbage, and other such plants. And an enquiry, concerning the materials that may be used in making paper.... Collected from the Dictionary of arts and sciences, and from various modern authors.... Philadelphia: Printed by Robert Bell, 1777. 4 p.l., 159, [1] p., fold. plate.
Mp-15597.
Copies: DLC; MWA; NN; PPL; RPJCB.

1098. [Bordley, John Beale] 1727–1804.
A summary view of the courses of crops, in the husbandry of England & Maryland; with a comparison of their products; and a system of improved courses, proposed for farms in America. Philadelphia: Printed by Charles Cist, 1784. 22 p.
Mp-18373.
Copies: CSmH; CtY; DLC; DNAL; ICJ; MH-BA; MWA; MeB; MiU-C; NN; NcD; OClWHi; PPAmP; RPJCB; ViU.

1099. Philadelphia Society for Promoting Agriculture.
An address from the Philadelphia Society for Promoting Agriculture, with a summary of its laws, and premiums offered.... [Philadelphia]: 1785. 16 p.
Mp-19193.
Copies: DNLA: DeGE; MH; MWA (20 p.).

1100. Varlo, Charles, 1725?–1795?
A new system of husbandry. From many years experience, with tables shewing the expence and profit of each crop.... Of trench ploughing, shewing how to raise good crops without manure.... A newly invented thrashing floor.... By C. Varlo.... Philadelphia: Printed for the author, 1785. 2 v., front. (fold. table).
First published at York, England, 1770.
Mp-19338.
Copies: DNAL; DeGE; MBAt; MH; MWA; MiU-C; NN; PPL.

1101. Bowler, Metcalf, 1762–1789.
A treatise on agriculture and practical husbandry. Designed for the information of landowners and farmers. With a brief account of the advantages arising from the new method of culture practised in Europe.... Providence: Printed by Bennett Wheeler, for the author, 1786. [88] p.
Mp-19522.
Copies: DLC; DNAL; MHi; MWA; OClWHi; PU-V; RPJCB.

1102. [Thomson, Charles] 1729–1824.
Notes on farming. New York: Printed in the year 1787. 38 p.
Mp-20599.
Copies: CtY; MBAt; MWA; NHi.

1103. South Carolina Society for Promoting and Improving Agriculture.
Letters and observations on agriculture &c. Addressed to, or made by the South-Carolina Society for Promoting Agriculture, and other rural concerns. [Charleston]: Printed by Bowen & Co., 1788. 32 p.
Mp-21471.
Copies: MBAt; PHi.

1104. [Coxe, Tench] 1755–1824.
Observations on the agriculture, manufactures and commerce of the United States. In a letter to a member of Congress. By a citizen of the United States. New York: Printed by Francis Childs and John Swaine, 1789. 102 p.
Mp-21774.
Copies: CtY; DLC; ICU; MH-BA; MWA; MiU-C; PPAmP; RPJCB.

1105. Deane, Samuel, 1733–1814.
The New-England farmer; or, Georgical dictionary: containing a compendious account of the ways and methods in which the most important art of husbandry, in all its various branches, is, or may be, practised to the greatest advantage in this country.... Worcester, Mass.: Printed by I. Thomas, 1790. viii, 335 p.
Mp-22450.
Copies: DLC; DNAL; DeGE; ICJ; InU; MBAt; MH; MWA; MiU-C; NN; OHi; PU; RPJCB.

1106. Poellnitz, Friedrich Carl Hans Bruno, Baron von, 1734–1801.
Essay on agriculture. By F. A. B. Poellnitz. New York: Printed by Francis Childs and John Swaine, 1790. 24 p., 4 fold. plates.
Mp-22805.
Copies: MBAt; MWA; NNS.

1107. Logan, George, 1753–1821.
Fourteen experiments on agriculture (agreeably to the English mode of farming), the result of seven years close and attentive observation.... Philadelphia: Printed and sold by Eleazer Oswald, 1791.
E-23506.
No copy known.

1108. [Bordley, John Beale] 1727–1804.
Sketches on rotations of crops. Philadelphia: Printed by Charles Cist, 1792. 1 p. l., 47 p.
Signed: B.
Mp-24129.
Copies: DLC; DeGE; IObB; MBAt; MHi; MdBE; MiU; NNC; PPAmP; PPL.

1109. ——— [S.l.]: Printed for J. Cowan, in Talboton, 1792. 1 p.l., 47 p.
B-7935; Mp-46390.
Copies: MdHi.

1110. [Young, Arthur] 1741–1820.
Rural economy; or, Essays on the practical parts of husbandry.... To which is

1110 (*continued*)
added The rural Socrates, being memoirs of a country philosopher. By the author of the Farmer's letters. 3d ed. Burlington, [N.J.]: Printed by I. Neale, 1792. 299, [1] p.

The rural Socrates (p. [191]-299) was originally written in German by H. K. Hirzel.

Mp-25061.
Copies: DeGE; MBAt; MWA.

1111. Coxe, Tench, 1755–1824.
A plan for encouraging agriculture, and increasing the value of farms in the midland and more western counties of Pennsylvania, applicable to several other parts of that state and to many parts of the United States. [Philadelphia: 1793.] 11 p.

Mp-25355.
Copies: DLC; MWA; NN; RPJCB.

1112. Spurrier, John.
The practical farmer: being a new and compendious system of husbandry, adapted to the different soils and climates of America. Containing the mechanical, chemical and philosophical elements of agriculture. With many other useful and interesting subjects.... Wilmington, [Del.]: Printed by Brynberg and Andrews, 1793. [2], x, iii–x, [11]–360, xv p.

Mp-26198.
Copies: DLC; DNAL; DeGE; DeHi; DeWI; DeWin; MBAt; MWA; NN; PPL; PU; RPJCB.

1113. Young, Arthur, 1741–1820.
Notes on farming, collected from Young's Farmer's tour through England, with additions and observations by American farmers on various subjects, more particuarly adapted to their own soil. Catskill, [N.Y.]: Printed by M. Croswell, 1794. 3 p.l., [13]-110 p.

N-N0333908.
Copies: MH-BA; N.

1114. ———— Hudson: Printed by Ashbel Stoddard, 1794.

E-28144.
No copy known.

1115. [Webster, Noah] 1758–1843.
The farmer's catechism. Containing plain rules of husbandry.... Canaan, [N.Y.]: Printed and sold by Elihu Phinney, 1795. 8 p.

Mp-29855.
Copies: CtY; MWA.

1116. The American farmer's guide; or, A new and excellent treatise on agriculture. Wherein the planter and farmer will see such a judicious method for the management of stockraisiing [*sic*], manure-enriching and improving his lands that they shall produce the most plentiful crops.... Philadelphia: Reprinted for the Rev. Mason L. Weems, by Jacob Johnson & Co., [1796?]. 83, [2] p.

Dated by Evans: 1795.
Mp-31718.
Copies: DLC; DNAL; MWA; PHi; PPAmP; RPB; RPJCB.

1117. [Bordley, John Beale] 1727–1804.
Sketches on rotations of crops, and other matters. To which are annexed, Intimations on manufactures; on the fruits of agriculture; and on new sources of trade, interfering with products of the United States of America in foreign markets. Philadelphia: Printed by Charles Cist, 1796. [2], 76 p., illus.

Signed: J.B.B.
Mp-30103.
Copies: ICN; MWA.

1118. [Dabney, John] 1752–1819.
An address to farmers.... To which is added an appendix, containing the most approved methods for the management and improvement of tillage, mowing and pasture lands; and for the practice of the art of husbandry in general. (Extracted principally

from a variety of authors, who have written judiciously on these important subjects.).... Newburyport, [Mass.]: Printed by Blunt and March, for John Dabney, Salem, 1796. [2], 64 p.

Mp-30307.

Copies: CU; DLC; DNAL; DeGE; MH; MSaE; MWA; MiU-C; NHi; NN; RPJCB.

1119. [Bordley, John Beale] 1727–1804.

Queries selected from a paper of the Board of Agriculture in London, on the nature and principles of vegetation: With answers and observations. By J. B. B. [Philadelphia: Printed by Charles Cist, 1797.] [19] p.

Mp-31845.

Copies: DLC; DNAL; MBAt; MH; PPL.

1120. [Bordley, John Beale] 1727–1804.

Sketches on rotations of crops, and other rural matters. To which are annexed, Intimations on manufactures; on the fruits of agriculture; and on new sources of trade, interfering with products of the United States of America in foreign markets. Philadelphia: Printed by C. Cist, 1797. 1 p.l., 76 p., illus.

Signed: J.B.B.

Contains a printed sheet on ice houses, copyrighted by Charles Cist, 20 Jan., 1797.

Mp-31846.

Copies: CtY; DLC; DNAL; IU; MBAt; MWA; MdBP; NNC; NcAS; PPL; PU; RPJCB; ScC.

1121. Deane, Samuel, 1733–1814.

The Newengland farmer; or, Georgical dictionary. Containing a compendious account of the ways and methods in which the important art of husbandry, in all its various branches, is, or may be, practised, to the greatest advantage in this country.... 2d ed., corr., improved, and enl. by the author.... Worcester, Mass.: Printed at the press of Isaiah Thomas by Leonard Worcester, for Isaiah Thomas, 1797. viii, 396 [397] p.

Mp-32020.

Copies: DNAL; DeGE; MWA; MiU; MnU; NHi; NN; NjP; PU; RPJCB; ViU.

1122. Logan, George, 1753–1821.

Fourteen agricultural experiments, to ascertain the best rotation of crops; addressed to the "Philadelphia Agricultural Society...." Philadelphia: Printed by Francis and Robert Bailey, 1797. 41 p.

Mp-32382.

Copies: DLC; DNAL; MBAt; MWA; MiU-C; NN; PPL.

1123. Essays on the spirit of legislation, in the encouragement of agriculture, population, manufactures, and commerce.... Interspersed with various remarks in the practice of agriculture. Societies of agriculture, rewards.... Industry. Machines.... Translated from the original French, which gained the premiums offered by the Oeconomical Society of Berne, in Switzerland.... Newark, [N.J.]: Printed by Pennington & Dodge, for William Reid, 1798. [2], [7]–479, vii, [5] p.

By Jean Bertrand, Benjamin Carrard, and Gabriel Seigneux de Correvon, originally published in the *Mémoires* of the Ökonomische Gesellschaft des Kantons Bern. This ed. is reprinted from the London ed. of 1792.

Mp-33407.

Copies: DLC; KyU; MWA; NN.

1124. [Lambert, John.]

A short and practical essay on farming, being the experience of a farmer of about sixty years of age, near forty years of which were spent in England, Essex County, on land where farming is done in the greatest perfection and near seven years on three hundred and twenty acres of worn out land in Patty Grove and Alloway Creek in Salem County, West Jersey. Shewing the means whereby these worn out lands may be improved.... [Philadelphia?: 1798.] 8 p.

1124 *(continued)*
Mp-33974.
Copies: PPL.

1125. Bordley, John Beale, 1727–1804.
Essays and notes on husbandry and rural affairs. . . . Philadelphia: Printed by Budd and Bartram, for Thomas Dobson, 1799. viii, 591, 7 p., 4 plates.
Mp-35216.
Copies: CU; CtY; DLC; DeGE; IaAS; MH; MWA; MiU-C; NN; NNS; PPHor; PPL; RPJCB; ScC.

1126. Essays on the spirit of legislation, in the encouragement of agriculture, population, manufactures, and commerce. . . . Interspersed with various remarks on the practice of agriculture. Societies of agriculture. Rewards. . . . Industry. Machines. . . . Translated from the original French, which gained the premiums offered by the Oeconomical Society of Berne, in Switzerland. . . . Newark, [N.J.]: Printed by Pennington & Dodge, for William Reid, 1799. 479, vii, [6] p.
Essays by Jean Bertrand, Benjamin Carrard, and Gabriel Seigneux de Correvon.
Mp-35186.
Copies: AU; DLC; MWA; NjP.

1127. Massachusetts Society for Promoting Agriculture.
Papers on agriculture, consisting of communications made to the Massachusetts Society for Promoting Agriculture, with extracts from various publications; recommended to the attention of farmers, by the trustees of the Society. Boston: Printed by Young and Minns, 1799. 103 p., plate.
Mp-35802.
Copies: DLC; DeGE; MH; MWA; NN; RPJCB.

1128. Parkinson, Richard, 1748–1815.
The experienced farmer, an entire new work, in which the whole system of agriculture, husbandry, and breeding of cattle, is explained . . . and the best methods, with the most recent improvements, pointed out. . . . Philadelphia: Printed by Charles Cist, 1799. 2 v.
The text was printed in England, the title pages in Philadelphia.
Mp-36041.
Copies: DLC; DNAL; MB; MH; MWA; MdBE; NN; NjR; PPL; PPiU; RPJCB; ViW.

1129. Essays on the spirit of legislation, in the encouragement of agriculture, population, manufactures, and commerce. Translated from the original French, which gained the premiums offered by the Oeconomical Society of Berne, in Switzerland. . . . Newark, [N.J.]: Printed for William Reid, by Pennington & Gould, 1800. 479, [1], vii p.
Essays by Jean Bertrand, Benjamin Carrard, and Gabriel Seigneux de Correvon.
Mp-37382.
Copies: CSmH; DLC; DeGE; MB; MH; MWA; NjP; PHi; PU; RPJCB.

1130. Foster, Edmund, 1752–1826.
Husbandry, an ancient, honourable and useful employment. An oration, delivered before the Western Society of Middlesex Husbandmen, at their semiannual meeting at Littleton, on Monday, October 28, 1799. Amherst, N.H.: Printed by Samuel Preston, 1800. 15 p.
Mp-37436.
Copies: DLC; MBH; MHi; MWA; MeB; NhD; RPJCB.

1131. [Hirzel, Hans Kaspar] 1725–1803.
The rural Socrates; or, An account of a celebrated philosophical farmer, lately living in Switzerland, and known by the name of Kliyogg. . . . Hallowell, Me.: Printed by Peter Edes, 1800. xii, [2], 203, xiii p.
Probably edited by Benjamin Vaughan.

Mp-38923.
Copies: DLC; DeGE; InU; MBAt; MH; MWA; MeB; MeHi; MiU-C; NN; NjP; PPA; PPAmP; PPL; PU.

1132. Logan, George, 1753–1821.
A letter to the citizens of Pennsylvania, on the necessity of promoting agriculture, manufactures, and the useful arts.... Lancaster, [Pa.]: Printed by W. & R. Dickson, 1800. 28 p.
Contains the constitution of the Lancaster County Society for Promoting Agriculture, Manufactures, and the Useful Arts.
Mp-37840.
Copies: DLC; MWA; NN; PPL.

1133. ———— 2d ed. Philadelphia: Printed by Patterson & Cochran, 1800. 30 p.
Mp-37841.
Copies: DLC; DNAL; DeGE; MH; MWA; NHi; NN; PPAmP; PPL; RPJCB.

1134. Massachusetts Society for Promoting Agriculture.
Inquiries by the Agricultural Society. [Boston: Printed by Young & Minns, 1800.] 29 p.
Contains 50 inquiries.
Mp-37935.
Copies: CtY; DLC; DNAL; DeGE; MB; MH; MHi; MWA; N; NBu; NN; NjR; PPAmP; RPJCB.

1135. Bordley, John Beale, 1727–1804.
Essays and notes on husbandry and rural affairs. By J. B. Bordley.... 2d ed. with additions. Philadelphia: Printed by Budd and Bartram, for Thomas Dobson, 1801. viii, 536, 8, [22] p., plates.
S&S-210.
Copies: DLC; DNAL; DeGE; GU; ICJ; KEmT; MH; MWA; NN; NcD; OU; PPL; PU; ViW.

1136. Massachusetts Society for Promoting Agriculture.
Papers on agriculture, consisting of communications made to the Massachusetts Agricultural Society, with extracts from various publications; by the trustees of the Society. Boston: Printed by Young & Minns, 1801. 93, [2] p.
S&S-906, 911.
Copies: DeGE; ICU; MB; MH; MWA.

1137. Moore, Thomas, 1760–1822.
The great error of American agriculture exposed; and hints for improvement suggested.... Baltimore: The author; [Printed by Bonsal and Niles], 1801. iv, [5]–72 p.
S&S-951.
Copies: CSmH; DLC; DNAL; MH; MWA; NjP; PHi; PPAmP; Vi.

1138. Parkinson, Richard, 1748–1815.
Of turnip & pea fallows, with a design of a rotation of crops, recommended to the farmers and planters of the United States of America.... Washington City: Printed by C. Cist, 1801. 2 p.l., 16, iii, ii, [277]–284, [293]–296 p.
A supplement to his *The Experienced farmer.*
S&S-1093.
Copies: DLC; DNAL.

1139. Robbins, Asher, 1757–1845.
An address to the Society for the Promotion of Agriculture, and Other Useful Arts, in the state of Rhode-Island ... on the second day of September, A.D. 1802.... Newport, R.I.: Printed at the office of the Newport Mercury, [1802]. 16 p.
S&S-3000.
Copies: DLC; DNAL; MWA; RNHi; RP.

1140. Society for Promoting Agriculture in the State of Connecticut.
Transactions of the Society for Promoting Agriculture in the State of Connecticut.... New Haven: Printed by William W. Morse, 1802. 21, [1] p.
Prepared for publication by Charles Chauncey.

1140 (*continued*)
S&S-2081; ULS-p. 3978.
Copies: Ct; DLC; DNAL; DeGE; MH; MHi; MWA; NN; NjPT; NjR; OClWHi.

1141. Binns, John Alexander, 1761(ca.)–1813.
A treatise of practical farming; embracing particularly the following subjects, viz. The use of plaister of Paris ... and general observations on the use of other manures. On deep ploughing; thick sowing of grain; method of preventing fruit trees from decaying, and farming in general.... Fredericktown, Md.: Printed by John B. Colvin, 1803. 2 p.l., 72 p.
S&S-3834.
Copies: DLC; DNAL; DeGE; MH; MWA; NN; NcD; NjP; PPAmP; PPL; Tx; ViU.

1142. Gleanings from the most celebrated books on husbandry, gardening, the rural affairs.... From the London 2d edition of 1803. Interspersed with remarks and observations by a gentleman of Philadelphia. Philadelphia: Printed and sold by James Humphreys, 1803. viii, 365 p., fold. plate.
"A list of the books out of which, principally, were selected materials for the following work," p. [iv].
Notes attributed to J. B. Bordley.
S&S-4284.
Copies: CSmH; DLC; DNAL; DeGE; ICJ; MH; MWA; MiU-C; NjR; PPL; PU; ViU.

1143. Massachusetts Society for Promoting Agriculture.
Papers on agriculture, consisting of communications made to the Massachusetts Society for Promoting Agriculture. Published by order of the trustees. Boston: Printed by Young and Minns, 1803. 93, [2] p.
S&S-4617.
Copies: DeGE; MH; MWA; NN.

1144. Washington, George, 1732–1799.
Letters from His Excellency George Washington ... to Arthur Young ... and Sir John Sinclair ... containing an account of his husbandry, with his opinions on various questions in agriculture; and many particulars of the rural economy of the United States. Alexandria, [Va.]: Printed by Cottom and Stewart, 1803. 128 p.
Letters dated from August 5, 1786 to November 6, 1797.
S&S-5536.
Copies: CSmH; DLC; DNAL; MH; MWA; NN; ViHi.

1145. Binns, John Alexander, 1761(ca.)–1813.
A treatise on practical farming: embracing particularly the following subjects, viz. the use of plaister of Paris, with directions for using it: and general observations on the use of other manures. On deep ploughing; thick sowing of grain; method of preventing fruit trees from decaying; and farming in general. By John A. Binns.... 2d ed. Richmond: Printed by S. Pleasants, jun., 1804. 83 p.
S&S-5878.
Copies: CSmH; DNAL; Vi; ViU.

1146. Bullard, John.
A discourse on agriculture, delivered before the Western Society of Middlesex Husbandmen, at their first meeting ... at Littleton, October 19, 1803.... Cambridge, [Mass.]: Printed by W. Hilliard, 1804. 15 p.
S&S-5926.
Copies: DLC; MH; MHi; MWA; PMA.

1147. Massachusetts Society for Promoting Agriculture.
Papers on agriculture; consisting of communications made to the Massachusetts Society for Promoting Agriculture, with extracts from various publications. By the trustees of the Society. Boston: Printed by Young & Minns, 1804. 61, [1] p.
S&S-6753.
Copies: DeGE; MH; MWA; MeB; NN; OrCS.

1148. Roberts, Job, 1757–1851.

The Pennsylvania farmer; being a selection from the most approved treatises on husbandry, interspersed with observations and experiments.... Philadelphia: Jacob Johnson & Co., [etc.], 1804. v, [3], [5]–224 p.

S&S-7203.

Copies: DLC; DNAL; DeGE; MWA; NN; RPB.

1149. Brodin de la Jutais, Pierre, d. 1765.

L'abondance; ou, La véritable pierre philosophale, consistant en un moyen de multiplier abondamment les grains, les fruits, les fleurs, et tous les végétaux généralement; ouvrage publié à Paris en 1752, par le chevalier Brodin de la Jutais ... corrigé et augmenté de principes physiques par son gendre, Faming de la Jutais ... dedié à Son Excellence John [*sic*] Jefferson, président des Etats-Unis de l'Amérique. Philadelphie: 1805. 47 p.

Recommends the use of nitrogen fertilizers.

S&S-8084.

Copies: DLC; NIC; PPAmP.

1150. Massachusetts Society for Promoting Agriculture.

Inquiries by the Massachusetts Society for Promoting Agriculture. [Boston: 1805?] 31 p.

Contains 66 questions.

S&S-8875.

Copies: DNAL (dated: 1801?); MWA; NN.

1151. Massachusetts Society for Promoting Agriculture.

Papers on agriculture; consisting of communications made to the Massachusetts Society for Promoting Agriculture.... Boston: Printed by Young & Minns, 1804 [1805]. 111 p.

S&S-6752, 8876.

Copies: DeGE; MH; MHa; MWA.

1152. Wilder, John, 1758–1836.

An address, to the members of the Attleborough Society for the Encouragement of Agriculture, Arts, and Social Intercourse, delivered on Friday, 22d of February, 1805.... Providence: Printed by Heaton & Williams, [1805]. 12 p.

S&S-9730.

Copies: CSmH; CtY; MWA; NHi.

1153. Massachusetts Society for Promoting Agriculture.

Papers; consisting of communications made to the Massachusetts Society for Promoting Agriculture.... Boston: Printed for Young & Minns by Greenough & Stebbins, 1806. 90 p.

S&S-10833.

Copies: DeGE; MH; MWA; NN.

1154. Masschusetts Society for Promoting Agriculture.

Inquiries by the Massachusetts Society for Promoting Agriculture. [Boston: Adams and Rhoades?, 1807?] 31 p.

S&S-13038.

Copies: MH.

1155. Massachusetts Society for Promoting Agriculture.

Papers; consisting of communications made to the Massachusetts Society for Promoting Agriculture, and extracts. Published by the trustees of the Society. Boston: Printed by Adams & Rhoades, 1807. 86 p.

S&S-13037.

Copies: DLC; DeGE; MH; MHa; MHi; MNBedf; MWA; MWfo; MeB.

1156. Custis, George Washington Parke, 1781–1857.

An address to the people of the United States, on the importance of encouraging agriculture and domestic manufactures.... Together with an account of the improvements in sheep at Arlington, the native sheep of Smith's Island, and the plans proposed of extending this valuable race of animals, for the benefit of the country at large. By George W. P. Custis.... Alexandria, [Va.]: Printed by S. Snowden, 1808. 43 p., 1 l.

S&S-14815.

Copies: CtHT-W; DLC; DNAL; MBAt; PP; ViU.

1157. An essay on the climate of the United States: or, an Inquiry into the causes of the difference in climate between the Eastern side of the continent of North America and Europe, with practical remarks on the influence of climate on agriculture, and particularly the cultivation of the vine. Philadelphia: Hopkins and Earle; Fry and Kammerer, printers, 1809. 42 p.

S&S-17462.

Copies: DLC; DNLM; ICRL; MWA; MdBP; NNN; PHi; PPL.

1158. Massachusetts Society for Promoting Agriculture.

Georgick papers for 1809, consisting of letters and extracts communicated to the Massachusetts Society for Promoting Agriculture. Published by the trustees.... Boston: Printed by Russell and Cutler, 1809. 91 p., plate.

S&S-18054.

Copies: DLC; DeGE; ICU; MWA; MeB; NN.

1159. [Hirte, Tobias.]

Die americanische Goldgrube; das ist, die Kunst des Landmanns, seine zeitlichen Güter fünfzig—bis hundertfältig zu vermehren. Vorgestelle in nützlichen, patriotischen Anweisungen, den Ackerbau, den Weinbau, die Viehzucht, und ganz besonders die einträgliche Schaafzucht zu veredeln.... Libanon, Pa.: Gedruckt auf Kosten des Verfassers, bei Heinr. Sage, 1810. viii, 192 p.

S&S-19359.

Copies: DLC; MWA; NN; OO; PPG; PPL; PPeSchw.

1160. Massachusetts Society for Promoting Agriculture.

Papers for 1810, communicated to the Massachusetts Society for Promoting Agriculture. Published by the trustees.... Boston: Printed by Russell and Cutler, 1810. 112 p.

S&S-20697.

Copies: CSt; DLC; DeGE; MH; MHi; MWA; NN.

1161. Massachusetts Society for Promoting Agriculture.

Papers for 1811, communicated to the Massachusetts Society for Promoting Agriculture. Published by the trustees.... Boston: Printed by Munroe & French, 1811. 114 p.

Includes a new edition of Jared Eliot's *Essays on field-husbandry*, and Samuel Tenney's *Advice relative to the business of making an orchard.*

S&S-23349.

Copies: DLC; DNAL; DeGE; MHi; MWA; MWeyAA; NN.

1162. Watson, Elkanah, 1758–1842.

Mr. Watson's address, delivered to the members of the Berkshire Agricultural Society at the Town-House in Pittsfield, September 24, 1811.... [Pittsfield, Mass.: Printed by P. Allen, 1811.] 11 p.

S&S-24383; Sabin-102134.

Copies: MBAt; N; NHi; NN.

1163. [Taylor, John] 1753–1824.

Arator; being a series of agricultural essays, practical & political: in sixty one numbers. By a citizen of Virginia. Georgetown, Columbia: Printed and published by J. M. & J. B. Carter, 1813. 296 p.

S&S-29920.

Copies: CSmH; DLC; DNAL; DeGE; MB; MH; MWA; MdBS; MnU; PWW; ScU; ViU.

1164. Vose, John, 1766–1840.

An oration, pronounced at Londonderry, before the Rockingham Agricultural Society ... June 7, 1813.... Concord, [N.H.]: Printed by George Hough, 1813. 15 p.

S&S-30434.

Copies: DLC; MWA; NhHi.

1165. Adams, Rufus W.

The farmer's assistant, containing a correct description of the best methods of raising & keeping cows, & making butter and cheese; also, useful and interesting observations on the best methods of raising common and merino sheep.... Marietta, Ohio: Printed by S. Fairlamp, 1814. 48 p.

S&S-30631.
Copies: DLC.

1166. Nicholson, John, 1765–1820.
The farmer's assistant.... The subjects on which this work treats, are numerous, embracing every article relating to agriculture, arranged in alphabetical order. Albany: H. C. Southwick, 1814. vii, [2], 10–324 [338] p., diagrs.
S&S-32356.
Copies: DNAL; DeU; GS; MB; MDeeP; MWA; MiD; N; NIC; NN; NRU; PHi; PU.

1167. Taylor, John, 1753–1824.
Arator; being a series of agricultural essays, practical & political: in sixty-one numbers. 2d ed., rev. and enl.... Georgetown, Columbia: Printed and published by J. M. Carter, 1814. 2 p.l., [9]–279, [1] p.
S&S-32909.
Copies: CLU; DLC; DNAL; FTaSU; IU; LNHT; MB; MWA; NNC; NcRA; NjR; OC; PLF; ScU; Vi.

1168. Watson, Elkanah, 1758–1842.
Address of Elkanah Watson, Esq. delivered before the Berkshire Agricultural Society.... 7th October, 1814 ... together with the premiums awarded.... Pittsfield, [Mass.]: Printed by Phinehas Allen, 1814. [3], 16 p.
S&S-33594.
Copies: DNAL; MBAt; MWA; N; PHi.

1169. Brownell, Thomas Church, 1779–1865.
Annual address, on the theory of agriculture: delivered by appointment, before the Society for the Promotion of Useful Arts, at the Capitol, in the city of Albany, on the 14th February, 1815. By Thomas C. Brownell.... Albany: Printed by Websters and Skinners, for the Society for the Promotion of Useful Arts, 1815. 43 p.
S&S-34237.
Copies: CtHT; CtY; DNAL; DeGE; ICU; MWA; MiD-B; N; NNNAM; PPL; PU-S; RPB.

1170. ——— Albany: Printed by Websters and Skinners, 1815. 51 p.
Copies: DNAL.

1171. Melvill, Thomas, d. 1846.
Address of Thomas Melvill, jun.... delivered before the Berkshire Society for the Promotion of Agriculture and Manufactures, at Pittsfield, October 3, 1815, together with the premiums awarded on said day. Pittsfield, [Mass.]: Printed by P. Allen, 1815. 31 p.
S&S-35251.
Copies: DNAL; MBAt; MHi; MWA; NN; RPB.

1172. Nicholson, John, 1765–1820.
The farmer's assistant; being a complete treatise on agriculture in general.... Arranged in alphabetical order ... Albany: H. C. Southwick, and T. C. Fay; H. C. Southwick, printer, 1815. x, [iii]–vii, [2], 10–327, [1] p. illus.
S&S-35497.
Copies: MWA; MiU; NBu; NGH; NRHi; PU; VtMiS.

1173. Briggs, Isaac.
Statements and remarks, addressed to Thomas Newton, Chairman of the Committee of Commerce and Manufactures, on the subject of agriculture, manufactures, and commerce. Washington: Printed by W. A. Davis, 1816. 18 p.
S&S-37087.
Copies: DLC; DeGE; MB; MH-BA; MHi; MWA; MdHi; MnU; NN; NjR; PPi.

1174. Gold, Thomas, 1759–1827.
Address of Thomas Gold ... delivered before the Berkshire Association, for the Promotion of Agriculture and Manufactures, at Pittsfield, Oct. 3d, 1816.... Together with the report of the Committee on Agriculture.... Pittsfield, [Mass.]: Printed by Phinehas Allen, [1816]. 22 p.
S&S-37717.
Copies: CtY; ICN; MHi; MWA; N; NN.

1175. Humphreys, David, 1752–1818.

A discourse on the agriculture of the state of Connecticut, and the means of making it more beneficial to the state: delivered at New-Haven, on Thursday, 12th September, 1816, by David Humphreys.... New Haven: Printed by T. G. Woodward, 1816. 42 [44] p.

S&S-37892.

Copies: Ct; CtHC; CtHT; CtY; DLC; DeGE; MBAt; MBH; MHi; MWA; NN.

1176. Peters, Richard, 1744–1828.

A discourse on agriculture. Its antiquity; and importance, to every member of the community: the obligations all are under to encourage it; and the necessity of diffusing a knowledge of its principles, and systematical practice, as the first step towards its advancement. Delivered before the Philadelphia Society for Promoting Agriculture ... on their annual meeting, the 9th of January, 1816.... Philadelphia: Johnson and Warner, 1816. [5], 6–40 p.

Pages [39]–40 contain "the eleventh article of the Society."

S&S-38603.

Copies: DNAL; MWA; NNBG; PHi; PPL.

1177. ———— Philadelphia: Johnson and Warner, 1816. 37 p.

"Errata" slip mounted on p. 37.

Copies: DeGE.

1178. Gold, Thomas, 1759–1827.

Address of Thomas Gold, esq., president of the Berkshire Agricultural Society ... delivered before the Berkshire Association for the Promotion of Agriculture and Manufactures, at Pittsfield, October 2d, 1817.... Pittsfield, [Mass.]: Printed by Phinehas Allen, [1817]. 32 p.

S&S-40927.

Copies: DLC; DNAL; MBAt; MBH; MH; MPB; MWA; NHi; NN; NNBG.

1179. Mease, James, 1771–1846.

An address on the progress of agriculture, with hints for its improvement in the United States. Delivered before the Philadelphia Society for Promoting Agriculture, at their annual meeting, January 14th, 1817.... [Philadelphia]: Published by order of the Society, 1817. 45 p.

S&S-41392.

Copies: DNAL; DeGE; MB; MWA.

1180. Morris, Jacob, 1755–1844.

Address of Gen. Jacob Morris, president of the Otsego County Agricultural Society, delivered before said association established for the promotion of agriculture and manufactures, at Cooperstown, Oct. 14, 1817.... Together with the address of Elkanah Watson, Esq., previous to declaring the premiums reported. Cooperstown, [N.Y.]: Printed by H. and E. Phinney, 1817. 16 p.

S&S-41491.

Copies: DLC; MB; MH; MWA; MiD-B; N.

1181. Swan, James.

An address to the Senate and House of Representatives of the United States, on the question for an inquiry into the state of agriculture, manufactures and commerce. Boston: Printed by John Eliot, 1817. 24 p.

S&S-42237.

Copies: DLC; MHi; NcD.

1182. Taylor, John, 1753–1824.

Arator; being a series of agricultural essays, practical and political: in sixty-one numbers.... 3d ed., rev. and enl. Baltimore: Printed by J. Robinson, for John M. Carter, 1817. iv, [5]–220 p.

S&S-42250.

Copies: CSmH; DLC; DNAL; ICU; IaAS; MH; MWA; MdBE; NIC; NcU; PPL; PU; TxU; ViU.

1183. Agricultural Society of South Carolina.

Papers published by order of the Agricultural Society of South Carolina. Columbia, S. C.: Printed at the Telescope Press,

1818. 24 p.
S&S-45770.
Copies: N.

1184. Ellsworth, Henry Leavitt, 1791–1858.
Sketches of an address delivered before the Hartford County Agricultural Society on the twenty-fourth of March, 1818. By Henry L. Ellsworth.... Hartford: G. Goodwin & Sons, printers, 1818. 23 p.
S&S-43935.
Copies: CSmH; CtHi; CtY; DLC; DNAL; MBH; MWA; NAuT; NHi; NjPT.

1185. Garnett, James Mercer, 1770–1843.
An address delivered before the Virginia Agricultural Society, of Fredericksburg.... October 28, 1818.... [Fredericksburg, Va.]: Printed by Green & Harrow, [1818]. 15 p.
S&S-44128.
Copies: CSmH; ViHi.

1186. Hammond, Abijah.
An address delivered before the Agricultural Society of the County of Westchester, N.Y., in the Court-House, in the town of White-Plains, on the 4th of November, 1818 —together with the proceedings of the Society and the premiums offered for the ensuing year. Mount Pleasant, [N.Y.]: Stephen Addington, 1818. 13 p.
N-H0080528.
Copies: MHi; N.

1187. [Jeffreys, George W.]
Agricultural letters. [Raleigh?: 1818.] 113 p., illus.
Includes descriptions and illustrations of agricultural implements.
N-J0074011.
Copies: DLC.

1188. Le Ray de Chaumont, James Donatien, 1760–1840.
An address, delivered at the meeting at the Agricultural Society of Jefferson County, December 29, 1817. By J. Le Ray de Chaumont.... To which is subjoined, the articles of association and the by-laws of the society.... New York: Printed by William A. Mercein, 1818. 20 p.
S&S-43592.
Copies: DLC; MH; N; NN; PHi.

1189. ———— Watertown, N.Y.: Printed by John H. Lord, Jr., 1818. 16 p.
S&S-43593.
Copies: MB; NN.

1190. Logan, George, 1753–1821.
An address on the errors of husbandry in the United States. Delivered before the Philadelphia Society for Promoting Agriculture, at their annual meeting, January 14, 1818.... Philadelphia: Printed by Lydia R. Bailey, 1818. 11 p.
S&S-44608.
Copies: DLC; MWA; NNC.

1191. Lowell, John, 1769–1840.
An address delivered before the Massachusetts Agricultural Society at the Brighton cattle show, October 13, 1818. By J. Lowell.... Boston: Printed by Wells and Lilly, 1818. 22, [2] p.
S&S-44625.
Copies: CSt; DLC; DNAL; ICU; MB; MWA; MiD-B; NIC; NN; OCHP; PPL; WU-A.

1192. Madison, James, 1751–1836.
An address delivered before the Agricultural Society of Albemarle, on Tuesday, May 12, 1818.... Richmond: Printed by Shepherd & Pollard, 1818. 31 p.
S&S-44668.
Copies: DLC; MB; MWA; MdHi; N; PPAmP; PPL; ViU.

1193. Mease, James, 1771–1846.
Address on the subject of establishing a pattern farm in the vicinity of Philadelphia: delivered at the annual meeting of the "Philadelphia Society for Promoting Agriculture.... " Philadelphia: Printed by order of

1193 (*continued*)
the Society, 1818. 14 p.
S&S-44778.
Copies: DeGE; MB; MWA; NjR; PPAmP.

1194. Mertz, John Nicholas.
The golden rules of agriculture; containing an explanation of the true causes of the Hessian fly.... Baltimore: Printed by J. Robinson, for F. Lucas, 1818. 40 p.
S&S-44820.
Copies: MdHi.

1195. Philadelphia Society for Promoting Agriculture.
Address to the citizens of Pennsylvania on the importance of a more liberal encouragement of agriculture. Accompanied with inquiries on agricultural subjects proposed by the Philadelphia Society for Promoting Agriculture, with a view to form an expose of the state of agriculture in Pennsylvania. Philadelphia: Printed by order of the Society, 1818. 14 p.
S&S-45299.
Copies: AzU; CtU; DNAL; MBH; MH; MWA; NcRA; NjR; Ok; PU.

1196. Pickering, Timothy, 1745–1829.
Address to the Essex Agricultural Society, May 5, 1818. By the Hon. Timothy Pickering, President of the Society. Salem, [Mass.]: Printed by T. C. Cushing, 1818. 27 p.
S&S-45327.
Copies: DNAL; DeGE; MBAt; MWA; NjR.

1197. Senter, Nathaniel Greene M.
A lecture on the beauty, principles and importance of agriculture. Delivered at Waterville and Bangor, by Nath. G. M. Senter. Bangor, [Me.]: Press of James Burton, Jun.; J. Harris, printer, 1818. 23 p.
S&S-45684.
Copies: DLC; DNAL; InU.

1198. Sinclair, John, Sir, bart., 1754–1835.
The code of agriculture: including observations on gardens, orchards, woods, and plantations.... 1st American ed. with notes. Hartford: Hudson and Co., and Cooke and Hale, 1818. viii, [3]–424 p., 8 plates (part col.).
First published in London, 1817.
S&S-45720.
Copies: CtHT-W; CtHi; DeGE; ICS; IaAS; InCW; MB; MH; MWA; MiD; MiU-C; NIC; NN; NjR; O; Vi.

1199. Society of Virginia for Promoting Agriculture.
Memoirs of the "Society of Virginia for Promoting Agriculture": containing communications on various subjects in husbandry and rural affairs. Richmond: Printed by Shepherd & Pollard, 1818. xiii, [9]–132 p., illus.
Contains papers presented in 1811 and 1818.
S&S-45756.
Copies: CSmH; DLC; DNAL; MWA; N; PPAmP; Vi; ViHi; ViRUT; ViU; ViW.

1200. Taylor, John, 1753–1824.
Arator, being a series of agricultural essays, practical and political: in sixty-four numbers.... 4th ed., rev. and enl. Petersburg, [Va.]: Printed by Whitworth & Yancey, for John M. Carter, 1818. 239, [1] p.
S&S-45842.
Copies: CSmH; DLC; ICU; MB; MH; MWA; MdAS; MiU-C; OClWHi; TxU; Vi; ViU.

1201. ——— 5th ed., rev. and enl. Petersburg, [Va.]: Printed by Whitworth & Yancey, for John M. Carter, 1818. xi, [1], 13–239, [1] p.
S&S-45843.
Copies: DLC; DNAL; KyLoF; MB; MWA; NN; PPL; TNJ; Vi; ViHi; VtU.

1202. ———— 6th ed., rev. and enl. Petersburg, [Va.]: Printed by Whitworth & Yancey, for John M. Carter, 1818. ix, [3], [13]–239, [1] p.
S&S-45844.
Copies: CSmH; DNAL; DeGE; MB; MWA; NN; NcD; OClWHi; TU; Vi; ViU.

1203. Webster, Noah, 1758–1843.
An address delivered before the Hampshire, Franklin and Hampden Agricultural Society, at their annual meeting in Northampton, Oct. 14, 1818. . . . Northampton, [Mass.]: Printed by Thomas W. Shepard & Co., 1818. 28 p.
S&S-46740.
Copies: CSmH; CoD; DLC; DNAL; MA; MBAt; MH-AH; MPE; MWA; NN.

1204. Allen, Wilkes, 1775–1845.
An address, delivered before the Western Society of Middlesex Husbandmen, at their annual meeting, holden at Westford, September 25th, 1819. Concord, [Mass.]: Printed by Caleb Cushing, 1819.
Copies: MBAt; MBH; MWA; NN.

1205. Andrews, Ethan Allen, 1787–1858.
Remarks on the present state of agricultural science, and the general means of improving the art of husbandry in the county of Hartford, read before the Hartford County Agricultural Society, at their annual meeting, February, A.D. 1819. . . . Hartford: G. Goodwin & Sons, printers, 1819. 24 p.
Proceedings of the meeting [etc.], p. [17]–24.
S&S-47002.
Copies: Ct; CtHC; CtHi; CtY; DeGE; MBAt; MWA; NHi; WU-A.

1206. Butler, Frederick, 1766?–1843.
The farmer's manual; being a plain practical treatise on the art of husbandry, designed to promote an acquaintance with the modern improvements in agriculture, together with remarks on gardening, and a treatise on the management of bees. . . . Hartford: Samuel G. Goodrich; Clark & Lyman, print., Middletown, 1819. 1 p.l, 224 p.
S&S-47479.
Copies: CtY; DLC; DeGE; GU; MHi; MWA; NIC; NcRS; NjR; PP; RPB.

1207. Casey, Joseph p.
The farmers' and gardeners' hive, shewing the expense and profit attending the cultivation of three hundred acres of land . . . and the work necessary to be done on a farm and in a garden. . . . Also, a treatise on the cultivation of the peach. To which is added a number of recipes, to protect all sorts of fruit trees. . . . Baltimore: Printed by W. F. Redding, 1819. 24 p., front.
S&S-47521.
Copies: DNAL; ICU; MWA; MdHi; PPL.

1208. Coventry, Alexander.
Address to the Agricultural Society of the County of Oneida; delivered at Whitestown, on the 27th day of Sept., 1818. Utica, N.Y.: printed by William Williams, 1819. 30 p. 1819. 30 p.
S&S-47738.
Copies: CSmH; MBAt; MBH; MWA; NCH; NUt; NUtHi.

1209. Davie, William Richardson, 1756–1820.
An address delivered before the South Carolina Agricultural Society, at their anniversary meeting, held in Columbia, on the 8th of December, 1818. . . . Together with the report of the curators for the proceding year . . . Columbia, S.C.: Printed at the Telescope Press, 1819. 29, [3] p.
S&S-47792.
Copies: NcD; NcU; PPAmP.

1210. De Witt, Simeon, 1756–1834.
Considerations on the necessity of establishing an agricultural college. . . . Albany: Printed by Websters and Skinners, 1819. 42 p.
S&S-47825.
Copies: CtHT-W; DLC; MWA; MiU-C; NIC; NN; NbU; NjR; PPL; WU.

1211. [Haines, Charles Glidden] 1792–1825.

An examination into the expediency of establishing a board of agriculture in the state of New-York. Published by the New-York Corresponding Association for the Promotion of Internal Improvements. Brooklyn: E. Worthington, printer, 1819. 64 p.

S&S-48907.

Copies: CSmH; DLC; DNAL; ICMcC; MB; MWA; N; PPL; ViU.

1212. Hammond, Abijah.

An address delivered before the West Chester Agricultural Society, on the 27th October, 1819.... New York: 1819. 15 p.

S&S-48144.

Copies: MHi.

1213. Hopkins, Samuel Miles, 1772–1837.

Address to the Agricultural Society of the County of Genesee, delivered at Batavia, on the 11th day of October, 1819, by Samuel M. Hopkins, Esq., president of the Society.... Batavia, [N.Y.]: Printed by Miller & Blodgett, 1819. 24 p.

S&S-48272.

Copies: DLC; DNAL; MB; MWA; N.

1214. [Jeffreys, George W.]

A series of essays on agriculture and rural affairs in forty-seven numbers. By "Agricola," a North-Carolina farmer.... Raleigh: Printed and published by Joseph Gales, 1819. 2 p.l., [3]–223 p., 1 l., illus.

S&S-48370, 49397.

Copies: AFIT; CSmH; DLC; DeGE; ICJ; MWA; MiU-C; NN; NcAS; NcD; NcU; NjR; ViU.

1215. Lincoln, Levi, 1782–1868.

Address, delivered before the Worcester Agricultural Society, October 7, 1819, being their first anniversary cattle show and exhibition of manufactures.... Worcester, [Mass.]: Printed by Manning & Trumbull, 1819. 30 p.

S&S-48498.

Copies: CSmH; CtY; DLC; MB; MHi; MTemNHi; MWA; MiD-B; OClWHi; RPB.

1216. Massachusetts. Laws.

An act for the encouragement of agriculture and manufactures. [Boston: 1819.] 4 p.

S&S-48614.

Copies: MWA.

1217. Nicholas, John, 1761–1819.

Address, delivered before the Ontario Agricultural Society, at its first annual meeting, October 13, 1819.... Canandaigua, [N.Y.]: Printed by J. D. Bemis & Co., [1819]. 22 p.

S&S-48932.

Copies: MWA; NCanHi; NGH; NjP.

1218. Nicholson, John, 1765–1820.

The farmer's assistant; being a digest of all that relates to agriculture, and the conducting of rural affairs.... Published by Benjamin Warner.... Lancaster, [Pa.]: William Dickson, printer, 1819. 468 p., illus.

N-N0247377; S&S-48935.

Copies: DeU; NjR.

1219. Parkes, Samuel, 1759–1825.

A letter to the farmers and graziers of great Britain, to explain the advantages of using salt in the various branches of agriculture, and in feeding all kinds of farming stock. Philadelphia: M. Carey and Son, 1819. xi, 89 p.

Also published in London, 1819.

S&S-49014.

Copies: MWA; NHi; NjR; PP.

1220. Quincy, Josiah, 1772–1864.

An address delivered before the Massachusetts Agricultural Society, at the Brighton Cattle Show, October 12th, 1819.... [S.l.]: 1819. 15 p.

S&S-49229.

Copies: CSt; DGU; DLC; M; MNBedf; MWA; NjR.

1221. Rawle, William.

An address delivered before the Philadelphia Society for Promoting Agriculture: at its anniversary meeting, January 19, 1819.

Philadelphia: By order of the Society, 1819. 35 p.
S&S-49239.
Copies: DLC; DNAL; MH; NN; PP; PU.

1222. Tibbits, George.
Address, delivered before the Rensselaer Agricultural Society, at their first anniversary meeting, October 13, 1819.... Troy, [N.Y.]: Printed by Wm. S. Parker, 1819. 16 p.
S&S-49595.
Copies: NN; NT.

1223. [Armstrong, John] 1758–1843.
A treatise on agriculture; comprising a concise history of its origin and progress, the present condition of the art, abroad and at home, and the theory and practice of husbandry.... By a practical farmer. Albany: Printed by J. Buel, and for sale at the bookstores in Albany, by J. Wiley & Co., New York, [etc., 1820]. 168 p.
Also dated [1819] (cf. S&S-47013) on the basis of the copyright notice: 2d December in the 44th year of independence.
Note on p. [3]: Board of Agriculture, in general committee, January 24, 1820....
S-143.
Copies: DNAL; DeGE; MB; MH; MU; MWA; MiU-C; MoU; N; NIC; NN; NSyU.

1224. Atherton, Charles Humphrey, 1773–1853.
Address to the Hillsborough Agricultural Society, Weare, Oct. 25, 1820. By Charles H. Atherton. Amherst, N.H.: Printed by E. Mansur, 1820. 30 p.
S-160.
Copies: DLC; IU; MB; MH; MWA; MiD-B; Nh.

1225. Barnard, Hannah Jenkins, b. 1754.
Dialogues on domestic and rural economy.... Hudson, N.Y.: Printed for the author, 1820. 84 p.
S-325.
Copies: MWA; PHC.

1226. Burr, Richard, ed.
Authentic documents relative to the Black River and St. Lawrence countries; being extracts from addresses, reports, &c. made to the Agricultural Society of Jefferson County. Philadelphia: Richard Burr, 1820. 55 p.
N-B0988615; S-629, 1791.
Copies: NCaS; PPAmP.

1227. Carey, Mathew, 1760–1839.
Three letters on the present calamitous state of affairs. Addressed to J. M. Garnett, Esq., president of the Fredericsburg Agricultural Society.... Philadelphia: M. Carey & Son, 1820. liv, [13]–60 p.
S-683.
Copies: CtHT-W; CtY; DLC; M; MB; MH; MWA; MiD-B; NBLiHi; NHi; NN; NcD; P; PPWa.

1228. Fessenden, Thomas Green, 1771–1837.
The husbandman and housewife: a collection of valuable recipes and directions, relating to agriculture and domestic economy. By Thomas G. Fessenden. Bellows Falls, [Vt.]: Printed by Bill Blake & Co., 1820. 191, [15] p.
S-1227.
Copies: DLC; DNAL; ICMcC; MH; MWA; NhD; RPB; VtU.

1229. Gold, Thomas Ruggles, 1764–1827.
Address delivered before the Agricultural Society of the County of Oneida ... at Whitestown, October 18, 1820. Utica, [N.Y.]: Printed for the Society, by William Williams, 1820. 21 p.
S-1400.
Copies: MBH; MWA; NHi; PHi.

1230. Granger, Gideon, 1767–1822.
Address, delivered before the Ontario Agricultural Society, at its second annual meeting, October 3, 1820.... Canandaigua, [N.Y.]: Printed by J. D. Bemis & Co., [1820]. 23 p.
N-G0384434; S-1428.
Copies: MB; MWA; NCanHi; NHi; NN; NRHi.

1231. Le Ray de Chaumont, James Donatien, 1760–1840.

Remarks addressed to the Jefferson County Agricultural Society, at their semi-annual meeting in March, 1820.... New York: W. A. Mercein, 1820. 22, [2] p.

S-725.

Copies: MH; PHi; PPAmP.

1232. ________ Watertown, [N.Y.]: Seth Alden Abbey, 1820. 14 p.

S-726, 1935.

Copies: MWA.

1233. Lyman, Jonathan Huntington, 1783–1825.

An address, delivered before the Hampshire, Franklin and Hampden Agricultural Society ... October 25, 1820.... Northampton, [Mass.]: Printed by Thomas W. Shepard & Co., 1820. 31 p.

S-2035.

Copies: MBAt; MBH; MNF; MWA.

1234. Maxcy, Virgil, 1785–1844.

Address to the Agricultural society of Maryland. Delivered at their anniversary meeting ... at Annapolis ... the 15th December, 1819. By the Hon. Virgil Maxcy.... Annapolis: Printed by J. Green, 1820. 19 p.

S-2190.

Copies: DLC; DNAL; DeGE; MB; MBAt; MWA; MdBE; MdHi; NcD; PHi; PPAmP; WU-A.

1235. New York (State). Board of Agriculture.

Address of the General Committee of the Board of Agriculture of the State of New York to the county agricultural societies for 1820, with accompanying documents. Albany: Printed for S. Southwick at the Agricultural Press by J. O. Cole, 1820. 31, [1] p.

S-2478.

Copies: CSmH; DLC; DeGE; MB; MH; MWA; N; NBLiHi; NHi; NN; ViU.

1236. Newberry, H.

Sketches of an address read before the Hartford County Agricultural Society, at their annual meeting, October 5, 1820, by H. Newberry. Hartford: G. Goodwin & Sons, printers, 1820. 15 p.

S-2548.

Copies: Ct; CtHC; CtHi; CtSoP; DNAL; DeGE; MBH; MWeA.

1237. Nicholson, John, 1765–1820.

The farmer's assistant; being a digest of all that relates to agriculture, and the conducting of rural affairs; alphabetically arranged, and adapted for the United States.... 2d ed., corr. and enl. Philadelphia: Benjamin Warner; William Dickson, printer, Lancaster, (Pa.), 1820. 2 p. l., [9]–468 p., illus., diagrs.

S-2551.

Copies: C; DLC; DNAL; DeGE; ICU; In; MH; MWA; NIC; NcU; PU; ScU; TxU; Vi.

1238. Pickering, Timothy, 1745–1829.

A discourse, read before the Essex Agricultural Society, in Massachusetts, February 21, 1820, suggesting some improvements in the agriculture of the county.... Salem, [Mass.]: John D. Cushing, printer, 1820. 27 p.

S-2770.

Copies: CSmH; CtHT-W; DLC; DNAL; DeGE; ICN; MB; MH; MWA; NIC; NN; NNC; NcD; PPL; PU; WHi.

1239. Quincy, Josiah, 1772–1864.

An address delivered before the Massachusetts Agricultural Society, at the Brighton Cattle Show, October 12, 1819. Boston: Wells & Lilly, 1820. 46 p.

Possibly the same as no. 1220.

S-2928.

Copies: MeB (not located).

1240. Rose, Robert H.

An address delivered before the Agricultural Society of Susquehanna County, at its organization, December 6, 1820.... Mon-

trose, [Pa.]: Printed by Adam Waldie, 1820. 22 p.
S-3054.
Copies: MWA.

1241. Tilghman, William, 1756–1827.
An address delivered before the Philadelphia Society for Promoting Agriculture; at its anniversary meeting, January 18, 1820. Philadelphia: Published by order of the Society; William Fry, printer, 1820. 36 p.
S-3441.
Copies: DLC; DNAL; MB; MBAt; MH-BA; MWA; MdHi; NN; NcD; NjR; OkS; P; PPAmP; PPL; PU; ScCleU; TxCM.

1242. ——— [Philadelphia: 1820.] [117]–148 p.
Offprint from the *Memoirs of the Philadelphia Society for Promoting Agriculture*
Copies: DeGE.

1243. Walker, William, 1751?–1831.
Address of Hon. William Walker ... delivered before the Berkshire Association for the Promotion of Agriculture and Manufactures ... October 5, 1820.... Pittsfield, [Mass.]: Printed by Phinehas Allen, 1820. 24 p.
S-4152.
Copies: MB; MPB; MWA; N; NjR.

1244. Allen, Jonathan, b. 1773.
Address of Jonathan Allen ... delivered before the Berkshire Association for the Promotion of Agriculture and Manufactures, at Pittsfield, October 4, 1821. Together with the reports of the committees of departments. Pittsfield, [Mass.]: Printed by P. Allen, 1821. 40 p.
S-4450.
Copies: DNAL; M; MB; MBAt; MWA.

1245. Barnum, A. W.
Address delivered to the Vergennes Agricultural Society, at their annual cattle show and fair, held in Vergennes, September 27, 1821. Middlebury, [Vt.]: Printed by Copeland & Allen, 1821. 18 p.
S-4611.
Copies: MB; VtHi; VtMiS.

1246. Beatson, Alexander, 1759–1833.
A new system of cultivation, without lime, or dung, or summer fallows, as practised at Knowle-Farm, in the County of Sussex. Philadelphia: M. Carey, 1821. iv, [3]–5, 77 p., plates.
First published in London, 1820.
S-4642.
Copies: DeGE; MWA; NjR; PPAmP; PPL; PU-V.

1247. Bigelow, Lewis, 1785–1838.
Address, delivered before the Worcester Agricultural Society, October 12, 1820, being their anniversary cattle show and exhibition of manufactures.... Worcester: Printed by Manning & Trumbull, 1821. 32, [2] p.
S-4707.
Copies: DLC; DNAL; ICN; MBAt; MH; MHi; MWA; MdBJ; MiD-B; NN; NjR; RPB.

1248. Brackett, Adino N.
Address delivered before the Coos Agricultural Society, October 17, 1821. Concord, [N.H.]: Printed by Hill and Moore, 1821. 16 p.
S-4824.
Copies: ICN; LNHT; MWA; Nh.

1249. Broom, James Madison, 1778–1850.
An address delivered before the Philadelphia Society for Promoting Agriculture, at its anniversary meeting, January 16, 1821, by James M. Broom.... Philadelphia: Printed for Littell & Henry, by Clark & Raser, 1821. 24 p.
S-4844.
Copies: DNAL; DeGE; MB; MBAt; MWA; MiD-B; NN; NjP; PHi; PPAN; PPAmP; PPPrHi; PU.

1250. Butler, Frederick, 1766?–1843.
Farmer's manual; being a plain prac-

1250 (*continued*)
tical treatise on the art of husbandry, designed to promote an acquaintance with the modern improvements in agriculture, together with remarks on gardening, and a treatise on the management of bees. . . . Weathersfield, [Conn.]: Printed by Clark and Lyman, for the author, 1821. 1 p.l., 224 p.

S-4882.

Copies: CSmH; Ct; CtY; DLC; DNAL; ICJ; MB; MFiHi; MWA; MiU-C; NBuG; NN; PMA; PPL; RNR.

1251. Genet, Edmond Charles, 1763–1834.

Address on the means of opening new sources of wealth for the northern states. Delivered on the 19th October, 1821, before the Agricultural Society of the County of Rensselaer, and the public attending their proceedings. . . . Troy, N.Y.: Printed by F. Adancourt, 1821. 56 p., plate.

Concerns particularly the merino sheep, and the cultivation of flax and hemp. Includes a description of James Lee's machine to clean flax and hemp.

S-5434.

Copies: DLC; DeGE; MB; MH-BA; MiU-C; N; NN.

1252. Hallett, Jacob W.

Address before the Ontario Agricultural Society, at its third annual meeting, October 16, 1821. . . . Canandaigua, [N.Y]: Printed by J. D. Bemis & Co., 1821. 16 p.

S-5522.

Copies: N.

1253. Hammond, Abijah.

An address, delivered before the Agricultural Society, of the County of Westchester, New-York, October 26, 1820. Mount Pleasant, [N.Y.]: Marshall, printer, 1821. 23 p.

S-5530.

Copies: NHi.

1254. Marshall, Thomas, 1784–1835.

Rotation of crops. A paper laid before the Agricultural Society of Virginia. No. 1. Oak Hill, Fauquier, October 14, 1818. . . . Baltimore: American farmer, 1821. Broadside 27 x 23 cm.

Offprint from *American farmer,* v. 3, Sept. 21, 1821.

N-M0250169.

Copies: ViU.

1255. Moulton, Joseph White, 1789–1875.

An address delivered at St. Paul's Church, Buffalo, on the anniversary celebration of the Niagara and Erie Society for Promoting Agriculture and Domestic Manufactures, the 30th day of October, 1821. By Joseph W. Moulton. . . . Buffalo: Printed by D. M. Day, and H. A. Salisbury, 1821. 36 p.

S-6128.

Copies: CSmH; DNAL; DeGE; MH; MWA; N; NBuG; NBuHi; NT; Nh.

1256. New York (State). Board of Agriculture.

Memoirs of the Board of Agriculture of the State of New-York. Published by authority. Vol. I. Albany: Printed by S. Southwick, 1821. xlviii, [18], 364, 109, [10] p.

S-6251.

Copies: DLC; DeGE; ICJ; MH; MWA; MiEM; MnHi; NIC; NNA; PPAmP; PPF; WU-A.

1257. Nichols, Andrew, 1785–1853.

An address, to the Essex Agricultural Society, at the first cattle show, at Topsfield, Oct. 5, 1820. . . . Salem, [Mass.]: Printed by John D. Cushing, 1821. [3], 36–56 p.

N-N0244353; S-6294, 6295; Tapley-p. 464.

Copies: MSaE; MWA; NN.

1258. Plumer, William, 1789–1854.

An address delivered before the Rockingham Agricultural Society, October 18th, 1821. . . . Exeter, [N.H.]: Printed by J. J. Williams, 1821. 24 p.

S-6492.

Copies: CtSoP; DLC; DNAL; MBAt; MH; MHi; MSa; MWA; NN; NhHi.

1259. Remarks on a petition from certain inhabitants of the county of Westchester, to the Legislature, praying for the repeal of the acts for improving the agriculture of this state. By a Westchester farmer. [S.l.: 1821.] 8 p.

Dated: February, 1821.
Sabin-102957.
Copies: NHi.

1260. Russell, Jonathan, 1771–1832.

Address delivered before the Worcester Agricultural Society, September 27, 1821, being their anniversary cattle show and exhibition of manufactures.... Worcester, [Mass.]: Printed by Manning and Trumbull, 1821. 24, [2] p.

S-6685.
Copies: DLC; IaAS; MB; MDeeP; MHi; MWA; MWH; MiD-B; N; OClWHi; OO; RPB.

1261. Taylor, John, 1753–1824.

A letter on the necessity of defending the rights and interests of agriculture, addressed to the delegation of the United Agricultural Societies of Virginia.... Petersburg, [Va.]: Intelligencer Press, 1821. 10 p.

S-6941.
Copies: DLC; NcD; OkU.

1262. A treatise on soils and manures, as founded on actual experience, and as combined with the leading principles of agriculture in which the theory and doctrines of Sir Humphry Davy, and other agricultural chemists, are rendered familiar to the experienced farmer. By a practical agriculturist.... Philadelphia: B. Warner, 1821. 92 p.

First published in London, 1818.
S-6995.
Copies: DLC; ICJ; MWA; NjP; P; PPC; PU.

1263. Abbot, Abiel, 1770–1828.

An address delivered before the Essex Agricultural Society at the agricultural exhibition in Danvers, October 17, 1821.... Andover, [Mass.]: Printed for the Society by Flagg and Gould, 1822. 55 p.

The Trustees' account of the agricultural exhibition at Danvers, October 16 and 17, 1821, p. [17]–55.

S-7727, 8635.
Copies: DLC; ICN; MB; MH-AH; MWA; MnHi; NN; NjR; PPL; RPB.

1264. Albany County Agricultural Society.

Agricultural tracts, no. 1, for October, 1822. Albany: D. Steele & Son; Packard & Van Benthuysen, printers, 1822. 48 p.

No more published?
S-7767.
Copies: N; NN.

1265. Biddle, Nicholas, 1786–1844.

Address delivered before the Philadelphia Society for Promoting Agriculture, at its annual meeting on the fifteenth of January, 1822.... Printed by order of the Society. Philadelphia: Clark & Raser, printers, 1822. 39 p.

S-8059.
Copies: CtY; DLC; DNAL; DeGE; ICU; IaAS; MB; MH; MWA; NbU; NjP; NjR; P; PPAmP; PPL; PU; ScC.

1266. Burrall, Thomas D.

Address, delivered before the Ontario Agricultural Society, at its fourth annual meeting, October 22, 1822.... Canandaigua, [N.Y.]: Printed by J. D. Bemis & Co., 1822. 23 p.

S-8221.
Copies: DLC; DNAL; N; NCanHi.

1267. Colman, Henry, 1785–1849.

An address delivered before the Massachusetts Agricultural Society at the Brighton Cattle Show, October 17th, 1821.... Boston: Wells and Lilly, 1822. 17 p.

S-8396.

1267 *(continued)*
Copies: CtY; DLC; MBH; MH; MHi; MeB.

1268. Deane, Samuel, 1733–1814.
The New-England farmer; or, Georgical dictionary. Containing a compendious account of the ways and methods in which the important art of husbandry, in all its various branches, is, or may be, practised, to the greatest advantage, in this country.... 3d ed., corr., impr., greatly enl.... Boston: Wells and Lilly, 1822. xi, 532 p.
S-8520.
Copies: CU; CtHT-W; DNAL; DeGE; FWpR; ICJ; InLP; MH; MWA; Me; MiU; MnU; NB; NBu; NNBG; PP; PPL; PU; RNR; ViL; WU-A.

1269. Denny, Nathaniel P.
Address delivered before the Worcester Agricultural Society, September 25, 1822, being their anniversary cattle show and exhibition of manufactures.... Worcester, [Mass.]: Printed by W. Manning, [1822]. 24, [2] p.
S-8538.
Copies: DeGE; MH; MWA; MWHi; MiD-B; NjPT; NjR; OClWHi; RPB.

1270. Eastman, Moses.
A discourse delivered before the Hillsborough Agricultural Society, at Amherst, N.H., September 25, 1822. Amherst: Printed by Richard Boylston, for the Society, 1822. 24 p.
S-8584.
Copies: DLC; ICJ; MBAt; MBC; Nh; NhHi; NjR; OClWHi.

1271. Helme, R. H.
An address delivered before the Agricultural Society of North Carolina, at Raleigh, December 16, 1822, by Dr. R. H. Helme.... Raleigh: Printed by J. Gales & Son, [1822]. 12 p.
N-H0263834.
Copies: NNC; NcU.

1272. M'Kay, Samuel M.
An address delivered before the Berkshire Association for the Promotion of Agriculture and Manufactures, at Pittsfield, October 3, 1822.... Pittsfield, [Mass.]: Printed by Phinehas Allen, 1822. 34 p.
S-9319.
Copies: M; MB; MH; MPB; MWA.

1273. Martin, Anthony B.
Remarks on agriculture and a method of improving soils, by creating artificial manure, founded on chymical and scientific principles.... Baltimore: J. D. Toy, printer, 1822. 19 p.
S-9367.
Copies: DNAL; Vi.

1274. Moore, Humphrey, 1778–1871.
An address delivered at Hopkinton, before the Hillsborough Society for the Promotion of Agriculture and Domestic Manufactures, October 17, 1821.... Amherst, N.H.: Printed by E. Mansur, 1822. 14 p.
S-9531.
Copies: DNAL; MBH; MH; MWA; MiD-B; NHi; NjR.

1275. New Hampshire. Board of Agriculture.
The New-Hampshire agricultural repository. Published by the state Board of Agriculture, under the patronage of the legislature of the state of New-Hampshire. Concord: Printed by Hill and Moore, 1822. 135 p.
At head of title: No. 1.
S-9634.
Copies: DLC; DNAL; IaAS; MB; MBH; MH; MWA; MiD-B; Nh; NhHi; NjR.

1276. New York (State). Board of Agriculture.
Report of the General Committee to the Board of Agriculture; read and accepted, January 23d, 1822. Albany: Printed by Packard & Van Benthuysen, 1822. 10 p.
N-N0188582; S-9684.
Copies: CSmH; MB.

1277. Pickering, Timothy, 1745–1829.
An address delivered before the Massachusetts Agricultural Society at the Brighton cattle show, October 9th, 1822. [Boston: 1822.] 26 p.
S-9934.
Copies: MH.

1278. Thompson, Otis, 1776–1859.
An address, delivered before the Bristol County Agricultural Society on their first anniversary at Taunton, October 30, 1822. [S.l.: 1822.] 16 p.
S-10438.
Copies: CSmH; MTaHi; Nh; RPB.

1279. Adams, Josiah, 1781–1854.
An address delivered before the Society of Middlesex Husbandmen and Manufacturers, on their anniversary at Concord, October 2, 1823. Concord, [Mass.]: Printed by William Gould, 1823. 13 p.
S-11533.
Copies: MH-BA; MWA; NN.

1280. Adams, William H.
Address, delivered before the Ontario Agricultural Society, at its fifth annual meeting, October 28, 1823. Canandaigua, [N.Y.]: Printed by J. D. Bemis and Co., [1823]. 16 p.
S-11534.
Copies: NCanHi; NN.

1281. Agricultural Society of South Carolina.
Report of the committee, appointed by the South-Carolina Agricultural Society, to consider what beneficial effects would result to the agricultural interests of the state, by importing foreign seeds, plants, and implements of husbandry. To which is added, Gen. Thomas Pinckney's letter on the water culture of rice.... Charleston: Printed by Archibald E. Miller, 1823. iv, [5]–24 p.
S-14170.
Copies: DLC; MHi; MWA; NcD; PU.

1282. Appeal to the President, Senate, House of Representatives and country on its agriculture, manufactures, commerce, and finance. By a Pennsylvanian. Philadelphia: 1823. 79 p.
By Mathew Carey; the same as No. 1284.
N-A0357837.
Copies: PHi.

1283. Bates, Isaac Chapman, 1780–1845.
Address to the Hampshire, Franklin and Hampden Agricultural Society, Northampton, Mass. October 23, 1823. Northampton: Printed by Thomas W. Shepard, 1823. 35 p.
S-11762.
Copies: CtY; DLC; ICN; MB; MWA; PPL; RPB.

1284. [Carey, Mathew] 1760–1839.
The crisis: a solemn appeal to the President, the Senate ... on the destructive effects of the present policy of this country, on its agriculture, manufactures, commerce and finances.... By a Pennsylvanian. Philadelphia: H. C. Carey & I. Lea, Nov. 24, 1823. viii, [9]–79 p.
S-12056.
Copies: CtY; DLC; ICU; MH; MWA; MdBP; MiU-C; NN; NNC;NbU; Nh; NjP; PPAmP; PPL; RP; ScU.

1285. ——— 2d ed ... Philadelphia: H. C. Carey & I. Lea, Dec. 17, 1823. viii, [9]–79 p.
S-12057.
Copies: CtY; DLC; DeGE; MH-L; MnU; NIC; NNC; PPL; PU; ViW.

1286. Eaton, Peter, 1765–1848.
An address delivered before the Essex Agricultural Society, at the agricultural exhibition in Topsfield, Oct. 2, 1822.... Salem, [Mass.]: Printed for the Society by John D. and Thomas C. Cushing, Jr., 1823. 60 p.
Also contains reports of the Society, p. [15]–60.
S-12425.
Copies: DLC; DNAL; MBAt; MDeeP; MHa; MLow; MNe; MPeaHi; MSaE; MWA; NN.

1287. Fiske, Oliver, 1762–1836.
Address delivered before the Worcester Agricultural Society, October 8, 1823, being their anniversary cattle show and exhibition of manufactures.... [Also, reports of committees.] Worcester, [Mass.]: Printed by W. Manning, [1823]. 36 p.
S-12554.
Copies: DLC; MH; MHi; MWA; MiD-B; MnHi; OClWHi; OO.

1288. Maryland. Laws.
An act to provide for the public instruction of youth throughout this state, and to promote the important interests of husbandry and agriculture. [Annapolis: 1823.] 24 p.
S-13225.
Copies: DLC; MBAt.

1289. Morris, James Pemberton.
Address delivered before the Agricultural Society of Bucks County, on the 28th of April, 1823.... Doylestown, [Pa.]: Printed by W. T. Rogers, 1823. 12 p.
S-13401.
Copies: PHi.

1290. ——— Philadelphia: Clark & Raser, printers, 1823. 12 p.
S-13402.
Copies: DLC; PHi.

1291. New York (State). Board of Agriculture.
Memoirs of the Board of Agriculture of the state of New York ... Vol.II. Albany: Printed by Packard & Van Benthuysen, 1823. 532, [10] p.
S-13551.
Copies: M; MWA; MoSM; NNA; NNBG.

1292. Pennsylvania Agricultural Society.
Strictures upon Arator's attack. [Philadelphia: 1823.] 4 p.
Signed: A farmer.
S-13719.
Copies: PPL.

1293. Peters, Richard, 1744–1828.
Address delivered before the Philadelphia Society for Promoting Agriculture, at its annual meeting, on the twenty-third of January, 1823, by Richard Peters, its President.... Philadelphia: Clark & Raser, printers, 1823. 1 p.l., 25 p.
S-13738.
Copies: AzU; DLC; DeGE; MH; MdU; NcD; NjR; PPAmP; PPL; PU-V; ScClU; TxCM.

1294. Sedgwick, Theodore, 1780–1839.
An address delivered before the Berkshire Association for the Promotion of Agriculture and Manufactures.... Oct. 2, 1823. Pittsfield, [Mass.]: Printed by P. Allen, 1823. 32 p.
S-14076.
Copies: M; MH-BA; MWA.

1295. Stimson, Earl.
Practical husbandry. An address, delivered at the fifth anniversary meeting of the Saratoga County Agricultural Society, held at Ballston Spa, on the eighth day of October, 1823. Ballston Spa, [N.Y.]: Printed for the Society by J. Comstock, 1823. 14, [1] p.
S-14219.
Copies: MWA; N; NRU; PPL.

1296. Whipple, Thomas.
An address delivered before the Grafton Agricultural Society, at their annual meeting, October 1822. Haverhill, N.H.: Printed by Sylvester T. Gross, 1823. 13 p.
S-14886.
Copies: Nh.

1297. Adams, Daniel, 1773–1864.
The agricultural reader, designed for the use of schools.... Boston: Richardson & Lord; printed by J. H. A. Frost, 1824. 264 p.
S-15021.
Copies: DeGE; IU; InCW; MB; MH; MWA; MeHi; NNC; NhHi; OHi; PPL; PU; TxU.

1298. Agricultural Society of South Carolina.

Original communications made to the Agricultural Society of South-Carolina; and extracts from select authors on agriculture. Published by order of the Society. Charleston: Printed by A. E. Miller, 1824. x, 316 p., diagrs., 2 plates.

S-18058.

Copies: DLC; DNAL; GHi; MWA; ScC.

1299. Carey, Mathew, 1760–1839.

Address delivered before the Philadelphia Society for Promoting Agriculture, at its meeting, on the twentieth of July, 1824.... Printed by order of the Society. Philadelphia: Printed by J. R. A. Skerrett, 1824. xii, [9]-80 p.

S-15646.

Copies: DLC; DNAL; DeGE; MB; MH-BA; MWA; MiU-C; MnU; PPAmP; PPL; PU; ViW.

1300. ———— 2d ed. Philadelphia: Printed by J. R. A. Skerrett, 1824. 66 p.

S-15647.

Copies: MH-BA; MiEM; PU.

1301. ———— 3d ed. ... Philadelphia: Printed by J. R. A. Skerrett, Aug. 14, 1824. 66 p.

N-C0133041.

Copies: MWA (2 title pages, dated Aug. 14 and 17); MeWC.

1302. ———— 4th ed., rev. and corr. Philadelphia: J. R. A. Skerrett, 1824. xiii, [15]–108 p.

S-15648.

Copies: CtY; DLC; DNAL; DeGE; ICU; InU; LU; MH-BA; MWA; MdHi; MiD-B; NIC; NbU; NjPT; OCHP; PPL; TxCM; WU.

1303. Chipman, Samuel.

Address, delivered before the Ontario Agricultural Society, at its fifth [sixth] annual meeting, October 26, 1824.... Canandaigua, [N.Y.]: Printed by J. D. Bemis and Co., [1824]. 16 p.

S-15742.

Copies: CSmH; MWA; N.

1304. Cuming, Francis Higgins, 1799–1862.

An address delivered in Rochester before the Monroe County Agricultural Society, October 20, 1824.... To which are annexed some remarks by James Sperry.... Rochester, N.Y.: Printed by Levi W. Sibley, 1824. 16 p.

N-C0831107.

Copies: N; NRU.

1305. Drowne William 793–1874.

Compendium of agriculture; or, The farmer's guide, in the most essential parts of husbandry and gardening; compiled from the best American and European publications, and the unwritten opinions of experienced cultivators.... By William Drown [*sic*], with the aid and inspection of Solomon Drown [*sic*], M.D. Providence: Printed by Field & Maxcy, 1824. xi, [13]–288 p.

S-16006.

Copies: CtY; DLC; DNAL; DeGE; ICU; MB; MWA; MiU-C; NNBG; NbCrD; NcRS; NhHi; NjR; OU; PPL; RPE.

1306. Goodwin, Isaac, 1786–1832.

Address delivered before the Worcester Agricultural Society, October 13, 1824, being their anniversary cattle show and exhibition of manufactures. Worcester, [Mass.]: Printed by Wm. Manning, [1824]. 32 p.

S-16325.

Copies: DLC; M; MB; MBAt; MWA; MWH; MiD-B; PHi; RPB.

1307. Howes, Frederick, 1782–1855.

An address to the Essex Agricultural Society, in Massachusetts, at Topsfield, October 6, 1823.... Salem: Printed for the Society by W. & S. B. Ives, 1824. 23 p. (*Transactions of the Essex Agricultural Society,* v. 1, no. 6)

S-16602.

Copies: CSmH; CtY; DLC; DNAL; MB; MBAt; MSaE; MWA; NjR.

1308. Le Ray de Chaumont, James Donatien, 1760–1840.

An address, delivered before the Jefferson County Agricultural Society, at the annual fair, 1824. Watertown, [N.Y.]: Printed by W. Woodward, 1824. 16 p.

N-L0272970; S-16883 (variant title; 24 p.).

Copies: MWA; NN; ViU (N-L0272970).

1309. [Putnam, Henry] 1778–1827.

Touches on agriculture; including a treatise on the preservation of the apple-tree, together with family recipes, experiments on insects, &c. &c. By the author of the Description of Brunswick, and other towns in Maine. Portland, [Me.]: A. W. Thayer, printer, 1824. 43 p.

S-17738.

Copies: DNAL; ICJ; MB; MWA; MeHi; MeLB; MeP; MeU.

1309a. Spalding, Thomas, 1774–1851.

Address of Hon. Thomas Spalding before the Union Agricultural Society, Darien, Georgia, May 13th, 1824. Prefaced by the constitution of the Society. Griffin, [Ga.]: The Goen Printing Co., [1824]. 16 p.

N-S0793172.

Copies: DNAL; KyU; ViU.

1310. Barbour, James, 1775–1842.

Address of James Barbour, Esq., president of the Agricultural Society of Albemarle, at their meeting on the 8th of November 1825.... Charlottesville, Va.: Printed by C. P. M'Kennie, 1825. 14 p.

S-19605.

Copies: CSmH; ViU.

1311. Briggs, Charles, 1791–1873.

A discourse delivered at Concord, October the fifth, 1825.... Concord, [Mass.]: Printed for the Society by John C. Allen, 1825. 34 p.

"Published by request of the Society of Middlesex Husbandmen and Manufacturers."

S-19865.

Copies: DLC; DNAL; DeGE; MH; MLow; MWA; MiD-B; MiU; PHi.

1312. Lathrop, Leonard Elijah.

The farmer's library; or, Essays designed to encourage the pursuits and promote the science of agriculture. By Leonard E. Lathrop.... Rutland, [Vt.]: Printed by William Fay, 1825. xii, [13]-215, [1] p.

S-21166.

Copies: CtHT-W; CtW; CtY; DLC; DNAL; MWA; PPL; Vt; VtMiS.

1313. Lorain, John.

Nature and reason harmonized in the practice of husbandry. By the late John Lorain.... Philadelphia: H. C. Carey & I. Lea, 1825. x, 563 p.

S-21248.

Copies: CtHT; DLC; DNAL; InLP; MH; MU; MdU; MiU-C; MnM; NIC; NjP; NjR; PPAmP; PPL; PU; Vi.

1314. Muse, Joseph Ennals.

An oration upon the dignity and utility of the art and science of agriculture, delivered before the Dorchester Agricultural Society, and a numerous audience, at their first exhibition and fair, held in Cambridge, November 10, 1825. Baltimore: Printed by John D. Toy, 1825. 22 p.

S-21579.

Copies: DLC; IObB.

1315. North Carolina. Board of Agriculture.

Papers on agricultural subjects, and professor Olmsted's Report on the geology of North Carolina. Published by order of the Board of Agriculture. Raleigh: Printed by J. Gales & Son, 1825-27. 3 v.

Volumes 1–2 are paged continuously. Vol. 3 has the title: *Report on the geology of North Carolina conducted under the direction of the Board of Agriculture.* Part III is by Elisha Mitchell.

S-21702.

Copies: CSmH; DI-GS; DLC; NcD.

1316. North Carolina. Board of Agriculture.

Proceedings of the Board of Agriculture of North-Carolina, at their session in 1824. Raleigh: J. Gales & Son, printers, 1825. 8 p.

S-21703.
Copies: NcD; NcU.

1317. Putnam, John, 1778–1827.
Touches on agriculture, including a treatise on the preservation of the apple tree. Together with family recipes, experiments on insects, &c.... 2d ed., enl. Salem, [Mass.]: John D. Cushing, printer, 1825. 64 p.
S-22013.
Copies: CtSoP; MB; MBAt; MBH; MH; MSaE; MWA; NjR.

1318. Riggs, Caleb Smith, 1763–1826.
Address, delivered before the Essex County Agricultural Society, at their annual fair, held in Orange—October, 1824. Newark, N.J.: Printed by W. Tuttle & Co., 1825. 16 p.
S-22099.
Copies: CSmH; NjR.

1319. Tracy, William S.
An address to the Agricultural Society of Portage County, delivered at their late annual fair and cattle-show. Ravenna, [Ohio]: Pub. by the order of the Society; J. B. Butler, printer, 1825. 12 p.
S-22497.
Copies: CtY; MiD; OClWHi.

1320. Vaux, Roberts, 1786–1836.
Address delivered before the Philadelphia Society for Promoting Agriculture, at its annual meeting on the eighteenth of January, 1825.... Philadelphia: Printed at the Port folio office, 1825. 28 p.
S-23147.
Copies: DLC; DNAL; DeGE; IU; MH; MiD-B; N; NjR; P; PPL; PU.

1321. Walker, Lyman B.
An address delivered before the Strafford Agricultural Society, at their annual fair and cattle show, holden on the 19th and 20th October, 1825, at Sandwich. Sandbornton, [N.H.]: Strafford Gazette Press; D. V. Moulton, 1825. 27 p.
S-23196.
Copies: CLCM; ICJ; MH; NN; Nh; NhHi; PPL.

1322. Williams, Thomas, 1779–1876.
An address before the Attleborough Agricultural Society, 4th July, 1825. Providence: Printed by Barnum Field & Co., 1825. 22 p.
S-23333.
Copies: CtSoP; M; MAtt; MB; MHi; MWA; RHi.

1323. Agriculture, commerce, and manufactures. Boston: [E. Bellamy, printer], 1826. 22 p.
Extracted from *Niles' Register*, November 11, 1826, and other sources.
S-23436.
Copies: DLC; DeGE; ICU; IU; MB; MH-BA; MHi; MWA; NN; PHi; WHi.

1324. Doolittle, Mark, d. 1855.
An address, at Northampton, October 11, 1826, to the Society of the counties of Hampshire, Franklin and Hampden, for the promotion of agriculture and the mechanic arts.... Northampton, [Mass.]: Printed by T. Watson Shepard, 1826. 24 p.
S-24359.
Copies: CtSoP; MA; MB; MBNEH; MHi; MNF; MWA; N; NN; RPB.

1325. Goodrich, Charles Augustus, 1790–1862.
An address delivered before the Hartford County Agricultural Society, October 12, 1826, by Rev. Charles A. Goodrich. Hartford: Printed by Goodwin & Co., 1826. 20 p.
S-24719.
Copies: CSmH; Ct; CtY; DNAL; MU; MWA; MiD-B; N; RPB; WU.

1326. Lathrop, Leonard Elijah.
The farmer's library; or, Essays designed to encourage the pursuits, and promote the science of agriculture. By Leonard

1326 (*continued*)
E. Lathrop, Esq.... 2d ed., corr. & enlarged. Windsor, [Vt.]: Printed by Wyman Spooner, 1826. 1 p.l., [v]–x, [13]–300 p.
S-25066.
Copies: DLC; DNAL; DeGE; ICU; IaAS; MU; MWA; NhD; PPL; VtU; WaU.

1327. New York (State). Board of Agriculture.
Memoirs of the Board of Agriculture of the state of New York.... Vol. III. Albany: Printed by Packard & Van Benthuysen, 1826. 558, [15] p.
S-25553.
Copies: DNAL; ICJ; ICU; M; MBH; MWA; NSyOHi.

1328. Patterson, Angus.
An address to the Farmers' Society of Barnwell District. Delivered on the second day of January, 1826. Charleston, [S.C.]: A. E. Miller, 1826. 18 p.
N-P0142750.
Copies: DNAL.

1329. Carey, Mathew, 1760–1839.
Address delivered before the Philadelphia Society for Promoting Agriculture, at its meeting on the twentieth of July, 1824.... 5th ed., rev. and corr. Philadelphia: Mifflin and Parry, printers, November 1827. 71 p.
S-28388.
Copies: CSmH; CtY; DLC; ICU; In; MH-BA; MWA; NN; OO; PPL; PU; ScHi; TxU; Vi; WHi.

1330. Gadsden, James, 1788–1858.
Oration delivered to the Florida Institute of Agriculture, Antiquities and Science, at its first public anniversary ... 4th January, 1827. Tallahassee: Printed at the office of the Florida intelligencer, 1827. [16] p.
S-28988.
Copies: CtY; DLC; FU; PP; PPAmP.

1331. Hitchcock, Edward, 1793–1864.
Scientific agriculture. An address delivered before the Hampshire, Franklin, & Hampden Agricultural Society; at Northampton, October 24, 1827. Published by the Society. Amherst, Mass.: J. S. & G. Adams, printers, [1827?]. 24 p.
S-29230.
Copies: CSmH; CtY; MB; MH; MWA; N; RPB.

1332. Niles, Hezekiah, 1777–1839.
Agriculture of the United States; or, An essay concerning internal improvement & domestic manufactures, shewing their inseperable [*sic*] connection with the business and interests of agriculture ... by H. Niles, of Baltimore. [Baltimore?: 1827.] 16 p.
With additions; first published in *Niles' Register,* March 24, 1827.
S-30062.
Copies: DLC; DeGE; MBAt; MdBP; PHi; ViU.

1333. Pennsylvania Agricultural Society.
Hints for American husbandmen, with communications to the Pennsylvania Agricultural Society. By order of the directors. Philadelphia: Printed by Clark & Raser, 1827. [178] p., fold. front., 5 plates.
Cover-title: *Hints for American husbandmen, with original notices and extracts on selecting and breeding neat cattle and sheep.*
S-30185.
Copies: DLC; DeGE; ICU; IU; M; MH; NNC; NcRA; P; PHi; PPL; PPi; PU; RPA.

1334. Stout, Z. Barton.
An address, delivered before the Ontario Agricultural Society ... in Canandaigua, October 2, 1827.... Canandaigua, [N.Y.]: Printed by Bemis, Morse and Ward, 1827. 26 p.
S-30723.
Copies: DLC; NCanHi; NIC; NRHi; NRU.

1335. Tufts, George Aaron.
Address, delivered before the Worcester Agricultural Society, October 12, 1825; being their seventh anniversary cattle show and

exhibition of manufactures. Worcester, [Mass.]: Charles Griffin, printer, 1827. [3], 16 p.

S-30850.

Copies: DLC; M; MB; MH; MH-BA; MHi; MWA; MiD-B; NHi; OClWHi; RPB.

1336. Washburn, Emory, 1800–1877.

Address delivered before the Worcester Agricultural Society, October 11, 1826; being their eighth anniversary cattle show and exhibition of manufactures.... Worcester: Charles Griffin, printer, 1827. 16 p.

S-31666.

Copies: CSmH; DLC; DeGE; MB; MH; MWA; MiD-B; NN; OClWHi; RPB.

1337. Washington, William.

An address delivered in Charleston before the Agricultural Society of South Carolina, at its anniversary meeting, on the 21st of August, 1827.... Charleston: Printed by A. E. Miller, 1827. 20 p.

S-31668.

Copies: NcD; NcU; ScC.

1338. Whiting, John.

An address delivered before the Berkshire Association for the Promotion of Agriculture and Manufactures, at Pittsfield, Oct. 5, 1826. Pittsfield, [Mass.]: Printed by Phinehas Allen, 1827. 23 p.

S-31738.

Copies: M; MPB.

1339. Breck, Samuel, 1771–1862.

Address delivered before the Blockley and Merion Agricultural Society, on Saturday, September 20th, 1828, on the death on [*sic*] their late president, the Hon. Richard Peters.... Philadelphia: Printed by Lydia R. Bailey, 1828. 27 p.

Includes references to the construction of the permanent bridge over the Schuylkill River and to Peters' contributions to improvements in agriculture.

S-32465.

Copies: DNAL; DeGE; DeWI; ICU; MH; MWA; MiD-B; PHC; PP; PPAmP; PPL; RPB.

1340. Horry, Elias, 1743–1834.

An address delivered in Charleston before the Agricultural Society of South Carolina, at its anniversary meeting on Tuesday, the 19th August, 1823. Charleston: Printed by A. E. Miller, 1828. 40 p.

S-33593.

Copies: CSmH; ICN; MHi; NcD; PPAmP; ScC; ScCC; ScU; TxU.

1341. Kersey, Jesse, 1768–1845.

Lectures on agriculture, delivered before the Downingtown Society for the Acquisition and Promotion of Natural Knowledge. West Chester, Penn.: Printed by Simeon Siegfried, 1828. [3], 80 p.

S-33770.

Copies: DeGE; P; PPL; PPi.

1342. Lathrop, Leonard Elijah.

The farmers' library: or, Essays designed to encourage the pursuits, and promote the science of agriculture.... 3d ed., corr. and enl. Rochester, N.Y.: Printed by Marshall & Dean, 1828. 344 p.

S-33819.

Copies: CSmH; MNe; NGH; NIC; NNC; NRU; NcD; PHi; PPAmP; PPL; PU.

1343. Muse, Joseph Ennals.

An address upon the dominant errors of the agriculture of Maryland, delivered ... before the Dorchester Agricultural Society ... at the third exhibition and fair, held in Cambridge, Oct. 29, 1827.... Baltimore: Printed by J. D. Toy, 1828. 25 p.

S-34354.

Copies: CtY; DNAL; MdBP; MdHi; NN; PPL.

1344. Seabrook, Whitemarsh Benjamin, 1795–1855.

An address delivered at the first anniversary meeting of the United Agricultural Society of South Carolina, in the hall of representatives at Columbia on 6th Dec., 1827. Published at the request of the Society. Charleston: A. E. Miller, 1828. 40 p.

S-35135.

1344 (*continued*)
Copies: CLU; CtY; DNAL; MH; NcU; PPAmP; RPB; ScC; ViU.

1345. Townsend, John.
An address delivered on Edisto Island, before the Agricultural Society of St. John's, Colleton, at its anniversary meeting, on Wednesday, the 9th of July, 1828.... Charleston, [S.C.]: Printed by A. E. Miller, 1828. 51 p.
S-35545.
Copies: NHi; PHi.

1346. Briggs, George Nixon, 1796–1861.
An address delivered before the Berkshire Association for the Promotion of Agriculture and Manufactures, at Pittsfield, October 2, 1828.... Pittsfield, [Mass.]: Printed by S. W. Bush, 1829. 21, [2] p.
S-37942.
Copies: DNAL; MPB.

1347. Este, David Kirkpatrick.
Address, to the Agricultural Society of Hamilton County, at the annual exhibition and fair, June 19, 1829.... Cincinnati: Hamilton County Agricultural Society; Wm. J. Ferris, printer, 1829. 12 p.
S-38491.
Copies: MWA; OCHP.

1348. Foster, Festus, 1776–1846.
An address delivered before the Hampshire, Franklin, and Hampden Agricultural Society, at Northampton, Oct. 29, 1829.... Northampton, [Mass.]: T. Watson Shepard, printer, 1829. 15, [1] p.
S-38610.
Copies: CSmH; CtSoP; MBH; MWA; N; NN; NjPT; NjR; RPB.

1349. Le Ray de Chaumont, James Donatien, 1760–1840.
An address delivered before the Jefferson County Agricultural Society, 15th September, 1829, at Watertown, N.Y. ... To which is subjoined, notes, and an abstract of the report of the Viewing Committee. [Watertown?, N.Y.]: 1829. 21 p.
S-39272.
Copies: DLC; DNAL; MB; MH; N; NN; PHi; TxU.

1350. Massachusetts. General Court. House. Committee on Agriculture.
Report on so much of the Governor's message as relates to agriculture, on petitions of agricultural societies for extension of grants, and on culture of silk-worm. [Boston: 1829.] 7, 1 p. (House. Doc. 17)
Hasse (Massachusetts)-p.117.
Copies: M.

1351. North Carolina. Board of Agriculture.
Papers on agricultural subjects; on the silk worm and making silk; on the vine and making wine, and Professor Mitchell's Report on the geology of North-Carolina. Pub. by order of the Board of Agriculture. Raleigh: Printed by J. Gales & Son, 1829. 119 p.
S-39861.
Copies: NcD; NcU.

1352. Pickering, Timothy, 1745–1829.
Address to Essex Agricultural Society, with premium list of 1828 and premiums offered 1829. Salem, [Mass.]: 1829. 77 p.
N-P0346857.
Copies: MWA.

1353. Allen, Samuel Clesson, 1772–1842.
An address delivered at Northampton, before the Hampshire, Franklin & Hampden Agricultural Society, October 27, 1830.... Northampton, [Mass.]: T. Watson Shepard, printer, 1830. 32 p.
C-63.
Copies: CSmH; CtY; DLC; MBAt; MBH; MH; MWA; MiD-B; N.

1354. Cuthbert, James.

An address delivered in Charleston, before the Agricultural Society of South-Carolina, at its anniversary meeting, on Tuesday, 18th August, 1830.... Charleston: Printed by A. E. Miller, 1830. 16 p.

Discusses the value of cattle as source of fertilizer.

C-1065.

Copies: CSmH.

1355. Hints addressed to the farmers of Essex. Salem, [Mass.]: Printed by W. & S. B. Ives, 1830. 22 p.

Hints to increase farm production, particularly through proper use of fertilizers.

C-1848.

Copies: DLC.

1356. Le Ray de Chaumont, Vincent, 1790–1866.

Address delivered before the Jefferson County Agricultural Society, at the annual cattle show, 28th September, 1830. By V. Le Ray de Chaumont.... [Watertown?, N.Y.]: 1830. 28 p.

Report of the Viewing Committee, p. 22–26.

C-2206.

Copies: DLC; MB; MiD-B; N; NN; PPAmP; TxU.

1357. [Marcet, Jane Haldimand] 1769–1858.

Conversations on vegetable physiology; comprehending the elements of botany, with their applications to agriculture.... Adapted to the use of schools by Rev. J. L. Blake.... Boston: Crocker & Brewster; New York: J. Leavitt, 1830. 372 p., 4 plates.

First published in London, 1829.

C-2390.

Copies: CSt; IU; MB; MH; MdU; NNC; NcU; OU; PPC; PU; TxU.

1358. ——— New York: Printed for G. & C. Carvill by Sleight and Robinson, 1830. xii, [13]–354 p., 4 plates.

C-2391.

Copies: CtY; MB; MH; MWA; MiU; MnHi; NNNAM; NRU; NcD; NjR; OClWHi; PU; TxU.

1359. Sedgwick, Theodore, 1780–1839.

An address, delivered before the Berkshire Agricultural Society, October 7, 1830.... Pittsfield, [Mass.]: Printed by P. Allen and Son, 1830. 20 p.

C-3438.

Copies: DLC; DNAL; MBC; MH; MPB; PPAmP.

1360. The western agriculturist, and practical farmer's guide. Prepared under the superintendence of the Hamilton County Agricultural Society. Cincinnati: Robinson and Fairbank, 1830. 367 p., illus., plates.

Published in 1832 under the title: *The Farmer's guide, and western agriculturist.*

C-5411.

Copies: DLC; ICN; IU; InU; KyLxT; MH; NcD; C; OC; OCHP; OHi.

3. Agricultural exhibitions

1361. Agricultural Society of Jefferson County, N.Y.

Proceedings of the first cattle show and fair of the Agricultural Society of Jefferson County, September, 1818. Watertown, [N.Y.]: Printed by John H. Lord, Jr.,[1818]. 24 p.

S&S-44459.

Copies: CSmH; MBAt; MWA; NN; NNG.

1362. Philadelphia County Fair Association.

List of premiums, rules and regulations

1362 (*continued*)
of the 7th annual fair of the Philadelphia County Fair Association ... 1818. [Philadelphia: Byberry, 1818.] 64 p.
S&S-45289.
Copies: OClWHi.

1363. Agricultural Society of Jefferson County, N.Y.
Proceedings of the second cattle show and fair ... September, 1819. Watertown, N.Y.: Printed by Seth Alden Abbey, for the Society, [1819]. 33 p.
S&S-46943.
Copies: CSmH; MWA; N; NHi; NN.

1364. Essex Agricultural Society.
Premiums offered by the Essex Agricultural Society, December 7, 1819. Salem, Mass.: Printed by T. C. Cushing, [1819]. 7 p.
S&S-47912.
Copies: MSaE.

1365. Worcester County Agricultural Society.
Worcester County cattle show and exhibition of manufactures. The Trustees of the Agricultural Society of the County of Worcester, desirous that the County should derive the earliest practicable advantages ... announce to the publick their determination to have a cattle show and exhibition of manufactures, at Worcester, on Thursday the seventh day of October next ... and they propose the following premiums to be awarded for excellence on that occasion.... Worcester, [Mass.]: March 24, 1819. Broadside.
S&S-50169.
Copies: MH; MWA.

1366. Massachusetts Society for Promoting Agriculture.
Cattle show, exhibition of manufactures, ploughing match, and public sale of animals and manufactures, at Brighton ... the 17th and 18th of October, 1820.... [Boston]: 1820. Broadside 50 x 41 cm.
Copies: MHi.

1367. Ontario County Agricultural Society.
Premiums for 1820.... [Canandaigua, N.Y.: 1820?] Broadside 44.5 x 27.5 cm.
Includes premiums for domestic manufactures and plows.
S-2607.
Copies: NCanHi.

1368. Essex Agricultural Society.
Essex cattle show. Rules and regulations.... Ploughing match and sale of animals, manufactures, &c. at Danvers Plains, Oct. 16 and 17, 1821. [Salem?, Mass.: 1821.] Broadside.
S-5263.
Copies: MSaE.

1369. [Essex Agricultural Society.]
The trustees account of the cattle show, and other exhibitions, at Topsfield, Oct. 5, 1820. Salem, [Mass.]: Printed by John D. Cushing, 1821. 56 p.
An address to the Essex Agricultural Society by Andrew Nichols, p. [33]–56.
S-5262.
Copies: DLC; DNAL; KHi; MB; MH; MSaE; MWA; MiD; MnHi; NjR; PPL.

1370. Massachusetts Society for Promoting Agriculture.
Cattle show, exhibition of manufactures ... and public sale of animals and manufactures, at Brighton, Mass.... the 10th and 11th of October, 1821.... Boston: Russell and Gardner, printers, 1821. Broadside 63 x 51 cm.
S-5989.
Copies: MHi; MWA.

1371. Hillsborough Society for the Promotion of Agriculture and Domestic Manufactures.
Hillsborough cattle show and fair, at Amherst ... 24th and 25th days of September, 1822. [Amherst?, N.H.: 1822.] Broadside 51 x 35 cm.
Signed: Charles Barrett, chairman,

February 22, 1822.
Copies: MWA.

1372. Massachusetts Society for Promoting Agriculture.
Cattle show, exhibition of manufactures ... and public sale of animals and manufactures, at Brighton, Massachusetts ... the 9th and 10th of October 1822.... Boston: Printed by Russell and Gardner, [1822]. Broadside 61 x 52 cm.
Copies: MWA.

1373. Cattle show and fair, for the exhibition and sale of all kinds of live stock, agricultural implements. To be held on Thursday & Friday, the 7th & 8th of June, 1824. At the Maryland Tavern, four miles from Baltimore, on the Frederick turn pike road.... Charles Ridgely, of Hampton, chairman of committee of arrang. Baltimore: Printed by Robinson, [1824]. 1 p.
S-15702.
Copies: NcU.

1374. Essex Agricultural Society.
Cattle show, ploughing match, and exhibition of manufactures, at Topsfield, on Wednesday, October 5th, 1825. List of premiums. [Salem, Mass.]: Printed at the Salem Gazette office, [1825]. Broadside.
S-20422.
Copies: MHi.

1375. Massachusetts Society for Promoting Agriculture.
Cattle show, exhibition of manufactures ... and public sale of animals and manufactures, at Brighton, Massachusetts ... 19th and 20th of October, 1825.... Boston: Printed by John B. Russell, [1825]. Broadside 62 x 52 cm.
Copies: MWA.

1376. Essex Agricultural Society.
Account of premiums awarded in 1825, and a list of premiums offered in 1826; accompanied with remarks & explanations for the information of the farmers of the county. Salem, [Mass.]: W. & S. B. Ives, printers, 1826. 33 p. (Its *Transactions*, v. 1, no. 7)
S-24443.
Copies: DLC; MB; MBAt; MH; MHi; MPeaHi; MWA; MnHi; NjR; OClWHi.

1377. Massachusetts Society for Promoting Agriculture.
Cattle show, exhibition of manufactures ... and public sale of animals and manufactures, at Brighton, Massachusetts ... 17th of October, 1827.... [Boston]: Printed by John B. Russell, [1827]. Broadside 64 x 52 cm.
Copies: MWA.

1378. Ontario County Agricultural Society.
Premiums for 1827.... Canandaigua, [N.Y.]: Printed by Bemis, Morse & Ward, [1827]. Broadside 43.5 x 35 cm.
Includes premiums for plows, threshing machine, and a machine for dressing flax.
S-30122.
Copies: NCanHi.

1379. Massachusetts Society for Promoting Agriculture.
Cattle show, exhibition of manufactures ... and public sale of animals and manufactures, at Brighton, Massachusetts ... 15th of October, 1828.... Boston: John B. Russell, printer, [1828]. Broadside 62 x 52 cm.
Copies: MWA.

1380. Rhode Island Society for the Encouragement of Domestic Industry.
Cattle show, exhibition of manufactures ... sale of animals, manufactures, at Pawtucket, R.I.... 21st and 22d day of October, 1828.... [S.l.]: 1828. Broadside 52 x 25 cm.
N-R0227776.
Copies: MWA; RPB.

1381. Essex Agricultural Society.
Account of premiums awarded in 1828, and a list of premiums offered in 1829; with Col. Pickering's address; and a list of the

1381 (*continued*)
members of the Society.... Salem, [Mass.]: W. & S. B. Ives, printers, 1829. 77, 4 p. (Its *Transactions,* v. 1, no. 8)
S-38490.
Copies: CSmH; DLC; DeGE; MBAt; MH; MoSM; NCH; NjR.

1382. Massachusetts Society for Promoting Agriculture.
Cattle show, exhibition of manufactures ... and public sale of animals and manufactures, at Brighton, Massachusetts ... 14th of October, 1829.... [Boston]: Press of Isaac R. Butts, [1829]. Broadside 64 x 52 cm.
Copies: MWA.

1383. Essex Agricultural Society.
Reports of committees and premiums awarded in 1829; and a list of premiums offered in 1830; with remarks and hints to farmers, &c.... Salem, [Mass.]: W. & S. B. Ives, printers, Observer office, 1830. 88 p. (Pamphlet no. IX, 1829)
C-1278.
Copies: MBAt; MH; MHi; MWA.

4. Agricultural implements and machines

1384. A short treatise on rice machinery; with a schedule of the expense. Also, the construction of pumps. With proper tables, &c. By J. B. Esq., of South-Carolina. [Charleston: Printed by Bowen and Markland?], 1786.
E-19483.
No copy known.

1385. Tyler, Benjamin.
The United States of America. To all to whom these letters patent shall come: whereas, Benjamin Tyler, a citizen of New Hampshire in the United States, hath alledged that he has invented a new and useful improvement, in the mode of cleaning wheat, rye, buck-wheat, and all other kinds of grain. [S.l.]: 1796. Broadside.
Copies: DNAL; DP.

1386. Jefferson, Thomas, 1743–1826.
A supplementary note on the mould board described in a letter to Sir John Sinclair, of March 23, 1798, inserted in the American Philosophical transactions, vol. 4, and in Maese's [Mease's] Domestic encyclopaedia voce Plough.... [S.l.: 1804?] Broadside.
Wrongly dated by Evans as 1798; MHi dates it on good authority as 1805.
E-33934; S&S-6552.
Copies: DLC; MH; MHi; NN.

1387. Adams, Paul.
Patent shelling machine.... To all people to whom these greeting shall come. Paul Adams, of Newsbury ... send greetings. Whereas, by virtue of an act of Congress entitled "An act to promote the progress of useful arts, &c." Paul Pilsbury of Newsbury, aforesaid, hath obtained letters patent, for a machine for shelling Indian corn on a new and improved method.... [Newsbury, Mass.]: 1805. Broadside 23 x 18 cm.
Part of the text is filled in by hand and signed the 29th day of October, 1805: Paul Adams.
Copies: MHi; MWA.

1388. Button, Richard.
Richard Button's flax & hemp machine.... Cazenovia, [N.Y.]: Printed by Dockstader & Bunce, 1807. Broadside, illus.

S&S-12256.
Copies: NN.

1389. Evans, Evan.
Patents for cyder, or tobacco press, straw cutter, flour press. [Philadelphia: 1807?] Broadside, illus. 35 x 25 cm.
Contains three engravings by F. Shallus; those for the straw cutter and flour press are marked respectively Fig. 6 and 7.
Announcement of Oliver Evans' Mars Works is dated: Philadelphia, April 30, 1807.
Privately held.

1390. Evans, Evan.
Cyder or tobacco press ... [and] Evan Evans's straw-cutters.... [Washington, D.C.: ca. 1813.] Broadside, illus. 27 x 25 cm.
Contains 3 engravings by F. Shallus.
On verso: Licence given by Evan Evans, of Washington, to Timothy Kirk at Lancaster Pa., to use his patented presses for packing flour. Dated in ink: 5th August, 1813.
Copies: DeGE.

1391. Evans, Evan.
An explanation of the improvements on straw cutters; and flour, cider and tobacco presses. As also, improvements on agriculture.... [Washington]: Printed for the author, 1813. 47, [1] p., 2 fold. plates.
S&S-28456.
Copies: DGU; DLC; DNAL; DeGE; PU; PU-V.

1392. Gale, Luther.
C. Tobey's patent bark mill. Important notice to tanners and others.... Now know ye that I, Luther Gale ... have purchased the sole and exclusive right and title of C. Tobey's patent bark mill.... New York: John C. Totten, printer, [1813]. Broadside 45 x 28 cm.
Dated: New York, December 30, 1813.
S&S-51301.
Copies: DeGE.

1393. United States. Dept. of State.
[Letters patent to Nathan Read for the invention of a machine for threshing wheat and other small grains. Washington: 1817.] Broadside.
Signed by James Monroe, President of the United States, and John Quincy Adams, Secretary of State, December 23, 1817.
S&S-42646.
Copies: MSaE.

1394. Gale, Luther.
Notice. Whereas I have the exclusive right, so secured to me by patent, for a mill of cast iron, to grind plaister, corn and cob, and I am credibly informed that these mills are made and for sale in Albany; as this is an infringement on my right, this is to notify the owners of mills, that ... they must expect to pay me the right of using them.... Lenox, [Mass.]: 1818. Broadside 15 x 20 cm.
Dated: Lenox, May 9, 1818.
Copies: DeGE.

1395. Herring, H., firm, Baltimore.
Jacob Bromwell's patent self feeding wheat fan.... Baltimore: [ca. 1818.] [2] p., illus.
Offered for sale by the firm of H. Herring, Baltimore.
Copies: DeGE.

1396. Bromwell, Jacob.
Jacob Bromwell's patent self feeding wheat fan.... Permission is hereby granted to use one of Jacob Bromwell's patent wheat fans.... [Baltimore: 1819.] 2 p.
S&S-47434.
Copies: DLC.

1397. Wood, Jethro.
Specification of letters patent for an improvement in a plow. New York: 1819. Broadside.
S&S-50148.
Copies: MSaE.

1398. Notice. Important to farmers, tavern-keepers, cartmen, travellers, and economists in general ... a straw cutting machine.... [New York: 1820.] Broadside.
Dated: New-York, Feb. 21, 1820

1398 (*continued*)
S-2570.
Copies: NN.

1399. United States. Congress. House. Committee on Agriculture.
Report of the Committee on Agriculture on the petition of Anthony Dey and James Macdonald. March 12, 1822.... [Washington: 1822.] 3 p. (17th Cong., 1st sess. House. Rept. 67)
Concerning a machine for breaking and cleaning hemp and flax, invented by J. Macdonald.
S-11144.
Copies: DLC; DeGE; PU.

1400. United States. Congress. House. Committee on the Judiciary.
Report of the Committee on the Judiciary, on the petition of Anthony Dey and James Macdonald, and a remonstrance of David Melville. April 29, 1822.... [Washington: 1822.] 2 p. (17th Cong., 1st sess. House. Rept. 102)
Refusal to extend the patent for a machine for breaking and cleaning hemp and flax; patented by J. Macdonald.
S-11200.
Copies: DLC; DeGE; DeU; MWA; NjP.

1401. Ohio. General Assembly. House. Committee to Examine Hines & Bain's Machine.
Report of the Committee appointed by the House of Representatives of Ohio, to examine Hines & Bain's machine and report their opinion of its merits.... Columbus: David Smith, printer, 1824. 7 p.
Concerns the flax-dressing machine, patented by Joseph Hines and William Bain on August 12, 1826.
S-17481.
Copies: O (not located).

1402. Warren, Edmund.
To all to whom these presents shall come.[S.l.: 1824?] [2] p.
Concerns his improved threshing machine, patented on March 6, 1824.
Copies: NN.

1403. United States. Laws.
A bill to extend the patent of Jonathan Lucas, Jr., for a further period of fourteen years. April 24, 1826.... [Washington: 1826.] 1 p. (s.137)
Concerns rice hulling, cleaning and polishing machines.
S-26826.
Copies: DNA.

1404. United States. Laws.
A bill authorizing the Secretary of State to issue a patent to Elizabeth H. Bulkley, widow of Chauncey Bulkley, deceased. January 7, 1828.... [Washington: 1828.] 1 p. (S.47)
Concerns a corn sheller and a cast iron hoe.
S-35732.
Copies: DNA.

1405. Davis, Gideon.
Directions how to stock and use Gideon Davis's improved barshare or fallow plough. Gideon Davis, manufacturer. Georgetown, D.C.: [1829]. 2 p.
S-38344.
Copies: DLC.

5. Mills and milling

1406. Georgia. Laws.
An act to regulate the toll to be taken at mills. [Augusta: 1786.] 4 p.
The item also includes three other acts.
B-6292; Mp-44896.
Copies: GU-De.

1407. Evans, John.

To the millers.... Lancaster, [Pa.]: Printed by Steemer, Albright & Lahn, [1787]. Broadside.

Signed and dated: John, Theophilus & Oliver Evans.... Redclay Creek, Dec. 19, 1787.

Advertises their merchant mill near Wilmington, Del., equipped with elevators and hopperboys, recently invented by Oliver Evans.

Copies: MHi.

1408. Burnes (Daniel) and Son, Philadelphia.

Burr mill-stones, made and sold at Daniel Burnes and son's store.... Philadelphia: Joseph Jones, 1788. Broadside 18.5 x 22.5 cm.

Dated: Philadelphia, August 11, 1788.

B-6680; Mp-45234.

Copies: NHi.

1409. Parker, James.

To be sold at public vendue, on Thursday the 14th day of January next, the noted grist-mill in Alexandria ... New Jersey.... New Brunswick: Abraham Blauvelt, [1789]. Broadside 36 x 22.5 cm.

Dated: December 19, 1789.

B-7020.

Copies: NjR.

1410. Evans, Oliver, 1755–1819.

Improvements on the art of manufacturing grain into flour or meal.... [Philadelphia?: Printed by William Young?, 1791.] Broadside, illus., 54 x 47.5 cm.

Also assumed to have been printed at Wilmington, Delaware.

B-7695; Mp-46164.

Copies: NN.

1411. Evans, Oliver, 1755–1819.

To the public.... [Philadelphia: 1795.] Broadside 52 x 42 cm.

Advertising for his forthcoming *Young mill-wright and miller's guide,* with an outline of contents.

Mp-28643.

Copies: PHi.

1412. Evans, Oliver, 1755–1819.

The young mill-wright & miller's guide. In five parts—embellished with twenty five plates.... Philadelphia: Printed for, and sold by the author, 1795. [8], 160, ii, [3]–178, x, [11]–90, 10, [11] p., 26 plates (2 fold.).

Part 5 is by Thomas Ellicott.

Mp-28644.

Copies: CtY; DLC; DeGE; ICJ; MBAt; MH-BA; MWA; MiU-C; NN; PPAmP; PPF; PPL; PSC; PU.

1413. Article of agreement, made and concluded upon this first day of April, one thousand seven hundred and ninety-six, with the subscribers, for the rules and wages of mill-wrighting.... [Wilmington?, Del.: 1796.] Broadside 34 x 21 cm.

Signed: Joseph Dutton, Brandywine; James Corbett, Wilmington; John Harlan, Nottingham, etc.

Gives rates for making specified parts for flour-mills.

Copies: DeGE.

1414. Hughlett, William

Delaware State Mills. Being desirous of driving my mills to the extent, it requires a great deal of wheat, corn and rye, to keep them in operation.... [Wilmington?, Del.: 1796.] Broadside 24 x 16 cm.

Signed: (in autograph) Wm Hughlett, February 13th 1796.

B-9578; Mp-47810.

Copies: PPL.

1415. Evans, Oliver, 1755–1819.

Oliver Evans, No. 126, South Second-Street, Philadelphia, keeps for sale at the Old Windmill, plaister of Paris ... stucco work ... bolting cloth ... patent burr millstones ...

1415 (*continued*)
Philadelphia: Printed by Henry Sweitzer, [1797]. Broadside.
In English and German.
Copies: MiU-T.

1416. Evans, Oliver, 1755–1819.
Patent burr mill-stones, made by Oliver Evans, No. 126, South Second-Street, Philadelphia. Perfectly regular and true, balanced on their centers.... [Philadelphia]: Printed by H. Sweitzer, [1797?]. Broadside.
Copies: MiU-T.

1417. To the farmers and others, who bring wheat to the city of New-York for sale. We, the subscribers, residing in the said city, engaged in the manufactory of flour ... interested ... in improving the quality ... address you.... New York: Isaac Collins, [1798]. Broadside 44 x 27 cm.
Dated: New-York, June 1st, 1798.
B-10544; Mp-48633.
Copies: N.

1418. Coates, Moses.
A new invention and improvement on saw mills, for accelerating the sawing of timber into boards, scantling, &c. &c. Accompanied with an explanatory [*sic*] plate. Lancaster, [Pa.]: Printed by Robert Bailey, 1802. 8 p., fold. plate.
S&S-2042.
Copies: DeGE (xerox copy); PPAmP.

1419. United States. Congress. House. Committee of Commerce and Manufactures.
Report from the Committee of Commerce and Manufactures, to whom was referred on the twenty first ultimo, the petition of Oliver Evans of the city of Philadelphia. 22d January, 1805.... Washington: Printed by William Duane & Son, 1805. 4 p.
Reported by Jacob Crowninshield.
Concerns extension of Oliver Evans' patent on flour-mills.
S&S-9603.
Copies: DLC; MBAt; NN; PPAmP.

1420. Habbel's new-invented water-wheel, for corn mills.... Rutland, 3d November, 1806. [Rutland, Vt.: Wm. Fay, 1806.] Broadside 39 x 33 cm.
S&S-10509.
Copies: VtHi.

1421. Evans, Oliver, 1755–1819.
The young mill-wright's & miller's guide; in five parts.... Octoraro, [Pa.]: Printed by Francis Bailey, 1807. viii, 364 p., plates (part fold.).
S&S-12525.
Copies: CSmH; DeGE; MWA; MiD; N; OU; PSC; PU.

1422. United States. Congress. House. Committee to Whom was Referred the Petition of Oliver Evans.
Report of the Committee, to whom was referred, on the eleventh ultimo, the petition of Oliver Evans, of the city of Philadelphia. December 2d, 1807. Read, and together with a bill for the relief of Oliver Evans, committed to a Committee of the Whole House. City of Washington: A. & G. Way, printers, 1807. 5, [3] p.
Concerns Oliver Evans's patent on flour-mills.
S&S-14076.
Copies: DLC; DeGE; MWA; NN.

1423. [Evans, Oliver] 1755–1819.
Law case. The following decision on the patent right of Mr. Oliver Evans, of this city, was given at the last session of the Circuit Court of the United States, held by judges Washington and Peters. Oliver Evans vs. John Weiss. This was an action on the case for a violation of the plaintiff's patent right.... [Philadelphia: 1808.] 2 p.
Concerns his reissued patent for mill machinery.
Copies: PHi.

1424. Evans, Oliver, 1755–1819.
To the millers of the United States. Take notice—that Congress did on the 21st day of January, 1808, pass the following act: An act for the relief of Oliver Evans....

[Philadelphia: 1809.] Broadside 25 x 20.5 cm.
Dated: December 30, 1809.
Contains the act authorizing the reissue of Evans's patent on mill machinery.
Copies: PHi.

1425. ———— [Philadelphia: 1810.] [3] p.
"An estimate of the profits arising from the use of Evans's improvements in the manufacture of flour and meal," on p. [3], signed: Oliver Evans, Philadelphia, April 20, 1810.
Copies: DeGE; MWA.

1426. Evans, Oliver, 1755–1819.
Instructions. The subscriber gives to his several agents the following instructions for regulating the prices of his licenses to use his several patented improvements during his patent terms.... [Philadelphia: 1811.] Broadside 33.5 x 21 cm.
Dated: Philadelphia, Jan. 1, 1811.
Concerns his mill machinery.
Copies: PPF.

1427. United States. Congress. House. Select Committee to Whom was Referred the Petition of John Brumback and Others.
Report of the Select Committee, to whom was referred on the nineteenth ultimo, the petition of John Brumback and others, of the state of Virginia. January 19, 1811.... Washington: A. and G. Way, printers, 1810 [1811]. 7 p.
Reported by S. L. Mitchell.
Concerns the infringement of Oliver Evans's patents on flour-mills.
S&S-24311.
Copies: DLC; DeGE; MWA.

1428. Evans, Oliver, 1755–1819.
Brief explanation of the screw mill for breaking hard substances, or Indian corn to be ground with the cobs.... [Philadelphia: ca. 1813.] Broadside, illus. 31 x 27 cm.
Includes licence, granted by Oliver Evans to Timothy Kirk, Lancaster Co., Pennsylvania, filled in in ink and dated August 5th, 1813.
Copies: DeGE.

1429. Evans, Oliver, 1755–1819.
Licence. Know all men, that I, Oliver Evans, of the City of Philadelphia, have received of.... [Philadelphia?: ca. 1813.] Broadside 23 x 25 cm.
Licence to use Evans's patented improvements of flour mills, given to Timothy Kirk, Lancaster, Pa.Filled out in ink and dated August 5, 1813.
On verso: Reproduction of the eighth plate from Evans's *Millwright and miller's guide,* representing his automatic flour-mill.
Copies: DeGE.

1430. Evans, Oliver, 1755–1819.
Notice is hereby given you, that the following act has been passed by Congress: An act for the relief of Oliver Evans.... [Philadelphia?: ca. 1813.] Broadside 40 x 25 cm.
Request to pay for the licence to use Evans's patented improvements of flour-mills. Delivered to Martin Rittenhouse, Wissahickon, Philadelphia Co., Pa., and dated June 8, 1813.
Copies: DeGE; PPF.

1431. ———— [Philadelphia?: ca. 1813.] Broadside 25 x 21 cm.
Copy of a notice of debt for the use of Evans's patented improvements of the flour-mills, delivered to William Reynolds, Juniata, Pa., and dated 12 August, 1814.
Copies: DeGE; PPF.

1432. Evidence accompanying the petition of John Worthington, and others. January 11, 1813. Printed by order of the Senate of the United States. Washington City: Printed by Roger C. Weightman, 1813. 7 p.
Contains the statement of Samuel Stroud and William Marshall, of Stanton, Delaware, concerning Oliver Evans's patented flour-mill.
S&S-30180.
Copies: DLC; DeGE; MWA.

1433. Memorial to Congress of sundry citizens of the United States, praying relief from the oppressive operation of Oliver Evans'

1433 (*continued*)
patent; vesting in him an exclusive right to certain machinery, now in general use, in the manufacture of flour; with evidence to shew that the said Evans was neither the original inventor nor first applier of said machinery, and consequently not entitled to the reward due to an actual inventor under the patent laws of the United States. Baltimore: Printed by J. Robinson, 1813. 50 p.

Prepared by Isaac M'Pherson.

S&S-29147.

Copies: DLC; DeGE; MWA; Md; MdBP; PPL.

1434. To the Senate and House of Representatives of the United States of America. [Baltimore: 1813.] 4 p.

Dated at end: Baltimore, January 1, 1813.

A petition protesting the renewal of Oliver Evans's patent on the flour-mill.

Copies: DeGE.

1435. United States. Congress. Senate. Committee to Whom Were Referred the Memorials of John Worthington, and Others.

Report of the Committee to whom were referred the memorials of John Worthington, and others, and Oliver Evans. February 22d, 1813.... Washington City: Printed by Roger C. Weightman, 1813. 4 p.

Concerns the patents issued to Oliver Evans for improvements on the flour-mills.

S&S-30357.

Copies: DLC; DeGE; MWA.

1436. [Evans, Oliver] 1755–1819.

A trip made by a small man in a wrestle with a very great man. Answer to the Baltimore millers' memorial and Thomas Jefferson's letter. [S.l.: 1814.] 25 p.

Answer to Jefferson's letter, published in *Niles' Register,* together with the memorial of Isaac M'Pherson, Elisha Tyson, Thomas Ellicott and others, concerning Evans's patent on flour-mills.

S&S-31434.

Copies: DLC; DGU; DeGE (microfilm); MBAt; NN; ViU.

1437. United States. Congress. House. Committee to Whom was Referred a Petition of Inhabitants of Baltimore County.

Report of the Committee to whom was referred, on the fourth instant, a petition of sundry inhabitants of Baltimore County, in the state of Maryland, complaining against the oppressive exercise, &c. of the patent right granted to Oliver Evans for the invention of certain mill machinery.... Washington: A. & G. Way, printers, 1814. 4 p. (13th Cong., 2d sess. House. Doc. 24)

S&S-33477.

Copies: DLC; DeGE; MWA.

1438. United States. Congress. Senate. Committee to Whom was Referred the Memorial of Isaac M'Pherson and Others.

Report of the Committee to whom was referred the memorial of Isaac M'Pherson and others, on the subject of the patent granted to Oliver Evans, under the act of twenty-first January, 1808, entitled "An act for the relief of Oliver Evans." February 25, 1814.... Washington City: Printed by Roger C. Weightman, 1814. 6 p.

Concerns the flour-mill patented by Evans.

S&S-33481.

Copies: DLC; DeGE; MWA; MiD-B.

1439. United States. Congress. House. Select Committee to Whom was Referred the Petition of John and Benjamin Tyler.

Report of the Select Committee, to whom was referred, on the 19th December, 1815, the petition of John and Benjamin Tyler. January 9, 1816. Made, and accompanied with a bill, to extend to Benjamin and John Tyler, of New-Hampshire, the patent right, &c. [Washington: 1816.] 2 p. (14th Cong., 1st sess. House. Doc. 15)

Concerns the improvement of flour and other mills.

S&S-39635.

Copies: DLC; DeGE; MWA.

1440. United States. Laws.

A bill to extend to Benjamin Tyler and John Tyler, of the state of New Hampshire,

the patent right to certain improvements in the construction of corn and other mills. January 9, 1816. [Washington: 1816.] 2 p. (H.R. 28)
S&S-39334.
Copies: DLC.

1441. Evans, Oliver, 1755–1819.
The young mill-wright's & miller's guide. In five parts; embellished with twenty-five plates ... 3d ed. Philadelphia: Printed by M. Carey, 1818. viii, [9]–364 p., 26 plates (2 fold.).
Part 5, by Thomas Ellicott, p. [282]–354.
S&S-43978.
Copies: DeGE; MWA; PSC.

1442. [Evans, Oliver] 1755–1819.
Address to the constituted authorities of the United States, on: 1. The inherent exclusive right of inventors to the labors of their minds and bodies.... 2. The recognition of those rights by the constitution of the United States.... 3. The acts of Congress to secure these rights.... 4. The patents granted under these acts to secure these rights.... By an advocate for patentees, inventors of improvements in the useful arts.... City of Washington: Printed by Jonathan Elliot, 1819. 55 p.
N-E0205387.
Copies: DP; PPAmP.

1443. Brush, John C.
A small tract, entitled A candid and impartial exposition of the various opinions on the subject of the comparative quality of the wheat and flour in the northern and southern sections of the United States.... To which are added instructions showing how to improve the character of the northern flour ... in a letter from John C. Brush ... to Samuel Latham Mitchill.... Washington City: Printed by Jacob Gideon, Junior, 1820. 52 p.
S-609.
Copies: CSmH; DLC; DeGE; MWA; NN; PPL; ScU.

1444. [Coles, Benjamin U.]
A memoir on the subject of the wheat and flour of the state of New-York. Read at a sitting of the board of managers of the New-York County Agricultural Society, February 17, 1820.... New-York: Printed by William A. Mercein, for Kirk and Mercein, 1820. 46 p., illus.
S-814, 2524.
Copies: CtY; MB; MH-BA; NIC; NN; NNS; ScU.

1445. Evans, Oliver, 1755–1819.
The young mill-wright and miller's guide. In five parts.... Embellished with twenty-five plates ... 4th ed. Philadelphia: M. Carey & Son, 1821. viii, [9]–374 p., 26 plates (part fold.).
Part 5: *The Practical mill-wright, or instructions for building mills* by Thomas Ellicott.
Essay on saw-mills by William French, p. 371–372.
S-5277.
Copies: CSf; DeGE; DeWI; ICJ; MH; NN; NNC; OBg; PPF; PSC; ScCleU; ViU; WU.

1446. Hart, Webb.
To mill owners. I have discovered a new and valuable improvement in the application of power, which may be applied to mill stones, or any other machinery ... Webb Hart. Petersburg, Va.: Pescud, printer, [1824]. Broadside.
S-16458.
Copies: DNA.

1447. Evans, Oliver, 1755–1819.
The young mill-wright and miller's guide, illustrated by twenty-eight descriptive plates. By Oliver Evans. 5th ed. With additions and corrections by the professor of mechanics, in the Franklin Institute of the state of Pennsylvania [Thomas P. Jones]; and a description of an improved merchant flour-mill, with engravings. By C. & O. Evans.... Philadelphia: H. C. Carey and I. Lea, 1826. viii, [9]–383 p., 28 plates (part fold.).
" ... Practical mill-wright, or Instructions for building mills, with all their pro-

1447 (*continued*)
portions; suitable to all falls, of from three to thirty-six feet. Received from Thomas Ellicott." p. [275]–344.
S-24450.
Copies: AU; CtY; DeGE; NWM; PSC; RP; ScU.

1448. [Cowell, Benjamin] 1781–1860.
Ancient documents relative to the old grist-mill, with some remarks on the opinions of Messrs. Hunter and Greene, counsel employed by the town to examine the same. [Providence]: Herald office, 1829. 32 p.
S-38292.
Copies: CSmH; DLC; MH-L; MWA; RHi; RNR; RPB; RPaw.

1449. Evans, Oliver, 1755–1819.
The young mill-wright and miller's guide. Illustrated by twenty-eight descriptive plates.... The 6th ed. With additions and corrections by the professor of mechanics in the Franklin Institute of the state of Pennsylvania; and a description of an improved merchant flour-mill, with engravings, by C. & O. Evans.... Philadelphia: Carey & Lea, 1829. viii, [9]–383 p., 28 plates (part fold.).
S-38503.
Copies: GHi; MdBE; MiU; NbU; NcD; OHi.

1450. Hunter, William, 1774–1849.
The legal opinion of the Hon. William Hunter, on the question of the town's interest in the ancient grist mill. [Providence]: Printed at the Patriot office, 1829. 15 p.
S-39053.
Copies: DLC; MWA; OCLaw; RHi; RPB.

1451. Paul, Joseph M., firm, Philadelphia.
Mills at Eayrestown, for sale or to let.... [Philadelphia]: Printed by J. Rakestraw, [ca. 1830]. Broadside 29 x 23 cm.
Includes a description of the mills.
C-2959.
Copies: DeGE.

6. Food and beverage manufacture and preservation

1452. Chapman, Thomas.
Cyder-maker's instructor, sweet maker's assistant, and victualler's and housekeeper's director. In three parts. Part I. Directs the grower to make his cyder in the manner foreign wines are made.... Part II. Instructs the trader or house-keeper to make raison wines.... Part III. Directs the brewer to fine his beer and ale in a short time.... London printed, and Philadelphia: Reprinted by And. Steuart, [1762]. 23, [1] p.
Mp-9085.
Copies: PHi.

1453. ———— To which is added, a method to make yest[*sic*] to ferment beer.... London printed. Boston: Reprinted and sold by Green & Russell, 1762. 28 p.
Mp-9084.
Copies: DLC; MHi; MWA; MeB; PPL; RPJCB.

1454. Remarks on the manufacturing of maple sugar; with directions for its further improvement. Collected by a society of gen-

tlemen, in Philadelphia, and published for the general information and benefit of the citizens of the United States. Philadelphia: Printed by James & Johnson, 1790. 24 p.
Mp-22832.
Copies: DLC; DeGE; MH; MWA; PHi; PPF; RPJCB.

1455. ——— Albany: Printed and sold by Charles R. & George Webster, 1791. 24 p.
E-23732.
Copies: DeGE.

1456. ——— Philadelphia, printed. New York: Reprinted by William Morton, 1791. 23 p.
Title varies slightly.
Mp-23731.
Copies: MWA; PHi; RPJCB.

1457. The art of cheese making, taught from actual experiments, by which more and better cheese may be made from the same quantity of milk. Concord: Printed and sold by George Hough, 1793. 16 p.
Mp-25123.
Copies: NhHi.

1458. Society for Promoting the Manufacture of Sugar from the Sugar Maple-Tree.
Constitution of the Society for Promoting the Manufacture of Sugar from the Sugar Maple-Tree and furthering the interest of agriculture in the state of Pennsylvania. Established the 18th of September, 1792. Philadelphia: Printed by Robert Aitken & Son, 1793. 8 p.
B-8476; Mp-46876.
Copies: MHi; PPAmP.

1459. [Callender, James Thomson] 1758–1803.
A short history of the nature and consequences of excise laws; including some account of the recent interruption to the manufactories of snuff and refined sugar.... Philadelphia: Printed for the booksellers, Dec. 7, 1795. 116 p.
Mp-28383.
Copies: DLC; DeGE; MBAt; MH; MWA; MiU-C; NHi; NN; NjP; PPL; RPJCB; ScC; ViU.

1460. ——— Philadelphia: Printed for Thomas Stephens, 1795. 116 p.
Mp-28384.
Copies: DLC; MBAt; MdHi; MiU-C; NjP; PHi; ViU.

1461. Child, Samuel.
Every man his own brewer; a small treatise, explaining the art and mystery of brewing porter, ale, and table-beer; recommending and proving the ease and possibility of every man's brewing his own porter, ale and beer.... 2d American ed., improved and calculated according to the measures and current money of the United States. Philadelphia: Printed for T. Condie, 1796. 34 p.
First published in London, 1768.
Mp-30189.
Copies: DLC; MWA; P; PPL.

1462. Strother, John.
A treatise on the distillation of ardent spirits from materials of the growth of the United States. [Richmond?]: 1796.
E-31251.
No copy known.

1463. Twamley, Josiah.
Dairying exemplified; or, The business of cheese-making: laid down from approved rules, collected from the most experienced dairy-women, of several counties. Digested under various heads: from a series of obser-

1463 (*continued*)
vations, during thirty years practice in the cheese trade. By J. Twamley. The first American edition, from the second British, corrected and improved. Providence, (Rhode Island): Printed by and for Carter and Wilkinson, 1796. iv, [5]–78, [5] p.
Mp-31324.
Copies: DLC; MWA; RHi.

1464. The art of cheese-making, taught from actual experiments, by which more and better cheese may be made from the same quantity of milk. Boston: Reprinted by B. Edes, 1797. 16 p.
Kress-B3328; N-A0437300.
Copies: MH.

1465. Die wahre Brantewein-Brennerey; oder, Brantwein-, Gin- und Cordialmacher-Kunst; wie auch die ächte Färbe-Kunst, wie man alle Couleuren auf Seide, Leinen und Wolle färben kan. York, [Pa.]: Gedruckt bey Salomon Mayer, 1797. 2 vols. in 1 (40, 30 p.).
Second volume has the title: *Besondere Kunst, auf eine leichte und wohlfeile Art zu färben; allerley Farbe, so whol auf Wolle, Halbwolle, Leinen und Seide....* Von Henrich Hartmann, berühmten Schönfärber.
Mp-33851.
Copies: DLC; ICJ.

1466. The art of cheese-making, taught from actual experiments; by which more and better cheese may be made from the same quantity of milk. [Litchfield, Conn.: Printed by Thomas Collier, 1798.] 16 p.
E-33315.
No copy known.

1467. ———— Windham, Conn.: Printed and sold by John Byrne, 1798. 8 p.
Mp-33316.
Copies: MWA (imperfect).

1468. Johnson, Joshua.
The art of cheese-making, reduced to rules, and made sure and easy, from accurate observation & experience.... Albany: Printed for the author, by Charles R. and George Webster, 1801. 12 p.
S&S-739.
Copies: MiU-C; N; NHi; NNS.

1469. Krafft, Michael August.
The American distiller; or, The theory and practice of distilling, according to the latest discoveries and improvements, including the most improved methods of constructing stills, and of rectification. By Michael Krafft.... Philadelphia: Thomas Dobson; Archibald Bartram, printer, 1804. [12], [9]–219 p., 2 fold. plates.
S&S-6606.
Copies: CU-A; DLC; DNAL; DeGE; ICJ; MH; MWA; NN; NNNAM; PHC; PHi; PU; ViU.

1470. The complete family brewer; or, The best method of brewing or making any quantity of good strong ale and small beer.... Philadelphia: B. Graves, 1805. 22, 12 p.
An Account of the operations in brewing, as well as the vessels made use of therein, 12 p. at end.
S&S-8227.
Copies: P.

1471. Boucherie, Anthony.
Petition of Anthony Boucherie of the city of Philadelphia. December 14, 1807. Referred to Mr. Holland, Milnor, and Clark. December 18, 1807. Bill presented. City of Washington: A. & G. Way, printers, 1807. 6 p.

Describes a new method to make sugar from sugar cane.
S&S-12186, 14035.
Copies: CtY; DLC; DS; DeGE; MBAt; MWA; NN; PPL; ViU.

1472. Nutt, Frederic.
The complete confectioner; or, The whole art of confectionary made easy: containing ... the art of making the various kinds of biscuits, drops ... as also the most approved method of making cheeses, puddings, cakes ... 4th ed., with considerable additions. London printed. New York: Reprinted for Richard Scott, 1807. 91 p., plate.
S&S-13270.
Copies: CtY; DLC; MB; MH; MWA; MnHi; NN.

1473. M'Harry, Samuel.
The practical distiller; or, An introduction to making whiskey, gin, brandy, spirits ... from ... rye, corn, buckwheat, apples, peaches, potatoes, pumpions and turnips. And sundry extracts of approved receipts for making cider, domestic wines, and beer.... Harrisburgh, (Penn.): John Wyeth, 1809. 184 p.
S&S-17955.
Copies: CSmH; DeGE; MBAt; MWA; MiU-C; PPL.

1474. Gillespie, Robert.
A new plan for distilling with greater facility than has ever been heretofore practised in any country. By R. Gillespie. Baltimore: Printed by Benjamin Edes, 1810. 15 p., illus.
S&S-20219.
Copies: MB.

1475. Chaptal de Chanteloup, Jean Antoine Claude, comte, 1756–1832.
A treatise upon wines. By M. Chaptal.... Translated from the French by John Sargent. Charleston, [S.C.]: Printed by John H. Sargent, 1811. viii, 176 p.
S&S-22513.
Copies: MB; ScU.

1476. Appert, Nicolas, 1750–1841.
Art of preserving all kinds of animal and vegetable substances for several years.... Translated from the French, from the 2d London ed. of 1812. New York: D. Longworth, 1812. 103 p.
N-A0359334; S&S-24624.
Copies: CtY; DLC; ICJ; ICU; MWA; NBuG; NNC; RPB.

1477. Pennsylvania Vine Company.
At a meeting of the stockholders of the Pennsylvania Vine Company ... the managers by their secretary M. Cloud, made a report. [Philadelphia: 1812.] 10 p.
Dated: Philadelphia, April 25th, 1812.
An account of the operation of the vineyard.
S&S-26428a.
Copies: PPL.

1478. Adams, Rufus W.
A dissertation, designed for the yeomanry of the western country, containing a correct description of the best method of making butter and cheese.... Marietta, Ohio: Printed at the office of the American friend [by Thomas G. Ransom, for D. Everett, T. Buell & D. H. Buell, 1813?] 36 p.
S&S-27655.
Copies: MWA; OCHP.

1479. Hall, Harrison, 1785–1866.
Hall's Distiller, containing 1. Full and particular directions for mashing and distilling all kinds of grain.... 2. A notice of the different kinds of stills.... 3. A treatise on fermentation.... 4. Directions for making

1479 *(continued)*
yeast.... 5. The Rev. Mr. Allison's process of rectification.... 6. Instructions for making all kinds of cordials, compound waters, &c.... Adapted to the use of farmers as well as distillers.... Philadelphia: Printed by John Bioren, 1813. x, 244 p., fold. front., plates.

"List of patents granted by the United States, for improvements in distillation, on stills, and refining liquors," p. 238–244.

S&S-28692, 28693.

Copies: CtY; DLC; DeGE; MH; MWA; MiU-C; NN; NcD; NjR; PPAmP; PPL; RPB; ViU.

1480. Coppinger, Joseph.

The American practical brewer and tanner.... New York: Printed by Van Winkle and Wiley, 1815. vii, [11]–246, [2] p., illus.

S&S-34445.

Copies: CtY; DLC; DeGE; ICJ; MBAt; MH; MWA; MiU-C; MoSU; NBuG; NNCC; NcD.

1481. Gillespie, Robert.

A short sketch of Gillespie's improvements on distillation. [S.l.: 1815?] 16 p.

N-G0218513; S&S-37709 (dated [1816]).

Copies: DLC; DNA; MB.

1482. United States. Treasury Dept.

Letter from the Secretary of the Treasury, transmitting a report made in pursuance of a resolution of the House of Representatives of March 9, 1816, requiring the Secretary of the Treasury to report to the next session of Congress, whether any, and if any, what alterations are necessary to equalize the duty on the capacity of stills, boilers, and other instruments used in distillation. January 13, 1817. Washington City: Printed by William A. Davis, 1817. 58 p.

S&S-42630.

Copies: DLC; G; MWA.

1483. Degelos, P. A.

Statement of the amount of sugar and molasses made in Louisiana in 1818. Taken by P.A. Degelos. [S.l.]: Printed by Benjamin Levy, [1818]. Broadside 50 x 42 cm.

Hummel-809.

Copies: NcD.

1484. Hall, Harrison, 1785–1866.

The distiller.... Adapted to the use of farmers and distillers.... The 2d ed., enlarged and improved. Philadelphia: Printed by J. Bioren for the author, 1818. xii, 338 p., 1 l., illus., plates.

Lists of patents granted by the United States, for improvements in distillation, on stills, and refining of liquors, p. 329–335.

S&S-44227.

Copies: DLC; DNAL; DeGE; ICJ; KyLx; MWA; NN; OU; PP; PPF.

1485. Boucherie, Anthony.

The art of making whiskey, so as to obtain a better, purer, cheaper and greater quantity of spirit.... Also the art of converting it into gin ... translated from the French by C.M. Lexington, Ky.: Printed by Worsley & Smith, 1819. 43 p.

S&S-47403.

Copies: ICU; PPiU.

1486. The progress of the dairy; descriptive of the method of making butter and cheese.... New York: Samuel Wood & Sons, and Samuel S. Wood & Co., Baltimore, 1819. 25 p., illus.

S&S-49194.

Copies: MWA; NjP.

1487. Accum, Friedrich Christian, 1769–1838.

A treatise on adulterations of food, and culinary poisons. Exhibiting the fraudulent sophistications of bread, beer, wine, spiritous liquors, tea, coffee, cream, confectionery, vinegar, mustard, pepper, cheese, olive oil, pickles, and other articles employed in domestic economy. And methods of detecting them.... Philadelphia: Ab'm Small, 1820. xii, [13]–269 p.

S-8.

Copies: CSmH; CU; CoFS; CtY; DNAL; ICJ; IaMp; InU; KU; MB; MH; MWA; NN; NjP; OC; P; RPM; ViU.

1488. The grocer's guide; being a directory for making and managing all kinds of foreign liquors and domestic compounds, principally selected from Davies' Innkeeper and butler's guide. New York: Printed and published by Broderick and Ritter, 1820. 202 p.

John Davies's *Inn-keeper and butler's guide* was first published at Liverpool, 1805.

S-1463.

Copies: CtY; ICU; MWA.

1489. Whitlow, Charles.

On preserving potatoes for sea stores of foreign consumption. Boston: Wells & Lilly, 1821. 5 p.

S-7651.

Copies: MeB (not located).

1490. Adlum, John, 1759–1836.

A memoir on the cultivation of the vine in America, and the best mode of making wine.... Washington: Printed by Davis and Force, 1823. 142 p.

S-11545.

Copies: CoFS; DLC; DNAL; GHi; MBH; MNBedf; MiU; NNBG; NNS; NRU; PHi; PPA.

1491. Chaptal de Chanteloup, Jean Antoine Claude, comte, 1756–1832.

Treatise on the culture of the vine, and of the art of making wine & vinegar. By Chaptal, Abbe Rozier and Parmentier. Translated from the French by M. Riviere. Nashville: Printed by George Wilson, 1823. 90, [2] p.

Original French edition, Paris, 1801.

S-12099.

Copies: T; TKL; TNJ.

1492. Cutbush, James, 1788–1823.

Lectures on the adulteration of food and culinary poisons, the detection of poisons in general, and of adulterations in sundry chemical preparations, &c. used in medicine and the arts, with the means of discovering them: and rules for determining the purity of substances. Delivered in the United States Military Academy.... Newburgh, N.Y.: Printed by Ward M. Gazlay, 1823. xii, [3]–464 p.

S-12312.

Copies: CSmH; CtY; DP; DeGE; KU; LNOP; NNNAM; NT; PPAN.

1493. An essay on the importance and the best mode of converting grain into spirit, as a means of promoting the wealth and prosperity of the western country. Read before the Kentucky Institute at its meeting in Lexington on Wednesday evening, July 30, 1823, and published by its direction. To which is added, a recipe for making three gallons of spirit to the bushel of grain.... A recipe for clarifying spirits—A recipe for making colouring for spirit—A recipe for making yest. Lexington, Ky.: Printed at W. W. Worsley's book-selling, book-binding and book-printing establishment, 1823. 20 p.

S-12465.

Copies: CtY; ICU.

1494. Cobbett, William, 1763–1835.

Cottage economy: containing information relative to the brewing of beer, making of bread, keeping of cows, pigs, bees, ewes, goats, poultry and rabbits ... to which are added instructions relative to the selecting, the cutting, and the bleaching of the plants of English grass and grain, for the purpose of making hats and bonnets ... 1st American from the 1st London ed. New York: S. Gould and Son, 1824. 125 p., front.

First edition, London, 1822.

S-15794.

Copies: CSmH; CtW; CtY; DeGE; MB; MH; MWA; NRom; NjR.

1495. Cooper, Thomas, 1759–1839.

A treatise of domestic medicine.... Also The art of preserving all kinds of animal and vegetable substances for many years, by M. Appert. Reading, [Pa.]: G. Getz, 1824. [5], 128 p., 3 plates.

The art of preserving ... by M. Appert, first published by order of the French Minister of the Interior, 1810.

S-15867.

Copies: CSmH; DLC; DeGE; MWA; MdU; NIC; NN; NcD; PPC; PPL; PU-S; ScU.

1496. Adlum, John, 1759–1836.
Adlum on making wine. Georgetown, D.C.: James C. Dunn, printer, 1826. 36 p.
S-23425.
Copies: DLC; MH; OO; NjR; PPAmP; PPL.

1497. Dufour, John James.
The American vine-dresser's guide, being a treatise on the cultivation of the vine, and the process of wine making; adapted to the soil and climate of the United States.... Cincinnati: Printed by S. J. Browne, 1826. 2 p.l., [3]–314, [2] p., illus.
S-24373.
Copies: CSmH; CtY; DLC; In; InHi; KyU; MB; MWA; MiU-C; NIC; OC; OCHP; PPL; TKL.

1498. Loubat, Alphonse.
The American vine dresser's guide.... New York: G. & C. Carvill; Joseph Desnoues, printer, 1827. 137, iii p.
In English and French.
S-29533.
Copies: DNAL; GHi; IU; KyBgW; MB; MWA; MiD; MoU; NNC; NjP; P; RNR; Vi.

1499. Adlum, John, 1759–1836.
A memoir on the cultivation of the vine in America, and the best mode of making wine. 2d ed.... Washington: Printed for the author, by Wm. Greer, 1828. 179, 1 p.
S-31902.
Copies: CSmH; DLC; ICJ; MN; MNe; MWA; MdBS; MoSHi; NPV; NbU; NcA-S; PPAmP; PPL; PU-V; T; ViU.

1500. Adlum, John, 1759–1836.
Petition of John Adlum, stating that he had prepared a treatise on the cultivation of the vine, and the manufacture of wine therefrom; and praying that encouragement be extended to him, by rendering said treatise useful to the United States, and of advantage to himself. April 30, 1828.... Washington: Duff Green, 1828. 3 p. (20th Cong., 1st sess. Senate. Doc. 185)
S-36718.
Copies: DLC; DeU; MWA; NjR.

1501. [Embree, Davis.]
Circular on steam mills and distilleries, containing a description of a patent for the use of the scape steam of an engine in distilling, &c. Cincinnati: Looker & Reynolds, printers, 1828. 22 p.
S-33066.
Copies: PPF.

1502. Kecht, J. S.
Der verbesserte praktische Weinbau in Gärten und vorzüglich auf Weinbergen. Mit einer Anweisung den Wein ohne Presse zu keltern. Von J. S. Kecht. Den amerikanischen Weinbauern gewidmet von Heinrich B. Sage. Reading, [Pa.]: Gedruckt bey G. Adolph Sage, 1828. viii, 84 p., 2 plates.
S-33750.
Copies: DLC; IU; P; PPG; PPL.

1503. Kurtzgefasste Anweisung über die Anlegung von Weinbergen, die Behandlung des Mostes und des Weines und andere dahin gehörigen Gegenständen mit besonderer Rücksicht auf die Vereinigten Staaten. Von einem praktischen teutschen Weingärtner. Baltimore: Johann T. Hanzsche, 1828.
S-33804.
No copy known.

1504. Beastall, William.
A useful guide, for grocers, distillers, hotel & tavernkeepers, and wine and spirit dealers ... being a complete directory for making and managing all kinds of wines and spirituos liquors.... New York: The author; C. Bartlett, printer, 1829. xii, 340 p.
S-37732.
Copies: CSmH; DLC; MB; MWA; NN; NNNAM; NcU.

1505. Loubat, Alphonse.
The American vine dresser's guide.... 2d ed. New York: G. & C. Carvill, 1829. 142 p.
In English and French.

S-39324.
Copies: CtY; PPL; ViU.

1506. Archbald, William A.
Prospectus d'une mode perfectionné de fabrication pour le sucre brut, d'aprés les brevets de William A. Archbald. [New Orleans: 1830.] [2] p.
C-182.
Copies: PPL.

7. Fertilizers

1507. Logan, George, 1753–1821.
Agricultural experiments, on gypsum, or plaister of Paris: with some observations on the fertilizing quality, and natural history of that fossil.... Philadelphia: Printed by F. and R. Bailey, 1797. 18 p.
N-L0449754.
Copies: ICU; PPL; ViU.

1508. Peters, Richard, 1744–1828.
Agricultural enquiries on plaister of Paris. Also, facts, observations and conjectures on that subtance [*sic*], when applied as manure. Collected, chiefly from the practice of farmers in Pennsylvania, and published as much with a view to invite, as to give information. Philadelphia: Printed by Charles Cist and John Markland, 1797. 1 p.l., vii, [9]–111, [3] p.
Mp-32673.
Copies: DLC; DNAL; DeGE; MBAt; MWA; NN; PPL; PU.

1509. Anderson, James, 1739–1808.
An essay on quick-lime, as a cement and as a manure.... Boston: Printed by S. Etheridge for J. Nancrede, 1799. 115 p.
Usually found in: Marshall, Charles, *An introduction to the knowledge and practice of gardening,* Boston, 1799.
First published in England.
Mp-35770.
Copies: DLC; DeGE; ICJ; BM; MH; MWA; MeB; MiU-C; NN; NjP; PPL; RPJCB; ScU.

1510. Kirwan, Richard, 1733–1812.
The manures most advantageously applicable to the various kinds of soils, and the causes of their beneficial effect in each particular instance.... From the 6th London ed. Philadelphia: Printed by Kimber, Conrad and Co., 1807. 94 p.
S&S-12870.
Copies: DLC; ICJ; MB; MWA; MiU-C; PPL; Vi.

1511. Redd, George.
A late discovery, extremely interesting to planters and farmers, relative to fertilizing poor and exhausted lands, upon a cheap and easy plan; with some remarks and observations on orcharding and gardening. Winchester, [Va.]: Printed by J. A. Lingan, 1809. 16 p.
S&S-18480.
Copies: DLC; DNAL; MiU-C; ViHi.

1512. Peters, Richard, 1744–1828.
Agricultural inquiries on plaister of Paris. Also facts, observations and conjectures on that substance, when applied as manure.... Philadelphia: Reprinted by Jane Aitken, 1810. vi, [11], [ix]–xvi, [17]–129, ix p.
S&S-21050.
Copies: CSmH; DLC; DeGE; MWA; PPAmP.

1513. Redd, George.

Eine neue Entdeckung, äusserst wichtig für Pflanzer und Landbauer, betreffend die Fruchtbarmachung des magern und erschöpften Landes. . . . Nebst einigen Bemerkungen über den Obst- und Garten-Bau, von Georg Redd. Aus dem Englischen übersetzt. Lancaster, pa.: Gedruckt für den Verfasser, von W. Hamilton und Cie., 1810. 16 p.

Concerns mainly the use of salt as a fertilizer.

S&S-21180.

Copies: PHi (at PPL); PLF.

1514. Sullivan, Richard, 1779–1861.

Remarks, by Richard Sullivan, on salt, as a manure, and as promoting the health and fattening of domestic animals. Boston: Wells & Lilly, 1824. 10 p.

S-18125.

Copies: MeB (not located).

8. Culture of crops

1515. Budd, Thomas, d. 1698.

Good order established in Pennsylvania & New Jersey in America, being a true account of the country; with its produce and commodities there made. And the great improvements that may be made by means of publick store-houses for hemp, flax, and linen-cloth; also, the advantages of a publick-school, the profits of a publick-bank, and the probability of its arising, if those directions here laid down are followed. With the advantages of public granaries. . . . [Philadelphia]: Printed [by W. Bradford] in the year 1685. 39, [1] p.

Mp-386.

Copies: DLC; ICN; MiU-C; N; PPL; RPJCB.

1516. [Arguments offered to the inhabitants of New England, and the other neighbouring provinces, to encourage the sowing of hemp. Boston: 1705.] Broadside.

B-269.

No copy known.

1517. Hall, Richard.

Observations made by Richard Hall, of the city of Dublin, hemp and flax dresser; on the methods used in Holland, in cultivating and raising hemp and flax. And likewise, his remarks on Mr. Slator's book. Printed at Dublin in the year 1724. And now published for the benefit of the inhabitants of New-England. . . . Boston, N.E.: Printed by S. Kneeland and T. Green, for Daniel Henchman, 1735. 1 p.l., 39 p., illus.

Mp-3911.

Copies: DLC; MWA; ViU.

1518. Slator, Lionel.

Instructions for the cultivating and raising of flax and hemp: in a better manner, than that generally practis'd in Ireland. . . . Printed at Dublin in the year 1724. And now published for the benefit of the inhabitants of New-England, and recommended to their perusal. Boston, N.E.: Printed by S. Kneeland and T. Green, for Daniel Henchman, 1735. iv, 45, [3] p.

Mp-3959.

Copies: DLC; MWA; ViU.

1519. A scheme (by striking twenty thousand pounds paper money) to encourage the raising of hemp and the manufacturing of iron in the Province of New York, with some observations, shewing the necessity and advantages thereof. [New York: 1737.] 6 p.

Published as a supplement to the *New York weekly journal,* for March 28, 1737.

Mp-4194.

Copies: MWA.

1520. Royal Dublin Society.

Extracts from the essays of the Dublin Society; relating to the culture and manufacture of flax. [Annapolis: Printed by Jonas Green, 1748.] 1 p.l., ii, 34, [1] p., illus.

"Reprinted from the *Maryland gazette,*" cf. Evans.

Mp-6127; N-R0481609.

Copies: MBAt (incomplete); NNC.

1521. Society for Encouraging Industry and Employing the Poor, Boston.

Advertisement. Boston, January 15, 1752. For the encouragement of the raising and well-curing and dressing of flax within this province, a praemium ... is hereby offered.... The claims of the several persons to the four last praemiums to be determined at the Manufactory-House in Boston, on the first Thursday in April, 1753.... [Boston: 1752.] Broadside 27 x 21 cm.

B-1629; Mp-40671.

Copies: DLC; MH; MHi; MWA; NN (photocopy).

1522. Woodmason, Charles.

A letter from a gentleman of South-Carolina, on the cultivation of indigo.... [Charleston: 1754.] 25 p.

Caption title from p. 3.

Dated at end: Charles-Town, South-Carolina, October 1st, 1754.

B-1696.

Copies: NNC (p. 1–2 lacking).

1523. Method of raising hemp, in Lancaster County, Pennsylvania. Hartford: Printed [by Thomas Green] and sold at the Heart and Crown, 1765. 8 p.

Mp-10074.

Copies: CtY; RPJCB.

1524. Quincy, Edmund, 1703–1788.

A treatise of hemp-husbandry; being a collection of approved instructions, as to the choice and preparation of the soils, most proper for the growth of that useful and valuable material, and also as to the subsequent management thereof, agreable to the experience of several countries wherein it has been produced, both in Europe and America.... Boston: Printed by Green & Russell, by order of the Honorable House of Representatives, 1765. 32, [1] p., plate.

Plate engraved by Paul Revere, from a drawing by Gershom Flagg.

Mp-10151.

Copies: DLC; DNAL; DeGE; MB; MBAt; MHi; MWA; PPL; RPJCB.

1525. Marcandier.

An abstract of the most useful parts of a late treatise on hemp, translated from the French of M. Marcandier, Magistrate of Bourges, and inscribed by the editor at London, to the laudable Society for Promoting Arts, Manufactures, &c. Being much recommended to the growers and manufacturers of that valuable material, from some modern discoveries and experiments of a method of preparation.... Together with some observations upon the prospect of singular advantage which may be derived to Great-Britain and her colonies from their early adopting the method prescribed. To which is added, some account of the use of the horse-chestnut; and a plan of the Pennsylvania hemp brake. Boston: Printed and sold by Edes & Gill, 1766. [4], 30 p., 1 l. (plan).

At head of title: Published for the general benefit of Great-Britain and the North American colonies.

An abstract of the full London ed., 1764.

Mp-10362.

Copies: CtY; DLC; DNAL; DeGE; MBAt; MH; MHi; MWA; MiU-C; NN; PPL; RPJCB.

1526. An essay on the culture and management of hemp, more particularly for the purpose of making coarse linens.... By a farmer. Annapolis: Printed by Anne Catherine Green, and Son, 1775. 52 p.

1526 (*continued*)
Mp-14022.
Copies: DLC.

1527. [Loocock, Aaron.]
Some observations and directions for the culture of madder. [Charles-Town, S.C.: Printed by Peter Timothy, 1775.] [8] p.
Signed at end: Aaron Loocock.
Mp-14166.
Copies: DLC.

1528. [Bordley, John Beale] 1727–1804.
Necessaries; best product of land; best staple of commerce. Philadelphia: Printed by J. Humphreys, junr., 1776. 17, [2] p.
Written in opposition to silk-culture and wine-making in Pennsylvania.
Mp-14896.
Copies: MHi; MWA; NN; PPAmP; PPL; RPJCB.

1529. An essay on the culture and management of hemp.... By a farmer. [Baltimore: M. K. Goddard, 1776?]
Advertised Jan. 10, 1776, as "Published and sold at the Printing-Office."
B-4209 (cf. E-14022).
No copy known.

1530. Directions for ploughing, harrowing, sowing, pulling, watering, grassing, and full management of white flax, till brought to the hackle. Taken from a treatise on agriculture, entitled The Yorkshire farmer. To which is added, a table of expense and profit of an acre of white flax, were it raised in Ireland according to the English method. Hartford: Printed by Hudson and Goodwin, 1781. 7 p.
B-5273; Mp-43970.
Copies: NN.

1531. Maryland. Laws.
An act to regulate the inspection of tobacco. Passed December 21, 1789. Annapolis: Printed by Frederick Green, printer to the state, [1789]. 28 p.
B-6980; Mp-45511.
Copies: DLC.

1532. Pennsylvania. Laws.
In General Assembly, Saturday, December 21, 1793. A supplement to.... "An act to enable the Governor ... to incorporate a company ... for promoting the cultivation of vines...." Philadelphia: T. Bradford, [1793]. 2 p.
B-8443; Mp-46847.
Copies: MWA.

1533. Amoureux, M.
A short and practical treatise on the culture of the wine-grapes in the United States of America.... George-Town: Printed by Hanson & Briggs, [1794]. Broadside 38 x 22 cm.
B-8571; Mp-46960.
Copies: DLC.

1534. Russell, Samuel.
Statement of snuff and tobacco manufactures.... Samuel Russell, agent for the snuff and tobacco manufacturers ... of New-York. The general facts ... in the above ... apply in ... Pennsylvania. Thomas Leiper, tobacconist. [Philadelphia: 1795.] Broadside 39 x 24 cm.
Dated: Philadelphia, December 7, 1795.
B-9284.
Copies: NHi (not located).

1535. Moseley, Benjamin, 1742–1819.
A treatise concerning the properties and effects of coffee. 1st American ed. ... Philadelphia: Samuel H. Smith, 1796. 45 p.
First published in London, 1775.
B-9611; Mp-47837.
Copies: DLC; NNNAM; PHi; PPC.

1536. Massachusetts Society for Promoting Agriculture.
On the culture of potatoes: extracted from communications made to the Board of Agriculture in Great-Britain. Published by order of the trustees of the Massachusetts Society for Promoting Agriculture.... Boston: Printed by Young & Minns, 1798. 25 p.
Mp-34084.
Copies: DLC; DeGE; DeU; MBAt; MH; MWA; RPJCB.

1537. A memorial on the practicability of growing vineyards in the state of South-Carolina. Read by the author to the Agricultural Society at their extra meeting on Wednesday, the 7th of November, 1798, and published under their patronage. Charleston: Printed by W. P. Young and T. C. Cox, 1798. [2], 21 p.

Mp-34093.

Copies: PHi; ScC.

1538. [Bordley, John Beale] 1727–1804.

Hemp. [Philadelphia: Printed by Charles Cist, 1799.] 9 p.

Signed on p. 7: B., January, 1799.

Mp-35217.

Copies: DLC; DeGE; PPL; PU.

1539. Dupré, Lewis.

Observations on the culture of cotton, by Lewis Du Pre. Georgetown, [S.C.]: Printed by Elliott & Burd, 1799.

E-35427.

No copy known.

1540. Ambler, Jacquelin, 1742–1798.

A treatise on the culture of lucerne. By Jacqulin [*sic*] Ambler.... Richmond: Printed by T. Nicolson, [1800?]. 9 p.

"Observations on Mr. Ambler's notes on the culture of lucerne," by R. Parker, p. 7-9.

B-10961.

Copies: DLC; DNAL.

1541. Sibbald, George.

Notes and observations, on the pine lands of Georgia, shewing the advantage they possess, particularly in the culture of cotton. Addressed to persons emigrating, and those disposed to encourage migration to this state. Together with a plan of emigration, for their immediate settlement. To which is added, a geographical sketch of the state of Georgia, with a comparative view of the population of 1791, and 1801, and the exports of the years 1791 & 1800.... Augusta: Printed by William J. Bunce, 1801. 71, [6] p.

S&S-1322.

Copies: DLC; DNAL; MH; MWA; MdBP.

1542. Bryarly, Wakeman, d. 1821.

An inaugural essay, on the Lupulus communis, of Gaertner; or the common hop.... Philadelphia: Printed for the author, by John H. Oswald, 1805. 34 p.

N-B0904652.

Copies: CtY-M; DLC; DNLM; MWA; PHi; PPC; PPL; PU.

1543. Fothergill, Anthony, 1732?–1813.

Remarks on the smut and mildew of wheat; with hints on the most probable means of prevention; addressed to the Agricultural Society of Philadelphia, by A. Fothergill.... Philadelphia: Printed by Jane Aitken, 1808. [1], 20 p.

S&S-15032.

Copies: PPiU.

1543a. Pinckney, Thomas, 1750–1828.

A letter from Thomas Pinckney, Esq. to the Agricultural Society of South Carolina, on the water culture of rice. 2d ed. Charleston, S.C.: Printed for the Society by E. Morford, Willington and Co., 1811. 14 p.

S&S-23695.

Copies: ScCC.

1544. Sternberg, P.

A treatise on the raising and culture of wheat.... Albany: Printed for the author, [1811]. 60 p.

S&S-23979.

Copies: MBAt; MWA; N.

1545. United States. Navy Dept.

Letter from the Secretary of the Navy to the Chairman of the Committee of Commerce and Manufactures, on the propriety of encouraging the culture of hemp by protecting import duties, or by prohibiting the importation of that article into the United States and its territories. February 21, 1811.... Washington: A. & G. Way, printers, 1811. 11 p.

Includes observations relative to the culture and dressing of hemp.

S&S-24162.

Copies: DLC; MWA; NjR.

1546. Clarke, Adam, 1760?-1832.

A dissertation on the use and abuse of tobacco, wherein the advantages and disadvantages attending the consumption of that entertaining weed, are particularly considered.... 1st American ed., with additions and improvements.... Burlington, N.J.: Printed and published by David Allinson & Co., 1812. iv, [5]–24 p.

First published in London, 1797.

S&S-25085.

Copies: CSmH; CtY-M; DLC; DNLM; MH; MWA; NcD-MC; OMC; PPPrHi; RPB.

1547. ———— New York: Printed by Paul and Thomas for E. Sargeant, and Griffin and Rudd, 1812. 24 p.

S&S-25086.

Copies: MWA.

1548. ———— 2d American ed. ... Newbury-port, [Mass.]: Thomas & Whipple, 1812. 24 p.

S&S-25087.

Copies: DLC; DNLM; ICJ; MH; MWA; NN; OC.

1549. Pennsylvania. Commissioners for the Culture of Vine.

Culture of the vine. An address of the Commissioners for the Cultivation of the Vine to the public. [Philadelphia?: 1815?] [2] p.

Signed: Timothy Pickering, Samuel Miles, Israel Whelen, Peter S. Duponceau ... Benjamin S. Barton [and others].

Printed not later than 1815.

Copies: PPAmP.

1550. Barton, William Paul Crillon, 1787–1856.

Some account of a plant used in Lancaster County, Pennsylvania, as a substitute for chocolate; and which appears to be the Holcus bicolor of Willdenow's Species plantarum. Read to the Philadelphia Linnean Society, October 17th, 1816.... Philadelphia: Printed by T. H. Palmer, 1816. 8 p.

S&S-36888.

Copies: DLC; MB; NjP; PPL.

1551. Spalding, Thomas, 1774–1851.

Observations on the method of planting and cultivating the sugar-cane in Georgia and South Carolina, together with the process of boiling and granulating.... Charleston: J. Hoff, 1816. 35 p.

S&S-38987.

No copy known.

1552. [Coxe, Tench] 1755–1824.

A memoir, of February, 1817, upon the subject of the cotton wool cultivation, the cotton trade, and the cotton manufactories of the United States of America. [Philadelphia: 1817.] 16 p.

S&S-40584.

Copies: DLC; MB; MH-BA; MWA; NcD; PPAmP.

1553. Watterston, George, 1783–1854.

A memoir of the history, culture, manufactures, uses, &c. of the tobacco plant.... Washington City: Printed by Jonathan Elliot, 1817. 12 p.

S&S-42843.

Copies: DLC; DNAL; MB; MH-BA; PPL.

1554. Coxe, Tench, 1755–1824.

An addition, of December, 1818, to the memoir of February and August 1817, on the subject of the cotton culture, the cotton commerce, and the cotton manufacture of the United States.... [Philadelphia?: 1818.] 16 p.

S&S-43761.

Copies: DLC; DNAL; MB; MH-BA; MWA; PHi; PPAmP.

1555. ———— [Philadelphia?: 1818.] 16 p.

With added printed footnotes.

S&S-43762.

Copies: MWA.

1556. [Jeffreys, George W.]

On orchard grass, as adapted to the soil and climate of North-Carolina. [Raleigh?:

1818?] 18 p.

At head of title: For the use of the Board of Agriculture.

"Directions for the management of silk worms, reeling, and preparing silk, and for the cultivation of the white mulberry. (Communicated by Mr. Gideon B. Smith, Baltimore)," p. 5–11; "On the cultivation of vine," p. 12–18.

N-J0074013.

Copies: DLC.

1557. Clarke, Adam, 1760?–1832.

Dissertation on the use and abuse of tobacco. Wherein the advantages and disadvantages attending the consumption of that entertaining weed are particularly considered.... New York: Printed by C. S. Van Winkle, 1819. 24 p.

First published in London, 1797.

S&S-47628.

Copies: CtHi; DLC; DNLM; IU; NHi; NNG; NNS; OClWHi; TxU.

1558. Cobbett, William, 1763–1835.

List of field seeds and garden seeds, contained in one of the boxes.... Also, hints respecting the sewing of each sort of seed, and the cultivation of the plants.... [New York]: Printed by Clayton & Kingsland, 1819. 30 p.

N-C0502607; S&S-47651.

Copies: MH; MWA.

1559. Carey, Mathew, 1760–1839.

A view of the ruinous consequences of a dependance on foreign markets for the sale of the great staples of this nation, flour, cotton and tobacco. Addressed to the Congress of the United States.... Philadelphia: M. Carey & Son, 1820. x, 42 p.

S-684.

Copies: CSt; DLC; ICU; KU; MH-BA; MWA; MdBP; MiD-B; NCH; NN; PPL; RPB; TxU; WHi.

1560. Chazotte, Peter Stephen.

Facts and observations on the culture of vines, olives, capers, almonds, &c. in the Southern States, and of coffee, cocoa, cochineal, in East Florida.... Philadelphia: J. Maxwell, 1821. 24 p.

S-4966.

Copies: DLC; FU; IaH.; LU; MBH; MH; MWA; MiU; NIC; OCHP; P; PPAN; PPAmP; Vi.

1561. Pomeroy, Samuel Wyllys.

Essay on flax husbandry. Boston: Wells & Lilly, 1821. 21 p.

S-6502.

Copies: MeB.

1562. Cropper, Benson and Company, Liverpool.

Circular on the cultivation of cotton. Liverpool, 9 mo. 27, 1822.... [Philadelphia: Printed at the office of the Aurora, 1822?] 16 p.

S-8471.

Copies: CtY; MWA; PPAmP; PPL.

1563. Dey, Anthony.

New York, 20th March, 1822. In answer to correspondents on the subject of hemp and flax, and the machine to dress the same in an unwretted (commonly called unrotted) state, I deem it advisable to send them all that I have written on the subject.... [New York: 1822.] 15 p.

N-D0233668; S-8516 (author's name misspelled Day).

Copies: DLC.

1564. Minor, Peter.

Notes on the cultivation and management of tobacco, from the plant bed to the prize; according to the most approved practices in Albemarle, and the adjacent counties in Virginia.... Baltimore: Printed by J. Robinson, 1822. 24 p., illus.

S-9504.

Copies: DLC; DNAL; KyU; N; ViU.

1565. United States. Congress. House. Committee on Public Lands.

Report of the Committee on the Public Lands on the petition of Peter S. Chazotte and others, in behalf of the American Coffee Land Association. February 20, 1822.

1565 (*continued*)
[Washington: 1822.] 35 p. (17th Cong., 1st sess. House. Rept. 47)

"Facts and observations on the culture of vines, olives, capers, almonds, &c. in the southern states, and of coffee, cocoa, and cochineal in East Florida," by Peter Stephen Chazotte, p. 13–30.

S-11210.

Copies: DLC; DeGE; DeU; MWA; PU.

1566. ———— December 12, 1822, reprinted.... [Washington: 1822.] 35 p. (17th Cong., 2d sess. House. Rept. 14)

S-11211.

Copies: DLC; DeU; MWA; NjP.

1567. United States. Treasury Dept.

Report of the Secretary of the Treasury, on the petition of Lefebvre Desnoettes, and others, French emigrants in Alabama, engaged in the cultivation of the vine and olive, praying a modification of the condition of their grant. March 20, 1822.... Washington: Printed by Gales & Seaton, 1822. 9 p. (17th Cong., 1st sess. Senate. Doc. 70)

S-11233.

Copies: DLC; DeGE; DeU; MWA; NjP.

1568. United States. Treasury Dept.

Letter from the Secretary of the Treasury, transmitting a list of the names of persons who entered into contract with the Treasury Department for the cultivation of the vine and olive, and of the tracts of land allotted to each. February 16, 1825.... Washington: Printed by Gales & Seaton, 1825. 14 p. (18th Cong., 2d sess. House. Doc. 87)

S-22930.

Copies: DLC; DeGE; DeU; MWA; NjP.

1569. Hutton, Isaac G.

The vigneron; an essay on the culture of the grape and the making of wine.... Washington: The author, 1827. 60 p.

S-29293.

Copies: DLC; NBuG.

1570. Seabrook, Whitemarsh Benjamin, 1795–1855.

A report accompanied with sundry letters on the causes which contribute to the production of fine sea-island cotton; read before the Agricultural Society of St. John's, Colleton, on the 14th March, 1827. Charleston: A. E. Miller, 1827. 36 p.

S-30554.

Copies: ICU; NcD; OCHP; ScC.

1571. United States. Treasury Dept.

Report from the Secretary of the Treasury (in compliance with a resolution of the Senate of 20th May, 1826) in relation to grants of land to French emigrants, to encourage the cultivation of the vine and olive. December 24, 1827.... Washington: Printed by Duff Green, 1827. 52 p. (20th Cong., 1st sess. Senate. Doc. 11)

Copies: DeU; MWA.

1572. Agricultural Society of South Carolina.

The report of the Committee of the Agricultural Society of South-Carolina, on the mill, constructed by Mr. John Ravenel, for pounding and preparing rice for market. Read before the Agricultural Society ... Oct. 20, 1829.... Together with Mr. Ravenel's letter. Charleston: Printed by A. E. Miller, 1829. 27 p.

S-37387.

Copies: MHi; NHi.

1573. Connecticut. General Assembly. House. Committee on the Culture of Hemp.

Report of the Committee on the Culture of Hemp, &c. Printed by order of the House of Representatives. Hartford: Printed at the Times office, by J. Russell, 1829. 16 p.

Includes a statement by William K. Scott on the culture of hemp.

N-C0636482.

Copies: CtY; NN.

1574. Fairbanks, Erastus.

Compilation of articles relating to the culture and manufacture of hemp in the United States; selected principally from

newspapers and journals devoted to the interest of agriculture. St. Johnsbury, [Vt.]: Printed at the Farmer's herald office, by Jewett and Porter, 1829. 39 p.
S-38524.
Copies: PHi.

1575. Thiébaut de Berneaud, Arsène, 1777–1850.
The vine-dressers' theoretical and practical manual; or, The art of cultivating the vine; and making wine, brandy, and vinegar ... by Thiebaut de Berneaud.... From the 2d French ed., by the translator of Le solitaire.... New York: P. Canfield, 1829. 158, iii p., fold. plate.
S-40634.
Copies: CU; DLC; MBL; MWA; OCl; PPC.

1576. Prince, William Robert, 1795–1869.
A treatise on the vine; embracing its history from the earliest ages to the present day, with descriptions of above two hundred foreign and eighty American varieties; together with a complete dissertation on the establishment, culture, and management of vineyards.... By William Robert Prince, aided by William Prince.... New York: T. & J. Swords,[etc.]; Philadelphia: J. Dobson, [etc., etc.], 1830. viii, [9]–355 p., front.
C-3177.
Copies: CSmH; CtHT-W; DLC; DeGE; GU; IU; KyBgW; LNB; MH; MWA; MdU; MeU; NIC; NbU; NcU; NjR; PPL; RPB.

1577. Rafinesque, Constantine Samuel, 1783–1840.
American manual of the grape vines and the art of making wine: including an account of 62 species of vines, with nearly 300 varieties. An account of the principal wines, American and foreign. Properties and uses of wines and grapes. Cultivation of vines in America; and the art to make good wines. With 8 figures. By C. S. Rafinesque.... Philadelphia: Printed for the author, 1830. 1 p.l., [5]–64 p., 1 l., illus.
C-3237.
Copies: CtHT-W; CtY; DLC; DNAL; KU; MBH; MH; NYBT; OCLloyd; PHi; PPL; VtU; WHi.

9. Animal husbandry and products; hunting and fishing

1578. Wily, John.
A treatise on the propagation of sheep, the manufacture of wool, and the cultivation and manufacture of flax, with directions for making several utensils for the business. Williamsburg: Printed by J. Royle, 1765. 52 p.
Mp-10214.
Copies: MBAt; NcD; Vi.

1579. Articles of agreement relative to the whale-fishery. Stores agreed upon as necessary for whaling voyage, viz.... The following charges to be deducted from the whole stock, viz.... After which the following division of the neat stock is to be made, viz.... To the performance of which we bind ourselves each to the other in the penal sum of two hundred pounds. Witness our hands at Boston, New-England, the tenth day of February, 1769. [Boston: 1769.] Broadside 21 x 32 cm.
Mp-11162.
Copies: MB; MWA.

1580. The sportsman's companion; or, An essay on shooting.... New York: Printed by Robertson, Mills & Hicks, 1783. 89, [2] p.

1580 (*continued*)
Mp-23785.
Copies: CtY.

1581. [Jefferson, Thomas] 1743–1826.
Observations on the whale-fishery. [New York: 1788.] 18 p.
Mp-21345.
Copies: MBAt.

1582. [Bordley, John Beale] 1727–1804.
Purport of a letter on sheep, written in Maryland, March the 30th, 1789. [Philadelphia: Printed by Daniel Humphreys, 1789.] 6 p.
Mp-21699.
Copies: MBAt; MWA; MdBJ-G; PPL; RPJCB.

1583. The sportsman's companion; or, An essay on shooting; illustriously shewing in what manner to fire at birds of game. And, directions to gentlemen for the treatment and breaking their own pointers and spaniels.... By a gentleman, who has made shooting his favorite amusement upwards of twenty-six years. 2d ed. Burlington, [N.J.]: Printed by Isaac Neale, 1791. viii, [9]–89, [2] p.
Mp-23786.
Copies: DLC; MWA; MiU-C; NN.

1584. United States. Dept. of State.
Report of the Secretary of State, on the subject of the cod and whale fisheries, made conformably to an order of the House of Representatives of the United States, referring to him the representation of the General Court of the Commonwealth of Massachusetts on those subjects; February 1, 1791. Philadelphia: Printed by Francis Childs and John Swaine, 1791. 28 p.
Signed: Thomas Jefferson, Secretary of State. February 1st, 1791.
Mp-23911.
Copies: DLC; MBAt; MWA; NN; RPJCB.

1585. ———— Philadelphia: Printed by J. Fenno, 1791. 34 p., tables.
Mp-23912.
Copies: DLC; DeGE; MBAt; MWA; NN; RPJCB.

1586. A complete guide for the management of bees, through the year. By a farmer of Massachusetts. Illustrated with a copperplate. Worcester, Mass.: Printed by Isaiah Thomas, & Leonard Worcester for Isaiah Thomas, 1792. 46 p., illus., plate.
Mp-24207.
Copies: CtY; DLC; MHi; MWA; RPJCB.

1587. The sportsman's companion; or, An essay on shooting: illustriously shewing in what manner to fire at birds of game.... By a gentleman, who has made shooting his favorite amusement upward of twenty-six years.... 3d ed. Philadelphia: Printed for the purchasers, [1792]. 36 p.
E-24809; N-S0829554.
Copies: DLC.

1588. United States. Dept. of State.
Report of the Secretary of State, on the subject of the cod and whale fisheries, made conformably to an order of the House of Representatives of the United States, referring to him the representation of the General Court of the Commonwealth of Massachusetts on those subjects; Feb. 1, 1791. Philadelphia: Printed by Francis Childs and John Swaine, 1792. 45 p., fold. table.
Thomas Jefferson, Secretary of State.
Mp-24924.
Copies: DLC; DeGE; MHi; PPAmP.

1589. Mills, John, d. 1784?
A treatise on cattle; showing the most approved methods of breeding, rearing and fitting for use, asses, mules, horned cattle, sheep, goats, and swine.... Carefully collected from the best authorities, and interspersed with remarks.... Boston: Printed and sold by William Spotswood, 1795. iv, 215, [9] p.
First published in London, 1776.
Mp-29083; N-M0598883.
Copies: DNAL; MH; MWA; MeB; MiU-C; NN; PPL; RPJCB.

1590. Moreau de Saint Méry, Médéric Louis Élie, 1750–1819.
Essai sur la manière d'ameliorer l'education des chevaux en Amérique. Philadelphie: De l'imprimerie de Moreau de Saint Méry, 1795. iv, 40 p.
Mp-29108.
Copies: DLC; PPAmP; PPL.

1591. Moreau de Saint Méry, Médéric Louis Élie, 1750–1819.
An essay on the manner of improving the breed of horses in America. Philadelphia: Printed & sold by Moreau de Saint-Méry, 1795. iv, 39 p.
Mp-29109.
Copies: DLC; NN; PPL.

1592. [Bordley, John Beale] 1727–1804.
Cattle pastured and soiled in summer: kept and fattened in winter.... [Philadelphia: Charles Cist, 1798.] 12 p.
B-10242; Mp-48372.
Copies: PPL.

1593. A short history of bees. In two parts.... Philadelphia: J. Johnson, 1803. xiv, 90 p.
S&S-50392.
Copies: NjR.

1594. United States. Congress. House. Committee on Fisheries.
Report of the Committee appointed on so much of the message of the President of the United States, of the fifteenth ultimo, "as relates to the fostering of the fisheries of the United States." 12th February, 1803. Washington: Printed by William Duane & Son, 1803. 9 p.
S&S-5409.
Copies: DLC; MWA; NN.

1595. Culley, George, 1735–1813.
Observations on live stock: containing hints for choosing and improving the best breeds of the most useful kinds of domestic animals.... To which is added, some observations respecting the cattle imported into America, in 1792, by Robert Heaton.... New York: D. Longworth, 1804. 103 p.
First published in Dublin, 1789.
S&S-6100.
Copies: DNAL; MH; MWA.

1596. Minor, Thaddeus.
The experienced bee-keeper; or, A short treatise on the management of bees.... Litchfield, [Conn.]: Printed at T. Collier's office, 1804. 21 p.
S&S-6786.
Copies: CtY; DLC.

1597. Souder, David.
The rural economist's assistant in the management of bees, principally taken from the German.... Lancaster, [Pa.]: William Greear, 1807. 55 p.
S&S-13618.
Copies: P; WU-A.

1598. Vanière, Jacques, 1664–1739.
The bees, from the Latin of J. Vaniere: being the fourteenth book of his Praedium rusticum. By Arthur Murphy.... Middletown, [Conn.]: Printed by I. Riley, New York, 1808. 59 p.
S&S-16614, 50894.
Copies: DNAL; MWA; NN; NjP.

1599. Livingston, Robert R., 1746–1813.
Essay on sheep; their varieties–account of the merinoes of Spain, France, &c. Reflections on the best method of treating them, and raising a flock in the United States; together with miscellaneous remarks on sheep and woollen manufactures.... New York: Printed by T. and J. Swords, 1809. 186 p.
S&S-17926.
Copies: CU; DLC; DNAL; DeGE; MWA; MiU-C; NIC; NN; NjP; PPL; Tx.

1600. Antidote to the merino-mania now progressing through the United States; or, The value of the merino breed, placed by observation and experience, upon a proper basis.... Philadelphia: Printed and sold by J.

1600 (*continued*)
& A. Y. Humphreys, 1810. iv, [5]–52 p.
S&S-19372.
Copies: DLC; DeGE; IU; MH-BA; MWA; MdHi; NIC; NN; NbU; PPAmP; PPL.

1601. Daubenton, Louis Jean Marie, 1716–1799.
Advice to shepherds and owners of flocks, on the care and management of sheep. Translated from the French of M. Daubenton. 1st ed. Boston: Printed by J. Belcher, 1810. 136 p.
Translation of his *Instruction pour les proprietaires de troupeaux,* Paris, 1802, by James Bowdoin.
S&S-19917.
Copies: DLC; MBAt; MBCo; MeB.

1602. Livingston, Robert R., 1746–1813.
Essay on sheep; their varieties–account of the merinos of Spain, France &c. Reflections on the best method of treating them, and raising a flock in the United States: together with miscellaneous remarks on sheep, and woollen manufactures.... 2d ed., much enlarged.... New York: Collins and Perkins, 1810. 164 p., 1 l., 3 plates.
"Printed by order of the Legislature of the state of New York."
S&S-20581.
Copies: CU; CtY; DLC; InU; MH-BA; MWA; MiGr; PU; ViU.

1603. ________ Trenton: James Oram, 1810. 149, [3] p., plates.
Includes a 2 page "Supplement": *New-Jersey sheep shearing.*
S&S-20580.
Copies: MWA; PPL; PU.

1604. Mackenzie, George Steuart, Sir, 1780–1848.
A treatise on the diseases and management of sheep.... And an appendix, containing documents exhibiting the value of the merino breed of sheep, and their progress in Scotland.... New York: I. Riley, 1810. viii, [3], 12–100, 45 p., 5 plates.
First published at Edinburgh, 1809.
S&S-20622.
Copies: DNAL; MA; MWA; NIC; NN; PU-V.

1605. Merino sheep. Authentic documents, relative to the quality and breed of merino sheep, imported in the ship Mount Hope, belonging to Gen. Elias Hasket Derby of Salem.... [S.l.: 1810.] Broadside 40 x 32 cm.
Copies: MWA.

1606. Richardson, Luther, 1774–1811.
An essay on sheep. Boston: Printed by Russell and Cutler, 1810. 24 p.
S&S-21220.
Copies: MeB.

1607. Bard, Samuel, 1742–1821.
A guide for young shepherds; or, Facts and observations on the character and value of merino sheep: with rules and precepts for their management, and the treatment of their diseases, as well as of sheep in general. Collected from the latest and best writers on these subjects, and confirmed by the experience of the author and his friends.... New York: Collins & Co., 1811. vi, [7]–112 p., illus.
S&S-22298.
Copies: DLC; DeGE; MWA; NIC; NN; NNNAM; PPL; ViU; WyU.

1608. Daubenton, Louis Jean Marie, 1716–1799.
Advice to shepherds and owners of flocks, on the care and management of sheep.

Translated from the original French of M. Daubenton, by a gentleman of Boston. To which are added, explanations of the plates, and a table of contents. Boston: Printed by Joshua Belcher, 1811. vi, 154 p., 4 plates.

Translation of his *Instruction pour les proprietaires de troupeaux,* Paris 1802, by James Bowdoin.

S&S-22662.

Copies: CU; DLC; DeGE; IU; InU; MH; MWA; MeB; PPL; RPB.

1609. Merino Society of the Middle States of North America.

The constitution of the Merino Society of the Middle States of North America; premiums proposed for 1812 and 1813. To which is prefixed, some Introductory observations on the origin, design, and probable advantages of the Merino Society—some of the objections against Merino flocks, subjected to the test of recent facts and present experience in the United States. Philadelphia: Printed for the Merino Society, 1811. 16 p.

Signed: President James Caldwell.... Vice-President E. J. Dupont [and others].

S&S-51117.

Copies: DeGE; MB; MWA; PHi; PPAmP.

1610. Tessier, Alexandre Henri, 1741–1837.

A complete treatise on merinos and other sheep, with plates. Recently published at Paris, by order of the government, compiled by Mr. Tessier.... Translated from the French.... New York: Printed at the Economical School office [by Joseph Desnoues], 1811. v, [1], 2, 9–158, 17, [1] p., 3 plates.

First published as *Instruction sur les bêtes à laine, et particulièrement sur la race des merinos....* Paris, 1810.

S&S-24023.

Copies: DLC; DNAL; DeGE; MWA; NjR; PPL.

1611. Doddridge, Joseph, 1769–1826.

A treatise on the culture of bees, in which is given, a description of the bee-house and hive, the natural history of bees, directions for managing them at the time of swarming, rules for taking the honey without destroying the bees and keeping them through the winter.... St. Clairsville, Ohio: Printed and published by Alexander Armstrong, 1813. 32 p.

S&S-28342.

Copies: MHi; MU; OClWHi; PMA; PPAN.

1612. Livingston, Robert R., 1746–1813.

Essay on sheep; their varieties—account of the merinoes of Spain, France, &c. Reflections on the best method of treating them, and raising a flock in the United States; together with miscellaneous remarks on sheep and woollen manufactures. Concord, N.H.: D. Cooledge, 1813. 143 p.

S&S-28966.

Copies: DNAL; DeGE; IU; LNHT; MBC; MWA; MdBP; MiU-C; NjR; PMA; WU-A.

1613. Bakewell, Robert, 1768–1843.

Observations on the influence of soil and climate upon wool; from which is deduced, a certain and easy method of improving the quality of English clothing wools, and preserving the health of sheep; with hints for the management of sheep after shearing: an inquiry into the structure, growth, and formation of wool and hair; and remarks on the means by which the Spanish breed of sheep may be made to preserve the best qualities of its fleece unchanged in different climates. By Robert Bakewell. With occasional notes and remarks by the Right Hon. Lord Somerville. Philadelphia: Printed by William Fry for Kimber and Conrad, 1814. 120 p.

First published in London, 1808.

S&S-30728.

1613 (*continued*)
Copies: DLC; DeGE; MB; MH; MWA; NHi; PPL.

1614. Wiestling, John S.
Der vollständige Bienen-Wärter; oder, Nützliche Anweisungen zur Bienen-Zucht.... Harrisburg, [Pa.]: Gedruckt bey John S. Wiestling, 1819. 50 p., illus.
S&S-50110.
Copies: MWA.

1615. Blake, Edward.
United States of America. To all to whom these letters patent shall come: Whereas Edward Blake, a citizen of the United States, hath alledged, that he has invented a new and useful improvement in the bee hive.... [Washington?: 1820.] Broadside.
Accompanied by a small sheet with woodcut illustrations of the bee hive.
Copies: DP.

1616. Goodwin, Joseph.
A new system of shoeing horses, abridged from the work of Joseph Goodwin ... containing a comparison between the English and French methods.... Boston: Printed by Wells and Lilly, 1821. vi, 139 p., 5 plates.
First published in London, 1820.
S-5469.
Copies: DLC; MBAt; MWA; MeU; NN.

1617. New York Association for the Improvement of the Breed of Horses.
Articles and rules.... New York: Printed by J. Seymour, 1823.
S-13561.
Privately held; photocopy of title page at NN.

1618. Grove, H. D.
An essay on sheep.... Boston: Printed at the New England farmer office, by W. Nichols, 1824. 12 p.
S-16356.
Copies: DLC; MWA.

1619. Boyd, James.
Patent sponge boots, for horses feet. Boston, June 1, 1825. Boston: True & Greene, printers, [1825], 1 p.
S-19842.
Copies: DLC.

1620. The American shooter's manual, comprising such plain and simple rules as are necessary to introduce the inexperienced into a full knowledge of all that relates to the dog, and the correct use of the gun; also a description of the game of this country. By a gentleman of Philadelphia County. Philadelphia: Carey, Lea & Carey, 1827. 2 p.l., [ix]–xii, [13]–249, [2], [5] p., illus.
Data about the quality of gunpowder, including that manufactured by E. I. du Pont, p. 232–238.
Supposedly written by Jesse Y. Kester.
S-27838.
Copies: CSmH; CtY; DLC; DNAL; DeGE; MWA; MiU-C; MnSM; MoSHi; NWM; NcWsW; PHi; PU; WyU.

1621. Du Couëdic de Villeneuve, Pierre Louis de, 1743–1822.
The pyramidal bee-hive: a plain and natural method of preserving and perpetuating the population of bees, and of receiving annually, from each family, a box full of wax and pure honey.... By P. Ducouédic.... Abridged, and translated from the French, by Silas Dinsmoor. Philadelphia: Carey, Lea & Carey, 1829. 103 p., front.
Translation of his *La Ruche pyramidale,*

Paris, 1812.
S-38426.
Copies: CU-A; DLC; KMK; MB; MU; NBuG; PPAmP.

1622. Thacher, James, 1754–1844.
A practical treatise on the management of bees; and the establishment of apiaries.... Boston: Marsh & Capen, 1829. 162, [2] p.
S-40627.
Copies: CSmH; DLC; MB; MBH; MH; MWA; MoS; NGH; NIC; NNC; Nh; NjP; OCY; OClW; WU-A.

1623. [Milnor, William] 1769–1848.
An authentic historical memoir of the Schuylkill Fishing Company of the state in Schuylkill.... By a member. Philadelphia: Judah Dobson, 1830. viii, 127 p. illus.
C-2578.
Copies: CtY; DLC; MB; MBAt; MH; MWA; MiU-C; PP; PPAmP; PPL; PU; RNR; ViU; WHi.

1624. Repplogel, George.
A book comprising a new system for the preservation of bees. Discovered through a series of practical experience. Greensburgh, Pa.: Printed for the author, 1830. 24 p.
C-3291.
Copies: PHi.

10. Forestry and forest products

1625. New husbandry to New-England; or, An experienced way to raise quick hedges and clover grass, and the way to make syder. [Philadelphia: Printed by William Bradford, 1692.]
E-626.
No copy known.

1626. Bridger, John.
Information and directions for the making of tar, and choice of trees for the same, as in Finland, &c. Boston, in New England: Printed by B. Green, 1707. [2] p.
Signed at end: J. Bridger.
B-296; Mp-39443.
Copies: MHi; NHi.

1627. Massachusetts (Colony). Laws.
An act for rendering more effectual the laws already made relating to shingles and for regulating the assize of staves, hoops and clapboards.... [Boston: 1762.] Broadside 35 x 22.5 cm.
At head of title: Province of the Massachusetts-Bay; dated: June 12, 1762.
B-2295; Mp-41281.
Copies: MHi; MWA; PHi.

1628. Marshall, Humphry, 1722–1801.
Arbustrum americanum: the American grove; or, An alphabetical catalogue of forest trees and shrubs, natives of the American United States.... Also, some hints of their uses in medicine, dyes, and domestic economy.... Philadelphia: Printed by J. Crukshank, 1785. xx, 174 p.
Mp-19068.
Copies: CU; CtY; DLC; DNAL; DeGE; MBAt; MWA; MiU; NjP; PPL; RPJCB.

1629. Rush, Benjamin, 1745–1813.

An account of the sugar maple-tree, of the United States, and of the methods of obtaining sugar from it, together with observations upon the advantages both public and private of this sugar. In a letter to Thomas Jefferson, Esq. Secretary of State of the United States.... Read in the American Philosophical Society, on the 19 of August, 1791, and extracted from the third volume of their transactions now in the press.... Philadelphia: Printed by R. Aitken & Son, 1792. 16 p.

Mp-24761.

Copies: DLC; DNAL; MWA; PHi; RPJCB.

1630. Bach (Robert) & Company.

Essence of spruce. [New York: 1804?] Broadside 22 x 9 cm.

Advertisement with directions for making spruce beer.

N-B0015101.

Copies: DNLM.

1631. Independent Company of New-York Foresters.

Constitution of the Independent Company of New-York Foresters. New York: Printed by L. Nichols, 1804. 8 p., illus.

S&S-6533.

Copies: NN.

1632. Main, Thomas.

Directions for the transplantation and management of young thorn or other hedge plants.... With some practical observations on the method of plain hedging.... City of Washington: A. & G. Way, printers, 1807. 37, [1] p.

S&S-12976, 12977.

Copies: DLC; DNAL; MB; MWA; MdHi; PU; ScU; ViU.

1633. United States. Laws.

An act authorizing and requiring the Secretary of State to issue letters patent to Andrew Kurtz. In the Senate of the United States. March 7th, 1816.... [Washington: 1816.] 2 p. (H.R. 113)

Concerns the manufacturing of charcoal.

S&S-39148.

Copies: DLC.

1634. [Akerly, Samuel] 1785–1845.

On the cultivation of forest trees; in a letter addressed to Jonathan Thompson, Esq.... [New York: 1823.] 2 p.

Signed: New-York, 25 July, 1823. Samuel Akerly.

S-11555.

Copies: MB; MWA.

1635. Akerly, Samuel, 1785–1845.

Remarks on the cultivation of the locust tree (Robinia pseudo-acacia) written in 1817, by Dr. S. Akerly. [New York?: 1823?] 3 p.

Postscript dated 1823.

S-11556.

Copies: MB.

1636. Guillet, Peter.

Timber merchant's guide. Also, a table, whereby, at one view, may be seen the solid and superficial measure of any square or unequal hewed logs or plank, from one to forty-seven inches. Also, plates representing the figures of the principal pieces of timber, used in building a seventy-four gun ship of the line, in standing trees. By Peter Guillet, l'aîné.... Baltimore: J. Lovegrove, 1823. 24 p., 30 col. plates, tables.

S-12738.

Copies: DLC; DeGE; IaAS; MH-BA; MWA; MdBP; MeBa; MiU-C; NIC; NNC; NcD; OCl; PP; PPWI; RPAt.

11. Horticulture

1637. Squibb, Robert.
The gardener's calendar, for South-Carolina, Georgia, and North-Carolina. Containing an account of all work necessary to be done in the kitchen and fruit gardens every month of the year, with instructions for performing the same; also particular directions relative to soil and situation, adapted to the different kinds of plants and trees most proper for cultivation in these states.... Charleston: Printed by Samuel Wright and Co., for R. Squibb, 1787. [8], 183, [8] p.
Mp-20722.
Copies: NcD.

1638. [Randolph, John] 1727–1784.
A treatise on gardening. By a native of this state. Richmond: Printed by Thomas Nicolson, 1793.
Assigned to John Taylor by Evans.
E-26275; Sabin-96739.
No copy known.

1639. A new gardener's calendar; or, Every man a complete gardener: containing all the necessary information relating to gardens, orchards, &c. grafting, transplanting, &c. destroying vermin, judging the weather, managing bees, &c. New London, [Conn.]: Sold by Charles Holt, 1798.
E-34191.
No copy known.

1640. Marshall, Charles, d. 1818.
An introduction to the knowledge and practice of gardening, by Charles Marshall.... 1st American from the 2d London ed., considerably enl. and improved. To which is added An essay on quick-lime, as a cement and as a manure, by James Anderson.... Boston: Printed by S. Etheridge, for J. Nancrede, 1799. 2 v.
"An essay on quick-lime ... " by James Anderson, 115 p. at end of vol. 2.
Mp-35770.
Copies: CtY; DLC; DNAL; DeGE; InU; MBAt; MWA; MiU-C; NN; PU; RPJCB; ViU.

1641. Forsyth, William, 1737–1804.
A treatise on the culture and management of fruit trees; in which a new method of pruning and training is fully described.... By William Forsyth.... To which are added, an introduction and notes, adapting the rules of the treatise to the climates and seasons of the United States of America, by William Cobbett. New York: Ezra Sargeant & Co., 1802. xii, 259 p., 13 plates.
S&S-2235.
Copies: CtHT-W; CtY; IU; MU; MWA; NNBG.

1642. ———— Philadelphia: Printed for J. Morgan, 1802. xii, 259 p., 13 plates.
S&S-2236.
Copies: CSt; DLC; DNAL; MWA; MeB; MiU-C; NjP; PPL; ViU.

1643. Forsyth, William, 1737–1804.
An epitome of Mr. Forsyth's treatise on the culture and management of fruit trees. Also, notes on American gardening and fruits: with designs for promoting the ripening of fruits ... and further, of economical principles in building farmers' habitations. By an American farmer. Philadelphia: Printed by T. L. Plowman, for John Morgan, 1803. [4], 186, [6] p., 15 plates.
Edited by William Cobbett.
S&S-4217.
Copies: CtY; DLC; InU; KU; MH; MWA; MiU-C; MnU; MeB; NN; NcD; OO; PPL; PU; TxU.

1644. Forsyth, William, 1737–1804.

A treatise on the culture and management of fruit trees; in which a new method of pruning and training is fully described.... By William Forsyth.... To which are added an introduction and notes, adapting the rules of the treatise to the climates and seasons of the United States of America. By William Cobbett. Albany: D. & S. Whiting [etc., etc.], 1803. 1 p.l., [v]–xii, [13]–280 p., 13 plates.

S&S-4218.

Copies: CSmH; CSt; CU; CtW; DLC; ICJ; InU; KU; LU; MH; MWA; MdBP; MeU; MiU-C; MoSW; NN; NcD; NjR; PU-V; RPB; TxU; ViU.

1645. Forsyth, William, 1737–1804.

An epitome of Mr. Forsyth's Treatise on the culture and management of fruit-trees. Also notes on American gardening and fruits; with designs for promoting the ripening of fruits, and securing them as family comforts: and further of economical principles in building farmers' habitations. By an American farmer. Philadelphia: Wm. Poyntell & Co., 1804. [4], 186, [6] p., 15 plates (13 fold.).

Edited by William Cobbett.

S&S-6318.

Copies: CU; DeGE; InU; MH-A; MdBP; MWA; NIC; NjP; OU; PP; ViLxW.

1646. Gardiner, John.

The American gardener, containing ample directions for working a kitchen garden.... and copious instructions for the cultivation of flower gardens, vineyards, nurseries, hop-yards, green houses and hot houses. By John Gardiner & David Hepburn.... City of Washington: Printed by Samuel H. Smith, for the authors, 1804. 204, [4] p.

S&S-6366.

Copies: DGU; DLC; ICJ; MBH; MWA; NN; PSt; ViU.

1647. MacMahon, Bernard, 1775?–1816.

American gardener's calendar; adapted to the climates and seasons of the United States. Containing a complete account of all the work necessary to be done ... with ample practical directions for performing the same.... Philadelphia: Printed by B. Graves, for the author, 1806. v, 648, [18] p., fold. table.

S&S-10771.

Copies: CU; DNAL; DeGE; ICJ; KyLx; MWA; NIC; NN; NcU; NjR; OU; PCC; PPAmP; PPL; PU; ViU.

1648. Forsyth, William, 1737–1804.

A treatise on the culture and management of fruit trees, in which a new method of pruning and training is fully described.... To which is added, an introduction and notes, adapting ... the treatise to the climates and soils of the United States of America. New York: Ezra Sargeant & Co., 1807. xii, 259 p., 13 plates.

S&S-12584.

Copies: NNS.

1649. Squibb, Robert.

Gardener's calender for North-Carolina, South-Carolina and Georgia.... With amendments and additions. Charleston: Printed and sold by W. P. Young, 1809. viii, [3]–162 p.

S&S-18679.

Copies: ABS; DLC; NIC; NNBG; NcD.

1650. Landreth, David, 1802–1880.

A catalogue of greenhouse plants, hardy trees, evergreen shrubs, flowering shrubs, bulbous rooted, and herbaceous plants.... With a collection of the most esteemed varieties of fruit trees. Sold by David and Cuthbert Landreth, nursery and seedsmen, Federal Street, near the Arsenal, Philadelphia. Philadelphia: Fry and Kammerer, printers, 1811. 47 p.

S&S-23177.

Copies: MB; PPAmP.

1651. Thorburn, Grant, 1773–1863.

The gentleman and gardener's kalendar, for the middle states of North America.... New York: Printed for Hardcastle & Van Pelt, 1812. 114 p., front.

S&S-26881.
Copies: DNAL; MB; NHi.

1652. Squibb, Robert.
The gardener's calendar for North-Carolina, South-Carolina, and Georgia.... Charleston, S.C.: W. P. Young, 1813. viii, [3]–162 p.
S&S-29844.
Copies: DNAL; NcD; NcU.

1653. Coxe, William.
A view of the cultivation of fruit trees, and the management of orchards and cider, with accurate descriptions of the most estimable varieties of native and foreign apples, pears, peaches, plums, and cherries, cultivated in the middle states of America.... Philadelphia: M. Carey and Son, 1817. iv, [5]–253, [15] p. diagrs., plates.
S&S-40585.
Copies: CoFS; DNAL; DeGE; MH; MWA; MiU-C; NIC; NcWsW; Nj; NjR; OClW; PPL; PU; ViU.

1654. Elliott, William R.
Practical and comprehensive treatise on fruit and floral culture, and a few hints on landscape gardening. By William R. Elliott. [Pittsburg: Joseph Horne & Co., 1817?] 100 p., illus.
Wrongly dated; probably printed ca. 1870.
S&S-40750.
Copies: NIC; PP; PPi.

1655. Thorburn, Grant, 1773–1863.
The gentleman & gardener's kalendar, for the middle states of North America.... 2d ed., corr. and improved. New York: Printed by E. B. Gould, 1817. 117 p.
S&S-42288.
Copies: DLC; N; NN.

1656. Gardiner, John.
The American gardener, containing ample directions for working a kitchen garden ... and copious instructions for the cultivation of flower gardens, vineyards, nurseries, hop yards, green houses and hot houses. By John Gardiner and David Hepburn.... A new ed., much enlarged. To which is added, A treatise on gardening, by a citizen of Virginia. Also, a few hints on the cultivation of native vines, and directions for making domestic wines. Georgetown, D.C.: Joseph Milligan, 1818. ix, [2], [13]–347, [1] p.
A Treatise on gardening is by John Randolph.
S&S-44127.
Copies: CSmH; DLC; DNAL; LNHT; MWA; NIC; NcD; NhD; ViU.

1657. Cobbett, William, 1763–1835.
The American gardener; or, A treatise on the situation, soil, fencing and laying-out of gardens; on the making and managing of hot-beds and green-houses; and on the propagation and cultivation of ... vegetables, herbs, fruits and flowers.... Claremont, N.H.: Manufacturing Co., Simeon Ide, ag't, [1819]. 230 p., illus.
S&S-47650.
Copies: CU; DNAL; KyLxT; KyU; MWA; NbU; NjP; NjR.

1658. MacMahon, Bernard, 1775?–1816.
The American gardener's calendar; adapted to the climates and seasons of the United States. Containing a complete account of all the work necessary to be done ... with ample practical directions for performing the same.... 2d ed., improved. Philadelphia: T. P. M'Mahon, 1819. 618 p., fold. table.
S&S-48550.
Copies: CSt; DNAL; DeGE; InCW; MWA; NNC; NjR; PSC; ViW.

1659. The practical American gardener; exhibiting the time for every kind of work in the kitchen garden, fruit garden, orchard, nursery, shrubbery, pleasure ground, flower garden, hop yard, green house, hot house, and grape vines. For every month in the year. By an old gardener. Baltimore: F. Lucas, Jr., 1819. xii, [5]–424 p.
S&S-49175.
Copies: DNAL; MWA; MdHi.

1660. The young florists' companion.... Hartford: Printed by Roberts & Burr, for S. G. Goodrich, 1819. 31 p.
S&S-50188.
Copies: CSmH.

1661. MacMahon, Bernard, 1775?–1816.
The American gardener's calendar; adapted to the climates and seasons of the United States. Containing a complete account of all the work necessary to be done ... With ample practical directions for performing the same. 4th ed., improved. Philadelphia: T. P. M'Mahon, 1820. 618 p., fold. table.
S-2066.
Copies: CoFS; MH-A; MLy; MWA; NIC; OClW.

1662. Prince, William, 1766–1842.
A treatise on fruit and ornamental trees and plants, cultivated at the Linnaean Botanic Garden, Flushing, Long-Island, near New-York. William Prince, proprietor. New York: Printed by T. and J. Swords, 1820. 81 p.
S-2871.
Copies: DNAL; MB; MH; MWA; NNC.

1663. Casey, Joseph P.
A treatise on the culture and growth of different sorts of flower roots, and of green house plants kept in rooms, &c.... Baltimore: Printed by John D. Toy, for the author, 1821. 99, [2] p.
S-4928.
Copies: DLC; IU; MBH; MWA; MdBE; MdBP; MdHi; NIC; PHi; ViW.

1664. Thacher, James, 1754–1844.
The American orchardist; or, A practical treatise on the culture and management of apple and other fruit trees.... Boston: 1821. 226 p.
Several differences exist in the preliminary matter from the 1822 edition, which is generally considered the first edition.
Privately held.

1665. Thorburn, Grant, 1773–1863.
The gentleman and gardener's kalendar; containing ample directions for the cultivation of the kitchen and flower garden, green house, nursery, orchard &c. 3d ed., corr. and enlarged. New York: Printed by B. Young, 1821. 132 p.
S-6970.
Copies: MH; NIC; NjR.

1666. The practical American gardener; exhibiting the time for every kind of work in the kitchen garden, fruit garden, orchard, nursery, shrubbery, pleasure ground, flower garden, hop yard, green house, hot house, and grape vines. For every month in the year. By an old gardener. Baltimore: Fielding Lucas, Jr., 1822. xii, [5]–424 p.
S-9995.
Copies: DeU; InLP; MSaP; MdBE; NIC; RPB; ScCM.

1667. Thacher, James, 1754–1844.
The American orchardist; or, A practical treatise on the culture and management of apple and other fruit trees.... To which is added the most approved method of manufacturing and preserving cider.... Boston: Joseph W. Ingraham, 1822. vi, [2], [9]–226 p.
S-10430.
Copies: CU; Ct; DLC; DeGE; ICJ; MB; MH; MWA; MeU; NIC; NNNAM; NcRA; NhPet; OO; OrCS; PU; RNR; RPB.

1668. Cobbett, William, 1763–1835.
The American gardener; or, A treatise on ... cultivation of ... vegetables, herbs, fruits and flowers. Baltimore: J. Robinson Circulating Library, and J. Robinson & Co., Frederick, Md., 1823. ix, 252 p., illus.
First published in London, 1821.
S-12191.
Copies: DNAL; MBH; MNBedf; MSaP; MdBE; MdBS; NIC; NNBG; Nh; ViW.

1669. Agricola, P., pseud.
The New-York gardener; or, Twelve letters, from a farmer to his son in which he

describes the method of laying out and managing the kitchen-garden. By P. Agricola [pseud.]. Albany: Daniel Steele & Son; Packard & Van Benthuysen, printers, 1824. 214, [2] p.
S-15040.
Copies: CSmH; DNAL; INS; MH; NCooHi; WU.

1670. Hosack, David, 1769–1835.
An inaugural discourse, delivered before the New-York Horticultural Society at their anniversary meeting, on the 31st of August, 1824.... New York: Printed by J. Seymour, 1824. 46 p.
S-16594.
Copies: CtY; DLC; DNAL; MB; MBAt; MWA; MiD-B; NIC; NNC; NcD; NjR; PMA; PPL; RPB; TNJ.

1671. Landreth, David, 1802–1880.
A catalogue of greenhouse plants, hardy trees, evergreen shrubs, flowering shrubs, bulbous rooted, and herbaceous plants ... with a collection of the most esteemed varieties of fruit trees, and a catalogue of garden seeds. Cultivated at the nurseries and gardens of David and Cuthbert Landreth, Federal Street, near the Arsenal. Philadelphia: Printed for D. & C. Landreth; I. Ashmead & Co., printers, 1824. 48 p.
N-L0069099.
Copies: MH-A.

1672. New York State Horticultural Society.
Constitution and by-laws of the New-York Horticultural Society, incorporated March 22, 1822. New York: Printed by J. Seymour, 1824. 28 p.
Founded in 1818.
S-17412.
Copies: MB; NNMus.

1673. New York State Horticultural Society.
Constitution of the New-York State Horticultural Society. [New York]: 1824. 15 p.
Founded in 1818.
S-17421.
Copies: DLC; MWA.

1674. New York State Horticultural Society.
Officers of the New-York Horticultural Society, elected on the 31st August, 1824. [New York: 1824.] [1] p.
In: Hosack, David, *An Inaugural discourse, delivered before the New-York Horticultural Society ... 31st of August, 1824.*
N-N0215343; S-17413 (erroneously giving 40 or 46 pages).
Copies: CtY; DLC; DNAL; MB; MBAt; MWA; MiD-B; NIC; NNC; NjR; PMA; PPL; RPB; TNJ.

1675. New York State Horticultural Society.
Prospectus of the New-York State Horticultural Society. [New York?]: 1824. 8 p.
S-17422.
Copies: PPL.

1676. Peirce, Joshua, 1795?–1869.
Catalogue of fruit and ornamental trees and plants, bulbous flower roots, green-house plants, &c. &c. cultivated at the Linnaean Hill nursery, on Rock Creek, near Washington, D.C. Joshua Peirce, proprietor. To which is added an essay on the planting and cultivation of orchards. Washington City: Printed at the Columbian office, 1824. 31, [1] p.
N-P0184814.
Copies: DLC.

1677. Thacher, James, 1754–1844.
The American orchardist; or, A practical treatise on the culture and management of apple and other fruit trees.... To which is added the most approved method of manufacturing and preserving cider, and also wine from apple juice and currants. Adapted to the use of American farmers ... 2d ed., much improved.... Plymouth, Mass.: Ezra Collier, 1825. v, [3], [9]–236 p.
S-22442.
Copies: DLC; ICJ; MBH; MWA; MeLB; Mi; NBLiHi; NN; NNNAM; NjP; PPL.

1678. Gardiner, John.
The American gardener: containing ample directions for working a kitchen garden.... A new ed., much enl. To which is added, A treatise on gardening, by a citizen of Virginia. Also a few hints on the cultivation of native vines, and directions for making domestic wines. 3d ed. Washington City: W. Cooper, jr., 1826. viii, [2], [11]-304 p.
A Treatise on gardening is by John Randolph.
S-24659.
Copies: CSmH; DLC; IU; KMK; NNBG; ViU.

1679. Mitchill, Samuel Latham, 1764–1831.
Address pronounced before the New-York Horticultural Society ... on the annual celebration, August 29, 1826. New York: Printed by E. Conrad, 1826. 32 p.
S-25404.
Copies: MB; NNAM; NNC; PPAN; RPB.

1680. [Randolph, John] 1727–1784.
Randolph's culinary gardener, enlarged and adapted to the present state of our climate, by an experienced gardener, a native of Virginia. Richmond: Collins & Co., 1826. 115 p.
Additions signed: M.
S-25899.
Copies: CSmH; DLC; DNAL; Vi; ViHi; ViSwC; ViW.

1681. Agricola, P., pseud.
The New York gardener; or, Twelve letters from a farmer to his son, in which he describes the method of laying out and managing the kitchen garden. By P. Agricola [pseud.]. White Creek, [N.Y.]: A. Crosby; G. M. Davison, printer, Saratoga Springs, 1827. 95, [1] p.
S-27789.
Copies: CSmH; MB; MBH; N; NNBG.

1682. Carter, Nathaniel Hazeltine, 1787–1830.
Address read before the New York Horticultural Society, at the anniversary celebration, of the 28th of August, 1827. New York: Pub. at the request of the Society; Sleight & George, printers, Jamaica, L.I., 1827. 22 p.
S-28415.
Copies: CSmH; IC; MBH; NHi; NN; NNMus; PPAmP.

1683. Peirce, Joshua, 1795?–1869.
Catalogue of fruit and ornamental trees and plants, bulbous flower roots, green-house plants, &c. cultivated at the Linnaean Hill, Rock Creek, near Washington, D.C. To which is added, an essay on the planting and cultivation of orchards. Joshua Peirce, proprietor. Washington: Printed by Gales & Seaton, 1827. 32 p.
N-P0184815.
Copies: DLC.

1684. Pennsylvania Horticultural Society.
Constitution of the Pennsylvania Horticultural Society at Philadelphia. Philadelphia: Clark & Raser, printers, 1827. 7 p.
S-30193.
Copies: PPL.

1685. Sagra, Ramon de la, 1798–1871.
Memorias para servir de introducción a la horticultura cubana. Por d. Ramon de la Sagra.... Nueva York: Lanuza, Mendia y. c., 1827. viii, 24 p.
S-30508.
Copies: CU-B; DNA; NN.

1686. Squibb, Robert.
The gardener's calendar for the states of North-Carolina, South-Carolina, and Georgia.... With appendix, containing a variety of particular information on husbandry and horticulture. Charleston, S.C.: P. Hoff, 1827. 3 p.l., 176 p., front.
S-30696.
Copies: CtW; DLC; GU-De; IaAS; MBH; NHi; NcAS; NcD; ScC; ScHi; ScU.

1687. Agricola, P., pseud.
The New York gardener; or, Twelve letters from a farmer to his son, in which he describes the method of laying out and managing the kitchen-garden. Saratoga Springs: Pub. by A. C. Crosby; G. W. Davison, printer, 1828. 96 p.
S-31907.
Copies: MBH; MiD-B.

1688. Fessenden, Thomas Green, 1771–1837.
The new American gardener; containing practical directions on the culture of fruits and vegetables; including landscape and ornamental gardening, grape-vines, silk, strawberries, &c. &c.... Boston: J. B. Russell, 1828. 306, [1], [5] p.
S-33168.
Copies: CSmH; CtY; DLC; DNAL; DeGE; InLP; LNHT; MH; MWA; MdBJ; MiU; NIC; NjP; PPL; RNR; TU.

1689. Green, Roland.
A treatise on the cultivation of ornamental flowers ... with directions for the general treatment of bulbous flower roots.... Boston: John B. Russell, press of I. R. Butts & Co., [etc.], 1828. 59, [1] p.
S-33414.
Copies: DLC; LNHT; MB; MBH; MH-A; MWA; NIC; NN; NNBG; PMA.

1690. Holley, Myron, 1779–1841.
An initiatory discourse, delivered at Geneva, 27th November, 1828, before an assembly from which, on that day, was formed the Domestic Horticultural Society of the western parts of the state of New York.... Geneva, [N.Y.]: Printed by James Bogert, 1828. 27 p.
S-33581.
Copies: CtY; DLC; MBH; MiU-C; NIC; NN; PPL.

1691. MacMahon, Bernard, 1775?–1816.
The American gardener's calendar, adapted to the climates and seasons of the United States, containing a complete account of all work necessary to be done in the kitchen-garden, fruit garden, orchard, etc., for every month of the year ... 7th ed., improved. Philadelphia: A. M'Mahon, 1828. 618 p., fold. table.
S-33959.
Copies: MH; ViW.

1692. New York State Horticultural Society.
Proceedings of the New-York Horticultural Society, at the celebration of its tenth anniversary, August 26, 1828. New York: Printed by E. Conrad, 1828. 24, [1] p.
S-34541.
Copies: MBH; MWA; MiD-B; PPL; WU-A.

1693. Prince, William, 1766–1842.
A short treatise on horticulture: embracing descriptions of a great variety of fruit and ornamental trees and shrubs, grape vines, bulbous flowers, green-house trees and plants, &c., nearly all of which are at present comprised in the collection of the Linnaean Botanic Garden, at Flushing, near New-York. New York: Printed by T. and J. Swords, 1828. ix, 196 p.
S-34887.
Copies: ArU; CU; DNAL; DeGE; IU; IaAs; InLP; KU; MH; MWA; MdU; MiU; MnU; NBLiHi; NIC; NNC; NcU; NjR; OHi; PPL; RP; TNJ; WU.

1694. Wilson, William.
Economy of the kitchen-garden, the orchard, and the vinery, with plain practical directions for their management.... New York: Anderson, Davis & Co., 1828. 206 p., plate.
S-37211.
Copies: CSmH; DNAL; MWA; Nh; NjR; PPL.

1695. Albany Horticultural Society.
Constitution and by-laws.... Albany: Printed by Websters and Skinners, 1829. 15 p.

1695 (*continued*)
Adopted January 15, 1829.
S-37405.
Copies: MB; NN.

1696. Bridgeman, Thomas, d. 1850.
The young gardener's assistant, containing a catalogue of garden and flower seeds, with practical directions under each head, for the cultivation of culinary vegetables. Some of which are not generally introduced into the United States. Brooklyn: Printed by Nichols and Matthews, 1829. 96 p.
S-37938.
Copies: CSmH; MBH; NBLiHi; NcD; Nh.

1697. Dearborn, Henry Alexander Scammell, 1783–1851.
An address delivered before the Massachusetts Horticultural Society, on the celebration of their first anniversary, September 19, 1829.... Boston: From the press of Isaac R. Butts, 1829. 28 p.
S-38354.
Copies: CSmH; DLC; DNAL; MBAt; MBH; MH; MWA; MiD-B; NCH; NjR; PPL; ViU.

1698. Massachusetts Horticultural Society.
[Proceedings on the establishment of the] Massachusetts Horticultural Society. Boston: Press of Isaac R. Butts, 1829. 51, [1] p.
S-39519.
Copies: DLC; IU; MB; MH; MWA; NN; OMC; PPL; RPB; ViU.

1699. Clinton, George William, 1807–1885.
An address, delivered at Canandaigua, June 22, 1830, before the Domestic Horticultural Society of the western parts of the state of New-York. Published at the request of the Society. Canandaigua, [N.Y.]: Printed by Morse, Ward and Co., 1830. 23 p.
C-925.
Copies: CtY; MB; MBH; NGH; NN; NRHi.

1700. Cook, Zebedee, 1786–1858.
An address, pronounced before the Massachusetts Horticultural Society. In commemoration of its second annual festival, the 10th of September, 1830. Boston: Printed by Isaac R. Butts, 1830. 54 p.
C-1006.
Copies: CU; DLC; DNAL; MH; MHi; MWA; MiD; NCH; NN.

1701. Fessenden, Thomas Green, 1771–1837.
The new American gardener; containing practical directions on the culture of fruits and vegetables; including landscape and ornamental gardening, grape-vines, silk, strawberries, &c., &c. 4th ed. Boston: Carter and Hendee; stereotyped at the Boston Type and Stereotype Foundry, 1830. [1], 16–306 p.
C-1369.
Copies: CU; IU; MNBedf; MWA.

1702. Francis, John Wakefield, 1789–1861.
An address delivered before the New York Horticultural Society, at their anniversary, on the eighth of September, 1829.... New York: Printed by E. Conrad, 1830. 31 p.
C-1432.
Copies: CtY; DLC; DNLM; MBH; MH; MHi; MiD-B; NN; NNNAM; PPL; RPB.

1703. MacMahon, Bernard, 1775?–1816.
American gardener's calendar; adapted to the climate and seasons of the United States. Containing a complete account of all the work necessary to be done.... 8th ed., improved. Philadelphia: Printed by John Young for the author, 1830. 618 p., fold. table.
C-2339.
Copies: MH; NIC; NNBG; ViW.

1704. Stout, Z. Barton.
An address, delivered before the Domestic Horticultural Society of the western part of New York, at its annual meeting, in Geneva, September 28, 1830. By Z. Barton

Stout.... Canadaigua, [N.Y.]: Printed by Morse & Harvey, 1830. 15, [1] p.

C-3614.

Copies: CSmH; NCanHi; PPAmP.

12. Silk culture

1705. Great Britain. Laws.

Anno regni Georgii II. regis Magnae Britanniae, Franciae & Hiberniae, vicesimo tertio. At the Parliament begun and holden at Westminster, the tenth day of November, anno Dom. 1747 ... and continued ... to the sixteenth day of November, 1749.... London: Printed by Thomas Baskett, 1750. New London: Reprinted by Timothy Green, 1751. [2], 45–64 p.

Contains: An act for encouraging the growth and culture of raw silk in His Majesty's colonies and plantations in America; and An act to encourage the importation of pig and bar iron from ... colonies in America, and to prevent the erection of any mill or other engine for the slitting or rolling of iron.

Mp-6682.

Copies: Ct; CtY.

1706. Sauvages, Pierre Augustin Boissier de, 1710–1795.

Directions for the breeding and management of silk worms. Extracted from the treatises of the Abbé Boissier de Sauvages, and Pullein. With a preface, giving some account of the rise and progress of the scheme for encouraging the culture of silk in Pennsylvania, and the adjacent colonies. Philadelphia: Printed by J. Crukshank, and I. Collins, 1770. xv, [1], 32 p.

Boissier's treatise translated and epitomized by J. Odell.

"An appendix: containing some extracts from Pullien [*sic*] ... ," p. [21]–32.

Mp-11574.

Copies: DLC; DeGE; MH; PPL.

1707. Ottolenghe, Joseph.

Directions for breeding silk-worms, extracted from a letter of Joseph Ottolenghe, Esq., late superintendent of the public filature in Georgia. Philadelphia: Printed by Joseph Crukshank, 1771. 8 p.

Mp-12172.

Copies: MiU-C; PHi; PPL.

1708. An essay on the culture of silk, and raising white mulberry trees, the leaves of which are the only proper food of the silkworm. By a citizen of Philadelphia. Philadelphia: J. Crukshank, 1790. 8 p.

Attributed to Pelatiah Webster.

Mp-22491.

Copies: CSmH; CtY; DLC; DeGE; MH; MWA; MdBJ; MnU; NN; PPL; RPJCB; ViU.

1709. Extract from an Essay on the culture of silk. Burlington, [N.J.]: Printed by Isaac Neale, 1792.

E-24306.

No copy known.

1710. Some modern directions for the culture and manufacture of silk, taken from a manuscript as it was wrote by a gentleman in Italy. Containing the most necessary instructions for culture and manufacture of silk, from the hatching of the worm, till the silk is prepared for the loom. Transcribed by a friend to the public. Windham, [Conn.]: Printed for Mess. Samuel Storrs, and Lemuel Eldredge, of Mansfield [by John Byrne], 1792. 38 p.

Mp-24802.

Copies: CSmH; MH; NN.

1711. A treatise on silk worms. In two parts. The first on their management within doors, and the second, concerning their treatment in open air, upon hedges. By a member of the Agricultural Society of the State of New-York. 1793. New York: Printed by T. Allen, 1793. 30, [2] p.

"An act to incorporate the Society instituted in the State of New York, for the Promotion of Agriculture, Arts and Manufactures," [2] p. at end.

Manuscript note on the title page of DeGE copy assigns the authorship to Peter de La Bigarre, [Pierre de La Bigarre].

Mp-26276.

Copies: DeGE; MH; MWA; NHi; NN.

1712. Sharpless, John Townsend.

An essay on the Bombyx mori, or silk worm.... [Philadelphia: 1826?] [21] p.

Extract from the *Franklin journal,* July–Sept., 1826.

S-26061.

Copies: PPL.

1713. United States. Congress. House. Committee on Agriculture.

Mulberry–silk worm. May 2, 1826, Mr. Van Rensselaer, from the Committee on Agriculture ... made the following report. [Washington: 1826.] 10 p. (19th Cong., 1st sess. House. Rept. 182)

S-27260.

Copies: DLC; DeU; MWA; NjP.

1714. United States. Treasury Dept.

The House of Representatives having, at the last session of Congress, passed a resolution that "The Secretary of the Treasury cause to be prepared a well-digested manual, containing the best practical information that can be collected, on the growth and manufacture of silk, adapted to the different parts of the Union ... " the following queries have been prepared.... Richard Rush, Treasury Department, 29th July, 1826. [Washington: 1826.] 3 p.

S-27020.

Copies: DLC; MWA.

1715. Georgia. General Assembly. Joint Committee on Agriculture.

Resolution [and report of the Joint Committee on Agriculture, concerning silk, the olive, etc. November, 1827]. [Milledgeville?, Ga.: 1827.] 5 p.

Bowker-p. 819.

Copies: GU.

1716. Buchanan, William B.

Observations on the silk worm. Baltimore: Printed by J. D. Toy, 1828. 20 p.

S-32508.

Copies: DeGE; MdBL; MdHi.

1717. Hazzi, Joseph, Ritter von, 1768–1845.

Letter from James Mease, transmitting a treatise on the rearing of silkworms by Mr. de Hazzi, of Munich.... April 21, 1828. Printed by order of the Senate of the United States. Washington: Printed by Duff Green, 1828. 106 p., fold. front., fold. table. (20th Cong., 1st sess. Senate. Doc. 175, pt. 2)

Added title page: *A Treatise on the culture of silk in Germany.*

S-36492.

Copies: CtY; DLC; KU; MHi; MWA; NIC; NcD-MC; NjP.

1718. Hazzi, Joseph, Ritter von, 1768–1845.

Silk-worms. Letter from James Mease, transmitting a treatise on the rearing of silk-worms, by Mr. de Hazzi, of Munich.... Washington: Printed by Gales & Seaton, 1828. 108 p., col. fold. front., fold. table. (20th Cong., 1st sess. House. Doc. 226)

Added title page: *A Treatise on the culture of silk in Germany, and especially in Bavaria....*

S-36885.

Copies: CSmH; DLC; DeGE; DeU; MB; MWA; NjR; PPF; PPL.

1719. Memorial of sundry inhabitants of the counties of Windham and Tolland, state of Connecticut, praying for the aid of government in the cultivation of the mulberry tree and of silk. February 25, 1828.... Washington: Printed by Gales & Seaton, 1828. 6 p.

(20th Cong., 1st sess. House. Doc. 159)
At head of title: Connecticut.
S-36227.
Copies: DLC; DeU; MWA; NjR.

1720. Pennsylvania Society for Promoting the Culture of the Mulberry Tree, and the Raising of Silk Worms.
At a meeting of the subscribers to the plan ... held at the Indian Queen, April 2, 1828.... The following constitution was read, and, after due consideration, adopted. [Philadelphia: 1828.] 4 p.
S-34705.
Copies: MWA; PPL.

1721. Pennsylvania Society for Promoting the Culture of the Mulberry Tree, and the Raising of Silk Worms.
Directions for the rearing of silk worms, and the culture of the white mulberry tree.... Philadelphia: Clark & Raser, printer, 1828. 25 p., fold. table.
S-34706.
Copies: DeGE; MB; MWA; P.

1722. United States. Treasury Dept.
Growth and manufacture of silk. Letter from the Secretary of the Treasury, transmitting the information required by a resolution of the House of Representatives, of May 11, 1826, in relation to the growth and manufacture of silk, adapted to the different parts of the Union. February 7, 1828.... Washington: Printed by Gales & Seaton, 1828. 8, viii, 220 p., illus., 5 plates. (20th Cong., 1st sess. House. Doc. 158)
A manual, compiled by Edmond Charles Genet.
"Translation from a German treatise on dying silk, by George William Helterhoff, published at Erfurt, 1809, p. [195]–213.
S-36324.
Copies: CtHT; DLC; DeGE; DeU; MNoaMV; MWA; NjR.

1723. United States. Treasury Dept.
Letter from the Secretary of the Treasury, transmitting the information required by a resolution of the House of Representatives, of May 11, 1826, in relation to the growth and manufacture of silk, adapted to the different parts of the Union. February 7, 1828.... Washington: Printed by D. Green, 1828. 224 p., illus., 5 plates. (20th Cong., 1st sess. Senate. Doc. 175)
A manual compiled by Edmond Charles Genet.
"Translation from a German treatise on dying silk, by George William Helterhoff, published at Erfurt, 1809," p. [203]–220.
S-36504.
Copies: DLC; DeGE; DeU; MBH; MNoanMV.

1724. Vernon, William Henry.
A methodical treatise on the cultivation of the mulberry tree, on the raising of silk worms, and on winding the silk from the cocoons. United to an accurate description of the winding mill. With plates. Abridged from the French of M. de La Brousse; with notes and an appendix. By William H. Vernon. Boston: Hilliard, Gray & Co.; Eastman & Bridgham, printers, Providence, 1828. vi, [9]–174 p., 4 fold. plates.
S-37016.
Copies: CSt; DLC; DeGE; ICJ; MBAt; MH; NIC; OCLloyd; PPAmP; PU; RHi.

1725. Pascalis-Ouvière, Félix, 1750?–1833.
Practical instructions and directions for silkworm nurseries, and for the culture of the mulberry tree ... by Felix Pascalis.... New York: J. Seymour, printer, 1829–30. 2 v., plate.
S-39972.
Copies: CtY; DLC; DNAL; MBH; MWA (vol. 1 only); NNC; NNNAM; OCHP; OHi; OMC; OrCS; ScCC; ViU.

1726. United States. Congress.
Summary of the proceedings in Congress, in 1828–29, in relation to the mulberry tree, with a view to the culture of silk in the United States.... New York: Printed by G. G. Sickles, 1829. 16 p.
S-40575.
Copies: CSmH; Nh.

1727. United States. Congress. House. Committee on Agriculture.

Jacob B. Clarke. January 19, 1829. Mr. Martindale, from the Committee on Agriculture ... made the following report: The Committee on Agriculture, to which was referred the petition of Jacob B. Clarke, for the grant of the public land at Greenbush, in the state of New York, to aid him in the rearing of the mulberry tree, preparatory to the culture of silk, report.... [Washington: Gales & Seaton, printers, 1829.] 6 p. (20th Cong., 2d sess. House. Rept. 49)

S-41177.

Copies: DLC; DeU; MWA; NjR.

1728. Harper, Joseph Morrill, 1789–1865.

Mr. Harper's report to the legislature of the state of New-Hampshire on the culture of silk. [Concord, N.H.: 1830.] 32 p.

C-1759; Kress-C2528.

Copies: CtY; DLC; DeGE; IaAS; MH-BA; MWA; Nh.

1729. Homergue, Jean d'.

Essays on American silk, and the best means of rendering it a source of individual and national wealth. With directions to farmers for raising silk worms. By John d'Homergue ... and Peter Stephen Duponceau.... Philadelphia: John Grigg, 1830. xxii, 120, [1] p., illus.

C-1156.

Copies: CtY; DLC; DNAL; DeGE; MB; MH-BA; MWA; MdBP; NN; NNC; NcD; PPAmP; PPL; PU; RPB; TxU; ViU.

1730. Pennsylvania Silk Society.

Address of the Pennsylvania Silk Society to the public. [Philadelphia: 1830.] 4 p., illus.

Signed: Benjamin R. Morgan, President, Mathew Carey, Secretary [and others], Philadelphia, Jan. 1, 1830.

Sometimes assigned to Mathew Carey.

C-3006.

Copies: DeGE; MWA; PPAmP; PU.

1731. United States. Congress. House. Committee on Agriculture.

D'Homergue upon American silk. Report of the Committee on Agriculture on the growth and manufacture of silk; to which is annexed, essays on American silk, with directions to farmers for raising silk worms, by John d'Homergue ... and Peter Stephen Duponceau.... Washington: Printed by D. Green, 1830. 88 p., illus. (21st Cong., 1st sess. House. Doc. 126)

C-4676.

Copies: DLC; DeGE; MNoanMV; MWA; NjR.

1732. United States. Congress. House. Committee on Agriculture.

Growth and manufacture of silk. March 12, 1830. Mr. Spencer, of New York, from the Committee on Agriculture ... made the following report. [Washington: 1830.] 11 p. (21st Cong., 1st sess. House. Rept. 289)

Letter from Peter S. Duponceau, p. 3–11.

C-4737.

Copies: DLC; DeU; MWA; NjR.

1733. United States. Congress. House. Committee on Agriculture.

P. S. Du Ponceau—flag of American silk.... Mr. Spencer, of New York, from the Committee on Agriculture, to which was referred the letter of Peter S. Du Ponceau, Esq., presenting to the House a flag of American silk and manufacture, made the following report. December 21, 1830.... [Washington: 1830.] 9 p. (21st Cong., 2d sess. House. Rept. 7)

Letters from P. S. Du Ponceau on the growing and manufacture of silk, p. 2–9.

C-5035.

Copies: DLC; DeGE; DeU; MWA; NjR.

1734. United States. Congress. House. Committee on Agriculture.

Rapport sur la culture et manufacture de la soie, fait ... par M. Spencer, au nom du Comité d'agriculture, le 12 mars, 1830. Philadelphia: Kay, 1830. 20 p.

C-5077.

Copies: PPAmP.

1735. United States. Laws.

A bill for promoting the growth and manufacture of silk. March 8, 1830.... [Washington: 1830.] 8 p. (H.R. 345)

C-3994.

Copies: DNA.

IV. CRAFTS AND TRADES

1. Carpentry

1736. The carpenters rules of work, in the town of Boston. Boston: Mills and Hicks, 1774. 11 p.
B-3699; Mp-42561.
Copies: MWA (photocopy); NHi.

1737. Philadelphia, June 20th, 1775. Proposals, for printing by subscription, the Gentleman and cabinet-maker's assistant. Containing a great variety of useful and ornamental household furniture.... [Philadelphia: R. Bell, 1775.] Broadside 42 x 25 cm.
The proposed work was not published.
Copies: PHi (photocopy; location of original unknown).

1738. Carpenters' company of the City and County of Philadelphia.
Articles of the Carpenters Company of Philadelphia, and their rules for measuring and valuing house-carpenters work. Philadelphia: Printed by Hall and Sellers, 1786. xv, 47 p., plates.
B-6260; Mp-44867.
Copies: MWiW-C; NNC; P; PPAmP.

1739. Jarves, ________.
Jarves, cabinet, chair and clock-case maker, from London, (No. 76) Newbury-Street, Boston, respectfully informs ... that he makes ... mahogany furniture.... [Boston: 1787?] Broadside 8 x 10 cm.
B-6506; Mp-45085.
Copies: MHi.

1740. United Society of House-Carpenters and Joiners.
Rules and regulations, formed by the United Society of House-Carpenters and Joiners of the towns of Lansingburgh and Troy.... [Lansingburgh?: 1790.] Broadside 39 x 28.5 cm.
Dated: Lansingburgh, 19th June, 1790.
B-7519; Mp-46014.
Copies: MWA (photocopy); N.

1741. Carpenters' Society of Baltimore.
Constitution of the Carpenters' Society of Baltimore. Baltimore: Printed by William Goddard and James Angell, 1791. 12 p.
E-22324.
Copies: MWA (photocopy); PPCC (not located for microprinting).

1742. Associated Body of House Carpenters, New York.
The constitution of the Associated Body of House Carpenters of the City of New-York. New York: 1792.
E-24606.
No copy known.

1743. Hartford Cabinet Makers Society.
At a meeting of the cabinet makers, held in the city, the following resolutions were agreed upon by us, whose names are hereunto affixed, and who have formed ourselves into a society for the purpose of regulating the prices of our work; on the principle of dealing in cash, and of establishing a uniformity in our trade, for the general interest of ourselves and customers. [Hartford: 1792.] 6+ [2?] p.
Dated: Hartford, August 1, 1792.
B-8015; Mp-46461.
Copies: CtHT-W (imperfect).

1744. A price-book, in alphabetical order, of sundry carpenters work, collected, calculated, and now, by the authority of an angry block-cornice architect, dictated for the carpenters of Baltimore Town; together with explanatory notes, &c. &c. and a number of additional prices, &c. &c. for curriers grainboards, salt-boles, brush-handles, woodhorses, three legged stools, &c. &c. Never attempted to be calculated by any person heretofore. Baltimore: Printed by David Graham; sold by John Dalrymple, John Scrogs, Christian Baum, in behalf of themselves and the Carpenters Society, 1792.
E-24064.
No copy known.

1745. The carpenter's rules of work, for the town of Boston. [Boston: 1794.]
E-26685.
No copy known.

1746. [Federal Society of Cabinet and Chairmakers, Philadelphia.]
The Philadelphia cabinet and chair-makers book of prices. Instituted April 14th, 1794. Philadelphia: 1794.
Entry from copyright notice, issued to John Lindsay for and on behalf of the Federal Society of Cabinet and Chairmakers.
E-27508.
No copy known.

1747. The carpenters' rules of work, in the town of Boston. Boston: Printed for W. P. Blake, 1795. 19, [xxi]–xxiii p.
Mp-28318.
Copies: DLC; MHi.

1748. Carpenters' Society of Baltimore.
Additional rules, &c.... [Baltimore: 1795.] 21 p.
Includes prices for carpentry work.
B-9045; Mp-47372.
Copies: MWA (photocopy); P; PPAmP.

1749. Carpenters' Society of Carlisle.
Constitution of the Carpenter's Society of Carlisle. Carlisle, [Pa.]: Printed by George Kline, 1795. 24 p.
The Society was concerned with the regulation and prices of carpentry work.
Mp-28393.
Copies: MWA.

1750. Federal Society of Cabinet and Chairmakers, Philadelphia.
The journeymen cabinet and chairmakers, Philadelphia, book of prices. 2d ed., corr. and enlarged. Philadelphia: Printed by Ormrod and Conrad, 1795. 82, [2] p., 20 tables.
First ed., 1794, issued under the title: *The Philadelphia cabinet and chair-makers book of prices.*
Mp-29308.
Copies: PHi.

1751. At a general meeting of the housejoiners and cabinet-makers in the county of Hampshire, held at Hatfield, on the second day of March, 1796, when they agreed on the following prises [*sic*] of work.... [Northampton, Mass.: Printed by William Butler, 1796.] Broadside 47 x 28.5 cm.
Mp-30547.
Copies: PPL.

1752. The journeymen cabinet and chairmaker's New-York book of prices. New York: 1796.
E-30880.
No copy known.

1753. Pain, William, 1730–1790.
The practical house carpenter; or, Youth's instructor: containing, a great variety of useful designs in carpentry and architecture; as centering the groins, niches, &c. Examples of roofs, sky lights, &c. The five orders laid down by a new scale, mouldings, &c.... To which is added, a list of the price of carpenters' work.... The first American from the fifth London edition, with additions. Boston: Printed and sold by William Norman, 1796. 16, 7, [1] p., 146 plates.
The Price of carpenters' work in the town of Boston, 7 p., is sometimes listed independently.
Mp-30940.

Copies: MB; MH; MSaE; MWA; NNMM; RPJCB.

1754. The Philadelphia cabinet & chain-makers' book of prices. Instituted March 4th, 1796. Philadelphia: Printed by Richard Folwell, 1796. 28 p.
Mp-30999.
Copies: PHi.

1755. Rules for house-carpenters work, in the town of Providence. Providence, Rhode Island: Printed by Carter and Wilkinson, 1796. 29, [3] p.
Mp-31062.
Copies: CSmH; MH-BA; RHi; RPB; RPJCB.

1756. Pain, William, 1730–1790.
The carpenter's pocket directory; containing the best methods of framing timber buildings of all figures and dimensions, with their several parts, as floors, roofs in ledgment, their lengths and backings; trussed roofs, spires, and domes, trussing girders, partitions, and bridges, with abutments; centering for arches, vaults, &c.... Philadelphia: J. H. Dobelbower, and J. Thackara, 1797. [28] p., 24 plates.
Originally published in England.
Mp-32627.
Copies: DeWint; MH; NN; NNC.

1757. Pain, William, 1730–1790.
The practical house carpenter; or, Youth's instructor: containing a great variety of useful designs in carpentry and architecture; as centering for groins, niches, &c. Examples for roofs, sky-lights, &c. The five orders laid down by a new scale.... Elevations and sections of houses for town and country, lodges, hot-houses, green-houses, stables, &c. Design for a church, with plan, elevation, and two sections.... The 6th ed., with additions. Philadelphia: Printed by Thomas Dobson, 1797. v, [30] p., 147 plates.
The 2d American edition.
Mp-32628.
Copies: DLC; DeWint; MWA; NN; PPF; PPL; PU; RPJCB.

1758. Regulations ascertaining the work and wages of house-joiners and cabinet makers; agreed upon at Hatfield, in the county of Hampshire, March 2nd, 1796.... Rutland, [Vt.]: Printed by Josiah Fay, 1797.
E-32237.
No copy known.

1759. Bill of rates of carpenter and house-joiner work, settled and agreed on, by the subscribers, Chambersburg, April 13th, 1799. To which is affixed, an appendix, by other subscribers, Chambersburg, December, 1799. Chambersburg, [Pa.]: Printed by Robert Harper, 1799. 14 p.
Mp-35292.
Copies: P.

1760. The rules of work, of the carpenters, in the town of Boston. Formed, and most accurately corrected, by a large number of the first workmen of the town.... [Charlestown?, Mass.]: Printed for the proprietors [by Samuel Etheridge], 1800. iv, [2], [7]–34 p.
Mp-37003, 38430.
Copies: DeGE; DeU; MB; MBAt; MH; MWA; NHi; NN; RPJCB.

1761. Carpenters' Company of the City and County of Philadelphia.
Articles of the Carpenter's Company of Philadelphia and the rules for measuring and valuing house-carpenter's work. [Philadelphia]: 1801.
S&S-282.
No copy known.

1762. House carpenters' book of prices, and rules, for measuring and valuing all their different kinds of work. Philadelphia: Printed by R. Folwell, 1801. 35, [1] p.
Prices quoted in pounds, shillings and pence.
S&S-679
Copies: CSmH; DLC; PHi; PPL.

1763. Company of Master Builders, New York.
House carpenters' book of prices, made

1763 *(continued)*
and adopted by the Company of Master Builders for the city and county of New-York; with rules of measuring and valuing all their different kinds of work. New York: Printed by Isaac Collins and Son, 1802. 80 p.
S&S-2059.
Copies: PPL.

1764. The New-York book of prices for cabinet and chair work. Agreed upon by the employers. September, 1802. New York: Southwick and Crooker, [1802]. 63 p.
S&S-2787.
Copies: NN.

1765. The builder's easy guide; or, Young carpenter's assistant: containing a great variety of useful designs in carpentry and architecture.... To which is added a list of the price of carpenter's work, in the town of Boston.... Boston: W. Norman, 1803. 8 p., illus., 48 plates.
Attributed to William Pain.
S&S-3902.
Copies: CtY (microfilm); NN (microfilm); NNS.

1766. The rules of work, of the house-wrights, in the town of Salem. Formed and accurately corrected by a committee chosen for that purpose. Salem, [Mass.]: Printed by Joshua Cushing, for the proprietors, 1804. 40 p.
N-H0546284; S&S-6506.
Copies: DLC; MSaE; MWA.

1767. Associated Housewrights in Boston.
The constitution of the Associated Housewright Society, of the town of Boston. Instituted October, 1804. Boston: Printed by Gilbert & Dean, 1805. 15 p.
S&S-7890.
Copies: CtY; DLC; MH; MWA.

1768. Biddle, Owen, 1774–1806.
The young carpenter's assistant; or, A system of architecture, adapted to the style of building in the United States.... Philadelphia: Printed and sold by B. Johnson, 1805. 64 p., 44 plates (part. fold.).
S&S-8018.
Copies: CtY; DeGE; DeWint; IU; MBAt; MH; MWA; MiU-C; NcU; NjR; PP; PPL; ViU.

1769. Carpenters' Company of the City and County of Philadelphia.
Articles of the Carpenters Company of the City and County of Philadelphia, and their rules for measuring and valuing house carpenters work, as revised and agreed on in the year 1805. Philadelphia: Printed by D. Humphreys, 1805. xvi, 84 p., plates.
S&S-8140.
Copies: DLC; IaHA; MWA; PHi.

1770. Company of Master Builders, New York.
Whereas experience has shewn that much irregularity and confusion has taken place for want of a uniform regulation, pointing out the proper time to be observed by carpenters and masons ... the following regulations should be observed ... New York: Printed by Isaac Collins & Son, [1805]. Broadside 30 x 23 cm.
Resolved by the Company of Master Builders, 11th March, 1805. Signed: Andrew Morrell, President.
S&S-8224.
Copies: CSt-L; DeGE; MWA; MWiW; NjR.

1771. General rules of work for house-wrights, in Newburyport. Newburyport, [Mass.]: Published by Angier March, for the Committee; Printed by W. & J. Gilman, 1805. 15, [1] p.
Committee which drafted the rules: Thomas Harvey, Andrews Palmer, Thomas Dodge [a.o.].
S&S-8508.
Copies: DeGE; MHi; MWA.

1772. The rules of work of the house-wrights in the town of Beverly. Formed and accurately corrected by a committee chosen for that purpose. Salem, [Mass.]: Printed by

Joshua Cushing for the proprietors, 1805. 40 p.
S&S-7983.
Copies: CtY; MWA.

1773. Biddle, Owen, 1774–1806.
The young carpenter's assistant; or, A system of architecture adapted to the style of building in the United States. Philadelphia: Johnson and Warner; printed by Robert and William Carr, 1810. 61 p., 44 plates.
S&S-19551.
Copies: DAIA; ICA; MB; MWA; NHi; NcD; PPPM; ViU.

1774. General Society of Journeymen House-Carpenters, New York.
List of prices of the General Society of Journeymen House-Carpenters, of the city of New-York. New York: Printed for the Society by J. K. Furman, 1810. 44 p.
S&S-20574.
Copies: MWA (xerox copy, imperfect); NRvS.

1775. Associated Housewrights in Boston.
The constitution of the Associated Housewright Society of the Town of Boston. As revised in October, 1812. Boston: Printed by J. T. Buckingham, 1812. 15 p.
N-A0461583.
Copies: DeWint; MH.

1776. The Carpenters' and Joiners' Society of Alexandria, D.C.
The builders' price book. Carefully revised and corrected.... Alexandria, D.C.: Printed by Cottom and Stewart, 1812. 23 p.
S&S-25028.
Copies: MWA.

1777. Practical House Carpenters' Society of the City and County of Philadelphia.
The constitution of the incorporated Practical House Carpenters' Society, of the City and County of Philadelphia: together with rules & regulations for measuring & valuing house carpenters' work, 1812. Incorporated June 29, 1811. [Philadelphia: 1812.] 80 p., plates.
S&S-26510.
Copies: DLC; MWA; OClWHi; PP; PPL.

1778. The book of prices adopted by the house carpenters of the borough of Pittsburgh, February 15, 1813. Pittsburgh: Printed by S. Engles & Co., 1813. 41 p.
S&S-27980.
Copies: NIC; OC; OClWHi.

1779. Biddle, Owen, 1774–1806.
The young carpenter's assistant; or, A system of architecture, adapted to the style of building in the United States.... Philadelphia: Johnson and Warner; William Brown, printer, 1815. 62 p., 44 plates (part fold.).
S&S-34117.
Copies: CtY; DeGE; MWA; NNC; NcD; P.

1780. Worcester South-West United Architectural Society, Worcester, Mass.
The carpenters' rules of work, for the Worcester South-West United Architectural Society.... [Worcester, Mass.?]: 1815. 36 p.
S&S-36646.
Copies: MWA; NNC.

1781. Biddle, Owen, 1774–1806.
The young carpenter's assistant; or, A system of architecture, adapted to the style of building in the United States.... Philadelphia: Benjamin Warner; William Dickson, printer, Lancaster, Pa., [1817]. 61 p. 44 plates.
S&S-40245.
Copies: MWA; NHi; PPL.

1782. New York revised prices for manufacturing cabinet and chair work, June, 1818. New York: Smith, 1818. 88 p.
S&S-45088.
Copies: MiGr.

1783. Nicholson, Peter, 1765–1844.
The carpenter's new guide: being a complete book of lines for carpentry and joinery. Treating fully on practical geometry, soffits, brick and plaster groins, niches of every description, sky-lights, lines for roofs and domes . . . 8th ed., from the sixth London ed. Philadelphia: M. Carey & Son; Griggs & Co., printers, 1818. vi, [3], 10–127, [1] p., 84 plates.
First published in London, 1792.
S&S-45104–5.
Copies: CtY; ICJ; MH; MWA; MeHi; MiU-C; MoS; NNC; NNS; NWM; OU.

1784. The book of prices adopted by the house carpenters of the town of Canton, April 10, 1819. Canton, Ohio: Printed by John Saxton, 1819. 8 p.
S&S-47506.
Copies: OClWHi.

1785. House carpenters' book of prices and rules, for measuring and valuing all their different kinds of work. (Adapted to federal currency.) Philadelphia: 1819. 50 p.
Issued also as an appendix to John Haviland's *The Builders' assistant,* v. 3, 1821.
S&S-48277.
Copies: CSmH; CtY; DSI; DeGE; MWA; MiU-C; MoS; NNC-A; PPCC; PPL.

1786. ——— Philadelphia: 1819. 33, [1] p.
N-H0545261.
Copies: CSmH; DLC; MWA; MiU-C; PPCC; PPL.

1787. Housewrights rules of work for the town of Portland, 1819. Portland, Me.: Printed by A. & J. Shirley, 1819. 24 p.
S&S-48278.
Copies: MeHi.

1788. [Master Carpenters' and Joiners' Society of Cincinnati.]
The book of prices of the house carpenters and joiners of Cincinnati, adopted Monday, January 4, 1819. Cincinnati: Printed by Mason & Palmer, 1819. vi, 32, [2] p.
S&S-47617.
Copies: OCU.

1789. Housewright Society of the Town of Boston.
Price book; or, Rules of carpenter work: recommended by the Housewright Society, of the town of Boston. Revised, January 1820. . . . Boston: Printed by E. G. House, for the Housewright Society, 1820. vi, [7]–56 p.
S-1682.
Copies: MH-BA.

1790. The Richmond and Alexandria builders' price book; containing the house carpenters' and joiner's book of prices, and rules for measuring and valuing all their different kinds of work, &c. To which is added, the ship joiners', stone masons', and plaisterers' prices, and customary way of measuring their work, &c. Richmond: P. Cottom, 1820. 59, [1] p.
S-3018.
Copies: DeGE; MH-BA; MWA; ViPet.

1791. Associated Housewrights in Boston.
The constitution as revised in October, 1822. Boston: Printed by E. G. House, 1823. 20 p.
N-A0461584.
Copies: MH.

1792. Master Carpenters' Society of the City of Washington.
The constitution and by-laws of the Master Carpenters' Society of the City of Washington. Washington: Printed by W. Duncan, 1823. 9 p.

N-M0323426.
Copies: DLC.

1793. The cabinet-maker's guide; or, Rules and instructions in the art of varnishing, dying [*sic*], staining, japanning, polishing, lackering and beautifying wood, ivory, tortoise-shell and metal. With observations on their management and application. A new ed., with considerable additions, including an appendix, containing several valuable tables. From the latest London ed. Greenfield, Mass.: Phelps, 1825. viii, 108 p.
N-C0005021.
Copies: CtY; MB; MWA; MiU-C; NHi; NN; PPF.

1794. ———— Concord, [N.H.]: Printed by J. B. Moore, 1827. vii, [9]–120 p.
S-28353.
Copies: CtY; DeGE; MWA; NN; NjR; PPL.

1795. Cunningham, Joseph L., auctioneer, Boston.
Catalogue of elegant cabinet furniture, mahogany and fancy chairs ... to be sold on Tuesday, March 13, 1827.... Boston: E. Bellamy, printer, 1827. 8 p.
S-28641.
Copies: MHi.

1796. Hale, Benjamin, 1797–1863.
Introduction to the mechanical principles of carpentry.... Boston: Richardson & Lord, [etc.], 1827. 182 p., illus.
S-29106.
Copies: CtY; DeGE; IU; MB; MH; MWA; MiU; NNC; NNE; NR; NcD.

1797. Nicholson, Peter, 1765–1844.
The carpenter's new guide; being a complete book of lines for carpentry and joinery. Treating fully on practical geometry, soffits, brick and plaster groins ... with a great variety of designs for roofs, trussed girders ... 9th ed. Philadelphia: J. Grigg; Clark & Raser, printers, 1827. vi, [7]–126 p., 84 plates.
First published London, 1782.
S-30060.
Copies: CtY; DeU; MNBedf; NHi; OOxM; PP; PU; ViU; WaU.

1798. Pennsylvania. Laws.
An act to incorporate the Carpenters Company. Philadelphia: Printed by T. Town, 1827. 30 p.
S-28414.
Copies: PHi; PPCC; WHi.

1799. The book of prices adopted by the house carpenters of the towns of Zanesville, & Putnam, March 22, 1828. Zanesville, [Ohio]: Printed by Peters & Pelham, 1828. 15, [2] p.
S-32412.
Copies: ICN.

1800. The Philadelphia cabinet and chair maker's union book of prices for manufacturing cabinet ware. Established January, 1828. By a committee of employers and journeymen. Philadelphia: Printed for the cabinet and chair makers, by William Stavely, 1828. 1 p.l., [5]–93, [1] p.
S-34751.
Copies: MiGr; PPL.

1801. Pennsylvania Society of Journeymen Cabinet-makers.
Constitution of the Pennsylvania Society of Journeymen Cabinet-makers, of the city of Philadelphia. Instituted Nov. 8, 1806. Incorporated May 20, 1825. Revised Oct. 10, 1829. Philadelphia: Printed by Garden and Thompson, 1829. 20 p.

1801 *(continued)*
S-40016.
Copies: PHi.

1802. Benjamin, Asher, 1773–1845.
The practical house carpenter. Being a complete development of the Grecian orders of architecture.... Boston: The author, [etc.], 1830. 119 p., 64 plates.
C-440.
Copies: CSmH; CtY; DLC; DeWint; InU; MBAt; MWA; MiU-C; NN; NNE; PPF; RPA; ViU.

1803. ———— 2d ed.... Boston: S. Walker, [etc.], 1830. 119 p., 64 plates.
N-B0310873.
Copies: N; NHi; NNC; OOxM; PP; ViU.

1804. The carpenter's standard price book. Carefully revised by a competent hand. Washington, [D.C.]: 1830. 35 p.
C-786.
Copies: ICU.

1805. [Master Carpenters' and Joiners' Society of Cincinnati.]
The book of prices of the house carpenters and joiners of the city of Cincinnati. Adopted: Monday, January 4, 1819. Cincinnati: Printed at the Chronicle office, 1830. 48, [2] p.
C-890.
Copies: ICU; OOxM.

1806. Nicholson, Peter, 1765–1844.
The carpenter's new guide: being a complete book of lines for carpentry and joinery. Treating fully on practical geometry, soffits, brick and plaster groins ... with a great variety of designs for roofs, trussed girders ... 10th ed. Philadelphia: J. Grigg, 1830. vi, [7]–126, [1] p., 84 plates.
First published ice London, 1792.
C-2849.
Copies: DLC; InTR; Md; NHi; NNC; NjR.

1807. The Pittsburgh cabinet makers' book of prices. Pittsburgh: Jos. Snowder, printer, 1830. 91, [1] p.
C-3096.
No copy known.

1808. Society of Master Carpenters and Joiners of Dayton.
The book of prices of the house carpenters and joiners of the town of Dayton. Adopted Friday, July 16, 1830. Dayton, Ohio: Printed at the Journal office, 1830. vi, 28, [1] p.
Preamble and constitution of the Society of Master Carpenters and Joiners of Dayton, p. iii–vi.
C-1106.
Copies: ODa.

2. Clothing and shoemaking

1809. A list of prices of boots and shoes, &c. as agreed to by the master cordwainers of the city and liberties of Philadelphia, at a meeting held the 8th November, 1790. [Philadelphia: 1790.] Broadside.

B-7461; Mp-45964.

Copies: DLC; PPL.

1810. Merkel, Lot.

Lot Merkel, No. 31, Hanover-Square, nearly opposite the bank, manufactures and has for sale a large and general assortment of muffs and tippets.... New York: Printed by A. M'Lean, [1791]. Broadside.

Dated: October 20, 1791.

Mp-23563.

Copies: NHi.

1811. Newark Stocking Manufactory.

A number of journeymen stocking makers are wanted, who will find constant employment and generous wages, at the stocking manufactory, lately established in the flourishing, pleasant and healthy town of Newark, in New-Jersey.... Michael Trappel, superintendent. [Newark: Printed by John Woods, 1792.] Broadside.

Dated: June 5, 1792.

Mp-24617.

Copies: NHi.

1812. True Assistant Society of Hatters, New York.

Rules and regulations, adopted by the True Assistant Society of Hatters in New-York, and ordered to be printed for their government. [New York: 1795.] 14 p.

B-9333; Mp-47635.

Copies: MWA.

1813. Bedford, John.

Method of making J. Bedford's patent nailed shoes. [S.l.: 1806?] Broadside 21 x 17 cm.

Copies: DeGE.

1814. Memorial of sundry manufacturers of hats, in the city of Philadelphia, December 10, 1805.... Washington, D.C.: Printed by A. & G. Way, 1806. 20 [19] p.

S&S-11622.

Copies: DLC; MWA; NN; NNC.

1815. The trial of the boot & shoemakers of Philadelphia, on an indictment for a combination and conspiracy to raise their wages. Taken in shorthand by Thomas Lloyd. Philadelphia: Printed by B. Graves for T. Lloyd and B. Graves, 1806. [4], 159 p.

Commonwealth of Pennsylvania vs. George Pullis, et al. of the Federal Society of Journeymen Cordwainers.

S&S-11472.

Copies: DLC; DeGE; IaU; MBAt; MH; MWA; MdBJ; NN; NNC-L; NjP; PP; PPL.

.1816. Queen, James.

The tailor's instructor; or, A comprehensive analysis of the elements of cutting garments, of every kind.... By James Queen and William Lapsley. Philadelphia: Printed for and sold by the authors, 1809. 157 p., 8 plates.

S&S-18470, 18732.

Copies: DLC; LNHT; MWA; NjN; PPFHi; ViW.

1817. Bedford, John.

To the Honorable the Senate and House of Representatives of the state of Pennsylvania. Gentlemen, as I have some doubts of having sufficiently explained in my

1817 (*continued*)
petition, the object thereof, and the advantages which would result to the community at large, and to that useful class of citizens, the country shoemakers, in particular.... [Philadelphia?: 1810?] Broadside 34 x 20 cm.
Dated on verso in ink: Jan-ry 13,1810.
Concerns his patent on shoemaking.
S&S-16972 (dated 1809; words in title arbitrarily selected).
Copies: DeGE.

1818. Journeymen Cordwainers of the City of New York.
Trial of the Journeymen Cordwainers of the City of New-York; for a conspiracy to raise their wages.... Reported by William Sampson.... New York: I. Riley, 1810. 2 p.l., 168 p.
S&S-20885.
Copies: DLC; MB; MH-L; MiU-L; NIC; NN; PPB; ViU-L.

1819. Patent anatomical dancing shoes.... Philadelphia: [1810]. Broadside.
Copies: PHi.

1820. Boston Association of Master Tailors.
The Boston Association of Master Tailors will consider the following as the regular prices to be paid for work, after Feb. 4, 1811.... [Boston: 1811.] Broadside 18 cm.
S&S-22436.
Copies: MH-BA; MHi; MWA.

1821. Report of the trial of the journeyman cordwainers of the borough of Pittsburgh: had at an adjourned Court of Quarter Sessions for the county of Allegheny, holden at Pittsburgh the first Monday of December, 1815. Taken by Charles Shaler, Esq. Pittsburgh, Pa.: Printed by Robert Ferguson & Co., 1816. 52 p.
S&S-38786.
Copies: NNLI; PP.

1822. Lancaster, Joseph, 1778–1838.
The Lancasterian system of tuition in needle-work, whereby one mistress can as easily instruct two hundred girls as twenty.... Baltimore: Wm. Ogden Niles, printer, 1821. 16 p., illus.
S-5789.
Copies: DHEW; MiDSH.

1823. Jones, Amanda.
Rules and directions for cutting men's clothes, by the square rule; by which, in a few hours, a person may acquire such a knowledge of the art, as will enable him to cut all sizes and fashions, with the greatest accuracy. Improved ed. Middlebury, [Vt.]: Amanda Jones; J. W. Copeland, printer, 1822. 23 p., diagrs., fold. table.
Cover title: *The Taylor's assistant and family's director, comprising rules and directions for cutting men's clothes by the square rule.*
S-9166.
Copies: DLC; DeGE; ICJ; MWA; NN; OHi; Vt; VtHi; VtMiS.

1824. Clark, Fielder.
An easy and correct method of cutting men's garments by geometrical rules. Woodstock, Vt.: Printed by David Watson, 1823. 21 p.
S-12155.
Copies: MWA.

1825. [Jones, Amanda.]
The tailor's assistant; comprising rules & directions. For cutting men's clothes, by the square rule ... 2d improved ed. Revised, corrected and published by William Summer. Whitehall, [N.Y.]: E. Adams, printer, 1823. 24 p. diagrs.
S-12964.
Copies: CSmH; DeGE; LNHT; MWA; MiU-C; PPL.

1826. Moxley, John.
Every one his own tailor. The improved compass rule, now called by the thirds, to cut garments. Danville, [Vt.]: E. Eaton, printer, [1823]. 16 p.
S-13416.
Copies: VtHi.

1827. Beneficial Society of Journeyman Hatters of Brooklyn.

The constitution and by-laws of the Beneficial Society of Journeyman Hatters of Brooklyn. Rev. and adopted November 1825. Brooklyn: Printed by Alden Spooner, 1825. 18 p.

S-19659.

Copies: NHi.

1828. [Murray, Hannah Lindley] 1777–1836.

The American toilet. New York: Printed and published at Imbert's lithographic office, [1825?]. 20 l. of illus.

Emblematic illustrations of ladies' toiletry and jewelry.

S-21541.

Copies: CtY; MWA; NBuG; NN; PP; PPL; RPB.

1829. [Standley, Samuel.]

An essay on the manufacture of straw bonnets, containing an historical account of the introduction of the manufacture, its effects upon the employments, dress, food, health, morals, social intercourse, &c. of the inhabitants of the several towns in which it has been carried on; with moral, political and miscellaneous remarks.... Providence: Barnum Field & Co., printers, 1825. 67 p., 1 l.

S-22358.

Copies: DLC; DeGE; M; MB; MF; MH; MWA; MiD; NN; RPB.

1830. The London and Paris union rule; or, The tailor's assistant and unerring instructor. Boston: J. D. Yates, 1826. 80 p.

S-25130.

Copies: MBAt.

1831. [Murray, Hannah Lindley] 1777–1836.

The American toilet. 2d ed. New York: Printed and published at Imbert's lithographic office, [1827]. 20 l. of illus.

Emblematic illustrations of ladies' toiletry and jewelry.

S-29838.

Copies: CSt; MBAt; NN; NjP; PP.

1832. The trial of twenty-four journeymen tailors charged with a conspiracy: before the Mayor's Court of the City of Philadelphia, September sessions, 1827. Reported by Marcus T. C. Gould, stenographer. Philadelphia: 1827. 167 p.

S-29056.

Copies: DLC; ICU; MB; MH-BA; MoU; NNC-L; NjR; PP; PPL; WHi.

1833. Union Association of Journeymen Hatters, Cincinnati.

The constitution and by-laws of the Union Association of Journeymen Hatters, of the city of Cincinnati. Cincinnati: S. J. Browne, printer, 1827. 18 p.

S-28497.

Copies: NN.

1834. Watts, Joseph.

The tailor's instructor; or, A comprehensive analysis of the elements of cutting garments of every kind. Hallowell, [Me.]: Printed at the Advocate office, 1828. 12 p.

S-37105.

Copies: DLC.

3. Dyeing and painting

1835. Beauvais-Raseau, ________ de

Indigorier; or, Art of making indigo. Translated from the French of Rafeau [*sic*]. Charles Town, [S.C.]: 1776.

French ed. published in Paris, 1770.

E-15040.

No copy known.

1836. Walker, James L.

Painting in general.... By James L. Walker.... Baltimore: Printed by Philip Edwards, [1792]. Broadside 27.5 x 21.5 cm.

Concerns painting of houses.

Mp-24979.

Copies: DLC.

1837. Ellis, Asa.

The country dyer's assistant, by Asa Ellis, Jun. Brookfield, Mass.: Printed by E. Merriam & Co. for the author, [1798]. viii, [9]–139, [4] p.

Mp-33670.

Copies: CtY; DLC; DeGE; ICJ; InU; MH; MWA; MiU-C; NHi; PPAmP; PPF; RPJCB; ViU.

1838. Percy, John.

The schedule referred to in these letters patent, and making part of the same, containing a description, in the words of the said John Percy himself, of his improvement, being a new mode of dying [*sic*] a blue colour. [S.l.: 1800.] Broadside 17 x 22 cm.

Dated: March 1, 1800.

B-11107; Mp-49133.

Copies: NN.

1839. Cadet de Vaux, Antoine Alexis François, 1743–1828.

A new and cheap composition for painting. By Antoine-Alexis Cadet de Vaux. London, printed. New London, [Conn.]: Reprinted by James Springer, 1802. 12 p.

S&S-1975.

Copies: CtY; MWA; N.

1840. Ellis, Asa.

Country dyer's assistant. Brookfield, Mass: Merriam & Co., 1805. 136 p.

S&S-8386.

Copies: WBaraHi (not located).

1841. Bemiss, Elijah.

The dyer's companion; in two parts. Part 1st containing upwards of one hundred receipts.... The 2d part containing directions for jacking and varnishing leather.... New London, [Conn.]: Printed by Cady & Eells for the author, [1806]. viii, [9]–118, 75, [7] p.

S&S-7971, 9955.

Copies: CSmH; DeGE; ICJ; MWA; Miu-C; NN; NcD; PPL; RP.

1842. Morse, ________.

The country dyer, and farmer's companion. By Morse and Lathrop. [S.l]: Printed for the authors, 1808. 40 p.

S&S-15650.

Copies: DeGE; ICJ; MWA.

1843. Kramb, Christian.

Auf Erfahrung gegründete Vorschriften, um Wolle, Leinen und Baumwolle zu farben.... Herausgegeben von Christian Kramb. Libanon, [Pa.]: Gedruckt bey Heinrich B. Sage, 1809. viii, 52 p.

S&S-17884.

Copies: DSI; NN; PHi.

1844. Fulhame, Mrs.

An essay on combustion, with a view to a new art of dying and painting. Wherein the phlogistic and antiphlogistic hypotheses are proved erroneous.... The 1st American ed. Philadelphia: Printed and sold by James Humphreys, 1810. xxiv, [25]–248 p.

First published in London, 1794.

S&S-20172.

Copies: CtY; DLC; DNLM; DeGE; MWA; MiD; NjR; OC; PPL; PU; RPB; ScU.

1845. Haigh, James.

The dier's assistant in the art of dying wool and woollen goods. Extracted from the philosophical and chymical works of ... Ferguson, Dufay, Hellot, Geoffery, Colbert; and that reputable French dier, Mons. de Julienne. Translated from the French. With additions and practical experiments. By James Haigh. . . . Philadelphia: J. Humphreys, 1810, xxi p., 1 l., [25]-311 p.

First published at Leeds, England, 1778.
S&S-20271.
Copies: DLC; DeGE; MWA; MiU-C; NIC; PPL; PU.

1846. Morse, ________.
The country dyer, and farmer's companion, by Morse & Lathrop. Bennington, Vt.: Printed by A. Haswell, 1810. 36 p.
S&S-19876.
Copies: CSmH.

1847. Mueller, Peter.
Elegant dying receipts, for colouring of silk, cotton, and wool. Frederick Town, [Md.]: Printed by C. T. Melsheimer, [181–]. 20 p.
C. T. Melsheimer printed at Fredericktown from 1810 to 1814.
Copies: MWA.

1848. The Best system of dying woolen, cotton and silk, of all colors.... Bennington, Vt.: Printed by B. Smead, 1811. 108 p.
S&S-22340.
Copies: CtY; MB; MWA.

1849. The domestic dyer, being receipts for dying cotton and linen, hot and cold. New England: Printed for domestic uses, 1811. 12 p.
S&S-22716.
Copies: CtY; DeGE; MWA; NN; Nh.

1849a. Reynolds, Hezekiah, 1756–1833.
Directions for house and ship painting; Shewing in a plain and concise manner the best method of preparing, mixing and laying the various colours now in use, designed for the use of learners. By H. Reynolds. New Haven: Printed by Eli Hudson, 1812. 22 p.
Copies: MWA.

1850. Haigh, James.
The dier's assistant in the art of dying wool and woollen goods. Extracted from the philosophical and chymical works of those most eminent authors Ferguson, Dufay, Hellot, Geoffery, Colbert; and that reputable French dier Mons. de Julienne. Translated from the French. With additions and practical experiments. By James Haigh.... Poughkeepsie, [N.Y.]: Paraclete Potter; P.&.S. Potter, printers, 1813. xviii p., 1 l., [25]–278 p.
S&S-28688.
Copies: CSmH; DLC; DSI; DeGE; ICJ; MWA; MiU-C; MnHi; WM.

1851. Bancroft, Edward, 1744–1821.
Experimental researches concerning the philosophy of permanent colours; and the best means of producing them, by dyeing, calico-printing &c.... Philadelphia: Thomas Dobson, 1814. 2 v.
First published in London, 1794.
S&S-30738.
Copies: CSmH; DLC; ICJ; IaU; KyLx; LNHT; MH; MWA; NN; NNM; NjP; NjR; P; PPL; PU; ViU.

1852. Bemiss, Elijah.
The dyer's companion, in two parts. Part first, containing a general plan of dying wool and woollen, cotton and linen cloths, yarn and thread. Also, directions for milling and finishing, stamping and bleaching cloths. Part second, contains many useful receipts on dying, staining, painting, &c. 2d ed., enl. and improved. New York: Evert Duyckinck, 1815. viii, [5]–307 p.
S&S-34038.
Copies: DLC; DSI; DeGE; MH-BA; MWA; NN; NjR; OHi; PU; RPB.

1853. Cooper, Thomas, 1759–1839.
A practical treatise on dyeing and callicoe printing: exhibiting the processes in the French, German, English, and American practice of fixing colours on woollen, cotton, silk and linen.... Philadelphia: Thomas Dobson; William Fry, printer, 1815. xv, 506 p.
S&S-34442.
Copies: DLC; DeGE; ICJ; KyU; MH; MWA; MeU; NBu; NIC; NTR; NcAS; PPL; PU.

1854. Rauch, John.
John Rauch's receipts on dyeing, in a series of letters to a friend. Contains correct and exact copies of all his best receipts on dyeing cotton & woollen goods.... Also—a true description of his invented substitute for woad.... New York: Printed for Joseph I. Badger & Co., proprietors in New Haven, 1815. iv, [5]–97 [98] p.
S&S-35749.
Copies: DLC; MWA; ViW.

1855. ———— New York: Printed for the proprietors, 1815. 97 p.
A new setting of the entire work.
Copies: MNoanMV.

1856. Rauch, John.
John Rauch's receipts on dyeing of cotton and woollen goods: containing correct and exact copies of all his best receipts on dyeing, obtained and improved by him, during twelve years practice, in the best manufactories in Switzerland, France, Germany and America.... New York: Printed for John Sterly Ermantinger, 1815. iv, [5]–97 [98] p.
S&S-35750.
Copies: DeGE; MWA; PU-S.

1857. Waite (Daniel) & Company.
The new American dier; or, An entirely new and superior method of dying woollen cloths. By Daniel Waite & Co. Brookfield, [Mass.]: Printed by E. Merriam & Co., 1815. viii, [9]–80, [3] p.
S&S-36465.
Copies: DSI; DeGE; MWA; MiU-C; NN.

1858. Lasteyrie du Saillant, Charles Philibert, comte de, 1759–1849.
A treatise on the culture, preparation, history, and analysis of pastel, or woad: the different methods of extracting the coloring matter, and the manner of using it, and indigo, in dyeing. By C. P. de Lasteyrie. To which is added, information upon the art of extracting indigo from the leaves of pastel.... Translated from the French, by H. A. S. Dearborn. Boston: Printed by Rowe & Hooper, 1816. xii, [13]–140 p.
S&S-38032.
Copies: CtY; DLC; DNAL; DeGE; ICJ; MA; MH; MWA; NBuG; NNC; NhHi; NjR; PU.

1859. Domestic dyer, being receipts for dying cotton & linen, hot and cold. 2d ed. Deering, N.H.: Printed & sold by Eben R. Chase, 1818. 46, [2] p.
S&S-43875.
Copies: MB; NhHi.

1860. Hohman, Johann Georg.
Die Land-und Haus-Apotheke; oder, Getreuer und gründlicher Unterricht für den Bauer und Stadtmann.... Nebst einem grossen Anhang von der aechten Färberey, um türkisch, roth, blau, satin-roth, patent-grün und viele andere Farben mehr zu färben. Erste amerikanische Auflage. Herausgegeben von Johann Georg Homan [*sic*].... Reading, [Pa.]: Gedruckt bey Carl A. Bruckman, 1818. 6 p.l., 169, [11] p.
S&S-44367.
Copies: CSmH; DLC; DNLM; MU; MWA; MiU-C; NcD-MC; PPL; PSt.

1861. Woodworth, Joseph.
The dyer's assistant. 1st ed. Middlebury, Vt.: Printed by J. W. Copeland, 1819. 14 p.
S&S-50159.
Copies: MWA.

1862. Uz, Augustus.
A treatise on wall-coloring; or, A complete family instructor, teaching the process of mixing, and manner of laying on any color whatever, so as to insure uniform good success.... Pittsburgh: Eichbaum and Johnston, 1820. 57 p.
S-4073.
Copies: P.

1863. Whipple, Oliver.
Economy and cheapness. The public are respectfully presented with two new compositions or coatings for buildings.... [S.l.:

1821?] Broadside 32 x 20 cm.

Dated and signed by hand: 4th Dec. 1821, Silvester G. Whipple, attr. for Oliver Whipple.

Last printed date in text: 1818.

Describes a patented white-wash for houses.

Copies: MWA.

1864. Partridge, William.

A practical treatise on dying of woollen, cotton, and skein silk, the manufacturing of broadcloth and cassimere, including the most improved methods pursued in the west of England, in which the various manipulations are accurately delineated. Also, a correct description of sulphuring woollens and chemical bleaching of cottons. New York: H. Wallis & Co., 1823. 288 p., fold. table.

S-13674.

Copies: DLC; DeGE; LNHT; MWA; NbU; OHi; PPL; PU; RPB.

1865. Bronson, J.

The family directory, in the art of dyeing; being an abridgment of The domestic manufacturer's assistant. By J. & R. Bronson. Utica: Printed by Colwell & Bennett, and sold wholesale and retail by H. & E. Phinney & Co., Utica and H. & E. Phinney, Cooperstown, 1826. vi, [7]-96 p.

S-23929.

Copies: NN; NUt.

1866. Imschwiller, P.

The family dyer, containing: a number of excellent dies [*sic*]. Carefully selected for the use of private families; in the English and German language. By P. Imswiler [*sic*], dyer. Der Familien Färber, enthaltend: eine Anzahl vortreffliche Farben.... York, [Pa.]: Printed for the publisher, 1826. 70 p.

S-24947.

Copies: DLC; DSI; MWA.

1867. Morse, ________ .

The country dyer, and farmer's companion. By Morse and Lathrop. 2d ed. Westfield, Mass.: 1829. 36 p.

Copies: MSo.

1868. Colors warranted or no pay. Mode for all kinds of coloring.... [S.l.: 1830?] Broadside 30 x 27.5 cm.

Receipts for coloring wool blue, green, yellow, red, black, brown and lilac.

Copies: MiU-C.

1869. Jones, Thomas.

The practical dyer, containing a collection of choice receipts for producing all the most common and permanent colours on cotton, woolen, silk, and linen.... New York: Printed for the publisher, 1830. 24 p.

C-2081.

Copies: MNoanMV; MWA.

1869a. Towers, William.

Every man his own painter: or, A complete guide to painting and graining, containing general instructions in the art of preparing, compounding, and applying all kinds of paints, and making various varnishes for house, chair, and furniture painting and graining: also, a treatise on the art of imitating fancy woods and marbles ... to which are added ... instructions and receipts for preparing and applying ... distemper colors for walls. By W. & T. J. Towers. Utica, [N.Y.]: Printed by J. Colwell, 1830. [5], 6–119 p.

N-T0293747.

Copies: DSI; ICU; MWA.

1870. Tucker, William.

The family dyer and scourer; being a complete treatise on the arts of dyeing and cleaning every article of dress, bed and window furniture, silks, bonnets, feathers, etc ... From the 4th London ed. Philadelphia: E. L. Carey and A. Hart, [etc., 1830]. xxiii, [25]–180 p., illus.

First published in London, 1817.

C-3776.

Copies: DLC; DeGE; MWA; OClWHi; PPL.

4. Printing

1871. [Markland, J.]
Typographia. An ode on printing. Inscrib'd to the Honourable William Gooch.... Williamsburg: Printed by William Parks, 1730. iv, 5–15 p.
Mp-3298.
Copies: RPJCB.

1872. Mein and Fleeming, Boston.
Boston, New-England. Specimen of Mein and Fleeming's printing types. [Boston: 1768.] Broadside 28 x 20 cm.
Also dated: 1766.
B-2687; Mp-41643.
Copies: MHi; NN; RPJCB (all photocopies; original privately held).

1873. Buell, Abel, 1742–1822.
Abel Buell, of Killingworth in Connecticut, jeweller and lapidary, begs leave to acquaint the public, and the printers of the several colonies, that he hath discoverd the art, and hath already entered upon the business of founding types.... [Boston: Edes & Gill, May, 1769.] Broadside 8.5 x 9.5 cm.
Proofsheet of an advertisement set in types of Buell's design and casting.
N-B0927735.
Copies: CtY.

1874. Buell, Abel, 1742–1822.
To the Honorable the General Assembly of the Colony of Connecticut, convened at New-Haven the second Thursday of October AD 1769: The memorial of Abel Buell of Killingworth humbly sheweth; that your memorialist ... has discover'd the art of letter-founding; and as a specimen of his abilities presents this memorial impress'd with the types of his own manufacture.... [Killingworth: Printed by Abel Buell, 1769.] Broadside 10 x 16 cm.
Mp-11192.
Copies: Ct; CtY; MiU-C.

1875. Dearborn, Benjamin, 1754–1838.
A scheme, for reducing the science of music to a more simple state, and to bring all its characters within the compass of a common fount of printing-types.... Portsmouth, N.H.: 1785. 16 p.
B-6055; Mp-44674.
Copies: NHi.

1876. Thomas, Isaiah, 1749–1831.
A specimen of Isaiah Thomas's printing types. Being as large and complete an assortment as is to be met with in any one printing-office in America.Chiefly manufactured by that great artist William Caslon, Esq. of London. Worcester, Mass.: Printed by I. Thomas, 1785. 42 [47] l.
Mp-19272.
Copies: CSmH; MB; MH; MWA; NN.

1877. Bache, Benjamin Franklin, 1769–1798.
A specimen of printing types belonging to Benjamin Franklin Bache's printing office, Philadelphia. [Philadelphia: 1790?] [4] p.
B-7227; Mp-45754.
Copies: CSmH; CtY; MHi; MWA; NN (all facsims.).

1878. Franklin Society, Philadelphia.
Constitution of the Franklin Society, of the city of Philadelphia; instituted March 8, 1788.... [Philadelphia]: Printed by Stewart & Cochran, 1792. 15 p.
A society of printers.
Mp-24687.
Copies: MB; NN.

1879. Philadelphia Company of Printers and Booksellers.

The constitution and proceedings, &c. of the Philadelphia Company of Printers and Booksellers. [Philadelphia]: Printed by Daniel Humphreys, 1793. 12 p.

B-8449.

Copies: PPL.

1880. Company of Philadelphia Printers, Philadelphia.

The constitution of the Company of Printers, of Philadelphia. Philadelphia: Printed by William W. Woodward, 1794. 8 p.

Mp-27505.

Copies: DLC.

1881. Philadelphia Company of Printers and Booksellers.

The constitution, proceedings, &c. of the Philadelphia Company of Printers & Booksellers.... [Philadelphia]: Printed by Daniel Humphreys, 1793 [1794]. 17 p.

Includes minutes of meetings on July 1, 1793, Jan. 6, and April 7, 1794 on p. 15–17, probably printed as an appendix in 1794.

B-8451, 8824-8826.

Copies: PPL.

1882. Kinsley, Apollos, 1766–1803.

The New star. [No.1] Tuesday, February 2, 1796. Hartford: Printed by Apollos Kinsley, 1796. [2] p.

Printed as a sample on Kinsley's newly invented printing press, which he patented in November, 1796.

E-30869.

Copies: MWA.

1883. Condie, Thomas, 1775?–1814.

Philadelphia: Sir, I take the liberty to inform you, that I have lately commenced the manufacture of printing ink.... [Philadelphia]: From Thompson's hot-press, printed September, 1799. Broadside 26 x 21 cm.

B-10747.

Copies: PHi.

1884. Carey, Mathew, 1760–1839.

Address to the printers and booksellers throughout the United States.... [Philadelphia: Mathew Carey, 1801.] Broadside 44 x 26 cm.

Signed and dated: Mathew Carey, Philadelphia, December 19, 1801.

Proposal for a book fair.

S&S-275.

Copies: MH; MWA.

1885. Clough, John.

An address, delivered on the fourth of July, 1801, before the Franklin Typographical Association of New York, and a select company.... New York: Printed by George F. Hopkins, [1801]. 16 p.

S&S-319.

Copies: MWA.

1886. Washington Printing and Bookselling Company.

To the public.... Constitution of the Printing and Book-Selling Company.... [Washington: 1801.] Broadside.

S&S-1611.

Copies: NN.

1887. Burdick, William.

An oration on the nature and effects of the art of printing. Delivered in Franklin-Hall, July 5, 1802, before the Boston Franklin Association.... Boston: Printed by Munroe & Francis, 1802. 31 p.

S&S-1965.

Copies: DLC; DeGE; InU; MBAt; MHi; MWA; NN; PPL; RPJCB.

1888. United States. Congress. House. Committee of Commerce and Manufactures.

Report of the Committee of Commerce and Manufactures, on the several memorials of the "Franklin Association," and other journeyman printers.... 21st February, 1803. [Washington: 1803.] 4 p.

Concerns the importation of type fonts.

S&S-5435.

Copies: MWA; NN.

1889. Dean, Henry, 1788–1849.

The printer's academical companion, containing an account of the origin and progress of printing, the true method of forming letters by scale ... as an assistant to engravers, painters, masons, &c. Salem, Mass.: Printed by Haven Pool, 1806. 4 p., 23 plates.

Usually with his: *Dean's analytical guide to the art of penmanship.*

S&S-10264.

Copies: ICN; MB; MH; MSaE; MWA; PU.

1890. [Perkins, Jacob] 1766–1849.

The permanent stereotype steel plate, with observations on its importance. Newburyport, Mass.: C. Stebbins, printer, 1806. 8 p.

Signed at end: Jacob Perkins, Newburyport, Mass., Jan. 1, 1806.

S&S-11137.

Copies: MB; MWA.

1891. [Brewster, Abel.]

The universal vitriolic test, for producing an uniform, safe, and intelligible kind of bank bills; with an explanation of its importance and utility. Hartford: Printed by Lincoln & Gleason, 1807. 12 p., fold. plate.

S&S-12205.

Copies: NHi.

1892. Russell, John, 1764?–1831.

An address, presented to the members of the Faustus Association, in Boston, at their annual celebration, Oct. 4, 1808.... Boston: [Printed by Russell & Cutler], 1808. 23 p.

Concerns the art of printing.

S&S-16115.

Copies: DLC; MB; MH; MWA; NN.

1893. Binny & Ronaldson, Philadelphia.

A specimen of metal ornaments cast at the letter foundery of Binny & Ronaldson. Philadelphia: Printed by Fry and Kammerer, 1809. [23] p., illus., 2 fold. plates.

S&S-17043, 17958.

Copies: CSmH; DLC; ICN; MWA (photocopy); NNC-Typ; PHi.

1894. Perkins, Jacob, 1766–1849.

Perkins' bank bill test; consisting of original impressions from the permanent stereotype steel plates of Massachusetts paper currency ... together with the standard check plate, which will apply equally to every bank in the United States which has adopted or which may adopt this principle. Newburyport, Mass.: Published by the proprietor; W. & J. Gilman, printers, 1809. [4] p., 3 plates.

S&S-18362.

Copies: DeGE; MHi; MWA; NhHi.

1895. [Brewster, Abel.]

A plan for producing an uniformity in the ornamental part of bank or other bills where there is danger of forgery, and for furnishing the public with a convenient and infallible test for the same. With a brief explanation of its importance, together with a new set of figures designed particularly for bank bills.... Philadelphia: Printed by T. Town, 1810. 16 p.

S&S-19634. 21095.

Copies: NN; PP; PPL; RPB.

1896. New York Typographical Society.

Constitution of the New-York Typographical Society, passed July, 1809. Revised, and ordered to be printed, July, 1810. New York: Printed by C. S. Van Winkle, 1810. 19 p., 1 l.

S&S-20908.

Copies: DLC.

1897. Norris (C.) and Company.

Specimen of types in C. Norris and Co's printing-office. In the building lately occupied by E. M. Blunt, State-Street. Newburyport, [Mass.]: May, 1810. 8 p.

S&S-20924.

Copies: MSaE; MWA (xerox copy).

1898. Thomas, Isaiah, 1749–1831.

The history of printing in America.

With a biography of printers, and an account of newspapers. To which is prefixed a concise view of the discovery and progress of the art in other parts of the world.... Worcester: From the press of Isaiah Thomas, Jun.; Isaac Sturtevant, printer, 1810. 2 v., fold. facsim., 4 plates (1 fold.)

S&S-21483.

Copies: CLU; CSmH; DLC; DeGE; MNt; MWA; NN; NhD; NjR; RPB; ScU.

1899. Asbridge, George.

An oration delivered before the New York Typographical Society, at their 2d anniversary on the 4th of July, 1811. New York: Printed by C. S. Van Winkle, 1811. 23 p.

S&S-22209.

Copies: DLC; RPB.

1900. ———— New York: Printed by C. S. Van Winkle, 1811. 28 p.

Ode, by Samuel Woodworth, p. 25–26; *Ode on the art of printing,* written by a member, p. 27–28.

N-A0446210.

Copies: NNC; RPB.

1901. Binny & Ronaldson, Philadelphia.

Specimen of printing types from the foundery of Binny & Ronaldson. [Philadelphia]: Fry and Kammerer, printers, 1812. [41] l., fold. plate.

N-B0494040; S&S-24877.

Copies: CSmH; ICN; MWA; NNC; PPAmP; PPL.

1902. White and Company, New York.

Specimen of printing types, from the foundry of White & Co. New York: Printed by C. S. Van Winkle, 1812. 14 l.

The company was owned by Elihu White.

S&S-27534.

Copies: MBC; NNC-Typ.

1903. Churchill, George.

An address, pronounced before the Albany Typographical Society, November 6, 1813, on the advantages resulting to mankind from the invention of the art of printing. Albany: 1813. 8 p.

S&S-28144.

Copies: CtY; DLC; MB, NN; NjPT; PPPrHi; WHi.

1904. Wells, John I.

Printing ink. Ground in patent mills, sold by the manufacturer, John I. Wells, Hartford, Connecticut.... [Hartford: 1813.] Broadside.

Copies: PHi.

1905. Albany Typographical Society.

Scale of prices, established, January 7th, 1815, by the Albany Typographical Society. [Albany: 1815.] Broadside 14 x 9 cm.

S&S-33815.

Copies: DLC.

1906. [Brewster, Abel.]

An appeal to banks in particular, and the public in general. Hartford: Printed by Sheldon & Goodwin, 1815. 40, [2] p.

Recommendations of techniques and improvements in printing bank notes to prevent counterfeiting.

S&S-34208.

Copies: CtSoP; CtY; DeGE; MB; MWA; MdBP; Mid-B; NNC; NjPT; PPL; RPB.

1907. Bruce (D. and G.), New York.

[Specimens of printing types by D. & G. Bruce, letter founders. New York: 1815.] [15] l.

S&S-34237a.

Copies: NNC.

1908. Columbia Typographical Society, Washington, D.C.

List of prices, of the Columbia Typographical Society. Adopted November 4, 1815. [Washington: 1815.] 1 p.

N-C0571549.

Copies: DLC.

1909. New York Typographical Society.
Scale of prices adopted October 7th . . . 1815. [New York]: 1815. Broadside.
S&S-35492.
Copies: MWA.

1910. Bruce (D. and G.), New York.
A specimen of printing types, by D. & G. Bruce, stereotype and letter founders. New York: [1816]. [22] l.
S&S-37107.
Copies: NNC.

1911. Chandler, Adoniram.
An oration, delivered before the New-York Typographical Society, on their seventh anniversary, July 4, 1816.... New York: Printed by J. Seymour, for the Society, 1816. 15 p.
N-C0294588; S&S-40439 (dated 1817).
Copies: DLC; NNC.

1912. Law, Andrew, 1749–1821.
Petition of Andrew Law. February 1st, 1816 ... to the honourable the Senate and House of Representatives of the United States of America, in Congress assembled. [Washington: 1816.] 4 p.
A petition for the renewal of a patent issued May 12, 1802, for printing music.
S&S-39580.
Copies: CtY; DLC; DeGE; MWA; NN; ViU.

1913. Ronaldson, James.
Specimen of printing type, from the letter foundry of James Ronaldson, successor to Binny & Ronaldson.... Philadelphia: [Printed by James Ronaldson], 1816. [6] p., [40] l.
S&S-38834.
Copies: DLC; MB; MDeeP; MWA; NN; PPL.

1914. ——— [Philadelphia]: 1816. [90] l.
Romaine-p. 285; S&S-38835.
Copies: MWA; NNC; NNGr.

1915. Rowe & Hooper, Boston.
Rowe & Hooper, printers, No. 78 State-Street, Boston. Printing neatly executed.... [Boston: 1816.] Broadside.
S&S-38840.
Copies: MHi.

1916. United States. Laws.
A bill to extend the patent granted to Andrew Law, for an improvement in the mode of printing music. March 20, 1816. [Washington: 1816.] 1 p. (H.R. 139)
S&S-39333.
Copies: DLC.

1917. Gaskill & Copper, Philadelphia.
Patterns of rolls, stamps, scrolls and connecting lines for ornamenting the covers of books. Philadelphia: Gaskill & Copper, 1817.1 p.l., 24 p.
S&S-40892.
Copies: PPF; PSt; PU-FA.

1918. White, Elihu, firm, typefounders.
A specimen of printing types, from the foundry of E. White. New York: J. Seymour, printer, 1817. [54] l.
S&S-42892.
Copies: MWA; NN; NNC.

1919. Bruce (D. and G.), New York.
A specimen of printing types, cast at the letter-foundry of D. & G. Bruce. New York: 1818. [33] l., illus.
S&S-43469.
Copies: DLC; MB; MWA; NNC.

1920. New York Typographical Society.
Constitution and by-laws of the New-York Typographical Society. Instituted July 4, 1809. Incorporated February 27, 1818. New York: Printed by Benjamin G. Jansen, 1818. 24 p.
N-N0216604.
Copies: MWA.

1921. Reich, Starr and Company, Philadelphia.
Specimen of printing types by Reich, Starr & Co., stereotype and letter founders.

Philadelphia: J. Maxwell, printer, 1818. [21] p.
S&S-45485.
Copies: MWA; NNC; PHi; PPiU.

1922. United States. Laws.
A bill authorizing the renewal of a patent heretofore granted to Andrew Law. February 24, 1818.... [Washington: 1818.] 1 p. (H.R. 101)
Concerns the printing of music.
S&S-45951.
Copies: DLC.

1923. Van Winkle, Cornelius S.
The printers' guide; or, An introduction to the art of printing: including an essay on punctuation, and remarks on orthography. By C. S. Van Winkle, 1818. xii, [13]–229, [1], [54] p., fold. front., diagrs.
"A specimen of printing types from the foundry of E. White, No. 11 Thames-Street, New York," [32] p.; and "A specimen of printing types cast at D. & G. Bruce's foundry, Chambers-Street, New York," [22] p. at end.
S&S-46630.
Copies: DeGE; MWA; NN; PPL.

1924. White, Elihu, firm, typefounders.
A specimen of printing types from the foundry of E. White.... New York: 1818. 32 p.
N-W0248770; Romaine-p. 287.
Copies: MWA; NN.

1925. Rogers (John K.) & Company, Boston.
Specimen sheets from the letter foundry of John K. Rogers & Co. Known as the Boston Type Foundry, commenced A.D. 1819. No. 6 Spring Lane, Boston. Boston: 1819.
Sheets not numbered.
S&S-493169
Copies: NNC.

1926. Boston Type and Stereotype Foundry.
A specimen of printing types cast at the Boston Foundry. [Boston]:1820. [36] l., fold. plate.
First specimen book printed by the Boston Type Foundry.
S-537.
Copies: MB; MH; NNC.

1927. Bowen, Abel, 1790–1850.
[Specimen impressions of Congreve patent bank checks and rewards of merit. Boston: 1820–1833.] 12 plates, mounted on cardboard.
S-567.
Copies: MB.

1928. Bruce (D. and G.), New York.
A specimen of printing types, cast at the letter-foundry of D. & G. Bruce. New York: 1820. [62] l., illus.
S-606.
Copies: CtY; DLC; MB; NNC.

1929. Chandler (A.) & Company, New York.
Specimen of ornamental types and embellishments, cast at the foundry of A. Chandler & Co.... New York: 1820. 24 l.
S-705.
Copies: DLC; PPAmP.

1930. Bruce (D. and G.), New York.
A specimen of printing types, cast at the letter-foundry of D. & G. Bruce. New York: [1821?]. [53] l., illus.
Specimen leaves dated from 1817 to 1821.
N-B0879512; S-4860.
Copies: NNC.

1931. Columbia Typographical Society, Washington, D.C.
Constitution of the Columbia Typographical Society: to which are annexed the by-laws and rules of order.... Established January, 1815. Washington: 1821. 16 p.
S-5031.
Copies: DLC.

1932. Fisher, A. M.
Wells' lever printing press. [New Haven: 1821?] 12 p., fold. plate.
Reprinted from *American journal of science and arts,* v. 3, 1821.
N-F0161399.
Copies: DLC; PPF.

1933. Prout, Paul.
Phoenix Ink Manufactory . . . continues to make and sell . . . printing inks. . . . [New York: 1821.] Broadside.
Dated: November 7, 1821.
Copies: NjP.

1934. White, Elihu, firm, typefounders.
A specimen of printing types from the foundry of E. White, No. 11 Thames St., New York. New York: Abraham Paul, printer, 1821. 78 l.
S-7647.
Copies: MWA; MiD; NNC.

1935. Chandler (A.) & Company, New York.
Specimen of ornamental type and printing ornaments cast at A. Chandler's foundry. . . . New York: 1822. [21] l.
S-8308.
Copies: NNC.

1936. The memorial of the subscribers, interested in printing, papermaking, bookselling. . . . [Philadelphia: 1822.] 12 p.
S-9466.
Copies: M.

1937. Puglia, James Philip.
Forgery defeated; or, A new plan for invalidating and detecting all attempts of the kind. . . . Philadelphia: J. F. Hortel, 1822. 30 p.
Concerns the detection of counterfeit bank notes.
S-10052.
Copies: DLC; MWA; PHi; PPAmP; WHi.

1938. Ronaldson, James.
Specimen of printing type from the letter foundry of James Ronaldson. Philadelphia: 1822. 54 l.
S-10135.
Copies: MB; NN; NNC; NNC-Typ; PHi; PPAmP; PPF; PPL.

1939. ——— Philadelphia: 1822. 48 l.
Copies: CSmH; NNC; PPL.

1940. ——— Philadelphia: 1822. 384 p.
Romaine-p. 285.
Copies: NNC.

1941. Howe (J.) & Company, Philadelphia.
Specimens of metal ornaments and job types cast, and for sale at the stereotype foundry of J. Howe, corner of Crown and Callowhill Streets, Philadelphia. Philadelphia: 1823. 43 l.
S-12879.
Copies: DLC.

1942. Kinsley, Alonzo W.
[Catalog of printing types. Albany: 1823.] 14 l., fold. plates.
Annenberg-p. 172.
Copies: NAll.

1943. Neall, Daniel.
To the printers in the United States. The above cut is a representation of the press lately invented and made by me. . . . Philadelphia: Printed on the vertical press, by the inventor, [1823]. Broadside, illus.
Dated: Philadelphia, 10 mo. 28th, 1823.
Copies: PPF.

1944. Bruce (D. and G.), New York.
[Specimen of printing types. New York: 1824.] 31 l.
S-15584.
Copies: NNC.

1945. Franklin Typographical Society, Boston.

Constitution and by-laws of the Franklin Typographical Society, instituted in Boston, Feb. 1824. Boston: 1824. 12 p.
Copies: MWA.

1946. United States. Laws.
An act to authorize the issuing of letters patent to Nathaniel Sylvester. May 14, 1824.... [Washington: 1824.] 1 p. (H.R. 230)
Concerns the prevention of counterfeiting.
S-18353.
Copies: DNA.

1947. Boston Type and Stereotype Foundry.
Specimen of modern printing types and stereotype cuts, from the Boston Type and Stereotype Foundry, by T. Harrington Carter & Co. Boston: 1825. 4 p., [38] l.
N-B0677375.
Copies: NNC.

1948. Durand (A.B.C.), Wright and Company.
Specimens of bank note engraving submitted to the banking institutions in the United States. New York: A. B. C. Durand, Wright, & Co., 1825. [3] p.
S-20348.
Copies: MBAt.

1949. Boston Master Printers Association.
Boston, May 25, 1826. The committee to whom was recommitted the 12th article of the Constitution for revision ... submit the following.... Boston: 1826. Broadside 27 x 18 cm.
A schedule of printers' charges.
N-B0676377.
Copies: MB.

1950. Boston Type and Stereotype Foundry.
Specimen of modern printing types and stereotype cuts from the Boston Type and Stereotype Foundry. Incorporated 1826.Late T. H. Carter & Co.; John Gorham Rogers, agent.... Boston: 1826. 1 v. (unpaged).
S-23899.
Copies: ICN; NN; NNC.

1951. Clark, Jefferson.
Address delivered at the anniversary celebration of the Franklin Typographical Society, January 17th, 1826.... Boston: Dutton & Wentworth, 1826. 31 p.
S-24118.
Copies: CSmH; CtY; DLC; IU; IaU; LNHT; MB; MBAt; MH; MWA; PPAmP.

1952. Columbia Typographical Society, Washington, D.C.
Constitution of the Columbia Typographical Society: to which are annexed the by-laws and rules of order. Adopted January, 1815—amended May, 1825. Washington: Printed by Way and Gideon, 1826. 20 p.
S-24168.
Copies: DLC.

1953. Hagar (William) and Company, New York.
New specimen of printing types, from the foundry of William Hagar & Co.... New York: Clayton & Van Norden, printers, 1826. [91] l., fold. plates.
S-24772.
Copies: NNC.

1954. Starr (R.) and Company, Albany.
Specimen of printing types, by R. Starr & Co., letter founders, No. 8 Liberty Street, Albany. Albany: Printed by Packard & Van Benthuysen, 1826. 64 l.
S-26133.
Copies: CSmH; NNC.

1955. White, Elihu, firm, typefounders.
A specimen of printing types from the foundry of E. White, No. 11 Thames Street, New-York. [New York]: 1826. 89 l.
S-27632.
Copies: NN; NNC.

1956. Boston Type and Stereotype Foundry.

1956 (*continued*)

Act of incorporation and by-laws of the Boston Type and Stereotype Foundry. Boston: T. R. Marvin, printer, 1827. 8 p.

N-B0677372.

Copies: DLC; RP.

1957. Boston Type and Stereotype Foundry.

Specimen of printing types and metal ornaments, from the Boston Type and Stereotype Foundry. . . . Boston: Dutton & Wentworth, 1827. 138 l.

N-B0677380.

Copies: NN.

1958. Bruce (D. and G.), New York.

A specimen of printing types, and ornaments, cast in the letter-foundry of George Bruce. New York: 1827. 1 v. (unpaged), illus.

N-B0879516.

Copies: DLC; NBuG; PPLas.

1959. Cincinnati Type Foundry.

[Specimen book of types. Cincinnati: 1827.] 94 l.

Romaine-p. 275.

Copies: NNC.

1960. Directions for transferring prints to the surface of wood. From the [Boston] Mechanicks magazine. . . . [Boston: 1827.] 8 p.

S-28707.

Copies: MB.

1961. Hagar (William) and Company, New York.

[Specimen of printing type. New York: 1827.] 88 l.

Annenberg-p. 144.

Copies: CHi.

1962. New England Type Foundry Company, Boston.

Specimen of printing types, stereotype cuts and ornaments, cast at the New England Type Fo[u]ndry. . . . [Boston: 1827?] 64 l.

Many leaves are dated 1826 or 1827.

N-N0147412.

Copies: MH.

1963. Van Winkle, Cornelius S.

The printer's guide; or, An introduction to the art of printing: including an essay on punctuation and remarks on orthography. 2d ed., with additions and alterations. New York: White, Gallaher, & White; [C. S. Van Winkle, printer], 1827. 240 p., plate.

S-31586.

Copies: DLC; MWA; Md; MdBP; NNC; OMC; P; TxU.

1964. Wells (O. and H.), firm, Cincinnati.

A specimen of printing type, from the foundry of O. & H. Wells, Walnut, between Third & Fourth Streets, Cincinnati. Cincinnati: 1827. [94] l.

S-31706.

Copies: NNC.

1965. Boston Type and Stereotype Foundry.

Specimen of printing types from the Boston Type and Stereotype Foundry. John G. Rogers, agent. . . . Boston: Dutton and Wentworth, printers, 1828. [243] p.

S-32449.

Copies: DLC; MAm; MB; NNC.

1966. Bruce (D. and G.), New York.

A specimen of printing types, and ornaments, cast in the letter-foundry of George Bruce. New York: 1828. 1 v. (unpaged), illus.

N-B0879517.

Copies: DLC.

1967. Johnson, John, 1777–1848.

An abridgment of Johnson's Typographia; or, The printers' instructor: with an appendix. Boston: C. L. Adams, print., 1828. 2 p. l., 316 p., illus.

Engraved title page.

S-33721.

Copies: DeGE; ICN; IaU; MB; MWA; MdBP; NNC; OU.

1968. Kinsley (A. W.) and Company, Albany.

A specimen of printing types, cast in the Franklin Letter Foundry of A. W. Kinsley & Co.... [Albany]: Webster & Wood, [1828?]. [108]l. (1 fold.).
S-33777.
Copies: MB.

1969. Starr, Little and Company, Albany.
[Specimen of printing types. Albany: 1828.] 75 l.
Romaine-p. 286.
Copies: NNC.

1970. Wells, Darius.
[Specimen book of wood type. New York: March, 1828.] 13 l.
Copies: NNC.

1971. Kinsley (A. W.) and Company, Albany.
A specimen of printing types cast in the Franklin Letter Foundry of A. W. Kinsley & Co. Albany: 1829. [147] l.
S-39207.
Copies: NNC.

1972. New England Type Foundry Company, Boston.
Specimen of printing types and metal ornaments, cast at the New England Type Foundry, by Greele & Willis.... Boston: Beals, Homer & Co., printers, 1829. 146 l. (part fold.).
N-N0147409.
Copies: ICJ.

1973. New York Typographical Society.
Answer of the Typographical Society to communications received from Robert Dale Owen. At general meeting of the New-York Typographical Society, held on the evening of the 5th of Dec. 1829. New York: John P. Haven, 1829. 4 p.
Signed: Ado. Chandler, W. E. Dean, D. H. Reins [a.o.].
S-39839.
Copies: DLC.

1974. Wells (Oliver) and Company, Cincinnati.
A specimen of printing types and ornaments cast by Oliver Wells & Company.... Cincinnati: 1829. 138 l.
Romaine-p. 287; Sabin-102612.
Copies: DLC.

1975. White, Elihu, firm, typefounders.
A specimen of printing types and ornaments, from the letter-foundry of E. White. [New York]: Sleight & Robinson, 1829. 130 l.
N-0248773; Romaine-p. 287.
Copies: NNC.

1976. Howe (J.) & Company, Philadelphia.
Specimen of printing types, and ornaments, from the letter-foundry of J. Howe & Co.... Philadelphia: 1830. 147 l. (partly fold.), illus.
C-1930.
Copies: DLC; MWA; NNC.

5. Shorthand

1977. Sarjeant, Thomas.
An easy and compendious system of short-hand; adapted to the arts and sciences, and to the learned professions. By Thomas Sarjeant.... Philadelphia: Printed by Dobson & Lang for the editor, sold by T. Dobson, [etc.], 1789. 46, [2] p., 12 plates, port.
Half title: *Gurney's Brachygraphy improved.*
Mp-21869.
Copies: DLC; MWA; NBu; NN; ViU.

1978. ———— Abstracted from the larger ed. By Thomas Sarjeant.... Illustrated with ten copper-plates. Philadelphia: Printed by T. Lang, sold by T. Dobson, 1792. 21, [1] p., 9 plates, port.

Mp-24370.

Copies: DLC; MWA; N; NBuG; NN.

1979. [Mitchell, John.]

The elementary principles of short hand, exemplified in a variety of easy lessons.... By an eminent shorthand writer. New York: Sold by Berry, Rogers, and Berry, 1793.

Also published in London, 1793.

E-25828.

No copy known.

1980. The system of short-hand, practised by Mr. Thomas Lloyd, in taking down the debates of Congress; and now (with his permission) published for general use, by J. C. Philadelphia: Sold by H. and P. Rice, 1793. [2], 16 p., 3 plates.

Attributed to Thomas Lloyd and John Carey.

Mp-25252.

Copies: DLC; DeGE; MWA; NN; PPL.

1981. Sarjeant, Thomas.

Gurney's Easy and compendious system of short hand; adapted to the arts and sciences, and to the learned professions. Improved by Thomas Sargeant [*sic*]. 2d American ed. Lancaster: Printed for Mathew Carey, Philadelphia, by W. & R. Dickson, 1799. 38 p., 12 plates.

Mp-35578.

Copies: MWA; NN; RPJCB.

1982. Rees, Thomas.

A new system of stenography, or short hand.... Philadelphia: Reprinted by James Humphreys from the sixth London ed., 1800. 14 p., 2 plates.

Mp-38364.

Copies: MH; MWA; NN; P; PPL.

1983. [Taylor, Samuel] 1749–1811.

Stenography; or, The art of short hand perfected, containing rules and regulations, whereby the most illiterate may acquire the mode of taking down trials, orations, lectures, &c. Boston: S. G. Snelling, printer, 1809. 16 p., 4 plates.

Also attributed to Cornelius Mangan.

First published in London, 1802.

S&S-18730.

Copies: CSmH; MB; MH; MWA; NN.

1984. Lloyd, Thomas, 1756–1827.

The system of short-hand.... New Haven: Printed by E. Hudson, 1810. 8 p.

S&S-20584.

Copies: MWA; NN.

1985. Taylor, Samuel, 1749–1811.

An essay intended to establish a standard for an universal system of stenography, or, short-hand writing.... Albany: Reprinted from the London copy, for J. Cation, by Websters and Skinner, [1810]. 80 p., 11 plates.

S&S-21466.

Copies: DLC; DeGE; MB; MH; MWA; NN; ViRUT.

1986. [Taylor, Samuel] 1749–1811.

Stenography; or, The art of short hand perfected, containing rules and regulations where-by the most illiterate may acquire the mode of taking down trials, orations, lectures ... 2d ed. Boston: S. G. Snelling, printer, 1810. 16 p., 4 plates.

Also attributed to Cornelius Mangan.

S&S-21467.

Copies: MH; MWA; NN.

1987. ———— 3d ed., enl., corr. and impr., with an additional plate, by C. Mangan. Boston: E. Bliss, 1810: [3]–16, [1] p., 5 plates.

S&S-21468.

Copies: DeGE; MB; MNF; MWA; NBLiHi; NN; NjR.

1988. Sarjeant, Thomas.

Gurney's Easy and compendious system of short hand; adapted to the arts and sciences, and to the learned professions. Improved by Thomas Sergeant [*sic*]. 3d American ed. . . . Philadelphia: Printed for Mathew Carey, by Ann Cochran, 1813. 39 p., port., 12 plates.

S&S-29728, 51309.

Copies: DLC; DeGE; MWA; NN; PHi.

1989. Newell, John Lyman.

A new system of chirography, comprising rules and plates . . . to which are subjoined some useful expositions and dilucidations relative to the art of brachygraphy . . . engraved for the author by A. Reed. Hartford: Printed by Hamlen & Newton, 1818. 12 p., 4 plates.

S&S-45095.

Copies: CtHi; CtY; ICN; MWA; MiU-C; NhD; ViU.

1990. [Taylor, Samuel] 1749–1811.

Stenography; or, The art of short hand perfected . . . 4th ed., enlarged, corr. and improved with an additional plate, by C. Mangan. Boston: R. P. & C. Williams; Thomas G. Bangs, printer, 1818. 16 p., 5 plates.

Also attributed to Cornelius Mangan.

S&S-45847.

Copies: MWA; NN.

1991. Lloyd, Thomas, 1756–1827.

Lloyd's stenography, publicly practised by him for nearly half a century, with his latest improvements patented. . . . Philadelphia: T. Lloyd, 1819. 3 p.l., 31 p., 5 plates.

N-L0428727.

Copies: DFo; DLC; NN; PPL; TU.

1992. Rees, Thomas.

A new system of stenography, or short-hand. . . . Philadelphia: M. Carey & Son, 1819. 15 p., plate.

N-R0116136.

Copies: CtY; DLC; NN.

1993. [Taylor, Samuel] 1749–1811.

Stenography; or, The art of short hand perfected. . . . 5th ed., enl., corr. and improved with an additional plate, by C. Mangan. Boston: R. P. & C. Williams, 1819. 16, 8 p., 5 plates.

Also attributed to Cornelius Mangan.

S&S-49560.

Copies: DLC; MB; MHi; MWA; NN; PHi; PPL; PU.

1994. Gould, Marcus Tullius Cicero, 1793–1860.

Stenographic synopsis; or, An authentic key to the most approved system of short-hand writing. . . . Albany: Packard & Van Benthuysen, 1820. Broadside 34 x 46 cm.

Also another issue, printed at Utica, N.Y., by W. Williams, with a slightly different subtitle.

N-G0355190; S-1418, 1419.

Copies: DLC; NN.

1995. Bailey, Phinehas, 1787–1861.

A system of stenography, or short hand writing. Selected from the most approved authors, with new improvements. Haverhill, [Mass.]: Burrill & Hersey, 1821. 18 p.

S-4519.

Copies: CSmH; MB; MH; NN; RPB.

1996. ———— Newburyport, Mass.: Printed by Allen & Stickney, 1821. 16 p.

S-4520.

Copies: MWA; NN.

1997. Walker, Ambrose.

A complete system of stenography. Published and taught by A. Walker. Philadelphia: 1821. 8 p., fold. table.

S-7551.

Copies: DLC.

1998. Bailey, Phinehas, 1787–1861.

An improved system of stenography; containing analogous abbreviations, adapted to the convenience of instructors and prac-

1998 (*continued*)
titioners.... 3d ed., enlarged and improved. Poultney, Vt.: Printed by Smith and Shute, 1822. 44 p., 5 plates.
S-7864.
Copies: CSmH; DLC; MWA; NN; VtHi.

1999. Dodge, Jonathan.
A complete system of stenography, or short-hand writing.... Containing ten copper plate engravings. By J. Dodge.... [New London, Conn.]: Printed for the author, 1823. 12 p., 10 [9] plates.
Added engraved title page, regarded as plate I, has imprint: Providence, R.I.: Engraved for the author.
S-12388.
Copies: Ct; CtY; DLC; MWA; MiU-C; NBu; NN; PPL; RHi; RPB.

2000. ———— 2d ed. [New London, Conn.]: Printed for the author [by S. Green], 1823. 12 p., 10 [9] plates.
Copies: DeGE; MWA.

2001. Gould, Marcus Tullius Cicero, 1793–1860.
The analytic guide, and authentic key to the art of short hand writing.... By M. T. C. Gould.... 2d ed. Albany: Printed by Packard and Van Benthuysen, 1823. xii, [13]-36 p., 12 plates.
S-12710.
Copies: CtHT-W; CtY; DeGE; ICN; MB; MWA; MiMus; NBu; NN; NWM; PU; ViW.

2002. Hewett, Daniel, d. 1846.
New and complete system of short hand or stenography.... Philadelphia: J. R. M. Bicking, printer, 1823. 8 p., 2 plates.
S-12820.
Copies: DLC; MBAt; N.

2003. Barnum, H.L.
An authentic key to the art of short-hand writing.... Baltimore: J. D. Toy, 1824. 28 p., 7 plates.
S-15273.
Copies: CtY; ICN; NN.

2004. Gould, Marcus Tullius Cicero, 1793–1860.
The analytic guide, and authentic key to the art of short hand writing.... 3d ed. New Haven: T. G. Woodward and Co., printers, 1824. xi, 13–35, [1] p., 12 plates.
S-16327.
Copies: CtHC; DLC; GDC; MH; MWA; NN; NNC-T; OClWHi.

2005. ———— 4th ed. New York: Collins and Co., 1824. 36 p., 14 plates.
S-16328.
Copies: CtHT-W; CtY; NN; PHi; RNR.

2006. Hewett, Daniel, d. 1846.
The self-taught stenographer; or, a new and complete system of short hand.... Washington City: Printed by J. Wilson, 1824. 16 p., 2 plates.
S-16518.
Copies: DLC; NN.

2007. ———— 2d ed., improved and enlarged. Washington City: Printed by J. Wilson, 1824. 16 p., 2 plates.
S-16519.
Copies: DLC; ICN; NN.

2008. Wrifford, Allison, 1780?–1844.
Wrifford's Mercantile penmanship, reduced to a regular system of chirographic drawing, the letters classed according to similarity and simplicity. Engraved by Peter Maverick. New York: 1824. 44 p., 7 plates.
S-19338.
Copies: MWA; NhD.

2009. ———— Plymouth, [Mass.]: Luke Loomis, [1824?]. 44 p., 13 plates.
S-19339.
Copies: NNC.

2010. [Durgin, Clement.]
A system of stenography, or short hand writing. Selected from the most approved authors. Boston: Printed by Charles Spear,

1825. 12 p.
S-20349.
Copies: DeGE; MWA; NN; WHi.

2011. Hewett, Daniel, d. 1846.
A new and complete system of short hand, or stenography.... Philadelphia: J. R. M. Bicking, 1825. 8 p., 2 plates.
S-20873.
Copies: NN.

2012. Adams, James.
Teacher's assistant and scholar's guide to shorthand writing.... Richmond: Printed for the author, 1826. 32 p.
S-23418.
Copies: In.

2013. Gould, Marcus Tullius Cicero, 1793–1860.
The analytic guide and authentic key to the art of shorthand writing.... Compiled from the latest European publications, with sundry improvements ... 5th ed.... Baltimore: Printed by William Wooddy, 1826. 36 p., 14 plates.
N-G0355163.
Copies: MdBP; NNC.

2014. Taylor, Samuel, 1749-1811.
An essay, intended to establish a standard for an universal system of stenography, or shorthand writing. Hallowell: Printed by Calvin Spaulding, 1826. 15 p., 5 plates.
S-26183.
Copies: MBC; MH; MWA; MeHi.

2015. Gould, Marcus Tullius Cicero, 1793–1860.
The analytic guide, and authentic key to the art of short hand writing ... 6th ed.... Philadelphia: Published by the author; I. Ashmead, printer, 1827. 33 p., 14 [11] plates.
Plates no. 2, 3, and 6 are omitted from this edition.
N-G0355164.
Copies: CtY.

2016. Bailey, Phinehas, 1787–1861.
An improved system of stenography; containing analogous abbreviations, adapted to the convenience of instructors and practicioners ... 4th ed., enlarged and improved. Royalton, Vt.: W. Spooner, 1829. 44 p., 5 plates.
S-37544.
Copies: MH; MWA; NN.

2017. Gould, Marcus Tullius Cicero, 1793–1860.
The art of short-hand writing: compiled from the latest European publications with sundry improvements, adapted to the present state of literature in the United States ... 7th ed.... Philadelphia: Carey, Lea & Carey, 1829. 51 p., 17 plates.
S-38802.
Copies: CL; CtHT; CtW; DLC; KHi; MH; MdBP; NN; NNC-T; NjP; PSC-Hi.

2018. Gould, Marcus Tullius Cicero, 1793–1860.
American Stenographic Academy, No. 6, North Eighth Street, Philadelphia. [Philadelphia: 1830.] ix, [3] p.
C-1646.
Copies: PPL.

2019. Gould, Marcus Tullius Cicero, 1793–1860.
The art of short-hand writing; compiled from the latest European publications, with sundry improvements, adapted to the present state of literature in the United States. By M. T. C. Gould.... Stereotype ed., with seventeen new engravings. Philadelphia: 1830. 47 p., 17 plates.
C-1647.
Copies: CSmH; DLC; DeGE.

2020. ———— Rev. stereotype ed., with new engravings. Philadelphia: The author, 1830. 60 p., 17 plates.
C-1648.
Copies: MB; MBC; MH; NjR; OHi; P; WGr.

6. Other trades and crafts

2021. [Buell, Abel] 1742–1822.
The sequel of arts and sciences. New Haven: Printed for and sold by the proprietors, A. B. and J. R., [1773]. Broadside 43.5 x 35 cm.
The border, engraved by Abel Buell, contains views of the arts of painting, engraving, printing, hat making, glass grinding, and mechanics.
Copies: CtHi; NHi.

2022. The artists assistant in drawing, perspective, etching, engraving, mezzotinto-scraping, painting on glass, in crayons, in water-colours, and on silks and sattins. The art of japanning, &c.... The 6th ed., improved. Philadelphia: Printed by Benjamin Johnson, for Benjamin Davies, 1794. 3 p.l., vi–vii, [8]–70 p., 4 fold. plates.
Assigned to Carington Bowles by the British Museum. Frist published in London.
Mp-26577.
Copies: CtY; NN; RPJCB.

2023. A list of articles with their prices annexed in repairing clocks and watches, agreed upon by the clock and watchmakers of Philadelphia. Philadelphia: 1795.
E-29310.
No copy known.

2024. Greenwood, Isaac, 1730–1803.
An exposition of the emblems contained in a certificate for the members of the Providence Association of Mechanics and Manufacturers.... The above concise explanation [of the emblems representing trades and crafts] ... submitted to the members ... by Isaac Greenwood, delineator. [Providence]: Carter and Wilkinson, [1796?]. Broadside 25 x 34 p.
B-9561; Mp-47795.
Copies: MWA (photocopy); RHi.

2025. Bricklayers' Company of the City and County of Philadelphia.
Charter articles, and bye laws, of the Bricklayers Corporation. Founded Anno Domini 1790. Incorporated by the name and style of the Bricklayers Company, of the City and County of Philadelphia, Anno Domni [*sic*] 1797. Northern Liberties, [Philadelphia]: Printed by B. Scheffler & Co., 1798. 23 p.
E-32678 (erroneously dated 1797).
Copies: PHi (at PPL).

2026. Bricklayers' Company of the City and County of Philadelphia.
A bye law for the regulation of measuring and valuing stone mason and bricklayer's work, passed by the Bricklayer's Company of the City and County of Philadelphia, on the 11th day of March, 1799. Philadelphia: Printed by R. Folwell, 1801.
S&S-264.
Copies: DLC (title page only).

2027. Burrowes, Stephen.
Sir, I take the liberty of transmitting to you, the information contained in the following advertisement ... [on] patent elastic saddles, by Stephen Burrowes, at his manufactory.... Philadelphia: 1801. Broadside.
S&S-262.
Copies: DLC.

2028. Leavenworth, William.
Clocks, made and sold by William

Leavenworth, and warranted, if cased and well used.... New Haven: Printed at the Herald-office, [1806]. Broadside 27 x 23 cm.

Dated: Waterbury, (Con.) Feb. 10, 1806.

Copies: DeGE.

2029. Terry, Eli, 1772–1852.

Clocks, made by Eli Terry, for Levi G. and E. Porter, and warranted, if cased and well used.... Waterbury, Conn.: [ca. 1808]. Broadside 28 x 20 cm.

Copies: DeGE.

2030. The rules of work of the masons, in the town of Boston. Revised and corrected, and the prices fixed, by a large and respectable committee of the first workmen in the town. Approved ... May 1, 1809. Boston: Printed by Josiah Hall, 1809. 22 p.

S&S-18546.

Copies: MB.

2031. Truman, James.

James Truman has invented a new and useful improvement, being a method of washing & cleaning clothes by steam ... December 3, 1811. [Philadelphia?: 1811.] Broadside.

Catalog of Craig W. Ross, Medina, N.Y., Jan., 1974.

Present location unknown.

2032. Bill of prices agreed on by the bricklayers of Cincinnati, March 1, 1814.... Cincinnati: Printed by Looker & Wallace, [1814]. Broadside.

S&S-30919.

Copies: NHi.

2033. Hoadley, Silas.

Clocks with improvements of boxing the pivots with ivory, etc. Plymouth, Conn.: [ca. 1815]. Broadside.

Romaine-p. 104.

Copies: MWA.

2034. Coach, Chaise and Harness Manufacturing Association, Boston.

Rules and regulations of the Coach, Chaise and Harness Manufacturing Association. Instituted March, 1816. Boston: Printed by T. G. Bangs, 1816. 10 p.

List of members on p. 9–10, followed by additional names in manuscript covering two and one half pages.

Copies: DeGE.

2035. Lafoy, John B. M. D.

The complete coiffeur; or, An essay on the art of adorning natural, and of creating artificial beauty.... By J. B. M. D. Lafoy.... New York: Stereotyped for the proprietors, 1817. 2 v., plates.

The second volume is in part an abridged French version, with an engraved title: *Essai sur l'art de parer la beauté naturelle et de créer la beauté factice.*

S&S-41222.

Copies: CSmH; DLC; MH; MWA; NN; NNS; PU; RPB.

2036. New York (State). Laws.

An act, to amend an act, entitled "An act, to regulate the culling of staves and heading." Passed April 14, 1817. New York: Printed by George Forman, 1819. 7 p.

Copies: NHi.

2037. Newell, Theodore.

American patent improved orrery. [Albany?: 1819.] Broadside 40 x 32 cm.

Copies: DeGE.

2038. Parker, John R., firm, Boston.

A catalogue of music and musical intruments, comprising every species of musical merchandise.... Boston: Printed by Thomas Badger, Jr., 1820. 53, [1] p.

S-2648.

Copies: MWA.

2039. New York Friendly Association of Master Book-Binders.

New York Friendly Association of Master Book-Binders' list of prices. A.D. 1822. New York: Pub. by the Society, 1822. Broadside.

S-9697.

Copies: MWA.

2040. Peale, Rubens, 1784–1865.
Outlines of the life of Theodore Newell, and history of his inventions. Compiled by Rubens Peale and Joseph Lancaster, from Theodore Newell's own account, and published solely for his benefit. Baltimore: Printed by John D. Toy, 1824. 12 p.
Concerns mainly Newell's orrery.
Copies: DeGE (xerox copy); PPAmP.

2041. Warren, John Collins, 1778–1856.
Description of an Egyptian mummy ... with an account of the operation of embalming in ancient and modern times.... [Boston: 1824.] 34 p., illus.
Dated variously [1821], [1823?], and [1824].
S-7586, 14831, 19191.
Copies: DLC; MB; MWA; NN; ScU.

2042. Whole art of bookbinding, containing valuable receipts for sprinkling, marbling, colouring, &c. The 1st American, from the 3rd London ed., with considerable additions. Richmond: P. Cottom, 1824. 60 p., fold. table.
First ed. published in London, 1811.
S-19276.
Copies: CSmH; CtHT-W; DLC; MWA; MiU; NN; NhD; PHi; PPF; PPL; ViHi.

2043. Association of Journeymen Book Binders, Philadelphia.
List of prices of the Philadelphia Association of Journeymen Book Binders. [Philadelphia: 1825.] Broadside.
Dated: May 4, 1825.
S-21227.
Copies: PPL.

2044. Francoeur, Louis Benjamin, 1773–1849.
An introduction to linear drawing. Translated from the French of M. Francoeur, and adapted to the use of publick schools in the United States. By William B. Fowle.... Boston: Cummings, Hilliard, and Co., 1825. vii, 64 p. diagrs., fold plate.
S-20546.
Copies: DLC; IObB; MB; MH; MWA; NN; NNC-T; OO; PPL; RPB.

2045. Hubard, James, 1809–1862.
A list of musical compositions performed by the panharmonicon, with a description of the instrument, and a catalogue of the Papyrotomia, or Hubard Gallery of Cuttings; to which is prefixed a brief memoir of Master Hubard. 4th ed. Boston: Printed by Munroe and Francis, 1825. 22, [2] p., illus.
History and description of the panharmonicon, p. 20–22, not present in previous editions.
S-20970.
Copies: DLC; MWA; NN; PPL; TxU.

2046. Francoeur, Louis Benjamin, 1773–1849.
An introduction to linear drawing; translated from the French of M. Francoeur; with alterations and additions to adapt it to the use of schools in the United States. To which is added, the elements of linear perspective.... By William B. Fowle ... 2d ed. Boston: Hilliard, Gray & Wilkins, 1828. vi, [2], 86, [2] p., diagrs., 2 fold. plates.
S-33233.
Copies: CtY; DLC; MWA; MWHi; MiU; NNC; OCl; TxDaM.

2047. ——— 3d ed. Boston: Hilliard, Gray, Little and Wilkins, 1830. vi, [2], 86 p., diagrs., 3 fold. plates.
C-1433.
Copies: AU; CtY; DLC; MB; MH; MdBL; MoS; NNC; NbHi; OrCS.

V. MEDICAL TECHNOLOGY

2048. Day (John) and Company, Philadelphia.

Catalogue of drugs, chymical and Galenical preparations, shop furniture, patent medicines and surgeons' instruments, sold by John Day & Co., druggists and chymists in Second Street, Philadelphia. Philadelphia: Printed by John Dunlap, 1771. 33 p.

E-12024.

Copies: PPAmP.

2049. Buchan, William, 1729–1805.

A letter to the patentee [G. Holland] concerning the medical properties of the fleecy hosiery.... The 1st American ed., with additional notes and observations by the editor [F. Wetherill]. London printed. New York: Reprinted by G. Forman, for F. Wetherill, 1794. 34, [1] p.

Mp-26707.

Copies: DNLM.

2050. ———— The 2d American ed., with additional notes and observations by the editor. London printed. Providence: Reprinted by Carter & Wilkinson, for F. Wetherill, 1795. 34, [1] p.

Mp-28367.

Copies: MWA; RHi.

2051. ———— The 3d American ed., with additional notes and observations, by the editor. London printed. Newark, New Jersey: Reprinted by John Woods, for M. Trappal, 1796. 23, [1] p.

B-9498; Mp-47742.

Copies: MHi; MiU-C.

2052. Perkins, Elisha, 1741–1799.

Certificates of the efficacy of Doctor Perkins's patent metallic instruments. New London: From S. Green's Press, [1796]. 16 p.

Dated: Plainfield, August 12, 1796.

B-9656; Mp-47879.

Copies: DNLM; MBCo; MH; MWA; NNNAM; PPL; RPJCB.

2053. ———— Newburyport, [Mass.]: Printed by Edmund M. Blunt, [1796]. 24 p.

Preface dated: October, 1796.

Sabin-60942.

Copies: CtHi; MBAt; MHi; MWA; NN; PPL.

2054. Perkins, Elisha, 1741–1799.

To all people to whom these presents shall come, greetings: Whereas I, Elisha Perkins ... obtained a patent ... in the method of removing pains and inflammations from the human body by application of metallic substances.... [Philadelphia?: 1796.] Broadside 31 x 17 cm.

Licence granting the right to use Perkins's metallic tractors, including reprint of his patent, dated February 19, 1796.

B-9657; Mp-47880.

Copies: CtHi; MBCo; MSaE; NNNAM; PPL.

2055. Perkins, Elisha, 1741–1799.

Certificates of efficacy of Doctor Perkins's patent metallic instruments. Newburyport: Printed by Edmund Blunt, [1797]. 24 p.

Probably the same as No. 2053.

Mp-32669.

Copies: DNLM; MBAt; MHi; MWA.

2056. Perkins, Elisha, 1741–1799.
Evidences of the efficacy of Doctor Perkins's patent metallic instruments. New London: From S. Green's Press, [1797]. 32 p.
Preface dated: January 2, 1797.
Mp-32668.
Copies: CtHi; MWA; NHi; NN; PPL.

2057. ________ Philadelphia: Printed by Richard Folwell, [1797]. 36 p.
"To the public," signed: Elisha Perkins, Plainfield, June 1797.
Mp-32670.
Copies: CtHi; DLC; DeGE; MBAt; MWA; NN; PPL.

2058. ________ Richmond: Printed by John Dixon, [1797]. 40 p.
B-10066; Mp-48225.
Copies: CSmH.

2059. Vaughan, John, 1775–1807.
Observations on animal electricity. In explanation of the metallic operation of Doctor Perkins.... [Wilmington, Del.]: From the office of the Delaware gazette, by W. C. Smyth, 1797. viii, [9]–32 p.
Mp-33112.
Copies: DLC; DNLM; DeGE; MH; MWA; NBM; NN.

2060. Lockette, Henry Wilson.
An inaugural dissertation on the warm bath.... Philadelphia: Carr & Smith, 1801. [6], 53, [1] p., plate.
The plate describes the steam bath designed by C. W. Peale.
S&S-836.
Copies: DLC; MWA; PHi; PPAmP; PPL; PU.

2061. Gale, T.
Electricity; or, Ethereal fire, considered: 1st. Naturally.... 2d. Astronomically.... 3d. Medically.... Comprehending both the theory and practice of medical electricity.... By T. Gale.... Troy, [N.Y.]: Printed by Moffitt & Lyon, 1802. 285, [3] p.
Includes descriptions of equipment and apparatus.
S&S-2299.
Copies: CSt; DLC; DeGE; IU; MWA; MiU; NIC; NN; NcD; PPAmP; PPC; PPL.

2062. Hebert, H. O.
Hebert's air pump vapour bath, a machine for conveying a vapour bath to diseased limbs, and for taking off the pressure of the atmosphere.... [Philadelphia: 1802.] 8 p., plate.
Dated at end: Philadelphia, 5th October, 1802.
S&S-2391
Copies: DLC; PHC.

2063. Niderburg, S. N.
Improved Galvanismus, and its medical application. By S. N. Niderburg, M.D. and member of the faculty of Vienna. New York: Printed by George Forman, for the author, 1803. 15 p.
S&S-4770.
Copies: DNLM; DeGE; MWA; NN.

2064. [Fessenden, Thomas Green] 1771–1837.
Terrible tractoration!! A poetical petition against galvanising trumpery, and the Perkinistic institution. In four cantos. Most respectfully addressed to the Royal College of Physicians, by Christopher Caustic [pseud.] ... 1st American from the 2nd London ed., rev. and corr. by the author, with additional notes. New York: Printed for Samuel Stansbury, 1804. xxxv, 192 p., 4 plates.
S&S-6297.
Copies: CSmH; CtY; DLC; DNLM; InU; KyU; MH; MWA; NN; NcU; NjP; OCl; PPAmP; PPL; RPB; TU; ViU.

2065. Gale, T.
Brief instructions for administering medical electricity.... Hartford: Printed for the author, 1805. v, [7]–21 p.

S&S-8499.
Copies: DLC; NN.

2066. Claiborne, Devereux Jarrat, 1785–1871.

An inaugural essay, on the use of artificial drains, in the prevention and cure of disease.... By Devereux J. Claiborne. Philadelphia: Printed by Thomas J. Stites [Stiles], 1806. 31 p.

S&S-10150.
Copies: CtY; DLC; MBCo; MWA; MiD-B; NNN; PHC; PPL; PU; Vi.

2067. [Fessenden, Thomas Green] 1771–1837.

The modern philosopher; or, Terrible tractoration! In four cantos, most respectfully addressed to the Royal College of Physicians, London. By Christopher Caustick [pseud.] ... 2d American ed., rev., corr. and much enlarged by the author. Philadelphia: From the Lorenzo press of E. Bronson, 1806. xxxii, 271 p., plates.

Published in 1804 under the title *Terrible tractoration.*

N-F0108757.
Copies: CSmH; DLC; DNLM; IaU; KyU; MWA; MWH; MiU; NN; NcD; OU; OrU; PHi; PU; ViU.

2068. ———— Alexandria, Va.: Printed by E. Bronson for Robert Gray, 1806. 303 p., plates.

S&S-10408.
Copies: MW; NIC; NN; NjP; OCHP; PMA; TNP.

2069. Hall, Richard Willmott, 1785–1847.

An inaugural essay on the use of electricity in medicine.... Philadelphia: Printed by Thomas and George Palmer, 1806. 35 p.

Includes description of apparatus.

S&S-10513.
Copies: CtY; DLC; DeGE; MWA; MdBJ; NNN; PPC; PPL; PU.

2070. Moore, Samuel W., 1786–1854.

An inaugural dissertation on the medical virtues of the white oxide of bismuth; with some preliminary observations on the chemical properties of that metal. Submitted to the public examination of the Faculty of Physic under the authority of the trustees of Columbia College, in the state of New-York ... for the degree of Doctor of Physic, on the 13th day of November, 1810. By Samuel W. Moore.... New York: Printed by T. & J. Swords, 1810. 39 p.

S&S-20770.
Copies: CtY; DLC; DeGE; NNNAM; PPAmP; PPL.

2071. Jennings, Samuel Kennedy, 1771–1854.

A plain, elementary explanation of the nature and cure of disease, predicated upon facts and experience; presenting a view of that train of thinking which led to the invention of the patent, portable warm and hot bath. By Samuel K. Jennings. Washington City: Printed by Roger C. Weightman, 1814. 120 p.

S&S-31825.
Copies: DLC; DNLM; DeGE; ICRL; MB; MBCo; MWA; OU; PPL; TxU; ViU.

2072. ———— Richmond: Printed by Ritchie and Truehart, 1814. 90, [6] p.

S&S-31826.
Copies: CSmH; DNLM; MBCo; MWA; MdBP; MoSW-M; NjP; PPC; PPL; Vi.

2073. [Jennings, Samuel Kennedy] 1771–1854.

Letters and certificates, recommending the patent steam bath. [Lynchburg, Va.: Haas & Lamb, printers, 1815.] 23 p.

Invented by Samuel K. Jennings.

N-J0082024.
Copies: MBCo.

2074. Jennings, Samuel Kennedy, 1771–1854.

Plain instructions for using the patent portable warm and hot bath. By Samuel K. Jennings. Richmond: Printed at the Argus

2074 (*continued*)
book-press, [1815]. 24 p.
S&S-35014.
Copies: ICJ.

2075. Coffin, John Gorham, 1770–1829.
Remarks on the preservation of the teeth, with directions for using Charles White's tooth powder. [Boston: Rowe & Hooper, 1816.] 7 p.
"Recommendations on the dentrifice," p. 6–7.
S&S-37285.
Copies: DNLM.

2076. Jennings, Samuel Kennedy, 1771–1854.
Letters and certificates, recommending the patent portable warm and hot bath. To which are added explanations of the nature of the remedy and instructions for its application; designating some of the cases in which it will be particularly beneficial, for the use of families. By Samuel K. Jennings, patentee. Norfolk, [Va.]: Printed by S. Dillworth & C. Keemle, for the author, 1816. 60 p.
S&S-37952.
Copies: CtY-M; DLC; DNLM; MWA; PPL; ViU.

2077. Brown, Thomas, b. 1766.
The ethereal physician; or, Medical electricity revived ... with some observations on the nature of electric fluid ... by Thomas Brown. No. 1–[2]. To which is added A brief account of its medical practice, by Jesse Everett. Albany: Printed for the author, by G. J. Loomis & Co., 1817–1823. 2 v., front.
No. 2, printed by E. and E. Hosford, has the title: *The Ethereal physician; or, The medical powers of electricity demonstrated.*
List of persons who purchased machines from the authors, p. [152].
S&S-40338; S-12009.
Copies: CtY; DLC; MA; MWA; NNNAM; PP; PPC.

2078. Brown, Thomas, b. 1766.
Medical electricity, recommended as a remedy for many diseases.... Albany: E.& E. Hosford, 1817. Broadside 60 cm.
S&S-40339.
Copies: CtY-M; DLC.

2079. Tucker, Benjamin.
Medical electricity. Benjamin Tucker, of the city of Philadelphia, having made the science of electricity an object of study for several years; and, for a considerable time past, having attended to the medical application of the fluid, by the agency of powerful machines, with jars so modified as to prevent the pungency of the shock.... Philadelphia: 1819. Broadside 31 x 20 cm.
Dated: Philadelphia, 10th Mo. 1819.
Copies: The only known copy is privately held.

2080. Hull, Amos Gerald, b. 1775.
Practical elucidation of the nature of hernia; together with some remarks on the unfit instruments hitherto used for its confinement, with an explanation of the experienced utility and effectual cures of the instrument now recommended.... New York: Seymour, 1823. iv, 20 p.
S-12887.
Copies: DLC; DNLM; MBAt; MBCo; NIC; NNNAM; OC.

2081. Hull, Amos Gerald, b. 1775.
Esplicacion practica de la naturaleza de la hernia, con algunas observaciones sobre la impropriedad de los instrumentos usados hasta el dia para sujetarlas, y una relacion de la utilitad esperimentada y curas hechas con el braguero que ahora se recomienda. Escrito en ingles por su inventor, dr. Amos G. Hull.... Traducido al castellano por Manuel Dominguez.... New York: H. Spear, 1824. v, [7]–27 p., front.
S-16613.
Copies: DLC; NNNAM.

2082. Hull, Amos Gerald, b. 1775.
Practical elucidation of the nature of hernia, together with some remarks on the unfit instruments hitherto used for its confinement. With an explanation of the experienced utility and effectual cures of the

instrument now recommended. By A. G. Hull.... New York: Printed by H. Spear, 1824. iv, [5]–28 p., front.
S-16611.
Copies: DLC; DNLM; MBCo; NNNAM.

2083. ________ New-York: Printed by H. Spear, 1824. 24 p.
S-16612.
Copies: CtY.

2084. Kitchiner, William, 1775?–1827.
The economy of the eyes: precepts for the improvement and preservation of the sight. Plain rules which will enable all to judge ... what spectacles are best calculated for their eyes; observations on opera glasses ... and an account of the pancratic magnifier, for double stars, and day telescopes.... Boston: Wells and Lilly; Philadelphia: E. Littell, 1824. viii, 224 p., illus., plates.
Also published in London, 1824.
S-16815.
Copies: CtY; DLC; DNLM; MB; MBCo; MH; MiU; NNC-M; PPL; ViU.

2085. Everett, Jesse.
A synopsis of rules and directions for exciting electricity ... as a remedy for diseases.... New York: Printed by P. and H. Van Pelt, 1825. 54 p.
S-20445.
Copies: NNNAM.

2086. Hull, Amos Gerald, b. 1775.
Practical elucidation of the nature of hernia, together with some remarks on the unfit instruments hitherto used for its confinement, with an explanation of the experience, utility and effectual cures, of the instrument now recommended. By A. G. Hull ... 3d ed.... New York: H. Spear, printer, 1826. iv, [5]–32 p., front.
S-24910.
Copies: DNLM; MBCo; MHi; MdBM; NBM; NN; NNNAM.

2087. Whitridge, J. B.
Remarks on fractures, embracing a description of an improved apparatus for their treatment.... Boston: Wells & Lilly, 1827. 21 p., illus.
S-31744.
Copies: DNLM; NNNAM.

2088. Casey, James K.
Specification of a patent for an apparatus denominated the dormant balance, for the cure of crooked or infected spine in the human body. Granted to James K. Casey, Esq. of New York.... Philadelphia: 1829. 6 p.
S-38079.
Copies: MBAt; NNNAM.

2089. Collin, Victor.
Manual for the use of the stethoscope. A short treatise on the different methods of investigating the diseases of the chest. Translated from the French of M. Collin ... by W. N. Ryland ... From the 3d London ed.... Boston: Benjamin Perkins & Co., 1829. xv, [1], 133, [2] p., 4 plates.
S-38225.
Copies: CSt-L; CtY; DLC; DNLM; ICJ; MB; MdBM; MdU-H; MeB; MnU; MoU; NIC; NjR; OClM; PPC; PPL.

VI. MILITARY TECHNOLOGY

1. General works

2090. [Boone, Nicholas] 1679–1738.
Military discipline. The compleat souldier, or expert artillery-man: containing the several postures and exercises of the musket & fire-lock.... Being a collection from Col. Elton, Bariff, and others. Boston: Printed for and sold by Nicholas Boone, 1701. 96 p.
"First book written by an American author on military science...." cf. G. E. Littlefield, *Early Boston booksellers, 1642–1711,* p. [192].
Mp-971.
Copies: DLC; MHi; MWA.

2091. Boone, Nicholas, 1679–1738.
Military discipline. The compleat soldier, containing I. The exercise of musket, as now in use. II. The Duke of Monmouth's exercise of the musket. . . . The 2d ed. with additions. Being a collection from several authors by N. Boone. Boston: printed by B. Green for Benj. Eliot, 1706. [4], 124 p.
MP-1245.
Copies: ICN; MHi; MWA; NN.

2092. Boone, Nicholas, 1679–1738.
Military discipline. The newest way and method of exercising horse & foot. With some account of field officers, and a description of the arms of horse and foot. Boston in N.E.: Printed by John Allen, for Nicholas Boone, 1718. 1 p.l., 80 p.
B-39.
Copies: DLC.

2093. Bland, Humphrey, 1686–1763.
An abstract of military discipline. More particularly with regard to the manual exercise, evolutions and firings of the foot. From Col. Bland. Boston: Printed for Rogers & Fowle for D. Henchman, 1743. [2], 62 p.
Description of techniques to fire the flintlocks. Abstracted from Col. Bland's *A Treatise on military discipline*, first published in London, 1727.
MP-5133.
Copies: DLC; MB; MWA.

2094. ———— Boston: Printed by B. Green and Co. for D. Henchman: 1744. [4], 70 p.
Mp-5344.
Copies: DLC; NN; PPL.

2095. ———— Boston: Printed for D. Henchman, 1747. [4], 70 p.
Mp-5910.
Copies: MWA; NN; PPL.

2096. ———— Boston, printed. New York: Reprinted and sold by Henry De Forest, 1754. 47 p.
Mp-7154.
Copies: MiU.

2097. Blakeney, William, 1672–1761.
The new manual exercise, by General

2097 (*continued*)

Blakeney. To which is added, The evolutions of the foot, by General Bland. Corrected, with additions. Philadelphia: Printed by William Bradford, 1755. 22, [2] p.

Includes instructions to use and fire the flintlock.

Mp-40738.

Copies: MWA; PPL.

2098. Bland, Humphrey, 1686–1763.

An abstract of military discipline, from Col. Bland; more particularly with regard to the manual exercise, evolutions, and firings of the foot. Boston: Printed by Fowle for D. Henchman, 1755. 3 l., 67 p.

Description of techniques to fire the flintlock.

E-7360.

Copies: MHi; MWA.

2099. The gentleman's compleat military dictionary. Containing the military art; explaining the terms and phrases us'd in the field and garrison; the terms relating to artillery; the works and motions of attack and defence ... 18th ed. Boston: Edes and Gill; Green and Russell, 1759. [111] p., plate.

Originally published in England.

Mp-8355.

Copies: MH; MHi; MWA; RPJCB.

2100. [Harvey, Edward.]

The manual exercise as ordered by His Majesty in 1764. Together with plans and explanations.... Boston, New England: Printed by T. and J. Fleet, [1774]. 39, [1] p. 2 fold. plates.

Compiled by Edward Harvey, Adjutant General of Gt. Britain. Recommended for use by the Massachusetts Provincial Congress at Cambridge, Oct. 29, 1774.

Mp-13319.

Copies: CSmH; DLC; InU; MB; MBAt; MH; MWA; NN; RPJCB.

2101. ———— Boston: I. Thomas, [1774]. 31 p., 2 fold. plates.

Also published in 1774 at New Haven (Mp-13322), Newburyport (Mp-13321), Norwich [Conn.] (Mp-42620, 42621) and Providence (Mp-13323).

Mp-13320.

Copies: DLC; MWA; MWiW-C; MiU-C; ViW.

2102. Hanson, Thomas, d. 1777.

The Prussian evolutions in actual engagements.... With some additions ... explained with thirty folio copperplates. To which is added, The Prussian manual exercise; also The theory and some practices of gunnery.... Philadelphia: Printed for the author, by J. Douglass M'Dougall, [1775]. [2], [2], [5], iv, 64, [2], 56 p., 30 fold. plates.

Second vol. has the title: *Book the second, containing The theory and practices of gunnery, with some methods of the English larbartrary*[*sic*].

Mp-14098.

Copies: DLC; DNW; MBAt; MH; MWA; MiU-C; NN; NWM; NjP; PHi.

2103. [Harvey, Edward.]

The manual exercise, as ordered by His Majesty, in 1764. Together with plans and explanations. . . . Wilmington, [Del.]: Printed by James Adams, 1775. 37 p.

Also published in 1775 at Lancaster (Mp-14102), New York (Mp-1403), Norwich (Mp-14104), and Philadelphia (Mp-14105, 14106).

Mp-14109.

Copies: DeWI; NHi; PPL.

2104. Clairac, Louis André de la Mamie de, ca. 1690–1750.

L'ingenieur de campagne; or, Field engineer. Written in French by the Chevalier de Clairac, and translated by Major Lewis Nicola. To which is added, by way of an appendix, a short treatise on sea batteries, shewing their defects, and an attempt to remedy them. Likewise, an explanation of all the technical terms used in the work. By the translator. Illustrated with a variety of copper-plates. Philadelphia: Printed and sold by R. Aitken, 1776. 3 p.l., 256, [2] p., 39 fold. plates.

Mp-14678.

Copies: CtY; DLC; IaU; MB; MBAt; MiU-C; NN; NcU; PPL; RPJCB; ViAl.

2105. [Harvey, Edward.]
The manual exercise, as ordered by His Majesty, in 1764. Together with plans and explanations.... Philadelphia: J. Humphreys, R. Bell and R. Aitken, 1776. 35 p.
Mp-14797.
Copies: DLC; MH; MiU-C; NN; PPL.

2106. Simes, Thomas.
The military guide for young officers.... London, printed. Philadelphia: Reprinted by J. Humphreys, R. Bell, and R. Aitken, 1776. 2 v., illus., plates.
Vol. 2 has the title: *A New military, historical, and explanatory dictionary....*
First published in London, 1772.
Mp-15083.
Copies: CSmH; CtY; DLC; MB; MWA; MdAN; MiU-C; N; NcD; NhD; PPL; PU; ViU.

2107. Wolfe, James, 1727–1759.
General Wolfe's instructions to young officers: also his orders for a battalion and an army, together with the orders and signals used in embarking and debarking an army, by flat-bottom'd boats, &c.... Philadelphia: Printed and sold by Robert Bell, 1778. 142, [2] p.
First published in London, 1768.
Mp-16174.
Copies: DLC; MH; MWA; MiU-C; NNC; PPL; ViU.

2108. Hoyt, Epaphras, 1765–1850.
A treatise on the military art.... By E. Hoyt.... Vol. I. Part I. & II. Brattleborough, [Vt.]: Printed by Benjamin Smead, for the author, 1798. v, [7]–212, [2] p., 10 fold. plates.
Concerns the cavalry. Includes techniques to use firearms.
Mp-33907.
Copies: CtY; DLC; MBAt; MWA; MiU-C; NN; RPJCB; VtU.

2109. Tielke, Johann Gottlieb, 1731–1787.
Instructions for officers who wish to become field engineers; or, who are desireous of becoming useful in the field. Illustrated by examples from the last war, and provided with the necessary plans. By J. G. Tielke.... Translated from the German, by Charles Smith.... New York: Charles Smith, 1798.
E-34663.
No copy known.

2110. United States. President (John Adams).
Message from the President of the United States, transmitting two reports; one from the Acting Secretary of War, the other from the Secretary of the Treasury, with details of the expenditure of monies appropriated.... February 27, 1801.... [Washington: 1801.] 35 p.
Contains detailed reports on the purchase and manufacture of arms, and the establishment of armories and foundries for the period from May 17, 1798 to Feb. 23, 1801.
S&S-1518.
Copies: DLC; MWA; NN.

2111. United States. President (Jefferson).
Message from the President of the United States, accompanying certain documents, exhibiting 1. A return of ordnance, arms, and military stores, the property of the United States. 2. Returns of muskets and bayonets fabricated at the armouries of the United States, at Springfield and Harper's Ferry, and of the expenditures at those places. 3. An estimate of expenditures, which may be necessary for fortifications and barracks ... 2d February, 1802.... [Washington: 1802.] 15 p.
S&S-3346.
Copies: DLC; DeGE; MWA; NN.

2112. New York (City). Ordinances.
A law for preventing and extinguishing fires and to regulate the keeping and transportation of gun-powder in the City of New-York. Passed 10th November, 1806. New-York: Printed by James Cheetham, 1806. 12 p.

2112 (*continued*)
S&S-11000.
Copies: N; NHi.

2113. United States Military Philosophical Society.
Extracts from the minutes of the United States Military Philosophical Society, at a stated meeting held October 6, 1806. [N.p.: 1806?] [14] p.
Concerned with fortifications, artillery, and other matters of military science.
S&S-11747.
Copies: DLC; MWA.

2114. United States Military Philosophical Society.
West Point, December 20, 1806. Sir, in conformity to the enclosed resolutions ... I have the honour of laying before you a summary view of their Transactions during the year 1806.... [New York?: 1806.] [12] p.
Signed: Jonathan Williams, president.
S&S-11748.
Copies: CSmH.

2115. Duane, William, 1760–1835.
The American military library; or, Compendium of the modern tactics. Embracing the discipline, manoeuvres, and duties of every species of troops. Infantry, rifle corps, cavalry, artillery of position, and horse artillery; a treatise on defensive works in the field, the exercise in sea coast batteries, and regular fortifications.... Philadelphia: Printed by and for the author, 1807–1809. 2 v., illus.
S&S-12468.
Copies: CSmH; DLC; MH; NWM; OO; PPL; WHi.

2116. [Godefroy, Maximilian.]
Military reflections, on four modes of defence, for the United States, with a plan of defence, adapted to their circumstances, and the existing state of things.... By Maxan. ***, ex-officer of the Etat Major in the service of S.M.T.C. Translated by Eliza Anderson. Baltimore: Printed by Joseph Robinson, 1807. 42 p.
S&S-12671.
Copies: CSt; CtY; DLC; MB; MWA; MdHi; NHi; NN; NNS; NWM; NcD; PPL.

2117. Lacroix, Irénée Amelot de.
Military and political hints, by Col. Ir. Amelot de Lacroix.... Digested and translated by Samuel Mackay. To which is added The artillerist, translated by the same. Boston: Etheridge and Bliss; Greenough and Stebbins, printers, 1808. 121, 54 p.
The French artillerist, with a special title page, is 54 p. at end.
S&S-15387.
Copies: DLC; DNW; MW; MWA; MiD; MiU-C; NcD; PPL.

2118. United States Military Philosophical Society.
Extracts from the minutes of the United States Military Philosophical Society, at an occasional meeting held at Washington, January 30, 1808. [S.l.: 1808.] 23 p.
Address by Jonathan Williams, President, p. 5-13.
S&S-14101, 16607.
Copies: DLC; DeGE; MBAt; MWA.

2119. Duane, William, 1760–1835.
The American military library; or, Compendium of the modern tactics. Embracing the discipline, manoeuvres, and duties of every species of troops ... a treatise on defensive works in the field, the exercise in sea coast batteries, and regular fortifications.... Philadelphia: Printed by and for the author, 1809. 2 v. and 1 v. of plates (part fold.).
S&S-17406.
Copies: DLC; DeGE; GU; KyU; MH; MWA; N; NWM; OCHP; OClWHi; PHi; PPL; ViU; WHi.

2120. Duane, William, 1760–1835.
A military dictionary; or, Explanation of the several systems of discipline of different kinds of troops, infantry, artillery and cavalry; the principles of fortification, and all the modern improvements in the science of tactics ... the technical terms and phrases of the art of war.... Philadelphia: Printed and published by William Duane, 1810. xi, 748 p., diagrs.

S&S-20000.
Copies: CU-B; DLC; DeGE; ICJ; MB; MWA; MnU; NN; NjR; PPL; ViU.

2121. The military instructor, Compiled for the use of the militia of Tennessee. Containing Steuben's regulation for the order and discipline of the troops of the United States ... with some remarks to give the reader an idea of the present military art.... Nashville: Printed by Thomas G. Bradford, 1810.
N-M0577845; S&S-20737; Sabin-94784.
Copies: THi (not located).

2122. United States. War Dept.
A compendious exercise for the garrison and field ordnance, as practised in the United States. December 31st, 1810. Printed by order of the Senate of the United States. Washington: R. C. Weightman, 1810. 48 p.
S&S-21650.
Copies: DLC; MWA; MiGr; PPL.

2123. United States Military Philosophical Society.
Extracts from the minutes of the United States Military Philosophical Society at an occasional meeting held at New-York, December 28th, 1809. [S.l.: 1810?] 22 p.
Address by Jonathan Williams, President, p. [3]–13.
S&S-14101.
Copies: DLC; MWA.

2124. Hoyt, Epaphras, 1765–1850.
Practical instructions for military officers: comprehending a concise system of military geometry, field fortification and tactics of riflemen and light infantry.... Greenfield, [Mass.]: Printed by J. Denio, 1811. iv–vii, 479 p., 14 plates (1 fold.)
S&S-23058.
Copies: CoHi; CtHi; DLC; InLP; MH; MWA; MiU-C; MoU; NHi; NcU; NjR; PPA; RNR; TxU; ViU; VtU.

2125. Maltby, Isaac, 1767–1819.
The elements of war.... Boston: Printed by T. R. Wait and Co., 1811. xxiv, 208 p., 18 plates.
S&S-23280.
Copies: DLC; ICN; MWA; Nh; ViFbE; ViW.

2126. [Armstrong, John] 1758–1843.
Hints to young generals. By an old soldier.... Kingston, [N.Y.]: J. Buel, [1812]. 36, [2], 37–71 p., illus.
S&S-24629.
Copies: DLC; DNW; MWA; NN; ViW.

2127. Duane, William, 1760–1835.
A hand book for infantry; containing the first principles of military discipline ... for the use of the military force of the United States. Philadelphia: Printed for the author, 1812. v, 112 p., plates.
Special emphasis on the parts and handling of flintlocks.
S&S-25296.
Copies: MWA; NGH; NWM; NcU; ViU.

2128. ——— 2d ed. Philadelphia: Printed for the author, 1812. v, 112 p., plates.
S&S-25297.
Copies: MWA; NNC; NcD.

2129. Duane, William, 1760–1835.
A hand book for riflemen; containing the first principles of military discipline ... and the latest improvements in the modern art of war.... Philadelphia: Printed for the author, 1812. v, 108 p., illus., 12 plates.
S&S-25298.
Copies: DNW; MWA; NjP; PHi; ViU; ViW.

2130. The military instructor, containining Baron Steuben's directions respecting the arms of officers and men, for forming company and its exercise.... 2d ed. Nashville: Printed by Thomas G. Bradford,[1812]. xvi, 116 p.
N-M0577846; S&S-26075.
Copies: THi.

2131. United States. Laws.
A bill for the organization of a Corps of Artificers. March 23d, 1812. [Washington: 1812.] 3 p. (H.R. 90)
S&S-26987.
Copies: DLC.

2132. Duane, William, 1760–1835.
A hand book for infantry; containing the first principles of military discipline ... for the use of the military forces of the United States ... 3d ed.... Philadelphia: Printed for the author, 1813. v, 112 p., plates.
Special emphasis on the parts and handling of flintlocks.
S&S-28362.
Copies: MWA; PPAmP.

2133. ———— 4th ed.... Philadelphia: Printed for the author, 1813. vii, 112 p., plates.
S&S-28363.
Copies: DNW; MBAt; MWA; NNC; ViU.

2134. ———— 5th ed.... Philadelphia: Printed for the author, 1813. vii, 112 p., plates.
S&S-28364.
Copies: DLC; DNW; MH; MWA; NN; NWM; NcD; PHi; PPAmP.

2135. Duane, William, 1760–1835.
A handbook for riflemen; containing the first principles of military discipline ... and the latest improvements on the modern art of war. 2d ed.... Philadelphia: Printed for the author, 1813. vi, 108 p., illus., plates.
N-D0388826; S&S-28365.
Copies: DLC; MBAt; MWA; NWM; NcU; Nh.

2136. ———— 3d ed.... Philadelphia: Printed for the author, 1813. vi, 108 p., illus., plates.
S&S-28366.
Copies: Ct; DNW; MWA; NjR.

2137. Maltby, Isaac, 1767–1819.
The elements of war ... 2d ed. Boston: Printed by T. B. Wait & Co., 1813. xxvi, 208 p., 18 plates.
S&S-29038.
Copies: CSt; DLC; ICJ; MWA; NBu; NN.

2138. The military monitor; or, Advice to the officers and soldiers of the American army; with a few additional hints to the officers of the ordnance, and to the Secretary at War.... Baltimore: J. C. O'Reilly & Co., 1813. 183 p., folded plate.
S&S-29164.
Copies: AAP; CSmH; Ct; DLC; MB; MdBP; NcD; OFH.

2139. New Brunswick, N.J. Common Council.
Gun-powder. An ordinance, to prevent the storage, or otherwise keeping, within half a mile of the line of buildings of this city, certain quantities of gunpowder.... [New Brunswick, N.J.: A. Blauvelt, printer, 1813.] Broadside 33.5 x 21.5 cm.
Signed: James Schureman, Mayor.
Copies: DeGE.

2140. United States Military Philosophical Society.
[Minutes of a meeting] held at Washington Hall, in the city of New York, on the first day of November, 1813. [New York: 1813.] 4 p.
S&S-30385.
Copies: MHi.

2141. Duane, William, 1760–1835.
A hand book for infantry; containing the first principles of military discipline ... for the use of the military force of the United States ... 6th ed.... Philadelphia: Printed for the author, 1814. vii, 112 p., plates.
Special emphasis on the parts and handling of flintlocks.
S&S-31361.
Copies: DNW; MWA; NN.

2142. ———— 8th ed.... Philadelphia: Printed for the author, 1814. vii, 112 p., plates.
S&S-31362.
Copies: CtY; DLC; MWA; MWM; MWiW.

2143. ———— 9th ed.... Philadelphia: Printed for the author, 1814. vii, 112 p., plates.
S&S-31363.
Copies: CtY; DLC; DNW; FU; MH; MWA; MiU; NIC; NcU; OHi; PHi; ViW.

2144. Machiavelli, Niccolò, 1469–1527.
The art of war. In seven books. Written by Nicholas Machiavel.... To which is added, Hints relative to warfare, by a gentleman of the state of New-York. Albany: Printed by H. C. Southwick, 1815. 349 p., illus.
Hints, relative to torpedo warfare, by a gentleman of the state of New-York, p. [323]–349, usually attributed to Robert Fulton.
S&S-35164.
Copies: CSmH; CtY; DLC; ICJ; MH; MWA; MeU; MiD; NN; NWM; NjP; PPL; ViU; WHi.

2145. Maltby, Isaac, 1767–1819.
The elements of war ... 3d ed., rev. and corr. Hartford: Printed by P. B. Gleason and Co., 1815. xxvii, 208 p., 18 plates.
S&S-35176.
Copies: CSmH; CtHi; DLC; MB; MWA; NBu; PMA; RPB.

2146. United States. Congress. House. Committee on Naval Affairs.
Report of the Committee on Naval Affairs, on the petition of Uriah Brown. January 27, 1815.... Washington City: Printed by Roger C. Weightman, 1815. [3] p.
Unfavorable report on Brown's proposal to use inflammable fluid for defense.
S&S-36387.
Copies: DLC; MNBedf; MWA; MeB; O.

2147. [Gay de Vernon, Simon François, baron] 1760–1822.
A treatise on the science of war and fortification: composed for the use of the Imperial Polytechnick School, and military schools; and translated for the War Department, for the use of the Military Academy of the United States ... by John Michael O'Connor.... New York: Printed by J. Seymour, 1817. 2 v. and atlas of maps and plates.
French ed. was published in 1805.
S&S-40893.
Copies: DGU; DLC; InU; MBAt; MH; MWA; MiU; NN; NWM; NjR; OClWHi; OrU; PU; RP.

2148. [Levy, Aaron.]
Hand-book for the drill and instruction of the companies attached to the field pieces.... New York: Printed by Van Winkle, Wiley & Co., 1817. 34 p.
Preface signed: Aaron Levy.
S&S-41252.
Copies: MWA.

2149. Massachusetts. Adjutant-General's Office.
A system of artillery discipline; to which is prefixed, the soldier's drill.... Published according to a resolve of the legislature of Massachusetts. Boston: Munroe and Francis, 1817. 60 p., illus., 5 fold. plates.
S&S-41371.
Copies: InU; M; MH; MWA; MeHi.

2150. Swett, Samuel, 1782–1866.
An abstract of the Baron de Rogniat's considerations on the art of war ... By S. Swett.... Boston: Cummings and Hilliard, 1817. 24 p.
S&S-42242.
Copies: DLC; MBNEH; MH; MSb; MSo; MWA; MdAN; WHi.

2151. United States. War Dept.
The infantry exercise of the United States army, abridged for the use of the militia of the United States. Poughkeepsie, [N.Y.]: P. Potter, 1817. 132 p., illus.
N-U0262091.
Copies: CSt; CtY; DLC; DN; DNW.
Also: 2d ed., corr. and improved. Poughkeepsie, [N.Y.]: P. Potter, and Sheldon Potter, Philadelphia, 1817. 156 p., plates.
S&S-42595.
Copies: CSmH; CtHT-W; MWA; N; NN.

2152. Massachusetts. Laws.
An act, further regulating the storage, safe keeping, and transportation of gunpowder, in the town of Boston, together with the rules and regulations of the firewards, relative to the same. Boston: Pub. by order of the firewards, 1821. 1 p.l., 8 p.
S-5965.
Copies: DLC; DeGE.

2153. United States. War Dept.
Letter from the Secretary of War, transmitting ... reports of the newspapers, journals, and other periodical publications, charts, and instruments, maps, and prints, taken at the public expense in the War Department, also a catalogue of the books, which have been purchased on public account, within the past six years. February 11, 1823.... Washington: Printed by Gales & Seaton, 1823. 35 p. (17th Cong., 2d sess. House. Doc. 68)
S-14620.
Copies: DLC; DeGE; DeU; MWA; NjP.

2154. Cutbush, James, 1788–1823.
Proposals, for publishing by subscription a new work, entitled A system of pyrotechny, comprehending the theory and practice, with the application of chemistry, designed for exhibition and for war, in four parts.... [West Point, N.Y.: 1824.] 1 p.
S-15912.
Copies: DLC.

2155. Dyckman, J. G.
The American militia officer's manual, being a plain and concise system of instruction for infantry, field & horse artillery, cavalry and riflemen.... New York: N. B. Holmes, 1824. xix, 216 p.
N-D0471912.
Copies: MB.

2156. The soldier's manual for cavalry, artillery, light infantry and infantry.... Philadelphia: James H. Nesmith, 1824. 73 p., plate.
Manual for the use of firearms and other equipment.
S-18046.
Copies: PPL.

2157. Willcox, Richard.
A new and complete system of naval and military tactics, formed by a combination of ... discoveries in chemistry, pyrotechnics, metallurgy, and mechanics. New York: D. Fenshaw, 1824. 16 p., col. diagrs.
Prospectus of a proposed work.
S-19282.
Copies: MB; NHi; NN.

2158. Cutbush, James, 1788–1823.
A system of pyrotechny, comprehending the theory and practice, with application of chemistry; designed for exhibition and for war. In four parts: containing an account of the substances used in fire works; the instruments, utensils, and manipulations; fireworks for exhibition; and military pyrotechny.... Philadelphia: C. F. Cutbush, 1825. xliv, 612 p., plates.
S-20239.
Copies: CU; CtY; DNLM; DeGE; ICJ; IU; KU; MB; MH; MWA; MdAN; MdBP; MiU; MnU; NCH; NjR; OMC; PP; PPAmP; PU; TNJ; TxH; ViRA; WM.

2159. Dyckman, J. G.
An American militia officer's manual, being a plain and concise system of instruction for infantry, field & horse artillery, cavalry and riflemen ... 2d ed., rev., corr. and enlarged. New York: N. B. Holmes; C. S. Van Winkle, printer, 1825. 216 p.
S-20357.
Copies: CSfCW; DLC; MH; MWA; MiU; NBLiHi; NGH; NWM; NjR; OU; PPT.

2160. ——— New York: N. B. Holmes; C. S. Van Winkle, printer, 1825. 179 p.
S-20358.
Copies: MB; MH; NN; P; PHi; PPT.

2. Arms and armories

2161. New York (City). Ordinances.
An ordinance . . . for building a battery. [New York: Printed by William Bradford, 1695.]
E-737.
No copy known.

2162. Elton, Richard.
The exercise of the musket, and forming of battalions. Boston: Printed by B. Green and J. Allen, 1702. 22 p.
E-1047.
No copy known.

2163. The new exercise of firelocks and bayonets. . . . Publish'd by a lover of the art military. New London: Printed and sold by Timothy Green, 1717. [4], 27, 1 p.
Mp-1914.
Copies: CtY.

2164. The manual exercise: teaching the military exercise of the firelock. The 2d ed. [Williamsburg, Va.]: W. Parks, [1739].
B-1050.
No copy known.

2165. Bruff, Charles Oliver.
All those gentlemen who are forming themselves into companies in defence of their liberties; and others that are not provided with swords, may be suited therewith by applying to Charles Oliver Bruff, in Maiden Lane. . . . [New York: 1775.] Broadside 21.5 x 17 cm.
B-3935; Mp-42779.
Copies: NN (photocopy); RPJCB.

2166. Connecticut (Colony). General Assembly.
At a General Assembly of the Governor and Company of the Colony of Connecticut, holden at Hartford . . . on the 26th day of April, Annoque [sic] Domini, 1775. Resolved that the three thousand stands of arms to be procured for the use of this Colony, be of the following dimensions. . . . [Hartford?: 1775.] Broadside 20.5 x 17 cm.
B-3948; Mp-42799.
Copies: Matt B. Jones, Boston (1933); NN (photocopy).

2167. Massachusetts (Colony). General Court. House.
Whereas it is of the utmost importance to the welfare and happiness of these colonies, that the manufacturing of fire arms and provisions of military stores, be effectually promoted and encouraged. . . . [Watertown: 1775.] Broadside.
Dated: In Council, November 4, 1775.
Mp-14210.
Copies: MHi; MSaE; NN.

2168. New York (Colony). Provincial Congress.
Resolutions of the Provincial Congress, of the Colony of New-York, for the encouragement of manufactories of gun powder, musket barrels, musket locks, and salt. New York: Printed by John Holt, 1776. 7 p.
Mp-14938.
Copies: DLC; NHi.

2169. Muller, John, 1699–1784.
A treatise of artillery: Containing I. General constructions of brass and iron guns used by sea and land, and their carriages. II. General constructions of mortars and howitzes [*sic*], their beds and carriages. III. Dimensions of all carriages used in artillery. IV. Exercise of the regiment at home and service abroad in a siege or battle. V. Its

2169 (*continued*)
march and encampment, ammunition, stores, and horses. VI. Lastly, the necessary laboratory work for fire-ships, &c. To which is prefixed, an introduction, with A theory of powder applied to fire–arms. With large additions, alterations, and corrections, according to the 2d London ed.... Philadelphia: Printed by Styner and Cist, for John Norman, 1779. 4 p.l., x1, 215 p., 29 plates (28 fold.), 3 fold. tables.

First published in London, 1757.
Mp-16378.
Copies: CSmH; CtY; DLC; ICJ; MB; MH; MeB; MiU-C; NN; NNE; PPL.

2170. Cursory observations relative to the mounting of cannon in a new way, and to naval war. New York: Printed by Eleazer Oswald, for Elizabeth Holt, 1785. 23 p.

Sometimes attributed to James Jay.
Mp-18979.
Copies: NN; NNC; PPL; RPJCB.

2171. United States. Laws.

An act to provide for erecting and repairing arsenals and magazines, and for other purposes.... [Philadelphia: Printed by Childs and Swaine, 1794.] Broadside.

Approved April 2, 1794.
Mp-27842.
Copies: MWA; RHi; RPJCB.

2172. Stevens, William.

A system for the discipline of the artillery of the United States of America; or, The young artillerist's pocket companion. In three parts. Part I. Containing: The formation of a corps of artillery.... Part II. Containing: The theory and practice of heavy artillery.... Part III. Containing: Laboratory duty.... In three volumes. By William Stevens, an officer in the American artillery.... New York: Printed by William A. Davis, for the author, 1797. 14, [13]–260 p., 24 plates (engraved by William Rollinson).

Only vol. I has been published.
Mp-32878.
Copies: DLC; DeGE; MWA; NN; RPJCB.

2173. [Pawtucket cannon factory. 5 December, 1798. Providence: 1798.] Broadside.

B-10457.
No copy known.

2174. United States. Treasury Dept.

Articles of agreement made on the fourth day of September, one thousand seven hundred and ninety-eight, between Oliver Wolcott, Secretary of the Treasury of the United States, of the one part, and And. Welton, of Waterbury in the state of Connecticut, of the other part.... [Philadelphia: 1798.] 2 p.

Contract for one thousand muskets of the Charleville model, with specifications.
E-34849.
No copy known.

2175. United States. War Dept.

Letter from the Secretary of War, inclosing a statement of the number of cannon purchased for the use of the frigates, the revenue cutters and fortifications, since January, 1794; with the sizes of the cannon, and what portion of them have been cast in the United States, and at what foundaries they were cast; with a statement of the prices for each size. In pursuance of a resolution of this House, of the 28th of March, 1798. 12th April, 1798, ordered to lie on the table.... Philadelphia: Printed by Way & Groff, [1798]. 7 p.

Mp-34889.
Copies: MBAt; MWA; RPJCB.

2176. Scheel, Heinrich Otto von, 1745–1807.

Tables and plates, refered to in A new system of artillery. Translated from the French of M. de Scheel [by Jonathan Williams. Philadelphia: Printed for the War Office, by John Ward Fenno, 1800]. 1 v. of tables and plates.

Part of his *Treatise of artillery.*
Mp-38467.
Copies: DLC; DeGE; MB; MBAt; MWA; MWiW; N; NHi; PPAmP.

2177. Scheel, Heinrich Otto von, 1745–1807.

A treatise of artillery; containing a new system; or, The alterations made in the French artillery, since 1765. Translated from the French of M. de Scheel [by Jonathan Williams]. Philadelphia: Printed for the War Office, by John Ward Fenno, 1800. [4], [3]–154, 6, [1] p. and atlas of tables and plates.

First published at Copenhagen, Denmark, 1777.

Mp-38467.

Copies: DLC; MB; MBAt; MWA; MWiW; N; NHi; PPAmP.

2178. Smith, Amasa, 1756?–1847.

A short compendium of the duty of artillerists: shewing the method of exercise with light fieldpieces; of ascertaining the true line of direction and elevation.... Also, an easy method of finding the distance of an object, by a plain table.... Also, observations on experimental gunnery.... Worcester, Mass.: Isaiah Thomas, Jun., 1800. [12], [13]–56 p., illus.

Mp-38510.

Copies: DGU; DLC; DeGE; MSaE; MWA; NN.

2179. United States. War Dept.

Letter from the Secretary at [*sic*] War, accompanying a report exhibiting the expenses of the National Armory at Springfield, Massachusetts. 7th January, 1800.... Published by order of the House of Representatives. [Philadelphia: 1800.] 8 p.

James M'Henry, Secretary of War.

Mp-38755.

Copies: DLC; DeGE; MBAt; MWA; NN; PPL; RPJCB.

2180. United States. War Dept.

Message from the President of the United States accompanying certain documents [from the War Department] exhibiting 1. A return of ordnance, arms and military stores, the property of the United States. 2. Returns of muskets and bayonets fabricated at the armouries of the United States.... 3. An estimate of expenditures which may be necessary for fortifications and barracks ... 2d February, 1802.... [Washington: 1802.] 16 p.

Henry Dearborn, Secretary of War Department.

S&S-3346.

Copies: DLC; DeGE; MWA; NN.

2181. Memorial of sundry gun manufacturers of the borough of Lancaster, in the state of Pennsylvania. 4th February, 1803.... Washington City: Printed by William Duane & Son, 1803. 6 p.

Signed: Jacob Dickert, Peter Gonter, Atram Henry [and others].

S&S-5349.

Copies: DLC; DeGE; MWA; NN.

2182. Adye, Ralph Willett, d. 1808.

The bombardier, and pocket gunner ... 1st American from the 2d London ed. Boston: Printed for E. Larkin, by William Greenough, Charlestown, 1804. 7 p.l., vi, 277 p., diagrs., tables.

First issued in London, 1798.

S&S-5666.

Copies: CSmH; DLC; ICJ; MHi; MWA; MiU; NN; NcD; PPL.

2183. United States. Superintendent of Military Stores, Philadelphia.

Letter from Callender Irvine, relative to the public buildings on the Schuylkill. Accompanying a letter from the Secretary of War, received the 15th of February, 1805. [Washington: 1805.] 8 p.

Concerns the United States arsenal in Philadelphia.

S&S-9537.

Copies: DLC.

2184. United States. Congress. House. Committee to Whom was Referred the Petition of Andrew Joseph Villard.

Report of the Committee, to whom was referred, on the eighteenth of December last, the petition of Andrew Joseph Villard. March 17, 1806.... City of Washington: A. & G. Way, printers, 1806. 10 p.

Concerns a new method of mounting

2184 (*continued*)
cannon for the defense of ports and harbors.

S&S-11730.

Copies: DLC; DeGE; MWA; NN.

2185. United States. Congress. House. Committee to Whom was Referred the Petition of Andrew Joseph Villard.

Report of the Committee to whom was referred on the 18th of December last, the petition of Andrew Joseph Villard, together with a report of a select committee made thereon, on the 17th of March, 1806. January 5, 1807. . . . Washington City: Printed by A. & G. Way, 1807. 11 p.

Concerns a method of mounting cannon.

S&S-14075.

Copies: DLC; MWA; MiD-B; NN.

2186. United States. Congress. House. Committee to Whom was Referred the Petition of Andrew Joseph Villard.

Report of the Committee to whom was referred, on the twenty-ninth ultimo, the petition of Andre [*sic*] Joseph Villard, presented on the 18th of December, 1805. November 19, 1807. . . . City of Washington: A. & G. Way, printers, 1807. 13 p.

Concerns a new method of mounting cannon for the protection of harbors and ports.

S&S-14080.

Copies: DLC; DS; DeGE; MWA; NN.

2187. United States. President (Jefferson).

Message from the President of the United States, communicating information of the effect of gun-boats in the protection and defence of harbours . . . February 10, 1807. . . . City of Washington: A. & G. Way, printers, 1807. 11 p.

Accompanying documents, p. [7]–11.

S&S-13982.

Copies: DLC; MWA; N; NNS.

2188. Virginia. Manufactory of Arms.

The superintendant [*sic*] of the Virginia Manufactory of Arms to the Governor of the Commonwealth. [Richmond: 1807?] 12 p.

Signed: John Clarke, December 31st, 1806.

S&S-14136.

Copies: DLC; Vi.

2189. Kościuszko, Tadeusz Andrzej Bonawentura, 1746–1817.

Manoeuvres of horse artillery, written at Paris in the year 1800, at the request of General Wm. R. Davis. . . . Translated, with notes and descriptive plates, by Jonathan Williams. Published by direction of the [United States Military Philosophical] Society. New York: Sold by Campbell & Mitchell, 1808. 77 p., plans.

S&S-15380.

Copies: CtY; DLC; DNW; ICN; MWA; NN; NWM; NcU; PPAmP; PU; ViU; ViW.

2190. Lacroix, Irénée Amelot de.

The French artillerist, containing brief instructions for the different objects of artillery. Selected from French authorities . . . translated by Samuel Mackay. Boston: Etheridge and Bliss; Greenough and Stebbins, printers, 1808. 54 p.

S&S-15386.

Copies: DLC; DNW; MB; MWA; MiD; MiU-C; NcD; PPL.

2191. Philadelphia. Purveyor of Public Supplies.

As a considerable number of contracts for muskets, with bayonets, has been made, I take an early occasion to address a letter of information and remarks to all the contractors, which may be useful. Philadelphia: 1808. Broadside 34 x 20.5 cm.

At head of title: Purveyor's Office, Philadelphia, December 9, 1808.

Copies: DeGE.

2192. Banks, Henry, fl. 1781–1826.

A compend[i]ous view of the establishment & operations of the manufactory of arms, and of the late public investigation, from the commencement to the expulsion of the officers in February, 1809. Richmond: Printed for the author, [1809]. 120 [220] p.

S&S-16906.
Copies: PPL; ViHi; ViU; ViW.

2193. Tousard, Louis de, 1749–1817.
American artillerist's companion; or, Elements of artillery. Treating all kinds of firearms in detail ... preceded by an introductory dissertation on cannon.... Philadelphia: C. and A. Conrad and Co., [etc.], 1809–1813. 2 v. and atlas of 70 plates.
The atlas has the imprint: Philadelphia: Bradford and Inskeep, 1813.
S&S-18770.
Copies: DLC; MBAt; MWA; MiD-B; NN; NjP; NjR; PPL; PU; RNR; VtU.

2194. Fulton, Robert, 1765–1815.
Concluding address of Mr. Fulton's lecture on the mechanism, practice and effects of torpedoes. Delivered at Washington, February 17, 1810. Printed at the desire of the audience. [Washington, D.C.: 1810.] 10 p.
S&S-20176.
Copies: DLC; DeGE (xerox copy); LNHT; MB; PPAmP.

2195. Fulton, Robert, 1765–1815.
Torpedo war, and submarine explosions. By Robert Fulton.... New York: Printed by William Elliot, 1810. 57, [3] p., 5 plates.
S&S-20177.
Copies: CSmH; DLC; DeGE; ICJ; MB; MWA; MdAN; MiU; NN; NNC; PPAmP; PPL; RPJCB.

2196. United States. Congress. Senate.
Mr. Bradley's motion. February 20th, 1810. Printed by order of the Senate of the United States. Washington City: Printed by R. C. Weightman, 1810. [4] p.
A motion to employ torpedo or submarine explosions for the defense of the ports and harbors of the United States.
S&S-21759.
Copies: DLC; MWA; NN.

2197. United States. Congress. Senate. Committee to Inquire into the Expediency of Employing the Torpedo or Submarine Explosions.
Report of the Committee, to whom was referred by the Senate, a resolution to inquire into the expediency of employing the torpedo or submarine explosions. February 26th, 1810.... Washington City: Printed by Roger C. Weightman, 1810. 4 p.
Reported by Mr. Bradley; concerns the experiments by Robert Fulton.
S&S-21839.
Copies: DLC; DeGE; NN; NNS.

2198. United States. Congress. House. Committee on the Manufacture of Cannon and Small Arms.
Report of the Committee, appointed on so much of the message of the President, as relates to the manufacture of cannon and small arms, and the providing the munitions of war. December 16th, 1811.... Washington City: Printed by R. C. Weightman, 1811. 6 p.
S&S-24266.
Copies: DLC; DeGE; MWA; ViU.

2199. United States. Navy Dept.
Letter from the Secretary of the Navy, transmitting sundry documents exhibiting certain preliminary experiments which have been made in the city and harbor of New York, in conformity with the act of Congress, entitled "An act making appropriation for the purpose of trying the practical use of the torpedo or sub-marine explosion." February 14, 1811.... Washington: A. and G. Way, printers, 1811. 55 p., illus.
Letter from Robert Fulton to the Secretary of the Navy, p. 43–55.
Concerns experiments with torpedoes (mines) constructed by Robert Fulton.
S&S-24168; Sabin-26196.
Copies: CSmH; DLC; DeGE; MH; MWA; RPB.

2200. United States. War Dept.
Letter from the Secretary of War, transmitting, in obedience to the fifth section of the "Act to provide for the erecting and repairing

2200 (*continued*)
of arsenals and magazines and for other purposes," an account of the expenses of the national armories at Springfield and Harper's Ferry, and of the arms made and repaired at each, during the year 1810. March 2, 1811.... Washington: A. & G. Way, printers, 1811. 7 p.
S&S-24198.
Copies: DLC; MWA; RPB.

2201. Kościuszko, Tadeusz Andrzej Bonawentura, 1746–1817.
Exercise for garrison and field ordnance, together with manoeuvres of horse artillery, as altered from the manual of General Kosciusko, and adapted to the service of the United States. By an officer of the United States artillerists. In two parts.... New York: Printed by Pelsue & Gould, 1812. 128 p., plates.
S&S-25803.
Copies: DLC; DNW; MWA; NN; PPAmP.

2202. United States. Congress. House. Committee on the petition of Edward Clark.
Report of the committee to whom was referred the petition of Edward Clark, relative to his projected floating battery. April 13th, 1812.... Washington City: Printed by R. C. Weightman, 1812. [4] p.
S&S-27337.
Copies: DLC; MWA; N.

2203. United States. Laws.
A bill for the better regulation of the ordnance. April 23rd, 1812.... [Washington: 1812.] 4 p. (H.R. 91)
S&S-26984.
Copies: DLC.

2204. United States. War Dept.
Letter from the Secretary of War, transmitting a statement of the expenditures on account of the national armories, and of the number of arms manufactured and repaired at the same, in the year 1811. February 19th, 1812.... Washington City: Printed by R. C. Weightman, 1812. [4] p.
S&S-27196.
Copies: DLC; MWA.

2205. Invincible floating batteries. New York: W. Elliot, 1813. 8 p., plate.
N-I0143408.
Copies: DLC.

2206. Smith, Amasa, 1756?–1847.
A short compendium of the duty of artillerists: shewing the method of exercise with light fieldpieces ... also, an easy method of finding the distance of an object, by a plain table.... Also, observations on experimental gunnery ... 2d ed. Boston: Isaiah Thomas, Jun.; W. S. & H. Spear, printers, 1813. xv, [2], 18–60 p., illus.
S&S-29794.
Copies: MB; MH; MWA.

2207. United States. Laws.
An act authorising the president of the United States to cause to be built, equipped, and employed, one or more floating batteries for the defence of the waters of the United States ... February 16th, 1814. [Washington: 1814.] 1 p.
S&S-32988.
Copies: DLC.

2208. United States. Laws.
A bill authorising the President of the United States to cause to be built, equipped, and employed one or more floating batteries for the defence of the waters of the United States. February 10th, 1814. [Washington: 1814.] 1 p. (S. 20)
S&S-33031.
Copies: DLC.

2209. United States. Laws.
A bill to insure uniformity in the manufacture of arms, and for the better regulation of the public armories. February 3d, 1814. [Washington: 1814.] 1 p. (S. 18)
S&S-33199.
Copies: DLC.

2210. Hints relative to torpedo warfare, by A gentleman of the State of New-York. [Albany: Printed by Henry C. Southwick, 1815.] [323]–349 p., illus.

Attributed sometimes to Robert Fulton. Published in: Machiavelli, Niccolò, *The Art of war*.... Albany, 1815, with a separate title page.

S&S-35164.

Copies: DLC; DeGE; MStoc; MWA; MiU; NN; NWM; NjP.

2211. Stevens, William.

A system for the discipline of the artillery of the United States of America; or, The young artillerist's pocket companion.... Albany: Printed by Websters and Skinners, 1815. 14, [13]–260 p., 24 plates.

S&S-36018.

Copies: DLC.

2212. United States. War Dept.

Letter from the Secretary of War, transmitting a statement of the amount of expenditures on account of the national armories, at Harper's Ferry and Springfield during the year 1814 ... February 24, 1815.... Washington: Printed by A. & G. Way, 1815. 4 p.

S&S-36312.

Copies: DLC; MNBedf; MWA; O.

2213. Hall, John Hancock, 1781–1841.

Remarks upon the patent improved rifles, made by John H. Hall, of Portland (District of Maine). Portland: Printed by F. Douglas, 1816. 16 p.

S&S-37775.

Copies: NWM.

2214. Smith, Amasa, 1756?–1847.

A short compendium of the duty of artillerists; shewing the method of exercise with light fieldpieces.... Also, observations on experimental gunnery.... Boston: Isaiah Thomas, Jr., 1816. 60 p., illus.

Probably the same as No. 2206.

S&S-38945.

Copies: NWM.

2215. United States. Laws.

A bill authorizing the establishment of a national armoury. January 13, 1817. [Washington: 1816, i.e. 1817.] 2 p. (H.R. 51).

S&S-42363.

Copies: DLC.

2216. A description of the United States armory at Springfield (Mass.). [S.l.: 1818.] 4 p., fold. table.

S&S-43844.

Copies: MWA.

2217. Observations on the intended establishment of a naval arsenal on the waters of the Chesapeake Bay. Annapolis: Printed by J. Chandler, 1818. 15 p.

S&S-45132.

Copies: CtY; DLC; MdHi.

2218. Smith, Amasa, 1756?–1847.

A short compendium of the duty of artillerists: shewing the method of exercise with light fieldpieces.... Also, observations on experimental gunnery.... Boston: Isaiah Thomas, 1818. 60 p., illus.

S&S-45725.

Copies: CtHT-W; MB.

2219. United States. Congress. House. Committee on Claims.

Report of the Committee of Claims, on the case of Samuel Hughes. February 3, 1818....[Washington: 1818.] 6 p. (15th Cong., 1st sess. House. Doc. 75)

Claim for compensation for a gun foundry, destroyed by the British in 1813.

S&S-46482.

Copies: DLC; DeGE; DeU; G; MWA; NjP.

2220. United States. Ordnance Dept.

Message from the President of the United States, transmitting, in compliance with a resolution of 2d inst., a statement of all the arms and accoutrements which have been manufactured at the different armories of the United States ... February 24, 1818....

2220 (*continued*)
Washington: Printed by E. De Krafft, 1818. 8 p., 3 fold. tables. (15th Cong., 1st sess. Senate. Doc. 118)

Transmitting statements from the Ordnance Dept.

S&S-46431.

Copies: DLC; DeU; MWA.

2221. United States. War Dept.

Letter from the Secretary of War, transmitting ... a report of the number and conditions of the contracts which have been made for the supply of arms for the United States, since the adoption of the Constitution ... February 11, 1818.... Washington: Printed by E. De Krafft, 1818. 6 p. (15th Cong., 1st sess. House. Doc. 93)

S&S-46389.

Copies: DLC; DeU; G; MWA.

2222. United States. War Dept.

Report of the Secretary of War, as to the establishment of a national armory on the western waters. December 7, 1818.... Washington City: Printed by E. De Krafft, 1818. 6 p. (15th Cong., 2d sess. House. Doc. 17)

S&S-46578.

Copies: DLC; DeGE; G; MWA; NjP; P; PPAmP; R.

2223. United States. War Dept.

Letter from the Secretary of War, transmitting statements of contracts made by the Commissary General of Subsistence, the Ordnance Department, the Commissary General of Purchases, and the Engineer Department, in the year 1818. February 25, 1819.... Washington: Printed by E. De Krafft, 1819. 40 p., fold. tables. (15th Cong., 2d sess. House. Doc. 148)

S&S-49861.

Copies: DLC; DeGE; DeU; MWA; NjP; O; R.

2224. Beck, Paul, 1760?–1844.

Memorial of Paul Beck, Jr. and Thomas Sparks, manufacturers of shot. December 8, 1820.... Washington: Printed by Gales & Seaton, 1820. 6 p. (16th Cong., 2d sess. Senate. Doc. 17)

S-3871.

Copies: DLC; DeGE; DeU; G; MWA; NjP; R.

2225. Lallemand, Henri Dominique, 1777–1823.

A treatise on artillery: to which is added, A summary of military reconnoitring, of fortification, of the attack and defense of places, and of castrametation. Translated from the manuscript of the author, by James Renwick. New York: Printed by C. S. Van Winkle, 1820. 2 v., plans, plates.

S-1897.

Copies: CSmH; CtY; DLC; LNHT; MB; MH; MiD-B; NN; NWM; NjR; OCHP; P; PPM; PU; TxSaSM; ViU; WU.

2226. Smith, Robert.

Circular. Sir, you must know, that it is above ten years since I left the District of Columbia and went out to Pittsburgh, and must inform you that I contributed, for the people of the United States, important services. [Carlisle, Pa.: 1820.] 2 p.

Dated: Carlisle, Pa., November 3d, 1820.

Concerns his invention of case shot.

S-3247.

Copies: DLC; PPAmP.

2227. Smith, Robert.

To the Honorable the Senate, and House of Representatives of the United States in Congress assembled. The petition of Robert Smith, a native of the state of Pennsylvania. [S.l.: 1820.] Broadside 23 x 19 cm.

Dated: November 5th, 1820.

Requests compensation for his inventions of shot and cannon.

Copies: PPAmP.

2228. Marshall, George, d. 1855.

Marshall's practical marine gunnery; containing a view of the magnitude, weight, description & use, of every article used in the sea gunner's department, in the navy of the United States. Norfolk: Printed by Thomas

G. Broughton, and published by C. Hall, 1822. ix, [13]–136 p., fold. form.
S-9364.
Copies: CtY; DLC; DN; DeGE; MB; MH; NN; NcD; PPL.

2229. United States. Laws.
A bill to establish on the western waters a national armory. March 19, 1822. [Washington: 1822.] 1 p. (S. 75)
S-10755.
Copies: DNA.

2230. United States. Ordnance Dept.
Message from the President of the United States, transmitting (in pursuance of a resolution of the House of Representatives) a report of the ordnance and ordnance stores, now on hand in the several fortifications and arsenals of the United States; also, estimates of the amount required to provide all which are deemed necessary for the public service. February 7, 1822.... Washington: Printed by Gales & Seaton, 1822. [15] p., 5 fold. tables. (17th Cong., 1st sess. House. Doc. 54)
S-11031.
Copies: DLC; DeGE; MWA; PU.

2231. United States. Laws.
A bill to establish a national armory on the western waters. January 31, 1823.... [Washington: 1823.] 1 p. (S. 46)
S-14483.
Copies: DNA.

2232. United States. Ordnance Dept.
In Senate of the United States, January 20, 1823. Ordered, that the report of the ordnance Department, which accompanied the President's message of the 13th instant, on the subject of establishing a national armory on the western waters, be printed for the use of the Senate. [Washington: 1823.] 4 p. (17th Cong., 2d sess. Senate. Doc. 18)
S-14512.
Copies: DLC; DeU; NjP.

2233. United States. War Dept.
Letter from the Secretary of War, transmitting a statement of the expenditures at the national armories, and of the arms made and repaired therein, during the year 1822. February 15, 1822.... Washington: Printed by Gales & Seaton, 1823. 4 p., fold. table. (17th Cong., 2d sess. House. Doc. 69)
S-14615.
Copies: DLC; DeU; MWA; NjP.

2234. United States. War Dept.
Message from the President of the United States, transmitting a report of the Secretary of War of the expenses of the Ordnance Department for the years 1817, 18, 19, 20, 21 and part of 1822.... January 6, 1823.... Washington: Printed by Gales & Seaton, 1823. 3 v. (117 p. and fold. tables). (17th Cong., 2d sess. House. Doc. 111)
Includes lists of armaments, description of buildings, and equipment of the armories.
S-14647.
Copies: DLC; DeU; MWA; NjP.

2235. United States. War Dept.
Message from the President of the United States, transmitting the information in relation to the expenditures on the Ordnance Department, the quantity and description of arms now belonging to the United States, as required by the resolution of the 17th of December last. February 17, 1823.... Washington: Printed by Gales & Seaton, 1823. 13 p., 13 fold. tables. (17th Cong., 2d sess. House. Doc. 73)
Transmitting a report from the War Department.
S-14661.
Copies: DLC; DeU; MWA; NjP.

2236. United States. War Dept.
Message from the President of the United States, transmitting a statement of the Secretary of War, in relation to contracts for cannon, cannon shot, muskets, and other small arms, which have been entered into since 1st January, 1820 ... January 6, 1824.... Washington: Printed by Gales & Seaton, 1824. 43 p. (18th Cong., 1st sess. House. Doc. 23)
S-18910.
Copies: DLC; DeGE, DeU; MWA; NjP.

2237. Armory on western waters. Petition of sundry inhabitants of Pendleton County, in the state of Kentucky, upon the subject of a national armory in the West. December 19, 1825.... Washington: Printed by Gales & Seaton, 1825. 8 p. (19th Cong., 1st sess. House. Doc. 12)

Includes information about the water power available at the Horse-Shoe Bend on the Licking River.

S-22673.

Copies: DLC; DeGE; DeU; MWA; NjP.

2238. [Ashcroft Rifle Company.]

Directions for using the arms known by the name of Hall's rifle. Harpers-Ferry, Va.: J. S. Gallaher, 1825. 17 p.

S-19465.

Copies: MB.

2239. United States. Commissioners on a National Armory on the Western Waters.

Message from the President of the United States, transmitting a report of the commissioners appointed under the act of the third March, 1823, to establish a national armory on the western waters. January 18, 1825.... Washington: Printed by Gales & Seaton, 1825. 161 p., tables. (18th Cong., 2d sess. House. Doc. 55)

Commissioners: McKee, Lee and Talcott.

S-22994.

Copies: DLC; DeU; MWA; NjP.

2240. Documents in relation to Tygart's Valley River, as a site for an armory. March 17, 1826. Referred to the Committee of the Whole on Bill No. 116, for the location of an armory on the western waters. Washington: Printed by Gales & Seaton, 1826. 8 p. (19th Cong., 1st sess. House. Doc. 136)

S-26958.

Copies: DLC; DeGE; DeU; MWA; NjP.

2241. Tennessee. General Assembly.

National armory. Resolution of the General Assembly of the state of Tennessee, in relation to the establishment of a national armory on the western waters. January 16, 1826. Read, and referred to the Committee on Military Affairs. Washington: Gales & Seaton, 1826. 3 p. (19th Cong., 1st sess. House. Doc. 44)

S-27265.

Copies: DLC; DeU; MWA; NjP.

2242. United States. Congress. House. Committee on Military Affairs.

Arsenal—St. Louis. March 21, 1826. Mr. Hamilton, from the Committee on Military Affairs ... made the following report. [Washington: 1826.] 4 p. (19th Cong., 1st sess. House. Rept. 129)

S-26372.

Copies: DLC; DeU; MWA; NjP.

2243. United States. Congress. House. Committee on Military Affairs.

Arsenal Watervliet—New York. March 31, 1826. Mr. Hamilton, from the Committee on Military Affairs ... made the following report. [Washington: 1826.] 5 p. (19th Cong., 1st sess. House. Rept. 148)

S-26373.

Copies: DLC; DeU; MWA; NjP.

2244. United States. Laws.

A bill to authorize the Secretary of the War Department to purchase a site for an arsenal, at St. Louis, in the state of Missouri, and to provide for the erection of an arsenal on the same. March 21, 1826.... [Washington: 1826.] 1 p. (H.R. 170)

S-26776.

Copies: DNA.

2245. ———— April 22, 1826.... [Washington: 1826.] 1 p. (S. 136)

S-26777.

Copies: DNA.

2246. United States. Laws.

A bill to establish an arsenal in the vicinity of Augusta, and to authorize the sale or removal of the arsenal heretofore estab-

lished at that place. March 7, 1826. [Washington: 1826.] 1 p. (S. 84)
S-26814.
Copies: DNA.

2247. United States. Ordnance Dept.
Arsenal at Augusta, Georgia. Message from the President of the United States, transmitting a report from the Ordnance Department, in relation to the site of the arsenal of the United States, at Augusta, in the state of Georgia. February 17, 1826.... Washington: Printed by Gales & Seaton, 1826. 8 p. (19th Cong., 1st sess. House. Doc. 96)
S-26371.
Copies: DLC; DeU; MWA; NjP.

2248. United States. War Dept.
National armories. Letter from the Secretary of War, transmitting a statement of the expenditures at the national armories, and the arms, &c. made therein during the year 1825. February 22, 1826.... Washington: Printed by Gales & Seaton, 1826. 6 p., fold. table. (19th Cong., 1st sess. House. Doc. 103)
S-27263.
Copies: DLC; DeU; MWA; NjR.

2249. United States. War Dept.
National armory—Ohiopile Falls. Letter from the Secretary of War, transmitting the information required by a resolution of the House of Representatives, of the 9th instant, in relation to the establishment of a national armory, on the western waters. February 13, 1826.... Washington: Printed by Gales & Seaton, 1826. 7 p. (19th Cong., 1st sess. House. Doc. 88)
S-27264.
Copies: DLC; DeU; MWA; NjP.

2250. United States. War Dept.
Report from the Secretary of War, relative to additional means of defence for the Mississippi River, in compliance with a resolution of the Senate, of April 28, 1826. May 8, 1826.... Washington: Printed by Gales & Seaton, 1826. 11 p. (19th Cong., 1st sess. House. Doc. 90)
Concerns the use of batteries on steam vessels.
S-27332.
Copies: DLC; DeU; MWA; NjP.

2251. United States. War Dept.
Site for an armory—Louisville, Kentucky. Letter from the Secretary of War, transmitting copies of a correspondence with the president, directors, and company of the Louisville and Portland Canal Company in Kentucky, upon the subject of a site for a national armory. January 18, 1826.... Washington: Printed by Gales & Seaton, 1826. 15 p. (19th Cong., 1st sess. House. Doc. 46)
Includes a statement on water power for an armory.
S-27396.
Copies: DLC; DeGE; DeU; MWA; NjP.

2252. United States. Laws.
A bill authorizing the establishment of an arsenal in the town of Augusta, in Maine. January 16, 1827.... [Washington: 1827.] 1 p. (S. 57)
S-30938.
Copies: DNA.

2253. United States. Ordnance Dept.
Expenditures at national armories. Letter from the Secretary of War, transmitting a statement of the expenditures at the national armories and the arms, &c. made therein, during the year 1826. February 13, 1827.... Washington, D.C.: Printed by Gales & Seaton, 1827. 7 p. (19th Cong., 2d sess. House. Doc. 99)
Signed: G. Bomford, Brev. Col. on Ordnance Service.
S-31253.
Copies: DLC; DeGE, DeU; NjP.

2254. Brown, Uriah.
Uriah Brown. Communication of Uriah Brown. February 18, 1828.... Washington: Printed by Gales & Seaton, 1828. 5 p. (20th Cong., 1st sess. House. Doc. 150)
Concerns the invention and use of Greek fire.
S-36956.
Copies: DLC; DeU; NjR.

2255. Maryland. Governor (J. Kent).

Communication from the executive enclosing reports of the several armorers in this state. Annapolis: Printed by Jeremiah Hughes, 1828. 36 p.

S-37301.

Copies: MdBP (not located).

2256. Memorial of inhabitants of Blairsville, upon the subject of an armory, &c. March 24, 1828.... Washington: Printed by Gales & Seaton, 1828. 3 p. (20th Cong., 1st sess. House. Doc. 214)

At head of title: Pennsylvania.

S-36706.

Copies: DLC; DeU; MWA; NjR.

2257. United States. Congress. House. Committee on Naval Affairs.

Uriah Brown. May 10, 1828. Mr. Dorsey, from the Committee on Naval Affairs, to which was referred the petition of Uriah Brown, relative to a system of coast and harbor defence, by means of "impregnable and invincible fire ships," made the following report.... [Washington: 1828.] 8 p. (20th Cong., 1st sess. House. Rept. 252)

S-36957.

Copies: DLC; DeU; MWA; NjR.

2258. United States. Congress. Senate. Committee on Military Affairs.

In Senate of the United States. January 23, 1828. Mr. Harrison made the following report: The Committee on Military Affairs, to whom was referred "A bill authorizing the establishment of an arsenal at or near Pensacola, in Florida," make the following report. [Washington: 1828.] 8 p. (20th Cong., 1st sess. Senate. Doc. 59)

S-36349.

Copies: DLC; DeU; MWA; NjR.

2259. United States. Laws.

A bill authorizing the establishment of an arsenal at or near Pensacola, in Florida, January 11, 1828.... [Washington: 1828.] 1 p. (S. 55)

S-35722.

Copies: DNA.

2260. United States. Laws.

A bill to establish an armory on the western waters. February 7, 1828.... [Washington: 1828.] 1 p. (H.R. 150)

S-36129.

Copies: DNA.

2261. United States. War Dept.

Superintendent, armory—Harper's-Ferry. Letter from the Secretary of War, transmitting the information ... in relation to coal and gunstocks, purchased since 1st January, 1820, for the use of said armory. January 22, 1828.... Washington: Printed by Gales & Seaton, 1828. 21 p. (20th Cong., 1st sess. House. Doc. 78)

S-36909.

Copies: DLC; DeGE; DeU; MWA; NjR.

2262. United States. War Dept.

Armories—United States. Letter from the Secretary of War, transmitting a statement of the expenditures at the national armories, and of the arms, &c. made therein, during the year 1828. March 2, 1829.... [Washington: Gales & Seaton, printers, 1829.] 4 p. (20th Cong., 2d sess. House. Doc. 139)

S-40804.

Copies: DLC; DeU; MWA; NjR.

2263. United States. War Dept.

A system of exercise and instruction of field-artillery, including manoeuvres for light or horse-artillery. Published by the Department of War, under ... the act of Congress on the 2nd of March, 1829. Boston: Hilliard, Gray, Little and Wilkins, 1829. 78 p., 56 plates.

Includes a gunner's manual with descriptions of cannon and artillery equipment.

S-41324.

Copies: Ct; DeGE; GMM; ICU; IaHi; KyDC; MWA; Nh; OClWHi; OMC; WaPS.

2264. The archer's manual; or, The art of shooting with the long-bow, as practised by the United Bowmen of Philadelphia. Philadelphia: R. H. Hobson, 1830. 66 p., plates.
C-185.
Copies: CtY; DLC; ICHi; LNHT; MB; MWA; MdBP; PHi.

2265. United States. Laws.
A bill to authorize the Secretary of War to purchase sites and for the erection of a powder magazine, arsenal, and other buildings, in the vicinity of Detroit ... February 22, 1830.... [Washington: 1830.] 2 p. (S. 105)
C-4472.
Copies: DNA.

2266. United States. War Dept.
Armories and arms. Letter from the Secretary of War, transmitting a statement of expenditures at the United States' armories, and the arms manufactured therein, during the year 1828. March 26, 1830.... [Washington: 1830.] 7 p. (21st Cong., 1st sess. House. Doc. 84)
C-3921.
Copies: DLC; DeU; MWA; NjR.

3. Fortifications

2267. New York (Colony). Laws.
An act for the better fortifying of this colony and other purposes herein mentioned. [New York: William Bradford, 1741.] 11 p.
B-1148; Mp-40250.
Copies: MWA; N.

2268. New York (Colony). Laws.
An act to apply the sum of six hundred and seventeen pounds, thirteen shillings and four pence half penny for repairing Fort-George.... [New York: Printed by William Bradford, 1742.] 8–11 p.
B-1182; Mp-40282.
Copies: MWA; N.

2269. Gridley, Richard, 1711–1796.
A plan of the city and fortress of Louisburg; with a small plan of the harbour. Done in metzotinto on royal paper, from the original drawing of Richard Gridley, Esq.; commander of the train of artillery at the siege of Louisburg. Boston: Sold by J. Smibert, 1746. Broadside 46 x 56 cm.
Mp-5783.
Copies: MHi; MWM; RPJCB.

2270. Scheme of a lottery to raise £ 10,000 old-tenor, for ... fortifications for the defense of ... Rhode-Island. [Providence?: 1756.] Broadside 26 x 20 cm.
B-1840; Mp-40862.
Copies: PHi.

2271. United States. Congress. House. Committee on the Defense of Ports and Harbors.
Report of the Committee directed to report on such of the ports and harbors as require to be put in a state of defense. February 28, 1794. [Philadelphia: Printed by F. Childs and J. Swaine, 1794.] 7 p.
"An estimate of the probable expense of fortifying the following harbors in the manner specified ...," p. 2–7, submitted to the House of Representatives by H. Knox, Secretary of War.
Mp-27905.
Copies: MBAt; MWA; RPJCB.

2272. United States. Laws.

An act supplementary to "An act to provide for the defence of certain ports and harbors in the United States...." [Philadelphia: Printed by Childs and Swaine, 1794.] Broadside.

"Approved—May the ninth, 1794."

Mp-27851.

Copies: MWA; RHi.

2273. United States. Congress. House. Committee on Fortifications.

Report of the Committee appointed to take into consideration the state of the fortification of our harbours; the measures which have been pursed [*sic*] for obtaining proper sites for arsenals, and for replenishing our magazines with military stores ... 9th May, 1796.... [Philadelphia: Printed by Francis Childs, 1796.] [4] p.

Mp-31394.

Copies: MBAt; MWA.

2274. United States. War Dept.

Report from the Department of War, relative to the fortifications of the ports and harbours of the United States. Published by order of the House of Representatives. [Philadelphia: 1796.] 9 p.

Signed: Timothy Pickering, War-Office, January 16, 1796.

Mp-31468.

Copies: MBAt; MWA; RPJCB.

2275. United States. Congress. House. Committee on Fortifications.

Report of the committee appointed on the sixteenth of December last, to enquire into the actual state of the fortifications of the ports and harbours of the United States ... 11th February, 1797, committed to a Committee of the Whole House.... [Philadelphia: Printed by William Ross, 1797.] 4 p.

Mp-32988.

Copies: MBAt; MWA.

2276. United States. Treasury Dept.

Letter from the Secretary of the Treasury, accompanying his report, and two estimates of the Secretary at War, of the sums required to be appropriated for the use of the Naval Department, and the fortifications erecting for the defence of the ports and harbours of the United States. 19th January 1797 ... [Philadelphia: Printed by William Ross, 1797.] 7 p.

Mp-33082.

Copies: DLC; MBAt; MWA; MiU-C; RPJCB.

2277. United States. Treasury Dept.

Letter and report of the Secretary of the Treasury, accompanied with sundry statements relative to the military and naval establishments [1791–1796] and to the fortification of the ports and harbours of the United States, in pursuance of three resolutions of the House of Representatives, of the 3d of March, 1797. 7th February, 1798.... [Philadelphia]: Printed by W. Ross, [1798]. [56] p.

Mp-34852.

Copies: DLC; MH; MiU-C; NN.

2278. [Belair, Alexandre Pierre Julienne de] 1747–1819.

The elements of fortification. Translated from the French. [Philadelphia]: Printed for the War Office, by John Ward Fenno, 1799. 42 p., 6 fold. plates.

Translated by Jonathan Williams.

Mp-35175.

Copies: MWA; NcD.

2279. ——— 2d ed. [Philadelphia]: Printed for the War-Office, by C. P. Wayne, 1801. 49 p., 7 fold. plates, 2 fold. tables.

S&S-440.

Copies: CtY; DLC; MB; NWM; PPAmP.

2280. Quincy, Josiah, 1772–1864.

A speech, delivered in Congress, on the 15th April, 1806. While the House, in Committee of the Whole, were discussing the bill for fortifying the ports and harbours of the United States.... Boston: Printed by Russell & Cutler, 1806. 22 p.

S&S-11235.

Copies: MH; MWA; NN.

2281. United States. Congress. Senate. Committee on Defence of Sea-Port Towns and Harbors.

In Senate of the United States. December 15, 1806. Mr. Mitchill from the Committee appointed on the 5th inst. to take into consideration that part of the message of the President of the United States ... which relates to the protection of the seaports, harbors, and waters of the United States, made a report.... [Washington: 1806.] 4 p.

S&S-11582.

Copies: DLC; MWA; NN.

2282. United States. Congress. Senate. Committee to Whom was Referred the Memorial of the Merchants of New York.

In Senate of the United States. January 28, 1806. Mr. Mitchill, from the Committee to whom was referred the memorial of the merchants of New York.... [Washington: 1806.] 6 p.

Concerns the defense of the New York harbor, describing the state of fortifications and their advantages.

S&S-11566.

Copies: DLC; MWA; NN.

2283. United States. Laws.

A bill for making further appropriations for fortifying the ports and harbors of the United States, and for building gun boats. [Washington: 1806.] 2 p. (S.11)

S&S-11524.

Copies: DLC.

2284. United States. President (Jefferson).

Message from the President of the United States, transmitting an act of the state of South Carolina, ceding to the United States various forts and fortifications, for sites for the erection of forts, in the said state. February 13, 1806.... City of Washington: A. & G. Way, printers, 1806. 8 p.

S&S-11652.

Copies: DLC; MWA; NN; NNS.

2285. United States. War Dept.

Message from the President of the United States, transmitting a report made to him by the Secretary of War, relative to fortifications erected at the several ports and harbours of the United States and their territories; and a statement exhibiting the amount of money disbursed on account of the Navy, in each year, since the establishment of the Navy Department; in pursuance of a resolution of the House, of the twenty fourth ultimo. February 18, 1806.... City of Washington: A. & G. Way, printers, 1806. 8 p., 5 fold. tables.

S&S-11646.

Copies: DLC; DeGE; MWA; NN; NNS.

2286. New York (City). Committee on Fortifying the Harbor of New York.

Report of the Committee of the Corporation, on the subject of fortifying the harbour of New-York. New York: Printed for the Corporation; Henry C. Southwick, printer, 1807. 14 p.

Selah Strong, Jacob Mott, and John D. Miller were members of the committee.

S&S-13219.

Copies: DeGE; N.

2287. United States. Congress. House.

Documents accompanying the bill for the defence of the mouth of the Mississippi, and for the protection of New Orleans, and its dependencies, presented the 15th January, 1807. City of Washington: A. & G. Way, printers, 1807. 7 p.

S&S-13890.

Copies: DLC; MWA; NNS; NhHi.

2288. United States. Congress. House. Committee on Repairs of Fortifications.

Report of the Committee appointed on the third ultimo, on so much of the message of the President of the United States as relates to the repairs of fortifications, and to the farther protection of our ports, towns and rivers. January 12, 1807.... Washington City: A. & G. Way, printers, 1807. 6 p.

S&S-14051.

Copies: DLC; DeGE; MWA; NN.

2289. United States. Congress. Senate. Committee on Defence of Sea-Port Towns

2289 (*continued*)
and Harbors.

Further report of the Committee on that part of the message of the President of the United States, which relates to the defence of our sea-port towns and harbors. December 3, 1807.... Washington: Printed by R. C. Weightman, 1807. 31 p.

Contains a statement from the War Department with a list of ports and harbors and their defenses; H. Dearborn, Secretary of War.

S&S-13907.

Copies: DLC; DeGE; MWA; MiU-C; N; NHi; NN; NNS.

2290. United States. Congress. Senate. Committee on Defence of Sea-Port Towns and Harbors.

In Senate of the United States. March 2, 1807. Mr. Bradley, from the Committee to whom was referred the bill making further appropriations for fortifying the ports and harbours of the United States, and for building gun boats, reported the following amendment. [Washington: 1807.] 1 p.

S&S-13923.

Copies: DLC.

2291. United States. Congress. Senate. Committee on Defence of Sea-Port Towns and Harbors.

Report, in part, of the Committee on that part of the message of the President of the United States, which relates to the defence of our sea-port towns and harbors. November 20th, 1807. Printed by order of the Senate. Washington City: Printed by R. C. Weightman, 1807. 5 p.

S&S-14045.

Copies: DLC; MWA; MiU-C; NN; NNS.

2292. United States. Laws.

A bill for the defence of the mouth of the Mississippi, and for the protection of New Orleans, and its dependencies. January 15, 1807.... [Washington: 1807.] 1 p. (H.R. 35)

S&S-13814.

Copies: DLC.

2293. United States. Laws.

A bill supplementary to an act, entitled, "An act for fortifying the ports and harbours of the United States, and for building gun-boats." Dec. 1, 1807.... [Washington: 1807.] 1 p. (H.R. 10)

S&S-13854.

Copies: MH.

2294. United States. War Dept.

Letter from the secretary of War, enclosing his report on "the state of the fortifications of the respective ports and harbours of the United States, with a statement of the monies appropriated for fortifications and remaining unexpended, and an estimate of the sums necessary for completing such fortifications as may be deemed requisite for their defence...." December 8, 1807.... City of Washington: A. & G. Way, printers, 1807. 8 p.

S&S-13968.

Copies: CSt; DLC; DS; MBC; MWA; MdBL; NN.

2295. Williams, Jonathan, 1750–1815.

Letter from Col. Jonathan Williams, on the subject of fortifying and protecting the harbour of New-York. New York: Printed for the Corporation; H. C. Southwick, printer, 1807. 16 p.

S&S-14223.

Copies: CSmH; DeGE; MH; MWA; NN; NWM.

2296. Williams, Jonathan, 1750–1815.

Plan of Col. Jonathan Williams, for fortifying the narrows, between Long and Staten Islands. New York; Printed for the Corporation of the City of New York, by H. C. Southwick, July 31, 1807. 15 p.

S&S-14224.

Copies: CtY; DLC; DeGE; MBAt; MWA; NN; NWM.

2297. United States. President (Jefferson).

Message from the President of the United States, respecting the execution of the act for fortifying the ports and harbours of the United States. March 25, 1808.... City of

Washington: A. & G. Way, printers, 1808. 4 p.
S&S-16470.
Copies: DLC; MWA; NN.

2298. United States. War Dept.
Message from the President of the United States, transmitting a report of the Secretary of War, upon the subject of the execution of the last session of Congress for fortifying the ports and harbors of the United States. January 7, 1809. City of Washington: A. & G. Way, printers, 1809. 19 p.
Also includes estimates for chains across New York harbor.
S&S-18970.
Copies: DLC; MWA; MiU-C; NN; PHi.

2299. United States. War Dept.
Report of the Secretary of War ... February 9th, 1809. Washington City: Printed by R. C. Weightman, 1809. 4 p.
Concerns the fortifications at or near Wilmington, New Castle, Port Penn, and Lewes in Delaware.
S&S-19069.
Copies: DLC; NN.

2300. United States. War Dept.
Report of the Secretary of War ... [on] the sums necessary to complete the fortifications commenced or contemplated, and also the amount of any deficiency of former appropriations for this object. June 6, 1809.... City of Washington: A. & G. Way, printers, 1809. [2] p.
S&S-19067.
Copies: DLC; MWA; NN.

2301. United States. War Dept.
Letter from the Secretary of War, to the Committee appointed on that part of the message of the President of the United States, of the fifth ultimo, relating to land forces and fortifications, transmitting an estimate of monies required on account of fortifications for the year 1811. January 18, 1811. Washington City: A. & G. Way, printers, 1810 [1811]. 10 p.
S&S-24193.
Copies: DLC; MWA; NN; NjR.

2302. A brief description of the towns, rivers, and fortifications contiguous to the seat of war.... New York; Printed [by Spooner & Sleight, Brooklyn] for John Tiebout, 1812. 24 p.
S&S-24930.
Copies: MWA.

2303. New York (State). Commission to Fortify the Narrows.
[Report of work done and proposals for additional fortifications.] New York: 1812. Broadside 20 x 33 cm.
Report dated: New York, March 15, 1812.
N-N0190706.
Copies: InU.

2304. Papers on the defence of Boston and other places. [Boston: 1813.] 16 p.
N-P0068748.
Copies: DLC; MWA; MiU-C; PMA.

2305. New York (State). Commissioners of Fortifications.
Message ... with the report of the Commissioners of Fortifications. [Albany?: 1814.] 7 p.
"In Assembly, March 24, 1814."
S&S-32316.
Copies: MB; MiU; N; NN.

2306. A narrative of a tour of observation, made during the summer of 1817, by James Monroe, President of the United States, through the north-eastern and north-western departments of the Union: with a view to the examination of their several military defences.... Philadelphia: S. A. Mitchell & H. Ames; Clark & Raser, printers, 1818. xii, [13]–228, xxxvi p.
Not the same as Waldo's *Tour of James Monroe,* Hartford, 1818.
S&S-44962, 46674.
Copies: DLC; ICN; Ia; InU; MH; MWA; Mi; MiD-B; MiU; MoSM; NNA; NcD; NjR; OCHP; PPL; ScCC; TxU; ViU; WHi.

2307. United States. Engineer Dept.

Message from the President of the United States, transmitting ... a report of the Secretary of War, of the expenditures on fortifications from the year 1816 to the year 1819 ... their contemplated magnitude and present condition ... January 20, 1820.... Washington: Printed by Gales & Seaton, 1820. 15 p. (16th Cong., 1st sess. House. Doc. 51)

Transmits a report from the Engineer Department.

S-3919.

Copies: DLC; DeGE; DeU; MWA; R.

2308. United States. War Dept.

Letter from the Secretary of War, transmitting (pursuant to a resolution of the House of Representatives) copies of contracts made for building, and repairing fortifications &c. since the year 1815 ... April 6, 1820.... Washington: Printed by Gales & Seaton, 1820. 48 p., fold. table. (16th Cong., 1st sess. House. Doc. 108)

S-3850.

Copies: DLC; DeGE; DeU; MWA; O; R.

2309. United States. Engineer Dept.

Letter from the Secretary of War, transmitting (in pursuance of a resolution of the House of Representatives, of the 9th instant) information in relation to the progress of the Board of Engineers, in the selection of sites for fortifications, &c. February 15, 1821.... Washington: Printed by Gales & Seaton, 1821. 18, [2] p., 2 fold. tables. (16th Cong., 2d sess. House. Doc. 98)

S-7356.

Copies: DLC; DeGE; DeU; MWA; NjP.

2310. United States. Congress. House. Committee on Military Affairs.

Report of the Committee on Military Affairs on the message of the President of the United States, of the 26th ult., on the subject of fortifications generally, and particularly of those on Mobile Point and Dauphine Island. April 13, 1822.... [Washington: 1822.] 5 p. (17th Cong., 1st sess. House. Rept. 94)

S-11157.

Copies: DLC; MWA; PU.

2311. United States. Congress. House. Committee on Military Affairs.

Report of the Committee on Military Affairs, to which was referred so much of the memorial of the Legislature of the state of Alabama, and of certain citizens thereof, as relates to fortifications. February 28, 1822.... [Washington: 1822.] 9 p. (17th Cong., 1st sess. House. Rept. 51)

S-11159.

Copies: DLC; DeGE; MWA; PU.

2312. United States. Engineer Dept.

Letter from the Secretary of War, to the chairman of the Committee on Military Affairs upon the subject of fortifications.... March 7, 1822.... Washington: Printed by Gales & Seaton, 1822. 8 p., fold. table. (17th Cong., 1st sess. House. Doc. 87)

Transmits a report from the Engineer Department.

S-10972.

Copies: DLC; DeGE; DeU; MWA; NjP.

2313. United States, Engineer Dept.

Report of the Engineer Department, in relation to a required appropriation of $500,000 for fortifications. March 7, 1822. Laid before the House of Representatives by the Chairman of the Committee of Ways and Means.... Washington: Printed by Gales & Seaton, 1822. 5 p. (17th Cong., 1st sess. House. Doc. 89)

S-11217.

Copies: DLC; DeGE; MWA; NjP.

2314. United States. President (Monroe).

Message from the President of the United States, upon the subject of fortifications on Dauphine Island and Mobile Point. March 26, 1822.... Washington: Printed by Gales & Seaton, 1822. 24 p., fold. map. (17th Cong., 1st sess. House. Doc. 99)

S-11052.

Copies: DLC; DeGE; DeU; MjP.

2315. ———— March 27, 1822.... Washington: Printed by Gales & Seaton, 1822. 9 p. (17th Cong., 1st sess. Senate. Doc. 74)
S-11053.
Copies: DLC; DeGE; DeU; MWA; NjP.

2316. United States. War Dept.
Letter from the Secretary of War to the chairman of the Committee on Military Affairs, accompanied with a statement of the expenditures on the fortification at the Pea Patch. As, also, the sum originally estimated for that work, &c. April 20, 1822.... Washington: Printed by Gales & Seaton, 1822. 8 p., fold. table. (17th Cong., 1st sess. House. Doc. 115)
Includes a statement from the Engineer Department.
S-10968.
Copies: DLC; DeGE; DeU; MWA; NjP.

2317. United States. War Dept.
Letter from the Secretary of War to the chairman of the Committee on Military Affairs, in relation to the application of money appropriated for fortifications. April 19, 1822.... Washington: Printed by Gales & Seaton, 1822. 7 p. (17th Cong., 1st sess. House. Doc. 114)
Includes a report from the Engineer Department.
S-10969.
Copies: DLC; DeGE; DeU; MWA; NjP.

2318. United States. Congress. House. Committee on Military Affairs.
Report of the Committee on Military Affairs, to which was referred the memorial of the General Assembly of the state of Alabama, on the subject of the fortifications at Mobile Point and Dauphin Island. February 25, 1823.... [Washington: 1823.] 1 p. (17th Cong., 2d sess. House. Rept. 103)
S-14694.
Copies: DLC; DeU; MWA; NjP.

2319. United States. Engineer Dept.
Report of Maj. Gen. Alex. Macomb, Chief Engineer, to the Secretary of War, in relation to the nature and extent of the work done, and the materials delivered at the several fortifications, viz: Forts Delaware, Washington, Monroe, Calhoun, and Rigolets, under the appropriation for the year 1822. January 23, 1823.... Washington: Printed by Gales & Seaton, 1823. 17 p. (17th Cong., 2d sess. House. Doc. 39)
S-14670.
Copies: DLC; DeU; MWA; NjP.

2320. United States. Laws.
A bill making an appropriation for the gradual armament of the new fortresses of the United States. January 31, 1823. [Washington: 1823.] 1 p. (S.45)
S-14430.
Copies: DNA.

2321. United States. Navy Dept.
Message from the President of the United States, transmitting information in relation to the fortifying of Thompson's Island, usually called Key West, &c. &c. January 20, 1823.... Washington: Printed by Gales & Seaton, 1823. 6 p. (17th Cong., 2d sess. House. Doc. 29)
Transmitting the report of the Secretary of the Navy, Smith Thompson.
S-14654.
Copies: DLC; DeU; MWA; NjP.

2322. United States. War Dept.
Letter from the Secretary of War, transmitting a report and statement respecting the appropriations for fortifications, in the years 1820, 1821, & 1822. February 17, 1823.... Washington: Printed by Gales & Seaton, 1823. 6 p., fold. table. (17th Cong., 2d sess. House. Doc. 72)
S-14610.
Copies: DLC; DeU; MWA; NjP.

2323. Farrow, Nimrod.
Memorial of Nimrod Farrow, for himself and Richard Harris. To the Senate and House of Representatives of the United States, in Congress. [Washington?: 1824.] 66 p., fold. map.
Concerns the fortifications on Dauphin

2323 (*continued*)
Island.
N-F0042246.
Copies: ICN; MB; ViU.

2324. United States. Laws.
An act for making appropriations for certain fortifications of the United States for the year one-thousand eight hundred and twenty four. April 24, 1824.... [Washington: 1824.] 1 p. (H.R. 88)
S-18328.
Copies: DNA.

2325. United States. President (Monroe).
Message from the President of the United States, transmitting ... information of the number and position of the permanent fortifications of the United States, &c. March 4, 1824.... Washington: Printed by Gales & Seaton, 1824. 8 p., fold. table. (18th Cong., 1st sess. House. Doc. 104)
S-18918.
Copies: DLC; DeU; MWA; NjP.

2326. United States. Congress. House. Committee on Military Affairs.
Report of the Committee on Military Affairs on the resolution instructing them to inquire into the expediency of purchasing additional land for the completion of the site of Fort Washington, accompanied with a bill providing for the accomplishment of that object. February 17, 1825.... [Washington: 1825.] 8 p. (18th Cong., 2d sess. House. Rept. 74)
S-23060.
Copies: DLC; DeU; MWA; NjP.

2327. United States. Engineer Dept.
Fortifications, harbor of Pensacola. Letter from the Secretary of War, transmitting information in relation to a survey and estimate of the harbor of Pensacola, with a view to fortifications. December 23, 1825.... Washington: Printed by Gales & Seaton, 1825. 8 p. (19th Cong., 1st sess. House. Doc. 14)
S-22861.
Copies: DLC; DeU; MWA; NjP.

2328. United States. Laws.
A bill making appropriations for certain fortifications of the United States, for the year one thousand eight hundred and twenty-five. January 19, 1825.... [Washington: 1825.] 1 p. (H.R. 280)
S-22779.
Copies: DNA.

2329. ________ for the year eighteen hundred and twenty-six. December 21, 1825.... [Washington: 1825.] 1 p. (H.R. 10)
S-22780.
Copies: DNA.

2330. United States. War Dept.
Letter from the Secretary of War, in relation to the fulfilment of the contract for erecting fortifications on Dauphin Island, and of damages sustained by R. Harris and N. Farrow, the contractors. February 9, 1825.... Washington: Printed by Gales & Seaton, 1825. 29 p. (18th Cong., 2d sess. Senate. Doc. 29)
Report of Thomas Swann, commissioner appointed by the War Dept., p. 7–14.
S-22947.
Copies: DLC; DeGE; DeU; MWA; NjP.

2331. United States. War Dept.
Letter from the Secretary of War, transmitting a report in relation to the works carrying on at the Pea Patch, in the Delaware River; and also, the proceedings of a court martial ordered for the trial of Major S. Babcock. The testimony taken in the case, and the judgment pronounced by the court thereupon. January 19, 1825.... Washington: Printed by Gales & Seaton, 1825. 78 p., 2 fold. tables. (18th Cong., 2d sess. House. Doc. 63)
S-22948.
Copies: DLC; DeU; MWA; NjP.

2332. United States. War Dept.
Letter from the Secretary of War, transmitting, in compliance with a resolution of the House of Representatives of the 16th instant, all the information contained in the department, in relation to contracts for the

supply of stone for fortifications on the waters of the Chesapeake Bay, since the 1st of January, 1824. February 21, 1825.... Wahington: Printed by Gales & Seaton, 1825. 13 p. (18th Cong., 2d sess. House. Doc. 98)
S-22958.
Copies: DLC; DeU; MWA; NjP.

2333. United States. Congress. House. Committee on Military Affairs.
Fort Washington. March 20, 1826. Mr. Hamilton, from the Committee on Military Affairs ... made the following report. [Washington: 1826.] 5 p. (19th Cong., 1st sess. House. Rept. 125)
S-26992.
Copies: DLC; DeU; MWA; NjP.

2334. United States. Congress. House. Committee on Military Affairs.
Fortifications—Pensacola Bay. January 23, 1826.... Report of the Committee on Military Affairs, on the resolution instructing them to inquire into the expediency of constructing fortifications at the entrance of Pensacola Bay, for the protection of the Navy Yard at that place. [Washington: 1826.] 1 p. (19th Cong., 1st sess. House. Rept. 46)
S-26996.
Copies: DLC; DeU; MWA; NjP.

2335. United States. President (John Q. Adams).
System of fortifications. Message from the President of the United States, transmitting the information required by the resolutions of the House of Representatives of 6th May, 1824, and the 5th January last, in relation to a system of fortifications, contemplated by the Board of Engineers. March 1, 1826.... Washington: Printed by Gales & Seaton, 1826. 9 p., 8 tables. (19th Cong., 1st sess. House. Doc. 78)
S-27419.
Copies: DLC; DeU; MWA; NjP.

2336. United States. War Dept.
Fortifications at Charleston, S.C. Letter from the Secretary of War, transmitting information ... upon the subject of the works intended for the defence of Charleston, in the state of South Carolina. December 22, 1826.... Washington: Printed by Gales & Seaton, 1826. 4 p. (19th Cong., 2d sess. House. Doc. 22)
S-26993.
Copies: DLC; DeU; MWA; NjP.

2337. ________ transmitting the information called for by the resolution of the House of Representatives of the 20th instant, respecting the fortifications contemplated for the defence of Charleston, S.C. January 24, 1826.... Washington: Gales & Seaton, 1826. 4 p. (19th Cong., 1st sess. House. Doc. 55)
S-26994.
Copies: DLC; DeU; MWA; NjP.

2338. United States. War Dept.
Fortifications on Pensacola Bay. Letter from the Secretary of War, transmitting the information required by a resolution of the House of Representatives of the United States of the 8th instant, in relation to the commencement of the fortifications at the entrance of Pensacola Bay. February 20, 1826.... Washington: Printed by Gales & Seaton, 1826. 4 p. (19th Cong., 1st sess. House. Doc. 98)
S-26995.
Copies: DLC; DeU; MWA; NjP.

2339. United States. Engineer Dept.
Fortifications for Pensacola. Letter from the Secretary of War, transmitting a report respecting plans and estimates of fortifications contemplated for Pensacola. January 3, 1827.... Washington: Printed by Gales & Seaton, 1827. 6 p. (19th Cong., 2d sess. House. Doc. 30)
S-31261.
Copies: DLC; DeU; MWA; NjP.

2340. United States. Engineer Dept.
Fortifications—Pensacola Harbor. Letter from the Secretary of War, transmitting a report from the Engineer Department in relation to the contemplated fortification of Pensacola Harbor. February 9, 1827....

2340 (*continued*)
Washington: Printed by Gales & Seaton, 1827. 4 p. (19th Cong., 2d sess. House. Doc. 88)
S-31260.
Copies: DLC; DeU; MWA; NjP.

2341. United States. Laws.
A bill making appropriations for certain fortifications of the United States, for the year 1827. January 19, 1827.... [Washington: 1827.] 1 p. (H. R. 385)
S-31087.
Copies: DNA.

2342. United States. War Dept.
Fortifications on Staten Island. Message from the President of the United States, with sundry documents, relating to a proposition by the state of New York, to sell to the government of the United States the fortifications erected by said state on Staten Island. January 15, 1827.... Washington: Printed by Gales & Seaton, 1827. 12 p. (19th Cong., 2d sess. House. Doc. 49)
S-31262.
Copies: DLC; DeU; MWA; NIC; NjP.

2343. Farrow, Nimrod.
Farrow and Harris. Letter from the Secretary of War transmitting a memorial from Nimrod Farrow upon the subject of the claim of Farrow and Harris. December 29, 1828 ... Washington: Printed by Gales & Seaton, 1828. 11 p. (20th Cong., 2d sess. House. Doc. 36)
Concerns work done on the fortifications on Dauphin Island.
S-36300.
Copies: DLC; DeU; MWA; NjR.

2344. United States. Laws.
A bill making appropriations for certain fortifications of the United States, for the first quarter of the year one thousand eight hundred and twenty-nine. April 17, 1828.... [Washington: 1828.] 1 p. (H.R. 272)
S-35997.
Copies: DNA.

2345. ———— for the year one thousand eight hundred and twenty eight. January 21, 1828.... [Washington: 1828.] 1 p. (H.R. 96)
S-35998.
Copies: DNA.

2346. United States. War Dept.
Barrataria. Documents accompanying the amendment proposed by the Senate to the Bill (H.R. No. 96) making appropriations for certain fortifications of the United States, for the year 1828. March 7, 1828.... Washington: Printed by Gales & Seaton, 1828. 6 p. (20th Cong., 1st sess. House. Doc. 186)
S-35704.
Copies: DLC; DeU; MWA; NjP.

2347. United States. War Dept.
Farrow and Harris. Documents &c. relating to the claim of Farrow and Harris. December 18, 1828.... [Washington: Gales & Seaton, printers, 1828.] 5 p. (20th Cong., 2d sess. House. Doc. 21)
Concerns work done on the fortifications on Dauphin Island.
S-36299.
Copies: DLC; DeU; MWA; NjR.

2348. United States. War Dept.
Farrow and Harris. Letter from the Secretary of War transmitting further information in relation to the settlement of the claim of Farrow and Harris against the United States. December 19, 1828.... Washington: Printed by Gales & Seaton, 1828. 2 p. (20th Cong., 2d sess. House. Doc. 27)
Concerns work done on the fortifications on Dauphin Island.
S-36301.
Copies: DLC; DeU; MWA; NjR.

2349. United States. Laws.
An act making additional appropriations for certain fortifications of the United States, for the year one thousand eight hundred twenty-nine. Feb. 23, 1829.... [Washington: 1829.] 1 p. (H.R. 398)
S-40759.
Copies: DNA.

2350. United States. Laws.
A bill making appropriations for the erection and completion of certain forts, barracks, and quarters; for the completion of the military road from the Matanawcook River to Mars Hill, and for other purposes. Feb. 7, 1829.... [Washington: 1829.] 3 p. (H.R. 424)
S-40990.
Copies: DNA.

2351. United States. War Dept.
Message from the President of the United States with a report from the Secretary of War, transmitting the inspection returns of Brevet Major General Gaines, for 1826 and 1827, relative to the organization of the army and militia, &c. February 27, 1829.... [Washington: 1829.] 79 p. (20th Cong., 2d sess. Senate. Doc. 101)
Includes reports on fortifications, military roads, and canals.
Copies: DeU; MWA.

2352. United States. Laws.
A bill for the security of the Pea Patch Island; for the construction of a new water tank; and for graveling the parade at Fort Delaware. April 7, 1830.... [Washington: 1830.] 1 p. (H.R. 409)
C-4320.
Copies: DNA.

2353. United States. Laws.
A bill making appropriations for certain fortifications for the year one thousand eight hundred and thirty. January 13, 1830.... [Washington: 1830.] 1 p. (H.R. 110)
C-4363.
Copies: DNA.

2354. United States. Laws.
A bill to authorize the Secretary of War to purchase an additional quantity of land for the fortifications at Fort Washington, upon the river Potomac. February 4, 1830.... [Washington: 1830.] 2 p. (H.R. 192)
C-4471.
Copies: DNA.

2355. United States. Treasury Dept.
Disbursements for certain specific purposes. Letter from the Secretary of the Treasury, transmitting, in compliance with a resolution of the House of Representatives of the 26th of May last, a statement of disbursements made since the year 1789, for fortifications, light-houses ... and internal improvements, &c. December 21, 1830.... [Washington: 1830.] 32 p. (21st Cong., 2d sess. House. Doc. 11)
C-4679.
Copies: DLC; DeU; MWA; NjR.

2356. United States. War Dept.
Fortifications. Letter from the Secretary of War, transmitting the information required by a resolution of the House of Representatives of the 27th ultimo, in relation to the appropriations for fortifications since 1815, &c.&c. March 9, 1830.... [Washington: 1830.] 5 p. (21st Cong., 1st sess. House. Doc. 75)
C-4717.
Copies: DLC; DeU; MWA; NjR.

2357. United States. War Dept.
Fortifications—Pea Patch Island. Message from the President of the United States, transmitting a report from the War Department, in relation to the fortifications at Pea Patch Island, &c.&c. March 31, 1830.... [Washington: 1830.] 3 p. (21st Cong., 1st sess. House. Doc. 88)
C-4718.
Copies: DLC; DeU; MWA; NjR.

VII. CIVIL ENGINEERING

1. General works

2358. Sganzin, Joseph Mathieu, 1750–1837.

An elementary course of civil engineering. Translated from the French of M. I. Sganzin.... From the 3d French ed. with notes and applications adapted to the United States. Boston: Hilliard, Gray, Little, and Wilkins, 1827. viii, 220 p., 6 fold. plates.

S-30576.

Copies: DLC; DeGE; IC; ICJ; LNHT; MA; MB; MWA; MiD; MnHi; NjP; NjR; P; PPL; PU; VtU.

2359. ________ 2d ed. Boston: Hilliard, Gray, Little, and Wilkins, 1828. viii, 232 p., 6 fold. plates.

S-35156.

Copies: DLC; GU; LNL; MB; MWA; MdBJ; MiU; MnS; NNE; OCl; PU.

2360. Eaton, Amos, 1776–1842.

Art without science; or, Mensuration, surveying and engineering, divested of the speculative printiples and technical language of mathematics ... 2d ed., much enlarged. Albany: Printed by Websters & Skinners, 1830. 96 p., illus.

C-1228.

Copies: CSmH; CtY; DeGE; MMeT; MiU; NCH; NNC; Nh; OkU; PU; RPB.

2. Surveying

2361. New Jersey (Colony). Surveyor General.

General instructions to the deputy surveyors of the western division of New Jersey. [Trenton?: 1721?] 5 l., illus.

N-N0152517.

Copies: DLC.

2362. [Chandler, William] 1698–1754.

A journal of the survey of the Narragansett Bay, and parts adjacent. Taken in the months of May and June, A.D. 1741. By order of the Honourable Court of Commissioners appointed by His Majesty King George the Second. [Newport: Ann Franklin, 1741.] Broadside 39.5 x 31.5 cm.

Signed: W.C.

Mp-40232.

Copies: MHi; MWA (photocopy); RHi; RPB.

2363. New Jersey (Colony). Surveyor General.

General instructions by the Surveyor General, to the deputy surveyors of the eastern division of New-Jersey. [Philadelphia?: B. Franklin?, 1747.] 5, [1] p.

Attributed also to James Parker's press in New York.

James Alexander, Surveyor General.

B-1354; Mp-40434.

Copies: CSmH; DLC; NN; NjR; PHi; PU; RPJCB.

2364. New Jersey (Colony). Surveyor General.

General instructions by the Surveyor General, to the deputy surveyors of the western division of New-Jersey. [Philadelphia?: B. Franklin?, 1747.] 5, [1] p.

Attributed also to James Parker's press in New York.

James Alexander, Surveyor General.

B-1355; Mp-40435.

Copies: PHi; PU.

2365. Gordon, John, b. 1700.

A supplement to the mathematical traverse tables in epitome, intitled A sure guide to all practical mariners, surveyors and other gentlemen that deals [*sic*] in land.... [Philadelphia: Printed by W. Dunlap, 1760]. 13 l., tables.

B-2133; Mp-41127.

Copies: NN; NjR; PPAmP (defective copy).

2366. Milton, Abraham.

The farmer's companion, directing how to survey land after a new and particular method. By Abraham Milton, farmer of Kent County, in Maryland. Annapolis: Printed for the author, 1761. 34 p., fold. plate.

B-2231.

Privately held.

2367. ——— instructing how to run land without a compass, and to plot the same in an easy manner. Also a supplement thereto, directing how any person may tell the time of day by a walking stick, or a piece of board, and thereby set off any course of compass.... Philadelphia: Printed by Andrew Steuart, 1761.

E-8929.

No copy known.

2368. Moody, Thomas.

A compendium of surveying; or, The surveyor's pocket companion. Shewing, by a small table, how to work trigonometry, without the help of logarithms, sines, tangents or secants; likewise by the same table, a short easy way to find the area of a triangle ... and to ascertain the true content of any piece or tract of land ... Burlington, [N.J.]: Printed by Isaac Collins, for the author, 1771. iii, 48 p.

Mp-12129.

Copies: DLC; NHi; PPL.

2369. [Gale, Samuel.]

Philadelphia, March 12th, 1772. The author of the Complete surveyor, to the public in general, and to the subscribers in particular.... [Philadelphia: John Dunlap, 1772.] Broadside 37 x 27.5 cm.

Advertising the intended publication of *The Complete surveyor* by Samuel Gale. The work was not printed.

B-3477; Mp-42367.

Copies: PHi.

2370. Carter, John.

The young surveyor's instructor; or, An introduction to the art of surveying.... Philadelphia: Printed and sold by W. and T. Bradford, 1774. iv, 64 p.

B-3709; Mp-42570.

Copies: DLC.

2371. Gibson, Robert.

A treatise of practical surveying; which is demonstrated from its first principles. Wherein every thing that is useful and curious in that art, is fully considered and explained.... The whole illustrated with copper-plates. 4th ed. With alterations and amendments, adapted to the use of American surveyors. Philadelphia: J. Crukshank, 1785. viii, 272, [1], 90 p., 12 plates.

First American edition.

The tables (90 p. at end) have special title page: *Tables of difference of latitude and departure....*

Mp-19026.

Copies: CtY; DeGE; ICJ; MH; MWA; MiU-C; NIC; NcD; OC; PPL; TxU.

2372. Wall, George, 1745?-1804.

A description, with instructions for the use, of a newly invented surveying instrument, called the trigonometer. For the making and vending of which, the inventor hath obtained an exclusive right for twenty-one years, from the Legislature of Pennsylvania. To which are added, I. A method of finding the content of any piece of land, by a scale of equal parts and compasses only.... II. A table of natural sines, calculated to five places of decimals.... III. Rules, respecting the variation of the needle. IV. Theorems for solving all cases in plane trigonometry.... V. An easy method of finding the exact course and distance between two objects, which are invisible the one from the other. The whole illustrated with cuts, and an elegant engraving of the instrument. By George Wall, Junior, of Bucks County, Pennsylvania. Philadelphia: Printed by Zachariah Poulson, Junr., 1788. 32 p., illus., plates.

Mp-21568.

Copies: MBAt; MiU-C; NN; PHi; PPL.

2373. Gibson, Robert.

A treatise of practical surveying; which is demonstrated from its first principles. Wherein every thing that is useful and curious in that art, is fully considered and explained. Particularly three new and very concise methods for determining the areas of right-lined figures arithmetically, or by calculation.... The 5th ed.... With alterations and amendments, adapted to the use of American surveyors. Philadelphia: Printed by Joseph Crukshank, 1789. viii, 272, [1], 90 p., 12 fold. plates.

Second American edition.

Mp-21856.

Copies: CU; DLC; KU; MH; MWA; NN; PPAmP; PPL. PU; RPJCB.

2374. Hampton, Samuel.

Longitude; or, The art of measuring at sea or land made easy.... Also, a standard for invariable measures and weights. [Philadelphia?]: 1790.

E-22554.

No copy known.

2375. [Robertson, John] 1712–1776.

Tables of difference of latitude and departure: constructed to every quarter of a degree of the quadrant, and continued from one to the distance of one hundred miles or chains. Philadelphia: Printed and sold by Joseph Crukshank, 1790. 1 p.l., 90 p.

Also found bound with: Gibson, Robert, *A Treatise of practical surveying*, 1792.

Mp-22856.

Copies: MWA.

2376. Gibson, Robert.

A treatise of practical surveying; which is demonstrated from its first principles. Wherein every thing that is useful and curious in that art, is fully considered and explained. Particularly three new and very concise methods for determining the areas of right-lined figures arithmetically, or by calculation, as well as the geometrical ones heretofore treated of.... The 6th ed. Phil-

2376 (*continued*)
adelphia: Printed by Joseph Crukshank, 1792. vii, 288, [1], 90 p., 13 fold. plates.
Third American edition.
Mp-24357.
Copies: MH; MWA; MiU; NIC; NNE; OU; PHi; RPJCB; ViU.

2377. Clendinin, John.
The practical surveyor's assistant. Calculated by John Clendinin. Philadelphia: Printed for the author, by Benjamin Johnson, 1793. v [iv], 20, 10 p.
B-8307; Mp-46718.
Copies: DGU; DeGE; MWA; NN; RPJCB.

2378. Clendinin, John.
The practical surveyor's assistant. In two parts. Part the I. being a table of difference of latitude and departure, fitted to every degree of the quadrant, and continued from one tenth of a perch to a mile. Part the II. a like table fitted to every quarter of a degree of the quadrant, and continued from one tenth of a perch to four hundred and fifty perches. Calculated by John Clendinin, land surveyor. Philadelphia: Printed by Benjamin Johnson, for the author, 1793. 3 p.l., 45, 45 p.
Includes "Subscriber's names."
Copies: DeGE; PPAmP; RPJCB.

2379. ———— Philadelphia: Printed for the author, 1793. [2], 45, [1], 45 p.
Second printing, without the list of subscribers.
Mp-25305.
Copies: DGU; MB; MWA; MiU; NN; NNC; PP; PPAmP; ViU.

2380. Love, John.
Geodaesia; or, The art of surveying, and measuring land made easy.... The 12th ed., adapted to American surveyors. New York: Samuel Campbell, 1793. [viii], 190, 15, 3, 36, 8 p., illus., diagrs., tables.
First published in London, 1688.
Mp-25731.
Copies: DAU; DeGE; MH; MWA; NN; PPL.

2381. [Gibson, Robert.]
Mathematical tables: difference of latitude and departure: logarithms, from 1 to 10,000. Artificial sines, tangents, and secants. Philadelphia: Printed and sold by Joseph Crukshank, 1794. [4], 90 p.
Bound with his *A Treatise of practical surveying*, 1792.
Mp-24357.
Copies: MWA.

2382. Ellicott, Andrew, 1754–1820.
Several methods by which meridianal lines may be found with ease and accuracy: recommended to the attention of the surveyors in the United States.... Philadelphia: Printed by Thomas Dobson, 1796. 32 p., 2 plates (1 fold.).
Mp-30385.
Copies: MBAt; MdHi; NN; NNC; PPAmP; PPL; RPJCB.

2383. Gibson, Robert.
A treatise of practical surveying; which is demonstrated from its first principles. Wherein every thing that is useful and curious in that art, is fully considered and explained. Particularly three new and very concise methods for determining the areas of right-lined figures, arithmetically or by calculation, as well as the geometrical ones heretofore treated of.... The 7th ed.... With alterations and amendments, adapted to the use of American surveyors. Philadelphia: Printed for Joseph Crukshank, 1796. viii, 288, [1], 90, [4], 60 p., 13 fold. plates.
B-9556; Mp-47790.
Copies: CtY; DLC; DeGE; MWA; MiU; MnU; NjP; OSW; P; PPL; PU; TKL.

2384. Love, John.
Geodaesia; or, The art of surveying, and measuring land made easy.... The 13th ed., adapted to American surveyors. New York: Printed and sold by Samuel Campbell, 1796. [2], [6], [8], 9–189, [16], 3, [36], 7 p., diagrs., tables.
First published in London, 1688.
Mp-30704.
Copies: DLC; MWA; MiU-C; NN; NNC; NNE.

2385. Moore, Samuel, 1737?–1810.
An accurate system of surveying; in which is contained, I. Decimal fractions.... II. The extraction of the squareroot. III. Plane trigonometry.... IV. An exact method to cast up the contents of lands. V. Field surveying. The whole being performed without the use of scale and compasses, or a table of logarithms. In which is given, some account of the variation of the needle, and the causes of its attraction. Litchfield, [Conn.]: Printed by T. Collier, 1796. x, [3], 131 [132], [1] p., illus.
Mp-30810.
Copies: CtHi; CtY; DLC; DeGE; MH; MWA; NN; PPL; RPJCB.

2386. [Redfield, Nathan.]
A treatise of surveying, containing both the theory and practice of that art.... [Hartford?]: 1796.
E-31076.
No copy known.

2387. [Robertson, John] 1712–1776.
Tables of difference of latitude and departure.... Philadelphia: Printed for, and sold by Joseph Crukshank, 1796. [1], 90 p.
Also found bound with Gibson, Robert, *A Treatise of practical surveying*, 1796.
Mp-47790.
Copies: CtY; DLC; DeGE; MWA; MiU; MnU; NjP; OSW; P; PPL; PU; TKL.

2388. Gibson, Robert.
A treatise of practical surveying; which is demonstrated from its first principles. Wherein every thing that is useful and curious in that art, is fully considered & explained.... The whole illustrated with copper-plates. The 8th ed. ... with alterations and amendments, adapted to the use of American surveyors. New York: Printed by William A. Davis & Co., for Gaine & Ten Eyck, [etc.], 1798. 452 p., 13 plates.
Mp-33795.
Copies: DLC; DSI; MWA; MiU-C; MnU; NIC; NN; RPB; ViU.

2389. Dewey, Solomon, 1750–1813.
A short and easy method of surveying; adapted to every capacity; designed chiefly for schools and common practice: to which is added the square root.... Hartford: Printed by Elisha Babcock, 1799.[4], 5–24 p.
Mp-35406.
Copies: CtHi; DLC; DeGE; MWA.

2390. Jess, Zachariah.
A compendious system of practical surveying, and dividing of land: concisely defined, methodically arranged, and fully exemplified. The whole adapted for the easy and regular instruction of youth, in our American schools. Compiled by Zachariah Jess.... Wilmington, [Del.]: Printed by Bonsal and Niles for the compiler, 1799. v, [1] p., 1 l., 12 [212] p., diagrs.
Mp-35670.
Copies: CU; DLC; DeGE; MH; MHi; MWA; NHi; NN; NcD; NjP; OClWHi; PPAmP; PPL; PU; RPJCB.

2391. Little, Ezekiel, 1762–1840.
The usher. Comprising arithmetic ... description and use of Coggeshall's sliding rule, some uses of Gunter's scale .. surveying; the surveyor's pocket companion....

2391 (*continued*)
Exeter, [N.H.]: Printed by H. Ranlet, 1799. 240 p.

Mp-35734.

Copies: CtY; DAU; IU; MWA; MiU-C; NN; PU; RPJCB.

2392. [Robertson, John] 1712–1776.

Tables of difference of latitude and departure: constructed to every quarter of a degree.... Wilmington, [Del.]: Printed by Bonsal and Niles, for Zachariah Jess, 1799. [1], 91, [3], 59, [1] p.

Usually bound with Jess, Z., *Compendious system of practical sruveying.*

Mp-36236.

Copies: DLC; DeGE; MH; MWA; NN; PPAmP; PPL; PU; RPJCB.

2393. Cook, David.

Cook's American arithmetic; being a system of decimal arithmetic; comporting with the federal currency of the United States of America. To which is annexed, by way of a second part, The American surveyor. The whole on an entire new plan. To which is added, a draught of instruments, adapted to carry the above into effect.... By David Cook, jun. New-Haven: Printed by Thomas Green and Son, 1800. 107, 9 p., 3 plates.

Mp-37246.

Copies: CtY; DLC; MWA.

2394. [Eaton, Amos] 1776–1842.

Art without science; or, The art of surveying unshackled with the terms and science of mathematics. Designed for farmers' boys. Hudson, N.Y.: Printed by Sampson, Chittenden & Croswell, 1802. 36 p., illus.

Attributed to Amos Eaton by Penrose R. Hoopes.

S&S-1770.

Copies: DeGE; MWA; NN.

2395. Gibson, Robert.

A treatise of practical surveying ... 2d New York ed. With alterations and amendments, adapted to the use of American surveyors. New York: Printed for E. Duyckinck, by L. Nichols, 1803. vii, [8]–452 p., 13 plates.

S&S-4277.

Copies: DLC; DeGE; InU; MB; MWA; NN; NNC; NNE; NjP.

2396. ———— The 8th ed.... With alterations and amendments, adapted to the use of American surveyors. Philadelphia: Printed and sold by Joseph & James Crukshank, 1803. viii, 288, 152 p., 13 fold. plates.

Tables of difference of latitude and departure, 152 p. at end.

S&S-4278.

Copies: DLC; IEN; IU; MWA; NNE; OCU; PHi; PPi; ViU.

2397. Flint, Abel, 1765–1825.

A system of geometry and trigonometry; together with a treatise on surveying.... Hartford: Printed for Oliver D. Cooke, by Lincoln & Gleason, 1804. viii, [9]–168 p., 4 fold. plates.

S&S-6311.

Copies: CSmH; CSt; CoCC; CtY; DAU; FMU; InU; MB; MH; MWA; MiU; NN; NNC; NbOU; NjR.

2398. Gibson, Robert.

A treatise of practical surveying ... 3d New-York ed. With alterations and amendments, adapted to the use of American surveyors. New York: Printed by Lewis Nichols for Evert Duyckinck, 1806. vii, [8]–452 p., 13 fold. plates.

S&S-10482.

Copies: MFiHi; MWA; MiU; NjP; ViU.

2399. United States. Congress. House. Committee of Commerce and Manufactures.

Report of the Committee of Commerce and Manufactures, who were directed, by a resolution of the House, of the fifth instant, to inquire into the expediency of authorizing the Secretary of the Treasury, to cause a survey to be made of the shoals of Cape Hatteras, Cape Look-Out and the Frying-Pan. February 27, 1806. Accompanying a bill directing the

Secretary of the Treasury to cause the coast of North Carolina, between Cape Hatteras and Cape Fear, to be surveyed.... City of Washington: A. & G. Way, printers, 1806. 6 p.
S&S-11716.
Copies: DLC; MWA; NN; PPAmP.

2400. Conway, Miles W.
Geodoesia; or, A treatise of practical surveying.... Made for the use of the Western surveyors in particular.... Lexington, [Ky.]: Printed by Daniel Bradford, 1807. 60 p., diagrs.
S&S-12342.
Copies: DLC.

2401. A traverse table to every degree and quarter degree of the compass or horizon, for the use of surveyors. New Brunswick, N.J.: Printed & published by William Elliot, 1807. [144] p.
S&S-13726.
Copies: CtY; DeGE; MWA; MiDSH; NN; NjR.

2402. United States. Commissioners for Surveying the Coast of North Carolina.
To the Honorable Albert Gallatin, Esq., Secretary of the Treasury of the United States. Sir, the Commissioners for Surveying the Coast of North Carolina have the honor to report: that on the twenty-eighth of May last, they met at Shell Castle.... January 27, 1807. [Washington: 1807.] 8 p.
S&S-14095.
Copies: MWA; NcU.

2403. Flint, Abel, 1765–1825.
A system of geometry and trigonometry; together with a treatise on surveying.... Likewise, rectangular surveying.... To the whole are added several mathematical tables, necessary for solving questions in trigonometry and surveying ... 2d ed. Compiled from various authors. Hartford: Printed for Oliver D. Cooke, by Lincoln & Gleason, 1808. viii, [9]–168 p., 4 fold. plates.
S&S-15023.
Copies: CSmH; CtHT; CtY; DLC; DeGE; MH; MWA; MeB; MiU-C; NN; OClWHi; OrP; PPF; RPB; ViU.

2404. Gibson, Robert.
A treatise of practical surveying ... 9th ed., with alterations and amendments, adapted to the use of American surveyors. Philadelphia: Printed by Joseph Crukshank, 1808. viii, 288, 152 p., 13 plates.
Mathematical tables, 152 p. at end.
S&S-15102.
Copies: CtY; MWA; MiU; NNC; PPL; PU; Vi.

2405. Gibson, Robert.
The theory and practice of surveying: containing all the instructions for the skilful [*sic*] practice of this art. By Robert Gibson.... The whole corr., newly arranged, and greatly enlarged ... by D. P. Adams.... New York: E. Duyckinck, 1811. 2 p.l., 324, 184 p., 13 fold. plates.
S&S-22908.
Copies: CtY; DeGE; MB; MH; MWA; MiU; NN; OClW; PU; ViU.

2406. United States. Treasury Dept.
Message from the President of the United States, transmitting a report of the Secretary of the Treasury, on the execution of the act providing for the survey of the coast of the United States. February 4th, 1811. Printed by order of the Senate of the United States. Washington: Printed by R. C. Weightman, 1811. 58 p., 2 plates.
S&S-24220.
Copies: DLC; MWA; RPB.

2407. United States. Treasury Dept.
Report of the Secretary of the Treasury, agreeably to the resolution of the Senate, of the 20th Dec. 1810, on the survey of the coasts of the U.S. February 4th, 1811. Printed by order of the Senate of the United States. Washington City: Printed by R. C. Weightman, 1811. 21 p.
Instructions for a survey of the coast of the Gulf of Mexico west of the Mississippi River.
S&S-24306.
Copies: DLC; MWA; RPB.

2408. Gibson, Robert.

The theory and practice of surveying . . . newly arranged, improved, and enlarged . . . by James Ryan. New York: E. Duyckinck, 1812. v, [2], 360, 184 p., 13 plates.

S&S-51202.

Copies: CtY; DLC; MH; MiU.

2409. Flint, Abel, 1765–1825.

A system of geometry and trigonometry: together with a treatise on surveying. . . . Likewise, rectangular surveying. . . . To the whole are added several mathematical tables, necessary for solving questions in trigonometry and surveying. . . . Compiled from various authors, by Abel Flint, A.M. 3d ed. Hartford: Oliver D. Cooke; Samuel T. Armstrong, printer, Boston, 1813. viii, [9]–168 p., 4 fold. plates.

S&S-28521.

Copies: CSt; CtHi; DeGE; ICU; MBAt; MH; MWA; MiU; MnU; MoS; PU; ViU; VtU; WM.

2410. Gibson, Robert.

The theory and practice of surveying; containing all the instructions requisite for the skilful [*sic*] practice of this art. By Robert Gibson. . . . Whole corrected, newly arranged and greatly enlarged with useful selections, and a new set of accurate mathematical tables. By D. P. Adams. . . . New York: Evert Duyckinck; G. Long, Printer, 1814. [4], 324, 184 p., 13 plates.

S&S-31580.

Copies: CSmH; IU; MB; MH; MWA; MiU; NbOU; RNR; ViU.

2411. Gummere, John, 1784–1845.

Solutions to the miscellaneous questions in Gummere's Surveying. By the author. Philadelphia: Kimber & Richardson, 1814. 15 p., fold. plate.

S&S-31624.

Copies: DeGE; KU; MWA.

2412. Gummere, John, 1784–1845.

A treatise on surveying, containing the theory and practice: to which is prefixed, a perspicuous system of plane trigonometry. . . . Philadelphia: Kimber & Richardson; W. Brown, printer, 1814. v. [6], 10–202, 152 p., 8 plates.

Mathematical tables, 152 p. at end.

S&S-31625.

Copies: DAU; InCW; MH; MNBedf; MWA; MiU; NN; OUrC; PPAmP; PPL; PSC-Hi; PU.

2413. Jess, Zachariah.

A compendious system of practical surveying, and dividing of land: concisely defined, methodically arranged, and fully exemplified. The whole adapted for the easy and regular instruction of youth, in our American schools. Compiled by Zachariah Jess. 2d ed., improved. Philadelphia: Johnson and Warner; Robert Porter, printer, Wilmington, Delaware, 1814. v, [3], 227 p., illus.

S&S-31827, 31828.

Copies: CtY; DLC; DeGE; DeHi; IU; MB; MWA; NIC; OClW; PHi; PPFHi; TNJ; TxU.

2414. [Robertson, John] 1712–1776.

Tables of difference of latitude and departure, constructed to every quarter of a degree, and continued from one to the distance of one hundred miles or chains. Philadelphia: Johnson & Warner, 1814. 1 p. l., 154 p.

S&S-32906.

Copies: DLC; DeGE; MWA.

2415. Gibson, Robert.

A treatise on practical surveying; which is demonstrated from its first principles. . . . Rev., corr. and adapted to the use of schools, American surveyors, &c. by John D. Craig. Baltimore; F. Lucas and J. Cushing, 1816. 320 [330], 152 p., illus.

Mathematical tables, 152 p. at end.

S&S-37705, 37706.

Copies: ICU; MB; MWA; MiD-B; ViU.

2416. United States. Treasury Dept.

Message from the President of the United States, transmitting a report of Secretary of the Treasury, relative to the measures which have been taken to complete an accurate survey of the coast of the United States; in compliance with a resolution of the Senate, of the 26th ult. April 4, 1816.... Washington: Printed by William A. Davis, 1816. 10 p., 3 fold. tables.

List of mathematical instruments and books acquired by F. R. Hassler in Europe, p. 6–10.

S&S-39560.

Copies: DLC; DNA; DeGE; MWA.

2417. Gummere, John, 1784–1845.

A treatise on surveying, containing the theory and practice; to which is prefixed a perspicuous system of plane trigonometry ... 2d ed., improved. Philadelphia: John Richardson; Griggs & Co., printers, 1817. vi, [7]–206, 152 p., 8 fold. plates.

Mathematical tables, 152 p., at end.

S&S-40968.

Copies: DLC; DeGE; MWA; NNC; PHC; PPAN; PWW.

2418. Fairlamb, Jonas Preston.

A new and concise method of completely obviating the difficulty occasioned in surveying land by local attractions of the magnetic needle, in most parts of the United States, by back observation.... Wilmington, Del.: Printed by W. A. Miller, 1818. 15 p., 2 fold. plates.

S&S-43987.

Copies: DLC.

2419. Flint, Abel, 1765–1825.

A system of geometry and trigonometry; together with a treatise on surveying ... 4th ed., rev. and corr. Hartford: Cooke & Hale, 1818. viii, [9]–168 p., 4 plates.

S&S-44054.

Copies: CSt; CtY; DLC; MH; MWA; MiU; NN; NNC; NbOU; NCU; NJP; OC; RNR; ScCleU; VtMiS.

2420. Gibson, Robert.

A treatise on practical surveying, which is demonstrated from its first principles.... Rev., corr. and adapted to the use of schools, American surveyors, &c. by John D. Craig. Baltimore: F. Lucas and J. Cushing, 1818. 326, 152 p., illus.

Mathematical tables.... Baltimore: 1819. 152 p. at end.

S&S-44156.

Copies: CoHi; InCW; MH; MWA; MiU; NIC; NNC; PPiPT; ViU.

2421. United States. President (Monroe).

Message from the President of the United States, transmitting copies of the reports in relation to the surveys and examinations made by naval officers in co-operation with officers of the Corps of Engineers. February 11, 1818.... Washington: Printed by E. De Krafft, 1818. 26 p., fold. table. (15th Cong., 1st sess. Senate. Doc. 104)

Concerns the eastern entrance to Long Island Sound, the harbors of Newport and Hampton Roads, and sites for a navy yard north of the Delaware River.

S&S-46429.

Copies: DLC; DeGE; DeU; MWA; N; R.

2422. United States. Treasury Dept.

Message from the President of the United States, transmitting, in obedience to a resolution of the Senate ... a report from the Secretary of the Treasury, respecting the progress made under the act to provide for surveying the coast of the United States. March 16, 1818.... Washington: Printed by E. De Krafft, 1818. 21 p. (15th Cong., 1st sess. Senate. Doc. 143)

Reports from F. R. Hassler on the progress of the survey of the coast: p. 8–20.

S&S-46433.

Copies: CSmH; DLC; DeGE; DeU; MWA; NjP.

2423. [Gibson, Robert.]

Mathematical tables: difference of latitude and departure; logarithms, from 1 to 1000; and artificial sines, tangents, and secants. Baltimore: Printed by T. H. Palmer for F. Lucas, Jun., 1819. 152 p.

Also issued as part of certain editions of his *A Treatise of practical surveying.*

S&S-48644.

Copies: CoHi; CtY; LNHT; MWA; NIC; NNC; NNE.

2424. Gummere, John, 1784–1845.

A treatise on surveying, containing the theory and practice: to which is prefixed a perspicuous system of plane trigonometry ... 3d ed., improved. Philadelphia: John Richardson, and Kimber and Sharpless, 1820. vi, [7]–206, 152 p., 8 fold. plates.

Mathematical tables, 152 p. at end.

S-1470, 1471 (called 2d ed.).

Copies: ABS; KWiF; MH; MWA; MiHM; MoMM; NN; NNC; PPL; PSC; TU; ViW.

2425. Gibson, Robert.

The theory and practice of surveying.... By Robert Gibson.... Newly arranged, improved and enlarged ... by James Ryan.... New York: E. Duyckinck, 1821. v, [2], 360, 184 p., 14 fold. plates.

S-5453.

Copies: CSmH; CtY; DLC; FTaSU; ICU; IaGG; KHi; MB; MdU; MiU; MoSU; NCaS; NbOU; NcD; ODW; PHi; ViU.

2426. Messinger, John, 1771–1846.

A manual, or hand book, intended for convenience in practical surveying; containing some explanation of the tables of natural numbers.... St. Louis: Printed by William Orr, 1821. 37, [7] p., illus.

S-6026.

Copies: CtY; IU.

2427. Gibson, Robert.

A treatise on practical surveying; which is demonstrated from its first principles ... 3d ed., rev., corr. and adapted to the use of schools, American surveyors, &c., by John D. Craig. Baltimore: F. Lucas, Jun., and Cushing and Jewett, 1822. 326, 152 p., diagrs.

Mathematical tables, 152 p. at and.

S-8825.

Copies: CtY; ICU; InLP; LNB; MdBP; MoS; NNC; NNE; NSyU.

2428. Hassler, Ferdinand Rudolph, 1770–1843.

Papers on various subjects connected with the survey of the coast of the United States. Extracted from the American philosophical transactions, vol. 2, new series. Philadelphia: Printed by Abraham Small, 1824. 200 p., 8 plates.

S-16480.

Copies: MA; NIC; NN; NjR.

2429. Villee, H. W.

Der Bauer als Landmesser; oder, Die practische Feldmesskunst; enthaltend: eine gründliche Unterweisung, Ländereyen, Höhen, Mauerwerke, Baumstämme und alle übrige körperliche Dinge zu messen und zu berechnen ... die spezifische Schwere der Producte, Wein- und Branntwein-Proben, etc.... Reading, [Pa.]: Gedruckt für den Verfasser, bey Johann Ritter, 1824. xi, [1], 154, [2] p., 2 fold. plates.

S-19146.

Copies: DLC; DeGE; MH; MWA; PLF; PPG; PPL; PPeSchw; PU.

2430. Flint, Abel, 1765–1825.

A system of geometry and trigonometry, together with a treatise on surveying ... 5th ed., with important additions, by George Gillet.... Hartford: Oliver D. Cooke & Co., 1825. 104, 137 p., diagrs.

S-20536.

Copies: CSt; CU; CtY; DeGE; GU; ICJ; IaK; MH; MU; NIC; NN; OMC; PU; TNJ; WaU.

2431. Gummere, John, 1784–1845.

Mathematical tables: differences of latitude and departure; logarithms.... Stereotype ed., carefully rev. and corr. Philadelphia: John Richardson; stereotyped by J. Howe, 1825. 152 p.

Usually issued as part of his *A Treatise on surveying.*

S-20770, 21372.

Copies: CtY; DLC; PPL.

2432. Gummere, John, 1784–1845.

A treatise on surveying, containing the theory and practice: to which is prefixed a perspicuous system of plane trigonometry. 4th ed., improved. Philadelphia: Kimber & Sharpless, [etc.], 1825. 216, 152 p., 8 fold.

plates.

Mathematical tables, 152 p. at end.

S-20771.

Copies: CtY; DLC; ICU; IaU; MWA; MdBL; MoS; NIC; NcU; NjR; OO; PPL; PU; WM; WvU.

2433. Gummere, John, 1784–1845.

A lecture, on land surveying, delivered November 23, 1825, in the chapel of Nassau Hall, before the Literary and Philosophical Society of New Jersey. Princeton: Princeton Press, printed for the Society, by D. A. Borrenstein, 1826. 14 p., diagr.

S-24764.

Copies: CtY; NjP; PHC; PU.

2434. Gummere, John, 1784–1845.

Solutions to the miscellaneous questions in Gummere's Surveying.... Philadelphia: Kimber & Sharpless, 1826. 15 p., 2 plates.

S-24765.

Copies: CtY; MWA; OO; PPL.

2435. Hassler, Ferdinand Rudolph, 1770–1843.

Corrections on the papers on the coast survey; published in the Philosophical transactions of Philadelphia. New York: F. R. Hassler; James Bloomfield, printer, 1826. 13 p.

S-24811.

Copies: MBAt; NIC; NjR; PU.

2436. Roberdeau, Isaac, 1763–1829.

Observations on the survey of the sea coast of the United States. [Washington: 1827.] 24 p.

At head of title: The following paper was read before the Columbia Institute on the 9th of December, 1826.

S-30467.

Copies: DLC; MWA; NN; PPAmP; PPL.

2437. Gibson, Robert.

The theory and practice of surveying; containing all the instructions requisite for the skilful [*sic*] practice of this art.... Newly arranged, improved, and enlarged, with useful selections, and a new set of accurate mathematical tables. By James Ryan.... New York: Printed by J.& J. Harper, for E. Duyckinck, Collins and Hannay, and O. A. Roorbach, [etc.], 1828. v, [7]–251, 184 p., front., 14 fold. plates.

S-33349.

Copies: DeGE; LNHT; MNBedf; OO; PPF; Vi.

2438. Gummere, John, 1784–1845.

A treatise on surveying, containing the theory and practice: to which is prefixed, a perspicuous system of plane trigonometry ... 5th ed., improved. Philadelphia: Kimber & Sharpless; Adam Waldie & Co., printers, 1828. 216, 152 p., 8 fold. plates.

Mathematical tables, 152 p. at end.

S-24766 (dated 1826), 33446.

Copies: CSt; DLC; In; MnU; NIC; NP; OCU; OOxM; PHC; PHi; PSC; ScC.

2439. United States. Navy Dept.

Survey, the coasts of the United States, submitted by Mr. Carter, from the Committee, on Naval Affairs, to accompany bill (No. 288) to amend the acts to provide for surveying the coasts of the United States. May 1, 1828.... Washington: Printed by Gales & Seaton, 1828. 12 p. (20th Cong., 1st sess. House. Doc. 264)

A report from the Navy Dept., including lists of surveying instruments.

S-36919.

Copies: DLC; DeGE; DeU; MWA; NjR.

2440. Hale, James.

Elements of geometry and trigonometry, with an easy and concise system of land surveying. Bellows Falls, Vt.: James I. Cutler, 1829. 115 p., illus.

S-38880.

Copies: CtY; DeGE; MH; MiU; VtHi.

2441. Davies, Charles, 1798–1876.

Elements of surveying. With the necessary tables.... New York: Printed and Published by J. & J. Harper; sold by Collins and Hannay, [etc.], 1830. 147, 62, 91 p., 9

2441 (*continued*)
fold. plates.
C-1094.
Copies: CSt; CtY; DLC; DeGE; MH; MiU; NN; NcD; NjR; PU.

2442. Flint, Abel, 1765–1825.
A system of geometry and trigonometry with a treatise on surveying.... By Abel Flint ... with improvements by George Gillet ... 6th ed., rev. and enlarged by the addition of copious notes and illustrations, and a concise treatise on logarithms, by Frederic A. P. Barnard.... Hartford: Cooke and Co., 1830. iv, [7]–112, 10, 62, 100 p.
C-1396.
Copies: Ct; DLC; DeGE; ICJ; MH; OClWHi; TxU.

3. City planning and improvements

2443. Ellicott, Andrew, 1754–1820.
A plan of the city of Washington, in the Territory of Columbia, ceded by the states of Virginia and Maryland to the United States of America, and by them established, as a seat of their government, after the year MDCCC. Philadelphia: Engraved by J. Thackara and J. Vallance, 1792. Map 51 x 68 cm.
Based on the plan begun by P. C. L'Enfant.
Mp-24296.
Copies: CtY; DLC; MWA; MiU-C; NNE; PPAmP; RPJCB.

2444. ________ Boston: Engraved by Saml. Hill, [1792]. Map 43.5 x 52.5 cm.
B-7987; Mp-46437.
Copies: DLC; MWA; MiU-C; RPJCB.

2445. L'Enfant, Pierre Charles, 1755–1825.
Plan of the city intended for the permanent seat of the government of the United States, projected agreeably to the direction of the President of the United States, in pursuance of an act of Congress, passed on the 16th of July, 1790 "establishing the permanent seat on the banks of the Potowmack." By Peter Charles L'Enfant. Philadelphia: Printed by John Dunlap, 1792.
The Evans entry (E-24471) erroneously refers to the manuscript map now at the Library of Congress, cf. Richard W. Stephenson, "The Delineation of a Great Plan," *The Quarterly Journal of the Library of Congress,* v. 36, no. 3 (1979) p. 217.

2446. [Ellicott, Andrew] 1754–1820.
Plan of the city of Washington in the territory of Columbia.... Rollinson sculp.t. New York: I. Reid, L. Wayland and C. Smith, [1795?]. Map 41.5 x 57.5 cm.
B-9088; Mp-47413.
Copies: CSmH.

2447. Essai sur la ville de Washington. Par un citoyen des Etats Unis. New York: De l'imprimerie de J. Delafond, 1795. 19 p.
Concerns the plan proposed by L'Enfant.
Mp-28637.
Copies: DLC; MWA; NN.

2448. United States. Commissioners Appointed Under the Act for Establishing the Temporary and Permanent Seat of the Government of the United States.
Message from the President of the United States, inclosing a memorial of the Commissioners appointed under the "Act for establishing the temporary and permanent seat of the government of the United States," representing the situation and circumstances of the city of Washington. 23d February,

1798.... Philadelphia: Printed by J. Gales, [1798]. 14 p.
Mp-34807.
Copies: DLC; MWA.

2449. Birch, William Russell, 1755–1834.
The city of Philadelphia, in the state of Pennsylvania, North America, as it appeared in the year 1800; consisting of twenty eight plates. Drawn and engraved by William Birch & son. Springland Cot, near Neshaminy Bridge and the Bristol Road, Pa.: 1800. 3 p.l., 29 col. plates.
The individual plates dated from 1798 to 1800.
Mp-38259.
Copies: DLC; ICN; MWA; NHi; RNHi.

2450. Hints towards promoting the health and cleanliness of the city of New York. New York: Printed by T. & J. Swords, 1802. 16 p., illus.
S&S-2403.
Copies: DNLM; MB; NSm; NjP.

2451. New York (City). Common Council.
Report of the Committee of the Common Council upon the subject of new streets. New York: Printed by James Cheetham, 1803. 14 p.
S&S-4753.
Copies: DLC.

2452. United States. Congress. House. Committee on Claims.
Report of the Committee of Claims ... on the petition of Peter Charles L'Enfant ... 17th February, 1803. Washington: Printed by William Duane & Son, 1803. 7 p.
Concerns the planning and laying out of the city of Washington.
S&S-5413.
Copies: DLC; MWA; NN; RPB.

2453. United States. Dept. of State.
Letter from the Secretary of State, transmitting sundry documents relative to the claim of Peter Charles L'Enfant, for planning and laying out the city of Washington ... 3d January, 1803.... Washington City: Printed by William Duane & Son, 1803. 11 p.
S&S-5326.
Copies: DLC; DeGE; MWA; NN.

2454. New York (City). Street Commissioner.
Observations on the improvements of the city of New-York south of Grand-Street. Submitted to the board by the Street Commissioner. New York: Printed by James Cheetham, 1806. 9 [8] p.
Last page numbered "9".
John S. Hunn, Street Commissioner.
S&S-11001.
Copies: MWA; N; NHi.

2455. Boston. City Council.
Report. The following report ... relative to the continuation of Front Street, is printed.... Boston: 1809. Broadside.
Dated: November 6, 1809.
S&S-17064.
Copies: MH; MHi; MWA (xerox copy); PHi.

2456. The rectifier; or, A survey of the streets of the city of Philadelphia. [By] A citizen. Philadelphia: Printed for the author, by Ann Cochran, 1813. [3]–21 p.
Attributed to Paul Beck by the Historical Society of Pennsylvania.
S&S-29628.
Copies: PHi; PPM.

2457. United States. Commissioner of the Public Buildings.
Letter from the Commissioner of the Public Buildings, transmitting an estimate for enclosing and improving the Capitol square. February 3, 1817.... Washington City: Printed by William A. Davis, 1817. 5 p. (14th Cong., 2d sess. House. Doc. 64)
Commissioner: Samuel Lane.
S&S-42615.
Copies: DLC; DeGE; G; MWA; MeB; MiD-B.

2458. [Beck, Paul] 1760?–1844.

A proposal for altering the eastern front of the city of Philadelphia, with a view to prevent the recurrence of malignant disorders, on a plan conformable to the original design of William Penn. By a citizen of Philadelphia. Philadelphia: Printed and published by William Fry, 1820. 11 p., fold. plan.

Signed: Paul Beck, Jr.

Statement by William Strickland, architect, p. 10–11.

S-358.

Copies: DLC; DeGE; DeWi; MB; MWA; PHC; PPAmP; PPL; PU; RPB.

2459. The stranger's guide. Containing a list of the streets ... wharves, roads, institutions, public buildings ... in the city of Philadelphia, and the villages of Mantua and Hamilton.... [Philadelphia]: R. Desilver, [182–]. 64 p. fold. plan.

S-3352.

Copies: DLC.

2460. Maryland. Commissioners on the Plan of the City of Baltimore.

Plan of the city of Baltimore, compiled from actual survey made under the direction of the Commissioners appointed by the Legislature of Maryland and by Lewis Brantz, Esq., under the authority of the Mayor and City Council of Baltimore. Engraved by B. T. Welch & Co. Baltimore: F. Lucas, Jr., 1822. Map.

S-9304.

Copies: MdBE; NjP; PHi.

2461. United States. Laws.

A bill to authorize the paving of Pennsylvania Avenue. Feb. 26, 1822. In Senate of the United States. [Washington: 1822.] 2 p. (S. 55)

S-10737.

Copies: DNA.

2462. Maryland. Commissioners on the Plan of the City of Baltimore.

Plan of the city of Baltimore, as enlarged & laid out under the direction of the Commissioners appointed by the General Assembly of Maryland in Feb'y, 1818.... By ... T. H. Poppleton, surveyor to the board. New York: Harrison, 1823. Plan 110 x 126 cm.

S-13235.

Copies: MB; MdBE; MdBP.

2463. Boston. City Council. Joint Committee on Streets.

[Report of the Joint Committee on a prospective plan and elevation of all the streets in the city. Boston: 1824.] 4 p.

S-15505.

Copies: MBAt.

2464. New York (City). Board of Aldermen. Committee on Streets.

Report of the Street Committee on the regulations of streets and avenues, from First-Street to Fourteenth-Street, and from Broadway to the East River. New York: P. & H. Van Pelt,[1826]. 12 p.

N-N0169081; S-25540.

Copies: NN.

2465. New York (City). Board of Aldermen. Committee on Streets.

Report of the Street Committee, on the subject of regulating the ground between North & Fourteenth Streets, the Bowery & East River. New York: Printed by W. Davis, 1826. 54 p.

N-N0169082.

Copies: CtY.

2466. The picture of New-York, and Stranger's guide to the commercial metropolis of the United States. New York: A. T. Goodrich, [1828]. viii, 492 p., illus., plates, fold. plan

Attributed to Samuel Latham Mitchill.

S-34796.

Copies: CSmH; DLC; IaDaM; NN; NNC; NNS; PHi; TxU.

4. Architecture and construction of buildings

2467. Massachusetts (Colony). Laws.
At a Special General Court held at Boston, Novemb. 7, 1683. [An order regulating the construction of new buildings.] [Cambridge: Printed by S. Green, 1683.] Broadside 29 x 17 cm.
Mp-347.
Copies: MBAt.

2468. Massachusetts (Colony). Laws.
At a General Court on adjournment. Held at Boston, Feb. 13, 1684. That for encouragement to build in Boston with brick and stone according to a law made the last session; as an addition to the said law.... [Cambridge: S. Green, 1684.] Broadside 23 x 14.5 cm.
Mp-365.
Copies: MBAt.

2469. [Franklin, Benjamin] 1706–1790,
Some account of the Pennsylvania Hospital; from its first rise to the beginning of the fifth month, called May, 1754. Philadelphia: Printed by B. Franklin, and D. Hall, 1754. 40 p.
Mp-7197.
Copies: DLNM; DeGE; MHi; MWA; MiU-C; NcD-M; PPL.

2470. ———— Philadelphia: Printed by B. Franklin and D. Hall, 1754. 77 p., 4 plates.
"Continuation of the account of the Pennsylvania Hospital; from the first of May, 1754 to the fifth of May, 1761.... Philadelphia: Printed by B. Franklin and D. Hall, 1761," p. [39]–77.
N-F0340828.
Copies: CU; CtHT-W; DLC; DNLM; ICJ; MB; MWA; NN; PPL; PU; RPJCB.

2471. Swan, Abraham.
The British architect; or, The builder's treasury of stair-cases. Containing, I. An easier, more intelligible and expeditious method of drawing the five orders, than has hitherto been published, by a scale of twelve equal parts.... II. Likewise stair cases ... shewing their most convenient situation.... With a great variety of curious ornaments.... III. Designs of arches, doors, windows.... IV. A great variety of new and curious chimney-pieces.... V. Corbels, shields and other beautiful decorations. VI. Several useful and necessary rules of carpentry ... The whole being illustrated with upwards of one hundred designs and examples.... Philadelphia: Printed by R. Bell, for John Norman, 1775. vi, 17, [1] p., 60 plates.
First published in London, 1745.
B-4124; Mp-42944.
Copies: DLC; DeWint; MWA; MiU-C; N; NHi; NNC; PHi; PPF; RPJCB.

2472. Swan, Abraham.
A collection of designs in architecture, containing new plans and elevations of houses for general use, with a great variety of sections of rooms; from a common room, to the most grand and magnificent. Their decorations, viz. bases, surbases, architraves, freezes, and cornices, properly inriched with foliages, frets and flowers.... To which are added, curious designs of stone and timber bridges.... Likewise some screens and pavilions. In two volumes.... Designed by Abraham Swan ... and engraved by John Norman. Vol. I. Philadelphia: Printed by R. Bell, 1775. [2], viii p., 10 plates.
No more published.
Originally published in London, 1757.
Mp-14481.
Copies: NN.

2473. [Norman, John] 1748–1817.

The town and country builder's assistant.... Explaining short and easy rules ... for drawing and working, the five orders of columns entire: or any part of an order, without regard to the module or diameter. And to enrich them with their rusticks, flutings, cablings, dentules, modillions, &c. Also to proportion their doors, windows, intercolumnations, portico's and arcades.... The whole illustrated by upwards of 200 examples, engraved on 60 folio copper-plates. By a lover of architect. Boston, N.E.: Engraved, printed and sold by J. Norman, [1786]. 12, [1] p., 60 plates.

Mp-20027.

Copies: DLC; MB; MHi; MWA; NN; RPJCB.

2474. Pain, William, 1730–1790.

The practical builder; or, Workman's general assistant ... with plans and elevations of gentlemen's and farm-houses, barns, &c. ... [First American from the] fourth [London] ed., rev. and corr. by the author.... Boston: Printed and sold by John Norman, 1792. 8 p., 83 plates.

First published in London, 1774.

Mp-25956.

Copies: CtY; DeWint; MH; MWA; NN; PPL; RPJCB; ViU.

2475. Clark, Joseph, d. 1798.

Polymathy; or, The American builder. A work calculated equally to edify and entertain the gentleman farmer, surveyor, builder, and mechanic. Annapolis: Printed by F. and S. Green, 1794.

E-26770.

No copy known.

2476. [Norman, John] 1748–1817.

The town and country builders' assistant. Absolutely necessary to be understood by builders and workmen in general.... The whole illustrated by upwards of 200 examples on 60 copper-plates. By a lover of architecture. Philadelphia: Printed by Benjamin Johnson, 1794.

E-27810.

No copy known.

2477. Pain, William, 1730–1790.

The builder's pocket-treasure. In which not only the theory, but the practical parts of architecture are carefully explained. Illustrated with useful designs of frontispieces, chimney-pieces, &c. Bases, capitals and entablatures at large. Architrave frontispieces, cornices and moldings for the inside rooms, &c. The construction of stairs, with their ramp and twist rails.... A new ed. London, printed. Boston: Reprinted and sold by William Norman, 1794. iv, 20 p., 15 plates (55 illus.) on 22 leaves.

Misdated by Evans: 1793; first published in London, 1763.

Mp-25955.

Copies: CtY; MB; MnU; NHi; NIC; NNC-A; NNMM; PPL; RPJCB.

2478. Swan, Abraham.

The British architect; or, The builder's treasury of stair-cases. Containing, I. An easier, more intelligible, and expeditious method of drawing the five orders.... II. Likewise stair cases ... shewing their most convenient situation, and the form of their ascending in the most grand manner.... III. Designs of arches, doors, and windows. IV. A great variety of new and curious chimney-pieces.... V. Corbels, shields, and other beautiful decorations. VI. Several useful and necessary rules of carpentry.... Boston: Printed by John W. Folsom, for John Norman, 1794. iv, 12 p., 60 plates.

Mp-27761.

Copies: MB; MWA; NHi; NNMM; PPF.

2479. Philadelphia. Ordinances.

An ordinance to prevent the erection of wooden buildings within certain limits of the city of Philadelphia.... [Philadelphia: 1796.] Broadside.

Dated: Sixth day of June, 1796.

Mp-30995.

Copies: PHi.

2480. United States. Commissioners Appointed Under the Act for Establishing the

Temporary and Permanent Seat of the Government of the United States.

Letter from the Secretary of the Treasury, transmitting the copy of a letter from the Commissioners appointed under the act "for establishing the temporary and permanent seat of the government of the United States"; with sundry documents.... [Philadelphia: 1796?] 26 p., folded table.

"29th of December, 1796, ordered to lie on the table."

Report on the progress and expenditures in the construction of government buildings.

Mp-33086.

Copies: DLC; MWA; RPJCB.

2481. United States. Commissioners Appointed Under the Act for Establishing the Temporary and Permanent Seat of the Government of the United States.

Message from the President of the United States, accompanying a memorial of the Commissioners appointed by virtue of the Act for establishing the temporary and permanent seat of the government of the United States. 8th January, 1796.... [Philadelphia: 1796.] 13 p.

Report of the House committee on the memorial, p. 8–13.

Reports on the progress made in erecting the government buildings and in laying out the city.

Mp-31406.

Copies: CSmH; MBAt; MWA; NN; PPL; RPB.

2482. Benjamin, Asher, 1773–1845.

The country builder's assistant: containing a collection of new designs of carpentry and architecture, which will be particularly useful to country workmen in general. Illustrated with new and useful designs of frontispieces, chimney pieces, &c. Tuscan, Doric, Ionic, and Corinthian orders, with their bases, capitals, and entablatures; architraves for doors, windows, and chimneys ... the construction of stairs.... Greenfield, [Mass.]: Printed by Thomas Dickman, 1797. [32] p., 30 plates.

Mp-31797.

Copies: IU; MH; MWA; NNC; OO; RPJCB.

2483. United States. Commissioners Appointed Under the Act for Establishing the Temporary and Permanent Seat of the Government of the United States.

Letter from the Secretary of the Treasury, transmitting a copy of a letter from the commissioners of the city of Washington, in the District of Columbia; including sundry documents marked A,B,C,D and E ... 3d of June. Ordered to lie on the table.... [Philadelphia: 1797.] 10 p.

Report on the expenditures on work and materials for the government buildings.

Mp-33085.

Copies: MBAt; MWA.

2484. Benjamin, Asher, 1773–1845.

The country builder's assistant, fully explaining the best methods for striking regular and quirked mouldings; for drawing and working the Tuscan, Doric, Ionic and Corinthian orders with their pedestals, bases, capitals and entablatures, architraves for doors, windows and chimneys.... The construction of stairs and their ramp and twist rails.... Plans and elevations of houses.... Boston: Printed by Spotswood and Etheridge, for the author, 1798. [36] p., 37 plates.

Mp-33399.

Copies: CtY; DLC; DeWint; ICJ; KU; MBAt; MH; MWA; MiU-C; NHi; NNC; PPL; RPJCB.

2485. Bordley, John Beale, 1727–1804.

Country habitations. [Philadelphia?: 1798.] 13 p., plate.

Mp-33435.

Copies: DLC; PPL.

2486. Latrobe, Benjamin Henry, 1764–1820.

Designs of a building proposed to be erected at Richmond in Virginia, to contain a theatre, assembly-rooms, and an hotel, by B. Henry Latrobe Boneval.... Begun Decr. 2d, 1797, finished Janr. 8th 1798. [Richmond?: 1798.] 2 p.l., 12 plates (part col., incl. plans).

2486 (*continued*)
N-L0122662.
Copies: DLC.

2487. Latrobe, Benjamin Henry, 1764–1820.

Designs of buildings erected or proposed to be built in Virginia, by B. Henry Latrobe Boneval. From 1795–1799. Philadelphia: 1799. 1 p.l., 2 p., 25 plates (part col., incl. plans).

N-L0122664.
Copies: DLC.

2488. United States. Commissioners Appointed Under the Act for Establishing the Temporary and Permanent Seat of the Government of the United States.

Letter from the Commissioners of the city of Washington, to the President of the United States, with an account of the present state of the public buildings in that city.... Published by order of the House of Representatives. [Philadelphia: 1799.] 9 p., 2 fold. tables.

Dated: 5th December, 1799.

Report on the progress and expenditures in the construction of government buildings. Reports by James Hoban, p. 5–9.

Mp-36525.
Copies: DLC; MBAt; MWA; NN; PPL; RPJCB.

2489. United States. Commissioners Appointed Under the Act for Establishing the Temporary and Permanent Seat of the Government of the United States.

Letter from the Secretary of the Treasury, transmitting copies of two letters from the Commissioners of the city of Washington, with sundry documents. Exhibiting the view of the receipts and expenditures ... and the progress and state of all the business and funds ... 5th December, 1799.... [Philadelphia]: Printed by W. Ross, [1799]. 17 p.

Includes reports of James Hoban with detailed descriptions of the progress in the construction of public buildings.

Mp-36539.
Copies: DLC; MB; MWA; NjP; PPL; RPJCB.

2490. Benjamin, Asher, 1773–1845.

The builder's assistant; fully explaining the best methods for striking regular and quirked mouldings ... 3d ed. Greenfield, [Mass.]: Printed by Thomas Dickman, 1800. [36] p., 38 plates (2 folded).

Mp-36941.
Copies: MWA; NHi; PP.

2491. Langley, Batty, 1696–1751.

The builder's jewel, the youth's instructor and workman's remembrancer.... Illustrated by upwards of 200 examples, engraved on 100 copper plates. The 1st American ed. Charlestown, [Mass.]: Printed by S. Etheridge, for Samuel Hill [Hall], [1800]. 46 p., 100 plates.

First published in London, 1741.
Mp-37778.
Copies: Iu; MH; MHi; MWA; MiU-C; NHi; NN; NNC; NNNM; PP; RPJCB.

2492. United States. Commissioners Appointed Under the Act for Establishing the Temporary and Permanent Seat of the Government of the United States.

Message from the President of the United States, transmitting a representation of the Commissioners of the city of Washington.... 30th Jan., 1801. [Washington: 1801.] 51 p.

Concerns the construction of buildings.
S&S-1514.
Copies: DLC; MWA; NN; RPB.

2493. Washington Building Company.

The constitution; or, Articles of agreement of the Washington Building Company.... Washington: Printed by Way and Groff, 1801. 16 p.

S&S-1609.
Copies: MB.

2494. Latrobe, Benjamin Henry, 1764–1820.

Design of a city hall proposed to be built in New York. By. B. H. Latrobe.... Philadelphia: 1802. 1 p.l., 27 plates (part col., incl. plans).

N-L0122663.
Copies: DLC.

2495. Moore, Thomas, 1760–1822.
An essay on the most eligible construction of ice-houses. Also, a description of the newly invented machine called the refrigerator. By Thomas Moore. Baltimore: Printed by Bonsal & Niles, 1803. 28 p.
S&S-4677.
Copies: DLC; DeGE; MH; MWA; MdHi; NN; PPAmP; ViU.

2496. [Evans, John.]
The builder's universal price book; containing a correct list of prices to the works of the several artificers concerned in building.... With observations on erecting sound and permanent buildings. Washington: Printed for the author, 1804. vi, [3], 10–44 p.
"Advertisement" signed: John Evans.
S&S-6263.
Copies: DLC; MH-BA; MWA; OHi.

2497. United States. Congress. House. Committee on Public Buildings.
Report of the Committee to enquire into the expediency of making provision ... for the completion of the public buildings belonging to the United States, near Philadelphia. 24th December, 1804.... [Washington: 1804.] 5 p.
Concerns the United States arsenal.
S&S-7595.
Copies: DLC; NN.

2498. United States. Congress. House. Committee on Public Buildings.
Report of the Committee to whom was referred ... the message of the President of the United States, communicating a report of the Surveyor of the Public Buildings at the city of Washington, 6th March, 1804 ... [Washington: 1804.] 14 p.
Letter from B. Henry Latrobe, Surveyor of Public Buildings, p. [7]–14.
S&S-7623.
Copies: DLC; NN; RPB.

2499. United States. Surveyor of Public Buildings.
Message from the President of the United States, communicating a report of the Surveyor of the Public Buildings at the city of Washington, on the subject of the said buildings ... 6th December, 1804. Washington City: Printed by William Duane & Son, 1804. 13 p.
Report signed: B. Henry Latrobe, Surveyor of the public buildings.
S&S-7546.
Copies: DLC; MWA; NN; PPAmP; ViU.

2500. United States. Surveyor of Public Buildings.
Message from the President of the United States, transmitting a report of the Surveyor of the Public Buildings at the city of Washington. 22d February, 1804. [Washington: 1804.] 10 p.
B. H. Latrobe, Surveyor.
S&S-7553, 7554.
Copies: DLC; NN; ViU.

2501. Benjamin, Asher, 1773–1845.
The country builder's assistant: fully explaining the best methods for striking regular and quirked mouldings: for drawing and working the Tuscan, Doric, Ionic, and Corinthian orders ... plans and elevations of houses.... Also, the tracing of groins, angle brackets, circular soffits in circular walls, &c. Correctly engraved on thirty-seven copperplates with a printed explanation to each. ... Greenfield, Mass.: Printed by John Denio, 1805. [36] p., 38 plates.
S&S-7974.
Copies: DeGE; DeWin; IU; MWA; NN; NNC; PP.

2502. Johnson, Stephen William.
To the public.... Proposals for publishing by subscription, Rural practices; or, A treatise on the mode of erecting dwelling–houses, public or private buildings ... in pisé ... With a treatise on turnpike roading.... By S. W. Johnson.... New Brunswick, N.J.: 1805. Broadside, illus. 33 x 19 cm.
Copies: DeGE.

2503. [Thornton, William] 1759–1828.
To the members of the House of Representatives of the United States. [Washington: 1805.] 12 p.
Signed: W. Thornton, and dated: City of Washington, Jan. 1, 1805.
A reply to B. H. Latrobe's criticisms on Thornton's plan for the Capitol.
S&S-9477, 9636.
Copies: DLC; MBAt; MH; MiU-C.

2504. United States. Surveyor of Public Buildings.
Letter from the Surveyor of the Public Buildings at the city of Washington, to the chairman of the committee appointed the seventeenth ultimo, on a message from the President of the United States, on the subject of said buildings. Accompanying a bill making an appropriation for compleating the south wing of the Capitol ... 9th January, 1805. [Washington: 1805.] 5 p.
Surveyor: B. H. Latrobe.
S&S-9561.
Copies: DLC; MWA; NcD.

2505. United States. Surveyor of Public Buildings.
Message from the President of the United States, communicating a report of the Surveyor of the Public Buildings at the city of Washington, on the subject of the said buildings ... December 27, 1805. Washington: A. & G. Way, printers, 1805. 12 p.
Surveyor: B. H. Latrobe.
S&S-9566.
Copies: DLC; MWA; NN; NjP; ViU.

2506. Andrews, George.
To builders. Composition ornaments for sale.... Boston: 1806. Broadside.
Dated: June 16, 1806.
S&S-9845.
Copies: MWA; NN.

2507. Benjamin, Asher, 1773–1845.
The American builder's companion; or, A new system of architecture; particularly adapted to the present style of building in the United States of America.... By Asher Benjamin ... And Daniel Raynerd.... Boston: Etheridge and Bliss; S. Etheridge, printer, 1806. viii, [9]–70 p., 44 plates.
S&S-9956.
Copies: CSmH; DLC; DeWint; IU; MB; MH; MWA; MiU-C; NWM; PPL; WU.

2508. Johnson, Stephen William.
Rural economy: containing a treatise on pisé building; as recommended by the Board of Agriculture in Great Britain, with improvements by the author; on buildings in general; particularly on the arrangement of those belonging to farms: on the culture of the vine; and on turnpike roads.... By S. W. Johnson. New Brunswick, N.J.: Printed by W. Elliot for I. Riley & Co., New York, 1806. 3 p.l., [v]–viii, 246, [4] p., plates.
S&S-10644.
Copies: CSmH; DLC; DeGE; ICA; KU; KyLx; MBAt; MH; MWA; MiU; NN; NjR; PPL; ViU.

2509. Latrobe, Benjamin Henry, 1764–1820.
A private letter to the individual members of Congress, on the subject of the public buildings of the United States at Washington, from B. Henry Latrobe, surveyor of the public buildings. Washington City: Printed by Samuel H. Smith, 1806. 32, 6 p., fold. plan.
S&S-10705.
Copies: DLC; KyLx; MBAt; MH; MWA; MdHi; NN; PHi; ViU.

2510. Massachusetts. State Prison. Board of Visitors.
An account of the Massachusetts State Prison. Containing a description and plan of the edifice; the law, regulations, rules and orders: with a view of the present state of the institution. By the Board of Visitors. Charlestown: Printed by Samuel Etheridge, 1806. 47, [1] p., 2 fold. plates.
The plates represent the floor plan and the front elevation of the building, accompanied with a description in the text.
S&S-10817.

Copies: DeGE; MWA; MWiW; MiU-C; MnHi; NN; NcD; ViW.

2511. United States. Surveyor of Public Building.
Message from the President of the United States, communicating a report of the Surveyor of the Public Buildings at the city of Washington ... December 15, 1806.... City of Washington: A. & G. Way, printers, 1806. 11 p.
B. H. Latrobe, Surveyor.
S&S-11631.
Copies: CtY-M; DLC; MWA; MiU-C; NN; ViU.

2512. United States. Surveyor of Public Buildings.
Letter and report from B. H. Latrobe, Surveyor of the Public Buildings ... on the subject of the said buildings. 28th October, 1807. City of Washington: A. & G. Way, printers, 1807. 7 p.
S&S-13938.
Copies: DLC; MBAt; MWA; NN.

2513. Birch, William Russell, 1755–1834.
The country seats of the United States of North America, with some scenes connected with them.... Designed and published by W. Birch.... Springland near Bristol, Pa.: 1808. [2] p., 4 l., 19 plates.
N-B0498565.
Copies: DLC; MWA (20 plates).

2514. United States. Surveyor of Public Buildings.
Message from the President of the United States, communicating the report of the Surveyor of the Public Buildings at the city of Washington. March 25, 1808.... Washington City: A. & G. Way, printers, 1808. 16 p.
B. H. Latrobe, Surveyor.
Copies: DLC.

2515. ———— April 5, 1808.... Washington City: A. & G. Way, printers, 1808. 16 p.
S&S-16467.
Copies: DLC; MWA; NN; PPAmP.

2516. United States. Surveyor of Public Buildings.
Message from the President of the United States, transmitting a report of the Surveyor of the Public Buildings of the United States, in the city of Washington. December 1, 1808.... City of Washington: A. & G. Way, printers, 1808. 11 p.
B. H. Latrobe, Surveyor.
S&S-16477.
Copies: DLC; MWA; NN.

2517. ———— December 1, 1808. Revised.... Washington: A. & G. Way, printers, 1808. 11 p.
S&S-16478.
Copies: DLC; MWA.

2518. Architectural Library of Boston.
Catalogue of the Architectural Library of Boston. Boston: 1809. 5 p.
Bound with its Constitution.
Copies: MWA.

2519. Architectural Library of Boston.
The constitution of the proprietors of the Architectural Library of Boston. Instituted 15th November, 1809. Boston: Printed by T. Kennard, 1809. 9, 5 p.
Catalogue of the Architectural Library of Boston, 5 p. at end.
Copies: MWA.

2520. United States. Congress. House. Committee Appointed to Confer with the Surveyor of the Public Buildings.
Report of the Committee appointed to confer with the Surveyor of the Public Buildings ... February 18th, 1809.... Washington City: Printed by R. C. Weightman, 1809. 6 p.
B. H. Latrobe, Surveyor.
S&S-19043.
Copies: DLC; MWA; NN; NNS.

2521. United States. Surveyor of Public Buildings.
Letter from Mr. B. Henry Latrobe on

2521 (*continued*)
repairs in the temporary Senate chamber, June 13, 1809.... Washington City: Printed by R. C. Weightman, 1809. 5 p.
S&S-18918.
Copies: DLC; MWA; NNS.

2522. United States. Surveyor of Public Buildings.
Message from the President of the United States, transmitting a report of the Surveyor of the Public Buildings. December 21st, 1809.... Washington City: Printed by R. C. Weightman, 1809. 11 p.
B. H. Latrobe, Surveyor.
S&S-18971.
Copies: DLC; MWA; NNS; ViU.

2523. United States. Surveyor of Public Buildings.
Report of Mr. B. Henry Latrobe, on public buildings, June 13th, 1809. Washington City: Printed by R. C. Weightman, 1809. 5 p.
S&S-19037.
Copies: DLC; NNS.

2524. To the ... Senate of South-Carolina. The memorial of sundry inhabitants of the city of Charleston.... [S.l.: 1810.] Broadside 32 x 19 cm.
Asks for legislation requiring buildings to be constructed of brick and stone.
Hummel-2025.
Copies: ScAr.

2525. United States. Congress. House. Committee on Public Buildings.
Report of the Committee, on the President's message. communicating a report of the Surveyor of Public Buildings. January 11, 1810. Accompanying a bill making further appropriations for completing the Capitol.... Washington City: Printed by R. C. Weightman, 1810. [4] p.
S&S-21833.
Copies: DLC.

2526. United States. Surveyor of Public Buildings.
Report of the Surveyor of the Public Buildings. January 11, 1810. Accompanying a bill making further appropriations for completing the Capitol.... Washington City: Printed by R. C. Weightman, 1810. 2 p.
B. H. Latrobe, Surveyor.
S&S-21862.
Copies: MWA (xerox copy); NNS.

2527. Benjamin, Asher, 1773–1845.
The American builder's companion; or, A system of architecture, particularly adapted to the present style of building ... 2d ed., corr. and enl. Charlestown, [Mass.]: Printed by Samuel Etheridge, Jnr., 1811. 104 p., 59 plates.
S&S-22332.
Copies: DAIA; DeWint; IU; MB; MBAt; MWA; MiU-C; NNC; NcD; OC; ViU.

2528. Latrobe, Benjamin Henry, 1764–1820.
Anniversary address, pronounced before the Society of Artists of the United States ... on the eighth of May, 1811. [Philadelphia: 1811.] 32 p.
Contains references to architecture and buildings in America.
S&S-23189.
Copies: DLC; DeGE; MWA; MdHi; NjR; PPL.

2529. Massachusetts. State Prison.
Rules and regulations for the government of the Massachusetts State-Prison.... with a description of the edifice.... Boston: J. Belcher, 1811. 23 p.
S&S-23335.
Copies: DLC; MWA; NN.

2530. United States. President (Madison).
Message from the President of the United States, transmitting reports from the Superintendent of the City, and the Surveyor of the Public Buildings ... February 26, 1811.... Washington: A. & G. Way, printers, 1811. 14 p.

B. H. Latrobe, Surveyor.
S&S-24230.
Copies: DLC; MWA.

2531. ________ of the expenditure of the money, appropriated on the first of May, 1810, for completing the Capitol, and for other purposes ... January 15, 1811. Washington: A. & G. Way, printers, 1811. 16 p., fold. table.
S&S-24231.
Copies: DLC; MWA; RPB.

2532. United States. Surveyor of Public Buildings.
Message from the President of the United States, transmitting a report of the Surveyor of the Public Buildings, relative to the progress and present state of them. January 15, 1811. Washington: A & G. Way, printers, 1811. 11 p.
B. H. Latrobe, Surveyor.
S&S-24223.
Copies: DLC; MWA; NjR; RPB; ViU.

2533. Dyster, Joseph Joshua.
A prospectus of the newly-discovered American, Dysterian, iron architecture; as applicable to the erection of iron bridges, and various other architectural purposes: such as domes, saloons, halls, vaulted ceilings, arched cellars, gateways, fire-proof buildings, &c.... Philadelphia: [ca. 1812]. Broadside 46 x 29 cm.
Signed: Joseph Joshua Dyster, Philadelphia.
Copies: DeGE; PPAmP.

2534. United States. Laws.
A bill making an appropriation for the purpose of discharging all the outstanding claims for the construction and repair of the Capitol and the President's House; for the compensation of the late Surveyor of the Public Buildings, and for furniture for the different apartments of the Capitol. [Washington: 1812.] 2 p.
No. 113, April 24th, 1812.
S&S-27031.
Copies: DLC.

2535. Benjamin, Asher, 1773–1845.
The rudiments of architecture: being a treatise on practical geometry, on Grecian and Roman mouldings.... Also, on the origin of building, on the five orders of architecture, on their general and particular parts and embellishments; with examples for cornices, base and surbase mouldings, architraves, and stairs.... Boston: Printed for the author by Munroe and Francis, 1814. 96 p., 32 plates (1 double).
S&S-30848.
Copies: DLC; DeGE; ICA; MB; MBAt; MH; MWA; NN; NNC; NjP; OO; RP.

2536. Browne, Peter Arrell, 1782–1860.
A summary of the law of Pennsylvania, securing to mechanics, and others, payment for their labour, and materials, in erecting any house, or other building, within the city and county of Philadelphia, the borough of Erie, the borough of Lancaster, and the borough of Pittsburgh.... Philadelphia: J. Maxwell, printer, 1814. 26 p.
S&S-31017.
Copies: DLC; MH-L; NNC; PHi; PPB.

2537. United States. Congress. House. Committee on Rebuilding or Repairing the Public Buildings.
Report of the Committee appointed on the twentieth ult. to inquire into the expediency of rebuilding, or repairing the public buildings in the city of Washington ... November 21, 1814.... Washington City: Printed by Roger C. Weightman, 1814. 4 p.
S&S-33403.
Copies: DLC; MWA; O.

2538. View of the New-York state prison in the city of New-York. By a member of the institution.... New York: Printed by T. and J. Swords, 1815. 89 p., front.
S&S-36443.
Copies: DLC; MB; MBAt; NHi; NN; PPAmP; PU.

2539. Washington. Ordinances.
Regulations of buildings, etc. [Washington: 1815.] 15 p.

2539 (*continued*)
S&S-36485.
Copies: DLC.

2540. Benjamin, Asher, 1773–1845.
The American builder's companion; or, A system of architecture, particularly adapted to the present style of building ... 3d ed., corr. and enlarged.... Boston: R. P. & C. Williams, for the author; Thomas G. Bangs, printer, 1816. iv, [5]–104 p., 59 plates.
S&S-36927.
Copies: DLC; DeWint; MWA; NN; NWM; PMA.

2541. Massachusetts. State Prison.
Description and historical sketch of the Massachusetts State Prison.... Charlestown, [Mass.]: S. Etheridge, Jr., 1816. 38 p.
Prepared by Gamaliel Bradford, warden.
S&S-38194.
Copies: CSt; DLC; ICJ; MH; MWA; NcD.

2542. United States. Treasury Dept.
Message from the President of the United States, transmitting a report of the Secretary of the Treasury of certain expenses which have been incurred for public edifices and improvements in the city of Washington ... March 11, 1816.... Washington: Printed by William A. Davis, 1816. 6 p.
Includes specifications of public works for which money had been appropriated.
S&S-39558.
Copies: DLC; MWA.

2543. [Franklin, Benjamin] 1706–1790.
Some account of the Pennsylvania Hospital; from its first rise to the beginning of the fifth month, called May, 1754. Philadelphia: Printed at the office of the United States' gazette, 1817. 145 p.
"Continuation of the account of the Pennsylvania hospital ... to the fifth of May, 1761 ... ," p. [71]–145.
S&S-40847.
Copies: CtY-M; DLC; DNLM; GU; ICJ; MH; MWA; MdBJ; MnU; NcD-M; NjP; PPL; PU; ViU.

2544. Gram, Olaves Peter, b. 1779.
Letter to a friend, concerning the building of the Capitol in the city of Washington.... [Washington: 1817.] 6 p.
S&S-40946.
Copies: DLC.

2545. Latrobe, Benjamin Henry, 1764–1820.
Memorial of B. Henry Latrobe, surveyor of the Capitol of the United States. February 26, 1817. [Washington: 1817.] 5 p. (14th Cong., 2d sess. House. Doc. 98)
S&S-42647.
Copies: DLC; DeGE; MWA.

2546. United States. Congress. House. Committee on Public Buildings.
Report of the Committee on the Public Buildings. Made February 18, 1817.... [Washington: 1817.] 40 p. (14th Cong., 2d sess. House. Doc. 89)
S&S-42720.
Copies: DLC; DeGE; MWA.

2547. United States. Surveyor of Public Buildings.
In Senate of the United States. February 14, 1817.... Copy of the reports of Messrs. Latrobe and Hoban, on the public buildings. [Washington: 1817.] 19 p. (14th Cong., 2d sess. Senate. Doc. 101)
S&S-42539.
Copies: DLC; DeGE; MWA.

2548. Haviland, John, 1792–1852.
The builders' assistant, containing the five orders of architecture ... and a variety of mouldings, modillions & foliage ... with working drawings ... for the use of builders, carpenters, masons, plasterers, cabinetmakers, and carvers ... with 150 copperplates ... Philadelphia: J. Bioren, [etc.], 1818–1821. 3 v. and atlas of 150 plates.
House carpenters' book of prices and rules, for measuring and valuing all their different kinds of work.... Philadelphia: 1819. 50 p. at end of v. 3.
The first American architectural book to describe the Greek as well as the Roman

orders.

S&S-44282.

Copies: CSmH; CtY; DSI; DeGE; MWA; MiU-C; MoS; NNC-A; PPCC; PPL.

2549. United States. Commissioner of the Public Buildings.

Message from the President of the United States, transmitting a report of the Commissioner of the Public Buildings. November 26, 1818.... Washington: Printed by E. De Krafft, 1818. 15 p. (15th Cong., 2d sess. House. Doc. 8)

Commissioner: Samuel Lane. Includes also a report of C. Bulfinch, Architect of the Capitol, on the progress made in the construction.

S&S-46418.

Copies: DLC; DeGE; DeU; G; MWA; NjP; O; R.

2550. United States. Commissioner of the Public Buildings.

Message from the President of the United States, transmitting a statement of the expenditures upon the public buildings, and an account of their progress for the year 1818. February 16, 1818.... Washington: Printed by E. De Krafft, 1818. 18 p. (15th Cong., 1st sess. Senate. Doc. 109)

Commissioner: Samuel Lane.

S&S-46422.

Copies: DLC; DeGE; MH; MWA; O.

2551. United States. Congress. House. Committee on Public Buildings.

Report of the Committee on the expenditures of the public buildings. April 3, 1818.... [Washington: 1818.] 23 p. (15th Cong., 1st sess. House. Doc. 184)

Includes an account of work completed.

S&S-46547.

Copies: DLC; DeGE; DeU; G; MWA; NjP.

2552. United States. Congress. House. Committee on Public Buildings.

Report of the Committee on the Public Buildings, &c. accompanied with a bill to provide for erecting additional buildings for the accomodation of the several executive departments. January 26, 1818.... [Washington: 1818.] 11 p., fold. tables. (15th Cong., 1st sess. House. Doc. 62)

Includes estimates of proposed work.

S&S-46511.

Copies: DLC; DeGE; DeU; MWA; NjP.

2553. Coppinger, Joseph.

On the construction of flat roofed buildings, whether of stone, brick, or wood, and the mode of rendering them fire proof.... New York: Printed for the author, by C. S. Van Winkle, 1819. 23 p.

S&S-47729.

Copies: DLC; MSaE; NN.

2554. Latrobe, Benjamin Henry, 1764–1820.

Memorial of Benjamin H. Latrobe, late Surveyor of the Public Buildings, in the city of Washington, in vindication of his professional skill. January 5, 1819. Printed by order of the Senate of the United States. Washington: Printed by E. De Krafft, 1819. 8 p. (15th Cong., 2d sess. Senate. Doc. 53)

S&S-49866.

Copies: DLC; DeGE; DeU; G; MWA; O.

2555. Pennsylvania. Commissioners for Erecting a State Capitol.

Pennsylvania. Capitol. The Commissioners for erecting a state capitol on the public grounds at Harrisburg, invite architects to exhibit to them plans and elevations.... [Harrisburg: 1819.] Broadside 21 x 42 cm.

Dated: Harrisburg, January 1819.

S&S-49041.

Copies: DLC.

2556. United States. Commissioner of the Public Buildings.

Message from the President of the United States, transmitting a report of the Commissioner of the Public Buildings. December 28, 1819.... Washington: Printed by Gales & Seaton, 1819. 4 p. (15th Cong., 1st sess. House. Doc. 15)

2556 (*continued*)
Commissioner: Samuel Lane.
S&S-49885.
Copies: DLC; DeU; MWA; NjP; O; R.

2557. United States. Congress. House. Committee on Public Buildings.
Report of the Committee on the Public Buildings, accompanying "A bill making appropriations for the public buildings ... and furnishing a supply of water for the use of certain public buildings." January 7, 1819.... [Washington: 1819.] 9 p. (15th Cong., 2d sess. House. Doc. 77)
S&S-49930.
Copies: DLC; DeU; G; MWA; NN.

2558. Benjamin, Asher, 1773–1845.
The American builder's companion; or, A system of architecture, particularly adapted to the present style of building ... 4th ed., corr. and enlarged.... Boston: R. P. & C. Williams, 1820. iv, [5]–104 p., 61 plates.
S-388.
Copies: DLC; LNHT; MWA; NN; NhHi; OCl; OEac; PU; RP; ViW.

2559. Benjamin, Asher, 1773–1845.
The rudiments of architecture: being a treatise on practical geometry, on Grecian and Roman mouldings.... Also, on the origin of building, on the five orders of architecture, on their general and particular parts and embellishments; with examples for cornices, base and surbase mouldings, architraves, and stairs ... 2d ed., enl., with a plan and elevations of a church. Boston: R. P. & C. Williams, 1820. 96 p., 34 plates (3 fold.)
S-389.
Copies: DeGE; IU; KBB; LNHT; MBAt; MH; MWA; MiD-B; NjR; PU; RPJCB.

2560. United States. Commissioner of the Public Buildings.
Message from the President of the United States, transmitting a report of the Commissioner of the Public Buildings. November 27, 1820.... Washington: Printed by Gales & Seaton, 1820. 7 p. (16th Cong., 2d sess. House. Doc. 3)
Samuel Lane, Commissioner.
S-3905.
Copies: DLC; DeGE; DeU; G; MWA; R.

2561. United States. Congress. House. Committee on Public Buildings.
Additional report of the Committee on the Public Buildings. February 16, 1820.... [Washington: 1820.] 11 p., fold. table. (16th Cong., 1st sess. House. Rept. 69)
S-3523.
Copies: DLC; DeGE; DeU; MWA; NjP.

2562. United States. Congress. House. Committee on Public Buildings.
Report of the Committee on so much of the President's message as relates to the public buildings. January 6, 1820.... [Washington: 1820.] 16 p. (16th Cong., 1st sess. House. Rept. 33)
S-4013.
Copies: DLC; DeGE; DeU; MWA; NjP.

2563. United States. Congress. House. Committee on Public Buildings.
Report of the Committee on the expenditures upon the public buildings, March 21, 1820.... [Washington: 1820.] 5 p., fold. tables. (16th Cong., 1st sess. House. Rept. 79)
S-4006.
Copies: DLC; DeGE; DeU; MWA; NjP.

2564. Description of Fulton Market House, intended to be built on the block of ground bounded by Fulton, Front, Crane-Wharf, and South-Street. [S.l.]: Printed by George L. Birch, 1821. 8 p.
Sabin-54243.
Copies: MB; MBAt.

2565. Haviland, John, 1792–1852.
Description of Haviland's design for the new penitentiary now erecting near Philadelphia.... Philadelphia: Robert Desilver, 1821. 12 p.
Erroneously dated; should be 1824.

S-5561.
Copies: PPL.

2566. Olmsted, Denison, 1791–1859.
Remarks on the preparation of mortar. By D. Olmsted.... Addressed to the Agricultural Society of North-Carolina. Read in Society and ordered to be printed, December 14, 1821. [Raleigh?: 1821.] 13 p.
N-00080482.
Copies: DLC; DeGE; NcD; NcU.

2567. United States. Congress. House. Committee on Public Buildings.
Report of the Committee on the Public Buildings, with a bill making appropriations for the public buildings. January 30, 1821.... [Washington: 1821.] 9 p. (16th Cong., 2d sess. House. Rept. 50)
S-7446.
Copies: DLC; DeU; MWA; NjP.

2568. United States. Congress. House. Committee on Public Buildings.
Report of the Committee on the Public Buildings, with a bill making appropriations for the public buildings. March 25, 1822 ... [Washington: 1822.] 10 p., fold. table. (17th Cong., 1st sess. House. Rept. 76)
Account of work done, and estimates for work and materials for 1822.
S-11206.
Copies: DLC; DeGE; DeU; MWA; NjP.

2569. Washington. Ordinances.
An act to regulate the manner of erecting, and the materials to be used, in the erection of houses in the city of Washington, with the Building regulations, &c. Published by order of the Council. [Washington: 1822.] 8 p.
N-W0100685.
Copies: DLC.

2570. United States. Congress. House. Committee on Public Buildings.
Report on the public buildings.... February 18, 1823. [Washington: 1823.] 12 p. (17th Cong., 2d sess. House. Rept. 93)
Copies: DeU; MWA.

2571. Haviland, John, 1792–1852.
A description of Haviland's design for the new penitentiary, now erecting near Philadelphia, accompanied by a birdseye view.... Philadelphia: Robert Desilver, 1824. 12 p., illus.
N-H0189684; S-5561 (erroneously dated 1821).
Copies: PHi; PPAmP; PPL.

2572. United States. Congress. House. Committee on Public Buildings.
Report of the Committee on the Public Buildings, in relation to the operation of said buildings during the last year, and their present state. February 12, 1824. [Washington: 1824.] 12 p. (18th Cong., 1st sess. House. Rept. 60)
S-19046.
Copies: DLC; DeU; MWA; NjP.

2573. Hardie, James, 1760(ca.)–1826.
A census of the new buildings erected in this city, in the year 1824, arranged in distinct classes, according to their materials and number of stories.... New York: Printed by Samuel Marks, 1825. 48 p.
S-20808.
Copies: CtY; DLC; InHi; NBLiHi; NNC; NNNAM; NNUT; PHi; PPL.

2574. Olmsted, Denison, 1791–1859.
Remarks on the preparation of mortar, by D. Olmsted.... Addressed to the Agricultural Society of North-Carolina. Read in Society and ordered to be printed, December 14, 1821. Reprinted by order of the Board of Agriculture, June, 1825. [Raleigh?]: 1825. 1 p.l., [61]–72 p.
Copies: DeGE.

2575. Benjamin, Asher, 1773–1845.
The American builder's companion; or, A system of architecture, particularly adapted to the present style of building.... 5th ed., corr. and enl.... Boston: R. P. & C. Williams, 1826. 106, [2] p., 62 plates.
S-23743.
Copies: CU; CtY; DeWint; MBAt; MNBedf; MWA; NNC.

2576. United States. Commissioner of the Public Buildings.

Public buildings. Message from the President of the United States, transmitting the annual statement of the Commissioner of the Public Buildings, of the expenditure of the same, and of the progress of the said buildings. December 8, 1826. . . . Washington: Gales & Seaton, 1826. 5 p. (19th Cong., 2d sess. House. Doc. 6)

S-27321.

Copies: DLC; DeU; MWA; NjP.

2577. United States. Congress. House. Committee on Public Buildings.

Public buildings. March 17, 1826. Mr. Van Rensselaer, from the Select Committee to which was referred, on the third January last, the subject of "finishing and furnishing the public buildings," made the following report. [Washington: 1826.] 23 p. (19th Cong., 1st sess. House. Rept. 122)

S-27319.

Copies: DLC; DeU; MWA; NjP.

2578. Benjamin, Asher, 1773–1845.

The American builder's companion; or, A system of architecture adapted to the present style of building . . . illustrated with seventy [*sic*] copperplate engravings . . . 6th ed., corr. and enl. Boston: R. P. & C. Williams, 1827. 114 p., 68 plates.

S-28099.

Copies: CSmH; CtY; DeWint; KyU; MA; MBMu; MH; MWA; NN; NNC; NcU; OO; PP; VtU.

2579. Bulfinch, Charles, 1763–1844.

Bulfinch on penitentiaries. Report of Charles Bulfinch on the subject of penitentiaries. Feb. 13, 1827. . . . Washington: Printed by Gales & Seaton, 1827. 8 p. (19th Cong., 2d sess. House. Rept. 98)

S-31201.

Copies: DLC; MH; MWA; NjP; PP; PPAmP.

2580. United States. Congress. House. Committee on Public Buildings.

Public buildings. February 7, 1827. Mr. Van Rensselaer, from the Committee on the Public Buildings . . . made the following report. [Washington: 1827.] 8 p. (19th Cong., 2d sess. House. Rept. 75)

An account of work, materials, and cost.

S-31458.

Copies: DLC; DeU; MWA; NjP.

2581. United States. Dept. of State.

Improvement in hall of House of Reps. Letter from the Secretary of State, transmitting a report of the board of inspection, appointed 19th of May last, in relation to improvements in the hall of the House of Representatives . . . February 12, 1827. Washington: Printed by Gales & Seaton, 1827. 10 p. (19th Cong., 2d sess. House. Doc. 93)

S-31282.

Copies: DLC; DeU; MWA; NjP.

2582. United States. Commissioner of the Public Buildings.

Annual report—Commissioner of Public Buildings. Message from the President of the United States, transmitting the annual report of the Commissioner of the Public Buildings. January 4, 1828. . . . Washington: Printed by Gales & Seaton, 1828. 8 p. (20th Cong., 1st sess. House. Doc. 41)

S-35681.

Copies: DLC; DeU; MWA; NjR.

2583. United States. Commissioner of the Public Buildings.

Message from the President of the United States, transmitting the annual report of the Commissioner of the Public Buildings . . . January 4, 1828. . . . Washington: Printed by Duff Green, 1828. 9 p. (20th Cong., 1st sess. Senate. Doc. 20)

Refers specifically to the heating and fire protection of the Capitol.

S-36616.

Copies: DLC; DeU; MWA; NjR.

2584. Lafever, Minard, 1797–1854.

The young builder's general instructor; containing the five orders of architecture, selected from the best specimens of the Greek and Roman . . . and a variety of mouldings, and fancy pilasters.... Newark, N.J.: Printed by W. Tuttle & Co., 1829. 175 p., 66 [67] plates.

S-39244.

Copies: CtY; DLC; ICA; NN; NNC; NcD; NjN; OO.

2585. [Eliot, William Havard] 1796–1831.

A description of Tremont House, with architectural illustrations. Boston: Gray and Bowen; printed by E. W. Metcalf and Co., Cambridge, 1830. 2 p.l., 36 p., 31 plates (1 col.).

C-1245.

Copies: CtY; DLC; ICJ; MB; MHi; MWA; MiU-C; NBLiHi; NBu; NIC; NNC; PU; RPA.

2586. Haviland, John, 1792–1852.

The practical builder's assistant; for the use of carpenters, masons, plasterers, cabinet-makers and carvers, with working drawings selected from ... examples, from the antique: together with ... original designs, with their plans, elevations and sections ... 150 engravings ... 2d ed. Baltimore: F. Lucas, jr., [1830?]. 3 v. and atlas of 150 plates.

C-1784.

Copies: CtY; DLC; ICJ; ICRL; MWA; MiD; NjP; OU; TxU.

2587. Shaw, Edward.

Civil architecture; or, A complete theoretical and practical system of building, containing the fundamental rules of the art of geometry and mensuration ... also the five orders of architecture: with a great variety of beautiful examples ... with many useful and elegant ornaments, and rules for projecting them.... Boston: Shaw and Stratton, 1830. 76 p., 25 plates.

Vol. I, parts 1 and 2; probably all that was issued of the first ed.; 2d ed. [1832].

Hitchcock-1143; N-S0477949.

Copies: MB; MWA; NWM.

2588. United States. Architect of the Capitol.

Memorial of Charles Bulfinch, on the subject of the hall of the House of Representatives. January 25, 1830.... [Washington: 1830.] 11 p. (21st Cong., 1st sess. House. Rept. 123)

C-4940.

Copies: DLC; DeU; MWA; NjR; PP.

2588a. A view and description of the Eastern Penitentiary of Pennsylvania. Philadelphia: Published for the Philadelphia Society for Alleviating the Miseries of Public Prisons, by C. G. Childs, 1830. p.l., 8 p., plan.

Signed: G. W. S. [George Washington Smith?]

C-3513; N-V0154006.

Copies: DLC; DNLM; KyHi; MWA; NNC; OClWHi; PHC; PHi; PPL; PU.

5. Bridges and tunnels

2589. [Walton, John.]

An account of the surprizing events of Providence, which hapned at the raising of a bridge in Norwich, June 28th, 1728. New London: Printed by T. Green, 1728. 10 p.

E-2981.

Copies: CtHi (not located).

2590. New-Haven lottery, for building a bridge over East-River.... [New Haven: 1780.] Broadside 41 x 17.5 cm.

Dated: New-Haven, July 31, 1780.

B-5138; Mp-43846.

Copies: CtHi.

2591. A list of fortunate numbers in the first class of New-Haven bridge lottery. [New Haven: 1781?] Broadside 35.5 x 21 cm.
B-5321; Mp-44009.
Copies: NHi.

2592. Proprietors of Charles River Bridge.
Charles River Bridge. At a meeting of Proprietors of Charles River Bridge—under the act of government for their incorporation, the Honorable John Hancock, John Lowell, and Nathaniel Gorham, Esq'rs; who were appointed a committee to prepare rules and regulations for the government of the Society, report.... [Boston: 1785.] 8, [6] p.
Subscribers' names: [6] p. at end.
B-6037; Mp-44662.
Copies: MB; MWA.

2593. Proprietors of Charles River Bridge.
Charles River Bridge. Boston, July 12, 1785. Voted that each proprietor be allowed fifteen days for payment from every notification of an assessment.... [Boston: 1785.] Broadside, folded into 4 p.
B-6038; Mp-44663.
Copies: MHi; MWA.

2594. Malden Bridge Corporation, Malden, Mass.
Boston, March 24, 1787. Subscribers to Malden-Bridge.... [Boston: Printed by S. Hall, 1787.] 13 p.
Contains a list of 41 subscribers, rules and regulations, and rates of toll.
Mp-20480.
Copies: MBAt.

2595. Pennsylvania. Laws.
State of Pennsylvania. In General Assembly, Wednesday, Nov. 14 ... 1787. The bill entitled, "An act to incorporate the subscribers to the plan for erecting a permanent bridge over the river Schuylkill," at the western extremity of the High-Street of the city of Philadelphia was read the second time.... [Philadelphia: 1787.] Broadside 43 x 25.5 cm.
B-6559; Mp-45127.
Copies: PHi.

2596. Bridge lottery. Scheme of a lottery to raise the sum of four hundred and seventy pounds ten shillings, for the salutary purpose of building a bridge over Conogocheague.... Elizabeth-(Hager's) Town: Printed by Stewart Herbert, [1790]. Broadside 20.5 x 17 cm.
B-7330; Mp-45853.
Copies: MdHi.

2597. New Jersey. Commissioners to Erect Bridges.
Contract for erecting bridges over the Hackinsack and Passaick rivers, between Powlas-Hook and Newark, in the state of New-Jersey. [Elizabeth-Town: Printed by Shepard Kollock?, 1791.] Broadside.
Dated: Newark, State of New-Jersey, April 22, 1791.
Mp-23626.
Copies: NHi.

2598. Omitted in numbering.

2599. Pennsylvania. Laws.
An act to authorize the Governor ... to incorporate a company for erecting a bridge over the river Delaware at the borough of Easton.... Philadelphia: Printed by T. Bradford, [1794]. 16 p.
At head of title: In General Assembly, Thursday, March 20, 1794.
B-8805.
Copies: PHi (at PPL).

2600. Proprietors of the Canal Bridge.
To the members of the Massachusetts legislature. Equal justice and protection.... The petition of A. Craigie, as president of the Proprietors of the Canal Bridge; now before the legislature.... [Boston: 1795?] Broadside 28.5 x 23 cm.
B-9042; Mp-47369.
Copies: NHi.

2601. Proprietors of the Kennebec Bridge.
Extracts from the act of incorporation, and bye-laws of the Proprietors of the Kennebec Bridge. Hallowell, [Me.]: Printed by Peter Edes, 1796. 16 p.
Mp-30004.
Copies: MeHi.

2602. Peale, Charles Willson, 1741–1827.
An essay on building wooden bridges. By Charles W. Peale.... Philadelphia: Printed by Francis Bailey, for the author, 1797. iv, [5]–16 p., 6 plates.
Mp-32642.
Copies: DeGE; MWA; PPAmP; PPL.

2603. Schuylkill Permanent Bridge Company.
Plan for the establishment of a company, to be incorporated by the Legislature of Pennsylvania ... by the name and title of "The Company for Erecting a Bridge over the River Schuylkill, at or near the City of Philadelphia." [Philadelphia: 1797.] [3] p.
Dated: Philadelphia, January 25, 1797.
Mp-32682.
Copies: PHi.

2604. To the honourable the Senate and House of Representatives of ... Pennsylvania. The memorial of the subscribers ... to promote ... a company for erecting a permanent bridge over the ... Schuylkill.... [Philadelphia: 1797.] Broadside 28 x 20 cm.
Dated: Philadelphia, January 25, 1797.
B-10128; Mp-48271.
Copies: PHi.

2605. Calculation, shewing in what time a permanent bridge over the river Schuylkill, supposed to cost two hundred thousand dollars, may be built and redeemed from the estimated combined product of duties on sales at auction within the city of Philadelphia and the product of bridge toll.... [Philadelphia]: Printed by Z. Poulson, Jun., 1798. 3 p.
Dated: January 8, 1798.
Mp-34360.
Copies: DLC; PHi; PPL.

2606. Calculation, shewing that within a certain period of time, a bridge, supposed to cost two hundred thousand dollars, may be erected and completed.... [Philadelphia]: Printed by Z. Poulson, Jun., 1798. 3 p.
Dated: January 8, 1798.
Mp-34361.
Copies: PHi.

2607. Lewis, William, 1751–1819.
Opinions of William Lewis and William Rawle, Esquires, on the rights of the Corporation of the City of Philadelphia to the Schuylkill-ferry, and of the rights of the Corporation for erecting a permanent bridge over that river. Philadelphia: Printed by John Ward Fenno, 1798. 18 p.
Mp-34001.
Copies: DLC; MWA; PHi.

2608. Pennsylvania. Laws.
An act of the Legislature of Pennsylvania and a patent to incorporate a company for erecting a permanent bridge over the river Schuylkill. Philadelphia: Printed by John Fenno, [1798]. 21 p.
Mp-34321.
Copies: DLC; MHi; PHi; PPAmP.

2609. Philadelphia. Councils. Joint Committee on the Bridge and Ferry over Schuylkill.
A communication from the Joint Committee of the Select and Common Councils, appointed to superintend the city interest in the bridge and ferry over Schuylkill, at the west end of High-Street. [Philadelphia: 1798.] 7 p.
Dated: July 30, 1798.
Mp-34348.
Copies: NN.

2610. West Boston Bridge Company.
Concise view of the facts, &c. [Boston: 1800.] [2]–8 p.
Mp-37021.
Copies: MHi.

2611. Connecticut. Laws.

The act of incorporation of the Milford and Stratford Bridge Company. New Haven: Printed by T. Green & Son, 1802. 8 p.

S&S-2068.

Copies: CtY; MB; MWA.

2612. Interesting essays, chiefly on the subject of the yellow fever, interspersed with some remarks on the progress of the preparations for a permanent bridge.... By a Philadelphian.... Philadelphia: Printed by Budd & Bartram, 1802. 51 p.

S&S-2452.

Copies: CtY-M; DLC; DNLM; MWA; PP.

2613. Proprietors of Neponset Bridge.

Acts of incorporation, and rules and regulations of the Proprietors of Neponset Bridge. Boston: Printed for the Proprietors, [1802]. 16 p.

Dated at end: July 26, 1802.

S&S-2730.

Copies: ICU; MBAt; MBBS; MWA; WHi.

2614. Connecticut. Laws.

Act of incorporation of the Milford and Stratford Bridge Company. Bridgeport, [Conn.]: Printed by Samuel Mallory, [1803]. 7, [1] p.

S&S-4653.

Copies: CtY.

2615. Massachusetts. Laws.

An act to incorporate the Proprietors of West-Boston Bridge. With the subsequent acts, relative to that institution. Boston: Printed by Russell and Cutler, 1803. 18 p.

S&S-4600.

Copies: MWA.

2616. Proprietors of Neponset Bridge.

Acts of incorporation, and rules and regulations of the Proprietors of Neponset Bridge. Boston: Printed for the Proprietors, [1803?]. 16 p.

"Additional act" is dated February 7, 1803.

S&S-4708.

Copies: CtY; ICN; MH-BA; MWA.

2617. Schuylkill Permanent Bridge Company.

Report of the building committee. [Philadelphia: 1803.] [4] p.

Dated: July 4th, 1803.

Copies: MWA.

2618. Collection of facts and documents relative to the project of a bridge from South Street in Boston to Dorchester Neck, and the annexation of that peninsula to the town of Boston.... Boston: Printed by E. Lincoln, 1805. 62, [1] p.

S&S-8207.

Copies: DLC; DeGE; MB; MH; MWA; MiD-B; PHi.

2619. Georgetown, D.C. Committee of the Corporation.

Memorial. To the Honorable the Committee of the Senate, for the District of Columbia. [Washington?: 1805.] 6 p.

Memorial of the Committee of the Corporation of Georgetown, asking for permission to build a free bridge over the Potomac River at or near the place called "The Three Sisters."

S&S-8511.

Copies: MWA.

2620. Lathrop, Joseph, 1731–1820.

A discourse delivered at Springfield, October 30, 1805. On occasion of the completion and opening of the great bridge over Connecticut River, between the towns of Springfield and West-Springfield.... Springfield, Mass.: H. Brewer, printer, [1805]. 16 p.

S&S-8754.

Copies: Ct; DeGE; MWA; MeB; NjPT; PPPrHi.

2621. ———— 2d ed. Springfield, Mass.: H. Brewer, printer, [1805]. 16 p.

S&S-8755.

Copies: CSmH; DLC; MH; MWA; NN.

2622. Schuylkill Permanent Bridge Company.

To the stockholders of the Schuylkill Permanent Bridge Company. [Philadelphia: Jane Aitken, printer, 1805.] 11 p.

Signed and dated on p. 6: Richard Peters, president, December 26th, 1805.

Statement on expenditures made in erecting the permanent bridge, p. 7–11.

S&S-9322.

Copies: PPAmP.

2623. Trenton Delaware Bridge. In compliance with the resolution of the board at their former meeting, the president laid before them the following statement.... Trenton: 1805. 3 p.

Signed and dated at end: By order of the board, John Beatty, president. Trenton, December 31, 1805.

N-T0326935.

Copies: PHi.

2624. United States. Congress. House. Committee on the District of Columbia.

Report of the Committee to whom was referred ... the memorials and petitions of sundry citizens and inhabitants of the District of Columbia, for and against the building of a bridge across the Potomac River ... 4th February, 1805.... [Washington: 1805.] 4 p.

S&S-9625.

Copies: DLC.

2625. An answer to a pamphlet entitled, "Considerations on the public expediency of a bridge from one part of Boston to the other."Boston: Printed by E. Lincoln, 1806. 31, 8 p.

S&S-9853.

Copies: DLC; DeGE; ICN; MB; MH-BA; MWA; NN; NNE.

2626. Delaware. Laws.

An act to incorporate a company to erect a draw bridge over the river Christiana at Wilmington.... [Wilmington, Del.: 1806.] 18 p.

S&S-10271.

Copies: DeHi.

2627. New Bedford Bridge Corporation.

By-laws.... [New Bedford, Mass.?: 1806?] Broadside.

S&S-10955.

Copies: MNBedf (not located).

2628. [Tudor, William] 1750–1819.

Considerations on the public expediency of a bridge from one part of Boston to the other. Boston: Printed by Manning & Loring, 1806. 33 p.

S&S-11484.

Copies: DLC; DeGE; MH; MWA; NNE.

2629. United States. Congress. House. Committee on the District of Columbia.

Report of the Committee to whom were referred on the fifth ultimo, the several memorials and petitions of citizens and inhabitants of the county of Alexandria, of the city of Washington, and of the towns of Georgetown and Alexandria ... January 21, 1806.... City of Washington: A. & G. Way, printers, 1806. 8 p.

Concerns a bridge over the Potomac River at Washington.

S&S-11734.

Copies: DLC; MWA; NN.

2630. United States. Laws.

An act authorizing the erection of a bridge over the river Potomac, within the District of Columbia.... [Washington: 1806.] 12 p.

At head of title: In Senate of the United States, 17th December, 1806.

S&S-11494.

Copies: DLC.

2631. Burr, Theodore.

In Senate of the United States, January 22, 1807. The following letter was read, and ordered to be printed. [Washington: 1807.] 4 p.

A letter from Theodore Burr, concerning the proposed bridge over the Potomac River.

S&S-13915.

Copies: DLC.

2632. Lyman, Joseph, 1749–1828.

A sermon preached at Hatfield, October 20, 1807. At the opening of Hatfield bridge.... Northampton, [Mass.]: Printed by William Butler, 1807. 16 p.

S&S-12954.

Copies: CSmH; CtY; DLC; DeGE; MA; MWA; MWiW; NGH; NN.

2633. Massachusetts. Laws.

An act for incorporating the Proprietors of Canal Bridge. [Boston: Printed by Belcher and Armstrong, 1807.] 10 p.

S&S-13011.

Copies: MB; MH.

2634. Memorial of the citizens of Georgetown ... addressed to the Senate of the United States. December 7, 1807.... Washington City: Printed by R. C. Weightman, 1807. 5 p.

In opposition to the proposed bridge over the Potomac River, which the memorialists consider harmful to the navigation.

S&S-13980.

Copies: DLC; NN.

2635. [Peters, Richard] 1744–1828.

A statistical account of the Schuylkill permanent bridge, communicated to the Philadelphia Society of Agriculture, 1806. Philadelphia: Printed by Jane Aitken, 1807. 3 p.l., [5]–84 p., front.

Also published in: Philadelphia Society for Promoting Agriculture, *Memoirs,* v. 1. Philadelphia, 1808.

S&S-13366.

Copies: CSmH; DLC; DeGE; IU; KyHi; MH; MWA; NNE; PPAmP; PPL.

2636. Schuylkill Permanent Bridge Company.

To the stockholders of the Schuylkill Permanent Bridge Company. [Philadelphia: 1807.] 6 p.

Dated: January 1st, 1807.

S&S-13552.

Copies: PHi; PPAmP.

2637. Statements by remonstrants against another bridge over Charles River.... [Boston: 1807.] Broadside.

Survey by Osgood Carleton, attested by John Gardner, Justice of the Peace, at Boston, February 10, 1807.

S&S-12177.

Copies: MHi; MWA (xerox copy).

2638. To the Honorable the Senate and House of Representatives of the State of Delaware. The memorial of the undersigned persons, citizens of New-Castle County, in the said State, respectfully represents, that your memorialists feel themselves imperiously called on in consequence of the renewed attempts of certain persons of the Borough of Wilmington ... to obtain an Act incorporating a company to erect a bridge over Christiana Creek, in the said borough.... [S.l.: 1807?] Broadside 30 x 23 cm.

Copies: DeA; DeGE (photocopy).

2639. To the Hon. the Senate, and the Honorable the House of Representatives, of the Commonwealth of Massachusetts.... The undersigned petitioners for a free dam, or bridge, from Wheeler's Point, to South Boston ... sumbit [*sic*] the following reasons why ... the petition should be granted.... [Boston: 1807?] 32 p.

S&S-13711.

Copies: MWA.

2640. United States. Congress. House. Committee on the District of Columbia.

Report of the Committee to whom was re-committed on the twenty third instant the engrossed bill authorizing the erection of a bridge over the river Potomac, within the District of Columbia. November 24, 1807.... City of Washington: A. & G. Way, printers, 1807. 4 p.

S&S-14074.

Copies: DLC; NN.

2641. United States. Congress. Senate.

Additional documents on the subject of bridges. Submitted to the Senate of the United States. January 12, 1807.... Washington

City: Printed by Duane & Son, 1807. 14 p.
Includes documents I–III.
S&S-13793, 13794.
Copies: DLC; NhHi.

2642. ——— January 14, 1807. Printed by order of the Senate. Washington City: Printed by Duane & Son, 1807. 23 p.
At head of title: No. IV to VII.
S&S-13792.
Copies: DLC; PPAmP.

2643. United States. Congress. Senate.
In Senate of the United States, January 5th, 1807. Ordered, that Mr. White, Mr. Smith of Vermont, and Mr. Gaillard, be a committee, to examine the papers relative to the contemplated bridge across the Potomac ... 6th January, 1807.... Washington City: Duane & Son, printers, 1807. 49 p.
Includes petitions of the residents of the District of Columbia, a letter from B. H. Latrobe on the probable effect of the bridge on the navigation of the river, and other documents.
S&S-13910.
Copies: DLC; NN; NNS; PHi; PPAmP.

2644. United States. Congress. Senate.
In Senate of the United States. January 21, 1807. Mr. Bayard submitted the following documents on the subject of bridges, which were read, and ordered that they be printed for the use of the Senate. [Washington: 1807.] 7 p.
S&S-13914.
Copies: DLC; MWA.

2645. United States. Congress. Senate.
In Senate of the United States. January 21, 1807. Mr. Giles laid before the Senate the following letter.... [Washington: 1807.] 2 p.
Concerns the bridge across the Potomac River.
S&S-13913.
Copies: DLC.

2646. United States. Congress. Senate.
In Senate of the United States. January 22, 1807. The following letter was read.... [Washington: 1807.] 4 p.
Concerns the bridge across the Potomac River.
S&S-13915.
Copies: DLC.

2647. Hatfield Bridge Lottery.
A list of prizes and fortunate numbers in the fourth class of Hatfield Bridge Lottery. Which commenced drawing in Boston, November 9, 1808, and was completed December 21, 1808. Boston: Printed by Oliver and Munroe, [1808?]. [14] p.
S&S-15204.
Copies: MWA.

2648. [Ludlow, Daniel.]
To the Honorable the Legislature of the state of New-York in Senate and Assembly convened. Albany: Printed by John Barber, [1808]. 11 p.
Concerns the Harlem River bridge; dated January 25, 1808.
S&S-15192, 15460.
Copies: M; N.

2649. Malden Bridge Corporation, Malden, Mass.
Rules and regulations of the Corporation of Malden Bridge, as revised and established, at the annual meeting, July, 1808. Charlestown, [Mass.]: Printed by Samuel Etheridge, 1808. 9 p.
Includes the rates of tolls for the Malden and Chelsea bridges.
S&S-15487.
Copies: DeGE; MBAt; MHi; NNC.

2650. Proprietors of the Kennebec Bridge.
By-laws of the Proprietors of the Kennebec Bridge. Augusta, Me.: 1808.
S&S-15354.
Copies: MH (not located).

2651. Willard, Samuel, 1776–1859.
A sermon, preached at Northampton, October 27, 1808, at the opening of Northampton Bridge.... Northampton, [Mass.]:

2651 (*continued*)
Bull and Butler, 1808. 19 p.
S&S-16740.
Copies: CtSoP; DLC; MB; MBAt; MH-AH; MWA; NHi.

2652. Finley, James.
A description of the chain bridge with data and remarks, illustrative of the power, cost, durability and comparative superiority of this mode of bridging. Uniontown, [Pa.]: 1809. 9 p.
S&S-17510.
Copies: CSmH.

2653. New Jersey. Laws.
An act to empower the Board of Chosen Freeholders of the county of Salem, to erect a bridge over Salem creek.... [Trenton: 1809.] 1 p.
S&S-18207.
Copies: NjR.

2654. Dyster, Joseph Joshua.
Patent tubular iron bridges, a circular setting forth its advantages. Philadelphia: 1810. Broadside.
S&S-20018.
Copies: MSaE; MWA (xerox copy).

2655. Considerations upon the propriety and expediency of erecting a bridge over the Cayuga Lake.... Canandaigua, N. Y.: Printed at the Messenger-office, 1811. 8 p.
S&S-22608.
Copies: N.

2656. Finley, James.
A description of the chain bridge; invented by Judge Finley.... With data and remarks illustrative of the power, cost, durability, and comparative superiority of this mode of bridging. Union Town, [Pa.]: Printed by W. Campbell, 1811. 13 p., front.
S&S-22821.
Copies: DLC; MSaE; PPL.

2657. Pennsylvania. Laws.
An act to authorize and direct the Governor to incorporate a company for erecting a permanent bridge over the river Schuylkill.... Philadelphia: Printed by Jane Aitken, 1811. 21 p.
S&S-23651.
Copies: MWA.

2658. Pope, Thomas.
A treatise on bridge architecture; in which the superior advantages of the flying pendent lever bridge are fully proved, with an historical account and description of different bridges.... New York: Printed for the author, by Alexander Niven, 1811. 9 p.l., ix–xxxii, 33–288 p., plates.
S&S-23721.
Copies: DLC; DeGE; ICJ; MB; MWA; NN; NNC; NNE; NjR; PPL; RPB.

2659. The rights and privileges of the Passaic and Hackinsack [*sic*] Bridge Company examined ... by Monitor. With a few introductory observations. By Justice.... Newark: Printed by W. Tuttle, 1811. 31 p.
S&S-23820; Sabin-58999.
Copies: NHi; NjR.

2660. United States. Congress. Senate.
Mr. Anderson's motion. January 23d, 1811. Printed by order of the Senate of the United States. Washington City: Printed by R. C. Weightman, 1811. [4] p.
Resolution that a committee be appointed to consider whether the United States should give aid for the building of a bridge at Havre de Grace, Maryland.
S&S-24236.
Copies: DLC.

2661. Hopkins, James.
Opinion of James Hopkins, Esq., relative to the capital stock of the Susquehanna Bridge Company, at or near the town of Columbia. Columbia, Pa.: Thomas A. Wilson, 1813. 15 p.
S&S-28771.
Copies: PPL.

2662. Schuylkill Permanent Bridge Company.

Report of the Committee of Accounts, with the report of the Schuylkill Permanent Bridge Committee on the present state of the bridge. Philadelphia: Printed by Jane Aitken, 1813. 13 p.

S&S-29496.

Copies: DeGE; PHi.

2663. Wernwag, Lewis.

[Plate, illustrating three bridges, at Newhope, dated 1813, over the Schuylkill River in Philadelphia, and over the Frankfort Creek at Neshaminy. Drawn and engraved by E. G. Gridley. S.l.: 1813?] Broadside 20 x 40 cm.

Copies: PPAmP.

2664. Lancaster and Schuylkill Bridge Company.

Report of the managers of the Lancaster and Schuylkill Bridge Company to the stockholders, March 3, 1814. Philadelphia: Printed at the office of the United States' gazette, 1814. 15 p.

S&S-31887, 51404.

Copies: DeGE; IEN; MWA; MdBP.

2665. New York (State). Legislature. Assembly.

Documents relative to the construction of a bridge across the Hudson River at or near the city of Albany. [Albany: 1814.] 16 p.

"In Assembly, March 11, 1814."

S&S-32300.

Copies: DeGE; N; NIC.

2666. Reasons why the Legislature of New-Jersey cannot deprive the proprietors of the bridges over the rivers Passaick and Hackensack of any of their legal rights. [Newark?, N.J.: 1814?] [3]–21 p.

S&S-32611.

Copies: DLC; Nj; NN.

2667. [Peters, Richard] 1744–1828.

A statistical account of the Schuylkill permanent bridge, communicated to the Philadelphia Society of Agriculture, 1806. Philadelphia: Johnson & Warner; William Greer, printer, 1815. 88 p. front.

S&S-35622.

Copies: CSmH; DLC; MH; MiU.

2668. United States. Laws.

A bill for the relief of the Anacosta Bridge Company. January 26, 1815.... [Washington: 1815.] 1 p. (H.R. 62)

S&S-36194.

Copies: DLC.

2669. A brief statement of facts, shewing the importance of a bridge over the river Susquehanna, at Connowingo Creek, and of the roads to the river Susquehanna, to the growth of the city of Baltimore, and to the comfort and convenience of its inhabitants: with a map, shewing the different routes to Lancaster and Philadelphia. Baltimore: Printed by J. Wane & Co., 1816. 12 p., fold. map.

Report by John Davis, p. 3–7.

S&S-37402.

Copies: DLC; MWA; MdHi.

2670. New Hope Delaware Bridge Company.

New-Hope-Delaware bridge. At a general meeting of the stockholders ... the following report was submitted. [S.l.: 1816.] 4 p.

Signed: W. Maris, president. January 4, 1816.

S&S-38403.

Copies: MdBP.

2671. North Carolina. Laws.

An act to to [*sic*] incorporate a company to build a bridge across the Yadkin River. Raleigh: Tho. Henderson, Jun., 1817. 8 p.

N-N0310993.

Copies: NcU (microfilm).

2672. Cotting, Uriah.

Observations concerning the canal bridge. Boston: 1818. 20 p.

S&S-43745.

Copies: MWA.

2673. Massachusetts. General Court. House. Committee on the Kennebec Bridge.

Report, in part, of the Committee to inquire into the conduct of the Proprietors of the Kennebec Bridge. The Committee appointed by the House. . . . [Boston: 1818.] 16 p.

S&S-44739.

Copies: MWA.

2674. Massachusetts. General Court. House. Committee on the Kennebec Bridge.

Report of the Committee appointed by the House of Representatives, to inquire and examine into the conduct of the Proprietors of the Kennebec Bridge. . . . Boston: 1818. 16 p.

S&S-44741.

Copies: MWA.

2675. United States. Laws.

A bill to incorporate a company to build a bridge over the eastern branch of the Potomac, from the southern termination of Eleventh Street East, in the city of Washington. March 3, 1818. . . . [Washington: 1818.] 1 p. (H.R. 114)

S&S-46102.

Copies: DLC.

2676. Observations concerning the canal bridge. [Boston?: 1819.] 6 p.

Concerns the bridge over the Charles River, Boston, with revenues up to May 24, 1819.

S&S-48960.

Copies: CSt; MWA; NN.

2677. Springfield Bridge Lottery.

Managers' list of prizes, in the 7th class, Springfield Bridge Lottery. . . . Boston: Printed by Sylvester T. Goss, 1819. [6] p.

S&S-49487.

Copies: MWA.

2678. Binney, Horace, 1780–1875.

Opinion of Horace Binney, Esquire, on a report of a committee of stockholders of the Columbia Bridge Company, made on the fourth day of May, 1820. [Columbia, Pa.: William Greer, 1820.] 12 p.

S-464.

Copies: PPAmP; PPL.

2679. Canal bridge! An answer to a pamphlet circulated among the members of the Legislature, at their last session, entitled, "Observations concerning the canal bridge." Boston: 1820. 11 p.

S-129.

Copies: DeGE; MiU.

2680. Columbia Bridge Company, Columbia, Pa.

A reply of the board of managers, to a report of a committee of the stockholders, of the Columbia, Pa., Bridge Company. Columbia, Pa.: Printed by William Greer, 1820. 46 p.

S-821.

Copies: CSmH; DLC; MH-BA; PHi (in PPL).

2681. Considerations addressed to the legislature on the subject of the proposed bridge from Charlestown to Boston. [Boston?: 1820.]

Inscribed: Justice.

Brit. Museum. *Gen cat.*, v. 120, col. 71; S-854.

Copies: BM.

2682. New Hope Delaware Bridge Company.

Charter of the New Hope Delaware Bridge Company. Incorporated A.D. 1812 by the states of New-Jersey and Pennsylvania. [New York]: Printed by J. & J. Harper, [1820]. 24 p.

Copies: PPL.

2683. Pennsylvania. General Assembly. Senate. Committee on the Affairs of the Columbia Bridge Company.

Report of a committee of the Senate, relative to the affairs of the Columbia Bridge Company, read Jan. 29, 1821. [Harrisburg: 1821.] 12 p.

S-6414.

Copies: PPAmP; PHi.

2684. [Pennsylvania and New Jersey Communication Company.]
Bridge documents, submitted to the consideration of those who have not before had an opportunity of investigating the subject to which they refer. Philadelphia: 1821. 16 p.
Dated: April 28, 1821.
N-P0214497; S-4834.
Copies: CtY; DeGE; PHi; PPL.

2685. Town, Ithiel, 1784–1844.
A description of Ithiel Town's improvement in the construction of wood and iron bridges: intended as a general system of bridge-building.... New Haven: Printed by S. Converse, 1821. 10 p., 3 plates.
S-6983.
Copies: CtHi; CtY; DeGE; ICJ; M; MB; MHi; MdBP; MdHi; MiU-C; NN; NbU.

2686. Schuylkill Permanent Bridge Company.
By-laws, rules, & regulations of the corporation ... to which are prefixed the act of the legislature authorizing the incorporation.... Philadelphia: Printed by William Fry, 1822. 40 p.
S-10189.
Copies: MWA; PHi; PPAmP.

2687. Citizens of the South-End! Boston: 1824. Broadside.
Dated: Boston, March 15, 1824.
Handbill, urging attendance at a meeting at Faneuil Hall to oppose the building of a bridge or dam at Wheeler's Point.
S-15498.
Copies: MB; MBAt.

2688. A concise view of some of the facts and arguments, respecting another bridge, to South Boston. Respectfully addressed to the citizens of Boston. [Boston: 1824?] 20 p.
Sometimes attributed to Lemuel Shaw.
Dated [1806?] by CtY and DLC.
S-17954.
Copies: CtY; DLC; MH; MWA.

2689. An appeal to the good sense of the legislature and the community, in favor of a new bridge to South Boston. By a citizen. Boston: Printed by True and Greene, 1825. 22 p.
S-19455.
Copies: CtSoP; CtY; DLC; DeGE; M; MB; MBAt; MH; MHi; MWA; MiD-B; MiU; NN; NcD.

2690. Lowber, John Cole, 1779–1834.
Report of the case of Alexander and others against the president, managers and company of the Schuylkill Navigation Company. Instituted to recover damages for the loss of the bridge at the Falls of Schuylkill, during the fresh of the 21st of February, 1822. Tried before the Court of Common Pleas of Philadelphia County, on the 18th day of February, 1824. Reported for the Watering Committee by John C. Lowber. Philadelphia: Printed by L. R. Bailey, 1825. 78 p.
S-21261.
Copies: DeGE; MH-L; MWA; PPAmP; PPB; PPL.

2691. Reasons, principally of a public nature, against a new bridge from Charlestown to Boston. Boston: Printed by Wells and Lilly, 1825. 32 p.
Attributed to Richard Sullivan, on basis of mss. notes on some copies.
S-22041.
Copies: DLC; DeGE; MB; MH; MH-BA; MWA; MiD-B; NN; NNE; PPL.

2692. Town, Ithiel, 1784–1844.
A description of Ithiel Town's improvement in the construction and practical execution of bridges, whether built entirely of wood, or of cast or wrought iron.... New Haven: T. G. Woodward & Co., printers, 1825. 1 p.l., 2 p.
S-22489.
Copies: MBAt; N; NN.

2693. United States. Laws.
A bill to authorize the construction of a bridge from the Navy Yard Island, in Portsmouth harbor, to Kittery, in the state of

2693 (*continued*)
Maine. February 17, 1825.... [Washington: 1825.] 1 p. (S. 56)
S-22803.
Copies: DNA.

2694. Boston. City Council.
The report of the Special Committee, and the report of the Joint Committee, on the subject of the free bridge; with the documents accompanying. Boston: True & Greene, printers, 1826. 23 p.
S-23868.
Copies: DLC; MBAt; MH-BA; MWo.

2695. Boston. Committee on the South Boston Bridge.
Report and accompanying documents on the petition of Lot Wheelwright and others for a free bridge to South Boston, Oct. 2, 1826. [Boston: 1826.] 16 p.
S-23860.
Copies: MBAt; MHi.

2696. Boston. Committee on the South Boston Bridge.
[Report of the committee instructed to inquire into the proceedings of the city authorities and the proprietors of the Boston South Bridge Corporation. Boston: 1826.] 16 p.
S-23865.
Copies: DLC.

2697. Massachusetts. General Court. Committee on Bridges.
[Report of the committee, of both houses of the legislature, on bridges, to whom were recommitted the petitions of John Skinner & al. and of Ezra Trull & al. on the principal facts, with an opinion on the questions of law arising in the case; also, how far the state has a reversionary interest in the Charles River bridge, which could be affected by the grant of a new bridge. Boston: 1826?] 14 p.
S-25270.
Copies: DLC; MWA.

2698. Massachusetts. Laws.
An act, establishing a free bridge in the city of Boston. [Boston: 1826.] 6 p.
S-25252.
Copies: DLC.

2699. United States. Congress. Senate. Committee on the District of Columbia.
In Senate of the United States, May 1, 1826. Mr. Eaton, from the Committee on the District of Columbia, laid the following documents on the table relative to the bill "authorizing the Corporation of Georgetown to erect a bridge over the river Potomac, within the District of Columbia.... " [Washington: 1826.] 7 p. (19th Cong., 1st sess. Senate. Doc. 86)
Includes a memorial from the Georgetown Bridge Company.
S-27071.
Copies: DLC; DeU; MWA; NjP.

2700. United States. Laws.
A bill to authorize the corporation of Georgetown to erect a bridge over the river Potomac, within the district of Columbia. March 22, 1826.... [Washington: 1826.] 5 p. (H.R. 174)
S-26399.
Copies: DNA.

2701. ________ May 1, 1826.... [Washington: 1826.] 5 p. (S. 147)
S-26400.
Copies: DNA.

2702. ________ December 19, 1826.... [Washington: 1826.] 5 p. (S.18)
S-26401.
Copies: DNA.

2703. Boston. City Council. Committee on the Memorial of the Boston Free Bridge Corporation.
City of Boston, in Common Council, June 25, 1827.... The committee of both branches of the City Council to whom was referred the petition of the Free Bridge Corporation, respectfully report.... [Boston:

1827.] 15 p.
N-B0669763, B0670367; S-28225.
Copies: M; MB; MH; NN.

2704. A description of the Thames tunnel. New York: Elliott & Palmer, [1827?]. 9 p.
S-28693.
Copies: DLC.

2705. Massachusetts. Laws.
An act to establish the Warren Bridge Corporation. [Boston: 1827.] 4 p.
S-29660.
Copies: MH.

2706. Parker River Bridge Corporation.
Act and by-laws and rules.... Newburyport, [Mass.]: Printed by W. & J. Gilman, 1827. 8 p.
S-30152.
Copies: MH-BA.

2707. Review of the case of the free bridge, between Boston and Charlestown; in which the expediency and constitutionality of that measure are considered; with a statement of the legislative proceedings upon it at the late January session. Including the public documents. Boston: Dutton & Wentworth, printers, 1827. vi, 106 p.
S-30433.
Copies: Ct; DeGE; ICN; M; MH; MWA; MiD-B; NN; NNE; NjR.

2708. United States. Laws.
A bill to amend "An act authorizing the erection of a bridge over the River Potomac, within the District of Columbia." January 30, 1827.... [Washington: 1827.] 1 p. (H.R. 406)
S-31119.
Copies: DNA.

2709. Boston. City Council. Joint Committee on the Boston Free Bridge.
City of Boston. In Common Council, June 30, 1828. The joint committee of the City Council, who were charged ... with the location and general superintendance of the Boston Free Bridge ... report. [Boston: 1828.] 8 p.
S-32415.
Copies: MBAt; MHi.

2710. Cartersville Bridge Company.
Documents laid before the Legislature, at the session of 1827–28, upon the subject of the application of the Cartersville Bridge Company, for a loan or lottery, to re-build the bridge; with the proceedings thereupon. Richmond: Printed by Samuel Shepherd & Co., 1828. 52 p.
S-32611.
Copies: ViHi; ViU.

2711. Harrison, Randolph, 1769–1839.
To the speakers and members of both houses of the General Assembly of Virginia. Randolph Harrison begs leave to offer a respectful memorial ... in relation to the Cartersville Bridge Company.... [Richmond?: 1828.] 4 p.
S-33484.
Copies: Vi.

2712. Proprietors of Charles River Bridge.
To the Honourable the Senate, and House of Representatives in General Court assembled. The memorial of the Proprietors of Charles River Bridge. [Boston?: 1828.] 4 p.
Signed: J. Belknap, chairman.
N-P0606221.
Copies: MH-BA.

2713. South Carolina. Laws.
Part of an "Act to establish certain roads, bridges and ferries.... " [Columbia, S.C.: 1828.] Broadside 25 x 20 cm.
Sections 33–39 of the act, authorizing the establishment of the Broad River Bridge Company.
Hummel-2079.
Copies: ScU.

2714. United States. Laws.
A bill authorizing the corporation of Georgetown to erect a bridge over the river Potomac, within the District of Columbia. March 17, 1828. [Washington: 1828.] 3 p.

2714 (*continued*)
(H.R. 223)
S-35721.
Copies: DNA.

2715. Wernwag, Lewis.
Memorial of Lewis Wernwag, to the Senate and House of Representatives, praying an extension of his patent for his improvement in bridge architecture. [S.l.]: Printed by John S. Gallaher, 1828. 12 p.
Preface dated "Harper's Ferry, Feb. 10, 1828."
S-37135.
Copies: NIC.

2716. Malden Bridge to the people. Boston: 1829. 20 p.
Concerns mainly the effects of the proposed Warren Bridge over the Charles River.
S-39421.
Copies: DLC; DeGE; ICJ; ICN; MBAt; MH; MH-BA; MWA; NN; NNE; NjR; WHi.

2717. United States. Laws.
A bill to alter the bridge and draw across the Potomac, from Washington City to Alexandria. Feb. 2, 1829.... [Washington: 1829.] 3 p. (S. 76)
S-41005.
Copies: DNA.

2718. Long, Stephen Harriman, 1784–1864.
Description of the Jackson Bridge, together with directions to builders of wooden or frame bridges. By S. H. Long.... Patented March, 1830. Baltimore: Printed by Sands & Neilson, 1830. 24 p., illus., 2 plates.
C-2266.
Copies: DBRE; DLC; DeGE; MWA; MdHi; MiU-T; NIC; NN; PPAmP.

2719. Pickering, Octavius, 1791–1868.
Case of the Proprietors of Charles River Bridge against the Proprietors of Warren Bridge, argued and determined in the Supreme Judicial Court of Mass.... Boston: Hilliard, Gray, Little & Wilkins, 1830. 203 p.
C-3070.
Copies: DLC; M; MBS; MH; MWA; MiD-B; NcD.

2720. Proprietors of Charles River Bridge.
The Proprietors of Charles River Bridge, in equity vs. the Proprietors of the Warren Bridge. [Boston: Samuel Condon, printer, 1830.] 206 p.
Case before the supreme Court of Massachusetts.
C-2475.
Copies: MH; MHi.

2721. ——— [Boston: 1830.] 72 p.
Sabin-101499.
Copies: MBAt.

2722. United States. Congress. House. Committee on Internal Improvements.
Wheeling and Belmont Bridge Company. May 11, 1830. Mr. Letcher, from the Committee on Internal Improvements, made the following report. [Washington: Printed by D. Green, 1830.] 1 p. (21st Cong., 1st sess. House. Rept. 399)
C-5234.
Copies: DLC; MWA; NjR.

2723. Unites States. Congress. House. Committee on the Judiciary.
Bridge accross the Ohio. April 29, 1830 ... Mr. Wickliffe, from the Committee on the Judiciary ... made the following report. [Washington: 1830.] 1 p. (21st Cong., 1st sess. House. Rept. 387)
C-4602.
Copies: DLC; MWA; NjR.

2724. United States. Laws.
An act to alter the bridge and draws across the Potomac, from Washington City to Alexandria. February 2, 1830. Mr. Powers, from the Committee for the District of Columbia, to which was referred the bill from the Senate, entitled "An act to alter the

bridge and draw across the Potomac, from Washington City to Alexandria," reported the same with amendments. [Washington: 1830.] 4 p. (S. 40)

C-3836.

Copies: DNA.

2725. Wheeling Bridge Company.

Virginia. Memorial of the Wheeling Bridge Company. March 12, 1830.... [Washington: 1830.] 5 p. (21st Cong., 1st sess. House. Rept. 349)

C-5226.

Copies: DLC; MWA; NjR.

6. Hydraulic engineering

2726. To the inhabitants of Philadelphia.... [Philadelphia: Printed by E. and T. Bradford, 1771.] Broadside.

Concerns the cost of maintaining public pumps, and the nightly watch and street lamps in Philadelphia.

Mp-12245.

Copies: PPL.

2727. Colles, Christopher, 1738–1816.

As the several inhabitants of this city are particularly interested in the following affair, it is therefore judged proper to lay the same before them. Copy of a proposal of Christopher Colles, for furnishing the city of New-York with a constant supply of fresh water.... New York: Printed by Hugh Gaine, [1774]. Broadside 38 x 24 cm.

B-3713; Mp-42574.

Copies: MWiW-C; NN (photocopy).

2728. New Jersey. Laws.

An act to enable the owners and possessors of meadow and marsh, lying on Newton-Creek, in the county of Gloucester, to make and maintain a bank, dam and necessary waterworks, to stop the tide out of the said creek, and to keep the watercourse thereof open and clear. [Trenton: 1786.] [6] p.

N-N0154716.

Copies: DLC.

2729. Salem and Danvers Aqueduct.

Rules and regulations of the proprietors of the Salem and Danvers Aqueduct, 1797. [Salem, Mass.: Printed by Thomas C. Cushing, 1797.] Broadside.

E-32799.

Copies: MSaE (not located).

2730. Philadelphia. Councils. Joint Committee on Water Supply.

Report of the Joint Committee of the Select and Common Councils, on the subject of bringing water to the city. Philadelphia: Printed by Zachariah Poulson, Junior, 1798. 12 p.

"The president, managers and company of the Delaware and Schuylkill Canal Navigation, to the Select and Common Councils of the city of Philadelphia," p. [8]–12, signed William Smith, chairman.

Mp-34350.

Copies: CSmH; DeGE; MHi; MWA; NN; PPL; RPJCB.

2731. Faris, William.

A treatise on the new invented patented machine.... To which is added, remarks on several subjects by the author. Baltimore: Printed by Warner & Hanna, for the author, 1799. 66 p.

A water elevator to operate an overshot wheel.

Copies: MWA.

2732. [Latrobe, Benjamin Henry] 1764–1820.

An answer to the Joint Committee of the Select and Common Councils of Philadelphia, on the subject of a plan for supplying the city with water, &c. [Philadelphia: 1799.] 7 p.

Signed: B. H. Latrobe, engineer, Philadelphia, March 2d, 1799.

"Account of steam engines, &c.," p. 5–7.

Mp-36086.

Copies: DLC; DeGE; MWA; NN; NNC; PHi.

2733. Latrobe, Benjamin Henry, 1764–1820.

Remarks on the Address of the Committee of the Delaware and Schuylkill Canal Company to the Committee of the Senate and House of Representatives, as far as it notices the "View of the practicability and means of supplying the City of Philadelphia with wholesome water," by B. Henry Latrobe, engineer, in a letter to John Miller, junr. Esquire, Chairman of the Committee of the Select Council. Philadelphia: Printed by Z. Poulson, 1799. 18 p.

Mp-35714.

Copies: DeGE; PHi; PPL.

2734. Latrobe, Benjamin Henry, 1764–1820.

View of the practicability and means of supplying the city of Philadelphia with wholesome water. In a letter to John Miller, Esquire, from B. Henry Latrobe, engineer. December 29th, 1798. Printed by order of the Corporation of Philadelphia. Philadelphia: Printed by Zachariah Poulson, junior, 1799. 20 p.

Mp-35715.

Copies: DLC; DeGE; MWA; N; NNE; PPAmP; PPL; PU; RPJCB.

2735. Manhattan Company, New York.

Report of the Manhattan Committee.... [New York]: Printed by John Furman, [1799]. 37 p.

Dated: Fourth month 22d, 1799.

Concerns supplying water for the city of New York by the Manhattan Company, including proposals from Nicholas Roosevelt for a steam engine, and other proposals and estimates for the system from Christopher Colles, William Weston, Joseph Browne, and Elias Ring.

Mp-35950.

Copies: N; NN; NNC.

2736. New York (City). Common Council.

Proceedings of the Corporation of New-York, on supplying the city with pure and wholesome water; with a Memoir of Joseph Browne, M. D. on the same subject. [New York]: Printed by John Furman, 1799. v, [3], [9]–29 p.

Mp-35941.

Copies: DLC; DeGE; MB; MWA; N; NHi; NN.

2737. New York (State). Laws.

An act of incorporation of the Manhattan Company. [New York]: Printed by John Furman, 1799. 12 p.

Caption title on p. 3: *An Act for supplying the city of New York with pure and wholesome water.* (Passed 2d April, 1799).

Mp-35917.

Copies: CSmH; MB; MH; NHi; NN.

2738. ———— [New York]: David, 1799. 8 p.

N-M0170066.

Copies: PPAmP.

2739. Philadelphia. Councils. Joint Committee on Water Supply.

Report of the Joint Committee of the Select and Common Councils, appointed to receive information on the subject of watering the city, and to employ agents, when necessary, for promoting that object.... Philadelphia: Printed by Zachariah Poulson, Junior, 1799. 7 p.

Mp-36089.

Copies: NHi; NN; PPAmP; PPL.

2740. Philadelphia. Councils. Joint Committee on Water Supply.

Report to the Select and Common Councils, on the progress and state of the water works, on the 24th of November, 1799. Printed by the order of the Corporation of Philadelphia. Philadelphia: Printed by Zachariah Poulson, junior, 1799. 44 p.

Mp-36091.

Copies: DLC; DeGE; MWA; MHi; NN; PPAmP; PPL; RPJCB.

2741. Philadelphia. Ordinances.

An ordinance providing for the raising of a sum of money for supplying the city of Philadelphia with wholesome water. Printed by order of the Corporation of Philadelphia. Philadelphia: Printed by Zachariah Poulson, junior, 1799. 7 p.

Mp-36087.

Copies: DLC; DNLM; NN; PPL.

2742. To the citizens of New York. Your attention and curiosity having been much excited, by an act passed at the close of the late session of the Legislature, incorporating Aaron Burr, Dr. Browne, John B. Church, John Watts, and others, into a company, stiled the "Manhattan Company.... " [New York: 1799.] Broadside 33 x 26 cm.

Dated: April 30, 1799 and signed "A citizen."

The Manhattan Company was established to supply water for the city of New York.

B-10912; Mp-48965.

Copies: NHi; NN.

2743. Weston, William, ca. 1752–1833.

Report of William Weston, Esquire, on the practicability of introducing the water of the river Bronx into the city of New-York. Done at the request of the Corporation of the said city. [New York]: Printed by John Furman, 1799. 16 p.

Mp-36709.

Copies: CtY; DLC; DeGE; MBAt; MH; MWA; NHi; NN; NNC; PPL.

2744. Philadelphia. Councils. Committee for the Introduction of Wholesome Water.

Report of the Committee for the Introduction of Wholesome Water into the city of Philadelphia. Printed by order of the Select and Common Councils. Philadelphia: Printed by Zachariah Poulson, junior, 1801. 16 p.

Signed: John Miller, junr., Chairman.

S&S-1145.

Copies: CSmH; DLC; DeGE; MWA; NN.

2745. Philadelphia. Councils. Common Council. Committee on Water Works.

Report of the Committee appointed by the Common Council to enquire into the state of the water works. Printed by order of the Common Council. Philadelphia: Printed by William Duane, at the Aurora printing office, 1802. 85 p.

Letter from Mr. Latrobe, Dec. 29, 1798, p. 6–21.

S&S-2888.

Copies: DLC; DeGE; MiD-B; NN; OClWHi; PPL.

2746. Philadelphia. Councils. Joint Committee on Water Works.

Report of the Joint Committee appointed by the Select and Common Councils for the purpose of superintending and directing the water works.... Philadelphia: Printed by Robert Cochran, 1802. 23 p.

S&S-2889.

Copies: PHi; PPAmP.

2747. Philadelphia. Watering Committee.

Report of the Watering Committee to the Select & Common Councils, November 1, 1803. Philadelphia: Printed by William Duane, 1803. 21 p.

S&S-4856.

Copies: MWA; PHi (at PPL); PPL.

2748. ———— November 1, 1804. Philadelphia: Printed by Robert Cochran, 1804. 15 p.

S&S-7041.

Copies: MWA; PPL.

2749. Boston. Committee upon the Mill Pond.

Report of the Committee upon the Mill Pond. Boston: Gilbert & Dean, printers, [1805]. Broadside.

Dated: Jan. 15, 1805.

S&S-8054.

Copies: MH.

2750. Salem and Danvers Aqueduct.

Rules and regulations of the proprietors of the Salem and Danvers Aqueduct, as established by their directors for the direction of those persons who have a right to take water from the same. [Salem, Mass.?: 1805.] Broadside 32 x 22 cm.

DeGE copy dated in ink: "First day of August 1805."

S&S-9301.

Copies: DeGE; MSaE; MWA; NN.

2751. New York (City). Comptroller's Office.

Comptroller's report, on the water right belonging to the Corporation. New York: Printed by James Cheetham, 1806. 9 p.

S&S-10994.

Copies: DLC; MWA; NHi.

2752. Philadelphia. Watering Committee.

Report of the Watering Committee to the Select and Common Councils, October 7th, 1805.... Philadelphia: Printed by William M'Corkle, 1806. 5, [15] p.

S&S-50710.

Copies: PPF; PPL.

2753. ———— November 13th, 1806.... Philadelphia: Printed by Jane Aitken, 1806. 6, [13] p.

S&S-11151.

Copies: PPF; PPL.

2754. Boston.

Contract between the town of Boston and the Mill Corporation. [Boston: Belcher & Armstrong, printers, 1807?] 12 p.

Concerns the use of water power.

S&S-12175.

Copies: MH; MWA; PPL.

2755. Boston.

Report [on] petition of James Ivers, Esq., and others, relative to the filling up of the Mill Pond, May 25, 1807.... Boston: Oliver & Munroe, 1807. Broadside.

S&S-12176.

Copies: MH.

2756. Pennsylvania. Laws.

An act to authorize Robert Kennedy, his heirs and assigns, to dig and support a mill-race in and adjacent to the river Schuylkill near its falls.... [Lancaster, Pa.]: William Greear, printer, [1807]. 3 p.

At head of title: No. 38. January 7, 1807.

S&S-13333.

Copies: CSmH; MWA (xerox copy).

2757. Philadelphia. Watering Committee.

Report of the Watering Committee to the Select & Common Councils, November 13th, 1807.... Philadelphia: Printed by Jane Aitken, 1807. 6 [8], 10 p.

S&S-13372.

Copies: MWA; PPF; PPL.

2758. Baltimore Water Company.

The president and directors of the Water Company, considering the necessity of mutual understanding between all persons supplied with water from their works ... have agreed, that the following rules and regulations must be strictly complied with.... [Baltimore]: G. Dobbin & Murphy, print., [1808?]. Broadside 34 x 21 cm.

S&S-14412.

Copies: MWA (xerox copy); MdHi.

2759. Philadelphia. Watering Committee.

Report of the Watering Committee to the Select and Common Councils, November 24th, 1808.... Philadelphia: Printed by Jane Aitken, 1808. 6, [10] p.

S&S-15908.

Copies: PPF; PPL.

2760. ———— November 2d, 1809.... Philadelphia: Printed by Jane Aitken, 1809. 5, [11] p.

S&S-18374.

Copies: MWA; PPF; PPL.

2761. Philadelphia. Watering Committee.

Watering Committee. January 16th, 1809. [Philadelphia: 1809.] [3] p.

Caption title.

Copies: PHi.

2762. Salem and Danvers Aqueduct.

Rules and regulations of the proprietors of the Salem and Danvers Aqueduct, as established by their directors, for the direction of those persons who have right to take water from the same. Salem, Mass.: 1810. Broadside.

S&S-21264.

Copies: MSaE; MWA (xerox copy).

2763. Clapp, Joseph, 1770– .

Description of Clapp's patented crank engine, for raising and throwing water.... Greenfield, [Mass.]: 1811. 8 p., plate.

N-C0447616.

Copies: CtY.

2764. Philadelphia. Watering Committee.

Report of the Watering Committee to the Select & Common Councils, November 7, 1811.... Philadelphia: Printed by Jane Aitken, 1811. 6, [9] p.

Copies: MWA; PPF; PPL.

2765. Latrobe, Benjamin Henry, 1764–1820.

Proposals for the establishment of a company for the supply of New Orleans with water. [Washington?: 1812.] 3 p.

Signed in mss.: B. H. Latrobe, Surveyor of the public buildings U.S.

Copies: LNA.

2766. Perkins, Jacob, 1766–1849.

An entire view and section of an improved fire-engine, patented by Jacob Perkins and Allen Pollock. [Philadelphia?: 1812?] Broadside.

Copies: PHi.

2767. Philadelphia. Watering Committee.

Report of the Watering Committee to the Select and Common Councils, November 5th, 1812.... Philadelphia: Printed by Jane Aitken, 1812. 6, [8] p.

Also a variant printing: Philadelphia: Printed by Ann Cochran, 1812. 6, [9] p. at MWA.

S&S-26441.

Copies: MWA; PPL.

2768. Philadelphia. Watering Committee.

Report of the Watering Committee upon the present state of the works for supplying the city with water, and several other plans proposed for that purpose. May 2, 1812. Philadelphia: Printed by Jane Aitken, 1812. 25 p., plan.

S&S-26442.

Copies: MHi; PHi; PPL.

2769. Boston. Committee on the Petition of Isaac P. Davis.

Report on the petition of Isaac P. Davis and others, for liberty to build a mill dam and turnpike road from the bottom of Beacon Street, and for other mill improvements.... [Boston: 1813.] 7 p.

Dated: 11th day of June, 1813.

S&S-27990.

Copies: DLC; MB; MH; MWA.

2770. Hough, Horatio Gates.

Diving; or, An attempt to describe upon hydraulic and hydrostatic principles, a method of supplying the diver with air under water.... Hartford: Printed by John Russell, jr., 1813. 7, [1] p., diagr.

S&S-28780.

Copies: CSmH; Ct; CtHi; DLC; DeGE; ICJ; MB; MWA; MiU-C; NN; PPF; PPL.

2771. Manhattan Company, New York.

Manhattan water-works. The president and directors of the Manhattan Company

2771 *(continued)*
have resolved the following resolutions.... [New York]: John Law, printer, [1813]. Broadside 22 x 15 cm.
Copies: MWA (photocopy); NHi.

2772. Philadelphia. Watering Committee.
Report of the Watering Committee to the Select and Common Councils, November 11th, 1813.... Philadelphia: Printed by Lydia R. Bailey, 1813. 8, [8] p.
S&S-29497.
Copies: MWA; PPF; PPL.

2773. Boston. Committee on the Petition of Isaac P. Davis.
Report of the town's Committee on the Petition of Isaac P. Davis, and others, for liberty to build a mill dam and turnpike road from the bottom of Beacon-Street, and for other mill improvements. [Boston: 1814?] 20 p.
S&S-30961.
Copies: DLC; MH-BA; MWA; NB; NN.

2774. ———— [With other documents relating to the establishment of the Boston and Roxbury Mill Corporation. Boston: 1814?] 28 p.
S&S-32059.
Copies: MH; MHi; MWA.

2775. Dearborn, Benjamin, 1754–1838.
A plan of those parts of Boston, and the towns in its vicinity, with the waters and flats adjacent, which are immediately or remotely connected with the contemplated design of erecting perpetual tide-mills. Boston: 1814. Map 47 x 36 cm.
S&S-31304.
Copies: MH; MHi; MSaE; MWA.

2776. Massachusetts. General Court. Committee on the Petitions of Isaac P. Davis.
The committee of both houses, to whom was referred the several petitions of Isaac P. Davis and others, praying that they may be incorporated by the name of the Boston and Roxbury Mill Corporation ... exhibit ... a description of the works contemplated by the ... petitioners.... [Boston: 1814.] 16 p.
The *Act of incorporation* is on p. [11]–16.
S&S-32039.
Copies: DLC; M; MH; MWA.

2777. United States. Laws.
A bill to incorporate a company for the purpose of supplying Georgetown with water. March 11th, 1814. [Washington: 1814.] 10 p. (S. 36)
S&S-33182.
Copies: DLC.

2778. Philadelphia. Watering Committee.
Report of the Watering Committee to the Select and Common Councils. Read January 11, 1815.... Philadelphia: Printed at the office of the United States gazette, 1815. 10, [8] p.
Copies: PPF; PPL.

2779. ———— Read January 25, 1816.... Philadelphia: Printed by William Fry, 1816. 32 p.
Copies: PPF; PPL.

2780. Marshall, John, 1755–1835.
Opinion of Judge Marshall, in relation to the Potomac Company; and agreement by the Potomac Company with General Henry Lee, relative to the use of the surplus water at Great Falls of Potomac. Georgetown, D.C.: Printed by W. A. Rind and Co., 1817. 4 p.
Contains only the "Opinion of Judge Marshall," dated January 31, 1792.
S&S-41350.
Copies: DLC; DWP.

2781. Philadelphia. Watering Committee.
Report of the Watering Committee to the Select and Common Councils. Read January 23, 1817. Published by order of the Councils. Philadelphia: Printed by William Fry, 1817. 18 p.
S&S-41782.
Copies: PPF; PPL.

2782. [Austin, Benjamin] 1752–1820.
Remarks on a late proposition from the Boston and Roxbury Mill Corporation.... [Boston: Rowe, 1818.] 12 p.
Signed: An impartial citizen.
Concerns the Boston mill dam.
S&S-43117.
Copies: MB; MHi; MWA.

2783. [Cotting, Uriah.]
The Boston and Roxbury Mill Corporation. The subscriber and his associates now offer to the public the project of their great improvement, viz. the contemplated mill power and toll road.... [Boston: 1818.] 24 p.
Signed and dated on p. 20: Uriah Cotting, Boston, Jan. 1818.
S&S-43744.
Copies: CtY; DLC; DeGE; MB; MH-BA; MWA; MnU; PPL.

2784. Description of the common sewers to be built. New York: Thomas P. Low, 1818. 11 p.
Dated: July 30, 1818.
Copies: NN.

2785. Massachusetts. Laws.
An act more effectually to provide for the upholding of mills and manufactories worked by water, and for repealing certain acts, and parts of acts relating thereto. [Boston: February 1818.] 7 p.
N-M0310889.
Copies: KPT.

2786. ——— [Boston: May 1818.] 8 p.
N-M0310890.
Copies: KPT.

2787. Philadelphia. Watering Committee.
Report of the Watering Committee to the Select and Common Councils. Read January 22, 1818. Philadelphia: Printed by William Fry, 1818. 17 p.
S&S-45285.
Copies: MWA; PPF; PPL.

2788. ——— November the twelfth, 1818.... Philadelphia: Printed by William Fry, 1818. 42 p.
S&S-45286.
Copies: PHi; PPL.

2789. Philadelphia. Watering Committee.
An additional report on water power, by the Watering Committee: with communications on the subject from Messrs. Ariel Cooley, Lewis Wernwag, Thomas Oakes, William Briggs, and William Lehman. And other documents.... Philadelphia: Printed by William Fry, 1819. 15 p.
S&S-49075.
Copies: MWA; P; PPL.

2790. Philadelphia. Watering Committee.
Report of the Watering Committee of the agreements with the Schuylkill Navigation Company and White and Gillingham, relating to the water power of the river Schuylkill. Published by order of Councils. Philadelphia: Printed by W. Fry, 1819. 23 p.
S&S-49076.
Copies: DeGE; MWA; PPL.

2791. Philadelphia. Watering Committee.
Report of the Watering Committee to the Select and Common Councils. Read January 14, 1819.... Philadelphia: Printed by William Fry, 1819. 17 p.
Copies: PPL.

2792. Taylor, William, 1763–1838.
Mill dam survey. [Boston: 1819.] Map 130 x 188 cm. (manuscript).
S&S-49567.
Copies: MB.

2793. Taylor, William, 1763–1838.
Plan of the aqueduct from Jamaica Pond to William's store. [Boston: 1819.] Map 130 x 188 cm. (manuscript).
S&S-49568.
Copies: MB.

2794. Hale, Elisha.
A description and general view of

2794 *(continued)*
Elisha Hale's horizontal water wheel. Patented July 5, 1820. Baltimore: Printed by Joseph Robinson, 1820. 8 p., diagrs.
S-1489.
Copies: DLC; N.

2795. Philadelphia. Watering Committee.
Report of the Watering Committee, on the propriety of raising the dam at Fair Mount. Philadelphia: Printed by Lydia R. Bailey, 1820. 8 p.
S-2734.
Copies: DeGE; NN; PPL.

2796. Philadelphia. Watering Committee.
Report of the Watering Committee, on the subject of obtaining water power from the river Schuylkill. Philadelphia: Printed by Lydia R. Bailey, 1820. 50 p.
S-2735.
Copies: DeGE; MWA; PPL; PU; RP.

2797. Philadelphia. Watering Committee.
Report of the Watering Committee to the Select and Common Councils. Read February 10, 1820. Published by order of the Councils. Philadelphia: Printed by Lydia R. Bailey, 1820. 18 p.
S-2737.
Copies: MWA; PPF; PPL.

2798. Philadelphia. Watering Committee.
The Watering Committee respectfully state to Councils.... [Philadelphia: 1820.] 2 p.
Dated: December 19, 1820.
Copies: PHi.

2799. United States. Commissioner of the Public Buildings.
Message from the President of the United States, communicating (in pursuance of a resolution of the Senate of the 6th inst.) information concerning the purchase of fire engines. December 13, 1820.... Washington: Printed by Gales & Seaton, 1820. 9 p. (16th Cong., 2d sess. Senate. Doc. 20)
Transmitting a report of Samuel Lane, Commissioner of Public Buildings.
S-3918.
Copies: A-Ar; DLC; DeGE; DeU; G; NjP; R.

2800. Merrick (S. V.) & Company.
A description of the patent improved fire engines, and other hydraulic machines, invented by Jacob Perkins, and manufactured by S. V. Merrick & Co., successors to Perkins and Bacon, at the factory, back of St. James's Church, North Seventh-Street, Philadelphia. With information respecting their use. [Philadelphia: 1821.] 8 p., 2 plates.
N-M0472894; S&S-35612 (incorrectly dated [1815?]).
Copies: DLC; DeGE; NN; PPAmP; PPL.

2801. Philadelphia. Ordinances.
A supplement to an ordinance, entitled, "An ordinance empowering the Mayor to raise money, to be applied to laying down iron pipes of conduit, from the water works at Fair Mount." [Philadelphia: 1821.] Broadside.
Copies: PPL.

2802. Philadelphia. Watering Committee.
Report. [Philadelphia: 1821.] [1], 2, 2 p.
Dated: November 22, 1821.
Copies: PHi.

2803. Philadelphia. Watering Committee.
Report of the Watering Committee to the Select and Common Councils. Read January 18, 1821.... Philadelphia: Printed by Lydia R. Bailey, 1821. 20 p.
S-6448.
Copies: MWA; PHi; PPL.

2804. For sale. On the 8th of the Sixth month (June) next.... Mill site for sale, June 8, 1822, on Octorara Creek, in Caecil County, and state of Maryland.... The site is within fifty rods of the boatable waters of the Susquehanna Canal.... The stage road from Port Deposit to Lancaster passes along the boundary line of this property.... Wilming-

ton, [Del.]: Printed by J. Wilson, 1822. Broadside 35 cm.
S-8721.
Copies: DeHi.

2805. [Kenworthy, William.]
The economical hydrostatic lift. [Baltimore: 1822.] 12 p.
Reprinted from the *American farmer,* v. 4, no. 10, 31st May, 1822.
S-9198.
Copies: DLC; ICJ.

2806. Philadelphia. Ordinances.
Acts of Assembly and ordinances relating to the water works. [Philadelphia: 1822.] 20 p.
S-9839.
Copies: DLC; DeGE; PPAmP.

2807. Philadelphia. Ordinances.
An ordinance, for raising supplies to defray the expenses for completing the general survey, and the expense attending the repairs of public pumps, the public highways and expenses for 1822. [Philadelphia: 1822.] 1 p.
S-9903.
Copies: PHi.

2808. Philadelphia. Ordinances.
A supplement to an ordinance, entitled "An ordinance empowering the Watering Committee to purchase of Josiah White and Joseph Gillingham their rights to water power of the river Schuylkill, and also to raise money on loan for the purpose of erecting a dam and other works at or near Fair Mount." [Philadelphia: 1822?] 1 p.
Copies: PPL.

2809. Philadelphia. Watering Committee.
Report of the Watering Committee, read in Select Council, November the twelfth, 1818. Printed by order of Select Councils. Philadelphia: Printed by William Fry, 1822. 42 p., illus.
Concerns mainly the need for a new water main and a new system of piping.
S-9888.
Copies: DLC; DeGE; PPL.

2810. Philadelphia. Watering Committee.
Report of the Watering Committee, to the Select and Common Councils. Read January 24, 1822. Philadelphia: Printed by Lydia R. Bailey, 1822. 22 p.
S-9889.
Copies: MWA; PPF; PPL.

2811. Philadelphia. Watering Committee.
Report of the Watering Committee, to the Select & Common Councils of the city of Philadelphia, relative to the Fair Mount water works. Read January 9, 1823. Philadelphia: Printed by Lydia R Bailey, 1823. 22 p., plates (part fold.).
S-13744.
Copies: CtHC; DLC; DeGE; IU; MH; MWA; MdHi; NRom; PP; PPL; PU.

2812. ——— Philadelphia: Printed by Lydia R. Bailey, 1823. 24 p., plates.
A variant issue of the above.
S-13745.
Copies: MWA; PPL.

2813. Philadelphia. Watering Committee.
Report of the Watering Committee, to the Select & Common Councils. Read in Select Council, January 8, 1824.... Philadelphia: Printed by Lydia R. Bailey, 1824. 10 p.
S-17604.
Copies: DeGE; MWA; PPF; PPL (2 copies, each from a variant issue).

2814. Philadelphia. Watering Committee.
The Watering Committee, having always in view the wishes of Councils.... [Philadelphia: 1824.] [2], [2] p.
Copies: PHi.

2815. Haight, Nicholas.
Practical treatise on the water delivered by the Manhattan Company; with practical observations on waters surrounding the city of

2815 *(continued)*
New York. New York: P. & H. Van Pelt, 1825. 14 p.
S-20780.
Copies: DNLM; NB.

2816. Philadelphia. Watering Committee.
Report of the Watering Committee, on the disposal of the surplus water power of the river Schuylkill, and the construction of a canal between the Schuylkill and Delaware. Published by order of the Councils. Philadelphia: Printed by Lydia R. Bailey, 1825. 24 p.
"Papers accompanying the report" include an examination of canal routes by Benjamin Wright and Canvass White.
S-21868.
Copies: DBRE; DLC; DeGE; MB; MWA; PPL; RP.

2817. Philadelphia. Watering Committee.
Report of the Watering Committee, to the Select and Common Councils. Read January 13, 1825.... Philadelphia: Printed by Lydia R. Bailey, 1825. 18 p.
S-21869.
Copies: PHi; PPL.

2818. ——— Philadelphia: Printed by Lydia R. Bailey, 1825. 12 p.
Copies: MWA; PPL.

2819. Treadwell, Daniel, 1791–1872.
Report made to the mayor and aldermen of the city of Boston, on the subject of supplying the inhabitants of that city with water. Boston: Printed by order of the Common Council; True & Greene, printers, 1825. 32 p.
S-22503.
Copies: DLC; M; MB; MBAt; MH; MHi; MWA.

2820. Baldwin, Loammi, 1780–1838.
Letter to the Hon. John Pickering, president of the Salem Mill Dam Corporation, upon the estimates, &c. of that work. Salem, [Mass.]: Printed by Andrews & Foote, 1826. 11 p.
S-23581.
Copies: CtSoP; DLC; MB; MH; MSa; MWA; MiU-C; NjPT; NjR; PPL.

2821. Cincinnati.
[Report of the committee to investigate the present state of the Cincinnati Water Works, for the purpose of fixing a value on the property . . . Cincinnati: 1826.] Broadside.
S-24107.
Copies: OCHP (not located).

2822. [Griffith.]
Essay on the art of boring the earth for the obtainment of a spontaneous flow of water. With hints towards forming a new theory for the rise of waters. New Brunswick, [N.J.]: Rutgers Press, printed and published by Terhune & Letson, 1826. 46, [1] p.
"Erratum," 1 p. at end.
DeGE and PPL copies contain manuscript inscriptions assigning the authorship to Griffith.
S-24441.
Copies: CtY; DeGE; MB; MH; MWA; MdHi; NN; NjR; OClWHi; OkU; PPL; TxDaM.

2823. New York Water Works Company.
Report to the directors of the New-York Water Works Company, by their engineer. [New York: 1826.] 11 p.
Signed: Canvass White, engineer. New-York, Jan. 9th, 1826.
N-N0217708; Shaw-680.
Copies: DNLM; NNC.

2824. Philadelphia. Watering Committee.
Report of the Watering Committee, to the Select and Common Councils. Read January 12, 1826.... Philadelphia: Printed by Lydia R. Bailey, 1826. 19 p.
S-25734.
Copies: MWA; PPL.

2825. Quin, Henry.
A newly improved moving lock; termed the hydraustatic lift, with illustrations of the

principles, and a full description of the plan, together with observations relative to the utility of; and contrasted with the latest and original plans.... Easton, [Pa.]: Printed by Henry & William Hutter, 1826. 10 p.
Copies: DeGE.

2826. Salem Mill-Dam Corporation.
An act to incorporate the Salem Mill-Dam Corporation, together with the by-laws. May 20, 1826. Salem, [Mass.]: Printed by Andrews & Foote, 1826. 16 p.
S-26008.
Copies: MHi; MPeaHi; MiU-C.

2827. Salem Mill-Dam Corporation.
Report and resolutions of the directors of the Salem Mill-Dam Corporation. Presented October 5, 1826. [Salem, Mass.: 1826.] 8 p.
S-26009.
Copies: DLC; MH-BA; PPL.

2828. Philadelphia. Watering Committee.
Report of the Watering Committee to the Commissioners of the District of Spring Garden, January 22, 1827. Philadelphia: 1827. 12 p.
S-30694.
Copies: PHi (not located).

2829. Philadelphia. Watering Committee.
Report of the Watering Committee, to the Select and Common Councils. Read January 11, 1827. Published by order of the Councils. Philadelphia: Printed by Lydia R. Bailey, 1827. 24 p.
S-30215.
Copies: MWA; PHi; PPL.

2830. Philadelphia. Watering Committee.
Report of the Watering Committee, to the Select and Common Councils of the city of Philadelphia, relative to the Fair Mount Water Works. Read January 9, 1823. 2d ed., with additions. Philadelphia: Printed by Lydia R. Bailey, 1827. 27 p., 3 plates (2 fold.).
Appendix: Extract from the annual report of the Watering Committee, dated January 3, 1824 ... , p. [23]–27.
S-30216.
Copies: DeGE; PHi; PPL.

2831. Renwick, James, 1790–1863.
Report on the water power at Kingsbridge, near the city of New York, belonging to the New-York Hydraulic Manufacturing and Bridge Company.... New York: Printed by Samuel Marks, No. 63 Vesey-Street, 1827. 12 p., 3 fold. plans.
S-30424.
Copies: CSmH; DLC; DeGE; MWA; NHi; NN; NNC; NNNAM.

2832. Salem Mill-Dam Corporation.
Report of the directors of the Salem Mill Dam Corporation, at a meeting of the stockholders, held May 7th, 1827. Salem, [Mass.]: Printed by W. & S. B. Ives, Observer office, 1827. 9 p.
S-30515.
Copies: CtSoP; DLC; MH; MH-BA; MWA; PPL.

2833. [Colden, Cadwallader David] 1769–1834.
An answer to Mr. John L. Sullivan's report to the Manufacturing Society in New-Jersey, in a letter to the directors of the Morris Canal Company, by the president of that company. New York: William A. Davis, printer, 1828. 39 p.
Signed: Cadwallader D. Colden.
Concerns the use of water power from the Passaic River.
S-32763.
Copies: DLC; NN; NNC; NjR.

2834. Cooper, Phelps and Campbell, Windsor, Vt.
Fire engines. Windsor (Vermont) March 14, 1828.... Cooper, Phelps & Campbell. [Windsor, Vt.: 1828.] Broadside 38 x 27.5 cm.
S-32834.
Copies: MB.

2835. Harrison, H.

Hydraulics; or, Improvement of the Mississippi; offering a plan for the reduction of its overflow, and of that of its alluvial tributary streams. The expense arising from the execution of this plan, the apportionment of that expense, and the distribution of it among the different parties who may receive permanent benefits from the effectuation of it. To which is subjoined, a brief enumeration of some of the most important consequences which will result. Cincinnati: Geo. T. Williamson, printer, 1828. 67, [1] p.

S-33482.

Copies: CSmH; CtY; ICJ; MB; MHi; MWA; MnHi; Nh; O; PPL.

2836. Philadelphia. Watering Committee.

Report of the Watering Committee, to the Select and Common councils. Read January 10, 1828. Published by order of the Councils. Philadelphia: Printed by Lydia R. Bailey, 1828. 27 p.

S-34745.

Copies: MWA; PHi

2837. Philadelphia. Watering Committee.

Report of the Watering Committee, to the Select and Common Councils of the city of Philadelphia, relative to the dam at Fair Mount. Read and adopted September 25, 1828. Philadelphia: Printed by L. R. Bailey, 1828. 22 p.

S-34746.

Copies: DeGE; MWA; PHi; PPL.

2838. [Sullivan, John Langdon] 1777–1865.

Refutation of Mr. Colden's "Answer" to Mr. Sullivan's report to the Society for Establishing Useful Manufactories in New-Jersey, upon the intended encroachments of the Morris Canal Company in diverting from their natural course the waters of the Passaic. [Boston?: 1828.] 56 p.

Signed: J. L. Sullivan.

S-35437.

Copies: CtY; DBRE; DLC; MH; NN; NNE.

2839. [Sullivan, John Langdon] 1777–1865.

Report, on the origin and increase of the Paterson manufactories, and the intended diversion of their waters by the Morris Canal Company: also on post rail roads, as the means of cheap conveyance throughout New-Jersey, of bringing Susquehanna coal to the iron mines and forges, and to supply Paterson and New-York: also on a method of supplying the city of New-York with water from the great falls of the Passaic. Paterson: Printed by Day & Burnett, 1828. 60, [2] p., fold. map.

Signed: Jno. L. Sullivan.

S-35438.

Copies: CSmH; DBRE; DLC; DeGE; MWA; NN; NNE; NjR.

2840. Treadwell, Daniel, 1791–1872.

To S. C. Phillips, Benja. Hawkes and Perley Putnam, Esq's, committee of the directors of the Salem Mill Dam Corporation. [Boston: 1828.] 16 p.

Circular letter, signed: Daniel Treadwell. Boston, 1828.

S-35554.

Copies: DLC; MH; NjR.

2841. Lehman, George, d. 1870.

Four views of Fairmount water-works. Drawn and acquatinted by George Lehman. Philadelphia: C. G. Childs, R. H. Hobson, 1829. 4 plates 25x30 cm

Cover title.

S-39268.

Copies: DLC; PPAmP.

2842. New Jersey. Court of Chancery.

Law opinion, in the case of the Paterson Manufacturing Society, against the Morris Canal and Banking Company. Paterson, N.J.: D. Burnett, printer, 1829. 27 p.

Caption [p.3]: In Chancery, New Jersey. Between the Society for Establishing Useful Manufactures and others, complainants, and the Morris Canal and Banking Company, defendants.

Concerns the use of water power.

S-39764.
Copies: DeGE; MdHi; NjR.

2843. New York (State). Commissioners Appointed in Relation to the Draining of Cayuga Marshes.
Report of the Commissioners appointed in relation to the draining of Cayuga marshes. [Albany: 1829.] 4 p. (Assembly. Doc. 153)
Dated: In Assembly, March 11, 1829.
N-N0190898.
Copies: MWA; NN.

2844. Philadelphia. Watering Committee.
Report of the Watering Committee, to the Select and Common Councils. Read Jan. 22, 1829. Pub. by order of the Councils. Philadelphia: Printed by Lydia R. Bailey, 1829. 37 p.
S-40040.
Copies: MWA; PHi.

2845. Washington. Mayor.
Draining low grounds—Washington. Message from the President of the United States, transmitting the information required by a resolution of the House of Representatives ... for draining the low ground on and near the public reservations in the city of Washington. April 3, 1826.... [Washington: Gales & Seaton, printers, 1829.] 12 p. (20th Cong., 2d sess. House. Doc. 55)
Transmitting a letter from the mayor's office.
S-26960 (erroneously dated 1826).
Copies: DLC; DeU; MWA; NjR.

2846. Baltimore. City Council. Water Committee.
Report of the Water Committee to the city Council of Baltimore. [Baltimore: 1830?] 15 p.
The Committee was appointed in January 1829, to report to the next annual session of the City Council. The report is concerned with best modes of furnishing water to the city.
S-37569.
Copies: MdBP.

2847. Baltimore Water Company.
Memorial of the Water Company, to the City Council of Baltimore. [Baltimore: 1830.] 8 p.
N-B0079400.
Copies: DLC.

2848. Camaron, George.
Water for the Capitol. February 2, 1830.... Letter from George Camaron, upon the subject of furnishing a supply of water for the use of the Capitol. [Washington: 1830.] 1 p. (21st Cong., 1st sess. House. Rept. 145)
C-5232.
Copies: DLC; DeU; MWA; NjR.

2849. Lynchburg, Va. Watering Committee.
Report of the Watering Committee to the Common Council of the Corporation of Lynchburg, relative to the water-works; read and approved March 24, 1830, and ordered to be printed. Lynchburg, Va.: Printed by Fletcher & Toler, 1830. 12 p.
Report of Albert Stein, civil engineer, p. 4–11.
C-2300.
Copies: NN; NcD; Vi.

2850. Mills, Robert, 1781–1855.
Letter from Robert Mills to the chairman of the Committee on the Public Buildings, upon the subject of providing a supply of water for the use of the Capitol. April 5, 1830.... [Washington: Printed by D. Green, 1830.] 4 p. (21st Cong., 1st sess. House. Rept. 344)
C-5233.
Copies: DLC; MWA, NjR.

2851. New York (City). Common Council.
Report of the Special Committee to whom was referred the resolutions of alderman Stevens, relating to the supply of water by the Manhattan Company. Common Council, May 31, 1830. [New York: 1830.] 6 p.
N-N0174297.
Copies: DNLM.

2852. New York (State). Legislature. Assembly. Committee on Canals.

Report on petitions and remonstrances respective Fort Miller Dam. Albany: 1830. 3 p. (Assembly. Doc. 226)

Hasse (New York)-p. 146; Whitford-p. 1271.

Copies: MWA; N; NN.

2853. New York (State). Legislature. Senate.

Report on memorial respecting incorporation for furnishing water supply for the city of Albany. Albany: 1830. 2 p. (Senate. Doc. 202)

Hasse (New York)-p. 551.

Copies: MWA; N.

2854. New York (State). Legislature. Senate. Select Committee on the Construction of a Dam Across the Oneida River.

Report of the Select Committee, relative to the construction of a dam across the Oneida River, at Oak Orchard reef. [Albany: 1830.] 5 p. (Senate. Doc. 137)

Dated: February 13, 1830.

N-N0201478; Whitford-p. 1247.

Copies: MWA; NN.

2855. Philadelphia. Watering Committee.

Report of the Watering Committee, to the Select and Common Councils. Read February 11, 1830.... Philadelphia: Printed by Lydia R. Bailey, 1830. 30 p.

C-3028.

Copies: CtY; DBRE; DLC; ICJ; MH; MHi; MWA; MiU-T; NN; PPAmP; PPL.

2856. Skinner, Ichabod Lord, 1767–1852.

Skinner's grand basin—water for public buildings, &c. March 8, 1830.... [Washington: 1830.] 5 p. (21st Cong., 1st sess. House. Rept. 281)

Concerns the creation of sufficient water supply for the federal buildings in Washington.

C-5166.

Copies: DLC; DeU; MWA; NjR; ViU.

2857. Sullivan, John Langdon, 1777–1865.

A description of a sub-marine aqueduct, to supply New-York with water from New-Jersey; connected with a commercial canal, and rail-way, for the direct western trade of this city. By J. L. Sullivan.... New York: G. & C. & H. Carvill, 1830. 29 p.

C-3632.

Copies: MWA; MiU-T; NN; NNE; PPL; ScU; WHi.

2858. United States. Congress. House. Committee on Internal Improvements.

Skinner's grand central basin. April 22, 1830 ... Mr. Craig, of Virginia, from the Committee on Internal Improvements ... made the following report. [Washington: 1830.] 1 p. (21st Cong., 1st sess. House. Rept. 376)

C-5168.

Copies: DLC; MWA; NjR.

2859. United States. Congress. House. Committee on the District of Columbia.

Skinner's grand central basin. April 20, 1830 ... Mr. Taliaferro, from the Committee for the District of Columbia, to which had been referred, in part, the letter from Mr. Skinner, upon the subject of a central canal basin in the city of Washington, made the following report. [Washington: 1830.] 1 p. (21st Cong., 1st sess. House. Rept. 375)

C-5167.

Copies: DLC; MWA; NjR.

VIII. MECHANICAL ENGINEERING

1. Tools and machines

2860. Boston, May 13, 1756. To be seen (for a short time) at the house of Mr. William Fletcher, merchant, New-Boston; that elaborate and matchless pile of art, called the Microcosm, or the World in miniature.... [Boston: Printed by Edes and Gill, 1756.] Broadside.

The "Microcosm" was a very complicated automaton, described as containing "upwards of twelve hundred wheels and pinions." It was built by Henry Bridges in London.

B-1794; Mp-40825.

Copies: MWA.

2861. Bogle, Samuel.

By the medium of the curious numerical machine, invented by the ingenious Mr. Cox, are to be dsposed [*sic*] of the following two hundred and fifty lots of ... plated goods, just imported by Samuel Bogle.... [Philadelphia: 1773.] Broadside 41 x 25 cm.

Dated at head of text: Philadelphia 9th of December, 1773.

B-3621; Mp-42409.

Copies: PPL.

2862. Martin, John.

The new invented friction roller, made and sold by John Martin, at his manufactory.... New York: Printed by W. Morton, 1788. Broadside.

E-21219.

No copy known.

2863. Rumsey, James, 1743?–1792.

The explanations and annexed plates, of the following improvements in mechanics, viz. 1st, a new constructed boiler. 2d, a machine for raising water. 3d, a grist-mill, &c. 4th, a saw-mill, &c. are respectfully dedicated to the patrons of the plans, by the inventor. Philadelphia: Printed by Joseph James, 1788. 8 p., [3] l., 3 fold. plates.

Includes: *The Explanations and annexed plates* (8 p., 2 plates); *Explanation of a steam engine, and the method of applying it to propel a boat* (2 l., plate); and *A Specification of a boiler for generating steam* (1 l.).

Mp-21439.

Copies: NN; PPAmP.

2864. Leslie, Robert, d. 1803.

A catalogue of plans of machines, and mechanical apparatus now collected: of which, together with many others, models will be made as soon as possible, and exhibited by Robert Leslie, at his museum in Philadelphia.... [Philadelphia: 1789.] Broadside 44 x 26.5 cm.

Copies: DeGE.

2865. Pennsylvania. Laws.

An act for vesting in James Rumsey, Esquire, the exclusive right and privilege of making, using, and vending divers engines, machines and devices, by him invented, or improved, for a term of years therein mentioned. [Philadelphia: 1789.] [3] p.

The inventions and improvements claimed are: pipe-boiler, steamboat, improvement upon Savery's steam engine, and im-

2865 (*continued*)
provement of Dr. Barker's mill.
Mp-22051.
Copies: RPJCB.

2866. Pierson, Josiah G.
New invented washing-mill, made by Josiah G. Pierson, No. 10 White-Hall-Street.... [New York]: T. Greenleaf, print., [1792]. Broadside 27.5 x 23 cm.
Dated: New-York, October 6, 1792.
B-8109; Mp-46547.
Copies: DLC.

2867. ———— [New York]: T. Greenleaf, printer, [1792]. Broadside 27.5 x 22.5 cm.
Dated: New-York, November 3, 1792. Includes instructions for use.
B-8110.
Copies: CSmH.

2868. Du Buc Marentille, Abraham.
All people wrecked at sea saved. Description of the machines invented for that purpose, and for which patents have been obtained. By M. A. Du Buc Marentille. Elizabethtown, [N.J.]: Printed by Shepard Kollock for the author, 1803. 18 p.
S&S-4116.
Copies: CtY; DLC; DeGE; MWA; NN; NjR.

2869. The mystery of the invisible lady unfolded and explained.... Salem, [Mass.]: Printed for the purchasers, 1805. Broadside 51 x 22 cm.
Explains how a "philosophical machine" responds to questions spoken into its trumpet with answers spoken in a female voice.
Copies: MWA.

2870. Save labour. The subscriber, having purchased of the original patentee, the right of making, using and vending his patent auger, for boring post-holes, wells, pumps, &c. &c. in the state of New-Jersey, hereby offers to dispose of the same.... [Burlington?, N.J.: ca. 1805.] Broadside 30 x 19 cm.
Probably referring to a patent for an auger for boring post holes, issued in 1804 to Christopher Hoxie from Hudson, N.Y.
Copies: DeGE.

2871. Baldwin, Isaac.
[Receipt given as agent for one patent washing machine. Salem, Mass.: 1808.] Broadside.
Dated: March 15, 1808.
S&S-14400.
Copies: MSaE; MWA (xerox copy).

2872. An accurate delineation of Readheffers [*sic*] perpetual motion. [Philadelphia]: Sold by W. Y. Birch, No. 37 S. 2d Str., [ca. 1810]. Broadside, illus. 21 x 35 cm.
The inventor's full name: Charles Redheffer.
Engraving of the machine with explanatory notes.
W. Y. Birch was active at this address from 1805 to 1810.
Copies: DeGE.

2873. United States. Congress. House. Select Committee to Whom was Referred the Memorial of Benjamin Tyler.
Report of the Select Committee, to whom was referred on the seventeenth instant, the memorial of Benjamin Tyler, Jun. and John Tyler. January 28, 1811.... Washington: A. & G. Way, printers, 1811. 4 p.
Concerns the renewal of their patent on a fly-wheel.
S&S-24312.
Copies: DLC; MWA.

2874. Denison, Erastus.
The United States of America. To all to

whom these letters patent shall come. Whereas Erastus Denison, a citizen of the United States, hath alleged that he has invented a new and useful improvement in the washing machine.... [Washington: 1812.] Broadside 32 x 20 cm.

Dated: Washington, 30 December, 1812.

Copies: DeGE.

2875. Colles, Christopher, 1738–1816.

Description of the numerical telegraph. For communicating unexpected intelligence by figures, letters, words, and sentences, with directions for writing the correspondence either public or private, and shewing the manner of working the machine with perfect accuracy and despatch. Brooklyn: Printed by Alden Spooner, 1813. 19 p., front.

S&S-28182.

Copies: MdBP; MiU-C; N; NBLiHi; NHi.

2876. Perkins, Jacob, 1766–1849.

Petition of Jacob Perkins, of Newburyport, in Essex County, and State of Massachusetts, for a renewal of his patent right to a nail cutting machine. June 15, 1813. Referred to a select committee.... Washington: A. & G. Way, printers, 1813. 6 p.

S&S-30313.

Copies: DLC; DeGE; MWA.

2877. Odiorne, G.T.

To the public ... nail machines. Odiornes vs. Perkins.... Boston: 1815. Broadside 30 x 18 cm.

Signed: G. T. & E. Odiorne, Boston, Sept. 26th, 1815.

Disputes the claim of Jacob Perkins concerning a nail cutting machine supposedly invented by Jacob Reed.

Copies: MWA.

2878. Pollock, Allen.

Description of Allen Pollock's patent scale beam. [Boston: 1815.] 1 p., plate.

Copies: PPAmP.

2879. Reynolds, Thomas.

Licence. Know all men, that I, Thomas Reynolds, of Chester County, Pennsylvania, have received of ________ in full for licence hereby granted ________ to construct, make, and use one of my patented machines, called the crane screw & bale, for raising mill stones.... [S.l.: ca. 1815.] Broadside, illus. 12.5 x 19.5 cm.

Patent granted to Thomas Reynolds on Nov. 8, 1814.

Copies: DeGE.

2880. United States. Congress. Senate. Committee on Claims.

In Senate of the United States, January 12, 1818. The Committee of Claims, to whom was referred the petition of William Faris, report.... [Washington: 1818.] 4 p. (15th Cong., 1st sess. Senate. Doc. 55)

Concerns a patent for propelling both carriages and boats, issued 1797.

S&S-46150.

Copies: DLC; DeGE; DeU. MWA.

2881. Brown, William.

Description of a machine, or mill for breaking plaster of Paris and other hard substances; invented by William Brown, of the city of Baltimore and state of Maryland, and patented. [Baltimore: 1820?] Broadside, illus. 40 x 25 cm.

Receipt, at the bottom of the sheet, dated in ink: 1st mo. 4th, 1821.

Copies: DeGE.

2882. United States. Congress. Senate. Committee on the Judiciary.

In Senate of the United States. February 9, 1821. Mr. Smith, from the Committee on the Judiciary, to which was referred

2882 (*continued*)
the petition of Thomas Oxley, made the following report.... [Washington: 1821.] 1 p. (16th Cong., 2d sess. Senate. Doc. 86)

Concerns the patents for machines to clear land.

S-7266.

Copies: DLC; DeU; MWA; NjP.

2883. United States. Laws.

A bill authorizing the Secretary of State to issue a patent to Thomas Oxley. February 4, 1821.... [Washington: 1821.] 1 p. (H.R. 240)

Concerns equipment and method for land clearing.

S-7061.

Copies: DLC.

2884. Wellford, Robert.

Carpenter's friend. The subscribers having obtained a patent from the United States, are now enabled to ease the very great labor of Youth as well as of old age, by furnishing them with an improved plane with friction metallic and other rollers, for planing floors. Philadelphia: Manning, printer, 1821. Broadside, illus. 21 x 17 cm.

Dated: May 14, 1821, and signed: Robert Wellford & James H. Deas.

Copies: PPF.

2885. United States. Congress. Senate. Committee on the Judiciary.

In Senate of the United States, February 11, 1822. The Committee on the Judiciary, to which was referred the petition of Abel Pratt, made the following report.... [Washington: 1822.] 1 p. (17th Cong., 1st sess. Senate. Doc. 39)

Refusal to extend the patent on machinery for making combs, issued to Phineas Pratt in 1805.

S-10851.

Copies: DLC; DeGE; MWA; NjP.

2886. Allen, Stephen, 1767–1852.

Reports on the stepping or discipline mill, at the New-York Penitentiary: together with several letters on the subject. By the Mayor. New York: Printed by Van Pelt and Spear, 1823. 32 p., front.

S-11582.

Copies: CSmH; DLC; DeGE; MBAt; NCH; NN.

2887. Hardie, James, 1760 (ca.)–1826.

The history of the tread-mill, containing an account of its origin, construction, operation, effects as it respects the health and morals of the convicts, with their treatment and diet; also, a general view of the penitentiary system, with alterations necessary to be introduced into our criminal code, for its improvement. New York: Printed by Samuel Marks, 1824. viii, [9]–70 p., front.

S-16406.

Copies: CSmH; DeGE; MB; MH; MWA; NBLiHi; NN; NjPT; NjR; PU.

2888. Mills, Robert, 1781–1855.

Improvements in the rotatory engine; or, Several modes by which a direct circular motion may be obtained; using either as an agent steam, air, water, gunpowder or any other propelling power. Columbia, S.C.: 1824. 11 p., 12 plates.

S-17165.

Copies: PPF.

2889. United States. Laws.

An act to authorize the issuing of letters patent to Samuel Brown. February 11, 1824.... [Washington: 1824.] 1 p. (H.R. 56)

Concerns a gas engine.

S-18354.

Copies: DNA.

2890. West, John.

To all these whom these letters patent

shall come: Whereas, John West, a citizen of the United States, hath alleged that he has invented a new and useful improvement in the washing machine.... [Washington, D.C.: 1824.] [4] p.

Dated: December 14, 1824.
Copies: MWA.

2891. Owings, John.

Memorial of John Owings, of Baltimore, the inventor and patentee of a certain machine, therein described; praying the purchase of it by Congress. February 14, 1825.... Washington: Printed by Gales & Seaton, 1825. 5 p. (18th Cong., 2d sess. Senate. Doc. 31)

Concerns an earthmoving machine.
S-22969.
Copies: DLC; DeGE; DeU; MWA; NjR.

2892. [Bradford, Gamaliel] 1795–1839.

The history and analysis of the supposed automaton chess player, of M. de Kempelen, now exhibiting in this country, by Mr. Maelzel; with lithographic figures, illustrative of the probable method by which its motions are directed.... Boston: Hilliard, Gray and Co., 1826. 24 p., plate.

S-23918.
Copies: CSmH; DLC; M; MB; MBAt; MHi; MWA; NN; NjR; OCl; PP; PPAmP; PPL; ViU; WHi; WU.

2893. Maelzel, Johann Nepomuk, 1772–1838.

Maelzel's exhibition.... Automaton chess player. Automaton trumpeter. Automaton rope dancers. [Boston: W. W. Clapp, printer, 1826.] Broadside.

S-25183.
Copies: DLC; NN.

2894. Nicholson, John.

The operative mechanic, and British machinist; being a practical display of the manufactories and mechanical arts of the United Kingdom.... First American from the second London ed., with additions.... Philadelphia: H. C. Carey & I. Lea, 1826. 2 v., plates (part fold.).

S-25596.
Copies: CtY; DP; DeGE; FWpR; GU; ICU; IaU; InCW; LNL; MH; MWA; MdBP; MeB; MiU; NN; NjR; OClWHi; P; PPL; PU; RPA; ScCleU; Vi; VtU.

2895. [Jones, Thomas P.] 1774–1848.

Observations upon the automaton chess player of von Kempelen, and upon other automata and androides, now exhibiting in the United States, by Mr. Maelzel.... Philadelphia: J. Dobson, 1827. 12 p., illus.

S-29382.
Copies: DLC.

2896. Documents for and against Charles Redheffer, as it respects his alleged invention of perpetual motion. Collected from authentic sources. Philadelphia: 1828. 1 p.l., 24 p.

Copies: DeGE.

2. Heating, lighting, and ventilation

2897. [Franklin, Benjamin] 1706–1790.

An account of the new invented Pensylvanian fire-places: wherein their construction and manner of operation is particularly explained; their advantages above every other method of warming rooms dem-

2897 (*continued*)
onstrated; and all objections that have been raised against the use of them, answered and obviated. With directions for putting them up, and for using them to the best advantage. And a copper-plate, in which the several parts of the machine are exactly laid down, from a scale of equal parts. Philadelphia: Printed and sold by B. Franklin, 1744. [2], 37, [1] p., illus., fold. plate.

Mp-5395.

Copies: CSmH; CtY; DLC; DeGE; ICJ; InU; MH; MWA; MiU-C; P; PPL; PPiU; RPJCB; ViU.

2898. Curtenius (Peter T.) and Company, New York.

New York air furnace. Peter T. Curtenius, and Co. have repaired the New-York air furnace, and have procured the best workmen, together with the necessary apparatus to carry on the manufacture of cast-iron, in the completest and best manner, so that the ware they make will be equal to any imported from Europe, and the price less.... New York: Printed by J. M'Lean, [1787]. Broadside, illus. 33 x 22 cm.

B-6478; Mp-45058.

Copies: NHi; NN (photocopy).

2899. Rumford, Sir Benjamin Thompson, count, 1753–1814.

Essays, political, economical, and philosophical. By Benjamin, Count of Rumford.... The 1st American from the 3d London ed.... Boston: Printed by Manning and Loring, for David West, 1798–1804. 3 v., illus., plates (part fold.).

Contains essays on fireplaces, fuel, heating of houses, and application of steam.

Mp-34656, 36251; S&S-7212.

Copies: CtY; DLC; MBAt; MH-BA; MWA; NN; PPF; PPL; PU; RPJCB.

2900. Robotham, Richard.

Description of the air-pump ventilator, for the ventilating of ships, mines, prisons, hospitals, &c. Hudson, N.Y.: 1801. Broadside, illus. 29 x 27 cm.

S&S-1260.

Copies: DeGE; MHi.

2901. DeWitt, Benjamin, 1774–1819.

Letter on chimney fire-places, addressed to Robert R. Livingston.... Albany: Printed by S. Southwick, 1810. 12 p.

S&S-19958.

Copies: DeGE; N; NN.

2902. Pettibone, Daniel.

Description of the improvements of the rarifying air-stove, for warming and ventilating hospitals, churches, colleges, courts of justice, dwellinghouses.... Philadelphia: Printed for the author, 1810. 33 p.

S&S-21052.

Copies: CtY; DeGE; MWA; NHi; PPAmP; PPL.

2903. Pettibone, Daniel.

Pettibone's economy of fuel; or, Description of his improvements of the rarefying air-stoves, grates, tubes, pipes, cylinders, or open stoves, or common fire places ... for warming and ventilating hospitals, churches, colleges, courts of justice, dwelling houses ... gun-powder, or other manufactories, banking-houses ... &c. &c. with or without the application of steam ... 2d ed., improved, with engravings. Also an extract from James Sharp's London edition of his improvements on that of Dr. Benjamin Franklin. Philadelphia: Printed by A. Dickinson for the author, 1812. 2 p. l., [3]–62 p., illus.

S&S-26439.

Copies: DLC; DeGE; LNHT; M; MBAt; MWA; N; NN; PPL.

2904. Pettibone, Daniel.

The economy of fuel; or, Russian stove

noticed.... Philadelphia: 1813. 2 p.
Circular letter.
S&S-29493.
Copies: PPL (not located 1975).

2905. Melville, David, 1773–1856.
Melville's patent improved gas-apparatus.... Any person or company wishing to be furnished with the improved gas apparatus are requested to apply to Winslow Lewis, Boston, or David Melville, Newport, R.I. who are the sole proprietors of the patent right. [Boston: 1814?] Broadside, illus. 30.5 x 25 cm.
Copies: PPAmP.

2906. Jenkins, John, d. 1823.
A new and useful improvement, offered to the publick, for constructing fire places & chimnies, with horizontal funnels, on a plan, which will save fuel, insure warm rooms, and preserve health, at a small expense. By J. Jenkins.... Warren, (R.I.): Printed by S. Randall, for the author, 1815. iv, [5]–24 p., illus.
N-J0078699.
Copies: CtY; DeGE; MH-BA.

2907. Cooper, Thomas, 1759–1839.
Some information concerning gas lights.... Philadelphia: John Conrad & Co.; J. Maxwell, printer, 1816. vii, [9]–190 p., 1 l., illus., 2 plates.
S&S-37340.
Copies: CtY; DLC; DeGE; ICJ; InLP; MB; MH; MWA; NN; NNC; NcD; NjR; PPL; PU.

2908. United States. Congress. House. Committee on Public Buildings.
Report of the Committee on the Public Buildings, relative to Daniel Pettibone's petition. February 19, 1817.... [Washington: 1817.] 10 p. (14th Cong., 2d sess. House. Doc. 90)
Concerns the heating of the Capitol.
S&S-42721.
Copies: DLC; DeGE; MWA.

2909. Johnson, Joseph, 1776–1862.
An address to the Literary and Philosophical Society of South-Carolina; read before them, at their anniversary, May the 8th, 1822.... Charleston, S.C.: Printed for the Society by J. Hoff, 1822. 11 p., illus.
Includes a description and illustration of a fire place invented by Johnson.
S-9156.
Copies: DLC; MWA; PPAmP; ScC.

2910. Boston Gas Light Company.
Extracts from the bye-laws of the company, together with some statements respecting oil gas, lately received from the Liverpool Company. [Boston: 1823.] 6, 2 p.
S-11942.
Copies: M.

2911. Facts, in relation to the introduction of gas light into the city of Philadelphia. With a plan of a company to be formed for that purpose. [Philadelphia: 1824.] 12 p.
Attributed to Philadelphia Gas Light Company.
S-17611.
Copies: DeGE; MWA; P; PPAmP; PPL; PU; RPB.

2912. New York Gas Light Company.
The New York Gas Light Company vs. Samuel Leggett and others. [New York: 1825.] 15 p.
Copies: MWA.

2913. Bull, Marcus.
Experiments to determine the compar-

ative value of the principal varieties of fuel used in the United States, and also in Europe; and on the ordinary apparatus used for their combustion. Philadelphia: J. Dobson; New York: G. & C. Carvill, [etc., etc.], 1827. x, 103, [1] p., front., tables.

S-28317.

Copies: CSt; CtY; DLC; DeGE; GHi; IU; KSalW; MH; MWA; NN; NNBG; NNE; NjR; OCl; P; PPL; RPB; ViU.

2914. [Bigelow, Jacob] 1787–1879.

A short reply to a pamphlet published at Philadelphia entitled "A defence of the Experiments to determine the comparative value of the principal varieties of fuel used in the United States, and also in Europe.... By Marcus Bull...." By one of the committee of the American Academy. Boston: Hilliard, Gray, Little and Wilkins; New York: G. and C. Carvill, [etc., etc.], 1828. 12 p.

S-32377.

Copies: CSmH; DLC; DeGE; MB; MH; MHi; MdHi; NN; PPAmP; PPL; RPB; ScU.

2915. Bull, Marcus.

An answer to "A short reply to 'A defence of the Experiments to determine the comparative value of the principal varieties of fuel, &c.' by one of the Committee of the American Academy."—(Published in Boston).... Philadelphia: J. Dobson, [etc.]; London: J. Miller, [etc., etc], 1828. 16 p.

S-32510.

Copies: CtY; DLC; DeGE; MB; MH; MWA; MdHi; NN; NjR; PPAmP; PPL.

2916. Bull, Marcus.

A defence of the Experiments to determine the comparative value of the principal varieties of fuel used in the United States, and also in Europe. containing a correspondence with a committee of the American Academy of Arts and Sciences; their report, and remarks thereon; and animadversions on the manner in which the trust confided to the Academy by Count Rumford has been managed.... Philadelphia: Judah Dobson; New York: G. & C. Carvill, [etc.], 1828. 51 p.

S-32511.

Copies: CtY; DLC; DeGE; IaDaM; MBAt; MH-BA; MWA; MdHi; NN; NjR; PPAmP; PPL; RPB; VtU.

2917. [Fessenden, Thomas Green] 1771–1837.

Patent lamp tea-kettle, boiler, &c. To be used with oil or alcohol. Boston: Press of Isaac R. Butts, 1829. 12 p.

S-38566.

Copies: DLC; MBAt.

2918. Bull, Marcus.

Experiments to determine the comparative quantities of heat evolved in the combustion of the principal varieties of wood and coal used in the United States for fuel; and also, to determine the comparative quantities of heat lost by the ordinary apparatus made use of for their combustion. Read April 7, 1826. [Philadelphia: 1830.] 63 p., plate.

From the *Transactions of the American Philosophical Society,* new ser., v. 3, 1830.

N-B0945467.

Copies: IaU; MH-A; MdBJ; NIC; NN.

2919. Vinton, Thomas.

Specification of a patent for an improved cooking grate, intended chiefly for cooking by means of anthracite coal. Granted to Thomas Vinton ... October 31, 1829. With remarks by the editor of the Journal of the Franklin Institute. From the Journal of the Franklin Institute, for March, 1830. Philadelphia: 1830. 7 p.

C-5175.

Copies: MB.

3. Steam engineering

2920. Rumsey, James, 1743?–1792.
Explanation of a steam engine, and the method of applying it to propel a boat. Invented by James Rumsey, of Berkeley County, Virginia. Philadelphia: Printed by Joseph James, 1788. 2l., plate.
Mp-21439; N-R0511465.
Copies: NN; PPAmP.

2921. Rumsey, James, 1743?–1792.
Philadelphia, 1788. At a meeting of the American Philosophical Society, on the 18th of April, 1788, a letter was received from Mr. James Rumsey.... [Philadelphia: 178.] [2] p.
Excerpts from the minutes of the American Philosophical Society, concerning steam engines, with an added note on John Voight's pipe boiler. Signed by James Rumsey and dated: Philadelphia, May 10, 1788.
B-6660; Mp-45218.
Copies: DLC; RPJCB.

2922. Rumsey, James, 1743?–1792.
A plan, wherein the power of steam is fully shewn, by a new constructed machine, for propelling boats and vessels of any burthen, against the most rapid streams or rivers, with great velocity. Also, a machine, constructed on similar philosphical [*sic*] principles, by which water may be raised for grist or saw-mills, watering of meadows, &.&.... [Richmond?: Printed by Bartgis and Willcocks?, 1788.] 20, [2] p.
Postscript on water mills and the author's apology of mistakes and errata, [2]p. at end.
Mp-21440.
Copies: DLC; MBAt; MWA; NHi; PPAmP.

2923. ———— [Philadelphia: 1788.] 20 p.
Reprint of the Virginia edition, without the postscript. Issued with John Fitch's *The Original steam-boat supported*
N-R0511467.
Copies: DLC; DeGE; NN; NNC.

2924. Rumsey, James, 1743?–1792.
A short treatise on the application of steam, whereby is clearly shewn, from actual experiments, that steam may be applied to propel boats or vessels of any burthen against rapid currents with great velocity. The same principles are also introduced with effect, by a machine of a simple and cheap construction, for the purpose of raising water sufficient for the working of grist-mills, saw-mills, &c., and for watering meadows and other purposes of agriculture.... Philadelphia: Printed by Joseph James, 1788. 26 p.
Corrected reissue of the work which was originally published under the title: *A Plan wherein the power of steam is fully shewn.*
The *Proofs, &c,* p. 11–26, were also issued as a separate pamphlet: [Richmond?: 1788.] 16 p. (copies at CSmH and ViU).
Mp-21442.
Copies: DLC; DeGE; MB; MWA; NN; PPAmP; PPL; RPJCB.

2925. ———— Philadelphia: Printed by Joseph James, 1788. 26 p.
Hastily printed first state, containing misprints in title and text.
Mp-21441.
Copies: DLC; MH; NN.

2926. ———— Philadelphia: Joseph James, 1788. 25 p.
Title page and text reset, differing from Mp-21441 and 21442.
B-6796; Mp-45353.
Copies: CSmH.

2927. Philadelphia. Councils. Joint Committee on Water Works.

Report. The Joint Committee, to whom was referred the memorial and remonstrance of Nicholas J. Roosevelt . . . offer the following report. [Philadelphia: 1801?] 3 p.

Concerns the steam engines installed by N. J. Roosevelt for the Philadelphia water works.

Signed: John Vallance, J. Worrell, Wm. Dalzell, Sam. Wetherill, Benjamin Say.

S&S-1147.

Copies: PPAmP.

2928. Proposals for establishing a company for the purpose of employing the surplus power of the steam-engine erected near the river Schuylkill: under the title and firm of the Philadelpia Rolling Company. [Philadelphia: 1801?] 3, [1] p.

Attributed to Benjamin H. Latrobe and to Nicholas J. Roosevelt.

"Agreement . . . to pay into the hands of the Treasurer of the Philadelphia Rolling Company of the sum . . . towards the formation of capital stock . . . ," [1] p. at end.

N-P0312588 (dated [1800]).

Copies: PHi; RPJCB.

2929. Evans, Oliver, 1755–1819.

The subscriber, with diffidence presumes to lay before the Honourable Senators and Representatives in Congress, individually . . . the following short description of the principles of steam engines. . . . [S.l.: 1804.] [4] p.

Signed on page 3: Oliver Evans.

S&S-6265.

Copies: DLC; MWA.

2930. Evans, Oliver, 1755–1819.

The abortion of The young steam engineer's guide: containing an investigation of the principles, construction and powers of steam engines. A description of a steam engine on new principles. . . . A description of a machine, and its principles, for making ice and cooling water in large quantities . . . by the power of steam. . . . A description of four other patented inventions. . . . Philadelphia: Printed for the author by Fry and Kammerer, 1805. xii, 139 p., 4 plates.

"Appendix: containing a concise account of the invention and improvement of steam engines. Extracted from Dobson's edition of the Encyclopedia. With notes and additions by the author. To which is added Mr. John Stevens' letters on the author's improvements, with the answers thereto," p. [81]–139.

S&S-8409.

Copies: CSt; DLC; DeGE; ICJ; LNHT; MH-BA; MWA; MiU-T; PHi; PPL; PSC; ViW.

2931. [Evans, Oliver] 1755–1819.

Memorial &c. to the honourable the senators and representatives in Congress, the memorial of Oliver Evans respectfully showeth, that since the age of twenty, your memorialist has devoted much time, labour, and study, to discover and invent many new and useful improvements. . . . [Washington?: 1811.] 14 p.

Concerns the renewal of his patents on the steam engine and other inventions.

N-E0205395.

Copies: DLC; MB; PPAmP.

2932. Evans, Oliver, 1755–1819.

To the honorable the senators and representatives in Congress individually. [S.l.: 1813.] 2 p.

Concerns the extension of his patent on the steam engine.

Copies: DLC; PPF.

2933. Evans, Oliver, 1755–1819.

To the Congress of the United States, each honorable member individually. [Washington: 1814.] 4 p.

Dated at end: Washington, February 5, 1814.

Concerns his patent on the steam engine.

S&S-31433.

Copies: DLC; MB.

2934. [Evans, Oliver] 1755–1819.

To the honorable the senators and

representatives in Congress. Each individual member is humbly requested to read the following with attention.... [Washington: 1814.] 8 p.

Concerns the extension of his patent on the steam engine.

Copies: MB; MiU-T; PHi.

2935. United States. Laws.

A bill to amend the act, entitled "An act for the relief of Oliver Evans." [Washington: 1814.] 2 p. (S. 27)

Dated: February 25, 1814.

S&S-33149.

Copies: DLC.

2936. United States. Laws.

A bill to extend the time of Oliver Evans' patent for steam engines. February 17th, 1814.... [Washington: 1814.] 1 p. (H.R. 58)

S&S-33177.

Copies: DLC.

2937. ———— November 9, 1814.... [Washington: 1814.] 1 p. (S. 8)

S&S-33179.

Copies: DLC.

2938. Johnson, John.

Memorial of John Johnson, Junior. April 24, 1820.... Washington: Printed by Gales & Seaton, 1820. 4 p. (16th Cong., 1st sess. House. Doc. 117)

Recommends the reduction of duty on English pig iron, which, mixed with the American iron, is used in castings to produce parts of steam engines.

S-3866.

Copies: DLC; DeGE; G; MWA; NjP; O.

2939. Evans, Oliver, 1755–1819.

The young steam engineer's guide; containing an investigation of the principles, construction and powers of steam engines. A description of the steam engine on new principles.... A description of a machine, and its principles for making ice and cooling water in large quantities.... A description of other patented inventions. Illustrated with two engravings.... Philadelphia: H. C. Carey & I. Lea, [1824?]. xii, 73, [81]-139 p., 2 plates.

"Appendix: Containing a concise account of the invention and improvement of steam engines. Extracted from Dobson's edition of the Encyclopedia. With notes and additions by the author. To which is added Mr. John Stevens' letter on the author's improvements, with the answer thereto," p. [81]–139.

Issue of part of *The Abortion of the young steam engineer's guide,* 1805, with a new title page, p. 73 reprinted, and p. 74– [80] (Article XVIII–"Useful invention by different persons," with plates [III]–IV) omitted.

This publication has been also dated as [1804?], cf. S&S-6266, and [1805] cf. S&S-8410. Both are erroneous, because the publisher did not use the name "H. C. Carey & I. Lea" until 1822.

N-E0205426.

Copies: CtY; DLC; DeGE; ICJ; MH; MWA; NN; NNE; PP; PPAmP; PSC; PU; ViU.

2940. Lardner, Dionysius, 1793–1859.

Popular lectures on the steam engine, in which its construction and operation are familiarly explained; with an historical sketch of its invention and progressive improvement. By the Rev. Dionysius Lardner.... With additions by James Renwick.... New York: E. Bliss, 1828. xi, 171 p., plates.

First published in London, 1828.

S-33817.

Copies: CSmH; CtHT; DBRE; GDC; IaHi; MH; MWA; MiU-C; NCH; NNC; PPL; TxU; Vi.

2941. [Lester, Ebenezer A.]

Lester's pendulum steam engine. [Boston: 1830.] 16 p.

Signed: Ebenezer A. Lester. Boston, August 2, 1830.

C-2212.

Copies: DLC; MH; MH-BA; MWA.

2942. Renwick, James, 1790–1863.
Treatise on the steam engine.... New York: G. & C. & H. Carvill, 1830. viii, [9]–328 p., illus., 9 fold. plates.
C-3280.
Copies: CSmH; CtY; DBRE; DLC; DeGE; ICJ; LNHT; MH-BA; MWA; NIC; NN; NjR; OCl; P; PPF; PPL; PSC; RNR; Vi.

2943. [Whiting, Henry], 1788–1851.
The age of steam, by Anonymous, Esq. Detroit: [Printed at the Journal office], 1830. 16 p.
C-5441.
Copies: MiD-B.

2944. ——— Detroit: 1830. 16 p.
A variant issue.
C-5442.
Copies: MiD-B.

IX. MANUFACTURING

1. General works

2945. Barnard, Thomas, 1716–1776.

A sermon preached in Boston, New-England, before the Society for Encouraging Industry and Employing the Poor, September 20, 1758.... Boston: S. Kneeland, 1758. 1 p.l., 25, [2] p.

"Appendix, by the committee of the Society, &c.," [2] p. at end.

Urges the employment in manufactories.

Mp-8078.

Copies: DeGE; MB; MBAt; MH-BA; MWA.

2946. At a meeting of the freeholders and other inhabitants of the town of Boston, legally assembled at Faneuil-Hall, on Wednesday the 28th of October, 1767. The town then took into consideration the petition of a number of inhabitants, "That some effectual measures might be agreed upon to promote industry, oeconomy, and manufactures; thereby to prevent the unnecessary importation of European commodities, which threaten the country with poverty and ruin...." [Boston: Printed by Edes and Gill?, 1767.] Broadside 33 x 21 cm.

Mp-10564.

Copies: MB; MH; MWA.

2947. The commercial conduct of the province of New-York considered, and the true interest of that colony attempted to be shewn. In a letter to the Society of Arts, Agriculture and Oeconomy. New York: Printed for the benefit of the Society of Arts, Agriculture and Oeconomy, of New-York, 1767. iv, [5]–20 p.

Urges the establishment of manufacturing enterprises, particularly for the production of linen cloth.

Mp-10584.

Copies: CtY; DLC; ICN; MBAt; MH; MHi; MiU-C; NN; PPL.

2948. The committee appointed by the inhabitants of the city of New-York, on Wednesday the 29th of December last, to consider of the expediency of entering into measures to encourage industry and frugality, and employ the poor, do report.... [New York: January 1768.] [2] p.

Mp-11008.

Copies: MWA; NN.

2949. Coxe, Tench, 1755–1824.

An address to an assembly of the friends of American manufactures, convened for the purpose of establishing a Society for the Encouragement of Manufactures and the Useful Arts, read in the University of Pennsylvania on Thursday the 9th of August, 1787.... Philadelphia: Printed by R. Aitken, 1787. 30 p.

Mp-20305.

Copies: CSmH; CtY; DLC; DeGE; ICU; MBAt; MH-BA; MHi; MWA; PPL.

2950. [Coxe, Tench] 1755–1824.

An enquiry into the principles on which a commercial system for the United States of America should be founded; to which are added some political observations connected with the subject. Read before the Society for Political Enquiries, convened at the house of his Excellency Benjamin Franklin, Esquire, in Philadelphia, May 11th, 1787. Philadelphia: Printed and sold by R. Aitken, 1787. 52 p.

Urges the encouragement of manufactures and the improvement of agriculture.

Mp-20306.

Copies: DLC; DeGE; MHi; MWA; PPL.

2951. Pennsylvania Society for the Encouragement of Manufactures and the Useful Arts.

The plan of the Pennsylvania Society for the Encouragement of Manufactures and the Useful Arts. Philadelphia: Printed by R. Aitken & Son, 1787. 12, [1] p.

Mp-20637.

Copies: DLC; MBAt; MH; MWA; PPL.

2952. Germantown Society for Promoting Domestic Manufactures.

The constitution of the Germantown Society for Promoting Domestic Manufactures. Philadelphia: Printed by Eleazer Oswald, 1790. 7 p.

Mp-22534.

Copies: DLC; NN; PHi.

2953. [Coxe, Tench] 1755–1824.

A brief examination of Lord Sheffield's Observations on the commerce of the United States. In seven numbers. With two supplementary notes on American manufactures. Philadelphia: From the press of M. Carey, 1791. vii, 135 p.

Mp-23295.

Copies: CtY; DLC; DeGE; ICU; MH; MHi; MWA; NHi; NN; PPAmP; RPJCB.

2954. New Jersey. Laws.

An act incorporating the Society for Establishing Useful Manufactures. Passed at Trenton, 22d November, 1791. Burlington: Printed by Isaac Neale, 1791.

E-23609.

No copy known.

2955. Society for Establishing Useful Manufactures.

The establishment of manufactures in the United States, when maturely considered, will be found to be of the highest importance to their prosperity.... [S.l.: 1791?] 10 p.

Last page refers to a New Jersey act of incorporation of such a society and ends: "In testimony whereof ... in the year one thousand seven hundred and ninety-one."

B-7277; E-23609; Mp-45794.

Copies: NHi.

2956. United States. Treasury Dept.

Report of the Secretary of the Treasury of the United States, on the subject of manufactures. Presented to the House of Representatives, December 5, 1791. [Philadelphia]: Printed by Childs and Swaine, [1791]. 2 p.l., 58 p.

Alexander Hamilton, Secretary of the Treasury.

Mp-23914.

Copies: DLC; DeGE; MBAt; MWA; NN; PPL; RPJCB.

2957. [Logan, George] 1753–1821.

Five letters, addressed to the yeomanry of the United States: containing some observations on the dangerous scheme of Governor Duer and Mr. Secretary Hamilton, to establish national manufactories. By a farmer. Philadelphia: Printed by Eleazer Oswald, 1792. 28 p.

Mp-24480.

Copies: DLC; DeGE; MBAt; MH; MWA; NN; PPAmP; PPL; RPJCB.

2958. [Bordley, John Beale] 1727–1804.

Intimations; on manufactures;—on the fruits of agriculture;—and on new sources of trade, interfering with products of the United states of America in foreign markets. [Philadelphia: 1794.] 8 p.

Signed: B., January, 1794.

Attributed to J. B. Bordley.

Mp-26681.

Copies: CtY; DeGE; MH-BA; MWA; NNC; PPF; PPL; RPJCB.

2959. Humphreys, David, 1752–1818.

A poem on industry. Addressed to the citizens of the United States.... Philadelphia: Printed for Mathew Carey, 1794. iv p., 1 l., [7]–22 p.

Mp-27145.

Copies: DLC; DeGE; MBAt; MH; MWA; NN; PPL; RPJCB.

2960. Scheme of a lottery to raise 39,900 dollars.... Newark, N.J.: Printed by John Woods, 1794. Broadside.

For the benefit of the Society for Establishing Useful Manufactures.

Mp-27675.

Copies: NjHi.

2961. Brissot de Warville, Jacques Pierre, 1754–1793.

The commerce of America with Europe, particularly with France and Great-Britain.... Shewing the importance of American Revolution ... and pointing out the actual situation of the United States of North-America, in regard to the trade, manufactures and population, by J. P. Brissot de Warville and Etienne Claviere. Translated from the last French edition, revised by Brissot and called the second volume of his View of America.... New York: T. and J. Swords, 1795. xxxv, 228 p., port.

Mp-28343.

Copies: CLU; CtY; DLC; ICJ; IaU; KyLx; MH-BA; MWA; MdBP; MiU; NN; NcU; NjP; PPL; PU; ViU.

2962. Lottery, for raising six thousand six hundred and sixty-seven dollars and fifty cents.... [Newark]: John Woods, 1795. Broadside 33.5 x 20 cm.

Dated: Patterson, November 24, 1795.

For the benefit of the Society for Establishing Useful Manufactures.

B-9173; Mp-47607.

Copies: DLC.

2963. Burrill, George Rawson, 1770–1818.

An oration delivered before the Providence Association of Mechanics and Manufacturers, at their annual election, April 11, 1796. By George R. Burrill, A.B. Providence: Printed and sold by Bennett Wheeler, 1796. 2 p. l., [5]–18, [5] p.

Mp-30145.

Copies: CtY; DLC; DeGE; MH-BA; MHi; MWA; NHi; NN; PPAmP; RPJCB.

2964. Memorial of sundry artisans and manufacturers of the city of Philadelphia ... 9th December, 1803.... [Washington: 1803.] 20 p.

In support of domestic manufactures, listing types of industries deserving promotion.

S&S-5348, 4638 ([Philadelphia]: Printed by B. Graves, 1803).

Copies: DLC; DeGE; MH; MWA; NIC.

2965. An essay on the manufacturing interest of the United States; with remarks on some passages contained in the report of the Committee of Commerce and Manufactures. By a member of the Society of Artists & Manufacturers of Philadelphia. To which is annexed, the memorial of that Society to Congress. Philadelphia: Printed by Bartholomew Graves, 1804. 32, xvii p.

Attributed to Tench Coxe.

S&S-6092.

Copies: DLC; DeGE; MB; MH-BA; MWA; NIC; NN; PPAmP; PPL; RPB; TU.

2966. Memorial of the artists and manufacturers of Philadelphia. [Philadelphia: 1804.] xvii p.

Issued as an appendix to *An Essay on the manufacturing interest of the United States.*

S&S-5728, 6092.

Copies: DLC; DeGE; MB; MH-BA; MWA; NIC; NN; PPL; RPB; TU.

2967. Pennsylvania Society for the Encouragement of Manufactures and the Useful Arts.

A communication from the Pennsylvania Society for the Encouragement of Manufactures and the Useful Arts. Philadelphia: Printed for the Society, by Samuel Akerman, 1804. 28 p.

"Circular communication" signed: Tench Coxe, President.

S&S-7024.

Copies: CSmH; DeGE; MH-BA; MWA; NN.

2968. United States. Congress. House. Committee of Commerce and Manufactures.

Report of the Committee of Commerce and Manufactures to whom were referred on the 8th, 14th, 19th and 26th of December last, the petitions & memorials of sundry manufacturers of corks, of sundry calico printers & dyers, and of sundry manufacturers of paper, within the United States. 25th January, 1804.... [Washington: 1804.] 10 p.

S&S-7608.

Copies: DLC; DeGE; NN; RPB.

2969. Philadelphia Society for the Encouragement of Domestic Manufactures.

Constitution of the Philadelphia Society for the Encouragement of Domestic Manufactures. Adopted March 14, 1806. Philadelphia: Printed by D. Hogan, 1806. 10 p.

Stephen Girard, President.

S&S-11157.

Copies: DeGE; MWA; PPL.

2970. Philadelphia Premium Society.

The directors of this Society have resolved to offer premiums for the following articles manufactured in, or of materials, the growth of the United States.... [Philadelphia: 1808.] 8 p.

Dated: July 25, 1808.

Promotion of American manufactures by offering premiums for best domestic textiles.

S&S-15913.

Copies: PHi.

2971. Princess Anne Society for the Encouragement of Domestic Manufactures, Kempsville, Va.

Articles of association of the Princess Anne Society of patriotism for the encouragement of domestic manufactures, and of rearing the best race of sheep. Norfolk, [Va.]: Printed by J. O'Connor, 1808. 6 p.

Copies: MNoanMV.

2972. Union Manufacturing Company.

Report of the committee, and constitution of the Union Manufacturing Company, of Maryland. Baltimore: Printed by Niles and Frailey, 1808. 31 p.

S&S-16355.

Copies: DLC.

2973. Seybert, Adam, 1773–1825.

An oration, delivered on the 19th of May, 1809, at the meeting of the manufacturers and mechanics of the city of Philadelphia. Philadelphia: Printed by order of the Committee, [1809]. 16 p.

A survey of the development of industrial arts.

S&S-18591.

Copies: DLC; DeGE; MB; MWA; MdHi; NN; NjR; P.

2974. United States. Treasury Dept.

Report of the Secretary of the Treasury on the subject of manufactures, made the fifth of December, 1791. December 7th, 1809. Washington City: Printed by R. C. Weightman, 1809. 109 p.

Alexander Hamilton, Secretary of the Treasury.

S&S-19064.

Copies: DLC; MWA; MiGr; RPB; WHi.

2975. The memorial of the mechanics and manufacturers of Lexington, Kentucky, to Congress. [Lexington, Ky.]: T. Smith, printer, December 21, 1810. 8 p.

Urges support of the domestic manufacturing industry.

N-L0327158.

Copies: CSmH.

2976. United States. Treasury Dept.

Letter from the Secretary of the Treasury, transmitting a report, in part, on the subject of American manufactures; prepared in obedience to a resolution of the House of Representatives. April 19th, 1810.... Washington City: Printed by Roger C. Weight man, 1810. 62 p., 4 fold. tables.

Albert Gallatin, Secretary of the Treasury.

S&S-21698.

Copies: DLC; DeGE; MWA; MiGr; N; NNS.

2977. United States. Treasury Dept.

Report from the Secretary of the Treasury on the subject of American manufactures, prepared in obedience to a resolution of the House of Representatives. April 19, 1810.... Boston: Farrand, Mallory, [etc.], 1810. 62 p., fold. tables.

Albert Gallatin, Secretary of the Treasury.

S&S-21795.

Copies: DeGE; MH-BA; MHi; MWA.

2978. United States. Treasury Dept.

Report of the Secretary of the Treasury on American manufactures, prepared in obedience to a resolution of the House of Representatives. April 17, 1810. Brooklyn, [N.Y.]: Printed by Thomas Kirk, [1810]. 48 p.

Albert Gallatin, Secretary of the Treasury.

S&S-21859.

Copies: DLC; DeGE; MHi; MWA; MiU-C; NN; PPAmP; PPL; RPB.

2979. Memorial of the mechanics and manufacturers of Lexington, Kentucky, to Congress, January 22d, 1811. Printed by order of the Senate of the United States. Washington City: Printed by R. C. Weightman, 1811. 14 p.

Calls for support of domestic manufactures.

S&S-24204.

Copies: DLC; MWA; PHi.

2980. Beck, Theodoric Romeyn, 1791–1855.

Annual address, delivered by appointment, before the Society for the Promotion of Useful Arts, at the Capitol, in the city of Albany, on the 3d of February, 1813.... Albany: Printed by Websters and Skinners, 1813. 2 p.l., [9]–44 p.

Appendix (p. [43]–44) contains "observations on the views and objects of the Mineralogical Committee of the Society for the Promotion of Useful Arts."

Includes observations on the production of raw materials and on the state of manufactures.

S&S-27855.

Copies: DeGE; MWA; NCH; NjR.

2981. Massachusetts. Laws.

An act, to incorporate the Massachusetts Manufacturing Company. [Boston: 1813.] 8 p.

S&S-29073.

Copies: MB.

2982. United States. Census Office. 3d Census, 1810.

A series of tables of the several branches of American manufactures exhibiting them in every county of the Union, so far as they are returned in the reports of the marshals, and of the secretaries of the territories ... in the autumn of the year 1810. [Philadelphia: 1813.] 169 p.

Assembled by Tench Coxe.

S&S-28241, 30375.

Copies: ICU; IU; MB; MH; MWA; OClWHi; PPAmP; PPL; TxU.

2983. United States. Treasury Dept.

Tabular statements of the several branches of American manufactures, exhibiting them I. By states, territories and districts, II. By counties, cities, and towns, so far as they are returned in the reports of the marshals, and of the secretaries of the territories ... in the autumn of the year 1810; together with similar returns of certain doubtful goods, productions of the soil and agricultural stock.... Philadelphia: Printed by A. Cornman, junr., 1813. 46, 169 [17]) p.

Tables prepared by Tench Coxe for the Treasury Dept.

S&S-30381.

Copies: DLC; MB; MH; MH-BA; MWA; PPAmP; PPL.

2984. Worthington Manufacturing Company.

The articles of association of the Worthington Manufacturing Company. Worthington, [Ohio]: Printed at the office of the Western intelligencer, 1813. 8 p.

S&S-30581.

Copies: CSmH; OClWHi.

2985. Young, William.
The remonstrance agreed unto at a meeting of improvers and manufacturers in the county of New-Castle. Wilmington, [Del.]: 1813. 8 p.
S&S-30596.
Copies: PHi.

2986. Massachusetts. Laws.
An act to incorporate Isaac P. Davis and others, by the name of the Boston and Roxbury Mill Corporation. [Boston: 1814.] 6 p.
N-B0674911.
Copies: DLC.

2987. Society for the Promotion of Industry, New York.
Plan for the Society for the Promotion of Industry. New York: Printed for the Society by J. Seymour, 1814. 12 p.
S&S-32812.
Copies: MH; NbU.

2988. United States. Laws.
A bill incorporating the Columbian Manufacturing Company of Alexandria, in the District of Columbia. March 31st, 1814.... [Washington: 1814.] 7 p. (H.R. 97)
S&S-33123.
Copies: DLC.

2989. ———— October 19, 1814.... [Washington: 1814.] 7 p. (H.R. 6)
S&S-33124.
Copies: DLC.

2990. United States. Laws.
A bill to incorporate the stockholders of the Columbian Manufacturing Company. March 31, 1814.... [Washington: 1814.] 8 p. (S.44)
S&S-33194.
Copies: DLC.

2991. United States. Treasury Dept.
A statement of the arts and manufactures of the United States of America, for the year 1810: digested and prepared by Tench Coxe, Esquire, of Philadelphia. Philadelphia: Printed by A. Cornman, junr., 1814. 6 p.l, v–[lxiv], 46, 169 [171] p.
S&S-31257.
Copies: DLC; M; MH; MWA; PPL; ViAl.

2992. United States. Treasury Dept.
Extract of a report to the House of Representatives on the subject of manufactures, by Alexander Hamilton.... New York: Printed by George Largin, 1815. 32 p.
S&S-36262.
Copies: DLC.

2993. New York (State). Legislature. Assembly. Committee on Manufactures.
Report of a committee ... 1816, to inquire into the state of manufactures.... New York: Van Winkle, Wiley & Co., [1816]. 16 p.
S&S-38442.
Copies: PPiU.

2994. Society for the Promotion of Industry, New York.
Plan for the Society for the Promotion of Industry; with the first report of the board of managers, and the names of subscribers to the institution. New York: Printed for the Society by J. Seymour, 1816. 27 p.
S&S-38968.
Copies: DLC; NjR.

2995. United States. Treasury Dept.
Extract of a report to the House of Representatives on the subject of manufactures, by Alexander Hamilton, Secretary of the Treasury. In obedience to an order of the House, of the 15th day of January, 1790. New York: Printed by George Largin, 1816. 32 p.
S&S-39366.
Copies: DLC; DeGE; MBAt.

2996. American Society for the Encouragement of Domestic Manufactures.
Address of the American Society for the Encouragement of Domestic Manufactures, to the people of the United States. New York:

Van Winkle, Wiley, & Co., printers, 1817. 32 p.

S&S-39988.

Copies: DLC; DeGE; InU; MB; MH-BA; MWA; MdHi; MiU-C; MnU; NjPT; PPL; ScU.

2997. American Society for the Encouragement of Domestic Manufactures.

Memorial of the American Society for the Encouragement of Domestic Manufactures. January 29, 1817.... [Washington: 1817.] [5] p. (14th Cong., 2d sess. House. Doc. 59)

Signed: Daniel D. Tompkins, president.

S&S-42649.

Copies: CtY; DLC; DeGE; MWA; PPL; ViU.

2998. American Society of the City of New York for Promoting Domestic Manufactures.

Extracts from an address of the American Society of New York for Promoting Domestic Manufactures, to the people of the United States. Published by order of the Philadelphia Society for the Promotion of American Manufactures. [Philadelphia?: 1817.] 12 p.

S&S-39991.

Copies: MWA.

2999. Briggs, Isaac, 1763–1825.

An address delivered before the Oneida Society for the Promotion of American Manufactures, in their annual meeting in Whitesboro on the 21st of October, 1817. Utica, [N.Y.]: Printed by William Williams, 1817. 8 p.

S&S-40319.

Copies: DLC; MH-BA; MdHi; NN; NNC.

3000. Connecticut Society for the Encouragement of American Manufactures.

Address of the Connecticut Society for the Encouragement of American Manufactures. Middletown, Conn.: T. Dunning, printer, 1817. 24 p.

Includes the constitution of the Society.

S&S-40556.

Copies: CSmH; CtHi; CtY DLC; DeGE; ICU; MB; MH-BA; MWA; NNC; PPL.

3001. Memorial of sundry citizens of Baltimore praying of the general government aid and protection to commerce and manufactures. February 23, 1816 [*sic*]. Referred to the Committee of Commerce and Manufactures. [Washington: 1817.] 13 p. (14th Cong., 2d sess. House. Doc. 96)

"Constitution of the American Society of Maryland for the Promotion of Domestic Manufactures and National Industry," p. [1]-3.

S&S-42648.

Copies: CtY; DLC; DeGE; MWA.

3002. Pennsylvania Society for the Promotion of Public Economy.

Report of the Committee on Domestic Economy to the Pennsylvania Society for the Promotion of Public Economy, read at its meeting, on November 10, 1817. Philadelphia: Printed for the Society, 1817. [2], [17]–27 p.

Report by Benj. Ives Gilman, Chairman, Committee on Domestic Economy, on a newly imported brewing machine and on sweeping of the chimneys.

S&S-41767.

Copies: DeGE; MH.

3003. Report of the committee appointed by the citizens of Pittsburgh, at a meeting held at the court-house on the 21st of December, 1816, to inquire into the state of the manufactures in the city and its immediate vicinity. Pittsburgh, Pa.: 1817. 16 p.

S&S-41834.

Copies: MB; NHi; PPi; WHi.

3004. ———— Made February 17, 1817. In the House of Representatives of the United States.... [Washington: 1817.] 24 p. (14th Cong., 2d sess. House. Doc. 86)

S&S-42684.

Copies: DLC; DeGE; MWA; PHi; PPi.

3005. Society for the Encouragement of American Manufactures.

Address of the Society for the Encouragement of American Manufactures. February 18, 1817. Middletown, [Conn.]: 1817. 24 p.

Sabin-15649.

Copies: MWA.

3006. Society of the State of Delaware for the Promotion of American Manufactures.

Constitution of the Society of the State of Delaware for the Promotion of American Manufactures. [Wilmington?, Del.: 1817.] 4 p.

S&S-42155.

Copies: DeGE.

3007. Society of the State of Delaware for the Promotion of American Manufactures.

To the Honorable the Senate and House of Representatives of the United States, in Congress assembled. The memorial and petition of the citizens of the state of Delaware and its vicinity, associated under the name of the Society for the Promotion of American Manufactures. [Wilmington?, Del.: 1817?] 8 p.

Signed: E. I. Dupont, President; Selleck Osborn, Secretary.

Dated on the basis of a reprint of the memorial in the *American Watchman,* Jan. 25, 1817.

S&S-42157.

Copies: DeGE.

3008. The soundness of the policy of protecting domestic manufactures; fully established by Alexander Hamilton, in his report to Congress on the subject, and by Thomas Jefferson, in his letter to Benjamin Austin. Philadelphia: Printed by J. R. A. Skerrett, 1817. 24 p.

Caption title: *Extract from the report of Alexander Hamilton, Esquire, Secretary of the Treasury, January, 1790.*

S&S-42169.

Copies: DeGE; MH-BA; MWA.

3009. ———— To which are added, Extracts from the address of the American Society for Promoting Domestic Manufactures, established in New York.... Philadelphia: Printed by J. R. A. Skerrett, for the Philadelphia Society for the Promotion of American Manufactures, 1817. 24, 12 p.

S&S-42170.

Copies: CSmH; CtY; DeGE; MBAt; MH-BA; MMeT; MWA.

3010. Worthington Manufacturing Company.

Articles of association.... Columbus, Ohio: P. H. Olmsted, 1817. 23 p.

S&S-42977.

Copies: OClWHi.

3011. Melish, John, 1771–1822.

The necessity of protecting and encouraging the manufactures of the United States, in a letter to James Monroe, Esq., President of the United States. Philadelphia: J. Melish, 1818. 30 p.

S&S-44795.

Copies: DeGE; MBAt; MH-BA; MWA; MiD-B; NcD; p.

3012. United States. Congress. House. Committee of Commerce and Manufactures.

Reports of the Committee of Commerce and Manufactures, upon the subject of domestic manufactures.... February 13, 1818.... [Washington: 1818.] 11 p.

S&S-46599.

Copies: DLC; DeGE; DeU; MWA; NjP; R.

3013. American manufactures. An adjourned meeting of the citizens of the city and county of Philadelphia, friendly to American manufactures, was held in the county court house, on Saturday the 2d of October, 1819. Matthew Lawler, Esq., Chairman, C. Raguet, Esq., Secretary. The committee appointed for the purpose, presented the following report. [Philadelphia: 1819.] 8 p.

Signed: Mathew [*sic*] Lawler, Chairman, Condy Raguet, Secretary, October 2, 1819.

Kress-C381.
Copies: DeGE; MH-BA; MWA; PU.

3014. [Carey, Mathew] 1760–1839.
Addresses of the Philadelphia Society for the Promotion of National Industry.... Philadelphia: M. Carey and Son, 1819. xvi, [9]–280 p.
Includes no. I–XIII and Supplement.
By Mathew Carey, except no. XII and XIII, which were by Dr. Samuel Jackson.
The Supplement consists of: "Report of the committee appointed by the citizens of Pittsburg ... to inquire into the state of manufactures in the city and its immediate vicinity," a memorial of the citizens of Baltimore, and "The memorial and petition of the citizens of the State of Delaware and its vicinity, associated under the name of The Society for the Promotion of American Manufactures," signed: E. I. Dupont, President, Selleck Osborn, Secretary.
S&S-47512.
Copies: DLC; DeGE; DeWI; MH-BA; MWA; MdHi; MnU; NNC; NcD; NjP; PPL; PU; RPB; TU; ViU; WaU.

3015. ———— [2d ed.]. Philadelphia: M. Carey & Son, 1819. xi, [9]–248 p.
N-C0133077.
Copies: CtY.

3016. ———— [3d ed.]. Philadelphia: M. Carey & Son, 1819. xvi, [9]–280 p.
Sabin-62041.
Copies: NcD; WaU.

3017. ———— 4th ed. Philadelphia: M. Carey & Son, 1819. xi, [9]–248 p.
S&S-47513.
Copies: CtY; DLC; DeGE; IEN; IdU; MWA; MiDSH; NjP.

3018. ———— Boston: W. W. Clapp, 1819. 116 p.
The Supplement is not included in this reprint.
Kress-C248; Sabin-62040.
Copies: MH.

3019. [Carey, Mathew] 1760–1839.
National interests & domestic manufactures. Address of the Philadelphia Society for the Promotion of Domestic Industry, to the citizens of the United States. Boston: Printed by W. W. Clapp, 1819. vii, [8]–116 p.
Includes no. I–XIII.
Republished by "several gentlemen of Boston," cf.p. [ii].
Written by Mathew Carey, except no. XII and XIII, which were by Dr. Samuel Jackson.
S&S-49096.
Copies: CSmH; DLC; DeGE; MH-BA; MWA; NN; NjP; PHC; PPAmP; PPL; PU; WHi.

3020. Convention of the Friends of National Industry, New York, 1819.
Memorial of a convention of the Friends of National Industry, assembled in the City of New York, to take into consideration the prostrate situation of our manufactures, and to petition Congress for their relief and protection.... December 20, 1819. Referred to the Committee on Manufactures. Washington: Printed by Gales & Seaton, 1819. 10 p. (16th Cong., 1st sess. House. Doc. 9)
Signed: W. Few, President, Mathew Carey, Secretary.
S&S-49865.
Copies: DLC; DeGE; G; MH; MWA; NjP; O; R.

3021. Convention of the Friends of National Industry, New York, 1819.
The proceedings of a convention of the Friends of National Industry, assembled in the city of New-York, November 29, 1819, consisting of delegates from the states of Massachusetts, Rhode Island, Connecticut, New-York, New-Jersey, Pennsylvania, Delaware, Maryland, and Ohio. New York: Printed by C. S. Van Winkle, 1819. 22 p.
Memorial to Congress, p. [13]–22.
S&S-48048.
Copies: DeGE; MB; MH-BA; PHi.

3022. Domestic industry: a poem, in a dialogue between a farmer and a manufacturer.... [Philadelphia?: 1819?] 28 p.
S&S-47841.
Copies: NHi; RPB.

3023. Memorial of sundry manufacturers residing in the state of Maryland, December 22, 1819, referred to the Committee of Manufactures. Washington, D.C.: Printed by Gales & Seaton, 1819. 5 p.
S&S-49870.
Copies: DLC; MH; MWA; MdBE; NhHi; NjP; O; R.

3024. New York (State). Legislature. Assembly.
List of all the incorporations in the state of New-York, except religious incorporations, with a recital of all their important particulars and peculiarities. Reported to the Assembly, pursuant to a resolution thereof. By Aaron Clark, clerk of Assembly. Albany: J. Buel, printer to the state, 1819. 106 p.
S&S-47623, 48896.
Copies: DLC; DeGE; NIC; NN; NNLI; NNS.

3025. Pennsylvania Society for the Encouragement of American Manufactures.
Constitution, adopted Sept. 4, 1819.... [Philadelphia: 1819.] 3 p.
N-P0216294.
Copies: MH-BA; MHi; PHi.

3026. Society of New-Castle County.
Report of the Committee appointed ... [18 May 1819] to prepare a plan for an association to promote the industry and domestic economy of the state.... [Wilmington?, Del.: 1819.] 11, [8] p.
Proposed constitution is on p. 9–10.
S&S-49453.
Copies: DeGE.

3027. Baldwin, Henry, 1780–1844.
Speeches of Henry Baldwin, esq. in the House of Representatives, on the bills reported by him as chairman of the Committee of Manufacturers.... Pittsburgh: Printed by Butler & Lambdin, for R. Patterson & Lambdin, 1820. 126 p., port.
S-192.
Copies: CSmH; MH-BA; OCU; PHi; PPi; T.

3028. Boston and Roxbury Mill Corporation.
An act to establish the Boston and Roxbury Mill Corporation. [Boston: 1820.] 19 p., fold. map.
At head of title: Commonwealth of Massachusetts. In the year of our Lord one thousand eight hundred and fourteen.
Includes also later additions to this act, *Grant of the town of Boston* and *By-laws of the Corporation,* with additions and alterations up to March 8th, 1820.
S-2146.
Copies: DLC; DeGE; MB; MHi; MWA; NjR.

3029. [Carey, Mathew] 1760–1839.
Addresses of the Philadelphia Society for the Promotion of National Industry ... 5th ed. Philadelphia: J. Maxwell, 1820. xii, [13]–299, [1] p.
Includes no. I–XIII and New series, I–II.
Written by Mathew Carey, except no. XII and XIII, which were by Dr. Samuel Jackson.
"A sermon delivered at Litchfield ... December 2, 1819. By Lyman Beecher," p. 261–286.
S-2750.
Copies: CtY; DLC; DeGE; ICN; KU; LU; MH-BA; MWA; NN; NcD; NjR; OO; PPi; RPB; Vi.

3030. Carey, Mathew, 1760–1839.
The new olive branch; or, An attempt to establish an identity of interest between agriculture, manufactures, and commerce.... Philadelphia: M. Carey & Son, 1820. 4 p.l., [iii]–x, [17]–248 p.
"This work may be considered as a second edition, much enlarged and improved, of the *Three letters to Mr. Garnett.*" Introd., p. [vii].

S-680.
Copies: CSmH; CtY; DLC; DeGE; ICN; MH; MWA; MiD; MnSM; NN; NcD; Nh; NjP; OClW; PPL; PU; RPB; T; WHi; WU.

3031. Convention of the Friends of National Industry, New Jersey, 1820.

Petition of a convention of the Friends of National Industry, in New Jersey. February 7, 1820. Read and referred to the Committee on Commerce and Manufactures, and ordered to be printed for the use of the Senate. Washington: Printed by Gales & Seaton, 1820. 5 p. (16th Cong., 1st sess. Senate. Doc. 40)

S-3930.
Copies: DLC; DeGE; G; MWA; NNC; NjP; O.

3032. National Institution for Promoting Industry in the United States.

An address to the people of the United States, drawn up by order of the National Institution for the Promotion of Industry, established in June, 1820, by delegates from New-York, New-Jersey, Pennsylvania, Connecticut, Massachusetts, Rhode-Island, Ohio, and Delaware. New York: Printed by J. Seymour, 1820. 22 p.

S-2391.
Copies: DLC; IU; MB; MH; MWA; MiD-B; N; NN; OClWHi; P; PPAmP; PPM; RPB; WHi.

3033. National Institution for Promoting Industry in the United States.

Circular and address of the National Institution for Promoting Industry in the United States, to their fellow-citizens. New York: Printed by J. Seymour, 1820. 28 p.

S-2392.
Copies: CSt; CtHT; DLC; DeGE; LU; MH; MWA; MiD-B; N; NN; OO; P; TKL.

3034. New York County Agricultural Society.

Memorial of the New York County Agricultural Society, for the protection of national industry. December 15, 1820. Referred to the Committee on Manufactures. Washington: Printed by Gales & Seaton, 1820. 7 p. (16th Cong., 2d sess. House. Doc. 28)

S-3895.
Copies: DLC; DeGE; G; MWA; NN.

3035. The paradox solved; or, A financial secret worth knowing; with the means of dissipating dearth and desolation, by the due encouragement of national industry, in a letter to the President of the United States. By the author of Statisticus. Baltimore: Printed by J. Robinson, 1820. 24 p.

S-2644.
Copies: DeGE; M; MB; MWA; MdHi; N; NN; NcD; NhHi; PPPrHi.

3036. Pennsylvania Society for the Encouragement of American Manufactures.

Memorial of the Pennsylvania Society for the Encouragement of American Manufactures. April 28, 1820. Printed by order of the Senate of the United States. Washington: Printed by Gales & Seaton, 1820. 15 p. (16th Cong., 1st sess. Senate. Doc. 126)

S-3896.
Copies: DLC; G; MWA; NjP; PHi.

3037. Pennsylvania Society for the Encouragement of American Manufactures.

Memorial. To the Senate and House of Representatives of the United States: The memorial of the Pennsylvania Society for the Promotion of American Manufactures. [Philadelphia: 1820.] 14 p.

Signed: Thomas Leiper, Vice President, James Mease, Secretary pro tempore. Philadelphia, April 3d, 1820.

S-2711, 6461.
Copies: DLC; DeGE; MWA; N; PPAmP; PPL; PU.

3038. Pennsylvania Society for the Encouragement of American Manufactures.

To the Honourable the Senate and House of Representatives of the United States in Congress assembled. The memorial of the

3038 (*continued*)
Board of Manufactures.... [Philadelphia: 1820.] [3] p.
Dated: Philadelphia, April 25th, 1820.
S-2712.
Copies: NN.

3039. Rhode Island Society for the Encouragement of Domestic Industry.
The state of Rhode-Island justly claims the honour of being one of the earliest seats of mechanic arts, and manufactures. . . . [Providence?]: 1820. Broadside 42 x 37 cm.
"The foregoing were adopted as the primary laws of the Society ... 25th day of February, 1820."
Copies: MWA.

3040. Wolcott, Oliver, 1760–1833.
Remarks on the present state of currency, credit, commerce, and national industry.... New York: C. Wiley and Co.; C. S. Van Winkle, printer, 1820. 43 p.
S-4315.
Copies: CtY; DLC; DeGE; ICU; MB; MH; MWA; MdBJ; NN; PPAmP; PPL; RPB.

3041. Bellamy, Edward S.
Domestic manufactures, the sources of real independence: an oration, delivered before the New York Typographical Society, on the celebration of their twelfth anniversary, July 4, 1821. New York: Printed by J. Seymour, 1821. 20 p.
S-4657.
Copies: DLC; MBAt; NjPT; PHi.

3042. Carey, Mathew, 1760–1839.
Address to the farmers of the United States, on the ruinous consequences to their vital interests, of the existing policy of this country.... By M. Carey. Philadelphia: M. Carey & Son, 1821. iv, [5]–84 p.
Observations on the decline of manufactures in the United States.
Second ed. (S-4914) was not independently published.
S-4913.
Copies: DeGE; MBAt; MWA; MdHi; P; PPL; PPM.

3043. Carey, Mathew, 1760–1839.
The new olive branch; or, An attempt to establish an identity of interest between agriculture, manufactures, and commerce ... 2d ed. Philadelphia: M. Carey & Sons, 1821. 130 p.
S-4916.
Copies: CSt; CoFS; ICN; ICU; M; MBAt; MH-BA; MWA; MdHi; MiU-C; NNA; NjR; P; PHi; PPAmP; Pi; PU; ScC.

3044. Knight, Perry G.
A guide for the home-domestick manufacturer; containing some useful observations on the most essential points of manufacturing woolen goods.... By Perry G. Knight.... Providence: Printed by Miller and Hutchens, 1821. 12 p.
S-5779.
Copies: RHi; RPB.

3045. National Institution for Promoting Industry in the United States.
An address to the people of the United States, drawn up by order of the National Institution for the Promotion of Industry, established in June, 1820, by delegates from New-York, New-Jersey, Pennsylvania, Connecticut, Massachusetts, Rhode-Island, Ohio, and Delaware. Concord, [N.H.]: Printed by Hill & Moore, 1821. 48 p.
S-6180.
Copies: DLC; ICJ; MH-BA; NhHi; OClWHi.

3046. National Institution for Promoting Industry in the United States.
Memorial of the National Institution for the Promotion of Industry, to the Hon. Senate and House of Representatives of the state of New York. Jan. 13, 1821. New York: [1821]. 8 p.
N-N0055579.
Copies: PHi.

3047. Newport Association of Mechanics & Manufacturers, Newport, R.I.
The constitution and by-laws of the Newport Association of Mechanics & Man-

ufacturers.... Newport, R.I.: Printed by William Simons, 1821. 15 p.

S-6292.

Copies: MWA; RNHi.

3048. Remarks on the United States factories. Washington City: Printed at the Gazette office, 1821. 8 p.

S-6609.

Copies: MB; MWA.

3049. Town of Boston. At a legal meeting of the inhabitants of the town of Boston, held by adjournment on the 22d October, 1821, a report was read by the chairman of the committee for erecting a House of Industry.... [Boston: 1821.] 15 p.

S-4786.

Copies: DLC; M; MB; MBAt; MHi; MWA; NIC.

3050. United States. Congress. House. Committee on Manufactures.

Report of the Committee on Manufactures, on the various memorials praying for, and remonstrating against, an increase of the duties on imports. January 15, 1821.... [Washington: 1821.] 82, 54 p., 4 fold. tables. (16th Cong., 2d sess. House. Rept. 34)

"Statements, accompanying the report ...," 54 p. at end.

Consideration of ways to further American manufactures.

S-7429, 7486.

Copies: DLC; DeGE; MWA; NjP.

3051. United States. Treasury Dept.

Official reports on publick credit, a national bank, manufactures, and a mint. By Alexander Hamilton. Philadelphia: W. M'Kean, 1821. vii p., 1 l., 325 p., port.

A reprint, with a new title page and different frontispiece, of v. 1 of *Hamilton's Works,* New York, 1810.

S-5523.

Copies: DLC; DeGE; ICJ; LNHT; MB; MWA; MdBP; MiU-C; NIC; NN; NcU; NjP; OCHP; PU; RPB.

3052. American Society for the Encouragement of Domestic Manufactures.

Memorial of the American Society for the Encouragement of Domestic Manufactures. January 19, 1822.... Washington: Printed by Gales & Seaton, 1822. 5 p. (17th Cong., 1st sess. Senate. Doc. 15)

S-11001.

Copies: DLC; DeU; MWA; NNLI; NjP.

3053. ——— January 21, 1822.... Washington: Printed by Gales & Seaton, 1822. 5 p. (17th Cong., 1st sess. House. Doc. 29)

S-11002.

Copies: DLC; DeU; MWA; NjP.

3054. Burges, Tristam, 1770–1853.

Address to the Rhode-Island Society for the Encouragement of Domestick Industry, delivered at Pawtuxet, October 17, 1821.... Providence: Miller and Hutchens, printers, 1822. 29 p., 1 l.

S-8216.

Copies: CSmH; DLC; DeGE; MBAt; MH; MWA; PPAmP; PPL; RHi; RP; RPB.

3055. [Carey, Mathew] 1760–1839.

Addresses of the Philadelphia Society for the Promotion of National Industry.... 6th ed. Philadelphia: 1822. [7]–221 p.

No. 12 and 13 by Dr. Samuel Jackson.

Issued, with separate title, in his *Essays on political economy.*

N-C0133081.

Copies: CSmH; DLC; DeGE; ICJ; MWA; MdBP; NN; PP; WHi.

3056. Carey, Mathew, 1760–1839.

Essays on political economy; or, The most certain means of promoting the wealth, power, resources, and happiness of nations: applied particularly to the United States.... Philadelphia: H. C. Carey & I. Lea, 1822. [6], [v]–x, [7]–546, [10] p.

Contents.—Addresses of the Philadelphia Society for the Promotion of National Industry.—The new olive branch.—Address to Congress.—Address to the farmers of the

3056 (*continued*)
United States.—The farmer's & planter's friend.—Strictures on Mr. Cambreleng's work, entitled, "An examination of the new tariff."

S-8269.

Copies: CSmH; CtY; DLC; DeGE; GDC; ICJ; IU; In; LU; MH; MWA; MdBJ; NN; Nh; OO; PPL.

3057. Pennsylvania Society for the Encouragement of American Manufactures.

Memorial of the Board of Manufacturers of the Pennsylvania Society for the Encouragement of American Manufactures. March 1, 1822.... Washington: Printed by Gales & Seaton, 1822. 9 p. (17th Cong., 1st sess. House. Doc. 81)

S-11003.

Copies: DLC; DeGE; MWA; NjP.

3058. Pennsylvania Society for the Encouragement of American Manufactures.

To the Honourable the Senate and House of Representatives of the United States, in Congress assembled. The memorial of the Board of Manufactures of the Pennsylvania Society for the Encouragement of American Manufactures.... [Philadelphia: 1822.] 4 p.

Signed: Philadelphia, December 30th, 1822. William Tilghman, president.

S&S-45302 (erroneously dated [1818]).

Copies: MB; MWA.

3059. Robbins, Asher, 1757–1845.

Address to the Rhode-Island Society for the Encouragement of Domestick Industry, delivered at Pawtuxet, October 16th, 1822.... Providence: Miller & Hutchens, printers, 1822. 54 p.

S-10117.

Copies: DLC; DNAL; DeGE; MH-BA; MWA; MiD-B; NN; PPL; RHi; RNHi; RP; RPB.

3060. National industry.... [New York: Dec. 2, 1823.] 8 p.

S-13468.

Copies: MWA; MdBP.

3061. The rights of men and things. Attend East! attend West! attend North! and particularly attend South! Niles is our second Franklin, and deserves well of his country. From Niles' weekly register, February 1st, 1823. [Wilmington, Del.: Printed at the Gazette office, 1823.] 8 p.

In support of domestic manufactures.

Copies: DeGE; MWA.

3062. ———— Poughkeepsie, [N.Y.]: Printed at the office of the Dutchess observer, 1823. 8 p.

S-13953.

Copies: CSmH; MH-BA.

3063. United States. Census Office. 4th Census, 1820.

Digest of accounts of manufacturing establishments in the United States, and of their manufactures. Made under direction of the Secretary of State, in pursuance of a resolution of Congress, of 30th March, 1822. Washington: Printed by Gales & Seaton, 1823. 2 p.l., [123] p.

S-14502.

Copies: DLC; DeGE; MWA; MiU-C; PPL.

3064. United States. Dept. of State.

Letter from the Secretary of State, transmitting a supplementary return to the Digest of the accounts of the manufacturing establishments, and their manufactures. February 27, 1823.... Washington: Printed by Gales & Seaton, 1823. 12 p., fold. tables. (17th Cong., 2d sess. House. Doc. 90)

Signed: John Quincy Adams.

S-14565.

Copies: DLC; DeGE; DeU; MWA; MiU-C; PPL; PU.

3065. [Carey, Mathew] 1760–1839.

Twenty-one golden rules to depress agriculture, impede the progress of manufactures, paralize commerce, impair national resources, produce a constant fluctuation in the value of every species of property, and blight and blast the bounties of nature ... in a word, to cripple a great nation.... By a

Pennsylvanian. Philadelphia: H. C. Carey &. I. Lea, 1824. 56 p.
S-15677.
Copies: CtY; DLC; DeGE; MB; MBAt; MChB; MH; MHi; P; PPAmP; PPL; PU; ScU.

3066. ———— Salem, [Mass.]: Reprinted by W. Palfray, Jun., 1824. 60 p.
S-15678.
Copies: CtY; DLC; ICU; MB; MH-BA; MWA; MdBP; MiD-B; MiU-C; PPL.

3067. Clay, Henry, 1777–1852.
Mr. Clay's speech in support of an American system for the protection of American industry, delivered March 30th and 31st, 1824. [Washington?: 1824.] 32 p.
S-15777.
Copies: CSmH; DLC; DeGE; MA; MH-BA; MWA; MdBJ; MiD-B; MiU-C; OCHP; PHi.

3068. Clay, Henry, 1777–1852.
Speech in support of an American system for the protection of American industry; delivered in the House of Representatives, on the 30th and 31st of March, 1824, by Henry Clay. Washington City: Printed at the Columbian office, 1824. 39 p.
S-15778.
Copies: CLU; CSmH; DLC; DeGE; MH-BA; MWA; MdBJ; MiD-B; MiU-C; OCHP; PHi.

3068a. ———— Philadelphia: Joseph R. A. Skerrett, 1824. 34 p.
N-C0472521.
Copies: MWA; PHi.

3069. ———— Lexington, Ky.: Printed at the Kentucky Reporter Office, 1824. 40 p.
S-15779.
Copies: CSmH; KyU; OC.

3070. Dover Manufacturing Company, Dover, N.H.
The acts of the legislature of New Hampshire incorporating the Dover Manufacturing Company, and the by-laws of the corporation. Dover. [N.H.]: John Mann, 1824. 26 p.
S-16001.
Copies: MBAt; MHi.

3071. Pennsylvania Society for the Encouragement of American Manufactures.
Memorial of the board of managers of the Pennsylvania Society for the Encouragement of American Manufactures, agreed to March 10th, 1824. Philadelphia: Printed by Joseph R. A. Skerrett, 1824. 24 p.
Signed: William Tilghman, president, Mathew Carey, Secretary.
S-17578.
Copies: DeGE; OClWHi; PPL.

3072. Pennsylvania Society for the Encouragement of American Manufactures.
Memorial of the Pennsylvania Society for the Encouragement of American Manufactures. March 15, 1824. Read, and referred to the Committee of the Whole House.... Washington: Printed by Gales & Seaton, 1824. (18th Cong., 1st sess. House. Doc. 116)
Signed: Wm. Tilghman, president; Mathew Carey.
S-18883.
Copies: DLC; DeGE; DeU; MWA; NjP.

3073. Society of Political Economists, Philadelphia.
Domestic industry. [Philadelphia: 1824.] 3 p.
Dated: Sept. 15, 1824.
S-18040.
Copies: PU.

3074. United States. Dept. of State.
Report of the Secretary of State, of such articles manufactured in the United States as would be liable to duties if imported from foreign countries; as, also, the amount of capital invested in each county, respectively. With a schedule of factories incorporated by state laws, from 1800 to 1820, inclusive. Prepared in obedience to a resolution of the Senate, of the 1st March, 1823. January 27,

3074 *(continued)*
1824.... Washington: Printed by Gales & Seaton, 1824. 199 p. (18th Cong., 1st sess. Senate. Doc. 45)
S-19061.
Copies: DLC; DeGE; DeU; MWA; NjP.

3075. United States. Treasury Dept.
Report of the Secretary of the Treasury, on the subject of manufactures, made the 5th of December, 1791.... Philadelphia: Printed by J. R. A. Skerrett, 1824. vii, [9]–143 p.
Alexander Hamilton, Secretary of the Treasury.
Preface by Mathew Carey.
S-19066.
Copies: DLC; DeGE; MB; MHi; MWA; MdBL; NB; NCH; OClWHi; PHi; PPL.

3076. Morgan, James.
The manufacturer's book of wages, and workpeople's companion, in two parts. Part first. Exhibiting at one view, any amount made by time.... Part second. Shewing any amount made by the piece.... Baltimore: Printed for the publisher, by R. J. Matchett, 1825. [46] p.
S-21520.
Copies: DLC; DeGE; MdBP; PPL.

3077. Tibbits, George, 1763–1849.
A memoir on the expediency and practicability of improving or creating home markets for the sale of agricultural productions and raw materials, by the introduction or growth of artizans and manufacturers; read before the Board of Agriculture of the state of New York, March 8, 1825.... Albany: Packard, & Van Benthuysen, 1825. 47 p.
S-22468.
Copies: ICN; MH; NHi; NCH.

3078. Delaware Co., Pa. Committee on Manufactories, Mills and Unimproved Mill Seats.
Report of the Committee of Delaware County, on the subject of manufactories, unimproved mill seats, &c. in said county, 1826. Chester, Pa.: Printed by Joseph M. G. Lescure, 1826. 27 p.
Report of the Committee signed on p. 4 by G. G. Leiper and W. Martin.
The statistical report on mills prepared by Benjamin Pearson, p. [3].
S-24330.
Copies: DeGE; MWA; PPFM.

3079. Drowne, Solomon, 1753–1834.
Annual address, delivered before the Rhode-Island Society for the Encouragement of Domestic Industry, October 15, 1823. Providence: Published by order of the Society; Carlile and Brown, printers, 1826. 18 p.
S-24367.
Copies: DeGE; MWA; RHi; RPB.

3080. Hirst, William Lucas, 1802?–1876.
Anniversary discourse pronounced before the Philadelphia Forum, November 4, 1826.... Philadelphia: H. C. Carey and I. Lea, 1826. 42 p.
Includes observations on the progress of manufactures and inland navigation.
S-24853.
Copies: DeGE; MHi; PHi; PPAmP; PPB; PPL.

3081. Hunter, William, 1774–1849.
Annual address, delivered before the R. I. Society for the Encouragement of Domestic Industry, October 20, 1824. Providence: Pub. by order of the Society; Carlile and Brown, printers, 1826. 43 p.
S-24915.
Copies: DLC; MB; MH-BA; MHi; MWA; RP; RPB.

3082. New England Society for the Promotion of Manufactures and the Mechanic Arts.
Act of incorporation, and by-laws of the New England Society for the Promotion of Manufactures and the Mechanic Arts. Incorporated March 3, 1826. Boston: Printed by Beals and Homer, 1826. 16 p.
S-25504.
Copies: DeGE; MH-BA; MHi; MWA; MiD-B; RPB.

3083. Pennsylvania Society for the Promotion of Manufactures and the Mechanic Arts.

Constitution of the Pennsylvania Society for the Promotion of Manufactures and the Mechanic Arts. [Phildelphia: 1826.] Broadside 28 x 18 cm.

Dated: December 14, 1826.

S-25717.

Copies: MWA; PU.

3084. Report of the committee appointed to enquire into the practicability and expediency of establishing manufactures in Salem. Salem, [Mass.]: Printed by Warwick Palfrey, jun., 1826. 31 p.

S-26005.

Copies: CtHT-W; CtSoP; DLC; ICJ; MB; MSa; MWA; MdBJ; MeHi; NNC; NhHi; PPL; PU; WHi.

3085. Richmond, William Ebenezer, 1786–1873.

Annual address delivered before the Rhode Island Society for the Encouragement of Domestic Industry, October 19, 1825. Providence: Carlile and Brown, printers, 1826. 24 p.

S-25943.

Copies: NN; RHi; RNHi; RP; RPB.

3086. Account of the dinner given to Professor List by the Pennsylvania Society for the Encouragement of Manufactures and the Mechanic Arts, at the Mansion House, Philadelphia, Nov. 3, 1827. [Philadelphia: 1827.] 15 p.

Friedrich List's speech, p. [5]–15.

S-27767.

Copies: DLC; DeGE; MB.

3087. Cocheco Manufacturing Company, Dover, N.H.

The acts of incorporation, and by-laws of the Cocheco Manufacturing Company. Together with extracts from the statute laws of New Hampshire relative to manufacturing corporations. Boston: Printed by Beals & Homer, 1827. 26 p.

S-28529.

Copies: MBAt; MHi.

3088. Convention of Delegates of the State of Pennsylvania, Held for the Promotion of the State Agricultural and Manufacturing Interests, Harrisburg, 1827.

Address of a Convention of Delegates of the State of Pennsylvania, for the State Agricultural and Manufacturing Interests, at Harrisburg, June 27, 1827. [Harrisburg: 1827.] 8 p.

Written by Charles Jared Ingersoll.

S-28587.

Copies: DeGE; MWA; NN; P.

3089. ———— [Boston: 1827?] 8 p.

With a manuscript note: "The address of J.I. as republished at Boston."

Copies: DeGE.

3090. Convention of Delegates of the State of Pennsylvania, Held for the Promotion of the State Agricultural and Manufacturing Interests, Harrisburg, 1827.

Address of the Convention of 27th June 1827 to their fellow citizens of Pennsylvania. [Harrisburg?: 1827.] 12 p.

Written by Charles Jared Ingersoll.

Copies: DeGE; MWA.

3091. Convention of Delegates of the State of Pennsylvania, Held for the Promotion of the State Agricultural and Manufacturing Interests, Harrisburg, 1827.

Journal of a Convention of Delegates of the State of Pennsylvania, held ... at the Capitol in Harrisburg on Wednesday, the 27th of June, A.D. 1827. [Harrisburg]: Atkinson & Alexander, printers, 1827. 16 p.

S-28588.

Copies: Ct; DLC; DeGE; ICU; MWA; N; OCHP.

3092. Harrisburg Convention, 1827.

National Convention. Harrisburg, July 30 [to August 1st], 1827. [Harrisburg: 1827.] 8 p.

Running title: Journal of the National Convention.

3092 (*continued*)

Proceedings of the first three days of the convention, held to promote domestic industries.

Signed: Mathew Carey, Chairman.

N-H0135776.

Copies: DeGE.

3093. Harrisburg Convention, 1827.

[Proceedings of the] General Convention of Agriculturists and Manufacturers, and Others Friendly to the Encouragement and Support of the Domestic Industry of the United States. Monday, July 30, 1827. [Baltimore: 1827.] 76 p.

Running title: Proceedings of the General Convention at Harrisburg.

S-29147.

Copies: CSmH; CtY; DLC; DeGE; ICJ; MBAt; MH; MWA; NNC; NjP; P; PPL; TxU; WHi.

3094. [Ingersoll, Charles Jared] 1782–1862.

At a ... meeting of farmers & manufacturers & friends of farmers & manufacturers, of the city & county of Philadelphia, held at the house of John Neville, in Germantown ... May 31, 1827.... The address ... for the promotion of manufactures & the mechanic arts.... [S.l.: 1827.] 4 p.

S-29326.

Copies: MWA.

3095. Jones, Thomas P., 1774–1848.

An address on the progress of manufactures and internal improvement, in the United States; and particularly, on the advantages to be derived from the employment of slaves in the manufacturing of cotton and other goods. Delivered in the hall of the Franklin Institute, November 6, 1827. Philadelphia: J. Dobson, 1827. 18 p.

S-29381.

Copies: CLU; DLC; ICU; LU; MBAt; MiU-C; NNC; NNNAM; NcU; PHi; PPAmP.

3096. [Lee, Henry] 1782–1867.

Report of a committee of the citizens of Boston and vicinity, opposed to a further increase of duties on importations. Boston: Press of N. Hale, 1827. iv, [5]–196 p.

Signed by Nathaniel Goddard and fourteen others; written by Henry Lee, one of the signers. The report gives a survey of the state of American manufactures.

S-28239.

Copies: CSmH; CtY; DLC; DeGE; ICN; MH; MWA; MdHi; MiD-B; NNS; NcD; NjR; PPAmP; RPB; ScU; WHi.

3097. Nashua Manufacturing Company.

The acts of incorporation, and by-laws of the Nashua Manufacturing Company. Together with extracts from the statute laws of New Hampshire, relative to manufacturing corporations. Boston: Printed by Beals and Homer, 1827. 24 p.

S-29883.

Copies: MCM; MHi; MWA.

3098. New Jersey Society for the Promotion of Manufactures and the Mechanic Arts.

Circular. Paterson, 8th December, 1827.... [Paterson, N.J.: 1827.] Broadside 25 x 18 cm.

Advocating the promotion and protection of domestic flax, hemp, woolen and cotton manufactures.

N-N0157270.

Copies: CtY.

3099. Pennsylvania Society for the Promotion of Manufactures and the Mechanic Arts.

Farmers' and manufacturers' meeting. At a meeting of the Pennsylvania Society for the Promotion of Manufactures and the Mechanic Arts, held at Philadelphia, on the 14th day of May, 1827. Charles Jared Ingersoll, Vice-President, in the chair, and Redwood Fisher, Esq., Secretary.... Boston: 1827. Broadside 64 x 56 cm.

Dated: Boston, June 1, 1827.

S-30195.

Copies: MWA; NjR.

3100. Pennsylvania Society for the Promotion of Manufactures and the Mechanic Arts.

Members of the Pennsylvania Society for the Promotion of Manufactures and the Mechanic Arts. [Phi]adelphia: 1827.] [2] l.

Cover title.

Includes the constitution of the Society, and the list of officers elected Jan. 19, 1827.

Handwritten list of members at end.

S-31871.

Copies: DeGE.

3101. Pennsylvania Society for the Promotion of Manufactures and the Mechanic Arts.

To the citizens of the United States. [Philadelphia: 1827.] 12 p.

Caption title from p. 3; the text on p. [1] begins: At a meeting of the Pennsylvania Society for the Promotion of Manufactures and Mechanic Arts, held in Philadelphia, on the 14th day of May, 1826....

Signed: Charles J. Ingersoll [and others].

An appeal for the support of industry, particularly woolen manufactures.

S-30197–8.

Copies: DeGE; MB; MH-BA; PPL.

3102. Tibbits, George, 1763–1849.

Essay on the expediency and practicability of improving or creating home markets for the sale of agricultural productions and raw materials, by the introduction or growth of artizans and manufacturers. Being an appendix and illustration of a memoir on the same subject, read before the Board of Agriculture of the State of New York, March 8, 1825.... Philadelphia: Printed by J. R. A. Skerrett, 1827. 56 p.

S-30801.

Copies: DeGE; KU; MB; MH-BA; MWA; MdHi; MiD-B; PPL; PU-W; ScCC; ViL.

3103. Tibbits, George, 1763–1849.

A memoir on the expediency and practicability, of improving or creating home markets for the sale of agricultural productions and raw materials, by the introduction or growth of artizans and manufacturers. Read before the Board of Agriculture of the State of New York, March 8, 1825 ... 3d ed. To which are added four appendices.... Albany: Printed; Philadelphia: Reprinted by J. R. A. Skerrett, 1827. 48 p.

S-30802.

Copies: DeGE; ICU; IU; MB; MH; MWA; MiD; N; NcD; PPL; WHi.

3104. Tillinghast, Joseph Leonard, 1790–1844.

Address delivered before the Rhode Island Society for the Encouragement of Domestic Industry at their anniversary, October 9, 1826.... Providence: Printed by Smith & Parmenter, 1827. 26 p.

S-30803.

Copies: MHi; NN; PFAmP; RP; RPB.

3105. United States. Treasury Dept.

Alexander Hamilton's report on the subject of manufactures, made in his capacity of Secretary of the Treasury, on the fifth of December, 1791. 6th ed. To which are prefixed two prefaces by the editor [Mathew Carey]. Philadelphia: Printed by W. Brown, 1827. 80 p.

S-30909.

Copies: CSmH; Ct; DLC; DeGE; MB; MH; MWA; OClWHi; PHi; PPL; RPJCB.

3106. Ware Manufacturing Company.

By-laws of the Ware Manufacturing Company, adopted March 17, 1827. Together with the acts of incorporation, and the laws relating to manufacturing corporations. Boston: Printed by Isaac R. Butts and Co., 1827. 23 p.

N-W0077710; Sabin-101413.

Copies: MB.

3107. Carey, Mathew, 1760–1839.

A valedictory address to the members of the acting committee of the Pennsylvania Society for the Promotion of Manufactures and the Mechanic Arts.... [Philadelphia:

3107 (*continued*)
1828.] [2] p.
Dated: July 23, 1828.
S-32601.
Copies: PHi.

3108. Considerations which demand the attention of farmers, mechanics, and friends of the American system.... By a citizen of the state of New York. New York: Sickels, printer, 1828. 16 p.
Promotion of domestic manufactures.
Also printed in New York by J. Seymour.
N-C0648882; S-32816.
Copies: CSmH; CtY; DLC; DeGE; IU; MH-BA; MWA; MiU-C; NN; NNC; NcD; NjR; OHi; TxU.

3109. [Hale, Nathan] 1784–1863.
The American system; or, The effects of high duties on imports designed for the encouragement of domestic industry; with remarks on the late annual Treasury report. Boston: From Nathan Hale's press, 1828. 86 p.
S-33454.
Copies: CtY; DLC; DeGE; ICJ; IU; MH; MWA; MiD-B; NNS; Nh; OO; PHi; RPB; WHi.

3110. Indiana. General Assembly.
Resolution of the General Assembly of Indiana in favor of the encouragement of domestic manufactures and internal improvements. February 20, 1828.... Washington: Printed by Duff Green, 1828. 3 p. (20th Cong., 1st sess. Senate. Doc. 113)
S-36801.
Copies: DLC; DeU; MWA; NjR.

3111. [Lee, Henry] 1782–1867.
Report of a committee of the citizens of Boston and vicinity opposed to a further increase of duties on importations. New York: Printed by Clayton & Van Norden, 1828. 196 p.
The report, written by Henry Lee, presents a general survey of the state of American manufactures.
S-32428.
Copies: CSmH; CtY; DLC; IU; InU; MH-BA; MWA; NIC; NjP; PPL; PU; ScU; Vi.

3112. ——— Boston. Printed 1827. Philadelphia: Reprinted by S. Parker, 1828. 180 p.
S-32429.
Copies: MWA; P; PPL.

3113. ——— Boston. Printed. Charleston: Reprinted by A. E. Miller, 1828. 122 p.
S-32430.
Copies: CtY; NcD; RPB; ScC.

3114. Memorial of agriculturists and manufacturers of the state of New York, friendly to the encouragement and protection of American industry. January 3, 1828.... Washington: Printed by Gales & Seaton, 1828. 13 p. (20th Cong., 1st sess. House. Doc. 33)
At head of title: New York.
S-36655.
Copies: DLC; DeGE; MWA; NNC; NjR.

3115. Ohio. General Assembly.
Resolutions of the General Assembly of Ohio dissenting from the opinions contained in certain resolutions of the Legislature of South Carolina, denying the power of the general government over internal improvements and protection of domestic manufactures. February 26, 1828.... Washington: Printed by Duff Green, 1828. 5 p. (20th Cong., 1st sess. Senate. Doc. 123)
S-36817.
Copies: DLC; MWA; NjR.

3116. Pennsylvania Society for the Promotion of Manufactures and the Mechanic Arts.
Observations on the report of the Committee of Ways and Means, made at Washington, 12th March, 1828. Philadelphia: 1828. 27 p.
S-34708.
Copies: DLC; DeGE; MH-BA; MWA; NNNAM; OClWHi; PPL.

3117. Pittsburgh Manufacturing Association.

Memorial on behalf of the Pittsburgh Manufacturing Association. January 3, 1828.... Washington: Printed by Duff Green, 1828. 5 p. (20th Cong., 1st sess. Senate. Doc. 16)

S-36603.

Copies: DLC; DeU; MH; MWA; NjR.

3118. Review of the Report of a committee of the citizens of Boston and vicinity, opposed to a further increase of duties on importations. Philadelphia: John Young, 1828. 87 p.

Includes data on American industrial production and urges further promotion of domestic manufactures.

S-35014, 35015.

Copies: DLC; DeGE; ICJ; ICU; MBAt; MH; MH-BA; MWA; MoSHi; Nh; OO; PPAmP; PPL; PU; TKL.

3119. South Carolina. General Assembly.

Memorial of the legislature of the state of South Carolina, remonstrating against the passing of laws by Congress, increasing the duties upon importations for the encouragement of domestic manufactures; against the exercise of the general power to construct roads and canals, either with or without the consent of the states ... January 14, 1828.... Washington: Printed by Gales & Seaton, 1828. 19 p. (20th Cong., 1st sess. House. Doc. 65)

S-36892.

Copies: DLC; DeU; MWA; NjR.

3120. United States. Congress. House. Committee on Manufactures.

Duties on imports. January 31, 1828. Mr. Mallary, from the Committee on Manufactures, made the following report.... [Washington: 1828.] 160 p. (20th Cong., 1st sess. House. Rept. 115)

"Minutes of evidence taken before the Committee on Manufactures," concerning the state of manufacturing industries, p. 19–160.

S-36271.

Copies: DLC; DeU; MWA; NjR.

3121. United States. Congress. House. Committee on Manufactures.

Minutes of evidence, taken before the Committee on Manufactures, first session of the Twentieth Congress. Ordered to be printed, 31st January, 1828. [Washington: 1828.] [2], 19–160 p. [(20th Cong., 1st sess. House. Rept. 115)]

Concerns the state of manufacturing industries.

S-36626 (part of S-36271).

Copies: DLC; DeGE; DeU; MWA; PHC; RPB.

3122. Canton Company of Baltimore.

Prospectus of the Canton Company of Baltimore, to which is prefixed a map of the company's ground; together with the act of incorporation, and an explanation of the designs of the company.... Baltimore: Printed by Benjamin Edes, 1829. 35 p., fold. map.

The map shows the Baltimore and Ohio, and the Baltimore and Susquehanna railroads.

S-38046.

Copies: DBRE; DLC; DeGE; MB; MH-BA; MWA; MdBP; MdHi; NN; NcD; PHi; PPAmP; PPL.

3123. Pitman, John, 1785–1864.

Annual address delivered before the Rhode Island Society for the Encouragement of Domestic Industry, Oct. 24, 1828. Providence: The Society; Carlile and Parmenter, printers, 1829. 28 p.

S-40090.

Copies: NN; RHi; RPB.

3124. Tibbits, George, 1763–1849.

Essay on the expediency and practicability of improving or creating home markets for the sale of agricultural productions and raw materials by the introduction or growth of artisans and manufacturers. Being an appendix and illustration of a memoir on the same subject, read before the Board of Agriculture of the state of New York, March

3124 *(continued)*
8, 1825. Philadelphia: Clark & Raser, printers, 1829. 2 p.l., 55 p.
S-40650.
Copies: CSmH; ICJ; IU; MH-BA; MWA; P; PU; ScHi; ScU.

3125. Essays on the protecting system.... I. Three of the addresses of the Philadelphia Society for the Promotion of National Industry.—Published in Philadelphia in the year 1819. II. Essay on the maxims of free trade, promulgated by Adam Smith.—From the London Quarterly review. III. Essay on free trade. —From Blackwood's magazine for May, 1825. IV. Extract from a message of Governor Wolcott, of Connecticut. Philadelphia: Printed by Mifflin & Parry, 1830. 72 p.
The addresses of the Philadelphia Society for the Promotion of National Industry are attributed to Mathew Carey.
C-1277.
Copies: DLC; DeGE; LU; MH-BA; MWA; MdHi; PHi; PPAmP; PPPrHi; PU.

3126. Exeter Manufacturing Company.
An act of the legislature of New-Hampshire incorporating the Exeter Manufacturing Company, and the by-laws of the corporation. Exeter: C. Norris, printer, 1830. 16 p.
N-N0149209.
Copies: NNC.

3127. Massachusetts. Laws.
An act defining the general powers and duties of manufacturing corporations. Passed February 23, 1830. [Boston: 1830.] 13 p.
C-2462.
Copies: MHi (not located).

3128. [Publicola], pseud.
Thirteen essays on the policy of manufacturing in this country. From the New York morning herald. Philadelphia: Printed by Clark & Raser, 1830. 1 l., 30 p.
Signed: Publicola.
Attributed to Mathew Carey by Sabin, and Halkett and Laing.
C-3223.
Copies: DLC; DeGE; KU; MB; MH; MWA.

3129. Rush, Richard, 1780–1859.
Interesting correspondence. Letter from Mr. Richard Rush, on the policy of the American system, in an answer to the following [letter from the owners of a line of packets between Providence, Rhode Island, and Philadelphia].... Washington: Printed at the office of the National Journal, 1830. 8 p.
Includes observations on the growth of manufactures and mineral industries.
C-3368.
Copies: DLC; DeGE; MWA; PPL.

3130. ________ [Providence?]: 1830. 18 p.
C-3367.
Copies: RHi; RPB.

3131. United States. Congress. House. Committee on Commerce.
Commerce and navigation. February 8, 1830. Mr. Cumbreleng, from the Committee on Commerce, submitted the following report.... [Washington: 1830.] 63 p. (21st Cong., 1st sess. House. Rept. 165)
Offers data on the state of American manufactures, commerce, and navigation.
C-4643.
Copies: DLC; DeU; MWA; NjR.

3132. United States. Congress. House. Committee on Commerce.
Report of the Committee on Commerce and Navigation ... February 8, 1830. Mr. Cambreleng, from the Committee on Commerce, submitted the following report. [2d ed.] [New York: Printed by J. M. Danforth ... for the merchants of New-York, 1830.] 63 p.
Contains data on the state of American manufactures.
C-5105.
Copies: DeGE; ICU; MBAt; MH-BA; NNC; PPL.

2. Exhibitions

3133. Ardenond, Mr.
A new and interesting mechanical exhibition, Concert Hall, Boston, June 3, 4, 1819. [Boston: 1819.] Broadside.
S&S-47011.
Copies: MHi.

3134. Committee to Procure an Exhibition.
Sir, Being appointed a committee to procure an exhibition ... at Washington ... of ... articles of American manufacture ... to enlist the feelings and judgment of Congress ... in favour of domestic industry ... solicit your cooperation.... [Philadelphia: 1824.] [2] p. (folded sheet).
Dated at end: Nov. 23d, 1824. Signed by Redwood Fisher and four others.
S-15834.
Copies: DeGE.

3135. Franklin Institute, Philadelphia.
Letter from the corresponding secretary of the Franklin Institute of the State of Pennsylvania, for the Promotion of Mechanic Arts. [Philadelphia: 1824.] 15 p.
Dated: Philadelphia, December 15, 1824, and signed Peter A. Browne.
Includes a list of manufactured items which have attracted the greatest attention at the exhibitions, and observations on some useful inventions.
S-16183.
Copies: DeGE; MWA; PPAmP.

3136. Franklin Institute, Philadelphia.
Report of the second annual exhibition of the Franklin Institute of the State of Pennsylvania.... Philadelphia: Printed for the Institute, 1825. 52 p.
S-20556.
Copies: DeGE; MHi; MWA; PHi; PPF.

3137. ——— Philadelphia: Printed for the Institute, 1825. 15 p.
S-20557.
Copies: PPAmP.

3138. Franklin Institute, Philadelphia.
Proposals of the Franklin Institute, for the exhibition of October, 1826. Addressed to the manufacturers of the United States. [Philadelphia: 1826.] 8 p.
N-F034299.
Copies: DLC; DeGE.

3139. Franklin Institute, Philadelphia.
Address of the Committee of Premiums and Exhibitions of the Franklin Institute of the State of Pennsylvania, and a list of the premiums offered to competitors at the exhibition to be held in October, 1827. Philadelphia: J. Harding, printer, 1827. 12 p.
S-28913.
Copies: MH; OClWHi.

3140. [American Institute of the City of New York.]
American Institute. [Report] June 5th, 1828. [New York: 1828.] 6 p.
List of premiums in fabrics, clothing, and other domestic products to be awarded under the superintendence of the American Institute, p. 4-6.
N-A0250964.
Copies: NSchU.

3141. Franklin Institute, Philadelphia.
Address of the Committee of Premiums of the Franklin Institute of the State of Pennsylvania. With a list of premiums offered to competitors at the exhibition to be held in October, 1828. Philadelphia: J. Harding, printer, 1828. 8 p.

3141 (*continued*)
S-33243.
Copies: DLC; MH; MdBJ; NN; RP.

3142. M'Henry, James, 1785–1845.
An ode, written by request, on the opening of the exhibition at the Franklin Institute of Philadelphia, October 1828. [Philadelphia: 1828.] Broadside.
S-33947.
Copies: PPL.

3143. American Institute of the City of New York.
Reports of the American Institute, in the City of New York on the subject of fairs. 2d ed. New York: Printed by J. Seymour, 1829. 23 p.
S-37445.
Copies: DLC; MB; MWA; MnHi; NN; Nh; PPL.

3144. American Institute of the City of New York.
Report of the third annual fair of the American Institute of the City of New-York, held at Masonic Hall, October, 1830; with a list of premiums, and a catalogue of all the articles exhibited, by whom made and sold, with the duties imposed on similar imported commodities. New York: Printed by J. Seymour, 1830. 32 p.
C-101.
Copies: MH; MiU-C; NCH; NbU; OCHP; PHi.

3145. Franklin Institute, Philadelphia.
Address of the Committee on Premiums and Exhibitions of the Franklin Institute of the State of Pennsylvania. With a list of the premiums offered to competitors at the exhibition to be held in September, 1830. Philadelphia: J. Harding, printer, 1830. 8 p.
C-1443.
Copies: DLC; MH-BA; MWA; PHi; PPAmP.

3146. Franklin Institute, Philadelphia.
Report of the Committee on Premiums and Exhibitions of the Franklin Institute of the State of Pennsylvania ... on the sixth exhibition, held September, 1830. Philadelphia: Jesper Harding, printer, 1830. 16 p.
"Extracted from the Journal of the Franklin Institute for October and November, 1830."
Copies: DeGE; MWA.

3. Chemicals

3147. Great Britain. Laws.
Anno regni Georgi II. regis Magnae Britanniae, Franciae & Hiberniae, vicesimo quarto. At the Parliament begun and holden at Westminster, the tenth day of November, anno Dom. 1747 ... and ... continued ... to January, 1750.... London: Printed by Thomas Baskett, 1751. New London: Reprinted by Timothy Green, 1751. [2],13–16 p.
Caption title on p. 13: An act for encouraging the making of pott ashes and pearl ashes in the British plantations in America.
Mp-6685.
Copies: Ct; CtHi.

3148. [Stephens, Thomas] d. 1780.
Method and plain process for making pot-ash, equal if not superior to the best foreign pot-ash. Boston: Sold by Edes & Gill, 1755.

Also published in London, 1755.
E-7575.
No copy known.

3149. [Mascarene, John] 1722–1779.
The manufacture of pot-ash in the British North-American plantations recommended. Boston: Printed by Z. Fowle for T. Leverett, 1757. [4], 11 p.
Mp-7940.
Copies: CtY; MH; MWA; NN; PPL.

3150. Directions for making calcined or pearl-ashes, as practiced in Hungary, &c. With a copper-plate drawing of a calcining furnace. Boston: Printed and sold by John Mein, 1766. 18, 2 p., plate.
Mp-10285.
Copies: MB; PPL.

3151. [Stephens, Thomas] d. 1780.
The method and plain process for making pot-ash, equal if not superior to the best foreign pot-ash. Published in consequence of the late encouragement granted by Parliament for that purpose. Interspersed with proper cuts, &c. to afford the greater idea of the works. New York: Printed and sold by W. Weyman, 1767.
First published in London, 1755.
E-10689.
No copy known.

3152. Palmer (Joseph) and Company.
Sperma-ceti candles warranted pure; are made by Joseph Palmer & co. at Germantown near Boston, & to be sold at their store in Boston, New England.... [Boston: 1770?] Broadside 15.5 x 18.5 cm.
Engraved text is by Nathaniel Hurd; upper portion is in English, lower portion is in French, a whale is in the center oval.
B-3240; Mp-42144.
Copies: MHi; MWA (xerox copy).

3153. Rhea, John.
John Rhea desires to return his thanks to all those persons, who have so chearfully [*sic*] countenanced his proposed plan of a pott and pearl ash works.... [Philadelphia: 1772.] Broadside 32 x 18 cm.
Dated: Philadelphia, February 18, 1772.
Mp-12536.
Copies: PPL.

3154. To the inhabitants of the city of Philadelphia, and parts adjacent. Gentlemen, whereas Mr. John Rhea of this city, merchant.... [Philadelphia: 1772.] Broadside 26 x 21 cm.
Dated: Philadelphia, 15th February, 1772.
A protest of Philadelphia soap boilers against John Rhea's proposal to manufacture potash.
Mp-12577.
Copies: MWA; MiU-C; NN; PPL.

3155. Massachusetts (Colony). General Court. House.
In the House of Representatives, October 31, 1775. Resolved that Dr. Whiting, a member of the committee appointed by this Court to improve the art and business of manufacturing salt petre, be directed immediately to repair again to the Colony of Connecticut.... [Watertown: Printed by Benjamin Edes, 1775.] Broadside.
Mp-14208.
Copies: MHi; MSaE; NN.

3156. United States. Continental Congress, 1775.
Several methods of making salt-petre; recommended to the inhabitants of the United Colonies by their representatives in Congress. Philadelphia: Printed by W. and T. Bradford, 1775. 12 p.
Sometimes attributed to W. Shewell, or Sewell.
Includes contributions from Benjamin Franklin and Benjamin Rush.
Mp-14584.
Copies: DLC; MBAt; NN; PPL; PU; RPJCB.

3157. ——— by the Honorable continental Congress. and re-published by order

3157 (*continued*)
of the General Assembly of the Colony of Massachusetts-Bay. Together with the resolve of the said Assembly, and an appendix, by Doctor William Whiting. Watertown: Printed and sold by Benjamin Edes, 1775. 20 p.

Mp-14585.

Copies: DLC; MB; MBAt; MH; MWA; RPJCB.

3158. Connecticut (Colony). General Assembly.

At a general Assembly of the Governor and Company of the English Colony of Connecticut, in New-England, holden at New-Haven, by special order of the Governor, on the 14th day of December, 1775. Resolved by this Assembly, that the selectmen for the time being, of each town of this Colony, be, and they are hereby appointed inspectors of nitre, or salt-petre, made and manufactured in this Colony.... [New London?: 1776.] [3] p.

Includes "An act for the more effectual carrying into execution the several acts relative to the making of salt-petre and gun-powder . . . ," dated "Hartford, on the second Thursday of May, 1776."

B-4204; Mp-43014.

Copies: NN.

3159. Connecticut (Colony). Laws.

Salt-petre and gun-powder. An act and law, made and passed by the General Court, or Assembly of His Majesty's English Colony of Connecticut, in New-England, in America, holden at New-Haven, by special order of the Governor of said Colony, on the fourteenth day of December, A.D. 1775. An act for encouraging the manufacture of salt-petre and gun-powder. [New London: Printed by T. Green, 1776.] [2] p.

Mp-14694.

Copies: CtHi; NN.

3160. Connecticut. Laws.

At a General Assembly of the Governor and Company of the State of Connecticut, holden at New-Haven . . . October 1776. An act for regulating the inspecting and vending of gun-powder.... An act in addition to an Act . . . relative to the making of salt-petre and gun-powder within this colony. New London: Printed by T. Green, 1776. [2] p.

Mp-14706.

Copies: CtHi; NNB.

3161. New York (Colony). Committee of Safety.

Essays upon the making of salt-petre and gun-powder. Published by order of the Committee of Safety of the Colony of New-York. New York: Printed by Samuel Loudon, 1776. 39 p.

Compiled by Thomas Tredwell, a member of the committee.

Mp-14930.

Copies: CtY; DLC; DeGE; MBAt; NN; PPL; RPJCB.

3162. Pennsylvania (Colony). Committee of Safety.

The process for extracting and refining salt-petre, according to the method practiced at the provincial works in Philadelphia. Published by order of the Committee of Safety. Philadelphia: Printed by William and Thomas Bradford, 1776. 8 p.

Attributed to William Lewis by Sabin.

Mp-14827; Sabin-40837.

Copies: PHi; PPL.

3163. United States. Continental Congress, 1776.

In Congress. August 28, 1776. Resolved, that there be an inspector or inspectors sufficiently qualified to judge the goodness of gunpowder. By order of Congress. John Hancock, President.... [Philadelphia: Printed by John Dunlap, 1776.] Broadside.

Mp-15166.

Copies: DLC; PPL.

3164. Massachusetts. General Court. House.

State of Massachusetts-Bay. In the House of Representatives, August 11, 1777. Whereas it has been represented, that some of the powder manufactured at the several powder mills in this state, has proved defective; and as there is a sufficient quantity of imported powder for the present demands: therefore, resolved.... [Boston: 1777.] Broadside 21 x 17 cm.

Mp-15435.

Copies: DLC; MBAt; MHi.

3165. [Lavoisier, Antoine Laurent] 1743–1794.

The art of manufacturing alkaline salts and potashes.... Translated from the French by Charles Williamos.... To which will be speedily added, an Appendix, stating sundry experiments intended to be made on the different species of woods, shrubs and plants of North America.... [Philadelphia: 1785?] vi, 50 p., 4 fold. plates.

Translation of *L'Art de fabriquer le saline et la potasse,* Paris, 1779.

The dating of this publication is debated: The Library of Congress assigns it to [181–?], and the Kress Library to [1820?]. The most likely date [1785?] is based on D. I. Duveen & H. S. Klickstein, *A Bibliography of the Works of Antoine Laurent Lavoisier,* London, 1954, p. 234.

B-6091; Mp-44708; S&S-20528.

Copies: DLC; MB; MH-BA; MdBP; PHi; PPAN; PPAmP.

3166. Joyce and Snowdon's American ink powder for records.... [New York: 1790.] Broadside.

Mp-22594.

Copies: NHi.

3167. An address to the manufacturers of pot and pearl ash, with an explanation of Samuel Hopkins's patent method of making the same; together with drawings of a furnace, &c. &c. To which are prefixed, the Act of Congress for promoting the progress of useful arts. Also, a copy of the patent granted to Samuel Hopkins. New York: Printed by Childs and Swaine, 1791. 27 p., plate.

Signed on p. 26: Samuel Hopkins, per his attorney, William Shotwell.

B-7629; Mp-46109.

Copies: NN (plate wanting).

3168. Greene, Caleb.

Greene's fine American ink-powder for records.... Sold wholesale and retail by Caleb Greene, New-Bedford, Massachusetts. [New Bedford: 1792?] Broadside 22 x 14 cm.

B-8009; Mp-46455.

Copies: MWA.

3169. Townsend, David.

Principles and observations applied to the manufacture and inspection of pot and pearl ashes.... Published according to act of Congress. Boston: Printed by Isaiah Thomas & Ebenezer T. Andrews, 1793. 48 p.

Mp-26270.

Copies: CtY; DLC; DeGE; MBAt; MHi; MWA; NN; PPAmP; RPJCB.

3170. Pennsylvania. Laws.

An act to regulate the exportation of pot and pearl ash.... [Philadelphia: T. Bradford, 1794.] 6 p.

Dated: In General Assembly, Tuesday, January 21, 1794.

B-8806; Mp-47165.

Copies: MWA.

3171. Aber, Israel.

The art of manufacturing saltpetre by cheap, easy, and expeditious methods. Compiled from different authors—reduced to practice, &c. By Israel Aber, junior.... Newark: J. Woods, printer, 1796. 23 p.

3171 (*continued*)
Mp-29946.
Copies: DLC.

3172. Frobisher, William.
A new and advantageous process for the manufacturing of pot-ash. By Wm. Frobisher. [Boston: Young & Minns?, 1797.] 17 p., fold. plate.
Caption title from p. 4.
Patent, issued to William Frobisher, and signed by President George Washington, and Secretary of State Timothy Pickering at Philadelphia 17th Nov. 1796, on p. [1], with caption title: The United States of America. To all whom these letters patent shall come.
Frobisher's offer to sell the rights to use his patent dated at Boston, May 5 [1797].
Copies: DeGE.

3173. [Mitchill, Samuel Latham] 1764–1831.
The case of the manufacturers of soap & candles, in the city of New-York, stated and examined. To which are prefixed, the laws of the state of New-York, concerning infectious diseases. With an addition, in form of an appendix, containing several documents and papers, relative to these subjects. Published by the Association of Tallow Chandlers and Soap Makers. New York: Printed by John Buel, for the Association, 1797. 62 p.
Mp-32564.
Copies: DLC; DNLM; DeGE; MBAt; MH; MHi; NHi; PPL.

3174. Constant, Joseph.
Patent cement. The patentee of the cement for preserving wood and brick from decay.... Directions.... [New York]: 1800. Broadside.
Dated: March 15, 1800.
The patent was issued to Silas Constant on February 4th, 1800.
Mp-37241.
Copies: NHi.

3175. Du Pont, Victor Marie, 1767–1827.
Riflemen, Attention; Gunpowder of the first quality ... manufactured by E. I. Dupont de Nemours & Co.... to be had ... by applying to the subscriber.... Wilmington, [Del.]: James Wilson, 1806. Broadside 27 x 21.5 cm.
Signed and dated at end: V. Dupont. Angelica, Allegany [*sic*] county, state of New York, November 1, 1806.
S&S-50670.
Copies: DeGE.

3176. [Carmichael, John.]
A treatise on soap-making, containing an account of the alkaline materials, test for discovering the presence of an alkali, &c., with full directions for manufacturing yellow, pure, white, and perfumed hard soap; also, complete instructions for the making of green or soft soap, with other requisites necessary to finish the soap-boiler. To which is added, abstracts of the excise laws relative to hard and soft soap-makers. By a manufacturer. New York: Collins & Perkins, 1810. viii, [xix]–xxi, [13]–75 p.
Kress-B5755; S&S-19724.
Copies: DP; MH; MWA; PPL.

3177. Du Pont de Nemours (E.I.) and Company, Wilmington, Del.
Villany [sic] detected. [Wilmington, Del.: James Wilson, 1812.] 16 p.
Written probably by E. I. du Pont. Concerns Thomas Ewell's attempts to secure Company's techniques and employees for his own powder factory near Washington.
Prefatory note dated June 16, 1812.
S&S-51190.
Copies: DeGE.

3178. Du Pont de Nemours (E.I.) and Company, Wilmington, Del.
Du Pont's powder. The subscribers having made extensive improvements ... to their

powder mills are now prepared to supply the demands of their customers at short notice and at reduced prices. Caution. Deception having been practiced ... in packing inferior powder in old kegs of Du Pont's brand, and selling or hawking it about the country as Du Pont's powder, the subscribers inform the public ... that no powder can be warranted as genuine, but such as is sold by their agents.... Wilmington, Del.: [ca. 1819]. Broadside 25 x 20 cm.

Copies: DeGE.

3179. United States. War Dept.

Letter from the Secretary of War, transmitting (in obedience to a resolution of the House of Representatives) copies of the respective contracts heretofore made with James Johnson, Alexander M'Rea, Elias Earle, and Peter Townsend, for the delivery of gunpowder, &c, &c. March 1, 1820. Washington: Printed by Gales & Seaton, 1820. 19 p., fold. table. (16th Cong., 1st sess. House. Doc. 90)

Includes specifications for gunpowder and arms.

S-3846.

Copies: DLC; DeGE; DeU; G; MWA; O; R.

3180. Boston. Firewards.

Rules and regulations, made and established by the Firewards of Boston relative to the sale, storage, safekeeping and transportation of gunpowder, in the said town.... [Boston: Russell and Gardner, 1821.] [3] p.

S-4783.

Copies: M.

3181. Boston. Ordinances.

Act regulating the storage of gunpowder; with rules and regulations of the fire-wards relating to the same. Boston: 1821. 8 p.

S-4769.

Copies: MHi (not located).

3182. Degrand, Peter Paul Francis, d. 1855.

[Printed letter, beginning] His Excellency John Brooks, Medford, Mass. Sir, at the request of Mr. E. I. du Pont, whose superior gunpowder is doubtless known to you, I beg leave to submit to you a few remarks relative to the inspection law of this state.... [Boston?:1821.] Broadside 35 x 23 cm.

Signed: P. P. F. Degrand, and dated Boston, July 25, 1821.

S-5166.

Copies: DeGE.

3183. Grotz, Christopher.

The art of making fireworks, detonating balls, &c. Containing plain and easy directions for mixing and preparing the ingredients and making ... devices in this ingenious art.... New York: S. King; W. Grattan, printer, 1821. 26 p., col. front.

First published in London, 1819.

S-5510.

Copies: CtY; PNR; PP.

3184. Degrand, Peter Paul Francis, d. 1855.

[Printed letter, beginning] His Excellency John Brooks, Medford, Mass. Sir, at the request of Mr. E. I. du Pont, whose superior gunpowder is doubtless known to you, I beg leave to submit to you a few remarks relative to the inspection law of this state.... [Boston?: 1822?] 2 l. 38 x 24 cm.

Signed: P. P. F. Degrand, Boston, July 25, 1821.

Leaf [2] contains an announcement by Robert M. Barnard, agent, Boston Powder Company, and E. Copeland, Jr., concerning Du Pont gunpowder, dated Boston, 15th Nov. 1822, and results of experiments made with Du Pont and English gunpowder.

Copies: DeGE.

3185. Grotz, Christopher.
The art of making fireworks, detonating balls, &c.... New York: S. King, 1822. 26 p., plate.
S-8903.
Copies: MH; PP.

3186. Carpenter, George Washington, 1802–1860.
Remarks on the use of piperine, with the formula for its manufacture.... [Philadelphia: 1825?] 12 p.
S-19984.
Copies: PPL.

3187. Morrison, John C.
Directions for using the chloride of lime, or bleaching powders, manufactured by the New-York Chemical Manufacturing Company, and sold by John C. Morrison, agent.... [New York: ca. 1827.] 12 p., illus.
Copies: DeGE; MWA.

3188. Labarraque, Antoine Germain, 1777–1850.
Instructions and observations concerning the use of the chlorides of soda and lime.... Translated by Jacob Porter.... New Haven: Printed by Baldwin and Treadway, 1829. 32 p.
S-39241.
Copies: CtY; DLC; DNLM; MB; MH; MWA; N; NcD-MC; P.

3189. Du Pont de Nemours (E.I.) and Company, Wilmington, Del.
[Circular explaining the invalidity of a recent test of powder of various manufacturers, and announcing the company had begun to manufacture orange stain powder.] [Wilmington?]: 1830. 1 p.
Dated at end: Wilmington, Nov. 8th 1830.
C-1210.
Copies: DeGE.

3190. Henry, Robert R.
Letter to the New-York Chamber of Commerce on "discretionary power," showing "it cannot be entrusted to any one without danger of abuse," consequently the necessity and propriety of their prompt interference to have that "dangerous power" taken by law from the Inspector General of Pot and Pearl Ashes, his deputies, and all other inspectors.... New York: Printed for the author, 1830. 124 p.
C-1825.
Copies: DLC; MWA; NN; PPAmP; PPL; ScU.

3191. Hopkins, William Fenn.
Gunpowder, by Lieut. Hopkins, professor of chemistry, U.S.M.A., West Point, N.Y. West Point, N.Y.; [183–?]. 12 p.
C-1915.
Copies: NWM.

3192. New York (State). Laws.
An act to incorporate the Chemical Manufacturing Company, passed February 24, 1823. [With "An act, entitled An act to amend an act to incorporate the New-York Chemical Manufacturing Company, passed April 1, 1824.]" New York: 1830. 22 p.
Goldsmiths-26183.

4. Glass and ceramics

3193. Newcastle lottery, instituted by the friends of the American China Manufactory, for the purpose of raising the clear sum of one thousand pounds, towards the encouragement of said manufactory.... [Wilmington?, Del.: Printed by James Adams? 1771.] Broadside 41 x 24.5 cm.

Dated above the text: Christiana Bridge, July 13, 1771.

Mp-12140 (as a Philadelphia imprint).

Copies: PPL.

3194. Pettie's-Island cash lottery, in three classes.[Philadelphia: 1773.] Broadside.

Lottery, to raise two thousand pounds for the American glass manufactory at Manheim, Pennsylvania.

Mp-12938.

Copies: PPL.

3195. Pettie's Island cash lottery for disposing of sundry very valuable buildings and lots of land ... also sundry well assorted packages of American flint glass therein specified.... [Philadelphia: 1774.] Broadside.

At head of title: Supplement to the American Flint Glass Manufactory.

Dated: January 17, 1774.

Mp-13533.

Copies: PPL.

3196. Amelung, John Frederick.

Remarks on manufactures, principally on the new established glass-house, near Frederick-Town, in the state of Maryland. By John F. Amelung. [Frederick-Town]: Printed [by Matthias Bartgis] for the author, 1787. 13 p.

Mp-20189.

Copies: MBAt; MdBJ; NCorniC; ScC.

3197. Albany glass-house, September 12, 1792. The subscribers, proprietors of the Albany glass-house ... inform the public, that their works are completed, and that they have on hand an assortment of glass.... [Albany: 1792.] Broadside 19 x 27.5 cm.

Advertisement of the glass-house and the firm of M'Clallen, MacGregor and Company.

B-7916; Mp-46373.

Copies: NHi; NN (photocopy).

3198. Rules, to be observed by the hands employed in the brick-making business, revised and amended at a meeting of the following master brickmakers, in and near the city of Baltimore, on the 23d day of February, 1802. [Baltimore: 1802]. Broadside.

S&S-1804.

Copies: MdHi.

3199. United States. Congress. House. Committee of Commerce and Manufactures.

Report of the committee of Commerce and Manufactures, to whom were referred, on the thirty-first ultimo, a petition of the president and directors of the Rensselaer Glass Factory in the state of New York, and a petition of the proprietors of a window-glass manufactory, in the same state. January 12, 1807.... City of Washington: A. & G. Way, printers, 1807. 4 p.

S&S-14062.

Copies: DLC; MWA.

3200. Crolius, Clarkson.

Stone ware for sale by Clarkson Crolius, at his manufactory, No. 4, Cross-Street.... New York: 1809. Broadside, illus. 22 x 17 cm.

Dated: New-York, March 28, 1809.

Copies: MWA.

3201. Tilley, J.

Wonderful mechanism.... J. Tilley, fancy glass blower–from London, respectfully informs the Ladies and Gentlemen, and the public in general, that he will exhibit the whole art of fancy glass blowing, spinning and reeling hot glass round a wheel....

3201 (*continued*)
[Salem, Mass.: 1821.] Broadside 19 x 16 cm.
S-7686a.
Copies: MSaE.

3202. Prices current of Crown window glass. Manufactured by the New England Crown Glass Company, Boston.... Boston: True and Greene, printers, 1827. Broadside 33 x 20 cm.
Dated: Oct. 1, 1827.
Copies: MWA.

5. Leather

3203. McBride, David, 1726–1778.
An improved method of tanning leather: from the Philosophical transactions.... Whereby the leather is not only improved in its quality, but tanned in much less time, and with a smaller quantity of bark, than in any method hitherto known or practised.... Philadelphia: Printed by Robert Smith, 1786. [1], [5]–19 p.
Dated at foot of p. 19: "Dublin, May the 1st, 1777."
Mp-19763.
Copies: MHi.

3204. Parker, Samuel.
The United States of America. To all to whom these letters patent shall come: whereas Samuel Parker, a citizen of the United States, hath alleged that he has invented a new and useful improvement in tanning.... [Washington: 1808.] [2] p.
Includes "The schedule, referred to in these letters patent, and making part of the same, containing a description in the words of the said Samuel Parker himself, of his improvements in tanning."
Copies: DeGE.

3205. United States. Congress. Senate. Committee on the Judiciary.
In Senate of the United States. February 13, 1821. Mr. Mills, from the Committee on the Judiciary, made the following report. [Washington: 1821.] 2 p. (16th Cong., 2d sess. Senate. Doc. 95)
Concerns the patent for currying and finishing leather, issued to Samuel Parker.
S-7275.
Copies: DLC; DeU; MWA; NjP.

3206. United States. Laws.
A bill to extend the term of Samuel Parker's patents for his improvement in currying and finishing leather of all kinds. In Senate of the United States, Feb. 13, 1821. [Washington: 1821.] 1 p. (S. 55)
S-7186.
Copies: DNA.

3207. Ball, Lemuel.
Tanning and currying in their various branches, explained and made easy, from actual experience. Sangerfield, [N.Y.]: Printed by Joseph Tenny, 1827. 29 p.
S-27937.
Copies: NNE.

3208. Shove, E.
Improvement in the art of tanning. Improvements in the manufacture of leather

have not heretofore resulted in much real advantage to the community, as more attention has been paid to the appearance than to the value of the article ... E. Shove and T. Hunt, patentees.... [S.l.]: 1829. 1 p.

S-40422.

Copies: DLC.

3209. United States. Congress. Senate. Committee on the Judiciary.

In Senate of the United States. January 11, 1830.... Mr. Rowan, from the Committee on the Judiciary, made the following report: The Judiciary Committee, to whom was referred the petition of David Keizer, stating that he had discovered the art of rendering leather impervious to water ... report. [Washington: 1830.] 1 p. (21st Cong., 1st sess. Senate. Doc. 21)

Copies: DeU; MWA.

6. Metals and metal products

3210. Pennsylvania (Colony). Lieutenant Governor.

By the Honourable James Hamilton, Esq., Lieutenant Governor, and Commander in Chief, of the Province of Pennsylvania, and counties of Newcastle, Kent and Sussex, on Delaware, a proclamation.... Philadelphia: Printed by B. Franklin, Printer to the Province, 1750. Broadside.

Announcing the act of Parliament "to encourage the importation of pig and bar iron from His Majesty's colonies in America, and to prevent the erection of any mill, or other engine, for slitting or rolling of iron, or any plating forge to work with a tilt hammer or any furnace for making steel in any of the said colonies."

Mp-6582.

Copies: PHi.

3211. Franklin, Samuel.

Samuel Franklin, rasor-maker from London; living at the sign of the Rasor and Crown ... Boston: makes and grinds, rasors, lancets, scissars, penknives, scaits ... all sorts of instruments for drawing teeth, and sundry other sorts of cutlary.... [Boston: 1760?] Broadside 21 x 15.5 cm.

B-2129.

Copies: MWo.

3212. Eliot, Jared, 1685–1763.

An essay on the invention, or art of making very good, if not the best iron, from black sea sand.... New York: Printed and sold by John Holt, 1762. [4], [3]–34 p.

Mp-9109.

Copies: CSmH; CtY; DeGE; MHi; MWA (photocopy); NHi; PPL; ViU.

3213. Salisbury (S. and S.), Boston.

Hard-ware goods. S. and S. Salisbury continue importing from London, Bristol, Birmingham and Sheffield ... assortment of hard-ware goods.... [Boston: 1767?] Broadside 31.5 x 19 cm.

B-2104.

Copies: MWA.

3214. Cranch, Robert Garland.

Robert G. Cranch, sadlers ironmonger and bridle-cutter, on the Exchange, Boston, has imported in the last ship from Bristol, bought of the manufacturers, and sells, whole-sale and retail on the lowest terms.... Boston: Printed by I. Thomas, [1774]. Broadside.

List of hardware for sale.

Mp-13233.

Copies: MB.

3215. Zane, Isaac.

To be sold or rented, Marlbro' Iron-Works, situated about 12 miles from the borough of Winchester, in the state of Virginia, at a distance of 65 to 90 miles from several sea port towns on Potowmack and Rappahannock Rivers, consisting in part, of the following particular buildings, land and appurtenances.... [Frederick: Printed by Matthias Bartgis, 1788.] Broadside 30.5 x 19.5 cm.

B-6873; Mp-45418.

Copies: RPJCB.

3216. Omitted in numbering.

3217. Zane, Isaac.

To be sold or rented, Marlbro' Iron-works, situate about 12 miles from the borough of Winchester, in the state of Virginia ... Isaac Zane.... [Winchester, Va.: 1791.] Broadside 30.5 x 19.5 cm.

Also assumed to be printed in Philadelphia.

Description of the ironworks property.

B-7910; Mp-46366.

Copies: MWA; PHi.

3218. United States. Laws.

A bill to enable the President ... to purchase or lease one or more foundries. [Philadelphia]: John Fenno, [1798]. Broadside 34.5 x 21 cm.

B-10633; Mp-48702.

Copies: MWiW; MiU-C.

3219. Fenton & Lyon, New Haven.

Church bells; of a superior workmanship, far exceeding any from Europe ... are made by Fenton & Lyon, in New Haven & West Stafford.[New Haven: 1800.] Broadside, illus. 22 x 14 cm.

Dated: New-Haven, January 20th, 1800.

Mp-38058.

Copies: CtY; NN.

3220. [Manufacturers' catalog of metal fittings. S.l.: 180–?]. 65 plates (part double, part fold.).

Contains more than 700 designs for cloak pins, curtain pins, furniture ornaments, door plates, and frames.

N-M0192941.

Copies: NNC.

3221. Dean, George.

Hard ware; a large assortment, for sale by George Dean at his store in Essex-Street, Salem.... Also, a few travelling desks; and an assortment of Boston glass, as cheap as at the manufactory. [Salem, Mass.: 1801.] Broadside.

S&S-390.

Copies: MSaE.

3222. Massachusetts. Laws.

An act for incorporating the Salem Iron Factory Company. Salem, [Mass.]: Printed by Joshua Cushing, 1804. 40 p.

S&S-6735.

Copies: CSmH; MWA; NN.

3223. Salem Iron Factory Company.

Be it known, that ________ of ________ in the county of ________ is entitled to one share ... of the property of the Salem Iron Factory Company.... [Salem, Mass.: 1804.] Broadside, illus. 33 x 22 cm.

Stock certificate, with an engraving of the factory by George Graham.

Copies: DeGE.

3224. Secrets relative to metals, respectfully submitted to the mechanics of the United States, by an ingenious artist.... Philadelphia: Printed and sold by T. T. Stiles, 1805. 60 p.

S&S-9333.

Copies: DeGE.

3225. New Hampshire Iron Factory Company.

An act to incorporate certain persons, by the name of the New-Hampshire Iron Factory Company; together with bye-laws of said company. Salem, [Mass.]: Printed by Pool & Palfray, 1808. 16 p.

S&S-15729.
Copies: MH-BA; MSaE.

3226. United States. Congress. House. Committee of Commerce and Manufactures.
Report of the Committee of Commerce and Manufactures, to whom were referred ... the petitions and memorial of Paul and Joseph W. Revere, of Boston ... of sundry copper smiths in the city and liberties of Philadelphia, and of sundry manufacturers of copper, in the city of New York. January 21, 1808.... City of Washington: A. & G. Way, printers, 1808. 6 p.
S&S-16571.
Copies: DLC; DS; DeGE; NN; NNS.

3227. Letter to the directors of the New-Hampshire Iron Factory.... [Salem, Mass.: T. C. Cushing], 1810. 12 p., plates.
N-L0293878; S&S-20861.
Copies: MSaE; NN.

3228. New Hampshire Iron Factory Company.
A report on the affairs of the New-Hampshire Iron Factory Co. and the present state of its works, with a view of its future prospects; made in consequence of an investigation of its concerns; at the request of the directors of said company. Salem, Mass.: Printed by Thomas C. Cushing, 1810. 27 p.
S&S-20862.
Copies: MSaE; NhHi.

3229. Andrews, Benjamin.
List of wares, supplied in large and small quantities, by Benjamin Andrews, No. 4, Union-Street, Boston; chiefly consisting of American manufactured hardwares.... [Boston: 1811.] 39 p.
Catalogue No. 1, October 1, 1811.
S&S-22198.
Copies: WHi.

3230. New Hampshire Iron Factory Company.
Report to the stockholders of the New Hampshire Iron Factory Company. [Salem, Mass.: Printed by W. Palfray, Jr., 1816.] 8 p.
Sabin-52873.
Copies: MB.

3231. Jones (T.K.) & Company, auctioneers, Boston.
Sheriff's sale. Catalogue of hard ware, to be sold at auction, on Tuesday, July 14, 1818, at stores No. 24 & 25, State Street. T. K. Jones & Co., auctioneers. [Boston]: Printed by Russell & Gardner, 1818. 66 p.
Copies: MSaE.

3232. Pettibone, Daniel.
The petition of Daniel Pettibone, to the Senate and House of Representatives of the United States, for the renewal of his patent for welding cast steel to iron in such a manner as to be applied to edge tools or other purposes where steel is required. Philadelphia: Printed by Thomas Town, 1818. 48 p.
S&S-45277.
Copies: DLC; DeGE; MWA; N; NhHi; OClWHi; PPAmP; PPL.

3233. Kemble, William.
West Point foundery and boring mill for the manufacture of iron and brass in all its branches: cannon, shot, shells, steam engines, cylinders cast and bored, sugar mills, mill work, rolling and slitting mill rollers.... List of wheel patterns now on hand, and daily increasing. [New York]: C. S. Van Winkle, printer, [1820?]. Broadside, illus. 40 x 26 cm.
Signed: William Kemble, agent.
Copies: DeGE (photocopy); NHi.

3234. ———— [New York]: 1821. Broadside, illus. 42 x 26 cm.
Signed: William Kemble, No. 17 Whitehall-street, New-York, June 1, 1821.
S-5748.
Copies: DeGE.

3235. United States. President (Monroe).
Message from the President of the United States, transmitting copies of letters received at the Department of State on the subject of the duties discriminating between

3235 (*continued*)
imported rolled and hammered iron. May 3, 1822.... Washington: Printed by Gales & Seaton, 1822. 12 p. (17th Cong., 1st sess. House. Doc. 125)

Includes data about iron manufacturing techniques.

S-11030.

Copies: DLC; DeGE; MWA; NjP.

3236. Committee of Iron Masters and Manufacturers of Iron, Philadelphia.

[Memorial, beginning] Sir, we beg leave to hand you herewith, a Memorial recently adopted at a meeting of iron masters and others interested in the different branches of that manufacture.... [Philadelphia: 1823.] [2] p.

Signed: Jonah Thompson, Mark Richards, David C. Wood, Committee of Iron Masters and Manufacturers of Iron.

Copies: DeGE.

3237. Irving, Ebenezer.

Catalogue of hardware, cutlery, saddlery, &c. for sale at auction by Ebenezer Irving, on Saturday, May 8, 1824 ... No. 145 Pearl-Street.... New York: H. & J. C. Spear, printers, 1824. 32 p.

Copies: MSaE.

3238. ———— Friday, Sept. 18, 1824.... [New York]: H. Spear, printer, 1824. 86 p.

Copies: MSaE.

3239. ———— Wednesday, Oct. 13, 1824.... [New York]: H. Spear, printer, 1824. 85 p.

Copies: MSaE.

3240. ———— Friday, Nov. 5, 1824.... [New York]: H. Spear, printer, 1824. 35 p.

Copies: MSaE.

3241. Salem Lead Manufacturing Company.

An act to incorporate the Salem Lead Manufacturing Company, and the by-laws of the corporation. Salem, [Mass.]: Ferdinand Andrews, printer, 1824. 14, 2 p.

S-17044.

Copies: MHi; MSaE.

3242. James & McArthur, Vincennes, Ind.

The iron store. James & McArthur are now receiving, direct from their furnaces, and will constantly keep on hand, a large and general assortment of ... iron & castings.... Vincennes, Ind.: 1825. Broadside 18 x 17 cm.

Dated: May 11, 1825.

S-21046.

Copies: In.

3243. Oliver, William B.

[Circular of furnishings and tinware, sperm oil, planished ware, patented lamps, lanterns, mills for various purposes, bellows, tankards, flatware, safes, stoves, coal hods, pokers, Britannia ware, and bathing tubs, to sell or let. Boston: ca. 1825.] Broadside.

Romaine-p. 216.

Copies: MWA.

3244. Wood, James, 1771–1851.

Improvement in the manufacture of shovels and spades. Specification forming part of letters patent, dated February 10, 1825. [Washington: 1825.] [2] p., illus.

At head of title: United States Patent Office.

Copies: DeGE (photocopy).

3245. Whitwell, Bond and Company, Boston.

Catalogue of hard ware and cutlery, to be sold by auction on Wednesday, June 21, 1826 ... at store corner of Union and Ann streets.... Whitwell, Bond & Co., auctioneers. Boston: Printed at Ezra Lincoln's office, 1826. 39 p.

Copies: MSaE.

3246. Mitchell & Freeman, Boston.

Prices of hollow ware, sold by Mitchell & Freeman, manufacturers' agents.... [Boston]: John G. Scobie, printer, [1827]. Broadside 35 x 20 cm.

Copies: MWA.

3247. Catalogue of assignee's sale of hard ware, to be sold by auction ... July 2, 1828 ... at No. 57, Water Street. [Boston: Beals, Homer & Co., printers, 1828.] 60 p.
Copies: MSaE.

3248. Coolidge, Poor & Head, auctioneers, Boston.
Catalogue of cutlery, hard ware, and plated saddlery, to be sold by public auction ... June 30 and July 1, 1828, at store No. 4, Dock Square, Boston ... Coolidge, Poor & Head, auctioneers. Boston: Beals, Homer & Co., printers, 1828. 55 p.
Copies: MSaE.

3249. Memorial of manufacturers of hardware, smiths and iron-founders of the city of Philadelphia. March 17, 1828.... Washington: Printed by Gales & Seaton, 1828. 8 p. (20th Cong., 1st sess. House. Doc. 202)
At head of title: Pennsylvania.
S-36709.
Copies: DLC; DeGE; DeU; MWA; NjR.

3250. New Jersey. Laws.
Act to incorporate the New Jersey Iron Company. [Trenton: 1829.] 7 p.
Bowker-p. 164.
No copy known.

3250a. Maryland Iron Company.
A bill, passed by the legislature of Maryland, in their session of 1828 ... to incorporate the Maryland Iron Company. Also, the proceedings by the commissioners under the said law; and the election of the president and directors. Philadelphia: Printed by Joseph Rakestraw, 1830. 11 p.
Copies: DeGE.

3251. Petition of sundry remanufacturers of iron, residing in Philadelphia, praying that the duty on certain descriptions of iron may be so modified as to afford to them an adequate protection in their business. March 2, 1830.... [Washington: 1830.] 7 p. (21st Cong., 1st sess. Senate. Doc. 75)
Covering letter signed: John Sarchet.
Includes observations on the types of iron and the use and need for them.
C-5060.
Copies: DLC; DeGE; DeU; MWA; NjR.

3252. Re-manufacture of iron. Petition of inhabitants of the city and county of Philadelphia, concerned in the re-manufacture of iron. March 2, 1830.... [Washington: 1830.] 6 p. (21st Cong., 1st sess. House. Rept. 266)
C-5082.
Copies: DLC; DeU; MWA; NjR.

3253. United States. Congress. Senate. Committee on Manufactures.
In Senate of the United States, May 18, 1830 ... Mr. Dickerson made the following report: The Committee on Manufactures, to whom were referred the petitions of the remanufacturers of iron, in all branches of smiths' work, in the city and county of Philadelphia, beg leave to report.... [Washington: 1830.] 4 p. (21st Cong., 1st sess. Senate. Doc. 142)
Includes observations on the types and quality of domestic and imported iron.
C-4792.
Copies: DLC; DeGE; DeU; MWA; NjR.

3254. Warren, Barry & Park, auctioneers, Boston.
Catalogue of hard ware, cutlery, &c. in store No. 17 Dock Square, to be sold at public auction, on Friday, April 16th, 1830, by order of the assignee ... Warren, Barry & Park, auct'rs. Boston: Beals & Homer, printers, 1830. 32 p.
Copies: MSaE.

7. Paper

3255. Bell, Robert, 1732?–1784.

Observations relative to the manufactures of paper and printed books in the province of Pennsylvania. [Philadelphia: Printed by Robert Bell, 1773.] Broadside 41 x 24 cm.

Dated: 25th January, 1773.
Mp-12668.
Copies: MWA (photocopy); MiU-C; PPL; PPRF.

3256. Boies and Clark, Milton, Mass.

Paper mills, in compliance with the recommendation of the late Provincial Congress, and to encourage the paper manufactory, we now propose to give one shilling ... per pound, for all white linnen.... Boies and Clark ... Milton, January 10th, 1775. [Boston: 1775.] Broadside 29 x 19 cm.

B-3930; Mp-42774.
Copies: Amos Hollingsworth (1949); MWA (photocopy).

3257. McLean (Hugh) and Company.

In the House of Representatives, February 16, 1776. Whereas the colony cannot be supplied with paper.... Paper-mills, at the slitting-mill, in Milton. In compliance with the foregoing resolve, and to encourage the paper manufacture, we now propose to give three coppers per pound for all white linen ... Hugh McLean and Co.... Salem, [Mass.]: Printed by E. Russell, [1776]. Broadside, illus. 39.5 x 25.5 cm.

B-4243; Mp-43056.
Copies: MWA.

3257a. Gilpin (Joshua) and Company.

Delaware paper manufactory. The subscribers having agreed to establish a paper manufactory at Brandywine Creek, in the county of New-Castle, about two miles from Wilmington.... Wilmington: Printed by Jacob A. Killen, [1787?]. Broadside 36 x 22 cm.

Copies: DeGE (xerox copy; original privately held).

3258. Clough, Ebenezer.

Boston paper staining manufactory, Ebenr. Clough, paper stainer, Charles River Bridge, Boston, manufactures & keeps constantly for sale, a great variety of paper hangings, etc.... [Boston: 1795.] Broadside.

Engraved by Samuel Hill, showing the processes of paper making and staining.

E-28428.
Copies: RPJCB.

3259. Bärtgis, Matthias, d. 1825.

A paper mill, printing office, and mill-seats for sale. The subscriber will sell his paper-mill on Tuscarora Creek, five and half miles from Frederick Town, Frederick County, state of Maryland, with sixty-nine acres of land.... Any person wishing to contract for the above property, may come and view the place and know the terms, by applying to Matthias Bartgis [*sic*], Pleasant Dale Paper Mill.... [Fredericktown, Md.: 1816.] Broadside.

Dated: April 26, 1816.
S&S-38551.
Copies: NHi.

3260. Society of Paper Makers of the States of Pennsylvania and Delaware.

Resolutions of the Society of Paper Makers of the States of Pennsylvania and

Delaware, in regard to "want of care in sending papers to market." [Philadelphia?: 1816.] 3 p.
Sabin-60617.
No copy known.

3261. Certificates respecting the machine paper, made at the Brandywine Paper Mills. [Philadelphia]: J. Maxwell, 1818. 8 l.
Copies: PPAmP.

3262. Society of Paper Makers of the States of Pennsylvania and Delaware.
To the Senate and House of Representatives of the United States, in Congress assembled, the Memorial of the Society of Paper Makers, of the States of Pennsylvania and Delaware.... [Philadelphia: 1820.] 15 p.
S-3261.
Copies: DLC; IU; MB; MWA; MiD-B; PHi; PPAmP.

3263. Gilpin and Company.
Pro patria. Gilpin & Co., Brandywine. Fine, quarto, post, hot pressed.... [Philadelphia?: ca. 1830.] Broadside 25 x 20 cm.
Engraved by J. W. Steel, advertising paper manufactured by Gilpin & Co.
C-1590.
Copies: DeGE.

8. Textiles

3264. Manufactory Company, Boston.
The manufactory scheme. In consequence of the scheme, which is inserted in the book of records of the Manufactory Company.... [Boston: 1740.] [4] p.
Dated at end: Suffolk, ss. Boston, December 4th, 1740.
Offer to sell stocks in the textile manufactory.
B-1087; Mp-40197.
Copies: DLC; MH-BA; MSaE; MWA; MiU-C; RPJCB.

3265. Manufactory Company, Boston.
Province of the Massachusetts-Bay in New-England. In order to redress the distressing circumstances which the trade of this province labours under.... [Boston: 1740.] Broadside.
Dated: Boston, March 10th, 1739, 40.
Concerns textile manufacturing, offering for sale stock in a new manufactory.
B-1088; Mp-40198.
Copies: MAr; MSaE.

3266. ———— [Boston: 1740.] Broadside.
Dated: September 9th, 1740.
B-1089; Mp-40199.
Copies: MSaE; RPJCB.

3267. Cox, Richard, Sir, 2d bart., 1702–1766.
A letter from Sir Richard Cox, Bart. to Thomas Prior, Esq.; shewing, from experience, a sure method to establish the linnen-manufacture, and the beneficial effects it will immediately produce.... Dublin printed. London reprinted & Boston, N.E.: Reprinted and sold by J. Draper, 1750. 36, [2] p.
Mp-6481.
Copies: CtY; ICN; LNHT; MB; MH-BA; MHi; MWA; NN; NNC; RPJCB.

3268. Cooper, Samuel, 1725–1783.
A sermon preached in Boston, New-England, before the Society for Encouraging Industry, and Employing the Poor; August 8, 1753.... Boston: Printed by J. Draper, for D. Henchman, 1753. 2 p.l., 38 p.

3268 (*continued*)
Recommends the establishment of linen manufactories.
Mp-6984.
Copies: CSmH; CtY; DLC; DeGE; ICN; IU; InU; MH; MWA; MiU; NN; NNC; NcD; OkU; PHi; RPJCB.

3269. Industry & frugality proposed as the surest means to make us a rich and flourishing people, and the linen manufacture recommended as tending to promote these among us.... Boston: Printed by Thomas Fleet, 1753. 15 p.
Mp-7027.
Copies: DLC; MB; MBAt; MH; MWA; MiU-C; NN; RPB.

3270. Society for Encouraging Industry and Employing the Poor, Boston.
The report of the committee to the Society for Encouraging Industry ... at their quarterly meeting ... February, A.D. 1752.... By the report ... in the looms, nine hundred and fifty yards of linen ... February the 12, 1752. [Boston: 1753?] Broadside 31 x 21 cm.
B-1630; Mp-40672.
Copies: MWA (xerox copy); NHi.

3271. Society for Encouraging Industry and Employing the Poor, Boston.
Whereas it is found by experience that this Province is not adapted for raising sheep, by reason of our long and tedious winters ... a number of gentlemen have formed themselves into a voluntary society, by the name and title of, The Society for Encouraging Industry and Employing the Poor. [Boston: 1754.] 12 p.
Dated at end: Boston, July 1754.
Includes a prospectus of a linen manufactory.
Mp-7155.
Copies: DLC; MHi; MWA.

3272. Philadelphia Linen Manufactory.
Whereas [Isaac Norris, Benjamin Franklin, Joseph Fox, et al.] have entered into an agreement of copartnership, for erecting and carrying on a linen manufactory, in or near the city of Philadelphia.... [Philadelphia: 1764.] Broadside 22.5 x 13 cm.
Names filled in by hand; dated and signed: Thirteenth day of August 1764, Sam. Preston Moore.
Copies: PHi; PPAmP; RPJCB.

3273. Philadelphia Linen Manufactory.
Whereas the number of poor in and around this city, is at present great ... therefore, in order to alleviate their wants, by finding them employment, it is proposed to erect a linen manufactory, in or near this city; and for that end the following articles are agreed on.... [Philadelphia: 1764.] 2 p.
Mp-9870.
Copies: PPL.

3274. Philadelphia Linen Manufactory.
A number of the inhabitants of this city, desirous of encouraging the poor, by finding them employ, did in the year 1764 agree to form themselves into a Company, and to raise a common stock to be laid out in the purchase of hemp, flax, &c. and in employing people to manufacture the same into coarse linen.... [Philadelphia: 1765.] 2 p.
Includes articles of agreement.
Mp-10134.
Copies: PPL.

3275. Proposals for carrying on a manufacture in the town of Boston, for employing the poor of said town. [Boston: 1768.] [2] p.
Dated: March 18, 1768.
Signed: John Barret, Ezekiel Goldthwait, Melatiah Bourn, Edward Payne, and Henderson Inches.
A prospectus for linen manufacture, also for the production of duck or sail cloth, in Boston.
Mp-10828; N-B0142894.
Copies: MB; MWA (photocopy); ViU.

3276. Woodbridge, Russell.
Advertisement. The subscriber hereby informs the public, that he has lately set up the business of weaving stockings, in East Hartford.... Hartford: 1771. Broadside 25.5 x 19 cm.

Dated: December 5, 1771.
The weaver, William White, requests their customers not to "over twist their thread and worsted."
Copies: CtHi.

3277. Wetherill, Samuel, 1764–1829.
Philadelphia. To the good women in this city and country adjacent, who want employment in the business of spinning.... [Philadelphia]: Printed by J. Dunlap, [ca. 1780]. Broadside, illus. 28 x 22 cm.
Signed: Samuel Wetherill, Jun.
Copies: DeGE.

3278. Mill-lottery. Boston.... According to an act of the General Assembly, passed February twenty-sixth, 1783. ... [Charlestown: 1783?] Broadside.
B-5748; Mp-44406.
Copies: MB.

3279. Hague, John.
American manufactures. The subscriber begs leave to inform the citizens of the United States, that he makes ... carding and spinning machines for wool and cotton.... Philadelphia: Printed by B. Towne, 1788. Broadside 27 x 22 cm.
Dated: Philadelphia, March 28, 1788.
An advertisement by John Hague, Manchester, England.
B-6714; Mp-45269.
Copies: DLC; RPJCB.

3280. Scheme of a lottery, for the purpose of extending and improving the woollen manufactory in the city of Hartford.... Class the second.... [Hartford: 1791.] Broadside 34.5 x 21 cm.
Dated: Hartford, February 17, 1791.
B-7907; Mp-46363.
Copies: CtHi.

3281. Hargrove, John.
The weavers draft book and clothiers assistant; humbly dedicated to Harry Dorsey Gough, Esq.e by his much obliged servant, John Hargrove. Baltimore: Printed, & sold by I. Hagerty, 1792. 28 p.
E-24996.
Copies: MNoanMV; MWA.

3282. Broom, Jacob, 1752–1810.
By legislative authority: a lottery! For raising four thousand dollars, clear of all expenses for the purpose of aiding the subscriber in the re-establishment of his cotton manufactory.... [Wilmington?, Del.: Printed by Samuel and John Adams? 1797.] Broadside 31 x 24 cm.
Dated: Wilmington, Delaware, July 24, 1797.
E-31875.
Copies: PHi.

3283. Slater (Samuel) and Company, North Providence, R.I.
Samuel Slater and Co's list of wholesale prices of cotton yarn.... North Providence, [R.I.]: 1802. Broadside 20 x 12 cm.
Dated: August 1, 1802.
Copies: MWA.

3284. Gibbs, Edward A.
Carding machine. The subscribers ... inform ... that they have erected a machine for breaking and carding wool ... at Mr. Joseph Bowman's mill ... in Rockingham County ... New-Market, [Va.]: Printed by Ambrose Henkel, 1809. Broadside 34 x 25 cm.
Signed: Edward A. Gibbs, Samuel Bowman, March 16, 1809.
Hummel-3190.
Copies: ViW.

3285. Sigler, Michael.
Carding machines. The subscribers have erected two new wool carding machines at Mr. Michael Sigler's plaster-mill ... 7 miles below New-Market on the stage road.... New Market, [Va.]: Printed by Ambrose Henkel & Co., 1810. Broadside 27 x 23 cm.
Signed and dated M. Sigler, Wm. Gibbs. April 26th, 1810.
Hummel-3203.
Copies: ViW.

3286. Hicks, Joseph.

Wool-carding machine. The subscriber ... begs leave to inform ... that his ... machine is now in opperation [*sic*] ... at Hoover's Mills on South-Fork, in Pendleton County.... New Market, [Va.]: Printed by Ambrose Henkel & Co., 1811. Broadside 22 x 25 cm.

Dated and signed: June 1st, 1811. Joseph Hicks.

Hummel-3246.

Copies: ViW.

3287. Read, Daniel.

Letters patent.... Whereas Daniel Read, a citizen of the United States, has alleged that he has invented a new and useful improvement for spinning wool, cotton, &c. called the pleasant spinner.... Utica, [N.Y.]: I. Merrell, p'r., [1811]. 2 l.

Dated: September 11, 1811; signed: Daniel Read.

Includes the description of the spinning device.

Copies: MWA.

3288. Brown, Edward.

Patent spinning machine. The subscriber informs the public, that he has the right for making and vending a machine for spinning wool and cotton for the state of Massachusetts, except Worcester and Hampshire counties, of which John Brown of Providence is the original patentee.... Cambridge, [Mass.]: 1812. Broadside 13.5 x 21.5 cm.

Dated: Sept. 22, 1812.

Copies: MNoanMV.

3289. Barrett, O.

Domestic rover and spinner, for cotton and wool. Hartford: John Russell, Jr., [1813]. Broadside, illus. 44.5 x 22 cm.

Dated: June, 1813.

"Good and warranted machines ... will be furnished ... by speedy application to Jeremiah Bolls, or to Oliver Farnsworth, agent of O. Barrett jun., patentee."

N-B0144495.

Copies: CtY.

3290. Loehner, George.

Wool carding machine. The subscriber ... has erected a new carding machine, at his mill on Shenandoah River.... New Market, [Va.]: Printed by Ambrose Henkel & Co., [1813]. Broadside 22 x 28 cm.

Dated and signed: May 15th, 1813. George Loehner.

Hummel-3268.

Copies: ViW.

3291. Machines, for picking & carding wool ... now put into operation at Jacob Kipf's ... near John Zerkel's saw-mill ... about one mile from New-Market.... New Market, [Va.]: Printed by Ambrose Henkel & Co., [1813]. Broadside 32 x 26 cm.

Dated and signed: June 23, 1813. Jacob Kipf, Geo. Kipf, Thomas Tousey.

Hummel-3261.

Copies: ViU.

3292. Miller, Philip W.

The young weaver's assistant, or A dissertation on the most important branches of weaving.... Boston: Printed by Munroe and Francis, 1813. 11, [1] p.

S&S-29165.

Copies: MWA; Nh.

3293. Umstead, John.

Hillsborough Manufacturing Company. At a meeting of a number of inhabitants of the county of Orange ... on the 28th day of May, 1813 ... for the purpose of establishing a cotton and wool manufactory in the town of Hillsborough.... [Charleston, S.C.: 1813.] Broadside 31 x 20 cm.

Signed and dated: John Umstead, Charleston, 1813.

Hummel-1404.

Copies: NcD.

3294. Walker, James.

Wool carding machines. The subscriber ... is erecting a ... set of wool carding machines, at Mr. Peter Mauck's hemp-mill ... Tom's Brook, Shenandoah County.... New Market, [Va.]: Printed by Ambrose Henkel & Co, [1813]. Broadside 30 x 25 cm.

Dated and signed: June 23d, 1813. James Walker.

Hummel-3269.

Copies: ViW.

3295. France, Joseph.

The weaver's complete guide; or, The webb analyzed. To which is annexed, The weaver's complete draught-book, containing seventy-three different draughts, from two to sixteen treadles.... [Providence?: 1814.] 26 p., 6 fold. plates.

S&S-31508.

Copies: DLC; MB; MWA; PPL; RHi.

3296. Grantzdorf, George.

Wool carding machines. The subscribers respectfully inform the public, that they have erected new and complete set of wool carding machines ... on Stony Creek.... New Market, [Va.]: Ambrose Henkel, [1814]. Broadside 19 x 23 cm.

Dated and signed: May 9th, 1814. George Grantzdorf, Philip Grantzdorf.

Hummel-3280.

Copies: ViU (photocopy).

3297. Ogden, Samuel.

Thoughts, what probable effect the peace with Great Britain will have on the cotton manufactures of this country: interspersed with remarks on our bad management in the business: and the way to improvement ... pointed out. Providence: Goddard & Mann, 1815. 39, [1] p.

S&S-35524.

Copies: MH-BA; MWA; NHi; RPB.

3298. Shakers.

Fulling mill. The society called Shakers, in Logan County, Ky. continue their fulling mill in operation.... Russellville, Ky.: Printed at the office of the Weekly messenger, 1815. Broadside, 22.5 x 25.5 cm.

Signed: John McComb, Saml. G. White, agents. South Union, Jasper Spring, Sept. 12th, 1815.

S&S-35910.

Copies: DLC; KyBgW; OClWHi.

3299. To the Honorable the Senate and House of Representatives of the United States, in Congress assembled. The petition of the citizens of the United States, engaged in manufactories on the Brandywine, and its vicinity.... [Wilmington?, Del : 1815.] 12 p.

Probably prepared by a standing Committee, consisting of E. I. du Pont, Joseph Bringhurst and Louis M'Lane to promote the manufacture of woolen and cotton goods.

S&S-34204.

Copies: DeGE; DeHi; MB; MH-BA; MWA; PPPrHi.

3300. To the Honourable the Senate and House of Representatives of the United States, in Congress assembled. The petition of the undersigned citizens of the United States, being a committee appointed by ... the cotton manufacturers residing in Providence and its vicinity, respectfully sheweth.... [Providence: 1815.] 7 p.

Describes the depressed state of cotton manufactures in Providence and vicinity.

Copies: MWA.

3301. Combs, Gilbert.

Picking & carding machines. The subscribers are now adding new cards to their machines ... on Buffaloe and Tinker Creek.... New Market, [Va.]: Printed in S. Henkel's office, [1816]. Broadside 19 x 24 cm.

Dated and signed: June 17th, 1816. Gilbert Combs, Thomas Tousey.

Hummel-3301.

Copies: ViW.

3302. Felch, Walton.

The manufacturer's pocket-piece; or, The cotton-mill moralized—a poem with illustrative notes.... Medway, [Mass.]: Samuel Allen, 1816. vi, [7]–23 p., table.

S&S-37573.

Copies: CtY; DLC; MB; MNoanMV; PPB.

3303. ———— [Newburyport, Mass.: Printed by William B. Allen & Co., 1816.] vi, [7]–23 p.

S&S-37574.

Copies: MWA.

3304. Hovey, William.
Wool carding and spinning. . . . Worcester: 1816. Broadside 15 x 17.5 cm.
Signed: Worcester, September 30, 1816. William Hovey, George March.
Announcing "to the publick that they have now in operation machinery for carding and spinning wool at the factory of William Hovey. . . ."
Copies: OClWHi.

3305. United States. Congress. House. Committee of Commerce and Manufactures.
Report of the Committee of Commerce and Manufactures on the memorials and petitions of the manufacturers of wool. March 6, 1816. . . . [Washington: 1816.] 2 p.
Reported by T. Newton.
S&S-39596.
Copies: DLC; MWA.

3306. United States. Congress. House. Committee of Commerce and Manufactures.
Report of the Committee of Commerce and Manufactures, to which was referred the memorials and petitions of the manufacturers of cotton wool. February 13, 1816. . . . [Washington: 1816.] 14 p. (14th Cong., 1st sess. House. Doc. 49)
Views the growth of cotton manufacturing, its present state and the amount of capital invested in it.
S&S-39597.
Copies: DLC; DeGE; MWA.

3307. Bronson, J.
The domestic manufacturer's assistant, and family directory, in the arts of weaving and dyeing: comprehending a plain system of directions, applying to those arts and other branches nearly connected with them in the manufacture of cotton and woollen goods. . . . By J. & R. Bronson. Utica: Printed by W. Williams, 1817. 204 p., illus.
S&S-40323.
Copies: CSmH; CtY; DLC; DeGE; LNHT; MWA; N; NGH; NNS; NcRS; OU; PPL; PPPM; RP.

3308. To the Honorable the Senate and House of Representatives of the United States in Congress assembled: the Memorial and petition of . . . manufacturers of woollen articles in the states of New Jersey, Pennsylvania and Delaware. . . . [S.l.: ca. 1817.] 6 p.
S&S-42300.
Copies: DeGE.

3309. Blakeman, Philo.
The weaver's assistant, explaining in a familiar manner the first principles of the art of weaving. . . . To which is added, observations on draft painting. . . . Bridgeport, Conn.: Printed by N. L. Skinner, 1818. 67 p.
S&S-43393.
Copies: CtHT-W; CtHi; CtY; ICU; MWA.

3310. Sykes, Stead & Company.
Sykes, Stead & Co. return their grateful acknowledgements. . . . We have leased for 5 years a mill . . . known by the name of Rockford Factory . . . we intend to carry on the woollen manufacturing business. . . . [Wilmington, Del.: 1818.] Broadside 25.5 x 20 cm.
Printed circular letter, dated at end: Rockford Factory. April 8, 1818. The company was later known as Joseph Sykes & Company.
S&S-45832.
Copies: DeGE.

3311. United States. Congress. House. Committee of Commerce and Manufactures.
Reports of the Committee of Commerce and Manufactures, upon the subject of domestic manufactures, made to the House of Representatives of the United States, at the first session of the Fourteenth Congress. February 13, 1818. . . . [Washington: 1818.] 11 p. (15th Cong., 1st sess. House. Doc. 105)
Concerns mainly woolen manufactures.
S&S-46599.
Copies: DLC; DeGE; DeU; MH-BA; MWA; NjP; PPL; R.

3312. Dunn, John D.
Wool carding and cloth dressing. The subscribers intend carrying on the above

business at the stand commonly known as Roosevelt's Mills in Johnsburg. Ashes and most kinds of country produce received in payment.... Caldwell, [N.Y.?]: Storer's print, [1820]. Broadside.

Dated: September 1, 1820, and signed: John D. Dunn and Caleb Bullock.

S-1057.

No copy known.

3313. A memorial to the Honorable the Senate and House of Representatives of the United States of America. By a company of Swiss settlers. Philadelphia: Printed by J. F. Hurtel, 1820. 27 p.

Proposal to establish in the United States a Swiss colony and manufactory of cotton and other fabrics.

"Memorial to the President of the United States," p. [18]–27, signed Charles Henry Du Pasquier.

S-2238.

Copies: CtY; DLC; MB; MH-BA; MWA; PHi; PPAmP; PPL; TxU.

3314. [Chase, Ebenezer.]

The weaver's guide. A choice selection of drafts, compiled from the newest fashions.... Andover, N.H.: Printed and published by Ebenezer Chase, 1821. [8] l. of diagrs.

Copies: MNoanMV.

3315. Brandywine Manufacturing Company.

An act to incorporate the Brandywine Manufacturing Company. [Wilmington, Del.: R. Porter, printer, 1822.] 8 p.

A cotton manufacturing company.

S-8173.

Copies: DeGE; PHi; PPAmP.

3316. Newmarket Manufacturing Company.

An act to incorporate the Newmarket Manufacturing Company, and the by-laws of the corporation. Salem, [Mass.]: Cushing and Andrews, printers, 1823. 11 p.

A textile manufacturing company.

S-13583.

Copies: MH-BA; MSaE; Nh.

3317. Slater, Samuel, 1768–1835.

Memorial of Samuel Slater, and others, a committee appointed by the manufacturers of cotton, and others interested in the manufacture of that article, in the state of Rhode Island. January 6, 1824.... Washington: Printed by Gales & Seaton, 1824. 13 p. (18th Cong., 1st sess. Senate. Doc. 13)

S-18843.

Copies: DLC; DeGE; DeU; MWA; NjP.

3318. Phenix Manufacturing Company, Paterson, N.J.

Patent Phenix mill canvass. [Paterson, N.J.: Day & Burnett, printers, 1826.] 10 p.

Price list and testimonials from the U.S. Navy Yard, and from ship builders and captains.

S-27751.

Copies: DeGE; MH-BA; PPL.

3319. Petition of the manufacturers of Paterson, in the state of New Jersey, December 31, 1827.... Washington: Printed by Gales & Seaton, 1827. 12 p. (20th Cong., 1st sess. House. Doc. 27)

At head of title: New Jersey.

Includes a survey of the manufacture of sail cloth, canvas, and other items from flax and hemp.

S-36653 (dated 1828).

Copies: DLC; DeU; MWA; NjR.

3320. Phenix Manufacturing Company, Paterson, N.J.

Phenix mill duck. Certificates of the superiority of the duck manufactured at Paterson, N.J. by the Phenix Manufacturing Company; with calculations comparing its cost with that of the foreign imported article. Paterson: Printed by Day & Burnett, 1827. 18 p.

Copies: DeGE.

3321. Proceedings of a meeting held in this city on Tuesday, the 5th inst. to take into consideration the present state of the wool-

3321 (*continued*)
growing and wool-manufacturing interests, and such other manufactures as may require encouragement, and to appoint delegates to attend a general convention ... to be held at Harrisburg on the 30th day of July next. [Boston?: 1827.] 16 p.

Dated: Boston, June 6, 1827.
S-30352.
Copies: MB; MH-BA.

3322. Proceedings of the meeting of manufacturers and growers of wool, in the state of Vermont. Held at Woodstock, December 13, 1826. [S.l.: 1827?] 12 p.

Memorial to the Honourable the Senate and House of Representatives of the United States of America, p. [3]–12.
S-29620.
Copies: DNA; MB; MH-BA; NNC.

3323. Exeter Manufacturing Company.

An act of the legislature of Newhampshire, incorporating the Exeter Manufacturing Company, and the by-laws of the corporation. Exeter, [N.H.]: C. Norris, printer, 1828. 16 p.

A cotton manufacturing company.
S-33109.
Copies: MH-BA.

3324. M'Ilvain, James, 1769–1850.

Memorial of James M'Ilvain, praying that the aid of government may be extended to the growers of wool, and the manufacturers of woollen goods. February 20, 1828.... Washington: Printed by Gales and Seaton, 1828; Wilmington, Del.: Reprinted by order of a number of Friends of National Industry, at the Advertiser Office, [1828]. 8 p. (20th Cong., 1st sess. House. Doc. 151)

S-33949, 36429 (10 p.)
Copies: DeGE; MWA.

3325. North Carolina. General Assembly. Senate. Select Committee on Cotton and Woolen Manufactories.

Report on the subject of cotton and woollen manufactories, and on the growing of wool in North Carolina.... Raleigh: Lawrence and Lemay, printers, 1828. 16 p.

Signed: Chas. Fisher, chairman.
Thornton (Off.)-1394.
Copies: NcD; NcU.

3326. ——— Raleigh: Printed by Lawrence and Lemay, printers to the state; Philadelphia: Reprinted by W. Brown, 1828. 16 p.

S-34582.
Copies: DLC; DeGE; MH; MH-BA; MWA; PPAmP.

3327. Whitwell, Bond and Company, Boston.

American goods. Catalogue of cotton, woollen, and other dry goods, of the manufacture of the United States, to be sold by auction, Thursday, August 27, 1829 ... by Whitwell, Bond & Co., & Coolidge & Haskell.... Boston: Printed by Beals, Homer and Co., 1829. 22 p.

S-41515.
Copies: MH-BA.

9. Other manufactures and processes

3328. United States. Dept. of State.

The Secretary of State, to whom was referred, by the House of Representatives of the United States, the petition of Jacob Isaack, of Newport, in Rhode-Island, has examined into the truth and importance of the allegations therein set forth, and makes thereon the following report.... [Philadelphia: 1791.] Broadside 26 x 20 cm.

Signed: Th. Jefferson, Philadelphia, November 21st, 1791.

Concerns a method to turn salt water into fresh water.

Mp-23919.

Copies: DLC; PPL.

3329. Williams, John Foster, 1743–1814.

Capt. J. F. Williams' apparatus for extracting fresh water from salt water. [Boston: 1793.] [2] p., illus.

Printed on verso of a "Shipping paper" made out for the schooner *Polly*.

B-8558; Mp-46947.

Copies: NN.

3330. United States. Congress. House. Committee on the State of the Mint.

Report on the state of the Mint, and what further measures are necessary to render the institution more beneficial. Published by order of the House of Representatives. Philadelphia: Printed by Francis Childs, 1795. 15 p.

Includes the description of equipment and machinery.

Mp-29774.

Copies: DeGE; MBAt; MWA; RPJCB.

3331. United States. Mint.

Letter from the Secretary of State, enclosing the reports of the late and present director of the Mint, exhibiting the state of that establishment, and shewing the necessity of some further legislative provisions to render it more efficient and secure. 14th December, 1795.... Philadelphia: Printed by Francis and Robert Bailey, 1795. 14 p.

The report is mostly concerned with technical difficulties in operating the Mint.

Mp-31433.

Copies: DeGE; MBAt; MWA.

3332. Williamson, Benjamin.

We, Benjamin Williamson, as the owner, and Jonathan Edwards, the manufacturing worker of a snuff mill moved by water, and having four mortars and four pestles, situated in the township of Springfield in the county of Delaware and state of Pennsylvania, do hereby make application to work the same for one year from the first day of April 1795.... [Philadelphia: 1795.] Broadside.

Addressed to John Nicholson, collector of revenue, district of Pennsylvania.

Mp-29286.

Copies: MWA.

3333. Smith & Bartlett, Boston.

Catalogue of drugs and medicines, instruments and utensils, dyestuffs, groceries and painters' colours, imported, prepared and sold by Smith & Bartlett at ... No. 61. Cornhill, Boston. Boston: Printed by Manning & Loring, 1796. 22 p.

Mp-29537.

Copies: DLC.

3334. United States. Mint.

Orders and directions for conducting the Mint of the United States, established by Elias Boudinot, director of said Mint. November 2, 1795. Philadelphia: Printed by John Fenno, 1796. 36 p.

Mp-31465.

Copies: DLC; PHi; PPL.

3335. United States. Mint.

Letter from the Secretary of State, inclosing a report of the Director of the Mint, suggesting the expediency of some alterations in its establishment ... 20th December, 1796, referred to Mr. Page ... 8th December, 1797, committed to a committee of the whole House.... [Philadelphia: Printed by William Ross, 1797.] [2], [125]–[141] p.

Mp-33068.

Copies: MBAt; MWA; RPJCB.

3336. United States. Congress. House. Committee of Revisal and Unfinished Business.

Report of the Committee of Revisal and Unfinished Business, who were instructed ... to enquire and report the probable amount for which the property occupied by the Mint,

3336 (*continued*)
in Philadelphia, could be sold; and the probable expense of the establishment on the principle of having it worked by steam. 25th January, 1803.... Washington City: Printed by William Duane & Son, 1803. 7 p.
S&S-5446.
Copies: DLC; MWA; NN.

3337. McAllister, John, firm, Philadelphia.
Description of the whips manufactured by John McAllister, No. 48 Chestnut Street.... Philadelphia: [1805]. 4 p.
S&S-50566.
Copies: PPL.

3338. ________ [Philadelphia: 1809.] [2] p.
S&S-17953.
Copies: DeGE; MWA; PPL.

3339. Lind, James, 1716–1794.
An essay on diseases incidental to Europeans, in hot climates.... To which is added ... a simple and easy way to render sea water fresh ... 1st American, from the 6th London ed.... Philadelphia: Printed by W. Duane, 1811. viii, 268 p.
First London ed. published in 1768.
S&S-23228.
Copies: CSmH; DLC; MWA; MdBJ; MoSW-M; NNNAM; VtU.

3340. Jackson, William.
Hair mattrasses.... [New York: 1820.] Broadside 33 x 25 cm.
Dated: New-York, June 1820.
Copies: MWA.

3341. Memorial of sundry umbrella-makers of Philadelphia, praying that protection be extended to the manufacture of cotton umbrellas. February 11, 1828.... Washington: Printed by Duff Green, 1828. 3 p. (20th Cong., 1st sess. Senate. Doc. 97)
S-36568.
Copies: DLC; DeGE; MWA; NjR; PPL.

3342. Bockius, Jacob, firm, Philadelphia.
List of prices at Jacob Bockius's brush manufactory, no. 204, North Second Street, Philadelphia. Philadelphia: Printed by D. Schneck, [ca. 1830]. Broadside 29 x 21 cm.
C-570.
Copies: DeGE.

X. MINING AND MINERAL PRODUCTION

1. General works

3343. Plattes, Gabriel, fl. 1638.

A discovery of subterranean treasure: containing useful explorations, concerning all manner of mines and minerals, from the gold to the coal; with plain directions and rules for the finding of them in all kingdoms and countries. In which the art of melting, refining, and assaying of them is plainly declared.... Whereunto is added a real experiment whereby every ignorant man may presently try whether any piece of gold that shall come to his hands be true or counterfeit.... Also a perfect way to try what colour any berry, leaf, flower ... will give; with a perfect way to make colours that they shall not stain nor fade.... Philadelphia: Printed and sold by Robert Bell, 1784. 37, [1] p.

Originally published in London, 1639.
Mp-18732.
Copies: DLC; DeGE; MWA; NHi; NN; NNE; PPL; RPJCB.

3344. Plattes, Gabriel, fl. 1638.

A discovery of subterranean treasure, viz. all manner of mines & minerals, from the gold to the coal, with plain directions and rules for finding them in all kingdoms and countries. Also the art of melting, refining and assaying them made plain and easy.... To which is added, a real experiment to try whether a piece of gold be true or counterfeit.... Also a sure way to try what colour any berry, leaf, flower, stalk, root, fruit, seed, bark or wood will give; together with directions for making colours that shall not stain or fade.... Philadelphia: Printed, 1792. 24 p.

Mp-24697.
Copies: DLC; MH; MHi; MWA; PPL.

3345. Henfrey, Benjamin.

Philadelphia, Feb. 4th, 1795. Sir, having from early age been concerned in working mines, and smelting and refining ores in Europe, and also several years engaged in the same line in the United States, I wish to become useful in promoting the opening of mines in this country ... Benjamin Henfrey. From the Gap-Copper-Mines, Lancaster County. Syllabus on mineralogy.... [Philadelphia: Thomas Dobson; Printed by Snowden and M'Corkle, 1795.] Broadside 19.5 x 23.5 cm.

Mp-28822.
Copies: DLC.

3346. Proposals for establishing an association for working mines, and manufacturing metals.... Philadelphia: Printed by Samuel H. Smith, 1796. 22 p.

B-9666; Mp-47886.
Copies: DLC.

3347. Roosevelt, Nicholas J., 1767–1854.

Papers, relative to an application to Congress, for an exclusive right of searching for and working mines, in the North-West and South-West Territory. By N.I. [*sic*] Roosevelt & J. Mark, and their associates. [Philadelphia]: Printed by Samuel Harrison Smith, [1796]. 28 p.

Mp-31473.
Copies: CSmH; DLC; MBAt; MWA; NN; PPAmP.

3348. Keating, William Hypolitus, 1799–1840.

Considerations upon the art of mining. To which are added reflections on its actual

3348 (*continued*)
state in Europe, and the advantages which would result from an introduction of this art into the United States.... Philadelphia: M. Carey and Sons, 1821. 87 p.

"Read before the American Philosophical Society, July 20th, 1821."

S-5745.

Copies: CU; CtY; DLC; DeGE; IU; MH; MWA; MdHi; MiHM; NN; PPL; RPJCB.

3349. New York (State). Laws.

An act to incorporate the North American Mining Company.... [Albany: 1822.] 21 p.

Passed March 1st, 1822.

Sabin-55554.

Copies: N.

3350. Batopilas Mining Company.

Articles of association of the Batopilas Mining Company. [S.l.: 1826.] 8 p.

Mines located in Mexico.

S-23715.

Copies: PPAmP; PPL.

3351. Interesting documents relative to the trade in coal, iron, lime, salt, gypsum, and other minerals, on the proposed Chenango Canal. [S.l.: 1827?] 11 p.

Consists of reports by James Geddes, Owen Forman, and Thomas Meredith.

N-I0105214.

Copies: DLC.

3352. Temascaltepec Mining Company, Baltimore.

Documents laid before a meeting of the stockholders of the Temascaltepec Mining Company of Baltimore, convened by public notice, August 24, 1827. Baltimore: Printed by Thomas Murphy, [1827?]. 33 p.

S-30773.

Copies: MH-BA; MWA; MdHi.

3353. United States. Laws.

A bill for authorizing a patent to be issued to Moses Shaw. March 26, 1830 ... Mr. Storrs, from the Committee on the Judiciary, reported the following bill. [Washington: 1830.] 1 p. (H.R. 393)

Concerns a method of blasting rocks.

C-3990.

Copies: DNA.

2. Metals

3354. Elixir magnum: the philosophers stone found out. Being a certain method to extract silver and gold out of the earth in great plenty. By way of address to the legislative powers of all the British colonies in North America. Philadelphia: Printed and sold by James Chattin, 1757. 29 p.

Mp-7887.

Copies: MWA; PPL.

3355. Gap Mine Company.

Plan of association of the Gap-Mine-Company. Agreed to by the subscribers at general meeting held at the mines, on the 14th day of June 1797. Philadelphia: Printed by Francis Bailey, 1797. [1], ix p.

Copies: PHi.

3356. Henfrey, Benjamin.

A plan with proposals for forming a company to work mines in the United States; and to smelt and refine the ores, whether of copper, lead, tin, silver, or gold.... Philadelphia; Printed by Snowden & M'Corkle, 1797. v, [7]–34 p., fold. tables.

Concerns the Gap Copper Mines.

Mp-32245.

Copies: DLC; DeGE; MBAt; MHi; MWA; NHi; NN; PPAmP; PPL.

3357. Hornblower, Josiah, 1729–1809.

Letter from Mr. Hornblower to Mr. Kitchell, on the subject of Schuyler's copper mine in New Jersey. April 18th, 1800. Printed by order of the House of Representatives of the United States. [Philadelphia?: 1800.] 4 p.

Mp-37648, 38754.

Copies: CU; CtY; DLC; DeGE; MBAt; MWA; NN; NjP; NjR; PPL.

3358. [Latrobe, Benjamin Henry] 1764–1820.

American copper-mines. [Philadelphia?: 1800]. 8 p.

Signed: B. H. Latrobe.

Caption title on p. [3]: To the chairman of the Committee of Commerce and Manufactures, to whom has been referred the petition of N.I. [*sic*] Roosevelt and his associates, praying for an act of incorporation of a mine and metal company.

Concerns the Schuyler Copper Mine in New Jersey.

Incorrectly dated by Evans as 1798.

Mp-37785.

Copies: DLC; DeGE; MWA; MiU-C; NN; NNC; NjP; PPL.

3359. United States. Congress. House. Committee on Copper Lands.

Report of the Committee appointed on the 5th instant, to enquire into the expediency of authorizing the President of the United States to appoint an agent to purchase of the Indians a tract of land on the south-side of Lake Superior which shall include the great copper bed.... [Philadelphia: 1800.] 4 p.

Dated: 17th March, 1800.

Reported by William Cooper.

Mp-38819.

Copies: DLC; MBAt; MWA; NN; RPJCB.

3360. United States. Congress. House. Committee on a Mine and Metal Company.

Report of the Committee to whom was referred the bill, intitled "An act to incorporate the persons therein named ... as a mine and metal company...." February 28th, 1801. [Washington: 1801.] [3] p.

Concerns the Schuyler Copper Mine in New Jersey.

S&S-1547.

Copies: NN.

3361. United States. Laws.

An act to incorporate the persons therein named, and their associates, as a mine and metal company. [Washington: 1801.] 4 p.

Passed the House of Representatives, January 30th, 1801; concerns the Schuyler Copper Mine in New Jersey.

S&S-1453.

Copies: DLC.

3362. United States. Laws.

A bill, to incorporate the persons therein named, and their associates, as a mine and metal company. [Washington: 1801.] 3, [1] p.

Dated: January 19, 1801.

Concerns the Schuyler Copper Mine in New Jersey.

S&S-1472.

Copies: DLC; MWA.

3363. The Soho Company.

The Soho Company, incorporated the 27th November, 1801, by the Assembly of the State of New-Jersey. New York: Printed by T. & J. Swords, 1802. 15 p.

"The Schuyler Copper-mine," p. 6–11.

S&S-3097.

Copies: DeGE; NHi; NjR.

3364. United States. Dept. of State.

Message from the President of the United States, accompanying a report of the Secretary of State, and sundry letters respecting the copper mines on the south side of Lake Superior, transmitted in pursuance of a resolution of the House of the 23d instant. 31st March, 1802.... [Washington: 1802.] 12 p.

S&S-3338.

Copies: DLC; MWA; NN.

3365. [Austin, Moses] 1761–1821.

A summary description of the lead mines in upper Louisiana. Also, an estimate of their produce for three years past. City of Washington: A. & G. Way, printers, 1804. 22 p.

Signed and dated at end: Moses Austin, February 13, 1804.

S&S-5744.

Copies: CSmH; DLC; MB; MWA; NNM; OClW; PPRF.

3366. United States. President (Jefferson).

Message from the President of the United States to both houses of Congress. 8th November, 1804.... Washington City: Printed by William Duane & Son, 1804. [2], 8, [11]–13, [7], 8–22 p.

"A summary description of the lead mines in Upper Louisiana. To which is added an estimate of their produce for three years past," by Moses Austin, at end.

S&S-7551.

Copies: DLC; ICN; MH; MWA; NN; RPB; ViU.

3367. North Carolina Gold-Mine Company.

North Carolina Gold-Mine Company. [Washington: 1806.] 20 p.

Prospectus, with rules and regulations, and other documents.

S&S-11040.

Copies: DLC; MH.

3368. Woodhouse, James, 1770–1809.

Woodhouse's reply to Seybert's strictures on his essay concerning the Perkiomen zinc mine. [Philadelphia: 1808.] 13 p.

Caption title.

S&S-16761.

Copies: DeGE; PPAmP; PPL.

3369. Binny & Ronaldson, Philadelphia.

[Circular letter concerning the mining of antimony and bismuth to make type metal. Philadelphia: 1809.] Broadside 31 cm.

Dated: Philadelphia, February 17, 1809.

N-B0494033.

Copies: DLC.

3370. United States. Congress. Senate. Committee on the Bill to Incorporate the Louisiana Lead Company.

In Senate of the United States. March 12, 1812. Mr. Bradley, from the Committee to whom was referred the bill to incorporate Moses Austin and others, reported the following amendments. [Washington: 1812.] 13 p.

Concerns the incorporation of the Louisiana Lead Company.

S&S-27264.

Copies: DLC.

3371. United States. Laws.

An act to incorporate Moses Austin, John Rice Jones, Henry Austin and others into a company, by the name of the Louisiana Lead Company ... April 7th, 1812. [Washington: 1812.] 8 p.

S&S-26938.

Copies: DLC.

3372. United States. Laws.

A bill concerning the lead mines in the territory of Missouri. June 29th, 1812.[Washington: 1812.] 3 p. (H.R. 142)

S&S-26969.

Copies: DLC.

3373. ———— December 24th, 1812. [Washington: 1812.] 4 p.

S&S-26970.

Copies: DLC.

3374. Fairfax, Ferdinando, 1766–1820.

Description of Ferdinando Fairfax's Shannondale iron estate, with a plan of a company for improving the same. Washington: J. Crossfield, 1815. 20 p.

S&S-34668.

Copies: DLC.

3375. United States. Laws.

A bill concerning the lead mines in the county of Washington, in the territory of Missouri. February 11, 1815. [Washington:

1815.] 8 p. (H.R. 80)
S&S-36171.
Copies: DLC.

3376. United States. Laws.
A bill authorizing the opening and working copper mines on Lake Superior, and for other purposes. February 6, 1816. [Washington]: 1815 [1816]. 2 p. (S. 20)
S&S-51635.
Copies: DLC.

3377. United States. Laws.
A bill authorizing the opening and working copper mines on Lake Superior, and for other purposes. January 28, 1817.... [Washington: 1817.] 2 p. (H.R. 88)
S&S-42364.
Copies: DLC.

3378. Notice on the malleable zinc in sheets; a discovery made not long since.... [New York]: 1819. Broadside 33 x 20 cm.
N-N0335269.
Copies: MH-BA.

3379. Schoolcraft, Henry Rowe, 1793–1864.
A view of the lead mines of Missouri; including some observations on the mineralogy, geology, geography, antiquities, soil, climate, population, and productions of Missouri and Arkansaw, and other sections of the western country. Accompanied by three engravings.... New York: Charles Wiley & Co.; J. Seymour, printer, 1819. 299 [297] p., plates.
S&S-49372.
Copies: DLC; DeGE; ICN; InI; MA; MH-BA; MWA; MWiW; MiDSH; MnHi; NN; NNE; NWM; NcD; PPL.

3380. United States. War Dept.
Message from the President of the United States, transmitting a report of the Secretary of War, on the number, value, & position, of the copper mines on the southern shore of Lake Superior. December 11, 1822.... Washington: Printed by Gales & Seaton, 1822. 33 p.(17th Cong., 1st sess. Senate. Doc. 5)
Report by Henry R. Schoolcraft, Indian Agent at the Sault St. Marie.
S-11023.
Copies: DLC; DeGE; DeU; MWA; NjP.

3381. United States. War Dept.
Message from the President of the United States, transmitting the information required by a resolution of the House of Representatives, respecting the lead mines of Missouri. May 7, 1822.... Washington: Printed by Gales & Seaton, 1822. 19 p. (17th Cong., 1st sess. House. Doc. 129)
The information supplied by the War Department includes documents from the Ordnance Department.
S-11050.
Copies: DLC; DeGE; MWA; NjP.

3382. ———— Washington: Printed by Gales & Seaton, 1822. 19 p. (17th Cong., 1st sess. Senate. Doc. 94)
S-11048.
Copies: DLC; DeGE; MWA; NjP.

3383. United States. Laws.
A bill to authorize the President of the United States to cause the lead mine and salt spring lands, belonging to the United States, to be exposed to public sale. January 21, 1823. [Washington: 1823.] 2 p. (S.36)
S-14467.
Copies: DNA.

3384. Michigan (Territory). Governor (L. Cass).
Letter from Governor Cass, of Michigan, on the advantage of purchasing the country upon Lake Superior where copper

3384 (*continued*)
has been found. Addressed to the Hon. Thomas H. Benton, of the Senate. January 21, 1825.... Washington: Printed by Gales & Seaton, 1825. 4 p. (18th Cong., 2d sess. Senate. Doc. 19)

S-22905.

Copies: DLC; DeU; MWA; NjP.

3385. United States. Laws.

A bill concerning the copper mines on the south side of Lake Superior. January 21, 1825.... [Washington:1825.] 1 p. (S.33)

S-22692.

Copies: DNA.

3386. United States. Laws.

A bill to authorize the President of the United States to cause the reserved lead mines, in the state of Missouri, to be exposed to public sale. March 3, 1826.... [Washington: 1826.] 1 p. (S.80)

S-26761.

Copies: DNA.

3387. ________ December 14, 1826.... [Washington: 1826.] 1 p. (S.7)

S-26762.

Copies: DNA.

3388. United States. President (John Q. Adams).

Lead mines – Illinois and Missouri. Message from the President of the United States, transmitting the information required by a resolution of the House of Representatives, of 8th May last, in relation to the lead mines belonging to the United States, in the states of Illinois and Missouri. December 8, 1826.... Washington: Printed by Gales & Seaton, 1826. 13 p. (19th Cong., 2d sess. House. Doc. 7)

S-27158.

Copies: DLC; DeU; MWA; NjP.

3389. United States. President (John Q. Adams).

Message from the President of the United States, transmitting, in compliance with a resolution of the Senate, of the 15th Dec., 1825, reports &c. relative to lead mines and salt springs. February 8, 1826.... Washington: Printed by Gales & Seaton, 1826. 23 p. (19th Cong., 1st sess. Senate. Doc. 38)

S-27221.

Copies: DLC; DeGE; DeU; MWA; NjP.

3390. United States. Treasury Dept.

Statement in relation to the lead mines and salt springs in Missouri. Laid before the Senate by the chairman of the Committee on Public Lands. December 27, 1826.... Washington: Printed by Gales & Seaton, 1826. 6 p. (19th Cong., 2d sess. Senate. Doc. 9)

Transmits a report from the Treasury Department.

S-27402.

Copies: DLC; DeU; MWA; NjP.

3391. United States. War Dept.

Letter from the Secretary of War, in relation to leases of lead mines and salt springs. February 22, 1826.... Washington: Printed by Gales & Seaton, 1826. 19 p. (19th Cong., 1st sess. Senate. Doc. 45)

Includes information on the methods to obtain and melt ore, and on the manufacture of shot.

S-27173.

Copies: DLC; DeU; MWA; NjP.

3392. United States. Laws.

A bill to authorize the President of the United States to cause the reserved lead mines, in the state of Missouri, to be exposed to public sale, and those on the Upper Mississippi to be leased. December 14, 1827. [Washington: 1827.] 1 p. (S.17)

S-31140.

Copies: DNA.

3393. United States. Congress. House. Committee on Public Lands.

Sales of lands—lead mines, Missouri. January 30, 1828. Mr. Jennings, from the Committee on the Public Lands ... made the following report. Washington: Printed by Gales & Seaton, 1828. 2 p. (20th Cong., 1st sess. House. Rept. 114)

S-36864.

Copies: DLC; DeU; MWA; NjR.

3394. United States. Laws.

An act to authorize the President to appoint a superintendent and receiver at the Fever River lead mines, and for other purposes. April 9, 1828 ... Mr. Duncan, from the Committee on the public Lands, to which was referred the bill from the Senate, entitled "An act to authorize the President to lease lots in the town of Galena, and lead mine on the Upper Mississippi," reported the same, with the following amendments, viz. amend the title to read.... [Washington: 1828.] 5 p. (S. 51)

S-35631.

Copies: DNA.

3395. United States. Laws.

A bill to authorize the President of the United States to cause the reserved lead mines, in the state of Missouri, to be exposed to public sale, and for other purposes. December 8, 1828.... [Washington: 1828.] 1 p. (S. 5)

S-36092.

Copies: DNA.

3396. United States. Laws.

A bill to authorize the sale of public lands containing lead mines, and which have been reserved from sale, in the state of Missouri. January 30, 1828.... [Washington: 1828.] 1 p. (H.R. 125)

S-36100.

Copies: DNA.

3397. United States. President (John Q. Adams).

Agent—lead mines, Missouri and Illinois, &c. Message from the President of the United States, transmitting the information required by a resolution of the House of Representatives of the 10th inst. in relation to instructions given for the government of the agent of the United States' Superintendent of the Lead Mines in Missouri and Illinois ... December 23, 1828.... [Washington: Gales & Seaton, printers, 1828.] 8 p. (20th Cong., 2d sess. House. Doc. 30)

S-35640.

Copies: DLC; DeU; NjR.

3398. United States. War Dept.

Lead mines. Letter from the Secretary of War, transmitting the information required by a resolution of the House of Representatives of the 2d inst. in relation to the lead mines of the United States. January 7, 1828.... Washington: Printed by Gales & Seaton, 1828. 10 p. (20th Cong., 1st sess. House. Doc. 45)

S-36489.

Copies: DLC; DeU; MWA; NjR.

3399. United States. War Dept.

Report from the Secretary of War, in relation to the lead and lead mines in Missouri. April 8, 1828.... Washington: Printed by Duff Green, 1828. 7 p., 2 fold. tables. (20th Cong., 1st sess. Senate. Doc. 169)

S-36755.

Copies: DLC; DeU; MWA; NjR.

3400. Chandler, R. W.

Map of the United States' lead mines on the Upper Mississippi River. Drawn and published by R. W. Chandler of Galena. Ebr. Martin, engraver. Cincinnati: 1829. Map 38 x 44 cm.

S-38102.

Copies: ICN; IaCrM; MB; OCHP; PPL.

3401. Silver mines in Mexico. [New York: Wm. A. Mercein, printer, 1829?] 16 p.
S-40425.
Copies: CU-B.

3402. United States. Laws.
A bill to encourage the discovery and working of certain copper mines, supposed to be on the south side of Lake Superior. December 16, 1828 ... Mr. Benton ... obtained leave to bring in the following bill ... January 29, 1829. Reported with an amendment. [Washington: 1829.] 1 p. (S. 28)
S-41031.
Copies: DNA.

3403. West-Chester Copper Mine Company.
West-Chester Copper Mine Company. [New York: 1829.] 10 p.
Prospectus; dated on p. 5: New-York, 25th September, 1829.
Sabin-102951.
Copies: NN.

3404. North Carolina. General Assembly. House. Committee on the Mecklenburg Gold Mining Company.
Report on incorporating the Mecklenburg Gold Mining Company. Raleigh: Printed by Lawrence & Lemay, printers to the state, 1830. 8 p. (Doc. 15)
C-2866.
Copies: NcU.

3405. North Carolina. General Assembly. Select Committee on the Establishment of a Mint.
Report on the establishment of a mint in the state of North Carolina. Raleigh: Printed by Lawrence & Lemay, printers to the state, 1830. 8 p.
Concerns also the mining of gold in North Carolina.
C-2868.
Copies: NcU.

3406. United States. Laws.
A bill authorizing the Secretary of State to issue a patent to John Powell. December 27, 1830 ... Mr. Buchanan, from the Committee on the Judiciary, reported the following bill. [Washington: 1830.] 1 p. (H.R. 515)
Concerns a method of washing gold ores.
C-3965.
Copies: DNA.

3407. United States. Laws.
A bill to authorize the President of the United States to appoint a superintendent and receiver at the Fever River lead mines ... January 27, 1830.... [Washington: 1830.] 3 p. (H.R. 166)
C-4466.
Copies: DNA.

3. Coal

3408. Lehigh Coal Mine Company.
Jacob Weiss, of Northampton County ... having discovered a certain coal mine.... [Philadelphia: 1792.] Broadside 28 x 15 cm.
Dated: February 13, 1792.
B-8043; Mp-46490.
Copies: PHi.

3409. An enquiry into the chymical character and properties of that species of coal, lately discovered at Rhode Island, together with observations on the useful application of it to the arts and manufactures of the eastern states. Boston: Printed by Snelling and Simons, 1808. 21 p.
Sometimes attributed to William Meade.
S&S-14946.
Copies: CtY; DLC; MH; MWA; NCH.

3410. Rhode Island. Laws.
An act to incorporate the Rhode Island Coal Company.... Boston: Printed by John Eliot, Jun., 1809. 24 p.
S&S-18512.
Copies: MHi; MWA; RHi.

3411. [Colden, Cadwallader David] 1769–1834.
Observations on the intended application of the North-American Coal and Mining Company, to the Legislature of the state of New-York. [New York: 1814.] 23 p.
Dated: February, 1814.
Signed: Cadwallader D. Colden, Jeremiah F. Randolph, and Hector Craig.
Includes a letter from Benjamin Silliman indicating favorable results of his analysis of a specimen of coal, from the vein to be developed by the company.
S&S-31192.
Copies: CtY; DeGE; MB; MH-BA; N.

3412. [Fulton, Robert] 1765–1815.
Plan for supplying the city of New-York with fuel, by the New-York Coal Company. New York: Printed by T. & J. Swords, 1814. 16 p.
Signed on p. 11: Robert Fulton.
S&S-31545.
Copies: DeGE; MB; MHi; MiU; NN; NNC.

3413. North American Coal and Mining Company.
Memorial of the North-American Coal and Mining Company, to the Legislature of the state of New York, and the report of the Select Committee thereon. Printed by order of the House.... [Albany?: 1814?] 16 p.
S&S-32314.
Copies: MB; NHi; NNMus.

3414. North American Coal Company.
An act to incorporate the North American Coal Company. Passed April 6, 1814. Together with the by-laws of the company. [New York?: 1814.] 11 p.
N-N0308114.
Copies: CtY; PPAmP.

3415. Rhode Island Coal Company.
Observations on the Rhode Island coal and certificates with regard to its qualities, values, and various uses. [Boston: 1814.] 16 p.
S&S-32643; Sabin-56563.
Copies: M; MH; PPAmP.

3416. Miner, Charles, 1780–1865.
Lehigh coal. Certificates from a number of persons shewing the use and value of the Lehigh stone coal. With some prefatory remarks. Wilkes Barre, Pa.: Printed by C.

3416 (*continued*)
Miner, [1815?]. 8 p.
S&S-35283.
Copies: PHi.

3417. Lehigh Coal and Navigation Company.
Statement of the Lehigh Navigation and Coal Mine Company. With the terms of subscription for stock. Philadelphia: William Brown, printer, 1818. 12 p.
S&S-44556.
Copies: DLC; PPAmP; PPL.

3418. Lyman & Ralston, Boston.
Notice. The subscribers, agents of the Lehigh Coal Company of Pennsylvania, take this method of informing the public that they are preparing to execute orders for coal ... No. 71, Broad Street, and 44 Market Street ... [Boston?: 1820?] Broadside 24 x 19 cm.
S-2038.
Copies: MHi.

3419. [New York and Schuylkill Coal Company.]
A cursory review of the Schuylkill coal, in reference to its introduction into New-York, and the other Atlantic cities. New York: Printed by G. F. Hopkins, 1823. 24 p.
"An act to incorporate the New-York and Schuylkill Coal Company," p. [22]–24.
S-12308.
Copies: CtY; DeGE; MH-BA.

3419a. [Schuylkill Coal Company.]
Remarks and observations, showing the justice and policy of incorporating "The Schuylkill Coal Company," respectfully addressed to the public, and particularly to the members of the Legislature. [Philadelphia: 1823.] 8 p.
Signed and dated at end: Manuel Eyre, Joshua Lippincott, Caspar W. Morris [et al.] Philadelphia, January, 1823.
Copies: DeGE.

3420. Coal corporations. [S.l.: 1824.] 21 p.
Argument for the incorporated company as the best means of developing Pennsylvania coal fields.
S-15791.
Copies: MH-BA.

3421. Lehigh Coal and Navigation Company.
Facts illustrative of the character of the anthracite, or Lehigh coal, found in the great mines at Mauch Chunk, in possession of the Lehigh Coal and Navigation Company, with certificates from various manufacturers, proving its decided superiority over every other kind of fuel. Philadelphia: S. W. Conrad, 1824. 12 p.
Kress-C1277; S-16880.
Copies: MH-BA.

3422. An address to the inhabitants of Rhode-Island, on the subject of their coal mines. New York: J. Seymour, printer, 1825. 16 p.
S-19375.
Copies: CtY; MH; MWA; PPAN; RP; RPB.

3423. Lehigh Coal and Navigation Company.
Facts illustrative of the character of the anthracite, or Lehigh coal, found in the great mines at Mauch Chunk, in possession of the Lehigh Coal and Navigation Company, with certificates from various manufacturers, proving its decided superiority over every other kind of fuel. Boston: Printed by T. R. Marvin, 1825. 22, [1] p.
Issued by Lyman & Ralston, Boston agents.
S-21187.
Copies: DLC; DeGE; MH; MWA; PPL.

3424. ———— with certificates from various manufacturers, and others, proving its decided superiority over every other kind of fuel. New York: Printed by Gray & Bunce, 1825. 18 p.
S-21188.
Copies: CtY; DeGE; ICJ; MB; MH; MdBP; PHi; PPL.

3425. Rhode Island Coal Company.
A report of the committee of the Rhode Island Coal Company. New York: Printed by J. Seymour, 1825. 16 p.
S-22085.
Copies: CtY; NNM; WHi.

3426. Vanuxem, Lardner, 1792–1848.
Experiments on anthracite, plumago, &c.... Philadelphia: J. Harding, printer, 1825. 13 p.
Extract from the *Journal of the Academy of Natural Sciences of Philadelphia,* June, 1825.
S-23145.
Copies: NNM; PPAN; PPAmP; PPL; PU.

3427. Lehigh Coal and Navigation Company.
Report of the engineers of the Lehigh Coal & Navigation Company, who were appointed, on the nineteenth instant, a committee on the subject of an expose of the property of the Company. Printed by order of the Board of Managers.... [Philadelphia]: S. W. Conrad, printer, 1826. 12 p., fold. diagr.
Signed: Josiah White, Erskine Hazard, engineers.
S-25081.
Copies: CSmH; CtY; DBRE; DLC; DeGE; MB; MH-BA; MiU-T; NN; P; PPF; PPL; ViU.

3428. Lyman & Ralston, Boston.
Directions for the use of the Lehigh coal in grates and stoves.... Boston: 1826. Broadside 30 x 18 cm.
Dated: Boston, Sept. 1, 1826.
Copies: DeGE.

3429. New York and Schuylkill Coal Company.
History of the coal lands and other real estate, owned by the New-York and Schuylkill Coal Company. New York: Geo. F. Hopkins, printer, 1826. 21 p.
S-25568.
Copies: MB; MH-BA; MWA; NN; NjR.

3430. [Vaux, James.]
Remarks upon the use of anthracite, and its application to the various purposes of domestic economy. [Philadelphia: 1826.] [5] p.
At head of title: From the *Franklin journal.*
Signed and dated: J.V., 11 Mo. 1, 1826.
N-V0062632; S-37001 (dated 1828).
Copies: NHi; NN; PPL.

3431. Lehigh Coal and Navigation Company.
Facts illustrative of the character of the anthracite, of Lehigh coal, found in the great mines at Mauch Chunk, in possession of the Lehigh Coal and Navigation Company, with certificates from various manufacturers, and others, proving its decided superiority over every other kind of fuel. Philadelphia: Printed by S. W. Conrad, 1827. 20 p., fold. diagr.
A printed letter from the Fall River Manufactory, Mass., dated February 8, 1828, attached to p. 19.
S-29471.
Copies: DLC; DeGE; MH-BA; MHi; PHi; PPAmP; PPL.

3432. National Anthracite Coal Company of Pennsylvania.
Maps of the lands of the National Anthracite Coal Co., & of the anthracite coal region of Pennsylvania. New York: 1827.
S-29887.
Copies: PPL (not located 1975).

3433. North American Coal Company.
A brief sketch of the property belonging to the North American Coal Company; with some general remarks on the subject of coal and coal mines. New York: Printed by William Davis, Jun., 1827. 14 p., 2 fold. maps.
S-30069.
Copies: DLC; DeGE; MH-BA; MWA; OC; OCHP; P; PHi; PPAmP; PPL.

3434. ———— New York: Printed by G. F. Hopkins, 1827. 23 p., 2 fold. maps.
S-31870.
Copies: DLC; DeGE; MH-BA; MWA; NN; PPAmP; PPL.

3435. Pennsylvania. Laws.
A supplement to the act entitled "An act to incorporate the Lehigh Coal and Navigation Company, passed thirteenth of February one thousand eight hundred and twenty two." [S. l.: 1827.] Broadside.
Copies: PPL.

3436. [Sedgwick, Henry Dwight] 1785–1831.
Circular. [Boston: W. L. Lewis, printer, 1827.] 20 p.
A circular to prospective stockholders in the Rhode Island Coal Company, describing the quality of anthracite coal, particularly in powering steamboats.
S-30566.
Copies: DLC; MB; MH; MWA.

3437. Lycoming Coal Company.
A brief description of the property belonging to the Lycoming Coal Company, with some general remarks on the subject of the coal and iron business. Poughkeepsie, [N.Y.]: Printed by P. Potter, 1828. 32 p., 2 fold. maps.
"An act authorising the governor to incorporate the Lycoming Navigation, Rail Road and Coal Company," p. 27–32.
S-33932.
Copies: CSmH; DIC; DLC; DeGE; MH; MWA; N; NHi; NN; NNE; PU.

3438. Hughes, George Wurtz, 1806–1870.
Report made to the committee of the citizens of Elmira (Tioga County), on the Tioga Coal & Iron Mines, situated on the Tioga River. By George W. Hughes, engineer. Albany: Printed by John B. Van Steenbergh, 1829. 8 p.
S-39042.
Copies: DeGE; MB; N.

3439. Delaware Coal Company.
Act of incorporation and by-laws. Wilmington: R. Porter & Son, 1830. 13 p.
C-1128.
Copies: IU; N; NN.

3440. Report on the coal formation of the valleys of Wyoming and Lackawana. With miscellaneous remarks and communications. New Haven: H. Howe, 1830. 29 p.
C-3289.
Copies: CtY; PHi.

4. Salt

3441. Connecticut (Colony). General Assembly.
At a Gereral Assembly ... holden at Hartford ... May, 1776. [New London: Printed by T. Green, 1776.] Broadside.
A resolution granting bounty for manufacture of salt.
Mp-14700.
Copies: CtHi; MWA (photocopy).

3442. [Paine, Robert Treat] 1731–1814.
The art of making common salt. Particularly adapted to the use of the American colonies. With an extract from Dr. Brownrigg's treatise on the art of making bay-salt. Detached from the Pennsylvania magazine for March 1776. Philadelphia: Printed by R. Aitken, 1776. 7 p., plate.

Mp-14651.
Copies: MBAt; MWA; PPL.

3443. ———— Boston: Reprinted by J. Gill, 1776. 15 p., plate.
Mp-14652.
Copies: DLC; MH; MHi; MWA; NHi.

3444. DeWitt, Benjamin, 1774–1819.
A memoir on the Onondaga salt springs and salt manufactories in the western part of the state of New-York; describing the situation and appearance of the springs, the nature and contents of the waters, and the process by which the salt is procured, &c.... Read before the Society for the Promotion of Agriculture, Arts and Manufactures of the State of New-York, held at the city of Albany, February 28, 1798. Albany: Printed by Loring Andrews, 1798. 28 p.
Mp-33632.
Copies: MH; MWA; NHi; NN; PPAmP.

3445. Fennell, James, 1766–1816.
Description of the principles and plan of proposed establishments of salt works; for the purpose of supplying the United States with home made salt. Philadelphia: Printed by John Bioren, 1798. 12,[17]–60 p.
Mp-33729.
Copies: DLC; NN; PPAmP; PPL.

3446. Fennell, James, 1766–1816.
Philadelphia, July 6th, 1798. To Mr. William Corbet.... [Philadelphia: 1798.] Broadside 45 x 31 cm.
In defense of his plan for making salt from sea water.
B-10298; Mp-48421.
Copies: PHi.

3447. New York (State). Superintendent of the Onondaga Salt Springs.
Annual report of the Superintendent of the Onondaga Salt Springs ... 1798. [Albany: Printed by Loring Andrews?, 1798?]
E-34272; Sabin-57362 (note).
No copy known.

3448. New York (State). Superintendent of the Onondaga Salt Springs.
To the representatives of the people of the state of New York.... The Superintendent of the Onondagua [*sic*] Salt-Works, pursuant to law, respectfully reports.... [Albany: 1800.] [4] p., 3 fold. tables.
E-38095.
Copies: N (not located).

3449. United States. Treasury Dept.
Letter from the Secretary of the Treasury, accompanying his reports on the memorial of David Jones, and the petitions of Obadiah Scott, and of George Gilbert and Jacob Gilbert. 25th March, 1800.... [Philadelphia: 1800.] 13 p.
Concerns salt mines in Ohio.
Mp-38763.
Copies: CtY; DLC; MB; MBAt; MWA; N; NN; RPJCB.

3450. United States. Laws.
A bill concerning the salt springs on the waters of the Wabash River. [Washington]: Duane & Son, printers, [1803]. Broadside.
Dated: 19th February, 1803.
S&S-5214.
Copies: DLC.

3451. [Collins, Thomas.]
A bill is now before the legislature of Pennsylvania, reported at the last session, for granting to Thomas Collins a loan of $5,000 for 5 years, to enable him to carry on his salt works in Butler County. [Pittsburgh: 1810.] Broadside 31 x 19 cm.
Dated above the text: Pittsburgh, December 17, 1810.
Copies: DeGE.

3451a. Proposals for establishing in New-Jersey a company for the purpose of carrying on the manufactory of sea-salt, the alkaline, neutral & purgative salts, the white and yellow paints, on the sea-shore thereof. Together with prefatory remarks and explanations on that subject. Addressed in a letter to Dr. William Campfield of Morris-town. [Morristown, N.J.?]: Printed by Henry P.

3451a (*continued*)
Russell, 1810. 15, [2] p., diagr.
Signed on p. 15: The patentee.
N-P0605321.
Copies: CtY.

3452. United States. Congress. House. Committee on Public Lands.
Report of the Committee on the Public Lands, on certain resolutions of the legislature of the Illinois Territory, relative to the salt springs. March 2d, 1813.... Washington City: Printed by Roger C. Weightman, 1813. [4] p.
S&S-30346.
Copies: DLC; MWA.

3453. United States. Laws.
A bill concerning the salt springs on the waters of the Wabash River. February 26th, 1813.... [Washington: 1813.] 1 p.
S&S-30046.
Copies: DLC.

3454. United States. Laws.
A bill to encourage the making of salt, at the United States' saline, in the Illinois Territory, by the partial introduction of negroes therein. February 6th, 1813.... [Washington: 1813.] 2 p. (H.R. 63)
S&S-30142.
Copies: DLC.

3455. Van Rensselaer, Jeremiah.
An essay on salt, containing notices of its origin, formation, geological position and principal localities, embracing a particular description of the American salines, with a view of its uses in the arts, manufactures and agriculture.... By Jer. Van Rensselaer.... New York: O. Wilder and J. M. Campbell, 1823. 80 p.
S-14770.
Copies: DI-GS; MB; MWA; NCH; NIC; NNM; NNNAM; NjR; OkU; PPAN; PPAmP; PPL.

3456. United States. Congress. House. Committee on Public Lands.
Report of the Committee on the Public Lands, on a resolution and memorial of the legislature of the state of Ohio, in relation to a sale of a tract of land, called the Scioto Salt Springs, with a bill to authorize the sale thereof. May 14, 1824.... [Washington: 1824.] 2 p. (18th Cong., 1st sess. House. Rept. 119)
S-19048.
Copies: DLC; DeU; MWA; NjP.

3457. United States. War Dept.
Message from the President of the United States, transmitting a report of the Secretary of War, and of the Commissioner of the General Land Office, in relation to the salt springs, and lead and copper mines. March 30, 1824.... Washington: Printed by Gales & Seaton, 1824. 147 p., fold. tables. (18th Cong., 1st sess. House. Doc. 128)
The report on the lead mines is from the Ordnance Department; the rest of the report is from the General Land Office.
S-18906.
Copies: DLC; DeGE; DeU; MWA; NjP.

3458. A few observations and remarks on that part of the executive message relating to the Onondaga salt springs which has been referred to committees by the Senate and Assembly. Albany: Office of the Daily advertiser, 1825. 10 p.
S-20515.
Copies: N.

3459. New York (State). Commissioners on the Onondaga Salt Springs.
Report of the Commissioners appointed to perform certain duties relative to the salt springs, in the county of Onondaga. Together with the bill introduced by said commissioners, to regulate the manufacture of salt in the town of Salina. Made to the Legislature, March 18, 1825. Albany: Printed by Croswell & Van Benthuysen, 1825. 82 p., fold. plate.
Signed: John Van Fossen, Reuben Whallon, Simeon Ford.
S-21668.
Copies: DeGE; ICU; MB; MH-BA; MWA; NNC; NjR; PHi; PPiU; WHi.

3460. Beck, Lewis Caleb, 1798–1853.

An account of the salt springs at Salina, in Onondaga County, state of New York; with a chemical examination of the water and of several varieties of salt manufactured at Salina and Syracuse.... New York: Printed by J. Seymour, 1826. 36 p.

S-23731.

Copies: CtY; DLC; MB; MH-AH; MWA; NN; NNNAM; NSyU; Nh; NjR.

3461. Pennsylvania Society for the Promotion of Internal Improvements in the Commonwealth.

At a meeting of the acting Committee of "The Pennsylvania Society for the Promotion of Internal Improvement," the following original paper was read by one of the members, and ordered to be published. Pennsylvania salt. [Philadelphia: 1826.] 4 p.

Dated Jan. 10, 1826.

Recommends the substitution of canals or railroads for turnpike roads to reduce the price of salt.

S-25711; Thomson-122.

Copies: DeGE; MWA; PU.

3462. United States. Congress. House. Committee on Public Lands.

Salt springs in Illinois. February 10, 1827. Mr. Scott, from the Committee on the Public Land, to which was referred the memorial of the Legislature of the state of Illinois, upon the subject of lands reserved for the use of salt springs, made the following report. [Washington: 1827.] 10 p. (19th Cong., 2d sess. House. Rept. 100)

S-31503.

Copies: DLC; DeU; MWA; NjP.

3463. United States. Laws.

An act to authorize the President of the United States to cause the reserved salt springs in the state of Missouri to be exposed to public sale. January 8, 1827.... [Washington: 1827.] 1 p. (S.6)

S-30898.

Copies: DNA.

3464. Memorial of the manufacturers of salt in the county of Kanawha, Virginia, against the repeal of the duty on imported salt. January 21, 1828.... Washington: Printed by Duff Green, 1828. 21 p., fold. table. (20th Cong., 1st sess. Senate. Doc. 47)

Provides an overview on the state of salt industry.

S-36592.

Copies: DLC; DeU; MWA; NjR.

3465. Memorial of the salt manufacturers in the town of Salina, Onondaga County, New York, against the repeal of duty on imported salt. January 21, 1828.... Washington: Printed by Duff Green, 1828. 4 p. (20th Cong., 1st sess. Senate. Doc. 48)

S-36600.

Copies: DLC; DeU; MWA; NjR.

3466. Teissier, J. A.

Projected salt works in Boston Bay. Principal conditions which constitute the agreement of the company. [Boston: 1828.] 8 p., plan.

S-35475.

Copies: DLC; MB; MH-BA.

3467. Carpenter, George Washington, 1802–1860.

On the muriate of soda, or common salt, with an account of the salt springs in the United States. By George W. Carpenter.... [Philadelphia?: 1829.] 8 p.

N-C0153115.

Copies: DLC.

3468. Smyth, Alexander, 1765–1830.

The saltworks case. The argument of Alexander Smyth before the Supreme Court of the United States. Wythe, Va.: Printed at the Argus office, [1829]. 24 p.

Concerns real estate, including salt works, bequeathed by William King, who died in 1808.

S-40470.

Copies: CSmH.

3469. [Frean, William.]

Proposals for establishing a salt company at Key West; accompanied with certificates and documents from persons of great respectability; elucidating the great advantages to be derived by manufacturing salt by solar evaporation, at that island, &c. &c. Washington: Printed by E. De Krafft, 1830. 27, [1] p.

N-F0350855; Sabin-66030.

Copies: DLC; MWA; NN.

3470. Memorial of the manufacturers of salt in Kanawha County, Virginia: praying for a restoration of the duty on imported salt. Addressed to the Senate and House of Representatives of the United States. Kanawha C.H., Va.: Printed at the office of the Kanawha banner, 1830. 19, 7, 16 p.

An Act, incorporating the James River and Kanawha Company, 16 p. at end.

C-2095.

Copies: CSmH; DLC; MB; ScU; WHi.

3471. Memorial of the manufacturers of salt in the county of Kenhawa [*sic*], Virginia, against the repeal of the duty on imported salt. April 30, 1830.... [Washington: 1830.] 18 p. (21st Cong., 1st sess. Senate. Doc. 134)

Provides an overview on the state of salt industry.

C-4966.

Copies: DLC; DeU; MWA; NjR.

3472. To the Hon. Thos. H. Benton, Senator from the state of Missouri. [Shawneetown, Ill.]: 1830. Broadside 47 x 29.5 cm.

A letter, signed by Timothy Guard, John Grenshaw, Benjamin White, Hugh M. Weed, Equality (Gallatin County Saline), Ill., Nov. 15, 1830, concerning the salt works on Saline Creek, Illinois. Issued as the *Illinois gazette extra.*

Byrd (Illinois)-102; N-L0291545.

Copies: PPL.

3473. United States. Laws.

A bill authorizing the legislature of the territory of Arkansas to lease the salt springs in said territory ... February 12, 1830 ... Mr. Irvin, from the Committee on Public Lands, reported the following bill. [Washington: 1830.] 2 p. (H.R. 241)

C-3962.

Copies: DNA.

3474. United States. Laws.

A bill to authorize the state of Indiana to sell her reserved salt springs. May 27, 1830.... [Washington: 1830.] 1 p. (H.R. 492)

C-4478.

Copies: DNA.

3475. United States. Treasury Dept.

Salt works—United States. Letter from the Secretary of the Treasury, transmitting a report of the number and nature of the salt works established in the United States, &c., &c. February 8, 1830.... [Washington: 1830.] 77 p. (21st Cong., 1st sess. House. Doc. 55)

C-5143.

Copies: DLC; DeU; MWA; NjR.

5. Other minerals and mineral waters

3476. Harrison, George.

Philadelphia, July 14, 1746. Advertisement. At the marble shop ... Philadelphia, are performed and sold, chimney pieces, slabs for hearths.... [Philadelphia: B. Franklin, 1746.] Broadside 20 x 16 cm.

B-1315; Mp-40396.

Copies: PHi.

3477. Rush, Benjamin, 1745–1813.
Experiments and observations on the mineral waters of Philadelphia, Abington, and Bristol, in the province of Pennsylvania. Read June 18, 1773, before the American Philosophical Society, held at Philadelphia.... Philadelphia: Printed by James Humphreys, junior, 1773. 30 p.
Mp-12995.
Copies: DLC; DeGE; MWA; NN; PHi; PPL; PU.

3478. Bowdoin, James, 1726–1790.
A philosophical discourse, addressed to the American Academy of Arts and Sciences ... in Boston, on the eighth of November M, DCC, LXXX, after the inauguration of the president into office.... Boston: Printed by Benjamin Edes and Sons, 1780. 35 p.
Includes a discussion of newly discovered deposits of talc, and its uses.
Mp-16720.
Copies: DLC; DeGE; MHi; MWA; MiU; NN; NcD; NjPT; OClWHi; PPL; RPJCB.

3479. Rush, Benjamin, 1745–1813.
Directions for the use of the mineral water and cold bath, at Harrogate, near Philadelphia. Philadelphia: Printed by Melchior Steiner, 1786. 12 p.
Mp-19971.
Copies: DLC; MWA; PPL.

3480. Rouelle, John.
A complete treatise on the mineral waters of Virginia: containing a description of their situation, their natural history, their analysis, contents, and their use in medicine.... Philadelphia: Printed for the author, by Charles Cist, and to be sold by Thomas Dobson, 1792. 4 p.l., xix, 68 p.
Mp-24757.
Copies: DLC; DNLM; MBAt; MH; MHi; MWA; NN; PPAmP; PPL.

3481. Seaman, Valentine, 1770–1817.
A dissertation on the mineral waters of Saratoga; containing a topographical description of the country, and the situation of the several springs; an analysis of the waters ... also, a method of making an artificial mineral water, resembling that of Saratoga.... New-York: Printed by Samuel Campbell, 1793. x, [11]–40 p.
Mp-26149.
Copies: DLC; DeGE; MH; MWA; NHi; NN; PPL.

3482. Hirte, Tobias.
Warnung, vor falschen Seneca-Oel. Indianisch-French-Crieck-Seneca-Oel. Ein vortrefliches ... Medicament, ist zu haben in Philadelphia, bey Tobias Hirte.... Kurze Nachricht von dessen Nutzen und Gebrauch.... [Philadelphia: 1794?] Broadside 36 x 23 cm.
Ten testimonials on the beneficial use of Seneca oil, dated from 1792 to 1794.
B-8692; Mp-47072.
Copies: MWA (photocopy); PPL.

3483. Ross, Aaron.
Stone cutting. The subscriber begs leave to inform the public ... he proposes carrying on the stone cutting business.... New Brunswick: A. Blauvelt, [1797]. Broadside 19 x 23.5 cm.
Signed and dated: Aaron Ross. New-Brunswick, May 13, 1797.
B-10091.
Copies: NjR.

3484. Stewart, Jesse.
Genuine French Creek Seneca oil, an excellent and approved medicine, may be had in Springfield, of Mr. Jesse Stewart, nigh Mr. Roll's tavern. A short account of this oil, its virtues and uses.... [Springfield, N.J.?: 1797?] Broadside.
B-10114; Mp-48258.
Copies: NNNAM; NjR (photocopy).

3485. Baltzell, John.
An essay, on the mineral properties of the Sweet Springs of Virginia, and conjec-

3485 (*continued*)
tures respecting the processes of their production by nature, together with hints relative to an artificial formation of similar medicinal waters. To which are added, a few concise strictures on a treatise composed by John Rouelle.... Baltimore: Printed by Warner & Hanna, 1802. 30 p.
S&S-1809.
Copies: DI-GS; DNLM; MdBJ; N; NNNAM; PPL.

3486. Howe, Timothy.
A history of the medicinal springs at Saratoga and Ballston: being a brief account of the situation, composition, operation, and effects, of these celebrated waters.... Brattleboro, Vt.: 1804. 23 p., plate.
S&S-6513.
Copies: DNLM; MWA; NhHi.

3487. United States. Congress. House. Committee of Commerce and Manufactures.
Report from the Committee of Commerce and Manufactures, to whom was referred, on the twelfth instant, the petition of the president and directors of the New York and Duchess County Slate Companies, in behalf of themselves and their associates. 15th November, 1804.... Washington City: Printed by William Duane & Son, 1804. 5 p.
S&S-7586.
Copies: DLC; DeGE; PPL.

3488. Law, Thomas, 1756–1834.
Ballston Springs.... New York: Printed by S. Gould, 1806. 48, 46 p.
"Additions to Ballston Springs," 46 p. at end.
S&S-10707.
Copies: DLC; DNLM; MBAt; MWA; MiU-C; NN.

3489. Seaman, Valentine, 1770–1817.
A dissertation on the mineral waters of Saratoga. Including an account of the waters of Ballston. 2d ed., enl.... New York: Printed and sold by Collins & Perkins, 1809. xxi, [23]–131 p., front. (fold. map).
S&S-18584.
Copies: DLC; DeGE; MBCo; MWA; MnHi; NN.

3490. Hosack, David, 1769–1835.
Analysis of Ballston water. Observations on the use of the Ballston mineral waters, in various diseases. [Albany: 1810?] 8 p.
S&S-20385.
Copies: CtY; MB.

3491. Mitchill, Samuel Latham, 1764–1831.
A concise description of Schooley's Mountain in New-Jersey, with some experiments on the water of its chalybeate spring. New York: J. Seymour, 1810. 1 p.l., 6–14 p.
S&S-20756.
Copies: MB; MH; NN; NjP; PPL.

3492. MacNeven, William James, 1763–1841.
Chemical examination of the mineral water of Schooley's Mountain. Read before the Literary and Philosophical Society of New-York.... New York: Van Winkle and Wiley, 1815. 21 p.
S&S-35172.
Copies: CtHT-W; DNLM; MBCo; NNNAM; PPAN.

3493. Meeker, John.
An inaugural dissertation on the principal mineral waters of the states of New York and New-Jersey.... New York: 1815. 37 p.
Thesis, Rutgers University.
S&S-35241.
Copies: MWA; NNNAM.

3494. Meade, William.
An experimental enquiry into the chemical properties and medicinal qualities of the principal mineral waters of Ballston and Saratoga, in the state of New-York. With directions for the use of those waters in the various diseases to which they are applicable; and observations on diet and regimen. To which is added an appendix, containing a chemical analysis of the Lebanon spring in the state of New York.... Philadelphia: Har-

rison Hall; William Fry, printer, 1817. xv, 195 p., illus., plates.
S&S-41391.
Copies: DLC; DeGE; MH; MWA; MnHi; NNNAM; PPA; PPAmP; PPC; PPF; PPL.

3495. The prices for the Kingsbridge marble, 1817. [New York]: George Long, printer, [1817]. 8 p.
S&S-41885.
Copies: MWA.

3496. Root, Erastus, 1773–1846.
An inaugural dissertation on the chemical and medicinal properties of the mineral spring in Guilford. Read before the Second Medicinal Society of the State of Vermont, on the 8th day of January, 1817.... Brattleborough, [Vt.]: Simon Ide, printer, 1817. 15 p.
S&S-42010.
Copies: NNBG.

3497. Steel, John Honeywood, 1780–1838.
An analysis of the mineral waters of Saratoga and Ballston.... Together with observations on the geology and mineralogy of the surrounding country ... by Doctor John H. Steel.... Albany: E. & E. Hosford, 1817. 94 p., table.
S&S-42216.
Copies: DNLM; MWA; MWiW; NNNAM; OC.

3498. Meade, William.
A chemical analysis of the waters of New Lebanon in the state of New York, with observations on their medicinal qualities and principally as a bath. Burlington, N.J.: D. Allinson, 1818. 45 p.
S&S-44776.
Copies: DNLM; MBAt; MBCo; N; PPAN; PPAmP.

3499. Sears, Reuben, 1778–1846.
A poem on the mineral waters of Ballston and Saratoga, with notes illustrating the history of the springs and adjacent country.... Ballston Spa, [N.Y.]: Published by the author; J. Comstock, printer, 1819. 108 p.
S&S-49390.
Copies: CSmH; DLC; MB; MBAt; MH; MWA; NN; NNC; RPB.

3500. Steel, John Honeywood, 1780–1838.
An analysis of the mineral waters of Saratoga and Ballston, with practical remarks on their use in various diseases. Containing observations on the geology and mineralogy of the surrounding country, with a geological map. 2d ed., enl. and improved. By Doctor John H. Steel.... Albany: D. Steele, Packard & Van Benthuysen, printers, 1819. 118, [1] p., fold. map, fold. table.
S&S-49503.
Copies: DLC; DeGE; MBAt; MH, MWA; NN; NNNAM; ScCC.

3501. Medical Society of Orange County, New York.
Report of the committee appointed by the Orange County Medical Society, to analyze the waters of Chechunk Spring, in the town of Goshen. Goshen, [N.Y.]: Printed by T. B. Crowell, 1820. 24 p.
Signed: David R. Arnell, Samuel S. Seward, James Heron.
S-2201.
Copies: DeGE; MBCo; MWA; NHi; NNNAM; NbU-M; PPL.

3502. Louisiana. Governor (T. B. Robertson).
Message from His Excellency, Thomas B. Robertson, governor of the state of Louisiana, to the Senate of the said state; on the subject of the alluvial soil known by the name of Batture of the subrub [*sic*] St. Mary; together with several documents accompanying said message.... New Orleans: J. C. de St. Romes, state printer, 1821. 29 p.
N-L0509205; S-5864.
Copies: LN; LU.

3503. Keating, William Hypolitus, 1799–1840.
Account of the Jeffersonite, a new mineral discovered at the Franklin Iron Works,

3503 (*continued*)
near Sparta, N.J. by L. Vanuxem and W. H. Keating. Described by William H. Keating.... Philadelphia: J. Harding, 1822. 12 p.
S-9181.
Copies: NGeno; NNC; PPAmP; ScU.

3504. Church, William.
An analysis of the waters of the Bedford mineral springs, &c.... Bedford, Pa.: Printed by C. M'Dowell, [1825?]. 8 p.
S-20075.
Copies: DLC; MBAt; PPHa.

3505. ———— Bedford, Pa.: N. T. Chapman, Jr., printer, [1825?]. 8 p.
N-C0413066.
Copies: DLC; MWA; ViU.

3506. Dawson, G.
Frankfort mineral springs.... Frankfort, Ky.: 1825. Broadside 16 x 19 cm.
Dated: August 4, 1825.
Hummel-695.
Copies: KyU.

3507. Prices of cut stone and marble, agreed on and established by all the master stone cutters in the city of New-York, Feb. 10th, 1825. New York: Printed by H. Sage, 1825. 8 p.
Copies: DeGE.

3508. Steel, John Honeywood, 1780–1838.
An analysis of the mineral waters of Saratoga and Ballston ... 2d ed., enl. and improved. Saratoga Springs: G. M. Davison, 1825. 118 p., map.
S-22368–9.
Copies: CtSoP; MH; MWA; NNNAM.

3509. Armstrong, John, 1784–1829.
A treatise on sulphureous waters.... Staunton, Va.: Printed at the Spectator office, 1826. v, [7]–24 p.
S-23543.
Copies: DLC; DNLM.

3510. Mitchill, Samuel Latham, 1764–1831.
A chymical examination of the mineral water of Schooley's Mountain springs; together with a physical geography of the first range of mountains extending across New Jersey, from the Hudson to the Delaware. By Dr's Samuel L. Mitchill and James McNeven.... Morristown, [N.J.]: Printed by Jacob Mann, 1828. 23 p.
Contains also a description of the mineral resources of the area, including mica, flintstone, feldspar, and iron ore.
S-34168.
Copies: DNLM; NBLiHi; PHi.

3511. ———— New York: Sickels, printer, 1828. 24 p.
S-34169.
Copies: DLC; NNNAM; NjR; PPL.

3512. Northampton Slate Quarry Company.
Memorial of the Northampton Slate Quarry Company in Pennsylvania, praying that further protecting duty be laid on imported slate for roofing. March 19, 1828.... Washington: Printed by Duff Green, 1828. 3 p. (20th Cong., 1st sess. House. Doc. 146)
S-36598.
Copies: DLC; DeU; MWA; NjR.

3513. Steel, John Honeywood, 1780–1838.
An analysis of the mineral waters of Saratoga and Ballston, with practical remarks on their use in various diseases; containing observations on the geology and mineralogy of the surrounding country, with a geological map ... 2d ed., improved. Saratoga Springs: 1828. 118 p., map.
S-37339.
Copies: MBC; MWA; RPM.

XI. SEA TRANSPORTATION

1. Ships and shipbuilding

3514. Elford, William, Sir, 1749–1837.
Plan of an African ship's lower deck. Philadelphia: M. Carey, [1789]. Broadside.
Mp-21807.
Copies: NN.

3515. United States. War Dept.
Letter and reports of the Secretary at War, to the committee appointed to enquire into the naval equipment of the United States ... 2d March, 1796. [Philadelphia: 1796.] 11 p., fold. table.
Reports concerning the construction of frigates.
Mp-31467.
Copies: DLC; MBAt; MWA; MiU-C; RPJCB; ViU.

3516. United States. Congress. House. Committee on Naval Affairs.
Report of the Committee appointed to enquire into the state of the naval equipment ... 25th January, 1797 ... [Philadelphia: Printed by William Ross, 1797.] 6 p.
Progress report on the construction and equipment of frigates.
Mp-32986.
Copies: MBAt; MWA.

3517. United States. War Dept.
Letter from the Secretary at War, to the chairman of the Committee on the Naval Equipment; inclosing sundry statements relative to the subject. Accompanying a report of the Committee appointed on the 16th ultimo, to enquire into the state of the naval equipment. 25th January, 1797, committed to a committee of the whole House.... [Philadelphia: Printed by William Ross, 1797.] 11 p., fold. table.
Mp-33094.
Copies: DLC; MBAt; MWA; MiU-C; NN.

3518. United States. War Dept.
Letter from the Secretary of War, transmitting sundry statements relative to the frigates, United States, Constitution and Constellation, June 17, 1797.... Philadelphia: Printed by Zachariah Poulson, junior, 1797. 10 p., fold. table.
Report on the progress and expenditures in the construction of frigates.
Mp-33096.
Copies: CSmH; MBAt; MWA; PPL; RPJCB.

3519. United States. Navy Dept.
Letter from the Secretary of the Navy, inclosing sundry statements relative to the vessels of war ... 26th December, 1798.... Published by order of the House of Representatives. Philadelphia: Printed by J. Gales, [1799]. 30 p., fold. table.
Includes an estimate of expenses for the construction and equipment of vessels.
Mp-36528.
Copies: MBAt; MWA; NHi; PHi; RPJCB.

3520. Guest, Henry.
New-Brunswick, New-Jersey, February, 1800. Sir, Your very obliging letter of 25th November ... to the superintendants of the constructions of vessels of war.... [New Brunswick, N.J.: 1800.] [2] p.
B-11043; Mp-49077.
Copies: PPL.

3521. Guest, Henry.

Observations on the sheathing of vessels, seasoning of timber, &c. In a series of letters. By Henry Guest, of New Brunswick, New Jersey. [New Brunswick, N.J.]: Printed for the author, by A. Blauvelt, 1801. 23 p.

Copies: NHi.

3522. [Latrobe, Benjamin Henry] 1764–1820.

Message from the president of the United States, transmitting plans and estimates of dry dock, for the preservation of our ships of war. 28th December, 1802.... Washington City: Printed by William Duane & Son, 1802. 25 p., fold. table.

Transmitting a letter from B. H. Latrobe to the Secretary of the Navy, and an extract from a letter from Latrobe to the President.

S&S-3361.

Copies: DLC; DeGE; MWA; NN; RPB.

3523. United States. Navy Dept.

Letter from the Secretary of the Navy, to Samuel L. Mitchill, chairman of the select committee appointed by the House of Representatives ... to consider so much of the message of the President of the United States ... as relates to naval preparations and the establishment of scites [*sic*] for naval purposes: together with certain documents relating thereto ... January 30, 1802 ... Washington City: Printed by Wm. Duane, 1802. 88 p.

Concerns mainly the navy yards.

S&S-3309.

Copies: DLC; MWA; NHi; NN.

3524. United States. Congress. House. Committee on Docks.

Report of the Committee appointed on the 20th instant, to enquire into the usefulness and propriety of constructing a dock or docks, at either of the public navy yards, or elsewhere, within the United States, for the building and repair of ships of war. 28th January, 1803.... Washington City: Printed by William Duane & Son, 1803. 4 p.

S&S-5457.

Copies: DLC; DeGE; MWA; NN.

3525. United States. Congress. House. Committee on Navy Yards.

Report of the Committee appointed on the 17th ultimo, to whom was referred so much of the message of the President of the United States, as relates to our navy yards and the building of docks. 10th January, 1803.... [Washington: 1803.] 4 p.

S&S-5406.

Copies: DLC; DeGE; MWA; NN.

3526. United States. Navy Dept.

Letter from the Secretary of the Navy, enclosing information respecting timber deposited in the navy yards of the United States, for the purpose of building ships of war ... 7th January, 1803. Washington City: Printed by William Duane & Son, 1803. 7 p.

S&S-5329.

Copies: DLC; MWA; NN; RPB.

3527. New York Journeymen Shipwrights' Society.

Constitution of the New York Journeymen Shipwrights' Society, adopted January 5, 1804. New York: Printed by T. and J. Swords, 1804. 12 p.

S&S-6924.

Copies: NHi.

3528. [Guest, Henry.]

Observations on sheathing vessels; seasoning timber; the proper time to fall timber; the nature and what force it is that causes the sap to rise; with a number of other valuable observations. [New Brunswick, N.J.]: Printed for the author, by A. Blauvelt, 1805. 61 p.

N-G0579582.

Copies: DLC; MH-A; PPAmP; PPF.

3529. United States. Navy Dept.

Message from the President of the United States, transmitting a report made to him by the Secretary of the Navy relative to the state of the frigates belonging to the United States ... January 5, 1807 ... City of Washington: A. & G. Way, printers, 1807. 7 p.

Progress report on the construction of the frigates.
S&S-13992.
Copies: DLC; MWA; N; NN.

3530. A memoir on the subject of a navigation act, including the encouragement of the manufactory of boats and sea vessels, and the protection of mariners. Philadelphia: Press of T. and G. Palmer, 1809. 16 p.
Sometimes attributed to Tench Coxe.
S&S-18065.
Copies: DLC; DeGE; ICU; MH-BA; MdBP; MiD-B; NN; NNC; NjP; PPAmP.

3531. Pakenham, Edward.
Captain Pakenham's invention of a substitute for a lost rudder, and to prevent its being lost. Also, a method of restoring the masts of ships.... New York: Edmund M. Blunt, 1811. 58, [2] p., illus.
First published in London, 1793.
S&S-23622.
Copies: MWA; NN.

3532. [Tudor, Frederic] 1783–1864.
Some memorandums, by which it is attempted to be shewn that an improved model may be adopted in the construction of ships, by a new application of well-known principles. [Boston?: ca. 1812.] 7 p., 4 fold. plates.
Copies: DeGE.

3533. United States. Navy Dept.
Letter from the secretary of the Navy, transmitting sundry statements of the expenditure of money at the several navy yards, pursuant to a resolution of the twenty-seventh of January last. May 27th, 1812.... Washington City: Printed by R. C. Weightman, 1812. 57 p.
S&S-27169.
Copies: DLC; MWA.

3534. Articles of agreement ... concluded upon this 12th day of April, one thousand eight hundred and thirteen, between William Flannagain and William Parsons ... master shipbuilders ... and James Beatty, Navy agent of the United States of America, at Baltimore.... [Baltimore]: S. Childs, [1813]. 10 p.
S&S-27755.
Copies: DNA.

3535. Articles of agreement, made and concluded upon this sixth day of May, eighteen hundred & thirteen, between Orlando B. Merrill, of Newbury, and William Cross, of Newburyport ... ship-builders on the one part—and Amos Binney, agent for the United States' Navy Department at Boston, on the other part, as follows: [a sloop of war (*Wasp*, the second) to be built on the Merrimack River, and to be delivered to the Navy by Sept. 6, 1813, with the following specifications]. [Boston?: 1813.] Broadside 52 x 44 cm. (printed in 5 columns).
N-B0493354.
Copies: MHi.

3536. Annesley, William.
Description of William Annesley's new system of boat and ship-building.... Albany: Printed by J. Buel, at the Argus office, 1816. 11 p., fold. plate.
S&S-36748.
Copies: MWA.

3537. Letters on the establishment of a navy yard, for the southern department of the United States; addressed to the commissioners of the Navy Board. Norfolk, [Va.]: Printed by O'Connor & Broughton, 1816. 47 p.
Contains five letters, first four signed: A friend of the Navy; the fifth is signed: A friend to the Navy.
S&S-38065.
Copies: DLC; MB.

3538. United States. Navy Dept.
Letter from the Secretary of the Navy, communicating the measures which have been taken to ascertain the most eligible site for a navy yard, in the vicinity of Chesapeake Bay in obedience to an order of the Senate of

3538 (*continued*)
December 21, 1815. January 25, 1816. Printed by order of the Senate. Washington: Printed by William A. Davis, 1816. 5 p.

Includes instructions for surveys.

S&S-39505.

Copies: DLC; MWA.

3539. Omitted in numbering.

3540. Busby, Charles Augustus, 1788–1834.

An essay on the propulsion of navigable bodies. By C. A. Busby, architect and engineer. Extracted from the American monthly magazine, for June, 1818. With considerable additions. New York: Printed by Benjamin G. Jansen, 1818. 31 p., illus.

S&S-43500.

Copies: CtY; DLC; DeGE; MB; MBAt; MWA; MiU; NJQ; NN; PPAmP; PPL; RPB.

3541. United States. Navy Dept.

Message from the President of the United States, transmitting a report of the Secretary of the Navy, in compliance with a resolution of the Senate of the eighth of January last, of the proceedings which may have been had, under "An act entitled An act for the general increase of the Navy of the United States," specifying the number of ships, put on the stocks ... the quantity of materials procured for ship building ... the contracts which may have been entered into ... February 2, 1818 ... Washington: Printed by E. De Krafft, 1818. 27 p. (15th Cong., 1st sess. Senate. Doc. 94)

S&S-46418a.

Copies: DLC; DeGE; DeU; PPAmP; R.

3542. United States. President (Monroe).

Message from the President of the United States, transmitting reports of the proceedings which have been held under the "Act for the gradual increase of the Navy...." January 4, 1819.... Washington: Printed by E. De Krafft, 1819. 29 p. (15th Cong., 2d sess. House. Doc. 70)

Includes reports from the Navy Dept., and the Commissioners of the Navy, documenting the work, materials, and costs of naval construction.

S&S-49896.

Copies: DLC; DeU; G; MWA; NN; O; R.

3543. Watts, C.

Sir: you and your friends are invited to inspect, and take passage for Weehawken in a horse boat, (on an entirely new plan).... New York: 1819. 2 p.

Signed and dated: C. Watts, New-York, Sept. 27, 1819.

The boat is said to have been invented by the Rev. M. Isaacs, and propelled at a velocity of 5 miles per hour by 4 horses.

Copies: CtHi.

3544. United States. Congress. House. Committee on Naval Affairs.

Report of the Committee on Naval Affairs upon the subject of reducing the expenses of the naval establishment of the United States. March 7, 1820.... [Washington: 1820.] 13 p., fold. table. (16th Cong., 1st sess. House. Rept. 74)

Includes proposals for selection of materials for the construction of ships, and a review of the progress of construction.

S-3983.

Copies: DLC; DeU; MWA; NjP.

3545. United States. Laws.

A bill authorizing a machine to be built for hauling up ships. Jan. 29, 1821. Read twice, and committed to a committee of the whole House on the state of the union. [Washington: 1821.] 1 p. (H.R. 229)

S-7054.

Copies: DLC.

3546. United States. Navy Dept.

Documents accompanying the "Bill to amend the act, entitled An act for the gradual increase of the navy of the United States." January 5, 1821.... Washington: Printed by Gales & Seaton, 1821. 17 p. (16th Cong., 2d sess. House. Doc. 43)

The documents, supplied by the Navy

Dept., include specifications for materials, and suggestions for methods to make them more durable.
S-7208.
Copies: DLC; DeU; MWA; NjP.

3547. Law papers and documents relating to the management of the Old, or Fulton Ferry. Brooklyn, [N.Y.]: 1822. 80 p.
S-9239.
Copies: MH-BA; NB; NBLiHi; NN; NNC; NSm; PPL.

3548. Remarks on the project of establishing a line of packets between Boston and Liverpool. [S.l.: 1822?] 20 p.
S-2984, 10084.
Copies: DLC; MWA; PPL; WHi.

3549. Sullivan, John Langdon, 1777–1865.
A commentary on the new system of naval architecture of William Annesley, architect. By John L. Sullivan. Troy, N.Y.: Printed by William S. Parker, 1823. iv, 59, [1] p., 11 fold. plates.
S-14238.
Copies: DLC; DeGE; NCooHi; NN; PPL.

3550. United States. Congress. House. Committee on Naval Affairs.
Report of the Committee on Naval Affairs, upon the President's message in relation to the marine railway, or inclined plane. February 12, 1823.... [Washington: 1823.] 1 p. (17th Cong., 2d sess. House. Rept. 89)
S-14706.
Copies: DLC; DeU; MWA; NjP.

3551. United States. Navy Dept.
Message from the President of the United States, transmitting ... information in detail of the expenses of building each vessel of war, authorized by the act of the 2d of January 1813, and its supplements ... January 4, 1823.... Washington: Printed by Gales & Seaton, 1823. 61 p., fold. tables. (17th Cong., 2d sess. Senate. Doc. 12)
S-14651.
Copies: DLC; DeGE; DeU; MWA; NjP.

3552. United States. Navy Dept.
Message from the President of the United States, upon the subject of Commodore Rodgers' marine rail-way, or inclined plane. January 27, 1823.... Washington: Printed by Gales & Seaton, 1823. 17 p. (17th Cong., 2d sess. House. Doc. 42)
Transmitting a report from the Navy Department.
S-14662.
Copies: DLC; DeGE; DeU; MWA; NjP.

3553. Thomas, John.
Improvements in naval architecture.... [Washington: S. A. Elliot, printer, 1824.] 6 p.
"Extracted from the *Washington quarterly magazine of arts, science and literature,* for April, 1824."
Copies: DeGE.

3553a. Observations on Thomas's naval architecture. [S.l.]: 1825. 22 p.
N-00010263.
Copies: PPAmP.

3554. United States. Navy Dept.
Report of the Secretary of the Navy, in relation to the construction of a dry dock. January 5, 1825.... Washington: Printed by Gales & Seaton, 1825. 22 p. (18th Cong., 2d sess. Senate. Doc. 10)
S-23098.
Copies: DLC; DeGE; DeU; MWA; NjP.

3555. The question of a south ferry to Long Island, stated and argued, in several papers originally published in the New-York American. By a freeman. New-York: G. F. Hopkins, printer, 1826. 63 p.
S-25881.
Copies: MB; NBLiHi.

3556. United States. Navy Dept.
Docks for repairing ships of war. Letter from the Secretary of the Navy, transmitting the information required by a resolution of the House of Representatives of the 18th inst.

3556 (*continued*)
upon the subject of constructing docks for the preservation and repair of the vessels belonging to the navy of the United States. March 31, 1826.... Washington: Printed by Gales & Seaton, 1826. 28 p. (19th Cong., 1st sess. House. Doc. 143)

S-26955.

Copies: DLC; DeU; MWA; NjP.

3557. [Weeks, Ezra.]
A statement in relation to the concerns of the New-York Dry Dock Company. [New York: 1826?] 12 p.

Signed and dated at end: Ezra Weeks. New-York, Dec. 28, 1825.

Sabin-90672.

Copies: NN.

3558. Clark, Edward.
Comparative remarks on dry docks and marine rail-roads. [S.l.: 1827.] 2, 3 p.

N-C0452471.

Copies: MBBS.

3559. Dry dock—Portsmouth, New Hampshire. Memorial of the citizens of Portsmouth, N.H., respecting the construction of a dry dock at the Navy Yard at that place. February 12, 1827.... Washington: Printed by Gales & Seaton, 1827. 5 p. (19th Cong., 2d sess. House. Doc. 92)

Also issued as Senate doc. 57.

S-31238.

Copies: DLC; DeU; MWA; NjR.

3560. Memorial of the citizens of Portsmouth, N.H., respecting the construction of a dry dock, at the Navy Yard at that place. February 15, 1827. Printed by order of the Senate of the United States. Washington: Printed by Gales & Seaton, 1827. 5 p. (19th Cong., 2d sess. Senate. Doc. 57)

S-31401.

Copies: DLC; DeU; MWA; NjP.

3561. Observations on the floating dock, invented by Commodore Barron; as described in the Franklin Journal for January, 1827, and recommended by a select committee, to merchants of Philadelphia.... Philadelphia: Joseph Harding, printer, 1827. 8 p.

Signed at end: A shipwright.

S-30608.

Copies: PPAmP.

3562. Sullivan, John Langdon, 1777–1865.
A description of the American marine rail-way, as constructed at New York, by Mr. John Thomas, naval architect. With explanations of its principle, and manifestations of its safety for ships of war. By John L. Sullivan.... To which is annexed, the report of the Committee of inventions of the Franklin Institute, to whom the above subject was referred. With engravings.... Philadelphia: J. Harding, printer, 1827. 16 p., diagrs. on 2 plates.

"Extracted from the *Franklin journal* for February, 1827."

S-30738.

Copies: DLC; DeGE; M; MB; MWA; NbO; NcD; NhHi; PPAmP; PPi; PU.

3563. United States. Navy Dept.
Dry docks—Portsmouth, N.H., Charlestown, Mass., &c. Message from the President of the United States, transmitting a report of an examination and survey which has been made of a site for a dry dock at Portsmouth, N.H., Charlestown, Mass., Brooklyn, N.Y., and Gosport, Va. January 10, 1827.... Washington: Printed by Gales & Seaton, 1827. 46 p. (19th Cong., 2d sess. House. Doc. 125)

Transmitting a report from the Navy Department.

S-31237.

Copies: DLC; DeU; MWA; NN; NNE; NjP.

3564. United States. Navy Dept.
Letter from the Secretary of the Navy, transmitting information respecting the cost of, and expenditures at the several navy yards ... February 12, 1827.... Washington: Printed by Gales & Seaton, 1827. 12 p., fold. table. (19th Cong., 2d sess. Senate. Doc. 54)

S-31363.
Copies: DLC; DeU; MWA; NjP.

3565. Barron, James, 1769–1851.
A brief essay on the cause of dry-rot in public and private ships and its remedy. Norfolk, [Va.]: Printed by Shields & Ashburn, 1828. 14 p., illus.
N-B0148556.
Copies: NjR.

3566. United States. Congress. House. Committee on Naval Affairs.
Dry docks. January 25, 1828. Mr. Hoffman, from the Committee on Naval Affairs ... made the following report.... [Washington: 1828.] 9 p. (20th Cong., 1st sess. House. Rept. 100)
S-36267.
Copies: DLC; DeU; MWA; NjR.

3567. ———— December 17, 1828. —Reprinted by order of the House of Representatives.... [Washington: 1828.] 8 p. (20th Cong., 2d sess. House. Rept. 8)
S-36268.
Copies: DLC; DeU; MWA; NjR.

3568. United States. Laws.
A bill to authorize the construction of additional dry docks. January 25, 1828.... [Washington: 1828.] 1 p. (H.R. 111)
S-36077.
Copies: DNA.

3569. Barron, James, 1769–1851.
A brief essay on the causes of dry-rot in public and private ships and its remedy. Norfolk, [Va.]: Printed by Shields & Ashburn, 1829. 16 p., illus.
S-37712.
Copies: CtY; DLC; MB; MH; NN; PPAmP; PPF; PPPrHi.

3570. New York Dry Dock Company.
Annual report. [Albany: 1829.] 3 p.
At head of title: No. 193. In Assembly, March 30, 1829.
N-N0210826.
Copies: NN.

3571. United States. Navy Dept.
Report from the Secretary of the Navy, in compliance with a resolution of the Senate of the 20th of May last, in relation to the establishment of a navy yard, between Cape Hatteras and Florida. January 6, 1829. [Washington: 1829.] 15 p. (20th Cong., 2d sess. Senate. Doc. 28)
Copies: DeU; MWA.

3572. Memorial of ship carpenters, smiths, rope makers, and others, employed or connected in ship building, in the city of Philadelphia. April 19, 1830. Referred to the Committee of the Whole House.... [Washington: Printed by D. Green, 1830.] 5 p. (21st Cong., 1st sess. House. Rept. 369)
At head of title: Pennsylvania.
C-5049.
Copies: DLC; DeGE; MWA; NjR.

3573. United States. Laws.
A bill to encourage ship building in the United States. May 20, 1830.... [Washington: 1830.] 4 p. (H.R.476)
C-4499.
Copies: DNA.

3574. United States. President (Jackson).
Marine rail-way, &c. Message from the President of the United States in reply to a resolution of the House of Representatives, in relation to the erection of a marine rail-way at the navy-yard at Pensacola, &c.&c. May 13, 1830.... [Washington: 1830.] 10 p. (21st Cong., 1st sess. House. Doc. 101)
C-4929.
Copies: DLC; DeU; MWA; NjR.

2. Steamboats and steamboat lines

3575. American Philosophical Society, Philadelphia.

Philadelphia, 1788. At a meeting of the American Philosophical Society.... [Philadelphia: 1788.] 2 p.

Extract from minutes, containing a report of the committee to examine the drawings of J. Rumsey's inventions concerning steamboats.

B-6660; Mp-45218.

Copies: DLC; RPJCB.

3576. Barnes, Joseph.

Remarks on Mr. John Fitch's reply to Mr. James Rumsey's pamphlet by Joseph Barnes, formerly assistant, and now attorney in fact to James Rumsey. Philadelphia: Printed by Joseph James, 1788. xvi, 16 p.

Concerns John Fitch's pamphlet, *The Original steam-boat supported,* which appeared in May, 1788.

Mp-20954.

Copies: CSmH; DLC; DeGE; MB; MdBJ-G; MiU; NN; NNE; NjP; PPAmP; PPL; PU; RPJCB.

3577. Cruse, Englehart.

The projector detected; or, some strictures, on the plan of Mr. James Rumsey's steam boat; by Englehart Cruse, of Baltimore. Baltimore: Printed by John Hayes, 1788. 14 p.

Mp-21032.

Copies: DLC.

3578. [Fitch, John] 1743–1798.

The original steam-boat supported; or, A reply to Mr. James Rumsey's pamphlet, shewing the true priority of John Fitch, and the false datings, &c. of James Rumsey. Philadelphia: Printed by Zachariah Poulson, junr., 1788. 34 p.

Mp-21092.

Copies: CSmH; DLC; DeGE; MBAt; MH; MWA; MiU-C; NN; NjP; OrU; PPAmP; PPL; RPJCB; Vi.

3579. ———— Philadelphia: Printed by Zachariah Poulson, junr., 1788. 34 p.

Second edition with corrections.

Mp-21093.

Copies: NN.

3580. Rumsey, James, 1743?–1792.

The petition of James Rumsey, of Berkeley County, in the State of Virginia, most respectfully sheweth, that your petitioner has been several years employed, with unremitted attention, and at a great expense, in inventing, and bringing to perfection, sundry machines and engines; namely, one for propelling boats on the water, by the power of steam.... [Philadelphia?: 1788.] Broadside 38 x 24 cm.

DeGE copy dated on verso in ink: Nov. 19, 1788.

Copies: DeGE; RPJCB.

3581. Rumsey, James, 1743?–1792.

Proposals for forming a company, to enable James Rumsey to carry into execution ... his steam-boat.... In witness of the premises, we, the subscribers, have hereunto set our names, this ———— day of 1788. [Philadelphia?: 1788.] Broadside 34.5 x 28.5 cm.

B-6795; Mp-45352.

Copies: DLC; WHi.

3582. Barnes, Joseph.

Philadelphia, September 21, 1789. As John Fitch has procured a number of handbills, containing his petition and remonstrance to the House of Assembly, to be printed and delivered to some of the members of the House, as well as to others, I hope this will be deemed a sufficient apology for my throwing together the following remarks on some parts of the said petition and remonstrance.... [Philadelphia: 1789.] Broadside 42 x 26.5 cm.

B-6896; Mp-45435.

Copies: PHi; RPJCB (photocopy).

3583. Fitch, John, 1743–1798.

The following Petition and remonstrance was read in the House of Assembly on the fifteenth instant, and as it has been said to be "improper, indecent and insulting," the subscriber wishes to give the world an opportunity of judging on its merits, and therefore presents this printed copy of it.... [Philadelphia?]: 1789. Broadside 44 x 41 cm.

Dated: September 11, 1789.

The petition is addressed: To the honorable the representatives of the freemen of the Commonwealth of Pennsylvania in General Assembly, and is entitled: *The Petition and remonstrance of John Fitch, on behalf of himself and associates.*

Concerns the dispute with James Rumsey.

B-6946; Mp-45475.

Copies: DLC; DeGE; MWA.

3584. Fitch, John, 1743–1798.

To the honorable the representatives of the Commonwealth of Pennsylvania. The petition and remonstrance of John Fitch, of the city of Philadelphia.... Philadelphia: 1789. Broadside 34 x 21 cm.

Dated: September 26, 1789.

B-6947; Mp-45476.

Copies: DLC; DeGE; PHi.

3585. Fulton, Robert, 1765–1815.

The steam-boat.... [New York?: 1808?] Broadside.

S&S-15081.

Copies: NN.

3586. New York (State). Laws.

An act for granting and securing to John Fitch the sole right and advantage of making and employing, for a limited time, the steam-boat by him lately invented. Passed 19th March, 1787. [S.l.: 1811?] 7 p.

The act is followed by five others, dated 1798–1811, in favor of Robert Livingston and Robert Fulton.

N-N0196659.

Copies: CtY; NjP.

3587. Ogden, Aaron, 1756–1839.

Memorial of Aaron Ogden, of Elizabeth Town, in the state of New Jersey. December 23d, 1811. Printed by order of the Senate of the United States. Washington City: Printed by R. C. Weightman, 1811. 4 p.

Concerns possible conflict of his steamboat, to be built by Daniel Dod according to an improved design invented by Dod, with the patents granted to Robert Fulton.

S&S-24201.

Copies: DLC; DeGE.

3588. Steam boats. We the subscribers hereby certify that we have examined with care and attention the steam boat built here, under the patent of Mr. Daniel French.... Philadelphia: 1811. Broadside 32 x 38.5 cm.

Kress: B5923.

Copies: MH-BA.

3589. Townsend, Isaiah.

Memorial of Isaiah Townsend and others, of the city of Albany, in the state of New York. December 27, 1811. Printed by order of the Senate of the United States. Washington City: Printed by R. C. Weightman, 1811. 4 p.

Arguments against the Livingston-Fulton steamboat monopoly, particularly against extending the patents of Robert Fulton.

S&S-24202.

Copies: DLC; NN.

3590. United States. Laws.

A bill for the relief of Robert Fulton and his associates. In Senate of the United States. December 5th, 1811. [Washington: 1811.] 6 p. (S.6)

S&S-24106.

Copies: DLC.

3591. Examination of the Chancellor's opinion in the case of Rob. R. Livingston and Rob. Fulton, vs. James Van Ingen, Lansing & others. [Albany]: Printed at the office of the Albany register, 1812. 54 p.

John Lansing, Jr., Chancellor.

Concerns the steamboat monopoly.

N-E0217561; S&S-26269.

Copies: DLC; MWA; NN; PP; PPAmP.

3592. Livingston, Robert R., 1746–1813.

Livingston, Robert R., and Fulton, Robert, complainants-appellants: and Van Ingen, James and others respond. Case on the part of the appellants in the Court for the Trial of Impeachments and the Correction of Errors. Albany: Office of the Albany register, 1812. 20 p.

Concerns the steamboat monopoly.

S&S-26273.

Copies: N.

3593. New York (State). Court for the Trial of Impeachments and Correction of Errors.

The opinions of the judges of the Supreme Court, delivered in the Court of Errors, in the case of Robert R. Livingston and Robert Fulton, vs. James Van Ingen, and twenty others. Albany: Printed by S. Southwick, 1812. 12, 12, 23, [1] p.

Includes the opinions of Judge Yates, Judge Thompson, and Chief Justice Kent, concerning the exclusive right to navigate the waters of the state of New York by steamboats.

S&S-26283.

Copies: CSmH; DeGE; InU; MH-L; MWA; N-L; NN; NNC-L.

3594. United States. Laws.

An act extending the time of certain patents granted to Robert Fulton. January 2d, 1812.... [Washington: 1812.] 2 p.

S&S-26935.

Copies: DLC.

3595. United States. Laws.

A bill respecting the enrolling and licensing of steam boats. January 24, 1812.... [Washington: 1812.] 2 p. (S. 20)

S&S-27055.

Copies: DLC.

3596. Van Ingen, James.

Livingston, R. R. and Fulton, R., Van Ingen, James et al. In the Court for Trial of Impeachments and Correction of Errors. Between James Van Ingen, and others, respondents; R. R. Livingston and Robert Fulton, appellants. Case of the respondents ... to navigate the waters of New York.... [Albany: 1812.] 12 p.

S&S-26275.

Copies: MH-L.

3597. ________ Albany: [1812]. 13 p.

S&S-26276.

Copies: N.

3598. ________ Albany: Office of the Albany register, 1812. 22 p.

S&S-26277.

Copies: N.

3599. ________ Albany: Office of the Albany register, 1812. 20 p.

S&S-26278.

Copies: N.

3600. Fulton, Robert, 1765–1815.

Report of the practicability of navigating with steam boats, on the southern waters of the United States, from the Chesapeak [*sic*] to the River St. Mary's, forming part of a line of steamboat communications, now establishing, from the northern extremity of Lake Champlain to East-Florida, a distance of fifteen hundred miles. By Robert Fulton, from the surveys of John D. Delacy. New York: Printed by E. Conrad, 1813. 16 p.

S&S-28593.
Copies: DLC; DeGE (xerox copy); MB; MdBP; MiU-C; N; NNE; PPAmP; PPL.

3601. New Jersey. Laws.
An act concerning steam boats. [Trenton: 1813?] 6 p.
Passed November 3, 1813.
N-N0154656.
Copies: NjP.

3602. Fulton, Robert, 1765–1815.
Memorial & petition of Robert Fulton & Edward P. Livingston, in behalf of themselves, and the heirs of the late Robert R. Livingston. [Albany: 1814.] 15 p.
At head of title: In Assembly, March 18, 1814.
Concerns steamboats.
S&S-31544.
Copies: DeGE; MB; NN.

3603. Livingston, John Robert, 1755–1851.
The petition of John R. Livingston & Robert J. Livingston, to the legislature of New Jersey, respecting steam-boats; with the laws of the states of New-York and New-Jersey, and affidavits of sundry persons, respecting the improvements said to be made by Daniel Dod and others on the machinery of R. Fulton. New York: Printed by Pelsue & Gould, 1814. 40 p.
S&S-31944.
Copies: DeGE; NNC.

3604. ———— New York: Printed by Pelsue & Gould, 1814. 43 p.
Preceded by *The Memorial of John R. Livingston & Robert J. Livingston to the Legislature of New-York, respecting steam-boats,* 8 p.
N-L0420597.
Copies: DLC; MWA (43, [15] p.); NN; PBL; PPL.

3605. New York (State). Legislature. Assembly. Committee to Whom was Referred the Memorial and Petition of Aaron Ogden.
Report of the Committee to whom was referred the memorial and petition of Aaron Ogden, a citizen of the state of New-Jersey. [Albany?]: 1814. 13, [1] p.
Dated: In Assembly . . . March 8, 1814.
Signed: W. A. Duer, chairman.
An Act, supplementary to the several acts heretofore passed concerning steam-boats, 1 p. at end.
S&S-32326.
Copies: DeGE; MB; MiU; N; NN; NNC.

3606. New York (State). Legislature. Assembly. Select Committee on the Memorial and Petition of Aaron Ogden.
Report of the Select Committee of the House of Assembly 1814, to whom was referred the memorial and petition of Aaron Ogden, a citizen of the state of New-Jersey. . . . [Albany: 1814.] 8 p.
Concerns steamboats.
N-N0200471.
Copies: NNC.

3607. Prospectus of a route proposed to be established between the cities of New-York and Philadelphia. A steam boat, drawing two feet six inches water, to ply between New York and the mineral spring, at Tinton Falls. . . . [Philadelphia: 1814.] Broadside.
S&S-32572.
Copies: PPL.

3608. Thornton, William, 1759–1828.
Short account of the origin of steam boats, written in 1810, and now committed to the press by W. Thornton, of the city of Washington. Washington City: Printed by Rapine and Elliot, Capitol Hill, 1814. 12 p.
S&S-32927 (erroneously giving 470 p.); Sabin-95647.
Copies: DLC; MH; MWA; NHi; NN.

3609. ———— Washington: Printed at Elliot's patent press, 1814. 20 p.
S&S-32928.
Copies: DLC; MH; MdHi; NHi; NNC.

3610. Fulton, Robert, 1765–1815.
Robert Fulton, to Aaron Ogden, Esq.

3610 (*continued*)
[S.l.: 1815?] 7 p.
Concerns Fulton's patent rights on steamboats.
N-F0428610.
Copies: DLC; MB; NjP; ScU.

3611. [Stockton, Lucius Horatio] 1768–1835.
A history of the steam-boat case, lately discussed by counsel before the legislature of New Jersey, comprised in a letter to a gentleman at Washington. Trenton: Printed for the author, 1815. 48, 4 p.
S&S-36022.
Copies: DeGE; MH-L; MWA; NN; NjR; PPAmP.

3612. Clinton, De Witt, 1769–1828.
A discourse, delivered before the American Academy of the Arts ... 23d October, 1816. New York: T. & W. Mercein, 1816. 30 p.
Partly devoted to Robert R. Livingston, Robert Fulton, and the development of steamboats.
S&S-37271.
Copies: CtY; DeGE; MB; MWA; MWiW; NBLiHi; NNC.

3613. Fairfax, Ferdinando, 1766–1820.
Memorial of Ferdinando Fairfax, against the extention [*sic*] of the patents granted to Robert Fulton, for improvements in propelling vessels by steam. Washington City: Printed by Edward De Krafft, 1816. 8 p.
S&S-37554.
Copies: DLC; DeGE (xerox copy); MBAt; MHi; OClWHi; PPAmP.

3614. Colden, Cadwallader David, 1769–1834.
The life of Robert Fulton, by his friend Cadwallader D. Colden. Read before the Literary and Philosophical Society of New York. Comprising some account of the invention, progress, and establishment of steam boats.... With an appendix. New York: Kirk & Mercein, 1817. vi, 371, [1] p., front. (port.), fold. table.
S&S-40508.
Copies: DLC; DeGE; ICJ; KU; MBAt; MH-BA; MWA; MdBP; NN; NNE; NSyOHi; NjP; OCU; PHC; PHi; PSC; ScU.

3615. Duer, William Alexander, 1780–1858.
A letter, addressed to Cadwallader D. Colden, Esquire. In answer to the strictures, contained in his "Life of Robert Fulton," upon the report of the select committee, to whom was referred a memorial relative to steam navigation, presented to the Legislature of New York, at the session of 1814. With an appendix, containing the several laws concerning steam boats; the petitions presented for their modification; and the reports of select committees thereupon.... Albany: E. and E. Hosford, 1817. 127 p., illus.
S&S-40696.
Copies: CSmH; CtY; DLC; DeGE; IU; MBAt; MH-BA; MWA; MdHi; MiU; NIC; NN; NNE; NWM; NjP; PPL; PSC; RPB.

3616. Philadelphia. Councils. Joint Committee on Steamboats.
Report of the Joint Committee appointed by the Select and Common Councils on the subject of steamboats. [Philadelphia: 1817.] 13 p.
S&S-41781.
Copies: DLC; PHi; PPF; PPL; RP.

3617. [Sullivan, John Langdon] 1777–1865.
Sir, I beg leave to submit to you the following statement of facts ... relative to the repeal of the second section of an act passed in 1808, and an act passed in 1811, declaring to be forfeited to Robert R. Livingston and Robert Fulton, all boats of every description navigated by steam or fire within the state, with power to seize the same. [Albany: 1817.] 14 p.
Dated: Albany, 17th March, 1817.
Kress-B7061; N-S1049653, S1049675.
Copies: DeGE; MB; MH-BA.

3618. Vaux, Roberts, 1786–1836.
A correspondence, containing useful information for steam-boat passengers. [Philadelphia: 1817.] 8 p.
A circular addressed by Roberts Vaux to steamboat passengers concerning boiler explosions.
Copies: PHi.

3619. Colden, Cadwallader David, 1769–1834.
A vindication by Cadwallader D. Colden, of the steam boat right granted by the state of New York [to Livingston and Fulton] in the form of an answer to the letter of Mr. Duer, addressed to Mr. Colden. Albany: Printed by Websters and Skinners, 1818. 2 p. l., [3]–178 p.
S&S-43662.
Copies: CtY; DLC; DeGE; ICJ; MB; MH; MWA; MdBP; MiU-C; NN; NNE; PPL; RPB.

3620. An examination of Cadwallader D. Colden's book, entitled A life of Robert Fulton. By a friend of John Fitch, deceased. [S.l.: 1818.] 38 p.
Possibly a London imprint; Penrose R. Hoopes attributed the authorship to John L. Sullivan.
N-E0217469; Sabin-14278.
Copies: CtY; DLC; M; MBAt; MiU; NIC; NN; PHi; PPF.

3621. A review of the letter, addressed by William Alexander Duer, Esquire, to Cadwallader Colden, Esquire, in answer to strictures contained in his "Life of Robert Fulton," relative to steam navigation. With an appendix, containing the acts of the Legislature. New York: 1818. 27 p.
Sometimes attributed to J. L. Sullivan.
S&S-45525.
Copies: DLC; DeGE; MH-BA; MWA; NN; NNC; OCHP.

3622. Sullivan, John Langdon, 1777–1865.
Explanation, by John L. Sullivan, of the nature of certain grants to him for the use of steam boats on Connecticut River, and of the nature of his claims to admission into the state of New-York with his patented improvements in steam boats. In a letter addressed to the editor of the New-York evening post. [New York?]: 1818. 29 p., [1] l.
S&S-45817.
Copies: Ct; DLC; DeGE; MB; MH-BA; MWA; NHi.

3623. Thornton, William, 1759–1828.
Short account of the origin of steamboats, written in 1810, and now committed to the press by W. Thornton, of the city of Washington. Albany: Printed by E. & E. Hosford, 1818. 18 p.
S&S-45866.
Copies: CtY; MSaE; MWA; N; PHi.

3624. Colden, Cadwallader David, 1769–1834.
A vindication by Cadwallader D. Colden, of the steamboat right granted by the state of New-York [to Livingston and Fulton] in the form of an answer to the letter of Mr. Duer, addressed to Mr. Colden. New York: Printed by W. A. Mercein, 1819. 96 p.
S&S-47660.
Copies: DLC; IC; MWA; MiU; MoK; N; NN; PP; PPAmP; ViU.

3625. Connecticut. Laws.
An act to protect the citizens of Connecticut in their right to navigate boats and vessels moved by fire or steam. [Hartford: 1819.] 4 p.
Copies: CtHi.

3626. Connecticut River Steamboat Company.
Copy of a petition about to be presented to the Legislature of Connecticut. To the Honorable the Senate, and House of Representatives of the State of Connecticut, in General Court assembled. The petition of the Connecticut River Steam-Boat Company. [Hartford: 1819.] 7 p.
Signed: David Watkinson, Christopher Colt . . . [and] John L. Sullivan.
Concerns the use of steamboats, as patented by John L. Sullivan, between Con-

3626 (*continued*)
necticut and New York.
S&S-47712.
Copies: CSmH; CtSoP; DeGE; MWA.

3627. Duer, William Alexander, 1780–1858.
A reply to Mr. Colden's Vindication of the steam-boat monopoly. With an appendix, containing copies of the most important documents referred to in the argument.... Albany: Printed and published by E. and E. Hosford, 1819. 184, xxvii p.
S&S-47850.
Copies: CSmH; CtHC; CtY; DLC; DeGE; ICJ; MBNEH; MH-BA; MWA; MdHi; MiU; NIC; NNE; PPL; RPB.

3628. Facts and considerations on the question: Why is steam-boat navigation interesting to Connecticut? Hartford: G. Goodwin & Sons, printers, 1819. 8 p.
Probably written by John L. Sullivan.
S&S-47933.
Copies: CtHi; CtY; DeGE; MWA.

3629. Gibbons, Thomas.
In the Court for the Trial of Impeachments and the Correction of Errors, between Thomas Gibbons, appellant, and Aaron Ogden, respondent. Case on the part of the appellant. New York: Printed by Clayton & Kingsland, 1819. 83 p.
At head of title: State of New-York.
Concerns the monopoly for steam navigation, granted to Livingston and Fulton by the state of New York.
N-G0195670.
Copies: MiU.

3630. Sullivan, John Langdon, 1777–1865.
[Circular letter to the proprietors of the Connecticut River Steamboat Company.] Boston, September 21, 1819. [Boston: 1819.] Broadside 22.5 x 21 cm.
Copies: CtHi.

3631. [Sullivan, John Langdon] 1777–1865.
Note relating to the steam-boat bill. [Hartford: 1819.] 6 p.
Dated: Hartford, May 21, 1819.
Contains a letter, dated Oct. 31, 1818, from Samuel Morey to William A. Duer describing the former's steamboat experience, and J. L. Sullivan's explanation of his own controversy with Robert Fulton over the rights to the steam tow-boat.
Copies: CtHi.

3632. New steam boat. Proposals for raising by subscription 100,000 dollars, in shares of 100 dollars each, to build a packet, of the burthen of 350 tons, or more, to be propelled by steam, and to run from New-York to Charleston.... [New York?: ca.1820.] 2 l.
S-2466.
Copies: DeGE; NN.

3633. Sullivan, John Langdon, 1777–1865.
A demonstration of the right to the navigation of the waters of New York without the license of the owners of the monopoly of steam and fire granted to Robert R. Livingston and Robert Fulton.... By John L. Sullivan.... Cambridge, [Mass.]: Printed by Hilliard and Metcalf, 1821. 41 p.
S-6918.
Copies: DLC; MB; MBAt; MH-BA; N; NN.

3634. Sullivan, John Langdon, 1777–1865.
In the Supreme Court of the United States, between John L. Sullivan & others, appellants, and the Fulton Steam-Boat Company, respondents. By appeal from the Circuit Court of the United States for the Southern District of New-York, in the Second Circuit. Case. New York: Printed by J. Seymour, 1821. 16 p.
S-7301.
Copies: MH-L.

3635. United States. Quartermaster's Dept.
Documents in relation to the claim of

James Johnson for transportation on the Missouri and Mississippi rivers. March 1, 1821.... Washington: Printed by Gales & Seaton, 1821. 193 [293] p., fold. tables. (16th Cong., 2d sess. House. Doc.110)

Last page misnumbered: 193.

Concerns the use of steamboats in river navigation.

S-7211.

Copies: DLC; DeGE; DeU; MWA; NjR.

3636. [Colden, Cadwallader David] 1769–1834.

A brief exposition of the views of John L. Sullivan, Esq., who holds in virtue of an act of the legislature of Massachusetts, an exclusive right to the use of steam tow boats, on part of the water of that state for forty-two years, viz: from 1814 to 1856!!! in opposition to similar rights granted by the state of New-York to Messrs. Livingston & Fulton. New York: Printed by William A. Mercein, 1822. 28 p.

Signed: Cadwallader D. Colden.

Appendix (p. 17–28) contains a copy of Mr. Sullivan's petition to the Massachusetts legislature, and correspondence between Sullivan, Colden, and D. Lynch.

S-8381.

Copies: CtY; DeGE; MB; MH; MWiW-C; MiD-B; N; NN; NjR.

3637. New York and Brooklyn Steam Ferry Boat Company.

A statement of facts, with remarks, &c. In answer to a pamphlet, published at Brooklyn, in relation to the steam boat ferry. Brooklyn: Printed by A. Spooner, 1822. 45, [1] p.

S-9691.

Copies: MWA; NB; NBLiHi; NHi; NJQ; NN; NSmB.

3638. United States. Laws.

In Senate of the United States, April 9, 1822. The bill from the House of Representatives, entitled "An act granting certain privileges to steam ships and vessels owned by incorporated companies," being under consideration, Mr. Otis proposed to amend the same, by adding thereto the following section.... [Washington: 1822.] 1 p.

S-10887.

Copies: DNA.

3639. Commercial line of steam-boats, and safety barges.... W. C. Redfield, agent for the Steam Navigation Company.... New York: Clayton, and Van Norden, printers, 1823. Broadside.

Dated: New York: July 30th, 1823.

S-12219.

Copies: NN.

3640. Connecticut. Governor (O. Wolcott).

Message of the Governor of the state of Connecticut, to the House of Representatives.... Containing his reasons for returning the steam boat bill to the legislature.

[Hartford: 1823.] 10–19 p.

Dated: May, 1822, and appended to Judge Wm. Johnson's opinion, delivered Aug. 7, 1823.

S-12237.

Copies: CtY.

3641. New Lake Erie steam-boat Superior, Jedediah Rogers, master, will sail from Buffalo for Detroit, on or about the twenty-fourth day of April next, if the lake is clear of ice ... Albany: Packard & Van Benthuysen, printers, [1823]. Broadside 32 x 22 cm.

Dated: Albany, February 23, 1823.

N-N0158273.

Copies: MiD.

3642. The steam-boat Superior, Captain Milner, will commence running up and down daily, for the season, on Friday the 18th inst. leaving Wilmington at 7 o'clock in the morning, and Philadelphia at two o'clock in the afternoon. (Sundays excepted). Passage one dollar and twenty-five cents. Freight at same rate as charged by the packets. [Wilmington, Del.]: Printed at the Gazette office, [1823]. Broadside.

S-14194.

Copies: DeHi.

3643. Sullivan, John Langdon, 1777–1865.
The answer of Mr. Sullivan, to the letter, and mis-statements of the Hon. Cadwallader D. Colden, in his "Brief exposition" of himself as the advocate of monopoly.... The legality of an extension of the term of time of a patent when for the good of state; and the just views and claims of patentees in steam navigation fully explained. Troy, [N.Y.]: Printed by William S. Parker, 1823. 40 p.
"Mr. Duer's opinion on the claim of Mr. Sullivan as a patentee, to navigate with steam boats, the waters of the state of New-York," p. 29–34.
S-14236.
Copies: DLC; MBAt; MH; MWA; NHi; NN; NNC.

3644. Sullivan, John Langdon, 1777–1865.
The answer of Mr. Sullivan, to the letter and mis-statements of the Hon. Cadwallader D. Colden, the advocate of the monopoly of steam and fire in navigating the rivers, coasts, and lakes of New York.... Supported by opinion of counsel. 2d ed., with a summary history of the origin, progress, and laws of the monopoly.... Troy, [N.Y.]: Printed by William S. Parker, 1823. 47 p.
"Mr. Duer's opinion on the claim of Mr. Sullivan as a patentee, to navigate with steam boats, the waters of the state of New-York," p. 30–32.
S-14237.
Copies: CSmH; CtY; DLC; DeGE; MB; MH-BA; MWA; MiU-C; NN.

3645. Gibbons, Thomas.
The case of Gibbons against Ogden, heard and determined in the Supreme Court of the United States, February term, 1824, on appeal from the Court of Errors of the state of New-York, and involving the constitutionality of the laws of that state, granting to Livingston and Fulton the exclusive navigation of its waters by steam-boats. Reported by Henry Wheaton. New York: R. Donaldson, 1824. 244 p.
S-19257.
Copies: CtY; MH; MH-BA; NHi; NN; PPB.

3646. New York (State). Attorney General.
Report relative to steamboats navigating the Hudson River. Albany: 1824. 3 p. (Assembly. Doc. 202)
Whitford-p. 1339.
Copies: N.

3647. Stanford, John, 1754–1834.
Aetna. A discourse, delivered in the New York City Hospital ... May 23, 1824. Intended as a moral improvement of the calamity occasioned by the bursting of the boiler of the steam-boat Aetna; by which several lives were lost.... New York: Printed by E. Conrad, 1824. 21 p.
Includes an account of the accident.
S-18076.
Copies: DeGE; MH; MWA; MNtcA; NNNAM; NRAB; PPL; PPPrHi; RPB.

3648. United States. Congress. House. Committee on Commerce.
Report of the Committee on Commerce, accompanied by a bill for regulating of steam boats, and for the security of passengers therein. May 22, 1824.... [Washington: 1824.] 8 p. (18th Cong., 1st sess. House. Rept. 125)
S-18999.
Copies: DLC; DeU; MWA; NjP.

3649. United States. Laws.
A bill for the regulation of steamboats, and for the security of passengers therein. May 22, 1824.... [Washington: 1824.] 3 p. (H.R. 234)
S-18432.
Copies: DNA.

3650. United States. Supreme Court.
The opinion of the Supreme Court of the United States, in the case of Gibbons vs. Ogden, delivered by Chief Justice Marshall, March 2, 1824. Wi[t]h a preface, containing an historical sketch of the steam-boat controversy. Albany: Printed by J. B. Van Steenbergh, 1824. 28 p.
Case on appeal from the Court for the Trial of Impeachments and Correction of

Errors of the State of New York, involving the constitutionality of the laws of that state granting to Livingston and Fulton the exclusive navigation of its waters by steamboats.

S-18938.

Copies: DLC; DeGE; MB; MH-BA; MWA; N; NHu; NIC; NN.

3651. Arrangement of the fleet of steamboats. On Friday, 4th November, 1825.... [New York]: Clayton & Van Norden, printers, [1825]. Broadside 31 x 21 cm.

Concerns ceremonies on the opening of the Erie Canal.

Copies: MWA.

3652. Delaware. Laws.

An act to establish a company under the name of "The Philadelphia, Dover and Norfolk Steam-Boat & Transportation Company," passed by the Legislature of the State of Delaware. Dover: A. M. Schee, printer, 1825. 8 p.

"Passed at Dover, February 10, 1825."

S-20284.

Copies: DeGE; O; PHi.

3653. McNeill, William L.

Steam boats Henrietta & North Carolina. Rates of freight, including toll. Fayetteville: 1825. 1 p.

Dated: August 7, 1825.

S-21286.

Copies: NcU.

3654. The objects of the builders of the steamboat "Codorus." [York?, Pa.: 1825?] Broadside 33 x 21 cm.

The steamboat "Codorus" was launched at York Haven, Pa., on April 26, 1826, to test the navigability of the Susquehanna River.

Copies: DeGE.

3655. Redfield, William Charles, 1789–1857.

Commercial line of steam-boats and safety barges.... New York: Clayton & Van Norden, printers, 1825. Broadside, illus. 24.5 x 19 cm.

Signed at end: W. C. Redfield, agent for the Steam Navigation Company, New York, July 30th, 1825.

Privately held; photocopy at NN and PHi.

3656. United States. Treasury Dept.

Letter from the Secretary of the Treasury, transmitting information collected by the department, upon the subject of accidents on board of steam boats. January 31, 1825... Washington: Printed by Gales & Seaton, 1825. 21 p. (18th Cong., 2d sess. House. Doc. 69)

S-22941.

Copies: DLC; DeGE; DeU; MWA; NjP.

3657. Pennsylvania. Laws.

An act to incorporate the Pennsylvania, Delaware & Maryland Steam Navigation Company, passed March 6, 1826. [Harrisburg?: 1826.] 8 p.

S-25707.

Copies: PPL.

3657a. Kennebec Steam Navigation Company.

Regulations and by-laws of the Kennebec Steam Navigation Company. Boston: True and Greene, 1827. 7 p.

Western Hemisphere, Stoughton, Mass., Cat. 58, 1977, No. 150.

Copies: MWA.

3658. North River Steam Boat Line.

For Albany at 6 o'clock A.M. The low pressure steam-boat Albany ... leaves ... on Friday the 1st of June.... [New York: 1827.] Broadside 27 x 18 cm.

S-30090.

Copies: MHi.

3659. Fulton, Robert, 1765–1815.

Report of the practicability of navigating with steam boats, on the southern waters of the United States, from the Chesapeake to the River St. Mary's, forming part of a line of steam boat communications, now estab-

3659 *(continued)*
lishing, from the northern extremity of Lake Champlain to East-Florida, a distance of fifteen hundred miles. First published, December, 1813, by Robert Fulton, from the surveys of John D. Delacy. 2d ed. Philadelphia: Printed by Thomas Town, 1828. 12, [3] p.

"Address to the public," signed by M. Carey, L. Elmaker and G. Ralston, committee appointed at a meeting of Philadelphia citizens held for the purpose of considering the extension of stage and steamboat lines from Norfolk to Charleston to connect with the line already established from Philadelphia to Norfolk, [3] p. at end.

S-33312.

Copies: CSmH; DeGE; MH-BA; MWA; MnHi; NIC; NcU; PPAmP; PPL; ViU; WHi.

3660. Narragansett Bay Company.

Charter of the Narragansett Bay Company. Granted at October session, A.D. 1828. Providence: Printed by Jones & Simons, 1828. 8 p.

A company to transport freight and persons in boats, to be propelled by steam, animal power, or otherwise.

S-34361.

Copies: RHi.

3661. Philadelphia, Dover and Norfolk Steam-Boat and Transportation Company.

At a meeting of stockholders of the above Company, held August 28th, 1828 ... Philadelphia, M. Carey in the chair, Henry Korn, secretary, the following paper was read. [Philadelphia: 1828.] 4, 3 p.

Dated at head of title: Philadelphia, August, 1828.

"Address to the public," 3 p. at end, signed M. Carey, L. Elmaker, G. Ralston, Philadelphia, February 4th, 1828.

S-34757.

Copies: MWA; PPL.

3662. Remarks on our inland communications, with a view to steam-boat navigation. [Schenectady, N.Y.: 1828.] 12 p.

Dated at end: Schenectady, September, 1828.

S-34990.

Copies: NHi; NN.

3663. Renwick, James, 1790–1863.

An account of some of the steam-boats navigating the Hudson River in the state of New York. In a letter from Mr. Renwick, Professor ... in Columbia College, to Captain Edward Sabine, R.A., Secretary of the Royal Society. [S.l.: 1828.] 24 p.

S-34996.

Copies: NN.

3664. United States. Laws.

A bill to incorporate the Washington, Alexandria, and Baltimore Steam Packet Company. March 7, 1828.... [Washington: 1828.] 5 p. (H.R. 208)

S-36143.

Copies: DNA.

3665. United States. Laws.

A bill to incorporate the Washington, Alexandria and George Town Steam Packet Company. Jan. 21, 1829. Mr. Chambers, from the Committee for the District of Columbia, reported the following bill.... [Washington: 1829.] 5 p. (S. 66)

S-41041.

Copies: DNA.

3666. Sullivan, John Langdon, 1777–1865.

Proposition for an anthracite coal steam power boat and barge company, for passengers on the North River. [New York]: Printed by Clayton & Van Norden, [1830]. 12 p., plate.

"Specification and description of the anthracite coal furnace for steamboat boilers, made and claimed by J. L. Sullivan," p. [5]–12.

C-3635.

Copies: DLC; MWA; MiU-C; WHi.

3667. United States. Laws.

A bill to amend the several acts authorizing the registering and granting of licences to steam-boats, and to provide for better security of the lives of passengers on board of vessels propelled in whole or in part by steam. May 4, 1830.... [Washington: 1830.] 5 p. (H.R. 458)

C-4431.

Copies: DNA.

3. Naval stores

3668. Great Britain. Laws.

An act of Parliament relating to pitch and tar made in the plantations. New York: Printed by William Bradford, 1719.

E-2025.

No copy known.

3669. British Tar Company.

Account of, and directions for using the different kinds of coal tar and varnish, prepared by the British Tar Company, at their works in England and Scotland, with certificates from ship masters and others. New York: Printed by J. and A. M'Lean, 1788. 50 p.

"Coal tar certificates," p. 11–50.

Mp-20982.

Copies: DLC; DeGE; MHi; MWA; NN; PPL.

3670. Kennebeck-River; price current for lumber delivered at Bath, Longreach. Masts hewed in best manner ... yards hewed 8 square of spruce or pine ... bowsprits hewed 8 square of pine ... timber ... planks ... common lumber.... [Portland?, Me.: 1792?] Broadside 27 x 16.5 cm.

Dated: Bath, August 1792.

B-8030; Mp-46477.

Copies: MWA (photocopy); RPJCB.

3671. Cutting, Nathaniel.

General observations respecting a new mechanical process for manufacturing cordage for maritime and other usage; the property of Nat. Cutting, of Boston, in the United States. [Boston?: 1803?] 15 p.

S&S-4038.

Copies: DLC; MH-BA; MHi; PPAmP.

3672. ———— [Boston: 1804.] 15 p.

S&S-6107.

Copies: DLC; MB.

3673. Cutting, Nathaniel.

Supplement to General observations respecting a new mechanical process for manufacturing cordage for maritime and other uses.... [Boston: 1804?] 15 p.

DLC copy dated at end in mss.: August 1804.

N-C0853753.

Copies: DLC; MHi; PPAmP.

3674. Cutting, Nathaniel.

General observations respecting a new mechanical process for manufacturing cordage for maritime and other uses; the property of Nat. Cutting, of Boston, in the United States. [Boston?: 1805?] 15 p.

Dated at end: Paris, Aug. 18, 1803.

S&S-8279.

Copies: DLC; DeGE; NHi.

3675. Cutting, Nathaniel.

Supplement to General observations respecting a new mechanical process for manufacturing cordage for maritime and other uses.... [Boston?: 1805?] 15 p.

Dated at end: May 1805.

S&S-8280.

Copies: DLC; DeGE; MH-BA.

3676. United States. Navy Dept.
American hemp. The Secretary of the Navy will receive ... proposals for furnishing water rotted hemp.... Norfolk: Printed by J. O'Connor, 1808. Broadside 34 x 21 cm.
Dated: Navy Department, May 3, 1808.
Hummel-3183.
Copies: Vi.

3677. Eastern Sail Company, Salem, Mass.
Rules and regulations for the Eastern Sail Company in Salem. [Salem: 1813.] Broadside 34 x 22 cm.
Tapley-p. 432.
Copies: MSaE.

3678. Brown, Samuel, Sir, 1776–1852.
Reports and observations on the patent iron cables, invented by Captain Samuel Brown, R. N. Agent, Martin Walker, Philadelphia. [Philadelphia: S. Merritt, printer, 1815.] vi, [7]–69 p.
"Copies of sundry reports, respecting the advantages of iron cables," p. [7]–69.
N-B0865611.
Copies: DeGE; PHi; PPAmP; PPF; PPWa.

3679. New York Journeymen Sail Makers Society.
Constitution of the New-York Journeymen Sail Makers Society. Instituted December 29, 1818, revised July 1, 1819. New York: Benjamin G. Jansen, 1819. 12 p.
The Current Company, Bristol, R.I., Catalog no. 8, 1973, no. 235.
Copies: MWA.

3680. Memorial of sundry manufacturers of cordage, of the city of New York. December 19, 1823.... Washington: Printed by Gales & Seaton, 1823. 4 p. (18th Cong., 1st sess. House.Doc. 9)
Signed: Joshua Sands, Peter Schermerhorn & Sons, and others.
S-14629.
Copies: DLC; DeGE; MWA; NjP.

3681. Memorial of the manufacturers of cordage, of the city and county of Philadelphia. January 30, 1823. Read, and referred to the Committee of the Whole House on the State of the Union, to which is committed the bill for the more effectual encouragement of certain domestic manufactures. Washington: Printed by Gales & Seaton, 1823. 4 p. (17th Cong., 2d sess. House. Doc. 49)
Signed: B. Williamson [and others].
S-14634.
Copies: DLC; DeGE; MWA; NjP.

3682. United States. Navy Dept.
Letter from the Secretary of the Navy, communicating information of the quantity of cordage manufactured of hemp of domestic growth, which has been used in the service of the Navy since the year 1802 ... February 23, 1824.... Washington: Printed by Gales & Seaton, 1824. 7 p. (18th Cong., 1st sess. Senate. Doc. 39)
S-18787.
Copies: DLC; DeU; MWA; NjP.

3683. United States. Navy Dept.
Message from the President of the United States, transmitting a report from the Secretary of the Navy, in relation to American canvass, cables, and cordage ... January 10, 1825.... Washington: Printed by Gales & Seaton, 1825. 34 p. (18th Cong., 2d sess. Senate. Doc. 12)
S-22991.
Copies: DLC; DeU; MWA; PU.

3684. United States. Board of Navy Commissioners.
Live oak timber. Letter from the Secretary of the Navy, transmitting a report of the Board of Navy Commissioners, in relation to the probable cost of certain lots of live oak timber, and the expense of providing for the preservation of the same. February 10, 1826.... Washington: Printed by Gales & Seaton, 1826. 3 p. (19th Cong., 1st sess. House. Doc. 85)
S-27188.
Copies: DLC; DeU; MWA; NjP.

3685. United States. Navy Dept.
Cotton cordage and canvas. Letter from the Secretary of the Navy, transmitting the information required by a resolution of the House of Representatives, of the 25th ultimo, in relation to the experiments which have been made to test the utility of cotton for cordage and sail cloth. February 1, 1826.... Washington: Printed by Gales & Seaton, 1826. 4 p. (19th Cong., 1st sess. House. Doc. 66)
S-26933.
Copies: DLC; DeU; MWA; NjP.

3686. United States. Navy Dept.
American water rotted hemp. Letter from the Secretary of the Navy, transmitting the information required by a resolution of the House of Representatives, of 2d March, 1827, in relation to experiments on American water rotted hemp. December 24, 1827.... Washington: Printed by Gales & Seaton, 1827. 8 p. (20th Cong., 1st sess. House. Doc. 22)
S-30914.
Copies: DLC; DeU; MWA; NjR.

3687. United States. Navy Dept.
Live oak timber. Letter from the Secretary of the Navy on the subject of live oak timber. February 26, 1827.... Washington: Printed by Gales & Seaton, 1827. 9 p. (19th Cong., 2d sess. House. Doc. 114)
S-31375.
Copies: DLC; DeU; MWA; NjP.

3688. United States. Navy Dept.
American hemp, flax, and cotton. Letter from the Secretary of the Navy transmitting the information required by a resolution of the House of Representatives of the 12th May last, in relation to connecting with two or more navy yards of the United States as many establishments for purchasing, water-rotting, and preparing for manufacture, American hemp, flax and cotton, &c. &c. December 22, 1828.... [Washington: Gales & Seaton, printers, 1828.] 3 p. (20th Cong., 2d sess. House. Doc. 28)
S-35674.
Copies: DLC; DeU; MWA; NjR.

3689. United States. Navy Dept.
American water-rotted hemp, &c. &c. Reports from the Navy Department, in relation to experiments on American water-rotted hemp, when made into canvass, cables, and cordage. January 18, 1828.... Washington: Printed by Gales & Seaton, 1828. 35 p. (20th Cong., 1st sess. House. Doc. 68)
S-35675.
Copies: DLC; DeGE; DeU; MWA; NhHi; NjR.

3690. Elliot, Jonathan, 1784–1846.
Remarks and correspondence on the use of cotton canvass, as preferable to hemp, for the sails of ships of war or merchant vessels, supported by the testimony of naval officers, ship owners, captains, &c. City of Washington: Jonathan Elliot, 1829. 25 p.
Remarks by Jonathan Elliot; correspondence between Warren R. Davis and the Secretary of the Navy, John Branch; and testimonials, p. [9]–25.
N-R0162350.
Copies: DeGE (xerox copy); MdBP; PPAmP.

3691. United States. Navy Dept.
Cotton and hemp for sail and cordage. Letter from the Secretary of the Navy on the subject of experiments made to test the comparative fitness of cotton and hemp for the use of sail in the public vessels of the United States, &c., &c. February 2, 1829.... [Washington: Gales & Seaton, printers, 1829.] 3 p. (20th Cong., 2d sess. House. Doc. 102)
S-41097.
Copies: DLC; DeU; MWA; NjR.

3692. [Elliot, Jonathan] 1784–1846.
Remarks and correspondence on the use of cotton canvass as preferable to hemp, for the sails of ships of war, or merchant vessels, supported by the testimony of naval officers, ship owners, captains, &c. Baltimore: Sands & Neilson, 1830. 27 p.
Remarks by Jonathan Elliot, followed

3692 (*continued*)
by correspondence between Warren R. Davis and the Secretary of the Navy, John Branch; and testimonials.

C-3270.

Copies: MH.

3693. United States. Navy Dept.

American canvass, cables, and cordage. In the House of Representatives, April 22, 1830. Mr. Spencer, of New York, from the Committee on Agriculture, reported the following.... [Washington: 1830.] 26 p. (21st Cong., 1st sess. House. Rept. 381)

Reprint of part of the *Message of the President of the United States,* 5th Jan., 1830, transmitting a report of the Navy Dept.

C-3896.

Copies: DLC; DeGE; MWA; NjR.

4. Harbors, piers and wharves

3694. A new map of the harbour of New-York, from a late survey. [Containing the soundings and settings of the tydes, and the bearings of the most remarkable places, with the proper places for anchoring.] [New York: Printed and sold by William Bradford, 1735.] Map 31 x 25 cm.

Mp-3922.

Copies: CSmH; NHi.

3695. Wind Mill Island cash lottery in four classes ... June 27, 1774.... [Philadelphia: 1774.] Broadside.

A lottery to raise money for the purchase and improvement of public landings on the Delaware River front in Philadelphia.

Mp-13776.

Copies: PPL.

3696. Charleston, S.C. Harbor Master.

Charleston. October 4, 1783. No. 3, Broadstreet. By virtue of the powers which have been vested in me, by an ordinance lately passed by the City Council, entituled [*sic*] "An ordinance for defending the duties of the Harbour-Master of the Port of Charleston," I do hereby order and direct the pilots of this bar and harbour not to bring any vessel to anchor within the distance of fifty fathoms from the outermost vessel at any wharf.... [Charleston]: Printed by N. Childs & Co., [1783]. Broadside 30 x 20 cm.

The word "defending" is corrected in ink to "defining."

B-5696; N-C0313682.

Copies: CtY.

3697. Pennsylvania. Laws.

An act 1784 for the further regulation of the port of Philadelphia.... [Philadelphia: Thomas Bradford, 1784.] 8 p.

Dated: April 1, 1784.

B-5957.

Copies: PHi (at PPL).

3698. Pennsylvania. Laws.

In the House of Representatives, Tuesday, September 20, 1791, read the first time, An act to provide for compleating the repairs of the wharf, and for paving the footways round the State-House Square.... [Philadelphia: 1791.] Broadside.

B-7791; Mp-46251.

Copies: PPL.

3699. United States. Congress. House. Committee of Commerce and Manufactures.

Report of the Committee of Commerce and Manufactures, to whom was referred ... the memorials of the Philadelphia Chamber of Commerce, and of sundry inhabitants of the town of New-Castle ... and of other citizens of the state of Delaware; and to whom was also referred ... a motion "for the appropriation ... for the erection and repair of piers in the river Delaware." 19th March, 1802.... [Washington]: 1802. 14 p.

Report from the Secretary of the Treasury, Oliver Wolcott, p. [9]-12.

S&S-3407.

Copies: DLC; DeGE; MWA; NN.

3700. United States. Laws.

A bill relative to public piers in the river Delaware. 19th March, 1802.... [Washington]: Duane, printer, [1802]. Broadside.

S&S-3241.

Copies: DLC.

3701. Petition of the inhabitants of the island and town of Nantucket, in the state of Massachusetts, 11th February, 1803. Referred to the Committee of Commerce and Manufactures. 16th February, 1803. Report made ... 7th November, 1803. [Washington: 1803.] 8 p.

Concerns opening the channel in Nantucket harbor.

S&S-5392.

Copies: DLC; MWA.

3702. United States. Congress. House. Committee of Commerce and Manufactures.

Report of the Committee of Commerce and Manufactures, to whom were referred, on the 28th ultimo, a letter and report from the Secretary of the Treasury, accompanied with sundry papers, relative to a survey of the harbour in the island of Nantucket, and of the bar and shoals near the same. 7th November, 1803.... [Washington: 1803.] 6, [2], 8, [8], 5, [3], 5 p.

An Estimate of building two piers for the harbor in the island of Nantucket, 5 p. at end.

S&S-5443.

Copies: DLC; MWA.

3703. United States. Treasury Dept.

An estimate of building two piers, for the harbor in the island of Nantucket. Accompanying a letter and report from the Secretary of the Treasury, received the 28th of October, 1803.... [Washington: 1803.] 5 p.

S&S-5290.

Copies: DLC; MWA; MiU-C.

3704. United States. Treasury Dept.

Letter from the Secretary of the Treasury, accompanying a report and sundry papers relative to a survey of the harbor in the island of Nantucket, and of the bar and shoals near the same. 28th October, 1803. Read ... 7th November, 1803. Report made. [Washington: 1803.] 4 p.

S&S-5331.

Copies: DLC; MWA; MiU-C; ViU.

3705. Williams, John Foster, 1743–1814.

Letter from John Foster Williams and Lemuel Cox, accompanying a report & sundry papers relative to a survey of the harbor in the island of Nantucket, and of the bar and shoals near the same ... 7th November, 1803.... [Washington: 1803.] 5 p.

S&S-5322.

Copies: DLC; MWA; MiU-C.

3706. United States. Navy Dept.

In Senate of the United States. January 27, 1817.... Report of the Secretary of the Navy, communicating, in obedience to a resolution of the Senate, information relating to the selection of a site for a naval depot. [Washington: 1817.] 53 p. (14th Cong., 2d sess. Senate. Doc. 72)

Concerns a site in Chesapeake Bay.

S&S-42512.

Copies: DLC; DeGE; MWA.

3707. [Hughes, B. F.]

Annapolis considered as a suitable situation, for a great naval depot and arsenal of marine stores. [Annapolis: Printed by J. Chandler, 1818.] 12 p.

S&S-44390.

Copies: MdHi.

3708. New York (City). Ordinances.
Harbour regulations, relating to warden, harbour masters and pilots, conformable to an act of the Assembly, Feb. 19, 1819. New York: Printed by J. Seymour, for Wm. Hooker, 1819. 23 p.
S&S-48888.
Copies: MWA.

3709. Wadsworth, Alexander.
Chart of Boston harbor, surveyed in 1817.... Philadelphia: John Melish, 1819. Map 102 x 93 cm.
S&S-51954.
Copies: M.

3710. United States. Laws.
A bill authorizing the repair of a sea-wall on the Isle of Shoals, and for other purposes. In Senate of the United States, Jan. 18, 1821. [Washington: 1821.] 2 p. (S. 34)
S-7059.
Copies: DNA.

3711. Kearney, James.
Pensacola harbor and bar. Surveyed in 1822 by Major James Kearney, topog. engineer, assisted by Lieuts. Thompson, Turnbull and Butler. Washington: 1822. Map 150 x 120 cm.
N-K0063389.
Copies: MB.

3712. Buffalo. Harbor Committee.
Documents and brief remarks, in reply to the pamphlet written by General Porter, and published by direction of the Black Rock Harbor Company, at Albany, in Dec., 1822, and at Buffalo in Jan., 1823. By the Buffalo Harbor Committee. Buffalo: Printed by David M. Day, 1823. 47 p., fold. map.
S-12018.
Copies: N; NBu.

3713. Porter, Peter Buel, 1773–1844.
Memorial in behalf of the Black Rock Harbor Company, praying authority to construct harbors at Buffalo and Black Rock. Albany: 1823. 2 p. (New York (State). Legislature. Assembly. Doc. 167)
Whitford-p. 1257.
Copies: N.

3714. Wilkeson, Samuel, 1781–1848.
Memorial on the subject of the Black Rock and Buffalo harbors. Albany: 1823. 4 p. (New York (State). Legislature. Assembly. Doc. 141)
Whitford-p. 1257.
Copies: N.

3715. Philadelphia. Chamber of Commerce.
Memorial of the Chamber of Commerce, of the city of Philadelphia. January 12, 1824.... Washington: Printed by Gales & Seaton, 1824. 6 p. (18th Cong., 1st sess. House. Doc. 27)
Signed: Robert Ralston, President; Jn. Vaughan, Secretary.
Concerns the building of a harbor at the Capes of Delaware.
S-18865.
Copies: DLC; DeGE; MWA; NjP.

3716. United States. Congress. Senate.
In Senate of the United States, May 8. 1824. Documents relating to the bill authorizing the examination and survey of the harbor of Charleston and the coast of Florida. [Washington: 1824.] 5 p. (18th Cong., 1st sess. Senate. Doc. 76)
Documents supplied by the War Department.
S-18756.
Copies: DLC; DeU; MWA; NjP.

3717. United States. Laws.
A bill authorizing the examination and survey of the harbor of Charleston, in South Carolina, and of the coast of Florida. May 8, 1824. Mr. Lloyd, of Massachusetts, from the Committee on Naval Affairs, to whom the subject was referred, reported the following bill. [Washington: 1824.] 1 p. (S. 107)
S-18388.
Copies: DNA.

3718. United States. Laws.
A bill making appropriations for deepening the channel leading into the harbor of Presque Isle, and for repairing Plymouth Beach. April 29, 1824.... [Washington: 1824.] 1 p. (H.R. 209)
S-18561.
Copies: DNA.

3719. Boston pier, or Long-wharf, incorporated ... July 14, 1772. Additional act passed ... June 24, 1824. Also, rules and by-laws.... Boston: 1825. 25 p.
S-19822.
Copies: MBAt; MHi.

3720. Jones, William, 1760–1831.
Remarks on the proposed breakwater at Cape Henlopen. Communicated, by request of the Secretary of the Treasury, by William Jones.... To which are added, the report of the Board of Engineers, and Captain Bainbridge of the Navy; the memorial of the Chamber of Commerce of Philadelphia &c. &c. Philadelphia: Printed by order of the Chamber of Commerce of Philadelphia, 1825. 31 p. incl. 2 diagrs.
S-21072.
Copies: CtY; DLC; DeGE; MB; MH-BA; MWA; MiD-B; NbO; Nj; NjPT; P; PPAmP; PPL.

3721. Philadelphia. Chamber of Commerce.
Breakwater in Delaware Bay. Memorial of the Chamber of Commerce of the city of Philadelphia, upon the construction of a breakwater in the Bay of Delaware. December 12, 1825.... Washington: Printed by Gales & Seaton, 1825. 6 p. (19th Cong., 1st sess. Senate. Doc. 5)
Robert Ralston, President.
S-22849.
Copies: DLC; DeGE; DeU; MWA; NjP.

3722. Baltimore. Mayor.
Memorial of the Mayor and City Councils of Baltimore, praying the establishment of a navy yard at that city. February 13, 1826.Washington: Printed by Gales & Seaton, 1826. 5 p. (19th Cong., 1st sess. Senate. Doc. 37)
Supplies data on the facilities of Baltimore harbor, depth of water, and availability of materials and naval stores.
S-27210.
Copies: DLC; DeU; MWA; NjP.

3723. Boston. Committee on the Petition of the Mercantile Wharf Corporation.
[Report of the committee of the City Council on the several petitions of the Mercantile Wharf Corporation. Boston: 1826.] 8 p.
S-23866.
Copies: DLC; NN.

3724. Central Wharf and Wet Dock Corporation, Boston.
[Notice of] a meeting of the trustees.... Boston, January 28, 1826. [Boston: 1826.] Broadside 9 x 21 cm.
S-23871.
Copies: MHi.

3725. Jones, William, 1760–1831.
Remarks on the proposed breakwater at Cape Henlopen. Communicated, by request of the Secretary of the Treasury, by William Jones.... To which are added, the report of the Board of Engineers, and Captain Bainbridge of the Navy; the memorial of the Chamber of Commerce of Philadelphia, &c. &c. 2d ed. Philadelphia: Printed by order of the Chamber of Commerce of Philadelphia; William Fry, printer, 1826. 34 p., illus.
S-25005.
Copies: DLC; DeGE; MB; MWA; PHi; PPL.

3726. Proceedings of a town meeting ... Philadelphia, December 28, 1825, with a view to memorialising Congress for an appropriation for constructing a breakwater in the Delaware Bay; with the memorial.... Philadelphia: Wm. Fry, printer, 1826 7 p.
S-25840.
Copies: NbO; PHi; PPAmP; PPL.

3727. To the Senate and House of Representatives of the United States. The undersigned, citizens of the village of Cleaveland [*sic*], in the state of Ohio, and its vicinity, beg leave to invite the attention of your honorable body, to the subject of the further improvement of the harbor at the mouth of the Cuyahoga River in said village. [Cleveland: 1826.] Broadside 21.5 x 30 cm.

Dated: Cleveland, Ohio, December 1, 1826.

S-24128.

Copies: OClWHi.

3728. United States. Laws.

A bill for improving certain harbors, and the navigation of certain rivers and creeks, and for authorizing surveys to be made of certain bays, sounds, and rivers, therein mentioned. March 30, 1826.... [Washington: 1826.] 3 p. (H.R. 192)

S-26449.

Copies: DNA.

3729. United States. Laws.

A bill for removing obstructions and deepening the channel of the harbor of Mobile. March 6, 1826. [Washington: 1826.] 1 p. (S. 81)

S-26452.

Copies: DNA.

3730. United States. Navy. Dept.

Breakwater at the capes of the Delaware. Letter from the Secretary of Navy, in reply to a resolution of the House of Representatives of the 6th ultimo, requiring information in relation to the effect of a breakwater at the capes of the Delaware, as a shelter from storms ... March 14, 1826.... Washington: Printed by Gales & Seaton, 1826. 4 p. (19th Cong., 1st sess. House. Doc. 128)

S-26885.

Copies: DLC; DeU; MWA; NjP.

3731. United States. Navy Dept.

Report of the Secretary of the Navy, with the report of the officer appointed to examine the harbors of Charleston and St. Mary's on the expediency of establishing a navy yard at either of these places, &c. January 23, 1826.... Washington: Printed by Gales & Seaton, 1826. 44 p. (19th Cong., 1st sess. Senate. Doc. 27)

S-27335.

Copies: DLC; DeU; MWA; NN; NjP.

3732. United States. Navy Dept.

Survey of the harbor of Baltimore. Letter from the Secretary of the Navy, transmitting a report on the survey of the harbor of Baltimore. December 18, 1826.... Washington: Printed by Gales & Seaton, 1826. 7 p. (19th Cong., 2d sess. House. Doc. 13)

Transmitting a report from Master Commandant Alexander Claxton.

S-27416.

Copies: DLC; DeU; MWA; NjP.

3733. United States. War Dept.

Harbors—Marblehead & Holmes' Hole. Letter from the Secretary of War, transmitting the information required by a resolution of the House of Representatives, of 30th Jan. last, respecting the improvement of the harbors of Marblehead & Holmes' Hole. May 6, 1826.... Washington: Printed by Gales & Seaton, 1826. 10 p. (19th Cong., 1st sess. House. Doc. 172)

S-27012.

Copies: DLC; DeU; MWA; PU.

3734. United States. War Dept.

Survey of Marblehead and Holmes' Hole. Letter from the Secretary of War, in answer to the resolution of the House of Representatives, of 31st ultimo, in relation to a survey of the harbor of Marblehead and Holmes' Hole. February 6, 1826.... Washington: Printed by Gales & Seaton, 1826. 3 p. (19th Cong., 1st sess. House. Doc. 71)

S-27414.

Copies: DLC; DeU; MWA; NjP.

3735. United States. War Dept.

Survey of Nantucket. Letter from the Secretary of War, in reply to a resolution of the House of Representatives, of 31st ultimo in relation to a survey of the island of

Nantucket, &c. February 6, 1826.... Washington: Printed by Gales & Seaton, 1826. 3 p. (19th Cong., 1st sess. House. Doc. 72)

Concerns a survey to construct a harbor on the Nantucket Island.

S-27409.

Copies: DLC; DeU; MWA; NjP.

3736. United States. Engineer Dept.

La Plaisance Bay harbor. Letter from the Secretary of War, transmitting a report of a survey of La Plaisance Bay harbor in the Territory of Michigan, January 26, 1827.... Washington: Printed by Gales & Seaton, 1827. 10 p. (19th Cong., 2d sess. House. Doc. 142)

Transmitting a report of the Chief Engineer.

S-31358.

Copies: DLC; DeU; MWA; NjP.

3737. United States. Engineer Dept.

Survey of Saugatuck Harbor and River. Letter from the Secretary of War transmitting a report and map of the survey of Saugatuck Harbor and River. February 23, 1827.... Washington: Printed by Gales & Seaton, 1827. 10 p., map. (19th Cong., 2d sess. House. Doc. 110)

S-31519.

Copies: DLC; DeU; MWA; NjP.

3738. United States. Laws.

An act to authorize the improving of certain harbors, the building of piers, and for other purposes. February 28, 1827. [Washington: 1827.] 3 p. (H.R. 438)

S-30896.

Copies: DNA.

3739. United States. Laws.

A bill making an appropriation for the erection of a breakwater near the mouth of Delaware Bay. January 3, 1827. [Washington: 1827.] 1 p. (S.37)

S-31084.

Copies: DNA.

3740. ——— December 12, 1827. [Washington: 1827.] 3 p. (S.10)

S-31085.

Copies: DNA.

3741. United States. Navy Dept.

Report of the Secretary of the Navy ... on surveys of the harbors of Charleston and Beaufort, South Carolina, and of Savannah, and St. Mary's, Georgia. December 31, 1827.... Washington: Printed by Duff Green, 1827. 8 p. (20th Cong., 1st sess. Senate. Doc. 12)

N-U219053.

Copies: DeU; ScU.

3742. Baltimore. Ordinances.

Extracts from ordinances to preserve the navigation of the harbor of Baltimore. To appoint port wardens and harbor masters, and prescribe their duties.... [Baltimore: 1828.] Broadside 40 x 32 cm.

Approved March 26th, 1828.

Copies: MWA.

3743. Black Rock Harbor Company.

Memorial to the Canal Commissioners stating grounds on which they declined to repair a breach in the harbor. Albany: 1828. 10 p. (New York (State). Legislature. Assembly. Doc. 28)

"Plans and propositions for the construction of a harbor at Black Rock," p. 9–10.

Whitford-p. 1258.

Copies: N.

3744. Boston. City Council.

City of Boston. In Common Council, Oct. 6, 1828.... The committee of the City Council who were directed to consider the expediency of selling ... the right of the city to extend a wharf over the flats, eastward of Faneuil Hall Market, respectfully report. [Boston: 1828.] 11 p.

S-32416.

Copies: MBAt; MHi.

3745. Jones, William, 1760–1831.

Remarks on the proposed breakwater at

3745 (*continued*)
Cape Henlopen. Communicated, by request of the Secretary of the Treasury, by Wm. Jones.... To which are added, the report of the Board of Engineers, and Captain Bainbridge of the Navy; the memorial of the Chamber of Commerce of Philadelphia, proceedings of a town meeting, shipwrecks, loss and disaster, &c., &c. 3d ed. Philadelphia: Printed by order of the Chamber of Commerce of Philadelphia, 1828. 24 p., 1 l., illus.
S-33734.
Copies: CtY; DLC; MB; MWA.

3746. Pennsylvania. General Assembly.
Breakwater—Delaware Bay. Resolution of the Legislature of Pennsylvania on the construction of a breakwater in the Delaware Bay. February 11, 1828.... Washington: Printed by Gales & Seaton, 1828. 3 p. (20th Cong., 1st sess. House. Doc. 134)
S-36179.
Copies: DLC; DeGE; MWA; NjR.

3747. United States. Congress. House. Committee on Commerce.
Breakwater in Delaware Bay. January 30, 1828. Mr. Sutherland, from the Committee on Commerce ... made the following report. [Washington: 1828.] 8 p. (20th Cong., 1st sess. House. Rept. 118)
S-36180.
Copies: DLC; DeU; MWA; NjR.

3748. United States. Congress. Senate. Committee on Commerce.
In Senate of the United States. January 21, 1828. Mr. Woodbury, from the Committee on Commerce, to whom was referred a resolution of the 31st ult. concerning the expediency of an appropriation to deepen the harbor of St. Marks, in Florida, and to erect a light house near it; and also a resolution of the 15th inst. concerning the establishment of a port of entry at St. Marks, and the removal of obstructions in the Apalachicola and Ocilla rivers in Florida, made a report.... [Washington: 1828.] 7 p. (20th Cong., 1st sess. Senate. Doc. 50)
S-36344.
Copies: DLC; DeU; MWA; NjR.

3749. United States. Engineer Dept.
Harbor of Stonington. Letter from the Secretary of War, transmitting ... a report of a survey of the harbor of Stonington, with a view to the erection of a sea wall, for the protection of said harbor. February 23, 1828.... Washington: Printed by Gales & Seaton, 1828. 9 p. (20th Cong., 1st sess. House. Doc. 153)
S-36326.
Copies: DLC; DeU; MWA; NN; NjR.

3750. United States. Engineer Dept.
Report from the Secretary of War (in compliance with a resolution of the Senate of the 10th instant) with report and plan of the survey of the harbor of Marblehead, Massachusetts. January 15, 1828.... Washington: Printed by Duff Green, 1828. 8 p. (20th Cong., 1st sess. Senate. Doc. 40)
Transmitting a report of the Chief Engineer.
S-36753.
Copies: DLC; DeU; MWA; NjR.

3751. United States. Engineer Dept.
Report from the Secretary of War, with report and plan of the survey of the island of Nantucket ... January 18, 1828.... Washington: Printed by Duff Green, 1828. 20 p.(20th Cong., 1st sess. Senate. Doc. 44)
Transmitting a report from the Engineer Department concerning improvements to the harbor and breakwater.
S-36756.
Copies: DLC; DeU; MWA; NjR.

3752. United States. Engineer Dept.
Survey, harbor of Black Rock. Letter from the Secretary of War, transmitting a report of the survey of the harbor of Black Rock, in the state of Connecticut. January 10, 1828.... Washington: Printed by Gales & Seaton, 1828. 8 p. (20th Cong., 1st sess. House. Doc. 52)
Transmitting a report of the Engineer Department.

S-36913.
Copies: DLC; DeU; MWA; NjR.

3753. United States. Engineer Dept.
Survey of the island of Nantucket. Letter from the Secretary of War, transmitting a report and plan of the survey of the island of Nantucket. January 22, 1828.... Washington: Printed by Gales & Seaton, 1828. 18 p.(20th Cong., 1st sess. House. Doc. 77)
Transmitting a report of the Engineer Department, indicating needs to improve the harbor and breakwater.
S-36917.
Copies: DLC; DeU; MWA; NjR.

3754. United States. Engineer Dept.
Surveys of harbors—Rhode Island. Letter from the Secretary of War, transmitting reports of surveys of Church's Cove Harbor; of the shores north end of Goat Island, and the river and harbor of Warren, all in the state of Rhode Island. February 23, 1828.... Washington: Printed by Gales & Seaton, 1828. 10 p. (20th Cong., 1st sess. House. Doc. 154)
Transmitting a report of the Engineer Department.
S-36924.
Copies: DLC; DeU; MWA; NjR.

3755. United States. Laws.
An act to authorize the improvement of certain harbors, the building of piers, and for other purposes. May 5, 1828.... [Washington: 1828.] 3 p. (H.R. 228)
S-35629.
Copies: DNA.

3756. United States. Laws.
A bill making an appropriation for the erection of a breakwater, near the island of Nantucket. May 21, 1828.... [Washington: 1828.] 1 p. (S. 150)
S-35989.
Copies: DNA.

3757. United States. Laws.
A bill making an appropriation for the erection of a breakwater near the mouth of Delaware Bay. January 30, 1828.... [Washington: 1828.] 1 p. (H.R. 131)
S-35990.
Copies: DNA.

3758. United States. Laws.
A bill to authorize the improving of certain harbors, the building of piers, and for other purposes. March 18, 1828.... [Washington: 1828.] 3 p. (H.R. 228)
S-36080.
Copies: DNA.

3759. United States. War Dept.
Message from the President of the United States, with a report from the Secretary of War ... relative to the practicability and probable cost of constructing an artificial harbor or breakwater at the mouth of the Mississippi. December 8, 1828.... Washington: Printed by Duff Green, 1828. 4 p. (20th Cong., 2d sess. Senate. Doc. 5)
S-36620.
Copies: DLC; DeU; MWA; NjR.

3760. Massachusetts. Laws.
An act, to authorize the proprietors of Union Wharf to extend the same. [Boston?: 1829.] 3 p.
S-39488.
Copies: MWA; OClWHi.

3761. United States. Engineer Dept.
Harbor of St. Mark's, Florida. Letter from the Secretary of War, transmitting a survey and estimate for the improvement of the harbor and river of St. Mark's, in Florida. January 17, 1829.... [Washington: 1829.] [4] p.(20th Cong., 2d sess. House. Doc. 76)
Transmitting a report of the Engineer Department.
Also issued as Senate doc. 45.
S-41131.
Copies: DLC; DeU; NhR.

3762. United States. Engineer Dept.
Report from the Secretary of War, transmitting a report of the survey and es-

3762 (*continued*)
timate for improving the harbor and river of St. Marks, in Florida. January 12, 1829.... [Washington: 1829.] 4 p. (20th Cong., 2d sess. Senate. Doc. 45)

Transmitting a report of the Engineer Department.

Also issued as House doc. 76.

Copies: DeU; MWA.

3763. United States. Engineer Dept.

Report of the Secretary of War, in compliance to a resolution of the Senate, relative to the erection of the pier at Kennebunk. December 28, 1829.... [Washington: 1829.] 13 p. (21st Cong., 1st sess. Senate. Doc. 7)

Transmitting an account on the construction of the pier from the Engineer Dept.

Copies: DeU; MWA.

3764. United States. Engineer Dept.

Survey of the harbor of Nantucket. January 30, 1829.... [Washington: Gales & Seaton, printers, 1829.] 6 p. (20th Cong., 2d sess. House. Doc. 97)

S-41322.

Copies: DLC; DeU; MWA; NjR.

3765. New York (State). Laws.

An act to incorporate the Long Island Sound Harbour Company, passed April 16, 1830. New York: Printed at the office of the New-York Amulet, 1830. 12 p.

C-2270.

Copies: NN.

3766. United States. Engineer Dept.

Harbor of Westbrook—Connecticut. Letter from the Secretary of War, transmitting a survey and estimate for the improvement of the harbor of Westbrook, in the state of Connecticut. February 16, 1830.... [Washington: 1830.] 4 p. (21st Cong., 1st sess. House. Doc. 124)

Transmitting a report of the Engineer Department.

C-4739.

Copies: DLC; MWA; NjR.

3767. United States. Engineer Dept.

Letter from the Secretary or War, transmitting a survey and estimate for the improvement of the harbor of Chicago, on Lake Michigan. March 2, 1830.... [Washington: 1830.] 4 p. (21st Cong., 1st sess. House. Doc. 69)

Transmitting a report of the Engineer Department.

C-4910.

Copies: DLC; DeU; MWA; NjR.

3768. United States. Engineer Dept.

Survey—Flat Beach. Letter from the Secretary of War, transmitting a copy of a report of the engineer appointed to survey Flat Beach, on Tucker's Island, at Egg Harbor, in New Jersey, with an estimate of the expense of securing the harbor. March 16, 1830.... [Washington: 1830.] 3 p. (21st Cong., 1st sess. House. Doc. 81)

C-5188.

Copies: DLC; DeU; MWA; NjR.

3769. United States. Engineer Dept.

Survey—Narraganset Bay, &c. &c. Letter from the Secretary of War, transmitting surveys and reports of the waters of Narraganset Bay and the harbor of Newport, Rhode Island, with a view to the establishment of a naval depot. January 14, 1830.... [Washington: 1830.] 11 p. (21st Cong., 1st sess. House. Doc. 21)

C-5189.

Copies: DLC; DeGE; DeU; MWA; NjR.

3770. United States. Engineer Dept.

Survey of Sandy Bay, Massachusetts. Message from the President of the United States transmitting a report of a survey of Sandy Bay, in the state of Massachusetts. April 24, 1830.... [Washington: 1830.] 5 p. (21st Cong., 1st sess. House. Doc. 95)

Includes estimates for a harbor and a breakwater.

C-5191.

Copies: DLC; DeU; MWA; NjR; Vi.

3771. United States. Engineer Dept.

Survey of the harbor of St. Augustine.

Message from the President of the United States transmitting a copy of the report of a survey of the harbor of St. Augustine. April 22, 1830.... [Washington: 1830.] 11 p. (21st Cong., 1st sess. House. Doc. 93)
C-5192.
Copies: DLC; MWA; NjR.

3772. United States. Laws.
An act making appropriations for the improvement of certain harbors, and for removing obstructions at the mouths of certain rivers, for the year one thousand eight hundred and thirty. March 31, 1830.... [Washington: 1830.] 3 p. (H.R. 242)
C-3829.
Copies: DNA.

3773. United States. Laws.
A bill making appropriations for the improvement of certain harbors, and for removing obstructions at the mouths of certain rivers, for the year one thousand eight hundred and thirty. February 12, 1830.... [Washington: 1830.] 3 p. (H.R. 242)
C-4366.
Copies: DNA.

5. Lighthouses, buoys, and other navigational markers

3774. Pennsylvania (Colony). Laws.
Act for erecting a light-house at the mouth of the Bay of Delaware. Philadelphia: B. Franklin, 1763.
N-P0204361.
Copies: N.

3775. Say-Brook Barr [*sic*] lottery, to raise the sum of five hundred and thirty seven pounds ... fixing buoys and other marks ... mouth of Connecticut River.... [Hartford: 1773.] Broadside 32 x 18.5 cm.
Dated: Hartford, June 5, 1773.
B-3640; Mp-42500.
Copies: CtHi; MWA (photocopy).

3776. United States. Laws.
An act for the establishment and support of light-houses, beacons, and buoys. 1789, July 20th. Read the third time, and passed to House of Representatives.... New York: Printed by Thomas Greenleaf, [1789]. Broadside 33.5 x 20 cm.
B-7111; Mp-45624.
Copies: DLC; NhD.

3777. United States. Laws.
An act for the establishment and support of light-houses, beacons, buoys and piers.... New York: Printed by Thomas Greenleaf, [1789]. Broadside 33.5 x 20 cm.
Ordered to be printed by the Senate, July 24, 1789.
B-7112.
Copies: DLC; NhD.

3778. United States. Laws.
An act for the establishment and support of lighthouses, beacons, buoys, and public piers ... Approved August 7th, 1789.... [New York: Childs and Swaine, 1789.] Broadside 35.5 cm.
At head of title: Congress of the United States.
B-7176; Mp-45698.
Copies: NN.

3779. United States. Laws.
An act authorizing the Secretary of the Treasury to finish the light-house, on Portland-Head, in the district of Maine. Approved August the tenth, 1790. [New

3779 (*continued*)
York: Printed by Francis Childs and John Swaine, 1790.] Broadside 29 x 24.5 cm.
Mp-22955.
Copies: NN; RHi.

3780. United States. Laws.
An act supplementary to the Act for the establishment and support of light-houses.... [New York]: John Fenno, [1790]. Broadside 33.5 x 21 cm.
B-7537.
Copies: DLC.

3781. United States. Laws.
An act to amend the Act for the establishment and support of light-houses, beacons, buoys, and public piers. Approved July the twenty-second, 1790. [New York: Printed by Francis Childs and John Swaine, 1790.] Broadside 29 x 23 cm.
Mp-22967.
Copies: DLC; MWA; NN; RHi.

3782. Specifications for the building of a light house at Cape-Henry. [S.l.]: 1791. Broadside 46 cm.
Shaw-653.
Copies: NHi (not located).

3783. United States. Laws.
An act for finishing the light-house on Baldhead at the mouth of Cape Fear River in the state of North-Carolina. [Philadelphia: Childs and Swaine, 1792.] Broadside 31 x 19.5 cm.
Approved April 2, 1792.
B-8209; Mp-46628.
Copies: MWA; NN.

3784. United States. Laws.
An act to erect a light-house on Montaugh Point.... [Philadelphia]: John Fenno, [1792]. Broadside 33.5 x 21 cm.
B-8186; Mp-46638.
Copies: DLC; NhD.

3785. United States. Laws.
An act supplementary to an act for the establishment and support of light-houses, beacons, buoys and public piers.... Approved March 2, 1793. [Philadelphia: Printed by Childs and Swaine, 1793.] [2] p.
Mp-26326.
Copies: DLC; MWA; NN.

3786. United States. Laws.
An act for erecting a lighthouse on the island of Seguin in the district of Maine, and for erecting a beacon and placing three buoys at the entrance of Saint Mary's River, in the state of Georgia.... Approved May the nineteenth 1794.... [Philadelphia: Printed by Childs and Swaine, 1794.] Broadside.
Mp-27855.
Copies: MWA; RHi.

3787. United States. Laws.
An act to erect a lighthouse on the head-land of Cape Hatteras; and a lighted beacon on Shell Castle Island in the harbor of Occacock in the state of North-Carolina.... [Philadelphia: Printed by Childs and Swaine, 1794.] Broadside 31 x 19 cm.
"Approved—May the thirteenth 1794."
Mp-27853.
Copies: MWA; NN; RHi; RPJCB.

3788. United States. Laws.
An act to provide for the placing buoys on certain rocks off the harbor of New-London, and in Providence River, and other places. [Philadelphia: Printed by Childs and Swaine, 1794.] Broadside.
Mp-27846.
Copies: MWA; RHi.

3789. United States. Laws.
An act authorizing the erection of a lighthouse near the entrance of George-Town harbour, in the state of South-Carolina.... Approved, February the twenty first, 1795.... [Philadelphia: Printed by Francis Childs, 1795.] Broadside.
E-29696.
Copies: DLC; MWA.

3790. United States. Laws.
An act authorizing the erection of a light-house on Baker's Island.... [Philadelphia: 1796?] Broadside 34 x 20 cm.
B-9753; Mp-47956.
Copies: MWA; MWiW; MiU-C; PPRF; RPJCB.

3791. United States. Laws.
An act authorizing the erection of a light-house on Cape Cod.... [Philadelphia?: 1796.] 2 p.
B-9754; Mp-47957.
Copies: MWA; MWiW; MiU-C; PPRF; RPJCB.

3792. Boston Marine Society.
Directions for sailing by Cape-Cod light-house. Boston: Boston Marine Society, 1797. Broadside 20.5 x 33.5 cm.
N-B0676358.
Copies: InU.

3793. United States. Laws.
A bill for the erection of a light-house, and placing buoys.... [Philadelphia?: 1798.] Broadside 34 x 21 cm.
Dated: 13th February, 1798.
The lighthouse to be erected on "Eaton's Neck, on Nassau Island," and the buoys "near Sandy Hook."
B-10598; Mp-48679.
Copies: NHi.

3794. United States. Congress. House. Committee of Commerce and Manufactures.
Report from the committee of Commerce and Manufactures, to whom was referred ... the bill ... "An act for the rebuilding of light-house on Gurnet Point ... " 30th March, 1802.... [Washington: 1802.] 5 p.
S&S-3405.
Copies: DLC; NN.

3795. United States. Laws.
Amendments of the Senate to the bill, intituled, "An act for the rebuilding the light-house on Gurnet Point, at the entrance of Plymouth Harbour—for rebuilding the light-house at the eastern end of Newcastle Island—for erecting a light-house on Lynde's Point—and for other purposes." [Washington]: Printed by Duane, [1802]. 3 p.
S&S-3208.
Copies: DLC.

3796. United States. Laws.
A bill for the rebuilding the light-house on Gurnet Point, at the entrance of Plymouth harbour—for rebuilding the light-house at the eastern end of New Castle Island—for erecting a light house on Lynde's Point; and for other purposes. [Washington]: Duane, printer, [1802]. 3 p.
S&S-3226.
Copies: DLC; MWA.

3797. United States. Laws.
The bill sent from the Senate, intituled "An act to erect a light house on the south end of St. Simon's Island, in the state of Georgia, and for the placing a buoy or buoys on or near St. Simon's Bar," was read the third time. [Washington: 1804.] 3 p.
S&S-7460.
Copies: DLC.

3798. United States. Laws.
A bill to erect a light house at the mouth of the Mississippi River, and also a light-house at or near the pitch of Cape Lookout in the state of North Carolina. [Washington: 1804.] 2 p.
S&S-7476.
Copies: DLC.

3799. United States. Laws.
A bill to erect a light house on the south end of St. Simon's Island in the state of Georgia, and for the placing a buoy or buoys on St. Simon's Bar. [Washington: 1804.] 2 p.
S&S-7477.
Copies: DLC.

3800. United States. Laws.
A bill to provide for light houses and in the cases therein mentioned, and to establish

3800 (*continued*)
the district of St. Mary's for that of Nanjemoy. [Washington: 1804.] 2 p.
S&S-7484.
Copies: DLC.

3801. United States. Laws.
An act authorizing the erection of certain light-houses, and the fixing of stakes, buoys, and beacons, in certain places therein named. In Senate of the United States, January 21, 1807. [Washington: 1807.] 3 p.
S&S-13768.
Copies: DLC; MWA.

3802. United States. Laws.
A bill making appropriations for rebuilding two light houses on Plumb Island. June 12, 1809.... [Washington: 1809.] [1] p. (H.R. 22)
S&S-18835.
Copies: NN.

3803. United States. Laws.
A bill to erect a light house at the entrance of Scituate harbor, a stone column on a spit of sand at the entrance into Boston harbor, and a beacon on Beach Point near Plymouth harbor, in the state of Massachusetts; and a light at the entrance of Bayou St. John into Lake Ponchartrain. March 21st, 1810. [Washington: 1810.] 3 p. (H.R. 79)
S&S-21630.
Copies: DLC.

3804. United States. Laws.
A bill to authorise the Secretary of the Treasury, under the direction of the President of the United States, to purchase of Winslow Lewis, his patent right to the new and improved method of lighting light houses, and for other purposes. January 3d, 1812. [Washington: 1812.] 2 p.(H.R. 38)
S&S-27090.
Copies: DLC.

3805. United States. Treasury Dept.
Documents accompanying a bill to authorise the Secretary of the Treasury, under the direction of the President of the United States, to purchase of Winslow Lewis his patent-right to the new and improved method of lighting light-houses, and for other purposes. Presented third January, 1812. Washington City: Printed by R. C. Weightman, 1812. 32 p., fold. table.
S&S-27136.
Copies: DLC; DeGE; MWA; NjR.

3806. [Lewis, Winslow] 1770–1850.
Description of the light houses, on the coast of the United States. Boston: Printed by Thomas G. Bangs, [1817]. 16 p., title vignette.
S&S-41255.
Copies: DeGE; MH; MHi; MMAl; MSaE; MWA; NN.

3807. United States. Laws.
A bill for erecting a light-house on the west chop of Holmes' Hole harbour, in the state of Massachusetts. February 11, 1817.... [Washington: 1817.] 1 p. (H.R. 127)
S&S-42379.
Copies: DLC.

3808. United States. Laws.
A bill for erecting a lighthouse upon Galloo Island, near the outlet of Lake Ontario; for placing sundry spindles or buoys, and for other purposes. April 9, 1818.... [Washington: 1818.] 1 p. (H.R. 186)
S&S-45971.
Copies: DLC.

3809. Melville, David, 1773–1856.
An exposé of facts, respectfully submitted to the government and citizens of the United States, relating to the conduct of Winslow Lewis, of Boston, "Superintendant for Lighting the United States' Light Houses," an accredited agent of the Treasury Department. Addressed to the Hon. the Secretary of the Treasury.... Providence: Miller & Hutchens, printers, 1819. 48 p.
N-M0426727; S&S-51915.
Copies: DLC; DeGE; PPL.

3810. Description of the light houses. [S.l.: 182- .] [3]–24 p.

Concerns the light houses on Great Lakes and the eastern coast of the United States.

ND0195631.

Copies: NNC.

3811. United States. Fifth Auditor's Office.

Letter from the Secretary of the Treasury, transmitting a report of the Fifth Auditor of the Treasury of the progress made in the erection of the lighthouse at the mouth of the Mississippi, March 14, 1820. [Washington: 1820.] 4 p. (16th Cong., 1st sess. House. Doc. 97)

S-3793.

Copies: DLC; DeU; MWA; O; R.

3812. United States. Laws.

An act supplementary to an act, entitled "An act concerning navigation." Read and committed to the Committee of the Whole House to which is committed the bill from the Senate entitled An act to authorize the erection of a light house on one of the Isles of Shoals near Portsmouth, in New Hampshire. [Washington: 1820.] 1 p. (S. 98)

Dated: April 14, 1820.

S-3754.

Copies: DLC.

3813. United States. Register of the Treasury.

Letter from the Secretary of the Treasury, transmitting (in obedience to a resolution of the Senate of 4th May, 1820) a statement of the money which has been annually appropriated, and paid, since the year seventeen hundred and seventy-five, for surveying the seacoast ... erecting and keeping in repair, lighthouses, beacons, and buoys, and for the purchase of ground for light-houses. November 17, 1820.... Washington: Printed by Gales & Seaton, 1820. 7 p., 4 fold. tables. (16th Cong., 2d sess. Senate. Doc. 6)

Signed: Joseph Nourse, Register of the Treasury.

S-3813.

Copies: DLC; DeGE; DeU; G; MWA; NjP.

3814. Bellamy, Elisha.

For the benefit of coasters and seafaring gentlemen and all persons concerned in commerce, the following hand-bill is issued gratis by E. Bellamy, printer of the Boston Weekly report. New light houses.... Important and useful invention.... [of a] marine life preserver.... [Boston: 1821?] Broadside, illus. 54 x 23 cm.

S-4658.

Copies: MBAt; MWA.

3815. United States. Laws.

A bill to authorize the building of light houses on Cross and Pond Islands, in the harbor of Boothbay, and at the mouth of Oswego River, and placing buoys on the shoals of Nantucket, and Vineyard Sound, near the harbor of Wickford, and on the Altamaha River, and for other purposes. Feb. 16, 1821.... [Washington: 1821.] 1 p. (H.R 257)

S-7163.

Copies: DLC.

3816. United States. President (Monroe).

Message from the President of the United States, accompanied with a report upon the subject of the light-houses ordered to be built by the act of third March, 1821. December 17, 1821.... Washington: Printed by Gales & Seaton, 1821. 9 p. (17th Cong., 1st sess. House. Doc. 5)

Report of the Commissioners appointed to consider the erection of a sea wall between Smutty Nose Island and Cedar Island of the Isles of Shoals, p. 7–9.

S-7377.

Copies: DLC; DeGE; DeU; MWA; NjP.

3817. United States. Congress. Senate. Committee on Commerce and Manufactures.

In Senate of the United States, March 18, 1822. Mr. Dickerson, from the Committee on commerce and Manufactures, to whom was referred the bill to authorize the building a light house, at Stonington Point, in the state of Connecticut, reported the same, with following amendments.... [Washington: 1822.] 2 p. (S. 45)

3817 (*continued*)
S-10875.
Copies: DNA.

3818. United States. Laws.
A bill to authorize the building of a light-house at Stonington Point, in the state of Connecticut. February 18, 1822.... [Washington: 1822.] 2 p. (S. 45)
S-10730.
Copies: DNA.

3819. United States. Laws.
A bill to authorize the building a light-house at Cape Romain, in South Carolina, and placing floating lights in Delaware Bay. January 28, 1823.... [Washington: 1823.] 1 p. (S. 43)
S-14461.
Copies: DNA.

3820. United States. Laws.
A bill to authorize the building of lighthouses, light vessels, and beacons, therein mentioned, and for other purposes. February 24, 1823.... [Washington: 1823.] 1 p. (S. 67)
S-14462.
Copies: DNA.

3821. United States. Dept. of State.
Letter from the Secretary of State, transmitting (in obedience to a resolution of the House of Representatives of the 23d of December last) information upon the subject of the establishment of a light near the Bahama Banks, for the security of navigation of the Gulf of Florida. December 27, 1824.... Washington: Printed by Gales & Seaton, 1824. 11 p. (18th Cong., 1st sess. House. Doc. 21)
S-18784.
Copies: DLC; DeU; MWA; NjP.

3822. United States. Laws.
An act to authorize the building of lighthouses, light-vessels, and beacons, therein mentioned, and for other purposes. May 18, 1824.... [Washington: 1824.] 3 p. (H.R. 202)
S-18349.
Copies: DNA.

3823. United States. Fifth Auditor's Office.
Light-houses. Letter from the Secretary of the Treasury, accompanied by a report of the Fifth Auditor, upon the subject of light-houses. December 28, 1825.... Washington: Printed by Gales & Seaton, 1825. 15 p. (19th Cong., 1st sess. House. Doc. 19)
S-22965.
Copies: DLC; DeU; MWA; NjP.

3824. United States. Laws.
An act to authorize the building of light houses and light vessels, and beacons and monuments, therein mentioned; and for other purposes. February 28, 1825.... [Washington: 1825.] 3 p. (H.R. 336)
S-22640.
Copies: DNA.

3825. United States. Treasury Dept.
Report from the Secretary of the Treasury (in compliance with a resolution of the Senate of the 21st ultimo) relative to light houses and buoys, on the coast of Florida and Gulf of Mexico. January 17, 1825.... Washington: Printed by Gales & Seaton, 1825. 4 p. (18th Cong., 2d sess. Senate. Doc. 17)
S-23024.
Copies: DLC; DeU; MWA; NN.

3826. United States. Laws.
An act for authorizing the building of light houses and light vessels, erecting beacon lights, placing buoys, removing obstructions in the River Savannah, and for other purposes. May 8, 1826. [Washington: 1826.] 7 p. (H.R. 185)
S-26277.
Copies: DNA.

3827. United States. Treasury Dept.
Light house on Dutch Island, and beacon on Warwick Neck. Letter from the Secretary of the Treasury, in relation to the erection of a light house on Dutch Island, in Narraganset Bay, and a beacon light on Warwick Neck, in the state of Rhode Island.

Rendered in obedience to a resolution of the House of Representatives of the 30th December 1825. January 9, 1826.... Washington; Gales & Seaton, 1826. 4 p. (19th Cong., 1st sess. House. Doc. 30)

S-27183.

Copies: DLC; DeU; MWA; NjP.

3828. United States. War Dept.

Throg's Point, Long Island. Letter from the Secretary of War, transmitting the information ... in relation to the purchase of Throg's Point on Long Island sound. January 26, 1826 ... Washington: Gales & Seaton, 1826. 12 p. (19th Cong., 1st sess. House. Doc. 60)

Concerns the erection of a lighthouse.

S-27424.

Copies: DLC; DeU; MWA; NjP.

3829. United States. Laws.

An act to authorize the building of lighthouses and beacons, and for other purposes. February 28, 1827. [Washington: 1827.] 5 p. (H.R. 403)

S-30895.

Copies: DNA.

3830. ———— May 5, 1828. [Washington: 1828.] 5 p. (H.R. 177)

S-35628.

Copies: DNA.

3831. United States. Treasury Dept.

Light vessel–Brandywine Shoal. Letter from the Secretary of Treasury, communicating information in relation to the removal of the light vessel from Brandywine Shoal to Tuckanuck Shoal. January 2, 1828.... Washington: Printed by Gales & Seaton, 1828. 7 p. (20th Cong., 1st sess. House. Doc. 32)

S-36512.

Copies: DLC; DeGE; DeU; MWA; NjR.

3832. United States. Laws.

An act making appropriations for building light-houses and beacons, and placing buoys, and for improving harbors, and directing surveys. Feb. 24, 1829. Read twice, and referred to the Committee on Commerce. [Washington: 1829.] 9 p. (H.R. 395)

S-40764.

Copies: DNA.

3833. United States. Congress. House. Committee on Commerce.

Mr. Woodbury, from the Committee on Commerce to which was referred the bill from the House of Representatives, entitled "An act making appropriations for building lighthouses, light-boats, beacons, and monuments, placing buoys, and for improving harbors, and directing surveys," reported the same with the following amendments. April 20, 1830. [Washington: 1830.] 5 p. (H.R. 304)

C-5007.

Copies: DNA.

3834. United States. Laws.

An act making appropriations for building light-houses, light-boats, beacons, and monuments, placing buoys, and for improving harbors and directing surveys. April 7, 1830.... [Washington: 1830.] 12 p. (H.R. 304)

C-3824.

Copies: DNA.

3835. ———— May 14, 1830. Printed as proposed to be amended by the Senate.... [Washington: 1830.] 16 p. (H.R. 304)

C-3825.

Copies: DNA.

3836. ———— December 14, 1830.... [Washington: 1830.] 15 p. (H.R. 304)

C-3826.

Copies: DNA.

3837. United States. Treasury Dept.

Light-house—Point Lookout. Letter from the Secretary of the Treasury, transmitting the information required, in relation to the erection of a beacon-light or small light-house on Point Lookout. February 15, 1830.... [Washington: 1830.] 6 p. (21st Cong., 1st sess. House. Doc. 64)

C-4917.

Copies: DLC; DeU; MWA; NjR.

3838. United States. Treasury Dept.

Report of the Secretary of the Treasury, on the subject of procuring a site for, and building a light-house near the mouth of Duck Creek, adjoining Delaware Bay. January 14, 1830.... [Washington: 1830.] 4 p. (21st Cong., 1st sess. Senate. Doc. 24)

Copies: DeU; MWA.

6. Seamanship

3839. Boston Marine Society.

Laws of the Marine Society, at Boston in New-England, incorporated ... February 2, 1754. [Boston: 1774]. [2] p.

The society had both, professional and charitable functions. List of members and the time of their entrance on verso.

B-3773; Mp-42635.

Copies: MHi; MWA (xerox copy).

3840. Haselden, Thomas.

The seaman's daily assistant, being a short, easy and plain method of keeping a journal at sea; in which are contained, rules, shewing how the allowances for lee-way, variation, heave of the sea, set of currents, &c. are to be made, and to correct the dead-reckoning by an observation, in all cases; and also all the tables that are in any ways necessary ... London, printed. Philadelphia: Reprinted and sold by J. Crukshank, 1777. 4 p.l., 160 p. including tables.

Mp-15360.

Copies: CSmH; CtY; DLC; MWA; RPJCB.

3841. ———— Philadelphia: Printed by J. Crukshank, 1779.

E-16301.

No copy known.

3842. [Burges, Bartholomew].

The seaman's journal. Being an easy and correct method of keeping the daily reckoning of a ship during the course of her voyage. New-York: Printed and sold by Hugh Gaine, 1786.

E-19534.

No copy known.

3843. Franklin, Benjamin, 1706–1790.

Maritime observations: in a letter from Doctor Franklin, to Mr. Alphonsus Le Roy, member of several academies, at Paris. Philadelphia: Printed by Robert Aitken, 1786. [1], 294-329 p., fold. plate, fold. map.

Offprint from *Transactions of the American Philosophical Society,* vol. 2, 1786, with a special title page.

B-6285; Mp-44888.

Copies: CSmH; CtY; DLC; DeGE; InU; MB; MH; MWA; MiU-C; PPAmP.

3844. Burges, Bartholomew.

The seaman's journal, being an easy method of keeping the daily reckoning of a ship during the course of her voyage. Newport: Printed and sold by Peter Edes, 1788.

E-20988.

No copy known.

3845. Buckner, Samuel.

The American sailor: a treatise on practical seamanship, with hints and remarks relating thereto. Designed to contribute towards making navigation in general more perfect, and of consequence, less destructive to health, lives and property. Newport:

Printed by Peter Edes, [1790]. [8], 96 p.

A condensed version of William Hutchinson's *A Treatise of practical seamanship*, Liverpool, 1777.

Mp-22377.

Copies: MB; NN; RPJCB.

3846. Boston Marine Society.

Constitution and laws of the Boston Marine Society, instituted in 1742. Boston: 1792. 37 p.

The Society had professional and charitable functions.

Mp-24133.

Copies: MWA.

3847. Malham, John, 1747–1821.

The naval gazetteer; or, Seaman's complete guide. Containing a full and accurate account ... of the several coasts of all the countries and islands in the known world.... Also comprehending ample directions for sailing into or out of the different ports, straits, and harbours.... Illustrated with a correct set of charts from the latest and best surveys. 1st American ed.... Boston: W. Spotswood and J. Nancrede, 1797. 2 v., 17 fold. maps.

First published in London, 1795.

Mp-32415.

Copies: CSt; CtY; DLC; KyLx; LNHT; MH; MWA; MiU-C MdAN; NIC; NN; NcD; PPL; PU; RPJCB; ViU.

3848. Plan for establishing a general marine society throughout the United States, and systems of regulations therein; written at sea, in the year 1794; now printed for the benefit of that institution, by the author. Philadelphia: Printed by John Fenno, 1798. vi, 8–32 p.

Signed: S.K.

Mp-33944.

Copies: CtY; DLC; MBAt; MH; MWA; NHi; NN; PPL; RPJCB; ViU.

3849. [Clerk, John] 1728–1812.

A system of seamanship and naval tactics. Extracted from the Encyclopaedia, published by Thomas Dobson. Illustrated with copperplates. Philadelphia: T. Dobson, 1799. 192 p., 8 fold. plates.

Mp-36393.

Copies: DeGE; MB; MWA.

3850. Boston Marine Society.

The constitution and laws of the Boston Marine Society.... Boston: Printed by Munroe & Francis, 1802. 42 p.

S&S-1933.

Copies: MWA.

3851. [Moore, J. J.]

The mariner's dictionary; or, American seaman's vocabulary of technical terms, and sea phrases, used in the construction, equipment, management, and military operations of ships and vessels of all descriptions. Improved from an English work. Illustrated with copperplates, descriptive of all objects appertaining to seamanship. Washington City: W. Duane, 1805. iv p., 1 l., 256 p., 8 plates.

English edition, 1801, was issued under the title: *British mariner's vocabulary.*

Erroneously attributed also to John Hamilton Moore.

S&S-8832.

Copies: DLC; DeGE; MH; MWA; MiU-C; NN; NcD; NjP; PPL; PU.

3852. The American ship-master's daily assistant; or, Compendium of marine law, and mercantile regulations and customs; being a correct and useful guide to all men in business.... Containing also a great variety of useful tables and commercial forms.... Portland, [Me.]: D. Johnson, 1807. xiv, 608 p., plates.

A Collection of all the requisite tables for a complete navigator, with an explanation prefixed, p. [349]–608.

S&S-11980.

Copies: DLC; DN; DeGE; MB; MH-BA; MWA; MeHi; NN; OCLaw; ViU.

3853. Clerk, John, 1728–1812.

A system of seamanship and naval

3853 (*continued*)
tactics.... Illustrated with copper-plates. Philadelphia: Printed by Bartram & Reynolds, 1807. 192 p., 7 fold. plates.
S&S-12309.
Copies: DLC; MSaE; MWA; NN; NcD; NjP; PSC.

3854. Salem East-India Marine Society, Salem, Mass.
By-laws and regulations of the East India Marine Society.... Salem, [Mass.]: Printed by Pool and Palfray, 1808. 20 p.
Includes the list of members.
S&S-14906.
Copies: MSaE; MWA.

3855. Boston Marine Society.
Constitution and laws of the Boston Marine Society, instituted in ... 1742; incorporated in ... 1754.... Boston: Printed by J. Belcher, 1809. 47 p.
Includes members' names and directions for sailing by Cape Cod lighthouse.
S&S-17068.
Copies: DLC; MWA.

3856. [Blunt, Edmund March] 1770–1862.
Seamanship, both in theory and practice. To which is annexed, an Act to establish a board of wardens for the port of New-York, and for the regulation of the pilots and pilotage of the said port, passed April 9, 1811. To which is added, the regulations for pilotage, adopted by the said board.... New York: Printed and sold by Edmund M. Blunt, 1811. vii, [2], 10–146, [4] p., illus., plates.
N-B0566632.
Copies: CtY; DLC; MWA.

3857. [Blunt, Edmund March] 1770–1862, ed.
Seamanship, both in theory and practice. To which is annexed an essay on naval tactics & signals. Also, regulations for the government of the navy of the United States of America, with a system of naval gunnery, and an act concerning letters of marque and reprisals. To which is subjoined, a list of post captains in the navy of the United States.... New York: Printed and sold by Edmund M. Blunt, 1812. 205, [1] p., illus., fold. plates.
"Advertisement by the editor" signed: Edmund M. Blunt.
S&S-24893.
Copies: CtNlCG; DeGE; MSaE; MWA; NhD; PPL; ScC.

3858. Malham, John, 1747–1821.
The naval gazetteer; or, Seaman's complete guide. Containing a full and accurate account ... of the several coasts of all the countries and islands in the known world.... Illustrated with a correct set of charts from the best and latest surveys.... Baltimore: Edward J. Coale, 1812. 2 v., maps.
First published in London, 1795.
S&S-25932.
Copies: CSmH; DGU; DN; MHi; MSaE; NN.

3859. [Blunt, Edmund March] 1770–1862.
Seamanship, both in theory and practice. To which is annexed an essay on naval tactics and signals. Also, regulations for the government of the navy of the United States of America, with observations and instructions for the use of ... naval officers.... New York: E. M. Blunt, 1813. 285 p., illus., plates.
S&S-27969.
Copies: MWA; MnM; NN.

3860. Cape Fear Navigation Company.
Bye laws, rules and regulations of the Cape Fear Navigation Company, with the act of incorporation of 1815.... Raleigh: Printed by the Minerva press, 1817. 21 p.
N-C0115645.
Copies: NcU; T.

3861. Lever, Darcy, 1760?–1836.
The young sea officer's sheet anchor; or, A key to the leading of rigging, and to practical seamanship.... With an appendix, containing several figures illustrative of novelties and improvements in rigging ... 1st Philadelphia, from the 2d London ed. [Philadelphia]: M. Carey & Sons, [1819]. xii, [1], 124 p., plates.
S&S-48484.

Copies: CtHT; CtY; DeGE; MB; MSaP; MWA; NcU; NjP; PPL; RPJCB.

3862. Boston Marine Society.
Constitution and laws of the Boston Marine Society, instituted in the year 1742; incorporated in the year 1754.... Boston: Printed by J. T. Buckingham, 1820. 48 p.
S-539.
Copies: CtSoP; MHi.

3863. New York Nautical Institution and Shipmasters' Society.
A system of constitutional by-laws, for the present government of the New-York Nautical Institution and Shipmasters' Society. New York: Abraham Paul, 1820. 12 p.
Includes a list of members.
S-2533.
Copies: CSmH.

3864. New York Nautical Institution and Shipmasters' Society.
Constitution and by-laws, of the New-York Nautical Institution and Shipmasters' Society. New York: Printed by R. Tyrell, 1821. 22 p.
N-N0213170.
Copies: NN.

3865. Salem East-India Marine Society, Salem, Mass.
The East-India Marine Society of Salem. [Salem, Mass.: Printed by W. Palfray, jr., 1821.] 100 p.
Contents: Introduction. — Act of incorporation. — By-laws. — List of members. — Officers. — Catalogue of the journals of the voyages. — Catalogue of the articles in the museum.
S-6699; Tapley-p. 462.
Copies: CtHT-W; DLC; MB; MBAt; MH; MHi; MSaE; NNC; NSy; PPAmP; PPL.

3866. Blunt, Edmund March, 1770–1862.
Theory and practice of seamanship; together with a system of naval tactics: with the rules, regulations, and instructions, for the naval service ... 2d ed. New York: Published by Edmund M. Blunt and sold by Wm. Hooker, and Edmund and George W. Blunt; J. Seymour, printer, 1824. 420, [38] p., plates.
S-15477.
Copies: CtY; LNHT; MB; MBAt; MSaP; MdAN; MiD; NN; NNA; NjP; OCHP; PPAmP.

3867. Portland Nautical Society, Portland, Me.
Constitution and by-laws of the Portland Nautical Society. Instituted November 2, 1825. Portland, [Me.]: Shirley & Edwards, printers, 1825. 24 p.
S-21951.
Copies: MeHi.

7. Navigation

3868. The complete mariner; or, A treatise of navigation.... Williamsburg, [Va.]: Printed [by] E. L. James Hubard, [1732]. [56] p., illus.
Dated:February 18th, 1731 [1732].
Title page and eight lines on next page are printed; the remainder is in manuscript.
B-867; Mp-39987.
Copies: NN (photocopy); ViWC.

3869. Fisher, Joshua, 1707–1783.
To the merchants & insurers of the city of Philadelphia this chart of Delaware Bay

3869 (*continued*)
from the sea-coast to Reedy-Island, containing a full and exact description of the shores, creeks, harbours, soundings, shoals, sands, and bearings of the most considerable land-marks; with a tide-table ... is dedicated by a friend to trade and navigation. Joshua Fisher, Feb. 28, 1756. Philadelphia: Printed by John Davis, for and sold by the author, 1756. Map 60 x 96 cm.

Mp-7657.

Copies: MdHi; PHi; RPJCB.

3870. Boston Marine Society.

Directions for sailing in and out of Plymouth harbor; taken by Moses Bennet, William Rhodes, Thomas Allen, and Nathaniel Green, being a committee of the Marine Society of Boston, appointed for this survey, by the desire of a committee of the General Court of the Massachusetts-Bay, appointed to build the light-house on the Gurnet, near Plymouth Harbour, in said province, in July, 1768. [Boston: 1768.] Broadside 32.5 x 22 cm.

Mp-10882.

Copies: MH; MWA.

3871. Directions for the gulph and river of St. Lawrence; with some occasional remarks. Philadelphia: Printed by William and Thomas Bradford, 1774. [2], 36 p.

Mp-13250.

Copies: DLC; ICN; MWA; MiU-C; NN; PPL; PU; RPJCB.

3872. Fisher, Joshua, 1707–1783.

To the merchants & insurers of the city of Philadelphia this chart of Delaware Bay and River, containing a full and exact description of the shores, creeks, harbours, soundings, shoals, sands, and bearings of the most considerable land-marks with a tide table ... by a friend to trade and navigation, Joshua Fisher. [Philadelphia: 1775?] Map 48 x 69 cm.

N-F0164360; Wheat-476.

Copies: MiU-C; PHi.

3873. Directions to sail into and up Delaware Bay.... [Philadelphia: February 9, 1778.] Broadside.

Mp-15784.

Copies: PPL.

3874. Massachusetts. Laws.

An act for regulating pilotage in several ports in this Commonwealth.... In the House of Representatives, July 11, 1783.... [Boston: 1783.] Broadside.

Mp-19424.

Copies: MWA.

3875. Mendenhall, Thomas, 1759–1843.

Traverse tables; or, Tables of difference of latitude and departure, constructed to every quarter of a degree of the quadrant, and to answer for all distances from one-tenth of a perch, or mile, to twelve hundred. Particularly suited to the purposes of land surveying. Philadelphia: Printed and sold by J. Crukshank, 1784. viii, [18] p.

Mp-18609.

Copies: MWA.

3876. Boston Marine Society.

Directions for sailing in and out of Plymouth harbour; taken by Moses Bennet, William Rhodes, Thomas Allen, and Nathaniel Green, being a committee of the Marine Society of Boston, appointed for this survey by the desire of a committee of the General Court of the Massachusetts-Bay.... [Boston: 1785.] Broadside 45 x 35 cm.

First published in 1768. Reissued by the Commonwealth of Massachusetts.

B-6058; Mp-44677.

Copies: MB; MH.

3877. Directions to and from the light house on the north east point of Nantucket. [Boston: 1788.] Broadside 43 x 33 cm.

Mp-21058.

Copies: MB.

3878. Franklin, Benjamin, 1706–1790.

Remarks upon the navigation from Newfoundland to New-York, in order to avoid the Gulf Stream on one hand, and on the other, the shoals that lie ten leagues due south of Nantucket, and south west of the shoalest ground of St. George's Banks, twenty-eight leagues. By Dr. Franklin.... Directions for sailing between the middle ground and the Horse Shoe, at the entrance of Cape-Henry.... Directions for sailing on the coast of North Carolina.... Directions of the tides within the capes of Delaware Bay. [Boston: Printed by Benjamin Edes and Son, 1790.] Broadside (2 p.).

Mp-22509.

Copies: MWA; NHi.

3879. [Norman, John] 1748–1817.

The American pilot, containing the navigation of the sea coast of North America, from the streights [*sic*] of Belle Isle to Cayenne, including the island and banks of Newfoundland, the West-India-Islands, and all the islands on the coast; with particular directions for sailing to and entering the principal harbours, rivers, &c. Describing also the capes, headlands, rivers, bays, roads, havens, harbours, straits, rocks, sands, shoals, banks, depths of water and anchorage.... Boston: Printed and sold by John Norman, 1791. 2 p. l., 12 maps.

Mp-23637.

Copies: MH.

3880. ———— Boston: Printed and sold by J. Norman, 1792. 2 p.l., 11 maps.

B-8092; Mp-46532.

Copies: DLC; MiU-C.

3881. Williams, Jonathan, 1750–1815.

Memoir on the use of the thermometer in navigation; presented to the American Philosophical Society, held at Philadelphia, for promoting useful knowledge. By Jonathan Williams, Jun.... Extracted from the third volume of their Transactions now in press. Philadelphia: Printed by R. Aitken & Son, 1792. [1], 82–100 p., fold. chart.

Mp-25040.

Copies: BM; NHi.

3882. [Norman, John] 1748–1817.

The American pilot, containing the navigation of the sea coast of North America, from the straights of Belle Isle to Cayenne, including the island and banks of Newfoundland, the West-India-Islands, and all the islands on the coast; with particular directions for sailing to and entering the principal harbours, rivers, &c. Describing also the capes, headlands, rivers, bays, roads, havens, harbours.... Boston: Printed and sold by John Norman, 1794. 2 p.l, 11 maps.

Dated by Evans: 1793.

Mp-25925.

Copies: CtY; DLC; RPJCB.

3883. Patterson, Robert, 1743–1824.

A new table of latitude and departure, for every degree and five minutes of the quadrant, and for every point and quarter of the compass: with a full explanation, and exemplification of its use in the various cases of plane trigonometry, surveying, navigation, practical astronomy and dialing. By R. Patterson.... [Philadelphia]: Printed by Stewart & Cochran, 1794. [6], [7]–30 p.

B-8785; Mp-47152.

Copies: NN; PPL.

3884. Truxtun, Thomas, 1755–1822.

Remarks, instructions, and examples relating to the latitude & longitude; also, the variation of the compass.... To which is annexed, a general chart of the globe, where the route made by the author ... is marked ... for the purpose of showing the best tract of the sea.... Together with a short ... account of variable winds, trade winds, monsoons, hurricanes.... Philadelphia: Printed by T. Dobson, 1794. [1], 5–8, [9]–74, 31, xxiii p., fold. map, fold. plate.

Mp-27823.

Copies: DLC; MB; MBAt; MH; MWA; NN; PPAmP; PPL; RPJCB.

3885. [Price, Jonathan.]

A description of Occacock Inlet, and of its coasts, islands, shoals, and anchorages, with the courses and distances, to and from the most remarkable places, and directions to

3885 (*continued*)
sail over the bar and through the channels. Newbern, [N.C.]: Printed by Francois X. Martin, 1795. [2], 8 p.

Mp-29351.

Copies: NcA.

3886. The American coast pilot; containing, the courses and distance from Boston to all the principal harbours, capes and headlands included between Passamaquady and the capes of Virginia ... also, a tide table ... together with the courses and distance from Cape-Cod and Cape-Ann, to the shoal of Georges.... By Capt. Lawrence Furlong ... (approved by experienced pilots and coasters). The 1st ed. Newburyport: Printed by Blunt and March, 1796. 5 p.l., [3]–125 p.

Mp-30464.

Copies: CtNlCG; DLC; MB; MWA; MdAN; MsU; NcU; NjP; PPL; RPJCB.

3887. Pinkham, Paul.

A chart of George's Bank, including Cape Cod, Nantucket, and the shoals lying on their coast, with directions for sailing over the same &c. Surveyed by Capt. Paul Pinkham. [Newburyport]: Printed for Edmund M. Blunt, 1797. Map 55 x 82 cm.

Mp-32693.

Copies: DLC; MWA.

3888. The American coast pilot; containing the courses and distances between the principal harbours, capes and headlands, from Passamaquoddy through the Gulph of Florida, with directions for sailing into the same, describing the soundings, bearings of the light-houses and beacons from the rocks, shoals, ledges &c.... By Capt. Lawrence Furlong. Corr. and improved by the most experienced pilots in the United States ... 2d ed., largely improved. Newburyport, Mass.: Printed by Edmund M. Blunt, 1798. xvi, [17]–172, [8], [177]–239, [1] p.

Mp-33772.

Copies: DLC; FMU; GU-De; MB; MWA; MiU-C; NN; NcD; OO; RPJCB.

3889. [Blunt, Edmund March] 1770–1862.

The new theoretic and practical navigator, containing later improvements particularly in working lunar observations, correcting the watch, &c.... Revised and corrected by Nicholas Pike.... Newburyport, [Mass.]: Printed by Edmund March Blunt, 1798.

E-33433.

No copy known.

3890. [Norman, William.]

The American pilot, containing the navigation of the sea coast of North America, from the streights of Belle-Isle to Essequebo, including the island and banks of Newfoundland, the West-India islands, and all the islands of the coast. With particular directions for sailing to, and entering the principal harbours, rivers, &c.... Boston: W. Norman, 1798. [6] p., 11 maps.

B-10445; Mp-48553.

Copies: DLC; MB.

3891. Hauducoeur, C. P.

A map of the head of Chesapeake Bay and Susquehanna River shewing the navigation of the same with a topographical description of the surrounding country from an actual survey by C. P. Hauducoeur, engineer. [Philadelphia]: 1799. Map 55 x 75 cm.

Wheat-520.

Copies: CtY; DLC; MdHi; MH; MHi; PHi; PPL.

3892. Moore, John Hamilton, d. 1807.

The new practical navigator; being an epitome of navigation.... The 1st American, from the 13th English ed. of John Hamilton Moore, improved by the introduction of several new tables, and by large additions to the former tables, and revised and corrected by a skilful mathematician and navigator [Nathaniel Bowditch] ... Newburyport, [Mass.]: Printed by Edmund M. Blunt (proprietor), 1799. xii, [1], 14–304, 114, 507–574 p., 8 plates.

First London edition, 1772.

Mp-35834.

Copies: CtY; InU; MB; MH; MWA; MiU.

3893. [Williams, Jonathan] 1750–1815.
Thermometrical navigation. Being a series of experiments and observations, tending to prove, that by ascertaining the relative heat of the sea-water from time to time, the passage of a ship through the Gulph Stream, and from deep water into soundings, may be discovered in time to avoid danger.... Extracted from the American philosophical transactions, vol. 2 & 3. With additions and improvements.... Philadelphia: R. Aitken, 1799. xii, 98, [3] p., fold. map, tables.
Mp-36722.
Copies: CtY; DLC; DeGE; MB; MBAt; MH; MWA; NN; PPAmP; PPL; RPJCB.

3894. The American coast pilot, containing the courses and distances between the principal harbours, capes and headlands, from Passamaquoddy through the Gulf of Florida.... By Capt. Lawrence Furlong. Corr. and improved by the most experienced pilots in the United States ... 3d ed. Newburyport, Mass.: Printed by Edmund M. Blunt (proprietor), 1800. xvi, [17]–251 p.
Mp-37483.
Copies: CtY; DLC; MB; MH; MWA; N; NN; OU; PPAmP; PPL; RPJCB.

3895. Churchman, John, 1753–1805.
The magnetic atlas; or, Variation charts of the whole terraqueous globe.... The 3d ed. with additions.... New York: Printed for the author, and sold by Gaine & Ten Eyck, 1800. vii, [1], xxix, [31]–82 p., 2 fold. maps, fold. plate.
First published in London, 1794.
Mp-37183.
Copies: CtHT-W; MB; MiU-C; NjP; PPAmP.

3896. Moore, John Hamilton, d. 1807.
The new practical navigator; being an epitome of navigation.... The first American from the thirteenth English edition of John Hamilton Moore, improved ... and revised and corrected by a skilful mathematician and navigator [Nathaniel Bowditch] ... 2d ed. Newburyport, [Mass.]: Printed by Edmund M. Blunt, 1800. xii, [13]–570, [2] p., 8 plates.
Mp-37991.
Copies: CtY; DGU; MB; MH; MWA; MeHi; MiU; NN; NhD; NjP; OClW; RPJCB.

3897. The seaman's journal; being an easy and correct method of keeping the daily reckoning of a ship. New Haven: Isaac Beers & Co., 1800. [78] p.
Mp-38478.
Copies: CtY.

3898. Clarke, Henry, 1743–1818.
Clarke's Seaman's desiderata; or, Concise, practical rules for computing the apparent time at sea, the latitude ... and the longitude.... With additions and corrections, by J. G. New Brunswick, N.J.: Printed by A. Blauvelt, 1801. v, [6]–48, [28] p., 5 plates.
First published at Bristol, England, 1801.
S&S-314.
Copies: DLC; MBAt; NN; NWM; OOxM; PPAmP; PPL.

3899. Gould, Chester.
Gould's patent sea logs. [Philadelphia?: 1801.] 8 p.
A device to measure the distance covered by vessels at sea.
Copies: DeGE.

3900. [Huddart, Joseph] 1741–1816.
The oriental navigator; or, New directions for sailing to and from the East Indies ... to which is added, The India officers and traders guide ... with practical directions for the choice of diamonds; and the use of the Chinese touch needles in judging of the value of gold. Likewise, accurate tables of the weights, measures, coins, money.... Philadelphia: Printed by J. Humphreys, 1801. 566, [2] p.
First published in London, 1794.
S&S-682.
Copies: DLC; MB; MH; MSaE; MWA; MiU-C; NIC; NN; OkU; PPL; PU.

3901. Bowditch, Nathaniel, 1773–1838.

The new American practical navigator; being an epitome of navigation ... 1st ed. Newburyport, [Mass.]: Printed by Edmund M. Blunt, for Brown and Stransbury, New York, 1802. 622 p., fold. map, plates.

S&S-1936.

Copies: NN.

3902. ———— Newburyport, [Mass.]: Printed by Edmund M. Blunt, for Caleb Bingham, Boston, 1802. 589 p., fold. map, plates.

S&S-1937.

Copies: DLC; MBAt.

3903. ———— Newburyport, [Mass.]: Printed by Edmund M. Blunt (proprietor) for Cushing & Appleton, Salem, 1802. xvi, 17–246, 286, [533]–589 p., fold. map, plates.

S&S-1938.

Copies: PRB.

3904. ———— Newburyport, Mass.: Printed by Edmund M. Blunt (proprietor) for Thomas & Andrews, Boston, 1802. xvi, 17–246, [284], 533–589, [14] p., fold. map, plates.

S&S-1939.

Copies: MWA; RPB.

3905. ———— Newburyport, Mass.: Printed by Edmund M. Blunt, for James White, Boston, 1802. 591 p., fold. map, plates.

S&S-50279.

Copies: MSaP.

3906. ———— Newburyport, [Mass.]: Printed by Edmund M. Blunt, for Samuel Hall, Boston, 1802. 591 p., fold. map, plates.

S&S-50280.

Copies: MSaP.

3907. ———— Newburyport, Mass.: Printed by Edmund M. Blunt, for William R. Wilder, Newport, 1802. 591 p., fold. map, plates.

S&S-50281.

Copies: DLC; ICJ; RPB.

3908. [Freeman, James] 1759–1835.

A description of the eastern coast of the county of Barnstable from Cape Cod ... to Cape Malebarre ... pointing out the spots ... where shipwrecked seamen may look for shelter. Boston: Printed by Hosea Sprague, 1802. 15 p.

Indicates the locations of lighthouses and other aids to navigation.

S&S-2255.

Copies: DLC; MPlPS; MWA; NN.

3909. Puységur, Antoine Hyacinthe Anne de Chastenet, comte de, 1752–1807.

A treatise upon the navigation of St. Domingo; with sailing directions for the whole extent of its coasts, channels, bays and harbours.... By M. de Chastenet Puysegur. Translated from the French by Charles de Monmonier. Baltimore: Printed for the translator, by W. Pechin, 1802. 4 p.l., [5]–112 p.

S&S-2957.

Copies: DLC; MB; MWA; MdBP; PPL; ViW.

3910. Trescott, William.

Star cross; by which can be found the hour and minute of the night, with ease and exactness, without the assistance of either clock or watch.... Peacham, [Vt.]: Printed by Samuel Goss, for the author, 1802. [24] p.

S&S-3179.

Copies: DeGE; MH; MWA.

3911. Baron, George, b. 1769.

Exhibition of the genuine principles of common navigation with a complete refutation of the false and spurious principles ignorantly imposed on the public in the "New American practical navigator...." New York: Printed by Sage and Clough, for the author, and sold by him and by W. Falconer, 1803. 34 p., plate.

S&S-3759.

Copies: MB; NN; PPAmP; PPL.

3912. Hoskins, Francis.

Navigation made easy; or, Mariners' complete guide, containing an accurate account of a voyage from the city of Phila-

delphia to the city of Cork.... Philadelphia: 1803. 32 p.
S&S-4403.
Copies: PPAmP (not located).

3913. [Norman, William.]
American pilot; containing the navigation of the sea-coast of North-America, from the streights of Belle Isle to Essequibo, including the island and banks of Newfoundland, and the West-India Islands and all the islands on the coast. Boston: William Norman, 1803. [12] p., 18 charts (part folded).
S&S-3667.
Copies: MWA; MiU-C; PMA.

3914. The American coast pilot: containing the courses and distances between the principal harbours, capes and headlands, from Passamaquoddy, through the Gulph of Florida; with directions for sailing into the same, describing the soundings, bearings of the light-houses and beacons from the rocks, shoals, ledges, &c.... By Capt. Lawrence Furlong. Corr. and improved by the most experienced pilots in the United States ... 4th ed. Newburyport, Mass.: Printed by Edmund M. Blunt (proprietor), [etc.], 1804. xiv p., 1 l., [17]–386 p., 11 charts.
S&S-5697, 6361.
Copies: DLC; DeGE; MBAt; MWA; NN; NjP.

3915. [Bowditch, Nathaniel] 1773–1838.
An appendix to the New American practical navigator, containing tables for clearing the apparent distance of the sun and moon, from the effects of parallax and refraction by a new method.... Newburyport, [Mass.]: Printed by Edmund M. Blunt, 1804. [37] p.
S&S-5896.
Copies: DeGE; M; MWA.

3916. Malham, John, 1747–1821.
The naval atlas; being a correct set of charts from the latest and best surveys. 2d American ed., considerably improved and enlarged. Philadelphia: Printed and sold by W. Spotswood, 1804. 1 p.l., 17 double maps.
First published in London; 1st American edition not located.
N-M0146902; S&S-6703.
Copies: CtHT-W; NN.

3917. Stedman, Ebenezer.
Catalogue of sea-books, charts, pilots, &c. for sale by Ebenezer Stedman, at the Newburyport book-store.... [Newburyport, Mass.: 1804.] Broadside 44 x 27 cm.
S&S-7303.
Copies: MB.

3918. Burke, Edward D.
A key to practical navigation; being an illustration of all the rules necessary to be used in conducting a vessel from any one part on the globe, to any other.... By Edw. D. Burke.... Baltimore: Printed by Wayne & Murphy, 1805. 16 p.
S&S-8118.
Copies: NN.

3919. Naylor, Samuel.
Dunn's linear tables described and their utility verified; with precepts and examples for shortening calculations and preserving accuracy in the lunar method of finding longitude at sea.... Boston: Printed for the proprietor; A. Newell, printer, [1805]. 15 p., fold. table.
S&S-8961.
Copies: MB; MH; MWA.

3920. Naylor, Samuel.
A new formula for longitude, having Dunn's linear tables. Published according to act of Congress, by Samuel Naylor at Boston.... Boston: A. Newell, printer, [1805]. Broadside 32 x 20.5 cm.
N-N0083453.
Copies: MB.

3921. The American coast pilot; containing the courses and distances between the principal harbours, capes and headlands, from Passamaquoddy, through the Gulph of Florida; with directions for sailing into the same, describing the soundings, bearings of

3921 (*continued*)
the light-houses and beacons from the rocks, shoals, ledges, &c.... By Capt. Lawrence Furlong. Corr. and improved by the most experienced pilots in the United States.... 5th ed. Newburyport, Mass.: Printed by Edmund M. Blunt (proprietor), [etc.], 1806. xii p.,1 l., [15]–408 p., charts.
S&S-9833, 10462.
Copies: DLC; MH; MWA; MeHi; NN.

3922. Bowditch, Nathaniel, 1773–1838.
Chart of the harbours of Salem, Marblehead, Beverly and Manchester, from a survey taken in the years 1804, 5 and 6, by Nathaniel Bowditch, assisted by George Burchmore and William Ropes. 3d. [Salem, Mass.]: 1806. Map.
S&S-10021.
Copies: MB; MWA; PPAmP.

3923. Bowditch, Nathaniel, 1773–1838.
Directions for sailing into the harbours of Salem, Marblehead, Beverly and Manchester ... Newburyport, [Mass.]: Printed by Edmund M. Blunt, 1806. 30 p.
S&S-10022.
Copies: MBAt; MH; MSaP; MWA; NN; PPAmP.

3924. Garnett, John.
Tables requisite to be used with The nautical ephemeris, for finding the latitude & longitude at sea. First published in London by order of the Commissioners of Longitude. To which are now added, the most useful astronomical tables.... New Brunswick, N.J.: Printed for the editor, [etc.], 1806. viii, [136], 56, 56 p.
N-G0061923; S&S-10465.
Copies: DGU; DLC; MB; MWA; NN; NNC; NjP; NjR; OOxM; ViM.

3925. ———— With considerable additions and improvements by the editor of the American nautical almanac. New Brunswick, N.J.: Printed for the editor, [etc.], 1806. 1 p.l., [134], 56 p.
N-G0061924.
Copies: DLC.

3926. Bowditch, Nathaniel, 1773–1838.
The new American practical navigator; being an epitome of navigation ... 2d ed. with many improvements. Newburyport, (Mass.): Printed by Edmund M. Blunt, 1807. 679 p., fold. map, plates.
S&S-12190.
Copies: DLC; MNBedf; MWA; MWiW; NN.

3927. Burroughs, Richard, 1775–1865.
A treatise on trigonometry & navigation, containing an explanation of their principles and tables, and a new method of working trigonometry by memory.... Middlebury, [Vt.]: Printed by J. D. Huntington, 1807. xii, [3]–81 p., 8 plates.
S&S-12252.
Copies: CSmH; CtY; DLC; DeGE; MH; MWA; N; VtMiS; VtU.

3928. Garnett, John.
Description and use of a new nautical chart for working the different problems in navigation.... New Brunswick, N.J.: Printed for the author [by William Elliot], 1807. 14 p., 1 l., 2, [2] p.
S&S-12645.
Copies: MB.

3929. Mackay, Andrew, 1760–1809.
The complete navigator; or, An easy and familiar guide to the theory and practice of navigation.... By Andrew Mackay.... To which is added, a concise system of calculations for finding the longitude at sea, by the lunar observations. By P. Delamar. Philadelphia: B. B. Hopkins and Co.; printed by T. & G. Palmer, 1807. 4 p.l., [vii]–xxiii, 275, 40, 220 [222], [2] p., 7 plates.
First published in London, 1804.
S&S-12965.
Copies: DGU; DLC; DeGE; IU; MWA; MiD; NN; NNA; NjP; PPL.

3930. The American coast pilot: containing the courses and distances between the principal harbours, capes and headlands, from Passamaquoddy, through the Gulph of Florida.... By Capt. Lawrence Furlong.

Corr. and improved by the most experienced pilots in the United States ... 6th ed. Newburyport, [Mass.]: E. M. Blunt, 1809. xii, [2], [15]–389 p., 15 charts.
S&S-16836.
Copies: CtHT; CtW; DLC; MH; MSaE; MWA; NN; NcD; OClWHi.

3931. Mangan, Cornelius.
A new method of ascertaining the latitude in the northern hemisphere. Boston: Printed by John Eliot, Jr., for William T. Clap, 1810. 15 p.
S&S-20636.
Copies: CtY; DLC; MBAt; MMhHi; MSaE; MWA.

3932. The seaman's daily assistant; containing plane, traverse, parallel, middle latitude, and Mercator's sailings ... taken from the works of Robertson, Mackay, Moore, and Norie. New York: William Eliot, 1810. [5], 2–60, [100] p., plate, diagrs.
S&S-21314.
Copies: CtY; MB; MWA; NN; NPStA.

3933. Bowditch, Nathaniel, 1773–1838.
The new American practical navigator, being an epitomy of navigation ... 3d ed. New York: Printed for the proprietor, by Edmund M. Blunt and sold by Edward Little, [etc.], 1811. xii, [2], 276, [280], 619–653, [5] p., fold. map, plates.
S&S-22445, 22446.
Copies: MBAt; MH; MSaP; MWA.

3934. The American coast pilot: containing the courses & distances between the principal harbours, capes & headlands, from Passamaquoddy, through the Gulph of Florida.... By Capt. Lawrence Furlong. Corr. and improved by the most experienced pilots in the United States ... 7th ed. New York: Printed by Edmund M. Blunt, [etc., 1812]. 7, [ix]–xiv, [2], [17]–311, 63 p., charts.
S&S-25486.
Copies: DLC; MB; MH; MSaE; MWA; OCl.

3935. Lambert, Samuel, 1768–1832.
A new chart of Nantucket shoals & George's Bank, with the adjacent coast.... Salem, [Mass.]: 1813. Map 75 x 90 cm.
A sailing chart.
N-L0051113.
Copies: MB.

3936. Whitney, Thomas.
Variation of the compass. [Philadelphia: 1814.] Broadside, illus. 19 x 44 cm.
Dated: Philadelphia, March 1st, 1814.
Copies: DeGE.

3937. The American coast pilot, containing the courses & distances between the principal harbours, capes, & headlands, from Passamaquoddy, through the Gulph of Florida; with directions for sailing into the same.... By Edmund M. Blunt. Corr. and improved by the most experienced pilots in the United States. 8th ed. New York: E. M. Blunt, 1815. 6, [2], [ix]–xvi, 297, 86 p., charts.
S&S-33842.
Copies: DLC; MB; MWA; NN; NjR; ViU.

3938. [Blunt, Edmund March] 1770–1862.
Sailing directions for Blunt's chart of the Mississippi River.... [New York: Edmund M. Blunt, 1815.] 8 p.
S&S-34156.
Copies: MWA.

3939. Brosius, Francis Xavier.
A new, easy and concise method of finding the latitude by double altitudes of the sun. Dedicated to the Boston Marine Society.... Cambridge, [Mass.]: Printed by Hilliard and Metcalf, 1815. 51 p., diagrs.
S&S-34219.
Copies: CtY; MBAt; MH; MWA; NN.

3940. Moore, Joseph, 1763–1845.
Navigation improved; with a number of requisite tables to ascertain the latitude and longitude at sea; together with proper rules and examples for illustrating the same.... Salem, [Mass.]: Printed by Thomas C.

3940 (*continued*)
Cushing, 1815. 30 p.
S&S-35311.
Copies: CtY; MBAt; MBC; MH; MSaE; MWA; NN.

3941. [Norman, William.]
The American pilot, containing charts of the West-Indies, South Carolina, Cape Hatteras, Cape-Roman, Chesapeake Bay, New-York and Massachusetts Bay; Together with the coast of New-England, Nova-Scotia and Newfoundland. Boston: Andrew J. Allen & Co., 1816. 1 l., 11 maps.
S&S-36710.
Copies: MB.

3942. The American coast pilot, containing the courses and distances between the principal harbours, capes, and headlands, from Passamaquoddy, through the Gulf of Florida; with directions for sailing into the same.... By Edmund M. Blunt. Corr. and improved by the most experienced pilots in the United States ... 9th ed. New York: Printed for E. M. Blunt, 1817. 4, [v]–xvi, 328, 84 p., charts.
S&S-40276.
Copies: CtSoP; CtW; DLC; MH; MSaP; MWA; MsU; NN; TxU.

3943. Bowditch, Nathaniel, 1773–1838.
The new American practical navigator; being an epitome of navigation ... 1st stereotype (4th) ed. New York: E. M. Blunt and S. A. Burtus; J. Seymour, printer, 1817. xiv, [2], 274, 280, [555]–597 p., diagrs., 12 plates.
S&S-40296.
Copies: DLC; DeGE; MBAt; MNBedf; MSaE; MWA; NN; PPL.

3944. Day, Jeremiah, 1773–1867.
The mathematical principles of navigation and surveying, with the mensuration of heights and distances. Being the fourth part of a course of mathematics.... New Haven: Printed by Steele and Gray, 1817. [4], 108, [11] p., 3 plates.
S&S-40626.
Copies: CU; Ct; CtY; InU; KU; MB; MH; MWA; MeB; NN; NWM; OClW; PU; RNR; ViU.

3945. Burroughs, Richard, 1775–1865.
A treatise on trigonometry and navigation.... New York: Sold by John C. Totten, 1818. 81 p., 8 plates.
S&S-43496.
Copies: MWA; MiU; PPL.

3946. Lambert, Samuel, 1768–1832.
Navigation school. Charts, nautical books, instruments and stationery. School room on Essex-street, opposite Newbury-street, Salem. Samuel Lambert. [Salem, Mass.: 1818.] Broadside 22.5 x 13 cm.
Copies: MSaE.

3947. Blunt, Edmund March, 1770–1862.
Analysis of the authorities upon which the dangers have been inserted in Blunt's new chart of the Atlantic, or Western Ocean. To which is added, an accurate table of the variations of the magnetic needle ... and a series of experiments and observations, tending to prove, that by ascertaining the relative heat of the sea-water, by use of the thermometer ... the passage of a ship through the Gulph Stream and from deep water into soundings, may be discovered in time to avoid danger.... Compiled by Edmund M. Blunt.... New York: Printed for William Hooker, by J. Seymour, 1819. 76 p.
S&S-47382.
Copies: DeGE; MHi; MMeT; MSaE; MWA.

3948. Heartte, Issac T.
Ars navigandi; or, Tables of longitude, for correcting the effects of parallax and refraction, on the distances observed between the sun and moon, or moon and star.... Baltimore: Printed for the author by B. Edes, 1819. 39 p.
S&S-48202.
Copies: DLC; MB; MdHi; MdU.

3949. Rules and regulations for the pilots of Sandy-Hook. New York: Printed by Chambers & Co., 1819. Broadside.
S&S-49332.
Copies: DNA.

3950. Watkins, Samuel P.
A complete set of improved linear [lunar] tables, for clearing the effect of refraction on lunar distances. With several other tables, and two of the shortest, and most easy methods, for obtaining the longitude by lunar observations; with formulas.... Philadelphia: Printed by Adam Waldie, 1819. 28 p.
Copies: MWA.

3951. Daboll, Nathan, 1750–1818.
Daboll's practical navigator: being a concise, easy, and comprehensive system of navigation. Calculated for the daily use of seamen, and also for an assistant to the teacher: containing plane, traverse, parallel, middle latitude, and Mercator's sailing; with all the necessary tables.... Also, a new, scientific, and very short method of correcting the dead reckoning.... New London, (Con.): Printed and sold by Samuel Green, 1820. iv, [4], 106, [104] p., front.
S-935.
Copies: CSmH; CV; Ct; CtHi; CtY; DLC; DeGE; MH; MWA; MeU; RPJCB.

3952. Hooker, William.
Charts and nautical books, published by Edmund M. Blunt, for William Hooker, 202 Water, cor. of Fulton street, New York. To which is added, A catalogue of charts, nautical books, instruments, and stationary comprising a very general assortment, constantly for sale. New York: Printed by Clayton & Kingsland, Sept. 1820. 18 p.
S-1662.
Copies: DLC.

3953. Lambert, Samuel, 1768–1832.
Information useful for navigators.... Salem, [Mass.]: Sold by Cushing & Appleton, S. Lambert and J. W. Archer; T. C. Cushing, printer, 1820. [14] p., illus.
S-1900.
Copies: MSaE; MWA.

3954. ———— Salem, [Mass.]: Sold by Cushing & Appleton, S. Lambert and J. W. Archer; T. C. Cushing, printer, 1820 [1821]. [14] p., illus.
A reprint.
N-L0051111.
Copies: MB.

3955. Ward, Edward C.
Sailing directions and remarks, made on board the sloop Orbit, in June, July, and August, 1820, on a survey of the Bahama Banks &c. By E. C. Ward ... Francis Mallaby ... and others. Accompanied by a chart, extending from New York to Havanna. New York: Printed by Clayton & Kingsland, for Edmund M. Blunt, 1820. 16 p., map.
N-W0069897: S-4161.
Copies: MSaE; MWA; NN.

3956. Watkins, Samuel P.
A complete set of improved lunar tables, for clearing the effect of refraction on lunar distances. With several other tables, and two of the shortest and most easy methods, for obtaining the longitude by lunar observations; with formulas.... Philadelphia: T. Dobson and Son, for the author, 1820. 28 p.
S-4186.
Copies: DLC; MSaE.

3957. Bowditch, Nathaniel, 1773–1838.
The new American practical navigator; being an epitome of navigation ... 5th stereotype ed. New York: E. M. Blunt for William Hooker; J. Seymour, printer, 1821. xiv, [2], 272, 296, [555]–596 p., fold. map, plates.
S-4811.
Copies: CU; CtHT-W; DLC; DeGE; IU; MB; MWA; MeHi; NNFM; NRU; PHi; PPL; ViU; VtMiS.

3958. Stansbury, Daniel.
Tables to facilitate the necessary calculations in nautical astronomy ... to enable navigators to determine their longitude from lunar observations.... New York: Printed for the author, by J. Seymour, 1821. cxi, 539 p.
S-6857.
Copies: MB; MH; MdAN; NNA.

3959. The American coast pilot; containing the courses and distances between the principal harbours, capes, and headlands, on the coast of North and South America; with directions for sailing into the same.... By Edmund M. Blunt. Corr. and improved by information derived from official documents, actual observations, and the most experienced pilots. 10th ed. New York: Pub. by Edmund M. Blunt for William Hooker; J. Seymour, printer, 1822. xvi, 495 p., charts.
S-8115.
Copies: DLC; DeWI; MAtt; MWA; MsU; NN; NNA; NcD; NhHi; ScCleU.

3960. Arnold, Thomas.
The American practical lunarian, and seaman's guide.... To which are annexed, a compendium of marine law, and mercantile regulations and customs.... Philadelphia: Robert Desilver, 1822. xviii, 844, 71, [5] p., illus., diagrs., plates (2 double).
S-7840.
Copies: CSmH; DLC; DeGE; MH; MdAN; PHi; PPAmP; PPF; RPJCB; ViFbE.

3961. Lambert, Samuel, 1768–1832.
Navigation school. Charts, pilots, nautical books, instruments and stationary [*sic*]. School room on Essex Street, opposite Newbury St. — Salem — Samuel Lambert.... [Salem, Mass.: 1822.] Broadside 21 x 11.5 cm.
S-10179.
Copies: MSaE.

3962. Stansbury, Daniel.
Tables to facilitate the necessary calculations in nautical astronomy; more particularly designed to enable navigators to determine their longitude from lunar observations.... New York: Printed for the author, sold by Collins & Co., 1822. cxi, 339, [1] p.
S-10332.
Copies: DLC; NNG; PHC; ScCC.

3963. Watson, George.
Description and use of a diagram of navigation; by which all problems in plane, traverse, parallel, middle latitude and Mercator's sailing may be instantly and accurately resolved. Belfast, Me.: Printed by Fellowes & Simpson, 1822. 64 p.
S-11354.
Copies: DLC; MH; MSaE.

3964. Ward, Edward C.
New lunar tables ... illustrated by appropriate rules and examples. To which are added plain and easy directions for taking lunar observations, and some useful remarks on the sextant. By E. C. Ward.... Stereotype ed. New York: G. W. Blunt, 1823. 12, [11] p. diagrs.
N-W0069890; S-14821.
Copies: DLC; MB.

3965. Day, Jeremiah, 1773–1867.
The mathematical principles of navigation and surveying, with the mensuration of heights and distances. Being the fourth part of a course of mathematics ... 2d ed. New Haven: S. Converse, 1824. [4], 108, [11] p., 3 plates.
S-15943.
Copies: CLU; Ct; CtY; DLC; NCH; NNC; PU.

3966. Ward, Edward C.
New lunar tables.... Illustrated by appropriate rules and examples, with plain and easy directions for taking a lunar observation. And some useful remarks on the sextant. To which are added the method of finding the longitude at sea by a chronometer.... By E. C. Ward ... 2d stereotype ed. New York: E. & G. W. Blunt, 1824. 20, [2], [20] p., diagrs.
N-W0069892.
Copies: DLC.

3967. Lambert, Samuel, 1768–1832.
Nantucket South Shoal.... Charts, corrected as above, for sale by S. Lambert. [Salem, Mass.]: 1825. Broadside 29 x 20 cm.
S-21156.
Copies: MSaE.

3968. The mariners atmospherical register, or weather book: comprehended in thirteen sheets. Constructed for recording the variations of the marine barometer and thermometer, the latitude, longitude, direction of wind, and other useful remarks.... Philadelphia: H. S. Tanner, 1825. 2 l., 13 double plates.

N-M0223240; S-21328.

Copies: DLC; DP.

3969. American coast pilot; containing the courses and distances between principal harbours, and headlands, on the coast of North and South America.... Together with a tide table. By Edmund M. Blunt. Corr. and improved by information derived from official documents, actual observations, and the most experienced pilots. 10th ed. To which is annexed, The shipmaster's assistant ... by Joseph Blunt.... New York: Pub. by Edmund M. Blunt; John Grey & Co., printers, 1826. xvi, 497, 114, [56] p., maps.

Advertisements, [56] p. at end.

S-23846.

Copies: DN; MH; MWA; NN; PSC.

3970. Bowditch, Nathaniel, 1778–1838.

The new American practical navigator; being an epitome of navigation ... 6th stereotype ed. New York: E. M. Blunt, 1826. xv, [1], 288, 298, [571]–616 p., fold. map, plates.

S-23909.

Copies: MB; MBC; MH; MSaP; MWA; MoS; NGH; NcD; NjP; PP; PPL; RPB; VtMiS.

3971. The American coast pilot: containing the courses and distances between the principal harbours, capes, and headlands, on the coast of North and South America: with directions for sailing into the same.... By Edmund M. Blunt. Corr. and improved by information derived from official documents, actual observations, and the most experienced pilots. 11th ed. New York: E. and G. W. Blunt, 1827. xvi, 676, 4 p., illus., plates, charts.

S-28205.

Copies: CSmH; CtW; DLC; MB; MBAt; MWA; NN; NNA; NjP; NjR; PHi.

3972. [Boston Marine Society.]

Rules and regulations for the pilotage of the harbor of Boston. Approved by the Governor with advice of Council, February 12, 1830. Boston: Dutton and Wentworth, 1830. 15 p.

C-591; N-B0676368.

Copies: DLC; MWA.

3973. Clarke, Francis Gedney, 1792–1843.

The seaman's manual; containing a variety of matters useful to the navigator.... Portland, [Me.]: Shirley, Hyde & Co., 1830. 367 p., diagrs.

C-908.

Copies: DeGE; ICN; NH-BA; MSaP; MeHi; NN; NNC; RPJCB.

8. Signaling

3974. Pennsylvania. Navy Board.

Day and night signals; for the fleet belonging to the state of Pennsylvania.... Published by order of the Navy Board. Philadelphia: Printed by W. and T. Bradford, 1777. 17 l.

Mp-15513.

Copies: PHi; PPL.

3975. System of night signals. [N.p.: 1805.] 39 p.

S&S-50618.

Copies: PPL.

3976. Grout, Jonathan, 1772–1849.

Telegraph. Jonathan Grout, of the Commonwealth of Massachusetts, in pursuance of a patent from the government of the United States, having established a line of telegraphs, embracing Boston and Martha's Vineyard, respectfully submits to the consideration of the merchants and others concerned in navigation ... the following terms.... [Boston]: Oliver & Munroe, printers, 1806. Broadside 48 x 37 cm.

Dated: September 13, 1806.

Copies: MWA.

3977. United States. Congress. House.

Mr. Crowninshield's motion. December 22, 1807.... City of Washington: A. & G. Way, printers, 1807. 4 p.

A resolution authorizing the President to establish telegraph stations along the coast.

S&S-14010.

Copies: DLC; MWA; NN.

3978. Telegraphic dictionary. [Brooklyn: T. Kirk, printer, 1812.] 328 p., fold. plate.

S&S-26851.

Copies: DLC; DN; MWA; NN.

3979. United States. Navy Dept.

Signals for the use of the United States' Navy as adopted by order of the Navy Department, August, 1813. Washington City: Printed for the use of the Navy, 1813. 163 p.

S&S-30376.

Copies: DNA.

3980. [Sargent, Charles Lenox.]

A system of general signals for night and day, whereby merchant vessels may communicate at a distance by means of the common colours of the ship, and with four lanterns by night, without going out of their course. Boston: Wells & Lilly, 1817. 64 p., 4 col. plates.

S&S-42053.

Copies: M; MBAt; MH; PPAmP; WHi.

3981. Squire, Nathaniel.

The universal signal book, containing a complete code of signals for the use of merchants' ships of all nations. By N. Squire; with improvements adapted to the vessels of the United States, by James M. Elford. Charleston, S.C.: James M. Elford, 1818. 39 p., illus.

S&S-45786.

Copies: DLC; DN; MH; MSaP; NNA.

3982. Reid, Samuel Chester, 1783–1861.

The New-York telegraph and signal book. Signals established between the telegraph on Staten-Island, and the shipping belonging to the port of New-York. New York: Printed for the author, 1822. 71+p.

S-10077.

Copies: MH (imperfect: all after p. 71 is missing).

3983. Berrian, Richard.

The American telegraph and signal book.... Arranged in order of classes and grades, for the purpose of communicating any degree of information from one vessel to another, either by day or night, as far as the eye can see, or the ear hear; and to any part of the habitable globe if necessary. New York: Printed by J. W. Bell, 1823. 44 p., illus.

S-11802.

Copies: DLC.

3984. Elford, James M.

J. M. Elford's marine telegraph; or, Universal signal book.... Being, also, the key to the patent telegraph, invented by the author.... Published and sold by James M. Elford. Charleston, S.C.: Printed by Archibald E. Miller, 1823. 75 p., front.

S-12436.

Copies: DN; IU; MWA; NN; NNA; PPAmP; PU.

3985. Boston Marine Society.

Telegraphic circular. Boston: 1827. 1l., illus. 25 cm.

Report of the Committee of Observation of the Society on the system of marine signals or telegraphic flags, published by Capt. James M. Elford.

N-B0676369.
Copies: MB.

3986. Parker, John R.
Marine telegraph.... Boston: 1827. Broadside, illus. 44 x 35 cm.
Concerns the use of Elford's code of marine signals.
N-P0101679.
Copies: MB.

3987. [Parker, John R.]
Harbor telegraph signals. Printed for the use of the patrons to the telegraph only. [Boston: 1829.] 8 p.
S-39964.
Copies: MHi.

XII. INLAND TRANSPORTATION

1. General works

3988. Evans, Lewis, 1700–1756.

Geographical, historical, political, philosophical, and mechanical essays. The first, containing an analysis of a general map of the middle British colonies in America ... and the maritime and inland navigations of the several rivers and lakes contained therein.... Philadelphia: Printed by B. Franklin and D. Hall, 1755. iv, 32 p., map.

Mp-7411.

Copies: CtY; DLC; MH; MWA; MiU-C; MsU; NN; OCl; PPL; PU; RPJCB; ViU.

3989. ——— 2d ed. Philadelphia: Printed by B. Franklin and D. Hall, 1755. iv, 32 p., map.

Mp-7412.

Copies: CSmH; CtY; ICN; MB; MBAt; MHi; MWA; MdBJ-G; MiD; MnU; MoU; NN; NjP; PPL; PU; RPJCB; ViU.

3990. ——— Philadelphia: Printed by B. Franklin and D. Hall, and sold by J. and R. Dodsley, London, 1755. iv, 32 p., map.

Second issue.

Mp-7413.

Copies: NN.

3991. To the merchants and other inhabitants of Pennsylvania.... [Philadelphia: 1771.] Broadside 33 x 19 cm.

Dated: Philadelphia, Dec. 13, 1771, and signed: A friend to trade.

Recommends the construction of canals between the Schuylkill and Susquehanna rivers and the Chesapeake and Delaware bays, as well as turnpike roads in eastern Pennsylvania.

Mp-12246.

Copies: MWA (photocopy); MiU-C; PHi; PPL.

3992. To the public. Various are the plans that have been proposed by individuals, for opening a free and easy communication of trade.... [Philadelphia: 1772.] Broadside 27 x 20.5 cm.

Dated: Philadelphia, January 15, 1772.

Discussion of the feasibility of canals between the Susquehanna and Schuylkill rivers, the Delaware River and the Chesapeake Bay, and of turnpike transportation in eastern Pennsylvania.

Mp-12580.

Copies: DLC; PPL.

3993. Colles, Christopher, 1738–1816.

Proposals for the speedy settlement of the waste and unappropriated lands on the western frontiers of the state of New-York, and for the improvement of the inland navigation between Albany and Oswego. New York: Printed by Samuel Loudon, 1785. 14 p.

Mp-18960.

Copies: DLC; MWA; MiU-C; NHi; NN.

3994. Churchman, John, 1753–1805.

Churchman's map of the peninsula between Delaware and Chesapeake bays ... exhibiting the maritime ports of Pennsylvania, Maryland, Virginia and all the Delaware state; with the ground marked out as levelled for the proposed canals, together with the Susquehanna ... as far northward as to include all the natural obstructions to the navigation in that river.... Philadelphia: Printed and sold by Eleazer Oswald, 1787. Map.

3994 *(continued)*
E-20272.
No copy known.

3995. Adlum, John, 1759–1836.
A map exhibiting a general view of the roads and inland navigation of Pennsylvania, and part of the adjacent states. Respectfully inscribed to Thomas Mifflin, Governor, and the General Assembly of the Commonwealth of Pennsylvania, by John Adlum, and John Wallis. Philadelphia: 1791. Map 80 x 92 cm.
Four insets, with views of canal locks.
Evan's title beginning: *A Geographical and hydrographical map,* is taken from title registered for copyright.
Mp-23104.
Copies: CtY; DLC; MB; MiU-C; NHi; PPL.

3996. Howell, Reading.
A map of Pennsylvania, and the parts connected therewith, relating to the roads and inland navigation, especially as proposed to be improved by the late proceedings of Assembly. Copied from his larger map. By Reading Howell, J. Trenchard, engraver. Philadelphia: Printed for the author, 1791. Map 76 x 50 cm.
Mp-23454.
Copies: DLC.

3997. Howell, Reading.
A map of the state of Pennsylvania.... Also, by actual survey, the rivers ... the portages and communications ... iron works, manufactories, mineral, and other noted places. [Philadelphia: Printed for the author, 1791.] Map.
E-23453.
No copy known.

3998. Howell, Reading.
A map of Pennsylvania, & the parts connected therewith, relating to the roads and inland navigation, especially as proposed to be improved by the late proceedings of Assembly (copied from his larger map) by Reading Howell.... [Philadelphia: 1792.] Map 49 x 90 cm.
On margin: J. Trenchard sculp.
Mp-24412.
Copies: DLC; PHi; PPL.

3999. Howell, Reading.
A map of the state of Pennsylvania, with part of Lake Erie, and Presque Isle, also, the rivers Susquehanna, Ohio, Alleghany, Monongahela, and lesser streams, the mountains, roads, portages, etc. [Philadelphia: Printed for the author, 1792.] Map 96 x 147 cm.
Mp-24411.
Copies: DLC.

4000. Howell, Reading.
A map of the county of Philadelphia in Pennsylvania shewing the rivers Delaware and Schuylkill and the other waters, hills, townships lines, canals, roads, water works, public houses, gentlemen's seats &c.... Philadelphia: 1794. Map.
E-27141.
No copy known.

4001. Howell, Reading.
A map of the United States of America shewing or containing the boundaries or division lines of the different and respective states ... lakes, rivers & principal creeks, mountains, canals, roads, portages and communications, forts, cities, towns.... Philadelphia: 1794. Map.
E-27142.
No copy known.

4002. General statement of the contracts for opening and improving roads and rivers, under the acts of the thirteenth of April, 1791,—the tenth of April, 1792,—the eleventh of April, 1793,—the twenty-second of April, 1794,—the seventeenth of April, 1795 ... and the fourth of April, 1796. Philadelphia: Printed by Zachariah Poulson, Junior, 1797. 19 p.
B-9948; Mp-48133.
Copies: NN.

4003. Tatham, William, 1752–1819.
The political economy of inland navigation, irrigation and drainage; with thoughts on the multiplication of commercial resources; and on means of bettering the condition of mankind, by the construction of canals, by the improvement of their various capacities for commerce, transfer, agriculture, household supplies, and mechanical power.... Philadelphia: Printed for the author, 1799. xvi, 500, [3] p., 11 plates (3 fold.).
Actually printed in London.
Mp-36397.
Copies: PPAmP.

4004. [Carnac, Turner.]
Facts and arguments respecting the great utility of an extensive plan of inland navigation in America. By a friend to national industry.... Philadelphia: Printed by W. Duane, 1805. 61, [1] p., fold. map.
Author's name is frequently misspelled as Camac.
Also attributed to William Blodget.
S&S-8418.
Copies: CtY; DLC; DeGE; ICU; MH; MWA; MiU-C; NHi; NN; NjPT; PPAmP; PPL; RPB.

4005. Orleans (Terr.). Laws.
An act for improving the inland navigation of the Territory of Orleans. New Orleans: Printed by J. M. Bradford, 1805. 15 p.
S&S-9069.
No copy known.

4006. Tarascon, Berthoud, and Company.
An address to the citizens of Philadelphia, on the great advantages which arise from the trade of the western country to the state of Pennsylvania at large, and to the city of Philadelphia in particular.... By Messrs. Tarascon junr., James Berthoud and Co. Philadelphia: Printed for the addressers, 1806. 13 p.
Attributed to Louis Anastasius Tarascon.
Concerns communication with the Ohio Valley.
S&S-11431.
Copies: CSt; DLC; MWA; MiD-B; NN; PPAmP; PPL.

4007. Agricola, pseud.
A letter to the inhabitants of the city and state of New York; on the subject of the commerce of the western waters. By Agricola. New York: S. Gould, [1807]. 40 p.
S&S-11950.
Copies: CSmH; DLC; MWA; NBu; NN.

4008. Colles, Christopher, 1738–1816.
Proposal of a design for the promotion of the interests of the United States of America ... by means of inland navigable communications.... New York: Printed for the author, by S. Wood, 1808. 22 p., fold. map, fold. plate.
S&S-14731.
Copies: DLC; DeGE; MWA; MiU-C; N; NN; NjP; PPL.

4009. United States. Treasury Dept.
Report of the Secretary of the Treasury, on the subject of public roads and canals, made in pursuance of a resolution of Senate, of March 2, 1807. April 12, 1808. Printed by order of the Senate. Washington: Printed by R. C. Weightman, 1808. 123 p.
Signed: Albert Gallatin, Secretary of the Treasury.
Contains communications on the subject of canals by B. H. Latrobe and Robert Fulton.
S&S-16589.
Copies: CSmH; DBRE; DLC; DeGE; MWA; MdBJ; NN; NjP.

4010. ——— Published by order of the Senate. 2d ed. Philadelphia: Printed by W. Duane, 1808. 123 p.
S&S-15883, 16590.
Copies: DLC; DeGE; M; MH; MWA; NdU; PPL; PPiU; RNR; ViL.

4011. Porter, Peter Buel, 1773–1844.
Mr. P. B. Porter's speech on internal improvements. Delivered in the House of

4011 (*continued*)
Representatives on the eighth February, 1810. [Washington: 1810.] 18 p.
S&S-21123.
Copies: DLC; MWA; MiU; NBu; NHi; PHi; PPAmP; PPL.

4012. Remarks, &c. [Project for inland navigation via the Middlesex Canal and Merrimack River, etc.] [Boston: 181– .] [3]–22 p.
S&S-21190.
Copies: DLC.

4013. United States. Laws.
A bill for the improvement of the United States by public roads and canals. February 23d, 1810.... [Washington: 1810.] 9 p. (H.R. 55)
S&S-21587.
Copies: DLC; NN.

4014. Duane, William John, 1780–1865.
Letters addressed to the people of Pennsylvania respecting the internal improvement of the Commonwealth by means of roads and canals.... Philadelphia: Printed by Jane Aitken, 1811. 1 p.l., 125 p.
S&S-20556, 22727.
Copies: DLC; DeGE; DeU; ICU; MBC; MH-BA; MHi; MWA; NIC; NN; OClWHi; P; PPL; PWb; RPB; ScU.

4015. New York (State). Commissioners on Internal Navigation.
Report of the commissioners appointed by joint resolutions of the honorable Senate and Assembly of the state of New-York, the 13th & 15th March, 1810, to explore the route of an inland navigation from Hudson's River to Lake Ontario and Lake Erie. Albany: Printed by S. Southwick, 1811. 35 p.
Signed: Gouv. Morris, S. V. Rensselaer, W. North, De Witt Clinton.
S&S-23551.
Copies: DLC; DeGE; MWA; NUtHi; PU; ScU.

4016. ———— New York: Printed for Prior and Dunning; Southwick and Pelsue, printers, 1811. 38 p.
Copies: DLC; DeGE; MWA.

4017. New York (State). Laws.
An act to provide for the improvement of the internal navigation of the state. Passed April 8th, 1811. [Albany: 1811.] Broadside 28 x 21 cm.
Copies: MWA; NHi.

4018. The right of a state to grant exclusive privileges, in roads, bridges, canals, navigable waters, &c. vindicated; by a candid examination of the grant from the state of New-York to, and contract with Robert R. Livingston and Robert Fulton, for the exclusive navigation of vessels, by steam or fire, for a limited time, on the waters of said state, and within the jurisdiction thereof. New York: Printed by E. Conrad, 1811. 44, 9 p.
"Supplementary remarks," 9 p. at end.
S&S-23819.
Copies: CSmH; DLC; DeGE; NIC; NN; PP; PPAmP; PPB.

4019. New York (State). Commissioners on Internal Navigation.
Report of the Commissioners, appointed by an act of the Legislature of the state of New-York, entitled "An act to provide for the improvement of internal navigation of the state," passed April 8th, 1811, for the consideration of all matters relating to the said inland navigation. Albany: Printed by S. Southwick, 1812. 40 p.
Documents, p. [23]–40.
S&S-26285.
Copies: CSmH; CtY; DLC; DeGE; ICN; MH-BA; MWA; MdBJ; NHi; NN; NNC; NjP; OClWHi; PPAmP; PPL.

4020. [Paleske, Charles G.]
Substance of an address intended to be delivered on the 25th January, 1812, before the Special Committee of the House of Representatives, on the remonstrance signed by Ebenezer Hazard, Turner Camac [Carnac]

and others, and also the Sub-Committee on Roads and Inland Navigation.... [Lancaster?, Pa.: 1812.] 7 p.
Dated: Lancaster, Jan. 23, 1812.
S&S-26372.
Copies: MH-BA; P; PHi.

4021. United States. Laws.
A bill to establish a system of navigation for the United States. December 24, 1813.... [Washington: 1813.] 29 p. (S. 4)
S&S-30144.
Copies: DLC.

4022. New York (State). Commissioners on Internal Navigation.
In Assembly, March 8, 1814. Report of the Commissioners appointed by the Legislature of New-York, to provide for the internal improvement of the state. [S.l.: 1814.] 8 p.
S&S-32325.
Copies: MWA; N; NHi.

4023. New York (State). Commissioners on Internal Navigation.
Report of the commissioners appointed by the Legislature of New York, on the 8th of March, 1814, for the internal improvement of the State. [S.l.: 1814.] 12 p.
Signed on p. 4: Gouv. Morris, S. Van Rensselaer, De Witt Clinton, [a.o.].
"The advantages of the proposed canal from Lake Erie to Hudson's River, fully illustrated in a correspondence between Gouverneur Morris and Robert Fulton," p. 5–12.
S&S-51427.
Copies: DeGE; MB; MWA; N; NBu; PPAmP; PPL.

4024. North Carolina. General Assembly. House. Committee on Inland Navigation.
Report of the Committee on Inland Navigation. Submitted to the legislature of North-Carolina, November 30, by Archibald D. Murphey, their chairman. Raleigh: Printed by Tho. Henderson, 1815. 16 p.
Concerns the improvement of North Carolina rivers for navigation.
S&S-35506.
Copies: NcU; NcWsW; NjP; ViU.

4025. United States. Laws.
A bill to establish a system of navigation for the United States. [Washington: 1815.] 31 p. (S. 42)
S&S-36235.
Copies: DLC.

4026. Baldwin, Benjamin F.
Report on inland navigation [in North Carolina]. Raleigh: 1816. Broadside 43 x 28 cm.
N-B0062341.
Copies: MB.

4027. Murphey, Archibald De Bow, 1777–1832.
To the freeholders of Orange County. [S.l.]: The author, [1816]. 3 p.
Dated: June 27, 1816.
Plea for "a liberal system for internal improvements."
N-M0894078.
Copies: NcU.

4028. New York (State). Laws.
An act to provide for the improvement of the internal navigation of this state ... April 17, 1816. [Albany: 1816.] 12 p.
Sabin-53463.
Copies: MBAt.

4029. North Carolina. General Assembly. House. Committee on Inland Navigation.
Mr. Murphey's report to the legislature of North-Carolina, on inland navigation, December, 1816. Raleigh: Printed by Tho. Henderson, Jun., [1816] 19 p.
S&S-38483.
Copies: MB; MWc; NN; NcU; PHi.

4030. United States. Congress. Senate. Committee on Roads and Canals.
In Senate of the United States, February 6th, 1816. The committee appointed on so much of the President's message, as relate to roads and canals, report.... [Washington:

4030 (*continued*)
1816.] 9 p. (14th Cong., 1st sess. Senate. Doc. 34)

A letter from the Treasury Dept., signed A. J. Dallas, p. 8–9.

S&S-39392.

Copies: DLC; DeGE; MWA.

4031. United States. Laws.

A bill making appropriations for the construction of roads and canals. February 6, 1816. [Washington: 1816.] 2 p. (S. 22)

S&S-39271.

Copies: DLC.

4032. United States. Laws.

A bill to establish a system of navigation for the United States. March 8, 1816. [Washington: 1816.] 33 p. (S. 55)

S&S-39326.

Copies: DLC.

4033. United States. Treasury Dept.

Report of the Secretary of the Treasury, on the subject of public roads and canals, made in pursuance of a resolution of Senate of March 2d, 1807. January 15, 1816. Printed.... Washington: Printed by W. A. Davis, 1816. 129 p. (14th Cong., 1st sess. Senate. Doc. 19)

Albert Gallatin, Secretary of the Treasury.

Contains communications on the subject of canals by B. H. Latrobe and R. Fulton.

S&S-39632.

Copies: DBRE; DLC; DeGE; IU; MH-BA; MWA; MiU-T; NN; NhHi.

4034. Virginia. Board of Public Works.

Annual report of the president and directors of the "Board of Public Works," to the Legislature of Virginia.... [Richmond: Shepherd and Pollard], 1816. 34 p., chart.

Shaw-664.

Copies: DLC; NN; Vi; ViHi; ViW.

4035. Virginia. General Assembly. House of Delegates. Committee on Roads and Internal Navigation.

Report on internal improvement, presented by the Committee of Roads and Internal Navigation ... 28th Dec. 1815.... [Richmond: 1816.] 12 p.

Presented by James Robertson, April, 1816.

N-V0186321.

Copies: NcD; PPAmP; ViW.

4036. Calhoun, John Caldwell, 1782–1850.

Speech of the Honorable John C. Calhoun, in the House of Representatives, February 4th, 1817. [Washington: 1817.] 7 p.

Concerns internal improvements.

S&S-40381.

Copies: DeGE; MeHi; NN; NcD.

4037. Shelton, William.

The means of improving Richmond, and the state of Virginia.... Richmond: Printed by Shepherd and Pollard, 1817. 76 p.

Concerns roads, river navigation, canals, and manufactories.

S&S-42108.

Copies: Vi.

4038. A treatise on internal navigation. Explaining the principles by which canals and their appendages are laid out, constructed and kept in repair.... To which is annexed, the report of Albert Gallatin on roads and canals. Ballston Spa, [N.Y.]: Printed by U. F. Doubleday, 1817. xii, 152, 115 p.

Attributed to Joshua Secor, Samuel Johnson, and Samuel Young.

S&S-42319.

Copies: DBRE; DLC; DeGE; ICJ; MH; MWA; N; NHi; NN; NNC; NNE.

4039. United States. Congress. House. Committee on Roads and Canals.

Report, in part, of the Committee on so much of the President's message as relates to roads, canals, and seminaries of learning. December 15, 1817.... [Washington: 1817.] 11 p. (15th Cong., 1st sess. House. Doc. 11)

S&S-42676.

Copies: DLC; DeGE; DeU; G; MWA; NjP.

4040. United States. Congress. House. Committee on Roads and Canals.

Report of the Committee on so much of the President's message as relates to roads and canals. Made February 7, 1817.... [Washington]: 1817. 18 p. (14th Cong., 2d sess. House. Doc. 72)

S&S-42715.

Copies: DLC; DeGE; MWA; N.

4041. United States. Congress. House. Select Committee on Roads and Canals.

Report of the Select Committee appointed on the 4th of December last, on so much of the message of the President of the United States as relates to roads and canals. February 20, 1817.... [Washington: 1817.] 8 p. (14th Cong., 2d sess. House. Doc. 92)

Memorial of the directors of the Chesapeake and Delaware Canal Company, signed by Kensey Johns, Jos. Gilpin, and James C. Fisher, p. 3-8.

S&S-42731.

Copies: DLC; DeGE; MWA.

4042. United States. Congress. Senate. Committee on Roads and Canals.

In Senate of the United States. February 14, 1817. The Committee appointed on so much of the President's message as relates to roads and canals, report in part.... [Washington: 1817.] 4 p. (14th Cong., 2d sess. Senate. Doc. 99)

S&S-42540.

Copies: DLC; DeGE; MWA.

4043. Breck, Samuel, 1771–1862.

Sketch of the internal improvements already made by Pennsylvania; with observations upon her physical and fiscal means for their extension; particularly as they have reference to the future growth and prosperity of Philadelphia.... Philadelphia: J. Maxwell, printer, 1818. 48 p., 3 fold. maps.

S&S-43442.

Copies: CSmH; DLC; ICN; MWA; MiU; NN; NjPT; NjR; PPL; ViU.

4044. ———— 2d ed., rev. and enlarged. Philadelphia: M. Thomas, 1818. 2 p.l., [3]–81, [1] p., map.

S&S-43443.

Copies: DLC; DeGE; ICJ; KU; MBAt; MH-BA; MWA; NN; PPAmP; PPL.

4045. [Hollins, William.]

Remarks on the intercourse of Baltimore with the western country. With a view of the communications proposed between the Atlantic and the western states. Baltimore: Printed and published by Joseph Robinson, 1818. iv, 29, [1] p., fold. map.

Also attributed to Charles W. Goldsborough and John Eager Howard.

S&S-44362.

Copies: CtY; DLC; DeGE; MH; MHi; MWA; MdBE; MiU; MoSM; PPAmP; PPB; PPL.

4046. Mercer, Charles Fenton, 1778–1858.

Speech of the Hon. Charles F. Mercer, of Virginia, on internal improvement, delivered in the House of Representatives of the United States, March 12, 1818. Philadelphia: Printed at the office of the Union, 1818. 49 p.

S&S-44810.

Copies: MH; MHi; MWA; PHi; ViU.

4047. North Carolina. Commissioners to Employ a Principal Engineer for the State.

Reports of sundry surveys made in obedience to certain resolutions of the General Assembly passed in the year 1817 and sumbitted to the General Assembly at their session in 1818. Raleigh: T. Henderson, 1818. 40 p.

Copies: DI-GS; NN; NcU.

4048. Observations on the subject of internal navigation. Addressed to the members of the Legislature of Pennsylvania. [Philadelphia: 1818.] 8 p.

Signed and dated at end: A friend to general improvement. Philadelphia, March 2d, 1818.

S&S-45133.

Copies: MH-BA; MWA; PHi.

4049. On the subject of internal improvement. [N.p.: 1818.] 10 p.

Signed: A friend to internal improvement.

S&S-45163.

Copies: MWA.

4050. Pennsylvania. General Assembly. House. Committee on Roads and Inland Navigation.

Report of Committee on Roads and Inland Navigation. Read March 11th, 1818; William Lehman, chairman. Harrisburg: Printed by James Peacock, [1818]. 10 p.

S&S-45253.

Copies: IU; MWA.

4051. Virginia. Board of Public Works.

Annual report of the president and directors of the Board of Public Works, to the General Assembly of Virginia.... Richmond: Printed by Thomas Ritchie, 1818. 78 p., map.

The second annual report of the Board, dated January 1818.

Also issued as part 2 of its: *A Collection of all laws and resolutions*.... Richmond: 1819.

Copies: MBAt; PPAmP; Vi; ViHi; ViW.

4052. ———— Richmond: Printed by Thomas Ritchie, 1818. 63 p.

Dated: December 22, 1818.

Also issued as part 3 of its: *A Collection of all laws and resolutions*.... Richmond: 1819.

Shaw-665.

Copies: NN; PPAmP; Vi; ViHi.

4053. Dana, Edmund.

Description of the principal roads and routes, by land and water, through the territory of the United States.... Cincinnati: Looker, Reynolds & Co., 1819. [49]–108 p., map.

Issued as part of his *A Description of the bounty lands,* with a special title page.

Copies with the map are very uncommon.

S&S-47780.

Copies: CtY; DLC; ICN; IU; MH; MWA; MiU-C; NIC; NN; OClWHi; WHi.

4054. [Murphey, Archibald De Bow] 1777–1832.

Memoir on the internal improvements contemplated by the legislature of North Carolina; and on the resources and finances of that state. Raleigh: Printed by J. Gales, 1819. 88 p.

S&S-48769.

Copies: CSmH; DLC; N; NcAS; NcD; NcU.

4055. North Carolina. Board of Internal Improvements.

Report of sundry surveys, made by Hamilton Fulton, Esq., state engineer, agreeably to certain instructions from Judge Murphey, chairman, &c., and submitted to the General Assembly, at their session in 1819. Raleigh: Printed by Thomas Henderson, 1819. 70 p.

S&S-51926.

Copies: DLC; N; NN; NcD; NcU.

4056. South Carolina. Civil and Military Engineer.

Report of the Civil and Military Engineer of the state of South-Carolina, for the year 1818. [Columbia, S.C.: Printed by D. Faust, 1819.] 23 p.

S&S-49463.

Copies: N; PPAmP; ScC.

4057. United States. War Dept.

Report of the Secretary of War, relative to roads and canals, in pursuance of a resolution of the House of Representatives, of the 4th April last. January 14, 1819.... Washington: Printed by E. De Krafft, 1819. 13 p. (15th Cong., 2d sess. House. Doc. 87)

S&S-49940.

Copies: DLC; DeGE; DeU; MWA; NN; O; OO; R; RPB; WHi.

4058. Virginia. Board of Public Works.

A collection of all laws and resolutions of the General Assembly of Virginia, relating to the Board of Public Works; the report of the commissioners appointed to view certain rivers within the commonwealth in 1812; the reports of 1814–15 from the Committee of

Roads and Internal Navigation to the legislature; and the several annual reports of the Board of Public Works to the General Assembly up to 1819. Richmond: Printed by Shepherd & Pollard, 1819. 1 p.l., 113, 78, 73, 16, 53, ii, 57–67, 8, viii p.
Copies: Vi; ViHi.

4059. Virginia. Board of Public Works.
Fourth annual report of the president and directors of the Board of Public Works.... Richmond: Printed by Thomas Ritchie, 1819. 41 p.
Also issued as part I of its: *The Fourth and fifth annual reports....* Richmond: 1821.
Copies: Vi.

4060. Virginia. Board of Public Works.
Supplemental report of the Board of Public Works to the General Assembly of Virginia. January, 1819. Richmond: Printed by Thomas Ritchie, 1819. 16 p.
Also issued as part 4 of its: *A Collection of all laws and resolutions....* Richmond: 1819.
Copies: Vi; ViHi.

4061. Virginia. Principal Engineer.
Letter from the engineer of the Board of Public Works to a member of the Board, January 1819. Richmond: Printed by Thomas Ritchie, 1819. 8 p.
Signed: J. M. [Thomas Moore].
Also issued as part 6 of: Virginia. Board of Public Works. *A Collection of all laws and resolutions....* Richmond: 1819.
Copies: Vi.

4062. Virginia. Principal Engineer.
To the Virginia Board of Public Works. [Richmond: 1819?] 67 p.
Signed on p. 53: Thomas Moore, Principal Engineer, Richmond, 12th Mo. 31st, 1818.
Also issued as part 5 of: Virginia. Board of Public Works. *A Collection of all laws and resolutions....* Richmond: 1819.
Copies: Vi; ViHi.

4063. Allen, S.
Grand state lottery, for the improvement of internal navigation. S. Allen, manager.... Draws October 1820.... Tickets for sale at Allen's Lucky Office, No. 18, South Third Street, Philadelphia.... [Philadelphia: 1820.] Broadside 21.5 x 14 cm.
S-82.
Copies: CSmH.

4064. Briggs, Isaac.
To the Board of Public Works of the Commonwealth of Virginia, Isaac Briggs respectfully offers the following report. [Richmond: 1820.] 63 p.
Dated: Richmond, 1st Mo. 15, 1820.
Also issued as part IV of: Virginia. Board of Public Works. *The Fourth and fifth annual reports....* Richmond: 1821.
Copies: Vi.

4065. Mills, Robert, 1781–1855.
A treatise on inland navigation, accompanied by a map.... Baltimore: F. Lucas, jun'r; J. D. Toy, printer, 1820. iv, [5]–103, [4] p., map.
S-2270.
Copies: CU-B; DLC; DeGE; KSalW; KU; MH-BA; MWA; MdBJ; MiD-B; MnST; NN; NNC; PPi; TxU; Vi; WHi.

4066. North Carolina. Board of Internal Improvements.
Annual report of the Board of Public Improvements of North Carolina to the General Assembly, November 27, 1820, together with Mr. Fulton's reports to the Board, on the public works projected and carrying on throughout the state during the present year. Raleigh: Printed by J. Gales, 1820. xi, 56 p.
S-2561.
Copies: DLC; NcD; NcU.

4067. Pennsylvania. General Assembly. House. Committee on Roads and Inland Navigation.
Report of the Committee on Roads and Inland Navigation. Wednesday, January 26, 1820. Harrisburg: James Peacock, printer,

4067 (*continued*)
[1820]. 8 p.
S-2697.
Copies: MWA; P; PHi; PPAmP.

4068. South Carolina. Board of Public Works.

Plans and progress of internal improvement in South Carolina: with observations on the advantages resulting therefrom to the agricultural and commercial interests of the state. Columbia, [S.C.]: Reprinted at the Telescope Press, 1820. 20 p.
S-3277.
Copies: MBAt; PPAmP.

4069. South Carolina. Board of Public Works.

Report of the Board of Public Works, to the Legislature of South-Carolina, for the year 1820. [Columbia, S.C.]: Printed by Daniel Faust, 1820. [7-14], 15–38 p.
S-3278.
Copies: MBAt.

4070. South Carolina. Civil and Military Engineer.

Report No. 1, of the Civil and Military Engineer, of the state of South-Carolina, for the year 1819. [Columbia, S.C.]: Faust, print. [1820]. 15 p.
S-3279.
Copies: MBAt; N; PPAmP; ScC; ViU.

4071. Virginia. Board of Public Works.

Fifth annual report of the president and directors of the Board of Public Works, to the General Assembly of Virginia.... Richmond: Printed by Thomas Ritchie, 1820. 139, vii p.

Also issued as part V of its: *The Fourth and fifth annual reports....* Richmond: 1821.
Copies: Vi.

4072. Virginia. Board of Public Works.

Supplement to the Fourth annual report ... January, 1820. Richmond: Printed by Thomas Ritchie, 1820. 15 p.

Also issued as part II of its: *The Fourth and fifth annual reports....* Richmond: 1821.
N-V0183301.
Copies: Vi; ViW.

4073. Comparative calculations, and remarks on internal improvements by roads, canals, and river navigation; illustrative of the advantages to be derived from the improvement of the river Lehigh. Philadelphia: Printed by William Brown, 1821. 12 p.
S-5045.
Copies: DeGE; MH-BA; MiU; PHi; PPAmP; PPL.

4074. Maryland. General Assembly. Senate. Committee on Internal Improvement.

Report of the Committee of the Senate of Maryland, on internal improvement, to whom was referred so much of the executive communication as relates to that subject. December session, 1821. Annapolis: J. Chandler, [1821?]. 8 p.
S-5956, 9381.
Copies: MdBE; MdHi; MiU.

4075. North Carolina. Board of Internal Improvements.

Annual report of the Board of Public Improvements of North-Carolina to the General Assembly, November 26, 1821.... With Mr. Fulton's reports to the Board. Raleigh: Printed by J. Gales, 1821. xxviii, 68 p.
S-6305.
Copies: MBAt; MHi; N.

4076. Pennsylvania. General Assembly. House. Committee on Roads and Inland Navigation.

Report of the Committee on Roads and Inland Navigation, read in the House of Representatives, January 30, 1821. [Harrisburg]: Hamilton & M'Ilwaine, printers, [1821]. 13 p.
S-6416.
Copies: DLC; P; PPAmP.

4077. South Carolina. Board of Public Works.

Report of the Board of Public Works, to the legislature of South Carolina, for the year

1821. [Columbia, S.C.: Faust, printer, 1821.] 22 p.
S-6827.
Copies: PPAmP.

4078. United States. Laws.
A bill for the promotion of internal improvements in the United States. January 10, 1821.... [Washington: 1821.] 1 p. (H.R. 193)
S-7080.
Copies: DLC.

4079. Virginia. Board of Public Works.
The fourth and fifth annual reports of the Board of Public Works, to the General Assembly of Virginia; also, a collection of all laws and resolutions of the General Assembly relating to the Board of Public Works, passed at the sessions of 1819–20 and 1820–21. Richmond: Shepherd & Pollard, printers, 1821. 41, 15, 114, 63, 139, vii p.
Contents: Part I. Fourth annual report.... Part II. Supplement to the Fourth annual report. Part III. Report of the Principal Engineer of the Board of Public Works on the subject of the survey of James and Kanawha rivers. Part IV. To the Board of Public Works.... Isaac Briggs ... offers the following report. Part V. Fifth annual report.
All parts have special title pages.
Copies: Vi.

4080. Virginia. Board of Public Works.
Sixth annual report of the president and directors of the Board of Public works, to the legislature of Virginia.... Richmond: Printed by Thomas Ritchie, 1821. 80 p.
Copies: Vi.

4081. Maryland. General Assembly. House. Committee on Internal Improvement.
Report to the House of Delegates of Maryland, by the Committee on Internal Improvement, to which was referred so much of the executive communication as relates to that subject. December session, 1821. Annapolis: Printed by J. Chandler, [1822]. 16 p.
N-M0289025.
Copies: MiU.

4082. Mills, Robert, 1781–1855.
Internal improvement of South-Carolina, particularly adapted to the low country.... Columbia: State Gazette office, 1822. 28 p.
S-9494.
Copies: DLC; IU; MiU; ScHi.

4083. North Carolina. Board of Internal Improvements.
Annual report of the Board of Internal Improvements of North Carolina, to the General Assembly, December 10, 1822, together with Mr. Fulton's reports to the Board on the public works projected and carrying throughout the state during the present year. Raleigh: Printed by J. Gales & Son, 1822. xii, 72 p.
S-9726.
Copies: NcU; NcWsW.

4084. On the internal improvement in Virginia; with Mr. Moore's report to the Board of Public Works, on the improvement of the navigation of Appomattox River; and on the subject of connecting Appomattox and Staunton rivers. Together with Robert Fulton's letter to the Secretary of the Treasury.... Richmond: Printed for the author, 1822. 59 p.
S-9785.
Copies: DLC; Vi; ViU.

4085. Pennsylvania. General Assembly. Senate. Committee on Roads, Bridges and Inland Navigation.
Report on roads, bridges and canals, read in the Senate, March 23, 1822. Mr. Raguet, Chairman. [Harrisburg]: C. Mowry, printer, [1822]. 18, 192, iii p. 3 fold. tables.
Documents, accompanying the Report of the Committee on Roads, Bridges and Inland Navigation ...,192, iii p., has a separate title page.
S-9841, 9852.
Copies: DLC; DeGE; MWA (without Documents); MiU; PHi; PPAmP; PPL; PPi.

4086. United States. Congress. House. Committee on Roads and Canals.
Report of the Committee on Roads and

4086 (*continued*)
Canals, on the subject of internal improvements, accompanied with a bill to procure the necessary surveys, &c. on the subject of roads and canals. January 2, 1822.... [Washington: 1822.] 100 p. (17th Cong., 1st sess. House. Rept. 8)

Contains also a report of the Secretary of the Treasury, and communications from B. H. Latrobe and R. Fulton.

S-11190.

Copies: DBRE; DLC; DeGE; IU; MH-BA; MWA; N; NN; PPL.

4087. United States. Congress. House. Committee on Roads and Canals.

Report of the Committee on Roads and Canals, upon the subject of internal improvements. April 26, 1822.... [Washington: 1822.] 8 p. (17th Cong., 1st sess. House. Rept. 98)

S-11192.

Copies: DLC; DeGE; DeU; MWA; NjP.

4088. United States. Laws.

A bill to procure the necessary surveys, plans and estimates, on the subject of roads and canals. Jan. 2, 1822.... [Washington: 1822.] 1 p. (H.R. 19)

S-10771.

Copies: DLC.

4089. ———— December 12, 1822.... [Washington: 1822.] 1 p. (H.R. 19)

S-10772.

Copies: DLC.

4090. United States. President (Monroe).

Extracts from the views of the President of the United States, on the subject of internal improvements, communicated to Congress, May 4, 1822. [S.l.: 1822.] 5 p.

Copies: PHi.

4091. United States. President (Monroe).

Message from the President of the United States, with his objections to the bill for the preservation and repair of the Cumberland Road; also, a paper, containing his Views on the subject of internal improvements. May 4, 1822.... Washington: Printed by Gales & Seaton, 1822. 60 p. (17th Cong., 1st sess. House. Doc. 127)

"Views of the President of the United States, on the subject of internal improvements." p. 7–60.

S-11056.

Copies: DLC; DeGE; MB; MWA; NjP.

4092. Virginia. Board of Public Works.

Seventh annual report of the president and directors of the Board of Public Works, to the legislature of Virginia.... Richmond: Printed by Thomas Ritchie, 1822. 68, [1] p.

Copies: Vi.

4093. Hemphill, Joseph, 1770–1842.

Mr. Hemphill's speech in the House of Representatives, on the 14th day of January, 1823, on the following bill: To procure the necessary surveys, plans and estimates, on the subject of roads and canals. [Washington]: Printed by D. Rapine, [1823]. 24 p.

S-12808.

Copies: ICU; MA; MWA; MdBP; NNC; OClWHi; PPAmP; PPL.

4094. Hemphill, Joseph, 1770–1842.

Mr. Hemphill's speech, on the following bill: To procure the necessary surveys, plans, and estimates, on the subject of roads and canals. [Washington: Printed at the Columbian office, 1823.] 16 p.

N-H0270436.

Copies: DeGE; IU; MB; NN; PHi.

4095. Mills, Robert, 1781–1855.

Internal improvement of South Carolina, particularly adapted to the low country. Columbia: State gazette office, 1823. 28 p.

N-M0599384.

Copies: DLC.

4096. North Carolina. Board of Internal Improvements.

Annual report of the Board for Internal Improvements of North-Carolina to the General Assembly, with a condensed view of their accounts from the organization of the Board, together with Mr. Fulton's reports on sundry public works. December 5, 1823. Raleigh: J. Gales & Son, state printers, 1823. 45 p. (Doc. 8)

Copies: DeGE; NcU.

4097. Pennsylvania. General Assembly. House. Committee on Roads and Inland Navigation.

Report of the Committee on Roads and Inland Navigation, read in the House of Representatives, Jan. 20, 1823. Mr. Lehman, chairman. Harrisburg: Printed by John S. Weistling [*sic*], 1823. 8 p.

S-13708.

Copies: PHi; PPAmP.

4098. Pennsylvania. General Assembly. House. Committee on Roads and Inland Navigation.

Report of the Committee relative to the subject of a board of commissioners for the purpose of improving the state. Read in the House of Representatives, Dec. 19, 1823. Mr. Lehman, chairman. [Harrisburg: John S. Wiestling, printer, 1823.] 8 p.

S-13709.

Copies: DeGE; P; PHi; PPAmP.

4099. United States. Laws.

A bill to procure the necessary surveys, plans, and estimates, upon the subject of roads and canals. December 15, 1823. [Washington: 1823.] 1 p. (H.R. 5)

S-14486.

Copies: DNA.

4100. An attempt to induce the people of Georgia, to commence the work of internal improvement, in a series of essays on the subject of canalling, together with a bill to be entitled "An act to establish a Board of Public Works." By a planter of Chatham County. Savannah: Printed at the Georgia job office by R. Fulton, 1824. 63, viii p.

S-15133.

Copies: DLC.

4101. Barbour, John Strode. 1790–1855.

Speech delivered in the House of Representatives of the United States, upon the bill for obtaining necessary surveys for roads and canals. [Washington?: 1824?] 9 p.

N-B0114115.

Copies: Vi.

4102. [Carey, Mathew] 1760–1839.

Address to those on whom Heaven has bestowed the goods of fortune, and, what is more valuable, hearts to make a proper use of them for the public benefit. . . . [Philadelphia: 1824.] 6 p.

Signed and dated: Philadelphia, October 27, 1824. Nicholas Biddle, Mathew Carey, John Sergeant [and others]. Attributed to Carey.

Proposal for a society for the promotion of national improvements.

S-15651.

Copies: MWA; NN; P; PPAmP; PPL; PU.

4103. Clay, Henry, 1777–1852.

Mr. Clay's speech on internal improvement. House of Representatives, January 16, 1824. [S.l.: 1824.] 24 p.

N-C0472519.

Copies: KyHi; NN.

4104. Holmes, John, 1773–1843.

Speech of Mr. Holmes, of Maine, on the subject of internal improvement. Delivered in the Senate of the United States, April 21–2, 1824. Washington: Printed by Gales & Seaton, 1824. 22 p.

S-16571.

Copies: DLC; MBAt; MWA; NN; OO; PPL.

4105. Kentucky. General Assembly. House. Committee on Internal Improvements.

Report on internal improvements [made] in the House of Representatives, November 22, 1824. [Frankfort?, Ky.: 1824.] 8 p.

N-K0099966.

Copies: KyU.

4106. Livingston, Edward, 1764–1836.
Speech of Mr. Livingston, of Louisiana, on the subject of internal improvement.... February 9, 1824. Washington: Printed by Gales & Seaton, 1824. 36 p.
S-16928.
Copies: DLC; LU; MH; MWA; NN; PPL.

4107. McDuffie, George, 1790–1851.
Speech of Mr. McDuffie, on internal improvements; with a few introductory remarks in answer to a pamphlet entitled "Consolidation." Columbia, [S.C.]: Printed by D. & J. M. Faust, 1824. 32 p.
S-16968.
Copies: DLC; NHi; NN; PPAmP; ViU.

4108. Memorial of sundry citizens of the county of Camden, in the state of Georgia, in relation to internal improvements. February 2, 1824. Referred to the Committee on Roads and Canals. Washington: Printed by Gales & Seaton, 1824. 7 p. (18th Cong., 1st sess. House. Doc. 54)
S-18849.
Copies: DLC; MWA; NjP; PHi.

4109. North Carolina. Board of Internal Improvements.
Annual report of the Board for Internal Improvements of North Carolina, to the General Assembly, with statements of their accounts ... November 23, 1824. Raleigh: Printed by J. Gales & Son, [1824]. 37 p. (No. 2)
S-17438.
Copies: MH; NcD; NcU; NcWsW.

4110. Pennsylvania Society for the Promotion of Internal Improvements in the Commonwealth.
Constitution of the Pennsylvania Society for the Promotion of Internal Improvements in the Commonwealth. [Philadelphia?:1824?] 4 p.
Dated: November 30, 1824.
S-17579.
Copies: DBRE; DLC; DeGE; MWA; MdBP; P; PPAmP; PPL; PU.

4111. Pennsylvania Society for the Promotion of Internal Improvements in the Commonwealth.
Sir, we beg to announce to you, the establishment, in this city, of an association, under the title of "The Pennsylvania Society for the Promotion of Internal Improvements in the Commonwealth.... " [Philadelphia: 1824.] Broadside 25 x 20 cm.
Signed: Mathew Carey, Joseph Hemphill, Richard Peters, Jr., Stephen Duncan, William Strickland, Acting Committee, December 18, 1824.
S-17580.
Copies: DLC; MWA.

4112. Pennsylvania Society for the Promotion of Internal Improvements in the Commonwealth.
Sir: We take the liberty to enclose the plan of the Society for the Promotion of National Improvements, to which we request your attention.... [Philadelphia: 1824.] 15 p.
Signed and dated: Nicholas Biddle, Mathew Carey [and others] Philadelphia, Oct. 19, 1824.
S-17581.
Copies: PPL.

4113. Randolph, John, 1773–1833.
Speeches of Mr. Randolph, on the Greek question; on internal improvement; and on the tariff bill. Delivered in the House of Representatives of the United States. Washington: Printed by Gales & Seaton, 1824. 57 p.
S-17752.
Copies: CSmH; DLC; MB; MH; MWA; MdHi; NCH; NNC; P; PPAmP; PPL; ScU; Vi.

4114. South Carolina. Superintendent of Public Works.
Report of the Superintendent of Public Works, to the legislature of South Carolina, for the year 1823. Columbia, S.C.: Printed by D. & J. M. Faust, [1824?]. 79, [80–112] p.
S-18055.
Copies: NHi; NN; NcD.

4115. To the people of New-Jersey. Your representatives having assembled in convention "to promote internal improvement," ... a few reflections upon the subject.... Belvidere, N.J.: Printed by Edmund P. Banks, [1824?]. Broadside.

S-18213.

Copies: NN.

4116. United States. Congress. Senate. Select Committee on Roads and Canals.

In Senate of the United States, March 23, 1824. Mr. Benton proposed the following amendment to the bill from the House of Representatives, entitled, "An act to procure the necessary surveys, plans, and estimates, upon the subject of roads and canals." [Washington: 1824.] 3 p. (H.R. 5)

S-18733.

Copies: DNA.

4117. Virginia. Board of Public Works.

Eighth annual report of the president and directors of the Board of Public Works, to the General Assembly of Virginia. Seventeenth December, 1823. Richmond: Printed by Shepherd and Pollard, 1824. 292 p.

S-19151.

Copies: PPL; Vi.

4118. Virginia. Board of Public Works.

The sixth, seventh, and eighth annual reports of the Board of Public Works, to the General Assembly of Virginia; also, a collection of acts and resolutions of the General Assembly, passed at the sessions of 1821–'22,1822–'23, 1823–'24, and other documents on subjects connected with the fund for internal improvement. Vol. III. Richmond: Printed by Shepherd & Pollard, 1824. 1 p.l., 80, 68, 292 p.

The reports also have separate title pages, dated respectively 1821, 1822, and 1824.

Copies: Vi.

4119. [Carey, Mathew] 1760–1839.

Internal improvement. Rail roads, canals, bridges, &c. [Philadelphia: 1825.] 28 p., illus.

Dated at end: Philadelphia, March 15, 1825.

Attributed also to William Strickland.

S-19977.

Copies: DBRE; DeGE; NN; P; PPAmP; PPL.

4120. [Carey, Mathew] 1760–1839.

Internal improvement. No. I. Railways and canals. [Philadelphia: 1825]. 5 p.

Signed and dated: Hamilton. Philadelphia, September 26, 1825.

S-19970.

Copies: MCM; MWA; PPL.

4121. [Carey, Mathew] 1760–1839.

Internal improvement. No. II. Canals and railways. [Philadelphia: 1825]. 4 p.

Signed and dated: Hamilton. Philadelphia, October 5, 1825.

S-19971.

Copies: MCM; MWA; P; PPL.

4122. [Carey, Mathew] 1760–1839.

Internal improvement. No. III. Canals and railways. [Philadelphia: 1825.] 5 p.

Signed and dated: Hamilton. Philadelphia, November 28, 1825.

S-19972.

Copies: MCM; MWA; P; PPL.

4123. ———— [Philadelphia: 1825.] 8 p.

Signed on p. 5: Philadelphia, Nov. 28, 1825. Hamilton.

"Internal improvement. No. IV. Canals and railways," p. [6]–8.

Copies: PPL.

4124. [Carey, Mathew] 1760–1839.

Internal improvement. No. IV. Canals and railways. [Philadelphia: 1825.] 5 p.

Signed at end: Philadelphia, November 28, 1825. Hamilton.

S-19973.

Copies: MWA; P; PPL.

4125. ———— [Philadelphia: 1825.] [6]–8 p.

Signed and dated: Hamilton. Philadel-

4125 (*continued*)
phia, December 7, 1825.
S-19974.
Copies: MCM; MWA; P; PPL.

4126. [Carey, Mathew] 1760–1839.
Internal improvement. No. V. Canals and railways. [Philadelphia: 1825.] 4 p.
Signed and dated: Hamilton. Philadelphia, December 20, 1825.
S-19975.
Copies: MCM; MWA; P; PPL.

4127. [Carey, Mathew] 1760–1839.
Internal improvement. No. VI. To the editors of the United States Gazette. [Philadelphia: 1825.] 4 p.
Signed and dated: Hamilton. Philadelphia, December 26, 1825.
S-19976.
Copies: MCM; MWA; PPL.

4128. Grahame, Thomas.
A letter, addressed to the proprietors and managers of canals and navigable rivers, on a new method for tracking and drawing vessels by a locomotive engine boat, with much greater speed, and at less than one third of the present expense. [Richmond]: Printed by Thomas Richie, printer for the Commonwealth, 1825. 18–27 p. (Virginia. General Assembly. House. Doc. 11)
First published in London, 1825.
S-20732.
Copies: DLC; NN; NNC; PPL; Vi.

4129. [Hollins, William.]
Views on the subject of internal improvement between the Atlantic and western states, 1825. Baltimore: Printed by W. Wooddy, 1825. 15 p.
Signed: William Hollins.
S-20929.
Copies: CtY; DLC; MWA; MdBP; NN; PPAmP; PPL.

4130. ———— and navigating the river Susquehanna with steam boats. No. 2. June 20, 1825. Baltimore: Printed by W. Wooddy, 1825. 21 p.
S-20930.
Copies: CtY; DLC; MdBP; MdHi; PHi; PPAmP; PPL.

4131. Internal improvement. At a numerous convention of delegates from the towns on the Kennebec River, assembled ... at Kimball's Hall in Waterville, for the purpose of adopting measures to remove the obstructions to the navigation of said river, and to consider the practicability and expediency of uniting the waters of the Kennebec and St. Lawrence. Waterville: 1825. Broadside 34.5 cm.
Dated: Waterville, December 28, 1825.
S-21025.
Copies: NHi.

4132. Internal Improvements Convention, Concord, N.H., 1825.
Facilitated carrying. A short statement on the proceedings at Concord, New Hampshire, on Friday, the third day of June, 1825, on the subject of internal improvements; and some remarks thereon. Boston: From the Commercial Gazette Press; A. Sampson, printer, 1825. 17 p.
Chairman of the meeting: Hon. William H. Kent; secretary: Josiah Quincy.
N-I0106026; S-20453.
Copies: CtY; DLC; DeGE; MB; MBAt; MH; MWA; MdBP; NN; NhHi; WHi.

4133. Maryland. State Convention on Internal Improvements, Baltimore, 1825.
The State Convention on Internal Improvements; containing the resolutions, &c. offered by the different gentlemen composing the Convention, together with an Appendix, containing the views of Thomas Ellicott, in support of the resolution offered by him to the Convention, and which were submitted to the Sub-Committee; by Thomas Phenix, assistant secretary to the Convention. Baltimore: Printed by W.O. Niles, 1825. 44, 19 p., fold. map.
S-21351.
Copies: DLC; DeGE; ICJ; MH-BA; MWA; MdBE; MdBJ-G; MdBL; MdHi; PPAmP; PPL; WHi.

4134. Maryland Society for Promoting Internal Improvement.

Constitution of the Maryland Society for Promoting Internal Improvement.... [Baltimore?: 1825.] [3] p.

S-21350.

Copies: MdHi.

4135. New York (State). Governor (Clinton).

Message of De Witt Clinton to the Legislature of the state of New York, January, 1825. New York: G. F. Hopkins, 1825. 26 p.

A substantial part of the message is concerned with internal improvements, particularly canals.

S-21666.

Copies: DLC; ICU; MWA; NIC; PPL.

4136. New York (State). Legislature. Joint Committee on Canals and Internal Improvements.

Report of the Joint Committee of the Senate and Assembly on Canals and Internal Improvements. Albany: Printed by Crosswell & Van Benthuysen, 1825. 33 p.

S-21670.

Copies: MWA; N; NN; PBL; RPB.

4137. North Carolina. Board of Internal Improvements.

Report of the Board of Internal Improvement to the General Assembly of North Carolina, in 1825. Raleigh: Printed by Bell & Lawrence, 1825. 25 p.

S-21704.

Copies: NcU.

4138. Pennsylvania. General Assembly.

Resolved by the Senate and House of Representatives of the Commonwealth of Pennsylvania in General Assembly met, that the Committee on Inland Navigation be instructed to report a bill constituting a board of some commissioners for internal improvement. [Harrisburg]: Mowry & Cameron, [1825.] 2 p.

Concerns the examination of three routes for water transportation through the state.

S-21834.

Copies: CSmH.

4139. Pennsylvania. General Assembly. House. Committee on Inland Navigation.

Report of the Committee on Inland Navigation read in the House of Representatives, March 15, 1825. Mr. Lehman, chairman. Harrisburg: Printed by John S. Wiestling, 1825. 3 p.

Concerns the navigation of the north branch of the Susquehanna, and a canal to be constructed from Susquehanna to Lehigh and Schuylkill rivers.

S-21829.

Copies: P; PPAmP.

4140. Pennsylvania Society for the Promotion of Internal Improvements in the Commonwealth.

Address to the citizens of Pennsylvania....[Philadelphia: 1825.] 2 p.

Signed: John Connelly, Vice President, John Y. Clark, Secretary, January 19, 1825.

The commerce of Pennsylvania compared to that of the state of New York after the opening of the Erie Canal.

S-21846.

Copies: DeGE: MCM; MWA; NN; PPAmP; PPL; PU.

4141. Pennsylvania Society for the Promotion of Internal Improvements in the Commonwealth.

At a meeting of the Acting Committee of the Pennsylvania Society for the Promotion of Internal Improvements, the following original paper was read by one of the members, and ordered to be published and put into general circulation ... No. I. The rivers of Pennsylvania. [Philadelphia: 1825.] 6 p.

Dated: Dec. 21, 1825.

S-21854.

Copies: DeGE; MBL; MWA; P; PHi.

4142. Pennsylvania Society for the Promotion of Internal Improvements in the Commonwealth.

Extracts from an address of the sub-

4142 (*continued*)
scribers, a committee of the Pennsylvania Society for the Promotion of Internal Improvements in the Commonwealth, to the citizens of the Commonwealth of Pennsylvania. [Philadelphia]: 1825. 8 p.

Dated: March 4, 1825.

S-21848.

Copies: NN; P.

4143. Pennsylvania Society for the Promotion of Internal Improvements in the Commonwealth.

Internal improvement. The Acting Committee . . . communicate to the public the following extract of their correspondence with Wm. Strickland . . . [Philadelphia: 1825.] 7 p.

Dated: September 23, 1825.

Includes a report of the Birmingham and Liverpool Railway Company, England.

S-21849.

Copies: MWA; NN; PPL; PU.

4144. Pennsylvania Society for the Promotion of Internal Improvements in the Commonwealth.

Internal improvement. Philadelphia, Dec. 21, 1825. The Acting Committee of "The Pennsylvania Society for the Promotion of Internal Improvement," communicate to the public the following extracts from their correspondence with William Strickland, Esq. and a communication received from him in reply of the same. [Philadelphia: 1825.] 7 p.

"Birmingham and Liverpool Railway," p. [5]–7.

Committee: Mathew Carey, Richard Peters, jun., Joseph Hemphill, and others.

S-19969, 21849, 21850.

Copies: DeGE; MWA; NN; PPL; PU.

4145. Pennsylvania Society for the Promotion of Internal Improvements in the Commonwealth.

To the citizens of the Commonwealth of Pennsylvania. The address of the subscribers, a committee of the "Pennsylvania Society for the Promotion of Internal Improvements in the Commonwealth." [Philadelphia: 1825.] 12 p.

Signed: Samuel Archer, Nicholas Biddle, John Connelly [and others]; dated at end: March 4, 1825.

"Constitution of the Pennsylvania Society for the Promotion of Internal Improvements in the Commonwealth," p. [11]–12.

Concerns a network of canals and railroads for Pennsylvania.

S-21857–8, 21861.

Copies: DBRE; MCM; MWA; NN; PPAmP; PPL; PU.

4146. ——— 2d ed. [Philadelphia: 1825.] 12 p.

S-21860.

Copies: PPL; PU.

4147. Tanner, Henry Schenck, 1786–1858.

The travellers guide. A map of the roads, canals and steam boat routes of the United States. . . . Philadelphia: H. S. Tanner, 1825.Map 45 x 51 cm.

S-22422.

Copies: ICU; IHi; IU; MH; MH-BA; MWA; MnSJ; NIC; PHi.

4148. United States. Board of Internal Improvement.

Message from the President of the United States, transmitting a report of the examination which has been made by the Board of Engineers, with a view to internal improvement, &c. February 14, 1825. . . . Washington: Printed by Gales & Seaton, 1825. 111 p. (18th Cong. 2d sess. House. Doc. 83)

Illustration of the Report of the Board of Internal Improvement, by J. L. Sullivan: p. [71]–111.

S-22996.

Copies: DLC; DeGE; DeU; MWA; NjP.

4149. ——— Washington: Printed by Gales & Seaton, 1825. 111 p. (18th Cong., 2d sess. Senate. Doc. 32)

Copies: DeGE; DeU; MWA; PPF.

4150. United States. Board of Internal Improvement.

Report of the Board of Engineers for internal improvements, as communicated, by message, from the President of the United States to Congress, February 14, 1825; and an illustration of the report by John L. Sullivan, a member of the Board. Washington: Printed by Gales & Seaton, 1825. 4, [2], 5–43 p.

Summary of the report.

N-U0065678; Parsons-p. 640.

Copies: DLC; ICJ; MB; MBAt; NN.

4151. United States. Congress. House.

Internal improvements. Mr. Bailey's resolutions proposing to amend the Constitution of the United States in relation to internal improvements. December 13, 1825.... Washington: [Printed by Gales & Seaton], 1825. 2 p. (19th Cong., 1st sess. House. Res. 4)

S-22893.

Copies: DLC; MWA; NjP.

4152. United States. Congress. House.

Internal improvements. Resolution submitted by Mr. Test. December 27, 1825.... Washington: [Printed by Gales & Seaton], 1825. 1 p. (19th Cong., 1st sess. House. Res. 7)

S-22894.

Copies: CtY; DLC; DeU; IU; MB; MH-BA; MWA; N; NN; NjP; P.

4153. United States. Congress. House. Committee on Roads and Canals.

Report of the Committee on Roads and Canals, upon the subject of internal improvements, accompanied by a bill "concerning internal improvements." February 26, 1825.... [Washington: 1825.] 16 p. (18th Cong., 2d sess. House. Rept. 83)

S-23075.

Copies: DLC; DeU; MWA; NN; NjP.

4154. United States. Engineer Dept.

Letter from the Secretary of War, transmitting copies of the reports and drawings of the superintending engineer, relating to the repair of Plymouth Beach. January 3, 1825.... Washington: Printed by Gales & Seaton, 1825. 11 p. (18th Cong., 2d sess. House. Doc. 27)

S-22956.

Copies: DLC; DeGE; DeU; MWA; NjP.

4155. [Armroyd, George.]

A connected view of the whole internal navigation of the United States, natural and artificial; present and prospective. With maps. Philadelphia: H. C. Carey & I. Lea, 1826. 192 p., 10 fold. maps.

S-23540.

Copies: CSmH; CSt; CtY; DBRE; DLC; DeGE; ICU; InU; MH-BA; MWA; MiD-B; NN; OClWHi; P; PPL; ScC; ViU; VtU; WHi.

4156. Carey (H. C.) & I. Lea, Philadelphia.

Just published by H. C. Carey & I. Lea, Philadelphia; in one volume, folio.... Reports upon canals, railways, roads, and other subjects, made by Wm. Strickland, Esq., engineer, agent for the Pennsylvania Society for the Promotion of Internal Improvement. on a tour through Great Britain. [Philadelphia: Carey & Lea, 1826.] 8 p.

S-23999.

Copies: MB; MH; PPL.

4157. [Carey, Mathew] 1760–1839.

Internal improvement. New series. No. I. [Philadelphia: 1826.] 8 p.

Dated on p. 4: Philadelphia, Jan. 19, 1826. Fulton.

Internal improvement. New series—No. II, p. [5]–8.

Dated on p. 6: Philadelphia, Jan. 31, 1826. Fulton.

S-24011–12.

Copies: DeGE; MCM; MWA; PPL; PU.

4158. Georgia. Board of Public Works.

Journal of the Board of Public Works of the state of Georgia at their first session held in Milledgeville, March 1826.... Milledgeville [Ga.]: 1826. 13 p.

S-24677.

Copies: WHi.

4159. Georgia. Board of Public Works.

Report of the Board of Public Works of the state of Georgia, to the General Assembly.... Milledgeville [Ga.]: Printed by Camak & Ragland, 1826. 14 p.

S-24680.

Copies: MHi.

4160. Maryland. General Assembly. House. Committee on Internal Improvement.

Report of the Committee on Internal Improvement, to the House of Delegates of Maryland, February 10th, 1826. With the accompanying bills. Annapolis: Printed by J. Hughes, 1826. 23 p.

N-M0289018.

Copies: DLC; MWA; MdBP.

4161. Massachusetts. Governor (L. Lincoln).

Speech of His Excellency Levi Lincoln, delivered before the legislature, June 6, 1826. Together with documents.... Boston: True and Greene, state printers, 1826. 75 p.

Discusses internal improvements and the relative merits of canals and railroads.

S-25272.

Copies: CSmH; DeGE; M; MBAt; MH; MWA; NcD; PMA.

4162. Pennsylvania. Commissioners for Promoting the Internal Improvement.

Report of the Commissioners for promoting the internal improvement of the state. No. II. Harrisburg: Printed by J. S. Wiestling, 1826. 19 p.

S-25696.

Copies: DBRE.

4163. Pennsylvania Society for the Promotion of Internal Improvements in the Commonwealth.

At a meeting of "The Acting Committee of the Pennsylvania Society for the Promotion of the Internal Improvement of the State," the following paper was read, discussed, agreed to, and directed to be printed. Signed by order, Mathew Carey, Vice-President.... The Union Canal. [Philadelphia: 1826.] 7 p.

Dated at head of title: Philadelphia, August 26, 1826.

Signed: P.

S-25712.

Copies: DeGE; MWA; PU.

4164. Pennsylvania Society for the Promotion of Internal Improvements in the Commonwealth.

At a meeting of "The Pennsylvania Society for the Promotion of Internal Improvement," the following address was reported by the acting committee.... To the citizens of the Commonwealth of Pennsylvania. [Philadelphia: 1826.] 3 p.

Dated: February 1, 1826.

S-25713; Thomson-119.

Copies: DeGE; MWA; NN; PHi.

4165. Pennsylvania Society for the Promotion of Internal Improvements in the Commonwealth.

At a meeting of "The Pennsylvania Society for the Promotion of Internal Improvement," the following original paper was read by one of the members, and ordered to be published. The Union Canal. [Philadelphia: 1826.] 6 p.

Dated: January 10, 1826.

S-25714.

Copies: P; PHi.

4166. Pennsylvania Society for the Promotion of Internal Improvements in the Commonwealth.

[Circular letter signed by Mathew Carey as vice president and Abraham Shoemaker as recording secretary, beginning:] Philadelphia, April 6, 1826.... [Philadelphia: 1826.] Broadside 25 x 18 cm.

On verso: "Constitution of the Auxiliary Internal Improvement Society of—."

Thomson-120.

Copies: MWA.

4167. Pennsylvania Society for the Promotion of Internal Improvements in the Commonwealth.

The first annual report of the Acting Committee of the Society for the Promotion

of Internal Improvement in the Commonwealth of Pennsylvania. Philadelphia: Printed by J. R. A. Skerrett, 1826. 45 p., 1 l.
S-25716.
Copies: DBRE; DLC; DeGE; ICJ; MH-BA; MWA; MiU-T; NN; NNC; P; PPL; PU.

4168. Pennsylvania Society for the Promotion of Internal Improvements in the Commonwealth.
The following letter and constitution are to be debated at half past five o'clock this evening, at the meeting of the Pennsylvania Society for the Promotion of Internal Improvements.... [Philadelphia: 1826.] [3] p.
Signed by Mathew Carey and others as Acting Committee.
Copies: MWA.

4169. Pennsylvania Society for the Promotion of Internal Improvements in the Commonwealth.
Sir, the Acting Committee of the "Pennsylvania Society for the Promotion of Internal Improvements in the Commonwealth," deeply anxious for the success of the grand measure of opening a communication between the eastern and western waters of the state.... [Philadelphia: 1826.] Broadside 25 x 20 cm.
Dated: Philadelphia, Jan. 10, 1826, and signed: Mathew Carey [and others].
Copies: MWA; PHi.

4170. Strickland, William, 1787–1854.
Reports on canals, railways, roads, and other subjects, made to "The Pennsylvania Society for the Promotion of Internal Improvement." By William Strickland ... while engaged in the service of the society. Philadelphia: H. C. Carey & I. Lea, 1826. vi, 51 p., illus., 57 plates (part col., part fold.), map.
S-26149.
Copies: CSmH; CtY; DLC; DeGE; ICU; InU; KU; LNL; MH; MWA; MdBJ; MiD-B; NBu; NN; NNC; NNE; NcU; Nj; OCl; PPL; RPB; T; Vi.

4171. United States. Congress. House.
Internal improvements. Mr. Mercer's resolution to provide a fund for internal improvement, and rules for its application. April 25, 1826.... [Washington: 1826.] 3 p. (19th Cong., 1st sess. House. Res. 45)
S-27099.
Copies: DLC; DeU; IU; MBAt; MH-BA; MWA; NN; NjP; P; PPAmP.

4172. United States. Engineer Dept.
Survey of the waters of Virginia & N. Carolina. Letter from the Secretary of War, transmitting a report of the Engineer Department, in relation to a survey of the waters of Virginia and North Carolina, in reference to a system of internal navigation. March 10, 1826.... Washington: Printed by Gales & Seaton, 1826. 9 p. (19th Cong., 1st sess. House. Doc. 125)
S-27417.
Copies: DLC; DeU; MWA; NjP.

4173. United States. Laws.
A bill concerning internal improvements. January 19, 1826. Reported by the Committee on Roads and Canals. [Washington: 1826.] 3 p. (H.R. 58)
S-26420.
Copies: DNA.

4174. United States. President (John Q. Adams).
Expenditures—internal improvements. Message from the President of the United States, transmitting the information required by a resolution of the House of Representatives of the 13th ult. in relation to expenditures incident, or relating to internal improvements, for the years 1824 & 1825. April 3, 1826.... Washington: Printed by Gales & Seaton, 1826. 7 p., fold. table. (19th Cong., 1st sess. House. Doc. 149)
Transmitting reports from the Treasury and War departments.
S-26984.
Copies: DLC; DeGE; DeU; MWA; NjP.

4175. Virginia. Board of Public Works.
The ninth and tenth annual reports of

4175 (*continued*)
the Board of Public Works to the General Assembly of Virginia; also, a collection of acts and resolutions of the General Assembly, passed at the sessions of 1824–'5 and 1825–'6, and other documents, on subjects connected with the fund for internal improvements. Vol. IV. Richmond: Printed by Shepherd & Pollard, 1826. 533 p.

S-27516.

Copies: CtY; DBRE; DLC; MCM; MWA; MiU-T; NN; PPL; Vi.

4176. At a very numerous and respectable meeting of farmers and manufacturers, and friends of farming and manufactures, of the city and county of Philadelphia, held at the house of John Neville, in Germantown, Thursday, May 31, 1827, Jacob Holgate, in the chair, Samuel Harvey, and Redwood Fisher, secretaries ... the following resolutions ... were presented with an appropriate address by Charles Ingersoll, Esq. and adopted by the meeting. [Philadelphia: 1827.] 4 p.

Promotes internal improvements and domestic manufactures.

S-27902.

Copies: DeGE; MWA; PU.

4177. [Baltimore. City Council. Committee on Internal Improvement.]

Report and resolution relative to internal improvement; and the Susquehanna Canal report. Adopted January 5th, 1827. Baltimore: Printed by Benjamin Edes, 1827. 29 p.

Signed: P. E. Thomas, Chairman. Attest. W. Meeteer, Sec.

S-27947.

Copies: DeGE; MdBE; MdBP; MdHi; PHi.

4178. [Carey, Mathew] 1760–1839.

To the citizens of the Commonwealth of Pennsylvania. Internal improvement. [Philadelphia: 1827.] 4 p.

Signed and dated: M. C., Philadelphia, March 6, 1827.

S-28407.

Copies: MWA; NN.

4179. ———— 2d ed. March 14, 1827. [Philadelphia: 1827.] 4 p.

Copies: PPL.

4180. ———— [Philadelphia: 1827.] 4 p.

At head of title: Third edition—March 17, 1827.

S-28408.

Copies: MWA; NN.

4181. Georgia. General Assembly. Senate. Committee on Internal Improvements.

Report on internal improvement. In Senate, 13th December, 1827.... [Milledgeville [Ga.]: 1827.] 8 p.

S-29013.

Copies: GU; WHi.

4182. Harrison, Randolph, 1769–1839.

Speech of Mr. Randolph Harrison, on the bill authorising [*sic*] a loan for internal improvement, and for other purposes. Richmond: Printed by Shepherd & Pollard, 1827. 16 p.

S-29149.

Copies: Vi; ViU.

4183. Internal improvement. At a meeting ... held at the Court House in the borough of Wilkesbarre, on Monday ... 15th January, 1827.... The object of the meeting being to consult upon ... making a complete ... navigation along the valley of the Susquehanna River.... [Wilkesbarre?: 1827?] Broadside.

S-29555.

Copies: PPL.

4184. Maryland. Board of Public Works.

Report of the Board of Public Works, to the General Assembly of Maryland.... Annapolis: Printed by Jeremiah Hughes, 1827. 10, [2] p.

S-29649.

Copies: MdBJ-G.

4185. Massachusetts. Governor (L. Lincoln).

Message of His Excellency Levi Lincoln, transmitted to both branches of the legislature January 3, 1827. Boston: True and Greene, state printers, 1827. 24 p.

Discusses the progress of internal improvements in Massachusetts.

S-29672.

Copies: DLC; DeGE; M; MBAt; MH; MHi; MWA; MiD-B; NjPT.

4186. Massachusetts. Governor (L. Lincoln).

Speech of His Excellency Levi Lincoln, before the two branches of the legislature, in convention. June 6, 1827. Boston: True and Greene, state printers, 1827. 16 p.

Discusses the newly constituted Board of Internal Improvements, and the progress of canals and railroad surveys in Massachusetts.

S-29684.

Copies: DeGE; M; MH; MHi; MWA; NcD; PHi.

4187. North Carolina. Board of Internal Improvements.

Report of the Board for Internal Improvements. Raleigh: Printed by Lawrence & Lemay, 1827. 28 p. (No. 7)

Letter of transmittal dated: 23d January, 1827.

S-30076.

Copies: NcU.

4188. ——— November, 1827. Raleigh: Printed by Lawrence & Lemay, 1827. 35 p.

S-30077.

Copies: DLC; NcU.

4189. North Carolina. Board of Internal Improvements.

Reports relative to the swamplands in North Carolina, by R. H. B. Brazier. Releigh: Printed by Lawrence & Lemay, 1827. 28 p.

S-30084; Thornton (Off.)—191.

Copies: DI-GS; NcU.

4190. North Carolina. Board of Internal Improvements.

Statement of expenditures on internal improvements. Raleigh: Lawrence & Lemay, printers, 1827. 5 p.

S-30086.

Copies: NcU.

4191. South Carolina. Superintendent of Public Works.

The report of the Superintendent of Public Works to the Legislature of South Carolina for the year 1827. Columbia, S.C.: Printed by Sweeny & Sims, at the Telescope Press, 1827. 18, 45, 6, 7, 8 p.

S-30668.

Copies: NN.

4192. Sullivan, John Langdon, 1777–1865.

Prospective economy, in the future public works of the state of New-York, describing the savings and effects on the Otsego, Chenango, Cayuga, Chemung and Allegany routes, by the use of the durable wooden lock, and single elevated railway. With general remarks on the consequences of the Pennsylvania improvements. By J. Langdon Sullivan.... Albany: Printed by Packard and Van Benthuysen, 1827. 30 p.

S-30740.

Copies: DBRE; DeGE; NHi.

4193. United States. Congress. House. Committee on Roads and Canals.

Surveys, with a view to making roads & canals. March 2, 1827. Mr. Mercer, from the Committee on Roads and Canals, made the following report. [Washington: 1827.] 31 p. (19th Cong., 2d sess. House. Rept. 102)

S-31523.

Copies: DLC; DeU; MWA; NjP.

4194. United States. Engineer Dept.

Letter from the Secretary of War, transmitting information respecting the surveys of roads and canals, and of their relative importance. February 19, 1827.... Washington: Printed by Gales & Seaton, 1827. 6 p. (19th Cong., 2d sess. House. Doc. 106)

S-31368.

Copies: DLC; DeGE; DeU; MWA; NjP.

4195. United States. Engineer Dept.
Surveys—roads and canals. Letter from the Secretary of War, transmitting (to the chairman of the Committee on Roads and Canals) a statement, exhibiting the number and designation of the examinations and surveys, which have been made with a view to the construction of roads and canals. February 7, 1827.... Washington: Printed by Gales & Seaton, 1827. 4 p., 4 fold. tables. (19th Cong., 2d sess. House. Doc. 83)
S-31521.
Copies: DLC; DeGE; DeU; MWA; NjP.

4196. Virginia. Board of Public Works.
Eleventh annual report of the president and directors of the Board of Public Works to the General Assembly of Virginia. January 17th, 1827. Richmond: Shepherd & Pollard, printers, 1827. 169 p.
Also issued together with the 12th and 13th annual reports, 1829.
Copies: Vi.

4197. [Carey, Mathew] 1760–1839.
To the citizens of the United States. Robert Fulton. [Philadelphia: 1828.] 2 p.
Signed: Philo Fulton. Philadelphia, March 17, 1828.
Concerned with internal improvements.
S-32598.
Copies: MWA; PPAmP; PPL; PU.

4198. Convention on Internal Improvement of the State, Charlottesville, Va., 1828.
Convention on Internal Improvement of the State, July 14th, 1828. Proceedings of the Charlottesville Convention. [Richmond: 1828.] 15 p.
S-32649.
Copies: NcU.

4199. Convention on Internal Improvement of the State, Charlottesville, Va., 1828.
To the honourable the speakers of the Senate and House of Delegates of Virginia, the memorial of the Convention assembled at Charlottesville [on July 14, 1828] on the subject of internal improvement. . . . [Charlottesville, Va.]: 1828. Broadside.
S-32650.
Copies: DLC; Vi.

4200. Georgia. General Assembly.
Legislature of Georgia on tariff and internal improvement. Letter from the Governor of Georgia, transmitting a report.... on the subject of the powers of the general government, for the purposes of encouraging domestic manufactures, and effecting a system of internal improvement. February 6, 1828.... Washington: Printed by Gales & Seaton, 1828. 14 p. (20th Cong., 1st sess. House. Doc. 120)
S-36490.
Copies: DLC; GU; MWA; NNG; NjR; ViU.

4201. Georgia. General Assembly.
Letter from the Governor of Georgia, with a report adopted by the Legislature of that state, on the tariff and internal improvement. February 6, 1828. Referred to the Committees on Manufactures and Roads and Canals.... Washington: Duff Green, 1828. 15 p. (20th Cong., 1st sess. Senate. Doc. 90)
S-36496.
Copies: DLC; MWA; NjR.

4202. Giles, William Branch, 1762–1830.
[Re-publication of opinions expressed by Wm. B. Giles on the subject of internal improvements. Richmond: T. W. White, printer, January 1, 1828.] 19 p.
Preface entitled "To the public" signed: Wm. B. Giles, January 1, 1828.
Concerns the public works in Virginia.
N-G0214723.
Copies: Vi; ViW.

4203. Maryland. General Assembly. House. Committee on Internal Improvement.
Report of the Committee on Internal Improvement. December session, 1827. [Annapolis: J. Green, 1828.] 22 p.
Dated: February 12, 1828.
Written by John V. L. McMahon.
S-34034.
Copies: DBRE (photocopy); NIC.

4204. Maryland. Laws.
A supplement to the act, entitled, An act for the promotion of internal improvement. [Baltimore: Wm. Wooddy, printer, 1828.] 16 p.
An Act passed by the Legislature of Pennsylvania to authorize the Baltimore and Ohio Rail Road Company to construct a rail road through Pennsylvania, p. [7]–14.
A Supplement to the act entitled, An act to incorporate the Baltimore and Ohio Rail Road Company, p. [15]–16.
S-34037.
Copies: CtY; MH-BA; MdBJ-G; NN; NcD.

4205. Massachusetts. Laws.
An act to provide for internal improvements by railroads.... [Boston: 1828.] 11 p.
N-M0311064.
Copies: DLC.

4206. Miner, Charles, 1780–1865.
Speech of Mr. Miner, on the constitutional power of Congress to make internal improvements. Delivered in the House of Representatives Monday, February 25, 1828. Washington: Printed by Gales & Seaton, 1828. 19 p.
S-34147.
Copies: DGU; DLC; DeGE; ICN; MBAt; MH; MWA; MoKU; PPL; ScU.

4207. North Carolina. Board of Internal Improvements.
Annual report of the Board for Int'l Improvements. November 29, 1828. Raleigh: Printed by Lawrence & Lemay, printers to the state, 1828. 43 p.
S-34569.
Copies: NcU.

4208. Pennsylvania. General Assembly. House. Committee on Internal Improvement.
Report of the Committee on Internal Improvement relative to the extension of the Pennsylvania Canal and the construction of a rail road. Read in the House of Representatives, February 4, 1828. Harrisburg: Printed by S. C. Stambaugh, 1828. 7 p.
S-34692.
Copies: DLC; P; PHi.

4209. Richard, Gaspard.
A petition & memorial, addressed to the Honorable Legislature of the state of New-York, and to the Congress of the United States. The plan and invention of Gaspard Richard. New York: Printed for the author, 1828. 20, 24 p.
On cover: No. 8. A new plan for internal improvement.
Concerns the financing of internal improvements.
S-35029.
Copies: DLC.

4210. Smith, William, 1762–1840.
Speech of the Hon. William Smith, of South-Carolina, in the Senate of the United States, on the bill making appropriations for internal improvements, delivered on the 11th April, 1828. Charleston: Printed and sold by A. E. Miller, 1828. 28 p.
S-35236.
Copies: CSmH; ICU; MiU; PPAmP; RPB; ScCC; ScU; TxU; WHi.

4211. United States. Congress. House. Committee on Roads and Canals.
Roads and canals. May 26, 1828 ... Mr. Mercer, from the Committee on Roads and Canals, made the following report.... [Washington: 1828.] 16 p. (20th Cong., 1st sess. House. Rept. 270)
S-36854.
Copies: DLC; DeGE; DeU; MWA; NjR.

4212. United States. Engineer Dept.
Internal improvement. Letter from the Secretary of War, transmitting the information required by a resolution of the House of Representatives of 19th Dec. 1826, in relation to the works of internal improvement which have been undertaken or projected by the general government, from the year 1824 to the year 1826, inclusive, &c. March 4, 1828. Washington: Printed by Gales & Seaton, 1828. 4 p., fold. table. (20th Cong., 1st sess. House. Doc. 172)

4212 (*continued*)
Report by Alex Macomb, Maj. Gen., Chief Engineer.
S-36418.
Copies: DLC; DeGE; DeU; MWA; NjR; PU.

4213. United States. Engineer Dept.
Internal improvement. Letter from the Secretary of War, transmitting the information required by a resolution of the House of Representatives of the 3d instant ... in relation to the amount deemed necessary in the execution and completion of each work of internal improvement specified in the report made on the 4th of March, 1828, &c. &c. April 29, 1828.... Washington: Printed by Gales & Seaton, 1828. 4 p., 2 fold. tables. (20th Cong., 1st sess. House. Doc. 261)
Transmitting the report of Alex. Macomb, Chief Engineer.
S-36419.
Copies: DLC; DeU; MWA; NjR.

4214. United States. Engineer Dept.
Internal improvements since 1824. Message from the President of the United States transmitting a report from the Secretary of War, in relation to harbors, roads, and other works of internal improvement, undertaken and projected since 30th April, 1824. December 8, 1828.... [Washington: Gales & Seaton, printers, 1828.] 5 p., 5 fold. tables. (20th Cong., 2d sess. House. Doc. 7)
Transmitting the report of the Engineer Department.
S-36420.
Copies: CtY; DLC; DeU; MB; MBAt; MH; MWA; NN; NjR.

4215. United States. Laws.
An act making appropriations for internal improvements. March 10, 1828.... [Washington: 1828.] 3 p. (H.R. 119)
S-35611.
Copies: DNA.

4216. Virginia. Board of Public Works.
Twelfth annual report of the Board of Public Works, to the General Assembly of Virginia. 24th January, 1828. Richmond: Printed by Thomas Ritchie, printer for the Commonwealth, 1828. 29 p.
Also issued together with the 11th and 13th annual reports, 1829.
S-37035.
Copies: CtY; DBRE; DLC; Vi.

4217. Wilde, Richard Henry, 1789–1847.
Speeches of Mr. Wilde, of Georgia, on internal improvements. Delivered in the House of Representatives of the United States, Feb. 29, and March 7, 1828. Washington: Printed by Gales & Seaton, 1828. 14 p.
S-37180.
Copies: DLC; MBAt; MH; MiD-B; NN.

4218. Cammeyer, William, firm, printers, Albany.
A new map of the Hudson River, the post roads between New York & Albany, the Northern and Western Canals, &c.,&c. Albany: W. Cammeyer, Junr., 1829. Col. map 122 x 11 cm., folded to 13 x 9 cm.
Inside back cover: Distances of places on the canals, from the sloop lock at Albany.
S-38037.
Copies: MB; N; NIC; NN; NNC.

4219. [Carey, Mathew] 1760–1839.
Internal improvement. [Philadelphia: 1829.] 4 p.
Signed and dated: Hamilton. Philadelphia, Nov. 12, 1829.
Copies: MWA; PU.

4220. ——— No. II. [Philadelphia: 1829.] 8 p.
Signed and dated: Hamilton. Philadelphia, December 5, 1829.
S-38062.
Copies: MBL; MHi; MWA; PPAmP; PPL; PU.

4221. An essay on inland navigation and internal improvement as now prosecuted in Pennsylvania. By a freeholder of Franklin County. Chambersburg, [Pa.]: Printed by

Geo. K. Harper, 1829. 15 p.
S-38488.
Copies: DLC; ICU; MdHi; MiU-T; NN; P.

4222. Finley, Anthony.
Pennsylvania. Philadelphia: A. Finley, 1829. Col. map 33 x 47 cm.
Shows routes of roads, canals, railroads and proposed railroads.
S-38574.
Copies: DLC; PHi.

4223. Maryland. General Assembly. House. Committee on Internal Improvement.
Report of the Committee on Internal Improvement to the House of Delegates, March 12, 1829, by Mr. Lee, chairman. Annapolis: Jeremiah Hughes, printer, 1829. 4, 47 p.
Includes the report of the Board of Engineers of the Baltimore and Ohio Railroad Company, dated Feb. 6, 1829, and its First annual report, dated Sept. 30, 1828.
S-39471; Thomson-318.
Copies: MWA; Md (not located).

4224. Massachusetts. Governor (L. Lincoln).
Message of His Excellency Levi Lincoln, transmitted to both branches of the legislature, January 7, 1829. Boston: True and Greene, state printers, 1829. 27 p.
Discusses the progress of internal improvements and the promotion of manufactures and agriculture in Massachusetts.
S-39499.
Copies: DLC; DeGE; M; MB; MBAt; MH; MHi; MWA; RPB; WHi.

4225. Massachusetts. Governor (L. Lincoln).
Speech of His Excellency Levi Lincoln, before the Honourable Council, and both branches of legislature, May 30, 1829. Boston: True and Greene, state printers, 1829. 22 p. (Senate. Doc. 1)
Parts of his speech refer to canals and railroads.
S-39512.
Copies: DBRE; DLC; DeGE; M; MB; MBAt; MH; MWA.

4226. North Carolina. Board of Internal Improvements.
Annual report of the Board for Int'l Improvements, Nov. 28, 1829. Raleigh: Printed by Lawrence & Lemay, printers to the state, 1829. 20 p. (No. 6)
S-39855.
Copies: NcU.

4227. North Carolina. Board of Internal Improvements.
Internal improvements. Raleigh: 1829. 20, [1] p.
S-39858.
Copies: NN; NcU.

4228. Pennsylvania. General Assembly. House. Committee on Inland Navigation and Internal Improvement.
Report of the Committee on Inland Navigation and Internal Improvement, relative to rail roads and canals. Mr. Lehman, chairman. Read in the House of Representatives, February 26, 1829. Harrisburg: Printed at the office of the Reporter, 1829. 8 p.
S-40003.
Copies: DLC; ICU; MWA; NNC; P; PHi.

4229. Richard, Gaspard.
Prospectus to form a new company for internal improvement. Invented by Gaspard Richard. New York: Printed for the author, 1829. 8 p.
A scheme to finance internal improvements.
S-40271.
Copies: DLC.

4230. South Carolina. Superintendent of Public Works.
Report of the Superintendent of Public Works. Columbia, S.C.: D. & J. M. Faust, 1829. 22 p.
Sabin-87727.
Copies: ScU.

4231. Tanner, Henry Schenck, 1786–1858.
Memoir on the recent surveys, observations, and internal improvements, in the United States, with brief notices of the new counties, towns, villages, canals, and railroads, never before delineated. By H. S. Tanner. Intended to accompany his new map of the United States. Philadelphia: The author, 1829. 1 p. l., 108 p.
S-40604.
Copies: A-Ar; CLSM; CSmH; CtY; DBRE; DLC; DeGE; FSaHi; ICN; In; KU; LNHT; MH; MWA; MdBP; MnHi; MoKU; NIC; NN; OClWHi; PPL; PPi; RPB; TNJ; ViU.

4232. United States. Board of Internal Improvement.
Map of the territory of Florida. From its northern boundary to lat. 27°, 30′ N, connected with the delta of the Mississippi. Annexed to the report of the Board of Internal Improvement dated Feb.y 19th 1829, relating to the canal contemplated to connect the Atlantic with the Gulf of Mexico; and describing the inland navigation parallel to the coast from the Mississippi to the Bay of Espiritu Santo; and from St. Mary's Harbour to St. Augustine. Drawn and compiled by W. H. Swift, Lt. Artillery. [Washington: 1829.] Map 67 x 160 cm.
N-U0065671.
Copies: DeGE; MWA; NIC; PPAmP; RPB.

4233. United States. Register of the Treasury.
Message from the President of the United States, with reports from the Secretaries of Treasury and War, showing the amount expended by the federal government of works of internal improvement, within the limits of several states, with an estimate of the amount necessary to complete any work begun and not yet completed. February 10, 1829.... [Washington: 1829.] 25 p. (20th Cong., 2d sess. Senate. Doc. 69)
Transmitting the reports of Joseph Nourse, Register of the Treasury.
Copies: DeU; MWA.

4234. Virginia. Board of Public Works.
The eleventh, twelfth and thirteenth annual reports of the Board of Public Works to the General Assembly of Virginia. Richmond: Printed by Samuel Shepherd & Co., 1829.[1], 606 p.
Individual reports have separate title pages.
S-41384.
Copies: MWA; PPL; Vi.

4235. Virginia. Board of Public Works.
Twelfth annual report of the president and directors of the Board of Public Works, to the General Assembly of Virginia. January 24th, 1828. Richmond: Printed by Samuel Shepherd & Co., 1828 [i.e. 1829]. [171]-410 p.
Issued together with the 11th and 13th annual reports, 1829.
S-37036.
Copies: CtY; DLC; IU; MiU-T; NN; Vi.

4236. Virginia. Board of Public Works.
Thirteenth annual report of the president and directors of the Board of Public Works, to the General Assembly of Virginia, January 23, 1829. Richmond: Printed by Samuel Shepherd & Co., 1829. 411–590 p.
Issued together with the 11th and 12th annual reports.
S-41387.
Copies: DLC; MiU-T; NN; PPL; Vi.

4237. [Williams, John S.]
Report of the engineer, to the Association for the Promotion of Internal Improvements in the State of Kentucky. [S.l.: 1829.] 8 p.
Signed: Jno. S. Williams, engineer. Maysville, Ky. Dec. 13th, 1829.
S-41541.
Copies: MB; MiU-T; NN.

4238. Alabama. Board of Internal Improvement.

Report of the Board of Interral [*sic*] Improvement, to the Hon. the Senate and House of Representatives of the state of Alabama. [S.l.: 1830.] 24 p.

Latest survey included is dated September 20, 1830. Includes a discussion of the relative merits of railroads over canals.

N-A0122697; Thomson-370.

Copies: MWA; NN.

4239. [Armroyd, George.]

A connected view of the whole internal navigation of the United States; natural and artificial, present and prospective: corrected and improved from the edition of 1826, and much enlarged.... The map ... is projected, on a reduced scale, from Tanner's ... map of the United States.... By a citizen of the United States. Philadelphia: Published by the author; printed by Lydia R. Bailey, 1830. viii, [9]–617 p., fold. map, fold. diagrs.

C-189.

Copies: CU; CtY; DLC; DeGE; ICN; IaU; InU; LNHT; MH; MWA; Md; MeU; MiU; MnU; MoSM; NN; NjP; OO; PPL; PU; RP; ViW.

4240. [Chapman, Isaac A.]

A sketch of the history of Wyoming.... To which is added, an appendix, containing a statistical account of the valley, and adjacent country. By a gentleman of Wilkesbarre. Wilkesbarre, Pa.: S. D. Lewis, 1830. 209 p.

The appendix offers information on roads, turnpikes, canals, natural resources such as coal and iron ore, and manufacturing enterprises.

C-826.

Copies: CSmH; CtHT-W; DLC; DeGE; ICN; Ia-HA; MH; MWA; MdBJ; MiD-B; MnHi; OU; PPL; PU.

4241. Coxe, Richard Smith, 1792–1865.

Address delivered by Richard S. Coxe, before the mechanics and other citizens of the city of Washington, friends of Henry Clay, on Monday, July 5, 1830. Washington: Printed by Peter Force, 1830. 15 p.

Includes observations on the progress in internal improvements and manufactures.

C-1031.

Copies: CSmH; DLC; DeGE; LNHT; MHi; MWA; MiU; NCH; OClWHi; OO; PFL; RPB.

4242. Massachusetts. Board of Internal Improvements.

To the Honourable House of Representatives.... [Boston: 1830.] 11 p. (House. Doc. 3)

Detailed account of expenditures, including those for reports, maps, and profiles.

N-M0306429; Thomson-417.

Copies: M; MB; MBAt; MH; MWA; NN.

4243. Massachusetts. Governor (L. Lincoln).

Message of His Excellency Levi Lincoln, communicated to the two branches of the legislature, January 6. 1830. Boston: Dutton and Wentworth, printers to the state, 1830. 40 p. (House. Doc. 1)

Parts of the message concern internal improvements in Massachusetts.

C-2473.

Copies: DeGE; M; MB; MBAt; MH; MHi; MWA; NNC.

4244. Massachusetts. Treasury Dept.

Statement of sundry expenditures for surveying railways and canals, since June session, 1824, including all monies paid as incident thereto. [Boston: 1830.] 8 p. (House. Doc. 14)

Caption title from p. [3].

Dated: Treasury Office, January 25, 1830.

C-2482.

Copies: DeGE; M; MB; MBAt; MH; MWA; NN.

4245. North Carolina. Board of Internal Improvements.

Annual report of the Board for Int'l Improvements. 1830. Raleigh: Printed by Lawrence & Lemay, printers to the state,

4245 (*continued*)
1830. 35 p.
C-2864.
Copies: NcU.

4246. Phelps, Humphrey.
Map of the state of New York, with the latest improvements.... New York: Humphrey Phelps, 1830. Map 42 x 50 cm.
Shows the proposed route of the Great Western Railway.
C-3025.
Copies: CSmH; MB; MH-BA; NN; PHi.

4247. Tanner, Henry Schenck, 1786–1858.
Memoir on the recent surveys, observations, and internal improvements, in the United States, with brief notices of the new counties, towns, villages, canals, and railroads, never before delineated. By H. S. Tanner. Intended to accompany his new map of the United States. 2d ed. Philadelphia: The author, 1830. 1 p., 108 p.
C-3663.
Copies: CSmH; CtY; DBRE; DLC; DeGE; ICJ; MH; MWA; NN; NcD; NjP; P; PPL; PU; TxU.

4248. Tanner, Henry Schenck, 1786–1858.
The traveller's pocket map of Ohio, with its canals, roads, & distances. [Philadelphia: H. S. Tanner, 1830.] Map 29 x 35 cm.
Engraved by E. B. Dawson.
C-3667.
Copies: OO.

4249. Tanner, Henry Schenck, 1786–1858.
The traveller's pocket map of Pennsylvania, with its canals, railroads & distances from place to place. Philadelphia: H. S. Tanner, 1830. Map.
C-3668.
Copies: DBRE; DLC; MB; MH.

4250. Tanner, Henry Schenck, 1786–1858.
The travellers pocket map of Tennessee with its roads and distances from place to place along the stage and steamboat routes. Philadelphia: H. S. Tanner, [1830]. Map 27 x 36 cm.
C-3669.
Copies: NNF; PHi.

4251. United States. Laws.
An act making appropriations for examinations and surveys, and also for certain works of internal improvement. April 2, 1830.... [Washington: 1830.] 2 p. (H.R. 279)
C-3828.
Copies: DNA.

4252. Virginia. Board of Public Works.
Fourteenth annual report of the Board of Public Works, to the General Assembly of Virginia. 22d January, 1830. Richmond: Printed by S. Shepherd & Co., 1830. 108 p.
With returns and reports of internal improvement companies.
C-5287.
Copies: Vi.

4253. ——— Richmond: Printed by T. Ritchie, 1830. 50 p.
C-5286.
Copies: Vi (not located).

4254. Young, James Hamilton.
The tourist's pocket map of Pennsylvania, exhibiting its internal improvements, roads, distances, &c. By J. H. Young. Philadelphia: S. Augustus Mitchell, [1830]. Map 29 x 36 cm.
C-5591.
Copies: MB; NN; P.

2. Navigation and improvement of rivers, lakes and inlets

4255. New York (Colony). Laws.
An act to regulate the ferry between the city of New York and the island of Nassau, and to establish the ferriage thereof. [New York: 1727?] 2 p.
N-N0187584.
Copies: NN.

4256. Remarks. The common rates of land carriage.... [Philadelphia: 1772.] Broadside, incl. map.
Dated: Philadelphia, January 20, 1772.
Concerns the use of the Susquehanna River for transportation.
B-3483; Mp-42370.
Copies: PPL.

4257. Maryland. Laws.
An act of the General assembly of the State of Maryland, entitled, An act for making the river Susquehanna navigable from the line of this state to tide water. Annapolis: Printed by Frederick Green, printer to the state, [1784]. 17 p.
B-5919; Mp-44553.
Copies: DLC; MdBJ-G (not located).

4258. Virginia. Laws.
An act for opening and extending the navigation of Potomack River.... Alexandria: Printed by George Richards and Company, [1784]. [4] p.
Signed: William Drew, G.S.
E-20103.
Copies: DLC; MH-BA.

4259. To the public. Pursuant to an act of the General Assembly passed at the last session, entitled "An act for opening and extending the navigation of Potowmack River...." [Richmond: Printed by James Hayes?, 1785.] Broadside 33.5 x 33 cm.
A report of the Commissioners of Virginia and Maryland, Jacquelin Ambler and John Beckley, managers. Dated: City of Richmond, Jan. 22, 1785.
B-6183; E-19274.
Copies: DLC (not located).

4260. Company for the Inland Navigation, from Santee to Cooper River.
Rules of the company for opening the inland navigation, between the Santee and Cooper Rivers, agreed to on March the 23d, 1786. To which is prefixed, the act of the Legislature, passed the 22d of March, for incorporating the company, and an ordinance, &c. Charleston: Printed by Bowen and Markland, 1786. [16] p.
B-6271; Mp-44875.
Copies: MiU-C; NN (photocopy).

4261. Georgia. Laws.
An act to improve the navigation of Brier Creek.... [Savannah?: J. Johnston?: 1786.] 4 p.
Dated: Augusta, 30th January, 1786.
"An act for improving the navigation of Savannah River," p. 2–4.
B-6290; Mp-44894.
Copies: GU-De.

4262. Leith and Shepherd, Detroit.
Transport on lakes Erie and Huron.... [S.l.: 1790.] Broadside 28 cm.
Dated: Detroit, 26th January, 1790.
B-7401; Mp-45899.
Copies: MWA (facsim.); N; NN (facsim.).

4263. Pennsylvania. Commissioners Appointed to Explore the Rivers Delaware, Lehigh and Schuylkill.
Copy of a report from Reading Howell,

4263 (*continued*)
Frederick Antes and William Dean, Esquires, Commissioners appointed to explore the headwaters of the rivers Delaware, Lehigh, and Schuylkill, and the north-east branch of Susquehanna. Also the report of the Commissioners appointed to examine the western waters of the state of Pennsylvania. Philadelphia: Printed by Francis Bailey, 1791. [33] p.

Mp-23678.

Copies: MWA; NN; NjR; P; PPL; PPiU; ScU.

4264. Pennsylvania. Commissioners Appointed to View and Explore the Rivers Susquehanna and Juniata.

Reports of sundry commissioners appointed to view and explore the rivers Susquehanna and Juniata; the river Delaware; the river Schuylkill, &c.... Philadelphia: Printed by Francis Bailey, 1791. [27] p.

Mp-23679.

Copies: MWA; NN; NNC; NjR; P; PPL; PPiU; ScU.

4265. Massachusetts. Laws.

An act incorporating the Proprietors of the Locks and Canals, on the Merrimack River.... Boston: Printed at the State Press, by Thomas Adams, 1792. 12 p.

Mp-24714.

Copies: MHi.

4266. New York (State). Committee to Explore the Western Waters.

The report of a Committee, appointed to explore the western waters in the state of New-York: for the purpose of prosecuting the inland lock navigation. Albany: Printed by Barber and Southwick, 1792. 24 p.

Members of the committee: Philip Schuyler, Goldsbrow Banyar, and Elkanah Watson.

A survey of the Mohawk River and Wood Creek for the Western Inland Lock Navigation Company.

Mp-24604.

Copies: MH-BA; MWA; NHi; NN.

4267. Northern Inland Lock Navigation Company.

A report of the committee appointed by the directors of the Northern Inland Lock Navigation Company, in the state of New-York, to examine Hudsons River. [New York]: From the press of W. Durell, [1792]. 20 p.

Dated: October 30, 1792; signed: Philip Schuyler, Abraham Ten Broeck, Jacobus Van Schenhoven, Stephen V. Rensselaer, Henry Quackenbosh.

Mp-24635.

Copies: CSmH; DeGE; MWA; NN.

4268. Proprietors of the Locks and Canals on Merrimack River.

Bye-laws of the Proprietors of Locks and Canals on Merrimack-River. Passed August 8, 1792. Newburyport, [Mass.]: Printed by John Mycall, 1792. 14 p.

Mp-24715.

Copies: MBAt; MHi.

4269. [Lear, Tobias] 1762–1816.

Observations on the river Potomack, the country adjacent, and the city of Washington. New York: Printed by Samuel Loudon and Son, 1793. 29 p., fold. map.

Concerns improvements in the river navigation and the construction of a canal.

Also attributed to Andrew Ellicott.

Mp-25711.

Copies: DLC; MHi; MWA; PPL; ViU.

4270. Virginia. Laws.

Acts of General Assembly, for clearing and improving the navigation of James River. Richmond: Printed by Thomas Nicolson, [1793]. 28 p.

Mp-26387.

Copies: DLC; MWA; NN; PPAmP.

4271. [Lear, Tobias] 1762–1816.

Observations on the river Potomack, the country adjacent, and the city of Washington. New York: Printed by Loudon and Brower, 1794. 30 p., fold. plan.

Concerns improvements in the river navigation and the construction of a canal.

Also attributed to Andrew Ellicott.
Mp-27209.
Copies: DLC; MBAt; MH; MWA; NN; NHi; PPL; RPJCB; ViU.

4272. Pennsylvania. Laws.
A supplement to the act entitled "An act to enable the Governor ... to incorporate a company for opening a canal and lock navigation on ... Brandywine Creek." [Philadelphia]: Zachariah Poulson, junr., [1794]. 2 p.
Signed into law March 19, 1794.
B-8810; Mp-47169.
Copies: MWA.

4273. Proprietors of the Locks and Canals on Merrimack River.
To the inhabitants of the towns bordering upon and near the Merrimac River. The Proprietors of the Locks and Canals of the said river, beg leave to represent, that: they have carefully examined the falls below Patuckett and in result are convinced, that to construct locks and canals by the said fall would involve an expence for which the trade of the river can never be expected to afford them a compensation. [Newburyport]: 1794. Broadside.
Dated: October 29, 1794.
Mp-27567.
Copies: MWA.

4274. [Condy, Jonathan Williams] 1770–1828.
A description of the river Susquehanna, with observations on the present state of its trade and navigation, and their practicable and probable improvement. Illustrated with a map, and an appendix. Philadelphia: Printed by Zachariah Poulson, junior, 1796. [4], 60 p., map.
Sometimes attributed to John Adlum.
Mp-30338.
Copies: CSmH; DLC; MBAt; MH; MWA; MiU-C; NN; NjPT; PPAmP; PPL; RPJCB; ScC.

4275. Maryland. Laws.
Acts passed at different sessions of the General Assembly of Maryland: concerning the navigation of the river Susquehanna, and establishment of Havre-de-Grace, &c. and naturalization laws of said state—and the United States, &c. Baltimore: Printed by W. Pechin, 1797. 37 p.
Includes the act chartering the Susquehanna Canal Company.
Mp-32429.
Copies: MH; PPAmP.

4276. The Ohio navigator. Comprising an ample account of that beautiful river from its head to its junction with the Mississippi ... containing accurate directions for its navigation ... Taken from the journals of gentlemen of observation who have frequently navigated it. Washington, [Ky.]: Printed by Hunter & Beaumont, 1798.
Zadok Cramer's *Ohio and Mississippi navigator* is reputedly based on this work.
E-33915. No copy known.

4277. ———— 2d ed. Washington, [Ky.]: Printed by Hunter and Beaumont, 1798.
E-33916.
No copy known.

4278. Pennsylvania. Laws.
An act to enable the Governor of this Commonwealth to incorporate a company for the purpose of improving the navigation of the river Lehigh.... [Philadelphia: Printed by Hall and Sellers, 1798.] 15 p.
Approved February 17, 1798.
Mp-34322.
Copies: DLC; PPAmP.

4279. Maryland. Laws.
A further supplement to the act ... for making the river Susquehanna navigable.... [Annapolis: F. Green, 1799.] Broadside 34 x 21 cm.
B-10858.
Copies: MdHi (not located).

4280. [Cramer, Zadok] 1773–1813.
The Ohio and Mississippi navigator....

4280 (*continued*)
A particular description of the several towns, posts, caves, ports, harbours &c. on their banks, and accurate directions how to navigate them ... Together with a description of Monongahela and Allegheny rivers. 3d corr. ed. Pittsburgh: Printed by John Scull, for Zadok Cramer, 1802. 40 p.
S&S-2096.
Copies: CtY; DLC; ICN; MWA; MiU-C; OCHP; PPi; TU.

4281. Notes on the navigation of the Mississippi; taken by a gentleman of talents and observation, and corrected after several voyages, in all stages of water.... Frankfort, Ky.: Printed by James M. Bradford, 1803. 56 p., illus.
S&S-4781.
Copies: WHi.

4282. Ogden, Aaron, 1756–1839.
Aaron Ogden and Thomas Gibbons, vs. Walter Rutherfurd. [Camden, N.J.: 1803.] 39 p.
Litigation concerning the sale of disputed property, including the Elizabeth Town Point Ferry.
N-O0040003.
Copies: MH-L; MiU; NjP.

4283. [Cramer, Zadok] 1773–1813.
The Ohio and Mississippi navigator.... Particular directions how to navigate them.... With a description of and directions how to navigate the Allegheny & Monongahela rivers.... A new and corr. ed. Pittsburgh: Zadok Cramer, 1804. 62 p.
S&S-6095.
Copies: MB; MWA.

4284. New Jersey. Laws.
Act of incorporation of the New Jersey Navigation Company. Trenton: 1804.
S&S-6902; Sabin-53036.
Copies: NHi (not located).

4285. Notes on the navigation of the Mississippi.... Taken by a gentleman of talents and observation.... Frankfort, Ky.: James M. Bradford, 1804. 56 p., 9 plates.
McMurtrie (Kentucky)-221; S&S-6946.
No copy known.

4286. James River Company.
The agents for the James River Company ... beg leave ... to present to each member of the legislature, a statement of the situation of the company from its commencement to the present day. [Richmond: 1805.] 11 p.
The report extends to January 15, 1805.
N-J0038198; S&S-8692.
Copies: DLC; ViU.

4287. Lehigh Coal and Navigation Company.
20,000 dollars!!! May be gained for 2 1/2 in the third class Lehigh Navigation unparalleled lottery ... Aug. 20, 1805. [Philadelphia: 1805.] Broadside.
S&S-8769a.
Copies: NN.

4288. [Cramer, Zadok] 1773–1813.
The navigator; or, The traders' useful guide in navigating the Monongahela, Allegheny, Ohio, and Mississippi rivers.... The 5th ed., much improved and enlarged.... Pittsburgh: From the press of Zadok Cramer, 1806. 94 p., maps.
S&S-10224.
Copies: CSmH; DLC; ICN; MB; MWA; NHi; OClWHi; OMC; PPiU.

4289. ——— Pittsburgh: From the press of Zadok Cramer, 1806. 156 p., maps.
S&S-10225.
Copies: CSmH; DLC; MWA; MiU; OClWHi; PPiU.

4290. Moore, Thomas.
Address to citizens of Georgetown and Washington, on improving the navigation of the river Potomac. Georgetown: 1806. 10 p.
S&S-10897.
Copies: DLC.

4291. United States. Laws.
A bill providing for the appointment of commissioners to ascertain the practicability of removing the obstructions in the navigation of the river Ohio, at the rapids. Feb. 26, 1807.[Washington: 1807.] 2 p.
S&S-13845.
Copies: DLC; MWA (photocopy).

4292. United States. Treasury Dept.
Letter from the Secretary of the Treasury, transmitting a report, prepared in obedience to "An act directing the Secretary of the Treasury to cause the coast of North Carolina between Cape Fear and Cape Hatteras, to be surveyed," passed on the 10th of April last. January 27, 1807.... City of Washington: A. & G. Way, printers, 1807. 4 p.
S&S-13959.
Copies: DLC; MWA; NcU.

4293. Catawba and Wateree Company.
Petition and memorial of the subscriber, in behalf of the company incorporated by law, for opening the navigation of the Catawba and Wateree rivers. [Charleston, S.C.: 1808.] 3, 12 p.
At head of title: To the Honorable the Senate and members of the House of Representatives of the state of South Carolina.
Signed: J. F. Grimke, president.
S&S-14652.
Copies: MH-BA; MWA; ScCC.

4294. [Cramer, Zadok] 1773–1813.
The navigator: containing directions for navigating the Monongahela, Alleghany, Ohio and Mississippi rivers; with an ample account of these much admired waters ... and a concise description of their towns, villages, harbours, settlements, &c. With accurate maps of the Ohio and Mississippi. To which is added an appendix, containing an account of Louisiana, and the Missouri and Columbia rivers ... 6th ed., improved and enlarged. Pittsburgh: Zadok Cramer; from the press of Cramer & Spear, 1808. 156 p., illus., maps.
S&S-14796.
Copies: CtY; DLC; DeU; ICN; MH-BA; MWA; MiU-C; MoSM; NN; NNC; OCHP; PPi; PPiU; RPJCB; ViU.

4295. United States. Congress. House. Committee on the Navigation of the Catawba and Wateree Rivers.
Report of the Committee to whom was referred, on the fourteenth November last, the petition of the company for opening the navigation of the Catawba and Wateree rivers. January 26, 1809. City of Washington: A. & G. Way, printers, 1809. 4 p.
S&S-19056.
Copies: DLC; MWA; NN.

4296. Sullivan, John Langdon, 1777–1865.
On inland navigation, by the Merrimack. [Concord?, N.H.]: Printed by J. C. Tuttle, [181–]. Broadside.
S&S-21439.
Copies: CSmH.

4297. United States. Laws.
A bill for the improvement of the navigation of the river Potomac. January 2d, 1810.... [Washington: 1810.] 3 p. (H.R. 25)
S&S-21586.
Copies: DLC.

4298. [Cramer, Zadok] 1773–1813.
The navigator: containing directions for navigating the Monongahela, Allegheny, Ohio, and Mississippi rivers; with an ample account of these much admired waters ... and a concise description of their towns, villages, harbours, settlements, &c. With accurate maps of the Ohio and Mississippi. To which is added, an appendix, containing an account of Louisiana and of the Missouri and Columbia Rivers ... 7th ed., improved and enlarged. Pittsburgh: Cramer, Spear & Eichbaum, 1811. x, [11]–295, [1] p., maps.
S&S-22636.
Copies: CHi; CtY; DLC; ICN; IaHi; KyLxT; LU; MH; MWA; MdHi; MiD-B; MoSHi; OClWHi; P; PPL.

4299. Moore, Thomas.

Ship navigation to Georgetown. [S.l.: 1811.] 56 p.

"Thomas Moore's comments on B. H. Latrobe's deposition and the subsequent documents," p. [29]–56.

N-M0750833.

Copies: MiU.

4300. [Tucker, George] 1775–1861.

A letter to a member of the General Assembly of North Carolina, on the navigation of the Roanoke and its branches. By a citizen of Pittsylvania. Richmond: Printed by John O'Lynch, 1811. 66 p.

N-T0371868; Thornton-2259.

Copies: NcU; PPiU; Vi; ViW.

4300a. Virginia. Laws.

An act for opening, improving and extending the navigation of the river Rappahannock, and all its improvable branches. Passed February 9, 1811. [Richmond: 1811?] 12 p.

N-V0187329.

Copies: ViU.

4301. Latrobe, Benjamin Henry, 1764–1820.

Opinion on a project for removing the obstructions to a ship navigation to Georgetown, Col., by B. H. Latrobe.... Washington: W. Cooper, printer, 1812. viii, 25 p.

N-L0122671; S&S-25830 (erroneously assigned to John Hazlehurst Boneval Latrobe).

Copies: DBRE; DLC; NN; OCHP; PPF.

4302. North Carolina. Laws.

Act to incorporate a company entitled The Roanoke Inlet Company, and for other purposes. [Raleigh?: 1812.] 669-672 p.

S&S-26310.

Copies: NcU.

4303. United States. Congress. House. Committee on the District of Columbia.

Report of the Committee for the District of Columbia, on the bill from the Senate, "for improving the navigation of the river Potomac, opposite the city of Washington," and sundry memorials in opposition to the passage of said bill. May 7th, 1812.... Washington City: Printed by R. C. Weightman, 1812. 4 p.

S&S-27297.

Copies: DLC.

4304. United States. Congress. Senate. Committee on Roads and Canals.

In Senate of the United States. March 26th, 1812. Mr. Goodrich, from the committee to whom was referred the bill for improving the navigation of the river Potomac opposite the city of Washington, reported that the bill be amended.... [Washington: 1812.] 3 p.

S&S-27271.

Copies: DLC.

4305. United States. Laws.

A bill for improving the navigation of the river Potomac, opposite the city of Washington. May 7th, 1812. [Washington: 1812.] 4 p. (H.R. 125)

S&S-26981.

Copies: DLC.

4306. Delacy, John D.

Having, as agent for Messrs. Livingston and Fulton, and under the immidiate [*sic*] direction [of] Mr. Fulton explored the sounds and waters south of the river Potomack, and Chesapeak Bay, I do certify there is [*sic*] but the three following obstructions to an inland communication between St. Mary's at East Florida, and the waters of the James River and Chesapeak Bay ... John D. Delacy. South-Carolina. July, 1813. [S.l.: 1813.] Broadside 17 x 23 cm.

S&S-28291.

Copies: MWA; PPAmP.

4307. [Cramer, Zadok] 1773–1813.

The navigator; containing directions for navigating the Monongahela, Allegheny, Ohio and Mississippi rivers ... and a concise description of their towns, villages, harbors, settlements &c. With maps of the Ohio and

Mississippi. To which is added an appendix, containing an account of Louisiana, and of the Missouri and Columbia rivers ... 8th ed. — improved and enlarged. Pittsburgh: Cramer, Spear and Eichbaum, 1814. 360 p., maps.
S&S-31260.
Copies: CLU; CSmH; CtY; DLC; DeGE; GU; ICN; In; MH; MWA; MiU-C; MnHi; NNA; OC; P; PPAmP; PPi.

4308. Badollet, John Louis, 1758–1837.
The report of John Badolett [*sic*], agent of information, appointed to explore the waters of Monongahela and Youghiogheny. Harrisburg: Printed by James Peacock, 1816. 24 p.
The report was printed in the *Appendix* of the *Journal of the House of Representatives*, Pennsylvania, 1816, p. 58–82. No separate publication has been located.
S&S-36784 (gives 84 p.).
Copies: DLC; PHi.

4309. James River Company.
Report in answer [to Resolutions of the Virginia Board of Public Works, requesting information relative to the several public works under their direction]. [Richmond: 1816.] 8 p.
N-J0038197.
Copies: Vi.

4310. North Carolina. Commissioners on Inland Navigation.
Report on inland navigation [Dec. 6, 1816, by P. Browne, chairman, Commissioners of Inland Navigation, enclosing a report of B. F. Baldwin, engineer, on a survey of Tar and Neuse rivers. Raleigh: 1816]. Broadside 43 x 22 cm.
S&S-36789.
Copies: MB; NcU.

4311. Pennsylvania. Commissioners Appointed to Explore the Rivers Susquehanna, Juniata and Schuylkill.
Appendix to the Journal of the 26th House of Representatives of Pennsylvania, 1815–16. Reports of Commissioners appointed to explore the rivers of the Susquehanna, Juniata and Schuylkill.... Report of Findley and Smith ... 1793. Report of John Badolett [*sic*].... Harrisburg: [1816?]. 84 p.
S&S-38577.
Copies: MWA; PHi.

4312. Schuylkill Navigation Company.
The Committee appointed by the Board of Managers of the Schuylkill Navigation Company, to view the improvements, by means of dams and locks, of the navigation of the Connecticut River, report.... [Philadelphia: 1816.] 12 p., illus.
Also includes an extract of Thomas Eddy's letter referring to navigational improvements on the Mohawk River.
S&S-38880.
Copies: Ct; DeGE; PPL.

4313. Virginia. Commissioners Appointed to View Certain Rivers.
Report of the Commissioners appointed to view certain rivers within the Commonwealth of Virginia, conformable to an act of said Commonwealth, passed the fourteenth day of February, 1814 ... and accompanied by a map. [Richmond]: Printed 1816. 39 p., map (map not present in copies seen).
N-V0183856; S&S-39685.
Copies: CSmH; DLC; MWA; NcD; NjR; ViHi; ViU.

4314. ———— [Richmond: Printed 1816.] 113 p.
Additional materials, p. [40]–113.
Also issued as Part I of: Virginia. Board of Public Works. *A Collection of all laws and resolutions....* Richmond: 1819.
Copies: Vi; ViHi.

4315. [Cramer, Zadok] 1773–1813.
The navigator, containing directions for navigating the Monongahela, Allegheny, Ohio and Mississippi rivers.... To which is added an appendix, containing an account of Louisiana, and of the Missouri and Columbia rivers ... 9th ed. Pittsburgh: Cramer, Spear and Eichbaum; Robert Ferguson & Co.,

4315 (*continued*)
printers, 1817. 307 p., maps.
S&S-40589, 40590.
Copies: CU; DLC; DeU; ICN; KyHi; MH; MWA; MiU; MoSU; NN; PPF; PPL.

4316. Kanawha County, W. Va.
At a meeting of the Standing Committee of Kanawha County, for the promotion of internal improvements, by opening the James and Great Kanawha rivers to the highest points of navigation, and connecting the same by an artificial road, held in the town of Charlestown, on Wednesday the 3d September, 1817.... [Charlestown?, W. Va.: 1817.] Broadside.
S&S-41179.
Copies: O-Ar.

4317. New York (State). Canal Commissioners.
Report relative to the navigation between Troy and Waterford. Albany: 1817. 4 p.
Submitted by William Weston and Benjamin Wright, Feb. 25, 1817.
Whitford-p.1236.
Copies: N.

4318. Schuylkill Navigation Company.
Address of the president and managers of the Schuylkill Navigation Company, to the stockholders, and to the publick in general. Philadelphia: Printed at the office of the United States' gazette, 1817. 8 p.
S&S-42066.
Copies: DBRE; DeGE; MWA; NN; PPL.

4319. United States. Engineer Dept.
Copy of Major Kearney's report, of November 5th, 1815. February 20, 1817. Printed by order of the House of Representatives. [Washington: 1817.] 14 p., illus. (14th Cong., 2d sess. House. Doc. 93)
Concerns a survey to connect the navigation of the Elizabeth River in Virginia to that of the waters of North Carolina.
Signed: James Kearney, Major, Topographical Engineers.
S&S-42459.
Copies: CtY; DLC; DeGE; MWA; NN; ViU.

4320. Brantz, Lewis.
Survey of the river Patapsco and part of the Chesapeake Bay, instituted by the insurance companies of Baltimore in 1817. [Baltimore]: 1818. Map 112 x 235 cm.
S&S-43441.
Copies: MdBE.

4321. A compendious view of the law authorizing the improvement of the navigation of the river Lehigh. Together with remarks on the facility of making the improvement, and advantages to be derived from it. Philadelphia: William Brown, printer, 1818. 14 p.
Josiah White, George F. A. Hauto, and Erskine Hazard were authorized by the Pennsylvania Legislature to make the Lehigh River navigable.
S&S-43686.
Copies: DeGE; MH-BA; MWA; PHi; PPAmP; PPL.

4322. [Cramer, Zadok] 1773–1813.
The navigator, containing directions for navigating the Monogahela, Allegheny, Ohio and Mississippi rivers.... To which is added an appendix, containing an account of Louisiana, and of the Missouri and Columbia rivers ... 10th ed. Pittsburgh: Cramer & Spear, 1818. 304 p., maps.
S&S-43766.
Copies: CoD; CtY; DLC; DeU; ICN; Ia-HA; KyLoF; LU; MWA; MdHi; NN; OCU; OHi; PPiU.

4323. Genet, Edmond Charles, 1763–1834.
Memorial on the alluvions or obstructions, at the head of the navigation of the river Hudson; the impossibility of removing them effectually; the practicability of a lateral canal along these impediments for vessels of all descriptions, as the only lasting remedy; with a view of the political advantages of that contemplated improvement, and of the accession of Albany to maritime commerce. By Edmund [*sic*] Charles Genet.... Albany:

Printed by I. W. Clark, 1818. 46 p.
S&S-44141.
Copies: CSmH; CtY; DeGE; ICJ; MH-BA; MWA; MiU; N; NNC; NjR; PBL; ViU.

4324. [Lehigh Coal and Navigation Company.]
Observations on the Lehigh navigation bill [by Josiah White and others]. [S.l.: 1818?] 4 p., 1 l.
N-L0218686.
Copies: MWA; ViU.

4325. New York (State). Legislature. Joint Committee on the Navigation of Hudson's River.
Report and bill relative to the improvement of the navigation of Hudson's River. Made to the Assembly, March 12, 1818. Albany: Printed by J. Buel, 1818. 24 p.
S&S-45049.
Copies: MB; N.

4326. Observations on the importance of improving the navigation of the river Schuylkill, for the purpose of connecting it with the Susquehanna, and through that river extending our communication to the Genesee lakes and the Ohio. [Philadelphia?: 1818.] 21, [1]p.
Attributed to William John Duane.
Appendix, [1] p. at end, signed: John L. Sullivan, Boston, May 4, 1818.
S&S-43889, 45129; S-2573 (dated 1820).
Copies: DLC; DeGE; ICJ; MH-BA; MWA; MiU-C; NN; PPAmP; PPL.

4327. Schuylkill Navigation Company.
Address of the president and managers of the Schuylkill Navigation Company, to the stockholders and to the publick in general. [Philadelphia: 1818.] 4 p.
S&S-45648.
Copies: PPAmP.

4328. Strotter, Seth.
Florida. Head of ship navigation on Mobile River. [Huntsville, Ala.: 1818.] Broadside 20.5 x 32 cm.
S&S-45813.
Copies: ICU.

4329. Virginia. General Assembly. House of Delegates. Committee to Inquire into the Complaints Against the James River Company.
Two reports of the Committee appointed to enquire into the complaints against the James River Company. With an accompanying document. Richmond: Thomas Ritchie, 1818. 19 p.
Copies: ViHi.

4330. Yadkin Navigation Company.
Bye laws, rules and regulations of the Yadkin Navigation Company, with the act of incorporation of 1816, and the recited elements of the Roanoke Charter, passed 1812. Fayetteville, N.C.: Printed by Carney and Dismukes, at the Carolina Observer office, 1818. 18 p.
S&S-46892.
Copies: NcD; NcU.

4331. Brantz, Lewis.
Survey of the river Patapsco and part of the Chesapeake Bay, instituted by the marine insurance companies of Baltimore and executed at their expense under the direction of Lewis Brantz.... Baltimore: F. Lucas, 1819. Map.
S&S-47423.
Copies: MdBE.

4332. Virginia. Commissioners for Examining the Obstructions to the Navigation of the Ohio River.
Reports of the Commissioners for Examining the Obstructions to the Navigation of the Ohio River. Richmond: Printed by Thomas Ritchie, 1819. 24 p.
Copies: ViHi.

4333. Virginia. Principal Engineer.

Report of the Principal Engineer of the Board of Public Works on the subject of the survey of James and Kanawha rivers, and the intermediate country, and an examination of the water courses ... with a view to a further improvement of their navigation.... To which is subjoined the report of the additional engineer, on the same subject. Richmond: Printed by Thomas Ritchie, 1819. 114, 63 p.

Also issued as parts III and IV of: Virginia. Board of Public Works. *The Fourth and fifth annual reports....* Richmond: 1821.

Bowker-P. 692; N-V0183287.

Copies: ICJ; Vi; ViW.

4334. Virginia. Principal Engineer.

Report of the Principal Engineer of the Board of Public Works, on the subject of the survey of James River and its branches.... Richmond: Thomas Ritchie, 1819. 1 p.l., 53, ii p.

Copies: ViHi.

4335. Bates, David Stanhope, 1777–1839.

Report on survey of Oswego River at or adjoining the falls. Albany: 1820. 7 p. (New York (State). Legislature. Assembly. Doc. 57)

Whitford-p. 1238.

Copies: N.

4336. Boyle, Alexander.

The Ohio pilot. Consisting of a set of charts of the Ohio River, in which are represented its channels, islands, ripples, rapids, shoals, bars, rocks &c. accompanied with directions for the use of navigators. Executed chiefly from the drawings and memorandums of Alexander Boyle.... Pittsburgh: Printed by Butler & Lambdin, for R. Patterson & Lambdin, 1820. [30] p., maps.

Attributed to J. C. Gilleland.

S-1387.

Copies: NN.

4337. Gilleland, James C.

The Ohio and Mississippi pilot, consisting of a set of charts of those rivers ... accompanied with directions for the use of navigators. To which is added a geography of the states and territories, west and south of the Allegheny Mountains. Pittsburgh: R. Patterson & Lambdin, 1820. 274 p., maps.

S-1386.

Copies: CtY; DLC; ICN; InHi; KyLoF; MB; MWA; MoSM; N; NBu; OCHP; OMC; P; PPi; PPiHi; ViFbE.

4338. M'Clure, David.

Report of the survey of a section of the river Delaware, from one mile below Chester, to Richmond, above Philadelphia. Taken by order of the Councils.... Philadelphia: Printed by L. R. Bailey, 1820. 47, [1] p., illus., fold. map.

S-2047.

Copies: CSt; DLC; DeGE; IU; MWA; N; NjR; P; PPAmP; PPL; PPM; RPB; ViL; ViRUT.

4339. New York (State). Canal Commissioners....

Report of the Canal Commissioners.... [Albany: 1820.] 26 p. (Assembly. Doc. 108)

Dated: In Assembly, February 23, 1820.

Includes a report by D. S. Bates on the survey of the Oswego and Seneca rivers and Onondaga lake, and by David Thomas on the Buffalo harbor.

S-2483.

Copies: DLC; MB; N; NN.

4340. New York (State). Commissioners for Improving the Navigation of the Hudson River.

Plans for improving the navigation of Hudson's River. [Albany: 1820.] 14 p.

Signed on p. 2: De Witt Clinton, Simeon De Witt, J. V. N. Yates [and others], Commissioners.

Report of Edmond C. Genet, to the Commissioners, p. 3–11.

S-2500.

Copies: DLC; DeGE; MH; MWA; N; NN; PPL.

4341. Virginia. Laws.

A bill, to amend the act, entitled, "An

act for clearing and improving the navigation of James River," and for uniting the eastern and western waters, by the James and Kanawha rivers. (Presented February 4th, 1820). [Richmond: 1820.] 12 p.
Copies: ViHi.

4342. Boyle, Alexander.
The Ohio pilot, consisting of a set of charts of that river, representing its channels, islands, ripples, rapids, shoals, bars, rocks, &c., accompanied with directions for the use of navigators ... 2d ed., corr. and improved. Pittsburgh: Printed by Butler and Lambdin for Patterson and Lambdin, 1821. [30] p., maps.
S-4817.
Copies: ODa; PPi.

4343. [Cramer, Zadok] 1773–1813.
The navigator, containing directions for navigating the Monongahela, Allegheny, Ohio and Mississippi rivers.... To which is added an appendix, containing an account of Louisiana, and of the Missouri and Columbia rivers ... 11th ed. Pittsburgh: Cramer & Spear, 1821. 283 p., maps.
S-5104.
Copies: AU; DLC; IU; MWA; OCl; OU; OrP; PPiU; WaU.

4344. Maryland. Governor (S. Sprigg).
Executive communication to the General Assembly of Maryland, at December session, 1821. Annapolis: Printed by J. Chandler, [1821?]. 32 p.
"Potomac navigation," p. [11]–17.
N-M0289289.
Copies: MiU.

4345. Pennsylvania. General Assembly. House. Committee on Roads and Inland Navigation.
In the House of Representatives, January 30th, 1821. The Committee, to whom was recommitted the bill "An act for the removal of obstructions to the navigation of the river Susquehanna at the Conewago falls, occasioned by the erection of James Hopkins' canal (or mill) dam," report.... [Harrisburg: 1821.] 2, 2 p.
Report concerning the James Hopkins's dam, 2 p. at end.
S-6417.
Copies: PPAmP.

4346. Schuylkill Navigation Company.
Address of the president and managers of the Schuykill Navigation Company, to the stockholders. Philadelphia: Printed by J. Frick, 1821. 8 p.
S-6731.
Copies: DBRE; DeGE; MB; PHi (at PPL); PPAmP.

4347. Virginia. Laws.
A bill to amend the act, entitled "An act to amend the act for clearing and improving the navigation of James River, and for uniting the eastern and western waters, by the James and Kanawha rivers." Presented January 25th, 1821. [Richmond: 1821.] 4 p.
S-7527.
Copies: DLC.

4348. Commissioners of Maryland and Virginia Appointed to Survey the Potomac River.
Message of the governor of Maryland, communicating the report of the commissioners appointed to survey the river Potomac. Published by authority. Annapolis: Printed by J. Hughes, 1822. 92 p.
S-8410.
Copies: CtY; DLC; DNA; DeU; MB; MBAt; Md; MdHi; ScU.

4349. Considerations on the practicability and importance of opening a navigation to the interior of the state, by the Housatonick River. [New Haven: A. H. Maltby & Co., printers, 1822.] 12 p.
S-8540.
Copies: CtY.

4350. Cumings, Samuel.
The western navigator; containing directions for the navigation of the Ohio and Mississippi.... Accompanied by charts of the

4350 *(continued)*
Ohio River ... and of the Mississippi River.... Philadelphia: E. Littell, 1822. 4, 232, [6] p. and atlas of 27 maps.

The atlas has the title: *The Western navigator; containing charts of the Ohio River ... and of the Mississippi River.... Accompanied by directions for the navigation of the Ohio and Mississippi....* vol.I.

The volume of text forms vol. 2 of the work.

Later editions are published under the title: *The Western pilot.*

Cumings' work is an amplification of Zadok Cramer's *The Navigator.*

"List of steam-boats ... in operation on the Mississippi and confluent waters ... 1822," v. 2, p. 69–70.

S-8481.

Copies: CtY; DLC; ICN; InHi; LNHT; MBAt; MH; MWA; MiU-C; Ms-Ar; NN; NjR; OClWHi; P.

4351. [Jones, William] 1760–1831.

Reflections upon the perils and difficulties of the winter navigation of the Delaware. Means by which these may be meliorated. Plans for the relief of vessels entangled in the ice. Improvement of the existing ice-harbours, and for the construction of intermediate harbours parallel to the direction of the stream and shore.... Philadelphia: Published by order of the Chamber of Commerce, 1822. 15 p., front.

Includes a description of a steam powered ice boat.

S-9170.

Copies: DLC; DeGE; MBAt; MWA; MiU-C; NN; NjPT; P; PPAmP; PPL; ScC.

4352. Lehigh Coal and Navigation Company.

An act to improve the navigation of the river Lehigh and an act to incorporate the Lehigh Coal and Navigation Company, together with the bye-laws. Philadelphia: Atkinson and Alexander, printers, 1822. 37 p.

S-9250.

Copies: IU; MB; MH-BA; PHi; PPL.

4353. Maryland. General Assembly. House. Committee on the Appointment of Commissioners to Inspect the Potomac River.

Report to the House of Delegates, by the Committee to whom was referred the executive communication relating to the appointment of commissioners to inspect the Potomac River, with accompanying documents; to which is added Mr. Millard's resolutions on internal improvement. December session, 1821. Annapolis: Printed by J. Chandler, [1822]. 15 p., illus.

N-M0289029; S-9383.

Copies: MdHi; MiU; NN.

4354. Schuylkill Navigation Company.

Report of the president and managers of the Schuylkill Navigation Company to the stockholders. Philadelphia: Printed by James Kay, Jun. & brother, 1822. 11 p.

Dated: January 7, 1822.

S-10188.

Copies: DeGE; MH-BA; NN; PPL.

4355. United States. Congress. House. Committee on the District of Columbia.

Report of the Committee of the District of Columbia, to whom were referred sundry memorials from the inhabitants of Pennsylvania, Maryland, and Virginia, praying the aid of the federal government towards the improvement of the navigation of the river Potomac. May 3, 1822.... [Washington: 1822.] 36 p. (17th Cong., 1st sess. House. Rept. 111)

S-11133.

Copies: DLC; DeGE; DeU; MWA; NjP.

4356. Workman, James.

Defence of the Orleans Navigation Company, before the Supreme Court; in the suit instituted against them by scire facias, pursuant to a resolution of the legislature of the state of Louisiana. By James Workman, of counsel for the defendants. New Orleans: Printed by Benjamin Levy & Co., 1822. 110 p.

Sabin-105479.

Copies: DGU; NHi.

4357. Clark, Edward.
Description of a plan for navigating the rapids in rivers, with an account of some experiments, instituted to establish its practicability.... Philadelphia: Printed by William Brown, 1823. 11 p., plate.
S-12154.
Copies: DLC; DeGE; MH; MWA; MdHi; NN; P; PPAmP; PPL; ViU.

4358. Commissioners of Maryland and Virginia Appointed to Survey the Potomac River.
Letter from the Governor and Council of Maryland, transmitting a report of the Commissioners appointed to survey the river Potomac. January 27, 1823. Printed by order of the Senate of the United States. Washington: Printed by Gales & Seaton, 1823. 53 p., 2 fold. tables. (17th Cong., 2d sess. Senate. Doc. 23)
Moses T. Hunter, Wm. T. T. Mason, Wm. Naylor, Athanas Fenwick, Elie Williams, commissioners.
S-12220, 14556.
Copies: CtY; DLC; DeGE; DeU; ICJ; MWA; MdBP; MiU; NN; NjP.

4359. ———— January 28, 1823. Washington: Printed by Gales & Seaton, 1823. 53 p., 2 fold. tables. (17th Cong., 2d sess. House. Doc. 46)
S-14557.
Copies: DLC; DeGE; DeU; MWA; NjP.

4360. Commissioners of Maryland and Virginia Appointed to Survey the Potomac River.
Report of the joint commissioners on the Potowmac River, with sundry documents. Richmond: Printed by T. Ritchie, 1823. 2 p.l., [77]–128 p., fold. table.
Usually found with: Virginia. Board of Public Works. *Seventh annual report....* Richmond: 1822.
N-J0138004.
Copies: DLC; Vi.

4361. Gregg, Andrew.
A letter from Andrew Gregg, Esq., relative to the navigation of the Susquehanna River, &c. &c.... Annapolis: Printed by J. Hughes, [1823]. [5], 6–17, [1] p. (Doc. B, accompanying the Executive communication, December session, 1823)
Heyl-93.
Copies: MdHi.

4362. James River Company.
Report from the James River Company, January 23, 1823. [Richmond: 1823.] 3 p.
N-J0038196.
Copies: MH-BA.

4363. Pennsylvania. Commissioners Appointed to Explore the River Susquehanna.
Report of the Commissioners appointed to explore the river Susquehanna, read in Senate, Dec. 21, 1823. [Harrisburg: 1823.] 7 p.
S-13705.
Copies: MWA; PPAmP.

4364. Schuylkill Navigation Company.
Report of the managers of the Schuylkill Navigation Company, to the stockholders and their proceedings thereon. Philadelphia, January 10th, 1823. Philadelphia: Printed for the Company, 1823. 11 p.
S-14054.
Copies: DeGE; ICJ; IU; PPAmP; PPL

4365. United States. Board of Internal Improvement.
Message from the President of the United States, transmitting a report of the Board of Engineers on the Ohio and Mississippi rivers, made in the year 1821. January 22, 1823. Read.... Washington: Printed by Gales & Seaton, 1823. 22 p. (17th Cong., 2d sess. House. Doc. 35)
Signed: Bernard, Brig. General; Joseph G. Totten, Maj. Eng'rs.
The report was accompanied by 65 maps, not printed.
S-14643.
Copies: DLC; DeGE; DeU; MWA; NjP.

4366. Omitted in numbering.

4367. Angell, Joseph Kinnicut, 1794–1857.
A treatise on the common law, in relation to water-courses. Intended more particularly as an illustration of the rights and duties of the owners and occupants of water privileges. To which is added an appendix, containing the principal adjudged cases. By Joseph K. Angell.... Boston: Wells and Lilly, 1824. xv, 96, 254 p.
S-15111.
Copies: Ct; DLC; DeGE; MBS; MH-L; MNe; MWCL; MWA; MdBP; MnU; NIC; NcD; PPB; RPB.

4368. [Cramer, Zadok] 1773–1813.
The navigator, containing directions for navigating the Ohio and Mississippi rivers.... To which is added an appendix, containing an account of Louisiana, and of the Missouri and Columbia rivers ... 12th ed. Pittsburgh: Cramer & Spear, 1824. 275 p., maps.
S-15888.
Copies: CSmH; DLC; ICN; ICU; InU; KHi; MH; MWA; MdBL; MnHi; MoHi; NN; O; OrP; PPiU; TxU; WaU.

4369. Cummins, Ebenezer Harlow, 1790?–1835.
The policy of Maryland, in relation to the Potomac and Susquehanna considered.... Baltimore: Printed by Thomas Murphy, 1824. 23 p.
S-15908.
Copies: DLC; MdBP; MdHi.

4370. Indiana. General Assembly. Senate. Committee on Roads and Canals.
Mr. Ewing, from the Committee, to whom was referred much of the Governor's message, as relates to the navigation of the Wabash River, begs leave to report. [Indianapolis]: 1824. 8 p.
Fatout-*Indiana canals*, p. 208.
Copies: In.

4371. Memorial to the Legislature of the state of Alabama, on the subject of connecting the waters of Tennessee and Alabama rivers, together with the act in relation thereto. Cahawba: Printed by William B. Allen & Co., 1824. 14 p.
Signed on p. 10: Jesse Beene, Samuel Pickens, Thomas White.
N-M0435785; S-15050.
Copies: DLC; MBAt; N.

4372. New York (State). Laws.
An act, for the further improvement of the navigation of the Hudson. Albany: 1824. 3 p. (Senate. Doc. 124)
Whitford-p. 1270.
Copies: N.

4373. Pennsylvania. Secretary of the Commonwealth.
Letter from the Secretary of the Commonwealth to the Legislature of the Commonwealth of Pennsylvania, accompanying the report of the Susquehanna Commissioners, on the subject of the navigation of the river from Columbia to the head of tide. Harrisburg: Printed by Charles Mowry, 1824. 8 p.
S-17557.
Copies: PHi; PPAmP.

4374. Proceedings of two meetings of citizens of Philadelphia respecting Col. Clark's plan for ascending rapids in rivers, and thereby improving the navigation of the river Delaware, beyond Trenton. To which is prefixed, a brief description of the plan, with a plate. Philadelphia: Printed by Joseph R. A. Skerrett, 1824. 15 p., plate.
Mathew Carey, chairman of the meetings.
S-17609.
Copies: DLC; DeGE (xerox copy); MdHi; PPAmP; PPL; RPB.

4375. Quin, Henry.
A new improvement in the art of stream navigation, whereby boats and crafts are to ascend and descend by the power of the water; with an illustration of the practica-

bility and advantages thereof; with draughts and explanations of the plan.... Philadelphia: Printed for the author, 1824. 11 p., plate.

"Improved saw mill," p. 11.

S-17745.

Copies: DLC; DeGE; MH-BA; MWA; NN.

4376. Roanoke Navigation Company.

Report of the directors to the stockholders in the Roanoke Navigation Company, on the subject of locking into the river at Weldon, etc.... Halifax, N.C.: Printed at the office of the "Free press," 1824. 14 p.

S-17830.

Copies: NcU.

4377. To the people of Charlotte, Prince Edward and Halifax. [S.l.: 1824?] 16 p.

Signed on p. 14: A., Dec. 10, 1824.

Concerns the Roanoke Navigation Company.

S-18211.

Copies: Vi.

4378. United States. Congress. House. Committee on Roads and Canals.

Report of the Committee on Roads and Canals, upon the subject of the navigation of the Ohio and Mississippi rivers, accompanied with a bill to improve the navigation of the Ohio and Mississippi rivers. February 28, 1824.... [Washington: 1824.] 4 p. (18th Cong., 1st sess. House. Rept. 75)

S-19040.

Copies: DLC; DeU; MWA; NjP.

4379. United States. Congress. House. Committee on the Navigation of the Ohio and Mississippi Rivers.

Report of the Committee to which was referred the message of the President of the United States, bearing date 22d January, 1823, upon the subject of the navigation of the Ohio and Mississippi rivers, accompanied with a bill appropriating money to assist the states of Kentucky and Ohio to open a canal around the falls at Louisville, Kentucky. March 31, 1824.... [Washington: 1824.] 10 p. (18th Cong., 1st sess. House. Rept. 98)

Report of the engineer, David S. Bates, p. 5–10.

S-19057.

Copies: DLC; DeGE; DeU; MWA; NjP.

4380. United States. Laws.

An act to improve the navigation of the Ohio and Mississippi rivers. May 12, 1824.... [Washington: 1824.] 3 p. (H.R. 139)

S-18367.

Copies: DNA.

4381. Association for Improving the Navigation of Connecticut River Above Hartford.

Two reports made by committees appointed by the directors of the Association for Improving the Navigation of Connecticut River Above Hartford; published by order of the board. Hartford, January, 1825. Hartford: P. B. Goodsell, printer, 1825. 22 p.

First report signed: David Porter, Alfred Smith; the second: David Porter, Wm. H. Imlay, Eliphalet Averill.

S-19468.

Copies: CtHi; CtSoP; DeGE; ICJ; MA; MH; MWA; MeHi; NN; PHi; VtMiHi.

4382. Connecticut. Laws.

Act to incorporate the Connecticut River Company. [Hartford?: 1825.] 12 p.

A company to improve the navigation of the Connecticut River.

S-20166.

Copies: CtSoP; MWA.

4383. Cumings, Samuel.

The western pilot, containing charts of the Ohio River, and of the Mississippi from the mouth of the Missouri to the Gulf of Mexico, accompanied with directions for navigating the same.... Cincinnati: Morgan, Lodge & Fisher, printers, 1825. 143 p., 43 maps.

S-20223.

Copies: DLC; F; ICU; IU; KyLoF; MB; MH; MiU; NN; NjR; OC; OClWHi; OMC; TKL.

4384. Journal of the proceedings of the convention, holden at Windsor, Vermont, February 16, 1825; for the purpose of taking preliminary measures to effect an improved navigation of the Connecticut River. Windsor, Vt.: W. Spooner, printer, [1825]. 12 p.

S-21080.

Copies: DLC; MH-BA; MWA; VtU.

4385. Lehigh Coal and Navigation Company.

Delaware navigation. The above is a sketch of the plan of improvement proposed by the Lehigh Coal and Navigation Company, to be adopted on the river Delaware.... [Philadelphia: 1825?] Broadside.

Copies: PPL.

4386. Memorial of sundry inhabitants of that part of Florida attached to the state of Louisiana. January 4, 1825. Read, and referred to the Committee of the Whole House, to which is committed the bill disapproving, in part, an act of the government of the late Territory of New Orleans, "incorporating the Orleans Navigation Company," passed on the 5th day of July, 1805. Washington: Printed by Gales & Seaton, 1825. 5 p. (18th Cong., 2d sess. House. Doc. 30)

Concerns navigation of the bayou or river St. Johns.

S-22971.

Copies: DLC; DeU; LN; MWA; NjP; ViU.

4387. Memorial of the citizens of Albany praying for a law to improve the navigation of the Hudson River. Albany: 1825. 2 p. (New York (State). Legislature. Assembly. Doc. 20)

Whitford-p. 1270.

Copies: N.

4388. Memorial of the inhabitants of Oswego relative to the improvement of the navigation of the Oswego River. Albany: 1825. 5 p. (New York (State). Legislature. Assembly. Doc. 159)

Whitford-p. 1239.

Copies: N.

4389. Orleans Navigation Company.

Reglemens établis par la Compagnie de navigation d'Orléans, relativement aux batimens, barques et bateaux qui naviguent dans le Bayou St. Jean, le canal Carondelet et le bassin. Nouvelle-Orléans: De l'imprimerie de Benjamin Levy, 1825. 12 p.

S-21754.

Copies: LU.

4390. Orleans Navigation Company.

Rules and regulations established by the Orleans Navigation Company, for the government and police of vessels and boats navigating the Bayou St. John, canal of Carondelet and basin. New Orleans: Printed by Benj. Levy, 1825. 11 p.

S-21755.

Copies: LU.

4391. Pennsylvania. Laws.

A supplement to the act, entitled "An act to improve the navigation of the river Lackawaxen." [Harrisburg]: Mowry & Cameron, printers, 1825. [4] p.

At head of title: Senate file, No. 117. Read January 21, 1825.

Copies: PPL.

4392. Yadkin Navigation Company.

Report of the president and directors of the Yadkin Navigation Company, rendered in pursuance of a resolution of the board of stockholders at a general meeting, on the tenth day of April, 1825. Salisbury, [N.C.]: Printed by Philo White, 1825. 63 p.

S-23387.

Copies: NcD; NcU.

4393. Commissioners of Alabama and Tennessee on the Obstructions to Navigation at the Muscle Shoals.

Report of the Commissioners of Alabama and Tennessee, respecting the obstruction to navigation at the Muscle Shoals, in Tennessee River, made to the Legislature of Tennessee at its session in November, 1826. Nashville: J. S. Simpson, 1826. 24 p.

S-26191.

Copies: PPL; T; THi.

4394. Connecticut River Company.
Report of the president and directors of the Connecticut River Company, with the report of H. Hutchinson, Esq. laid before the stockholders, at their annual meeting, January 3d, 1826. Hartford: Printed by Philemon Canfield, [1826]. 54 p.
Cover title: *Report on the improvement of Connecticut River.*
S-24206.
Copies: CLU; CtHi; CtSoP; CtY; DLC; DeGE; ICU; MB; MBAt; MH-BA; MWA; MiU-T; NN; NNE; NjPT; OClWHi; PPAmP.

4395. Facts connected with the application for power to improve the navigation of Connecticut River. [Hartford?: 1826?] 15 p.
S-24476.
Copies: Ct; CtSoP; CtY; MH-BA; MHi; MWA; NN; ViW.

4396. Gilleland, James C.
The Ohio pilot, consisting of a set of charts of that river, representing its channels, islands, ripples, rapids, shoals, bars, rocks, &c., accompanied with directions for the use of navigators. Pittsburgh: H. Holdship, printed by Johnston and Stockton, 1826. 32, [2] p., maps.
S-24690.
Copies: CtY; IHi.

4397. Louisiana. District Court.
Transcript, taken verbatim, from the authenticated record of the District Court of the state of Louisiana ... in the case of the state of Louisiana, versus the Orleans Navigation Company.... Washington City: Way & Gideon, printers, 1826. 24 p.
S-26232.
Copies: PPL.

4398. Mississippi. Legislature.
Memorial of the Legislature of Mississippi, praying Congress to enable the state to open a canal through the bar at East Pass, mouth of Pascagoula River. December 13, 1826.... Washington: Printed by Gales & Seaton, 1826. 6 p. (19th Cong., 2d sess. Senate. Doc. 4)
S-27209.
Copies: DLC; DeU; MWA; MiU; NjP.

4399. New Hampshire. Laws.
An act, to confirm an act of the General Assembly of the state of Vermont, entitled "An act to provide for improving the navigation in the valley of the Connecticut River." [S. l.]: 1826. 30 p.
Includes the text of the Vermont law, incorporating the Connecticut River Company to improve navigation of the Connecticut River.
S-25505.
Copies: NhHi.

4400. Orleans Navigation Company.
To the Congress of the United States; the memorial of the president and directors of the Orleans Navigation Company. [New Orleans?: 1826.] 16 p.
Dated: March, 1826.
N-00131195.
Copies: MB.

4401. Pennsylvania. Laws.
Acts of the Legislature of Pennsylvania, relating to the Schuylkill Navigation Company. Philadelphia: Printed by Lydia R. Bailey, 1826. 35 p.
S-25686.
Copies: IU; PPL.

4402. Pilots on the Ohio and Mississippi.
Memorial of sundry owners, masters, and pilots, of steam boats, navigating the Ohio and Mississippi Rivers. January 9, 1826. Referred to the Committee on Roads and Canals. Washington: Gales & Seaton, 1826. 4 p. (19th Cong., 1st sess. House. Doc. 36)
S-27305.
Copies: DLC; MWA; NjP.

4403. Schuylkill Navigation Company.
Report of the president and managers of the Schuylkill Navigation Company, to the stockholders. January 2, 1826. [Phila-

4403 (*continued*)
delphia]: Printed by Lydia R. Bailey, 1826. 12 p.

S-26023.

Copies: DLC; DeGE; MB; MH-BA; MWA; NN; PHi.

4404. United States. Congress. House. Committee on Roads and Canals.

Improvement of the navigation of the Kennebec River. March 14, 1826. Mr. Stewart, from the Committee on Roads and Canals, to which was referred the petition of inhabitants of the state of Maine, made the following report. [Washington: 1826.] 2 p. (19th Cong., 1st sess. House. Rept. 118)

S-27021.

Copies: DLC; DeU; MWA; NjP.

4405. United States. Congress. Senate. Committee on Commerce.

In Senate of the United States, March 6, 1826. Mr. Lloyd, from the Committee on Commerce, to whom was referred a resolution directing an inquiry into the expediency of making an appropriation for the removal of the obstructions and improving the navigation of the bay and harbor of Mobile, reported.... [Washington: 1826.] 2 p. (19th Cong., 1st sess. Senate. Doc. 54)

S-27048.

Copies: DLC; DeU; MWA; NjP.

4406. United States. Congress. Senate. Select Committee on Roads and Canals.

In Senate of the United States, March 20, 1826. Mr. Hendricks, from the Select Committee on Roads and Canals, to whom were referred a letter from the Secretary of War, on the subject of the navigation of the Ohio and Mississippi rivers, and the memorial of sundry owners, masters, and pilots, of steam boats navigating said rivers, made the following report. [Washington: 1826.] 3 p. (19th Cong., 1st sess. Senate. Doc. 62)

Concerns the removal of obstructions to navigation.

S-27056.

Copies: DLC; DeU; MWA; NjP.

4407. United States. Congress. Senate. Select Committee on Roads and Canals.

In Senate of the United States, May 1, 1826. Mr. Hendricks, from the Select Committee on Roads and Canals, to whom was referred the memorial of the legislature of the state of Mississippi, praying for an appropriation to aid in opening a passage for vessels over the bar, at the East Pass of the mouth of Pascagoula River, made thereon the following report. [Washington: 1826.] 4 p. (19th Cong., 1st sess. Senate. Doc. 84)

S-27072.

Copies: DLC; DeU; MWA; NjP.

4408. United States. Engineer Dept.

Letter from the Secretary of War, transmitting a report of the Chief Engineer, relative to the application of the appropriation for removing obstructions to the navigation of the Ohio and Mississippi rivers. January 5, 1826.... Washington: Printed by Gales & Seaton, 1826. 31 p. (19th Cong., 1st sess. Senate. Doc. 14)

S-27174.

Copies: DLC; DeU; MWA; NjP.

4409. United States. Laws.

A bill for removing the existing obstructions in the river Savannah. February 20, 1826. [Washington: 1826.] 1 p. (S. 66)

S-26453.

Copies: DNA.

4410. United States. Laws.

A bill making an appropriation for removing the bar at the east pass of the mouth of Pascagoula River, and for improving the harbor thereof. May 1, 1826.... [Washington: 1826.] 1 p. (S. 145)

S-26669.

Copies: DNA.

4411. United States. Laws.

A bill to amend and explain an act, approved May twenty-fourth, eighteen hundred and twenty-four, entitled "An act to improve the navigation of the Ohio and Mississippi Rivers." March 25, 1826.... [Washington: 1826.] 3 p. (H.R. 181)

S-26734.
Copies: DNA.

4412. United States. Laws.
A bill to improve the navigation of the Allegany [*sic*] River. April 4, 1826.... [Washington: 1826.] 1 p. (H.R. 206)
S-26834.
Copies: DNA.

4413. United States. President (John Q. Adams).
Sand bars of the Ohio River. Message from the President of the United States, transmitting the information required by a resolution of the House of Representatives of the 21st instant in relation to the measures which have been taken to improve the navigation over the sand bars in the Ohio River, &c., &c., March 31, 1826.... Washington: Printed by Gales & Seaton, 1826. 15 p. (19th Cong., 1st sess. House. Doc. 145)
S-27386.
Copies: DLC; DeU; MWA; NjP.

4414. United States. War Dept.
Survey—Connecticut River, &c. Letter from the Secretary of War, transmitting a report of a survey of Connecticut River, from Barnet, in Vermont, to Lake Connecticut. And, also, a canal route from Memphrymagog to Connecticut River. April 12, 1826.... Washington: Printed by Gales & Seaton, 1826. 65 p. (19th Cong., 1st sess. House. Doc. 154)
Surveyed by De Witt Clinton, engineer.
S-27411.
Copies: DBRE; DeU; MWA; NjP.

4415. Vermont. Laws.
An act to confirm an act of the General Assembly of the state of Vermont, entitled "An act to provide for improving the navigation in the valley of Connecticut River." [Boston: 1826.] 34 p.
At head of title: Commonwealth of Massachusetts. In the year of our Lord —twenty-six.
Contains only the reprint of the Vermont law.
S-25254.
Copies: MB.

4416. Virginia. Laws.
An act incorporating a company for the purpose of improving the navigation of James River. (Passed February 19th, 1816). Also an act to amend the act, entitled "An act incorporating a company for the purpose of improving the navigation of James River." (Passed February 18th, 1823). Richmond: Printed by Thomas W. White, 1826. 18 p.
Copies: ViHi.

4417. At a meeting of delegates from Newbern, Washington ... assembled at Washington on the 26th November, 1827, the following report was presented ... and the accompanying memorial subscribed.... [S.l.: 1827.] 4 p.
The memorial is headed: "To the Honorable the General Assembly of the State of North Carolina.... "
Concerns obstructions to navigation and deepening of the channel at Ocracoke, N.C.
Signed: William Gaston and others; dated November 27, 1827.
N-A0473137.
Copies: DLC.

4418. Facts relating to the improvement of Connecticut River.... [Boston?: 1827.] 4 p.
S-28833.
Copies: MH-BA.

4419. Massachusetts. Laws.
An act, to confirm an act of the General Assembly of the state of Vermont, entitled "An act to provide for improving the navigation in the valley of Connecticut River." Boston: 1827. 36 p. (House. Doc. 7)
Includes a reprint of the Vermont act, the Massachusetts act, and amendments adopted by the Connecticut legislature relating to the Vermont act.
S-29659; Thomson-158.
Copies: MB; MWA.

4420. North Carolina. General Assembly. House. Committee on the Memorial of Sundry Citizens of Newbern, Washington, Plymouth, Edenton, Murfreesborough, Elizabeth City and Hertford.

Report relative to Ocracoke Inlet, 1827. Raleigh: Lawrence & Lemay, state printers, 1827. 6 p. (Doc. 12)

S-30082.

Copies: NcU.

4421. Roanoke Navigation Company.

Report of the Roanoke Navigation Company, 1827. Raleigh: Printed by Lawrence & Lemay, printers to the state, 1827. 13 p. (No. 14)

S-30463.

Copies: NcU.

4422. Schuylkill Navigation Company.

Report of the president and managers of the Schuylkill Navigation Company, to the stockholders. January 1, 1827. [Philadelphia]: Printed by Lydia R. Bailey, 1827. 12 p.

S-30534.

Copies: DeGE; MWA; PHi; PPL.

4423. Shreve, Henry Miller, 1785–1851.

Letter from Henry M. Shreve to the Hon. C. A. Wickliffe, upon the subject of the navigation of the Mississippi River. December 14, 1827.... Washington: Printed by Gales & Seaton, 1827. 6 p. (19th Cong., 1st sess. House. Doc. 11)

S-31362.

Copies: DLC; DeU; MWA; NIC; NjR.

4424. Trcziyulny, Charles.

Report of Charles Trcziyulny, appointed to explore the river Susquehanna, in pursuance of an act of the General Assembly, with a view to its improvement, from New York to Maryland line. Printed for the Senate. Harrisburg: Cameron & Krause, 1827. 19 p., fold. table.

S-30826.

Copies: MdHi; P; PPAmP.

4425. [Trcziyulny, Charles.]

Report of the engineer appointed by the Commissioners for the Improvement of Navigation of the River Susquehanna, and the North Branch thereof, made in pursuance of resolutions of the legislature, passed at the session of 1826–7. Read in the House of Representatives, March 8, 1827. Harrisburg: Printed by J. S. Wiestling, 1827. 19 p.

Signed: Charles Trcziyulny.

S-30179.

Copies: DLC; NN; PPAmP; PPi.

4426. United States. Congress. House. Committee on Roads and Canals.

Inland channel — St. Mary's to St. John's River. February 3, 1827. Mr. Mercer, from the Committee on Roads and Canals, to which the subject had been referred, made the following report.... [Washington: 1827.] 11 p. (19th Cong., 2d sess. House. Rept. 70)

S-31327.

Copies: DLC; DeU; MWA; NjP.

4427. United States. Congress. House. Committee on Roads and Canals.

Navigation of Red River. March 1, 1827 ... Mr. Mercer, from the Committee on Roads and Canals, to which was referred the memorial of the General Assembly of Arkansas, in relation to obstructions to the navigation of Red River, made the following report. [Washington: 1827.] 8 p. (19th Cong., 2d sess. House. Rept. 96)

S-31435.

Copies: DLC; DeU; MWA; NjP.

4428. United States. Congress. House. Committee on the Judiciary.

Orleans Navigation Company. February 24, 1827 ... Mr. Webster, from the Committee on the Judiciary ... made the following report. [Washington: 1827.] 20 p. (19th Cong., 2d sess. House. Rept. 89)

"Memorial of the president and directors of the Orleans Navigation Company," p. 8–20.

Concerns the navigation of the Bayou St. John and the Carondelet Canal.

S-31445.
Copies: DLC; DeU; MWA; NjP.

4429. United States. Congress. Senate. Committee on Commerce.
In Senate of the United States. February 2, 1827, Mr. Johnson, of Louisiana, made the following report: The Committee of Commerce, to whom was referred the petition of the inhabitants of Wilmington, North Carolina.... [Washington: 1827.] 2 p. (19th Cong., 2d sess. Senate. Doc. 49)
Recommends the removal of obstructions for navigation in Cape Fear River.
S-31304.
Copies: DLC; DeU; MWA; NjP.

4430. United States. Congress. Senate. Select Committee on Roads and Canals.
In Senate of the United States, January 8, 1827. Mr. Hendricks, from the Select Committee on Roads and Canals, to whom was referred the memorial of the Legislature of the state of Mississippi, praying for an appropriation to aid in opening a passage for vessels over the bar, at the East Pass of the mouth of Pascagoula River, made thereon the following report.... [Washington: 1827.] 2 p. (19th Cong., 2d sess. Senate. Doc. 13)
S-31284.
Copies: DLC; DeU; MWA; NjP.

4431. United States. Engineer Dept.
Survey of Kennebec River. Letter from the Secretary of War transmitting a report of a survey of Kennebec River in the state of Maine. January 26, 1827.... Washington: Printed by Gales & Seaton, 1827. 6 p. (19th Cong., 2d sess. House. Doc. 103)
Transmitting a report of the Chief Engineer.
S-31518.
Copies: DLC; DeU; MWA; NjP.

4432. United States. Engineer Dept.
Surveys, harbor of Edgartown, Merrimack River, &c. Letter from the Secretary of War transmitting reports of a survey of the flats on the northwest side of the harbor of Edgartown; a survey of the mouth of Merrimack River; also, a survey of the harbor of Hyannis, in the Vineyard Sound, with a map of each; rendered in obedience to a resolution of the House of Representatives of the fourth of last month. February 9, 1827.... Washington: Printed by Gales & Seaton, 1827, 16 p., maps. (19th Cong., 2d sess. House. Doc. 140)
S-31522.
Copies: DLC; DeU; MWA; NjP.

4433. United States. Laws.
A bill for improving the inland navigation between the St. Mary's River and the entrance of the River St. John's, in Florida. February 3, 1827. [Washington: 1827.] 1 p. (H.R. 421)
S-30957.
Copies: DNA.

4434. United States. Laws.
A bill for improving the navigation on the Ohio River. February 6, 1827. [Washington: 1827.] 1 p. (S. 89)
S-30958.
Copies: DNA.

4435. Connecticut. General Assembly.
At a General Assembly of the state of Connecticut, holden at New-Haven in said state, on the first Wednesday of May, in the year ... one thousand eight hundred and twenty-eight. Upon the petition of the Connecticut River Company for an alteration of their charter.... [New Haven: 1828.] 2 p.
Concerns the improvement of the Connecticut River for navigation.
S-32804.
Copies: Ct.

4436. Connecticut River Company.
To the Honorable General Assembly of the state of Connecticut, to be holden at New-Haven in said state, on the first Wednesday of May, 1828. The petition of the Connecticut River Company.... [Hartford?: 1828.] 24 p.
Concerns building of locks at Enfield on the Connecticut River.
Supporting documents, p. [5]–24.
S-32814.
Copies: CtSoP.

4437. Memorial of inhabitants of Scituate, Pembroke, Hancock &c. in the state of Massachusetts, praying for the improvement of North River channel. May 5, 1828.... Washington: Printed by Gales & Seaton, 1828. 4 p. (20th Cong., 1st sess. House. Doc. 226)

S-36538.

Copies: DLC; MWA; NjR.

4438. Memorial of sundry citizens of Elizabeth City, and county of Pasquotank, in North Carolina, praying that an appropriation may be made, for the purpose of opening a water communication from Albemarle Sound to the Atlantic Ocean. January 21, 1828.... Washington: Printed by Duff Green, 1828. 4 p. (20th Cong., 1st sess. Senate. Doc. 49)

S-36555.

Copies: DLC; DeGE; DeU; MWA; NjR.

4439. Memorial of sundry inhabitants of Murfreesborough, N. Carolina, praying that a passage be made between Ocracock Inlet and the Atlantic Ocean, April 4, 1828.... Washington: Printed by Duff Green, 1828. 4 p. (20th Cong., 1st sess. House. Doc. 165)

S-36562.

Copies: DLC; DeU; MWA; NjR.

4440. North Carolina. Board of Internal Improvements.

Report on the improvement of the navigation of Neuse River. 1828. Raleigh: Printed by Lawrence & Lemay, printers to the state, 1828. 4 p. (Doc. 11)

Favoring the improvement of the Neuse River between the Smithfield and Cobbs' Mills.

S-34581.

Copies: NcU.

4441. North Carolina. General Assembly. Joint Select Committee on Improvement of Cape Fear River.

Report on an additional appropriation for the improvement of Cape Fear River below Wilmington, 1828. Raleigh: Lawrence & Lemay, printers, 1828. 6 p.

S-34580; Thornton (Off.)-1364.

Copies: NcU.

4442. Ohio. Laws.

Acts and proceedings of the government of the state of Ohio, relating to the navigation of the Muskingum River and its connection with the Ohio and Erie Canal. Zanesville [Ohio]: 1828. 30 p.

Includes: Report of the Canal Commissioners, Jan. 17, 1828; the Act to incorporate the Muskingum Navigation Company; the Act to authorize J. Taylor to erect a dam across the Muskingum River; and Reported estimate of expense of works authorized in Act for incorporating the Muskingum Navigation Company.

S-34597.

Copies: MWA; OMC; RPB.

4443. Petition of sundry residents of Hertford, N. Carolina, praying that an appropriation be made for opening an outlet for commerce to the ocean, at or near "Roanoke Inlet." March 3, 1828.... Washington: Printed by Duff Green, 1828. 4 p. (20th Cong., 1st sess. Senate. Doc. 129)

S-36720.

Copies: DLC; DeU; MWA; NjP.

4444. Schuylkill Navigation Company.

Report of the president and managers of the Schuylkill Navigation Company to the stockholders. January 7, 1828. [Philadelphia]: Printed by Lydia R. Bailey, 1828. 9 p.

S-35114.

Copies: DeGE; IU; MWA; PHi.

4445. Some remarks made before the Joint Committee on Rivers and Canals, relating to the proposed improvement of Connecticut River and the extension of the Hampshire and Hampden Canal. February 1828. Boston: Dutton & Wentworth, state printers, 1828. 24 p., 2 fold. plans.

Note on p. [2]: "The letters of Governor Clinton and Col. Baldwin, having been selected and published by the agents of the Hampshire and Hampden Canal Company, have led to the following publication."

S-35353.

Copies: CtHC; CtSoP; DBRE; DLC; DeGE; MH; MWA; MdBJ; MiU-T; NN; NhHi.

4446. United States. Board of Internal Improvement.

Survey of the Muscle Shoals. Letter from the Secretary of War, transmitting the information required by a resolution of the House of Representatives of the 16th January last, in relation to an examination of the Muscle Shoals in Tennessee River, with a view to removing the obstructions to the navigation thereof, and the construction of a canal around the same. May 14, 1828.... Washington: Printed by Gales & Seaton, 1828. 20 p. (20th Cong., 1st sess. House. Doc. 284)

Transmitting a report of the Board of Internal Improvement.

S-36918.

Copies: DLC; DeU; MWA; NjR.

4447. United States. Congress. House. Committee on Roads and Canals.

Navigation between St. Mary's and St. John's River, Florida. January 11, 1828. Mr. Mercer, from the Committee on Roads and Canals ... made the following report. [Washington: 1828.] 5 p. (20th Cong., 1st sess. House. Rept. 61)

S-36642.

Copies: DLC; DeU; MWA; NjR.

4448. United States. Congress. House. Committee on Roads and Canals.

Navigation of Pearl River. March 3, 1828. Mr. Gurley, from the Committee on Roads and Canals ... made the following report.... [Washington: 1828.] 2 p. (20th Cong., 1st sess. House. Rept. 175)

S-36643.

Copies: DLC; DeU; MWA; NjR.

4449. United States. Congress. House. Committee on Roads and Canals.

Navigation of rivers Teche, Vermillion, &c.—Lou[isiana]. March 12, 1828. Mr. Gurley, from the Committee on Roads and Canals ... made the following report. [Washington: 1828.] 5 p. (20th Cong., 1st sess. House. Rept. 188)

S-36644.

Copies: DLC; DeU; MWA; NjR.

4450. United States. Congress. House. Committee on Roads and Canals.

Obstructions in the river Ohio. March 27, 1828. Mr. Woods ... from the Committee on Roads and Canals ... made the following report. [Washington: 1828.] 4 p. (20th Cong., 1st sess. House. Rept. 213)

S-36673.

Copies: DLC; DeGE; DeU; MWA; NjR.

4451. United States. Congress. House. Committee on Roads and Canals.

Obstructions—the Ohio River. April 4, 1828 ... Mr. Woods, from the Committee on Roads and Canals ... made the following report. [Washington: 1828.] 2 p. (20th Cong., 1st sess. House. Rept. 269)

S-36672.

Copies: DLC; DeU; MWA; NjR.

4452. United States. Congress. House. Committee on Roads and Canals.

Pascagoula River. February 13, 1828. Mr. Gurley, from the Committee on Roads and Canals ... made the following report. [Washington: 1828.] 7 p. (20th Cong., 1st sess. House. Rept. 145)

S-36694.

Copies: DLC; DeU; MWA; NjR.

4453. United States. Congress. House. Committee on the Judiciary.

Orleans Navigation Company. February 11, 1828 ... Mr. P. P. Barbour, from the Committee on the Judiciary ... made the following report.... [Washington: 1828.] 22 p. (20th Cong., 1st sess. House. Rept. 138)

Concerns the navigation of the Bayou St. John in Louisiana.

S-36692.

Copies: DLC; DeU; MWA; NjR.

4454. United States. Congress. Senate. Committee on Commerce.

In Senate of the United States. April 4, 1828. Mr. Williams, from the Committee on Commerce, to which was referred the resolution of the Senate of the 1st inst., directing that Committee to "inquire into the expediency of making further appropriation for removing obstructions at the mouth of the Pascagoula River," laid the following letters from Silas Dinsmore and W. Barton, on the table.... [Washington: 1828.] 7 p. (20th Cong., 1st sess. Senate. Doc. 164)

S-36384.

Copies: DLC; DeU; MWA; NjR.

4455. United States. Congress. Senate. Committee on Commerce.

In Senate of the United States. April 28, 1828. Mr. Williams made the following report: The Committee on Commerce, who were instructed by a resolution of the Senate.... [Washington: 1828.] 8 p. (20th Cong., 1st sess. Senate. Doc. 179)

Recommends the removal of obstructions at the mouth of the Pascagoula River.

S-36389.

Copies: DLC; DeU; MWA; NjR.

4456. United States. Engineer Dept.

Hiwassee and Conesauga rivers. Letter from the Secretary of War, transmitting a communication from the Chief Engineer on the subject of a communication between the Hiwassee and Conesauga rivers, &c. December 11, 1828.... [Washington: Gales & Seaton, printers, 1828.] 10 p. (20th Cong., 2d sess. House. Doc. 15)

S-36332.

Copies: DLC; DeU; MWA; NN; NjR.

4457. United States. Engineer Dept.

Report of the Secretary of War, showing that progress has been made in deepening and clearing the channel, at a place called the Grand Chain. May 22, 1828.... Washington: Printed by Duff & Green, 1828. 7 p. (20th Cong., 1st sess. Senate. Doc. 203)

Transmits a report of the Chief Engineer concerning the clearing of the Ohio River at the Grand Chain.

S-36781.

Copies: DLC; DeU; MWA; NjR.

4458. United States. Engineer Dept.

Survey of Cape Fear River. Letter from the Secretary of War, transmitting a report and plan of the survey of Cape Fear River below the town of Wilmington. February 8, 1828.... Washington: Printed by Gales & Seaton, 1828. 14 p. (20th Cong., 1st sess. House. Doc. 127)

S-36914.

Copies: DLC; DeU; MWA; NjR.

4459. United States. Engineer Dept.

Survey of Ocracock Inlet. Letter from the Secretary of War transmitting a report in relation to the improvement of the navigation of Ocracock Inlet in the state of North Carolina. December 29, 1828.... [Washington: Gales & Seaton, printers, 1828.] 10 p. (20th Cong., 2d sess. House. Doc. 37)

S-36915.

Copies: DLC; DeU; MWA; NjR.

4460. United States. Engineer Dept.

Survey of Sandusky Bay, Ohio. Letter from the Secretary of War, transmitting a report and plan of the survey of Sandusky Bay in the state of Ohio. January 21, 1828.... Washington: Printed by Gales & Seaton, 1828. 5 p. (20th Cong., 1st sess. House. Doc. 72)

S-36916.

Copies: DLC; DeU; MWA; NjR.

4461. United States. Engineer Dept.

Survey—swash in Pamlico Sound. Letter from the Secretary of War, transmitting ... a report and plan of the survey of the swash in Pamlico Sound. January 18, 1828.... Washington: Printed by Gales & Seaton, 1828. 12 p. (20th Cong., 1st sess. House. Doc. 69)

Transmits a report from the Engineer Dept., including information about the machinery and expenses needed to remove the obstructions.

S-36912.
Copies: DLC; DeU; MWA; NjP.

4462. United States. Engineer Dept.
Surveys—Kennebec River, &c. Letter from the Secretary of War, transmitting a report of the surveys of the Kennebec River, and of contemplated routes for canals, connected with the waters of the said rivers. February 22, 1828. Referred to the Committee on Roads and Canals, March 4, 1828.... Washington: Printed by Gales & Seaton, 1828. 58 p. (20th Cong., 1st sess. House. Doc. 173)
Signed: John J. Abert, Lt. Col. T. E.
S-36922.
Copies: DLC; DeGE; MWA; DeU; NN; NjR.

4463. ———— Concord: Printed by Jacob B. Moore, 1828. 28 p.
S-36923.
Copies: DeGE; NN; NhHi.

4464. United States. Laws.
A bill for improving the inland navigation between the St. Mary's River and the entrance of the river St. John's, in Florida, and for other purposes. January 11, 1828.... [Washington: 1828.] 1 p. (H.R. 67)
S-35753.
Copies: DNA.

4465. United States. Laws.
A bill making a grant of land to aid in improving the navigation of rivers Teche, Vermillion, Plaquemine, and other water courses in the state of Louisiana. March 12, 1828.... [Washington: 1828.] 1 p. (H.R. 216)
S-35985.
Copies: DNA.

4466. United States. Laws.
A bill making an appropriation for removing the bar at the east pass of the mouth of Pascagoula River, and for improving the harbor thereof. February 13, 1828.... [Washington: 1828.] 1 p. (H.R. 165)
S-35988.
Copies: DNA.

4467. United States. Laws.
A bill making an appropriation for the improvement of the navigation of the Ohio and Mississippi rivers. March 27, 1828. [Washington: 1828.] 1 p. (H.R. 244)
S-35991.
Copies: DNA.

4468. United States. Laws.
A bill to aid the states of Mississippi and Louisiana in improving the navigation of Pearl River. March 3, 1828.... [Washington: 1828.] 1 p. (H.R. 201)
S-36046.
Copies: DNA.

4469. United States. War Dept.
Obstructions to navigation, Wabash Letter from the Secretary of War in reply to a resolution of the House of Representatives, on the 15th instant, requiring information in relation to the survey of the obstructions to the navigation of the Wabash River. December 18, 1828.... [Washington: Gales & Seaton, printers, 1828.] 1 p. (20th Cong., 2d sess. House. Doc. 26)
S-36674.
Copies: DLC; DeU; MWA; NjR.

4470. United States. War Dept.
Report of the Secretary of War, in compliance with the resolution of the Senate of the 13th instant, with statements showing the reasons that induced the department to postpone the expenditure of the appropriation of the last session, for the removal of obstructions at the mouth of Pascagoula Bay. March 19, 1828.... Washington: Printed by Duff Green, 1828. 9 p. (20th Cong., 1st sess. Senate. Doc. 147)
Includes reports from the Corps of Engineers.
S-36780.
Copies: DLC; DeU; MWA; NjR.

4471. Cumings, Samuel.
The western pilot, containing charts of the Ohio River, and of Mississippi from the mouth of the Missouri to the Gulf of Mexico, accompanied with directions for navigating

4471 (*continued*)
the same, and a gazetteer, or description of the towns on their banks, tributary streams, &c.... Cincinnati: N. & G. Guilford, 1829. 104 p., plates, 44 maps.
S-38315.
Copies: Ct; DLC; ICN; ICU; InHi; MB; MH; MWA; MiU-C; MnU; NN; NcD; OClWHi; PHi; T.

4472. Mebane, James.
Report of James Mebane, Esquire, concerning works on Cape Fear River, for the year 1829. Raleigh: Printed by Lawrence & Lemay, printers to the state, 1829. [8] p.
S-39540.
Copies: NcU.

4473. Missouri. General Assembly.
Memorial of the legislature of Missouri, praying that improvements be made in the navigation of the Mississippi River.... Feb. 19, 1829.... [Washington: 1829.] 2 p. (20th Cong., 2d sess. Senate. Doc. 88)
Copies: DeU; MWA.

4474. Navigation and advantages of the river Schuylkill for foreign trade, established by affidavits. [Philadelphia?: 1829?] 8 p.
S-39703.
Copies: DLC; DeGE; MH-BA; PHi; PPAmP.

4475. New York (State). Canal Commissioners.
Report relative to the navigation of the Hudson River. Albany: 1829. 4 p. (Assembly. Doc. 42)
Whitford-p. 1271.
Copies: N.

4476. North Carolina. General Assembly. Joint Select Committee on Albemarle Sound.
Report of the Joint Select Committee on so much of the Governor's message as relates to opening a communication from Albemarle Sound to the ocean. Raleigh: Printed by Lawrence & Lemay, 1829. 4 p.
S-39866.
Copies: NcU.

4477. North Carolina. Laws.
North Carolina. Act of the Legislature of the state of North Carolina, incorporating the Roanoke Inlet Company, and for other purposes. January 19, 1829.... Read, and referred to the Committee on Roads and Canals. Washington: Printed by Gales & Seaton [1829]. 5 p. (20th Cong., 2d sess. House. Doc. 82)
S-41265.
Copies: DLC; MWA; NjR; ViU.

4478. Ohio. Board of Canal Commissioners.
Report of the Canal Commissioners. [Jan. 17, 1829.] [Columbus: 1829.] 19, [1] p.
Transmitting the report of engineers on surveys and examinations of the Stillwater and Kilbuck creeks.
S-39905.
Copies: NN; PP.

4479. Schuylkill Navigation Company.
Report of the president and managers of the Schuylkill Navigation Company, to the stockholders. January 5, 1829. [Philadelphia]: Printed by Lydia R. Bailey, 1829. 11 p.
S-40365.
Copies: DeGE; MWA; PHi.

4480. United States. Congress. House. Committee on the Judiciary.
Orleans Navigation Company. Feb. 11, 1828. Read, and committed to a Committee of the Whole House to-morrow. Jan. 31, 1829. Reprinted by order of the House of Representatives. Mr. P. P. Barbour, from the Committee on the Judiciary, to which the subject had been referred, made the following report. [Washington: Gales & Seaton, printers, 1829.] 20 p. (20th Cong., 2d sess. House. Rept. 62)
Concerns navigation on the Bayou St. John in Louisiana.
S-41272.
Copies: DLC; DeU; MWA; NjR.

4481. United States. Engineer Dept.
Message from the President of the United States, transmitting copies of surveys made in pursuance of acts of Congress, of 30th April, 1824, and 2d March, 1829. December 16, 1829.... [Washington: 1829.] 34 p. (21st Cong., 1st sess. House. Doc. 7)
Includes the surveys made by the Engineer Department, of Deep Creek, Virginia, Pasquotank River, North Carolina, the mouth of the Mississippi River, and a railroad from Charleston to Hamburg, South Carolina.
S-41240.
Copies: CSt; CtY; DLC; DeU; IU; MB; MBAt; MH; MWA; NN; NjR; P; PPAmP.

4482. United States. Engineer Dept.
Navigation of the Ohio. Letter from the Secretary of War, transmitting the information required by a resolution of the House of Representatives in relation to the progress which has been made in moving obstructions to the navigation of the river Ohio. Jan. 17, 1829.... [Washington: Gales & Seaton, printers: 1829.] 2 p. (20th Cong., 2d sess. House. Doc. 77)
Transmitting a report of the Engineer Department.
S-41258.
Copies: DLC; DeU; MWA; NjR.

4483. United States. Engineer Dept.
Survey, Genesee River, L. Ontario, Oswego River, &c. Letter from the Secretary of War, transmitting a report of a survey of Genesee River, the south shore of Lake Ontario, between the Genesee and Oswego rivers, and Big Sandy Creek, in the state of New York. February 3, 1829.... [Washington: Gales & Seaton, printers, 1829.] 20 p. (20th Cong., 2d sess. House. Doc. 106)
Transmitting a report of the Engineer Department.
S-41320.
Copies: DLC; DeU; MWA; NjP.

4484. United States. Engineer Dept.
Survey—Kentucky River. Letter from the Secretary of War in reply to a resolution of the House of Representatives of the 24th ultimo, in relation to the progress which has been made in the survey of the Kentucky River, &c., &c. January 8, 1829.... [Washington: Gales & Seaton, printers, 1829.] 2 p. (20th Cong., 2d sess. House. Doc. 58)
Transmitting a report of the Engineer Department.
S-41319.
Copies: DLC; DeU; MWA; NjR.

4485. United States. Engineer Dept.
Survey of Roanoke Inlet, &c. Letter from the Secretary of War, transmitting a report of the survey of Roanoke Inlet and Sound, in the state of North Carolina. February 24, 1829.... [Washington: Gales & Seaton, 1829.] 24 p. (20th Cong., 2d sess. House. Doc. 128)
S-41321.
Copies: DLC; DeU; MWA; NjR.

4486. United States. War Dept.
Report of the Secretary of War, with one from the Engineer Department, on the practicability of an outlet from Albemarle Sound to the ocean, &c. February 23, 1829.... [Washington: 1829.] 24 p., fold. map. (20th Cong., 2d sess. Senate. Doc. 106)
Reports from Hamilton Fulton, engineer: p. 2–10.
N-U0264032.
Copies: DeU; MWA; NcD.

4487. Circular. Sir, the government of the United States having directed a survey and examination of Connecticut River, and also of the Farmington and the Hamphire and Hampden canals, it has become desirable to ascertain as far as may be, the quantity of goods ... to be transported to and from the town of ________ annually.... Northampton, Mass.: 1830. Broadside.
Dated: September 22d, 1830.
C-2468.
Copies: M.

4488. James River Company.

A collection of all acts and parts of acts, of the General Assembly of Virginia, from October, 1784, down to the session of 1829–30, inclusive, relative to the James River Company. Together with an appendix, containing the regulations of the president and directors of the said company, for the superintendence and preservation of the improvements under their direction. Richmond: Printed by S. Shepherd & Co., 1830. 104, 14, xxvii p.

C-5284; N-J0038195.

Copies: DLC; Vi.

4489. Journal of the convention, holden at Windsor, Vermont, September 29th & 30th, 1830. For the purpose of taking into consideration subjects connected with the improvement of the navigation of the Connecticut River.... Windsor, [Vt.]: S. Ide, printer, 1830. 19 p.

C-2085.

Copies: DLC; MB; MH-BA; MiD; Nh; Vt; VtU.

4490. Maryland. Commissioners on the Survey of the Pocomoke River.

Report of the Commissioners on the Survey of the Pocomoke River, &c. Annapolis: Printed by J. Hughes, 1830. 7 p.

C-2442.

Copies: MWA; MdHi.

4491. Maryland. Governor (T. K. Carroll).

(Document No. 1) Accompanying the executive communication of the 29th December 1830. Correspondence between the clerk of the Council, and the Secretary of War, relative to the survey of the Pocomoke. Annapolis: Printed by Jonas Green, 1830. 5 p.

C-2418.

Copies: MdHi.

4492. New York (State). Canal Commissioners.

Report of Canal Commissioners respecting improvement of navigation of Hudson River with steam dredging machine. Albany: 1830. 3 p. (Assembly. Doc. 335)

Hasse (New York)-p. 487; Whitford-p. 1271.

Copies: MWA; NN.

4493. New York (State). Legislature. Assembly. Select Committee on the Survey of the Susquehanna and Chemung Rivers.

Report of the Select Committee, in relation to the survey of the Susquehannah and Chemung rivers, and the connecting of the Otsego Lake with the Erie Canal, by canal or rail-road. [Albany: Printed by E. Croswell, 1830.] 5 p. (Assembly. Doc. 63).

Dated: January 27, 1830.

N-N0200488; Thomson-431.

Copies: CtY; DLC; MB; MWA; MiU-T; N; NN.

4494. New York (State) Legislature. Senate. Committee on Canals.

Report on the petition of O. W. Van Tuyl, president of the Neversink Navigation Company, praying for a further loan from the state, to complete the improvement of the navigation of the Neversink River. Albany: 1830. 6 p. (Senate. Doc. 358)

Whitford-p. 1274.

Copies: MWA; N.

4495. Ohio and Mississippi River. Memorial of sundry citizens of Louisville, Kentucky, praying for the improvements of the navigation of the Ohio and Mississippi rivers. [Washington: Printed by D. Green, 1830.] 2 p. (21st Cong., 1st sess. House. Rept. 337)

Dated: March 29, 1830.

C-5025.

Copies: DLC; MWA; NjR.

4496. Pennsylvania. Laws.

Acts of the Legislature of Pennsylvania, relating to the Schuylkill Navigation Company. Philadelphia: Printed by J. Kay, 1830. 43 p.

C-2980.

Copies: DLC; DeGE; MWA; NNC; PHi (at PPL); PPF.

4497. Petition of citizens of Louisville, Kentucky, praying for an appropriation for improving the navigation of the Ohio River. December 24, 1830.... [Washington: Printed by D. Green, 1830.] 3 p. (21st Cong., 2d sess. House. Rept. 11)

C-5059.

Copies: DLC; MWA; NjR.

4498. Roanoke Navigation Company.

Report of the Roanoke Navigation Company, 1830. Raleigh: Printed by Lawrence & Lemay, printers to the state, 1830. 4 p.

At head of title: Number 6, A.D. 1830.

C-3322.

Copies: NcU.

4499. Schuylkill Navigation Company.

Report of the president and managers of the Schuylkill Navigation Company, to the stockholders. January 4, 1830. Philadelphia: Printed by James Kay, jun. & Co., [1830]. 12 p.

C-3416.

Copies: DBRE; DIC; DLC; DeGE; ICJ; MH-BA; MWA; MiU-T; NN; OClWHi; PPAmP.

4500. Tennessee. General Assembly.

In Senate of the United States. January 11, 1830.... Resolutions of the legislature of Tennessee, to procure an appropriation by Congress to effect the examination and removal of the Harpeth Shoals on the Cumberland River. [Washington: 1830.] 2 p. (21st Cong., 1st sess. Senate. Doc. 16)

N-T0094754.

Copies: DeU; MWA; ViU.

4501. United States. Congress. House. Committee on Internal Improvements.

Survey—Cumberland River. February 10, 1830.... Mr. Blair, from the Committee on Internal Improvements ... made the following report. [Washington: 1830.] 1 p. (21st Cong., 1st sess. House. Rept. 181)

C-5187.

Copies: DLC; DeU; MWA; NjR.

4502. United States. Congress. House. Committee on Internal Improvements.

Survey of Connecticut River, &c. March 15, 1830.... Mr. Butman, from the Committee on Internal Improvements ... made the following report. [Washington: 1830.] 1 p. (21st Cong., 1st sess. House. Rept. 294)

Survey for a canal or railroad, and for improvement of navigation of the Connecticut River.

C-5190.

Copies: DLC; DeU; MWA; NN; NjR.

4503. United States. Congress. House. Select Committee on the Navigation of Roanoke Inlet.

Roanoke Inlet. March 19, 1830.... Mr. William B. Shepard, from the Select Committee ... made the following report. [Washington: 1830.] 4 p. (21st Cong., 1st sess. House. Rept. 317)

C-5133.

Copies: DLC; MWA; NjR.

4504. United States. Congress. House. Select Committee on the Navigation of the Ohio and Mississippi Rivers.

Navigation, Ohio and Mississippi rivers. (To accompany bill H.R. no. 440) April 23, 1830. Mr. Wickliffe, from the Select Committee to which the subject had been referred, made the following report. [Washington: Printed by D. Green, 1830.] 4 p. (21st Cong., 1st sess. House. Rept. 379)

Recommends the construction of steamboats for the removal of obstructions to river transportation.

C-5017.

Copies: DLC; DeU; MWA; NjR.

4505. United States. Congress. Senate. Committee on Commerce.

In Senate of the United States. May 8, 1830.... Mr. Woodbury made the following report: The Committee on Commerce, to whom were referred two letters from the Engineer Department on the subject of a survey of the Wabash River, ask leave to report. [Washington: 1830.] 1 p. (21st Cong.,

4505 (*continued*)
1st sess. Senate. Doc. 140)
C-4790.
Copies: DLC; DeU; MWA; NjR.

4506. United States. Congress. Senate. Committee on Public Lands.
In Senate of the United States. February 12, 1830. Mr. McKinley made the following report: The Committee on Public Lands, to which was referred the memorial of the legislature of the state of Alabama, asking a grant of land for the purpose of improving the navigation of the Coosa River and to connect its waters with the waters of the Tennessee River ... report. [Washington: 1830.] 7 p. (21st Cong., 1st sess. Senate. Doc. 55)
C-4759.
Copies: DLC; DeU; MWA; NjR.

4507. United States. Engineer Dept.
Cape Fear River. Letter from the Secretary of War, transmitting a report in relation to the works on Cape Fear River. February 1, 1830.... [Washington: 1830.] 3 p. (21st Cong., 1st sess. House. Doc. 44)
Transmitting a report from the Engineer Department.
C-4610.
Copies: DLC; MWA; NjR.

4508. United States. Engineer Dept.
Pensacola Bar. February 5, 1830.... Memoir and estimate on the improvement of the bar of Pensacola.... [Washington: 1830.] 4 p. (21st Cong., 1st sess. House. Doc. 48)
Report from the Engineer Department, transmitted to the House of Representatives by the Secretary of War.
C-5052.
Copies: DLC; MWA; NjR.

4509. United States. Engineer Dept.
Report on the survey of North River, Massachusetts. December 23, 1830.... [Washington: 1830.] 3 p. (21st Cong., 2d sess. House. Doc. 16)
A report from the Engineer Department, on the practicability of removing obstructions at the mouth of the North River.
Copies: DeU; MWA.

4510. United States. Engineer Dept.
Survey — Penobscot River. Message from the President of the United States, transmitting a copy of the report on survey made of the ship channel of Penobscot River, &c.&c. March 19, 1830.... [Washington: 1830.] 3 p. (21st Cong., 1st sess. House. Doc. 82)
C-5193.
Copies: DLC; DeU; MWA; NjR.

4511. United States. Engineer Dept.
Survey — River Thames, Conn. Letter from the Secretary of War, transmitting a copy of a report of the engineer appointed to make a survey of the river Thames in the state of Connecticut. March 9, 1830.... [Washington: 1830.] 7 p. (21st Cong., 1st sess. House. Doc. 125)
C-5194.
Copies: DLC; MWA; NjR.

4512. United States. Laws.
An act to grant to the state of Alabama certain lands for the purpose of improving the navigation of the Coosa River, and to connect its waters with those of the Tennessee River and Canal. April 12, 1830.... [Washington: 1830.] 3 p. (S.92)
C-3845.
Copies: DNA.

4513. United States. Laws.
A bill making further appropriations for the improvement of the navigation of the Ohio and Mississippi rivers, and their tributary streams, and for the deepening the channel at the mouth of the Mississippi River. April 23, 1830.... [Washington: 1830.] 2 p. (H.R. 440)
C-4380.
Copies: DNA.

4514. United States. Laws.
A bill to extend the time for commencing the improvement of the navigation of the

Tennessee River. February 17, 1830.... [Washington: 1830.] 1 p. (S. 97)
C-4514.
Copies: DNA.

4515. United States. Laws.
A bill to improve the Back Creek. February, 1830.... [Washington: 1830.] 1 p.(H.R. 293)
C-4521.
Copies: DNA.

4516. United States. Laws.
A bill to improve the navigation of the Monongahela and Alleghany Rivers. April 14, 1830.... [Washington: 1830.] 1 p. (S. 179)
C-4523.
Copies: DNA.

4517. United States. Laws.
A bill to provide for the improvement of the navigation of Coosa River, in the state of Alabama, and to connect the waters of said river Coosa with the waters of the Tennessee River. December 23, 1830.... [Washington: 1830.] 3 p. (H.R. 511)
C-4556.
Copies: DNA.

4518. United States. Treasury Dept.
Savannah River. Letter from the Secretary of the Treasury transmitting the information required by a resolution of the House of Representatives of the 6th ultimo, in relation to disbursements which have been made in removing obstructions in the Savannah River, under the authority of the act of 18th May, 1826, &c.&c. May 20, 1830.... [Washington: 1830.] 59 p. (21st Cong., 1st sess. House. Doc. 106)
C-5155.
Copies: DLC; MWA; NjR.

4519. United States. War Dept.
Letter from the Secretary of War, transmitting copies of the reports of H. M. Shreve and R. Delafield, on the improvement of the navigation of the Mississippi and Ohio rivers. December 16, 1830.... [Washington: 1830.] 12 p. (21st Cong., 2d sess. House. Doc. 9)
C-4911.
Copies: DLC; DeU; MWA; NjR.

4520. Virginia. Principal Engineer.
Report of the Principal Engineer. [Richmond: Printed by Thomas Ritchie, printer to the Commonwealth, 1830.] 7 p. (Virginia. House. Doc. 16)
Claudius Crozet, principal engineer.
Concerns the improvement of the James River and a proposed railroad.
N-C0811794; Thomson-397.
Copies: DBRE; DLC; M; NN; Vi.

4521. Wheeler, John Hill, 1806–1882.
Speech of Mr. John H. Wheeler ... on the subject of the Roanoke Inlet, delivered in a convention, assembled at Edenton, February 15, 1830; also, a report, made by Mr. Wheeler, on the same subject, to the House of Commons of the last legislature of North Carolina. [Raleigh: 1830.] 8 p.
Thornton (Off)-1703.
Copies: DLC.

3. Canals

4522. Chesapeak. [Philadelphia: 1768.] iv, [5]–18 p.
Proposals for the construction of a canal between the Chesapeake and Delaware bays.
Mp-10854.
Copies: DLC; PPL.

4523. New York (State). Laws.

An act for establishing and opening lock navigations within this state, passed the 30th of March, 1792. [New York: Francis Childs and John Swaine, printers to the state], 1792. 4 p.

B-8085; Mp-46526.

Copies: NHi; NN.

4524. Northern Inland Lock Navigation Company.

The Northern Inland-Lock Navigation Company will purchase the following articles, by contract.... Albany: Printed by Charles R. and George Webster, [1792]. Broadside 40.5 x 24.5 cm.

Signed: Ph. Schuyler, President. October 10, 1792.

Concerns bids for hewn timber of various dimensions.

Mp-24634.

Copies: NN.

4525. Schuyler, Philip John, 1733–1804.

To the stockholders in the Northern and Western Inland-Lock-Navigation Companies, in the state of New-York.... [Albany?: 1793.] Broadside 54.5 x 45 cm.

Dated: Stillwater, on Hudson's River, June 7, 1793.... Ph. Schuyler.

B-8474; Mp-46874.

Copies: DLC (photocopy); NN.

4526. Schuylkill and Susquehanna Navigation.

Plan for raising by loan a sum of money for the purpose of carrying on the work of the Schuylkill and Susquehanna Canal, till the end of the present year 1794. [Philadelphia: 1794.] Broadside 41 x 33.5 cm.

B-8862; Mp-47210.

Copies: DLC.

4527. Schuylkill and Susquehanna Navigation.

To the Senate and House of Representatives of the Commonwealth of Pennsylvania in General Assembly met.... [Philadelphia: Printed by Zachariah Poulson, Jr., 1794.] [2] p.

Letter from the president and report from the superintendent and the engineer, giving details of progress, costs, and estimates of uncompleted construction.

E-27678.

No copy known.

4528. Tatham, William, 1752–1819.

To His Excellency Henry Lee, Governor, and the respective gentlemen of the Council ... to the share-holders and others interested in the canals of Virginia.... [Richmond: 1794.] 16 p.

Signed: William Tatham. Richmond, Virginia, May 1794.

B-8882; N-T0050626.

Copies: CtY.

4529. A new map of that part of Massachusetts state, which includes the proposed canal, from Connecticut River, to the waters that surround the town of Boston. Boston: 1795. Map.

E-29183.

No copy known.

4530. Pennsylvania. Laws.

An act to enable the president and managers of the Schuylkill and Susquehanna Navigation and the president and managers of the Delaware and Schuylkill Canal Navigation to raise by way of lottery the sum of four hundred thousand dollars for the purpose of completing the works in their acts of incorporation mentioned. [Philadelphia]: Printed by Zachariah Poulson, junior, [1795]. 2 p.

"Approved April 17, 1795."

B-9224; Mp-47554.

Copies: PPL.

4531. Schuylkill and Susquehanna Navigation.

Canal lottery. Scheme of a lottery authorized by an act entitled "An act to enable the president and managers of the Schuylkill and Susquehanna Navigation, and the president and managers of the Delaware and Schuylkill Canal Navigation, to raise by way of lottery, the sum of four hundred thousand dollars, for the purpose of completing the works in their acts of incorporation mentioned. . . ." [Philadelphia: Printed by Zachariah Poulson, 1795.] Broadside 45 x 28.5 cm.

Dated: Philadelphia, May 18, 1795.
Mp-29473.
Copies: PHi.

4532. Schuylkill and Susquehanna Navigation.

An historical account of the rise, progress and present state of the canal navigation in Pennsylvania. With an appendix, containing abstracts of the Acts of the legislature since the year 1790, and their grants of money for improving roads and navigable waters throughout the state; to which is annexed, "an explanatory map." Published by direction of the president and manager of the Schuylkill and Susquehanna, and the Delaware and Schuylkill Navigation Companies. . . . Philadelphia: Printed by Zachariah Poulson, junior, 1795. 1 p.l., xvi, 80 p., fold. map.

"An act . . . to raise, by way of lottery, the sum of four hundred thousand dollars, for the purpose of compleating the works in their acts of incorporation. . . ." p. 79–80.

Ascribed to Robert Morris, or to William Smith.

Mp-29474.
Copies: DLC; DeGE; MBAt; MH; MWA; NN; PPAmP; PPL; RPJCB.

4533. ———— Philadelphia: Zachariah Poulson, 1795. 77 p., map.

B-9298; Mp-47605.
Copies: MWA; NHi; NN; RPJCB.

4534. Susquehanna Canal Company.

At a meeting of the governor and directors of the Maryland Susquehanna Canal, held at the canal, on Thursday, the 28th May, 1795. . . . Circular. [Baltimore?: 1795.] 2 p.

Dated: Maryland, Susquehanna Canal-office, May 29, 1795.
Mp-29032.
Copies: PHi.

4535. Delaware and Schuylkill Canal Navigation.

Resolutions of the stockholders of the Delaware and Schuylkill Canal Navigation, at a meeting held March 12, 1796; and the plan of a loan founded on the same. Philadelphia: 1796. 2 p.

Mp-30330.
Copies: PHi; RPJCB.

4536. New York (State). Legislature. Senate.

New-York. In Senate, February 10, 1796. Resolved, that the report of the Western [and Northern] Inland Lock Navigation Companies, be forthwith printed in the news-paper published by the printer to this state. By order, Abraham B. Bancker, Clerk. To the Honourable the Legislature of the state of New-York, in Senate and Assembly convened. The directors of the Western and Northern Lock Navigation Companies, respectfully report . . . Ph. Schuyler, president. [New York]: Printed by William Robins, [1796]. Broadside.

Mp-31622.
Copies: NN.

4537. Schuylkill and Susquehanna Navigation.

Canal lottery, no. two. Scheme of a lottery authorized by an act entitled "An act to enable the president and managers of the Schuylkill and Susquehanna Navigation, and the president and managers of the Delaware and Schuylkill Canal Navigation, to raise by way of lottery, the sum of four hundred thousand dollars, for the purpose of completing the works. . . ." [Philadelphia: Printed by Zachariah Poulson, Junior, 1796.] Broadside 41 x 24 cm.

Mp-31167.
Copies: PHi; PPL.

4538. Western Inland Lock Navigation Company.

Report of the directors of the Western and Northern Inland Lock Navigation Companies, in the state of New-York, to the Legislature: together with the report of Mr. William Weston, engineer. New York: Printed by George Forman, 1796. 20 p.

Mp-31623.

Copies: MWA; MiU-C; NHi; NN.

4539. Western Inland Lock Navigation Company.

To the Honourable the Legislature of the state of New-York, in Senate and Assembly convened. The directors of the Western and Northern Inland Lock Navigation Companies respectfully report.... [New York]: Printed by William Robins, [1796]. Broadside.

At head of title: New-York, In Senate, February 10, 1796 ...

Mp-31622.

Copies: NN.

4540. Western Inland Lock Navigation Company.

Western inland lock navigation. The directors of the Western Inland lock navigation Company have determined to construct a canal to connect the waters of the Mohawk River, with those of Wood Creek.... [New York: 1796.] Broadside.

Dated: New-York, January 18th, 1796.

Signed: Ph. Schuyler, President.

Mp-31621.

Copies: NN (defective).

4541. [Carroll, James.]

Reflections on the proposition to communicate, by a navigable canal, the waters of Chesapeake with those of Delaware Bay, addressed to the citizens of Maryland. To which is prefixed the bill, as published for consideration by order of the House of Delegates. Annapolis: Printed by Frederick Green, printer to the state, [1797?]. 50 p.

DeGE copy is dated in ink: 1796.

Mp-32744.

Copies: DeGE; MWA; PPAmP.

4542. Western Inland Lock Navigation Company.

Rates of toll. The directors of the Western Inland Lock Navigation Company, order, that the following rates of toll be received at the canal.... Albany: Printed by C. R. & G. Webster, [1797]. Broadside 35 x 21 cm.

Dated: New-York, March 1, 1797.

Mp-33207.

Copies: NN.

4543. Western Inland Lock Navigation Company.

Rates of toll to be received at Fort-Schuyler.... [Albany?: 1797.] Broadside 45 x 30 cm.

Dated: New-York, August 18, 1797.

Mp-33208.

Copies: NN.

4544. Conewago Canal Company.

Account of the Conewago-Canal, on the river Susquehanna. To which is prefixed, the act for incorporating the company. Philadelphia: Whitehall-Press, printed by William Young, 1798. [6], 18 p.

Signed: William Smith, Henry Miller. Includes a report by John Hall.

Mp-33551.

Copies: DLC; MH; MdHi; NN; PPAmP; PPL.

4545. Proprietors of the Middlesex Canal.

At a meeting of the Middlesex-Canal Corporation, held at Charlestown, on the 16th day of October, 1798.... [Boston: 1798.] Broadside 18 x 19 cm.

Mp-34106.

Copies: MB.

4546. Western Inland Lock Navigation Company.

At a meeting of a board of directors of the Western Inland Lock Navigation Company, held the 12th day of February, 1798. Resolved ... that the following rates of toll be established.... [Albany: Printed by Charles R. and George Webster, 1798.] Broadside.

A revised schedule of tolls to be collected at Fort Schuyler, dated February 12, 1798.

Mp-35011.
Copies: NN.

4547. Western Inland Lock Navigation Company.
Report of the directors of the Western Inland Lock-Navigation Company, to the Legislature, 16th February, 1798. Albany: Printed by Charles R. & George Webster, [1798]. 32 p.
Mp-35012.
Copies: CSmH; MH; MWA; N; NHi; NN; PPAmP.

4548. Blodget, Samuel] 1724–1807.
The subscriber has begun, and nearly accomplished a suit of locks on Amoskeig [*sic*] Falls in Merrimack River.... [Boston?: 1799.] Broadside 18 x 17 cm.
Subscription form for shares in the Amoskeag Canal Company.
B-10718.
Copies: MB.

4549. [Blodget, Samuel] 1724–1807.
To the public. The cutting a canal at Amoskeig Falls is of very great importance.... [Concord?, N.H.: 1799?] Broadside 42 x 33 cm.
Includes the text of a letter of Loammi Baldwin to Blodget, dated September 3, 1799, and the form for the subscription of shares in the Amoskeag Canal Company.
B-10719.
Copies: MB.

4550. Delaware and Schuylkill Canal Navigation.
Address of the committee of the Delaware and Schuylkill Canal Company, to the committees of the Senate and House of Representatives, on the memorial of said Company. Philadelphia: Printed by John Ormrod, [1799]. 32 p.
Signed on p. 28: William Sansom, Chairman of the Committee, January 19th, 1799.
Mp-35399.
Copies: DLC; DeGE; MH; MHi; NN; PPAmP; PPL.

4551. New Hampshire. Laws.
An act to authorize Samuel Blodget, Esquire, to set up a lottery for the purpose of locking Amoskeag Falls. [Portsmouth?, N.H.: 1799.] Broadside.
Mp-35887.
Copies: MB.

4552. New Hampshire. Laws.
An act to incorporate Samuel Blodget, Esquire, and others ... with the exclusive right ... of cutting a canal by Amoskeag Falls ... and locking the same. [Portsmouth?, N.H.: John Melcher?, 1799?] 3 p.
"Approved December 24, 1798."
B-10877.
Copies: MB.

4553. Proprietors of the Middlesex Canal.
Middlesex Canal navigation. The public are informed that a large boat, called the Washington ... will proceed from the head of the canal ... on every Thursday morning, and arrive at Charleston, the same day before night.... [Boston: 1799.] Broadside 26 x 21.5 cm.
Also dated [1804?] by MH-BA.
Mp-35819; N-P0606267.
Copies: MB; MH-BA.

4554. Remarks on a second publication of B. Henry Latrobe, engineer, said to be printed by order of the Committee of the councils (of the city) and distributed among the members of the Legislature. [Philadelphia: 1799.] 7 p.
Dated at end: January 26th, 1799.
Attributed to William Smith; concerns the Delaware and Schuylkill Canal.
E-36200 (microcarded with Mp-35399).
Copies: DLC; MBAt; MH; MWA; N; NN; PHi.

4555. Amoskeag Canal Lottery.
Amoskeag canal-lottery. The managers ... present the following scheme of the fouth [*sic*] class of said lottery.... [Exeter, N.H.: 1800.] Broadside 30 x 18 cm.
Dated and signed: Exeter, New-Hampshire, August 5, 1800. Ephraim

4555 (*continued*)
Robinson, Samuel Tinney [Tenney], Nathan'l Parker, managers.
B-10963; Mp-49004.
Copies: MB.

4556. Delaware and Schuylkill Canal Navigation.
Extracts from the resolutions of the stockholders of the Delaware and Schuylkill Canal; and the procedure of the managers thereon, in order to carry the same into effect. Philadelphia: Printed for the information of the stockholders, 1800. 12 p.
Dated: 25th January, 1800.
Mp-37313.
Copies: IU; MH-BA; NN; P; PP; PPL; RPJCB.

4557. Massachusetts. Laws.
Acts of the Legislature of Massachusetts, incorporating the Proprietors of the Middlesex Canal. [Boston?: Printed by Young and Minns?, 1800.] 14, [1] p.
Contains the acts of 1793, 1795, and 1800.
Mp-37911.
Copies: DLC; MH-BA; MWA; NNC.

4558. Amoskeag Canal Lottery ... May 23, 1801. Boston: 1801. Broadside.
Advertises the 6th–10th classes of lottery tickets, offered to finance the canal. Signed: Ephraim Robinson, Samuel Tenney and Nathaniel Parker.
Copies: MSaE; MWA.

4559. Francis, Thomas Willing, 1767–1815.
Letter to William Penrose, Esq., chairman of the committee to whom was referred the petition of Thomas W. Francis, for a toll on the Conewago Canal and locks on the river Susquehanna. Philadelphia: Printed by Z. Poulson, Jr., 1802. 16 p.
"Report on the Conewago Canal. To His Excellency Governor Mifflin. December 5, 1797," signed by John Hall, p. 10–14.
S&S-2247.
Copies: NN.

4560. Lehigh Coal and Navigation Company.
By authority. The second class of the Lehigh Navigation lottery.... Philadelphia: Printed by William W. Woodward, [1802]. Broadside.
Dated 22d February, 1802.
Describes the parts of the canal to be completed with the income from the lottery.
S&S-2514.
Copies: DLC.

4561. South Hadley Canal Lottery.
A list of the prizes and fortunate numbers in the second class of South Hadley Canal Lottery.... [Boston: Printed by Gilbert & Dean, 1802.] [4] p.
Dated: March 21, 1802.
Copies: MWA.

4562. Susquehanna Canal Company.
Report of the governor and directors to the proprietors of the Susquehanna Canal, at their semi-annual meeting, held in the city of Baltimore, October 25th, 1802. Baltimore; Printed by Warner & Hanna, 1802. 22 p.
Robert Gilmor, Governor of the Company.
"Report on the improvements effected in the navigation of the river Susquehanna.... " by B. H. Latrobe, p. 13–20.
S&S-3128.
Copies: DLC; DeGE; MdHi; NHi; NN; PPL.

4563. United States. Congress. House. Committee on the District of Columbia.
Report on the petition of sundry inhabitants of the city of Washington and District of Columbia ... 19th February, 1802.... [Washington: 1802.] 7 p.
Reported by Mr. Sprigg; concerns the incorporation of a company to open a canal between the Potomac River and the Eastern Branch through Tyber Creek and the low lands at the foot of Capitol Hill.
S&S-3440.
Copies: DLC; MWA; NN.

4564. United States. Laws.
A bill, for opening a canal to communicate from the Potomac River to the Eastern Branch thereof, through the city of Washington. [Washington]: Duane, printer, [1802]. 5 p.
Dated: 24th March, 1802.
S&S-5220 (dated [1803]).
Copies: DLC.

4565. Chesapeake and Delaware Canal Company.
The laws of the legislatures of Maryland, Delaware & Pennsylvania, respecting the Chesapeake and Delaware Canal. Philadelphia: Printed by Zachariah Poulson, Junior, 1803. 48 p.
S&S-4512.
Copies: DeWI; NHi; MdBE; NNC; PPAmP.

4566. Proprietors of the Middlesex Canal.
Rate of toll and freight by the Middlesex Canal packet-boats on the Middlesex Canal ... June, 1803. [Charlestown, Mass.: 1803?] Broadside 37 x 28 cm.
Signed: John L. Sullivan, agent of the Middlesex Canal Corporation.
N-P0606270.
Copies: MH-BA.

4567. South Hadley Canal Lottery.
A list of the prizes and fortunate numbers in the third class of South Hadley Canal Lottery.[Boston: Printed by Gilbert & Dean, 1803.] [4] p.
Dated: July 19, 1803.
Copies: MHi; MWA.

4568. South Hadley Canal Lottery.
Sir, the managers of South-Hadley Canal Lottery have fixed Wednesday the 15th day of June next to commence drawing the third class.... Springfield, [Mass.]: 1803. Broadside 25 x 20 cm.
Dated: Springfield, April 15th, 1803.
Copies: MWA.

4569. Catawba Canal Company, in South Carolina. The charter granted by the Legislature of South Carolina to the company for opening the Catawba and Wateree rivers, by clearing away the shoals and removing obstructions, may be justly considered as the most valuable ever granted in the United States. Charleston: 1804. Broadside 34 x 21 cm.
Dated: 1st September, 1804.
Hummel-2002.
Copies: NcC.

4570. Chesapeake and Delaware Canal Company.
First general report of the president and directors of the Chesapeake and Delaware Canal Company, June 4, 1804. [Philadelphia: J. W. Scott, printer, 1804.] 17, [2] p.
S&S-6009.
Copies: DLC; DeGE; DeU; ICJ; MH; MWA; MdBP, MiU; MnHi; NN; NWM; OU; PP; PPL; PU.

4571. [Law, Thomas] 1756–1834.
Observations on the intended canal in Washington City. City of Washington: 1804. 24 p.
Appendix, p. 12–24, includes: Act of Congress [to incorporate the Washington Canal Company] approved May 1, 1802; Rates of wharfage; An estimate of the quantity of materials necessary for a two story brick and frame house ... ; and A rough estimate of the probable proceeds of the canal, and extracts from the letters of M. Templeman and Col. Stoddert.
S&S-6630, 6955.
Copies: DLC; OCl.

4572. New Hampshire. Laws.
An act, for incorporating the proprietors of Milford Canal. Amherst, N.H.: Joseph Cushing, printer, March, 1804. 7 p.
N-N0149191.
Copies: MiU-C.

4573. Observations on the proposed canal. Wilmington, Del.: J. Wilson, printer, 1804. Broadside 46 x 39 cm.

4573 (*continued*)
Signed: A friend to inland navigation, June 2, 1804.
Concerns the Chesapeake and Delaware Canal.
Copies: MWA.

4574. Proprietors of the Middlesex Canal.
By the Committee of Operations of the Middlesex Canal. May 7th, 1804.... [Boston?: 1804.] [2] p.
Signed: James Sullivan, chairman.
Kress-B4824.
Copies: MH-BA.

4575. Proprietors of the Middlesex Canal.
Rules, regulations, and by-laws adopted by the Proprietors of the Middlesex Canal. Boston: Printed by Manning & Loring, 1804. 19 p.
S&S-6777.
Copies: DeGE; M; MH; MHi.

4576. South Hadley Canal Lottery.
A list of the prizes and fortunate numbers in the fifth class of South Hadley Canal Lottery.... [Boston: Gilbert & Dean, 1804.] [4] p.
Dated: Boston, July 4, 1804.
S&S-7296.
Copies: MHi; MWA.

4577. Allen, Ira, 1751–1814.
Particulars of the capture of the ship Olive Branch laden with a cargo of cannon and arms.... With an appendix proposing a ship canal, from Lake Champlain to the river St. Lawrence ... v. 2. Philadelphia: Printed for the author, 1805. 551 p.
Vol. 1, published in 1798, is concerned only with the capture of the ship *Olive Branch.*
S&S-7852.
Copies: DLC; MWA; NN; PPL.

4578. Chesapeake and Delaware Canal Company.
Facts and observations respecting the Chesapeake and Delaware Canal, and its extension into Pennsylvania. [S.l.: 1805?] 8 p.
S&S-10126 (dated 1806?).
Copies: DeGE; DeU; MWA; NNC.

4579. ———— [S.l.: 1805.] 26 p.
S&S-8176.
Copies: DBRE; NNC; PPAmP.

4580. Chesapeake and Delaware Canal Company.
General report of the president and directors of the Chesapeake and Delaware Canal Company. [Philadelphia: 1805.] 26 p.
S&S-8177.
Copies: PPL.

4581. Chesapeake and Delaware Canal Company.
The memorial and petition of the president and directors of the Chesapeake and Delaware Canal Company, to the Honorable the Senate and House of Representatives of the United States, in Congress assembled: [Philadelphia?: J. W. Scott, printer, 1805.] 26 p.
Signed and dated at end: Kinsey [*sic*] Johns, Joshua Gilpin, Robert H. Goldsborough, December 1, 1805.
"Observations respecting the Chesapeake and Delaware Canal," p. [9]–26.
S&S-8178, 11620.
Copies: CtY; DLC; DeGE; MWA; MiU-C; NN; PPAmP.

4582. Chesapeake and Delaware Canal Company.
The memorial and petition of the president and directors of the Chesapeak and Delaware Canal Company, to the Senate and House of Representatives of the Commonwealth of Pennsylvania. [Philadelphia?: 1805.] 14 p.
Dated: Feb. 1, 1805.
Signed: George Fox, Joshua Gilpin, George Roberts.
S&S-8179.
Copies: CSt; DLC; DeGE; MWA; NN.

4583. Chesapeake and Delaware Canal Company.

Second general report of the president and directors of the Chesapeake and Delaware Canal Company, June 3, 1805. [Philadelphia: 1805.] 11, [2] p.
S&S-8180, 50621.
Copies: DLC; DeGE; DeU; ICJ; MWA; PPL.

4584. Chesapeake and Delaware Canal Company.
To the Honorable the Senate and House of Representatives of the Commonwealth of Pennsylvania. Presented February 1st, 1805. The memorial and petition of the president and directors of the Chesapeake and Delaware Canal Company. [Philadelphia?: 1805.] 12 p.
Signed: George Fox, Joshua Gilpin, George Roberts.
S&S-8181.
Copies: DeU; MWA; NN; NNC; PPAmP.

4585. Hovey, Benjamin, 1760–1814.
Communication from General Benjamin Hovey to his associates, relative to opening a canal navigation near the rapids of Ohio River, with sundry documents therein referred to, marked from No. 1 to 7. [Washington?: 1805.] 12 p.
S&S-8650, 9516.
Copies: CtY; DLC; MWA; MiU-C; OC.

4586. Indiana (Territory). Laws.
An act to incorporate the Indiana Canal Company. [Vincennes, Ind.: Elihu Stout, 1805.] [4] p.
S&S-8676.
Copies: WHi.

4587. Ohio Canal Company.
Proceedings of the managers of the Ohio Canal Company, at Louisville, on Wednesday, the 11th day of September, 1805. Lexington, [Ky.]: Printed by Daniel Bradford, 1805. 19 p., map.
Report of the surveyor, Jared Brooks, p. [5]–19, includes communications from General Benjamin Hovey.
S&S-9056.
Copies: DLC; MiU-C.

4588. [Allen, Ira] 1751–1814.
A concise summary of the second volume of the Olive Branch.... To which is subjoined General Allen's circular letter, on the subject of a ship canal of commerce.... Philadelphia: Printed and sold by Thomas T. Stiles & Solomon Wieatt, [1806]. 15 p.
S&S-9821.
Copies: MH; MHi; MWA; MiU-C; NNS; Vt.

4589. [Blodget, Samuel] 1724–1807.
Statements concerning the Blodget Canal. at Amoskeag Falls, on Merrimac River.... [S.l.]: Printed for the information of the public, June, 1806. 8 p.
S&S-10005.
Copies: MB; MH-BA; MWA.

4590. Chesapeake and Delaware Canal Company.
The memorial and petition of the president and directors of the Chesapeak and Delaware Canal Company, to the Honorable the Senate and House of Representatives of the Commonwealth of Pennsylvania, December 24, 1806. [Philadelphia?: 1806.] 7 p.
Signed: George Fox, Joshua Gilpin.
S&S-10128.
Copies: CtY; DeGE; DeU; MWA; NN; NNC; OOxM.

4591. ———— to the Senate and House of Representatives of the Commonwealth of Pennsylvania, January 1, 1806. [Philadelphia: 1806.] 20 p.
S&S-10127.
Copies: DeU; MWA.

4592. Chesapeake and Delaware Canal Company.
Third general report of the president and directors of the Chesapeake and Delaware Canal Company, June 2, 1806. [Philadelphia: 1806.] 9, [2] p.
S&S-10129.
Copies: DLC; DeGE; DeWI; MWA; PPL.

4593. Chesapeake and Delaware Canal Company.

To the Honorable the Senate and House of Representatives of the Commonwealth of Pennsylvania. Presented February 1st, 1805. The memorial and petition of the president and directors of the Chesapeake and Delaware Canal Company. [Philadelphia?: 1806.] 12, 6, 10 p.

Includes also: *To the Honorable the Senate and House of Representatives of Pennsylvania. Presented January 1st, 1806. The memorial and petition of the president and directors of the Chesapeake and Delaware Canal Company,* 6 p.; and *Facts and observations respecting the Chesapeak and Delaware Canal, and its extension into Pennsylvania,* 10 p.

Copies: DeGE; DeU.

4594. ———— Presented January 1st, 1806. The memorial and petition of the president and directors of the Chesapeake and Delaware Canal Company. [Philadelphia?: 1806.] 10 p.

Signed on p. 6: George Fox, Joshua Gilpin.

Facts and observations respecting the Chesapeak and Delaware Canal and its extension into Pennsylvania, p. 7–10.

S&S-10130.

Copies: DLC; MWA; NN; NNC.

4595. Proprietors of the Middlesex Canal.

Report of a committee appointed by the directors to report the business done in the year 1805. [Charlestown?, Mass.: 1806.] 4 p.

Signed by James Sullivan and Loammi Baldwin.

N-P0606278.

Copies: DLC.

4596. Proprietors of the Middlesex Canal.

The subscribers, being a committee appointed by the directors of the Middlesex Canal Corporation, to examine, calculate, and report the quantity of business done on the canal in the course of the year 1805. [Boston: 1806.] 4 p.

Dated: February 6, 1806; signed by James Sullivan and Loammi Baldwin.

S&S-10869.

Copies: MB; MH-BA; MWA.

4597. United States. Congress. House. Committee on the Memorial of the Legislature of Kentucky.

Report of the Committee appointed on the tenth ultimo, on the memorial of the legislature of Kentucky, March 19, 1806.... Washington: A. & G. Way, printers, 1806. 4 p.

Concerns a canal to bypass the rapids of the Ohio River.

S&S-11696.

Copies: DLC; MWA; NN.

4598. United States. Congress. House. Committee on the Petition of the Chesapeake and Delaware Canal Company.

Report from the Committee to whom was referred, on the twenty-eighth January last, the petition of the president and directors of the Chesapeake and Delaware Canal Company. March 5, 1806.... City of Washington: A. & G. Way, printers, 1806. 4 p.

S&S-11683.

Copies: DLC; DeGE; MWA; NN.

4599. United States. Congress. Senate. Committee to Whom was Referred the Memorial of the Chesapeake and Delaware Canal Company.

In Senate of the United States. March 21, 1806. Mr. Logan, from the Committee to whom was referred, on the 28th of January last, the memorial of the president and directors of the Chesapeake and Delaware Canal Company, made report.... [Washington?: 1806.] 5 p.

S&S-11573.

Copies: DLC; DeGE; MWA; NN.

4600. [Allen, Ira] 1751–1814.

A concise summary of the second volume of the Olive Branch.... To which is subjoined General Allen's circular letter, on the subject of a ship canal of commerce.... Philadelphia: Printed for the author, 1807. 24 p.

S&S-11965.

Copies: CtY; DLC; MB; MWA.

4601. Chesapeake and Delaware Canal Company.

[Papers. Philadelphia: 1807.] 5, 6, 18 p.

No title page.

Contains the following publications: [1] *The Memorial and petition of the president and directors of the Chesapeake and Delaware Canal Company, to the Honorable the Senate and House of Representatives of the United States in Congress assembled* ... January 10, 1807; [2] Dec'r. 1, 1805. *To the Honorable the Senate and House of Representatives of the United States, in Congress Assembled, the memorial and petition of the president and directors of the Chesapeake and Delaware Canal Company,* and [3] *Observations respecting the Chesapeake and Delaware Canal* (dated December 1, 1805).

S&S-12290, 13979.

Copies: DLC; MWA; PPL.

4602. United States. Congress. Senate. Committee to Whom was Referred the Memorial of the Chesapeake and Delaware Canal Company.

In Senate of the United States. January 15, 1807. Mr. Bayard, from the committee to whom was referred, the memorial of the Chesapeake and Delaware Canal Company, praying the protection and aid of Congress; presented on the 28th of January, 1806, with the documents accompanying the same, and the report of a committee of the Senate thereon, reported.... [Washington: 1807.] 6 p.

Report of Mr. Logan, March 21, 1806, p. 2–6.

S&S-13912.

Copies: DLC; DeGE.

4603. United States. Congress. Senate. Select Committee on Roads and Canals.

In Senate of the United States. February 24, 1807. The Committee, to whom was referred the resolution proposing appropriation of a quantity not exceeding ________ acres of land, to assist in cutting out a canal at the rapids of the Ohio ... report. [Washington: 1807.] 4 p.

Refers to the survey of rapids by Jared Brooks.

S&S-13922.

Copies: DLC; MWA.

4604. Amoskeag Canal Lottery.

A list of prizes and fortunate numbers in the ... Amoskeag Canal Lottery ... completed May 12, 1808.... Boston: Printed by Oliver and Munroe, [1808]. [22] p.

S&S-14351.

Copies: MB.

4605. Paleske, Charles G.

Letter to the Delaware and Schuylkill Canal Navigation Company. [Philadelphia: 1808.] 3 p.

Dated: June 28, 1808.

N-P0032859.

Copies: PHi.

4606. [Paleske, Charles G.]

Observations on the application for a law to incorporate "The Union Canal Company," respectfully submitted to the members of both houses of the Legislature of Pennsylvania. [Philadelphia: Duane, printer, 1808.] 16 p.

Signed: On behalf of the Schuylkill and Susquehanna Navigation. Charles G. Paleske.

S&S-15833.

Copies: DLC; DeGE; MH-BA; MWA; NN; PHi; PP; PPAmP; PPL.

4607. Proprietors of the Middlesex Canal.

Extract from the act of the General Court, passed March 2, 1803 ... Middlesex Canal by-laws.... [Boston?: 1808.] Broadside 46 x 28.5 cm.

Signed at end: John L. Sullivan, agent of the Proprietors of Middlesex Canal. Canal-Office, Charlestown, April 4, 1808.

N-P0606263.

Copies: MB; MH-BA.

4608. Proprietors of the Middlesex Canal.

Notification. The Proprietors of the Middlesex Canal having caused booms to be fixed in the mill-pond at Charlestown, for the accomodation of all persons ... who may bring spars and timber down the canal, and

4608 (*continued*)
wish to leave them in a safe place for sale.... Charlestown, [Mass.]: 1808. Broadside.
Dated: April 27, 1808.
S&S-15597.
Copies: MHi; MWA (xerox copy).

4609. Proprietors of the Middlesex Canal.
Rate of toll on the Middlesex Canal ... until further notice. April 4, 1808. [Charlestown?, Mass.: 1808.] Broadside 44 x 28 cm.
N-P0606271; S&S-15598.
Copies: MB; MHi; NN.

4610. Proprietors of the Middlesex Canal.
Rules and regulations for the government of the Middlesex Canal—until further notice.... Charlestown, April 4, 1808. [Boston?: 1808.] Broadside 56.5 x 44.5 cm.
Signed: John L. Sullivan.
S&S-15599.
Copies: MBilHi; MH-BA; MHi.

4611. Tatham, William, 1752–1819.
A comparative view of the four projected coastwise canals, which are supposed by some to be in competition for the trade between Norfolk and North-Carolina.... Norfolk, Va.: Printed by J. O'Connor, 1808. 27 p.
S&S-16288.
Copies: DLC; MBAt; N; NHi; PPL.

4612. Tatham, William, 1752–1819.
A view of the proposed grand junction canal, designed to bring the commerce of North-Carolina to Norfolk.... Norfolk, Va.: Printed by J. G. O'Connor, 1808. 15 p.
S&S-16289.
Copies: DLC; MSaE; NN.

4613. United States. Treasury Dept.
Letters to the Honorable Albert Gallatin, Secretary of the Treasury of the United States, and other papers relative to the Chesapeake and Delaware Canal. Philadelphia: Printed by John W. Scott, [1808]. vii, [3]–46 p.
Report of the Secretary of the Treasury, p. [v]–vii; Letter from Joshua Gilpin, p. [3]–39; Letter from B. H. Latrobe, p. 40–46.
S&S-12663, 15120.
Copies: DBRE; DLC; DeGE; DeU; MB; MWA; MdBP; NN; PPAmP; PPL.

4614. Allen, Ira, 1751–1814.
Copies of letters to the Governor of Vermont & address to the legislature thereof, respecting a conspiracy against the author & respecting a ship canal from Lake Champlain to the river St. Lawrence, with the opinion of the attorney general on the case of the Olive Branch.... Philadelphia: The author, 1809. 61 p.
S&S-16821.
Copies: PHi; PPAmP.

4615. Baldwin, Loammi, 1780–1838.
Thoughts on the study of political economy, as connected with the population, industry and paper currency of the United States.... Cambridge, [Mass.]: Printed by Hilliard and Metcalf, 1809. 75 p.
Includes observations on the Middlesex Canal, turnpike roads, and inland navigation.
S&S-16891.
Copies: CSmH; CSt; DLC; DeGE; MBAt; MWA; NN; NcD.

4616. Chesapeake and Delaware Canal Company.
The memorial and petition of the president and directors of the Chesapeake and Delaware Canal Company. January 24th, 1809. Printed by order of the Senate. Washington City: Printed by R. C. Weightman, 1809. 29, [1] p.
"Observations respecting the Chesapeake and Delaware Canal," p. [10]–29.
Signed: Kinsey [Kensey] Johns, Joshua Gilpin, Robert H. Goldsborough.
S&S-18956.
Copies: DLC; DS; DeGE; MWA; MdBP; MdHi; N; NN.

4617. Chesapeake and Delaware Canal Company.

[Publications. Philadelphia: 1809.] 9 v. in 1.

Edited and provided with a prefatory note by Joshua Gilpin.

Contents: The memorial and petition ... [1805]; First general report, 1804; Second general report, 1805; Third general report, 1806; To the Honorable the Senate and House of Representatives of the Commonwealth of Pennsylvania.... Presented January 1st, 1806; The memorial and petition ... to ... the Senate and House of Representatives ... December 24, 1806; Facts and observations respecting the Chesapeak and Delaware Canal ... ; Letters to the Honorable Albert Gallatin and other papers.

S&S-17195.

Copies: DLC; PPAmP.

4618. Proprietors of the Middlesex Canal.

Manner of transacting the canal business, at landings no. 1 and 2, in Charlestown. For the information of those persons who may be concerned therein ... April, 1809. [Boston: 1809.] Broadside.

S&S-18084.

Copies: MHi; MWA (xerox copy).

4619. Proprietors of the Middlesex Canal.

Rate of dockage, wharfage, and storage, at the landing places of the Middlesex Canal.... [Boston]: 1809. Broadside.

S&S-18085.

Copies: MHi; MWA (xerox copy).

4620. Proprietors of the Middlesex Canal.

Report to the Proprietors of the Middlesex Canal, at their meeting January 25, 1809. [Boston?: 1809.] 8 p.

Signed: J. L. Sullivan, agent.

N-P0606279; S&S-18086, 18704.

Copies: DLC; MB; MH-BA; MHi; MWA; NN.

4621. Sullivan, John Langdon, 1777–1865.

Statement of facts relative to the Middlesex Canal, and the object of the proprietors' petition to the legislature.... [Boston]: 1809. Broadside 41 x 28 cm.

Dated, February, 1809, and signed: John L. Sullivan, agent of the Middlesex Canal Corporation.

Copies: MWA.

4622. Tatham, William, 1752–1819.

A view of the proposed grand junction canal.... Norfolk, Va.: Printed by J. D. O'Connor, 1809. 15 p.

S&S-18722.

Copies: DLC.

4623. United States. Congress. House.

Mr. Nicholson's motion. December 27th, 1809. Washington City: Printed by R. C. Weightman, 1809. [4] p.

Concerns the construction of canals and roads.

S&S-19008.

Copies: DLC; MH; ViU.

4624. Allen, Ira, 1751–1814.

Copies of letters to the Governor of Vermont, and address to the legislature thereof, respecting a conspiracy against the author; and respecting a ship canal from Lake Champlain to the river St. Lawrence.... Philadelphia: Printed for the author, January 1810. 32 p.

S&S-19336.

Copies: DLC; MH; MHi; PPAmP.

4625. Delaware and Schuylkill Canal Navigation.

Delaware and Schuylkill Canal Navigation, composed of four hundred and sixty-four shares.... Philadelphia: 1810. Broadside 41 x 51 cm.

Charles G. Paleske, Joseph Watson, committee; Wm. Govett, secretary.

Dated: Dec. 11th. 1810.

S&S-19948.

Copies: DeGE; MWA; PHi.

4626. [Eddy, Thomas] 1758–1827.

Observations on canal navigation. By an observer. [New York?: 1810]. 15 p , map.

S&S-20025.

Copies: DLC; NN.

4627. New Jersey. Legislature. Joint Committee on the Delaware and Raritan Canal.

Report. The Joint Committee to whom was referred the resolution of both houses, respecting a plan for the construction of two canals . . . to unite the waters of the Delaware with the waters of the bay of New-York . . . [Trenton: 181–]. Broadside.

S&S-20873.

Copies: NjR (not located).

4628. Proprietors of the Middlesex Canal.

Report to the Proprietors of the Middlesex Canal, at their meeting on the 3d February, 1810. [Boston: 1810]. 14 p.

Signed: John L. Sullivan, agent.

S&S-20733, 21440.

Copies: DLC; MH-BA; MWo; NN.

4629. Sampit and St. James Santee Canal Company, Charleston, S.C.

Rules and regulations of the Sampit and St. James Santee Canal Company, passed December 19, 1809. Charleston, S.C.: Printed by Gabriel M. Bounetheau, 1810. 12 p.

S&S-21276.

Copies: ScHi.

4630. Schuylkill and Susquehanna Navigation.

To the Senate and House of Representatives of the commonwealth of Pennsylvania. The memorial of the president, managers and company, of the Schuylkill and Susquehanna Navigation, and of the Delaware and Schuylkill Canal Navigation. [Philadelphia: 1810]. 3 p.

Signed: Charles G. Paleske.

A proposal to consolidate these companies as the Union Canal.

Kress-B5716; Sabin-60725.

Copies: MH-BA; PHi.

4631. To the General Assembly . . . Petition of citizens of the counties on the river Appomattox . . . [S.l.: 181–]. Broadside 32 x 22 cm.

Petition for a canal between the Staunton and Appomattox rivers.

Hummel-3199.

Copies: Vi; ViHi.

4632. Allen, Ira, 1751–1814.

Copies of letters to the Governor of Vermont and address to the legislature thereof, respecting a conspiracy against the author; and respecting a ship canal from Lake Champlain to the river St. Lawrence. . . . Philadelphia: Printed by John Binns for the author, [1811]. 61 p.

S&S-22175.

Copies: DLC; PPAmP.

4633. Delaware and Schuylkill Canal Navigation.

Memorial of stockholders in Delaware and Schuylkill Canal Navigation Company to Senate and House of Representatives. [S.l.: 1811.] 2 p.

Protests the incorporation of the Union Canal Company.

Hasse (Pennsylvania)–p. 482.

Copies: PHi.

4634. Eddy, John Henry.

Map of the western part of the state of New York shewing the route of a proposed canal from Lake Erie to Hudson's River. Newark, N.J.: P. Maverick, engraver, 1811. Map 52 x 28 cm.

S&S-22741.

Copies: MBAt; MH; MWA; NIC; NNR.

4635. New York (State). Commissioners on Internal Navigation.

Report of the commissioners appointed by joint resolutions of the Honorable Senate and Assembly of the state of New-York, the 13th & 15th March, 1810, to explore the route of an inland navigation from Hudson's River to Lake Ontario and Lake Erie. Albany: Printed by S. Southwick, 1811. 35 p.

Signed: Gouv. Morris, S. Van Rensselaer, W. North, De Witt Clinton [a.o.].

S&S-23551.

Copies: DLC; DeGE; ICN; MH; MWA; MiD-B; NHi; NN; PU; ScU.

4636. ________ New York: Printed for Prior and Dunning; Southwick and Pelsue, printers, 1811. 38 p.

Minor variations in title.

N-N0191020.

Copies: DLC; DeGE; MWA; MiU; NHi; NN; PP.

4637. New York (State). Laws.

Message from the President of the United States, transmitting the copy of an act of the Legislature of New York, relating to a canal navigation from the Great Lakes to Hudson River. December 23d, 1811. Read and referred to the committee appointed on the 7th instant, on the petition of the Union Canal Company of the state of Pennsylvania. Washington City: Printed by R. C. Weightman, 1811. 6 p.

S&S-24232.

Copies: DLC; MWA; NHi; NjP.

4638. Proprietors of the Middlesex Canal.

Middlesex Canal. At a meeting of the directors of the Middlesex Canal, on the 8th March, 1811, voted that Dr. Dexter, Mr. Joy and Mr. Weld be a committee to make a printed communication to the proprietors on the last year's transactions.... [Boston: 1811.] 15 p.

Caption title on p. [2]: *Report. Pursuant to the preceeding vote the Committee state to the Proprietors....*

Signed: J. L. Sullivan, agent.

S&S-23386, 23387.

Copies: DLC; MH-BA; MHi.

4639. Proprietors of the Middlesex Canal.

Regulation relative to rafts on Middlesex Canal. [Charlestown?, Mass.: 1811?] Broadside 18 x 20 cm.

Signed: John L. Sullivan, agent of the Corporation.

N-P0606274.

Copies: MH-BA.

4640. Union Canal Company of Pennsylvania.

An act to incorporate the Union Canal Company of Pennsylvania, with the bye-laws, rules, orders and regulations, enacted at a meeting of the stockholders, on the 24th July, 1811. Philadelphia: Printed by John Binns, 1811. 34 p.

S&S-24075.

Copies: DLC; MH-BA; MWA; PPAmP; PPL.

4641. [Paine, Edward.]

Remarks on the importance of the contemplated grand canal, between Lake Erie and the Hudson River. [New York?]: 1812. 31 p.

Assigned to Hugh Williamson by Sabin.

Petition to Simeon de Witt signed by Edward Paine and others.

Atticus, to the citizens of the state of New-York, by De Witt Clinton, p. [15]–27.

S&S-26370; Sabin-104454.

Copies: DLC; DeGE; MH-BA; MWA; NBu; NN; NNC; PPAmP; PPi.

4642. Proprietors of the Middlesex Canal.

Additional regulations made by the board of directors of the Middlesex Canal. March 20, 1812. [Charlestown, Mass.: 1812.] Broadside 23 x 29 cm.

Signed: John L. Sullivan, agent to the Middlesex Canal Corporation.

N-P0606258.

Copies: MH-BA.

4643. Proprietors of the Middlesex Canal.

Report of the agent of the Middlesex Canal Corporation to the proprietors on the business done in the year 1811.... [Boston: 1812.] 10 p.

Signed: J. L. Sullivan, agent.

Copies: MH-BA; MHi.

4644. Union Canal Company of Pennsylvania.

Address of the president & managers of the Union Canal Company, of the State of Pennsylvania. Philadelphia: Printed by J. Binns, 1812. 11 p.

Signed on p. 7: Charles G. Paleske, president; James Milnor, William Meredith, George Simpson, Joseph Watson, managers, Philadelphia, June 22, 1812.

4644 (*continued*)
S&S-51341.
Copies: DeGE; MWA; PPL.

4645. Union Canal Company of Pennsylvania.
Report and memorial of the president and managers, of the Union Canal Company, of Pennsylvania. Made and presented to the Legislature of the state of Pennsylvania, at their session 1812–13. Philadelphia: Printed by John Binns, 1812. 11 p.
The first annual report of the company.
S&S-26428.
Copies: DeGE; MH-BA; PHi; PPL.

4646. United States. Congress. House. Committee to Whom were Referred the Several Memorials of the Canal Companies.
Report of the Committee to whom were referred the several memorials of the canal companies, therein mentioned. February 20, 1812.... Washington: A. & G. Way, printers, 1812. 8 p.
Reported by H. M. Ridgely.
Concerns memorials from the Chesapeake and Delaware Canal Company, the Union Canal Company, and the commissioners of the state of New York, respecting the canal from the Great Lakes to the Hudson River.
Answers from the Treasury Department signed: Albert Gallatin.
S&S-27343.
Copies: DLC; DeGE; MWA; N; NHi.

4647. Sullivan, John Langdon, 1777–1865.
Remarks on the importance of inland navigation from Boston by the Middlesex Canal and Merrimack River in the present and probable future state of foreign commerce.... Boston: Printed by John Eliot, 1813. 22 p.
S&S-29899.
Copies: CSt; CtY; DBRE; DeGE; MB; MBAt; MH; MHi; MWA; NHi; NN; WHi.

4648. Union Canal Company of Pennsylvania.
Report of the president and managers to the stockholders of the Union Canal Company of Pennsylvania. Together with the memorial, presented to the Legislature of the state of Pennsylvania, at their session 1813–14. Philadelphia: Printed by John Binns, 1813. 18 p.
S&S-29485.
Copies: MH-BA; PHi; PPF.

4649. United States. Congress. House. Committee on Memorial of the Chesapeake and Delaware Canal Company.
Report of the Committee to whom was referred the Memorial of the president and directors of the Chesapeake & Delaware Canal Company. December 20, 1813.... Washington: A. & G. Way, printers, 1813. 7 p. (13th Cong., 2d sess. House. Doc. 7)
S&S-30356.
Copies: DLC; DeGE; MH-BA; MWA.

4650. United States. Congress. Senate. Committee on Roads and Canals.
Report of the Committee, to whom were referred the message of the President of the United States.... together with the legislative act of Pennsylvania, relating to the Chesapeake and Delaware Canal Company. July 12, 1813.... Washington City: Printed by Roger C. Weightman, 1813. 18 p.
S&S-30363.
Copies: DLC.

4651. Catawba and Wateree Company.
The petition and memorial of J. F. Grimke, president of the Catawba and Wateree Company, in behalf of the stockholders of the said company. [Charleston?: 1814?] 15 p.
Concerns the operation and financing of canals.
N-C0206769.
Copies: CtY.

4652. Morris, Gouverneur, 1752–1816.
The advantages of the proposed canal from Lake Erie, to Hudson's River, fully illustrated in a correspondence between the Hon. Gouverneur Morris, and Robert Fulton,

Esq. [New York?: 1814.] 13 p.
S&S-32170; S-8794.
Copies: CSmH; CtY; DeGE; MBAt (dated [1822]); MH; MWA; NHi; NN; PPAmP; PPL; ViU.

4653. [Paine, Edward.]
Remarks on the importance of the contemplated grand canal between Lake Erie and the Hudson River. [New York?]: 1814. 30 p.
Assigned to Hugh Williamson by Sabin.
Includes *Atticus, to the citizens of the state of New York,* by De Witt Clinton.
S&S-32417; Sabin-104454.
Copies: MHi; N; NHi; WHi.

4654. Proprietors of the Middlesex Canal.
To members of the General Court.... [Boston: 1814?] 6 p.
Arguments for granting a lottery in Massachusetts for the benefit of the Middlesex Canal.
N-P0606288.
Copies: MB.

4655. Union Canal Company of Pennsylvania.
Report of the president and managers to the stockholders of the Union Canal Company of Pennsylvania. [Philadelphia: 1814.] 2 p.
Signed: Charles G. Paleske, President; Philadelphia, November 15, 1814.
S&S-32982.
Copies: DeGE; PHi; PPAmP; PPL.

4656. New York (State). Laws.
An act for opening the navigation between the head waters of the Seneca Lake —the Chemung River. Passed March 31, 1815. Geneva, N.Y.: 1815. 3 p.
S&S-35459.
Copies: PHi.

4657. Pennsylvania. Laws.
An act to authorize the Governour to incorporate a company to make a lock navigation on the river Schuylkill. Philadelphia: Printed at the Office of the United States' gazette, 1815. 28 p.
S&S-35592.
Copies: CSt; DBRE; DeGE; MWA; PHi.

4658. Proprietors of the Middlesex Canal.
A report made to the directors of the Middlesex Canal, October 14, 1815. [Boston: 1815.] [3] p.
S&S-35274.
Copies: MH; MH-BA.

4659. Union Canal Company of Pennsylvania.
Report of the president and managers to the stockholders of the Union Canal Company of Pennsylvania. [Philadelphia: 1815.] [2] p.
Signed: Charles G. Paleske, president, Philadelphia, November 21, 1815.
Copies: DeGE; MH-BA; PHi; PPL.

4660. Clinton, De Witt, 1769–1828.
The memorial of the subscribers, in favor of a canal navigation between the great western lakes and the tidewaters of the Hudson.... [Albany: 1816.] 28 p.
At head of caption title: In Assembly. Feb. 21, 1816. To the Legislature of the State of New York.
N-C0491124; S&S-38427.
Copies: N.

4661. Clinton, De Witt, 1769–1828.
New-York, 13th January, 1816.... [New York: 1816.] 1 p.
Circular letter, on behalf of a committee appointed by the citizens of New York to correspond with other counties and states on the subject of a canal.
S&S-37270.
Copies: NN.

4662. [Clinton, De Witt] 1769–1828.
Remarks on the proposed canal, from Lake Erie to the Hudson River. By Atticus [pseud.]. New York: Printed by Samuel Wood & Sons, 1816. 14 p.
S&S-38779.
Copies: DLC; DeGE; ICN; MBAt;

4662 (*continued*)
MH-BA; MWA; MiU; N; NBu; NHi; NN; NNC; PPAmP.

4663. Commissioners Appointed from North Carolina and Virginia to Investigate Cutting a Canal.

Sir, the undersigned, Commissioners of the states of North-Carolina and Virginia, appointed under the laws of our respective states, entitled "An act to incorporate a company for the purpose of cutting a navigable canal...." [S.l.: 1816.] Broadside 25 x 19.5 cm.

Signed: Wm. E. Brodnax, Eaton Pugh [and others].

N-C0590473.

Copies: MiU-C.

4664. Commissioners Appointed from North Carolina and Virgina to Investigate Cutting a Canal.

To the Legislature of the state of North-Carolina. We, the undersigned commissioners, appointed by acts of the Legislature of North-Carolina ... and by the Legislature of Virginia on the 31st day of January, 1812.... [S.l.: 1816?] Broadside 34 x 21 cm.

Dated in manuscript: June 27, 1816.

N-C0590474.

Copies: MiU-C.

4665. The Grand Canal defeated, by a Democratic Senate. To the electors of the Western District.... [Buffalo: 1816?] Broadside 25.5 x 50 cm.

S&S-51591.

Copies: NBuHi.

4666. Memorial of the citizens of New-York in favour of a canal navigation between the great western lakes and the tide-waters of the Hudson. New York: Printed by Samuel Wood & Sons, 1816. 18 p., map.

Dated: February 21, 1816.

S&S-38439.

Copies: DBRE; DeGE; MH-BA; MWA; N; NHC; NN; NNC; NRU; PPF.

4667. New York (State). Canal Commissioners.

Memorial of the Commissioners of the state of New York, in behalf of said state: praying the aid of the general government in opening a communication between the navigable waters of Hudson River and the lakes. December 11, 1816.... Washington: Printed by William A. Davis, 1816. 13 p. (14th Cong., 2d sess. House. Doc. 6)

Signed: De Witt Clinton.

S&S-39544.

Copies: DLC; DeGE; IU; MWA; MiD-B; NNC; NhHi; TU.

4668. New York (State). Legislature. Joint Committee on Canals and Roads.

In Assembly, March 21, 1816. The Joint Committee on that part of His Excellency the Governor's speech, which relates to canals and roads, report.... Albany: 1816. 12 p.

Mainly concerned with canals.

S&S-38428.

Copies: MWA; N.

4669. Pennsylvania. Laws.

A supplement to the "Act entitled Act to authorize the Governor to incorporate a company to make a lock navigation in the river Schuylkill." [Harrisburg?]: John S. Wiestling, printer, 1816. 2 p. 33.5 x 21 cm.

Copies: DeGE.

4670. Proprietors of the Middlesex Canal.

The following report was made to the Proprietors of the Middlesex Canal at their annual meeting, January 31, 1816. [Boston?: 1816.] [3] p.

Signed: J. L. Sullivan, agent.

Copies: MH-BA; MHi.

4671. Proprietors of the Middlesex Canal.

Report of the committee of the board of directors of the Middlesex Canal ... to whom were referred the complaints made relative to certain privileges said to be granted in passing the locks on Merrimack River, to the Merrimack Boating Company.... [Boston: 1816.] 11 p.

Signed: Benjamin Weld, Joseph Hall, Josiah Quincy.
S&S-38247.
Copies: MH-BA; NNC.

4672. Remarks on the proposed canal from Rome on the Mohawk River, to Onondaga Lake; contained in a letter from a gentleman at Rome, who surveyed and took the levels of the country from Rome to Cayuga Lake, by order of the commissioners. New York: Printed by Samuel Wood & Sons, 1816. 6 p.
Kress-B6784.
Copies: DLC; MH-BA; MWA; N.

4673. [Roberts, William.]
A statement of facts, shewing why the bill reported by the D.C. committee, in favor of the Potomac Company ... should not be passed. Georgetown, Col.: Printed by W. A. Rind and Co., 1816. 13 p., plan.
Concerns the canal within the District of Columbia.
S&S-38822.
Copies: DLC.

4674. A serious appeal to the wisdom and patriotism of the Legislature of the state of New-York; on the subject of a canal communication between the great western lakes and the tide waters of the Hudson. By a friend to his country. [S. l.]: Printed for the author, 1816. 37 p.
Attributed to Ferris Pell.
S&S-38912.
Copies: DeGE; MB; MH; MWA [Pell, Ferris]; N; NBu.

4675. To the Legislature of the state of New York. The memorial of the subscribers, in favor of a canal navigation between the great western lakes and the tide waters of the Hudson.... [Albany: 1816.] Broadside.
At head of title: In Assembly, Feb. 21, 1816.
S&S-38427.
Copies: MWA; N.

4676. ——— [Albany: 1816.] 28 p.
Copies: MWA.

4677. Union Canal Company of Pennsylvania.
Report of the managers of the Union Canal Company of Pennsylvania, to the stockholders, at their general meeting, on the nineteenth day of November, 1816. [Philadelphia: 1816.] [2] p.
Signed: W. Meredith, vice-president, Philadelphia, 19th November, 1816.
Copies: DeGE; PHi; PPL.

4678. United States. Congress. Senate.
In Senate of the United States. March 30th, 1816. Mr. Mason, of Virginia, submitted the following motion for consideration. [Washington: 1816.] 2 p.
Motion to instruct the Committee on Roads and Canals to inquire into the expediency of subscribing to the stock of the Great Coastwise Canal and River Navigation Company.
S&S-39437.
Copies: DLC; MWA.

4679. United States. Congress. Senate. Committee on Roads and Canals.
In Senate of the United States. February 6th, 1816. The Committee, to whom was referred so much of the President's message as relates to roads and canals, having in their general report recommended to Congress the Chesapeake and Delaware Canal as an object of first importance ... report concerning the same. [Washington: 1816.] 39 p. (14th Cong., 1st sess. Senate. Doc. 38)
Memorial and petition of the president and directors of the Chesapeake and Delaware Canal Company, Dec. 1st, 1805, signed by Jos. Gilpin and Kensey Johns, p. 11–17.
Observations respecting the Chesapeake and Delaware Canal, Dec. 1st, 1805, signed by Kensey Johns, Joshua Gilpin, Robert H. Goldsborough, p. 18–39.
S&S-39394.
Copies: DLC; DeGE; MWA; NN.

4680. Chesapeake and Delaware Canal Company.

In Senate of the United States. Memorial of the directors of the Chesapeake and Delaware Canal Company. January 9, 1817. Read and ordered to be printed for the use of the Senate. [Washington: 1817.] 6 p.

S&S-42592.

Copies: DLC; DeGE; MWA; MiD-B.

4681. Granger, Gideon, 1767–1822.

Speech of Gideon Granger, Esq., delivered before a convention of the people of Ontario County, N.Y., Jan. 8, 1817, on the subject of a canal from Lake Erie to Hudson's River. . . .Canandaigua, [N.Y.]: Printed by J. D. Bemis, [1817]. 24 p.

S&S-40948.

Copies: DLC; MB; MH-BA; MWA; N; NHi; NN; NjPT; PHi; PPF.

4682. New Jersey. Commission on the Delaware and Raritan Canal.

Report of the Commissioners, appointed by an act of the Legislature of the state of New-Jersey, for ascertaining the most eligible route for, and the probable expense of a canal to connect the tide waters of the Delaware with those of the Raritan. Passed February 13, 1816. Trenton: Printed by J. J. Wilson, 1817. 31 p.

S&S-41595.

Copies: CtY; DeGE; NN; Nj; NjR; PPAmP; PPF.

4683. ———— Trenton: Printed by J. J. Wilson, 1817. 30 p.

N-N0153238.

Copies: DLC; NjP.

4684. ———— New-York: Printed by Van Winkle, Wiley & Co., 1817. 40 p.

S&S-41594.

Copies: DBRE; MBAt; MH-BA; MWA; NN.

4685. New York (State). Canal Commissioners.

Application to Congress. . . . [Albany: 1817?] 52 p., table.

At head of title: Appendix.

Concerns mainly the Erie Canal; contains also correspondence with the governor and legislature of Ohio, information respecting the Middlesex Canal, etc.

N-N0189713.

Copies: NBuG.

4686. New York (State). Canal Commissioners.

The official reports of the Canal Commissioners of the state of New-York, and the acts of the Legislature respecting navigable communications between the great western and northern lakes and the Atlantic Ocean; with perspicuous maps and profiles. Published at the request of the Board of Canal Commissioners. New-York: T. & W. Mercein, printers, 1817. 174, 12 p., fold. map.

Map has the title: *Map and profile of the proposed canal from Lake Erie to Hudson River in the state of New York.*

S&S-41628.

Copies: CtY; DLC; ICJ; NBu; NN; NNE; NcD.

4687. ———— Newburgh, [N.Y.]: B. F. Lewis, & Shelton & Kensett; Ward M. Gazlay, printer, 1817. 112 p., fold. map.

"Canal estimates or contract prices," a manuscript in an unknown hand, inserted in the DeGE copy.

S&S-41629.

Copies: CSmH; DLC; DeGE; ICJ; MWA; MiU-C; NN; NNE; NjR; PPF; PPL; ViU.

4688. New York (State). Canal Commissioners.

Report of the Commissioners of the state of New-York, on the canals from Lake Erie to the Hudson River, and from Lake Champlain to the same. Presented to the Legislature, 17th February, 1817. Albany: Printed by J. Buel, printer to the state, 1817. 90, 52 p.

Appendix, entitled *Application to Congress,* is not always present.

S&S-41630.

Copies: CSmH; DLC; DeGE; In; MH-BA; MWA; NHi; NN; NUtHi; PPF.

4689. New York (State). Canal Commissioners.

To the Legislature of the state of New York, pursuant to the act, entitled "An act respecting navigable communications between the great western and northern lakes, and the Atlantic Ocean," passed April 15, 1817. [Albany: 1817.] 34 p.

N-N0189751.

Copies: DLC.

4690. New York (State). Laws.

Laws of the state of New-York, respecting navigable communications between the great western and northern lakes and the Atlantic Ocean. New York: Printed by T. & W. Mercein, 1817. 14 p.

S&S-41625.

Copies: CSmH; MH-L; MWA;N; NNC; NbU; PPF.

4691. New York (State). Legislature. Joint Committee on Canals.

Report of the Joint Committee of the Legislature of New-York, on the subject of the canals, from Lake Erie to the Hudson River, and from Lake Champlain to the same. In Assembly, March 19, 1817.... Albany: Printed by Websters and Skinners, 1817. 24 p.

Includes *An Act respecting navigable communications between the great western and northern lakes and the Atlantic Ocean.*

S&S-41631.

Copies: MH-BA; MWA; N; NN; PBL; PPF; RPB; WHi.

4692. Proprietors of the Middlesex Canal.

Report to the board of directors of the Middlesex Canal. By a committee. April, 1817. Boston: Printed by John Eliot, [1817]. 20 p.

S&S-41432.

Copies: MH-BA; MHi; MLow; MWA; MWo.

4693. To the speakers and members of both houses of the Virginia legislature. The petition of sundry inhabitants ... respectfully sheweth, that all of the county above the falls contiguous to James River, is greatly interested in the establishment of a canal on the south side thereof.... [Petersburg?, Va.: 1817?] Broadside.

S&S-42304, 45876.

Copies: Vi (not located).

4694. Union Canal Company of Pennsylvania.

Report of the president and managers to the stockholders of the Union Canal Company of Pennsylvania. [Philadelphia: 1817.] [2] p.

Signed: Joseph Watson, President, Philadelphia, November 18, 1817.

Copies: DeGE; PHi; PPL.

4695. Washington Canal Company.

Report of the president and directors of the Washington Canal Company ... January 31, 1817. [Washington: 1817.] 2 p. (14th Cong., 2d sess. House. Doc. 63)

S&S-42729.

Copies: DLC; DeGE; MWA; ViU.

4696. The following is a statement of such facts as came under the cognizance of the individuals who explored a more southern route for the Grand Canal, through the counties of Ontario, Genesee, and Niagara.... [Buffalo?: 1818.] Broadside 43 x 54 cm.

S&S-51773.

Copies: NBuHi.

4697. [Haines, Charles Glidden] 1792–1825.

Considerations on the great western canal, from the Hudson to Lake Erie: with a view of its expence, advantages, and progress. Re-published by order of the New-York Corresponding Association, for the Promotion of Internal Improvements. Brooklyn: Spooner & Worthington, printers, 1818. 58 p., 1 l.

"The Northern Canal, from the Hudson to Lake Champlain," p. [51]–58.

N-H0035275.

4697 (*continued*)
Copies: DeGE; MH; MWA (81p.); MiU-C; NBu; NN; OCU; PPF; PU; RPB.

4698. ________ 2d ed. Published by order of the New-York Corresponding Association for the Promotion of Internal Improvements. Brooklyn: Spooner & Worthington, printers, 1818. 60 p.
S&S-44216.
Copies: CSmH; DLC; DNLM; MB; MH; MWA; NN; OClWHi.

4699. ________ Brooklyn: Spooner & Worthington, printers, 1818. 81 p.
"The Northern Canal, from the Hudson to Lake Champlain," p. [61]–68; "Mr. William Darby's letter," p. [69]–80.
N-H0035277.
Copies: CU; CtY; DeGE; ICJ; MH-BA; MWA; N; NN; NNC; PPAmP; PPL.

4700. Indiana. Laws.
An act of the General Assembly of the state of Indiana, incorporating the Jeffersonville Ohio Canal Company, approved January 28, 1818. Salem, Ind.: Printed by Patrick & Booth, 1818. 12 p.
S&S-44461.
Copies: ICU; N.

4701. Jeffersonville Ohio Canal Company.
To the Hon. the Secretary of the Treasury. Sir, In conformity with your circular of the 26th May last, on the subject of roads and canals, the president and directors of the Jeffersonville Ohio Canal Company, beg leave to submit the following report ... [Jeffersonville, Ind.: 1818]. Broadside 20 x 23 cm.
S&S-44462.
Copies: O-Ar.

4702. Massachusetts. Laws.
An act to establish the Navigation Canal Corporation. [Boston: 1818]. 8 p.
Established to build a canal across Cape Cod.
S&S-44717.
Copies: MB; MH-BA.

4703. New York (State). Canal Commissioners.
Report of the Commissioners of state of New-York, on the canals from Lake Erie to the Hudson River, and from Lake Champlain to the same. Presented to the Legislature, January 31st, 1818. Albany: Printed by J. Buel, printer to the state, 1818. 45. p., fold. table.
S&S-45050
Copies: DLC; MWA; NCanHi; NHi; NNE; NNS; PPM.

4704. Pennsylvania. Commissioners Appointed to Explore the route of the Seneca Lake and Tioga River Canal.
Report of the Commissioners, appointed to explore the route of the intended canal for uniting the waters of the Seneca Lake and Tioga River, in the state of New-York. Read February 7th, 1818. [Philadelphia: 1818]. 24 p.
Copies: MWA.

4705. Proprietors of the Middlesex Canal.
Abstract of the report of the agent of the canal ... 1818. [Boston?: 1818]. [3] p.
Signed: J. L. Sullivan, agent.
Copies: MH-BA.

4706. Proprietors of the Middlesex Canal.
To the board of directors of Middlesex Canal. Boston, Nov. 21, 1818. [Boston 1818]. 8 p.
Signed: John L. Sullivan.
S&S-44837.
Copies: MH-BA.

4707. Sullivan, John Langdon, 1777–1865.
Letters, first published in the Boston daily advertiser, in answer to certain inquiries relative to the Middlesex Canal ... Boston: Printed at the office of the Boston daily advertiser, 1818. 12 p.
S&S-45818.
Copies: DLC; MB; MH-BA; MWA; N.

4708. Union Canal Company of Pennsylvania.

Account of the receipts and expenditures of the Union Canal Company of Pennsylvania (Oct. 11, 1812, to Dec. 8, 1818). Harrisburg: 1818. 8 p.

Title taken from p. 3.

S&S-49659 (dated 1819).

Copies: PHi.

4709. Union Canal Company of Pennsylvania.

Report of the president & managers of the Union Canal Company of Pennsylvania; to the stockholders ... Philadelphia: Printed by John Bioren, 1818. 14 p.

Signed: Joseph Watson, president, Philadelphia, November 17, 1818.

S&S-45919.

Copies: DeGE; MH-BA; MWA; N; P; PHi: PP; PPL; PU; WU.

4710. [Beach, Samuel Bellamy] 1780–1866.

Considerations against continuing the great canal west of the Seneca: addressed to the members-elect of the Legislature of the state of New York. By Peter Ploughshare [pseud.] ... Utica, [N.Y.]: William Williams, 1819. 29 p.

S&S-47178.

Copies: DLC; M; MWA; NCH; NCanHi; NN; NNS.

4711. Flint, James.

Report of the Jeffersonville and Ohio Canal. [S.l.: 1819]. Broadside.

S&S-47996.

Copies: MWA.

4712. New York (State). Canal Commissioners.

Annual report of the Canal Commissioners, communicated to the Legislature, Jan. 25, 1819. Albany: Printed by J. Buel, printer to the state, 1819. 46 p.

S&S-48887.

Copies: DLC; DeGE; MB; MWA; NCanHi; NRom.

4713. New York (State). Governor (Clinton).

Speech of Governor Clinton, delivered this day to the Legislature of the state of New-York. [Albany: 1819]. 16 p.

At head of title: Albany Register —Extra. Tuesday, January 5, 1819.

Includes discussion of the progress of canal building in the state of New York.

S&S-48902.

Copies: MiU-C; N.

4714. New York (State). Laws.

Acts of the Legislature of the state of New-York, respecting navigable communications between the great western and northern lakes and the Hudson River. Albany: Printed by Websters and Skinners, 1819. 20 p.

Concerns the Erie and Champlain canals.

S&S-48886.

Copies: DLC; DeGE; MBAt; MWA; N; NHi; NN.

4715. New York (State). Legislature. Joint Committee Relative to Internal Improvements.

Report of the Joint Committee of the Senate and Asembly, relative to the internal improvements of the state; to whom was refered [*sic*] the annual report of the Canal Commissioners. Albany. Printed by J. Buel, 1819. 14 p.

Dated: 12th February, 1819.

Includes reply of the Canal Commissioners to the inquiry of the Committee, signed De Witt Clinton, Samuel Young and Ephraim Hart.

N-N0201038; S&S-48899.

Copies: MB; N; NHi; NN.

4716. Pennsylvania. General Assembly. Joint Committee on the Schuylkill and Susquehanna Canal Company.

Report of the Joint Committee appointed by the Senate and House ... for the purpose of examining the works erected by the former Schuylkill and Susquehanna Canal Company, &c. [Harrisburg: 1819]. 12 p.

4716 (*continued*)
S&S-49054.
Copies: PPAmP.

4717. Pennsylvania. Governor (W. Findlay).
Documents accompanying the Governor's message of December 10, 1819. [Harrisburg: 1819]. 14 p.
Concerns the Susquehanna Canal Company.
Copies: MWA.

4718. Union Canal Company of Pennsylvania.
Report of the president and managers, to the stockholders of the Union Canal Company of Pennsylvania: together with the memorial, presented to the Legislature of the state of Pennsylvania, at their session, 1818–19. And extracts from the report of the Joint Committee. Philadelphia: Printed by John Bioren, 1819. 12 p.
S&S-49660.
Copies: DeGE; MWA; PHi; PPL.

4719. United States. Engineer Dept.
Letter from the Secretary of War, transmitting topographical reports, made with a view to ascertain the practicability of uniting the waters of Illinois River, with those of Lake Michigan. December 28, 1819 ... Washington: Printed by Gales & Seaton, 1819. 10 p. (16th Cong., 1st sess. House. Doc. 17)
S&S-49863.
Copies: DLC; DeGE; DeU; MH-BA; MWA; NhHi; NjP; O; R.

4720. Foos, Joseph.
The highway of all nations. Columbus, [Ohio]: Printed by H. P. Olmsted, 1820. 24 p.
Concerns interoceanic canals.
S-1261
Copies: MBAt; N.

4721. Geddes, James, 1763–1838.
Map and profile of the Champlain Canal as made from Lake Champlain to the Hudson River and surveyed thence to the tide at Waterford. [Albany]: 1820. Map 46 x 67 cm.
N-G0095272.
Copies: MB.

4722. Memorial. To the Honorable the Legislature of the state of New York: the subscribers, inhabitants of the county of Niagara, most respectfully submit to your honorable body the following memorial ... [Buffalo?: 1820]. 8 p.
Dated at end: Niagara County, December, 1820.
Asking that the terminus of the Erie Canal be fixed at the mouth of the Buffalo Creek.
S-2549.
Copies: NBuHi.

4723. Morris Canal and Banking Company.
Statement, etc. [S.l.: 182–]. 7 p.
N-M0795909.
Copies: N.

4724. New Jersey. Laws.
An act to incorporate a company for making a canal or water communication in this state, between the tide waters of the Delaware and Raritan rivers. Trenton: Printed by Joseph Justice, 1820. 13 p.
S-2447
Copies: CSmH; DLC; MiU; NN; Nj; NjP; PPAmP.

4725. New York (State). Canal Commissioners.
Annual report of the Canal Commissioners, communicated to the Legislature, February 18, 1820. Albany: Printed by J. Buel, 1820. 29 p., map.
S-2480.
Copies: InHi; N; NHi; NNS; NRom.

4726. ——— Albany: Printed by Websters and Skinners, 1820. 23 p., map.
S-2479.
Copies: MWA; NHi.

4727. New York (State). Governor (Clinton).

Speech of His Excellency De Witt Clinton, delivered this day to the Legislature of New-York. [Albany: 1820.] 14 p.

"Albany Register—Extra. Tuesday, January 4, 1820."

Devoted partly to the Erie Canal and other internal improvements.

N-N0195227; S-2511.

Copies: MWA; N; NHi; NIC.

4728. New York (State). Governor (Clinton).

Speech of His Excellency Governor Clinton, delivered to the Legislature of the state of New-York, January 4, 1820. Albany: Printed by Jeremiah Tryon, 1820. 24 p.

Devoted partly to the Erie Canal.

S-2513.

Copies: DLC; MB; MWA; NBu; NHi; NbU.

4729. New York (State). Laws.

An act for the maintenance and protection of the Erie and Champlain canals, and the works connected therewith. [Albany: 1820.] 7 p.

S-2473.

Copies: N.

4730. New York (State). Laws.

Acts passed at the forty-third session of the Legislature ... respecting the canals from Lake Erie to the Hudson River, and from Lake Champlain to the same. Albany: Websters and Skinners, 1820. 20 p.

S-2477.

Copies: MB; MWA; N; NN.

4731. New York (State). Legislature. Assembly. Committee on Canals.

Report of the Committee on Canals. March 14, 1820. Albany: Printed by J. Buel, 1820. 33 p.

Concerns the disbursement of money for the eastern section of the Erie Canal and the southern section of the Champlain Canal.

S-2506.

Copies: CSmH; CtY; InHi; MB; MHi; N; NHi; NNS.

4732. Ohio. Board of Canal Commissioners.

Rules & specifications relating to the construction of the Miami Canal and estimating of work performed thereon ... [Columbus: 182–?]. 12 p.

N-00045789.

Copies: NN.

4733. Theory tested by experience: a comparison of the arguments in favor of the western section of the grand canal, with facts, by a citizen of Oswego, Oswego, N.Y.: Printed at the Palladium office, 1820. 24 p.

S-3423.

Copies: MWA; N.

4734. Union Canal Company of Pennsylvania.

Report of the president and managers, to the stockholders of the Union Canal Company of Pennsylvania. [Philadelphia: 1820]. 4 p.

Signed: Joseph Watson, President; Philadelphia, Nov. 21st, 1820.

S-3509.

Copies: DeGE; PHi; PPAmP; PPL.

4735. Virginia. Laws.

Acts of the General Assembly respecting the Dismal Swamp Canal Company. Norfolk, [Va.]: Thomas G. Broughton, 1820. 29, 1 p.

S-4112.

Copies: ICJ; ViHi; ViU.

4736. Watson, Elkanah, 1758–1842.

History of the rise, progress, and existing condition of the western canals in the state of New-York, from September 1788, to the completion of the middle section of the grand canal, in 1819. Together with the rise, progress, and existing state of modern agricultural societies, on the Berkshire system, from 1807 to the establishment of the Board of Agriculture in the state of New-York, January 10, 1820. Albany: D. Steele, 1820. 2 p.l., iv, [5]-104 p., 2 l., [111]-210, [2] p., maps, plates.

History of agricultural societies on the modern

4736 (*continued*)
Berkshire system ... has special title page.
S-4187.
Copies: CSmH; CtW; DLC; DNAL; DeGE; ICJ; MH; MWA; MiD; NN; Nh; NjP; OHi; PPL; PU; RPB; WHi.

4737. [Baldwin, Loammi] 1780–1838.
To the president and managers of the Union Canal Company of Pennsylvania. [1]–6. [Philadelphia?: 1821–1822]. 11, 9, 8, 13, 7, 4 p.
Six letters, signed: L. Baldwin, engineer, and dated respectively: Philadelphia, Nov. 19, 1821; Philadelphia, Nov. 29, 1821; Lebanon, Jan. 4, 1822; Lebanon, Aug. 2, 1822; Lebanon, Sept. 14, 1822; and Lebanon, Nov. 4, 1822.
S-4524, 7871.
Copies: DeGE; MH-BA; NN.

4738. [Clinton, De Witt] 1769–1828.
The canal policy of the state of New-York: delineated in a letter to Robert Troup, Esquire. By Tacitus [pseud.].... Albany: Printed and published by E. and E. Hosford, 1821. viii, [9]–54 p.
S-5006.
Copies: CtY; DLC; DeGE; ICJ; M; MH-BA; MWA; MdHi; N; NBu; NHi; NIC; NN; OCHP; PHi.

4739. Gilpin, Joshua, 1765–1840.
A memoir on the rise, progress, and present state of the Chesapeake and Delaware Canal, accompanied with original documents and maps.... Wilmington, [Del.]: Printed by R. Porter, 1821. 50, 72 p., maps (2 folded).
S-5461.
Copies: CU; DLC; DeGE; DeHi; DeWI; ICU; MH-BA; MWA; MdBE; MiD; MiU-C; NN; P; PPAmP; PPL; ViU.

4740. Lehigh Coal and Navigation Company.
An address of the Lehigh Coal and Navigation Company to their fellow citizens. Philadelphia: Printed by William Brown, 1821. 8 p.
S-5811.
Copies: MH-BA (not located).

4741. Memorial. To the Honorable the Legislature of the state of New-York, the citizens of the village of Buffalo most respectfully represent to your honorable body: that they have perused a memorial of the inhabitants of the county of Niagara, to your honorable body.... [Buffalo: 1821.] 10 p.
Concerns the terminus of the Erie Canal.
S-6022.
Copies: N.

4742. Mills, Robert, 1781–1855.
Inland navigation. Plan for a great canal between Charleston and Columbia, and for connecting our waters with those of the western country. Columbia, S.C.: Telescope Press, 1821. 93, [1] p., fold. map.
S-6051.
Copies: CtY; GU; MB; MWA; MiU-T; PPL; ScU.

4743. New York (State). Canal Commissioners.
Annual report of the Canal Commissioners of the state of New York ... presented to the Legislature 12th March, 1821. Albany: Printed by Cantine & Leake, 1821. 34 p.
S-6240.
Copies: N; PPL.

4744. New York (State). Legislature. Assembly. Committee on Canals.
Report on so much of the speech of the Governor as relates to the northern and western canals. Albany: 1821. 5 p. (Assembly. Doc. 131)
Whitford-p. 1193.
Copies: N.

4745. New York Corresponding Association for the Promotion of Internal Improvements.
Public documents relating to the New-

York canals, which are to connect the western and northern lakes, with the Atlantic Ocean; with an introduction. Printed under the direction of the New-York Corresponding Association, for the Promotion of Internal Improvements. New-York: W. A. Mercein, printer, 1821. lii, 16, 25–484 p., fold. table, fold. maps.

Introduction by Charles Glidden Haines.

Includes reports of the Canal Commissioners.

S-6265.

Copies: CtW; DLC; DeGE; ICJ; In; MH; MWA; MdHi; MiD-B; NIC; NNC; NcD; OClWHi; PPamP; PPL; Vi.

4746. Proprietors of the Middlesex Canal.

Rates of toll on the Middlesex and River canals . . . 1821. [Charlestown, Mass.?]: April, 1821. Broadside 34 x 41.5 cm.

S-6043.

Copies: MB; MH-BA.

4747. Proprietors of the Middlesex Canal.

To the board of directors of Middlesex Canal. . . . [Charlestown, Mass.: 1821.] [3] p.

Signed at end: In behalf of the rest of the proprietors, J. L. Sullivan.

On the rates of toll on the Middlesex Canal.

S-6044.

Copies: MB.

4748. Report, on the subject of a communication, between Canandaigua Lake and the Erie Canal; made on the 21st December, 1820, at a meeting of the citizens of Canandaigua and its vicinity. . . . Canandaigua, [N.Y.]: Printed by J. D. Bemis & Co., 1821. 23 p.

S-4908.

Copies: CSmH; MB; MWA; N; NHi; PHi.

4749. Troup, Robert, 1757–1832.

A vindication of the claim of Elkanah Watson, Esq. to the merit of projecting the Lake canal policy, as created by the Canal act of March, 1792. And also, a vindication of the claim of the late General Schuyler, to the merit of drawing that act, and procuring its passage through the legislature. . . . Geneva, [N.Y.]: J. Bogert, 1821. 1 p.l., 23, [1], 38 p.

S-7010.

Copies: CSmH; CtY; DLC; ICJ; MH; MH-BA; MWA; MiU; N; NBuG; NHi; NN.

4750. Union Boating Company, Concord, N.H.

To the public. . . . [Concord?,N.H.: 1821.] Broadside 40.5 x 34 cm.

Signed: Albe Cady, Isaac Hill, committee; Richard H. Ayer, agent. April 19, 1821.

A reply to a publication of the Proprietors of the Middlesex Canal, issued by J. L. Sullivan.

S-7027.

Copies: MB.

4751. Union Canal Company of Pennsylvania.

Report of the president and managers of the Union Canal Company of Pennsylvania, presented to the stockholders, at their annual meeting on the twentieth of November, 1821. Philadelphia: Printed by Jacob Frick & Co., 1821. 8 p., fold. table.

S-7028.

Copies: DeGE; ICJ; MH; MdBJ; NbO; PHi; PPL; WU.

4752. Union Canal Company of Pennsylvania.

A supplement to the Act to incorporate the Union Canal Company of Pennsylvania, and an extract from the "Act for the improvement of the state;" together with the by-laws, adopted by the stockholders on the 22d of May, 1821. Philadelphia: Printed by Jacob Frick and Co., 1821. 18, [1] p.

S-7029.

Copies: DLC; DeGE; PHi; PPAmP.

4753. Black Rock Harbor Company.

Documents, relating to the western termination of the Erie Canal; with explana-

4753 (*continued*)
tions and remarks. Published by direction of "The Black Rock Harbor Company." Black Rock, [N.Y.]: Printed by L. G. Hoffman, 1822. 60 p., fold. map.

By David Thomas, James Geddes, and others.

S-8092.

Copies: CSmH; DLC; N; NBu; NN.

4754. [Carey, Mathew] 1760–1839.

Sir, a portion of the result of the election, yesterday, of directors of the Chesapeake and Delaware Canal Company has excited general surprise.... [Philadelphia: 1822.] [3] p.

Signed: Philadelphia, July 16, 1822. M. Carey.

Copies: MWA.

4755. Connecticut. General Assembly. Committee on Canal by the Housatonic Valley.

Report of a committee of the legislature of Connecticut on the subject of a canal by the Ousatonic [*sic*] Valley, and charter of a company to construct the same. [New Haven: A. H. Maltby & Co., 1822]. 15 p.

S-8431.

Copies: CtY; N; NN.

4756. An examination of the line of the great Erie Canal, as adopted by the commissioners, from Schoharie Creek to its final destination at the tide waters of the Hudson River; contrasted with a more safe, direct, and practicable line, and exhibiting some of the reasons why the latter ought to be prefered [*sic*]. [S.l.]: Printed for the publisher, 1822. 32 p.

Caption title: An expose of errors, committed on the Erie Canal line.

Attributed to John Randel by Daniel H. Calhoun.

S-8650.

Copies: DLC; DeGE; N; NN; NNC.

4757. Gloucester Canal Corporation.

Act of incorporation and by-laws of the Gloucester Canal Corporation: instituted 1822. Salem, [Mass.]: Printed by W. Palfray, Jun., 1822. 15 p.

S-8840.

Copies: MSaE.

4757a. Great Western Canal. [S.l.: 1822.] 8 p.

Signed: W. G., and dated: Saratoga Springs, 20th June, 1822.

N-G0478370.

Copies: DLC; N; NBuG; NNC.

4758. Housatonic Canal Company.

Proposals for the stock of the Ousatonic [*sic*] Canal Company. [New Haven: A. H. Maltby & Co., printer, 1822?] 16 p.

Signed: Orange Merwin.

S-9069.

Copies: CtHi; CtY.

4759. Irvine, Baptis.

Commerce of southern America; its importance to us, with some remarks on a canal at Darien, and other cognate subjects.... Philadelphia: Printed for T. T. H., 1822. 59 p.

An early recommendation for what later became the Panama Canal.

S-9117.

Copies: DeGE; MiU-C; PPAmP; PPL.

4760. [Lacock, Abner] 1770–1837.

Great national object. Proposed connection of the eastern & western waters, by a communication through the Potomac country. [Washington: 1822.] 38 p.

Contains essays by Gen. Abner Lacock, reprinted from the *National intelligencer*, and a report by Charles Fenton Mercer, concerning the Chesapeake and Ohio Canal.

S-9224.

Copies: CtY; DLC; DeGE; MiU-C; N; OClWHi; PPL; ViU.

4761. Letters and extracts, relative to the New-Jersey canal. [S.l.: 1822.] 24 p.

Attributed to John Rutherford.

Hasse (New Jersey)-p. 187.

Copies: MWA; NN; NjHi.

4762. Maryland. General Assembly. House. Committee on the Susquehanna Canal Wing Dam.

Report of the Committee appointed to inspect the wing-dam at the head of the Susquehanna Canal, with the evidence &c. December session, 1821. Annapolis: Printed by Jehu Chandler, [1822]. 19 p.

S-9377.

Copies: MdHi.

4763. New York (State). Canal Commissioners.

The annual report of the Canal Commissioners of the state of New-York, presented to the Legislature, the 27th February, 1822. Albany: Printed by Cantine and Leake, 1822. 32 p.

S-9669.

Copies: DLC; N; NHi; NN.

4764. New York (State). Governor (Clinton).

Speech of His Excellency De Witt Clinton, to the Legislature of the state of New-York, at the opening of the session, 2d January, 1822. [Albany: 1822.] 8 p.

A substantial part is devoted to the progress of the construction of the Erie Canal.

S-9688.

Copies: DLC; IU; MB; MiU-C; N; NN; PHi; PPL.

4765. New York (State). Laws.

Acts passed at the forty-third session of the Legislature of the state of New-York, respecting the canals from Lake Erie to the Hudson River, and from Lake Champlain to the same. Utica: Printed by William Williams, 1822. 20, [3] p.

S-9668.

Copies: N; NcD.

4766. New York (State). Laws.

Revised statute. Relating to the navigation of the New-York state canals. Chapter IX. Title IX. Article seventh. [Albany?: 1822?] 14 p.

S-9686.

Copies: N.

4767. Ohio. General Assembly. House. Committee on Canals.

Report of the Committee on canals. Columbus: Printed at the office of the Columbus Gazette by P. H. Olmsted, 1822. 15 p.

S-9769.

Copies: DLC; MWA; N; OCHP; OClWHi.

4768. The President, Directors, and Company of the Farmington Canal.

The act of incorporation of the Farmington Canal Company, with the reports of the Hon. Benjamin Wright and Andrew A. Bartow, Esq. and of the committee of the legislature of Connecticut, on that subject. New Haven: S. M. Dutton, printer, 1822. 15 p.

S-8425, 8680.

Copies: CtHi; CtY; DeGE; MH-BA; NN.

4769. Randel, John.

Description of a direct route for the Erie Canal, at its eastern termination: with estimates of its expense, and comparative advantages. By John Randel, Jun. Albany: Printed by G. J. Loomis, 1822. 72 p.

S-10063.

Copies: CSmH; DeGE; MB; MH-BA; MWA; MWo; MeU; N; NBu; NN; NNE; NjR; PPL; WHi.

4770. Susquehanna Canal Company.

Counter report relative to the Susquehannah Canal wing-dam, accompanied by the report of the committee of the House of Delegates, to which was re-committed the bill relating to said wing-dam. December session, 1821. Annapolis: Jehu Chandler, [1822]. 13 p.

S-9372.

Copies: MdBJ-G.

4771. Susquehannah & St. Lawrence Canal. Report of a committee appointed for the purpose of ascertaining the most eligible route for a canal from Seneca Lake to the Erie Canal. Geneva, N. Y.: Printed by James Bogert, 1822. 22 p., fold. table.

4771 (*continued*)
S-10382.
Copies: N; NGH; NHi.

4772. Troup, Robert, 1757–1832.
A letter to the Honorable Brockholst Livingston, Esq., one of the justices of the Supreme Court of the United States, on the lake canal policy of the state of New-York.... With a supplement, and additional documents. Albany: Printed by Packard & Van Benthuysen, 1822. 38, 34, 42, 5 p.
S-10484.
Copies: CSmH; CtHC; DBRE; DLC; DeGE; ICJ; IU; MH; MWA; NN; NNE; NbU; NcD; NhHi; NjR; OCHP; PPL; PU; RPB; ScU; Vi; WU.

4773. Union Canal Company of Pennsylvania.
Report of the president and managers of the Union Canal Company of Pennsylvania to the stockholders. November 19, 1822. Philadelphia: Printed by order of the Company, [1822]. 8, [1] p.
S-10504.
Copies: DeGE; ICJ; MH-BA; MWA; MnHi; PHi; PPL; WU.

4774. United States. Congress. Senate. Committee on Public Lands.
In Senate of the United States, February 11, 1822. Mr. Thomas, from the Committee on Public Lands, to whom was referred the bill to authorize the state of Illinois to open a canal through the public lands to connect the Illinois River and Lake Michigan, reported the same with the following amendments.... [Washington: 1822.] 2 p. (S.26)
S-10852.
Copies: DLC.

4775. United States. Laws.
A bill to authorize the state of Illinois to open a canal through the public lands, to connect the Illinois River and Lake Michigan. January 14, 1822.... [Washington: 1822.] 2 p. (H.R. 34)
S-10739.
Copies: DLC.

4776. ———— January 24, 1822. In Senate of the United States. [Washington: 1822.] 2 p. (S.26)
S-10740.
Copies: DNA.

4777. Winchester, George.
Observations on the Susquehanna River & Maryland Canal, addressed to the legislatures of Pennsylvania & Maryland. Baltimore: Frederick G. Schaeffer, 1822. 29 p.
S-9751, 11459.
Copies: Md; MdBE; MdBP; PPAmP.

4778. Worcester Co., Mass. Committee on Canal Survey.
An account of the proposed canal from Worcester to Providence: containing the report of the engineer; together with some remarks upon inland navigation. Published by order of the Committee for the county of Worcester. Worcester: Printed by William Manning, 1822. 18 p.
Benjamin Wright, engineer.
S-11488.
Copies: DLC; DeGE; KHi; MH-BA; MWA; MiD; N; NBu; NN; NhHi; NjPT; PPAmP; RPB.

4779. Celebration of the passage of the first boat from the grand canal into the Hudson, at the city of Albany, on Wednesday, October 8, 1823. Order of arrangements. [Albany: 1823.] 1 p.
S-12083.
Copies: NN.

4780. Chesapeake and Delaware Canal Company.
A collection of the laws relative to the Chesapeake and Delaware Canal; passed by the legislatures of the States of Maryland, Delaware, and Pennsylvania, subsequent to the year 1798. Published by order of the President and Directors of the Chesapeake and Delaware Canal Company, June 1, 1823. Philadelphia: Printed by L. R. Bailey, 1823. 49 p.
S-12117.

Copies: DBRE; DeGE; DeWI; MWA; NN; NNC; PHi; PPAmP; PPF; PPL.

4781. Chesapeake and Delaware Canal Company.

Communication from the Chesapeake and Delaware Canal Company and a report and estimate of William Strickland, to the president and directors. Philadelphia: Printed by Joseph R. A. Skerrett, 1823. 8 p.

Report of W. Strickland, engineer, p. 5–8.

S-12118.

Copies: NN; PHi; PPAmP; PPL.

4782. Chesapeake and Delaware Canal Company.

Sir, we take the liberty of sending you a book for receiving subscriptions towards the completion of the Chesapeake and Delaware Canal. . . . Philadelphia: 1823. Broadside 25 x 30 cm.

Printed circular letter, dated April 14, 1823, and signed: Mathew Carey, Richard Peters, Jun., Samuel Spackman, Caleb Newbold, J. K. Kane, Sub-committee appointed to promote the completion of the Chesapeake and Delaware Canal.

Copies: DeGE.

4783. Chesapeake and Ohio Canal Convention, Washington, 1823, 1826.

Proceedings at a general convention of delegates, representing counties in Virginia, Maryland, Ohio, Pennsylvania, and the District of Columbia . . . 6th, 7th and 8th of Nov., 1823, on the subject of the Chesapeake & Ohio Canal. To which are added, an extract from the report of the commissioners appointed by the states of Virginia and Maryland to survey the Potomac Canal, and an act of Virginia incorporating the Potomac Canal Company. Washington: Printed at the office of the Washington Republican, 1823. 48 p.

S-12120.

Copies: DLC; MBAt; MWA; MdBP; NNC; NbU.

4784. Chesapeake and Ohio Canal Convention, Washington, 1823, 1825. Committee for the State of Maryland.

The memorial of a Committee for the State of Maryland, appointed (in pursuance of the resolution of a general convention of delegates from the states of Virginia, Pennsylvania, Ohio, Maryland, and the District of Columbia, held in the city of Washington on the 6th of November, 1823) "to prepare and present, in behalf of said Convention, a memorial to the General Assembly of Maryland, requesting their concurrence in the incorporation of a company, to form a connection of the Atlantic and Western waters, by a canal." [Baltimore?: 1823?] 17 p.

S-12119, 20040 (dated 1825?).

Copies: MdBP; MiU-C; NIC; ViU.

4785. Cumberland and Oxford Canal Corporation.

Report of the committee appointed to procure a survey for the route of the contemplated Cumberland and Oxford Canal; to which is added the report of the engineer, containing an estimate of the expence of making the canal. . . . Portland, [Me.]: Printed at the Mirror office, 1823. 15 p.

S-12299.

Copies: IU; MBAt; MH; MWo; MeHi; MeP; MeU; NN.

4786. Drawing of the Union Canal lottery, sixth class, new series. Philadelphia: 1823. [3] p.

Copies: MWA.

4787. Geddes, James, 1763–1838.

Canal report, made by James Geddes, esq., the engineer employed by the state of Ohio. By authority. Columbus: Printed at the office of the Columbus Gazette, by P. H. Olmsted, 1823. 14 p.

S-12649.

Copies: DLC; MH-BA; MWA; N; NN; OCHP.

4788. Maryland. Commissioners Appointed to Examine into the Practicability of a Canal from Baltimore to the Potomac.

4788 (*continued*)

Report of the Commissioners Appointed to Examine into the Practicability of a Canal from Baltimore to the Potomac, together with the engineer's report. Baltimore: Fielding Lucas, Jun.; John D. Toy, printer, 1823. 69 p.

Engineer's report signed: Isaac Briggs.

Main report signed: Athan's Fenwick, William Howard, commissioners.

S-13238.

Copies: DLC; MB; MH-BA; MWA; MdBP; MdHi; MiDSH; NN; NNC; PPL.

4789. Maryland. Laws.

An act to incorporate the Potomac Canal Company. [Annapolis: 1823.] 24 p. (House. Bill No. 57)

Heyl-154.

Copies: MdHi.

4790. Maryland. Laws.

A further supplement to an act, entitled, A supplement to an act entitled, An act to incorporate a company for the purpose of cutting and making a canal between the river Delaware and Chesapeake Bay. [Annapolis: 1823.] 3, [1] p. (House. Bill No. 63)

Heyl-178.

Copies: MdHi.

4791. Maryland. Susquehanna Canal Commissioners.

Report by the Maryland Commissioners on a proposed canal from Baltimore to Conewago.... Baltimore: F. Lucas, jun., 1823. 84, [1] p., illus., 2 fold. maps.

Signed: Theodorick Bland, George Winchester, John Patterson, Commissioners of Susquehanna Canal.

S-13236.

Copies: DLC; DeGE; DeU; MB; MWA; MdBP; MeHi; MiDSH; N; NN; NNC; NcD; PP; PPL; PU.

4792. Maryland. Susquehanna Canal Commissioners.

A topographical map of the route of a proposed canal and the country between Conewago and Baltimore. Drawn by direction of the Canal Commissioners by F. Lucas Jr. [Baltimore: 1823.] Map 48 x 40 cm.

S-14003.

Copies: DeGE; PPL.

4793. Maryland. Treasurer's Office.

Correspondence of the Treasurer of the Western Shore and H. D. Gilpin, relative to the Chesapeake and Delaware Canal.... Annapolis: J. Hughes, [1823?]. 20 p.

Copies: MdHi.

4794. Memorial of the inhabitants of Waterford, remonstrating against the passage of a bill now pending before the Hon. Assembly, for extending the Champlain Canal. Waterford, [N.Y.]: 1823. 8 p.

Dated: March 15, 1823.

S-14841.

Copies: N.

4795. Memorial of the trustees of the villages of Lansingburgh and Waterford concerning location of canals at said places. Albany: 1823. 2 p. (New York (State). Assembly. Doc. 143)

Whitford-p. 1194.

Copies: N.

4796. Mercer, Charles Fenton, 1778–1858.

Speech of Mr. C. F. Mercer, on the subject of the Chesapeake and Ohio Canal, delivered in the convention of delegates, held at the city of Washington, November 7, 1823. Washington: Printed by Gales & Seaton, 1823. 47 p.

S-13302.

Copies: DLC; MB; MBAt; MH; MWA; MdHi; NNC; NcD; OClWHi; PPAmP; PPi; ViU; WHi.

4797. Moulton, Joseph White, 1789–1875.

A remonstrance and memorial to the legislature, by Joseph W. Moulton ... relative to the western termination of the grand canal.... Albany: Printed by W. & J. G. White, 1823. 19 p.

S-13414.

Copies: MWA; MiU; N; NBu; NBuG; NN; PHi.

4798. New Jersey. Commission on the Delaware and Passaic Canal.

Report of the Commissioners appointed by the Legislature of the state of New Jersey, for the purpose of exploring the route of a canal to unite the river Delaware, near Easton, with the Passaic, near Newark, with accompanying documents. Morristown: Printed by J. Mann, 1823. [150] p. (various pagings), plan.

Report signed: George P. M'Culloch, Charles Kinsey, Thomas Capner.

Documents have the caption title: *Documents appertaining to the Commissioners' report, relative to the Morris Canal.*

S-13522.

Copies: DBRE; DLC; M; MB; MHi; MWA; MdBP; NNC; NNE; NbU; PPL.

4799. ———— Morristown: Printed by J. Mann, 1823. 87, [1] p., plan.

N-N0153236.

Copies: CtY; DeGE; ICJ; MH-BA; MWA; MiU-T; NN; PPL.

4800. New Jersey. Commission on the Delaware and Raritan Canal.

Legislature of New-Jersey. House of Assembly, Saturday, November 22, 1823. Mr. Griffith offered the following resolution: Resolved, that there be printed for this house, to be distributed at their discretion, five hundred copies of the Report of the Commissioners appointed by an act of the Legislature, passed February 14, 1816, for ascertaining the most eligible route of a canal to connect the tide waters of the Delaware with those of the Raritan River, including the Report of John Randel, jun., relative to said canal. [Trenton]: 1823. 24 p., fold. table.

Signed on p. 21: John Rutherford, John N. Simpson, George Holcomb. New Brunswick, January 22, 1817.

S-13517.

Copies: DeGE; MWA; NN; NjP.

4801. New York (State). Canal Commissioners.

The annual report of the Canal Commissioners of the state of New-York, presented to the Legislature, the 24th February, 1823. Albany: Printed by Cantine & Leake, printers to the state, 1823. 40 p.

S-13545.

Copies: InHi; MWA; N; NHi; WHi.

4802. New York (State). Canal Commissioners.

Report on the memorial of the mayor, aldermen, and commonalty of the city of Troy, concerning location of the canal of said city. Albany: 1823. 7 p. (Assembly. Doc. 207)

Whitford-p. 1236.

Copies: N.

4803. New York (State). Canal Commissioners.

Report relative to the construction of a harbor at Black Rock. Albany: 1823. 2 p. (Assembly. Doc. 167)

Concerns the terminus of the Erie Canal.

Whitford-p. 1257.

Copies: N.

4804. New York (State). Laws.

An act authorizing the construction of a basin in Albany at the termination of the Erie and Champlain canals. Albany: 1823. 4 p. (Assembly. Doc. 133)

Whitford-p. 1260.

Copies: N.

4805. New York (State). Laws.

An act authorizing the construction of a canal, from the Champlain Canal at Waterford to the Hudson at Troy. Albany: 1823. 2 p. (Assembly. Doc. 215)

Whitford-p. 1236.

Copies: N.

4806. New York (State). Laws.

An act authorizing William Waring and others to construct a canal through certain lands in the county of Jefferson, and

4806 (*continued*)
for other purposes. Albany: 1823. 2 p. (Assembly. Doc.)

Whitford-p. 1241.

Copies: N.

4807. New York (State). Laws.

An act to authorize the inhabitants of the village of Salina, to make a lateral canal in said village, and for the laying out an additional number of salt lots. Albany: 1823. 2 p. (Assembly. Doc. 138)

Whitford-p. 1231.

Copies: N.

4808. New York (State). Laws.

An act to incorporate the president, managers and company of the Delaware and Hudson Canal Company, passed April 23, 1823. Albany: 1823. 8 p. (Assembly. Doc. 222)

Whitford-p. 1254.

Copies: N.

4809. New York (State). Legislature. Assembly. Committee on Canals.

Report on the petition of the mayor, aldermen and commonalty of Troy relative to a canal between Troy and Waterford. Albany: 1823. 5 p. (Assembly. Doc. 156)

Whitford-p. 1236.

Copies: N.

4810. New York (State). Legislature. Assembly. Committee on the Ontario and Champlain Canal.

Report of the Committee appointed on the subject of the Ontario and Champlain Canal. Potsdam, N.Y.: Printed by F. C. Powell, [1823]. 8 p.

S-13640.

Copies: N.

4811. North Carolina. General Assembly. Joint Committee on Internal Improvements.

Report to the General Assembly, December 19, 1823. Raleigh: J. Gales, 1823. 8 p. (Doc. 9)

Concerns the construction of canals.

N-N0310664; Thornton (Off.)-1525.

Copies: NcU; ViU.

4812. Ohio. Board of Canal Commissioners.

Report of Canal Commissioners, made to the Twenty-first General Assembly, January 3, 1823.... Columbus: Printed at the Office of the Columbus Gazette, by P. H. Olmsted, 1823. 12 p.

S-13624.

Copies: DLC; MWA; N; OCHP; OClWHi; OHi.

4813. Ohio. Board of Canal Commissioners.

Second report of the Canal Commissioners, January 13, 1823.... Columbus: Printed at the office of the Columbus Gazette, by P. H. Olmsted, 1823. 4 p.

S-13627.

Copies: N; NN; OHi.

4814. The Ohio canal. [New York?: 1823.] 8 p.

Dated: New York, November, 1823, and signed "The voice of New-York."

S-14795.

Copies: N; NbU.

4815. Petition of the inhabitants of the northerly part of the state, praying for aid, to make a canal from Lake Ontario near Ogdensburgh to Lake Champlain, near Plattsburg. Albany: 1823. 3 p. (New York (State). Legislature. Assembly. Doc. 68)

Whitford-p. 1347.

Copies: N.

4816. The Potomac canal. Papers relating to the practicability, expediency, and cost of the Potomac canal; also, the act of the legislature of Virginia, incorporating the Potomac Canal Company. [Washington]: 1823. 39 p.

S-13824.

Copies: DLC; ICU; NBu; NNC; PPAmP.

4817. The President, Directors, and Company of the Farmington Canal.

Canal.... New Haven: S. M. Dutton, printer, 1823. Broadside 30 x 22 cm.

Announcement of opening subscriptions to the stock.

Copies: CtSoP.

4818. Proprietors of the Middlesex Canal.

To the members of the General Court.... [Boston?: 1823?] 4 p.

Application to raise money through lotteries.

S-13334.

Copies: MH-BA.

4819. Rhode Island. Laws.

An act to incorporate the Blackstone Canal Company.... [Providence: 1823.] [4] p.

S-11900.

Copies: RPB.

4820. Stewart, Andrew, 1791–1872.

Mr. Stewart's speech on the proposed connection of the eastern and western waters, by the Ohio and Potomac canal. Washington: Printed by E. De Krafft, 1823. 14 p.

N-S0933894.

Copies: ICU; PHi.

4821. Union Canal Company of Pennsylvania.

Letter and documents, in relation to the dissolution of the engagement of Loammi Baldwin, with the Union Canal Company. Harrisburg: Printed by Mowry & Cameron, 1823. 23 p.

Covering letter signed: Samuel Mifflin, president.

S-14330.

Copies: DBRE; DeGE; MB; NN; NNC; P; PPAmP; PPL.

4822. Union Canal Company of Pennsylvania.

Report of the president & managers of the Union Canal Company of Pennsylvania to the stockholders. Presented at the annual meeting, November 18, 1823. Philadelphia: Printed by Jacob Frick & Co., [1823]. 8, [1] p.

S-14331.

Copies: DeGE; MH-BA; N; NRom; OClWHi; PHi; PPL; WU.

4823. United States. Engineer Dept.

War Department ... Sir: In conformity with the instructions of this Department, the Board of Engineers have placed in your hands a copy of their report on the projected Morris Canal.... [S.l.: 1823.] 45 p., diagr.

Report signed by S. Bernard and Jos. G. Totten, from the Engineer Dept.

Letter of transmission dated: November 10, 1823.

Copies: DeGE; MWA; NN.

4824. Virginia. Commissioner of the James and Jackson's River Canal.

Commissioner's report on the James and Jackson's River Canal, with accompanying documents. [Richmond: 1823.] 16 p.

Signed: Randolph Harrison, 4th December, 1823.

N-V0183846.

Copies: NcD.

4825. Washington Canal Company.

Report of the Washington Canal Company, of their receipts and expenditures, during the year 1822. February 3, 1823 .. Washington: Printed by Gales & Seaton, 1823. 3 p. (17th Cong., 2d sess. House. Doc. 48)

S-14736.

Copies: DLC; MWA; NjP.

4826. Willis, Richard, d. 1830.

DeWitt Clinton's Grand Canal march. As performed by the West Point Band at the entrance of the first canal boat into the Hudson River on the 8th of October, 1823. New York: H. Sage, [1823]. [2] p.

S-14936, 14937 (reissued at New York: A. & W. Geib, [1823]).

Copies: DLC; MH; MWA; NN.

4827. [Carey, Mathew] 1760–1839.

Address to the stockholders of the Chesapeake & Delaware Canal on the subject of the route recommended by the board of engineers and adopted by the board of directors of that canal, by a Pennsylvanian. Philadelphia: H. C. Carey & I. Lea, 1824. 8 p.

S-15650.

Copies: MiU; NN; PPAmP; PPL.

4828. [Carey, Mathew] 1760–1839.

Canal policy—No. I. [Philadelphia: 1824.] 2 p.

Signed and dated: Fulton, Philadelphia, Dec. 8th, 1824.

Copies: PPAmP; PPL.

4829. ———— 2d ed. [Philadelphia: 1824.] 4 p.

Signed and dated on p. 2: Fulton, Philadelphia, Dec. 8, 1824.

Canal policy—No. II: p. 3–4; signed and dated at end: Fulton, Philadelphia, Dec. 13, 1824.

S-15654.

Copies: DeGE; MWA; MdHi; PPAmP; PPL; PU.

4830. ———— No. I [–II]. 3d ed. [Philadelphia: 1824.] 4 p.

Signed: Fulton, and dated respectively Philadelphia, Dec. 8th and Dec. 13th, 1824.

S-15653, 15655.

Copies: DeGE; MCM; MWA; PPL.

4831. Chesapeake and Delaware Canal Company.

Fifth general report of the president and directors of the Chesapeake and Delaware Canal Company. June 7, 1824. [Philadelphia?: 1824.] 3 p.l., [5]–19 p., 3 fold. maps.

S-15733.

Copies: DeGE; DeU; IU; MBAt; MWA; MdBE; MdBP; NIC; NbU; PU; PPL.

4832. Chesapeake and Delaware Canal Company.

Memorial of the president and directors of the Chesapeake and Delaware Canal Company. December 27, 1824.... Washington: Printed by Gales & Seaton, 1824. 6 p. (18th Cong., 2d sess. House. Doc. 23)

Signed: James C. Fisher, President.

S-18884.

Copies: DLC; DeGE; DeU; MWA; NjP.

4833. Erie Canal Transportation Company.

"Erie Canal Transportation Company." Four daily lines.... Albany: Packard & Van Benthuysen, 1824. Broadside.

S-16067.

Copies: NN.

4834. Harper, Robert Goodloe, 1765–1825.

Gen. Harper's speech to the citizens of Baltimore, on the expediency of promoting a connexion between the Ohio, at Pittsburgh, and the waters of the Chesapeake, at Baltimore, by a canal through the District of Columbia. With his reply to some of the objections of Mr. Winchester. Delivered at a meeting, held at the Exchange, on the 20th day of December, 1823. Baltimore: E. J. Coale, 1824. 78 p., fold. map.

S-16447.

Copies: CtY; DLC; DeGE; ICJ; MH; MWA; MdBP; MiD-B; MnHi; NN; Nh; NjP; OClWHi; PPL; ScU; Vi; WHi.

4835. Lackawaxen Coal Mine and Navigation Company.

Address to the public by the Lackawaxen Coal Mine and Navigation Company, relative to the proposed canal from the Hudson to the head waters of the Lackawaxen River. Accompanied by documents. New York: Printed by Wm. Grattan, 1824. 24 p.

S-16836.

Copies: CtY; DeGE; IU; MH-BA; MWA; NN; NRU.

4836. Map of the route of the proposed, and enacted, Hudson and Delaware Canal,

with the river Lackawaxen and its head waters, showing their relative situation to the coal mines. Philadelphia: Engraved by J. Yeager, 1824. Map 22.5 x 61 cm.
N-M0196961.
Copies: PBL.

4837. Maryland. Laws.
An act to provide for making the Baltimore Canal. January 24, 1824. [Annapolis]: J. Green, print., [1824]. 10 p.
Heyl-246.
Copies: MdHi.

4838. Maryland. Susquehanna Canal Commissioners.
To the Mayor and City Council of Baltimore. The Commissioners appointed by an act of the legislature of Maryland, passed at December session, 1821, Chap. 51, entitled "An act for the improvement of the navigation of the river Susquehanna," beg leave to report.... [Baltimore: 1824.] Broadside.
Signed: Isaac M'Kim, Wm. H. Winder, J. W. Patterson, Wm. Howard, Tho. Ellicott. Baltimore, January 20th, 1824.
Sabin-93938 (note).
Copies: MWA.

4839. Memorial of the counties of St. Lawrence, Franklin and Clinton, to the Legislature of the state of New-York, praying for an act, authorizing a survey of the route of a canal, to connect lakes Ontario and Champlain, commencing at the foot of sloop navigation of the river St. Lawrence. Presented January 28, 1824. Albany: W. M'Dougall's print, [1824]. 27 p.
Signed: B. Raymond.
S-17383.
Copies: MWA; NHi; NNC; PPL.

4840. ——— Presented January 27, 1824. [S.l.: 1824.] 27 p.
Copies: NHi.

4841. New Jersey. Commission on the Delaware and Raritan Canal.
Report by the New Jersey Commissioners upon the subject of a canal, from the Delaware to the Raritan. Made November 1824. Trenton: Printed by J. Justice, 1824. 26 p.
Commission created by an act passed December 8, 1823.
Signed on p. 21: Lucius Q. C. Elmer, George Holcombe, Peter Kean.
S-17344, 21632 (dated 1825).
Copies: CSmH; DLC; DeGE; MWA; NN; NNE; NjP; PPAmP; PPL.

4842. New Jersey. Laws.
An act to incorporate a company, to form an artificial navigation between the Passaic and Delaware rivers. [Trenton?: 1824.] 12 p.
At head of the title: By the legislature of the state of New Jersey.
Concerns the Morris Canal and Banking Company.
N-M0795892, N0154721 (dated 1826?).
Copies: CtY; NN.

4843. New Jersey. Legislature. Joint Committee on the Delaware and Raritan Canal.
Report of the Joint Committee, of the Council and Assembly of New-Jersey, upon the report of the commissioners, appointed by the Legislature, at their session of 1823, upon the subject of a canal between the Delaware and Raritan rivers.... Trenton: Printed by J. Justice, 1824. 7 p.
S-17345.
Copies: NN; NjP.

4844. New York (State). Canal Commissioners.
The annual report of the Canal Commissioners of the state of New-York, presented to the Legislature, the 20th February, 1824. Albany: Printed by Leake & Croswell, printers to the state, 1824. 47 p.
S-17365.
Copies: InHi; MWA; MWo; NNC; NbU.

4845. New York (State). Canal Commissioners.
Report of the acting Canal Commissioner of the state of New York relative to the

4845 (*continued*)
aqueduct across Genesee River. Albany: 1824. 6 p. (Assembly. Doc. 94)
Whitford-p. 1221.
Copies: N.

4846. New York (State). Laws.
Canal laws. Acts of the Legislature of the state of New-York, respecting the canals, from Lake Erie to the Hudson River, and from Lake Champlain to the same. Albany: Printed by Websters & Skinners, 1824. 38 p.
"Rules and regulations of the Canal Commissioners, March 21, 1825": p. 35–37; "Rates of toll on the Erie & Champlain canals": p. 37–38.
N-N0197254; S-17367.
Copies: DLC.

4847. New York (State). Legislature. Assembly. Committee on Canals.
Documents accompanying a report of the standing Committee on Canals and Internal Improvements, in relation to the western terminus of the Erie Canal. Albany: 1824. 30 p. (Assembly. Journal, 1824. Appendix D)
Whitford-p. 1258.
Copies: N.

4848. New York (State). Legislature. Assembly. Committee on Canals.
Report of the Canal Committee on the memorial for surveying a canal from Ogdensburgh to Lake Champlain. [Albany: 1824]. 5 p. (Assembly. Doc. 156)
Dated: March 19, 1824.
Whitford-p. 1347.
Copies: N; NHi.

4849. New York (State). Legislature. Assembly. Committee on Canals.
Report relative to the navigation of the Champlain Canal, from Fort Edward to Fort Miller. Albany: 1824. 3 p. (Assembly. Doc. 82)
Whitford-p. 1237.
Copies: N.

4850. New York (State). Legislature. Assembly. Committee on Canals.
Report relative to the report of the Acting Canal Commissioner in charge of constructing the canal across the Genesee River. Albany: 1824. 5 p. (Assembly. Doc. 193)
Whitford-p. 1221.
Copies: N.

4851. New York (State). Legislature. Assembly. Select Committee on Delaware and Hudson Canal.
Report on petition of inhabitants of Orange County, in relation to a canal, to unite the waters of the Delaware and Hudson rivers. Albany: 1824. 3 p. (Assembly. Doc. 150)
Whitford-p. 1254.
Copies: N.

4852. New York (State). Legislature. Senate. Committee on Canals.
Report on the petition for continuing the Champlain Canal from Fort Edward to Fort Miller. Albany: 1824. 3 p. (Senate. Doc. 155)
Whitford-p. 1237.
Copies: N.

4853. New York (State). Legislature. Senate. Select Committee on the Memorial of Maurice Wurts.
Report concerning the memorial of Maurice Wurts of Philadelphia, relative to Delaware and Hudson Canal. Albany: 1824. 2 p. (Senate. Doc. 153)
Whitford-p. 1254.
Copies: N.

4854. Ohio. Board of Canal Commissioners.
Report of the Board of Canal Commissioners, to the General Assembly of Ohio, with accompanying documents. Columbus: Printed by P. H. Olmsted, 1824. 33, 20 p.
S-17480.
Copies: DLC; MWA; MdBP; N; NN; OCHP; OCLaw; OClWHi.

4855. Ohio. Canal Engineer.

Governor's message, transmitting the report of the Canal Engineer, on the subject of a canal at the Falls of the Ohio; with accompanying documents. Columbus: P. H. Olmsted, printer, 1824. 19 p.

David Stanhope Bates, engineer.

"Mr. Kelley's report": p. 11–19.

S-17477.

Copies: DLC; MH-BA; N; NN; OCHP; OClWHi.

4856. Pennsylvania. Board of Canal Commissioners.

Annual report of the Board of Canal Commissioners of Pennsylvania, with accompanying documents. Printed by order of the House of Representatives, Jan. 17, 1824. Harrisburg: Printed by Henlock & Bratton, 1824. 80 p.

S-17554.

Copies: MiD-B; PPL; WHi.

4857. Pennsylvania. Commissioners on the Lake Erie and French Creek Canal.

Report of the Commissioners appointed by an act of the legislature of the 21st of March, 1823, to view and examine all the contemplated routes for connecting the waters of Lake Erie and French Creek by a canal and slack water navigation. Read in the Senate, January 30, 1824. Harrisburg: Printed by John S. Wiestling, 1824. 12 p.

Thomas Forster, James Herrington and William Marks, Jr., commissioners.

S-17559.

Copies: MWA; MiU; P; PHi (at PPL); PP; PPAmP.

4858. Pennsylvania. General Assembly. House. Committee on Juniata and Conemaugh Canal.

Report of the Committee to whom was referred the subject of the Juniata and Conemaugh Canal. Read in the House of Representatives, Feb. 10, 1824. Mr. Clarke, Chairman. [Harrisburg?]: J. S. Wiestling, printer, 1824. 13 p.

S-17562.

Copies: DeGE; ICJ; MWA; MiU-T; N; NN; P; PHi; PPAmP.

4859. Public proceedings on the removal of the Hon. De Witt Clinton from the office of Canal Commissioner. New York: Printed by Geo. F. Hopkins, 1824. 40 p.

S-17732.

Copies: MB; MWA; MiD-B; NBLiHi; NHi; NSyOHi; NbU; P.

4860. Shriver, James.

An account of surveys and examinations, with remarks and documents, relative to the projected Chesapeake and Ohio, and Ohio and Lake Erie canals.... Baltimore: F. Lucas, jr.; J. D. Toy, print., 1824. 116 p., illus., 2 maps.

S-17981.

Copies: CSmH; DLC; DeGE; ICJ; MB; MH-BA; MWA; MdBP; MiD-B; NIC; NNE; OClWHi; OMC; PPF; PPL; PPi; RPB; Vi; WHi; WvU.

4861. Sullivan, John Langdon, 1777–1865.

Suggestions on the canal policy of Pennsylvania, in reference to the effects of the inland navigation of the adjoining states, on the commerce of Philadelphia.... Philadelphia: Printed by J. Young, 1824. 49, [1] p.

S-18124.

Copies: CSmH; DLC; DeGE; ICJ; MWA; MdHi; NN; OClWHi; PPAmP; PPL.

4862. Tarascon, Louis Anastasius, 1757– .

An exposition of some of the reasons why measures should be taken for the construction of a canal round the falls of the river Ohio, on the Kentucky side; for the location of the western armory, near to it ... and for the opening of a waggon road, from the river Missouri to the navigable waters of the river Columbia.... By Lewis A. Tarascon. Louisville: Printed for the author, 1824. 7 p.

S-18144.

Copies: CSmH; InU.

4863. To the Honorable the Senate and Assembly of the state of New-York.... The memorial of the undersigned, inhabitants of

4863 (*continued*)
Long Island.... [New York?: 1824.] Broadside.

A memorial in support of an application about to be made by the Long Island Canal Company for an act of incorporation, accompanied by an advertisement of the Company, dated May 26, 1824, with map.
S-16933.
Copies: NN.

4864. To the stockholders of the Chesapeake & Delaware Canal.... [Philadelphia: 1824.] Broadside 21 x 33.5 cm.

Dated and signed: Philadelphia, March 26, 1824. A Philadelphia stockholder.
S-18217.
Copies: DLC; MWA.

4865. ———— [Philadelphia: 1824.] 2 p.
Signed: A Philadelphia stockholder, Philadelphia, May 10, 1824.
Copies: MWA.

4866. ———— No. I.... [Philadelphia: 1824.] Broadside 45 x 15 cm.
Signed: A stockholder; dated in text: November 25th, 1824.
Copies: MWA.

4867. Union Canal Company of Pennsylvania.

Annual report of the president and managers of the Union Canal Company of Pennsylvania, to the stockholders, November 16, 1824. Philadelphia: Printed by Lydia R. Bailey, 1824. 14 p.
S-18262.
Copies: DLC; DeGE; MB; MH-BA; MMal; MWA; MeHi; P; PHi; PPAmP; PPL.

4868. United States. Congress. House. Committee on Roads and Canals.

Report of the Committee on Roads and Canals, upon the subject of connecting the waters of the Tennessee, with those of the Alabama River, and also, upon connecting the waters of the Tennessee, with those of the Coosa River, by means of a canal. May 17, 1824.... [Washington: 1824.] 1 p. (18th Cong., 1st sess. House. Rept. 123)
S-19039.
Copies: DLC; DeU; MWA; MjP.

4869. United States. Congress. Senate. Select Committee on Roads and Canals.

In Senate of the United States, April 24, 1824. Mr. Van Dyke from the Select Committee on Roads and Canals, to whom was referred the memorial of the president and directors of the Chesapeake and Delaware Canal Company, reported.... [Washington: 1824.] 6 p. (18th Cong., 1st sess. Senate. Doc. 70)
S-18750.
Copies: DLC; DeGE; DeU; MWA; NjP.

4870. United States. Laws.

An act to authorize the state of Indiana to open a canal through the public lands, to connect the navigation of the rivers Wabash and the Miami of Lake Erie. May 14, 1824.... [Washington: 1824.] 1 p. (H.R. 6)
S-18360.
Copies: DNA.

4871. United States. Laws.

An act to authorize the Territory of Florida to open a canal, through the public lands, to unite the river St. John's with the bay of St. Augustine. May 14, 1824.... [Washington: 1824.] 1 p. (H.R. 188)
S-18361.
Copies: DNA.

4872. Views respecting the Chesapeak and Delaware Canal. By a citizen of Philadelphia. Philadelphia: Printed by T. Town, 1824. 1 p.l., 18 p.
S-19143.
Copies: DLC; DeGE; MB; MH-BA; NN; PPAmP; PPL.

4873. Washington Canal Company.

Annual report of the Washington Canal Company. April 21, 1824.... Washington: Printed by Gales & Seaton, 1824. 3 p. (18th

Cong., 1st sess. House. Doc. 138)
S-18382.
Copies: DLC; DeU; MWA; NjP.

4874. Wright, Benjamin, 1770–1842.
Report of Messrs. Benj. Wright and J. L. Sullivan, engineers, engaged in the survey of the route of the proposed canal from the Hudson, to the head waters of the Lackawaxen River. Accompanied by other documents, &c. Philadelphia: Printed by J. Young, 1824. viii, 70 p., fold. map, plates.
"An act to improve the navigation of the river Lackawaxen," p. [44]–56.
"An act to incorporate the president, managers and company, of the Delaware and Hudson canal company," p. [57]–70.
S-19340.
Copies: DLC; DeGE; ICU; IU; MCM; MH-BA; MWA; N; NNE; NRom; NcD; P; PPL; RPB.

4875. Wurts, Maurice.
Memorial of Maurice Wurts, of the city of Philadelphia, concerning the Delaware and Hudson Canal Company. Albany: 1824. 3 p. (New York (State). Legislature. Senate. Doc. 134)
Whitford-p. 1254.
Copies: N.

4876. Baltimore. Commission to Explore and Survey the Route for a Canal from Baltimore to Port Deposit.
Report of the Commissioners appointed by the mayor of the city of Baltimore, to explore and survey the rout [*sic*] for a canal and stillwater navigation from the city of Baltimore, to the head of tidewater at Port Deposit.... Baltimore: Published for the Commissioners by F. Lucas, jun'r; J. D. Toy, printer, 1825. 24 p., fold. map.
S-19508.
Copies: DLC; MB; MBAt; MH-BA; MdBP; MdHi; NNE; PHi; PPF.

4877. [Carey, Mathew] 1760–1839.
[Address in vindication of his efforts in behalf of the Chesapeake and Delaware Canal Company. Philadelphia: 1825.] 8 p.
Dated: May 9, 1825.
N-C0133044.
Copies: MH; MWA.

4878. [Carey, Mathew] 1760–1839.
An appeal to the justice and humanity of the stockholders of the Chesapeake and Delaware Canal. [Philadelphia: 1825.] 16 p.
S-19945.
Copies: MWA; PHi; PPAmP; PPL.

4879. ———— 2d ed., enlarged. [Philadelphia: 1825.] 20 p.
S-19946.
Copies: MB; MWA.

4880. ———— 3d ed. [Philadelphia: 1825.] 20 p.
S-19947.
Copies: MWA; PPAmP; PPL; PU.

4881. [Carey, Mathew] 1760–1839.
Canal policy—Fulton. No. V. [Philadelphia: 1825.] 4 p.
Signed: Fulton, March 19, 1825.
N-C0133141; S-15652.
Copies: DGU; DeGE; MCM; MWA; NN; PPL; PU.

4882. ———— 2d ed. [Harrisburg?: 1825.] 4 p.
Dated: Harrisburg, March 19, 1825.
S-19949.
Copies: MWA; PPL.

4883. [Carey, Mathew] 1760–1839.
Canal policy—No. III. [Philadelphia: 1825.] 8 p.
Signed on p. 4: Fulton, Jan. 13, 1825.
"The importance of the views of the canal policy of New York, presented by De Witt Clinton," p. 5–8.
Copies: DeGE; MWA; NN; PPL.

4884. ———— 2d ed. [Philadelphia: 1825.] 8 p.
Signed on p. 4: Jan. 13, 1825. Fulton.
Fulton—No. IV. 2d ed. Canals and railroads, p. [5]–8.

4884 (*continued*)
Signed at end: Philadelphia, Feb. 14, 1825. Fulton.
Copies: DeGE; PPL; PU.

4885. ——— 2d ed. Philadelphia, February 14, 1825. [Philadelphia: 1825.] 4 p.
S-19948.
Copies: MCM; MWA; MdHi; NN; PPL.

4886. [Carey, Mathew] 1760–1839.
Exhibit of the shocking oppression and injustice suffered for sixteen months by John Randel, Jun., Esq., contractor for the eastern section of the Chesapeake and Delaware Canal, from Judge Wright, Engineer in Chief, and the majority of the Board of Directors.... By the author of An appeal to the stockholders. Philadelphia: 1825. 23 p.
S-19950.
Copies: DLC; MWA; PPAmP; PPL.

4887. ——— 2d ed., improved. Philadelphia: 1825. 24 p.
S-19951.
Copies: MWA; MdHi; PHi; PPAmP.

4888. ——— 3d ed., improved. Philadelphia: 1825. 24 p.
S-19952.
Copies: DeGE; MWA; PU.

4889. [Carey, Mathew] 1760–1839.
Fulton.... No. IV. Canals and railroads. [Philadelphia: 1825.] 4 p.
Signed and dated: Fulton. Philadelphia, February 14, 1825.
No. IV of his *Canal policy* series.
S-19953.
Copies: MWA; NN.

4890. ——— 2d ed. [Philadelphia: 1825.] [5]–8 p.
S-19954.
Copies: DeGE; MWA; MdHi; NN; PPL; PU.

4891. [Carey, Mathew] 1760–1839.
Fulton, no. VI. Internal improvement. [Harrisburg: 1825.] 6 p.
Signed and dated: Fulton. Harrisburg, March 28, 1825.
Part of his *Canal policy* series.
S-19955.
Copies: DeGE; MCM; MWA; NN; PPL; PU.

4892. [Carey, Mathew] 1760–1839.
A last appeal to the stockholders of the Chesapeake and Delaware Canal, on the injustice exercised towards Mr. Randel, the contractor for the eastern section. [Philadelphia: 1825.] 8 p.
Signed and dated: M.C., October 12, 1825.
Copies: DeGE.

4893. ——— 2d ed. [Philadelphia: 1825.] 8 p.
S-19978, 19979.
Copies: MH; MWA; PHi; PU.

4894. [Carey, Mathew] 1760–1839.
Sir, at a town meeting on Friday last, John Sergeant, Esq.... [Philadelphia: 1825.] 8 p.
Dated: May 9, 1825.
Concerns the Chesapeake and Delaware Canal.
S-19980.
Copies: MWA; PU.

4895. Chesapeake and Delaware Canal Company.
Sixth general report of the president and directors of the Chesapeake and Delaware Canal Company, June 6, 1825. [Philadelphia: 1825.] 12 p., 2 fold. maps.
S-20038.
Copies: DeGE; DeU; MWA; OCHP; PPL.

4896. Chesapeake and Ohio Canal Company.
The acts of the states of Virginia and Maryland, and of the Congress of the United States, incorporating the Chesapeake and Ohio Canal Company. The proceedings of

the general special meeting of the Patowmac Company, declaring their assent thereto.... To which are added, extracts from the charter of the Patowmac Company shewing the property, rights, and privileges, now enjoyed by them ... also, a brief description of the works, executed by the Patowmac Company.... Washington: Printed by Gales & Seaton, 1825. 24 p.

S-20039.

Copies: CSmH; DLC; DeGE; MdBP; ViU.

4897. Colden, Cadwallader David, 1769–1834.

Memoir prepared at the request of a committee of the Common Council of the City of New York, and presented to the mayor of the city, at the celebration of the completion of the New York canals. By Cadwallader D. Colden. [New York]: Printed by order of the Corporation of New York, by W. A. Davis, 1825. v, 408, [2] p., port., maps, plates, facsims.

"Appendix, containing an account of the commemoration of the completion of the Erie Canal, by the Corporation of the City of New York.... New York: 1826," p. [103]–408. The appendix also has an engraved title page: *An Account of the grand canal celebration at New York. November the fourth 1825.*

S-20118.

Copies: CSmH; CtW; DLC; DeGE; ICJ; IU; MH-BA; MWA; MdBE; MiU-C; MnM; NNC; NjP; OC; PHi; PU; RNR; TNJ; Vi.

4898. ———— [New York]: Printed by order of the Corporation of the City of New York by W. A. Davis, 1825. 388 p., illus., maps.

S-20119.

Copies: CSmH.

4899. ———— New York: Printed by order of the Corporation of the City of New York by W. A. Davis, 1825. 130 p., illus., maps.

S-20120.

Copies: CSmH.

4900. Commencement of the Ohio Canal, at Licking Summit, July 4th, 1825. Lancaster, Ohio: Printed by John Herman, [1825]. 22 p.

Addresses by Thomas Ewing and De Witt Clinton.

S-20148.

Copies: DLC; N; NN; OClWHi; OFH.

4901. Considerations on the claims of the "southern tier of counties," addressed to the representatives of an intelligent public. Albany: Printed by J. B. Van Steenbergh, 1825. 31 p.

A plea for canal development in the southern counties of the state of New York.

N-C0648377; S-20177.

Copies: MiU-C; NBuG; NIC; NSy.

4902. Considerations on the proposed canal to connect the rivers Delaware and Raritan. Philadelphia: Printed by Joseph R. A. Skerrett, 1825. 14 p.

S-20178.

Copies: CSmH; DeGE; MWA; MiU; NN; NjP; PPAmP; PPL.

4903. Cuming, Francis Higgins, 1799–1862.

An address delivered at the laying of the cap-stone, of the ten combined locks at Lockport ... June 24th, 1825.... Lockport, [N.Y.]: Printed by Orsamus Turner, 1825. 13 p.

S-20221.

Copies: CtHW; M.

4904. Dismal Swamp Canal Company

An exhibition of the Dismal Swamp Canal, from its commencement; its progress, and its improvements contemplated by the president, directors and company. Submitted to the consideration of the members of the national legislature. Norfolk, [Va.]: T. G. Broughton, printer, 1825. 8 p.

S-20315.

Copies: ICJ; IU; MB; MBAt; MWA; NNC; NcD; OCHP.

4905. Facts and observations in relation to the origin and completion of the Erie Canal. New York: N. B. Holmes, 1825. 36 p., diagrs.

S-20456.

Copies: CtHT-W; DLC; DeGE; MH; MWA; MiU-C; NBu; NHi; NN; NNC; NNMus; NjR; PHi; PPAmP; PPF; ViU.

4906. Hutchinson, Holmes.

Documents relative to the survey and examination of the route of a proposed canal from Ogdensburgh to Lake Champlain. Albany: 1825. 34 p. (New York (State). Legislature. Assembly. Journal, 1825. Appendix D)

Whitford-p. 1347.

Copies: N.

4907. Illinois. Canal Commissioners.

Report of the Canal Commissioners of the state of Illinois, made to the General Assembly on the third of January, 1825. Accompanied by a map of the proposed canal: together with the law, incorporating a company to open a canal to connect the waters of Lake Michigan with those of the Illinois River.... Vandalia, Ill.: Printed by R. Blackwell, 1825. 26 p., fold. map.

The report was also issued separately, 19 p.

S-21000-21001.

Copies: CSmH; DLC; ICHi; ICJ; N; NN; PPAmP.

4908. Louisiana. Legislature.

Memorial and resolution of the Legislature of Louisiana, upon the subject of a canal from Lake Pontchartrain to the river Mississippi. January 21, 1825.... Washington: Printed by Gales & Seaton, 1825. 4 p. (18th Cong., 2d sess. House. Doc. 57)

S-22968.

Copies: DLC; DeU; MWA; NjP.

4909. Louisville and Portland Canal Company.

Charter, by-laws, &c., of the Louisville and Portland Canal Company. Louisville, Ky.: Printed by S. Penn, jr., [1825]. 10 p.

S-21259.

Copies: DLC; ICJ; ICU; KyLo; MHi; MWA.

4910. A map of the United States of America. Profile or vertical section of the Erie Canal; exhibiting its various locks, levels &c. Philadelphia: H. S. Tanner, 1825. Map.

S-21324.

Copies: MH; ViU.

4911. Maryland. Laws.

Letter from the Governor of the state of Maryland, transmitting a copy of an act of the legislature of said state, entitled "An act to confirm an act of the General Assembly of Virginia, entitled 'An act incorporating the Chesapeake and Ohio Canal Company.,'" February 2, 1825.... Washington: Printed by Gales & Seaton, 1825. 16 p. (18th Cong., 2d sess. Senate. Doc. 24)

Also issued as House Doc. 68.

S-22912.

Copies: DLC; DeU; MWA; NN; NjP; ViU.

4912. Maryland. Susquehanna Canal Commissioners.

Report of the Susquehanna Commissioners. Baltimore: Printed by Wm. Wooddy, 1825. 12, [1], 8 p.

S-21345.

Copies: MWA; MdBP; MdHi.

4913. Memorial of the citizens of the county of Oswego, and others, on the subject of the Oswego Canal. Albany: 1825. 5 p. (New York (State). Legislature. Assembly. Doc. 216)

Whitford-p. 1239.

Copies: N.

4914. Memorial of the counties of St. Lawrence, Franklin and Clinton, to the Legislature of the state of New York, for an act to authorize a survey of the route of a canal, from the river St. Lawrence to Lake Champlain. Albany: Printed by Packard & Van Benthuysen, 1825. 45 p.

S-21664.
Copies: DLC; MB; MH-BA; MWA; N.

4915. New Jersey. Laws.
An act to incorporate a company to form an artificial navigation between the Passaic and Delaware rivers. Passed 31st December, 1824. [Trenton: 1825?] 19 p.
Copies: MWA.

4916. New Jersey. Laws.
An act to incorporate the Delaware and Raritan Canal Company. [Trenton?: 1825?] 16 p.
Dated at end: December 30, 1824.
S-20289.
Copies: NN; NjR; PPAmP.

4917. ——— [Trenton?: 1825?] 11 p.
S-20290.
Copies: MWA; NN.

4918. New Jersey. Laws.
An act to incorporate the Morris Canal and Banking Company. [Trenton?: 1825?] 20 p.
Dated: December 31, 1824.
Copies: MWA.

4919. New York (State). Canal Commissioners.
The annual report of the Canal Commissioners of the state of New-York, presented to the Legislature, March 4, 1825. Albany: Printed by Croswell & Van Benthuysen, 1825. 40 p.
S-21650.
Copies: DLC; DeGE; MWA; MWo; N; NHi; NbU; PPL.

4920. New York (State). Canal Commissioners.
Communication on the survey of a route for a canal from the St. Lawrence River to Lake Champlain. Albany: 1825. 36 p. (Assembly. Doc. 183)
Whitford-p. 1347.
Copies: N.

4921. New York (State). Canal Commissioners.
Report concerning the erection of hydraulic works on the Erie and Champlain canals. Albany: 1825. 3 p. (Assembly. Doc. 61)
Whitford-p. 1329.
Copies: N.

4922. New York (State). Commissioners of the Canal Fund.
Annual report of the Commissioners of the Canal Fund. Albany: Printed by Croswell & Van Benthuysen, 1825. 20 p.
S-21651.
Copies: MH-BA; NBu; NHi; NNC; NRom; NbU.

4923. New York (State). Laws.
An act concerning the Erie and Champlain canals. Passed April 20, 1825. [Albany?: 1825.] 1 l.
S-21647.
Copies: NN.

4924. New York (State). Laws.
An act, explanatory of the act incorporating the Delaware and Hudson Canal Company, and of the several acts amendatory thereof, passed April 20. Albany: 1825. 2 p. (Senate. Doc. 215)
Whitford-p. 1254.
Copies: N.

4925. New York (State). Laws.
An act, to incorporate the Albany Basin Company. Albany: 1825. 5 p. (Assembly. Doc. 118)
Whitford-p. 1261.
Copies: N.

4926. New York (State). Laws.
Laws of the state of New York, in relation to the Erie and Champlain canals, together with the annual reports of the Canal Commissioners, and other documents, requisite for a complete official history of these works. Also, correct maps delineating the routes of the Erie and Champlain canals, and

4926 (*continued*)
designating the lands through which they pass. . . . Albany: Pub. by the authority of the state; E. & E. Hosford, printers, 1825. 2 v., plates, fold. maps.

S-21660.

Copies: CSmH; CU; Ct; DLC; ICJ; ICN; MH; MWA; MiD-B; MiU-T; MnHi; NN; NNC; NjP; PPL; RPB; Vi; WHi.

4927. Ohio. Board of Canal Commissioners.

Report of the Canal Commissioners, to the General Assembly of Ohio. . . . Columbus: Printed by P. H. Olmsted, 1825. 66 p.

Appendix: Report of David S. Bates, principal engineer.

Concerns the project to connect Lake Erie with the Ohio River by a canal.

S-21731.

Copies: CSmH; DLC; DeGE; ICJ; MWA; MiU-T; OCHP; OClWHi; PPL.

4928. Ohio. Board of Canal Commissioners.

Report of the Canal Commissioners to the General Assembly. Read and ordered to be printed, December 12, 1825. [Columbus]: George Nashee, state printer, 1825. 54 p.

S-21732.

Copies: CSmH; DLC; IU; MB; MWA; OClWHi.

4929. Ohio. Commissioners of the Canal Fund.

Report of the Commissioners of the Canal Fund. To the General Assembly of the State of Ohio. [Columbus: 1825.] 4 p.

Dated: December 9, 1825.

S-21733.

4930. Ohio. General Assembly.

Memorial of the legislature of the state of Ohio, in relation to the connection of Lake Erie, with the Ohio River, by means of canal navigation. February 22, 1825 . . . . Washington: Printed by Gales & Seaton, 1825. 6 p. (18th Cong., 2d sess. Senate. Doc. 36)

S-22977.

Copies: DLC; DeGE; DeU; MWA; MiU; NN; NjP.

4931. Ohio. General Assembly. House. Committee on Canals.

Report. The standing Committee on Canals, to whom was referred the memorial of sundry persons, styling themselves citizens of the Miami County and state of Ohio, charging Micajab T. Williams, one of the acting canal commissioners, with misconduct in the charge of his duties, and praying that he may be removed from office, &c. ask leave to report. [Columbus: 1825.] 3 p.

S-21736.

Copies: DLC.

4932. Ohio. Laws.

An act establishing an equitable mode of levying the taxes of this state; and An act to provide for the internal improvement of the state of Ohio, by navigable canals. Columbus: Printed by P. H. Olmsted, 1825. 27 p.

S-21721.

Copies: DLC; MWA; N; OCHP; OClWHi.

4933. Pennsylvania. Board of Canal Commissioners.

Report and correspondence of the Commissioners for promoting the internal improvements of the state. Harrisburg: John S. Wiestling, 1825. 71 p.

Dated: February 2, 1825.

S-21820.

Copies: CtY; NN; P; PPAmP; PPL.

4934. ——— Harrisburg: John S. Wiestling, 1825. 46 p.

Correspondence is omitted from this edition.

Copies: PHi.

4935. Pennsylvania. Board of Canal Commissioners.

Report of the Canal Commissionnrs [*sic*]. Report of the commissioners for promoting the internal improvement of the state.... [Harrisburg: 1825.] 46 p.

Signed: Jacob Holgate: James Clarke, Pennsylvania Canal Commissioners.

S-21821.

Copies: DBRE; DLC; DeGE; MiU; NN; P; PPAmP.

4936. Pennsylvania. Board of Canal Commissioners.

Report of the Canal Commissioners to the General Assembly . . . December 12, 1825. [Harrisburg:] George Hashee, state printer, 1825. 54 p.

With accompanying documents A–H.

S-21822.

Copies: P; PPAmP.

4937. Pennsylvania. General Assembly. House. Committee on Inland Navigation.

Report of the Committee on Inland Navigation, on the Delaware and Raritan Canal, read February 21, 1825. Harrisburg: Printed by John S. Wiestling, 1825. 8 p.

S-21828.

Copies: MiU; PPAmP.

4938. Pennsylvania. General Assembly. House. Committee on Inland Navigation.

Report of the Committee on Inland Navigation, relative to a water communication between the eastern and western waters. Read in the House of Representatives, Feb. 28, 1825. Mr. Lehman, chairman. Harrisburg: Printed by John S. Wiestling, 1825. 7 p.

S-21830.

Copies: DBRE; DLC; DeGE; NN; PHi; PPAmP.

4939. Pennsylvania. Laws.

An act making further provisions for the survey of routes for canals between the Delaware and Ohio. Harrisburg: Printed by Mowry and Cameron, 1825. 3 p.

S-21804.

Copies: CSmH.

4940. Pennsylvania. Laws.

An act relative to the Delaware and Raritan Canal, to be constructed in the state of New-Jersey. [Harrisburg]: Mowry & Cameron, printers, [1825]. 3 p.

At head of title: No. 395. Mr. Lehman, chairman. House of Representatives. Read February 21, 1825.

S-21805.

Copies: CSmH (2 variants, one with 9 sections, the other with 7 sections); PPL.

4941. Pennsylvania Canal Convention, Harrisburg, 1825.

Proceedings of the Pennsylvania Canal Convention, which met at the borough of Harrisburg, August 4, 1825. Harrisburg: Printed by Mowry & Cameron, 1825. 19 p.

Signed: Joseph Lawrence, President.

S-21842.

Copies: DeGE; P; PHi; PPAmP.

4942. Pennsylvania Canal Convention, Harrisburg, 1825.

To the citizens of the Commonwealth of Pennsylvania. [Harrisburg?: 1825.] 24 p.

Address on canals and other internal improvements, submitted by a committee of the Pennsylvania Canal Convention. Signed and dated: Mathew Carey, John Kennedy and S. D. Ingham, August 22, 1825.

S-21843, 21847, 21859.

Copies: CtY; DeGE; IU; MH-BA; MWA; NN; P; PHi; PPAmP; PPL; PU.

4943. Pennsylvania Society for the Promotion of Internal Improvements in the Commonwealth.

The subscribers, the Acting Committee of "The Pennsylvania Society for the Promotion of Internal Improvements in the Commonwealth," respectfully submit the following address on the subject of a canal to connect the waters of the Susquehannah with those of the Alleghany.... Philadelphia: 1825. 7 p.

Dated and signed Philadelphia, Jan. 13, 1825. Mathew Carey, Joseph Hemphill, Richard Peters, Jun. [et al.].

4943 (*continued*)
S-21855.
Copies: DeGE; PPL.

4944. ———— 2d ed. April 8, 1825. [Philadelphia: 1825]. 8 p.
S-21856.
Copies: PPL.

4945. [Pennsylvania Society for the Promotion of Internal Improvements in the Commonwealth.]
To the citizens of the Commonwealth of Pennsylvania. [Philadelphia: 1825.] 27 p.
Signed: Wm. Tilghman.
Concerns the canals in Pennsylvania.
Copies: PHi.

4946. Pennsylvania Society for the Promotion of Internal Improvements in the Commonwealth.
To the Senate and House of Representatives of the Commonwealth of Pennsylvania, the memorial of the subscribers ... respectfully represents that your memorialists ... are deeply impressed with the importance and usefulness of a canal to connect the waters of the Alleghany with those of the Susquehannah.... [Philadelphia: 1825.] 2 l.
Signed and dated: Mathew Carey, Joseph Hemphill, Richard Peters, Jun. [et al.], Acting Committee of the Pennsylvania Society for the Promotion of Internal Improvements, Philadelphia, Jan. 13, 1825.
Copies: DeGE.

4947. The President, Directors, and Company of the Farmington Canal.
Act of incorporation of the Farmington Canal Company; also the Act of incorporation of the Mechanics' Bank of New-Haven; together with sundry other documents and papers relative to the canal and bank. New-Haven: Sidney's Press, 1825. 22 p.
S-20165.
Copies: CSmH; CtHi; CtY; DLC; NN.

4948. Remarks upon the advantages of the Southwark Canal. [Philadelphia: ca. 1825.] [2] p.
Copies: DeGE.

4949. Reply to Mr. Carey's appeal. [Philadelphia: 1825.] 8 p.
Dated and signed at end: Philadelphia, October 13, 1825. A stockholder.
Concerns the Chesapeake and Delaware Canal.
S-22057.
Copies: PPL.

4950. South Carolina. Superintendent of Public Works.
Report of the Superintendent of Public Works on the Edisto Canal. [Columbia, S.C.: 1825.] 18 p.
Signed: A. Blanding, Superintendent Public Works, October 22, 1825.
S-22329.
Copies: DLC; ICN; NHi.

4951. Stevenson, John Andrew, 1761–1833.
The meeting of the waters of Hudson & Erie. New York: E. Riley, [1825]. 3 p.
A song, dedicated to the opening of the Erie Canal.
S-22377.
Copies: DLC; MH; MWA; MiU-C; NHi; NN.

4952. Stone, William Leete, 1792–1844.
Narrative of the festivities observed in honor of the completion of the grand Erie Canal. Prepared at the request of the committee of the corporation of the city of New York. New York: 1825. 289–408, 2 p., plates (part fold.), plans, facsims.
Usually found with C. D. Colden's *Memoir ... at the celebration of the New York canals....*
S-22383.
Copies: Ct; DLC; DeGE; M; MH-BA; MWA; MdBP; MiU-C; NBu; NP; NP; Nj; OC.

4953. Town meeting. Philadelphia, January 24, 1825. At a large and respectable meeting of the citizens of the city and county of Philadelphia, convened to consider the propriety of memorializing the Legislature to adopt the initiatory measures for opening a communication between the Susquehannah River and the western waters. [Philadelphia: 1825.] 7, 2, 6, 6, 7, 4, 3, 7, 8 p.

S-22491.

Copies: MWA.

4954. Union Canal Company of Pennsylvania.

Acts of the Legislature of Pennsylvania, relating to the Union Canal Company of Pennsylvania. Philadelphia: Printed for the Union Canal Company [by] L. R. Bailey, 1825. 72, 44 p.

Appendix "Bylaws, rules, orders and regulations," p. [61]–65.

List of stockholders at end.

S-22538.

Copies: DLC; DeGE; MWA; NNC; NNC-L; PPAmP; PPF; PPL.

4955. Union Canal Company of Pennsylvania.

Annual report of the president and managers of the Union Canal Company of Pennsylvania to the stockholders. November 15, 1825. Philadelphia: Printed by Lydia R. Bailey, 1825. 22 p.

S-22539.

Copies: DeGE; MWA; MeHi; P; PHi; PPL.

4956. United States. Comptroller of the Treasury.

Letter from the Comptroller of the Treasury to the chairman of the Committee on Commerce, upon the subject of licensing, enrolling, and exacting a tonnage duty on canal boats or vessels. Submitted by the chairman of the Committee on Commerce, with a bill concerning canal vessels or boats. January 7, 1825.... Washington: Printed by Gales & Seaton, 1825. 5 p. (18th Cong., 2d sess. House. Doc. 35)

Jos. Anderson, Comptroller.

S-22907.

Copies: DLC; DeGE; DeU; MWA; NjP.

4957. United States. Congress. House.

Canal in Florida. Mr. Webster's resolution. December 19, 1825. [Washington: 1825.] 1 p. (19th Cong., 1st sess. House. Res. 6)

S-22852.

Copies: DLC; DeGE; DeU; MWA; NjP.

4958. United States. Congress. House. Committee on Roads and Canals.

Return of the state of the Dismal Swamp Canal Company. November 30th, 1824. February 16, 1825. Referred to the Committee of the Whole House, to which is committed a bill authorizing a subscription of stock in the Dismal Swamp Canal Company. Washington: Printed by Gales & Seaton, 1825. 3 p. (18th Cong., 2d sess. House. Doc. 86)

S-23123.

Copies: DLC; DeU; MWA; NjP.

4959. United States. Congress. House. Select Committee on a Canal Between the Illinois River and Lake Michigan.

Report of the Select Committee to which was referred, on the 3d ultimo, a memorial of the General Assembly of Illinois, upon the subject of a canal communication between the Illinois River and Lake Michigan, accompanied with a bill to aid the state of Illinois in the accomplishment of the same. February 1, 1825.... [Washington: 1825.] 6 p. (18th Cong., 2d sess. House. Rept. 53)

S-23105.

Copies: DLC; DeU; MWA; NjP.

4960. United States. Congress. House. Select Committee on a Canal to Connect the Wabash River with Lake Erie.

Report of the Select Committee, to whom was referred the resolution of the 12th of January, on the subject of aiding the state of Indiana in opening a canal to connect the waters of the Wabash River, with those of Lake Erie, accompanied by a bill to accomplish that object. February 21, 1825.... [Washington: 1825.] 4 p. (18th Cong., 2d sess. House. Rept. 78)

S-23110.

Copies: DLC; DeU; MWA; NjP.

4961. United States. Congress. Senate. Select Committee on Roads and Canals.

In Senate of the United States, February 4, 1825. Mr. Brown, from the Select Committee on Roads and Canals, to whom was referred the memorial of the Legislature of Louisiana, requesting an appropriation of money for the purpose of cutting a navigable canal from Lake Pontchartrain to the river Mississippi, reported.... [Washington: 1825.] 3 p. (18th Cong., 2d sess. Senate. Doc. 26)

S-22871.

Copies: DLC; DeU; MWA; NjP.

4962. United States. Engineer Dept.

Dismal Swamp Canal. Letter from the Secretary of War, containing the information called for by the House of Representatives, in relation to the Dismal Swamp Canal. December 27, 1825.... Washington: Printed by Gales & Seaton, 1825. 8 p. (19th Cong., 1st sess. House. Doc. 15)

Transmits a report of the Engineer Dept.

S-22858.

Copies: DLC; DeGE; DeU; MWA; NjP.

4963. United States. Laws.

A bill concerning canal vessels or boats. January 7, 1825.... [Washington: 1825.] 5 p. (H.R. 256)

S-22687.

Copies: DNA.

4964. A view of the grand canal, from Lake Erie to the Hudson River; containing a particular description of its construction, and historical summary of the different towns, villages, &c. on the route of the canal. Also, a description of the Champlain Canal. New York: Printed for John Low, 1825. 28 p., illus.

Sometimes attributed to John Low.

S-23162.

Copies: MWA; N; NN; WHi.

4965. Washington Canal and Delaware State lottery ... draws in Washington City, Oct, 5, 1825.... Tickets ... for sale by S. Brewster, Jr.... Grand State lottery, which draws in Providence 14th October.... Norwich, [Conn.]: 1825. Broadside 28 x 19 cm.

Dated: Norwich, Oct. 1st, 1825.

Copies: DeGE.

4966. Washington Canal Company.

Annual report of the president and directors of the Washington Canal Company. February 24, 1825.... Washington: Printed by Gales & Seaton, 1825. 3 p. (18th Cong., 2d sess. House. Doc. 102)

S-22671.

Copies: DLC; DeU; MWA; NjR.

4967. Williams, Richard L.

The Grand Canal celebration ball. New York: E. Riley, 1825. 3 p.

Instrumental music without lyrics.

S-23332.

Copies: CSmH.

4968. Winnipiseogee Canal Committee.

Report of the committee appointed to procure a survey of the canal route from the tide waters of Piscataqua River by Alton Bay and the Winnipisiogee and Squam lakes to Pemigewasset River near Plymouth.... [Dover?, N.H.]: J. Mann, printer, [1825.] 13 p.

A report of James Baldwin, p. 5–13, is dated Boston, Sept. 12, 1825.

S-23355.

Copies: CtY; MBAt; MH-BA; MWA.

4969. Woodworth, Samuel, 1785–1842.

Ode for the canal celebration, written at the request of the printers of New York.... [New York]: Clayton, Van Norden, printers, [1825.] Broadside 26 x 27 cm.

"The foregoing ode was printed on a moveable stage, on the 4th day of November, 1825, during the procession in honour of the completion of the Grand Western Canal."

S-23377.

Copies: DLC; MWA; NHi.

4970. Worcester Co., Mass. Committee on Canal Survey.

Account of the proposed canal from Worcester to Providence, containing the report of the engineer: together with some remarks upon inland navigation. Published by order of the Committee for the county of Worcester. Providence: Printed by John Miller, 1825. 16 p.

Benjamin Wright, engineer.

S-19361.

Copies: DLC; DeGE; MWA; NNE; RP.

4971. [Baldwin, Loammi] 1780–1838.

Letters on the Union Canal of Pennsylvania. First published in the "Boston Daily Advertiser." [Boston: 1826.] iv, [5]–44 p.

S-4523 (erroneously dated 1821), 23582.

Copies: DeGE; MBAt; MH-BA; Nh; PHi; PPAmP.

4972. Binney, Horace, 1780–1875.

Opinions of counsel. [Jersey City: 1826.] 12 p.

Opinions on the charter of the Delaware and Raritan Canal Company; signed Horace Binney and Daniel Webster.

N-B0493484.

Copies: MiU; NN.

4973. [Carey, Mathew] 1760–1839.

Sir, although it is, in general, improper for a man to write or speak of himself.... [Philadelphia: 1826.] [4] p.

Signed: Mathew Carey, Philadelphia, February 15, 1826.

Concerns the Chesapeake and Delaware Canal.

Copies: MWA.

4974. Chesapeake and Delaware Canal Company.

Seventh general report of the president and directors of the Chesapeake and Delaware Canal Company. June 5, 1826. [Philadelphia: 1826.] 18 p., fold. map.

S-24078.

Copies: DeU; MWA; NN; PPL; PU.

4975. Chesapeake and Ohio Canal Convention, Washington, 1823, 1826.

Chesapeake and Ohio Canal. Memorial of the Central Committee of the Chesapeake and Ohio Canal Convention, and of the commissioners appointed by the states of Virgina and Maryland, and by the United States, to open books for subscription of stock to the said canal. April 3, 1826.... Washington: Printed by Gales & Seaton, 1826. 40 p. (19th Cong., 1st sess. House. Doc. 151)

S-26906.

Copies: DBRE; DLC; DeGE; DeU; MWA; NN; NjP.

4976. Chesapeake and Ohio Canal Convention, Washington, 1823, 1826.

Report of the central committee of the Chesapeake and Ohio Canal Convention, which met in the city of Washington, on the 6th December, 1826: acting under a resolution of the convention of delegates assembled in the capitol of the United States, on the 16th day of November, 1823. Washington: Printed by Gales & Seaton, 1826. 21 p.

S-24079.

Copies: DLC; DeGE; MiU-T; NN; OCl.

4977. The claims of the Delaware and Raritan Canal Company, to a repeal of the law of Pennsylvania, passed April 6th, 1825, entitled "An act relative to the Delaware and Raritan Canal, to be constructed in New Jersey," and to the assent of Pennsylvania ... to the use of the water of the Delaware for the supply of a canal.... Philadelphia: Printed by J. R. A. Skerrett, 1826. 27 p.

Signed: A citizen of Philadelphia.

S-24112.

Copies: CtY; DeGE; MB; MH-BA; MiU; NN; NjP; PPAmP; PPL.

4978. Delaware and Hudson Canal Company.

Annual report of the board of managers of the Delaware and Hudson Canal Co. to the stockholders for the year 1825. March 7th, 1826. New York: Nathan Lane, printer, 1826. 15 p.

S-24326.

4978 (*continued*)
Copies: DLC; ICJ; ICU; MA; MB; MCM; MH-BA; N; NHi; NN; NNE; NjP.

4979. Delaware and Raritan Canal Company.
Memorial of the Committee on behalf of the Delaware and Raritan Canal Company. Harrisburg: Printed by Cameron & Krause, 1826. 43 p., fold. map.
Signed by Garret D. Wall and James Neilson, agents and managers of the Delaware and Raritan Canal; addressed to the Senate and House of Representatives of Pennsylvania.
N-D0140268.
Copies: DI; DeGE; MWA (map wanting); PPAmP.

4980. ———— together with reports of the Canal Commissioners. Harrisburg: Printed by Cameron & Krause, 1826. 43, 40, 189 p., fold. map.
S-24328.
Copies: DBRE; DLC; MWA; MiU-T; NN; PHi;PPAmP.

4981. Delaware and Raritan Canal Company.
Report [of] the committee appointed by a resolution of the board of managers ... to take into consideration "the character of the company, the law of the legislature of Pennsylvania, and the report of the committee of the board of managers appointed ... to go to Harrisburg, and to report what measures ... it might be expedient to adopt." [Jersey City: 1826.] 7 p.
Signed: James Kent, Elisha Tibbets, L. Baker.
N-D0140281.
Copies: MiU; NN.

4982. Delaware and Rariton [*sic*] Canal. [Trenton?: 1826.] 12 p., fold. map.
S-24329.
Copies: DLC; DeGE; MWA; MiU-T; NN; PPAmP; PPL.

4983. Documents presented to the Legislature of the state of New-York at its last and present sessions, setting forth the expediency and public importance of incorporating the Long-Island Canal Company for the purpose expressed by the petitioners. New York: Printed by J. C. Spear, 1826. ix, 11–27 p., fold. map.
Dated: January, 1826.
S-24358, 25135.
Copies: CtSoP; DLC; MWA; MiU-C; N; NBLiHi; NIC; NJQ; NN.

4984. Florida (Terr.). Commissioners on a Canal Through the Peninsula of Florida.
Memorial of James Gadsden and E. R. Gibson, Commissioners appointed by the Legislature of Florida, to examine into the expediency of opening a canal through the peninsula of Florida. January 5, 1826.... Washington: Printed by Gales & Seaton, 1826. 4 p. (19th Cong., 1st sess. Senate. Doc. 15)
S-27201.
Copies: DLC; DeU; MWA; NjP.

4985. Hampshire and Hampden Canal Company.
Report of Jarvis Hurd, Esq., civil engineer, employed by the executive committee of the Hampshire and Hampden Canal Company, with an estimate of expense to complete the canal, from the termination of the Farmington Canal, on the line of the state at Southwick to the great bend of Connecticut River at Northampton. Northampton, [Mass.]: Printed by H. Ferry, 1826. 12, [9]-36 p.
S-24787.
Copies: DLC; DeGE; MAJ; MB; MDeeP; MWA; NN.

4986. [Holley, Myron] 1779–1841.
Memorial. To the Hon. the Legislature of the state of New-York, in Senate and Assembly convened.... [S.l.: 1826.] 37 p.
Signed on p. 21: Myron Holley, late canal commissioner.
Concerns his activities in the management of the Erie Canal.

S-24880.
Copies: DLC; M; N; NHi; PBL.

4987. Illinois. General Assembly.
Memorial of the General Assembly of the state of Illinois, on the construction of a canal, to unite the waters of Lake Michigan with the Illinois River. February 26, 1826.... Washington: Printed by Gales & Seaton, 1826. 4 p. (19th Cong., 1st sess. Senate. Doc. 49)
S-27207.
Copies: DLC; DeU; MWA; NjP.

4988. ———— February 30, 1826.... Washington: Printed by Gales & Seaton, 1826. 5 p. (19th Cong., 1st sess. House. Doc. 147)
Copies: DeU; MWA.

4989. Internal Improvement Convention, Baltimore, 1825.
Memorial of the Committee of the Convention on Internal Improvement, held in Baltimore on the 14th December, 1825. Addressed to the legislature of Maryland. Annapolis: J. Hughes, printer, 1826. 14 p.
Concerns the Chesapeake and Ohio Canal.
N-I0106001.
Copies: CtY; MWA.

4990. Jones, William, 1760–1831.
A report of a survey of the Genesee Valley Canal route. [Genesee, N.Y.: A. Harvey], 1826. 14 p.
S-25004.
Copies: NN.

4991. Maine. Laws.
An act to establish the Kennebec and Androscoggin Canal Association. [S.l.: 1826.] 15 p.
S-25191.
Copies: MeHi.

4992. Maryland. Governor (J. Kent).
Communication from the executive, transmitting information from the president and directors of the Chesapeake & Delaware Canal Company, relative to the situation, progress & prospects of said company.... Annapolis: Printed by J. Hughes, 1826. 6 p.
Letter: "To the Governor of the state of Maryland," p. 5–6, is signed by James C. Fisher, president, and H. D. Gilpin, secretary of the Chesapeake and Delaware Canal Company, and dated Philadelphia, 16th November, 1825.
Copies: DeGE.

4993. Maryland. Governor (J. Kent).
Communication from the Governor to the Legislature of Maryland, enclosing the law of the state of Pennsylvania, respecting the Chesapeake and Ohio Canal Company. February 16, 1826.... Annapolis: Printed by Jeremiah Hughes, 1826. 8 p.
N-M0289298; S-25235.
Copies: Md; NN.

4994. Maryland. Governor (S. Stevens).
Documents accompanying the message of the executive of Maryland, transmitting to the Senate and House of Delegates on the thirtieth December, eighteen hundred and twenty five, and referred to therein. Annapolis: Printed by J. Hughes, 1826. 18 p.
Concerns the Chesapeake and Ohio Canal Company.
N-M0289291.
Copies: ViU.

4995. Maryland. Governor (S. Stevens).
Message of the executive of Maryland, transmitted to the Senate and House of Delegates, Dec. 30th, 1825. Annapolis: Printed by J. Hughes, 1826. 11 p.
Concerns the Chesapeake and Ohio Canal.
N-M0289296.
Copies: ViU.

4996. Massachusetts. Board on Canal Route from Boston to the Connecticut and Hudson Rivers.
Report of the commissioners of the state of Massachusetts, on the routes of canals from Boston harbour to Connecticut and Hudson

4996 (*continued*)
rivers. Boston: True and Greene, state printers, 1826. 185, [1], 62 p., fold. map.

With the report of the engineer, Loammi Baldwin.

S-25269.

Copies: CSmH; CtY; DLC; DeGE; ICJ; ICU; MB; MH; MWA; MeU; NN; NNC; P; PPAmP; PPL; RPB; WHi.

4997. Memorial of the inhabitants of the counties of Montgomery and Hamilton, to the Legislature of the state of New York, relative to a canal from the Sacondaga River through Johnstown to the Erie Canal, accompanied by the report of the engineer on that subject. Johnstown, [N.Y.]: 1826. 6 p.

S-25410.

Copies: N.

4998. Morris Canal and Banking Company.

Annual report upon the Morris Canal, for the year ending July 1, 1826. Morristown, [N.J.]: 1826. 20 p.

Copies: DLC.

4999. ________ New York: Printed by W. Davis, 1826. 20 p.

N-M0795893.

Copies: DLC; NN.

5000. Morris Canal and Banking Company.

By-laws of the Morris Canal and Banking Company, passed at a meeting of the president and directors, convened for the purpose, at their banking house at the city of Jersey, the fifth day of December, 1826. [S.l.: 1826.] 13 p.

N-M0795894.

Copies: NjR.

5001. New Jersey. Legislature. Joint Committee on the Delaware and Raritan Canal.

Report of the Joint Committee of the Council and Assembly of the state of New Jersey, on the subject of the Delaware and Raritan Canal. November 29, 1826. Trenton: Printed by William Prall, 1826. 8 p.

S-25525.

Copies: DBRE; MiU; NN; NjP.

5002. New York (City). Common Council.

Appendix, containing an account of the commemoration of the completion of the Erie Canal, by the Corporation of the City of New York.... By order of the Common Council. New York: 1826. 4 p.l., [105]–288 p., maps, plates.

Added title page: *An Account of the grand canal celebration at New York, November the fourth, 1825.*

Issued with: Colden, C. D., *Memoir, prepared at the request ... of the Common Council,* 1825.

S-25534.

Copies: CtY; DLC; DeGE; M; MB; MWA; MiU-C; NBu; NN; NNCoCi; Nj; ViU.

5003. New York (State). Canal Commissioners.

The annual report of the Canal Commissioners of the state of New-York, presented to the Legislature, March 25, 1826. Albany: Printed by Croswell, Barnum & Van Benthuysen, 1826. 19 p.

S-25543.

Copies: MWA; N; NHi; NNC; NUtHi.

5004. New York (State). Canal Commissioners.

Report of the Canal Commissioners, made pursuant to the act, entitled "An act to provide for the survey of certain canal routes therein mentioned." Made to the Assembly, March 6, 1826. Albany: Printed by Croswell, Barnum & Van Benthuysen, 1826. 68 p.

S-25557.

Copies: MB; MH-BA; NHi; NNE; NRom; PBL.

5005. New York (State). Laws.

An act authorizing the construction of a canal from the head of Seneca Lake to Newtown. Albany: 1826. 2 p. (Assembly. Doc. 242)

Whitford-p. 1244.
Copies: N.

5006. New York (State). Laws.
An act authorizing the construction of a canal through the valley of the Chenango, to connect the waters of the Susquehanna with the Erie Canal. Albany: 1826. 2 p. (Assembly. Doc. 243)
Whitford-p. 1248.
Copies: N.

5007. New York (State). Laws.
An act to incorporate the Hudson Ship Canal Company. Albany: 1826. 9 p. (Senate Doc.)
Whitford-p. 1270.
Copies: N.

5008. New York (State). Legislature. Assembly. Committee on Canals.
Report of the Committee on Canals, on the subject of the sale of surplus waters. Together with the minutes of testimony taken by the committee. Albany: Printed by Croswell, Barnum & Van Benthuysen, 1826. 88 p.
S-25559.
Copies: N; NUtHi.

5009. New York (State). Legislature. Senate. Committee on Canals.
Report on Canal Commissioners' report. Albany: 1826. 5 p. (Senate. Doc. 265)
Whitford-p. 1194.
Copies: N.

5010. Ohio. Board of Canal Commissioners.
Report of the Canal Commissioners, December 27, 1826. [Columbus: 1826.] 20 p.
Bowker-p. 205.
Copies: MWA; OClWHi.

5011. Ohio. Commissioners of the Canal Fund.
Report of the Commissioners of the Canal Fund. To the General Assembly of the State of Ohio. [Columbus: 1826.] 6 p.
Dated: December 11, 1826.
Bowker-p. 205.
Copies: OClWHi.

5012. Ohio. General Assembly. House. Committee on Finance.
Report of the Committee of Finance, on the canal fund. [Columbus: 1826?] 4 p.
Report covers the year 1826.
S-25627.
Copies: DLC; PPAmP.

5013. Pennsylvania. Board of Canal Commissioners.
First report of Canal Commissioners. Read in the Senate, January 5, 1826. [Harrisburg: 1826.] 4 p.
Thomson-116.
Copies: ICJ; NN.

5014. Pennsylvania. Board of Canal Commissioners.
Report of the Canal Commissioners of Pennsylvania. Read in Senate, February 7, 1826. Harrisburg: Printed by Cameron & Krause, 1826. 18 p.
S-25695.
Copies: DBRE; DeGE; MWA; N; NN; P; PHi; PPAmP; PPL.

5015. Pennsylvania. Board of Canal Commissioners.
Report of the Commissioners, for promoting the internal improvement of the state. No. II. Harrisburg: Printed by J. S. Wiestling, 1826. 19 p.
Dated: February 3, 1826.
S-25696.
Copies: DBRE; DeGE; PHi.

5016. Pennsylvania. General Assembly. House. Committee on Inland Navigation.
Report of the Committee on Inland Navigation and Internal Improvement, relative to the Pennsylvania Canal. Read in the House of Representatives, January 16, 1826. Mr. Lehman, chairman. [Harrisburg: J. S. Wiestling, printer, 1826.] 7 p.
S-25698.
Copies: MWA; N.

5017. Pennsylvania. Laws.

An act to provide for the commencement of a canal, to be constructed at the expense of the state, and to be styled the Pennsylvania Canal. Harrisburg: 1826. 6 p.

Dated: February 5, 1826.

Copies: P.

5018. Pennsylvania. Laws.

Additional acts of the legislature of Pennsylvania, relating to the Union Canal Company of Pennsylvania. [Philadelphia: 1826.] 12 p.

Sabin-60750 (note).

Copies: PPL.

5019. Petition of the manufacturers of salt at Montezuma, for the construction of a canal from Erie Canal to the said salt works. Albany: 1826. 2 p. (New York (State). Legislature. Senate. Doc. 42)

Whitford-p. 1231.

Copies: MWA; N.

5020. Proceedings, in relation to the western termination of the Pennsylvania Canal. Pittsburgh: 1826. 8 p.

S-25788.

Copies: PPi (not located).

5021. Proprietors of the Locks and Canals on Merrimack River.

Proposals by the Proprietors of the Locks and Canals on Merrimack River, for the sale of their mill power and land at Lowell, in the county of Middlesex, in Massachusetts. [Boston: 1826?] [4] p.

Includes data on the height of falls.

S-25842.

Copies: CSmH; DLC.

5022. ———— Boston: Printed by Wells and Lilly, 1826. 18 p.

S-25843.

Copies: DLC; DeGE; ICN; MCM; MH-BA; MHi; MWA; MiD-B.

5023. Remarks on the recent letters of Mr. Sergeant and Mr. De Witt Clinton, on the subject of the Pennsylvania Canal. [Philadelphia: 1826?] Broadside 67 x 37 cm.

S-25919.

Copies: MWA.

5024. Remarks relative to a proposed union of the Erie Canal with Lake Ontario at Great Sodus Bay. By a border settler. Geneva, [N.Y.]: James Bogart, 1826. 23 p., chart.

Arguments in favor of making a cut from the Erie Canal to Great Sodus Bay, including the engineer's report.

S-25920.

Copies: CSmH.

5025. The Union Canal. [Philadelphia: Pennsylvania Society for the Promotion of Internal Improvements in the Commonwealth, 1826.] 7 p.

Signed at end: P.

At head of title: Philadelphia, August 26, 1826. At a meeting of "The Acting Committee of the Pennsylvania Society for the Promotion of the Internal Improvement of the State," the following paper was read, discussed, agreed to, and directed to be printed. Signed by order, Mathew Carey, Vice President.

S-25712.

Copies: DeGE; PU.

5026. Union Canal Company of Pennsylvania.

Annual report of the president and managers of the Union Canal Company of Pennsylvania, to the stockholders. November 21, 1826. Philadelphia: Printed by Lydia R. Bailey, 1826. 12 p.

S-26261.

Copies: DeGE; KyLxT; MH; MWA; MeHi; N; PHi; PPL; WU.

5027. United States. Board of Internal Improvement.

Message from the President of the United States, transmitting a report from the Secretary of War with that of the Board of Engineers for Internal Improvement, concerning the proposed Chesapeake and Ohio Canal. December 7, 1826.... Washington:

Printed by Gales & Seaton, 1826. 81 p., 20 fold. tables. (19th Cong., 2d sess. House. Doc. 10)
S-27218.
Copies: DLC; DeGE; DeU; MWA; NN; NNE; NjP; PPAN.

5028. United States. Congress. House. Committee on Roads and Canals.
Canal around the Muscle Shoals. May 15, 1826. Mr. Stewart, from the Committee on Roads and Canals, to which the subject had been referred, made the following report. [Washington: 1826.] 2 p. (19th Cong., 1st sess. House. Rept. 215)
S-26891a.
Copies: DLC; DeU; MWA; NjP.

5029. United States. Congress. House. Committee on Roads and Canals.
Canal—Boston harbor to Narraganset Bay. May 15, 1826 ... Mr. Stewart, from the Committee on Roads and Canals ... made the following report. [Washington: 1826.] 9 p. (19th Cong., 1st sess. House. Rept. 214)
S-26892.
Copies: DLC; DeGE; DeU; MWA; NjP.

5030. United States. Congress. House. Committee on Roads and Canals.
Canal—Lake Michigan to Illinois River. March 30, 1826. Mr. Stewart, from the Committee on Roads and Canals ... made the following report. [Washington: 1826.] 5 p. (19th Cong., 1st sess. House. Rept. 147)
S-26894.
Copies: DLC; DeU; MWA; NjP.

5031. United States. Congress. House. Committee on Roads and Canals.
Canal—Tennessee and Coosa Rivers. May 16, 1826 ... Mr. Stewart, from the Committee on Roads and Canals ... made the following report. [Washington: 1826.] 2 p. (19th Cong., 1st sess. House. Rept. 220)
S-26896.
Copies: DLC; DeU; MWA; NjP.

5032. United States. Congress. House. Committee on Roads and Canals.
Canal to connect lakes Champlain and Ontario. May 22, 1826 ... Mr Stewart, from the Committee on Roads and Canals ... made the following report. [Washington: 1826.] 3 p. (19th Cong., 1st sess. House. Rept. 230)
S-26897.
Copies: DLC; DeU; MWA; NjP.

5033. United States. Congress. House. Committee on Roads and Canals.
Chesapeake and Ohio Canal. May 22, 1826. Mr. Stewart, from the Committee on Roads and Canals ... made the following report. [Washington: 1826.] 122 p., 1 fold. plan. (19th Cong., 1st sess. House. Rept. 228)
S-26905.
Copies: DLC; DeGE; DeU; MWA; NjP.

5034. United States. Congress. House. Committee on Roads and Canals.
Chesapeake and Ohio Canal—extension of. May 16, 1826 ... Mr. Stewart, from the Committee on Roads and Canals ... made the following report. [Washington: Printed by Gales & Seaton, 1826.] 3 p. (19th Cong., 1st sess. House. Rept. 216)
S-26907.
Copies: DLC; DeGE, DeU; MWA; NjP.

5035. United States. Congress. House. Committee on Roads and Canals.
Survey—canal route—Olean Point to Erie Canal. April 4, 1826.... [Washington: 1826.] 4 p. (19th Cong., 1st sess. House. Rept. 151)
S-27410.
Copies: DLC; DeU; MWA; NjP.

5036. United States. Congress. House. Committee on Roads and Canals.
Survey of canal routes—Mississippi and Louisiana. April 14, 1826 ... Mr. Stewart, from the Committee on Roads and Canals ... made the following report. [Washington: 1826.] 4 p. (19th Cong., 1st sess. House. Rept. 162)
S-27412.
Copies: DLC; DeU; MWA; NjP.

5037. United States. Congress. Senate. Select Committee on Roads and Canals.

In Senate of the United States, April 19, 1826, Mr. Hendricks, from the Select Committee on Roads and Canals, made the following report: The Committee on Roads and Canals, to whom were referred several resolutions of the Senate, on the subject of providing for the surveys of canal routes, in the state of Indiana, and various routes for roads and canals, in the state of Maine.... [Washington: 1826.] 2 p.(19th Cong., 1st sess. Senate. Doc. 81)

S-27068.

Copies: DLC; DeU; MWA; NjP.

5038. United States. Congress. Senate. Select Committee on Roads and Canals.

In Senate of the United States, February 21, 1826. Mr. Hendricks from the Select Committee on Roads and Canals, to whom were referred a resolution of the Senate, and two memorials of the General Assembly of the state of Indiana, on the subject of a canal communication between the navigable waters of the Wabash and those of the Miami and Lake Erie, made the following report. [Washington: 1826.] 3 p. (19th Cong., 1st sess. Senate. Doc. 47)

S-27044.

Copies: DLC; DeU; MWA; NjP.

5039. United States. Congress. Senate. Select Committee on Roads and Canals.

In Senate of the United States, January 19, 1826. Mr. Hendricks, from the Select Committee on Roads and Canals, to whom was referred "A bill for the survey of a route for a canal between the Atlantic and the Gulf of Mexico," reported. [Washington: 1826.] 14 p. (19th Cong., 1st sess. Senate. Doc. 21)

S-27027.

Copies: DLC; DeU; MWA; NjP.

5040. United States. Engineer Dept.

Canal—Buzzard and Barnstable Bays. Letter from the Secretary of War, transmitting a memoir on the survey of the route of a canal, to connect Buzzard and Barnstable Bays, in the state of Massachusetts, with three sheets of drawings. March 2, 1826.... Washington: Printed by Gales & Seaton, 1826. 15 p., 3 fold. plates. (19th Cong., 1st sess. House. Doc. 174)

Memoir signed: Major P. H. Perrault, Top. engineer.

S-26893.

Copies: DLC; DeGE; DeU; MWA; NjP.

5041. United States. Engineer Dept.

Deep Creek and Castleman summits. Letter from the Secretary of War, transmitting the report of the engineer employed in exploring the country and running levels between the Deep Creek and Castleman's summits, on the contemplated route of the Chesapeake and Ohio Canal. April 29, 1826. Referred to the Committee on Roads and Canals. May 1, 1826, ordered to be printed. Washington: Gales & Seaton, printers, 1826. 12 p., table. (19th Cong., 1st sess. House. Doc. 169)

Signed: William G. McNeill.

S-26947.

Copies: DLC; DeGE; DeU; MWA; PU.

5042. United States. Engineer Dept.

Examination of the Muscle Shoals. Letter from the Secretary of War, transmitting the information required by a resolution of the House of Representatives of the 26th ultimo, in relation to the examination of the Muscle Shoals with a view to a canal to connect the waters of the Tennessee with those of the Coosa rivers. February 1, 1826.... Washington: Printed by Gales & Seaton, 1826. 4 p. (19th Cong., 1st sess. House. Doc. 67)

Transmitting a report of Alexander Macomb, Chief Engineer.

S-26977.

Copies: DLC; DeU; MWA; NjP.

5043. United States. Engineer Dept.

Juniatta, and Chesapeake and Ohio Canal. Letter from the Secretary of War, transmitting a report of the engineer, employed in exploring the country between

Bedford and Cumberland, with a view of uniting the Juniatta, by the route of the Juniatta and Wills Creek, with the Chesapeake and Ohio Canal. April 12, 1826, read ... April 15, 1826, referred to the Committee on Roads and Canals. Washington: Printed by Gales & Seaton, 1826. 13 p. (19th Cong., 1st sess. House. Doc. 158)

Signed: James Shriver, United States Assistant Civil Engineer.

S-27144.

Copies: DLC; DeGE; DeU; MWA; NjP.

5044. United States. Laws.

A bill concerning canals through Florida, and from thence to the Mississippi River. February 6, 1826.... [Washington: 1826.] 1 p.(H.R. 92)

S-26419.

Copies: DNA.

5045. United States. Laws.

A bill for the survey of a route for a canal between the Atlantic and the Gulf of Mexico. January 9, 1826.... [Washington: 1826.] 1 p. (S. 22)

S-26650.

Copies: DNA.

5046. United States. Laws.

A bill to aid in making a canal to connect the Michigan Lake with the Illinois River. May 12, 1826. [Washington: 1826.] 2 p. (H.R. 190)

S-26722.

Copies: DNA.

5047. United States. Laws.

A bill to aid the state of Ohio, in extending the Miami Canal from Dayton to Lake Erie. May 17, 1826.... [Washington: 1826.] 3 p. (H.R. 275)

S-26724.

Copies: DNA.

5048. United States. Laws.

Mr. Hendricks, from the Select Committee on Roads and Canals, to whom was referred the bill for the survey of a route for a canal between the Atlantic and the Gulf of Mexico, reported the same, with the following amendment. January 19, 1826.... [Washington: 1826.] 1 p. (S. 22)

S-27243.

Copies: DNA.

5049. United States. War Dept.

Buzzard's and Barnstable Bay canal. Letter from the Secretary of War, in reply to a resolution of the House of Representatives, of the 31st ultimo, in relation to the survey for a canal route between Buzzard's and Barnstable Bay. February 7, 1826.... Washington: Printed by Gales & Seaton, 1826. 3 p. (19th Cong., 1st sess. House. Doc. 77)

S-26890.

Copies: DLC; DeGE; DeU; MWA; NjP.

5050. United States. War Dept.

Canal—Mississippi to Lake Pontchartrain. Letter from the Secretary of War, transmitting information relative to making a canal between the Mississippi River and Lake Pontchartrain. December 18, 1826.... Washington: Printed by Gales & Seaton, 1826. 4 p. (19th Cong., 2d sess. House. Doc. 14)

S-26895.

Copies: DLC; DeU; MWA; NjP.

5051. Van Heuvel, Jacob Adrien, 1787–1874.

Memorial addresses to His Excellency Governor Clinton, relative to the northern boundary of the state, and the canal, proposed from the river St. Lawrence to Lake Champlain, dated January 12, 1825, by J. A. Vanden Heuvel, Esq. New York: Printed by J. Seymour, 1826. 15 p.

S-27496.

Copies: DLC.

5052. Virginia. Laws.

Letter from the Governor of the state of Virginia, transmitting a copy of an act passed by the legislature of that state, entitled "An act to amend the act, entitled An act incorporating the Chesapeake and Ohio Canal Company." February 9, 1826.... Washing-

5052 (*continued*)
ton: Printed by Gales & Seaton, 1826. 4 p. (19th Cong., 1st sess. House. Doc. 81)
S-27162.
Copies: DLC; DeU; MWA; NjP.

5053. Washington Canal Company.
Annual report of the president and directors of the Washington Canal Company. March 10, 1826.... Washington: Printed by Gales & Seaton, 1826. 3 p. (19th Cong., 1st sess. House. Doc. 127)
S-27463.
Copies: DLC; DeGE; DeU; MWA; NjP.

5054. Whippo, Charles T.
A report of a survey of the Tonewanda canal route from the Allegany River to the Erie Canal, by Charles T. Whippo, esq., civil engineer. And a memorial by the committee of the people to the legislature of New-York, in favor of its construction. [S.l.]: 1826. 12 p.
S-27630.
Copies: N.

5055. Winnipiseogee Canal Corporation.
Report of the committee appointed by the directors of the Winnipiseogee Canal, to prepare a statement of such facts and circumstances, in relation to the proposed canal, as they might deem proper to be laid before the publick; together with the charter of said canal corporation, and amendments thereto. Dover, N.H.: J. Dickman, 1826. 24 p.
S-27678.
Copies: CSmH; DLC; IU; MBAt; MCM; MH-BA; MWA; NN.

5056. Alspach, S.
Schuylkill Canal navigator.... Philadelphia: Printed by Joseph Rakestraw, for the author, 1827. 8 p.
S-27815.
Copies: CSmH; DeGE; MH; NN; P; PHi; PPL.

5057. Bernard, Simon, 1779–1839.
An estimate of the revenue which may be derived from the Morris Canal; extracted from a report made in 1823, to the War Department of the United States, by General Bernard, and Colonel Totten, of the Engineer Corps. With notes. New York: William Davis, jr., printer, 1827. 10 p.
S-28108.
Copies: MWA; MiU; NN; NNC; NjR.

5058. Binney, Horace, 1780–1875.
Union Canal Company. Opinion from Horace Binney, Esq. [Philadelphia: 1827.] 8 p.
S-28187.
Copies: PPAmP; PPL.

5059. [Bishop, Abraham] 1763–1844.
Farmington Canal. To the citizens of New-Haven. [New Haven?: 1827]. 14 p.
S-28190.
Copies: CtSoP.

5060. Chesapeake and Delaware Canal Company.
Eighth general report of the president and directors of the Chesapeake and Delaware Canal Company. June 4, 1827. [Philadelphia: 1827.] 28 p., fold. map.
S-28450.
Copies: DLC; DeGE; MWA; PPL.

5061. Chesapeake and Ohio Canal Company.
Laws relative to the Chesapeake and Ohio Canal. Washington: Printed by Way & Gideon, 1827. 28 p.
S-28451.
Copies: DBRE; DLC; MH-BA; MiD-B; NN; ViU.

5062. Chesapeake and Ohio Canal Convention, Washington, 1823, 1826.
Proceedings of the Chesapeake and Ohio Canal Convention, which assembled in the Capitol of the United States, in the city of Washington, on the sixth day of November, 1823, and re-assembled in the same city on the sixth day of December, 1826. Prepared and published pursuant to a resolution of the Convention. Washington City: Printed by

Way and Gideon, 1827. 112 p., fold. tables.
S-28452.
Copies: CSmH; DBRE; DLC; DeGE; ICJ; IU; MH-BA; MWA; MdBE; NN; NNE; OClW; PPAmP; PPi.

5063. [Clinton, De Witt] 1805– .
Report from the engineer appointed to survey certain canal routes. January 24, 1827. [S.l.: 1827.] 44 p., tables.
Signed: DeWitt Clinton.
Hasse (New York)-p. 88; Shaw–152.
Copies: NN.

5064. [Cruger, Alfred.]
A report of the proposed Saugatuck and New-Milford canal. [New York: 1827.] 11 p.
Signed: Alfred Cruger, engineer.
S-28626.
Copies: CtY; DLC.

5065. Delaware and Hudson Canal Company.
Considerations in support of the memorial of the Delaware & Hudson Canal Company. Albany: Croswell & Van Benthuysen, 1827. 16 p.
Signed: John Bolton, president.
N-D0140171; S-28582.
Copies: DLC; N; NRom.

5066. Enlargement of Erie Canal. Proposed loan anticipating the canal revenue, for expediting that magnificent work.... Rochester, December 31, 1826.... Committee of Correspondence. [Rochester: 1827?] Broadside.
S-28796.
Copies: NCanHi (not located).

5067. An exposition of the practicability of constructing a great central canal, from Lake Erie to the Hudson, through the southern tier of counties in the state of New-York; being a series of numbers originally published in the Western star, under the signature of Hercules. Westfield, N.Y.: Printed by Harvey Newcome, 1827. 31, iv p.
S-28827.
Copies: CSmH; CtY; MiU-T; NBu; NHi; NN.

5068. Facts and observations in relation to the origin and completion of the Erie Canal. 2d ed. Providence: F. Y. Carlisle and H. H. Brown, 1827. 32 p.
S-28831.
Copies: DLC; MB; MWA; MWo; Vi.

5069. Hampshire and Hampden Canal Company.
Report of Jarvis Hurd, Esq., civil engineer, employed by the Hampshire and Hampden Canal Company, to make a survey, and estimate the expense of a canal from the termination of the Hampshire and Hampden Canal, in Northampton, to the north line of Massachusetts, and thence to Brattleborough, in the state of Vermont. Boston: True & Greene, printers, 1827. 40 p.
S-29279.
Copies: CtY; MB; MBAt; MW; MWA.

5070. Hampshire and Hampden Canal Company.
To the honourable General Court of the Commonwealth of Massachusetts, now in session. The petition of the Hampshire and Hampden Canal Company, by their agents . . . respectfully sheweth. . . . [Boston: 1827.] 1 p.
Dated: Boston, January 13, 1827.
Prays for the extension of the canal from Northampton to Brattleborough, Vermont.
S-29125.
Copies: DLC.

5071. Hillhouse, James, 1754–1832.
Remarks of the Hon. James Hillhouse before the joint committee, on the petition of Samuel Hinkley and others, for the extension of the Hampshire and Hampden canal. Boston: True & Greene, printers, 1827. 11, [1] p.
S-29216.
Copies: CtHi; CtSoP; CtY; M; MA; MB; MBAt; MH-BA; MWA; NN.

5072. Illinois. General Assembly.
Memorial of the General Assembly of the state of Illinois, asking for a grant of land to aid the said state in opening a canal to connect the waters of Lake Michigan with the Illinois River. February 7, 1827.... Washington: Printed by Gales & Seaton, 1827. 4 p. (19th Cong., 1st sess. Senate. Doc. 46)
S-31404.
Copies: DLC; DeU; MWA; NjP.

5073. ——— Washington: Printed by Gales & Seaton, 1827. 4 p. (19th Cong., 2d sess. House. Doc. 81)
S-31405.
Copies: DLC; DeU; MH-L; MWA; NjP.

5074. Kentucky. Governor (J. Desha).
Governor's message.... [Frankfort?: 1827.] 16 p.
Signed at end: Joseph Desha; dated, December 6, 1827.
Refers partially to the Louisville Canal, built with private funds.
N-K0100060.
Copies: MiU-C.

5075. Maryland. Governor (J. Kent).
Message from His Excellency the Governor transmitting memorials respecting the Chesapeake and Ohio Canal, and the Chesapeake and Delaware Canal. Annapolis: Printed by Jeremiah Hughes, 1827. 8 p.
S-29646.
Copies: MdHi.

5076. Maryland. Laws.
An act to amend an act of incorporation of the Chesapeake and Ohio Canal Company. February 8, 1827. Referred to the Committee of the Whole House to which is committed the "Bill authorizing a subscription to the stock of the Chesapeake and Ohio Canal Company." Washington: Printed by Gales & Seaton, 1827. 3 p. (19th Cong., 2d sess. House. Doc. 85)
Contains an act of Maryland.
S-30894.
Copies: DLC; DeGE; DeU; MH-BA; MWA; NN; NjP.

5077. Massachusetts. General Court. House. Special Committee on the Hampshire and Hampden Canal Company.
The Special Committee of this House, to whom was committed the bill to extend the act to incorporate the Hampshire and Hampden Canal Company ... report. [Boston: 1827.] 4 p.(H.R. No. 64)
Dated: House of Representative, Feb. 27, 1827.
Copies: DeGE.

5078. Massachusetts. General Court. Joint Committee on Roads and Canals.
Report [of the Joint Committee on Roads and Canals, to whom was referred the petition of Samuel Hinkley and others, praying that they may be authorized to extend the Hampshire and Hampden Canal to the north line of the state]. [Boston: 1827.] 20 p. (House. Doc. 45)
S-2153 (dated 182–?), 29676.
Copies: M; MBAt; MWA; N; NN.

5079. Massachusetts. General Court. Senate. Committee on Roads and Canals.
Commonwealth of Massachusetts. The Committee on Roads and Canals, to whom was referred the petition of Aaron Tufts and others, praying that a survey may be made for a canal from Boston, by the waters of Charles River, to the Blackstone Canal.... report. [Boston: 1827.] 2 p. (H.R. No. 66)
Dated at end: In Senate, Feb. 27, 1827.
Copies: DeGE.

5080. Massachusetts. Laws.
An act in addition to an act entitled "An act to incorporate the Hampshire and Hampden Canal Company." [Boston: 1827.] 3 p. (H.R. No. 56)
Caption title from p. [2].
Order by the Massachusetts House of Representatives to print the act, dated Feb. 20, 1827, on p. [1].
Copies: DeGE.

5081. Massachusetts. Laws.
An act, in addition to an Act to incorporate the Blackstone Canal Company. [Bos-

ton: 1827.] 4 p. (H.R. No. 18)

Dated at head of title: Commonwealth of Massachusetts. In the year of our Lord, one thousand eight hundred and twenty-seven.

Copies: DeGE.

5082. Massachusetts. Laws.

An act, to authorize the Hampshire and Hampden Canal Company to extend a canal from Northampton to the north line of this Commonwealth. [Boston: 1827.] 20 p. (H.R. No. 53)

Dated at head of the title: In the year of our Lord, one thousand eight hundred and twenty-seven.

Copies: DeGE.

5083. Mills, John B.

A survey for a proposed canal along the Delaware River, from the mouth of the Lackawaxen River to the village of Deposite. Done in October and November, 1826, by John B. Mills.... Albany: Printed by Croswell & Van Benthuysen for the Chesapeake and Delaware Canal Company, 1827. 16 p.

Includes additions by Benjamin Wright.

S-29766.

Copies: CSmH; NNE.

5084. ———— New York: Printed by Elliott & Palmer, 1827. 17 p.

S-29767.

Copies: MWo; NN; NNE; NRom.

5085. Morris Canal and Banking Company.

A report to the directors of the Morris Canal and Banking Company; made by the president, Cadwallader D. Colden, May 1st, 1827: with a statement of the work then done upon the canal. A report from the chief engineer, and a communication from Governor Clinton. Printed by order of the board of directors. New York: William Davis, Jr., printer, 1827. 22 p., 1 l.

S-29813.

Copies: DLC; DeGE; MWA; NBuG; NN; NNC; NRU; NjR; PBL.

5086. Narrative of facts relative to the location of the western section of the Pennsylvania Canal, and the proceedings of the Canal Commissioners, prepared under a resolution of the Canal Committee of Pittsburgh. Pittsburgh: Printed at the office of the Statesman, 1827. 35 p.

S-30277.

Copies: PPL; PPiHi.

5087. New Jersey. Legislature. General Assembly. Committee on the Delaware and Raritan Canal.

Report of the Committee on the Delaware & Raritan Canal. [S.l.: 1827.] 8 p.

Signed: James Parker.

Hasse (New Jersey)-p. 188.

Copies: DeGE; NN.

5088. New York (State). Canal Commissioners.

Annual report of the Canal Commissioners of the state of New-York. Made to the Assembly, Feb. 10, 1827. Albany: Printed by Croswell and Van Benthuysen, 1827. 35 p.

S-29969.

Copies: MCM; N; NHi; NRom.

5089. New York (State). Laws.

An act to incorporate the Harlaem River Canal Company, passed April the 16th, 1827. New York: Dixon and Sickels, [1827]. 12 p.

Copies: NHi.

5090. New York (State). Laws.

Canal laws. The several laws relative to the canals of the state of New York. Taken from the laws, published by order of the Canal Board, April, 1827. Albany: Printed by Croswell & Van Benthuysen, 1827. 43 p.

S-29970.

Copies: IU; MWA (88 p.); N.

5091. New York (State). Laws.

Revised statute. Chapter IX, title IX, relating to the canals; passed at the extra session of the Legislature of the state of New-York, December 3, 1827. Albany:

5091 (*continued*)
Printed by E. Croswell, 1827. vii, 40 p.
Includes regulations of the Canal Board and Commission.
S-30023.
Copies: MH-L; NN; PPL.

5092. New York (State). Legislature. Assembly.
Reports of committees in relation to the Long Island Canal Company. Albany: 1827. 14 p.
S-30022.
Copies: MWA; N.

5093. New York (State). Legislature. Assembly. Committee on Canals.
Report of the Committee on Canals and Internal Improvements, in relation to the Chenango Canal. [Albany: 1827.] 7 p. (Assembly. Doc. 91)
Dated: February 16, 1827.
S-29990.
Copies: DLC.

5094. New York (State). Legislature. Assembly. Committee on Canals.
Report of the Committee on Canals on the engrossed bill from the Assembly, authorizing the construction of the canal, from Seneca Lake to Newtown. Albany: Printed by Croswell and Van Benthuysen, 1827. 55 p.
S-30017.
Copies: MPlPS; N.

5095. New York (State). Legislature. Assembly. Committee on Canals.
Report on the Genesee and Conewango canals. Albany: 1827. 3 p. (Assembly. Doc. 118)
Whitford-p. 1252.
Copies: MWA; N.

5096. New York (State). Legislature. Senate. Committee on Canals.
Report of the Committee on Canals, on the engrossed bill from the Assembly, entitled An act authorizing the construction of a canal through the valley of the Chenango. Made to the Senate, April 3, 1827. Albany: Printed by Croswell & Van Benthuysen, 1827. 40 p.
Copies: NHi.

5097. New York (State). Legislature. Senate. Committee on Canals.
Report of the Committee on Canals on the petition of David E. Evans, and others, made to the Senate, February 12, 1827. Albany: Printed by Croswell and Van Benthuysen, 1827. 36 p.
Concerns the request to construct a canal from the Erie Canal to the Alleghany River at Olean, by way of the Tonnewanda Creek.
S-30018.
Copies: DBRE; DLC; N; NBu; NN; NNE; NRom; PPL.

5098. New York (State). Legislature. Senate. Select Committee on the Incorporation of the Long Island Canal Company.
Reports of committees in relation to the Long Island Canal Company. Albany: Croswell & Van Benthuysen, 1827. 14 p.
Includes reports by Mr. Colden and Mr. Gardiner from the Select Committee, and by Mr. Wilkeson from the Canal Committee.
N-N0201485.
Copies: NN.

5099. Ohio. Commissioners of the Canal Fund.
Report of the Commissioners of the Canal Fund. To the General Assembly of the State of Ohio. [Columbus: 1827.] 7 p.
A corrected statement superseding the report dated December 11, 1826.
Copies: MWA; OClWHi.

5100. Ohio. Commissioners of the Canal Fund.
Report. To the General Assembly of the State of Ohio. [Columbus: 1827.] 5 p.
Dated: December 27, 1827.
Bowker-p. 205.
Copies: MWA; OClWHi.

5101. Ohio. General Assembly. House. Committee on Canals.

Report. The standing Committee on Canals to whom was referred the memorial of sundry citizens, residents of the Miami County, praying for an alteration of the law prescribing the mode for the assessment of damages for materials, &c. taken for the construction of canals, have had the same under consideration, and ask leave to report. [January 18, 1827. Columbus: 1827.] 3 p.

S-30109.

Copies: DLC; OClWHi.

5102. Ohio. General Assembly. Senate. Committee on Canals.

Report of the Committee on Canals, of the Senate, on the canal fund. [Columbus: 1827.] 6 p.

Dated: January 18, 1827.

Hasse(Ohio)-p. 381; N-00047592.

Copies: DLC; NN.

5103. Ohio. Laws.

Pennsylvania and Ohio canal law. [Columbus: 1827.] 12 p.

S-30108.

Copies: DLC.

5104. Pennsylvania. Auditor General.

Report of the Auditor General, giving a statement of the accounts of the commissioners and engineers, of the Pennsylvania Canal. Harrisburg: Printed by Cameron & Krause, 1827. 23 p.

S-30175.

Copies: PPAmP.

5105. Pennsylvania. Board of Canal Commissioners.

Communication from the Governor, accompanied with a report of the Board of Canal Commissioners of Pennsylvania, and documents accompanying the said report. Read in the House of Representatives, January 1, 1827. Harrisburg; Printed by John S. Wiestling, 1827. 96 p., 3 fold. tables.

Report submitted December 11, 1826.

S-30169.

Copies: DLC; ICJ; MB; MWA; N; P; PPAmP; PPF; PPL; PU.

5106. Pennsylvania. Board of Canal Commissioners.

Communications from the Governor, accompanied with a report of the Board of Canal Commissioners of Pennsylvania . . . No. 2. Read in the House of Representatives, February 7, 1827. Harrisburg: Printed by John S. Wiestling, 1827. 135, [1] p.

S-30170.

Copies: DLC; DeGE; MB; MH-BA; MWA; P; PPAmP; PPF; PPL.

5107. Pennsylvania. Board of Canal Commissioners.

First report of the Pennsylvania Canal Commissioners, accompanied with official documents. Harrisburg: Printed by Cameron & Krause, 1827. 96 p., 3 fold. tables.

Report submitted December 11, 1826.

S-30171.

Copies: CtY; DBRE; DLC; DeGE; MH-BA; N; NN; NNE; P; PHi; PPAmP; PPF; PPL.

5108. Pennsylvania. Board of Canal Commissioners.

Report of the Canal Commissioners of Pennsylvania to the Legislature. [Harrisburg: 1827.] 320 p., fold. tables.

S-30176.

Copies: CtY; ICJ; OCHP; PPF; PPL.

5109. Pennsylvania. Board of Canal Commissioners.

Second report of the Canal Commissioners together with reports and estimates of the engineers. Harrisburg: Printed by Cameron & Krause, 1827. 139 p.

Dated: Feb. 6, 1827.

S-30181.

Copies: CtY; MWA; MiU-T; N; NN; P; PHi; PPF; PPL.

5110. Pennsylvania. General Assembly. House. Committee on Inland Navigation and Internal Improvement.

Report of the Committee on Inland Navigation and Internal Improvement, relative to the further extension of the Pennsylvania Canal, accompanied with a bill. Mr.

5110 (*continued*)
Lehman, chairman. Read in the House of Representatives, February 28, 1827. Harrisburg: Printed by J. S. Wiestling, 1827. 18 p.
S-30178.
Copies: DLC; DeGE; IU; MH-BA; MWA; N; P; PHi; PPAmP.

5111. Profile[s] 1 [and 2] of the Farmington Canal and of the Hampshire and Hampden Canal. [S.l.: ca. 1827.] [3] p., fold. plate.
Title from the plate, which is accompanied by [3] pages of text, containing "Explanation of the profile."
Copies: DeGE.

5112. Quinebaug Canal Association.
Survey and charter of the Quinebaug Canal. Norwich, [Conn.]: Printed at the Courier office, 1827. 15 p.
The charter includes rules for the construction, use, and supervision of the canal.
N-Q0021215; S-30396.
Copies: CSmH; NN.

5113. Union Canal Company of Pennsylvania.
Annual report of the president and managers of the Union Canal Company of Pennsylvania to the stockholders. November 20, 1827. Philadelphia: Printed by Lydia R. Bailey, 1827. 12 p.
S-30863.
Copies: DeGE; MWA; NRom; PHi; PPL.

5114. United States. Congress. House. Committee on Naval Affairs.
Communication across the isthmus of Panama. January 24, 1827 ... Mr. Storrs, from the Committee on Naval Affairs, made the following report. [Washington: 1827.] 7 p. (19th Cong., 2d sess. House. Rept. 56)
S-31218.
Copies: DLC; DeU; MWA; NjP.

5115. United States. Congress. House. Committee on Roads and Canals.
Chesapeake and Ohio Canal. January 30, 1827. Mr. Mercer, from the Committee on Roads and Canals, to which was referred the memorial of the Chesapeake and Ohio Canal Convention, made the following report.... [Washington: 1827.] 122 p. (19th Cong., 2d sess. House. Rept. 90)
S-31212.
Copies: DLC; DeGE; DeU; MWA; NjP.

5116. United States. Congress. Senate. Select Committee on Roads and Canals.
In Senate of the United States, January 23, 1827. Mr. Hendricks, from the Select Committee on Roads and Canals, to whom were referred a resolution of the Senate and two memorials of the General Assembly of the state of Indiana, on the subject of a canal communication between the navigable waters of the Wabash and those of the Miami of Lake Erie, made the following report.... [Washington: 1827.] 6 p. (19th Cong., 2d sess. Senate. Doc. 29)
S-31297.
Copies: DLC; DeU; MWA; NjP.

5117. United States. Engineer Dept.
Canal—Lake Pontchartrain to the Mississippi. Letter from the Secretary of War, transmitting the information required by a resolution of the House of Representatives of the 13th December last, in relation to a canal from Lake Pontchartrain to the Mississippi River. March 3, 1827.... Washington: Printed by Gales & Seaton, 1827. 18 p., fold. table. (19th Cong., 2d sess. House. Doc. 133)
Transmitting a report of the Chief Engineer, with accompanying documents.
S-31206.
Copies: DLC; DeU; MWA; NjP.

5118. United States. Laws.
A bill to aid in making a canal to connect the Michigan Lake with the Illinois River. May 12, 1825 ... February 3, 1827. Re-printed by order of the House of Representatives. [Washington: 1827]. 2 p. (H.R. 190)
S-31113.
Copies: DNA.

5119. United States. Laws.
A bill to authorize the President of the United States to negotiate with the Chickasaw Indians for a canal route to connect the waters of the Tennessee and Tombeckby Rivers. January 13, 1827.... [Washington: 1827.] 1 p. (H.R. 376)
S-31142.
Copies: DNA.

5120. United States. Laws.
A bill to authorize the President of the United States to obtain from the Cherokee Indians a site for a canal to connect the western and southern waters. January 2, 1827.... [Washington: 1827.] 1 p. (H.R. 327)
S-31143.
Copies: DNA.

5121. United States. War Dept.
Canal—Barnstable to Buzzard's Bay. Letter from the Secretary of War, in reply to a resolution of the House of Representatives on the 2d of Jan. last, requiring an estimate of the expense of making a canal between Barnstable and Buzzard's bays. February 13, 1827.... Washington: Printed by Gales & Seaton, 1827. 3 p. (19th Cong., 2d sess. House. Doc. 97)
S-31205.
Copies: DLC; DeGE; DeU; MWA; NjR.

5122. Varlé, Charles.
Candid considerations, respecting the canal between the Chesapeake and Delaware Bays. Whereby the impracticabilities of proceeding successfully with the same ... is demonstrably proved. Baltimore: The author; Wm. Wooddy, printer, 1827. 19 p.
S-31587; Sabin-98628.
Copies: MBAt.

5123. Waterman, T. G.
Speech in the Senate on the bill authorizing the construction of the Chenango Canal, delivered April 9 and 10, 1827. Albany: 1827. 7 p.
Whitford-p. 1248.
Copies: N.

5124. [Carey, Mathew] 1760–1839.
Pennsylvania canals. [Philadelphia: 1828.] 3 p.
Signed and dated at end: Hamilton. Philadelphia, May 15th, 1828.
S-32587.
Copies: MWA; PPL; PU.

5125. [Carey, Mathew] 1760–1839.
When the following letter was written, it was without the slightest idea of its ever being published. It was originally addressed to a friend, then a member of Congress from Pennsylvania.... Dear Sir: According to your request, I now take up my pen to communicate to you my ideas ... on the policy of constructing canals.... [S.l.: 1828?] 8 p.
Dated on p. [1]: Meadville, Dec. 5, 1828; signed at end: H.
S-32602.
Copies: MWA; PU.

5126. Chemung Canal Company.
The following report.... Geneva, [N.Y.]: 1828. 9 p.
Dated: Jan. 1, 1828.
Copies: MWA.

5127. Chesapeake and Delaware Canal Company.
Ninth general report of the president and directors of the Chesapeake and Delaware Canal Company. June 2, 1828. [Philadelphia: 1828.] 15 p., fold. map.
S-32660.
Copies: DeU; MWA; NN; PPL; PU.

5128. Chesapeake and Ohio Canal Company.
Acts of the States of Virginia, Maryland, and Pennsylvania, and of the Congress of the United States, in relation to the Chesapeake & Ohio Canal Company; with the Proceedings of the Convention, which led to the formation of the said company. Also, The acts and resolutions of the States of Virginia and Maryland concerning the Potomac Company. To which are appended the by-laws, list of officers, &c. of the Chesapeake and Ohio Canal Company, with a copious

5128 (*continued*)
index. Washington: Printed by Gales & Seaton, 1828. 148, 13 p.
S-32662.
Copies: CSmH; CtY; DBRE; DLC; DeGE; ICJ; IU; LNHT; MH-BA; MWA; MiU-T; N; NN; NNE; NcD; Vi.

5129. Chesapeake and Ohio Canal Company.
Acts of Virginia, Maryland and Pennsylvania, and of the Congress of the United States in relation to the Chesapeake and Ohio Canal Company.... Washington: Printed by Gales & Seaton, 1828. 102, 6 p.
S-32661.
Copies: DLC.

5130. Chesapeake and Ohio Canal Company.
Chesapeake and Ohio Canal. Schedule of contracts let in August, 1828, with the estimated quantities of work to be performed, and the prices accepted. [Washington: 1828.] Broadside.
Copies: PPL.

5131. ———— in October, 1828, with the estimated quantity of work to be performed, and the prices accepted. [Washington: 1828.] 2 p.
Copies: PPL.

5132. Chesapeake and Ohio Canal Company.
Laws passed by the legislatures of Maryland, Virginia, and Pennsylvania and of the Congress of the United States, in relation to the Chesapeake and Ohio Canal. [Washington?: 1828.] 26 p.
S-32663.
Copies: DBRE; DLC; MH-BA; MWA; MiD-B.

5133. Chesapeake and Ohio Canal Company.
Memorial of the Chesapeake and Ohio Canal Company. December 5, 1828.... [Washington: 1828.] 32 p. (20th Cong., 2d sess. House. Doc. 12)
S-36202.
Copies: DLC; DeU; MWA; NjR; ViU.

5134. Chesapeake and Ohio Canal Company.
Proceedings of the president and directors of the Chesapeake and Ohio Canal Company, and of the corporations of Washington, Georgetown, & Alexandria, in relation to the location of the eastern termination of the Chesapeake and Ohio Canal. Printed by order of the general meeting of the stockholders; convened on the 10th September, 1828. Washington: Printed by Gales & Seaton, 1828. 31 p.
S-32664.
Copies: DLC; OU; ViU.

5135. Chesapeake and Ohio Canal Company.
Reports of agents appointed by the president and directors of the Chesapeake and Ohio Canal Company to explore the shores of the Potomac River, in search of materials suitable for the manufacture of water lime; and also of stone suitable for the construction of locks, aqueducts &c. [S.l.: 1828.] [4] p.
N-C0345907.
Copies: MiU-C.

5136. Chesapeake and Ohio Canal Company.
Rules adopted by the president and directors of the Chesapeake and Ohio Canal Company, for the government of the corps of engineers. Together with a distribution of the first division of the canal into residencies, with the engineers allotted to each. Also, a list of premiums for diligence and fidelity. Washington: Printed by Gales & Seaton, 1828. 10 p.
S-32665.
Copies: DLC; NN.

5137. Chesapeake and Ohio Canal Convention, Washington, D.C., 1823, 1826.
Chesapeake and Ohio Canal Company.

Memorial of the Central Committee of the Chesapeake and Ohio Canal Convention to the Congress of the United States. January 28, 1828. Washington: Printed by Gales & Seaton, 1828. 8 p. (20th Cong., 1st sess. House. Doc. 101)

Signed: C. F. Mercer, Chairman of Central Committee.

S-36201.

Copies: DLC; DeGE; DeU; MWA; NjR; ViU.

5138. Clinton, De Witt, 1769–1828.

Letters of Governor Clinton, and of Colonel L. Baldwin ... improved as evidence before the joint committee of the Legislature of Massachusetts, on the petition of Samuel Hinkley and others, for the extension of the Hampshire and Hampden Canal. February 1828. Boston: Dutton & Wentworth, state printers, 1828. 22 p.

S-32738.

Copies: CtSoP; DLC; M; MBAt; MH-BA; MWA; MWo; MiU-T; N; NN; NhHi.

5139. Clinton, Isaac, 1759–1840.

Speech of the Rev. Isaac Clinton, delivered at the court-house in the county of Lewis, before the Martinsburg Canal Convention, held on the 4th of December, 1827. Lowville, [N.Y.]: Printed by Wm. L. Easton, 1828. 8 p.

Concerns the Black River Canal.

S-32739.

Copies: CSmH.

5140. Conkling, Alfred, 1789–1874.

A discourse commemorative of the talents, virtues and services of the late De Witt Clinton; delivered at Schenectady, before the Society of the Phi Beta Kappa, July 22, 1828. Albany: Printed by Websters and Skinners, 1828. 31 p.

Partially concerned with the Erie Canal.

S-32802.

Copies: CtY; DLC; IU; MBAt; MWA; MiD-B; MnHi; NN; NNC; NhHi; NjR; OMC; PPL; RPB.

5141. Continuation of the numbers of "A Baltimorean," the first of which were published in the National Intelligencer. Canals. —Railroads. No. 7. October 21, 1828. [Baltimore: 1828.] [11] p.

S-32818.

Copies: DBRE; M; MB; MCM; MH-BA; MWA; NN; OClWHi.

5142. [Cruger, Alfred.]

Report on the proposed canal connecting the Altamaha and Ogeechee rivers, comprising the plan and estimates of the same. [New York?: 1828.] 40 p., fold. map.

Report of Benjamin Wright, p. 39–40.

S-32857.

Copies: GU; GU-De; MiU.

5143. Delaware and Hudson Canal Company.

Annual report of the board of managers of the Delaware and Hudson Canal Co. to the stockholders, for the year 1827. March 4th, 1828. New York: Nathan Lane, 1828. 20 p.

Thomson-205.

Copies: DBRE; DLC; ICJ; MCM; MH-BA; N; NHi; NN; NNE; NjP.

5144. Delaware and Hudson Canal Company.

Third general report of the president, managers and company, of the Delaware and Hudson Canal Company, to the stockholders. March 4, 1828. New York: Printed by Elliott and Palmer, 1828. 22 p.

S-32918.

Copies: CtY; DBRE; MiU-T; NN; PPL; WHi.

5145. Florida. Laws.

Chipola Canal Company—Act, Council, Florida. Copy of an act to incorporate the Chipola Canal Company, with powers to construct a canal or rail-way between the Chipola River and the eastern arm of St. Andrews Bay ... February 29, 1828.... Washington: Printed by Gales & Seaton, 1828. 7 p. (20th Cong., 1st sess. House. Doc. 166)

5145 (*continued*)
S-36203.
Copies: CtY; DBRE; DLC; DeU; MB; MBAt; MH; MWA; MiU-T; NN; NjR; P; PPAmP.

5146. [Geddes, James] 1763–1838.
Chesapeake and Ohio Canal. Letter from the Secretary of War, transmitting estimates of the cost of making a canal from Cumberland to Georgetown. March 10, 1828.... Washington: Printed by Gales & Seaton, 1828. 100 p., tables. (20th Cong., 1st sess. House. Doc. 192)
Report signed: James Geddes, Civil Engineer, Nathan S. Roberts, Civil Engineer.
S-36200.
Copies: DLC; DeGE; DeU; MWA; NjR; RPB; TxU.

5147. [Holley, Myron] 1779–1841.
Memorial. To the Hon. the Legislature of the state of New-York, in Senate and Assembly convened.... [New York?: 1828?] 21 p.
Signed and dated: Myron Holley, 9th January, 1828.
Concerns claims for services as Canal Commissioner.
S-33582.
Copies: NN.

5148. [Howard, William] 1793–1834.
Report on the survey of a canal from the Potomac to Baltimore. Baltimore: Printed by B. Edes, 1828. 14 p.
S-33595.
Copies: MdHi; NNC; ViU; WHi.

5149. Illinois. Laws.
A bill for "An act to provide for the construction of the Illinois and Michigan Canal." [Vandalia?: 1828.] 3 p. (H.R. no. 2)
S-33641.
Copies: I-Ar.

5150. Indiana. General Assembly. Senate. Committee on Canals and Internal Improvements.
Report of the Committee on Canals and Internal Improvements, relative to the Wabash and Miami Canal, by Mr. Ewing, December 25, 1828.... Indianapolis: Smith and Bolton, state printers, 1828. 15 p.
N-I0066423.
Copies: InU.

5151. Jocelyn, N. & S. S., publishers, New Haven.
Map exhibiting the Farmington & Hampshire & Hampden canals, together with the line of their proposed continuation through the valley of the Connecticut River to Canada. Engraved and published by N. & S. S. Jocelyn. New Haven: 1828. Map 121 x 39 cm.
S-33717.
Copies: CtY; ICJ; MB.

5152. Kilbourn, John, 1787–1833.
Public documents, concerning the Ohio canals, which are to connect Lake Erie with the Ohio River, comprising a complete official history of these great works of internal improvement. Columbus: Compiled and published by John Kilbourn; Olmsted, Bailache & Cameron, printers, 1828. [4], 320 p.
Consists of vol. 1, no. 1–20 of the *Civil engineer and herald of internal improvement,* Columbus, June 21–Nov. 1, 1828.
N-K0133860.
Copies: CtY; ICJ; ICN; MH-BA; MiU-C; N; NN; OCHP; OCl; OClWHi; OHi; OO.

5153. ———— Columbus: 1828. 403 p.
Originally published in the *Civil engineer and herald of internal improvement,* with additional documents.
S-33771; N-K0133859.
Copies: DLC; ICN; KyU; NBu; NBuG.

5154. Lehigh Coal and Navigation Company.
Report of the board of managers of the Lehigh Coal and Navigation Company, presented to the stockholders, January 14, 1828. Philadelphia: Printed by S. W. Conrad, [1828]. 12 p.

S-33839.
Copies: DLC; MH-BA; NN; PPAmP; PPL.

5155. Letter on the subject of raising money for the canal system of Pennsylvania, originally addressed to a member of Congress from Pennsylvania, and now submitted to the Legislature, [S.l.: 1828?] 8 p.

Dated: Meadville, Dec. 5, 1828, and signed: H.

Probably the same as No. 5125.

Sabin-60205. No copy known.

5156. Manley, James R., 1781–1851.

An eulogium on De Witt Clinton, late governor of the state of New-York. Delivered ... in the hall of Columbia College, on Friday, 11th July, 1828, New York: Printed by Gould & Jacobus, 1828. [4], 22 p.

Includes observations on the Erie Canal.

S-34003.
Copies: CSmH; DLC; ICU; IaU; MBAt; MH; MWA; MiD-B; MnHi; NN; NjR; OMC; PHi.

5157. Maryland. Governor (J. Kent).

A communication from the executive, relative to the Chesapeake and Delaware Canal. February 7, 1828. Annapolis: Printed by Jeremiah Hughes, Jan. [*sic*] 1828. 7 p., fold. map.

"A communication from the president and directors of the Chesapeake and Delaware Canal," p. [4]–7.

S-34027.
Copies: PPL (map wanting).

5158. Maryland. Laws.

Act—State of Maryland. An act of the General Assembly of Maryland, entitled A further supplement to the act, entitled "An act for the promotion of internal improvement." March 17, 1828.... Washington: Printed by Gales & Seaton, 1828. 5 p. (20th Cong., 1st sess. House. Doc. 205)

Concerns the Chesapeake and Ohio Canal.

S-35587.
Copies: DLC; DeU; MWA; NjR; ViU.

5159. Massachusetts. Board of Internal Improvements.

Report of the Board of Commissioners, of Internal Improvement, in relation to the examination of a route for a canal, from Boston to the Blackstone Canal, and thence to the line of Connecticut.... Boston: Dutton & Wentworth, printers to the state, 1828. 18 p. (Senate. Doc. 8)

S-34063.
Copies: DLC; DeGE; IU; MB; MH-BA; MWA; NN; NbO; PPF; RPB.

5160. Massachusetts. General Court. Joint Committee on Roads and Canals.

[Report on the petition of Samuel Hinkley and others to extend the Hampshire and Hampden Canal to the north line of the state, and on the petition for improvement of navigation of Connecticut River. Boston: 1828.] 13 p. (House. Doc. 41)

Hasse (Massachusetts)-p. 55
N-M0309807.
Copies: M; MWA; NNC.

5161. Massachusetts. Laws.

An act to authorize the Hampshire and Hampden Canal Company to construct a canal from Northampton to the north line of this commonwealth. [Boston: 1828.] 22 p. (Senate. No. 20)

Dated on p. 20: February 27, 1828.
S-33467.
Copies: DLC; DeGE. MWA.

5162. Massachusetts. Laws.

An act to incorporate the proprietors of the central locks and canals on Connecticut River. [Boston: 1828.] 11 p. (H.R. 45)

N-M0311046.
Copies: NNC.

5163. Mitchill, Samuel Latham, 1764–1831.

A discourse on the character and scientific attainments of De Witt Clinton, late

5163 (*continued*)
Governor of the state of New York; pronounced at the Lyceum of Natural History . . . 14th July, 1828. . . . New York: Printed by E. Conrad, 1828. 28 p.

Part of the discourse is devoted to the construction of Erie Canal.

S-34170.

Copies: CSmH; DLC; ICU; IU; MBAt; MH; MnHi; NNC; PPAmP; PPL; WHi.

5164. New Jersey. Legislature. Joint Committee on the Morris Canal.

Report of the Joint-Committee of Council and Assembly, appointed to view the Morris Canal and inclined planes. [Trenton?: 1828.] 27 p.

S-34457.

Copies: DBRE; MWA; Nj; NjR.

5165. New York (State). Canal Commissioners.

The annual report of the Canal Commissioners of the state of New York, made to the Legislature, the 19th January, 1828. Albany: Printed by Croswell and Van Benthuysen, 1828. 31, [1] p.

S-34477.

Copies: MWA; MWo; NRom.

5166. New York (State). Canal Commissioners.

Canal regulations, relative to the navigation and management of the canals. . . . [Albany]: April 1, 1828. Broadside 28 x 28 cm.

Signed by the Canal Commissioners: Samuel Young, Henry Seymour, Wm. C. Bouck.

Includes 12 printed regulations, to which have been added, on a separate sheet, regulations no. 13–14 in manuscript, signed: S. Young, Henry Seymour, Wm. C. Bouck.

Copies: DeGE.

5167. New York (State). Canal Commissioners.

Report relative to Buffalo and Black Rock harbors. Albany: 1828. 4 p. (Assembly. Doc. 175)

Whitford-p. 1258.

Copies: MWA; N.

5168. New York (State). Canal Commissioners.

Report respecting Chenango Canal. Albany: 1828. 38 p.

Submitted by N. S. Roberts and H. Hutchinson.

Probably the same as no. 5189.

Also issued as Appendix A, Assembly journal, Jan. 1829. 20 p.

Hasse (New York)-p. 135.

Copies: NN.

5169. New York (State). Comptroller's Office.

Report of the Comptroller, relative to the expenditure of moneys on the canals. . . . [Albany: 1828.] 163 p. (Assembly. Doc. 23)

Dated: In Assembly, January 12, 1828.

Concerns the charges of fraud against William L. Perce in the canal construction.

N-N0191311.

Copies: MB.

5170. New York (State). Laws.

An act authorizing the construction of the Chenango Canal. Albany: 1828. 2 p. (Assembly. Doc. 98)

Whitford-p. 1249.

Copies: MWA; N.

5171. New York (State). Laws.

An act to incorporate the Black River Canal Company. Passed March 20, 1828. [Albany: 1828.] 35 p., 2 maps.

Hasse (New York)-p. 130.

Copies: NHi; NN.

5172. New York (State). Legislature. Assembly. Committee on Canals.

Report on proposed amendment to the report relative to the proposed canal from the Susquehanna River at Binghampton to Erie Canal. Albany: 1828. 12 p. (Assembly. Doc. 93)

Whitford-p. 1249.
Copies: MWA; N.

5173. New York (State). Legislature. Assembly. Committee on Canals.

Report relative to the construction of a canal from Cayuga and Seneca Canal to East Cayuga. Albany: 1828. 2 p. (Assembly. Doc. 52)

Whitford-p. 1240.
Copies: N.

5174. New York (State). Legislature. Assembly. Committee on Canals.

Report relative to the construction of the Chenango Canal. Albany: 1828. 10 p. (Assembly. Doc. 92)

Whitford-p. 1249.
Copies: MWA; N.

5175. New York (State). Legislature. Assembly. Committee on Grievances.

Report relative to the Cayuga and Seneca Canal. Albany: 1828. 4 p. (Assembly. Doc. 236)

Whitford-p. 1240.
Copies: MWA; N.

5176. Ohio. Board of Canal Commissioners.

Report. To the General Assembly of the State of Ohio. [Columbus: 1828.] 28 p.

Dated on p. 23: Jan. 5th, 1828.
Bowker-p. 205.
Copies: MWA; OClWHi; PPAmP.

5177. Ohio. Board of Canal Commissioners.

Report of Canal Commissioners on the survey of the Muskingum River and in relation to the Pennsylvania and Ohio Canal. [Columbus: 1828.] 8 p.

Dated: January 17, 1828.
Hasse (Ohio)-p. 403.
Copies: MWA; NN; OHi.

5178. Ohio. Board of Canal Commissioners.

Report of the Board of Canal Commissioners. [Columbus: 1828]. 13 p.

Dated: January 17, 1828.
Concerns a survey made between Licking Summit and the Ohio River, by David S. Bates, engineer.
S-34602.
Copies: DLC; ICJ; MH-BA; NN; OMC.

5179. Ohio. General Assembly. House. Committee on Finance.

Report of the Committee on Finance on the report of the Commissioners of Canal Fund, and on so much of the report of the Canal Commissioners as relates to their accounts. . . . [Columbus: 1828]. 6 p.

Dated: February 6, 1828.
Hasse (Ohio)-p. 381.
Copies: OMC.

5180. ________ for the year ending Nov. 30, 1828. [Columbus: 1828]. 8 p.

Hasse (Ohio)-p. 381.
Copies: NN.

5181. Pennsylvania. Board of Canal Commissioners.

Documents accompanying the report of the Canal Commissioners, of the Commonwealth of Pennsylvania, for promoting the internal improvement of the state, with the exception of those relating to the surveys made the last season. Read in the Senate, January 4, 1828. Vol. 2. Harrisburg: Printed by S. C. Stambaugh, 1828. 180, 3 p., fold. tables.

Vol. 1 entitled: *Report of the Canal Commissioners.*
S-34675.
Copies: MH-BA; MiU-T; NN; P; PPAmP; PPF; PPL.

5182. Pennsylvania. Board of Canal Commissioners.

Message from the Governor, accompanied with the report of the Canal Commissioners. Read in the House of Representatives, December 13, 1828. Harrisburg: S. C. Stambaugh, printer, 1828. 326 p.

S-34681.
Copies: DLC; ICJ; MB; MWA; NN; P; PPAmP; PPF; PPL.

5183. ________ Read in the Senate, December 15, 1828. Harrisburg: S. C. Stambaugh, printer, 1828. 326 p.

5183 (*continued*)
S-34682.
Copies: DLC; DeGE; ICJ; MiU-T; P; PPL.

5184. Pennsylvania. Board of Canal Commissioners.
Report of the Canal Commissioners of the Commonwealth of Pennsylvania. Accompanied with documents. Read in the House of Representatives, January 4, 1828. Harrisburg: Printed by Samuel C. Stambaugh, 1828. 325p., fold. tables.
S-34684.
Copies: CtY; DBRE; DLC; DeGE; P; PPF; PPL.

5185. Pennsylvania. Board of Canal Commissioners.
Report of the Canal Commissioners, of the Commonwealth of Pennsylvania, for promoting the internal improvement of the state. With the documents relating to the surveys made the past season. Read in the Senate, January 4, 1828. Harrisburg: Printed by Samuel C. Stambauch [*sic*], 1828. 145, 2 p.
S-34685.
Copies: MH-BA; MWA; MiU-T; NN; P; PPAmP; PPF; PPL.

5186. Pennsylvania Canal Boat Company.
List of freights charged by the Pennsylvania Canal Boat Company to and from Middletown and Philadelphia. [Philadelphia: 1828.] 2 l.
S-34701.
Copies: DeGE.

5187. Pennsylvania Society for the Promotion of Internal Improvements in the Commonwealth.
Delaware Canal. [Philadelphia: 1828.] 12 p.
At head of title: "Published by the Society for the Promotion of Internal Improvement. From the Pennsylvania Gazette of March 4, 1828."
S-34707.
Copies: DLC; NN; PPAmP.

5188. Remarks on the Farmington Canal, &c. By an original stockholder. The construction of canals is the proper work of the sovereign power, and should never be committed to private stock companies. New Haven: 1828. 16 p.
S-34992.
Copies: CtHT-W; CtHi.

5189. Roberts, Nathan S., 1776–1852.
Reports relative to the Chenango Canal, by N. S. Roberts & H. Hutchinson, civil engineers. Made to the Senate, April 7th, 1828. Albany: Printed by Croswell and Van Benthuysen, 1828. 38 p.
S-35054.
Copies: MWA; NN.

5190. Shepard, Samuel W.
Erie & Junction canal directory; containing a list of the principal places on said canals, with their distance from each other, and from the several collectors' of ice. Little Falls, N.Y.: Griffing's Press, 1828. 1 p.
S-35170.
Copies: DLC; IU; NN.

5191. South Carolina. Superintendent of Public Works.
Report of the Superintendent of Public Works. [Columbia, S.C.]: Printed at the State Gazette office; D. & J. M. Faust, printers, [1828]. 15, 4 p.
Concerns the years 1819–1829, submitted in reply to a resolution. Report on a canal or railroad from Charleston to Hamburgh, dated March 15, 1828, 4 p. at end.
Signed: A. Blanding.
S-35359.
Copies: MCM; NN.

5192. South Carolina Canal and Railroad Company.
By laws of the South Carolina Canal and Rail Road Company, adopted by the stockholders May 13, 1828. [Charleston: J. S. Burges, printer, 1828.] 8 p.
S-35361.
Copies: DLC.

5193. ________ Together with the act of incorporation, granted by the state legislature. Charleston: Printed by James S. Burges, 1828. 16 p.

S-35362.

Copies: CSmH; DBRE; DLC; DNA; MB; MH-BA; NN; ScU; WM.

5194. South Carolina Canal and Railroad Company.

First semi-annual report, to the president and directors of the South-Carolina Canal and Rail-Road Company, by their committee of inquiry. Charleston: Printed by A. E. Miller, 1828. 24, [4] p.

S-35363.

Copies: CSmH; DLC; MB; MH-BA; MWA; NN; NNC; ScHi; ScU; WM.

5195. South Carolina Canal and Railroad Company.

Memorial of the South Carolina Canal and Rail Road Company. April 7, 1828. Referred to the Committee of Ways and Means. April 14, 1828. Bill reported, No. 268. Washington: Printed by Gales & Seaton, 1828. 4 p.(20th Cong., 1st sess. House. Doc. 246)

At head of title: South Carolina.

S-36602, 36889.

Copies: CtY; DBRE; DLC; DeU; IU; MB; MBAt; MH-BA; MWA; MiU-T; NN; NjP; P; PPAmP.

5196. Union Canal Company of Pennsylvania.

Annual report of the president and managers of the Union Canal Company of Pennsylvania, to the stockholders. November 18, 1828. Philadelphia: Printed by Lydia R. Bailey, 1828. 12 p.

Copies: DeGE; PHi; PPL.

5197. United States. Congress. House. Committee on Public Lands.

Lands to Ohio—for canals. February 11, 1828. Mr. Isacks, from the Committee on the Public Lands, to which the subject had been referred, made the following report.... [Washington: 1828.] 2 p. (20th Cong., 1st sess. House. Rept. 139)

S-36487.

Copies: DLC; DeU; MWA; NjR.

5198. United States. Congress. House. Committee on Roads and Canals.

Canal—Dayton to Lake Erie. January 18, 1828. Mr. Woods, of Ohio, from the Committee on Roads and Canals ... made the following report. [Washington: 1828.] 3 p. (20th Cong., 1st sess. House. Rept. 88)

S-36187.

Copies: DLC; DeU; MWA; NjR.

5199. United States. Congress. House. Committee on Roads and Canals.

Chesapeake and Ohio Canal. January 2, 1828. Mr. Mercer from the Committee on Roads and Canals, made the following report.... [Washington: Gales & Seaton], 1828. 136 p., fold. map, fold. plans. (20th Cong., 1st sess. House. Rept. 47)

S-36198.

Copies: DLC; DeGE; DeU; MWA; MiU-T; NN; NjR; PPF.

5200. United States. Congress. House. Committee on Roads and Canals.

Chesapeake and Ohio Canal, February 11, 1828. Read and referred to the Committee of the Whole House ... [Washington: 1828.] 146 p., fold. tables. (20th Cong., 1st sess. House. Rept. 141)

S-36199.

Copies: DLC; DeU; MWA; NjR.

5201. United States. Congress. House. Committee on Roads and Canals.

Chipola Canal Company. March 14, 1828. Mr. Mercer, from the Committee on Roads and Canals ... made the following report.... [Washington: 1828.] 8 p. (20th Cong., 1st sess. House. Rept. 191)

S-36204.

Copies: DLC; DeU; MWA; MiU-T; NN; NjR.

5202. United States. Congress. House. Committee on Roads and Canals.

Corporations—Washington, Georgetown, and Alexandria. January 30, 1828. Mr. Mercer, from the Committee on Roads and Canals to which was referred a memorial of the corporations of Washington, Georgetown and Alexandria, in the District of Columbia,

5202 (*continued*)
made the following report.... [Washington: 1828.] 32 p., 2 tables. (20th Cong., 1st sess. House. Rept. 112)

Concerns the subscription to the stock of the Chesapeake and Ohio Canal Company.

S-36242.

Copies: DLC; DeU; MWA; NjR.

5203. United States. Engineer Dept.

Canal—Baltimore to contemplated Chesapeake & Ohio Canal.... Letter from the Secretary of War, transmitting, pursuant to a resolution of the House of Representatives of the 12th ultimo, a report and plans of the survey of a route for a canal from the city of Baltimore to the contemplated Chesapeake and Ohio Canal. January 14, 1828.... Washington: Printed by Gales & Seaton, 1828. 31 p. (20th Cong., 1st sess. House. Doc. 58)

Signed: Wm. Howard, Assistant Civil Engineer.

S-36186.

Copies: DLC; DeGE; DeU; MWA; NjR.

5204. United States. Engineer Dept.

Chesapeake and Ohio Canal—extension of. Letter from the Secretary of War, transmitting a report, map, and estimate, of the Chesapeake and Ohio Canal to Alexandria, in the District of Columbia. April 21, 1828.... Washington: Printed by Gales & Seaton, 1828. 14 p., diagr. (20th Cong., 1st sess. House. Doc. 254)

S-36197.

Copies: DLC; DeGE; DeU; MWA; NjR.

5205. United States. Engineer Dept.

James River and Kenhawa Canal route. Letter from the Secretary of War, transmitting a report on the James River and Kenhawa Canal route. March 24, 1828.... Washington: Printed by Gales & Seaton, 1828. 68 p. (20th Cong., 1st sess. House. Doc. 216)

Report by Capt. William G. McNeill.

"Appendix, containing reports from Lieutenants Cook and Hazzard, illustrative of the location of the James and Kenhawa Canal; with reports from Lieutenants Dillahunty and Fessenden, on the supply of water," p. 33–68.

S-36433.

Copies: DBRE; DLC; DeGE; DeU; MWA; MiU-T; NN; NjR.

5206. United States. Laws.

An act to aid the state of Ohio in extending the Miami Canal from Dayton to Lake Erie. Approved May 4, 1828. An Act to grant land to Indiana for aiding said state in opening the Wabash and Erie Canal, March 2, 1827. [S.l.: 1828.] 4 p.

Hasse (Ohio)-p. 384.

Copies: NN; OMC.

5207. ———— May 20, 1828.... [Washington: 1828.] 3 p. (H.R. 94)

S-35624.

Copies: DNA.

5208. United States. Laws.

An act to amend and explain an act, entitled "An act confirming an act of the legislature of Virginia, incorporating the Chesapeake and Ohio Canal Company, and an act of the state of Maryland, for the same purpose." May 8, 1828.... [Washington: 1828.] 3 p. (H.R. 40)

S-35626.

Copies: DNA.

5209. United States. Laws.

A bill to incorporate the Alexandria Canal Company. December 29, 1828.... [Washington: 1828.] 19 p. (H.R. 341)

S-36140.

Copies: DNA.

5210. Vindex, pseud.

A review and refutation of the statements made in the late report of the Canal Commissioners, to the Legislature of this state. By Vindex [pseud.]. Albany: 1828. 8 + p.

Incomplete: all after p. 8 wanting.

Argument for the superiority of Buffalo over Black Rock as the western terminus of the Erie Canal.

S-37022.

Copies: MWA.

5211. Wright, Benjamin, 1770–1842.

Opinion of Judge Wright on certain

points in relation to the Chenango Canal. Albany: 1828. 10 p. (New York (State). Assembly. Doc. 106)
Whitford-p. 1249.
Copies: MWA; N.

5212. Binney, Amos, 1802–1878.
Memorial of Amos Binney, et al., Feb. 9, 1829. Committed to the Committee of the Whole House.... [Washington: Gales & Seaton, 1829.] 24 p. (20th Cong., 2d sess. House. Doc. 116)
Concerns the Chesapeake and Ohio Canal.
S-41229.
Copies: DLC; MWA; NjR; ViU.

5213. Blackstone Canal Company.
Letter concerning the issue of new stock; and vote of the company passed Nov. 14, 1829. Providence: 1829. 1 fold. sheet in 2 parts.
S-37857.
Copies: MW.

5214. Bradish, Luther, 1783–1863.
Substance of Mr. Bradish's remarks in the Committee of the Whole House, upon the bill, entitled, "An act to loan the credit of the state to the president, managers, and company of the Delaware & Hudson Canal Company." Albany: Printed by Croswell and Van Benthuysen, 1829. 38 p.
S-37925.
Copies: DBRE; DLC; NN; NRom.

5215. Chesapeake and Delaware Canal Company.
Chesapeake and Delaware Canal. Notice is hereby given, that this canal is now open for navigation.... [Philadelphia?: 1829.] Broadside.
Signed: Robert M. Lewis, President; H. D. Gilpin, Secretary.
Copies: DeHi.

5216. Chesapeake and Delaware Canal Company.
Memorial of the Chesapeake and Delaware Canal Company. January 12, 1829. Referred to the Committee on Roads and Canals. February 5, 1829. Bill reported, No. 417. [Washington: Gales & Seaton, 1829]. 1 p. (20th Cong., 2d sess. House. Doc. 111)
S-41076.
Copies: DLC; DeGE; DeU; MWA; NjR.

5217. Chesapeake and Delaware Canal Company.
Tenth general report of the president and directors of the Chesapeake and Delaware Canal Company. June 1, 1829. [Philadelphia: 1829.] 10 p., fold. map.
S-38125.
Copies: DLC; DeGE; MWA; PPL.

5218. Chesapeake and Ohio Canal Company.
Chesapeake and Ohio Canal. Schedule of contracts let in December, 1828, with the estimated quantity of work to be performed and the prices accepted. [Washington: 1829?] Broadside.
Copies: PPL.

5219. Chesapeake and Ohio Canal Company.
Communication from the Chesapeake and Ohio Canal Company, to the chairman of the Committee on Internal Improvement of the House of Delegates, March 7, 1829. Annapolis: Jeremiah Hughes, 1829. 3 p.
N-C0345829.
Copies: ViU.

5220. Chesapeake and Ohio Canal Company.
A communication from the president of the Chesapeake and Ohio Canal Company, to the executive of Maryland. February 28, 1829. Printed by order of the Senate. Annapolis: Jeremiah Hughes, printer, 1829. 7 p.
S-39447.
Copies: MWA; NN.

5221. Chesapeake and Ohio Canal Company.
Finances—Chesapeake and Ohio Canal Company. Letter from the President of the Chesapeake and Ohio Canal Company, transmitting information upon the subject of

5221 (*continued*)
the finances of said company, and its receipts and disbursements. Feb. 13, 1829. Read and committed to the Committee of the Whole House to which is committed the bill (no. 370) to amend the charter of the Chesapeake and Ohio Company.... [Washington: Gales & Seaton, printers, 1829.] 3 p. (20th Cong., 2d sess. House. Doc. 122)

S-41120.

Copies: DLC; DeGE; MWA; MiU; NN; NjR.

5222. Chesapeake and Ohio Canal Company.

First annual report of the president and directors of the Chesapeake and Ohio Canal Company, to the stockholders, in general meeting, June 1, 1829. Washington: Printed by Gales & Seaton, 1829. 21 p.

S-38126.

Copies: DLC.

5223. ———— Washington: Printed by Gales & Seaton, 1829. 42 p.

S-38128.

Copies: MH-BA; Md-LR; MdHi; NIC.

5224. Chesapeake and Ohio Canal Company.

First annual report of the president and directors of the Chesapeake and Ohio Canal Company, together with the proceedings of the stockholders, at their first annual meeting, begun on the first day of June, 1829, and held, by adjournments, on the 6th, 8th, 9th and 10th days of June; on the 31st day of July; and on the 11th, 25th, and 29th days of September, 1829. And also containing, in the appendix, the proceedings of the stockholders, at their special meeting in September, 1828. Washington: Printed by Gales & Seaton, 1829. 30, lix, 104 p.

Reports and letters from the engineers employed in the revised location of the western section of the Chesapeake and Ohio Canal, 104 p. at end.

Reports made by Benjamin Wright and Nathan S. Roberts.

S-38127.

Copies: CtY; DBRE; DLC; IU; MH-BA; MWA; MdBE; MdBJ; MdHi; MiU-T; NN; PPF; Vi.

5225. Chesapeake and Ohio Canal Company.

Fourth of July, 1829. Laying the cornerstone of one of the eastern locks of the Chesapeake and Ohio Canal.... By order of the committee, John Cox, Mayor. [Washington: 1829.] Broadside.

S-38129.

Copies: DLC.

5226. Chesapeake and Ohio Canal Company.

Letter from the president of the Chesapeake and Ohio Canal Company, transmitting information upon the subject of the finances of said company, and its receipts and disbursements. [Washington: Printed by Gales & Seaton], 1829. 3 p. (20th Cong., 2d sess. House. Doc. 122)

Charles Fenton Mercer, president.

N-M0458662.

Copies: ViU.

5227. Chesapeake and Ohio Canal Company.

Memorial of stockholders, &c. in the Chesapeake and Ohio Canal Company. February 12, 1829.... [Washington: Gales & Seaton, printers, 1829.] 4 p. (20th Cong., 2d sess. House. Doc. 120)

S-41077.

Copies: DLC; DeGE; DeU; MWA; MiU; NjR; ViU.

5228. Chesapeake and Ohio Canal Company.

Memorial of the Chesapeake and Ohio Canal Company, January 12, 1829.... [Washington: Gales & Seaton, printers, 1829.] 7 p. (20th Cong., 2d sess. House. Doc. 73)

Concerns the controversy with the Baltimore and Ohio Railroad Company.

S-41078.

Copies: DLC; DeGE; DeU; MWA; NjR; ViU.

5229. ———— [Washington?]:1829. 14 p.

N-C0345863.

Copies: ViU.

5230. ———— February 23, 1829.... [Washington: 1829.] 13 p. (20th Cong., 2d

sess. House. Doc. 127)
S-41079.
Copies: CtY; DBRE; DLC; DeU; MB; MBAt; MH; MWA; MiU-T; NN; NjP; PPAmP.

5231. Chesapeake and Ohio Canal Company.
Memorial of the president and directors of the Chesapeake and Ohio Canal Company, relative to differences between that company and the Baltimore and Ohio Railroad Company, as to precedence, in the selection and location of the route, at certain points. February 26, 1829. Referred to the Committee on Roads and Canals ... [Washington: Printed by Duff Green, 1829]. 14 p. (20th Cong., 2d sess. Senate. Doc. 99)
S-41233
Copies: CtY; DBRE; DLC; DeGE; DeU; M; MB; MH; MWA; MiU-T; N; NN; NjP; P; PPAmP.

5232. Delaware and Hudson Canal Company.
An analysis of the memorial of the Delaware and Hudson Canal Company, and the considerations offered in its support. [Albany: 1829]. 26 p.
Signed; John Bolton, president of the Delaware and Hudson Canal Company. Albany, Jan. 28, 1829.
S-38364.
Copies: DBRE; NN.

5233. Ferguson, James, 1797–1867.
Report of James Ferguson, Esq., engineer, relative to surveys and examinations of the French Creek section of the Pennsylvania Canal ... Harrisburg: Printed at the office of the Reporter, 1829. 8 p.
S-38565.
Copies: DLC.

5234. Hollins, William.
To the Honourable the Senate and House of Representatives of the United States of America, in Congress assembled. [A memorial concerning government subscriptions to the Chesapeake and Ohio Canal Company and to the Baltimore and Ohio Rail Road Company.] Baltimore: 1829. 1 l.
Dated: January 17, 1829.
N-H0466130.
Copies: CtY.

5235. Hosack, David, 1769–1835.
Memoir of De Witt Clinton: with an appendix, containing numerous documents, illustrative of the principal events of his life.... New York: Printed by J. Seymour, 1829. xxiv, [21]-530 p., 2 plates, fold. map.
Partially devoted to the construction of Erie Canal.
S-39031.
Copies: CSmH; CU; CtY; DLC; DNLM; DeGE; ICU; IaU; MH; MWA; MdHi; MiD-B; MiU-C; MnHi; MoSW; NIC; NN; NhHi; NjP; OClW; P; PPL; RPB; ViU; VtMiM; WHi.

5236. Illinois. General Assembly.
Memorial of the General Assembly of Illinois, praying that Congress will grant to that state a sufficient quantity of land to enable it to complete the canal between the Illinois River and Lake Michigan. February 16, 1829.... [Washington: 1829.] 2 p. (20th Cong., 2d sess. Senate. Doc. 80)
Copies: DeU; MWA.

5237. Indiana. Canal Commissioners.
Report of the Board of Commissioners of the Wabash and Miami Canal. December 15, 1829. [Indianapolis?: 1829.] 15 p.
Issued as *Appendix A* of the journals of the Senate and of the House of Representatives of Indiana.
Copies: In.

5238. Lehigh Coal and Navigation Company.
Report of the board of managers of the Lehigh Coal and Navigation Company, presented to the stockholders, January 12th, 1829. Philadelphia: Printed by Soloman W. Conrad, [1829]. 15 p.
S-39267.
Copies: IU; MBAt; MH-BA; MWA; NN; P; PPAmP; PPL.

5239. Louisville and Portland Canal Company.

5239 (*continued*)

Louisville and Portland Canal Company. To accompany Senate bill no. 14. Fourth annual report of the president and directors of the Louisville and Portland Canal Company. February 4, 1829.... Washington: [Gales & Seaton, printers, 1829]. 5 p. (20th Cong., 2d sess. House. Doc. 107)

S-41224.

Copies: DLC; DeU; MWA; NjR.

5240. Maryland. Governor (D. Martin).

A communication from the executive department of Maryland on the 16th of February, 1829, enclosing the report of the directors of the Chesapeake and Ohio Canal Company. Annapolis: Jeremiah Hughes, printer, 1829. 15 p., 11 fold. tables.

S-39443.

Copies: MWA; NN.

5241. Maryland. Governor (D. Martin).

A communication from the executive, enclosing the report of the Chesapeake and Delaware Canal Company. February 21, 1829. Annapolis: J. Hughes, printer, 1829. 8 p.

Report of the company signed: James Fisher, president.

Bowker-p. 623.

Copies: DeGE.

5242. Maryland. Governor (T. K. Carroll).

Document No. 9. Accompanying the executive message to the Legislature. December 31, 1829. Correspondence with the President of the United States and the Secretary of War, respecting survey of canal routes, &c. Annapolis: J. Hughes, printer, 1829. 7 p.

S-39458.

Copies: MWA; Md; MdHi.

5243. Maryland. Governor (T. K. Carroll).

Document No. 11. Accompanying the executive message, to the Legislature. December 31, 1829. Report respecting the Chesapeake and Ohio Canal. Annapolis: J. Hughes, printer, 1829. 4 p.

S-39459.

Copies: MdHi; ViU.

5244. Maryland. Governor (T. K. Carroll).

Document No. 14. Accompanying the executive communication of 31st December, 1829. Third annual report of the Chesapeake and Ohio Canal Company. Published by authority. Annapolis: J. Hughes, printer, 1829. 37, cxi p.

S-39461.

Copies: CtY; MWA; MdHi; MiU-T.

5245. Massachusetts. Board of Internal Improvements.

Report of the Board of Commissioners of Internal Improvements, in relation to the survey of a route for a canal from Boston to the Blackstone Canal, and thence to the line of Connecticut.... Boston: True and Greene, state printers, 1829. 51 p., fold. map. (Senate. Doc. 19)

"An estimate of the cost of constructing the canal," signed James Hayward, p. 47–51.

S-39502.

Copies: CSmH; DLC; DeGE; ICU; IU; KU; MB; MH; MWA; MeU; NN; PP; RPB; WHi.

5246. Massachusetts. General Court. Committee on Survey for Canal Route from Boston Harbor to Narragansett Bay.

[Report of the Committee appointed by an order of the legislature, in June, 1826, to aid and assist such engineers as might be appointed by the government of the United States, to make a survey for a canal route, from Boston harbour to Narragansett Bay.] [Boston: 1829.] 8 p. (House. Doc. 37)

Signed: Minott Thayer, Benj. Hobart, William Mason.

S-39505.

Copies: DLC; MB; MWA.

5247. Memorial of sundry holders of land, counter to the memorial of the Chesapeake and Ohio Canal Company [to the General Assembly of Maryland] February 9th, 1829. Annapolis: Jeremiah Hughes, printer, 1829. 7 p.

S-39555.

Copies: NNC; ViU.

5248. New Jersey. Governor (I. H. Williamson).

Governor's message, with accompanying documents, relative to the probable trade of the proposed Delaware and Raritan Canal. Read in the House of Assembly on Monday, 19th Jan., 1829.... [Trenton: 1829.] 16 p.
S-39762.
Copies: CtY; DBRE; DeGE; MWA; MiU-T; NjP.

5249. New Jersey. Legislature. General Assembly. Committee on the Delaware and Raritan Canal.
Report of the Committee, to whom was referred the subject of the Delaware & Raritan Canal, made to the House of Assembly on the fifteenth day of January, eighteen hundred and twenty-nine.... Trenton: Printed by J. Justice, 1829. 36 p.
S-39765.
Copies: CtY; DBRE; DLC; DeGE; ICJ; MH-BA; MWA; NN; Nj; NjP; NjR.

5250. ———— With other documents. [Trenton: 1829.] 32 p.
S-39766.
Copies: DBRE; MB; MH; MH-BA; MiU-T; NN; NjP; PPAmP.

5251. New York (State). Canal Board.
Report, on the petition of Paul Knight, referred to that board by the Honorable the Assembly. [Albany: 1829.] 5 p. (Assembly. Doc. 196)
Dated: In Assembly, March 30, 1829.
N-N0189708.
Copies: MWA; NN.

5252. New York (State). Canal Commissioners.
Annual report of the Canal Commissioners of the state of New York, made to the Assembly, Jan. 26, 1829. Albany: Printed by Croswell and Van Benthuysen, 1829. 28 p.
S-39786.
Copies: DLC; ICJ; MWA; N; NN; NRom; WHi.

5253. New York (State). Canal Commissioners.
Report of the Canal Commissioners, to whom was referred the petition of William Brayton and others, praying for a re-appraisal of damages done to their lands, by the construction of the Champlain Canal. [Albany: 1829.] 17 p. (Assembly. Doc. 156)
Dated: March 12, 1829.
Copies: MWA; NHi.

5254. New York (State). Canal Commissioners.
Report on the subject of lowering the Oneida Lake. Albany: 1829. [illegible] p. (Assembly. Doc. 120)
Whitford-p. 1247.
Copies: MWA; N.

5255. New York (State). Canal Commissioners.
Report relative to the survey of a canal route from Seneca Lake to Bath. Albany: 1829. 11 p. (Senate. Doc. 84)
Whitford-p. 1246.
Copies: MWA; N.

5256. New York (State). Commissioners of the Canal Fund.
Annual report of the Commissioners of Canal Fund to the Legislature of the state of New-York. [Albany: 1829.] 16 p. (Assembly. Doc. 174)
Dated: In Assembly, March 20, 1829.
Copies: MWA; NHi.

5257. New York (State). Laws.
An act authorizing the construction of the Penn-Yan Canal. Albany: 1829. 2 p. (Senate. Doc. 139)
Whitford-p. 1246.
Copies: MWA; N.

5258. New York (State). Legislature. Assembly. Committee on Canals.
Documents accompanying the report of the Committee on Canals ... respecting construction of Black River Canal. [Albany]: 1829. 23 p.
Hasse (New York) p. 130.
Copies: NN.

5259. New York (State). Legislature. Assembly. Committee on Canals.

5259 (*continued*)

Report of the Committee on Canals and Internal Improvements relative to the construction of the Black River Canal. [Albany]: 1829. 44, 35 p., maps. (Assembly. Doc. 146)

Signed: Benjamin Wright.

N-N0200263; Shaw-465.

Copies: MWA; N; NN; NRU.

5260. New York (State). Legislature. Assembly. Committee on Canals.

Report on the petitions for the construction of the Chenango Canal. Albany: 1829. 31 p. (Assembly. Doc. 24)

Whitford-p. 1249.

Copies: MWA; N.

5261. New York (State). Legislature. Senate. Committee on Canals.

Report of the Committee on Canals relative to the Crooked Lake Canal. [Albany: 1829.] 7 p. (Senate. Doc. 138)

N-N0201303.

Copies: MWA; NRU.

5262. Ohio. Board of Canal Commissioners.

Report. To the General Assembly of the State of Ohio. [Columbus: 1829.] 22 p.

Dated: January 6, 1829.

S-39904.

Copies: MWA; NN; OClWHi.

5263. Ohio. Board of Canal Commissioners.

Special report of Canal Commissioners relative to practicability of a canal from Dayton to Lake Erie by way of Fort Defiance. [Columbus: 1829.] 4 p.

Hasse (Ohio)-p. 354.

Copies: OMC.

5264. Ohio. Commissioners of the Canal Fund.

Report of the Canal Fund Commissioners. [Columbus: 1829.] 3 p.

Dated: December 7, 1829.

Bowker-p. 205.

Copies: OClWHi.

5265. Ohio. Governor (Trimble).

Canal lands. Special message [of the Governor, relative to the location of public lands granted by the United States to the state of Ohio to aid in the construction of canals. Columbus: 1829]. 3 p.

Dated: January 19, 1829.

Hasse (Ohio)-p. 363.

Copies: NN.

5266. Ohio Canal Company.

Ohio Canal. South of Deer-Creek. Notice to bidders. Each proposition must state distinctly.... The work must be completed by the first of April, 1831. Chillicothe, July 15, 1829. [Chillicothe, Ohio: 1829.] Broadside 26.5 x 37 cm.

Lists 29 projects of work on which bids will be taken.

S-39907.

Copies: WHi.

5267. Ohio Canal Company.

Rules and specifications, relating to the construction of the Ohio Canal, and the estimating of work performed thereon. Cleveland: Printed by John R. St. John, 1829. 14 p.

S-39908.

Copies: OClWHi; WHi.

5268. Ohio Canal Company.

Rules for the construction of masonry, on the Ohio Canal.... Chillicothe, Oct. 13, 1829. Chillicothe, [Ohio]: Printed at the office of the Scioto Gazette, [1829]. Broadside 29 x 38.5 cm.

S-39909.

Copies: WHi.

5269. Ohio Canal Company.

Sections of canal, to be let at Chillicothe, on July 15th, 1829.... Sections of canal to be let at Portsmouth, on July 20th, 1829. [Chillicothe?, Ohio: 1829.] Broadside 20 x 27 cm.

S-39910.

Copies: WHi.

5270. Ohio Canal Company.
Specifications for a lock. The walls of locks shall be of solid masonry, laid in water cement. Chillicothe, July 15, 1829. [Chillicothe, Ohio: 1829.] Broadside 19.5 x 29 cm.
S-39911.
Copies: WHi.

5271. Ohio Canal Company.
Specifications for a stone and wood lock. This lock is to consist of four pillars of stone masonry.... [Chillicothe, Ohio: 1829?] Broadside 18.5 x 22 cm.
S-39912.
Copies: WHi.

5272. Pennsylvania. Board of Canal Commissioners.
Message from the Governor, accompanied with the report of the Canal Commissioners, relative to the Pennsylvania Canal and Rail-Road. Read in the Senate—December 18, 1829. Harrisburg: Printed by Welsh & Miller, 1829. 14 p.
S-39999.
Copies: DBRE; DLC; MB; MHi; MWA; MiU-T; P; PHi.

5273. ——— Read in the House of Representatives, December 18, 1829. Harrisburg: Printed by Welsh & Miller, 1829. 13 p.
S-40000.
Copies: DLC; MB.

5274. Potts, Charles.
The canal engineer's and contractor's assistant, being an epitome of the theory of canal cutting; with rules for gauging water, locating curves, calculating tables, &c.... Together with a new and extensive table of excavation, adapted to the dimensions of the Pennsylvania Canal.... Philadelphia: W. Brown, printer, 1829. 40 p., incl. tables, 2 plates.
S-40137.
Copies: DLC; DeGE; NIC; NN; NjR; P; PHi.

5275. Renwick, James, 1790–1863.
A discourse on the character and public services of DeWitt Clinton, delivered before the alumni of Columbia College ... 6th May, 1829. New York: G. & C. & H. Carvill, 1829. 44 p.
S-40243.
Copies: CtHT; DLC; ICU; MH; MWA; NIC; NNC; NbU; NjR; PPL; RPB; WHi.

5276. Robinson, Moncure, 1802–1891.
Report of the engineer on the proposed line of a canal, from the north branch of the river Susquehanna, through the Nescopeck Valley, to the river Lehigh. By Moncure Robinson. Read in the House of Representatives, January 12, 1829. Harrisburg: Printed at the office of the Reporter 1829. 14 p.
N-R0337035; Sabin-72159.
Copies: PBL.

5277. South Carolina Canal and Railroad Company.
Memorial of the president and directors of the South Carolina Canal and Rail Road Company, praying that the United States may be authorized to subscribe two thousand five hundred shares of the capital stock of said company ... Feb. 19—Bill reported (No. 92).... [Washington: 1829.] 4 p. (20th Cong., 2d sess. Senate. Doc. 91)
S-41234.
Copies: CSt; CtY; DBRE; DLC; DeU; M; MB; MH; MWA; MiD; MiU-T; N; NN; NRU; NjP; P; PPAmP.

5278. South Carolina Canal and Railroad Company.
Report. Engineer's office.... To the president and directors of the South Carolina Canal and Railroad Co. Charleston: 1829. 16 p.
Signed: J. B. Petival.
Dated: October 10, 1829.
S-40491.
Copies: CSt; DBRE; MH.

5279. South Carolina Canal and Railroad Company.

Semi-annual report of the board of directors, and of the South Carolina Canal and Railroad Company. Charleston: Printed by A. E. Miller, 1829. 7 p.

S-40493.

Copies: CSmH; DLC; MH-BA; ScU; WM.

5280. Thomas, David, 1776–1859.

Survey of a canal route from the Seneca Lake to the village of Bath. Albany: 1829. 10 p. (New York (State). Legislature. Senate. Journal, 1829. Appendix D)

Whitford-p. 1246.

Copies: N.

5281. Tibbits, George, 1763–1849.

The finances of the Canal Fund of the state of New-York, and of their application, examined in a letter to the Hon. Stephen Allen and G. B. Throop, Esquires, from George Tibbets [*sic*], Esq. Albany: Printed by John B. Van Steenbergh, 1829. 24 p.

S-40651.

Copies: DLC; M; MB; MiD-B; NHi.

5282. Union Canal Company of Pennsylvania.

Annual report of the managers of the Union Canal Company of Pennsylvania, to the stockholders. November 17, 1829. Philadelphia: Printed by Thomas Desilver, 1829. 12 p.

S-40707.

Copies: DBRE; DLC; DeGE; IU; MH-BA; MWA; MiU-T; PHi; PPAmP; PPL; PU.

5283. United States. Board of Internal Improvement.

Canal—Atlantic and Gulf of Mexico. Message of the President of the United States, transmitting a report of the Board of Engineers, of a survey with a view to ascertain the most eligible route for a canal to connect the Atlantic with the Gulf of Mexico, &c. February 26, 1829.... [Washington: Gales & Seaton, printers, 1829.] 52 p. (20th Cong., 2d sess. House. Doc. 147)

S-41069.

Copies: DLC; DeU; MWA; NN; NjR.

5284. United States. Board of Internal Improvement.

Message from the President of the United States, in relation to the survey of a route for a canal between the Gulf of Mexico and the Atlantic Ocean. February 28, 1829.... [Washington: 1829.] 52 p. (20th Cong., 2d sess. Senate. Doc. 102)

Transmits a report of the Board of Internal Improvement.

N-U0065673.

Copies: DeGE; DeU; MWA; NN.

5285. United States. Congress. House. Committee on Public Lands.

Lake Pontchartrain Canal. (To accompany bill H.R. No. 445). February 24, 1829. Mr. Isacks, from the Committee on the Public Lands ... made the following report.... [Washington: Gales & Seaton, printers, 1829.] 3 p. (20th Cong., 2d sess. House. Rept. 94)

S-41209.

Copies: DLC; DeU; MWA; NjR.

5286. United States. Engineer Dept.

Canal—Connecticut River to Lake Memphremagog, &c. Letter from the Secretary of War, transmitting a report and survey of a canal from Connecticut River to Lake Memphremagog, and the report of a survey from the same river to Lake Champlain. Feb. 10, 1829.... [Washington: Gales & Seaton, printers, 1829.] 12 p. (20th Cong., 2d sess. House. Doc. 118)

S-41070.

Copies: DLC; DeU; MWA; NjR.

5287. United States. Laws.

Amendments to the "Bill to amend the charter of the Chesapeake and Ohio Canal Company and for other purposes." Mr. Mercer, from the Committee on Roads and Canals, submitted the following as amendments. [Washington: 1829]. 3 p. (H.R. 370)

S-40795.
Copies: DNA.

5288. United States. Laws.
A bill to amend the charter of the Chesapeake and Ohio Canal Company, and for other purposes. January 15, 1829 ... [Washington: 1829]. 3 p. (H.R. 370)
S-41012.
Copies: DNA.

5289. Williams, Charles.
To the Honorable the Senate and House of Representatives of the United States in Congress assembled ... Washington City: 1829. 1 p.
Memorial concerning the Chesapeake and Ohio Canal, dated Dec. 21, 1829.
S-41537.
Copies: DLC.

5290. The work on the Rappahannok Canal.... Fredericksburg, [Va.]: 1829. Broadside.
Dated: January 19, 1829.
Copies: MWA.

5291. Binney, Amos, 1802–1878.
Amos Binney. May 24, 1830. Read and referred to the Committee of the Whole House to which is committed the bill H.R. no. 441, confirming certain acts of the Legislature of Virginia, relating to the Chesapeake and Ohio Canal Company. Washington: Printed by Duff Green, 1830. 36 p. (21st Cong., 1st sess. House. Rept. 412)
Memorial by Amos Binney.
C-3897.
Copies: DLC; MWA; NjR.

5292. Chesapeake and Delaware Canal Company.
Eleventh general report of the president and directors of the Chesapeake and Delaware Canal Company. June 7, 1830. [Baltimore?: 1830.] 17 p.
C-839.
Copies: DLC; DeGE; MWA; NbU; PPL.

5293. Chesapeake and Ohio Canal Company.
Chesapeake and Ohio Canal Company. Memorial of the president and directors of the Chesapeake and Ohio Canal Company. May 24, 1830.... [Washington: Printed by Duff Green, 1830.] 108 p. (21st Cong., 2d sess. House. Doc. 53)
Appendix, consisting of reports and other documents: p. 11–108.
C-4628.
Copies: CSt; CtY; DLC; DeGE; DeU; IU; MH; MWA; MiU-T; NN; NjP; P; PPL; ViU.

5294. Chesapeake and Ohio Canal Company.
Correspondence between the Chesapeake and Ohio Canal Company and the Baltimore and Ohio Rail Road Company. In relation to the disputes between those companies concerning the right of way for their respective works along the Potomac River Baltimore: Printed by William Wooddy, 1830. 80 p., illus.
C-840.
Copies: CSt; DBRE; DLC; DeGE; MH; MWA; MdBJ-G; MdHi; MiU; NNE; PPF; PPi; ViU; WU.

5295. Chesapeake and Ohio Canal Company.
Letter from the president of the Chesapeake and Ohio Canal Company, with proceedings of a meeting of that company, to the Hon. E. F. Chambers, chairman of the Committee on the District of Columbia. May 24, 1830.... [Washington: 1830.] 3 p. (21st Cong., 1st sess. Senate. Doc. 144)
Concerns an agreement between the Chesapeake and Ohio Canal and Baltimore and Ohio Railroad companies on routes for the canal and railroad.
C-4899.
Copies: CSt; CtY; DLC; DeU; MB; MWA; NN; NjP; NjR; PPAmP.

5296. Chesapeake and Ohio Canal Company.
Memorial from the Chesapeake and

5296 (*continued*)
Ohio Canal Company to the legislature of Maryland. Annapolis: Jeremiah Hughes, 1830. 12 p.
C-841.
Copies: MdBP; ViU.

5297. Chesapeake and Ohio Canal Company.
Memorial of the president and directors of the Chesapeake and Ohio Canal Company. May 24, 1830. . . . [Washington: 1830.] 10 p.(21st Cong., 2d sess. House. Rept. 31)
C-4968.
Copies: CSt; CtY; DBRE; DLC; DeGE; DeU; IU; MB; MWA; MiU-T; N; NN; NjP; P; ViU.

5298. Chesapeake and Ohio Canal Company.
Second annual report of the president and directors of the Chesapeake and Ohio Canal Company, together with the proceedings of the stockholders, at their second annual meeting, begun on the seventh day of June, 1830. Washington: Printed by Gales & Seaton, 1830. 42 p., 8 fold. tables.
C-842.
Copies: CtY; DLC; LNHT; MWA; MdHi; MiU-T; NN; NNC; PPF.

5299. ———— begun on the 7th day of June, 1830, and held, by adjournment, on the 8th, 9th, 10th, and 12th days of June. Washington: Printed by Gales & Seaton, 1830. 38 p., 8 fold. tables.
C-843.
Copies: DLC; ICJ; NN.

5300. Elliot, Jonathan, 1784–1846.
Historical sketches of the ten miles square forming the District of Columbia . . . describing objects of general interest . . . the proposed route and plan of the Chesapeake and Ohio Canal, from Georgetown to Pittsburgh. . . . Washington: Printed by J. Elliot, jr., 1830. 554 p., front.
C-1246.
Copies: DLC; MB; MWA; MiU; NjP; OCl; Vi; ViU.

5301. Henry, John R.
Few statements showing the utility and importance of the Louisville & Portland Canal around the falls of the Ohio River. Philadelphia: Charles Alexander, printer, 1830. 23 p.
C-1818.
Copies: PPi.

5302. [Hillhouse, James] 1754–1832.
Farmington, and Hampshire and Hampden canals. Statement on the petition of the Farmington, and Hampshire and Hampden canals. (To accompany bill H.R. No. 276). February 22, 1830. . . . [Washington: 1830.] 7 p. (21st Cong., 1st sess. House. Rept. 221)
By James Hillhouse, agent for the Farmington, and Hampshire and Hampden canals.
C-4708.
Copies: DLC; DeGE; DeU; MWA; NjR.

5303. [Hillhouse, James] 1754–1832.
Farmington, &c. Canal Company. (To be annexed to Report no. 221). April 29, 1830. Answer to the remonstrance of the agent of the Connecticut River Company. [Washington: 1830.] 3 p. (21st Cong., 1st sess. House. Rept. 221)
By James Hillhouse, agent for the Farmington, and Hampshire and Hampden canals.
C-4710.
Copies: DLC; DeGE; MWA; NjR.

5304. [Hillhouse, William] 1757–1833.
New Hartford side cut. No.1. [New Haven: 1830.] 8 p.
Signed at end: William Hillhouse. New-Haven, June 28, 1830.
Concerns a branch from the Farmington Canal to New Hartford.
N-H0375409.
Copies: CtHT-W; CtSoP.

5305. Hollins, William.
Circular. To the editor of the Baltimore Gazette. . . . [Baltimore: 1830?] [3] p.

Concerns the canal from Baltimore to the Potomac River.
N-H0466124.
Copies: CtY.

5306. Illinois. Canal Commissioners.
Report of the Canal Commissioners of the state of Illinois to the General Assembly on Illinois and Michigan Canal. Report of the Commissioners approved under the act of Jan. 22, 1829. [Vandalia, Ill.: 1830.] 15 p.
Dated: Vandalia, Ill., December 27, 1830.
Copies: MWA.

5307. Indiana. Canal Commissioners.
Report of the Commissioners of Wabash and Erie Canal. December 18, 1830. [Indianapolis?: 1830.] 7 p.
Issued as Appendix B of the *Journal of the Senate* and Appendix C of the *Journal of the House of Representatives of Indiana.*
Copies: In.

5308. Lacock, Abner, 1770–1837.
Letter from Abner Lacock ... to the Hon. David Scott, Esq., president of the Board of Canal Commissioners, of Pennsylvania.... Pittsburgh: "Statesman" print, 1830. 1 p.l., 31 p.
Concerns the Pennsylvania Canal.
C-2151.
Copies: DLC; PPL.

5309. Lehigh Coal and Navigation Company.
Report of the board of managers of the Lehigh Coal and Navigation Company, presented to the stockholders, January 11th, 1830. Philadelphia: Printed by Timothy A. Conrad, 1830. 16 p.
C-2198.
Copies: DBRE; DIC; DLC; IU; MH-BA; MWA; MiU-T; NN; P; PPAmP; PPL.

5310. Louisville and Portland Canal Company.
Fifth annual report of the president and directors of the Louisville and Portland Canal Company. [Louisville: 1830.] 2 p.
C-2287.
Copies: MWA; WHi.

5311. Louisville and Portland Canal Company.
In the Senate of the United States. January 26, 1830.... Fifth annual report of the president and directors of the Louisville and Portland Canal Company. [Washington: 1830.] 5 p. (21st Cong., 1st sess. Senate. Doc. 36)
Copies: DeU; MWA.

5312. Maine. Laws.
An abstract of the laws and regulations relating to the Cumberland and Oxford Canal. Portland: Printed by Thomas Todd, 1830. 18, [2] p.
C-2358.
Copies: MeHi.

5313. Maryland. Governor (T. K. Carroll).
Communication from the executive department, enclosing the report and memorial of the president and directors of the Chesapeake and Ohio Canal Company, and a communication from the executive of Georgia, relative to the election of president and vice president, to the Legislature of Maryland. February 1st, 1830. Annapolis: Printed by J. Hughes, 1830. 10 p.
C-2413.
Copies: MdHi.

5314. Maryland. Governor (T. K. Carroll).
(Document No. 4) Accompanying the executive communication of the 29th December 1830. Communication from the Governor of Virginia, relative to the Chessapeake [*sic*] and Ohio Canal. Annapolis: Printed by Jonas Green, 1830. 7 p.
C-2422; N-V0186799.
Copies: MdHi; ViU.

5315. Monopoly is tyranny; or, An appeal to the people and legislature from the oppres-

5315 (*continued*)
sion of the Delaware & Hudson Canal Company. Dundaff, Pa.: Printed by S. Hamilton, 1830. 28 p.

N-M0708626; Whitford-p. 1255.

Copies: DLC; MiU-C; N.

5316. New York (State). Canal Board.

In Senate, March 19, 1830. Report of the Canal Board.... [Albany: 1830.] 14 p. (Senate. Doc. 291)

Concerns tolls on merchandise transported on the canals.

N-N0189689.

Copies: DLC.

5317. New York (State). Canal Board.

Report of the Canal Board, on the reference to that board of the petition of Andrew P. Tillman, by the Hon. the Assembly. March 4, 1830. [Albany: 1830.] 9 p. (Senate. Doc. 240)

N-N0189701.

Copies: NN.

5318. New York (State). Canal Commissioners.

Annual report of the Canal Commissioners of the state of New York on the Chenango Canal. Albany: 1830. 42 p. (Assembly. Doc. 47)

Whitford-p. 1249.

Copies: MWA; NN.

5319. New York (State). Canal Commissioners.

Col. Young's report. [S.l.: 1830.] 16 p.

Dated: February 23, 1830.

Reprint of *Report of one of the Canal Commissioners, in answer to the resolution of the Assembly of the 20th ult., requiring estimates of the costs and revenues of the Crooked Lake and Chemung canals,* presented to the Legislature of the state of New York by Samuel Young.

Copies: NHi.

5320. New York (State). Canal Commissioners.

Report of one of the Canal Commissioners, in answer to the resolution of the Assembly, of the 20th ult., requiring estimates of the costs and revenues of the Crooked Lake and Chemung canals. [Albany: 1830.] 19 p. (Assembly. Doc. 195)

Dated: February 23, 1830.

By Samuel Young, Canal Commissioner.

N-N0189726; Parsons-p. 620.

Copies: MWA; NN.

5321. New York (State). Canal Commissioners.

Report of the Canal Commissioners, in obedience to a resolution of the Honorable the Assembly, on the subject of fences and bridges on the canals. March 25, 1830. [Albany: 1830.] 3 p. (Senate. Doc. 334)

Hasse (New York)-p. 80.; N-N0189732.

Copies: MWA; NN.

5322. New York (State). Canal Commissioners.

Report of the Canal Commissioners, in relation to the lowering of Oneida Lake. January 26, 1830. [Albany: 1830.] 12 p. (Senate. Doc. 68)

N-N0189733; Whitford-p. 1247.

Copies: MWA; NN.

5323. New York (State). Canal Commissioners.

Report of the Canal Commissioners, on several petitions and a remonstrance on the subject of Fort Miller dam. February 1, 1830. [Albany: 1830.] 3 p. (Assembly. Doc. 86)

N-N0189736.

Copies: MWA; NN.

5324. New York (State). Canal Commissioners.

Report of the Canal Commissioners on the Chemung and Crooked Lake canals. [Albany: 1830.] 22, 39–41 p., 3 fold. tables. (Assembly. Doc. 97)

Dated: February 4, 1830.

Includes a report by Holmes Hutchinson, engineer.

N-N0189737; Whitford-p. 1244.

Copies: MB; MH-BA; MWA; N; NN; NRU.

5325. New York (State). Canal Commissioners.

Report of the Canal Commissioners, on the petition of Marcus Andrews. March 30, 1830. [Albany: 1830.] 4 p. (Assembly. Doc. 355)

N-N0189738.

Copies: MWA; NN.

5326. New York (State) Canal Commissioners.

Report of the Canal Commissioners, on the petition of Wm. James and others. February 1, 1830. [Albany: 1830.] 2 p. (Assembly. Doc. 83)

N-N0189739.

Copies: MWA; NN.

5327. New York (State). Canal Commissioners.

Report of the Canal Commissioners requiring estimates of costs and revenues of Crooked Lake and Chemung canals. Albany: 1830. 19 p. (Assembly. Doc. 195)

Hasse (New York)-p. 136.

Copies: MWA; NN.

5328. New York (State). Commissioners of the Canal Fund.

Annual report of the Commissioners of the Canal Fund of the state of New York.... Albany: Printed by Croswell and Van Benthuysen, 1830. 31 p. (Assembly. Doc. 152)

C-2776.

Copies: MCM; MWA; MiD-B.

5329. New York (State). Comptroller's Office.

Copy of the register of canal boats, as appears from the book of registry, of those boats in the Comptroller's Office, on the 1st day of January 1830. [Albany?: 1830.] 30 p.

C-2779.

Copies: N.

5330. New York (State). Comptroller's Office.

Report from the Comptroller, realtive to the amount of compensation paid to the canal appraisers, January 16, 1830. [Albany: 1830.] 5 p.(Assembly. Doc. 28)

N-N0191303.

Copies: MWA; NN.

5331. New York (State). Legislature. Assembly. Committee on Canals.

Report on lowering of Oneida Lake. Albany: 1830. 12 p. (Assembly. Doc. 68)

Hasse (New York)-p.154.

Copies: MWA; NN.

5332. New York (State). Legislature. Assembly. Committee on Canals.

Report on the Chemung Canal. Albany: 1830. 3 p. (Assembly. Doc. 326)

Whitford-p. 1244.

Copies: MWA; N.

5333. New York (State). Legislature. Assembly. Committee on Canals.

Report on the report of the Canal Commissioners on the Chenango Canal. Albany: 1830. 15 p. (Assembly. Doc. 299)

Whitford-p. 1249.

Copies: MWA; N.

5334. New York (State). Legislature. Senate. Committee on Canals and Internal Improvements.

Report of the Committee on Canals and Internal Improvements, on the petition of sundry inhabitants of the counties of Oneida, Madison and Onondaga, for an act authorising the lowering of the Oneida Lake. March 2, 1830. [Albany: 1830.] 8 p. (Assembly. Doc. 225)

N-N0201307.

Copies: MWA; NN.

5335. New York (State). Legislature. Senate. Select Committee on the Petition of Sundry Inhabitants of the County of Seneca.

Report of the Select Committee, on the petition of sundry inhabitants of the county of Seneca, and the report of the Canal Commissioners thereon. March 29, 1830. [Albany: 1830.] 1 l. (Assembly. Doc. 374)

N-N0201469.

Copies: MWA; NN.

5336. Ohio. Board of Canal Commissioners.

Report of the Board of Canal Commissioners for 1829. [Columbus: 1830.] 18 p.

Dated: January 9, 1830.

Copies: MWA (22 p.); OHi.

5337. Ohio. General Assembly. House. Committee on Canals.

Report of the Standing Committee on Canals on the report of W. Silliman, commissioner to treat with Indiana on subject of connecting by a navigable canal the waters of the Wabash with Lake Erie. [Columbus: 1830.] 3 p.

Hasse (Ohio)-p. 415.

Copies: OMC.

5338. Ohio. General Assembly. House. Committee on Finance.

Report of the Committee on Finance on the report of the Canal Fund, and so much of Canal Commissioners' as relates to their accounts for the year ending Dec. 1, 1829.... [Columbus: 1830.] 8 p.

Dated: January 23, 1830.

Hasse (Ohio)-p. 381.

Copies: NN.

5339. Pennsylvania. Board of Canal Commissioners.

Bericht der Canal Commissäre von Pennsylvanien, betreffend die pennsylvanische Canäle und Riegelstraszen. Gelesen im Hause der Representanten, Dez. 22, 1830. Harrisburg: Gedruckt bey Jacob Baab, 1830. 54 p.

C-2981.

Copies: PPL.

5340. Pennsylvania. Board of Canal Commissioners.

Report of the Canal Commissioners of Pennsylvania, relative to the Pennsylvania canals and rail-road. Read in the House of Representatives, Dec. 22, 1830. Harrisburg: Printed by Henry Welsh, 1830. 48 p.

C-2991.

Copies: DLC; DeGE; MBAt; MWA; NN; P; PPL.

5341. ——— Read in the Senate, Dec. 23, 1830. Harrisburg: Printed by Henry Welsh, 1830. 48 p.

C-2992.

Copies: MB; MH-BA; MWA; MiU-T; PPF.

5342. Pennsylvania. General Assembly. House. Committee on the Official Conduct of James Clarke.

Report of a committee of the House of Representatives, appointed upon the application of James Clarke, Acting Commissioner on the Juniata Division of the Pennsylvania Canal, to investigate his official conduct in relation to certain charges set forth against him in a petition which has been presented ... March 23, 1830. Harrisburg: Welsh and Miller, 1830. 14 p.

C-2990.

Copies: DLC; MWA; P; PHi; PPi.

5343. Pennsylvania. Laws.

An act to authorize a loan to defray the expenses of the Pennsylvania Canal and Rail Road.... [Harrisburg: 1830.] 8 p.

C-2979.

Copies: DBR; MB; MiU; NN; PHi.

5344. Proceedings of a public meeting, holden in Jefferson County, Ohio, on the subject of canal policy. [Steubenville?, Ohio]: 1830. 20 p.

Meeting held April 20, 1830.

C-2043.

Copies: MWA; NN.

5345. Proprietors of the Middlesex Canal.

Regulations, relative to the navigation of the Middlesex Canal. Boston: Printed at the office of the Mechanicks magazine, 1830. 18 p.

C-2564.

Copies: DLC; MH-BA; MHi; MWA.

5346. Sullivan, John Langdon, 1777–1865.

Report of the engineers to the commissioners of the states of Pennsylvania and New Jersey. Trenton: Printed by J. Justice, 1830. 27 p.

Report, submitted by J. L. Sullivan and H. G. Sargent, relates to the use of water from the Delaware River for the Delaware and Raritan Canal.

Bowker-p. 172; Hasse (New Jersey)-p. 188.

Copies: NjHi.

5347. Tennessee. General Assembly.

Canal—Tennessee River to Mobile Bay. February 25, 1830.... Memorial of the General Assembly of the State of Tennessee, upon the subject of a canal or rail road to connect the waters of Tennessee River and Mobile Bay. [Washington: 1830.] 3 p. (21st Cong., 1st sess. House. Rept. 228)

C-4609.

Copies: CSt; CtY; DBRE; DLC; DeU; MH; MWA; NN; NjR; P; PPAmP.

5348. Torrey, Jason.

To adventurers. The completion of the Delaware & Hudson Canal has opened a wide and promising field for enterprise, to the farmer, the merchant, and the mechanic. This canal forms an inland navigation for boats of twenty-five ton burthen, from tide on the Hudson River, at the village of Bolton, near Kingston ... to the village of Honesdale, in Wayne County, Pennsylvania.... Bethany, Pa.: W. Sasman, print., 1830. Broadside.

Signed: Jason Torrey, Bethany, June 17, 1830.

Facsimile in: Le Roy, Edwin D. *The Delaware and Hudson Canal.* [Honesdale, Pa.]: Wayne Co. Hist. Soc., 1950. Location of the original unknown.

5349. Tower, Reuben.

An appeal to the people of the state of New-York, in favor of the construction of the Chenango Canal, with statements and documents, to prove the claim of that part of the state for this improvement; and arguments on the subject of the supply of water, cost of construction, and revenue. By Col. R. Tower. Utica, [N.Y.]: Dauby & Maynard, printers, 1830. 32 p.

Appendix: Report of David S. Bates, engineer.

C-3745.

Copies: CSmH; CtY; DLC; ICJ; IU; M; MB; MWA; N; NBu; NN; NUt; NbU; RPB; WHi.

5350. Union Canal Company of Pennsylvania.

Annual report of the managers of the Union Canal Company of Pennsylvania, to the stockholders. November 16, 1830. Philadelphia: Printed by Lydia R. Bailey, 1830. 12 p.

C-3795.

Copies: DeGE; MWA; PHi; PPL; PPi.

5351. United States. Board of Internal Improvement.

Buzzard and Barnstable Bay Canal. Report and estimate on the Buzzard and Barnstable Bay Canal, Massachusetts. February 8, 1830.... Plan and estimate of the Buzzard and Barnstable Canal, drawn up by the Board of Internal Improvement, in conformity to a resolution of Congress, dated 2d January, 1827. [Washington: 1830.] 17 p. (21st Cong., 1st sess. House. Doc. 54)

C-4603.

Copies: DLC; DeGE; DeU; MWA; NjR.

5352. United States. Congress. House. Committee on Internal Improvements.

Canal—Tennessee and Coosa rivers. February 10, 1830 ... Mr. Blair, from the Committee on Internal Improvements ... made the following report. [Washington: 1830.] 2 p. (21st Cong., 1st sess. House. Rept. 180)

C-4608.

Copies: DLC; DeU; MWA; NjR.

5353. United States. Congress. House. Committee on Internal Improvements.

Chesapeake and Ohio Canal west of the

5353 (*continued*)
Alleghany. March 8, 1830 ... Mr. Hemphill, from the Committee on Internal Improvements ... made the following report. [Washington: 1830.] 2 p. (21st Cong., 1st sess. House. Rept. 280)
C-4629.
Copies: CSt; CtY; DBRE; DLC; DeU; MH; MWA; NN; NjR; P; PPAmP.

5354. United States. Congress. House. Committee on Public Lands.
Canal—Lake Michigan to Illinois River. Illinois to relinquish land and issue scrip. April 22, 1830. (To accompany Bill H.R. 255). Mr. Duncan, from the Committee on the Public Lands ... made the following report. [Washington: 1830.] 3 p. (21st Cong., 1st sess. House. Rept. 377)
C-4607.
Copies: DLC; MWA; NjR.

5355. United States. Congress. House. Committee on Roads and Canals.
Farmington and Hampshire Canal Company. April 2, 1830.... [Washington: 1830.] 2 p. (21st Cong., 1st sess. House. Rept. 341)
C-4709.
Copies: DLC; MWA; NjR.

5356. United States. Congress. House. Committee on Roads and Canals.
Sandy and Beaver Canal. April 12,1830.... [Washington: 1830.] 10 p. (21st Cong., 1st sess. House. Rept. 357)
C-5153.
Copies: DLC; MWA; NjR.

5357. United States. Congress. Senate. Committee on Public Lands.
In Senate of the United States. February 8, 1830 ... Mr. Barton made the following report: The Committee on Public Lands ... have had under consideration the expediency of modifying the act to aid the state of Ohio in completing the Miami Canal, from Dayton to lake Erie.... [Washington: 1830.] 2 p. (21st Cong., 1st sess. Senate. Doc. 49)
Poore-p. 217.
Copies: DeU; MWA.

5358. United State. Engineer Dept.
Letter from the Secretary of War to the chairman of the Committee on Internal Improvements, with the result of the examination of Back Creek, &c. April 26, 1830.... [Washington: 1830.] 5 p. (21st Cong., 1st sess. Senate. Doc. 131)
Transmits a report from the Chief Engineer S. H. Long, with estimates on work to improve the navigation of the Chesapeake and Delaware Canal.
C-4909.
Copies: DLC; DeGE; DeU; MWA; NjR.

5359. ———— December 31, 1830.... [Washington: 1830.] 5 p. (21st Cong., 2d sess. House. Doc. 13)
C-4908.
Copies: DLC; DeGE; DeU; MWA; NjR.

5360. United States. Laws.
An act to incorporate the Alexandria Canal Company. May 13, 1830.... [Washington: 1830.] 19 p. (H.R. 184)
C-3846.
Copies: DNA.

5361. ———— [Washington: 1830.] 12 p.
C-3847.
Copies: DLC.

5362. United States. Laws.
A bill amending and supplementary to the act to aid the state of Ohio in extending the Miami Canal from Dayton to Lake Erie, and to grant a quantity of land to said state, to aid in the construction of the canals authorized by law, and for making donations of land to certain persons in Arkansas Territory. February 8, 1830. Mr. Barton, from the Committee on Public Lands, reported the following bill.... [Washington: 1830.] 2 p. (S. 84)

C-3948.
Copies: DNA.

5363. United States. Laws.
A bill to authorize a change in the disposal of the land granted for the construction of the Illinois and Michigan Canal. February 16, 1830.... [Washington: 1830.] 4 p. (H.R. 255)
C-4437.
Copies: DNA.

5364. United States. Laws.
A bill to authorize a subscription of stock in the Savannah, Ogechee, and Alatamaha Canal Company. May 3, 1830 ... Mr. Blair of Tennessee, from the Committee on Internal Improvements, reported the following bill. [Washington: 1830.] 2 p. (H.R. 456)
C-4444.
Copies: DNA.

5365. United States. Treasury Dept.
Report of the Secretary of the Treasury ... relative to the affairs of the Potomac Company at the time of the surrender of its charter, and also, on the subject of the stock of the Chesapeake and Ohio Canal Company. [Washington: 1830.] 3 p. (21st Cong., 1st sess. Senate. Doc. 39)
N-U0253427.
Copies: DeU; MWA; ViU.

5366. United States. War Dept.
Canal, Atlantic to Gulf of Mexico. Letter from the Secretary of War, transmitting an estimate of the cost of completing a survey, &c. of a canal to connect the waters of the Atlantic and the Gulf of Mexico. January 28, 1830.... [Washington: 1830.] 3 p. (21st Cong., 1st sess. House. Doc. 41)
C-4605.
Copies: DLC; DeU; MWA; NjR; ViU.

5367. United States. War Dept.
Canal—Boston harbor to Narragansett Bay. February 5, 1830.... [Washington: 1830.] 1 p. (21st Cong., 1st sess. House. Doc. 62)
Letter from the War Department, signed: Jn. H. Eaton.
C-4606.
Copies: DLC; DeGE; DeU; MWA; NjR.

5368. Virginia. Laws.
Acts of Virginia—Chesapeake and Ohio Canal. Message from the President of the United States, transmitting two acts of the legislature of Virginia, respecting the Chesapeake and the Ohio Canal Company. March 9, 1830. Referred to the Committee on Internal Improvement. April 24, 1830. Bill reported, No. 441. Washington:[Printed by Duff Green], 1830. 4 p. (21st Cong., 1st sess. House. Doc. 94)
C-3858.
Copies: DLC; MWA; MiU; NjR; Vi; ViU.

4. Roads and land vehicles

5369. New Hampshire (Colony). Laws.
An act ... February 9th, 1760; An act in addition to ... an act, for the better regulating high ways. [Portsmouth: 1760.] 4 p.
B-2159.
Copies: NhHi (not located).

5370. Massachusetts (Colony). Laws.
The acts contained in this book were ordered to be left out of the last impression of

5370 (*continued*)
temporary laws and printed by themselves, viz.... Acts respecting highways, 39 to 43. Boston: Printed by Green and Russell, 1763. [2], 52 p.

The parts referring to highways are also concerned with bridges.

Mp-9432.

Copies: DLC; IU; MB; MWA; MiU-L; NN; RPJCB.

5371. Good public roads are of the greatest utility even in countries, where from the sterility of the soil, they do not produce sufficient food for the support of the inhabitants of it. [Philadelphia: 1773.] Broadside 26 x 21 cm.

Plan for the construction of a road from Philadelphia to Strasburg, south of Lancaster.

Mp-12797.

Copies: PPL.

5372. Pennsylvania. Laws.

An act for the regulation of waggons, carriages and pack-horses.... [Lancaster, Pa.: 1778.] 2 p.

Dated: January 2, 1778.

Mp-15957.

Copies: DLC; PPL.

5373. Pennsylvania. Laws.

A further supplement to the act entitled, "An act for the regulation of waggons, carriages and pack-horses...." Philadelphia: Printed by John Dunlap, [1778.] Broadside.

Mp-15962.

Copies: DLC; PPL.

5374. Observations by the committee of the land-holders, on the utility and importance of the roads, proposed to be laid open in Northampton and Luzerne.... Philadelphia: Printed by Joseph James, [1788.] Broadside 42 x 34 cm.

E-21369.

Copies: PHi; PPL.

5375. Colles, Christopher, 1738–1816.

Proposals for publishing a survey of the roads. [New York: 1789.] Broadside.

Mp-21740.

Copies: MWA.

5376. Colles, Christopher, 1738–1816.

A survey of the roads of the United States of America, by Christopher Colles.... [New York]: 1789. [2] p., 83 maps (plates 1–33; 40–86, and extra numbers 45*, 46*, 47*).

Mp-21741.

Copies: CSmH; CtY; DLC; MWA; MWiW-C; MiU-C; NN; NNC; PPL; RPJCB; ViU.

5377. Gray, George.

Sir, the subscribers have spared no expence.... A handsome stage waggon ... will run twice a day ... Gray's Ferry, May 20, 1789. [S.l.: 1789.] Broadside 25 x 20 cm.

B-6957; Mp-45487.

Copies: DLC (photocopy); NHi.

5378. Scheme of a lottery for assisting to pave the streets at Fell's-Point, etc. The following scheme has for its object the rendering the road between the town and Point, more easy and commodious for waggons, and to assist in paving the streets at the Point.... Baltimore: Printed by William Goddard, [1789.] Broadside 38 x 18 cm.

Dated: Baltimore, March 16, 1789.

B-7059.

Copies: MdBMA (not located).

5379. Post-road lottery. The subscribers, being appointed managers of a lottery ... for raising the sum of five hundred dollars, in specie, in order to repair the South Road ... leading from Providence in this state, to Voluntown in Connecticut.... [Providence]: Printed by Bennett Wheeler, 1791. Broadside 20.5 x 14 cm.

At head of title: From the United States chronicle.

Dated: Cranston, March 9th, 1791.

B-7798; Mp-46261.

Copies: DLC; MWA (photocopy); RPJCB.

5380. Scheme of a lottery to raise the sum of four thousand pounds, agreeably to an act of the Legislature of the state of New-Jersey, passed the 24th November, 1790, for the purpose of laying out and improving the roads and causeways between the town of Newark ... and Paules-Hook ferry.... Elizabeth Town, [N.J.]: Printed by Shepard Kollock, [1791.] Broadside.

Dated: March 8, 1791.

Mp-23754.

Copies: NHi.

5381. Pennsylvania. Laws.

An act ... for making an artificial road, from the city of Philadelphia, to the borough of Lancaster. [Philadelphia: Printed by Hall and Sellers, 1792.] [1], 13 p.

"Approved April the ninth, 1792."

Mp-24665.

Copies: MiU-C; P; PHi.

5382. United States. Laws.

Act to establish the post-office and post-roads within the United States. [Philadelphia: Printed by John Fenno, 1792.] 10 p.

Passed the House of Representatives, January 10, 1792.

Mp-24877.

Copies: DLC.

5383. ———— Approved February the twentieth, 1792.... [Philadelphia: Printed by Francis Childs and John Swaine, 1792.] 6 p.

B-8217; Mp-46640.

Copies: DLC.

5384. Eastern State Road lottery, granted by the Honourable General Assembly of the state of Connecticut, at thier [*sic*] session in May last, for the purpose of repairing the stage road, from Hartford, in said state, to Coventry in the state of Rhode Island.... [Norwich: Printed by John Trumbull], 1793. Broadside.

Dated: Norwich, June 1, 1793.

Mp-25429.

Copies: NHi.

5385. New York (State). Laws.

The road act: that is to say—An act for the better laying out, regulating and keeping in repair public highways.... Enacted into law ... the fourth day of May ... one thousand seven hundred and eighty four.... To which are added, two laws, amendatory of the same. Lansingburgh, [N.Y.]: Printed by William W. Wands, 1793. 4 p.

Mp-25899.

Copies: MWA.

5386. Maryland. Laws.

An act supplementary to an act ... for the more effectual paving of streets of Baltimore Town.... [Baltimore: Yundt and Patton, 1794.] Broadside 41 x 34 cm.

B-8735; Mp-47109.

Copies: MdBE.

5387. New Jersey. Laws.

An additional supplement to an act, entitled, "An act for regulating roads and bridges." [Trenton: Printed by Isaac Collins, 1794.] 4 p.

At head of title: State of New-Jersey.

Mp-27386.

Copies: NN.

5387a. Pennsylvania. Laws

An act for the improvement of the Wissahickon Road from the city of Philadelphia to the bridge over Wissahickon Creek. [Philadelphia: Printed by T. Bradford, 1794.] 7 p.

Dated: In General Assembly, February 6, 1794.

Copies: DeGE.

5388. Pennsylvania. Laws.

An act for the improvement of the Wissahickon Road from the city of Philadelphia, to the ten mile stone on the said road. [Philadelphia: Printed by Zachariah Poulson, junr., 1794.] [8] p.

Mp-27475.

Copies: PHi; PPL.

5389. Pennsylvania. Laws.

An act to enable the Governor of this Commonwealth to incorporate a company for making an artificial road from the borough of Lancaster to the river Susquehanna at or near Wright's Ferry.... [Philadelphia]: Printed by Zachariah Poulson, [1794.] 20 p.

B-8789.

Copies: PHi (at PPL).

5390. Philadelphia and Lancaster Turnpike Company.

The public are hereby informed that the president and the managers of the Philadelphia and Lancaster Turnpike Road ... have established toll gates.... [Philadelphia: 1794?] Broadside 40 x 33 cm.

Copies: DeGE.

5391. Scituate, R.I. Ordinances.

An act for mending and repairing the highways in the town of Scituate.... Providence: Printed by Bennett Wheeler, 1794. Broadside.

"At a town-meeting held in Scituate, in the County of Providence, on the 26th day of August, A.D. 1794."

Mp-47211.

Copies: MWA (xerox copy); RNHi.

5392. Pennsylvania. Laws.

An act to enable the president, managers and company of the Philadelphia and Lancaster Turnpike Road to increase the width of the said road in certain cases.... [Philadelphia: 1795.] Broadside 24 x 17 cm.

Approved April 17th, 1795.

B-9225; Mp-47551.

Copies: PHi.

5393. Pennsylvania. Laws.

[Act to incorporate the president, managers and company of the Wissahickon Toll Road.] [Philadelphia: 1795?] 6 p.

B-9229.

Copies: MWA (p. 3–6 only).

5394. Bradley, Abraham, d. 1838.

A map of the United States, exhibiting the post-roads, the situations, connections & distances of the post-offices, stage roads.... [Philadelphia]: 1796. Map on 3 sheets.

1st sheet: The nine northern states with parts of Virginia and the territory north of Ohio; 2d sheet: The southern states east of the Mississippi; 3d sheet: Northwestern territory. Includes a table of postal service.

Mp-30122, 30123.

Copies: MB; MH; MiU-C.

5395. Massachusetts. Laws.

Commonwealth of Massachusetts. In the year of our Lord, one thousand, seven hundred and ninety-seven. An act, in addition to several acts now in force, respecting highways. Boston: Printed by Young and Minns, state printers, [1797.] Broadside 45 x 34 cm.

Dated: February 28, 1797.

Mp-32445.

Copies: MWA; MHi; NN.

5396. New York (City). Ordinances.

City of New York ss. At a Common Council held on Thursday the 9th day of November, 1797.... A law to regulate carts and cartmen in the city of New-York.... [New York: Printed by George Forman, 1797.] Broadside.

Mp-32561.

Copies: DLC.

5397. New York (State). Laws.

The road-act, passed into a law, at Albany, March 21, 1797, at the second meeting of the twentieth session of the Legislature of the state of New-York.... Albany: Printed by Charles R. and George Webster, [1797.] 21 p.

Mp-32557.

Copies: MWA; NHi.

5398. Williamson, Charles, 1757–1808.

The legislature at their last session, having granted considerable aid to improve the great road leading from Fort-Schuyler, on the Mohawk-River, to the county of Ontario, and appointed a commissioner to carry the law into effect.... [Albany: Printed by

Loring Andrews & Co., 1797.] Broadside.
Dated: July 4, 1797.
Mp-33230.
Copies: NN.

5399. Connecticut. General Assembly.
Proceedings of the General Assembly of the State of Connecticut, relative to Hartford and New-Haven Turnpike Road. [Hartford: Printed by Hudson & Goodwin, 1798.] 12 p.
Members of the committee appointed to view the road: Heman Swift, Epaphroditus Champion, and Jonathan O. Moseley.
Mp-33560.
Copies: CtY; DeGE; MWA.

5400. New Jersey. Laws.
A bill, entitled "An act making provision for working and repairing highways." [Trenton: 1798.] 6 p.
At head of title: State of New Jersey.
B-10419; Mp-48532.
Copies: NN; Nj.

5401. New York (State). Laws.
An act to amend an act entitled An act to regulate highways. Albany: Printed by Loring Andrews & Co., 1798.
E-34207. No copy known.

5402. Pennsylvania. Laws.
A further supplement to an act entitled "An act for making an artificial road from the city of Philadelphia to the borough of Lancaster...." [Philadelphia: 1798.] Broadside 44 x 26 cm.
Approved April 4th, 1798.
Mp-34325; N-P0208799.
Copies: CtY; PHi.

5403. Bradley, Abraham, d. 1838.
Map of the United States, exhibiting the post roads, the situations, connections & distances of the post-offices, stage roads.... [Georgetown?: 1799.] Map 44 x 51 cm.
N-B0730329.
Copies: MB.

5404. New Hampshire. Laws.
An act to incorporate a company by the name of the Proprietors of the Third Turnpike Road in Newhampshire.... Approved, December 27, 1799.... [Portsmouth?: 1799.] Broadside 31 x 45.5 cm.
E-35888; N-N0149223.
Copies: MiU-T.

5405. Pennsylvania. Laws.
An act to extend for a limited time, an act entitled "A further supplement to the act, entitled an Act for making an artificial road from the city of Philadelphia to the borough of Lancaster; and for other purposes ... " [Philadelphia: 1799.] Broadside 28 x 20 cm.
B-10892; Mp-48953.
Copies: PHi; PPL.

5406. Sixth Massachusetts Turnpike Corporation.
By laws of the Sixth Massachusetts Turnpike Corporation, for establishing a turnpike road, from Amherst to Shrewsbury. Worcester: Printed by Mower & Greenleaf, [1799.] 18, [2] p.
Includes "An act for establishing a corporation by the name of the Sixth Massachusetts Turnpike Corporation."
Mp-35803.
Copies: MWA.

5407. Hartford and New Haven Turnpike Company.
Hartford and New-Haven turnpike road. [Hartford: Hudson and Goodwin?, 1800.] 5 p.
Signed: Heman Swift, Epaphroditus Champion, Jonathan O. Moseley, committee. Hartford, November 6, 1800.
"An act in addition to an act, entitled, 'An act concerning turnpike roads' (passed October session 1800)," p. 5.
Mp-37588.
Copies: CtY; DeGE; MH-BA; MiU.

5408. [Williamson, Charles] 1757–1808.
Observations on the proposed state road, from Hudson's River, near the city of Hudson, to Lake Erie, by the Oleout,

5408 (*continued*)
Catharine's, Bath, and Gray's settlement, on the western bounds of Steuben County. New York: Printed by T. & J. Swords, 1800. 18 p., fold. map.
Mp-39109.
Copies: DLC; M; MWA; MiU-C; NBu; NHi; NN; PPAmP; PPL.

5409. Connecticut. General Assembly. Committee to View the Road from New Haven to Hartford.
Hartford and New-Haven turnpike road. Minutes of committee in viewing the road from New-Haven to Hartford, October 1801. [S.l.: 1801.] 3 p.
Heman Swift, Epaphroditus Champion, Jonathan O. Moseley, members of the committee, appointed by the General Assembly of Connecticut.
N-H0147278.
Copies: CtY; DeGE.

5410. New York (State). Laws.
An act to establish a turnpike corporation for improving and making a road from Newburgh ... to Coshecton on the Delaware River. Passed March 20, 1801. Newburgh, [N.Y.]: Printed by Dennis Coles, 1801. 12 p.
S&S-1026.
Copies: MHi.

5411. New York (State). Laws.
The new road-act; being an act to regulate highways. Passed April 8th, 1801. Troy: Printed by R. Moffitt & Co., 1801. 24 p.
N-N0198578.
Copies: MiU.

5412. Pennsylvania. Laws.
An act to enable the Governor of this Commonwealth to incorporate a company to make an artificial road from the city of Philadelphia, through Germantown, to the ten mile stone on Chestnut Hill, and from thence to the new stone bridge over Perkiomen Creek, in the county of Montgomery. [Philadelphia: 1801.] 22 p.
S&S-35593 (dated 1815), 50241.
Copies: PU.

5413. United States. Congress. Senate. Committee on Post Offices and Post Roads.
Report of the Committee to whom was referred the bill, further to alter and to establish certain post roads. Printed by order of the Senate of the United States. February 27, 1801. [Washington?: 1801.] 4 p.
S&S-1549.
Copies: DLC; MWA; NN.

5414. United States. Laws.
A bill, to provide for the better establishment of the post roads of the United States. [Washington: 1801.] 6 p.
S&S-1473.
Copies: DLC.

5415. Albany and Schenectady Turnpike Company.
Charter of the Albany and Schenectady Turnpike Company. Granted by the Legislature of the State of New-York ... at Albany, the thirtieth day of March, in the year one thousand eight hundred and two. Albany: Printed by Charles R. and George Webster, [1802.] 15 p.
S&S-1735.
Copies: MWA; N.

5416. Bush, Philip.
Philip Bush's list of stages for the accomodation of travellers.... Frankfort, [Ky.]: James M. Bradford, printer, 1802. Broadside.
S&S-1972.
Copies: MH.

5417. Great Western Turnpike Road Company.
Charter of incorporation, granted to the first company of the Great Western Turnpike Road, by the Legislature of the state of New-York, by their acts of the 15th day of March, 1799, and of the 5th of April 1802. Albany: Printed by Charles R. and George Webster, 1802. 27 p.
S&S-2776.
Copies: MWA; N.

5418. Mohawk Turnpike and Bridge Company.

Charter of incorporation, of the Mohawk Turnpike and Bridge Company; granted by the Legislature of the state of New-York, by an act, passed the 4th of April, 1800; to which is added, an act, amendatory of the said act, passed 5th April, 1801. Albany: Printed by Charles R. and George Webster, 1802. 15 p.

S&S-2681.

Copies: N; NN.

5419. New Jersey. Laws.

An act for establishing a turnpike road from Hackensack to Hoboken, in the County of Bergen. November 30th, 1802. [Trenton?: 1802.]

Bowker-p. 164; S&S-2763.

No separate printing located.

5420. New York (State). Laws.

An act more effectually to provide for completing and keeping in repair the turnpike roads in this state. [Albany: 1802.] Broadside.

S&S-2773.

Copies: NN; NNMer.

5421. Salem Turnpike and Chelsea Bridge Corporation.

An act for incorporating certain persons for the purpose of laying out and making a turnpike road from Salem to Charles-River-Bridge. Salem, [Mass.]: Printed by Thomas C. Cushing, 1802. 38 p.

"Rules and regulations of the Salem Turnpike and Chelsea Bridge Corporation," p. [29]–38.

S&S-3044.

Copies: CSmH; DLC; DeGE; M; MWA; MiU-C; NN.

5422. United States. Congress. Senate. Committee on Post Offices and Post Roads.

In Senate of the United States. April 15th, 1802. The Committee to whom was referred the act further to alter and establish certain post roads, report the following amendments. [Washington: 1802.] 4 p.

S&S-3262.

Copies: DLC.

5423. United States. Congress. Senate. Committee on Transportation of Mail.

Report of the Committee of the Senate of the United States, appointed March 12, 1802, on the subject of transporting the mail . . . March 30, 1802. [Washington: 1802.] 35 p.

Report of the General Post Office, with accompanying documents, describing, among other things, the roads and availability of construction materials, p. [7]–35.

S&S-3413.

Copies: DLC; MWA; NN.

5424. United States. Laws.

A bill further to alter and establish certain post roads. [Washington]: Printed by Duane, [1802.] 6 p.

S&S-3232.

Copies: DLC.

5425. Cambridge and Concord Turnpike Corporation.

Act of incorporation of the Cambridge & Concord Turnpike, passed March 7, 1803. Also the bye laws of the same. Charlestown, [Mass.]: Printed by Samuel Etheridge, 1803 20 p.

S&S-4607.

Copies: MB.

5426. Massachusetts. Laws.

An act for incorporating certain persons for the purpose of laying out and making a turnpike road from Newburyport to Chelsea Bridge. [Boston: Printed by Russell & Cutler, 1803.] 20 p.

"Rules and regulations of the Newburyport Turnpike Corporation," p. [14]–20.

S&S-4595.

Copies: MWA; MiU.

5427. Pennsylvania. Laws.

An act to enable the Governor of this Commonwealth to make an artificial road, from the Rising Sun Tavern through Shoe-

5427 (*continued*)
maker's Town, to the Red Lion, on the Old York Road. Philadelphia: Printed by Tesson and Lee, 1803. 21 p.

Concerns the Willow Grove Turnpike.
S&S-5596.
Copies: PPL.

5428. Philadelphia and Lancaster Turnpike Company.

A bye-law of the president, managers and company of the Philadelphia and Lancaster turnpike road. [Philadelphia]: Printed by W. W. Woodward, [1803.] Broadside.

Dated: June 27, 1803.
S&S-4857.
Copies: PHi.

5429. Rhode Island. Laws.

An act for the better regulating the mending of the public highways in the town of Gloucester, in the county of Providence, &c. Providence: Printed at the Phoenix printing-office, 1803. Broadside 49 x 31 cm.

Copies: MWA.

5430. Third Company of the Great Western Turnpike.

Charter of incorporation of the Third Company of the Great Western Turnpike, granted by act of the Legislature of the state of New-York, fourth of April, 1803. Albany: Printed by Charles R. & George Webster, 1803. 16 p.

S&S-4313.
Copies: NN.

5431. United States. Congress. House. Committee on Post Office and Post Roads.

Report of the Committee ... who were directed ... to enquire by what means the mail may be conveyed with better dispatch than at present, between the city of Washington and Natchez and New Orleans. 13th December, 1803.... [Washington: 1803.] 5 p.

Includes the route of the new post road.
S&S-5472.
Copies: DLC; NN; RPB.

5432. United States. Laws.

A bill for the relief of the proprietors of the Salem turnpike road in Massachusetts. [Washington]: Printed by Duane & Son, [1803.] 2 p.

S&S-5235.
Copies: DLC.

5433. United States. Laws.

A bill to alter and establish certain post roads, and for other purposes. [Washington]: Printed by Duane & Son, [1803.] 3 p.

S&S-5238.
Copies: DLC.

5434. Bradley, Abraham, d. 1838.

Map of the United States, exhibiting the post-roads, the situations, connections & distances of the post-offices; stage roads, counties, ports of entry and delivery for foreign vessels, and the principal rivers. Philadelphia: [1804.] Map 86 x 93 cm.

S&S-5900.
Copies: DeGE; OCHP; ViU.

5435. Easton and Wilkesbarre Turnpike Company.

An ordinance, by the president, managers and company of the Easton & Wilkesbarre Turnpike Company. Also, the act of the legislature of Pennsylvania incorporating the said company. [S.l.: 1804.] 31 p.

N-E0010181.
Copies: PHi.

5436. Maryland. Laws.

[A bill, entitled, An act to incorporate a company to make a turnpike road to lead from the cross roads near Caton's limekiln to the city of Baltimore. Annapolis: F. Green, 1804.]

S&S-6719.
No copy known.

5437. Pennsylvania. Laws.

An act, to enable the Governor of this Commonwealth to incorporate a company for making an artificial or turnpike road, from the intersection of Bristol and Newtown roads, at the rock in Oxford through Bustleson [*sic*] and Smithfield, in the county of Philadelphia, to the Buck tavern in Southampton, in the county of Bucks. Passed the fifth day of March, 1804.... Philadelphia: Printed by Wm. Duane, 1804. 23 p.

S&S-7006.

Copies: DLC; DeGE; MiU; PP.

5438. Pennsylvania. Laws.

An act to enable the Governor of this Commonwealth to incorporate a company to make an artificial road from the top of Chestnut-Hill through Flower-Town to the Spring-House tavern in Montgomery County. [Lancaster: 1804.] 23 p.

Copies: PPL.

5439. Proprietors of the Second Turnpike Road in New Hampshire.

The act of incorporation, and bye laws of the Proprietors of the Second Turnpike Road in New-Hampshire. Published by order of the Proprietors. Amherst, N.H.: Printed by Joseph Cushing, 1804. 14 p.

S&S-7129.

Copies: MBAt; MWA; NN.

5440. Union Turnpike Road Company.

In the Court for the Trial of Impeachments and Correction of Errors. The president, directors and company of the Union Turnpike Road, adms. Thomas Jenkins. Case, on the part of defendants in error. Hudson, [N.Y.]: Printed by Ashbel Stoddard, 1804. 9 p.

S&S-7402.

Copies: DLC; NN.

5441. United States. Congress. House.

Mr. Dennis's motion. 7th February, 1804. Read, and ordered to be referred to the Committee of the Whole House, to whom was committed on the 25th ultimo, the bill to lay out and open a new public road in the county of Washington, in the District of Columbia. [Washington: 1804.] 3 p.

S&S-7567.

Copies: DLC; MWA; NN.

5442. United States. Congress. House.

Motion respecting the establishment of a post road from Knoxville, in the state of Tennessee to the settlements on the Tombigby River, in the Mississippi Territory, and from thence to New Orleans .. 7th December, 1804. Washington City: Printed by William Duane and Son, 1804. 4 p.

Resolution introduced by David Thomas.

S&S-7582.

Copies: DLC.

5443. United States. Congress. House. Committee on Post Office and Post Roads.

Mr. Anderson, from the Committee to whom was referred the bill further to alter and establish certain post roads, and for other purposes, reported the following amendments. [Washington: 1804.] 3 p.

S&S-7558.

Copies: DLC.

5444. United States. Congress. House. Committee on Post Office and Post Roads.

Supplementary report of the Committee to whom was recommitted ... their report relative to the means by which the mail may be conveyed with greater dispatch ... between the city of Washington & Natchez and New Orleans ... 12th January, 1804 ... [Washington: 1804.] 4 p.

S&S-7633.

Copies: DLC; MWA; NN.

5445. United States. Congress. House. Committee on Post Office and Post Roads.

10th December, 1804, referred to the Committee of the Whole House, to whom was referred on the 7th instant, a motion respecting the establishment of a post-road from Knoxville in the state of Tennessee, to the settlements on the Tombigby River, &c. [Washington: 1804.] 6 p.

"Report ... relative to a post route from the city of Washington, to Natchez and New-Orleans ... ," p. 5–6.

S&S-7634.

Copies: DLC; MWA.

5446. United States. Laws.

An act further to alter and establish certain post roads, and for other purposes. [Washington: 1804.] 5 p.

At head of title: In Senate of the United States. February 28, 1804.

S&S-7410.

Copies: DLC.

5447. United States. Laws.

A bill making provision for the application of the money heretofore appropriated to the laying out and making public roads, leading from the navigable waters emptying into the Atlantic, to the Ohio River. [Washington: 1804.] 3 p.

S&S-7453.

Copies: DLC.

5448. United States. Laws.

A bill to lay out and open a new public road in the county of Washington and District of Columbia. [Washington: 1804.] 3 p.

S&S-7480.

Copies: DLC.

5449. Great Northern Turnpike Company.

Charter of the Great Northern Turnpike Company, granted by the Legislature of the state of New-York, on the fourth day of April, 1805. Albany: Printed by Charles R. and George Webster, 1805. 15, [1] p.

Copies: MWA.

5450. New Jersey. Laws.

Bergen turnpike road act.... New York: Printed by George Forman, 1805. 24 p.

S&S-8998.

Copies: Nj (not located).

5451. Pennsylvania. Laws.

Law for authorizing the Governor to incorporate a company for making an artificial road in Wayne and Luzerne counties; also an ordinance by the president, managers, and company of the Coshecton and Great Bend Turnpike Road. Philadelphia: Printed by Jane Aitken, 1805. 30 p.

S&S-9125.

Copies: DLC; PPAmP.

5452. United States. Congress. Senate. Committee on the Admission of Ohio into the Union.

In Senate of the United States. December 19th, 1805. Mr. Tracy, from the committee to whom was referred the examination of the act, entitled "An act to enable the people of the eastern division of the territory northwest of the river Ohio, to form a constitution and state government, and for the admission of such state into the Union ... " report. [Washington: 1805.] 12 p.

Includes a list of roads proposed in and to the territory of Ohio, including a road from Cumberland, Maryland, to the Ohio river, with a description of the route.

S&S-9530.

Copies: DLC; MiU-C.

5453. United States. President (Jefferson).

Message from the President of the United States, communicating further information in relation to a public road from the City of Washington to New-Orleans. 23d February, 1805.... [Washington: 1805.] 11 p.

Includes a report from Isaac Briggs, surveyor.

S&S-9567.

Copies: DLC; MWA; NN.

5454. United States. President (Jefferson).

Message from the President of the United States, communicating information in part, on the subject of a post road, from the city of Washington to New-Orleans. 1st February 1805. Washington: Printed by William Duane & Son, 1805. 10 p.

Includes a survey of the proposed road by Isaac Briggs.

S&S-9568.

Copies: DLC; MWA; NN; RPB.

5455. Boston and Montreal Turnpike Company.

The act of incorporation and bye laws of the Boston and Montreal Turn Pike Company. Peacham, [Vt.]: Printed by Samuel Goss, 1806. 12 p.

S&S-10018.
Copies: RP; VtHi.

5456. Highland Turnpike Company.
To the Senate and House of Representatives of the United States in Congress assembled. The memorial of the president, directors and company of the Highland Turnpike in the state of New York, respectfully showeth.... [Washington: 1806.] 2 p.
Dated: December 22, 1806; signed: William Edgar, president.
Copies: DNA.

5457. Maryland. Laws.
An act to incorporate companies to make several turnpike-roads through Baltimore County, and for other purposes: passed on the 12th day of January 1805. Baltimore: Printed by John Hewes, 1806. 26 p.
S&S-10800.
Copies: MdBE; MdHi; PHi; PPL.

5458. New York (State). Laws.
An act to regulate highways, passed 8th April, 1801. To which is added, an appendix, containing sundry laws and extracts from laws, relative to roads. Albany: Printed by Websters and Skinners, 1806. 19 p.
S&S-11005.
Copies: PHi; PPL.

5459. Observations on the real, relative and market value of the turnpike stock of the state of New-York.... By a citizen. New York: Printed by S. Gould, 1806. 36 p.
S&S-11054.
Copies: MWA; NBLiHi; NN; RNR.

5460. Proprietors of the Coos Turnpike Road.
The act of incorporation, and bye-laws, of the Proprietors of the Coos Turnpike Road, in the state of New Hampshire. Concord: Printed by George Hough, for the proprietors, 1806. 23 p.
S&S-10209.
Copies: NhD.

5461. United States. Congress. House. Committee on the District of Columbia.
Report of the Committee to whom was referred on the seventeenth instant, the petition of sundry inhabitants of the counties of Alexandria and Washington, in the District of Columbia. December 23, 1806.... City of Washington: A. & G. Way, printers, 1806. 3, [3] p.
Concerns a turnpike road from Washington to Alexandria.
S&S-11731.
Copies: DLC; MWA; N; NN.

5462. United States. Post-Office Dept.
Letter from the Post-Master General transmitting a report, shewing the obstructions to the transmissions of the mail from Athens, in the state of Georgia, to New-Orleans; and giving an estimate of the probable expense of opening a horse road between the same points ... March 21, 1806. Washington: A. & G. Way, printers, 1806. 8 p.
S&S-11592.
Copies: DLC; MWA; MiU-C; ViU.

5463. New York (City). Street Commissioner.
Regulating and paving-streets. Proposals in writing will be received.... Street-Commissioner's Office, City Hall, June 17, 1807. [New York]: Printed by Henry C. Southwick, [1807.] Broadside.
S&S-13221.
Copies: NN.

5464. New York (State). Laws.
An act to regulate highways. Passed April 8th, 1801. Together with the amendments, passed 1802, and 1807. Poughkeepsie, [N.Y.]: Printed by Thomas Nelson & Son, 1807. 36 p.
N-N0196981.
Copies: NNC.

5465. Observations on the extension of Pratt-Street, respectfully addressed to the Legislature of Maryland. By an inhabitant of Baltimore. Baltimore: Printed by Joseph Robinson, 1807. 27 p.

5465 (*continued*)
S&S-13273.
Copies: MdBT; MdToH.

5466. Pennsylvania. Laws.
Act for making an artificial road from the city of Philadelphia to the borough of Lancaster. [Philadelphia?: 1807.] 22 p.
Two editions; 2d ed. with supplements.
S&S-13329.
Copies: PHi (at PPL).

5467. Susquehanna and Lehigh Turnpike Company.
General William M'Pherson, in compliance with the request contained in your note of the 7th inst., and with the wishes of the Secretary of the Treasury, the presidents and directors of "the Susquehannah and Lehigh," and of "the Susquehannah and Tioga Turnpike Road Companies," with peculiar pleasure communicate to you the following statement of their several undertakings in Pennsylvania.... Philadelphia: 1807. [6] p.
Dated: August 20th, 1807.
S&S-13664.
Copies: PPAmP.

5468. United States. Commissioners Appointed to Lay Out a Road from Cumberland, Maryland, to the State of Ohio.
Message from the President of the United States, transmitting a report of the commissioners appointed under the act passed the twenty-ninth day of March last, intituled "An act to regulate the laying out and making a road from Cumberland, to the state of Ohio." January 31, 1807.... City of Washington: A & G. Way, printers, 1807. 16 p.
S&S-13993.
Copies: DLC; MWA; N; NN; PHi.

5469. United States. Congress. House. Committee on the Memorial of the Highland Turnpike Company.
Report of the committee to whom was referred the petition and memorial of the president and directors of the Highland Turnpike Company in the state of New York. January 5, 1807.... City of Washington: A. & G. Way, printers, 1807. 4 p.
Concerns the turnpike from Kingsbridge across the Highlands to the town of Hudson.
S&S-14083.
Copies: DLC; MWA; NN.

5470. United States. Congress. Senate.
In Senate of the United States. January 22, 1807. Mr. White presented the following memorial.... Washington City: Duane & Son, printers, 1807. 4 p.
Concerns a turnpike road to be built between Alexandria and Washington.
S&S-13916.
Copies: DLC.

5471. United States. Laws.
A bill, for the establishment of a turnpike company in the county of Alexandria, in the District of Columbia. February 16, 1807.... [Washington: 1807.] 10 p. (H. 63)
S&S-13816.
Copies: DLC.

5472. United States. Laws.
A bill making a further appropriation towards defraying the expense of a road from Cumberland, in the state of Maryland, to the state of Ohio. Feb. 24, 1807.... [Washington: 1807.] 2 p.
S&S-13835.
Copies: DLC.

5473. United States. Laws.
A bill to alter and establish certain postroads, and for other purposes. February 16, 1807.... [Washington: 1807.] 5 p. (H.R. 61)
S&S-13857.
Copies: DLC; MWA.

5474. United States. Laws.
A bill to incorporate a company for making a turnpike road from the west end of Mason's causeway, to Alexandria. Jan. 12, 1807.... [Washington: 1807.] 13 p. (H.R. 30)
S&S-13867.
Copies: DLC; MWA (photocopy).

5475. Gentlemen, a meeting of a number of citizens and landholders of Wayne and Northampton Counties, and of citizens of Philadelphia and the Northern Liberties, was lately held at this place, on the subject of the improvement of the North Road. Philadelphia: 1808. [3] P.

Caption title.

Dated: Philadelphia, January 18, 1808.

Signed: Tench Coxe, Benjamin Chew, Junior, Horace Binney, John B. Wallace, George A. Baker.

Advocates the construction of the "North Road" to Canada.

Copies: DeGE.

5476. New York (State). Laws.

An act, to amend an act, entitled "An act to regulate highways." Passed April 11, 1808.... [Albany: 1808?] [2] p.

N-N0196777.

Copies: NNC.

5477. United States. Commissioners Appointed to Lay Out a Road from Cumberland, Maryland, to the State of Ohio.

Message from the President of the United States, communicating an additional report of the proceedings of the Commissioners appointed under an act of the 29th March, 1806, intituled "An act to regulate the laying out and making a road from Cumberland in the state of Maryland, to the state of Ohio." February 19, 1808.... City of Washington: A. & G. Way, printers, 1808. 8 p.

S&S-16460.

Copies: DLC; DS; DeGE; MWA; NN.

5478. United States. Commissioners Appointed to Lay Out a Road from Cumberland, Maryland, to the State of Ohio.

Message from the President of the United States, transmitting a report of the commissioners appointed under the act of March 29, 1808 [1806]; concerning a road from Cumberland to Ohio; being a statement of the proceedings under the said act, since their last report communicated to Congress. December 13, 1808. Received. December 14, 1808. Referred to Mr. Jeremiah Morrow [a.o.]. City of Washington: A. & G. Way, printers, 1808. 8 p.

Signed: Elie Williams, Thomas Moore. August 30, 1808.

S&S-16476.

Copies: DLC; DeGE; InHi.

5479. United States. Treasury Dept.

Document accompanying the bill in addition to the "Act to regulate the laying out and making a road from Cumberland, in the State of Maryland, to the State of Ohio." Presented the 8th of March, 1808. Washington: A. & G. Way, printers, 1808. 4 p.

Signed: Albert Gallatin.

S&S-16385.

Copies: DLC; DS; DeGE; MWA; NN.

5480. Dixville road lottery.... Boston: 1809. Broadside 32 x 20 cm.

S&S-17394.

Copies: MB.

5481. United States. Congress. House.

Documents accompanying a bill in addition to the "Act to regulate the laying out and making a road from Cumberland in the state of Maryland, to the state of Ohio." February 16, 1809.... Washington City: A. & G. Way, printers, 1809. 6 p.

S&S-18875.

Copies: DLC; MWA; NN.

5482. United States. Laws.

An act in addition to the "Act to regulate the laying out, and making a road from Cumberland in the state of Maryland, to the state of Ohio." December 13th, 1809.... [Washington: 1809.] 2 p.

S&S-18804.

Copies: DLC.

5483. Boston, Plymouth, Sandwich mail stage, continues to run as follows.... [Boston]: 1810. Broadside.

Signed: Boston, Leonard & Woodward, Nov. 24, 1810.

5483 (*continued*)
S&S-19603.
Copies: CSmH; MBBS; MWA (xerox copy).

5484. South Carolina. To the ... Senate. The memorial and petition of the subscribers.... [Charleston, S.C.: 181– .] Broadside 33 x 21 cm.

Requests the establishment of turnpike roads from the interior of the state to Charleston.

Hummel-2023.
Copies: ScAr.

5485. United States. Laws.

A bill to amend an act, entitled, "An act for the establishment of a turnpike company in the county of Alexandria, in the District of Columbia." January 18th, 1810.... [Washington: 1810.] 3 p. (H.R. 36)

S&S-21624.
Copies: DLC.

5486. United States. Laws.

A bill to establish post roads. March 22d, 1810.... [Washington: 1810.] 35 p. (H.R. 80)

Includes a list of proposed roads.
S&S-21632.
Copies: DLC.

5487. United States. Post-Office Dept.

Letter from the Post-Master General, inclosing a report and bill to establish post-roads; in obedience to a resolution of the House of the 28th of June last. February 22, 1810.... City of Washington: Printed by A. & G. Way, 1810. 61 p.

Includes a list of post roads to be established.
S&S-21676.
Copies: DLC; MWA; NNS; NhHi.

5488. Banks, Henry, fl. 1781–1826.

Sketches & propositions, recommending the establishment of an independent system of banking; permanent public roads ... with some observations necessary to illustrate these several topics.... Richmond: Manson, printer, [1811?]. 65 p.

S&S-22245.
Copies: DLC; ICRL; NN; PU; RPB; ViU.

5489. Mohawk Turnpike and Bridge Company.

Acts, relative to the Mohawk Turnpike and Bridge Company, and the Mohawk Turnpike Company, and also relative to the appointment of turnpike commissioners. Schenectady: Printed by William S. Buell, 1811. 20 p.

N-M0678041; S&S-23409.
Copies: NN.

5490. Pennsylvania. Laws.

An act to enable the Governor to incorporate a company for making an artificial road ... Vine and Tenth Streets, Philadelphia ... to the Perkiomen Bridge. [S.l.: 1811.] 17 p.

S&S-23818.
No copy known.

5491. Pennsylvania. Laws.

An act to incorporate a company for making an artificial road from or near the borough of Wilmington, in the county of New Castle, on the east side of the Brandywine Creek, in the route through West Chester, to the turnpike roads in the Great Valley, in the state of Pennsylvania. [S.l.]: 1811. 20 p.

Running title: Wilmington and Great Valley Turnpike.
S&S-22677 (listed as a Delaware law).
Copies: DeWI; PHi; PPL.

5492. United States. Congress. House. Committee on a Road from Vincennes to Dayton.

Report of the Committee appointed on the fifth instant, to inquire into the expediency of opening a road from Vincennes, in the Indiana Territory, towards Dayton, in the state of Ohio. Washington City: A. and G. Way, printers, 1810 [1811.] 4 p.

Dated: January 17, 1811.

S&S-21805, 24268.
Copies: DLC; ICJ; MWA; OCHP.

5493. United States. Congress. House. Committee on the Roads Contemplated by the Treaty of Brownstown.

Report of the committee appointed on the eleventh instant, on the expediency of laying out and making the roads contemplated by the treaty of Brownstown, concluded on the twenty fifth November, 1808. November 28, 1811.... Washington City: A. & G. Way, printers, 1811. 4 p.

Concerns the roads between the Michigan Territory and different settlements in Ohio.

S&S-24267.
Copies: DLC; MH-L; MWA; MiD-B.

5494. United States. Laws.

A bill to authorize the laying out a public road from the line established by the Treaty of Greenville, to the North Bend in the state of Ohio. November 15th, 1811.... [Washington: 1811.] 1 p. (H.R. 4)

S&S-24127.
Copies: DLC.

5495. United States. Laws.

A bill to authorise the surveying and making of certain roads, in the state of Ohio, as contemplated by the Treaty of Brownstown on the territory of Michigan. In Senate of the United States. November 18th, 1811. [Washington: 1811.] 2 p. (S. 3)

S&S-24129.
Copies: DLC.

5496. Virginia. Laws.

An act incorporating a company to establish a turnpike road from Fredericksburg to Swift-Run Gap in the Blue Ridge. Fredericksburg, [Va.]: Printed by T. Green, 1811. 21 p.

S&S-51142.
Copies: NcD; ViHi.

5497. Be it known ... that at the request of Joshua Ruffner ... commissioners of the turnpike road over the Massenutten Mountain, have this day ... examined, passed and received said road.... [New Market, Va.: Ambrose Henkel, 1812.] Broadside 17 x 20 cm.

Dated: 12th day of September, 1812.
Hummel-3247.
Copies: ViW.

5498. Pennsylvania. Laws.

An act to authorize the Governor to incorporate a company for making an artificial road from ... Belmont ... until it shall intersect the Easton and Wilkesbarre turnpike road in Northampton County. [Harrisburg: 1812.] 6 p.

S&S-26407.
Copies: PPAmP.

5499. Pennsylvania. Laws.

An act to enable the Governor to incorporate a company to make an artificial road from the town of Hanover, in the county of York, to the borough of Carlisle. Carlisle, [Pa.]: 1812. 16 p.

N-P0208908.
Copies: IU.

5500. United States. Congress. House. Committee on Post Office and Post Roads.

The Committee on the Post-Office and Post-Roads, to whom was referred a bill to alter and establish certain post-roads ... report the same with the following amendments. March 24th, 1812.... [Washington: 1812.] 5 p.

S&S-27123.
Copies: DLC.

5501. United States. Congress. House. Committee on the Cumberland Road.

Report of the Committee to whom were referred a message from the President of the United States, transmitting a report and letter relative to a road from Cumberland, in Maryland, to the State of Ohio. April 14, 1812.... Washington: A. & G. Way, printers, 1812. 4 p.

S&S-27342.
Copies: DLC; DeGE; MWA.

5502. United States. Laws.
A bill in addition to the act to regulate the laying out and making a road from Cumberland in the state of Maryland, to the state of Ohio. April 17, 1812.... [Washington: 1812.] 2 p. (S. 43)
S&S-27028.
Copies: DLC.

5503. United States. Treasury Dept.
Message from the President of the U. States, transmitting a report of the Secretary of the Treasury, containing a statement of the proceedings under the "Act to regulate the laying out and making a road from Cumberland in the State of Maryland to the State of Ohio." February 3d, 1812.... Washington: Printed by R. C. Weightman, 1812. 16 p.
S&S-27233.
Copies: DLC; DeGE; MWA; OClWHi.

5504. The carriage is, as represented in the engraving, standing on a segment of the same circle for which it was constructed.... [Philadelphia: 1813.] Broadside, illus. 33 x 22 cm.
A description of a carriage which can be moved by steam power.
Copies: MWA (photocopy); PPAmP.

5505. Mechanics.... "Specifications of a new, safe, easy and expeditious means of travelling and transporting merchandize, over land on prepared roads, or over ice, in carriages constructed for pleasure or burden...." New York: 1813. Broadside 44 x 28 cm.
Dated: January 15, 1813.
Describes a new improved carriage.
Copies: MWA.

5506. Road bill, shewing the stages on the different roads from Williamsport, Washington County, Maryland.... [S.l.: 1813.] Broadside.
S&S-29670.
Copies: PHi.

5507. Salem Turnpike and Chelsea Bridge Corporation.
An act for incorporating certain persons for the purpose of laying out and making a turnpike road from Salem to Charles River Bridge. Salem, [Mass.]: Printed by Joshua Cushing, 1813. 35 p.
"Rules and regulations of the Salem Turnpike and Chelsea Bridge Corporation," p. [27]–35.
S&S-29071.
Copies: DLC; DeGE; MH; MSaE; MWA.

5508. United States. Congress. Senate. Committee on Roads and Canals.
In Senate of the United States, July 2d, 1813. Mr. Brent, from the committee to whom was referred the bill, entitled "An act to incorporate a company for making a certain turnpike road in the county of Alexandria," reported the same with the following amendments. [Washington: 1813.] 1 p.
S&S-30273.
Copies: DLC.

5509. United States. Congress. Senate. Committee on Roads and Canals.
In Senate of the United States, July 2d, 1813. Mr. Brent, from the committee to whom was referred the bill, entitled "An act to incorporate a company for making a certain turnpike road in the county of Washington, in the District of Columbia," reported the same with the following amendments. [Washington: 1813.] 1 p.
S&S-30274.
Copies: DLC.

5510. United States. Engineer Dept.
Result of the survey of the main post road, from Robbinstown in Maine, to St. Mary's in Georgia. Made between June, 1812, and January, 1813. [S.l.: 1813.] 7 p.
S&S-30371.
Copies: MB; PPAmP.

5511. United States. Laws.
A bill supplementary to an act, entitled "An act to authorize the making a turnpike road from Mason's causeway to the town of

Alexandria." [Washington: 1813.] 6 p. (S. 30)
Dated: January 29, 1813.
S&S-30122.
Copies: DLC.

5512. United States. Laws.
A bill to authorise and empower the president and managers of the Washington Turnpike Company of the state of Maryland, when organized to extend and make their turnpike road to or from Georgetown, in the District of Columbia, through the said District, to the line thereof. February 16th, 1813.... [Washington: 1813.] 2 p. (S.25)
S&S-30134.
Copies: DLC.

5513. United States. Laws.
A bill to incorporate a company for making a certain turnpike road in the county of Alexandria. January 20th, 1813. [Washington: 1813.] 18 p. (H.R. 49)
S&S-30146.
Copies: DLC.

5514. United States. Laws.
A bill to incorporate a company for making a certain turnpike road in the county of Washington, in the District of Columbia. February 22d, 1813. [Washington: 1813.] 18 p. (H.R. 81)
S&S-30147.
Copies: DLC.

5515. United States. Treasury Dept.
Message from the President of the United States, transmitting a report of the Secretary of the Treasury, containing a statement of the Secretary of the Treasury ... of the proceedings under the "Act to regulate the laying out and making a road from Cumberland, in the state of Maryland, to the state of Ohio." March 2, 1813.... Washington City: Printed by A. & G. Way, 1813. 8 p.
S&S-30256.
Copies: DLC; MWA.

5515a. Virginia. Laws.
An act incorporating a company to establish a turnpike road, commencing at a point on the route laid down for the Leesburg turnpike road, and extending towards the Falls Bridge, as far as the boundary of the District of Columbia. [Richmond: 1813.] 6 p.
Passed February 18, 1813.
N-V0187335.
Copies: MiU; ViU.

5516. Melish, John, 1771–1822.
A description of the roads in the United States. Compiled from the most authentic materials.... Philadelphia: Printed by G. Palmer, 1814. iv, [5]–11, [41 p. in 82 columns].
S&S-32096.
Copies: DLC; IaHi; MB; MH; MEA; NN; OClWHi; PPAmP.

5517. New York (State). Laws.
Act of incorporation of the Waterford and Whitehall Turnpike Company. An act to establish a turnpike road from the village of Waterford through the town of Whitehall to Westhaven [with amendments of 1810 and 1814]. [Albany: 1814.] 15 p.
N-N0196739.
Copies: DLC; MWA; MiU; N; NHi.

5518. United States. Laws.
A bill to authorise the opening a road from Shawaneetown on the Ohio River, in the Illinois Territory to the United States saline, and from thence to the town of Kaskaskia. January 4th, 1814.... [Washington: 1814.] 1 p. (H.R. 12)
S&S-33160.
Copies: DLC.

5519. United States. Treasury Dept.
Message from the President of the United States, transmitting a report of the Secretary of the Treasury, containing a statement of the proceedings under the "Act to regulate the laying out and making a road from Cumberland in the state of Maryland to the state of Ohio." January 18, 1814.... Washington City: Printed by Roger C. Weightman, 1814. 6 p. (13th Cong., 2d sess. House. Doc. 36)

5519 (*continued*)
S&S-33313.
Copies: DLC; DeGE; MWA.

5520. Virginia. Laws.
An act to amend the act entitled "An act incorporating a company to establish a turnpike road from Fredericksburg to Swift-Run Gap...." [Fredericksburg?, Va.: 1814?] Broadside.
S&S-51447.
Copies: ViHi.

5521. Banks, Henry, fl. 1781–1826.
Observations and propositions to establish public roads from Richmond to North Carolina, Tennessee, Kentucky and Ohio. Richmond: 1815. 18 p.
S&S-33921.
Copies: PHi.

5522. Melish, John, 1771–1822.
A description of the roads in the United States.... Philadelphia: Printed by G. Palmer, 1815. iv, [44 p. in 88 col.], [1] p.
S&S-35245.
Copies: MWA.

5523. New York (State). Laws.
An act to amend an act, entitled "An act to incorporate the Highland Turnpike Company." April 15, 1815. [Albany: 1815.] Broadside 25.5 x 31 cm.
S&S-51518.
Copies: NN.

5524. Pennsylvania. Laws.
An act to enable the Governor to incorporate a company for making an artificial road.... Vine and Tenth Streets, Philadelphia ... to the Perkiomen Bridge.... [S.l: 1815.] 22 p.
S&S-35593.
No copy known.

5525. Spafford, Horatio Gates, 1778–1832.
Some cursory observations on the ordinary construction of wheel-carriages: with an attempt to point out their defects, and to show how they may be improved; whereby a saving may be made in the power applied, the motion be rendered more uniform and easy, and the danger of upsetting most effectually prevented.... Albany: Printed and published by E. & E. Hosford, 1815. 11 p., plate.
S&S-35985.
Copies: DBRE; DLC; DeGE; InLP; MH-BA; MWA; NNC; OMC; PMA; PPAmP.

5526. United States. Congress. House Committee on Repair of Road from Nashville to Natchez.
Report of the Committee appointed on the eighteenth ult., to inquire into the expediency of repairing and keeping in repair the road from Nashville to Natchez; as also, the road from Fort Hawkins, in Georgia, to St. Stephens, in the Mississippi Territory. February 9, 1815.... Washington City: Printed by Roger C. Weightman, 1815. 4 p. (13th Cong., 3d sess. House. Doc. 61)
S&S-36371.
Copies: DLC; DeGE; MNBedf; MWA; MeB; MiD-B.

5527. United States. Congress. Senate. Committee on Roads and Canals.
Report of the Committee to whom was referred the report of the Secretary of the Treasury, containing a statement of proceedings under the act to regulate the laying out and making a road from Cumberland, in the state of Maryland, to the state of Ohio. January 24, 1815. Printed by order of the Senate of the United States. Washington City: Printed by Roger C. Weightman, 1815. 6 p.
Reported by Mr. Smith.
S&S-36399.
Copies: DLC; DeGE; MWA; MiU-C; O.

5528. United States. Laws.
A bill in addition to the act to regulate the laying out and making a road from Cumberland in the state of Maryland, to the state of Ohio. January 24, 1815. [Washington: 1815.] 2 p. (S. 24)

S&S-36206.
Copies: DLC.

5529. United States. Laws.
A bill making an appropriation for repairing the road from Nashville to Natchez. February 9, 1815. [Washington: 1815.] 1 p. (H.R. 73)
S&S-36209.
Copies: DLC.

5530. United States. Treasury Dept.
Letter from the Secretary of the Treasury to the chairman of the Committee on the Cumberland Road showing the appropriations heretofore made therefor, &c., &c. February 2, 1815.... Washington City: Printed by R. C. Weightman, 1815. 6 p. (13th Cong., 3d sess. House. Doc. 57)
Includes estimates for continued construction and statements of contracts.
S&S-36297.
Copies: DLC; DeGE; MNBedf; MWA; MeB.

5531. United States. Treasury Dept.
Message from the President of the United States, transmitting a report from the Secretary of the Treasury, containing a statement of proceedings under the "Act to regulate the laying out and making a road from Cumberland, in the state of Maryland, to the state of Ohio." January 2, 1815.... Washington City: Printed by Roger C. Weightman, 1815. 7 p.
Contains a report from David Shriver, superintendent of Cumberland Road.
S&S-36325, 36326.
Copies: DLC; DeGE; MWA; O.

5532. Indiana (Territory). Laws.
An act supplementary to the several acts providing for the opening and repairing public roads and highways, passed at the second session of the Fifth General Assembly of Indiana Territory. [Lexington?: 1816.] 12 p.
S&S-37923.
No copy known.

5533. National road. [Philadelphia?: 1816?] 8 p.
A protest from Pennsylvania against the proposed route of the Cumberland Road. Includes a memorial of the General Assembly of Pennsylvania to the United States Congress.
S&S-38360.
Copies: MWA.

5534. United States. Congress. House. Committee on the Cumberland Road.
Report of the Committee to which was referred the message of the President of the United States, of the 12th instant, transmitting a report of the Secretary of the Treasury containing a statement relative to the Cumberland Road. March 23d, 1816... [Washington: 1816.] 9 p. (14th Cong., 1st sess. House. Doc. 75)
S&S-39619.
Copies: DLC; DeGE; MWA.

5535. United States. Congress. Senate.
In Senate of the United States. March 11th, 1816. Mr. Ruggles submitted the following motion for consideration. [Washington: 1816.] 2 p.
Concerns the location of the road from the rapids of the Miami River of Lake Erie to the western boundary of the Connecticut Reserve.
S&S-39423.
Copies: DLC; MWA.

5536. United States. Laws.
A bill making an appropriation for repairing the road from Duck River, in the state of Tennessee, to Madisonville, in the Mississippi Territory, and between Fort Hawkins, in the state of Georgia, and Fort Stoddard. April 15, 1816. [Washington: 1816.] 1 p. (H.R. 181)
S&S-39273.
Copies: DLC.

5537. United States. Laws.
A bill to authorize the surveying and making a road in the Territory of Illinois. [Washington: 1816.] 1 p.

5537 (*continued*)
S&S-39316.
Copies: DLC.

5538. United States. Laws.
A bill to establish post roads. March 30, 1816. [Washington: 1816.] 10 p. (H.R. 157)
S&S-39329.
Copies: DLC.

5539. United States. Treasury Dept.
Message from the President of the United States, transmitting a report of the Secretary of the Treasury on the subject of the Cumberland Road, with a statement of past appropriations and an estimate of required appropriations. March 13, 1816.... Washington: Printed by W. A. Davis, 1816. 11 p. (14th Cong., 1st sess. House. Doc. 67)
Includes a report from the Superintendent of the Cumberland Road.
S&S-39559.
Copies: DLC; DeGE; MWA.

5540. Maryland. Laws.
Acts on the subject of turnpike roads.... [Annapolis?: 1817.] 35 p.
N-M0289533.
Copies: DNAL.

5541. United States. Congress. Senate. Committee on Roads and Canals.
In Senate of the United States. January 8, 1817. The Committee on Roads and Canals to whom was referred the petition of a number of the inhabitants of Washington County, Pennsylvania, praying a change in part of the route of the western turnpike road, report. [Washington: 1817.] 2 p.
S&S-42484.
Copies: DLC; MWA.

5542. United States. Laws.
A bill to alter and establish certain post roads. February 22, 1817. [Washington: 1817.] 7 p., fold. table. (H.R. 155)
S&S-42422.
Copies: DLC.

5543. United States. Laws.
A bill to regulate the laying out and making a road from the Ohio River opposite where the Cumberland Road strikes that river, to the state of Indiana. January 25, 1817. [Washington: 1817.] 4 p. (H.R. 76)
S&S-42447.
Copies: DLC.

5544. [Hoge, John] 1760–1824.
To the stockholders of the Washington & Pittsburgh Turnpike Road Company. [Washington, Pa.: 1818?] 11 p.
Signed: Jno. Hoge.
N-H0448282.
Copies: DLC.

5545. Maryland. Laws.
Acts of Maryland and of Congress, on the subject of turnpike roads leading from the District of Columbia. Georgetown, D.C.: Printed by W. A. Rind & Co., 1818. 24 p.
S&S-44703.
Copies: DLC; MB.

5546. Philadelphia & Pittsburgh Transportation Company.
Address of the managers to citizens of Philadelphia. Philadelphia: 1818. 8 p.
S&S-45288.
Copies: PPAmP.

5547. Smyth, Charles.
Transportation to Pittsburgh, through the state of New-York.... Albany: 1818. Broadside (2 p.) 27 x 23 cm.
Dated: Albany, 16th November, 1818.
Table of routes, indicating types of transportation, time required and prices, on verso.
S&S-45739.
Copies: NHi.

5548. United States. Treasury Dept.
Letter from the Secretary of the Treasury, transmitting statements, in relation to the progress, &c. in the construction of the Cumberland Road. April 20, 1818.... Washington: Printed by E. De Krafft, 1818. 6 p., 2

fold. tables. (15th Cong., 1st sess. House. Doc. 201)

S&S-46361.

Copies: DLC; DeGE; DeU; MWA; MoU; NjP; O; R.

5549. United States. Treasury Dept.

Message from the President of the United States, transmitting, pursuant to a resolution of the House of Representatives of ninth Dec. last, information of the roads made, or in progress, under the authority of the Executive of the United States ... January 23, 1818. Washington: Printed by E. De Krafft, 1818. 11 p., 2 fold. tables. (15th Cong., 1st sess. House. Doc. 61)

Transmitting a report from the Treasury Department.

S&S-46439.

Copies: CSmH; DLC; DeGE; MWA; N; NhHi; NjP; R.

5550. Allard, Louis.

Proposals for the establishment of a road from Bayou St. John to Lake Pontchartrain.... New Orleans: 1819. Broadside 20 x 34 cm.

Dated: New Orleans, June 14, 1819.

Parallel text in French.

Hummel-808.

Copies: LN.

5551. Baltimore and Reisterstown Turnpike Road Company.

To the Honorable the General Assembly of Maryland. The memorial of the president and managers of the Baltimore and Reister's Town Turnpike Road Company, of the Baltimore and Frederick Town Turnpike Road Company.... [Baltimore: 1819.] 15, [1] p.

Heyl-233.

Copies: MdHi.

5552. Banks, Henry, fl. 1781–1826.

Observations shewing the benefits of opening a turnpike road from Lexington in Kentucky to James River in Virginia, with some remarks concerning the banking establishments of Kentucky and Virginia.... Frankfort, [Ky.]: Printed by Kendall and Russells, printers to the state, 1819. 56 p.

N-B0096142.

Copies: MiU-C.

5553. Dana, Edmund.

A description of the bounty lands in the state of Illinois; also, the principal roads and routes, by land and by water, through the territory of the United States.... Cincinnati: Looker, Reynolds & Co., printers, 1819. 108 p., map.

Description of the principal roads and routes.... with special title page, p. [49]–108.

Copies with the map are very uncommon.

S&S-47779.

Copies: CtY; DLC; ICN; IU; MH; MWA; MiU-C; NIC; NN; OClWHi; WHi.

5554. Maryland. Governor (C. Goldsborough).

Executive communication to the General Assembly of Maryland, at December session, 1818, on the subject of turnpike roads, made in pursuance of a resolve of December session, 1817, with the accompanying documents. Annapolis: Printed by Jehu Chandler, 1819. 45 p.

"Accompanying documents," p. 20–45.

S&S-48596.

Copies: DLC; DeGE; MWA; MdBE; MdBP; MdH.

5555. Petition to the General Assembly of Maryland, by citizens of the state, praying to be relieved from illegal exactions of toll, on the turnpike roads leading to Baltimore. Baltimore: Printed for the committee, 1819. 20 p.

S&S-48601.

Copies: MdHi.

5556. Tanner, Henry Schenck, 1786–1858.

The travellers guide through the state of New York, showing all the important roads, post-towns, &c. Philadelphia: [1819.] Map 52 x 65 cm.

S&S-49553.

Copies: ICN.

5557. Turnpike Company of the Commonwealth of Pennsylvania.

Report of the Turnpike Company, 1819. Harrisburg: 1819. 17 p.

S&S-49651.

No copy known.

5558. Heald, Henry.

Roads of New Castle County, surveyed and printed by Henry Heald. [Wilmington, Del.?]: 1820. Map 95x47 cm.

Inset: Statistical view of New Castle County.

N-H0218759.

Copies: DeGE; MiU-C (photocopy).

5559. Salem Turnpike and Chelsea Bridge Corporation.

Report of the directors of the Salem Turnpike and Chelsea Bridge Corporation. Salem, [Mass.]: 1820. 14 p.

S-3099.

Copies: MB; MBAt; MSaE.

5560. United States. Treasury Dept.

Message from the President of the United States, transmitting statements of expenditures in constructing the Cumberland Road, from the year 1806, to the year 1820. March 17, 1820.... Washington: Printed by Gales & Seaton, 1820. 9 p., fold. tables. (16th Cong., 1st sess. House. Doc. 99)

Transmits statements from the Treasury Department.

S-3923.

Copies: DLC; DeGE; DeU; G; MWA; O; R.

5561. Virginia. Laws.

An act incorporating a company to establish a turnpike road from Fredericksburg to Swift-Run Gap in the Blue Ridge. Fredericksburg, [Va.]: Printed by James D. Harrow, 1820. 22 p.

Copies: ViHi.

5562. McAdam, John Loudon, 1756–1836.

Remarks on the present system of roadmaking; with observations, deduced form [*sic*] practice and experience, with a view to a revision of the existing laws, and the introduction of improvement in the method of making, repairing, and preserving roads ... 1st American edition from the 3d London edition, carefully rev., with considerable additions, and an appendix.... Baltimore: Printed by John Robinson, 1821. 196 p.

S-5875.

Copies: DLC; MWA; MdBP; MdHi; NNC; NNE; NcU; P; PPA; PPAmP; PPL; PPi.

5563. United States. Treasury Dept.

Message from the President of the United States, transmitting a report of the Secretary of the Treasury, with copies of the instructions given to the commissioners for locating a road from Wheeling, in Virginia, to the Mississippi River; also, copies of the report of said commissioners of the progress made in the execution of the duties assigned to them. January 31, 1821.... Washington: Printed by Gales & Seaton, 1821. 10 p. (16th Cong., 2d sess. House. Doc. 82)

S-7385.

Copies: DLC; DeGE; DeU; MWA; NjP; TxU.

5564. Virginia. Laws.

An act giving the assent of the state of Virginia to an act of Congress for laying out and making a road from the Potomac River to the state of Ohio. January 9, 1821.... Printed by order of the House of Representatives of the United States. Washington: Printed by Gales & Seaton, 1821. [3] p. (16th Cong., 2d sess. House. Doc. 50)

The act was passed during the session of 1807.

S-7037.

Copies: DLC; DeGE; DeU; MWA; NjP.

5565. Hall, Lorenzo T.

The American directory and traveller's companion, containing a list of inn-keepers, on the principal mail stage roads, from

Darien (Geo.) to Boston (Mass.), showing the distances from each.... Boston: Printed by True & Greene, 1822. 36 p.
S-8919.
Copies: DLC; ScC.

5566. New York (State). Laws.
New road act, being a revised law of the state of New York, entitled An act to regulate highways. With a copious index, to enable persons at first sight to turn to any article of the law. Albany: Websters & Skinners, 1822. 20 p., 2 l.
N-N0198577.
Copies: NN.

5567. Pennsylvania. General Assembly. House. Committee on Domestic Manufactures.
Report of Mr. Stevenson, chairman of the Committee on Domestic Manufactures, read Tuesday, Dec. 24, 1822. Harrisburg: Printed by John S. Wiestling, 1822. iii, 8 p.
Recommendation to make certain turnpike roads in Pennsylvania toll free.
S-9848.
Copies: P; PPAmP; PPL.

5568. Pennsylvania. Laws.
Acts of Assembly, concerning the Lancaster, Elizabethtown, and Middletown turnpike road. Lancaster: Printed by John Bear, 1822. 22 p.
S-9838.
Copies: CSmH; P; PHi.

5569. Turnpike Company of the Commonwealth of Pennsylvania.
Statement of the Turnpike Company of the Commonwealth. [Harrisburg: 1822.] 8 p.
S-9859.
Copies: P (not found).

5570. United States. Commissioners Appointed to View and Inspect the Cumberland Road.
Letter from the Secretary of the Treasury transmitting a report of the Commissioners, appointed to view and inspect the Cumberland Road. January 15, 1822.... Washington: Printed by Gales & Seaton, 1822. 7 p. (17th Cong., 1st sess. House. Doc. 27)
S-10939.
Copies: DLC; DeGE; DeU; MWA; NjP.

5571. United States. Congress. House. Committee on Roads and Canals.
Report of the Committee on Roads and Canals, to which was referred a report of a select committee made the 12th of May, 1820, relative to carrying into effect the treaty of Brownstown, of 25th November, 1808, accompanied with "A bill for laying out and making a road from the Lower Rapids of the Miami of Lake Erie to the Connecticut Western Reserve, in the state of Ohio," &c. March 8, 1822.... [Washington: 1822.] 13 p. (17th Cong., 1st sess. House. Rept. 61)
S-11191.
Copies: DLC; MWA; PU.

5572. ——— Dec. 19, 1822. Ordered to be reprinted. [Washington: 1822.] 14 p. (17th Cong., 2d sess. House. Rept. 50)
S-11516.
Copies: DLC; DeGE; DeU; MWA; NjP.

5573. United States. Laws.
A bill for laying out, and making, a road, from the lower rapids of the Miami of Lake Erie to the western boundary of the Connecticut Western Reserve, in the state of Ohio, agreeable to the provisions of the treaty of Brownstown. March 8, 1822.... [Washington: 1822.] 1 p. (H.R. 112)
S-10566.
Copies: DLC.

5574. ——— Dec. 12, 1822. Ordered to be reprinted. [Washington: 1822.] 1 p. (H.R. 112)
S-10567.
Copies: DLC.

5575. United States. Laws.
A bill to keep in repair the Cumberland Road. January 8, 1822. Mr. Johnson, of

5575 (*continued*)
Kentucky, from the Committee on Roads and Canals, reported the following bill.... [Washington: 1822.] 3 p. (S. 12)
S-10764.
Copies: DNA.

5576. First Great South-Western Turnpike Road Company.
The memorial of the First Great South-Western Turnpike Road Company [to the Legislature of New York]. [S.l.]: 1823. 8 p.
N-F0152288; Sabin-47673.
Copies: N.

5577. Maryland. Governor (S. Stevens).
The Governor's communication of the 4th February, 1823, with accompanying documents. Annapolis: Printed by J. Hughes, 1823. 13 p.
Concerns the repair of the Cumberland Road.
N-M0289293.
Copies: ICN.

5578. Maryland. Laws.
An act for making a certain road in Baltimore County. [Annapolis: 1823.] 12 p.
Heyl-130.
Copies: MdHi.

5579. Maryland. Laws.
An act for the relief of the Westminster, Taney Town and Emmitsburg Turnpike Road Company. [Annapolis: 1823.] 5, [1] p. (House. Bill No. 56)
Heyl-230.
Copies: MdHi.

5580. United States. Laws.
A bill for the continuation of the Cumberland Road. December 18, 1823.... [Washington: 1823.] 2 p. (H.R. 14)
S-14375.
Copies: DNA.

5581. United States. Laws.
A bill for the preservation and repair of the Cumberland Road. December 16, 1823. [Washington: 1823.] 3 p. (H.R. 9)
S-14379.
Copies: DNA.

5582. United States. Laws.
A bill to authorize the laying out and opening certain public roads in the Territory of Florida. February 5, 1823.... [Washington: 1823.] 1 p. (H.R. 275)
S-14465.
Copies: DLC.

5583. United States. Laws.
A bill to authorize the surveying and making a road from Memphis, in the state of Tennessee to Little Rock, in the territory of Arkansas. December 24, 1823.... [Washington: 1823.] 1 p. (H.R. 29)
S-14471.
Copies: DNA.

5584. United States. Post Office Dept.
Letter from the Postmaster General, in relation to the state and condition of the Cumberland Road. (Made in obedience to a resolution of the House of Representatives of the 31st ult.) January 7, 1823. Washington: Printed by Gales & Seaton, 1823. 8 p. (17th Cong., 2d sess. House. Doc. 16)
S-14558.
Copies: DLC; DeGE; DeU; MWA; NjP.

5585. United States. Register of the Treasury.
Letter from the Secretary of the Treasury, transmitting statements of the expenditures in laying out and making the Cumberland Road. February 20, 1823.... Washington: Printed by Gales & Seaton, 1823. 6 p. (17th Cong., 2d sess. House. Doc. 77)
Signed: Joseph Nourse, Register.
S-14598.
Copies: DLC; DeGE; DeU; MWA; NjP.

5586. United States. War Dept.
Letter from the Secretary of War, transmitting a report of Major General Jacob Brown, upon the subject of the military road

leading from Plattsburg to Sackett's Harbour. January 22, 1823.... Washington: Printed by Gales & Seaton, 1823. 6 p. (17th Cong., 2d sess. House. Doc. 33)
S-14611.
Copies: DLC; DeU; MWA; NjP.

5587. New Jersey. Laws.
An act for facilitating the communication from Elizabeth-Town, in the county of Essex, through Morris-Town, in the county of Morris, and from thence into the county of Sussex, passed at Trenton, March 9, 1801. Elizabethtown: J. and E. Sanderson, printers, 1824. 12 p.
S-17340.
Copies: NjMo (not located).

5588. Ohio. Auditor.
Auditor's report in response to a resolution requesting report on progress relative to the road from the rapids of the Miami of Lake Erie to the Western Reserve. [Columbus: 1824.] 8 p.
The report is an abstract of two semiannual reports of the superintendent of the road, showing the details of contracts for making the road.
Hasse (Ohio)-p. 1030.
Copies: OClWHi.

5589. Pennsylvania. Laws.
The several acts of the General Assembly of the Commonwealth of Pennsylvania relating to the Lycoming and Potter Turnpike Road Company. [Philadelphia]: Printed for the company by J. Harding, 1824. 19 p.
S-17568.
Copies: PHi; PPL.

5590. Pennsylvania Society for the Promotion of Internal Improvements in the Commonwealth.
The subscribers, the Acting Committee of "The Pennsylvania Society for the Promotion of Internal Improvements in the Commonwealth," respectfully submit the following essay on the construction and reparation of roads to the consideration of their fellow citizens.... Philadelphia: 1824. 8 p.
Signed: Mathew Carey, Joseph Hemphill [a.o.], Philadelphia, Dec. 20, 1824.
S-17582.
Copies: DeGE; MBL; PPAmP; PPL.

5591. ———— 2d ed. Philadelphia: 1824. 8 p.
S-17583.
Copies: DeGE; MWA; PPL.

5592. Tarascon, Louis Anastasius, 1757– .
Lewis A. Tarascon, of Shippingport, Ky. To the people of the United States, on the propriety of establishing a wagon road from the river Missouri to the river Columbia, of the Pacific Ocean. [Louisville?: 1824?] Broadside 22.5 x 26.5 cm.
S-18145.
Copies: CSmH; DNA.

5593. Tarascon, Louis Anastasius, 1757– .
Petition of Lewis A. Tarascon (and others) praying the opening of a wagon road from the river Missouri, north of the river Kansas, to the river Columbia. December 13, 1824.... Washington: Printed by Gales & Seaton, 1824. 12 p. (18th Cong., 2d sess. Senate. Doc. 2)
S-18939.
Copies: DLC; DeU; ICN; MWA; NNC; NjP; PPL.

5594. United States. Congress. House. Committee on Roads and Canals.
Report of the Committee to which was referred that part of the President's message which relates to the repairs of the Cumberland Road. May 7, 1824.... [Washington: 1824.] 2 p. (18th Cong., 1st sess. House. Rept. 118)
S-19056.
Copies: DLC; DeGE; DeU; MWA; NjP.

5595. United States. Congress. Senate. Select Committee on Roads and Canals.
In Senate of the United Sates. March 18, 1824. Mr. M'Ilvine, from the Select

5595 (*continued*)
Committee on Roads and Canals, to whom was referred the memorial of J. L. Skinner, reported.... [Washington: 1824.] 5 p. (18th Cong., 1st sess. Senate. Doc. 55)

Concerns work done on the construction of the Cumberland Road.

S-18729.

Copies: DLC; DeU; MWA; NjP.

5596. United States. Engineer Dept.

Message from the President of the United States, transmitting a report of David Shriver, Superintendent of the Cumberland Road, stating the manner in which the appropriation made at the last session has been expended, &c. May 24, 1824.... Washington: Printed by Gales & Seaton, 1824. 10 p. (18th Cong., 1st sess. House. Doc. 154)

S-18900.

Copies: DLC; DeGE; DeU; MWA; NjP.

5597. United States. Laws.

A bill supplementary to "An act to incorpoarte a company for making certain turnpike roads in the District of Columbia," March 1, 1824. [Washington: 1824.] 1 p. (S. 62)

S-18589.

Copies: DNA.

5598. United States. Laws.

A bill to amend an act, entitled, "An act for the establishment of a turnpike company in the county of Alexandria, in the District of Columbia." Feb. 6, 1824.... [Washington: 1824.] 1 p. (S. 42)

S-18609.

Copies: DNA.

5599. United States. Laws.

A bill to authorize the laying out and opening of a road from Wheeling, in the state of Virginia, to the seat of government of the state of Missouri. Feb. 12, 1824.... [Washington: 1824.] 3 p. (H.R. 110)

S-18621.

Copies: DNA.

5600. United States. Laws.

A bill to authorize the President of the United States to cause to be made a military road from Fort St. Philip, on the river Mississippi, to the English Turn, as an auxiliary to the defence of New Orleans. January 23, 1824.... [Washington: 1824.] 1 p. (S. 25)

S-18626.

Copies: DNA.

5601. United States. Laws.

A bill to authorize the surveying and making of a road from a point in the northwestern boundary of the state of Ohio, near the foot of the rapids of the Miami of Lake Erie, to Detroit in the Territory of Michigan. Feb. 12, 1824.... [Washington: 1824.] 1 p. (H.R. 112)

S-18635.

Copies: DNA.

5602. United States. Laws.

A bill to authorize the surveying and opening of a road, from Detroit to Chicago, in the state of Illinois. April 26, 1824. [Washington: 1824.] 1 p. (H.R. 204)

S-18636.

Copies: DNA.

5603. United States. Post Office Dept.

Report of the Postmaster General, on the subject of the most practicable post route, from New Orleans to Washington City.... December 15, 1824.... Washington: Printed by Gales & Seaton, 1824. 5 p. (18th Cong., 2d sess. Senate. Doc. 4)

S-19059.

Copies: DLC; DeU; MWA; NjP.

5604. United States. War Dept.

Report of the Secretary of War, in relation to a military road, from St. Philip to the English Turn. Made in pursuance of a resolution of the Senate. January 13, 1824.... Washington: Printed by Gales & Seaton, 1824. 7 p. (18th Cong., 1st sess. Senate. Doc. 22)

S-19067.

Copies: DLC; DeU; MWA; NjP.

5605. Bradley, Abraham, d. 1838.
Map of the United States intending chiefly to exhibit the post roads and distances. Georgetown, D.C.: Engraved and printed by W. Harrison, [1825?]. Map 105 x 136 cm.
The map seems to cover the postal service for 1825. Contains Lieut. Talcott's route between St. Anthony's Falls and Council Bluff, 1820.
N-B0730333.
Copies: MB.

5606. National turnpike from Washington to New-Orleans. To the President of the United States.... [Winchester, Va.: 1825.] 8 p.
A memorial from the citizens of Frederick County, Virginia.
S-20559.
Copies: DLC.

5607. Ohio. Treasury Dept.
Receipts and expenditures, Miami and Western Reserve Road. December 22, 1825. [S.l.: 1825.] 7 p.
Hasse (Ohio)-p. 279; S-21729.
No copy known.

5608. Pennsylvania Society for the Promotion of Internal Improvements in the Commonwealth.
The subscribers, the Acting Committee of "The Pennsylvania Society for the Promotion of Internal Improvements in the Commonwealth," respectfully submit the following essay, on the construction and reparation of roads. [Philadelphia: 1825.] 8 p.
3d ed., July 31, 1825.
Signed: Mathew Carey, Joseph Hemphill, Richard Peters, Jun., [a.o.].
N-P0216329.
Copies: MH-BA; MWA; PPL.

5609. Petition of sundry inhabitants of the state of Missouri, upon the subject of a communication between the said state and the internal provinces of Mexico, with a letter from Alphonso Wetmore, upon the same subject. February 14, 1825.... Washington: Printed by Gales & Seaton, 1825. 8 p. (18th Cong., 2d sess. House. Doc. 79)
Concerns a road from Missouri to Mexico.
S-23016.
Copies: DLC; MWA; NIC; NjP; ViU.

5610. United States. Congress. Senate. Select Committee on Roads and Canals.
In Senate of the United States, February 2, 1825. Mr. Brown, from the Select Committee on Roads and Canals, to whom was referred so much of the memorial of the Legislative Council of the Territory of Michigan, as relates to roads in that territory, reported. [Washington: 1825.] 4 p. (18th Cong., 2d sess. Senate. Doc. 23)
S-22870.
Copies: DLC; DeU; MWA; NjP.

5611. United States. Laws.
An act for the continuation of the Cumberland Road. January 24, 1825.... [Washington: 1825.] 3 p. (H.R. 14)
S-22561.
Copies: DNA.

5612. United States. Laws.
An act supplementary to the act entitled "An act for the continuation of the Cumberland Road." February 28, 1825.... [Washington: 1825.] 1 p. (H.R. 370)
S-22637.
Copies: DNA.

5613. United States. Laws.
An act to authorize the laying out and opening of a public road from the St. Mary's River to the Bay of Tampa, in the Territory of Florida. February 14, 1825.... [Washington: 1825.] 1 p. (H.R. 269)
S-22642.
Copies: DNA.

5614. United States. Laws.
An act to authorize the surveying and making a road from Little Rock to Cantonment Gibson, in the Territory of Arkansas. January 27, 1825.... [Washington: 1825.] 1 p. (H.R. 270)

5614 (*continued*)
S-22648.
Copies: DNA.

5615. United States. Laws.
An act to authorize the surveying and opening of a road, from Detroit to Chicago, in the state of Illinois. February 2, 1825.... [Washington: 1825.] 2 p. (H.R. 204)
S-22649.
Copies: DNA.

5616. United States. Laws.
An act to provide an additional appropriation to complete the public road from Pensacola to St. Augustine, in the Territory of Florida. February 14, 1825.... [Washington: 1825.] 1 p. (H.R. 268)
S-22656.
Copies: DNA.

5617. United States. Laws.
A bill to authorize the President of the United States to cause a road to be marked out from the western frontier of Missouri, to the confines of New Mexico. January 11, 1825.... [Washington: 1825.] 1 p. (S. 23)
S-22812, 22813 (dated February 3, 1825)
Copies: DNA.

5618. United States. Post Office Dept.
Letter from the Postmaster General, in relation to an appropriation to render the post road through the Indian country between Columbus and Jackson, Mississippi, fit for the transportation of mail from the city of Washington to Natchez and New Orleans, &c. February 7, 1825.... Washington: Printed by Gales & Seaton, 1825. 3 p. (18th Cong., 2d sess. Senate. Doc. 28)
S-22913.
Copies: DLC; DeU; MWA; NjP.

5619. Virginia. Commissioner of the Kanawha Road and Navigation.
Report of the Commissioner of the Kanawha Road and Navigation, with accompanying documents. [Richmond: 1825.] 10 p.
William Anderson, Commissioner.
N-V0183848.
Copies: ViW.

5620. Fessenden, John Milton, 1802–1883.
Cumberland Road. Map of the country through which it is proposed to extend the National Road to the tide waters within the District of Columbia. Washington: 1826. Map 46 x 58 cm.
N-F0108664.
Copies: MB.

5621. National Road. Resolution of citizens of Adams County in support of a petition for a "branch of the Western National Road." [West Union, Ohio?: 1826.] Broadside.
S-23420.
Copies: O-Ar (not located).

5622. New York (State). Commissioners of the Great State Road.
In Assembly, March 29, 1826. Report of the Commissioners of the Great State Road. [Albany: 1826.] 12 p.
S-25549.
Copies: DLC.

5623. New York (State). Commissioners of the Great State Road.
Report of the Commissioners of the Great State Road between Lake Erie and Hudson River. [Albany: 1826.] 61 p.
Hasse(New York)-p.490.
Copies: NN.

5624. Remarks upon the utility of constructing a turnpike road from Flushing to Huntington, across the heads of the necks. New York: Printed by Samuel Wood & Sons, 1826. 11 p.
S-25921.
Copies: CtY; NBLiHi; NN.

5625. United States. Board of Internal Improvement.
Road from Washington to New Or-

leans. Letter from the Secretary of War, transmitting the report of the Board of Internal Improvement, upon the subject of a national road from the city of Washington to New Orleans. April 12, 1826.... Washington: Printed by Gales & Seaton, 1826. 28 p., fold. table. (19th Cong., 1st sess. House. Doc. 156)

Includes the report from the Engineer Dept.

S-27372.

Copies: DLC; DeGE; DeU; MWA; NjP.

5626. United States. Congress. House. Committee on Military Affairs.

Military road in Michigan. January 16, 1826. Mr. Vance, from the Committee on Military Affairs ... made the following report. [Washington: 1826.] 18 p. (19th Cong., 1st sess. House. Rept. 42)

Memoir from Lewis Cass, Governor of Michigan, p. 6–18.

S-27229.

Copies: DLC; DeGE; DeU; MWA; NjP.

5627. United States. Congress. House. Committee on Military Affairs.

Road—Cantonment Gibson to Natchitoches. February 8, 1826. Mr. Hamilton, from the Committee on Military Affairs, made the following report. [Washington: 1826.] 2 p.(19th Cong., 1st sess. House. Rept. 67)

S-27366.

Copies: DLC; DeU; MWA; NjR.

5628. United States. Congress. House. Committee on Post Office and Post Roads.

Road between Jackson and Columbus, Mississippi. Documents to accompany the bill from the Senate, (No.41), appropriating a sum of money for the repair of the post road between Jackson and Columbus, in the state of Mississippi. March 18, 1826.... Washington: Printed by Gales & Seaton, 1826. 5 p. (19th Cong., 1st sess. House. Doc. 134)

S-27365.

Copies: DLC; DeGE; DeU; MWA; NjP.

5629. United States. Congress. House. Committee on Post Office and Post Roads.

Road from Baltimore to Philadelphia. January 26, 1826. Mr. Ingham, from the Committee on the Post Office and Post Roads ... made the following report. [Washington: 1826.] 2 p. (19th Cong., 1st sess. House. Rept. 51)

S-27369.

Copies: DLC; DeGE; DeU; MWA; NjP.

5630. United States. Congress. House. Committee on Roads and Canals.

Daniel Loomis. March 21, 1826. Mr. Stewart, from the Committee on Roads and Canals, to which was referred the petition of Daniel Loomis, made the following report. [Washington: 1826.] 2 p. (19th Cong., 1st sess. House. Rept. 131)

Concerns repairs of the Cumberland Road.

S-26941.

Copies: DLC; DeGE; DeU; MWA; NjP.

5631. United States. Congress. House. Committee on Roads and Canals.

Daniel Steinrod. Jan. 27, 1826. Mr. Stewart, from the Committee on Roads and Canals, made the following report. [Washington: 1826.] 4 p. (19th Cong., 1st sess. House. Rept. 56)

Concerns repairs of the Cumberland Road.

S-26944.

Copies: DLC; DeU; MWA; NjP.

5632. United States. Congress. House. Committee on Roads and Canals.

Daniel Steinrod. May 2, 1826. Mr. Stewart, from the Committee on Roads and Canals, to which was recommitted the bill for the relief of Daniel Steinrod, made the following report. [Washington: 1826.] 19 p. (19th Cong., 1st sess. House. Rept. 183)

Concerns work done on the Cumberland Road.

S-26945.

Copies: DeGE; DeU; MWA.

5633. United States. Congress. House. Committee on Roads and Canals.

Moses Shepherd. March 22, 1826. Mr. Stewart, from the Committee on Roads and Canals, to which was referred the petition of Moses Shepherd, made the following report. [Washington: 1826.] 4 p. (19th Cong., 1st sess. House. Rept. 133)

Concerns masonry work done on the Cumberland Road.

S-27259.

Copies: DLC; DeGE; DeU; MWA; NjP.

5634. United States. Congress. House. Committee on Roads and Canals.

National Road—Cumberland to Washington. March 3, 1826 ... Mr. Stewart, from the Committee on Roads and Canals ... made the following report. [Washington: 1826.] 2 p. (19th Cong., 1st sess. House. Rept. 100)

S-27267.

Copies: DLC; DeGE; DeU; MWA; NjP.

5635. United States. Congress. House. Committee on Roads and Canals.

Preservation of the Cumberland Road. March 27, 1826. Mr. Stewart, from the Committee on Roads and Canals ... made the following report. [Washington: Printed by Gales & Seaton, 1826.] 16 p. (19th Cong., 1st sess. House. Rept. 143)

S-27311.

Copies: DLC; DeGE; DeU; MWA; NjP.

5636. United States. Congress. House. Committee on Roads and Canals.

Road from Buffalo to Hanover. May 17, 1826 ... Mr. Stewart, from the Committee on Roads and Canals ... made the following report. [Washington: 1826.] 1 p. (19th Cong., 1st sess. House. Rept. 223)

S-27370.

Copies: DLC; DeU; MWA; NjP.

5637. United States. Congress. House. Committee on Roads and Canals.

Road from Detroit to Lake Michigan. March 28, 1826. Mr. Stewart, from the Committee on Roads and Canals ... made the following report. [Washington: 1826.] 4 p. (19th Cong., 1st sess. House. Rept. 144)

S-27371.

Copies: DLC; DeGE; DeU; MWA; NjP.

5638. United States. Congress. House. Committee on Roads and Canals.

Roads—Black Swamp to Cumberland, & Pittsburg to rapids of Miami, of Lake Erie. May 15, 1826 ... Mr. Beecher, from the Committee on Roads and Canals ... made the following report. [Washington: 1826.] 6 p. (19th Cong., 1st sess. House. Rept. 212)

S-27378.

Copies: DLC; DeGE; DeU; MWA; NjP.

5639. United States. Congress. House. Committee on Roads and Canals.

Washington Turnpike Road Company. January 30, 1826. Mr. Stewart, from the Committee on Roads and Canals, made the following report. [Washington: 1826.] 5 p. (19th Cong., 1st sess. House. Rept. 57)

S-27464.

Copies: DLC; DeGE; DeU; MWA; NjP.

5640. United States. Congress. Senate. Committee on Roads and Canals.

In Senate of the United States. February 6, 1826. Mr. Hendricks, from the Committee on Roads and Canals, to whom were referred sundry documents relative to certain roads in the Territory of Florida, reported. [Washington: 1826.] 3 p. (19th Cong., 1st sess. Senate. Doc. 29)

S-27034.

Copies: DLC; DeGE; DeU; MWA; NjP.

5641. United States. Congress. Senate. Select Committee on Roads and Canals.

In Senate of the United States, March 7, 1826. Mr. Hendricks, from the Select Committee on Roads and Canals, to whom was referred the memorial of the General Assembly of the state of Indiana, on the subject of location of the Cumberland Road through the state, reported. [Washington: 1826.] 3 p.(19th Cong., 1st sess. Senate. Doc. 59)

S-27051.

Copies: DLC; DeU; MWA; NjP.

5642. United States. Engineer Dept.
Cumberland Road. Letter from the Secretary of War, transmitting the information required by a resolution of the House of Representatives, of the 28th instant, in relation to the state of the Cumberland Road. April 29, 1826.... Washington: Printed by Gales & Seaton, 1826. 19 p. (19th Cong., 1st sess. House. Doc. 168)
Transmits a report from the Engineer Department.
S-26937.
Copies: DLC; DeGE; DeU; MWA; PU.

5643. United States. Engineer Dept.
Road—Cumberland to District of Columbia. Letter from the Secretary of War, transmitting a report upon the subject of an extension of the National Road from Cumberland to the District of Columbia. December 18, 1826.... Washington: Printed by Gales & Seaton, 1826. 34 p., fold. table. (19th Cong., 2d sess. House. Doc. 16)
S-27367.
Copies: DLC; DeGE; DeU; MWA; NjP.

5644. United States. Engineer Dept.
Road—Detroit to Chicago. Letter from the Secretary of War, transmitting a report of the Chief Engineer in relation to a survey and location of a road from Detroit to Chicago. February 1, 1826.... Washington: Printed by Gales & Seaton, 1826. 6 p. (19th Cong., 1st sess. House. Doc. 68)
Alex. Macomb, Chief Engineer.
S-27368.
Copies: DLC; DeGE; DeU; MWA; NjP.

5645. United States. Engineer Dept.
Road from Wheeling to Missouri. Letter from the Secretary of War, accompanied by a report of the Chief Engineer, in relation to the survey, location, and progress, of the road from the right bank of the Ohio, opposite to Wheeling, to the state of Missouri. January 21, 1826.... Washington: Printed by Gales & Seaton, 1826. 32 p. (19th Cong., 1st sess. House. Doc. 51)
S-27373.
Copies: DLC; DeGE; DeU; MWA; NjP.

5646. United States. Engineer Dept.
Road, Miami of Lake Erie to Detroit. Letter from the Secretary of War, transmitting a report from the Chief Engineer, showing the proceedings under the Act of 26th May, 1824, for surveying and making a road from the foot of the rapids of the Miami of Lake Erie, to Detroit. January 10, 1826.... Washington: Printed by Gales & Seaton, 1826. 16 p. (19th Cong., 1st sess. House. Doc. 33)
S-27375.
Copies: DLC; DeGE; DeU; MWA; NjP.

5647. United States. Engineer Dept.
Road through Ohio, Indiana and Illinois. Letter from the Secretary of War, transmitting a report of the Chief Engineer, in relation to the road through Ohio, Indiana, and Illinois. December 20, 1826.... Washington: Printed by Gales & Seaton, 1826. 16 p.(19th Cong., 2d sess. House. Doc. 18)
S-27376.
Copies: DLC; DeGE; DeU; MWA; NjP.

5648. United States. Engineer Dept.
Road—Washington to Buffalo. Survey —Roanoke Inlet, &c. Letter from the Secretary of War, transmitting information in relation to routes for a national road from Washington City to Buffalo; also, information respecting the progress in fulfilment of the acts of the last session of Congress, authorizing a survey of Roanoke Inlet and Sound. December 18, 1826.... Washington: Printed by Gales & Seaton, 1826. 4 p. (19th Cong., 2d sess. House. Doc. 15)
Transmits a report of the Engineer Department.
S-27377.
Copies: DLC; DeU; MWA; NjP.

5649. United States. Laws.
An act for the preservation and repair of the Cumberland Road. May 18, 1826. [Washington: 1826.] 5 p. (H.R. 183)

5649 (*continued*)
S-26280.
Copies: DNA.

5650. United States. Laws.
An act to amend an act, entitled "An act to incorporate a company for making a certain turnpike road in the county of Alexandria," passed 18th July, 1818. May 5, 1826. [Washington: 1826.] 1 p. (H.R. 88)
S-26317.
Copies: DNA.

5651. United States. Laws.
An act to authorize the laying out and opening of certain roads in the Territory of Michigan. May 17, 1826. [Washington: 1826.] 1 p. (H.R. 62)
S-26320.
Copies: DNA.

5652. United States. Laws.
An act to authorize the surveying and making a road from Fort Smith, in the Territory of Arkansas, to Natchitoches, in the state of Louisiana. May 19, 1826. [Washington: 1826.] 1 p. (H.R. 94)
S-26322.
Copies: DNA.

5653. United States. Laws.
A bill authorizing the reopening and repairing the King's Road, from the Georgia line to New Smyrna, in the Territory of Florida. February 6, 1826. [Washington: 1826.] 1 p. (S. 42)
S-26410.
Copies: DNA.

5654. ———— for the completion of the road from Pensacola to St. Augustine, and for other purposes. December 26, 1826. [Washington: 1826.] 1 p. (H.R. 318)
S-26411.
Copies: DNA.

5655. United States. Laws.
A bill to establish sundry post roads. May 12, 1826.... [Washington: 1826.] 13 p. (H.R. 269)
S-26818.
Copies: DNA.

5656. United States. Laws.
A bill to regulate the laying out and making a road from the seat of government, in the city of Washington, to the city of New Orleans, for the more safe and expeditious transportation of the United States mail, and for other national purposes. April 25, 1826.... [Washington: 1826.] 3 p. (H.R. 226)
S-26870.
Copies: DNA.

5657. United States. President (John Q. Adams).
Road—Little Rock to Cantonment Gibson. Message from the President of the United States, transmitting the information required by a resolution of the House of Representatives in relation to the construction of a road from Little Rock to Cantonment Gibson. December 22, 1826.... Washington: Printed by Gales & Seaton, 1826. 4 p. (19th Cong., 2d sess. House. Doc. 20)
S-27374.
Copies: DLC; DeU; MWA; NjP.

5658. United States. War Dept.
Cost per mile—road to New Orleans. Letter from the Secretary of War, transmitting the information required by a resolution of the House of Representatives of the 24th ult. in relation to the probable cost per mile of making the road on each of the routes lately examined, between Washington City and New Orleans. May 13, 1826.... Washington: Printed by Gales & Seaton, 1826. 8 p. (19th Cong., 1st sess. House. Doc. 181)
S-26932.
Copies: DLC; DeU; MWA; PU.

5659. [Beck, Nicholas F.]
Considerations in favour of the construction of a great road, from Lake-Erie to the Hudson-River. By a citizen of New-York. Albany: Websters & Skinners, printers, 1827. 48 p.
S-28083.

Copies: CSmH; DLC; MB; MWA; N; NBu; NN; NjR; WHi.

5660. Columbus and Sandusky Turnpike Company.

Report of the committee of Columbus and Sandusky Turnpike Company instructed to tender surrender of company's charter. [Columbus: 1827.] 3 p.

Hasse (Ohio)-p. 1022.

Copies: OMC.

5661. Cumberland Road & domestic manufactures. Resolutions adopted at a meeting of inhabitants of Ohio County, in the state of Virginia, in relation to the Cumberland Road and the domestic manufactures of the county. February 19, 1827.... Washington: Printed by Gales & Seaton, 1827. 4 p. (19th Cong., 2d sess. House. Doc. 108)

S-31227.

Copies: DLC; DeGE; DeU; MWA; NjP.

5662. Kentucky. Laws.

An act to incorporate the Maysville and Lexington Turnpike Road Company. [Frankfort?, Ky.: 1827.] 16 p.

S-29707.

Copies: DNA.

5663. Michigan (Territory). Laws.

Actes, relatifs aux townships, aux grands chemins, aux elections, &c. Publiés conformement à certaines résolutions, passés pendant la seconde session du second Conseil législatif du Michigan. Detroit: Imprimé par Sheldon & Wells, 1827. 69 p.

S-29744.

Copies: MiD-B; WHi.

5664. Michigan (Territory). Laws.

Acts, relative to townships, highways, & elections. Published agreeably to certain resolutions, passed during the second session of the Second Legislative Council of the Territory of Michigan. Detroit: Printed by Sheldon & Wells, 1827. 55 p.

S-29745.

Copies: MiD-B.

5665. New York (State). Comptroller's Office.

A communication from the Comptroller, transmitting a report of Philip Church and Sylvanus Russell, Esqrs. relative to a road from Angelica to Hamilton; together with a petition of sundry persons. Made to the Senate Feb. 9, 1827. Albany: Printed by Croswell & Van Benthuysen, 1827. 21 p.

S-29988.

Copies: N; PPL.

5666. New York (State). Laws.

Revised statute, chapter XVI. Relating to highways, bridges and ferries, passed at the extra session of the Legislature of the state of New York, December 3 1827. Albany: Printed by E. Croswell, 1827. 31 p.

S-30026.

Copies: NIC.

5667. Ohio. General Assembly. House. Select Committee on the Columbus and Sandusky Turnpike Company.

Report of the Select Committee on Governor's message and on report of the company's committee relative to surrender of the charter of the company. [Columbus: 1827.] 7 p.

Hasse (Ohio)-p. 1022.

Copies: OMC.

5668. Ohio. Governor (Trimble).

Message of the Governor, relative to progress made by the Columbus and Sandusky Turnpike Company, and offer of surrender of charter of the company. [Columbus: 1827]. 4 p.

Dated: December 10, 1827.

Hasse (Ohio)-p. 1C21.

Copies: OMC.

5669. Turn-pike system. [Columbia, S.C.: 1827.] 16 p.

Plan in the form of a bill submitted by A. Blanding, Superintendent of Public Works, to the legislature "for chartering joint stock companies, for the improvement of the roads of the state."

S-30665; Sabin-87729.

Copies: NN.

5670. United States. Board of Internal Improvement.

Road-Baltimore to Philadelphia. Letter from the Postmaster General, transmitting a report of General Bernard, of surveys of routes for a post road from Baltimore to Philadelphia. February 12, 1827.... Washington: Printed by Gales & Seaton, 1827. 19 p., fold. tables, fold. map. (19th Cong., 2d sess. House. Doc. 94)

John McLean, Postmaster General.

Also issued as Senate Doc. 88.

S-31496.

Copies: DLC; DeGE; DeU; MWA; NjP.

5671. United States. Commissioner for Surveying and Locating the National Road from Wheeling to St. Louis.

National Road—Wheeling to Missouri. Letter from the Secretary of War, transmitting reports and drawings relative to the National Road, from Wheeling to the seat of government in the state of Missouri. February 3, 1827 ... Washington: Printed by Gales & Seaton, 1827. 26 p., table. (19th Cong., 2d sess. House. Doc. 74)

Transmits the reports from J. Knight, Commissioner.

S-31434.

Copies: DLC; DeGE; DeU; MWA; NjP.

5672. United States. Congress. House. Committee on Military Affairs.

Military road—Fishkill Landing to Cold Spring. February 6, 1827. Mr. Ward, from the Committee on Military Affairs ... made the following report. [Washington: 1827.] 4 p. (19th Cong., 2d sess. House. Rept. 72)

S-31427.

Copies: DLC; DeU; MWA; NjP.

5673. United States. Congress. House. Committee on Roads and Canals.

Road—La Plaisance Harbor, westward. January 10, 1827. Mr. Mercer, from the Committee on Roads and Canals ... made the following report. [Washington: 1827.] 4 p. (19th Cong., 2d sess. House. Rept. 39)

S-31497.

Copies: DLC; DeGE; DeU; MWA; NjP.

5674. United States. Congress. Senate. Select Committee on Roads and Canals.

In Senate of the United States, January 25, 1827. Mr. Hendricks made the following report: The Select Committee on Roads and Canals, to whom was referred the subject of a road from Lake Michigan, by the way of Indianapolis, to the Ohio River.... [Washington: 1827.] 2 p. (19th Cong., 2d sess. Senate. Doc. 31)

S-31298.

Copies: DLC; DeU; MWA; NjP.

5675. United States. Engineer Dept.

Road from Washington City to Buffalo. Letter from the Secretary of War, transmitting a report of surveys of proposed routes of a national road from the city of Washington to Buffalo, in the state of New York. February 17, 1827.... Washington: Printed by Gales & Seaton, 1827. 39 p., 3 fold. tables. (19th Cong., 2d sess. House. Doc. 105)

Survey signed: S. H. Long, Top. Eng.

S-31500.

Copies: DBRE; DLC; DeGE; DeU; MH-BA; MWA; NN; NjP.

5676. United States. Engineer Dept.

Road—Zanesville to Columbus. Letter from the Secretary of War, transmitting a report on the subject of the National Road between Zanesville and Columbus. January 3, 1827.... Washington: Printed by Gales & Seaton, 1826 [1827]. 7 p. (19th Cong., 2d sess. House. Doc. 31)

Transmits a report of the Engineer Department.

S-31499.

Copies: DLC; DeGE; DeU; MWA; NjP.

5677. United States. Laws.

An act to establish sundry post roads. February 22, 1827. [Washington: 1827.] 21 p. (H.R. 434)

S-30901.

Copies: DNA.

5678. United States. Laws.

An act to provide for the completion of the road from a point opposite to Memphis, in the state of Tennessee, to Little Rock, in the Territory of Arkansas. January 2, 1827. [Washington: 1827.] 1 p. (H.R. 299)

S-30903.

Copies: DNA.

5679. United States. Laws.

A bill for making a military road to facilitate the access to the military post of West Point, in the state of New York. February 6, 1827.... [Washington: 1827.] 1 p. (H.R. 425)

S-30959.

Copies: DNA.

5680. United States. Laws.

A bill for the continuation of the Cumberland Road. December 24, 1827.... [Washington: 1827.] 3 p. (S. 34)

S-30974.

Copies: DNA.

5681. United States. Laws.

A bill making an appropriation for the construction of the Cumberland Road, from Bridgeport to Zanesville, in the state of Ohio, and for continuing and completing the surveys of the Cumberland Road, from Zanesville to the seat of government in the state of Missouri. December 18, 1827. [Washington: 1827.] 1 p. (S. 26)

S-31083.

Copies: DNA.

5682. United States. Laws.

A bill to authorize the laying out and construction of a road from La Plaisance Harbor to the road leading from Detroit to Chicago, in the Territory of Michigan. January 10, 1827. [Washington: 1827.] 1 p.(H.R. 364)

S-31131.

Copies: DNA.

5683. United States. Laws.

A bill to authorize the President of the United States to cause a road to be opened from Fort Smith, in Arkansas, to Natchitoches, in Louisiana. January 4, 1827.... [Washington: 1827.] 1 p. (S. 39)

S-31139.

Copies: DNA.

5684. United States. Laws.

A bill to regulate the laying out and surveying a road, from the seat of government, in the city of Washington, to the eastern termination of the Cumberland Road, at Cumberland, in the state of Maryland. January 22, 1827.... [Washington: 1827.] 1 p.(H.R. 388)

S-31190.

Copies: DNA.

5685. United States. Laws.

Road—Washington City to New Orleans. February 22, 1827. Mr. Blair submitted the following, which he intends to offer as an amendment to the "Bill to regulate the laying out and making a road from the seat of government, in the city of Washington, to the city of New Orleans," &c. [Washington: 1827.] 2 p. (H.R. 226)

S-31498.

Copies: DNA.

5686. United States. President (John Q. Adams).

Message from the President of the United States, transmitting a detailed statement of the expenditures for the construction and repair of the Cumberland Road, in compliance with a resolution of the Senate of the 20th of May last. January 10, 1827.... Washington: Printed by Gales & Seaton, 1827. 7 p. (19th Cong., 2d sess. Senate. Doc. 14)

S-31415.

Copies: DLC; DeU; MWA; NjP.

5686a. Virginia. Commissioner of the Kanawha Road and Navigation.

Report of the Commissioner of the Kanawha Road and Navigation, and accompanying documents. Richmond: Printed by Thomas Ritchie, 1827. 20 p.

David S. Garland, Commissioner; J. C.

5686a (*continued*)
Warren, engineer.
Report covers March 1826 to January 20, 1827.
N-V0183849.
Copies: MB; ViU.

5686b. Virginia. Commissioner of the Kanawha Road and Navigation.
Report of the Commissioner of the Kanawha Road and Navigation, to the General Assembly of Virginia. [Richmond: 1827.] 19 p.
David S. Garland, Commissioner; John Staples, engineer.
Report covers March to November, 1827.
N-V0183850.
Copies: ViW.

5687. [Carey, Mathew] 1760–1839.
[Address to the public in favor of establishing a line of stages and steamboats from Norfolk to Charleston. Philadelphia: 1828.] [3] p.
Signed: M. Carey, L. Elmaker, G. Ralston.
N-C0133064
Copies: MH; MWA; PPAmP.

5688. Indiana. General Assembly.
Memorial of the General Assembly of Indiana, expressive of the advantages resulting from the Cumberland Road and the desire of that state for its completion. February 20, 1828.... Washington: Printed by Duff Green, 1828. 6 p. (20th Cong., 1st sess. Senate. Doc. 111).
S-36585.
Copies: DLC; DeU; MWA; NjR.

5689. Indiana. General Assembly.
National road. Memorial of the General Assembly of the state of Indiana, upon the subject of the Cumberland or western national road. February 25, 1828.... Washington: Printed by Gales & Seaton, 1828. 3 p. (20th Cong., 1st sess. House. Doc. 163)
S-36640.
Copies: DLC; DeU; MWA; NjR.

5690. Indiana. General Assembly.
Resolution of the General Assembly of Indiana, relative to the western mail route from Louisville, Kentucky, to St. Louis, Missouri ... February 20, 1828.... Washington: Printed by Duff Green, 1828.3 p. (20th Cong., 1st sess. Senate. Doc. 112)
S-36802.
Copies: DLC; MWA; NjR.

5691. Map of Baltimore, Ann Arundel, Montgomery, Frederick, and Washington counties, Md., showing drainage, roads, and important place names. [Baltimore: 1828?] Map.
Indicates routes surveyed and the location of a "rail road." The first survey was published in 1828.
Copies: DLC.

5692. Memorial of citizens of the city and county of Philadelphia, upon the subject of a post route between the cities of Philadelphia and Baltimore. March 17.... Washington: Printed by Gales & Seaton, 1828. 5 p. (20th Cong., 1st sess. House. Doc. 203)
At head of title: Pennsylvania.
S-36704.
Copies: DLC; DeGE; DeU; MWA; NjR.

5693. New York (State). Laws.
An act to incorporate the Watervliet Turnpike Company, passed March 31, 1828.... Albany: Printed by Websters and Skinners, 1828. 15 p.
S-34474.
Copies: NBC; OCHP.

5694. New York (State). Legislature. Assembly. Committee on Roads and Bridges.
Report of the Committee on Roads and Bridges, on the petition of inhabitants of the county of Columbia. [Albany: 1828.] 2 p.(Assembly. Doc. 202)
Dated: March 17, 1828.
S-34515.
Copies: MWA; NN.

5695. Ohio. Governor (Trimble).
Governor's special message on the correspondence with the Secretary of the Treasury, relative to grant for the construction of a road from Columbus to Sandusky. January 23, 1828. [Columbus: 1828.] 4 p.
Hasse (Ohio)-p. 1022.
Copies: OMC.

5696. Pennsylvania. General Assembly.
Cumberland Road. Message from the President of the United States. transmitting copies of resolutions of the legislature of Pennsylvania, in relation to the Cumberland Road. February 12, 1828.... Washington: Printed by Gales & Seaton, 1828. 7 p. (20th Cong., 1st sess. House. Doc. 152)
S-36249.
Copies: DLC; DeGE; DeU; MWA; NjR.

5697. Pennsylvania. Governor (J. A. Shulze).
Message from the President of the United States transmitting to Congress copies of a communication from the Governor of Pennsylvania, and resolutions of the Legislature of that Commonwealth relative to the permanent preservation and repair of so much of the Cumberland Road as passes through that state. February 25, 1828.... Washington: Printed by Duff Green, 1828. 7 p. (20th Cong., 1st sess. Senate. Doc. 120)
S-36617.
Copies: DLC; DeU; MWA; NjR.

5698. Pennsylvania. Laws.
A compilation and digest of the road laws, and the election laws, together with certain forms. Prepared and published under authority of an act of Assembly of Pennsylvania, passed the first day of February, Anno Domini 1827. Harrisburg: Printed at the office of the Reporter, 1828. 141, 16 p.
S-34674.
Copies: CU-L; DLC; MB; MH-L; P; PPB; PPi; PPiHi.

5699. Philadelphia City Tavern, January 31st, 1828. A meeting of a number of citizens was held ... [to appoint a committee to draft an] Address to the public [on the question of establishing a line of stages & steamboats from Norfolk to Charleston. Philadelphia: 1828.] 3 p.
Dated February 4th, 1828, and signed by the committee.
S-34754.
Copies: NH; MWA; PPAmP; PPL.

5700. Tanner, Henry Schenck, 1786–1858.
A new pocket atlas of the United States, with the roads and distances, designed for the use of travelers.... Philadelphia: The author, 1828. 2 p., 12 maps.
S-35460.
Copies: CLU; CtY; DLC.

5701. United States. Board of Internal Improvement.
Letter from the Postmaster General, transmitting a report of General Bernard, of surveys of routes for a post road from Baltimore to Philadelphia. [Washington: 1828.] 21 p., 6 fold. tables, fold. map. (20th Cong., 1st sess. Senate. Doc. 88)
Title from page [3]. Text on p. [1] begins: In Senate of the United States, February 5, 1828....
Report signed: S. Bernard, Brig. Gen., member of the Board of Inter. Imp. [and] William Tell Poussin, Capt. Top. Eng. and Assistant to the Board.
S-36360.
Copies: DLC; DeGE; DeU; MWA; NjR.

5702. United States. Commissioners Appointed to Locate the National Road.
Continuation of Cumberland Road. Letter from the Secretary of War, transmitting (in compliance with a resolution of the House of Representatives, of the 11th instant) a report of Commissioners, Appointed to Locate the National Road, under the act of March 3, 1825. February 18, 1828.... Washington: Printed by Gales & Seaton, 1828. 16 p., fold. table. (20th Cong., 1st sess. House. Doc. 149)
Signed: J. Knight, Commissioner.
S-36235.
Copies: DLC; DeGE; DeU; MWA; NjR.

5703. United States. Commissioners Appointed to Locate the National Road.

Report of the Secretary of War, in compliance with a resolution of the Senate, of the eighth instant, transmitting a report of the commissioner for locating the continuation of the Cumberland Road, under act of March 3, 1825. February 12, 1825 [*sic*].... Washington: Printed by Duff Green, 1828. 18 p., 2 fold. tables. (20th Cong., 1st sess. Senate. Doc. 99)

S-36779.

Copies: DLC; DeGE; DeU; MWA; NjR.

5704. United States. Congress. House. Committee on Claims.

Moses Shepherd. May 10, 1828 ... Mr. McIntire, from the Committee of Claims, to which was referred the bill from the Senate (No. 112) for the relief of Moses Shepherd, made the following report. [Washington: 1828.] 185 p. (20th Cong., 1st sess. House. Rept. 253)

Concerns the claims for compensation for masonry work on the Cumberland Road, with accompanying documents.

S-36634.

Copies: DLC; DeGE; DeU; MWA; NjR.

5705. United States. Congress. House. Committee on Roads and Canals.

Road—Baltimore to Philadelphia. May 13, 1828. Mr. Mercer, from the Committee on Roads and Canals ... made the following report. [Washington: 1828.] 2 p. (20th Cong., 1st sess. House. Rept. 254)

S-36841.

Copies: DLC; DeU; MWA; NjR.

5706. United States. Congress. House. Committee on Roads and Canals.

Road—Homochitto Swamp. January 24, 1828. Mr. Stewart, from the Committee on Roads and Canals ... made the following report. [Washington: 1828.] 10 p. (20th Cong., 1st sess. House. Rept. 97)

Refers to the removal of obstructions on the road from Natchez to New Orleans.

S-36842.

Copies: DLC; DeU; MWA; NjR.

5707. United States. Congress. House. Committee on Roads and Canals.

Road—Lawrenceburg to Fort Wayne, Indiana. April 15, 1828. Mr. Woods, of Ohio, from the Committee on Roads and Canals ... made the following report. [Washington: 1828.] 3 p. (20th Cong., 1st sess. House. Rept. 234)

S-36843.

Copies: DLC; DeGE; DeU; MWA; NjR.

5708. United States. Congress. House. Committee on Roads and Canals.

Road—Memphis to Tuscumbia. May 13, 1828. Mr. Mercer, from the Committee on Roads and Canals ... made the following report. [Washington: 1828.] 1 p. (20th Cong., 1st sess. Rept. 256)

S-36844.

Copies: DLC; DeU; MWA; NjR.

5709. United States. Congress. House. Committee on Roads and Canals.

Road—Penobscot River to New Brunswick. March 27, 1828. Mr. Davis, of Massachusetts, from the Committee on Roads and Canals ... made the following report. [Washington: 1828.] 2 p. (20th Cong., 1st sess. House. Rept. 214)

S-36845.

Copies: DLC; DeU; MWA; NjR.

5710. United States. Congress. House. Committee on Roads and Canals.

Road—Union Town, through Pittsburg [*sic*] to Lake Erie. May 21, 1828 ... Mr. Stewart from the Committee on Roads and Canals ... made the following report. [Washington: 1828.] 6 p. (20th Cong., 1st sess., House. Rept. 267)

S-36846.

Copies: DLC; DeGE; DeU; MWA; NjR.

5711. United States. Congress. House. Committee on Roads and Canals.

Road—Washington to New Orleans. January 2, 1828. Mr. Mercer, from the Committee on Roads and Canals ... made the following report. [Washington: Gales & Seaton, 1828.] 26 p., fold. table. (20th Cong.,

1st sess., House. Rept. 48)

Includes a report by S. Bernard, member of the Board of Internal Improvement.

S-36847.

Copies: DLC; DeGE; DeU; MWA; NjR.

5712. United States. Congress. House. Committee on Roads and Canals.

Road—Washington to northwestern parts of New York and Pennsylvania. January 21, 1828. Mr. Mercer, from the Committee on Roads and Canals ... made the following report. [Washington: 1828.] 43 p., 3 tables. (20th Cong., 1st sess. House. Rept. 91)

S-36849.

Copies: DLC; DeGE; DeU; MWA; NjR.

5713. United States. Congress. House. Committee on Roads and Canals.

Road—Wiggins' Ferry to lead mines on Fever River. March 4, 1828. Mr. Mercer, from the Committee on Roads and Canals ... made the following report. [Washington: 1828.] 3 p. (20th Cong., 1st sess. House. Rept. 177)

S-36850.

Copies: DLC; DeU; MWA; NjR.

5714. United States. Congress. House. Committee on Roads and Canals.

Roads in Louisiana. March 17, 1828. Mr. Gurley, from the Committee on Roads and Canals ... made the following report. [Washington: 1828.] 2 p. (20th Cong., 1st sess. House. Rept. 194)

Concerns the roads from Franklinton to St. Francisville, and from Covington to Opelousas.

S-36855.

Copies: DLC; DeU; MWA; NjR.

5715. United States. Congress. House. Committee on Roads and Canals.

Subscribe stock, Tenth Turnpike Road Company, N.H. April 14, 1828. Mr. Davis, of Massachusetts, from the Committee on Roads and Canals ... made the following report. [Washington: 1828.] 8 p. (20th Cong., 1st sess. House. Rept. 233)

Report on a petition to help to repair the Tenth Turnpike Road in New Hampshire.

S-36905.

Copies: DLC; DeU; MWA; NjR.

5716. United States. Engineer Dept.

Cumberland Road. Letter from the Secretary of War, transmitting a copy of the last annual report of the Superintendent of the Cumberland Road. December 11, 1828.... [Washington: Gales & Seaton, printers, 1828.] 9 p. (20th Cong., 2d sess. House. Doc. 14)

Casper W. Wever, Superintendent.

S-36248.

Copies: DLC; DeGE; DeU; MWA; NjR.

5717. United States. Engineer Dept.

Extend Cumberland Road. Letter from the Secretary of War transmitting a report of the engineer engaged to survey the route of the extension of the National Road from Cumberland to Washington City. December 29, 1828.... [Washington: Gales & Seaton, printers, 1828.] 7 p.(20th Cong., 2d sess. House. Doc. 39)

Report signed: And'w Alexander.

S-36297.

Copies: DLC; DeGE; DeU; MWA; NjR.

5718. United States. Engineer Dept.

National Road—Washington to New Orleans. Letter from the Secretary of War, transmitting a report of the Chief Engineer, accompanied by a report upon the reconnoissance of a route, across the Cumberland Mountain, of the National Road contemplated from Washington to New Orleans, and a map of the country between those two cities. January 30, 1828. Read ... February 7, 1828. Referred to the Committee of the Whole House ... Washington: Printed by Gales & Seaton, 1828. 19 p., 10 fold. tables. (20th Cong., 1st sess. House. Doc. 125)

Map not included.

S-36641.

Copies: DLC; DeGE; DeU; MWA; NN; NjR; PU.

5719. ———— Washington: Printed by Gales & Seaton, 1828. 16, 25–27 p., 10 fold. tables.

The text on pages 25–27 is identical to that on pages 17–19 of the other issue.

Copies: DeGE.

5720. United States. Engineer Dept.

Repair Cumberland Road. Letter from the Secretary of War, transmitting a report of the Superintendent of the Cumberland Road, relative to the mode of repairing the same. January 14, 1828.... Washington: Printed by Gales & Seaton, 1828. 8 p. (20th Cong., 1st sess. House. Doc. 57)

Signed: Casper W. Wever, Superintendent U.S. Road.

S-36741.

Copies: DLC; DeGE; DeU; MWA; NjR.

5721. United States. Engineer Dept.

Road—Washington to northwest part of New York. Letter from the Secretary of War, transmitting a report from the engineer appointed to examine a route for a national road from the city of Washington to the northwestern frontier of the state of New York. December 29, 1828.... [Washington: Gales & Seaton, printers, 1828.] 22 p., 2 fold. maps. (20th Cong., 2d sess. House. Doc. 38)

S-36848.

Copies: DLC; DeGE; DeU; MWA; NjR.

5722. United States. Engineer Dept.

Road—Zanesville, Ohio, to Florence, Alabama. Letter from the Secretary of War, transmitting a report of the reconnoissance of a route for a national road, from Zanesville, Ohio, to Florence, Alabama. March 18, 1828.... Washington: Printed by Gales & Seaton, 1828. 44 p. (20th Cong., 1st sess. House. Doc. 209)

Report prepared by Stephen Harriman Long.

S36851.

Copies: DLC; DeGE; DeU; MWA; NjR.

5723. United States. Laws.

An act to alter and establish post roads, and discontinue others. May 19, 1828. The following bill of the Senate, passed by the House of Representatives, with amendments, was referred to the Committee on the Post Office and Post Roads, and ordered to be printed.... [Washington: 1828.] 19 p. (S. 133)

S-35625.

Copies: DNA.

5724. United States. Laws.

Amendments: Mr. Mercer, from the Committee on Roads and Canals, to which was referred the bill from the Senate, entitled "An act to provide for opening and making a military road in the state of Maine," reported the same with the following amendments. April 12, 1828.... [Washington: 1828.] 1 p. (S. 113)

S-35673.

Copies: DNA.

5725. United States. Laws.

A bill authorizing a subscription of a stock for the establishment of a permanent post road between the cities of Baltimore and Philadelphia. May 13, 1828.... [Washington: 1828.] 3 p. (H.R. 301)

S-35718.

Copies: DNA.

5726. United States. Laws.

A bill for improving the post road leading from New York to Albany, in the state of New York, between Mesier's Mills, in the county of Dutchess, and Croton River, in the county of West Chester. May 21, 1828.... [Washington: 1828.] 3 p. (H.R. 303)

S-35754.

Copies: DNA.

5727. United States. Laws.

A bill for the preservation and repair of the Cumberland Road. January 2, 1828.... [Washington: 1828.] 5 p. (H.R. 42)

S-35774, 35775.

Copies: DNA.

5728. United States. Laws.

A bill to aid the state of Louisiana in

constructing a road from Franklinton to St. Francisville, and from Covington to Opelousas, in said state. March 17, 1828.... [Washington: 1828.] 1 p. (H.R. 224)

S-36044.

Copies: DNA.

5729. United States. Laws.

A bill to authorize the establishment of a national road from the junction of the Penobscot and Matanawcook rivers, in the state of Maine, to the boundary line between the United States and the Province of New Brunswick. March 27, 1828.... [Washington: 1828.] 1 p. (H.R. 243)

S-36078.

Copies: DNA.

5730. United States. Laws.

A bill to authorize the improvement of a road through the lands of the Chickasaw nation, in the direction between Memphis and Tuscumbia, in the state of Alabama. May 13, 1828.... [Washington: 1828.] 1 p. (H.R. 300)

S-36079.

Copies: DNA.

5731. United States. Laws.

A bill to authorize the laying out and construction of a road from La Plaisance Harbor to the road leading from Detroit to Chicago, in the Territory of Michigan. January 11, 1828.... [Washington: 1828.] 1 p. (H.R. 69)

S-36081.

Copies: DNA.

5732. United States. Laws.

A bill to authorize the laying out and making one or more roads from the city of Washington, towards the northwestern frontier of the states of New York and Pennsylvania. January 21, 1828.... [Washington: 1828.] 5 p. (H.R. 100)

S-36082.

Copies: DNA.

5733. United States. Laws.

A bill to authorize the subscription by the United States to the stock of the Tenth Turnpike Road Company, in the state of New Hampshire. April 11, 1828.... [Washington: 1828.] 1 p. (H.R. 269)

S-36070.

Copies: DNA.

5734. United States. Laws.

A bill to authorize the survey and opening of a road from St. Louis, on the River Mississippi, to Washington, in the Territory of Arkansas. April 5, 1828.... [Washington: 1828.] 3 p. (H.R. 256)

S-36102.

Copies: DNA.

5735. United States. Laws.

A bill to authorize the survey and opening of a road from Wiggins' Ferry, on the river Mississippi, to the lead mines on Fever River. March 4, 1828.... [Washington: 1828.] 3 p. (H.R. 205)

S-36103.

Copies: DNA.

5736. United States. Laws.

A bill to authorize the surveying and opening a road within the Territory of Michigan, from Detroit, westwardly, to Lake Michigan. January 11, 1828.... [Washington: 1828.] 3 p. (H.R. 68)

S-36104.

Copies: DNA.

5737. United States. Laws.

A bill to establish sundry post roads, and discontinue others. April 21, 1828.... [Washington: 1828.] 15 p. (H.R. 276)

S-36130.

Copies: DNA.

5738. United States. Laws.

A bill to provide for opening and making a military road in the state of Maine. March 4, 1828.... [Washington: 1828.] 1 p. (S. 113)

S-36151.

Copies: DNA.

5739. United States. Laws.

A bill to regulate the laying out and making of a national road from the city of Washington in the District of Columbia, to New Orleans, in the state of Louisiana. January 2, 1828.... [Washington: 1828.] 7 p. (H.R. 44)

S-36166.

Copies: DNA.

5740. United States. Laws.

A bill to repair the post road from Cincinnati to Portland, in the state of Ohio. March 17, 1828.... [Washington: 1828.] 1 p. (H.R. 225)

S-36168.

Copies: DNA.

5741. United States. Post-Office Dept.

Condition of the Cumberland Road. Letter from the Postmaster General, in reply to a resolution of the House of Representatives, requiring information in relation to the present state and condition of the Cumberland Road. May 10, 1828.... Washington: Printed by Gales & Seaton, 1828. 3 p. (20th Cong., 1st sess. House. Doc. 269)

S-36225.

Copies: DLC; DeU; MWA; NjR.

5742. United States. Post-Office Dept.

Road from Columbus to New Orleans. Letter from the Postmaster General, in reply to a resolution of the House of Representatives of the last session, in relation to the probable cost required for the repairs of the road from Columbus, passing through the Choctaw nation, to New Orleans ... Decem. 22, 1828.... [Washington: Gales & Seaton, printers, 1828.] 4 p. (20th Cong., 2d sess. House. Doc. 23)

S-36852.

Copies: DLC; DeGE; DeU; MWA; NjR.

5743. United States. Post-Office Dept.

Road through the Wyandot Reservation, &c. Letter from the Postmaster General, transmitting the information required by a resolution of the House of Representatives of the 8th inst. in relation to a road through the Wyandot Reservation, between Cincinnati and Portland in the state of Ohio, &c. &c. January 21, 1828.... Washington: Printed by Gales & Seaton, 1828. 3 p. (20th Cong., 1st sess. House. Doc. 73)

John McLean, Postmaster General.

S-36853.

Copies: DLC; DeGE; DeU; MWA; NjR.

5744. United States. War Dept.

Military road—defence of northwestern frontier—Michigan. Letter from the Secretary of War, transmitting the information required by a resolution of the House of Representatives, of the eighth instant, relative to the proceedings under the act of 2d March, 1827, for the defence of the northwestern frontier, &c. December 23, 1828.... [Washington: Gales & Seaton, printers, 1828.] 3 p. (20th Cong., 2d sess. House. Doc. 29)

Includes a report from Col. C. Gratiot of the Engineer Dept.

S-36622.

Copies: DLC; DeGE; DeU; MWA; NjR.

5745. Bussard, William.

Letter from William Bussard, surveyor, to the Hon. William Hendricks, chairman of the Committee on Roads and Canals, relative to the "Bill for subscription to stock in the Washington and Fredericktown Turnpike Road Company." February 16, 1829 ... [Washington: 1829.] 2 p. (20th Cong., 2d sess. Senate. Doc. 83)

Copies: DeU; MWA.

5746. Elmendorf, L.

Facts and observations on the merits of the memorial of the president and directors of the First Great South-Western Turn-Pike Road Company, for relief. Albany: 1829. 15 p.

S-38462; Sabin-22355.

No copy known.

5747. Lufborough, Nathan.
Letter from Nathan Lufborough, president of the Washington Turnpike Company, in relation to the bill before the Senate, for the subscription of stock in the said company ... Feb. 9, 1829. [Washington: 1829.] 6 p. (20th Cong., 2d. sess. Senate. Doc. 68)
Includes information about the route of the turnpike and cost estimates.
S-41218.
Copies: DeGE; MWA.

5748. Maryland. Governor (D. Martin).
A communication from the executive department to the House of Delegates ... enclosing a communication from the Washington Turnpike Company. Annapolis: Jeremiah Hughes, printer, 1829. 10 p.
S-39444.
Copies: MdBP.

5749. Memorial of sundry citizens of Montgomery County, Md. praying that the bill now before the Senate, for the subscription for stock in the Washington Turnpike Road Company, be postponed to the next session of Congress. Feb. 4, 1829.... [Washington: 1829.] 2 p. (20th Cong., 2d sess. Senate. Doc. 60)
S-41230.
Copies: DeGE; MWA.

5750. Missouri. General Assembly.
Memorial of the legislature of Missouri, for the speedy location and construction of the Cumberland Road to the City of Jefferson. February 19, 1829.... [Washington: 1829.] 2 p. (20th Cong., 2d sess. Senate. Doc. 87)
Copies: DeU; MWA.

5751. New York (State). Laws.
An act to lay out and improve a road through the Buffalo Indian reservation, in the county of Erie. [Albany: 1829.] 2 p. (Assembly. Rept. 28)
Dated: January 23, 1829.
S-39785.
Copies: CSmH; MWA.

5752. New York (State). Laws.
The revised act of the Legislature of the State of New York, respecting highways and bridges, which took effect on the first day of January, 1828. To which are annexed some of the necessary forms, for laying out roads, &c. under the said act. Rochester: E. Peck & Co., and Bemis & Ward, Canandaigua, 1829. 27 p.
S-39805.
Copies: NR.

5753. Ohio. Commissioners to View a Route for a Road from Gallipolis to Chillicothe.
Report of the Commissioners appointed to view a route for a road petitioned for by citizens of Gallia, Jackson and Ross counties, commencing at Gallipolis, thence to Jackson and thence to Chillicothe. [Columbus: 1829.] 2 p.
Hasse (Ohio)-p. 1015.
Copies: NN.

5754. Pennsylvania. General Assembly.
Cumberland Road. Message from the President of the United States, transmitting copies of resolutions of the legislature of Pennsylvania, in relation to the Cumberland Road. January 16, 1829. Reprinted by order of the House of Representatives. [Washington: Gales & Seaton, printers, 1829.] 2 p. (20th Cong., 2d sess. House. Doc. 75)
S-41102.
Copies: DLC; DeGE; DeU; MWA; NjR.

5755. Smith, Oliver Hampton, 1794–1858.
Speech of Mr. Smith, of Indiana, on the amendment of Mr. Buchanan to the Cumberland Road bill. Delivered in Committee of the Whole, on the floor of Congress. Wednesday, January 28, 1829. Washington: P. Force, 1829. 22 p.
Sabin-83692.
Copies: DLC.

5756. Stewart, Andrew, 1791–1872.
Speech of Mr. Stewart (of Pennsylvania) on the Cumberland road bill, delivered in the House of Representatives of the United

5756 (*continued*)
States, January 27, 1829. [Washington: 1829.] 24 p.
S-40537.
Copies: DLC; DeGE; MH-BA; PPL; ViU.

5757. United States. Commissioner Appointed to Lay out the National Road from Zanesville to St. Louis.
Road from Zanesville, Ohio, to St. Louis, Missouri. Message from the President of the United States, transmitting a report of the Commissioner appointed to lay out the National Road from Zanesville, in the state of Ohio, to the seat of government in the State of Missouri. February 7, 1829.... [Washington: Gales & Seaton, printers, 1829.] 12 p., fold. table. (20th Cong., 2d sess. House. Doc. 113)
Joseph Shriver, Commissioner.
S-41307.
Copies: DLC; DeGE; DeU; MWA; NjR.

5758. United States. Congress. House. Committee on Post Office and Post Roads.
Road—Mobile to New Orleans, &c. January 15, 1829 ... Mr. McKean, from the Committee on the Post Office and Post Roads, made the following report. [Washington: Gales & Seaton, printers, 1829.] 2 p. (20th Cong., 2d sess. House. Rept. 63)
S-41305.
Copies: DLC; DeU; MWA; NjR.

5759. United States. Engineer Dept.
Cumberland Road. Letter from the Secretary of War, transmitting letters, &c. from the Superintendent of the Cumberland Road. January 17, 1829.... [Washington: Gales & Seaton, printers, 1829.] 8 p. (20th Cong., 2d sess. House. Doc. 78)
Casper W. Wever, Superintendent.
S-41101.
Copies: DLC; DeGE; DeU; MWA; NjR.

5760. United States. Laws.
An act for the continuation of the Cumberland Road. Jan. 6, 1829. Read, and committed to the Committee of the Whole House on the state of the Union, to which is committed the bill for the preservation and repair of the Cumberland Road ... Mr. Mercer, from the Committee on Roads and Canals, to which was referred the bill from the Senate, entitled "An act for the continuation of the Cumberland Road," reported the same without amendment. [Washington: 1829.] 3 p. (S. 1)
S-40718.
Copies: DNA.

5761. United States. Laws.
An act for the preservation and repair of the Cumberland Road. February 19, 1829. [Washington: 1829.] 7 p. (H.R. 42)
S-40721.
Copies: DNA.

5762. United States. Laws.
An act making appropriations for continuing the road from Detroit to Chicago, for completing and repairing piers, for the improvement of certain harbors, and for making examinations and surveys. February 23, 1829.... [Washington: 1829.] 3 p. (H.R. 434)
S-40765.
Copies: DNA.

5763. United States. Laws.
A bill authorizing the construction of a road from the mouth of Cache or White River to Helena, on the Mississippi. Feb. 5, 1829 ... Mr. Stewart, from the Committee on Roads and Canals, reported the following bill. [Washington: 1829.] 1 p. (H.R. 418)
S-40824.
Copies: DNA.

5764. United States. Laws.
A bill for the construction of the Cumberland Road, westwardly of Zanesville. January 15, 1829.... [Washington: 1829.] 1 p. (S. 57)
S-40840.
Copies: DNA.

5765. United States. War Dept.
Road—Pensacola to Mobile Point.

Letter from the Secretary of War, transmitting the information required by a resolution of the House of Representatives, of the 24th ultimo, in relation to a military road recommended by the Quartermaster General in his report of 1826, &c. January 6, 1829.... [Washington: Gales & Seaton, printers, 1829.] 7 p. (20th Cong., 2d sess. House. Doc. 52)

S-41306.

Copies: DLC; DeU; MWA; NjR.

5766. Archer, William Segar, 1789–1855.

Speech of Mr. Archer of Virginia, on the bill proposing to construct a national road from Buffalo to New Orleans. Delivered in the House of Representatives of the United States, April, 1830. Washington: Printed by Duff Green, 1830. 20 p.

C-184.

Copies: DLC; LU; MWA; MdBP; PPAmP; PPL; Vi.

5767. Baltimore and Harford Turnpike Road Company.

Report of the Baltimore and Harford Turnpike Road Company, to the Legislature of Maryland. In obedience to an order of the House of Delegates of the 14th January, 1830. Annapolis: Jeremiah Hughes, printer, 1830. 7 p.

C-239.

Copies: MdHi.

5768. Baltimore and Havre-de-Grace Turnpike Company.

Report of the Baltimore and Havre-de-Grace Turnpike Company, to the Legislature of Maryland. Annapolis: Jeremiah Hughes, printer, 1830. [4] p., fold. leaf.

C-240.

Copies: MdHi.

5769. Barbour, Philip Pendleton, 1783–1841.

Speech of Mr. Philip P. Barbour, of Virginia on the national road bill delivered in the House of Representatives, March, 1830. Washington: Printed by Duff Green, 1830. 15 p.

Concerns the road from Buffalo to New Orleans, through Washington, D.C.

C-370.

Copies: DeGE; ICN; N; RPB; TxU; ViU.

5770. Carson, Samuel Price, 1798–1838.

Speech of Mr. Carson, of North Carolina, on the bill proposing to construct a national road from Buffalo to New Orleans. Delivered in the House of Representatives, March 29, 1830. Washington: Printed by Duff Green, 1830. 16 p.

N-C0166582.

Copies: DLC; DeGE; MdBP.

5771. [Cowan, John.]

Address to the free & independent citizens of Maryland; & particularly the citizens of the county & city of Baltimore, being a statement of facts & a brief exposition of chartered monopoly usurpation & outrage.... Baltimore: 1830. iv, [5]–43 p.

Concerns the Baltimore and Reisterstown Turnpike Road Company.

Copies: MdHi.

5772. Craig, Robert, 1792–1852.

Speech of Mr. Craig, of Virginia, on the bill to construct a national road from Buffalo to New Orleans. Delivered in the House of Representatives of the United States, March, 1830. Washington: Printed by Duff Green, 1830. 16 p.

N-C0769105.

Copies: DLC; DeGE; Vi.

5773. Crawford, Thomas Hartley, 1786–1863.

Speech of Mr. Crawford, of Pennsylvania, on the bill to construct a national road from Buffalo to New Orleans. Delivered in the House of Representatives of the United States, March 30, 1830. Washington: Printed by Duff Green, 1830. 8 p.

N-C0780985.

Copies: DLC; DeGE; MiU.

5774. Crockett, David, 1786–1836.

Speech of Mr. Crockett, of Tennessee,

5774 (*continued*)
on the bill proposing to construct a national road from Buffalo to New Orleans. Delivered in the House of Representatives of the United States, March, 1830. Washington: Printed by Duff Green, 1830. 8 p.
N-C0796579.
Copies: DLC; DeGE; MiU.

5775. [Gilpin Joshua] 1765–1840.
State of the application to Congress for aid to complete the post road from Philadelphia to Baltimore, and observations thereupon. [Philadelphia: 1830.] 3 p.
C-1588.
Copies: PPL.

5776. Hemphill, Joseph, 1770–1842.
Mr. Hemphill's speech on the bill to construct a national road from Buffalo, passing by the seat of the general government, to New Orleans. Delivered in the House of Representatives, U.S., 23d March, 1830. [Washington: W. Greer, printer, 1830.] 23 p.
C-1813.
Copies: DLC; DeGE; MWA; MdBP; NBu; PPAmP; PPL; PU.

5777. Indiana. General Assembly.
Cumberland Road in Indiana. Memorial of the legislature of Indiana, in relation to the Cumberland Road in that state. February 8, 1830. Read and referred to the Committee on Internal Improvements. [Washington: 1830.] 2 p. (21st Cong., 1st sess. House. Rept. 174)
Also issued as Senate Doc. 43.
C-4660.
Copies: DLC; DeGE; MWA; NjR.

5778. Indiana. General Assembly.
In Senate of the United States. January 20, 1830. Resolutions of Indiana and Illinois relative to the repair of the mail road leading to St. Louis, &c. [Washington: 1830.] 3 p. (21st Cong., 1st sess. Senate. Doc. 28)
Copies: DeU; MWA.

5779. Indiana. General Assembly.
Memorial of the General Assembly of Indiana, approving the system of internal improvements by the government, and praying for an appropriation to continue and construct the Cumberland Road in that state. February 4, 1830.... [Washington: 1830.] 2 p. (21st Cong., 1st sess. Senate. Doc. 43)
Also issued as House Doc. 174.
Copies: DeU; MWA.

5780. Irwin, Thomas, 1785–1870.
Speech of Mr. Irwin, of Pennsylvania, on the bill to construct a road from Buffalo, in the state of New York, to New Orleans, in the state of Louisiana. Delivered in the House of Representatives of the United States, April 1st, 1830. Washington: Printed at the office of Jonathan Elliot, 1830. 24 p.
N-I0164820.
Copies: NBu.

5781. Isacks, Jacob C., d. 1832.
Speech of Mr. Isacks, of Tennessee, on the bill proposing to construct a national road from Buffalo to New Orleans. Delivered in the House of Representatives of the United States, March 25, 1830. Washington: Printed by Duff Green, 1830. 16 p.
N-I0166594.
Copies: DLC; DeGE; MiU.

5782. Lea, Pryor, 1794–1879.
Speech of Mr. Pryor Lea, of Tennessee, on the bill proposing to construct a national road from Buffalo to New Orleans.... April, 1830. Washington: Printed by D. Green, 1830. 15 p.
C-2182.
Copies: DLC; ICU; MWA; NN; NNC.

5783. Maine. Agents to Open the Canada Road.
Reports of agents appointed to open the Canada Road. Portland: Day & Fraser, printers to the state, 1830. 8 p.
C-2373.
Copies: Me-LR; MeHi; MeWC.

5784. Maine. Governor (S. E. Smith).
Documents accompanying the Governor's message, relating to the road from Baring to Houlton. Portland: Day & Fraser, printers to the state, 1830. 16 p.
C-2360.
Copies: Me-LR; MeHi.

5785. Maryland. Laws.
An act to authorize the New-Castle and French-Town Turnpike Company to make a rail road from French-Town, on Elk River, to the Delaware line, in a direction towards New-Castle. [Annapolis?: 1830?] [21] p.
C-2837.
Copies: CSt.

5786. New Jersey. Laws.
Law of the New-Jersey Turnpike, with its several supplements. Somerville, N.J.: Printed at the Messenger office, 1830. 22 p.
Caption title: *An Act to incorporate the New-Jersey Turnpike Company.*
N-N0157518.
Copies: DLC.

5787. New York (State). Legislature. Assembly. Committee on Roads.
[Report on petition for grant from the state to construct a road from Lake George to Warrensburgh]. [Albany: 1830.] 2 p. (Assembly. Doc. 283)
Hasse (New York)-p. 491.
Copies: MWA; NN.

5788. New York (State). Legislature. Assembly. Committee on Roads.
[Report on petition of sundry inhabitants of St. Lawrence and Franklin counties for aid to construct a road through St. Regis reservation]. [Albany: 1830.] 2 p. (Assembly. Doc. 223)
Hasse (New York)-p. 491.
Copies: MWA; NN.

5789. New York (State). Legislature. Assembly. Committee on Roads.
Report on the petition of sundry inhabitants of the county of Chautauque [for a state road from Buffalo to Catalraugus Creek]. Albany: 1830. 2 p. (Assembly. Doc. 224)
Dated: Feb. 24. 1830.
Hasse (New York)-p. 491.
Copies: MWA; NN.

5790. New York (State). Legislature. Assembly. Committee on Roads.
[Report on the petition for aid for road improvement in Jefferson, Oswego, and Oneida counties]. [Albany: 1830.] 2 p. (Assembly. Doc. 350)
Hasse (New York)-p. 491.
Copies: MWA; NN.

5791. Observations on the road from Baltimore to Philadelphia [by] the agent requested ... by the borough of Wilmington and turnpike companies adjacent ... to represent their interests in an application to Congress for assistance to perfect the mail road ... from Baltimore to Philadelphia. [Philadelphia: 1830.] 12 p.
C-2887.
Copies: PPL.

5792. Polk, James Knox, 1795–1849.
Speech of Mr. James K. Polk, on the bill to construct a national road from Buffalo to New Orleans. Delivered in the House of Representatives, March 29, 1830. Washington: Printed by Duff Green, 1830. 20 p.
N-P0459958.
Copies: DLC; DeGE; IU; LNHT; MWA; MiU.

5793. Richardson, Joseph, 1778–1871.
Speech of Mr. Richardson, of Massachusetts House of Representatives, March 30, 1830. [Washington: 1830.] 12 p.
Proposes an amendment to the bill for the construction of a national road from Buffalo to New Orleans.
C-3309.
Copies: CSmH; MWA; NN; NjP.

5794. Standifer, James, d. 1837.
Speech of Mr. Standefer [*sic*], of Tennessee, on the bill proposing to construct a national road from Buffalo to New Orleans. Delivered in the House of Representatives of the United States, March 25, 1830. Washington: Printed by Duff Green, 1830. 8 p.
N-S0857332.
Copies: DLC; DeGE; NBu.

5795. Tennessee. General Assembly.
Turnpike road in Tennessee. Resolutions of the legislature of the state of Tennessee in relation to the construction of a turnpike road from the Virginia line to the Mississippi River. February 25, 1830. Read, and referred to the Committee of the Whole House ... to which is committed the bill (H.R. No. 285) authorizing a subscription of stock in the Maysville, Washington, Paris, and Lexington Turnpike Road Company. [Washington: Printed by D. Green, 1830.] 1 p. (21st Cong., 1st sess. House. Rept. 230)
C-5219.
Copies: DLC; MWA; NjR.

5796. United States. Congress. House. Committee on Internal Improvements.
Cumberland Road. May 24, 1830. Mr. Vinton, from the Committee on Internal Improvements, made the following report. [Washington: 1830.] 2 p. (21st Cong., 1st sess. House. Rept. 410)
C-4663.
Copies: DLC; MWA; NjR.

5797. United States. Congress. House. Committee on Internal Improvements.
Road—Uniontown, Pittsburg, Presque Isle. February 15, 1830 ... Mr. Hemphill, from the Committee on Internal Improvements ... made the following report. [Washington: 1830.] 2 p. (21st Cong., 1st sess. House. Rept. 196)
C-5132.
Copies: DLC; DeGE; DeU; MWA; NjR.

5798. United States. Congress. House. Committee on Post Office and Post Roads.
Road—Chicot County to Little Rock. February 22, 1830. Mr. Johnson of Kentucky, from the Committee on the Post Office and Post Roads, made the following report. [Washington: 1830.] 2 p. (21st Cong., 1st sess. House. Rept. 218)
C-5130.
Copies: DLC; DeGE; DeU; MWA; NjR.

5799. United States. Congress. House. Committee on Post Office and Post Roads.
Road—Memphis and Tuscumbia (with bill H. R. No. 354). March 15, 1830. Mr. Johnson, of Kentucky, from the Committee on the Post Office and Post Roads ... made the following report. [Washington: 1830.] 1 p. (21st Cong., 1st sess. House. Rept. 293)
C-5131.
Copies: DLC; DeU; MWA; NjR.

5800. United States. Congress. House. Committee on the District of Columbia.
Street—Capitol to executive offices. To accompany bill H.R. no. 224. February 10, 1830. Mr. Powers, from the Committee for the District of Columbia ... made the following report. [Washington: 1830.] 15 p. (21st Cong., 1st sess. House. Rept. 184)
Recommendation to macadamize Pennsylvania Avenue.
C-5181.
Copies: DLC; DeGE; DeU; MWA; NjR.

5801. United States. Congress. Senate. Committee on Roads and Canals.
Documents in relation to the continuation of the Cumberland Road in the states of Ohio and Indiana. (With Senate bill No. 100). February 22, 1830. Laid on table by Mr. Hendricks.... [Washington: 1830.] 9 p.

(21st Cong., 1st sess. Senate. Doc. 63)

Includes a report from the Engineer Department.

C-4682.

Copies: DLC; DeGE; DeU; MWA; NjR.

5802. United States. Congress. Senate. Committee on Roads and Canals.

In Senate of the United States, May 5, 1830 ... Mr. Hendricks made the following report: The Committee on Roads and Canals, to whom were referred a resolution of the Senate, and a memorial of many citizens of Allen County in the state of Indiana, asking a donation of land for the construction of a road from Lawrenceburg, by way of Fort Wayne, to the southern bend of the St. Joseph's of Lake Michigan, report. [Washington: 1830.] 5 p. (21st Cong., 1st sess. Senate. Doc. 137)

C-4788.

Copies: DLC; DeGE; DeU; MWA; NjR.

5803. United States. Engineer Dept.

Continuation of the Cumberland Road. Message from the President of the United States, transmitting a letter from the Chief of the Engineers with surveys of two routes, for the continuation of the Cumberland Road. February 12, 1830.... [Washington: 1830.] 12 p. (21st Cong., 1st sess.House. Doc. 59)

C-4655.

Copies: DLC; DeGE; DeU; MWA; NjR.

5804. United States. Engineer Dept.

Cumberland Road. January 27, 1830. Read, and referred to the Committee on Internal Improvements. Letter from the Secretary of War, transmitting a report relative to the repairs of the Cumberland Road. [Washington: 1830.] 21 p. (21st Cong., 1st sess. House. Doc. 36)

Transmits a report of the Engineer Department.

C-4661.

Copies: DLC; DeGE; MWA; NjR.

5805. United States. Engineer Dept.

Cumberland Road. Letter from the Secretary of War, transmitting, in pursuance of a resolution of the House of Representatives of the 16th instant, a report of the progress which has been made in the construction of the Cumberland Road ... December 21, 1830.... [Washington: 1830.] 14 p. (21st Cong., 2d sess. House. Doc. 12)

Transmits a report of the Engineer Department.

C-4662.

Copies: DLC; DeGE; DeU; MWA; NjR.

5806. United States. Engineer Dept.

In Senate of the United States. February 4, 1830 ... Report from the Secretary of War ... shewing the result of an examination, made by order of that department, of the route for a road from Washington to Fredericktown by way of New Market, and also by Rockville. [Washington: 1830.] 6 p. (21st Cong., 1st sess. Senate. Doc. 44)

Report signed by Lt. J. A. Dumeste, from the Topographical Bureau.

C-5102.

Copies: DeGE; DeU; MWA.

5807. United States. Engineer Dept.

In Senate of the United States. January 19, 1830. Report from the Secretary of War ... showing the progress made in opening the Cumberland Road continued through Indiana. [Washington: 1830.] 3 p. (21st Cong., 1st sess. Senate. Doc. 27)

Transmits a report from the Engineer Dept.

Copies: DeU; MWA.

5808. United States. Laws.

A bill amendatory of the act for the continuation of the Cumberland Road. May 24, 1830 ... Mr. Vinton, from the Committee on Internal Improvements, reported the fol-

5808 *(continued)*
lowing bill.... [Washington: 1830.] 2 p. (H.R. 483)
C-3947.
Copies: DNA.

5809. United States. Laws.
A bill for continuation of the Cumberland Road, in the states of Ohio, Indiana and Illinois. February 22, 1830.... [Washington: 1830.] 2 p. (S. 100)
C-4009.
Copies: DNA.

5810. ———— March 2, 1830.... [Washington: 1830.] 2 p. (H.R. 315)
C-4010.
Copies: DNA.

5811. United States. Laws.
A bill for the improvement of the mail road between Louisville and St. Louis. January 20, 1830.... [Washington: 1830.] 1 p. (S. 68)
C-4017.
Copies: DNA.

5812. United States. Laws.
A bill making appropriation to complete the road leading from Fooy's, opposite Memphis, in the state of Tennessee, to Little Rock, in the territory of Arkansas. January 13, 1830.... [Washington: 1830.] 2 p. (H.R. 111)
C-4357.
Copies: DNA.

5813. United States. Laws.
A bill to authorize a subscription for completing the mail road from Baltimore to Wilmington. March 15, 1830.... [Washington: 1830.] 3 p. (H.R. 352)
C-4438.
Copies: DNA.

5814. United States. Laws.
A bill to authorize the improvement of a road through the lands of the Chickasaw nation, in the direction between Memphis and Tuscumbia, in the state of Alabama, March 15, 1830.... [Washington: 1830.] 1 p. (H.R. 354)
C-4456.
Copies: DNA.

5815. United States. Laws.
A bill to authorize the survey and opening of a road from Mariana, in the territory of Florida, to Appalachicola Bay, in the aforesaid territory, and from Quincy to the same place. March 8, 1830.... [Washington: 1830.] 3 p. (H.R. 266)
C-4481.
Copies: DNA.

5816. United States. Laws.
A bill to authorize the survey and opening of a road from Washington, in the territory of Arkansas, to Jackson, in the aforesaid territory. March 3, 1830.... [Washington: 1830.] 3 p. (H.R. 321)
C-4482.
Copies: DNA.

5817. United States. Laws.
A bill to authorize the surveying and making of a road from La Plaisance Bay, in the territory of Michigan, to intersect the Chicago road. March 5, 1830.... [Washington: 1830.] 2 p. (H.R. 325)
C-4483.
Copies: DNA.

5818. United States. Laws.
A bill to construct a national road from Buffalo, in the state of New York, passing by the seat of the general government, in the

District of Columbia, to New Orleans, in the state of Louisiana. January 7, 1830.... [Washington: 1830.] 5 p.(H.R. 86)
C-4493.
Copies: DNA.

5819. United States. Laws.
A bill to construct a national road from Zanesville, in Ohio, to Florence, in Alabama. March 18, 1830.... [Washington: 1830.] 5 p. (H.R. 366)
C-4494.
Copies: DNA.

5820. United States. Laws.
A bill to provide for opening a road in Arkansas Territory, from Villemont, in the county of Chicot, to Little Rock. February 22, 1830.... [Washington: 1830.] 3 p. (H. R. 273)
C-4544.
Copies: DNA.

5821. United States. President (Jackson).
Maysville road. Message from the President of the United States, returning to the House of Representatives the enrolled bill entitled "An act authorizing a subscription of stock in the Maysville, Washington, Paris, and Lexington Turnpike Road Company," with his objections thereto. May 27, 1830.... [Washington: 1830.] 22 p. (21st Cong., 1st sess. House. Doc. 113)
C-4935.
Copies: DLC; DeGE; DeU; MWA; NjR.

5822. Washington and Baltimore Turnpike Road Company.
Letter from the president of the Washington and Baltimore Turnpike Road Company, enclosing an abstract, &c. Annapolis: Printed by J. Hughes, 1830. [4] p.
C-5347.
Copies: MdHi.

5823. Washington Turnpike Company.
Letter from the president of the Washington Turnpike Company, in relation to the recent survey of two routes for a road from Washington City to Frederick in Maryland. February 11, 1830.... [Washington: 1830.] 4 p. (21st Cong., 1st sess. Senate. Doc. 54)
Signed: Nathan Lufborough.
C-4900.
Copies: DLC; DeGE; DeU; MWA; NjR.

5824. [Williams, William] 1787–1850.
The stage, canal, and steamboat register, etc., etc. for 1830. Utica, [N.Y.]: Printed and published by William Williams, [1830]. 16 p., fold. map.
C-3582.
Copies: CSmH; NBu; NHi.

5. Railroads and equipment

5825. [Stevens, John] 1749–1838.
Hints, on the expediency of a rail road from Philadelphia to Pittsburg. Most respectfully addressed to the consideration of the members of the legislature of Pennsylvania. [S.l.: 18– .] 7 p.
N-H0386069; Thomson-8.
Copies: MiU-T; NN; PPAmP.

5826. [Stevens, John] 1749–1838.
Documents tending to prove the superior advantages of rail-ways and steam-carriages over canal navigation. New York: Printed by T. and J. Swords, 1812. 43 p.
Introduction signed: John Stevens.
S&S-26804.
Copies: CSmH; DLC; DeGE; MH-BA; MWA; N; NN; NNS; NjR; PPAmP; PPF.

5827. [Stevens, John] 1749–1838.
Further hints, on the expediency of a rail road from Philadelphia to Pittsburg. Most respectfully addressed to the consideration of the members of the legislature of Pennsylvania. [S.l.: 182– .] 9 p.
Sabin-61692, 91540; Thomson-36.
Copies: MWA; MiU-T; PPAmP.

5828. North Carolina. General Assembly. House. Committee on Internal Improvements.
Report in House of Commons, Dec. 9, 1824. [Raleigh: 1824.] 8 p. (No. 5)
Signed: Joseph A. Hill.
Concerns the construction of railroads.
S-17444.
Copies: NcD; ViU.

5829. Facts and arguments in favour of adopting railways in preference to canals, in the state of Pennsylvania. To which are added, a few remarks on the subject of internal improvements. 3d ed. Philadelphia: William Fry, printer, 1825. 66 p., illus.
Dated: July 20, 1825.
Attributed to George Washington Smith, and also erroneously to Mathew Carey and to William Strickland.
Parts of the text and the illustrations are identical with *Internal improvement. Rail roads, canals, bridges, &c.*, attributed to Mathew Carey.
S-20454.
Copies: CtY; DBRE; ICU; MH-BA; MWA; MiU; NN; NjP; PPAmP; PPL; PPi.

5830. ———— 4th ed. Philadelphia: William Fry, printer, August 1, 1825. 68 p., illus.
S-20455.
Copies: CtY; DLC; DeGE; IU; MH-BA; MHi; MWA; NN; P; PPL; PPi.

5831. Pennsylvania. Laws.
An act authorizing the governor to incorporate the Moosic Mountain Coal and Rail Road Company. [Harrisburg: John Wyeth, printer, 1825.] 14 p.
N-M0753925; Thomson-79.
Copies: MWA; NN.

5832. Pennsylvania Society for the Promotion of Internal Improvements in the Commonwealth.
Railways. The acting committee of "The Pennsylvania Society for the Promotion of Internal Improvement," have been, from the formation of the society, particularly desirous to lay before the public correct information on the subject of railways. [Philadelphia: 1825.] 6 p., illus.

Dated: Philadelphia, Feb. 25, 1825.
Acting committee: Mathew Carey, Joseph Hemphill, Richard Peters, Jun., and others.
S-21851.
Copies: DBRE; DeGE; MWA; MiU-T; NN; PPAmP.

5833. ———— 2d ed. Philadelphia: 1825. 6 p., illus.
Dated: March 22, 1825.
S-21852.
Copies: DBRE; NN; NNC; PPAmP.

5834. ———— 3d ed. Philadelphia: 1825. 6, [2] p., illus.
Dated: April 5, 1825.
S-21853.
Copies: MCM; PPL.

5835. Railways and canals. From the United States Gazette. No. 1–[2]. [Philadelphia: 1825.] Broadside 37 x 20 cm.
Signed: Audi alteram partem.
Two articles, written in response to Mathew Carey's *Internal improvement, no. 1. Railways and canals,* Sept. 26, 1825.
N-A0491202; S-19474.
Copies: MWA; NNC.

5836. Tredgold, Thomas, 1788–1829.
A practical treatise on rail-roads and carriages, showing the principles of estimating their strength, proportions, expense, and annual produce, and the conditions which render them effective, economical, and durable; with the theory, effect, and expense of steam carriages, stationary engines, and gas machines.... New York: E. Bliss and E. White, 1825. xi, 184 p., illus., 4 plates.
Also published in London in 1825.
S-22506.
Copies: DBRE; DLC; DeGE; ICJ; LNL; M; MNE; MWal; MWA; Md; MnSJ; N; NN; NNC; NNE; NcU; NdU; OClWHi; PPL; TNJ.

5837. Long, Stephen Harriman, 1784–1864.
Specification of certain improvements in the locomotive engine, and in the mode of transferring loaded carriages from one level to another, in their passage upon rail-ways. By S. H. Long.... Philadelphia: 1826. 12 p., 4 plates.
"Extracted from the *Franklin Journal* for September, 1826."
S-25133.
Copies: CtY; DLC; DeGE; IU; MWA; MiU-T; NN; PPAmP; PPF; PPL.

5838. Pennsylvania. Laws.
An act to incorporate the Columbia, Lancaster, and Philadelphia Rail-Road Company. Philadelphia: Pennsylvania Society for the Promotion of Internal Improvement, 1826. 10 p.
S-24167.
Copies: DeGE; MWA; PU.

5839. [Thomas, John.]
The railway docks at Brooklyn. [S.l.: 1826.] 8, 4 p.
Also attributed to John L. Sullivan.
N-1049665; S-26198.
Copies: NBLiHi; PPL

5840. Baltimore and Ohio Railroad Company.
By-laws. [Baltimore: 1827.] 2 p.
S-27951.
Copies: MBAt; NN.

5841. Baltimore and Ohio Railroad Company.
First annual report of the directors to the stockholders of the Baltimore & Ohio Railroad Co. [Baltimore: 1827.] 4 p.
Dated: 1st October, 1827.
S-27952.
Copies: CSmH; CtY; DBRE; IU; MH-BA; MdBJ-G; MdBP; MdHi; MiU-T; NN; NNE; NjP; WU.

5842. Baltimore and Ohio Railroad Company.

Regulations for the Engineer Department. Organization of the Engineer Department of the Baltimore and Ohio Rail Road Company, providing a system for the government of the same; and for securing a strict accountabilily [*sic*] in the fiscal operations of the Baltimore and Ohio Rail Road Company, and economy in the disbursement of its funds. [Baltimore: William Wooddy, printer, 1827?] 8 p.

S-27953.

Copies: CtY; MWA; NN.

5843. Baltimore and Ohio Railroad Company.

Report of the committee appointed by the Baltimore and Ohio Rail Road Company to examine the Mauch Chunck & Quincy rail roads. [Baltimore: Wm. Wooddy, printer, 1827.] 8 p.

S-27954.

Copies: CtY; DBRE; DLC; MBAt; MCM; MWA; MdBJ; MdHi; NN; NNE; OClWHi; PHi.

5844. [Carey, Mathew] 1760–1839.

Rail-roads. [Philadelphia: 1827.] 4 p.

Statement favoring the building of a railroad from the Delaware to the Raritan River in the state of New Jersey.

Signed: Fulton, Philadelphia, September 18, 1827.

S-28402.

Copies: MWA; NN; PHi; PPL; PU.

5845. ———— 2d ed. [Philadelphia: 1827.] 4 p.

Copies: PPL.

5846. ———— 3d ed. [Philadelphia: 1827.] 4 p.

Copies: PPL.

5847. [Hale, Nathan] 1784–1863.

Remarks on the practicability and expediency of establishing a rail road on one or more routes from Boston to the Connecticut River. By the editor of the Boston Daily Advertiser. Boston: William L. Lewis, printer, 1827. 71 p.

S-29108, 30421.

Copies: CSmH; CSt; CtY; DBRE; DLC; ICJ; MH; MWA; MiU-T; NN; NNC; NbO; PPL; RPB; WU.

5848. [Hollins, William.]

Rail roads in the United States of America; or, Protest and argument against a subscription on the part of the state of Maryland, to the Baltimore and Ohio Railroad Company. Addressed to the officers and representatives of the people in the several states of the Union, and the general government. By a citizen of Baltimore. Baltimore: Matchett, printer, 1827. 35 p.

S-29247.

Copies: DBRE; DLC; M; MBAt; MH-BA; MWA; Md; MdHi; MiU-T; N; PPL; PU.

5849. ———— Baltimore: Matchett, printer, 1827. 42 p.

Appendix: p. [37]–42; dated January 8, 1828.

N-H0466127; Thomson-147.

Copies: DBRE; MWA; MdBJ-G.

5850. Long, Stephen Harriman, 1784–1864.

Letter from Lt. Col. S. H. Long, U.S. topographical engineer, to Philip E. Thomas, Esq. on the subject of rail roads. Baltimore: Printed by Samuel Sands, 1827. 15 p.

S-29524.

Copies: DBRE; DLC; MdHi; PPAN; PPL.

5851. Maryland. Laws.

An act to incorporate the Baltimore and Ohio Rail Road Company. [Baltimore: William Wooddy, 1827.] 12 p.

S-27950.

Copies: CtY; DBRE; PPL.

5852. ———— passed at December session, 1826. Baltimore: Printed by William

Wooddy, 1827. 24 p.

"An act passed by the Virginia Legislature," p. 19–24.

S-29636.

Copies: CtY; DLC; MBAt; MH-BA; MWA; MdBJ; MdHi; NN; NNE; NNP; PHi.

5853. Massachusetts. General Court. House.

[Resolve that the Commissioners of Internal Improvements be directed to survey the railway from Boston, state line of Rhode Island. Boston: 1827.] 2 p. (House. Doc. 65)

S-29679.

Copies: M; MB; MBAt; MWA.

5854. Massachusetts. General Court. House. Select Committee on a Railway from Boston to the Hudson River.

Report of the Select Committee of the House of Representatives of Massachusetts, on the practicability and expediency of constructing a railway from Boston to the Hudson River, at or near Albany . . . January 19, 1827. Boston: True & Greene, 1827. 37, [1] p. (House. Doc. 13)

Signed: Abner Phelps, George W. Adams, Emory Washburn.

S-29677.

Copies: DBRE; DeGE; KU; M; MB; MBAt; MH; MWA; MeP; MiU-T; NN; NNC.

5855. Massachusetts. General Court. Joint Committee of Roads and Canals.

[Report of the Joint Committee on petition for survey for railway on one or more routes from Boston to Hudson River. Boston: 1827.] 2 p. (House. Doc. 1)

S-29675.

Copies: M; MWA.

5856. Memorial to the Baltimore and Ohio Rail Road Company. [Baltimore: B. Edes, printer, 1827.] 16 p.

Submitted by a committee appointed by the citizens of Kanawha County, Virginia, proposing a route for the railroad through the Kanawha Valley. Letter of transmittal signed: James C. McFarland; the memorial is signed by Philip R. Thompson, Lewis Summers, and others.

S-29570, 29726.

Copies: DBRE; MH-BA; MdBJ; MdBJ-G; MdHi; NN.

5857. Mohawk and Hudson Railroad Company.

Petition of the president and directors of the Mohawk and Hudson Rail Road Company. To the honourable the Legislature of the state of New-York. [New York: 1827.] 2 p.

At head of title: In Assembly, January 22, 1827.

Signed: S. Van Rensselaer, president, G. W. Featherstonhaugh, vice-president.

Later known as the Albany and Schenectady Railroad Company.

S-27798.

Copies: NN.

5858. Proceedings of sundry citizens of Baltimore, convened for the purpose of devising the most efficient means of improving the intercourse between that city and the western states. Baltimore: Printed by W. Wooddy, 1827. 38 p.

This meeting resulted in the formation of the Baltimore and Ohio Railroad Company.

S-27946.

Copies: CtY; DBRE; DLC; DeGE; MBAt; MWA; MdBJ-G; MiU-T; MnHi; NN; NNE; OMC; PPAmP; PPL; WU.

5859. Rogers, John.

Statements, calculations and hints, relative to rail-roads, and moving power locomotive engines to be used thereon.... Baltimore: Matchett, printer, 1827. vi, [7]–40 p., fold. front.

S-30478.

Copies: DBRE; DLC; MB; MCM; MH-BA; MWA; MdBP; MdHi; NN; PPL; WU.

5860. Single rail railway. [With lithograph plate by Pendleton.] [Boston: 1827.] 8 p., fold. plate.

5860 *(continued)*
Dated: April 30, 1827.
S-30618.
Copies: DBRE; DLC; IU; MB; MCM; MWA; MiU-T; NN; NNC.

5861. Suggestions for the improvement of the commerce of the state of South Carolina. As originally published in the Charleston Courier, the 27th October, 1827. Charleston: Printed by J. S. Burges, 1827. 12 p.
Recommends the construction of railroads.
S-30737.
Copies: DLC; MWA; NHi; ScCC.

5862. Sullivan, John Langdon, 1777–1865.
Mr. Sullivan's description of a cheap, durable rail road, and certain routes, in the states of New-York and New-Jersey. [S.l.: 1827.] 22 p., fold. plate.
S-30739.
Copies: MH; MH-BA; MiU-T; NN; PPL.

5863. Three papers on rail-roads. 1. Rail road to New Orleans for transporting the mail. 2. Timber rail-ways—economical plan. 3. General remarks on rail-roads. With plates of reference. From the American farmer. [S.l.: 1827?] 12 p., illus.
By Robert Mills and Stephen Harriman Long.
S-30799.
Copies: DBRE; DLC; DeGE; M; MB; MH-BA; MWA.

5864. Ward, Minus.
Remarks, propositions and calculations, relative to a rail-road and locomotive engines to be used upon the same, from Baltimore to the Ohio River.... Baltimore: Printed by John D. Toy, 1827. vi, [7]–50 p.
S-31657.
Copies: DBRE; DLC; DeGE; MCM; MH-BA; MWA; MdBJ-G; NN; NRom; PHi; WU.

5865. Baltimore and Ohio Railroad Company.
Baltimore and Ohio Rail Road Company. Documents accompanying a memorial of the president and directors of the Baltimore and Ohio Rail Road Company. May 19, 1828.... Washington: Printed by Gales & Seaton, 1828. 5 p. (20th Cong., 1st sess. House. Doc. 272)
S-35695.
Copies: CtY; DBRE; DLC; DeGE; DeU; MB; MH; MWA; NN; NjR.

5866. Baltimore and Ohio Railroad Company.
Baltimore and Ohio Rail Road—iron. Mr. Buchanan submitted the following letter, &c. upon the subject of obtaining a supply of iron for the Baltimore and Ohio Rail Road. May 16, 1828.... Washington: Printed by Gales and Seaton, 1828. 7 p. (20th Cong., 1st sess. House. Doc. 271)
Submits a letter from P. E. Thomas, president of the Baltimore and Ohio Railroad Company, and other documents.
S-35696.
Copies: CtY; DBRE; DLC; DeGE; DeU; MB; MWA; NN; NjR; P.

5867. Baltimore and Ohio Railroad Company.
Maryland. Memorial of the Baltimore and Ohio Rail Road Company, praying to import iron and machinery, free of duty. March 17, 1828.... Washington: Printed by Gales & Seaton, 1828. 4 p. (20th Cong., 1st sess. House. Doc. 206)
S-36533.
Copies: CtY; DBRE; DLC; DeGE; DeU; MBAt; MH; MWA; MiU-T; NN; NjP; PPAmP.

5868. Baltimore and Ohio Railroad Company.
Memorial of the Baltimore and Ohio Rail Road Company, praying that they may be permitted to import iron and iron machinery for said road free of duty. March 17, 1828.... Washington: Printed by Duff Green, 1828. 4 p. (20th Cong., 1st sess.

Senate. Doc. 140)
S-36570.
Copies: CtY; DBRE; DLC; DeU; MBAt; MH; MWA; MiU-T; NN; NjR; P; PPAmP.

5869. Baltimore and Ohio Railroad Company.
Rail Road Company, Baltimore and Ohio. December 23, 1828.... Memorial. To the Senate and House of Representatives of the United States.... The memorial of the president and directors of the Baltimore and Ohio Rail Road Company.... Washington: Printed by Gales & Seaton, 1828. 2 p. (20th Cong., 2d sess. House. Doc. 48)
S-36733.
Copies: CSt; CtY; DBRE; DLC; DeGE; ICJ; MB; MH; MWA; MiU-T; NN; NjR; P; PPL.

5870. Baltimore and Ohio Railroad Company.
Report of the engineers, on the reconnoissance and surveys, made in reference to the Baltimore and Ohio Rail Road. Baltimore: Printed by W. Wooddy, 1828. 188, [1] p.
Reports signed by Wm. Howard, S. H. Long, Wm. G. M'Neill, and J. Knight; appendix contains the descriptive memoirs of Lieuts. Barney and Trimble and Mr. Harrison.
S-32088.
Copies: CtY; DBRE; DIC; DLC; DeGE; ICJ; M; MBAt; MH; MWA; MdBE; MdBJ-G; MdHi; MnHi; NIC; NcD; PPF; PPL.

5871. Baltimore and Ohio Railroad Company.
Second annual report of the president and directors to the stockholders of the Baltimore & Ohio Rail Road Company. Baltimore: Printed by James Lucas, 1828. 11, [1], 43 [44] p.
Includes also: *First annual report of the Board of Engineers ...*, Sept. 30, 1828, with accompanying documents.
S-32089.
Copies: CtY; IU; MCM; MWA; MdBJ-G; NN; NNE.

5872. ———— Baltimore: Printed by William Wooddy, 1828. 11, 45 p., fold. map.
S-32090.
Copies: CtY; DBRE; DLC; MH-BA; MWA; MdHi; NNE; NjP; WU.

5873. Baltimore and Susquehanna Railroad Company.
Report and proceedings in relation to a rail road from Baltimore to the Susquehanna Baltimore: Printed by Benjamin Edes, 1828. 23 p.
S-32092.
Copies: DLC; MBAt; MCM; MWA; MdHi; NN; NNC.

5874. Baltimore and Susquehanna Railroad Company.
Report on the Baltimore and Susquehanna Rail Road. By General J. G. Swift, Chief Engineer. Also the annual report of the president and directors to the stockholders. Baltimore: Printed by Lucas & Deaver, 1823. 10, 24 p.
S-32093.
Copies: CtY; DBRE; DIC; DLC; IU; MH-BA; MWA; MdBP; MdHi; MiU-T; NIC, NN; NNE; NjR; PHi; PPL; WU.

5875. Barney, Joshua.
Map of the country embracing the various routes surveyed for the Balt. & Ohio Rail Road by order of the Board of Engineers. [Baltimore: 1828?] Map 27 x 61 cm.
Prepared to accompany: Baltimore and Ohio Railroad Company. *Report of the engineers, on the reconnoissance and surveys in reference to the Baltimore and Ohio Rail Road.* Baltimore: 1828.
Copies: DLC.

5876. [Caldwell, Joseph] 1773–1835.
The numbers of Carlton, addressed to the people of North Carolina, on a central rail-road through the state.... New York: G.

5876 (*continued*)
Long, 1828. 232 p.

A series of papers dealing with railroads in general, and the encouragement of railroad construction in North Carolina.

S-32549.

Copies: DBRE; MWA; MiU-T; NN; NcD; NcU; P; TxU; Vi.

5877. Charleston, S.C. Chamber of Commerce.

Report of a special committee appointed by the Chamber of Commerce, to inquire into the cost, revenue and advantages of a rail road communication between the city of Charleston and the towns of Hamburg & Augusta. Charleston: Printed by A. E. Miller, 1828. 32 p.

S-35358.

Copies: CSmH; DBRE; DLC; GU; MBAt; MH-BA; MWA; MiU-T; NN; RPB; ScCC; ScU; WU.

5878. Charleston, S.C. City Council.

Memorial of the city council of Charleston, praying that the South Carolina Canal and Rail Road Company be authorized to import, free of duty, iron and machinery, to be used in effecting the objects of their incorporation. April 7, 1828.... Washington: Printed by Duff Green, 1828. 4 p. (20th Cong., 1st sess. Senate. Doc. 166)

S-36583.

Copies: CtY; DBRE; DLC; DeU; MB; MBAt; MH; MWA; MiU-T; NN; NjR; P; PPAmP.

5879. A circumstantial estimate of the cost of a railway, followed by some calculations on its probable income. [Boston: 1828.] Broadside 77 x 16 cm.

First published in the *Chronicle and patriot*, February 9, 1828. Refers to the Western Railway.

S-32715.

Copies: MB.

5880. [Corri, Philip Antony] 1784–1832.

The Carrollton march; performed at the ceremony of commencing the Baltimore & Ohio Railroad, on the fourth of July, 1828. Dedicated by permission to the Hon. Charles Carroll, of Carrollton, by A. Clifton [pseud.] Baltimore: John Cole, [1828.] [2] p.

S-32839.

Copies: DLC.

5881. Delaware and Hudson Canal Company.

Delaware and Hudson Canal. Memorial of the Delaware and Hudson Canal Company. April 7, 1828. Referred to the Committee of Ways and Means. April 14, 1828. Bill reported no. 268. Washington: Printed by Gales & Seaton, 1828. 20 p. (20th Cong., 1st sess. House. Doc. 247)

The company requests that the iron for its railroad be admitted free of duty.

S-36254.

Copies: CtY; DBRE; DLC; DeGE; DeU; MB; MBAt; MH; MWA; NN; NjR; PPL; RPB.

5882. A detailed and correct account of the grand civic procession in the city of Baltimore on the fourth of July, 1828; in honor of the day, and in commemoration of the commencement of the Baltimore and Ohio Rail-Road; as published in the "American." Baltimore: Printed by Thomas Murphy, 1828. 46 p.

S-32969.

Copies: CtY; MWA; MdBJ-G; MdHi; NN; PHi.

5883. Long, Stephen Harriman, 1784–1864.

Rail road manual; or, A brief exposition of principles and deductions applicable in tracing the route of a rail road.... Baltimore: Wm. Wooddy, printer, 1828–1829. 2 vols., diagrs.

Part I, 109 p., published in 1829; part II, 62 p., published in 1828.

S-33897, 39316.

Copies: CSmH; DBRE; DLC; DeGE (part I only); MH-BA; MWA; MdBJ-G; MdHi; MiU-T; NN; NNC; NNE; NjP.

5884. Maryland. General Assembly. House. Committee on Internal Improvement.

Report of the Committee on Internal Improvement, to the House of Delegates. Annapolis: Printed by Jeremiah Hughes, 1828. 23 p.

Dated: February 12, 1828.

Devoted entirely to the Baltimore and Ohio Railroad.

S-34033.

Copies: MdBP; MdHi.

5885. Maryland. Laws.

An act to incorporate the Baltimore and Susquehanna Rail Road Company, passed December session, 1827. Baltimore: Printed by Benjamin Edes, 1828. 16 p.

S-32091.

Copies: DLC; MdBE; MdHi; NNE; PHi.

5886. Massachusetts. Board of Commissioners for the Survey of One or More Routes for a Railway from Boston to Albany.

Report of the Board of Commissioners for the survey of one or more routes for a railway from Boston to Albany. Boston: Dutton & Wentworth, printers to the state, 1828. 56 p. (Senate. Doc. 5)

Signed: Nahum Mitchell, Samuel F. McKay, commissioners; James F. Baldwin, engineer.

S-34066, 37149.

Copies: CSmH; DBRE; DLC; DeGE; IU; KU; MH; MWA; MiU-T; NN; NNC; NbO; NjP; OO; PPL; RPB.

5887. Massachusetts. Board of Internal Improvements.

Report of the Board of Commissioners, of Internal Improvement, in relation to the examination of sundry routes for a railway from Boston to Providence. With a Memoir of the survey. Boston: Dutton & Wentworth, printers to the state, 1828. 72 p., fold. map. (Senate. Doc. 4)

First issue, with "Erratum" on p. 72.

"Memoir of the survey," by James Hayward, p. 51–72.

S-34064.

Copies: CSmH; CtY; DLC; DeGE; MA; NN; PPL.

5888. ———— Boston: Dutton & Wentworth, printers to the state, 1828. 72 p., fold. map. (Senate. Doc. 4)

"In Senate, Feb. 2, 1828. Ordered that five hundred additional copies be printed."

"Erratum" has been corrected.

Copies: DeGE; M.

5889. ———— Boston: Dutton & Wentworth, printers to the state, 1828. 72 p., fold. map.

Issued without the document note.

S-34065.

Copies: CSmH; CSt; DBRE; DLC; IU; M; MB; MH-BA; MWA; N; NN; NRU; PPAmP.

5890. Massachusetts. Board of Internal Improvements.

Report [on the organization of the Board of Internal Improvements and on the prosecution of the surveys for proposed railroads in Massachusetts. Boston: 1828.] 15 p. (Senate. Doc. 6)

Signed on p. [11]: Levi Lincoln, Nathan Hale, George Bond [and others].

S-34070.

Copies: CtY; DBRE; DLC; M; MB; MBAt; MH; MWA; MiU-T; NN.

5891. Massachusetts. General Court. House. Committee on Roads and Railways.

The Committee on Roads and Railways, to whom was referred the report of the Board of Commissioners for the survey of one or more routes for a railway from Boston to Albany, and also the report of the Board of Commissioners of Internal Improvement in relation to the examination of sundry routes for a railway from Boston to Providence, report. [Boston: 1828.] 11 p. (House. Doc. 48)

Dated: House of Representatives, Feb. 15, 1828.

S-34068.

Copies: DBRE; DLC; DeGE; M; MBAt; MH; MWA; NN.

5892. Massachusetts. Governor (L. Lincoln).

[Governor's message and memorial of citizens of Troy in favor of a rail road from Boston to eastern shore of Hudson River.] [Boston: 1828.] 7 p. (House. Doc. 57)

Hasse (Massachusetts)-p. 238.

Copies: M; MWA.

5893. Massachusetts. Governor. (L. Lincoln).

[Governor's message transmitting accounts of the commissioners and engineer on the Western Railway and also of the Board of Internal Improvement, together with the record of the doings of the Executive Council, in auditing and allowing these accounts, and their advice thereon, to be prepared ... Feb. 13, 1828.] [Boston: 1828.] 26 p. (House. Doc. 52)

S-34056.

Copies: DBRE; M; MB; MBAt; MCM; MH; MH-BA; MWA; NN.

5894. Massachusetts. Governor (L. Lincoln).

Speech of His Excellency Levi Lincoln, delivered before the two branches of the legislature in convention, June 2, 1828. Boston: Dutton and Wentworth, 1828. 22 p. (Senate. Doc. 2)

Discussion of a railroad from Boston to the Hudson River, p. 7–12.

S-34073; Thomson-223.

Copies: DBRE; IU; M; MB; MBAt; MH; MWA; WHi.

5895. Massachusetts. Laws.

An act to incorporate the Berkshire and Hudson Rail Road Company. [Boston: Dutton & Wentworth, 1828.] 14 p. (House. Doc. 1)

S-34051.

Copies: DBRE; DLC; M; MBAt; MH; MWA; NN.

5896. Meineke, Christopher, 1782–1850.

Rail road march—for the Fourth of July—dedicated to the directors of the Baltimore & Ohio Rail Road. Composed & arranged for the piano forte.... Baltimore: Geo. Willig, 1828. 3 p.

Head piece: Engraved view of a railroad train.

S-34105.

Copies: MWA; MdBJ-G.

5897. ———— Baltimore: Geo. Willig, 1828. 3 p.

S-34106.

Copies: DLC; MNF; MiU-T; NN; ViU.

5898. Memorial of inhabitants of Philadelphia, praying that the Baltimore and Ohio Railroad Company may not be permitted to import iron free of duty. May 12, 1828.... Washington: Printed by Gales & Seaton, 1828. 6 p. (20th Cong., 1st sess. House. Doc. 270)

At head of title: Pennsylvania.

S-36708.

Copies: CtY; DBRE; DLC; DeU; MB; MBAt; MH; MWA; NjR; P; PPAmP.

5899. Memorial. To the Honorable the Legislature of the state of New-York, in Senate and Assembly convened.... [Albany?: 1828.] 8 p.

Text begins: The memorial of the undersigned, inhabitants of the County of ———— respectively showeth: that a memorial was presented to the Legislature of this state, in 1827, by Jacob Morris....

Asks for a charter for the Susquehanna Navigation and Rail Road Company, to build a railroad between the Erie Canal at Fort Plain and Otsego Lake.

Thomson-228A.

Copies: MWA; N; NN.

5900. New York (State). Inspectors of Auburn State Prison.

Memorial of the inspectors of the Auburn State Prison, &c. praying the construction of a rail-road from the said prison to the Erie Canal. To ... the Legislature of the state of New-York.... [Albany: 1828.] 5 p. (Assembly. Doc. 97)

Dated: February 16, 1828.

S-34503.
Copies: MWA; NN.

5901. New York (State). Laws.
An act authorizing the construction of a rail road from the state prison at Auburn, to the Erie Canal. [Albany: 1828.] 2 p. (Assembly. Doc. 117)
Dated: February 26, 1828.
S-34468.
Copies: MWA; NN.

5902. New York (State). Laws.
An act to facilitate the construction of a rail-road from the city of Boston to the Hudson River. [Albany: 1828.] [1] p. (Senate. Doc. 226)
Dated: March 24, 1828.
S-34470.
Copies: MWA; NN.

5903. New York (State). Laws.
An act to incorporate the Catskill and Ithaca Rail-Road Company. [Albany: 1828.] 4 p. (Assembly. Doc. 233)
Dated: In Assembly, March 27, 1828.
S-34471.
Copies: MWA; NN.

5904. New York (State). Laws.
An act to incorporate the "Great Sable Rail Road Company." [Albany: 1828.] 3 p. (Assembly. Doc. 114)
Dated: February 25, 1828.
S-34472.
Copies: MWA; NN.

5905. New York (State). Laws.
An act to incorporate the Hudson and Berkshire Rail-Road Company. [Albany: 1828.] 5 p.
S-34473.
Copies: MWA; NN.

5906. New York (State). Laws.
An act to incorporate the Ithaca and Owego Rail Road Company, passed January 28, 1828. [Albany?: 1828.] 19 p.
S-33680.
Copies: NIC; NN.

5907. New York (State). Laws.
An act to incorporate the Susquehannah Navigation and Rail-Road Company. [Albany: 1828.] 10 p. (Assembly. Doc. 170).
Dated: March 10, 1828.
Thomson-260.
Copies: MWA; N; NN.

5908. New York (State). Legislature. Assembly. Committee on Canals.
Report of the Committee on Canals and Internal Improvements, relative to the construction of a rail-road from the Auburn state prison, to the Erie Canal. [Albany: 1828.] 4 p. (Assembly. Doc. 116)
Dated: February 26, 1828.
S-34514.
Copies: MWA; NN.

5909. New York (State). Legislature. Assembly. Select Committee on a Rail-Road from Port Kent to the Sable River.
Report of the Select Committee, relative to the construction of a rail-road from Port Kent to the Sable River. [Albany: 1828.] 2 p. (Assembly. Doc. 113)
Dated: February 25, 1828.
S-34518.
Copies: MWA; NN.

5910. New York (State). Legislature. Assembly. Select Committee on a Rail-Road from the Hudson River to Ithaca.
Report of the Select Committee, relative to the construction of a rail-road from the Hudson River to Ithaca. [Albany: 1828.] 2 p. (Assembly. Doc. 232)
Dated: March 27, 1828.
S-34519.
Copies: MWA; NN.

5911. New York (State). Legislature. Assembly. Select Committee on the Susquehanna Navigation and Rail Road Company.
Report of the Select Committee relative to the incorporation of the Susquehannah

5911 (*continued*)
Navigation and Rail Road Company. [Albany: 1828.] 4 p. (Assembly. Doc. 169)
Dated: March 10, 1828.
S-34520.
Copies: MWA; NN.

5912. New York (State). Legislature. Senate. Select Committee on the Construction of a Rail-Road from the Auburn State Prison to the Erie Canal.
Report of the Select Committee, on the engrossed bill from the Assembly, relative to the construction of a rail-road from the Auburn State Prison to the Erie Canal. [Albany: 1828.] 3 p. (Senate. Doc. 251)
Dated: April 15, 1828.
S-34517.
Copies: MWA; NN.

5913. Pennsylvania. Board of Canal Commissioners.
Report of the Canal Commissioners, relative to a railroad from the mouth of the Swatara by Columbia to Philadelphia. [Philadelphia: 1828.] 18 p.
S-34686.
Copies: CSmH; MH-BA; P; PPAmP.

5914. Pennsylvania. Laws.
An act incorporating the Mill Creek & Mine Hill Navigation and Rail Road Company. [Harrisburg: 1828.] 10 p.
Passed February 7th, 1828.
Copies: P.

5915. Pennsylvania. Laws.
An act passed by the Legislature of Pennsylvania to authorize the Baltimore & Ohio Rail-Road to construct a rail road through Pennsylvania in the direction from Baltimore to the Ohio River. [Harrisburg: 1828.] [7] p.
S-34673.
Copies: DLC; MB.

5916. ———— [Baltimore]: Wm. Wooddy, printer, [1828]. 16 p.
S-32087.
Copies: MH-BA (not located).

5917. Rail-road meeting. Chatham County, N.C. August 1, 1828. Agreeably to previous notice, a number of citizens of the counties of Chatham, Randolph, Guilford and Orange.... [Hillsborough, N.C.: D. Heartt, printer, 1828.] 8 p.
Discussions on the need for a railroad from the coast, through Raleigh, to the western border of the state.
S-34951.
Copies: NcU.

5918. [Sedgwick, Theodore] 1780–1839.
Brief remarks on the rail roads, proposed in Massachusetts. By Berkshire [pseud.]. Stockbridge, [Mass.]: Printed by Charles Webster, 1828. 23 p.
S-32275.
Copies: DBRE; M; MB; MCM; MH-BA; MWA; NN.

5919. Swift, William Henry, 1800–1879.
Report of the survey of the route of the Ithaca and Owego Rail-Road.... Ithaca: Printed by Mack & Andrus, 1828. 16 p.
S-35453.
Copies: DBRE; MWA; NIC; NN; NNE.

5920. Troy, N.Y. Common Council.
Message of Levi Lincoln [Governor of Massachusetts, transmitting a memorial of the Troy (N.Y.) Common Council to the Legislature of the Commonwealth of Massachusetts in regard to the construction of a railroad from the city of Boston to the eastern shore of the Hudson River. Boston?: 1828]. 7 p. (Massachusetts. General Court. House. Doc. 57)
Memorial signed: Samuel McCoun, Mayor.
Letter of transmittal dated: Feb. 20th, 1828.
S-34061.
Copies: DBRE; M; MB; MBAt; MH-BA; MWA; MiU-T; NN.

5921. United States. Congress. Senate.

In Senate of the United States. May 6, 1828. The following letters were laid on the table by Mr. Smith, of Maryland.... [Washington: 1828.] 4 p. (20th Cong., 1st sess. Senate. Doc. 192)

Concerns the procurement of iron rails for the Baltimore and Ohio Railroad.

S-36392.

Copies: DLC; DeU; MWA; NjR.

5922. United States. Congress. Senate. Committee on Finance.

In Senate of the United States. April 2, 1828. Mr. Smith, from the Committee on Finance, to which was referred the memorial of the Baltimore and Ohio Rail Road Company, praying that they may be permitted to import iron and iron machinery for said road free of duty, reported a bill to admit iron and machinery necessary for railroads, duty free, and laid on the table the following document, which, with the memorial (Doc. 140) heretofore printed, were ordered printed. [Washington: 1828.] 5 p. (20th Cong., 1st sess. Senate. Doc. 161)

S-36383.

Copies: DLC; DeU; MWA; NjR.

5923. United States. Laws.

An act to admit iron and machinery necessary for rail roads, duty free. April 30, 1828. Committed to the Committee of the Whole House on the state of the Union, and printed by order of the House of Representatives. [Washington: 1828.] 1 p. (S. 132)

S-35623.

Copies: DNA.

5924. United States. Laws.

A bill to authorize a rail road within the District of Columbia. April 15, 1828.... [Washington: 1828.] 1 p. (S. 140)

S-36069.

Copies: DNA.

5925. Virginia. Laws.

A bill, concerning the Baltimore and Ohio Rail-Road. [Richmond: Printed by Thomas Ritchie, 1828.] 2 p.

S-37028.

Copies: DBRE; NN.

5926. Virginia. Principal Engineer.

Report of the Principal Engineer, on the rail-way from the coal-pits to James River. [Richmond: Printed by Thomas Ritchie, printer to the Commonwealth, 1828.] 6 p. (Virginia. House. Doc. 6)

Claudius Crozet, principal engineer.

Concerns the Chesterfield Railroad.

S-37034.

Copies: DBRE; MiU-T; NN; Vi.

5927. [Warfield, Charles.]

[Protest against the Baltimore and Ohio Railroad Company taking possession of part of his land for their railroad.] [Baltimore: 1828.] 8 p.

S-37085.

Copies: MdBJ-G.

5928. Baltimore and Ohio Railroad Company.

By-laws, and rules and regulations of the Baltimore and Ohio Rail Road Company. [Baltimore: 1829?] 1 p.l., 7 p.

S-37571.

Copies: DIC; MiU-T.

5929. Baltimore and Ohio Railroad Company.

Document no. 13. Accompanying the executive communication of 31st December, 1829. Report of the Baltimore and Ohio Rail Road Company. Published by authority. Annapolis: Printed by Jeremiah Hughes, Dec. 1829. 11 p.

Contains the *Third annual report of the Baltimore and Ohio Railroad Company,* dated 12th October, 1820 [1829], without engineer's report and maps.

S-39460; Thomson-279.

Copies: DeGE; MWA; MdHi; PPL.

5930. [Baltimore and Ohio Railroad Company.]

Experiments on rail roads, in England, illustrative of the safety, economy and speed

5930 (*continued*)
of transportation, which this system, as now improved, is capable of affording. Baltimore: Printed by W. Wooddy, 1829. 16, 2, 2 p.

Includes *The Memorial of the president and directors of the Baltimore and Ohio Rail Road Company* to the Senate and House of Representatives, 2 p.

S-37572.

Copies: CSmH; CtY; DBRE; DLC; DeGE; ICJ; MB; MH-BA; MWA; MdBP; MdHi; NN; NNC; NNE; NRU; OHi; PPi.

5931. Baltimore and Ohio Railroad Company.

Memorial of the Baltimore and Ohio Railroad Company. December 23, 1829.... [Washington: 1829.] 2 p. (21st Cong., 1st sess. House. Rept. 23)

S-41231.

Copies: CSt; CtY; DLC; DeU; IU; MB; MH; MWA; NjP; NjR; PPL.

5932. Baltimore and Ohio Railroad Company.

Memorial of the Baltimore and Ohio Rail Road Company, praying for the aid of the United States, by a subscription to the stock of said company. December 22, 1829.... [Washington: 1829.] 2 p. (21st Cong., 1st sess. Senate. Doc. 5)

S-41232.

Copies: CSt; CtY; DBRE; DLC; DeU; IU; MH; MWA; N; NN; NjP; P; PPAmP.

5933. Baltimore and Ohio Railroad Company.

Report of the Baltimore and Ohio Rail Road Company. Published by authority. Annapolis: Printed by Jeremiah Hughes, Dec. 1829. 11 p. (Maryland. General Assembly. House. Doc. 13)

State edition of the *Third annual report,* without engineer's report and maps.

Thomson-279.

See no. 5929.

5934. Baltimore and Ohio Railroad Company.

Report of the board of engineers to the Baltimore and Ohio Rail Road Company. Submitted to the Mayor and City Council of Baltimore. [Baltimore: 1829.] 12, [3] p.

S-37574.

Copies: MWA; MdBP; MdHi; NjP.

5935. Baltimore and Ohio Railroad Company.

Third annual report, of the president and directors, to the stockholders of the Baltimore and Ohio Rail Road Company. Baltimore: Printed by William Wooddy, 1829. 105 p., fold. map, fold. plate.

S-37575.

Copies: DBRE; DLC; ICJ; IU; MCM; MH-BA; MWA; MiU-T; NjP; PPAmP; PPF.

5936. ——— Baltimore: Printed by William Wooddy, 1829. 13 p., map.

Without accompanying documents.

S-37576.

Copies: LNL; MdHi; NNE; PPL; WU.

5937. Baltimore and Susquehanna Railroad Company.

Report of the city director of the Baltimore and Susquehanna Rail Road Company. [Baltimore?: 1829.] 7 p.

Signed: S. Brady.

N-B0078876.

Copies: NN.

5938. Bridgeman, George.

A description of Bridgeman's rotary railway. New York: Printed by G. F. Bunce, 1829. 8 p., illus.

Decribes a new type of wheel to improve the performance of vehicles.

S-37936.

Copies: CtY; MB; MH-BA; MWA.

5939. [Carroll, James.]

The claim of the western navigation of the city of Baltimore to a rail road deposit, stated and considered. Baltimore: Printed by

John D. Toy, 1829. 26 p., fold. map.
Signed: A friend of the rail road.
S-38075.
Copies: MWA; MdBJ-G; MdHi; NN.

5940. Fleming, Peter.
Report to the president and directors of the Mohawk and Hudson Railway Company. By Peter Fleming, civil engineer. New York: G. & C. Carvill, 1829. 15 p.
Survey of the projected railroad line between the Mohawk and the Hudson rivers.
S-38591.
Copies: DBRE; DLC; MCM; MH-BA; MWA; NHi; NN; PPAmP; PPL.

5941. [Hale, Nathan] 1784–1863.
Remarks on the practicability and expediency of rail roads from Boston to the Hudson River and from Boston to Providence. From the Boston Daily Advertiser. Boston: Press of the Boston Daily Advertiser; W. L. Lewis, printer, 1829. vii, 64 p.
S-38881.
Copies: DBRE; DLC; DeGE; ICJ; IU; MB; MH; MWA; MiU-T; NNC; PPL.

5942. [Hale, Nathan] 1784–1863.
To the commissioners appointed under the act of the Legislature of New York, relative to the construction of a rail road from the city of Boston to the Hudson River. [Boston?: 1829?] 20 p.
S-38882.
Copies: DBRE; DLC; MB; MH-BA; MWA; MiU-T; NN.

5943. Hopkins, William R.
Report relative to the rail road line from the west end of the Harrisburg bridge, through Cumberland and Franklin counties, to the borough of Chambersburg, together with an estimate of the cost, and from Chambersburg through Gettysburg to the borough of York. By W. R. Hopkins, engineer. Harrisburg: Printed at the office of the Reporter, 1829. 16 p.
Read in the House of Representatives, Jan. 19, 1829.
S-40006.
Copies: CSmH; DBRE; MWA; MiU; NN; PHi.

5944. Howard, William, 1793–1834.
Specification of a patent for an improved rail-way carriage. Patented November 22, 1828.... Philadelphia: 1829. 6 p., 3 plates.
Extracted from the *Journal of the Franklin Institute,* January, 1829.
S-39033.
Copies: MH-BA; MdHi; NN; PPF.

5945. Howard, William, 1793–1834.
Specification of a patent for locomotive engines or carriages propelled by steam, on rail, or other roads, denominated the Improved locomotive carriage. Granted to William Howard, Esq., United States civil engineer. Baltimore, December 10, 1828. With two copperplates. From the Journal of the Franklin Institute for May, 1829. Philadelphia: 1829. 4 p., 2 plates.
S-39032.
Copies: MB; NN; PPF.

5946. Howard, William, 1793–1834.
To the president and directors of the South-Carolina Canal and Rail Road Company.... [Charleston: 1829.] 4 p.
At head of title: South-Carolina Canal and Rail Road Company, Charleston, January 19, 1829. Resolved that two hundred copies of Dr. Howard's communication be published.
Signed: Wm. Howard, U.S. civil engineer.
N-H0556473; S-40494.
Copies: CSmH; MWA; NHi; NN; ViU; WM; WU.

5947. Jackson, William, 1783–1855.
A lecture on rail roads, delivered January 12, 1829, before the Massachusetts Charitable Mechanic Association. Boston: Printed by Crocker & Brewster, 1829. 32 p.
S-39109.
Copies: CSmH; CSt; CtY; DBRE; DLC; DeGE; ICJ; KU; MB; MBAt; MH; MWA; MdHi; MiU; NN; NNC; NjP; OClWHi; P; PHi.

5948. ________ Boston: H. Bowen, 1829. 36 p.
S-39110.
Copies: CtY; DLC; ICJ; MB; MH-BA; MWA; MiU-T; NN; PPAmP.

5949. ________ 2d ed. Boston: H. Bowen, 1829. 36 p.
S-39111.
Copies: DLC; DeGE; ICJ; MB; MBAt; MH-BA; MWA; NN; NjPT; PPAmP; PPL; VtU.

5950. ________ 3d ed. Boston: H. Bowen, 1829. 36 p.
S-39112.
Copies: DBRE; DIC; DeGE; MWA; N; NN; NNC; NNE.

5951. ________ 4th ed. Springfield, [Mass.]: A. G. Tannatt, printer, 1829. 36 p.
S-39113.
Copies: MSCV; MWA; NN.

5952. [List, Friedrich] 1789–1846.
Reports on the improvement of the Little Schuylkill. Reading, [Pa.]: 1829. 16 p.
Refers to the Little Schuylkill Railroad.
S-39298.
Copies: CSmH; DLC; NN.

5953. McCartney, the builder of the viaduct over the Patapsco, gives a frolic on the closing of the arches—after the men are made drunk, he orders them to kneel before him, and in that situation baptises them by pouring on the head a pint of whiskey. This impious violation of the holy rites of religion took place on the Baltimore and Ohio rail road, in the neighborhood of Ellicott's Mills, the office of his warm friend and supporter, the superintendent of the road, being in the distance. Rail-road scene. Printed all along shore. [S.l.: 1829.] Broadside.
In verse.
S-39354.
Copies: MdHi; NN.

5954. Maryland. Governor (D. Martin).
A communication from the executive department, to the House of Delegates, in obedience to their order of the 2d inst. enclosing a communication from the Baltimore and Ohio Rail Road Company. February 18, 1829. Annapolis: Jeremiah Hughes, printer, 1829. 7, [1] p.
S-39445; Thomson-319.
Copies: MWA; Md; NN.

5955. Massachusetts. Board of Internal Improvements.
Report of the Board of Directors of Internal Improvements of the state of Massachusetts, on the practicability and expediency of a rail-road from Boston to the Hudson River, and from Boston to Providence. Submitted to the General Court, January 16, 1829. To which are annexed, the reports of the engineers.... With plans and profiles of the routes. Boston: Press of the Boston daily advertiser, 1829. 76, 119 p., 6 fold. maps, tables.
Contents: Report of the Board of Directors.—Report of Mr. Baldwin on the rail-road surveys from Boston to the Hudson River.—Report of Mr. Hayward on the rail-road surveys between Boston and Providence.—Mr. Willard's report on the cost of stone for the rail-road.
S-39503.
Copies: CSmH; CSt; CtY; DBRE; DLC; DeGE; ICU; LNHT; MH; MWA; MeHi; MiU-T; MnU; NIC; NN; NNC; NbO; NjP; OO; PPL; VtU; WHi.

5956. Massachusetts. Board of Internal Improvements.
Report of the Commissioners of the Board of Internal Improvements, relative to a route for a rail road from Plymouth to Wareham, in the county of Plymouth.... [Boston: 1829.] 12 p. (Senate. Doc. 2)
"Report of the survey for a rail-way between Plymouth and Wareham," by James Hayward, p. [6]–12.
S-39504.
Copies: DBRE; DeGE; M; MB; MBAt; MH-BA; MWA; MiU-T.

5957. Massachusetts. Board of Internal Improvements.

[Report] to His Excellency Levi Lincoln, Governor of the Commonwealth of Massachusetts ... of a survey [of] ... a route for a rail road from the town of Taunton, to some point in the route previously surveyed between Boston and Providence. Boston: 1829. 8 p. (Senate. Doc. 10)

Dated and signed: December 15, 1829, Josiah J. Fiske, Willard Phillips, James Hayward, commissioners.

S-39501.

Copies: DBRE; DeGE; M; MB; MBAt; MH-BA; MWA; NN.

5958. Massachusetts. General Court. House.

Rail road resolutions. Mr. Bliss of S. offers the following amendment. February 12, 1829. [Boston: 1829.] 1 p. (House. Doc. 24)

S-39500.

Copies: M; MB; MBAt; MH; MWA.

5959. Massachusetts. General Court. House. Committee on Roads and Railways.

[Report of the Committee on a railroad from Boston to Providence and from Boston to the Hudson River.] [Boston: 1829.] 7 p. (House. Doc. 42)

S-39506; Thomson-321.

Copies: DBRE; M; MB; MH-BA; MWA; MiU-T; NN.

5960. Massachusetts. General Court. Joint Committee on Railways and Canals.

Commonwealth of Massachusetts. The Committee of both Houses on Railways and Canals, to whom was referred so much of his Excellency the Governor's message, as relates to the subject of railways, and also the report of the Directors of Internal Improvements on the practicability and expediency of constructing rail roads from Boston to Hudson River, and also from Boston to Providence, respectfully report.... [Boston: 1829.] 6 p. (House. Doc. 12)

Signed: John W. Lincoln.

S-39507.

Copies: DBRE; DEGE; M; MB; MBAt; MH; MWA; NN.

5961. Massachusetts. General Court. Senate. Committee on Roads and Canals.

In Senate, February 28, 1829. The Committee to whom was referred the report of the Committee of both houses on railways and canals, on so much of the message of ... the Governor as relates to the subject of railways, and on the report of the Board of Directors of Internal Improvements, on ... constructing railroads from Boston to Hudson River, and from Boston to Providence ... report. [Boston: 1829.] 12 p. (Senate. Doc. 24)

S-39508; Thomson-322.

Copies: DBRE; M; MB; MH-BA; MWA; MiU-T; NN; NNE.

5962. Massachusetts. Laws.

An act to establish the Massachusetts rail road corporations. [Boston: 1829.] 24 p. (H.R. 3)

An act incorporating the Massachusetts Rail Road Corporation and the Second Massachusetts Rail Road Corporation.

S-39489.

Copies: DBRE; DeGE; MB; MBAt; MH; MWA; NN.

5963. Massachusetts Rail Road Association.

Constitution of the Massachusetts Rail Road Association.... [Boston: 1829.] 3, [1] p.

Dated: Boston, April 2, 1829.

S-39524, 39525.

Copies: CSmH; MB; MH; MWA; NN.

5964. Massachusetts Rail Road Association.

Reports made to the directors of the Massachusetts Rail Road Association, on the practicability of conducting transportation on a single set of tracks. Printed for the use of the board, 1829. [Boston: 1829.] 37 p.

By Daniel Treadwell and others.

S-39526.

Copies: CSmH; DLC; MCM; MH-BA; MWA; NN; NjP.

5965. Mine Hill and Schuylkill Haven Railroad Company.

Act of incorporation and bylaws of the

5965 (*continued*)
Mine Hill and Schuylkill Haven Rail Road Company. Philadelphia: T. A. Conrad, printer, 1829. 16 p.
S-39601.
Copies: CtY; ICJ; MWA; NN.

5966. New York (State). Commissioners on the Construction of a Rail-Road from Boston to the Hudson River.
Report of the Commissioners, appointed under the act of April 15, 1828, relative to the construction of a rail-road from the city of Boston to the Hudson River. February 24, 1829.... [Albany: 1829.] 29, 47–64 p. (Senate. Doc. 113)
N-N0190911; Sabin-53901.
Copies: CSt; CtY; DLC; MH-BA; MWA; MiU; N; NN.

5967. New York (State). Legislature. Assembly. Select Committee on the Susquehanna and Chemung Navigation and Rail-Road Company.
Report of the Select Committee on sundry memorials from the citizens of the counties of Otsego, Delaware, &c. for a law incorporating the Susquehannah and Chemung Navigation and Rail-Road Company. [Albany: 1829.] 4 p.
N-N0200460; Thomson-340.
Copies: DBRE; NN.

5968. New York (State). Legislature. Senate. Committee on Roads and Bridges.
Report of the Committee on Roads and Bridges, relative to the construction of a rail-road from the village of Havana to the Chemung River. [Albany: 1829.] 3 p. (Senate. Doc. 149)
Dated: March 10, 1829.
N-N0201399; Thomson-339.
Copies: MWA; NN; NjP.

5969. Pennsylvania. Board of Canal Commissioners.
Letter from the president of the Board of Canal Commissioners, accompanied with a copy of the minutes of their proceedings and copies of certain documents ... relating to further surveys ... on the eastern end of the Pennsylvania Rail Road. Read in the House of Representatives, March 31, 1829. Harrisburg: Printed by S. C. Stambaugh, 1829. 17 p.
S-39997.
Copies: DBRE; P.

5970. ________ Read in the Senate, March 30, 1829. Harrisburg: Printed by S. C. Stambaugh, 1829. 17 p.
S-39998.
Copies: DeGE.

5971. Pennsylvania. Board of Canal Commissioners.
Report of the engineers, employed to resurvey the location of the contemplated rail road in the vicinity of Philadelphia. Also, a reply to a committee appointed at a meeting of the citizens of Philadelphia, to report in regard to the termination of the Pennsylvania Rail Road, by Joseph M'Ilvaine, Esq., Secretary to the Board of Canal Commissioners. Harrisburg: E. F. Cryder & Co., printers, 1829. 1 p.l., 26 p.
Engineers' report signed: Moncure Robinson, John Wilson, William R. Hopkins.
S-40005.
Copies: CtY; DBRE; DLC; DeGE; MWA; MiU-T; NN; P; PHi.

5972. Pennsylvania. General Assembly. House. Committee on Inland Navigation and Internal Improvement.
Report of the Committee on Inland Navigation and Internal Improvement, relative to the Baltimore and Susquehanna Rail Road. Read in the House of Representatives, February 3, 1829. [Harrisburg: 1829.] 4 p.
Thomson-347.
Copies: DLC; IU; MWA.

5973. Pennsylvania. General Assembly. Senate.
The preamble and resolution, offered by Mr. Logan, as a substitute for the report and resolution of the Committee on Roads, Bridges and Inland Navigation, on granting the Baltimore and Susquehanna Rail Road

Company the privilege of extending their rail road into the state of Pennsylvania. Read in the Senate, January 24, 1829. Harrisburg: Printed at the office of the Reporter, 1829. 12 p.

S-40001.

Copies: DBRE; DLC; MWA; NN; PHi.

5974. Pennsylvania. General Assembly. Senate. Committee on Roads, Bridges and Inland Navigation.

Columbia and Philadelphia Rail Road. No. 41. Senate file. Resolution relative to the Pennsylvania Rail Road reported by the Committee on Inland Navigation. [Harrisburg?: 1829.] 4 p.

N-P0208169; Thomson-346.

Copies: NNC.

5975. Pennsylvania. General Assembly. Senate. Committee on Roads, Bridges and Inland Navigation.

Report of the Committee on Roads, Bridges and Inland Navigation, on granting the Baltimore and Susquehanna Rail Road Company the privilege of extending their rail road into the state of Pennsylvania. Read in Senate, January 17, 1829. Harrisburg: Printed at the office of the Reporter, 1829. 11 p.

S-40004.

Copies: DLC; MWA; MiU; NN; PHi; PPAmP.

5976. Powel, John Hare, 1786–1865.

Mr. Powel's remarks on the termination of the Pennsylvania rail-way. March, 1829. [Philadelphia?: 1829.] 16, 8 p.

Termination of the Pennsylvania railway [statements of George Haines, J. H. Siddall, P. M. Price, and others], 8 p. at end.

C-3149 (dated [1830?]); Thomson-350.

Copies: DBRE; DIC; DeU; MH-BA; MWA; NN; P; PPAmP; PPL.

5977. [Redfield, William Charles] 1789–1857.

City railways. [S.l.: 1829?] 1 p.

Bound in: D. Olmstead, *Address on the scientific life and labors of William C. Redfield.* 1857.

S-40220.

Copies: PPAmP.

5978. [Redfield, William Charles] 1789–1857.

Sketch of the geographical rout of a great railway, by which it is proposed to connect the canals and navigable waters of the states of New-York, Pennsylvania, Ohio, Indiana, Illinois, Missouri; and the Michigan, North-West, and Missouri territories; opening thereby a free communication, at all seasons of the year, between the Atlantic states and the great valley of the Mississippi. New York: G. & C. & H. Carvill, 1829. 16 p., fold. map.

S-40221.

Copies: CSmH; CSt; CtY; DBRE; DLC; DeGE; IHi; MB; MH; MWA; MiU-T; NN; PPAmP; PPL; WU.

5979. Some observations on the Baltimore and Susquehanna Rail Road. By a citizen of York. [S.l.: 1829?] 7 p.

S-40481.

Copies: DBRE; DLC; NN.

5980. South Carolina Canal and Railroad Company.

Report on the Charleston and Hamburg rail-road, by William Howard, U. S. civil engineer. Charleston: 1829. 18 p., fold. map.

N-H0556469; S-40492.

Copies: DLC; DNA; MiU; NN; NcU; ScCC; WM.

5981. Sullivan, John Langdon, 1777–1865.

On the Baltimore rail-road carriage, invented by Ross Winans, Esq., and the manner of adapting it to streets; also on cheap railroads. By J. L. Sullivan, civil engineer. [Philadelphia: 1829.] 1 p.l., 9, [5] p., 2 plates.

Half-title: *Description from the Franklin Journal of a timber rail road artificially durable; and of the Baltimore rail road carriage.*

Appendix, [5] p., contains: *Specification of an improvement or combination in the art of*

5981 (*continued*)
constructing rail roads, whereby timber employed therein may be increased in durability, by J. L. Sullivan; and other communications on rail-roads.

S-40572, 40573.

Copies: DBRE; DeGE; MBAt; MCM; MWA; NN; PPAmP; PPF; PPL.

5981a. [Sullivan, John Langdon] 1777–1865.

Report and explanation of a survey for a rail-road from Paterson to New-York. Paterson, [N.J.]: D. Burnett, printer, 1829. 15 p., fold. map.

Signed: Jno. L. Sullivan.

Report made to the Society for Establishing Useful Manufactures on a route adopted by the Paterson and Hudson River Railroad Company, chartered in 1831.

N-S1049668.

Copies: CtY.

5982. To the citizens of Massachusetts: the undersigned, being a committee to receive subscriptions in money, to collect deeds of land ... &c. to aid the construction of a rail road from Boston to Brattleborough.... [Boston]: Printed by Dickinson, [1829.] Broadside 24 x 28 cm.

Dated: Boston, December 3, 1829, and signed: Joseph H. Dorr, Henry J. Oliver, and Andrew J. Allen.

M & S Rare Books, Inc. Weston, Mass. Catalog no. 13, no. 842.

Present location unknown.

5983. United States. Congress. House. Committee on Roads and Canals.

Baltimore and Ohio Rail Road Company. February 21, 1829 ... Mr. Woods, of Ohio, from the Committee on Roads and Canals, to which the subject had been referred, made the following report.... [Washington: Gales & Seaton, printers, 1829.] 3 p. (20th Cong., 2d sess. House. Rept. 92)

S-40812.

Copies: CtY; DLC; DeU; M; MB; MH; MWA; N; NN; NjP; P; PPAmP.

5984. United States. Congress. Senate. Committee on Roads and Canals.

In Senate of the United States. February 11, 1829. Mr. Hendricks made the following report. The Committee on Roads and Canals, to whom was referred the memorial of the president and directors of the Baltimore and Ohio Rail Road Company, report.... [Washington: 1829.] 3 p. (20th Cong., 2d sess. Senate. Doc. 73)

S-41156.

Copies: CSt; CtY; DBRE; DLC; DeU; M; MB; MWA; MiU-T; NN; NjR; P; PPAmP.

5985. United States. War Dept.

Hudson and Berkshire Rail Road. Letter from the Secretary of War, accompanied by the maps, reports, and proceedings of the survey of the Hudson and Berkshire Rail Road. January 21, 1829.... [Washington: Gales & Seaton, printers, 1829.] 26 p., fold. tables. (20th Cong., 2d sess. House. Doc. 89)

Maps not present in copies seen.

S-41137.

Copies: CtY; DBRE; DLC; DeU; MB; MH-BA; MWA; MiU-T; N; NN; NjP; NjR; P; PPAmP.

5986. Western rail-way. [Boston: 1829?] 2 p.

By The farmer's friend.

An argument in favor of a railroad from Boston to Albany.

S-38553.

Copies: CSmH; DBRE; MWA.

5987. Wright, James.

Specifications of patents for improvements in rail-roads and cars for turn-outs in rail-roads, and for a rail-way carriage for curved roads. Granted to James Wright, of Columbia, Pa. Philadelphia: J. Harding, printer, 1829. 7 p., illus.

"From the *Journal of the Franklin Institute* for October, 1829."

S-41600.

Copies: DeGE.

5988. Address to the citizens of Baltimore, relative to the contemplated extension of the rail road, down Pratt-Street, to the city block. Adopted at an adjourned meeting of citizens of the western and south-western section of the city, held at the Union Engine Committee room, Thursday evening, September 30, 1830. Baltimore: Printed at the office of the Baltimore Republican, 1830. 8 p.

C-28.

Copies: DLC; MdBJ-G; MdHi; NN.

5989. Baltimore and Ohio Railroad Company.

The Baltimore and Ohio Rail Road Company vs. the Chesapeake and Ohio Canal Company, and others. [Annapolis: 1830.] 71 p.

Proceedings before Chancellor Bland, concerning the right of way along the Potomac.

C-245.

Copies: PPF.

5990. Baltimore and Ohio Railroad Company.

Fourth annual report, of the president and directors, to the stockholders of the Baltimore and Ohio Rail Road Company. Baltimore: Printed by William Wooddy, 1830. 153 p., fold. tables.

C-241.

Copies: CSt; CtY; DBRE; DLC; DeGE; ICJ; LNHT; MBAt; MH; MWA; MdBJ-G; MdHi; MiU-T; NN; NjP; O; PPL.

5991. Baltimore and Ohio Railroad Company.

Information and directions, relative to laying a single track of wood rail way, on the Baltimore and Ohio Rail Road. [Baltimore: W. Wooddy, printer, 1830?] 8 p.

C-242.

Copies: MH-BA; NN.

5992. Baltimore and Ohio Railroad Company.

Letter and documents from the Baltimore and Ohio Rail Road Company, in refutation of the charges contained in a pamphlet laid before Congress by the Chesapeake and Ohio Canal Company. [Baltimore: 1830.] 8 p.

C-243.

Copies: CtY; DBRE; DLC; MWA; MiU-T; NN.

5993. ———— [Baltimore: 1830.] 7 p.

C-244.

Copies: CtY; DLC; MdHi; NN.

5994. Baltimore and Ohio Railroad Company.

Memorial of the Baltimore and Ohio Rail Road Company, for permission to extend a lateral branch of their road to some point within the District of Columbia. December 14, 1830. Referred to the Committee on Roads and Canals. Dec. 15, 1830. Ordered to be printed. [Washington: 1830.] 2 p. (21st Cong., 2d sess. Senate. Doc. 5)

C-4955.

Copies: CtY; DBRE; DLC; DeU; MB; MH; MWA; MiD; N; NN; NjP; PPAmP; PU.

5995. Baltimore and Susquehanna Railroad Company.

Address to the mayor and city council, on the trade of the Susquehanna, and the rail road to that river. Baltimore: Printed by Lucas & Deaver, 1830. 24 p.

Signed on p. 17: Geo. Winchester, President.

C-5507; Thomson-462.

Copies: CtY; DeGE; MH-BA; MdBP; MdHi; NNE; OCHP; PPL.

5996. Baltimore and Susquehanna Railroad Company.

Letter from George Winchester, Esq., president of the Baltimore and Susquehanna Rail Road Company. Annapolis: Jeremiah Hughes, printer, 1830. 6 p. (Maryland. House. Doc. 10).

Accompanying the executive message, December 31, 1829.

Letter to the governor of Maryland, urging him to aid in efforts to obtain a

5996 (*continued*)
charter for the railroad from the state of Pennsylvania.
C-2429; Thomson-463.
Copies: MWA; MdHi.

5997. Baltimore and Susquehanna Railroad Company.
Third annual report of the directors to the stockholders of the Baltimore and Susquehanna Rail Road Company. Together with the report of the engineers. 18th October, 1830. Baltimore: Printed by James Lucas & E. K. Deaver, 1830. 27 p.
C-246.
Copies: CSt; CtY; DBRE; DLC; DeGE; ICJ; MH-BA;MWA; MdHi; MiU-T; NN; NNE; PPL.

5998. [Carey, Mathew] 1760–1839.
Essay on rail roads. [Philadelphia: 1830.] 26 p., illus.
Signed on p. 6: Hamilton, Philadelphia, January 14th, 1830.
C-768.
Copies: DBRE; DLC; MB; MH-BA; MWA; MiU-T; NN; NNC; PHi; PPAmP; PU.

5999. ———— 2d ed. [Philadelphia; 1830.] 26 p., illus.
C-769.
Copies: DLC; MHi; MWA; PPL; PU.

6000. The case of the Albany and Schenectady Turnpike Company considered. [S.l.: 1830?] 2 p.
An opinion as to its rights to construct a railroad against the wish of the Mohawk and Hudson Railroad Company.
C-798.
Copies: NN.

6001. Clinton, De Witt, 1805– .
Correspondence on the importance & practicability of a rail road, from New York to New Orleans, in which is embraced a report on the subject by DeWitt Clinton.... New York: Vanderpool and Cole, printers, 1830. 2 p.l., 23 p.
C-924.
Copies: CtY; DBRE; DLC; DeGE; MB; MH-BA; MHi; MWA; MiU-T; NN; NNC; Nh; NjR; PPAmP; PPL; WU.

6002. Considerations on the practicability and utility of immediately constructing a central railway, from Pottsville to Sunbury and Danville, through the coal region of Mahanoy and Shamokin, with the proceedings of a meeting at Sunbury, Dec. 1830. Republished with additions by the Philadelphia Committee. December 23, 1830. [Philadelphia: 1830.] 1 p.l., 37 p.
C-997.
Copies: DBRE; DLC; DeGE; IU; MH-BA; MWA; MiU-T; N; NN; PPL; WU.

6003. Considerations on the practicability and utility of immediately constructing a rail way, from Pottsville to Sunbury and Danville, through the coal region of Mahonoy [*sic*] & Shamokin; with the proceedings of a meeting at Sunbury, December, 1830. [Milton, Pa.: Printed at the office of the "State's advocate," 1830.] 24 p.
C-996.
Copies: CSmH; CtY; DLC; DeGE; ICJ; MB; MH-BA; MWA; NN; PPAmP; PPL.

6004. The contemplated plan of the Franklin Rail Road Company, to commence with a route through the counties of Middlesex, Worcester and Franklin, to the line of Vermont; from thence to Lake Champlain and to Lake Ontario, with branches convenient to accommodate the inhabitants of Massachusetts. Boston: J. H. Eastburn, 1830. 7 p., map.
C-1000, 1445.
Copies: DBRE; DLC; ICJ; MBAt; MCM; MH-BA; MHi; MWA; NN; RPB.

6005. Earle, Thomas, 1796–1849.
A treatise on rail-roads and internal

communications. Compiled from the best and latest authorities, with original suggestions and remarks. Philadelphia: Sold by John Grigg, Towar, J. & D. M. Hogan; Miffin & Parry, printers, 1830. 120 p., illus., plates, 2 fold. maps.

C-1224.

Copies: CSt; CtY; DBRE; DLC; DeGE; DeU; ICJ; MB; MH-BA; MWA; MdBP; MiU-T; MnSJ; N; NN; NjP; OHi; PPF; PPL; WHi.

6006. Franklin Railroad Company.

Great publick improvement. Franklin Rail-Road Company. [Boston: Printed at the office of the Mechanicks magazine, and Journal of publick internal improvement, 1830?] 8 p.

C-1446.

Copies: CtY; DBRE; DLC; ICJ; MB; MH; MH-BA; MWA; MiU-C; NIC; NN; NNC; WHi.

6007. Great rail road meeting. Boston: 1830. Broadside 60 x 39 cm.

C-1663.

Copies: MB; NN.

6008. Hayward, James, 1786–1866.

[Report of the engineer appointed to make a survey for a rail road from the city of Boston to Lowell.] [Boston: Dutton and Wentworth, 1830.] 16 p. (Massachusetts. Senate. Doc. 3)

Signed: James Hayward.

N-H0212070.

Copies: CSmH; DBRE; DLC; M; MH-BA; MWA; MiU-T; NN; NNC.

6009. Hints for the representatives of the country. [Philadelphia: 1830.] 24 p.

Contents: Pennsylvania rail-way. The route on the western side of the Schuylkill is the cheapest and the best; Mr. Strickland's report; Engineers' and Canal Commissioners' opinions in favour of the western route [etc.].

C-1849.

Copies: DIC; DeGE; MH-BA; MiU-T; NN; NNC; P; PHC; PPL; PU.

6010. Improvements in Massachusetts. To the citizens in favor of a rail road from Boston to Vermont. Boston: John H. Eastburn, printer, [1830?]. Broadside.

C-1977.

Copies: MWA; NN (photocopy).

6011. Jones, Samuel T.

Specifications of two patents for a new rail-road car, denominated the pendulous rail-road car, and for a mode ... of adapting rail-way cars ... to run on ordinary roads ... and also of enabling carts ... to run securely upon rail-roads. Granted to Samuel T. Jones, of Philadelphia, on the 22nd day of February, 1830. To which are appended remarks by the editor of the Journal of the Franklin Institute.... Philadelphia: 1830. [2], 8 p., front.

From the *Journal of the Franklin Institute*, for March and April 1830.

C-2080.

Copies: DeGE; MWA; NbO; PPAmP; PPF; PPL.

6012. Kentucky. Laws.

An act to incorporate the Lexington and Ohio Rail Road Company. [S.l.: 1830.] 11 p.

C-2227.

Copies: DBRE; MWA; MiU-T; NN.

6013. Locomotive steam engine. Printed for Burd Patterson—Pottsville. From the London quarterly review. [S.l.: 1830?] 16 p.

Copies: DeGE.

6014. Long, Stephen Harriman, 1784–1864.

Narrative of the proceedings of the board of engineers, of the Baltimore and Ohio Rail Road Company, from its organization to its dissolution, together with an exposition of facts, illustrative of the conduct of sundry individuals. By Lt. Col. S. H. Long and Capt. Wm. Gibbs McNeill ... Baltimore: Printed by Bailey & Francis, 1830. 2 v., fold. table.

Part 2, printed by W. A. Francis, has the title: *Part II, in continuation of the Narrative.... To which is appended a brief review of the location of the road....*

6014 (*continued*)
C-2267.
Copies: CSt; CtY; DBRE; DLC; DeGE; MH; MWA; MdBJ-G; MiU-T; N; NNC; PPL.

6015. Louisiana. Laws.
An act to incorporate the Pontchartrain Rail Road Company. Published by order of the board of directors. New Orleans: Printed by Atherton T. Penniman & Co. for Thomas Stroud & Co., 1830. 16 p.
C-3128.
Copies: MB; MH-BA; MWA.

6016. McNeill, William Gibbs, 1800–1853.
To the stockholders of the Baltimore and Ohio Rail-Road Company. [Baltimore: 1830.] 36 p.
C-2344.
Copies: DBRE; NN.

6017. Maryland. Governor (T. K. Carroll).
(Document No. 10) Accompanying the executive message of the 31st December 1829. Letter from George Winchester, Esq., president of the Baltimore and Susquehanna Rail Road Company. Annapolis: Jeremiah Hughes, printer, 1830. 6 p.
C-2429.
See no. 5996.

6018. Massachusetts. Board of Internal Improvements.
Report of the Directors of Internal Improvement on the subject of rail roads. Transmitted to the legislature, January session, 1830. Boston: Dutton and Wentworth, printers to the state, 1830. 61 p. (Senate. Doc. 4)
C-2477.
Copies: CSmH; DBRE; DeGE; IU; MH-BA; MWA; MeHi; NIC; NN; NNC; NRU; PPL; RPB.

6019. Massachusetts. General Court. Joint Committee on Railways and Canals.
The Committee of both Houses on Railways and Canals, to whom was referred the petition of John F. Loring and others, for authority to construct a rail road from the city of Boston to the town of Lowell, and the remonstrance of the president, directors and company of the Middlesex Canal, against the granting of said petition, and to whom was also committed the message of His Excellency the Governor, transmitting the survey of a rail road to Lowell . . . report. [Boston: 1830.] 21 p. (Senate. Doc. 21)
Signed: John W. Lincoln, chairman.
"An act to establish the Boston and Lowell Rail Road Company," p. [13]–21.
C-2476.
Copies: DBRE; DeGE; M; MB; MBAt; MH; MWA; MiU-T; PPL.

6020. Massachusetts. General Court. Senate.
[Report on a survey for a railroad from the city of Boston to Lowell, by engineer James Hayward. Boston: 1830.] 16 p. (Senate. Doc. 3)
BRE-p. 184.
Copies: M; MWA; NNC.

6021. Massachusetts. Laws.
An act in addition to an act to incorporate the Franklin Rail Road Company. [Boston: Dutton and Wentworth?, 1830.] 3 p. (Senate. No. 4)
N-M0310872; Thomson-419.
Copies: DeGE; M; MB; MH; MWA; NN.

6022. Massachusetts. Laws.
An act to establish the Boston and Lowell Rail Road Corporation. [Boston: Dutton and Wentworth, 1830.] 10 p. (House. Doc. 5)
Thomson-380.
Copies: M; MB; MWA; NN.

6023. Massachusetts. Laws.
An act to establish the Boston, Providence and Taunton Rail Road Corporation. [Boston: 1830.] 11 p. (House. Doc. 57)
Dated at end: In Senate, March 3, 1830.

C-2463.
Copies: DLC; DeGE; M; MB; MBAt; MH; MWA; NNC; PPL.

6024. Massachusetts. Laws.
An act to establish the Massachusetts Rail Road Corporation. [Boston: 1830.] 11 p. (House. Doc. 56)
C-2464.
Copies: M; MB; MBAt; MH; MWA; NNC; PPL.

6025. Massachusetts. Laws.
An act to establish the Massachusetts rail road corporations. [Boston: 1830.] 21 p. (House. Doc. 4)
An act incorporating the Massachusetts Rail Road Corporation and the Second Massachusetts Rail Road Corporation.
C-2465.
Copies: DeGE; M; MB; MBAt; MH; MWA; NN; NNC; PPL.

6026. Massachusetts. Laws.
An act to incorporate the Franklin Railroad Company. [Boston: Dutton & Wentworth, 1830.] 17 p. (House. Doc. 35)
C-2466.
Copies: M; MB; MBAt; MH-BA; MWA; NN; NNC.

6027. Mount Carbon Railroad Company.
Act of incorporation and by-laws of the Mount Carbon Rail Road Company. Philadelphia: Printed for the Company [by] Thomas Town, 1830. 24 p.
C-2651.
Copies: DBRE; DeGE; ICU; IU; MH; MWA; MiU-T; NN; PPL.

6028. New Castle and Frenchtown Railroad Company.
Acts passed by the legislatures of Maryland and Delaware, authorizing the construction of a railroad from French-Town to New-Castle. [S.l.: 1830.] 23 p.
C-2714.
Copies: CSt; DBRE; NN; NNC.

6029. New York (State). Comptroller's Office.
Report of the Comptroller, to whom was referred the petition of William C. Young. February 18, 1830. [Albany: Printed by E. Croswell, 1830.] 8 p. (Assembly. Doc. 181)
Concerns the account of William C. Young as engineer of the railroad from Boston to the Hudson River.
Thomson-430.
Copies: CtY; DLC; MB; MWA; N; NN.

6030. New York (State). Laws.
An act to incorporate the Mohawk and Hudson Rail Road Company, passed April 17, 1826. And An act to amend the Act to incorporate the Mohawk and Hudson Rail Road Company, passed April 17, 1826. Passed March 28, 1830. New York: Ludwig & Tolefree, 1830. 12 p.
C-2609.
Copies: DBRE; N; NN.

6031. New York (State). Legislature. Assembly. Select Committee on Catskill and Ithaca Railroad Company.
Report of the Select Committee on the memorial of the commissioners of the Catskill and Ithaca Rail-road Company. April 7, 1830. [Albany: Printed by E. Croswell, printer to the state, 1830.] 2 p. (Assembly. Doc. 410)
N-N0201463 (listed as Senate doc. 410); Thomson-429.
Copies: CtY; DLC; MB; MWA; MiU-T; N; NN.

6032. New York (State). Legislature. Senate. Select Committee on Canjoharie and Catskill Railroad Company.
Report of the Select Committee, on the petition for the incorporation of the Canjoharie and Catskill Rail-Road Company. February 18, 1830. [Albany: Printed by E. Croswell, printer to the state, 1830.] 3 p. (Senate. Doc. 180)
Thomson-428.
Copies: CtY; DBRE; DLC; MB; MWA; MiU-T; N; NN.

6033. Pennsylvania. Laws.
The act, and supplements thereto, authorizing the incorporation of the Little Schuylkill Navigation, Rail Road and Coal Company. Reading: Douglas W. Hyde, 1830. 33 p.
Mainly concerned with the railroad.
C-2254.
Copies: ICJ; MB; MWA.

6034. Pennsylvania. Laws.
An act to incorporate the Tuscarora and Cold Run Tunnel and Railroad Company. Approved April 6, 1830. Pottsville, [Pa.]: Printed by Benjamin Bannan, 1830. 15 p.
C-3782.
Copies: CSmH; DBRE; MWA; PHi; PPM.

6035. Petersburg Railroad Company.
Petersburg Rail Road law. An act to incorporate the Petersburg Rail Road Company. (Passed February 10th, 1830).... [Petersburg, Va.: Pescud, printer, 1830.] 16 p.
C-3019.
Copies: Vi.

6036. Philadelphia. Watering Committee.
Report of the Watering Committee of the Select and Common Councils of Philadelphia, relative to the termination of the Columbia and Philadelphia Railroad. Philadelphia: Printed by L. R. Bailey, 1830. 18 p.
C-3027.
Copies: CSmH; DBRE; DeGE; MH; MWA; P; PPAmP; PPAmP; PPF; PPL; PU.

6037. [Proprietors of the Middlesex Canal.]
[Remonstrance of the president, directors, and company of the Middlesex Canal to granting the petition of John F. Loring, and others, for a rail road from Boston to Lowell. Boston: 1830.] 12 p. (Massachusetts. House. Doc. 40)
Signed: Wm. Sullivan, John Coolidge, Geo. Hallet, committee of the directors of the Middlesex Canal.
C-3274; Thomson-423.
Copies: DLC; M; MB; MBAt; MH-BA; MWA; NN.

6038. Read, John Meredith, 1797–1874.
Remarks of John M. Read, one of the deputation from the general town meeting of the citizens of the city of Philadelphia, held January 16, 1830, relative to the termination of the Columbia and Philadelphia Rail Road; before the Committee on Inland Navigation and Internal Improvement of the House of Representatives, at Harrisburg, on Wednesday the 17th February, 1830. Philadelphia: Printed by L. R. Bailey, 1830. 24 p., map.
C-3249.
Copies: CtY; DBRE; DLC; DeGE; ICU; MB; MH-BA; MWA; MiD-B; MnHi; NN; OClWHi; PPAmP; PPL.

6039. [Redfield, William Charles] 1789–1857.
Sketch of the geographical rout of a great railway, by which it is proposed to connect the canals and navigable waters, of New-York, Pennsylvania, Ohio, Indiana, Illinois, Michigan, Missouri, and the adjacent states and territories; opening thereby a free communication, at all seasons of the year, between the Atlantic states and the great valley of the Mississippi. 2d ed., with additions. New York: G. & C. & H. Carvill, 1830. 48 p., fold. map.
C-3253.
Copies: CU-B; CtY; DBRE; DLC; DeGE; ICN; MB; MBAt; MH; MWA; MiU-T; MnHi; N; NIC; NN; OClWHi; PPAmP; PPF; PPL; PU.

6040. Reports and documents relative to the termination of the Pennsylvania Rail Road. [Philadelphia: 1830.] 1 p.l., [5]–30 p.
Concerns the Philadelphia and Columbia Railroad.
C-3290.
Copies: DBRE; MWA; MiU; PHi; PPAmP; PU.

6041. Robinson, Moncure, 1802–1891.
Report of the engineer, on a survey of a rail-road from Petersburg to the Roanoke. [Petersburg, Va.: 1830.] 11 p.
Signed and dated at end: Moncure Robinson. April 8, 1830.
Report to the mayor and the Common Council of Petersburg, Virginia.
Copies: ViW (in Moncure Robinson Papers).

6042. Sullivan, John Langdon, 1777–1865.
Practical principles of rail-ways, proper for the climate of the United States, and adapted to the use of the Winans rail-way carriage.... [New York?: 1830.] 20 p.
Cover title: *American rail-way, and the Winans carriage.*
C-3633.
Copies: DBRE; DLC; NHi; NN; PHi.

6043. Sullivan, John Langdon, 1777–1865.
Profits of rail roads in the United States considered; with a description of the compound leverage carriage, and explanations of its increase of effect of power. Also, an inquiry into the proper method of building railroads, in reference to the climate of the middle and northern states. With some remarks on the partial application of the United States revenue to internal improvements. By J. L. Sullivan.... New York: Printed by Clayton & Van Norden, 1830. 20 p.
C-3634.
Copies: CSt; NN.

6044. United States. Congress. House. Committee on Internal Improvements.
Baltimore and Ohio Rail Road Company. February 19, 1830. Mr. Hemphill, from the Committee on Internal Improvements, to which had been referred the memorial of the Baltimore and Ohio Rail Road Company, made the following report.... [Washington: Printed by Duff Green, 1830.] 1 p. (21st Cong., 1st sess. House. Rept. 211)
C-3932.
Copies: CSt; CtY; DBRE; DLC; DeGE; DeU; MH; MWA; MiU-T; NN; NjR; P; PPAmP.

6045. United States. Laws.
A bill to authorize the extension, construction, and use of a lateral branch of the Baltimore and Ohio Rail Road, into and within the District of Columbia. December 8, 1830.... [Washington: 1830.] 6 p. (H.R. 517)
C-4454.
Copies: DNA.

6046. Walker, James Scott.
An accurate description of the Liverpool and Manchester rail-way, the tunnel, the bridges, and other works throughout the line; with a sketch of the objects which it presents interesting to the traveller or tourist.... Republished by the standing committee appointed by the inhabitants of Paterson, to forward the application for a rail-road from this place to the Hudson River, opposite New-York. Paterson, N.J.: D. Burnett, printer, 1830. 33, [1] p., fold. plate.
C-5316.
Copies: CSmH; CtY; DBRE; DIC; MWA; MiU-T; N; NN.

6047. Wallenpaupack Improvement Company.
Act of Assembly of Pennsylvania incorporating the Wallenpaupack Improvement Company, with the report of Henry G. Sargent, Esq., civil engineer. In relation to the proposed route, to open the coal region of the Lackawannock Valley to the New York market. Easton, Pa.: Josiah P. Hetrich, printer, 1830. 21 p.
Concerns a proposed railroad.
C-2978; N-P0208841.
Copies: MWA; MiU; PU.

6048. Western Railroad Corporation.
[Articles which will be brought over the Western Rail-way from the towns west of Boston. S.l: 1830.] Broadside.
C-5423.
Copies: DBRE, M.

6049. Wirt, William, 1772–1834.
Argument delivered at Annapolis, by William Wirt, Esquire ... before the Chancellor of Maryland, on the 19th and 20th

6049 (*continued*)
days of August, 1829, in support of a motion to dissolve an injunction obtained by the Baltimore and Ohio Rail-Road Company against the Chesapeake and Ohio Canal Company. Washington: Printed by Gales & Seaton, 1830. 2 p.l., [3]–206 p.
C-5515.
Copies: DIC; DLC; MH-BA; MWA; MdBJ-G; NN.

6050. ——— Washington: Printed by Gales & Seaton, 1830. 186 p.
Sabin-104870.
Copies: DLC; MH-BA; MiU-T.

6. Ballooning and flight

6051. Deeker, Joseph.
Air balloon. The subscriber, who is the proprietor of the Speaking Image ... is perfectly acquainted with the ... construction of air balloons.... [New York: 1789.] Broadside 34 x 20.5 cm.
Signed: Joseph Deeker. New-York, June 10, 1789.
B-6933; Mp-45463.
Copies: NHi.

6052. Blanchard, Jean Pierre, 1753–1809.
Journal de ma quarante-cinquième ascension, la première faite en Amérique, le 9 janvier, 1793.... Philadelphie: De l'imprimerie de Charles Cist, 1793. [27] p., front.
E-25207 (note).
Copies: DLC (photocopy); MH.

6053. Blanchard, Jean Pierre, 1753–1809.
Journal of my forty-fifth ascension, being the first performed in America, on the ninth of January, 1793.... Philadelphia: Printed by Charles Cist, [1793]. 27 p. front.
Mp-25207.
Copies: DLC; MBAt; MWA; NN; PHC; PPAmP; PPL.

6054. The forty-fifth aerial flight of the universally celebrated Mr. Blanchard, at Philadelphia.... [Philadelphia: 1793?] 7 p.
The balloon ascension took place on January 9, 1793.
B-8341; Mp-46753.
Copies: CSmH; DLC (photocopy).

6055. The principles, history & use, of air-balloons. Also, a prospectus of Messrs. Blanchard & Baker's intended aerial voyage from the city of New-York. New York: Printed by C. C. Van Alen for J. Fellows, 1796. [2], 46 p., plate.
Half title: *Blanchard's balloon.*
"List of Mr. Blanchard's aerial voyages," p. 40–41.
Includes a copy of the proclamation written by Washington and presented by him to Blanchard on his flight from Philadelphia, Jan. 9, 1793.
Mp-30099.
Copies: DLC; DeGE; MB; MWA; NHi; NN.

6056. Fowler, George.
A flight to the moon; or, The vision of Randalthus. By Geo. Fowler.... Baltimore: Printed by A. Miltenberger, 1813. 185 p.
Description of an imaginary voyage to the moon.
S&S-28550.
Copies: CSmH; DLC; IGK; MH; MWA; MdHi; MiDSH; MnU; N; NNC; PPFM; ViU.

6057. Walker, Thomas.
A treatise on the art of flying, by mechanical means; with a full explanation of the natural principles by which birds are enabled to fly; likewise instructions and plans for making a flying car with wings, in which a man may sit, and, by working a small lever, cause himself to ascend and soar through the air.... New York: Printed and sold by S. Wood, 1814. xiv, [15]–64 p., 2 plates.
First published at Hull, England, 1810.
S&S-33565.
Copies: DLC; DeGE; MH-BA; MWA.

6058. ———— New York: Printed by S. Wood & Sons, 1816. xviii, [19]–68 p., 2 plates.
S&S-39701.
Copies: DLC; RPB.

6059. The art of conjuring made easy; or, An entertaining selection of diverting tricks, deceptions and experiments ... together with the secret of making and filling of air and fire balloons.... New York: S. King, 1822. 24 p., col. front.
S-7841.
Copies: DLC; MB; MH.

6060. ———— New York: W. Borradaile, 1823. 24 p.
N-A0437315.
Copies: DLC; InU; MWA.

6061. Genet, Edmond Charles, 1763–1834.
Air navigation. Mr. Edmond Charles Genet and Mr. Eugene Robertson, well known as an aerostatic experimenter having formed a connection to promote the aerostatic science, and ascertain by actual experiment the practicability of navigating effectually, the air, by means of an aeronaut, constructed on the principles represented above.... They respectfully solicit the patronage of the citizens of the United States.... [New York: 1827.] 1 p.
S-29004.
Copies: DLC.

6062. Robertson, Eugène, 1800–1836.
Relation d'un voyage aérien, fait à New York, par M. Eugène Robertson, suivi d'un mémoire sur l'aerostation, depuis la découverte jusqu'à l'époque actuelle. Nouvelle-Orléans: A la libraire française et espagnole de A. L. Boimare, 1827. 59 p. port.
Translated from the English published originally in the *New-York Mirror and Ladies' Literary Gazette,* Dec. 23, 1826, and from a memoir by the author's father, Étienne G. Robertson.
N-R0325081.
Copies: CtY; DLC; LNHT; MWA.

6063. [Tucker, George] 1775–1861.
A voyage to the moon: with some account of the manners and customs, science and philosophy of the people of Morosofia, and other lunarians. By Joseph Atterley [pseud.].... New York: E. Bliss, 1827. 264 p.
An imaginary voyage.
S-30846.
Copies: CSmH; CtY; DLC; MB; MH; MWA; MnU; NGH; NNC; NcD; OMC; RPB.

6064. Durant, Charles Ferson, 1805–1873.
Grand aerostatic ascension, of Charles F. Durant, the 1st native citizen of the United States that ever attempted an ascension in a balloon ... on Thursday, September 9.... [New York]: J. C. Spear, printer, [1830.] Broadside, illus. 117 x 57 cm.
Listed by G. Cooper under the title: *Castle Garden grand aerostatic ascension.*
C-804; N-D0452196.
Copies: NN.

6065. Durant, Charles Ferson, 1805–1873.
Second grand ascension of Mr. Charles F. Durant will take place on Wednesday afternoon, Sept. 22.... [New York: 1830.] Broadside, illus. 75 x 58 cm.
N-D0452199.
Copies: NN.

INDEX

Index

NOTE: This Index lists names, subjects, and anonymous titles.
Index entries in capital letters are subject entries;
those in upper and lower case are name and title entries.
The numbers in this Index refer to the entry numbers used
in the checklist.

A

B

C

D

E

F

G

H

I

J

K

L

M

N

O

P

Q

R

S

T

U

V

W

Y

Z